NUMERICAL METHODS IN GEOMECHANICS / INNSBRUCK / 1988

PROCEEDINGS OF THE SIXTH INTERNATIONAL CONFERENCE ON NUMERICAL METHODS IN GEOMECHANICS / INNSBRUCK / 11-15 APRIL 1988

Numerical Methods in Geomechanics Innsbruck 1988

Edited by
G.SWOBODA
Institute of Structural Engineering, University of Innsbruck

VOLUME TWO:

4 *Ice mechanics*
5 *Rock hydraulics*
6 *Modeling of joints, interfaces and discontinuum*
7 *Modeling of infinite domains*
8 *Soil-structure interaction, piles*
9 *Earth structures, slopes, dams, embankments*

Published on behalf of the International Committee for Numerical Methods in Geomechanics by
A.A.BALKEMA / ROTTERDAM / BROOKFIELD / 1988

ORGANIZING COMMITTEES

International Conference Committee

Prof. G.Swoboda (Chairman) University of Innsbruck, Austria
Prof. C.S.Desai (Co-Chairman) University of Arizona, USA
Prof. H.Duddeck (Co-Chairman) Technical University of Braunschweig, FR Germany
Prof. W.Wittke (Co-Chairman), Technical University of Aachen, FR Germany
Prof. T.Adachi, Kyoto University, Japan
Prof. D.Aubry, Ecole Centrale des Arts, France
Dr. G.Beer, CSIRO, Division Geomechanics, Australia
Prof. Y.K.Cheung, University of Hong Kong, Hong Kong
Prof. G.Clough, Virginia State University, USA
Dr. J.T.Christian, Stone & Webster Engineer Corp., USA
Prof. W.D.L.Finn, University of British Columbia, Canada
Prof. G.Gioda, Politecnico di Milano, Italy
Prof. G.Gudehus, Technical University of Karlsruhe, FR Germany
Prof. Y.Ichikawa, Nagoya University, Japan
Prof. K.Kovári, ETH-Hönggerberg, Switzerland
Prof. Y.M.Lin, Northeast University, People's Republic of China
Prof. S.Ukhov, Moscow Civil Engineering Institute, USSR
Prof. F.Medina, Universidad de Chile, Chile
Prof. Z.Mróz, Polish Academy of Science, Poland
Prof. J.Prevost, Princeton University, USA
Prof. J.Smith, University of Manchester, UK
Prof. C.Tanimoto, Kyoto University, Japan
Prof. S.Valliappan, University of New South Wales, Australia
Prof. A.Varadarajan, ITT, Delhi, India
Prof. S.J.Wang, Academia Sinica, People's Republic of China
Prof. N.E.Wiberg, Chalmers University of Technology, Sweden
Prof. O.C.Zienkiewicz, University of Wales, UK

International Committee for Numerical Methods in Geomechanics

Prof. T.Adachi, Japan
Prof. D.Aubry, France
Prof. A.S.Balasubramaniam, Thailand
Prof. J.R.Booker, Australia
Dr. C.A.Brebbia, UK
Prof. Y.K.Cheung, Hong Kong
Dr. J.T.Christian, USA
Dr. A.Cividini, Italy
Prof. C.S.Desai, USA (Chairman)
Prof. J.M.Duncan, USA
Prof. Z.Eisenstein, Canada
Prof. A.J.Ferrante, Brazil
Prof. W.D.L.Finn, Canada
Dr. J.Geertsma, Netherlands
Prof. J.Ghaboussi, USA
Prof. K.Höeg, Norway
Prof. K.Ishihara, Japan
Prof. T.Kawamoto, Japan
Prof. K.Kovári, Switzerland
Prof. S.Prakash, India
Prof. J.M.Roesset, USA
Prof. I.M.Smith, UK
Prof. V.I.Solomin, USSR
Prof. G.Swoboda, Austria
Prof. S.Valliappan, Australia
Prof. C.Viggiani, Italy
Prof. S.Wang, People's Republic of China
Prof. N.E.Wiberg, Sweden
Prof. W.Wittke, FR Germany
Prof. O.C.Zienkiewicz, UK

The texts of the various papers in this volume were set individually by typists under the supervision of each of the authors concerned.

Published by

A.A.Balkema, P.O.Box 1675, 3000 BR Rotterdam, Netherlands

A.A.Balkema Publishers, Old Post Road, Brookfield, VT 05036, USA

For the complete set of three volumes ISBN 90 6191 809 X
For volume 1: ISBN 90 6191 810 3
For volume 2: ISBN 90 6191 811 1
For volume 3: ISBN 90 6191 812 X

Printed in the Netherlands

Numerical Methods in Geomechanics (Innsbruck 1988), Swoboda (ed.)
© 1988 Balkema, Rotterdam. ISBN 90 6191 809 X

Contents

6 *Modeling of joints, interfaces and discontinuum*

7 *Modeling of infinite domains*

8 *Soil-structure interaction, piles*

4 Ice mechanics

Numerical Methods in Geomechanics (Innsbruck 1988), Swoboda (ed.)
 ISBN 90 6191 809 X

Modelling of shallow slope creep by finite element method

D.F.E.Stolle
Department of Civil Engineering and Engineering Mechanics, McMaster University, Hamilton, Ontario, Canada

ABSTRACT: This paper provides a brief review of finite element modelling of ice creep and elucidates on a weighted residual finite element technique for obtaining solutions to large ice mass deformation problems requiring full three-dimensional stress analysis. The 3-D equilibrium equations are reduced to an equivalent two-dimensional form by using the shallow-ice approximation before discretization. The influence of stress variations, through depth of the ice mass, on flow field is accounted for by assuming appropriate stress distributions and then integrating the creep law to arrive at a relationship between vertically averaged stress and deformation rate measures. The approach presented herein can also be used to model lava and mud flows. Examples within a linear framework are presented to demonstrate the appropriateness of the approach for the class of problem addressed

1 INTRODUCTION

The theoretical framework for the modelling of large ice mass dynamics via continuum mechanics is well established; see for example Hutter [1] for an excellent review. Except for a few highly idealized problems, implementation of the theoretical concepts for the analysis of real glacier flow problems requires various degrees of simplification and the use of numerical techniques for obtaining solutions.

While the planar flow approximation which is most often adopted is applicable along the centerline of wide valley glaciers or certain sections of ice sheets, glacier flow is in general influenced by valley walls which have the effect of introducing out-of-plane shear stresses and the reducing basal shear stress as a result of reduced hydraulic radius, and ice sheets and glacier tongues tend to spread in both horizontal directions thereby requiring full 3-D analysis. The extension of the 2-D finite element methodology to 3-D flow is in principle straightforward, although numerical difficulties may arise due to the large system of equations which must be solve simultaneously. Consequently, only the simplest 3-D geometries may be investigated; even then a computer with large memory and fast computing capabilities is required. Further, while general formulations are able to give detailed information on stresses and velocities within an ice mass, significant data preparation is required for ice masses of complex geometry and accuracies provided by standard finite element formulations may be severely undermined by poor quality of data input.

This paper addresses large scale three-dimensional modelling of ice creep due to gravity loading with emphasis being placed on numerical modelling methodology via the finite element weighted residual approach. By taking into account typical large ice mass geometries, the 3-D problem is reduced to 2-D form, thereby allowing a significant reduction in data input preparation and computational effort. The emphasis in this paper is on boundary-valued problem modelling technique and not on material description. In order to simplify the presentation, linear isotropic ice rheology and basal sliding relationships are adopted, and it is assumed that the ice mass is isothermal and homogeneous. It must be stressed that the methodology presented in this paper can be applied using nonlinear constitutive relationships which may also depend on temperature. However the equations are naturally more complex; see, e.g. Ref. [2].

It is demonstrated that the finite element model presented herein gives reasonable large scale predictions when compared with solutions effected by using a standard finite element formulation.

2 REVIEW OF TWO-DIMENSIONAL APPROACHES

Comprehensive modelling of large ice masses requires, simultaneous consideration of mass, momentum and energy principles. However the short term flow behaviour of large ice masses can be studied independently of mass and energy considerations, provided that the spatial distribution of material properties and temperature, and basal boundary conditions are known a priori. Owing to the stress-strain-time behaviour of ice, residual stresses are quickly dissipated, unlike with other geologic materials. Consequently, it is not necessary to trace the entire deformation history when estimating the large scale stress and velocity fields within glaciers or ice sheets. At the same time it is important to recognize that the ice flow parameters may change with time due to strain-induced anisotropy (nonrandom fabric).

During the past several years two-dimensional finite element modelling has enjoyed an increase in popularity in applied glaciology for analysing ice flow. This is attributed to its ability to readily accommodate ice masses of arbitrary geometry with a wide range of material properties and complicated boundary conditions. Two distinct approaches to solving slope creep have been adopted.

2.1 Initial strain method

This procedure is attractive when modelling transient creep problems where ablation-accumulation influences can be neglected [3]. It may also be used to establish steady-state solutions via the method of successive approximations, provided that the boundary conditions and material properties of the problem being analyzed do not change [4]. In this approach, it is generally assumed that the material is elastic under instantaneous loading and creeps thereafter due to gravity. The principle of virtual work can be used to formulate the finite element equivalent of quasi-static equilibrium

$$\int_V \delta\varepsilon_{ij}\sigma_{ij}dV - \int_V \delta u_i b_i dV - \int_{S_t} \delta u_i t_i dS = 0 \qquad (1)$$

where $\delta\varepsilon_{ij}$ are virtual strains which are compatible with virtual displacements δu_i, b_i are body forces in domain V, t_i are surface tractions on part of boundary S_t, and σ_{ij} are stresses obtained by integrating stress rates $\dot{\sigma}_{ij}$ over time

$$\dot{\sigma}_{ij} = D_{ijk\ell}(\dot{\varepsilon}_{k\ell} - \dot{\varepsilon}^c_{k\ell}) \qquad (2)$$

in which $D_{ijk\ell}$ is the elastic constitutive matrix, $\dot{\varepsilon}_{k\ell}$ are total strain rates related to deformation gradients and $\dot{\varepsilon}^c_{k\ell}$ are initial (creep) strain rates. Index notation is assumed with repeated indices implying summation. Integration with respect to time is done incrementally. The unknowns which are being solved during each time interval are displacement increments, thus this method is sometimes referred to as displacement formulation.

When analysing problems where large deformations are expected, attention must be paid to the selection of proper conjugates for the large strain and stress tensors, and proper objective rates must be used for stress and strain rates (see, e.g. Ref. [5]). Further, it is advisable to adopt implicit time-marching schemes if the simulation time period is long or if numerical stability is of concern. The reader is referred to Ref. [6] for details on the initial strain finite element method and discretization.

2.2 Primitive variable

In this method, it is assumed that the influence of elastic strains on stresses is negligible and that ice creep can be modelled as a very viscous non-Newtonian fluid. This approach has been used extensively in the past and has led to steady-state models which were capable of predicting velocities that were in good agreement with measured ones; see, e.g. Ref. [7] and [8].

Since the ice is assumed to be incompressible, the constitutive relationship between total stresses σ_{ij} and rate of deformation d_{ij} is given by

$$\sigma_{ij} = -p\,\delta_{ij} + 2\eta\, d_{ij} \qquad (3)$$

where p is pressure, δ_{ij} is the Kronecker delta and η is a stress and temperature-dependent viscosity.

The equilibrium of the ice mass can be written in integral form by using the principle of virtual velocities

$$\int_V \delta d_{ij}\sigma_{ij}dV - \int_V \delta v_i b_i dV - \int_{S_t} \delta v_i t_i dS = 0 \qquad (4)$$

where δd_{ij} is virtual rate of deformation tensor which is consistent with virtual velocities δv_i and the other terms are the same as defined previously. A second virtual work rate equation is required in order to enforce incompressibility and is of the following form

$$\int_V \delta p\, v_{i,i}\, dV = 0 \qquad (5)$$

where δp is virtual pressure. The unknowns are in terms of velocities and pressure. Thus this method of analysis is sometimes referred to as the velocity-pressure formulation. The pressure degree of freedom may be eliminated via a penalty function approach incorporating reduced integration as discussed in Ref. [6]. While the estimates of the velocity field are generally good, such an approach may yield poor pressure predictions. Since the constitutive laws describing ice flow are generally nonlinear, Newton-Raphson or direct iterative solvers must be used to obtain solutions to boundary-valued problems. It should be noted that sliding at the ice-bedrock interface may be introduced via the boundary integral term in Equation (4) by relating the basal surface shear traction to sliding velocity.

3 THREE DIMENSIONAL MODEL

3.1 Constitutive and Sliding Relationships

As indicated previously, this paper is restricted to linear isotropic rheology. Adopting the usual assumption of incompressible creep, the flow law may be written as

$$S_{ij} = 2\eta\, d_{ij} \qquad (6a)$$

where η is the ice viscosity, and S_{ij} and d_{ij} are the deviatoric stress and rate of deformation tensors, respectively. Since linear rheology is being used we may also write the vertically averaged equation as

$$\bar{S}_{ij} = 2\eta\, \bar{d}_{ij} \tag{6b}$$

where $\overline{S}_{ij}$ and $\overline{d}_{ij}$ are the vertically averaged tensors. For nonlinear problems such vertically averaged relationships are naturally more complex as shown in Ref. [2] for 2-D modelling.

Owing to the temperature-dependence of creep in polythermal cold ice masses and sliding of ice over bedrock for temperate ice masses, high shear gradients tend to develop near to the ice-bedrock interface. It is assumed that these high gradients can be modelled by using a sliding relationship of the form

$$\tau_{b_i} = C\, v_i^b \tag{7}$$

where C is a function of ice temperature, bedrock roughness, properties of ice and shear stress, and τ_{b_i} and v_i^b are basal shear stress and sliding velocity in x_i-direction, respectively.

3.2 Kinematics

The usual approach of relating rate of deformation d_{ij} to velocity v_i is adopted herein,

$$d_{ij} = \frac{1}{2}(v_{i,j} + v_{j,i}). \tag{8}$$

in which $_{,j}$ represents differentiation with respect to x_j, etc.

The vertical axis is taken to be in the x_3-direction (see Figure 1) leaving x_1 and x_2 to define locations on the horizontal plane. Although we wish to retain as much mathematical rigour as possible to ensure proper modelling of the physical phenomenon, for application purposes, simplifying assumptions are introduced in order to develop a working model. Taking into account typical ice mass geometries, it is assumed that $v_{1,3} >> v_{3,1}$ and $v_{2,3} >> v_{3,2}$, yielding

$$\eta\, v_{1,3} = S_{13},\ \eta\, v_{2,3} = S_{23} \tag{9}$$

after substitution of Equation (6) into (8). Equation (9) can be integrated vertically provided that variations of the shear stresses with depth are defined. Based upon experience linear variation of shear with depth is a reasonable assumption

$$S_{13}(\theta) = (1-\theta)\,\tau_{b_1},\ S_{23}(\theta) = (1-\theta)\,\tau_{b_2} \tag{10}$$

where θ represents a dimensionless x_3 co-ordinate ranging from zero at bedrock to 1 at ice mass surface. It should be noted at this time that variations with depth of the other deviatoric stress components is also required when dealing with nonlinear ice rheology. Based on nonlinear 2-D modelling it has been found that a linear variation in the direct deviatoric stresses is appropriate for most problems with limited basal sliding; see, e.g. Ref. [2]. For problems with significant basal sliding, the direct deviatoric stresses are approximately uniform with depth.

Integration of Equation (9) subject to (7) and (10) yields

$$v_1(\theta) = \Omega(\theta)\, \bar{v}_1 \tag{11a}$$

$$v_2(\theta) = \Omega(\theta)\, \bar{v}_2 \tag{11b}$$

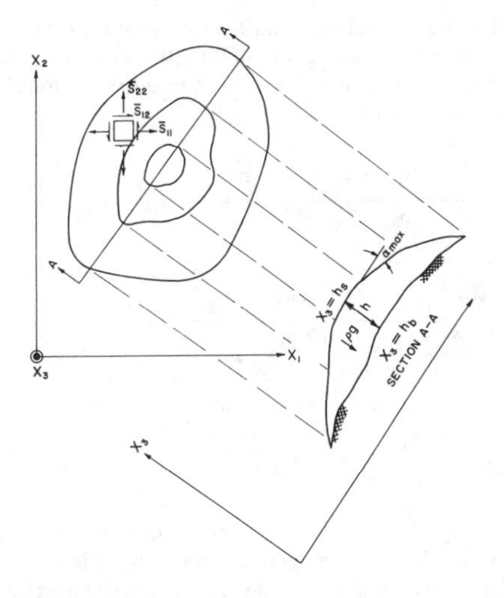

Figure 1. Co-ordinate system for 3-D glacier flow.

where $\bar{v}_1$ and $\bar{v}_2$ are average cross-sectional velocities and $\Omega(\theta)$ represents an interpolation function through the depth of the ice mass,

$$\Omega(\theta) = \left(1 + \frac{Ch}{\eta}\,\theta\left(1 - \frac{\theta}{2}\right)\right) / \left(1 + \frac{Ch}{3\eta}\right) \tag{12}$$

with $h = h_s - h_b$ being the thickness of the ice mass at the section of integration

The relationship between basal sliding velocity v_{b_i} and average velocity $\bar{v}_i$ can be obtained by letting $\theta = 0$. Since average longitudinal velocities vary in the horizontal plane, it is proposed that we may define the horizontal velocity variation within a finite element which lies in the horizontal plane as

$$v_i(x_1, x_2, \theta) = \Omega(x_1, x_2, \theta)\, N_j(x_1, x_2)\, \bar{v}_{i_j} \tag{13}$$

where $N_j(x_1, x_2)$ are interpolation functions corresponding to nodal average velocities $\bar{v}_{i_j}$. The function Ω now also depends upon x_1 and x_2 due to the variation of h in the horizontal plane. Based upon a sensitivity analysis it has been found that terms containing $\partial\Omega/\partial x_1$ and $\partial\Omega/\partial x_2$ may be neglected when evaluating average longitudinal rates of deformation $\bar{d}_{11}$ and $\bar{d}_{22}$, provided that ice mass thickness h does not change rapidly. It should be noted that this assumption is consistent with the shallow ice approximation adopted in this study The average rates of deformation in the horizontal plane are related to average velocity gradients via

$$\bar{d}_{ij} = \frac{1}{2}\int_0^1 \left\{ (\Omega N_k)_{,j} \, d\theta \, \bar{v}_{i_k} + (\Omega N_k)_{,i} \, \bar{v}_{j_k} \right\} d\theta \quad (14)$$

for i,j = 1,2. Further d_{33} can be evaluated by using the $d_{ii} = 0$ identity.

3.3 Equilibrium

Following an approach similar to that of Paterson [9] for studying the variations of longitudinal stresses along a glacier, we may express three-dimensional stress equilibrium as

$$\frac{\partial[h\,(2\,\bar{S}_{11} + \bar{S}_{22})]}{\partial x_1} + \frac{\partial[h\,\bar{S}_{12}]}{\partial x_2} = \tau_{b_1} - f_1 \quad (15a)$$

$$\frac{\partial[h\,\bar{S}_{12}]}{\partial x_1} + \frac{\partial[h\,(2\,\bar{S}_{11} + \bar{S}_{22})]}{\partial x_2} = \tau_{b_2} - f_2 \quad (15b)$$

with

$$h\,\bar{S}_{ij} = \int_{h_s}^{h_b} S_{ij}\,dx_3 = 2\eta \int_{h_b}^{h_s} d_{ij}\,dx_3 = 2\eta\,h\,\bar{d}_{ij} \quad (15c)$$

where h_b and h_s are bedrock and surface coordinates respectively, f_i are components of $\rho g h\,\alpha_{max}$ in x_i-directions, ρg is gravity loading ,α_{max} is the maximum angle which the surface makes with the horizontal plane and the other terms are the same as defined previously.

When developing Equation (15) the following assumptions are made: top surface is traction free; top and bottom boundaries do not deviate strongly from one another; ice is incompressible and isotropic; density is uniform with depth; and the force responsible for creep is along direction of maximum surface slope. For purposes of this study it is also assumed that the ice mass is isothermal and in steady state and that the short term influences of environmental changes at boundaries can be neglected.

The integrated equivalents for Equations (15a) and (15b) are obtained by multiplying these equations with weighting functions $\delta\bar{v}_1$ and $\delta\bar{v}_2$, respectively, which are consistent with the average longitudinal velocity fields, and then integrating by parts in order to achieve symmetry of the stiffness matrix,

$$\int_A \left\{ \delta\,\bar{v}_{1_{,1}}\,h\,(2\,\bar{S}_{11} + \bar{S}_{22}) + \delta\,\bar{v}_{1_{,2}}\,h\,\bar{S}_{12} \right\} dA + \int_A \delta\,\bar{v}_1 \tau_{b_1}\,dA = \int_A \delta\,\bar{v}_1 f_1\,dA \quad (16a)$$

$$\int_A \left\{ \delta\,\bar{v}_{2_{,1}}\,h\,\bar{S}_{12} + \bar{v}_{2_{,2}}\,h\,(2\,\bar{S}_{22} + \bar{S}_{11}) \right\} dA + \int_A \delta\,\bar{v}_2 \tau_{b_2}\,dA = \int_A \delta\,\bar{v}_2 f_2\,dA \quad (16b)$$

It should be noted that the boundary integrals have been neglected in Equation (16) since it is assumed that ice sheet thickness at the margins reduces to zero. After appropriate substitution of the interpolation functions into Equation (16), we may write the discretized form for equilibrium as

$$[\mathbf{K} + \mathbf{K}_\tau]\,\mathbf{u} = \mathbf{F} \quad (17a)$$

$$\left.\begin{aligned} \mathbf{K} &= \int_A \mathbf{B}^T h\,\mathbf{D}\,\mathbf{B}\,dA \\ \mathbf{K}_\tau &= \int_A \mathbf{N}^T\,\bar{\mathbf{C}}\,\mathbf{N}\,dA \\ \mathbf{F} &= \int_A \mathbf{N}^T \begin{Bmatrix} f_1 \\ f_2 \end{Bmatrix} dA \end{aligned}\right\} \quad (17b)$$

$$\left.\begin{aligned} \mathbf{u} &= \langle \bar{v}_{1_1}, \bar{v}_{2_1}, \bar{v}_{1_2}, \bar{v}_{2_2} \ldots \rangle^T \\ \mathbf{N} &= \begin{bmatrix} N_1 & 0 & N_2 & 0 & \ldots \\ 0 & N_1 & 0 & N_2 & \ldots \end{bmatrix} \end{aligned}\right\} \quad (17c)$$

$$\mathbf{B} = \begin{bmatrix} \frac{\partial}{\partial x_1} & 0 \\ 0 & \frac{\partial}{\partial x_2} \\ \frac{\partial}{\partial x_2} & \frac{\partial}{\partial x_1} \end{bmatrix} \begin{bmatrix} N_1 & 0 & N_2 & 0 & \ldots \\ 0 & N_1 & 0 & N_2 & \ldots \end{bmatrix} \quad (17d)$$

$$\mathbf{D} = 2\eta \begin{bmatrix} 2 & 1 & 0 \\ 1 & 2 & 0 \\ 0 & 0 & 1/2 \end{bmatrix} \qquad \bar{C} = C\left(1 + \frac{Ch}{3\eta}\right)^{-1} \quad (17e)$$

The matrix **K** is associated with the longitudinal straining within the ice mass while $\mathbf{K}_\tau$ is associated with basal shear. It should be noted that Equation (17) is similar in construction to the equations which describe stress equilibrium for planar problems. Consequently, finite element codes for studying planar problems can be easily modified to study three-dimensional ice flow. For details on the discretization of Equation (17) the reader is again referred to Ref. [6].

Low order 4-noded isoparametric elements were adopted for the simulations in this study. However, it should be noted that with use of higher order interpolation it should be possible to set up a curvilinear coordinate system within the isoparametric framework, thereby allowing one to take into account large scale curvature influences on equilibrium. This approach has been successful when degenerating the 20 node brick element to a form which is suitable for studying thick shell problems [10].

Through a minor modification of the load vector, creep of ice shelves may also be studied. While the emphasis in this paper has been on ice flow, the model in nonlinear form may also be used to study lava and mud flow phenomena. At this point it should be stressed that the model presented herein is applicable to large scale flow and by no means can one expect to obtain the same details provided by the more general finite element models. The main advantage of the approach presented in this study rests in reduced computational and data input efforts which are required for effecting solutions to glacier and ice sheet flow problems requiring 3-D analysis.

4. NUMERICAL EXAMPLES

Two idealized examples are presented to demonstrate the ability of the three dimensional model presented in this paper to simulate large scale ice creep. The geometries and grids for planar and axisymmetric ice masses, as viewed from above, are given in Figures 2(a) and 2(b), respectively. Identical profile geometry as shown in Figure 2(c) was used for both examples. Since these examples may be reduced to a two-dimensional form for analysis, it was possible to compare the solutions provided by the 3-D model with those generated by using a linear elastic 2-D 8-noded isoparametric finite element model. Owing to similarities between the linear flow law and the constitutive equation for elasticity, the ice viscosity (1500 kPa.yr) was introduced through the elastic shear modulus while incompressibility was approximated by using Poisson's Ratio of 0.49995. To avoid ill-conditioning of the matrix equation due to near incompressibility, 2×2 reduced integration was adopted. Basal sliding was suppressed in the 3-D model by letting C = 1000 kPa.yr.m^{-1}.

Figures 3 to 5 show predictions of horizontal surface velocity, vertical surface velocity and basal shear stress variations along the ice masses, respectively. It is clearly demonstrated that reasonably good agreement of solutions is achieved between the two- and three-dimensional models. It should be noted that the oscillations in the solutions given by the 8-noded isoparametric model are numerical and will disappear if a finer finite element mesh is used. Although 120 elements were used for the axisymmetric simulation via the 3-D model, which allowed similar node spacing in radial direction as was used for 2-D model, a grid-sensitivity analysis suggests that a coarser discretization would also give reasonable results. Of course one would generally not use the 3-D model for solving problems where a 2-D model is appropriate.

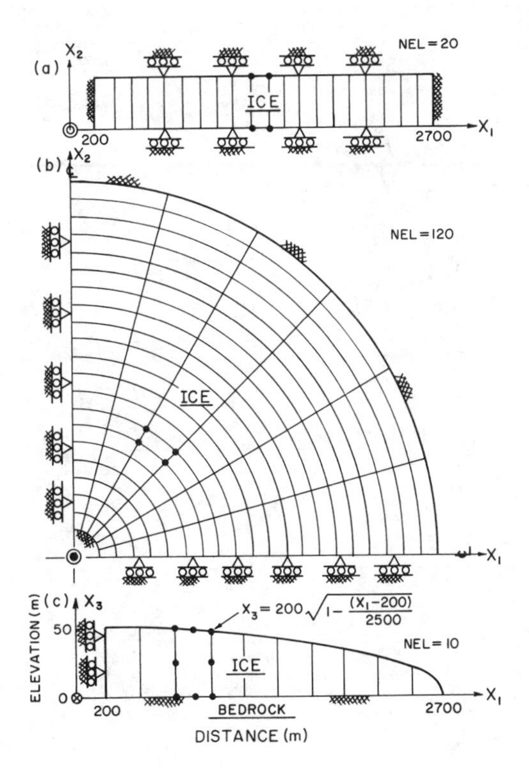

Figure 2. Finite element grids for (a) 3-D model plane strain simulation, (b) 3-D model axisymmetric simulation, and (c) 2-D isoparametric model simulation.

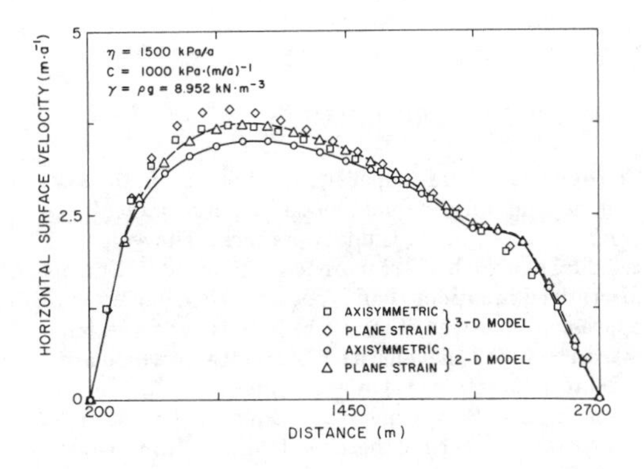

Figure 3. Horizontal surface velocity variation.

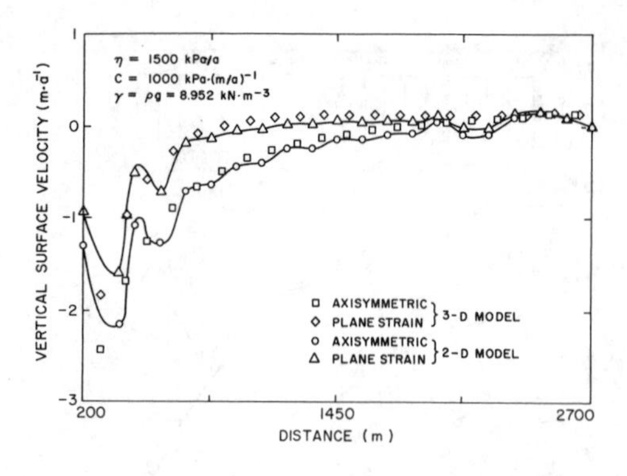

Figure 4. Vertical surface velocity variation.

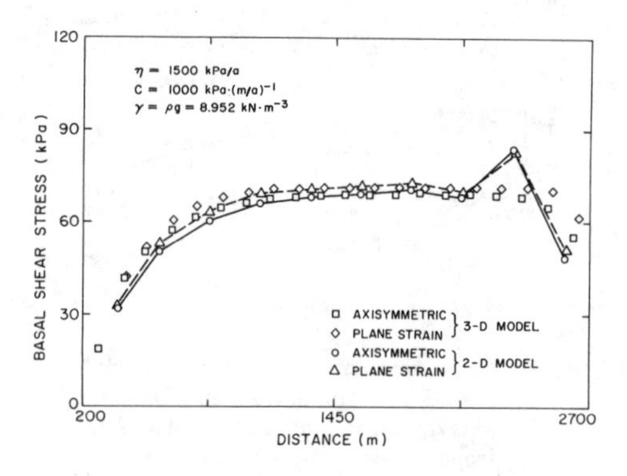

Figure 5. Basal shear stress variation.

5 CONCLUDING REMARKS

The emphasis in this paper has been placed on three-dimensional finite element modelling of large scale ice creep via a weighted residual approach. The weighted residual approach offers more flexibility for setting up discretized equations than is possible via a variational approach. However, it should be noted that Equation (4) can be used to develop a set of equations which are similar to Equation (17) by enforcing $\delta d_{ii} = 0$, observing that $S_{ii} = 0$ and using Equation (13) as an interpolation function; thereby yielding a Kantorovich type of approximation.

The advantages of the model proposed herein over the standard initial strain and primitive variable finite element approaches are the two-dimensional construction of finite element mesh, fewer nodal unknowns and the ability to give reasonable predictions when using coarse grids. Consequently, the computational and data preparation efforts are substantially reduced.

Of course the approximations involved in the reduction of the 3-D problem to 2-D form limits the types of problems which may be handled by the model proposed herein. If details on a small scale are important or if the geometry of the problem does not fit within the framework of shallow-ice approximation, a standard finite element, or perhaps boundary element, procedure must be used.

ACKNOWLEDGEMENTS

The financial support for this study provided by the Natural Sciences and Engineering Research Council of Canada is gratefully acknowledged.

REFERENCES

[1] H. Hutter, Theoretical Glaciology, D. Reidel Publishing Company, 1983

[2] D.F.E. Stolle, Two-dimensional line element for glacier flow problems, Engineering Analysis, 3, 161-165 (1986)

[3] D.F.E. Stolle and F.A. Mirza, Large ice mass surging via ice-bedrock interface mobilization, in A.P.S. Selvaduri and G.Z. Voyiadjis (eds.), Mechanics of Material Interfaces, Elsevier, 173-188 (1986)

[4] J.J. Emery, Simulation of ice creep, in B. Voight (ed.), Rockslides and Avalanches, Vol. 1, Natural Phenomena, Elsevier Scientific Publishing Company, 669-691 (1978)

[5] L.E. Malvern, Introduction to Mechanics of Continuous Medium, Prentice-Hall Inc. 1969

[6] O.C. Zienkiewicz, The Finite Element Method, Third Edition. Maidenhead, McGraw Hill, 1977

[7] R.L. Hooke, C.F. Raymond, R.L. Hotchkiss and R.J. Gustafson, Calculations of velocity and temperature in polar glacier using the finite element method, Journal of Glaciology, 24, 131-146 (1979)

[8] D.F.E. Stolle and M.S. Killeavy, Determination of particle paths using the finite element method, Journal of Glaciology 32, 219-223 (1986)

[9] W.S.B. Paterson, The physics of glaciers, Pergamon Press, 1969

[10] S. Ahmad, B.M. Irons and O.C. Zienkiewicz, Analysis of thick and thin shell structure by curved finite elements, International Journal for Numerical Methods in Engineering 2, 419-451 (1970)

Numerical Methods in Geomechanics (Innsbruck 1988), Swoboda (ed.)
© 1988 Balkema, Rotterdam. ISBN 90 6191 809 X

Numerical models of idealized ice shelf flow

M.A.Lange
Alfred Wegener Institute for Polar and Marine Research, Bremerhaven, FR Germany

D.R.MacAyeal
University of Chicago, Ill., USA

ABSTRACT: We present numerical model results on the flow of idealized, i.e. geometrically simple, fully specified ice shelves under a variety of boundary conditions. Using a time-dependent finite- element model, we construct their steady- state ice thickness- and velocity distributions by solving mass continuity, stress equilibrium and constitutive relations. The numerical domain is represented by rectangular elements of constant size. We discuss the following set of simulations: (i) one-ice stream models and (ii) two-ice stream models including a central ice rise, both with zero and constant ice stream inflow. The model results yield useful insight into the dynamics of natural ice shelves. They also yield estimates on minimum requirements for field measurements, which will allow detection of time- dependent ice stream flow rates.

1 INTRODUCTION

Ice shelves represent a major element in the mass balance of the Antarctic ice sheet. Bottom melting of- and iceberg calving from the major ice shelves represent the prime negative budget quantities of Antarctic ice. Because of their extensive catchment areas, ice shelves serve as sensitive indicators of possible, climatically induced long-term changes of the ice sheet as a whole. This requires assessment and monitoring of basic characteristics of ice shelves by field measurements.

However, ice shelves might be subject to time dependent changes in their boundary conditions on a shorter time scale, such as variations in ice stream influx. This might mask the effect of climatically induced alterations of the ice. In order to distinguish between long-term and short-term changes in its flow field, one needs to understand to what degree measureable quantities of an ice shelf such as ice thicknesses or ice velocities are affected by the different forcings. This requires the capability of predicting time dependent ice shelf flow as a function of varying boundary conditions.

However, ice shelves, even in their most simple form, represent highly non-linear systems, which cannot be adequately described by analytical methods. This is even more true, when time dependent boundary conditions are taken into consideration. Numerical simulations provide the as yet only meaningful tool to tackle these problems. However, while numerical models allow the simulation of ice shelf flow in response to specific, 'free' parameters, it is often difficult to adequately describe and represent the essential characteristics of a natural ice shelf (e.g. , its exact configuration) due to a lack of sufficient boundary conditions. This frequently renders the model results less reliable if not useless. Thus, the theoretical glaciologist is faced with the dilemma of having to use numerical simulations as the only feasible means of representing ice shelf flow on a natural scale, while on the other hand lacking appropriate data to control and verify his model results. At this point, we decided to use a different approach, by considering simple, well defined, 'ideal' ice shelves instead of modelling 'real' ice shelves. In doing this, we hope to distinguish the effects of particular boundary conditions on ice shelf flow and to better understand the dynamics of an ice shelf by gradually considering models with increasing complexity.

In this paper, we will present the first steps of this approach, where we have looked into the changes in the flow field of idealized ice shelves with minimum and maximum ice stream(s) inflow.

2 MODELLING TECHNIQUES

2.1 General

We compute steady-state ice shelf thickness and

velocity distributions using a time dependent finite- element nodel described by (2). The model solves mass continuity, stress equilibrium and constitutive equantions to determine the evolution of ice thickness H and horizontal velocity $\underline{u}$ (assumed independent of depth) under prescribed ice- volume flux, velocity and stress boundary conditions. We do not treat the thermodynamics of the ice shelf, i.e. its temperature- depth distribution, and its effect on ice rheology and use a constant stiffness parameter B in each model. To investigate the importance of ice rheology, we performed simulations with different values of the stiffness parameter, as well as with linear (Newtonian) and exponential ($n=3$) rheology. We do not account for firn densification and use a constant density for the entire domain ($=917$ kg/m^3). Except for ice stream inlets, ice velocities and volume fluxes along the inland ice boundaries are held at zero. Outward flux through the ice shelf front is constrained to balance the seaward advection. We run the models, starting with an arbitrary thickness distribution, until they reach steady state ice thicknesses after 1500 a (=annum) and analyze their thickness and velocity field at that stage.

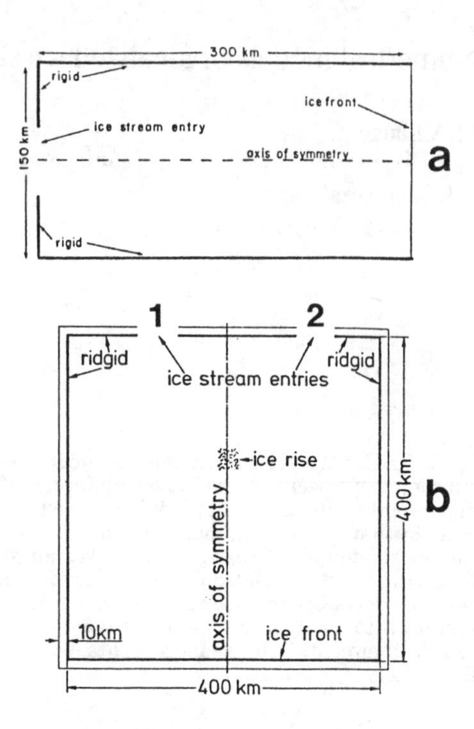

Fig. 1 Configurations of the one- ice stream- (a) and two- ice stream models. In b, the entire numerical domain, including a 10 km boundary, which is required in our calculations, is shown and will be displayed also in subsequent figures of results. The nominal domain is the 400 x 400 km box shown in heavy lines.

2.2 Model configuration

We use two different model configurations (Figure 1a, b) with one and two ice streams (in the following called model A and B, respectively; ice stream width = 50 km). Model A is fully symmetrical with respect to the centerline of the ice shelf, while in model B ice stream 2 is positioned closer to the ice shelf boundary than ice stream 1. In model B, an ice rise (20 x 20 km) symmetric to the centerline is added, to study its effect on ice shelf flow. We do not allow any changes in ice shelf configurations with time (i.e., no additional grounding, retreat of advance of the ice shelf). The numerical domain is represented by rectangular elements of 5 x 10 and 10 x 10 km for model A and B, respectively. We use a constant snow accumulation rate of 0.25 m/a (ice equivalent) and 0.1 m/a for model A and B, respectively and a 0 m/a bottom melting rate. The ice stream volume fluxes are fixed at either zero or 40 km^3/a (corresponding to 0 or 1000 m/a ice velocity, respectively). We are considering two different rheologies with the nominal value for $B = 1.6 \times 10^8$ Pa s$^{1/3}$ as well as $B = 0.8 \times 10^8$ Pa s$^{1/3}$ (model 1 and 2, respectively). For model A, we also carried out simulations with Newtonian rheology (model 3).

2.3 Ice thickness anomaly envelopes

Ice shelf thickness variations produced by ice stream discharge fluctuations that are perpetually bounded between extreme limits are also bounded. Instantaneous thickness deviations from the long term mean, referred to here as ice thickness anomalies, will thus fall within an envelope determinded by the maximum and minimum thickness attainable from ice stream forcing alone. In cases where particular details of ice stream transience are unknown, the spatial distribution of this envelope provides the best available measure of ice shelf response. This measure may be useful, e.g., to distinguish regions susceptible to changes in ice thickness caused by transient ice stream forcing from those affected solely by atmospheric and oceanic conditions. In the present study, we construct ice thickness anomaly envelopes from two steady state ice thickness distributions associated with the above given maximum and minimum ice stream discharge, respectively.

3 RESULTS AND DISCUSSION

Figure 2 gives thickness anomaly envelopes for the three one-ice stream models with varying rheology. Using model 1 ($B = 1.6 \times 10^8$ Pa s $^{1/3}$) as the baseline simulation, it is seen that a reduction of the stiffness parameter B by a factor of two (this corresponds to a warming of the ice by 12° C, i.e., an extreme climatic change; model 2) firstly results in an accelerated thinning of the ice shelf and restricts the region of greatest anomaly envelope amplitude closer to the ice stream outlet. With more ductile ice, ice thickness perturbations forced by the ice stream cannot persist as far downstream, because ice thickness gradients relax significantly faster by horizontal spreading. This result suggests that, if a climatic warming were to occur, ice stream influences on transient ice shelf phenomena will deminish relative to other oceanic or atmospheric influences.

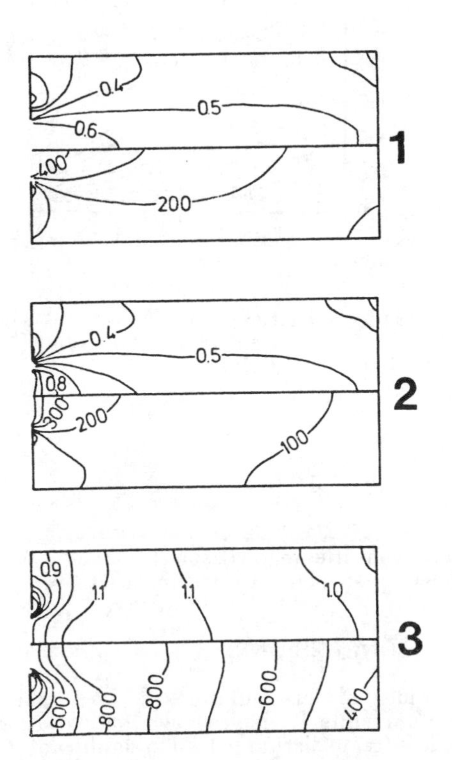

Fig. 2 Absolute (upper panels, in meters) and relative (lower panels) ice thickness anomaly envelopes for the one- ice stream models. 1 and 2 are models with exponential rheologies and 3 is a model with Newtonian rheology.

The effects of a Newtonian rheology as proposed by [1] is demonstrated by comparing model 1 and model 3. While we do not intend to support or refute the idea of a linearly viscous ice shelf, our results demonstrate significant qualitative differences, which should be observable by field measurements. The envelope amplitudes of model 3 are clearly different from those of models 1 and 2 in that their range is maximum downstream of the ice stream outlet and in that they are more uniformly distributed across the ice shelf widths. These differences are likely caused by high strain rates in model 3 as a result of greater friction generated at the margins of the ice shelf. With greater friction, greater ice thickness is required within the mid section of the ice shelf to produce the ice front volume flux needed to balance ice stream input.

Figure 3 gives the relative ice thickness anomalies for the baseline model (1) and the model with reduced ice stiffness (2) for the two-ice stream- ice rise model (model B). There is little qualitative difference between the one- ice stream- and the two- ice stream model. Again, reduced ice stiffness results in larger anomaly amplitudes, which are found directly at the ice stream outlet in model 2. Comparison between the two models for the maximum ice stream discharge case (1-2) demonstrates the significant thinning of the softer ice for otherwise identical conditions. This is also seen primarily directly at the ice stream outlets. Thus, changes in ice stream input will be observable only close to the entry of the ice streams in a warmer (i.e., softer) ice shelf, while further downstream the effects diminish.

The ice rise is seen only marginally in the ice thickness anomaly envelopes. Hence, ice stream transience will result in only limited ice thickness anomalies around restrictions to ice shelf flow. In contrast, the ice velocities around an ice rise will react much more strongly to variations in ice stream flow (Figure 4). While in the minimum- discharge case (a) the ice rise affects the ice flow only minimally, 'switching-on' the ice stream (b) produces a significant signal in the velocity field. Thus, variations in ice stream discharge could be monitored around an existing ice rise by observing changes in the adjacent ice velocities.

The closer proximity of ice stream 2 to the ice shelf boundary is clearly seen in the asymmetry in the velocity field (Fig. 4) and to a lesser degree in the ice thickness anomalies (Fig. 3). here again, the restriction to the ice shelf flow imposed on ice stream 2 by the rigid ice shelf boundary produces a stronger signal in response to ice stream transience in the velocity field compared to the ice thickness anomaly envelopes. Thus, partial grounding of an ice shelf, which results in changes of the distance between ice stream entrance and ice shelf boundary could be observed by a detailed survey

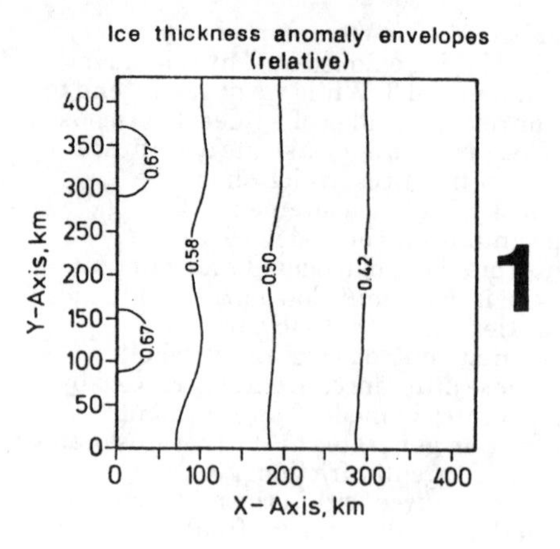

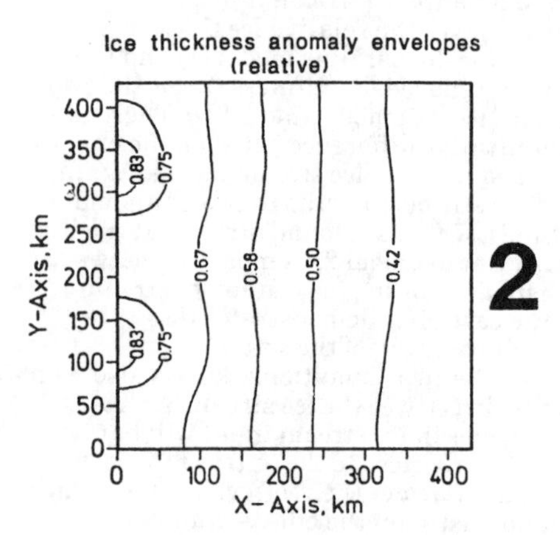

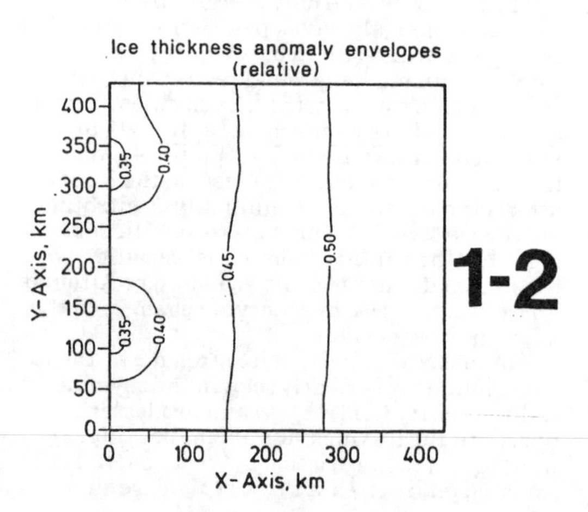

Fig. 3 Relative ice thickness anomaly envelopes for the two- ice stream models. 1 and 2 are models with exponential rheologies but reduced stiffnes parameter [2], 1-2 is a comparision between model 1 and 2 for the case of maximum ice stream discharge.

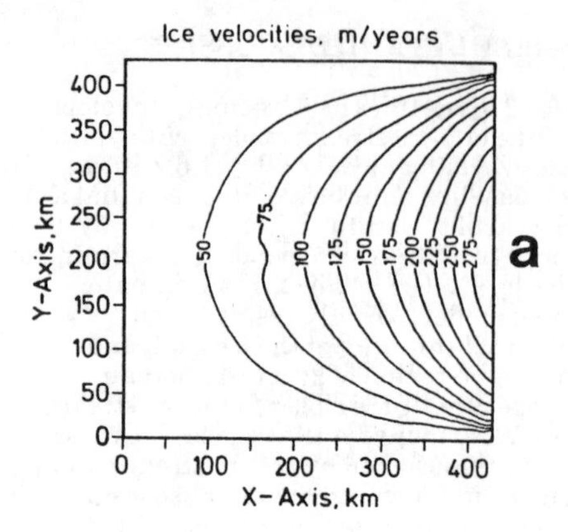

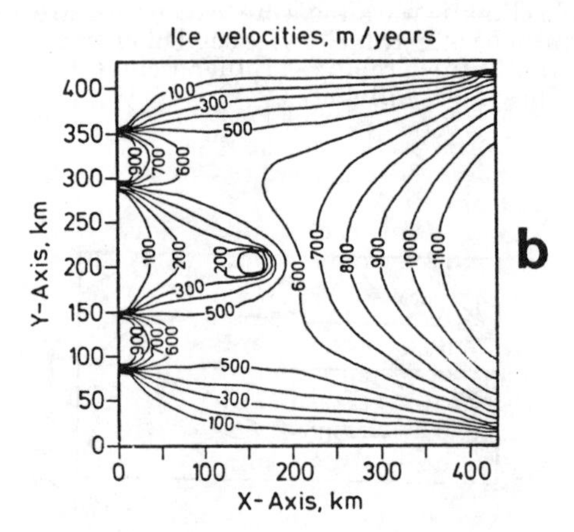

Fig. 4 Ice velocities of two- ice stream model (1) for minimum (a) and maximum (b) ice stream discharge.

of ice velocities (e.g. via satellite-imagery) downstream of the ice stream outlet.

4 CONCLUSIONS

Our ideal model simulations yield a number of useful results. We have shown that variations in ice stream discharge lead to significant changes in both the ice thickness- and the ice velocity field. Ice thicknesses anomaly envelopes, which are constructed based on steady- state ice thicknesses of models with minimum and maximum ice stream discharge bracket possible effects of ice stream transience on ice thickness. Using these anomaly envelopes, it is seen that ice rheologies influence significantly the changes in ice thicknesses. While this alone is well known, our models results allow quantification of this statement. Particularly notable is the comparison between exponential and linear ice rheology [1] (Fig.2), which should be observable by field measurements. Thus, given that changes in ice stream discharge occur, the pattern of ice thickness variations will shed light on the question of the prevailing rheology of an ice shelf under consideration.

Two- ice stream models (Fig. 3) produce qualitatively similar results compared with one- ice stream models. Using these simultations it is demonstrated that ice rises will be particularly sensitive indicators of ice stream transience when adjacent ice velocities are monitored. The ice thickness pattern around the ice rise will in contrast change only marginally in response to ice stream discharge fluctuations. This should be taken into account in field programs, where measurements around ice rises might be considered particularly suitable to indicate changing ice stream discharge.

Based on our results, we can estimate minimum requirements for field data in order for them to show the effect of ice stream transience on ice thicknesses and ice velocities. Variations in ice stream inflow on a timescale of a 300a between minimum and maximum discharge [3] will result in ice thickness changes, which range from 1 m/a to 2.5 m/a (close to the ice stream entry, or somewhat downstream, depending on ice rheology; cf. Fig. 2) as a maximum, to 0.3 m/a to 1.3 m/a (at the ice front) as a minimum. Thus, measurements of ice thicknesses (e.g., via remotely obtained ice shelf elevations) have to be at least of this accuracy, when observations are done yearly. Ice velocities vary in response to ice stream transience from <50 to 1000 m/a at the ice stream entrance to 275 to 1100 m/a at the ice front. Taking again the 300a time frame for the discharge variation into account, a maximum yearly change in ice velocities from between 3 m/a to 2.75 m/a has to be detected.

Acknowledgements: This research was supported by the Alfred-Wegener-Institute for Polar and Marine Research (Contr. No 81), the National Science foundation (DPP 85 09451) and the NATO Scientific Affairs Division (27-854/85).

REFERENCES

[1] Doake, C. S. M. and E. W. Wolff, Flow law in polar ice sheets, Nature, 314, 255-257, 1985.

[2] MacAyeal, D. R. and R. H. Thomas, The effects of basal melting on the present flow of the ross ice shelf, antarctica, J. Glaciology, 32, 72-86.

[3] Shabtaie, S. and C. R. Bently, West antarctic ice stream draining into the ross ice shelf: configuration and mass balance, J. Geophys. Res., 92, 1311-1336, 1987.

Numerical Methods in Geomechanics (Innsbruck 1988), Swoboda (ed.)
© 1988 Balkema, Rotterdam. ISBN 90 6191 809 X

Triaxial creep properties of frozen soil and their measurement

Song Ling
Northeast University of Technology, Shenyang, People's Republic of China
Zhang Xiang Dong
Fuxin Mining Institute, People's Republic of China

ABSTRACT: A 'low temperature box —— triaxial cell' testing equipment system have been made for carrying on compression strength and creep test in conditions of triaxial stresses. In the course of the testing the authors adopted the methods of artificially making specimens of frozen clay and used them to carry out the long-term creep tests at triaxial stress states so that the creep curves of the frozen clay were derived. According to the test results, three types of nonlinear creep equations are proposed to describe the nonlinear creep behaviour of frozen clay at different temperatures.

1. INTRODUCTION

Today investigating the mechanical properties dependent on time has become one of the most extensive and brisk topics in the field of cryoscopy, It is extremely important to make a study of the rheological property of frozen soil for defining the bearing capacity and the deformation of the frozen wall. Frozen soil is a complicated multiphase system. Its rheological characters are influenced by a lot of factors —— tempreture, soil property, ice (water) content and loads exerted on the frozen wall, for example. Up to the present time the rheological characters of forzen soil haven't been recognized wholly and integrally.

In the course of the experiment, the 'low temperature box —— triaxial cell' testing equipment system home-made by ourselves was used for carrying out creep testing up on the frozen clay in conditions of the triaxial stresses. The specimens of the frozen caly were made artificially.

Besides introdution of the light-duty testing equipment system in this paper, the points are put on the creep characters of forzen clay at triaxial stresses and method of measuring creep parameters.

2. TESTING EQUIPMENT SYSTEM

The 'low temperature box---triaxial cell' testing equipment system is composed of three parts (Fig.1).

2.1 Freezing part

The testing temperature are regulated by low temperature box at the lowest controlling temperature -40°C and controlling error ± 1°C.

2.2 Loading part

The triaxial cell is one of the main loading devices. It can be used to put axial load and confining pressure on the cylindrical specimen of ϕ50x100mm. Both axial force and confining pressure are produced by hydraulic power. Two oil circuits are used for joining the oil pump and hydraulic valves to the triaxial cell. The triaxial cell is designed to apply the axial force of up to 12T and confining pressureo of up to 20MPa. In order to carry on triaxial creep testing, the open-loop system of controlling by electrohydraulic proportional pilated pressure valve was used to regulate the oil pressure steplessly and stabilize it for a long time.

The open--loop system is showed in Fig.2

2.3 Measuring part

The cell's design also facilitates the measurement of axial force, confining pressure and strain under the conditions of triaxial stresses. Both axial force and confining pressure acted on the specimen are measured by pressure transducers and finally noted at the X-Y function recorder.

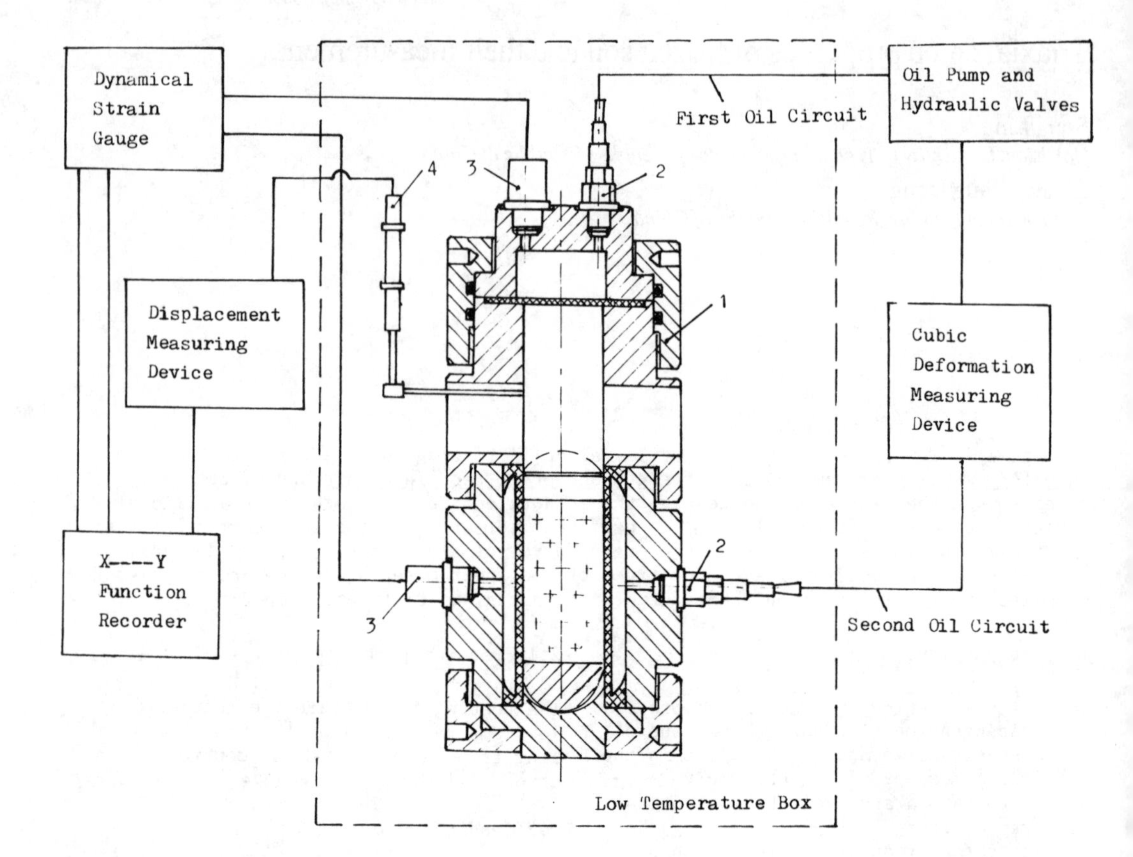

1. Triaxial cell
2. Fast adaptor
3. Pressure transducer
4. Displacement transducer

Fig. 1. Triaxial compressive testing equipment system

The axial deformations of specimen are determined by a displacement measuring device, and its creep curves are drawn on the function recorder. Besides the axial strains, the radial strains of specimen are also produced on the basis of triaxial stresses. It is very difficult to measure the radial strain on the specimen surface because forzen soil is different from other materials to varying degrees. The radial strain is very large in number and non-uniform when failure occurs. Its values may be calculated according to the formula (1), because the cubic strains of specimen are measured by using a cylinder.

$$\varepsilon_2 = \varepsilon_3 = (\varepsilon_v - \varepsilon_1)/2 \tag{1}$$

where ε_1 is axial strain of specimen, ε_v is cubic strain, ε_2 and ε_3 are average value of radial strain.

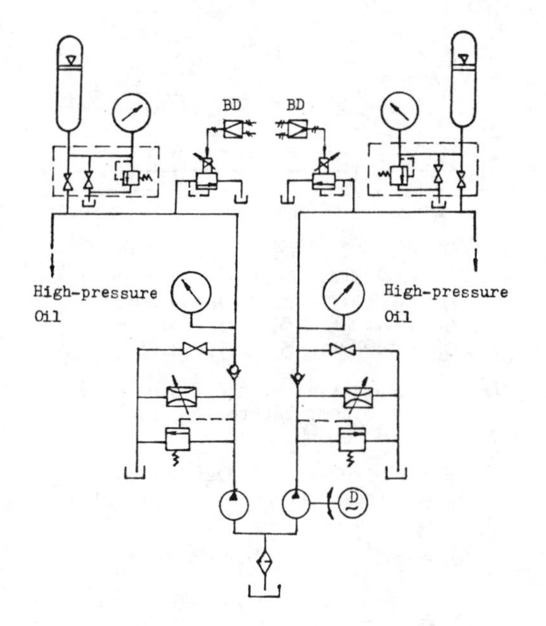

Fig. 2 Open-loop system controlled by means of electro-hydraulic proportional pilated pressure valve

3. TESTING CONDITIONS AND RESULTS

The frozen clay specimens was artificially made in the laboratory in order to carry on the traxial compression creep test. They were compacted to a dry density of 1.78 t/m^3, with moisture 27%. The method of of fast unidirectionally freezing was adopted to make frozen clay specimens, so that the crystal ice was fairly well-distributed in them and not to form a large ice-crystal.

The triaxial creep test had been carried out at different temperture (T=-5, -10, -14, -18°C) and in different states. The axial force and confining pressure were chosen on the basis of the real loading conditions of frozen wall. In the course of the creep testing, the mean normal stresses acted on all specimens are equal, but deviation stresses are different for each specimen in the condition of the same temperature.

The creep curves are illustrated in Fig. 3 to 6.

4. DEDUCING OF CREEP EQVATIONS

In order to fit perfectly to testing creep curves, some types of nonlinear creep equations have al-ready, been deduced to describe the creep behaviour of frozen soil in different conditions. In the course of deducing, the hypothesis was adopted, i.e. the cubic strain of frozen clay is nearly to zero (ε_v=0), it can be neglected so that creep equations are formed of deviation stress tensor (S_{ij}) and deviation strain tensor (e_{ij}). This hypothesis is suitable for forzen clay, which has been proved by tests.

4.1. Depenaence of deviation stress tensor and deviation strain tensor

At axial stress the creep equation can be shown as followed.

$$\varepsilon = A\sigma^B \varphi(t)$$

or

$$\dot{\varepsilon} = A\sigma^B \varphi'(t) \qquad (2)$$

In the conditions of triaxial stresses, the equation (3) is adopted. It is a simplified creep equation which was recommended by F.K.G. odquist and J. Hult.

$$\dot{\varepsilon}_{ij} = f_1(J_2) S_{ij} \qquad (3)$$

where $\dot{\varepsilon}_{ij}$ is a strain rate tensor, S_{ij} is a deviation stress tensor and J_2 is second invariant of deviation stresses,

$f_1(J_2)$ is a fuction which reflects the nonliear relations between deviation stresses, time and strain rates. The time factor in it can be separated for the sake of contrasting with equation (2), so that the equation (3) may be written as:

$$\dot{\varepsilon}_{ij} = f_2(J_2) S_{ij}\, \varphi'(t)$$

or

$$\varepsilon_{ij} = f_2(J_2) S_{ij}\, \varphi(t) \qquad (4)$$

where $f_2(J_2)$ only reflects the nonliear relations between invariant of deviation stressea and strain rates. Its specific form can be obtained through contrasting with axial creep function.

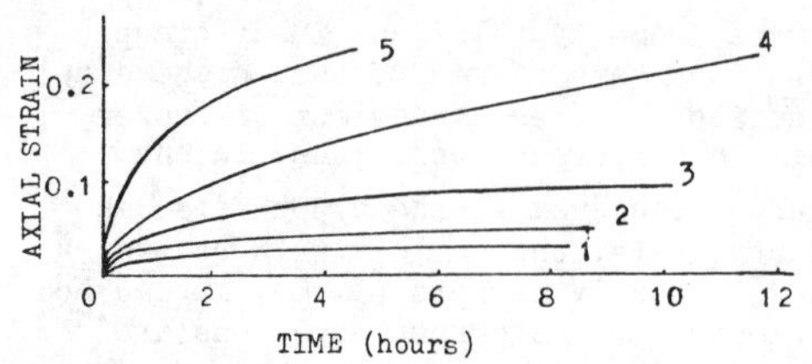

1. σ_1 = 5.5MPa, σ_2 = σ_3 = 3.25MPa;
2. σ_1 = 6MPa, σ_2 = σ_3 =3MPa;
3. σ_1 = 6.5MPa, σ_2 = σ_3 =2.75MPa;
4. σ_1 = 7MPa, σ_2 = σ_3 =2.5MPa;
5. σ_1 =7.5MPa, σ_2 = σ_3 =2.25MPa

Fig. 3 Creep curves of the frozen clay at the temperature T=-18°C

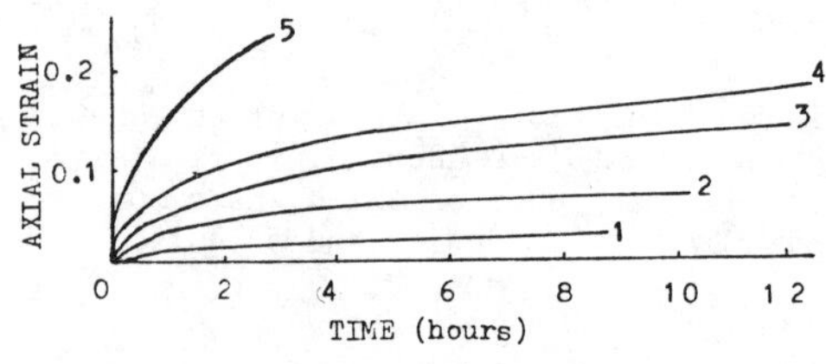

1. σ_1 = 5MPa, σ_2= σ_3 =3.5MPa;
2. σ_1 =5.5MPa, σ_2= σ_3 =3.25MPa;
3. σ_1 =6MPa, σ_2= σ_3 =3MPa;
4. σ_1 =6.5MPa, σ_2= σ_3 =2.75MPa;
5. σ_1 =7MPa, σ_2= σ_3 =2.5MPa

Fig. 4 Creep curves of the frozen clay at the temperature T=-14°C

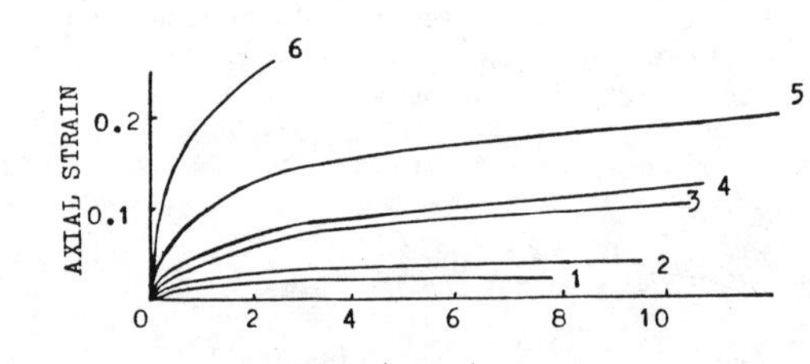

1. σ_1 = 4.5MPa, σ_2 = σ_3 = 3.75MPa;
2. σ_1 = 5MPa, σ_2 = σ_3 =3.5MPa;
3. σ_1 = 5.5MPa, σ_2= σ_3= 3.25MPa;
4. σ_1 = 5.2MPa, σ_2= σ_3= 2.85MPa;
5. σ_1 = 6MPa, σ_2= σ_3= 3MPa;
6. σ_1 = 6.5MPa, σ_2= σ_3= 2.75MPa

Fig. 5 Creep curves of the frozen clay at the tempeature T=-10°C

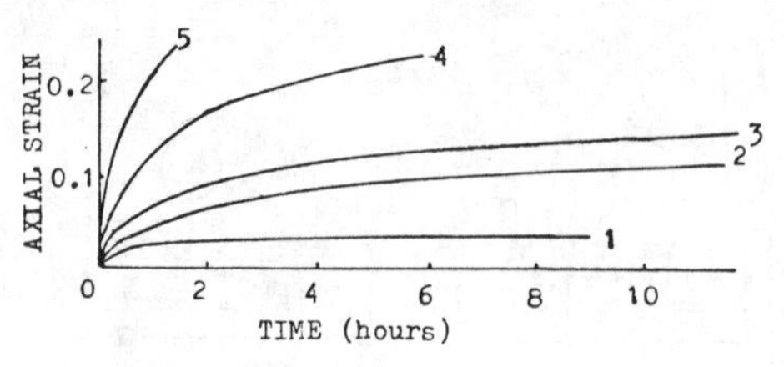

1. σ_1 = 4.5MPa, σ_2 = σ_3 = 3.75MPa;
2. σ_1 =4.75MPa, σ_2 = σ_3 = 3.625MPa;
3. σ_1 = 5MPa, σ_2 = σ_3 = 3.5MPa;
4. σ_1 = 5.25MPa, σ_2 = σ_3 = 3.37MPa;
5. σ_1 = 5.5MPa, σ_2 = σ_3 = 3.25MPa

Fig. 6 Creep curves of the frozen clay at the temperature T=-5°C.

At axial stress, the invariant J_2, deviation stress tensor e and strain tensor ε_{ij} become:

$$J_2 = \frac{1}{3}\sigma^2 \quad , \quad \sigma_m = \frac{1}{3}\sigma \quad ,$$

$$S_{ij} = \sigma_{ij} - \delta_{ij}\sigma_m = \begin{bmatrix} \frac{2}{3}\sigma & 0 & 0 \\ 0 & -\frac{\sigma}{3} & 0 \\ 0 & 0 & -\frac{\sigma}{3} \end{bmatrix}$$

$$\varepsilon_{ij} = e_{ij} = \begin{bmatrix} \varepsilon & 0 & 0 \\ 0 & -\frac{\varepsilon}{2} & 0 \\ 0 & 0 & -\frac{\varepsilon}{2} \end{bmatrix} \tag{5}$$

where σ_m is spherical tensor of stress.

Using formulas (5), the creep equation (3) become:

$$\varepsilon_{11} = \frac{3}{2}A(3J_2)^{\frac{B-1}{2}} S_{11}\varphi(t) \tag{6}$$

The equation (6) is a special case of equation (4). Both of them are identical when i and j are all one. Through contrasting with each other, the form of function $f_2(J_2)$ can be give:

$$f_2(J_2) = \frac{3}{2}A(3J_2)^{\frac{B-1}{2}} \tag{7}$$

Using above formula, the equation (4) become:

$$\dot{e}_{ij} = \dot{\varepsilon}_{ij} = \frac{3}{2}A(3J_2)^{\frac{B-1}{2}} S_{ij}\varphi(t) \tag{8}$$

or $$e_{ij} = \varepsilon_{ij} = \frac{3}{2}A(3J_2)^{\frac{B-1}{2}} S_{ij}\varphi(t)$$

4.2 Dependence of strain and time

In triaxial stress states, linear creep equation may be written as:

$$e_{ij} = \frac{S_{ij}}{2G}\left[1 + \int_0^t K(t)dt\right] \tag{9}$$

where K(t) is creep nucleus which reflect creep strain rate and G is shear modulus.

Neglecting instantaneous strain, the equation (9) becomes:

$$e_{ij} = \frac{S_{ij}}{2G}\int_0^t K(t)dt \tag{10}$$

Consulting the equation (8) and neglecting instantaneous strain, the nonliear creep equation of frozen soil at triaxial stress may be written as:

$$e_{ij} = \frac{3}{2}A(3J_2)^{\frac{B-1}{2}}S_{ij}\int_0^t K(t)dt$$

or

$$\dot{e}_{ij} = \frac{3}{2}A(3J_2)^{\frac{B-1}{2}}S_{ij}K(t) \tag{11}$$

4.3 Nonliear creep equations

If creep nucleus is different function forms, the nonliear creep equations will be different one from another. In this paper, three types of creep nucleus are adopted.

1. Creep nucleus is power function, i.e.:

$$K(t) = \alpha C t^{C-1} \tag{12}$$

Using above function, the equation (11) becomes:

$$e_{ij} = \bar{A}_1(3J_2)^{\frac{B_1-1}{2}}S_{ij}t^{C_1} \tag{13}$$

where $\bar{A}_1 = \frac{3}{2}A\alpha$, B_1=B and C_1=C; $\bar{A}_1$, B_1 and C_1 are three creep parameters of nonliear power function creep equation, their values may be determined by triaxial creep testing.

2. Creep nucleus is index function, i.e.:

$$K(t) = \delta_1 e^{-\delta t} \tag{14}$$

Using above formula, the equation (11) becomes:

$$e_{ij} = \bar{A}_2(3J_2)^{\frac{B_2-1}{2}}S_{ij}(1-e^{-C_2 t}) \tag{15}$$

where $\bar{A}_2 = \frac{3\delta_1 A}{2\delta}$, B_2=B and $C_2=\delta$; $\bar{A}_2$, B_2 and C_2 are three creep parameters of nonliear index function creep equation, their values may be determined by triaxial creep testing.

3. Creep nucleus is Logarithm function, i.e.:

$$K(t) = \frac{\nu}{1+t}\ln^{m-1}(1+t) \tag{16}$$

Using above formula, the equation (11) becomes:

$$e_{ij} = \bar{A}_3(3J_2)^{\frac{B_3-1}{2}}S_{ij}\ln^{C_3}(1+t) \tag{17}$$

where $\bar{A}_3 = \frac{3A\nu}{2m}$, B_3=B and C_3=m; $\bar{A}_3$, B_3 and C_3 are three parameters of nonliear logarithm function creep equation.

5. PROCESSING OF TESTING DATA AND ANALYSING OF THEIR RESULTS

According to the creep equations (13) (15) and (17), the testing data are processed so that the values of creep parameters could be difined. Because above-mentioned creep equations are similar in the form, the processing methods are idential. i.e. through derivation and simultaneously fetching logarithm on both sides of equation they may be change into standard duality linearity equations. Using duality linearity regression method, the values of creep parameters are gained by means of the computer. Results are presented in Tab. 1.

On the basis of the testing data, creep curves and regression results, five main conclusions have been yielded:

1. The cubic strain of the frozen clay is very small in the course of the creep (ε_v = 1.638x10^{-4} - 1.019x10^{-3}) and have no obvious variations with the passage of time, so that the creep equations of frozen clay are expressed in deviation stresses and strains.

Table 1. The values of frozen caly creep parameters.

Temperature	$-18^{\circ}C$	$-14^{\circ}C$	$-10^{\circ}C$	$-5^{\circ}C$
$\bar{A}_1$	$2.85x10^{-6}$	$5.81x10^{-5}$	$2.16x10^{-4}$	$2.05x10^{-3}$
B_1	2.77	2.1	1.86	1.45
C_1	0.408	0.4	0.424	0.401
$\bar{A}_2$	$9.45x10^{-7}$	$2.6x10^{-5}$	$2.42x10^{-4}$	$1.98x10^{-3}$
B_2	3.28	2.61	2.07	1.75
C_2	0.34	0.31	0.52	0.51
$\bar{A}_3$	$3.3x10^{-6}$	$6.87x10^{-5}$	$2.7x10^{-4}$	$2.65x10^{-3}$
B_3	2.81	2.15	1289	1.47
C_3	0.55	0.57	0.58	0.56

Table 2. The average values of c and the values of w,k, β and n

parameters	w	K	β	n	c
equation (13)	0.026	5.255	0.087	1.0006	0.410
equation (17)	0.017	5.360	0.091	1.0	0.564

2. Creep of frozen day seems to be characterized by a critical region which can be defined in terms of the ratio of axial stress to confining pressure. when this state is reached, the creep rate increases and the frozen day proceeds to failure at high ratio althrough the failure has not been measured. At lower ratio, attenuation creep sets in. This ratio relates to temperature, soil property and ice content of frozen soil. The testing results show that it increases along with dropping in temperature, for example, it is 1.2 at temperature of $-5^{\circ}C$ and 2.0 at temperature of $-18^{\circ}C$.
3. The testing results still show that the creep deformations of frozen clay are nonliear in the conditions of triaxial stresses.
4. The regression calculations show that the power and logarithm function creep equations perfectly coincide with testing curves. When the temperature of frozen clay varies from $-5^{\circ}C$ to $-10^{\circ}C$, the logarithm function creep equation is better, while power function creep equation is extremely well when the temperature varies from $-14^{\circ}C$ to $18^{\circ}C$.
5. Creep parameters $\bar{A}$ and B obviously relate to the temperature of frozen clay. They may be expressed in following formulas. Creep parameter C has no bearing on the temperature, its average values are presented in Tab.2.

$$\bar{A} = \frac{1}{w(1+|T|^{K})} \tag{18}$$

and

$$B = 1+\beta|T|^{n} \tag{19}$$

where T is temperature of frozen clay, w, K, β and n are testing constants, the values are presented in Tab 2.

6. CONCLUSIONS

The prelininary study has led to three main conclusions:

1. The 'low temperature box--triaxial cell' testing equipment system may be used to carry out the strength and creep testing upon the frozen soil in the conditions of the triaxial stresses. The testing conditions may be artificially regulated and automatically controlled-axial force, confining pressure and temperature, for example. In this testing equipment system, it is successful that the open-loop system of controlling by electro- hydroulic proportional valve has been used for egulating the oil pressure steplessly and stabilizing it for a long time, but further improvements are required.
2. The testing results show that the creep deformations of frozen clay are nonliear and cubic strains not only have rheologic property, but are approximately equal to zero ($\varepsilon_v = 0$).
3. The following equations are suggested to describe the creep characted of frozen clay at different temperatures.

$T = -5 \sim -10°C$

$$e_{ij} = \bar{A}_3 (3J_2)^{\frac{B_3-1}{2}} S_{ij} \ln^{C_3}(1+t) \quad (20)$$

$T = -14 \sim -18°C$

$$e_{ij} = \bar{A}_1 (3J_2)^{\frac{B_1-1}{2}} S_{ij} t^{C_1} \quad (21)$$

REFERENCES

(1) J.A. Franklin and E. Hoeck, Developments in Triaxial Testing Technique, Rock Mechanics 2, 223-228 (1970)

(2) A.M. Fish, Creep Strength, Strain rate, temperature and unfrozen water relationship in frozen soil, 4. Int. symp. on Ground Freezing, Vol 2, 29-36 (1985)

(3) R. Pusch, Creep of Frozen Soil, A Preliminary Physical Interpretation, 2. Int. Symp. on Ground Freezing, Vol. 1, 190-201 (1980)

(4) J. F. Ouvry, Results of triaxial Compression tests and triaxial Creep tests on an artificially Frozen Stiff clay, 4. Int. Symp. on Ground Freezing, Vol. 2, 207-212 (1985)

(5) J. C. Li and O. B. Andersland, Creep behavior of frozen sand under cyclic Loading Conditions, 2. Int. Symp. on Ground Freezing. Vol. 2, 223-234 (1980)

(6) T.H. W. Baker, Strain rate effect on the compressive Strength of frozen sand, sand, 1. Int. Symp. on GroundFreezing, Vol. 2, 73-79 (1978)

Numerical Methods in Geomechanics (Innsbruck 1988), Swoboda (ed.)
© 1988 Balkema, Rotterdam. ISBN 90 6191 809 X

A constitutive model for frozen sand

Toshihisa Adachi
Kyoto University, Japan
Fusao Oka
Gifu University, Japan

ABSTRACT: An elasto-viscoplastic constitutive model for frozen sand is proposed based on the elasto-viscoplasticity theory using the new time measure. The proposed model can describe the features of mechanical behavior such as rate sensititvity and strain-softening, under triaxial compression tests. The effect of temperature, confining pressure and the concentration of soil particles are also discussed.

1 INTRODUCTION

It has been experimentally found that frozen soil shows rate-sensitive, strain-hardening and strain-softening behaviors (Ladanyi 1981). These mechanical properties are closely related to soil concentration, temperature, strain rate and confining presssure. At low temperatures, such as are encountered when ground freezes, the material in frozen sand becomes brittle in nature. In this low temperature range, the typical stress-strain curve is of strain-hardening and strain-softening type, such that after it has reached peak strength, stress decreases and tends to become large strain strength. This brittleness grows evident with an increase in the strain rate and a decrease in temperature.

Since frozen soil is a mixture of ice and soil particles, its mechanical properties depend on the concentration of soil and the degree of the ice matrix. The shear strength is generally considered to be due to the strength of the ice, the frictional strength of soil particles and bonding. Peak strength is mainly due to the strength of the ice matrix, and the frictional strength is responsible for the residual or large strain strength. Peak strength and residual strength are, in general, affected by the strain rate.

Oka(1985) proposed a new type of viscoplasticity theory with memory and internal variables based on the generalized theory(Wang 1970 and Perzyna 1980). Using the strain measure instead of real time, Adachi and Oka(1985) have proposed an elasto-plastic constitutive model for soft rocks considering the strain-softening effect. Extending the proposed elasto-viscoplastic model, a viscoplastic constitutive model for frozen sand is constructed to describe the above mentioned properties of frozen sand.

Using the proposed model for frozen sand, we have simulated the triaxial test results with high and low concentrations of sand particles in the temperature range of -10 to -50°C(Shibata et al. 1985). Comparing the simulated and experimental results, discussion has been made on the effect of strain rate, temperature, confining pressure and soil concentration on deformation and strength properties.

2 VISCOPLASTIC CONSTITUTIVE EQUATIONS FOR FROZEN SAND

In this section, we will deal with only the infinitesimal strain field for simplicity. Introducing the strain measure instead of real time into the elasto-viscoplasticity theory(Oka 1986), Adachi and Oka (1985) constructed the elasto-plastic consti-tutive equation for soft rocks with strain hardening. The strain measure used in this theory is similar in concept to that proposed by Valanis(1971). In the present theory, we have adopted a new definition of time measure as follows:

$$dz=F(\text{strain rate})dt \quad (1)$$

where t is real time and rate dependency fuction F is the material function to be experimentally determined.

Using the above mentioned time measure, the stress history tensor is obtained by

$$\sigma_{ij}^* = \sigma^*_{ij}(\sigma_r^z(z-z')) \qquad (2)$$

$$\sigma_r^z = (\sigma_{ij}(z-z').\ 0<z'<z) \qquad (3)$$

In which σ_r^z is a reduced stress history.

Equation (2) shows that the stress history tensor is , in general, a functional of the reduced stress history.

In the present study, the stress history tensor is expressed by use of a single exponential type of kernel function as follows:

$$\sigma_{ij}^* = \frac{1}{\tau}\int_0^z \exp((z-z')/\tau)\sigma_{ij}(z')dz' \qquad (4)$$

in which z is a time measure defined by

$$dz = gdt \qquad (5)$$

$$g = (\dot{\varepsilon}_{ij}) \qquad (6)$$

If g is a homogeneous function of degree one, dz becomes the strain measure for rate independent materials, and when g is 1.0, the constitutive model leads to the viscoplastic theory proposed by Oka(1985).

In Eq.(4), τ is a material parameter which expresses the retardation of stress with respect to the time measure.

The total strain rate tensor is broken down
into the elastic strain rate tensor and the viscoplastic strain rate tensor.

$$d\varepsilon_{ij} = d\varepsilon_{ij}^e + d\varepsilon_{ij}^p \qquad (7)$$

The viscoplastic strain increment $d\varepsilon_{ij}^p$ is

assumed to be given by the non-associated flow rule.

$$d\varepsilon_{ij}^p = H\frac{\partial f_p}{\partial \sigma_{ij}}df_y \qquad (8)$$

in which f_p is the plastic potential function, f_y is the plastic yield function and H is the hardening and softening function.

The subsequent yield function is given by

$$\bar{\eta}^* - \kappa = 0 \qquad (9)$$

$$\bar{\eta}^* = (\eta_{ij}\eta_{ij})^{1/2} \qquad (10)$$

$$\eta_{ij} = s_{ij}^*/\sigma_m^* \qquad (11)$$

s_{ij}^* is the deviatoric part of the stress

history tensor and σ_m^* is the mean stress history.

Then, $df_y = d\bar{\eta}^*$

$$= [(s_{ij}^*/(\sqrt{2J_2}\sigma_m^*) - \sqrt{2J_2}/(\sigma_m^*)^2\delta_{ij}/3]d\sigma_{ij}^* \qquad (12)$$

$$d\sigma_{ij}^* = (\sigma_{ij} - \sigma_{ij}^*)dz \qquad (13)$$

In Eq.(9), κ is the strain-hardening and -softening parameter.

When $df_y>0$ and $f_y=0$, viscoplastic strain

occurrs, and when $df_y\leq 0$ or $f_y<0$, the material responds elastically.

The strain hardening and softening parameter is assumed to be given by the following strain-softening and -hardeing rule.

$$\bar{\eta}^* = \kappa = \frac{M_f^* G'\gamma^{p*}}{M_f^* + G'\gamma^{p*}} \qquad (14)$$

in which γ^{p*} is the second invariant of deviatoric plastic strain as:

$$\gamma^{p*} = (e_{ij}^* e_{ij}^*)^{1/2} \qquad (15)$$

By differentiating Eq.(14), function H in Eq.(9) is obtained.

Next, we will give the concrete form of the plastic potential function. The plastic potential function f is assumed to be as:

$$f_p = \eta^* + \tilde{M}^* \ln[(\sigma_m + b)/(\sigma_{mb} + b)] \quad (16)$$

$$\eta^* = (\eta_{ij}^* \eta_{ij}^*)^{1/2} \quad (17)$$

$$\eta_{ij}^* = s_{ij}/\sigma_m \quad (18)$$

in which s_{ij} is the deviatoric stress tensor, σ_m is the mean stress and $\tilde{M}^*$ is the parameter that controls the development of volumetric strain. Referring the previous work, we introduced the bounding surface which is influential for determining the value of $\tilde{M}^*$.

The boundary surface shown in Fig.1 is expressed by

$$f_b = \eta^* + M_m^* \ln[(\sigma_m + b)/(\sigma_{mb} + b)] \quad (19)$$

Inside the boundary surface($f_b < 0$), $\tilde{M}^*$ is given by

$$\tilde{M}^* = -\frac{\eta^*}{\ln[(\sigma_m + b)/(\sigma_{mb} + b)]} \quad (20)$$

On the other hand, outside the boundary surface $f_b = 0$, and the value of $\tilde{M}^*$ is constant.

$$\tilde{M}^* = M_m^* \ (f_b > 0) \quad (21)$$

The elastic strain increment is given by

$$d\varepsilon_{ij}^e = ds_{ij}/(2G) + d\sigma_m/(3K)\,\delta_{ij} \quad (22)$$

in which G is the elastic shear modulus and K is the elastic volumetric modulus.

3 APPLICATION TO FROZEN SAND UNDER TRIAXIAL CONDITION

The proposed constitutive model has been applied to the behavior of frozen sand at several temperatures. Toyoura sand was used for making the fozen sand. The triaxial compression test were performed for both a low and high concentration of

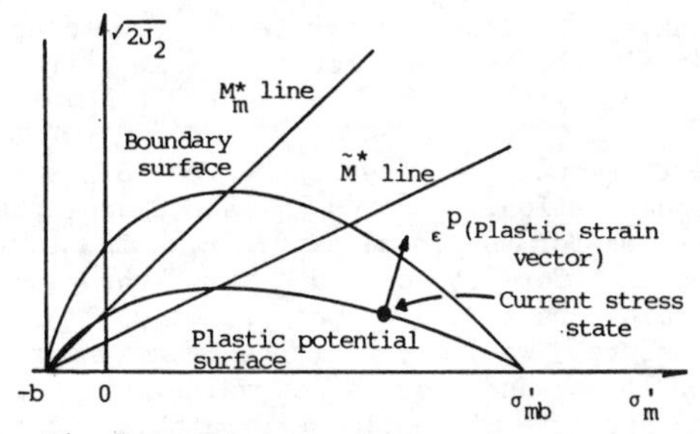

Fig.1 Boundary surface and plastic potential surface

soil particles at three axial strain rates. The material parameters are listed in Table 1.

For frozen sand, the following type of function F was adopted.

$$F = F(\dot{\varepsilon}_{11}) = F_0(\dot{\varepsilon}_{11}/\dot{\varepsilon}_0)^a \quad (23)$$

$$dz = Fdt \quad (24)$$

To begin with, we will discuss the results at the low temperature of -50 °C Figure 2 shows the typical deviator stress-strain and the volumetric deviatoric strain relations of dense sands. It is seen that the deviator stress increases until it reaches the peak value of strength, and after that stress decreases to the residual value of strength. This is a typical strain-hardening and -softening stress-strain relation. The volumetric strain-deviatoric strain relation shows good agreement with the experimental results.

Figures 3 and 4 are the results for lower strain rates with the same confining pressure at the same temperature. With a decrease in the strain rate, the difference between the peak and residual strengths becomes small. In the experimental results(S-3) at the lowest strain rate, the strength at large strain is larger than that in the other two tests(S-1,S-2). This phenomenon might be due to the time-hardening effect.

The effect of the value of confining pressure can be seen by comparing Cases S-1,-2, and C-1,-2 (Figs.2-3 and Figs.5-6). The test specimen under the higher confining pressure exhibits the higher strength at the large strain. On the other hand, peak strengths for these results are almost the same as for the same strain rate conditions.

Next, we will discuss the effect of soil concentration. Figures 7 and 8 are the results for the loose sand. From the test results for loose sand with the same strain rate under the same confining pressure, we note that the peak strength for loose sand shows a low value. However, the strength at the large strain is almost same. With respect to this point, the prediction shows a good agreement with these test results. For the specimen with a low concentration of soil particles, the strain rate effect is also observed.

Finally, let us note the effect of temperature for the behavior of frozen sand. In Figs. 9 and 10, the test results at -10°Care shown. These results show that at high temperature the strain-softening behavior cannot be observed and the peak strength is smaller than that of the frozen sand at a low temperature. With an increase in strain, the stress-strain curve is a monotonic increasing one.

CONCLUSION

An elasto-viscoplastic constitutive model for frozen sand was proposed and applied to the behavior of triaxial compression tests. From the comparison between experimental and predicted results, it is evident that the proposed theory can simulate the behavior of frozen sand such as strain-rate sensitivity, strain-hardening and -softening, the effects of temperature, confining pressure and the concentration of soil particles. Further research on behavior at a wide range of temperatures, strain rates, confining pressure and on the effect of grain crushing still remains to be done.

REFERENCE

(1) B.Ladanyi, Mechanical behaviour of frozen soils, Mechanics of structured Media, Part B,edt.by A.P.S.Selvadurai 203-245(1981)

(2) F.Oka,Elasto-viscoplastic constitutive equations with memory and internal variables,Computers & Geotechnics Vol. 1,No.1,59-69.

(3) T.Adachi and F.Oka,An elasto-plastic constitutive equation of geologic materials with memory, Proc. 5th ICONMG, Nagoya, 293-300(1985)

(4) C.-C.Wang,Generalized simple bodies, Arch. Rational Mech. Analysis,32,1-30, (1969)

(5) P.Perzyna, Memory effects and internal changes of a material,Int. J. Non-Linear Mech.,6,707-716(1971)

(6) T.Shibata,T.Adachi,A.Yashima,T.Takahashi,I.Yoshioka,Time-dependence and volumetric change characteristic of frozen sand, Proc. of 4th Int. Symp. on Ground Freezing,Sapporo,173-179. (1985)

(7) K.C.Valanis,A theory of viscoplasticity without a yield surface I, Arch. Mech Stos.,23-4,517-533(1971)

Table 1 Test Conditions

Test No.	Temperature °C	Confining Pressure Kgf/cm^2	Initial Void Ratio	Strain Rate $\dot{\varepsilon}_{11}$ %/min
S-1	-48.5	50.0	0.642	2.7
S-2	-51.5	50.0	0.654	0.29
S-3	-51.0	50.0	0.643	0.027
C-1	-49.0	100.0	0.644	2.7
C-2	-50.0	100.0	0.664	0.29
E-1	-50.0	50.0	0.919	2.7
E-2	-49.0	50.0	0.929	0.29
T-1	-10.0	50.0	0.651	2.7
T-2	-12.5	50.0	0.655	0.29

Table 2 Material parameters

Test No. of Series	M_f^*	M_m^*	Young's Modulus E (kgf/cm^2)	Elastic Bulk Modulus K(kgf/cm^2)	τ	a
S	1.7	1.6	100000	16000	130	0.920
C	1.5	1.4	800000	10000	75	0.920
E	1.53	1.42	80000	21060	110	0.920
T	1.50	1.40	20000	6000	40	0.720

$b=400(kgf/cm^2)$, $G'=200$ $\dot{\varepsilon}_0=0.270$ %/min and, $F_0=1$ and $\sigma'_{mb}=10^4 kgf/cm^2$ for all tests.

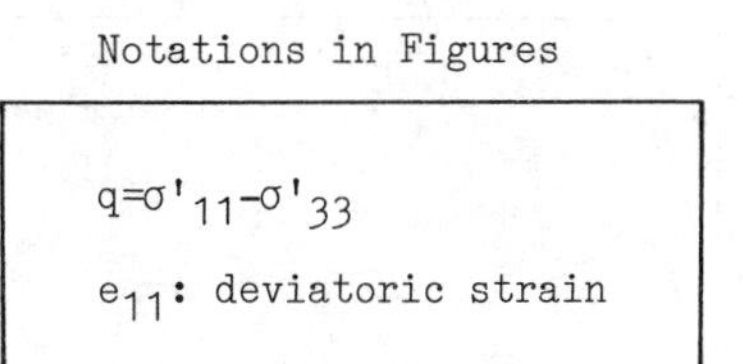

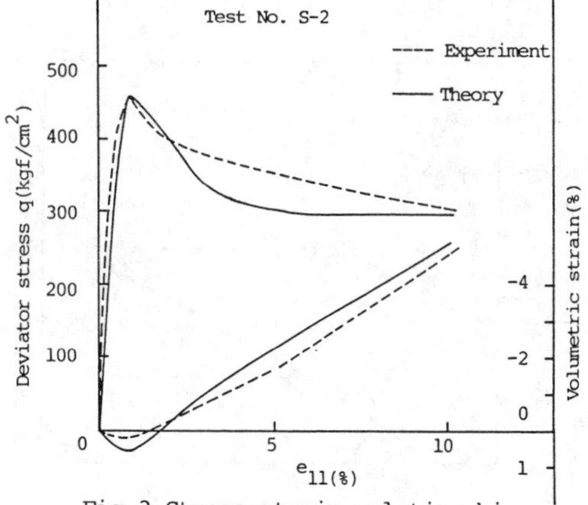

Fig.3 Stress-strain relationships

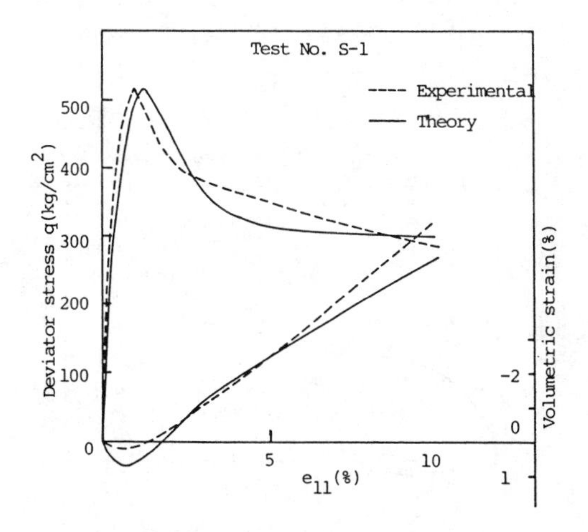

Fig.2 Stress-strain relationships

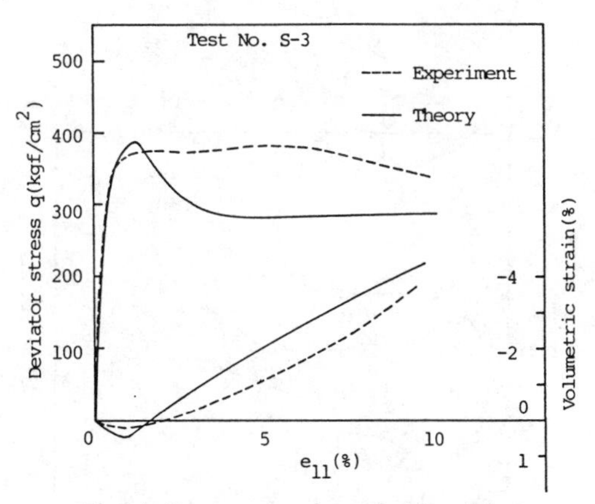

Fig.4 Stress-strain relationships

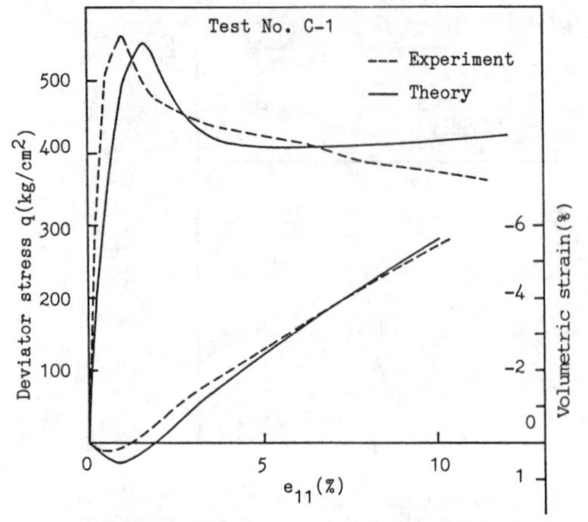

Fig.5 Stress-strain relationships

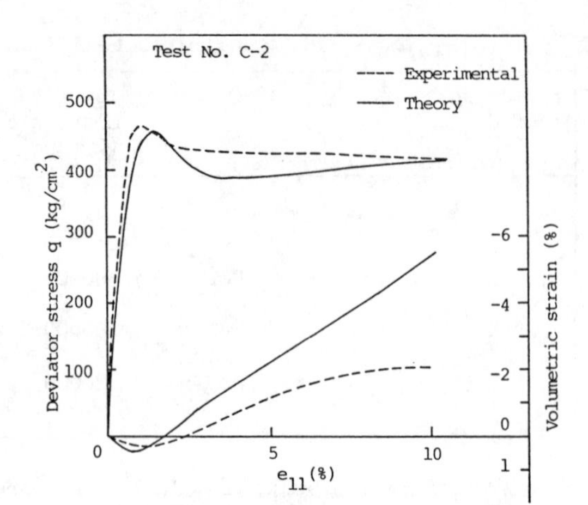

Fig.6 Stress-strain relationships

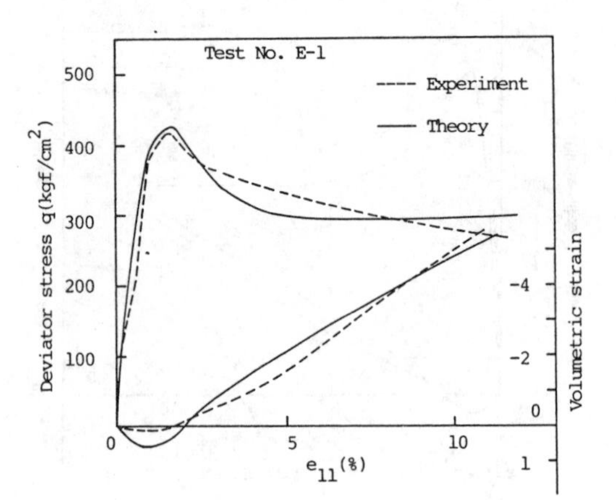

Fig.7 Stress-strain relationships

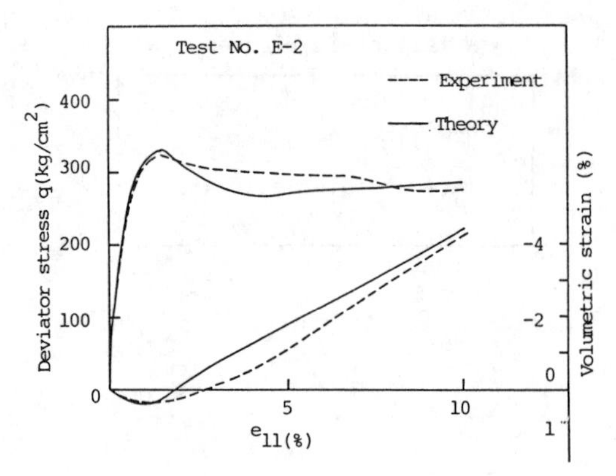

Fig.8 Stress-strain relationships

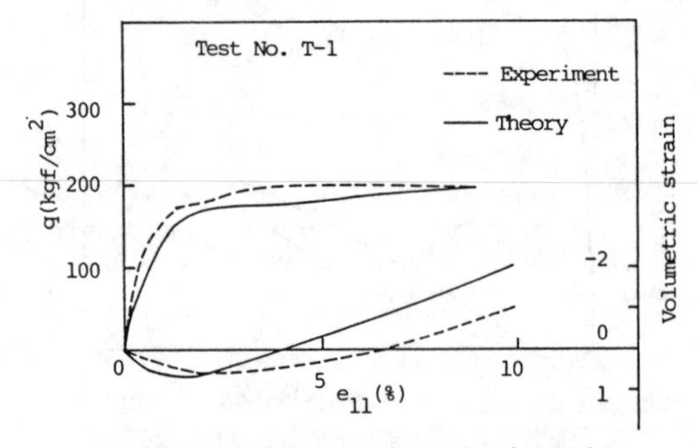

Fig.9 Stress-strain relationships

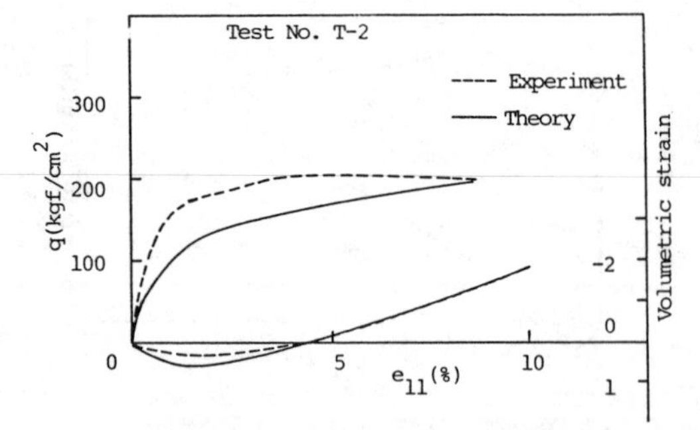

Fig.10 Stress-strain relationships

Numerical Methods in Geomechanics (Innsbruck 1988), Swoboda (ed.)
© 1988 Balkema, Rotterdam. ISBN 90 6191 809 X

Ground movements by tunnel driving in the protection of frozen shells

H.Meissner
University of Kaiserslautern, FR Germany

ABSTRACT: Using the artificial freezing method in tunnel driving, a reciprocal interaction exists between the subsoil and the frost shell, and it is presented from the initial heading state onwards. The settlements of the fround surface are described using results from numerical parametric studies. Consequently, settlements are a function of the freezing temperature, the tunnel and the frost shell dimensions, and the surface load. Comparisons with measured displacements, obtained by model tests are presented.

1. INTRODUCTION

The artificial freezing method in tunnel construction is a well known practice. In contrast to the progress of driving followed by trailing linings, the artificial freezing method consists of a reciprocal interaction between the subsoil and the frost shells, a phenomenon presented from the initial heading state onwards. Whenever a soil is frozen, there is a possibility of occurence of both heave and ice segregation. However, this phenomenon is not considered in these studies. The tunnels are driven in a saturated, non cohesive subsoil, assumed to be insensible to frost. Furthermore, the frost period in tunnel construction is lasting only for a short time of about 800 hours.

Computer simulations are performed to determine the ground movements due to both the excavation of the tunnel and the creep of the frost shell. In this paper, consideration will be given to the settlements of the ground surface. The bearing behaviour of the frost shells reported elsewhere. To verify reliability of the numerical model the results are compared with measured displacements obtained by model tests in the laboratory.

In the numerical model for unfrozen soil the plasticity material law is used covering various stress paths. The material law for frozen soil is described by an elastic viscoplastic law. Today, it is well established that this law must describe the entire stress-strain-time behaviour for frozen soil at least for a certain range (Ting, 1983, Orth, 1986). This paper extends earlier investigations (Meißner, 1985) by varying the tunnel dimensions as well as applying a surface load.

2. NUMERICAL MODEL

An underground section with the considered tunnel as well as the used discretization by a finite element mesh is presented in Fig. 1. Only plane deformation problems will be treated. The ranges of the different model parameters are chosen as follows:

Tunnel radius: $2\ m \leq r_a \leq 8\ m$

Overburden height: $12\ m \leq l_m \leq 28\ m$

wall thickness: $0{,}75\ m \leq d \leq 4\ m$

surface load: $0 \leq p_o \leq 655$ KPa

Creep time: $0 \leq t \leq 1000$ h

Average temperature: $-35°C \leq T_m \leq -10°C$

In all calculations the two distances l_u and b remain constant that is:

$l_u = 40$ m
and $b = 40$ m

The excavation is simulated by stepwise decrease of the balance loads P_n at the Tunnel surface, Fig. 1.

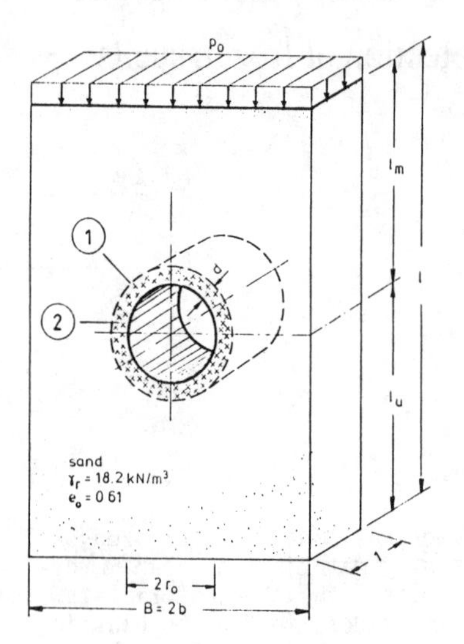

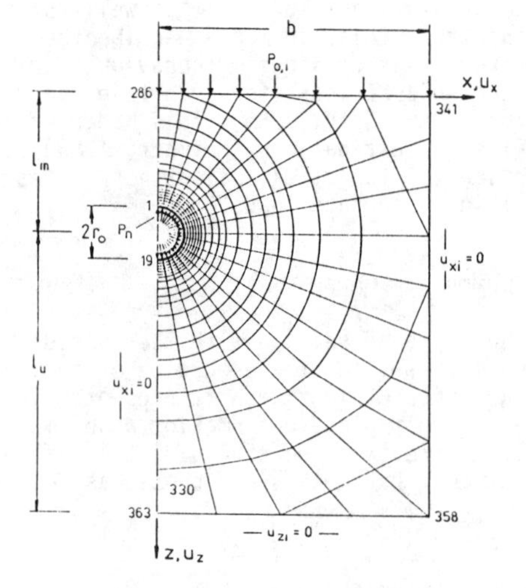

Fig. 1 Physical and numerical model of the tunnel.

2.1 Material law for unfrozen soil

The used plasticity law is reported in detail by Meißner (1983). It truly describes the stress strain behaviour of granular soils as investigated in biaxial and triaxial tests. Softening at large strains as well as changes of soil stiffness for bended stress paths from the earth pressure at rest are considered.

Figures are presented in the principal stress space or as one dimensional stress strain plots. The tensor of the plastic strain increments is assumed to be coaxial to Cauchy's stress tensor. Compressive stresses and contractive strains are negative.

The object of the plasticity model is to specify the relationship between the stress increments and the increments of strain as

$$d\sigma_{ij} = D_{ijkl} \; d\varepsilon_{ij} \tag{1}$$

in which $d\varepsilon_{ij}$ are the total strain increments and D_{ijkl} is the generally non symmetrical elasto-plastic matrix. The strains are separated into elastic and plastic components.

$$d\varepsilon_{ij} = d\varepsilon^e_{ij} + d\varepsilon^p_{ij} \tag{2}$$

Both, the elastic and the plastic components can be divided into

$$\begin{aligned} d\varepsilon^e_{ij} &= de^e_{ij} + dI_{\varepsilon e} \frac{\delta_{ij}}{3} \\ d\varepsilon^p_{ij} &= de^p_{ij} + dI_{\varepsilon p} \frac{\delta_{ij}}{3} \end{aligned} \tag{3}$$

where δ_{ij} is the Kronecker delta and dI_ε is the first invariant of the corresponding strain tensor. Further invariants used in following equations are defined as:

$$II_{ep} = e^p_{ij} \, e^p_{ji}$$

$$I_\sigma = \sigma_{ii}$$

$$II_s = s_{ij} \, s_{ji}$$

$$III_s = s_{il} \, s_{lj} \, s_{ji}$$

$$\cos 3\alpha_\sigma = \sqrt{6} \, \frac{III_s}{II_s^{3/2}}$$

in which the deviator stresses are

$$s_{ij} = \sigma_{ij} - I_\sigma \frac{\delta_{ij}}{3}$$

Let us denote the atmospheric pressure as p_a. Then the related first stress invariant is defined as

$$I_a = I_\sigma / p_a.$$

The stress can be expressed in the elastic strains using the symmetrical elasticity matrix C_{ijkl}.

$$d\sigma_{ij} = C_{ijkl} \, d\varepsilon_{kl} = C_{ijkl} \, (d\varepsilon_{kl} - d\varepsilon^p_{kl}) \tag{4}$$

To determine $d\varepsilon^p_{ij}$ we introduce the concept

of a yield function as well as flow rules.

Yield function

From laboratory tests we obtain a characteristic strain curve as shown in Fig. 2.

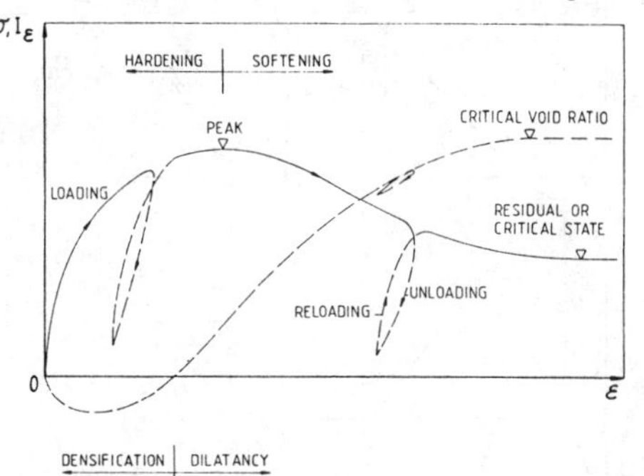

Fig. 2 Characteristic stress strain curves from triaxial tests

In the spatial stress space, test results can be presented as shown in Fig. 3. The yield surfaces separate loading stress paths from unloading/reloading stress paths. In the concept of strain hardening for each yield surface the deviator strain $II_{ep}^{1/2}$ is constant.

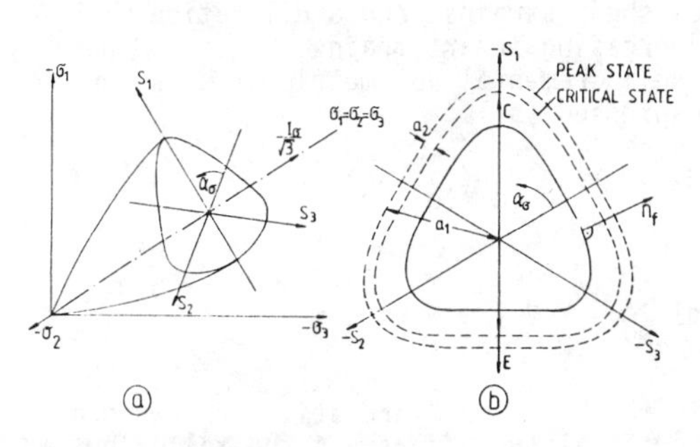

a_1: Hardening
a_2: Softening
C : Compression Tests
E : Extension Tests

Fig. 3 Sixfold yield surface in the spatial stress space

The general form of the yield function reads:

$$f = II_s^{1/2} - f_o\ (\sigma, \alpha_\sigma, \varepsilon_p, e_o) \quad (5)$$

where e_o is the initial void ratio and α_σ is the angle in the deviator plane as shown in Fig. 3. For f = 0 the yield condition is satisfied.

The unit vector at a point on the surface is denoted by n_f.

$$\{n_f\} = \frac{\partial f/\partial \sigma_{ij}}{|\partial f/\partial \sigma_{ij}|} \quad (6)$$

Considering the strain hardening model increasing values of $II_{ep}^{1/2}$ define the loading state whereas in the case of unloading, $II_{ep}^{1/2}$ remains constant. Based on a proposal of Stutz (1972), the following yield function is introduced in the plasticity law:

$$f = II_s^{1/2} - A\ I_\sigma\ (1 - \frac{B}{\sqrt{6}} \cos 3\alpha_\sigma)^{-m}$$

The material parameters A, B and m are determined from sand samples subjected to triaxial test conditions. The parameters can be expressed as:

$$A = -2 \cdot 10^{-4} (I_a II_{ep})^{0,5} a_1^{\ 2} (a_1\ II_{ep}^{0,5} + 1)^{-1} \quad (7)$$

where a_1 is given by

$$a_1 = \frac{10^3}{e_o\ I_a^{\ 0,5}} - 20$$

and

$$B = 1,90 + 3,6 \cdot 10^{-5}\ I_a^{\ 2} \quad (8)$$

The parameter m depends on B and we find:

$$m = \frac{0,1}{\ln\ (1 - B/\sqrt{6})} \quad (9)$$

At the peak points values of $II_{ep}^{1/2}$ are denoted by ξ_i, Fig. 5, and the corresponding A parameters take the values of Ap. From test results we find:

$$\xi = 0,06\ e_o \exp\ (0,12\ I_a^{\ 0,5}) \quad (10)$$

In good qualitative agreement with test results the parameter A for softening can then be calculated by the formula

$$A=A_p\ [1-(1-\frac{e_p}{e_k})(1-\exp(-75(II_{ep}^{1/2}-\xi)^2))]$$

$$II_{ep}^{1/2} \geq \xi \quad (11)$$

in which e_p is the actual void ratio at the peak state and e_k is the critical void ratio. Fig. 4 shows the decrease in the values of A with increasing $II_{ep}^{1/2}$. The relationship for e_p and e_k may be written as

$$e_p = e_o + (1+e_o)\frac{0{,}002}{e_o^4}(1-215\ \xi^2\ e_o^2) \quad (12)$$

and

$$e_k = e_p + (1+e_o)\ \frac{0{,}0004}{\xi\ e_o^4} \quad (13)$$

It should be noted that Eqs. 7 to 11 are valid for any given void ratio in the range from loose to dense.

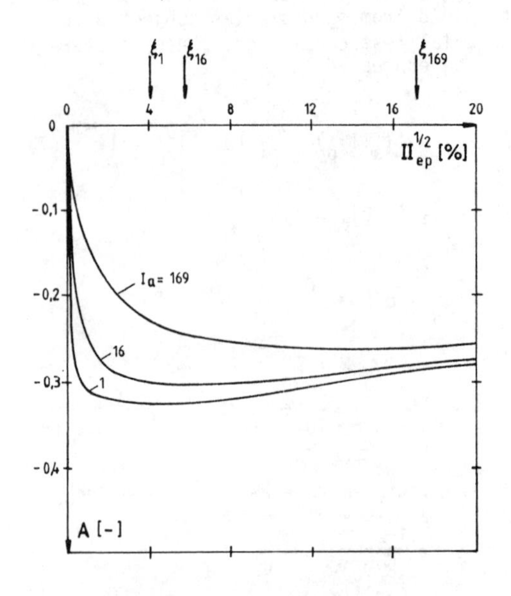

Fig. 4 Parameter A of the yield function, $e_o = 0{,}6$

Flow rule

The plastic volumetric strains of Eq. (3) can be divided into a consolidation volumetric strain and dilatancy strain.

$$dI_{\varepsilon p} = dI_{\varepsilon pc} + dI_{\varepsilon pd} \quad (14)$$

$dI_{\varepsilon pc}$ results from changes of the mean pressure dI_σ while $I_{\varepsilon pd}$ results from shear deformations. In Fig. 5 the two parts of the volumetric strains are shown.

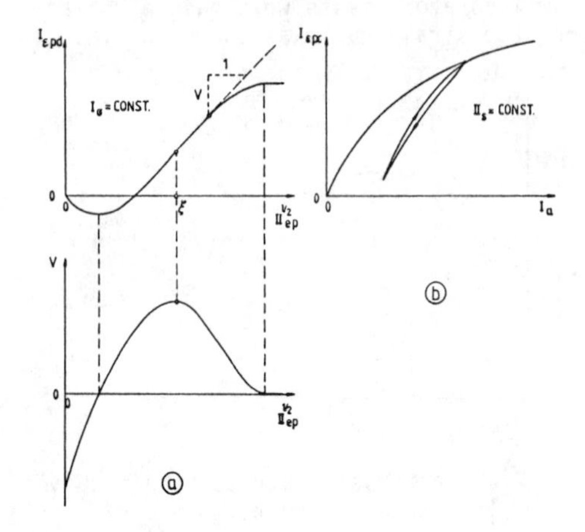

Fig. 5 Dilatancy and consolidation volumetric strains

Volumetric strain increments are often computed from spherical cap yield models and corresponding plastic potentials. Expressions for the incremental dilatancy volumetric strains, should of course describe both a densification, for small values of shear strains, and a dilatation with increasing shear strains, Fig. 2. Generally the incremental volumetric strains may be expressed as:

$$dI_{\varepsilon pc} = \lambda_c\ V_c\ (I_\sigma,\ e_o) \quad (15)$$

and

$$dI_{\varepsilon pd} = \lambda\ V\ (\sigma,\ \alpha_\sigma, e^p,\ e_o) \quad (16)$$

in which λ_c and λ are still unknown proportionality factors. In the following, we will assume that $dI_{\varepsilon pc}$ can be described using a variable bulk modulus K such that

$$dI_{\varepsilon e} + dI_{\varepsilon pc} = \frac{dI_\sigma}{K} \quad (17)$$

The parameter K may be written as

$$K = -\ p_a\ \frac{750}{e_o - 0{,}1}\ (1 + I_a)^{1/3} \quad (18)$$

Generally, in the quasi elastic state, as unloading and reloading, the bulk modulus is given by the relationship:

$$K = - p_a \frac{4000}{(1-o,o4(I_a - I_{au}))(e_o - 0,1)},$$

$$I_{au} \leq 100 \qquad (19)$$

where I_{au} is the value of I_a before unloading. The corresponding shear modulus reads:

$$G = \frac{ds_i}{2de_i^e} = - p_a \frac{200}{e_o^2}(1+0,01\ I_a - 1,5 \cdot 10^{-5} I_a^2) \qquad (20)$$

The function V in Eq. 16 is refered to as the dilatancy parameter defined by (Fig. 5a)

$$V = \frac{dI_{\varepsilon pd}}{dII_{ep}^{1/2}} \qquad (21)$$

Using the experimental results from both compression and extension tests we can express the dilatancy parameter by the formula

$$V = \frac{0,003}{\xi\ e_o^4} [1-25(II_{ep}^{1/2} - \xi)^2 \cdot \alpha]; \qquad (22)$$

$$\alpha = \begin{cases} \frac{1}{25\xi^2} + 17\ e_o^2 & \text{for } II_{ep}^{1/2} \leq \xi \\ 1 & \text{for } II_{ep}^{1/2} > \xi \end{cases}$$

Fig. 6 shows the variation of V with I_σ when the void ratio is assumed constant.

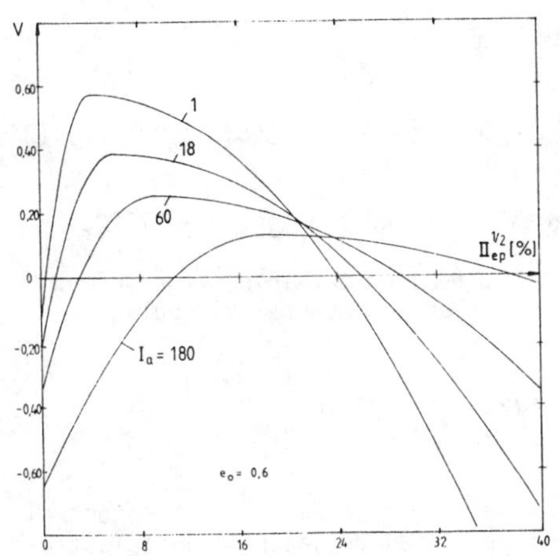

Fig. 6 Dilatation parameter V as function of $II_{ep}^{1/2}$

Generally, the direction of the deviator component of the plastic strain increment can be determined by a plastic potential. From tests with cubic samples it is known that up to values of $I_a \approx 2$ MPa the direction of de_p is independent of I_σ, $II_{ep}^{1/2}$ and e_o, Goldscheider (1975). From results of cubical triaxial tests the function g for the plastic potential may written as

$$g = II_s^{1/2} - x\ (1+0,47 \cos 3\ \alpha_\sigma)^{-o,2} \qquad (23)$$

where x is a factor such that g = o is satisfied. The derivation of g with respect to the deviator stress is in accord with the condition

$$de_{kk}^p = 0$$

$$de_{ij}^p = \lambda \left(\frac{\partial g}{\partial s_{ij}} - \frac{1}{3}\frac{\partial g}{\partial s_{kk}}\delta_{ij}\right) \qquad (24)$$

in which λ is the same proportionality factor as in Eq. (16). Again, an outward unit vector normal to the potential curve in the deviator plane can be introduced as shown in Fig. 7

$$\{n_g\} = \frac{de_{ij}^p}{|de_{ij}^p|} \qquad (25)$$

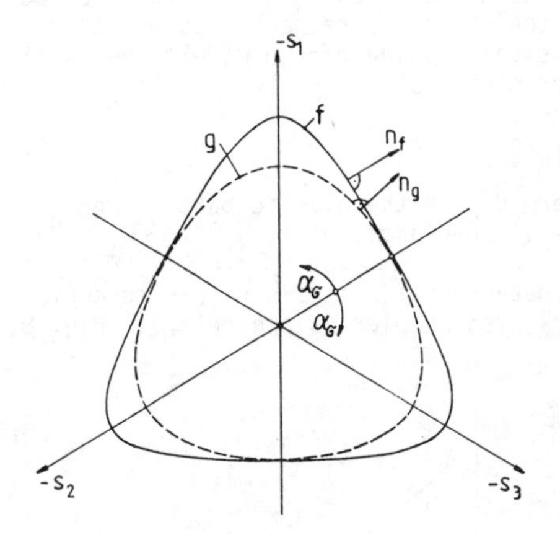

Fig. 7 Yield and potential curve in the Deviatoric plane

Thus, the total incremental plastic deformations take the form

$$d\varepsilon_{ij}^p = \lambda\ (n_{gij} + V \frac{\delta_{ij}}{3}) \qquad (26)$$

or

$$d\varepsilon^p_{ij} = \lambda \, h_{ij} \tag{27}$$

Now, the factor λ is obtained using the consistency condition. The well-known derivations yield:

$$\lambda = \frac{\{n_f\}^T \, [C] \, \{d\varepsilon\}}{\{n_f\}^T \, [C] \, \{h\}} \tag{28}$$

and

$$[D] = [C] - [C] \, \frac{\{h\} \, \{n_f\}^T \, [C]}{\{n_f\}^T [C] \, \{h\} + H} \tag{29}$$

where H is the plastic modulus defined as

$$H = - \frac{\{\frac{\partial f}{\partial \varepsilon^p}\}^T \{h\}}{|\partial f/\partial \sigma|} \tag{30}$$

Eq. (29) defines the plastic matrix D introduced in Eq. (1).

2.2 Material law for Frozen Soil

The equations of the model are explained in detail in an earlier paper (Meißner, 1985). The derivations of the formulae are based on results of frozen cylindric sand samples in triaxial tests. In this paper the significant expressions are reported in the necessary range to recognize the physical meaning of the model. The total incremental strain yields

$$d\varepsilon = d\varepsilon_e + d\varepsilon_{vp} \tag{31}$$

where $d\varepsilon_e$ is the elastic part and $d\varepsilon_{vp}$ the visco-plastic part.

A characteristic strain rate-time curve of loaded samples is presented in Fig. 8.

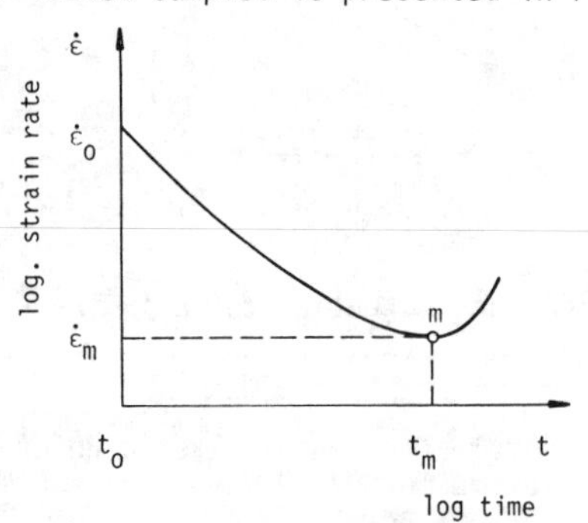

Fig. 8 Strain rate-time curve

The point m in Fig. 8 coincides with the point of inflection of the creep curve. In creep tests the volumetric strains are zero for a wide range of deformations. Thus,

$$\dot{\varepsilon}_{ij} = \dot{e}_{ij}$$

is valid.

Similar to an equation formulated Ting (1983) we find in good qualitative agreement with results form compression, extension as well as tension test, the function

$$II^{1/2}_{\dot{e}} = II^{1/2}_{\dot{e}m} \, \tau^{-1} \exp(\tau - 1) \tag{32}$$

where the parameter τ is defined as

$$\tau = \frac{t}{t_m}$$

Based on the results from tests with the conditions

$$-30°C \leq T \leq -2°C$$

and

$$1 \text{ MPa} \leq I_a \leq 15 \text{ MPa}$$

$$0{,}8 \text{ MPa} \leq II^{1/2}_s \leq 11{,}5 \text{ MPa}$$

The parameters t_m as well as $II^{1/2}_{\dot{e}m}$ may be written as

$$t_m = \Theta^{3,55} \left(\frac{II^{1/2}_s}{I_a}\right)^{-8,3} (650 I_a)^{-5,64} \quad [h] \tag{33}$$

and

$$II^{1/2}_{\dot{e}m} = \Theta^{-4,4} \left(\frac{II^{1/2}_s}{I_a}\right)^{11} (45{,}5 \, I_a)^{7} \quad [\%/h] \tag{34}$$

where $\Theta = T/T_{abs}$ and $T_{abs} = -273°C$.

With Eq. (32) wo obtain now the second invariant of incremental viscoplastic strains

$$\Delta \, II^{1/2}_{e,\,vp} = II^{1/2}_{\dot{e}} \cdot \Delta t \tag{35}$$

By assuming g from Eq. (23) as potential for incremental deviatoric viscoplastic strains we get

$$dv_{ij} = \frac{\partial g}{\partial s_{ij}} - \frac{\partial g}{\partial s_{kk}} \frac{\delta_{ij}}{3}$$

$$\Delta e^{vp}_{ij} = \frac{\Delta II^{1/2}_{e,\,vp}}{(dv_{ij}\, dv_{ji})^{1/2}} dv_{ij} \quad (36)$$

Other important components of the material law are the observerd quasi elastic strains occuring without delay of time. The elastic behaviour of the soil may be characterized by the shear modulus G and the Poisson's ratio ν. From test results we find the expressions

$$G = -\ 12{,}2\ p_a\ (20+3100\ \Theta + \frac{II_s^{1/2}}{p_a}) \cdot (1{,}63 - \frac{II_s^{1/2}}{I_a}) \quad (37)$$

$G \geq 5$ MPa

and

$$\nu = 0{,}45\ (0{,}1 + \frac{II_s^{1/2}}{I_a})^{o,2} \quad (38)$$

With this, all equations for the two material models are available.

In some tests the creep behaviour for frozen soil was observed after reducing the load. The creep rate was found in good quantitative agreement with values obtained for samples subjected initially to the reduced load. Consequently, in numerical analysis the creep rate depends only on the actual stress state. A remembrance of former stress states is swept out.

3. NUMERICAL RESULTS

By finite element analysis the effect of the different parameters presented in chapter 1 on the settlements of the ground surface was investigated. The destribution of the temperature in the frost shell is assumed to be

$$T_r = T_m \frac{\ln r/r_K}{\ln r_m/r_K},\ r_K = \begin{cases} r_o & \text{for } r < r_m \\ r_a & \text{for } r \geq r_m \end{cases} \quad (39)$$

where r_m is the radius of the freezing pipe circle and r_a the radius of the outer freezing front. The initial stress state in computations is the earth pressure at rest. Within 15 steps the balance loads at the tunnel surface are decreased to zero. The creep time is chosen to be about 800 hours. It should be noted that in all computations, the reference strain state is the state after the ground surface is subjected to p_o.

The concept to present the settlement curves is based on five distinct steps:

1 determination of the settlements s_{1o} of point 1 in Fig. 9. The overburden pressure p_o is zero, the time yield $t = 1$ h.
2 The ground surface is subjected to p_o. The settlements $s_{1p} = s_1 - s_{1o}$ are determined, where s_1 are the total settlements. The time is $t = 1$h.
3 The settlements in point 1 at times $t > 1$ h are approached by a function. Then, s_1 is denoted by s_{1t}.
4 The settlements s_{2t} at a distance of $3\ r_o$ from point 1 are expressed as function of s_{1t}.
5 Similar to a formula proposed by Kany (1974) the settlement curves are functions of s_{1t} and s_{2t}.

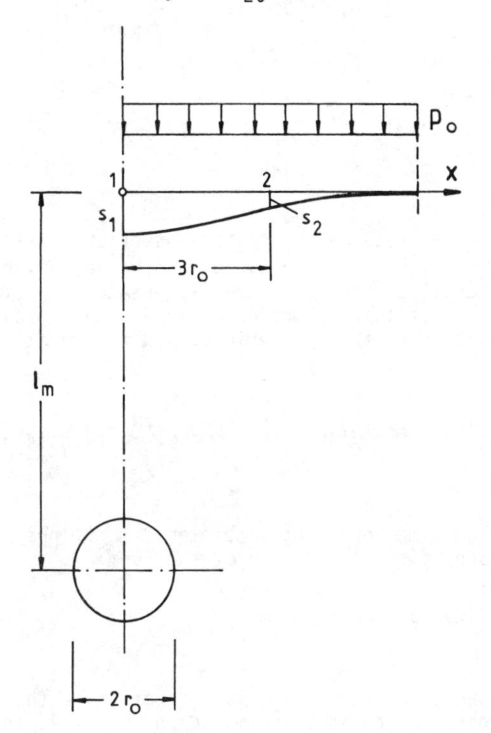

Fig. 9 Notations for the settlement curves

In our studies we use a certain medium sand. The void ratio remains constant. Therefore, the general expression for the settlements of the ground surface may be simplified and presented in dimensionless form as

$$\frac{s}{l_m} = f\left(\frac{d\gamma_s}{G}, \frac{r_o}{l_m}, \Theta, \frac{p_o}{G}, \frac{t}{t_1}\right), \qquad (40)$$

where t_1 = 1 h, γ_s: unit weight of grains; G: shear modulus of the frozen soil in Eq. (37) for $T=T_m$ and $II_s^{1/2} = 0$. f is a function which has to be determined by approximation of numerical results.

At first we consider the settlements s_{1o}. Then Eq. (40) becomes:

$$\frac{s_{1o}}{l_m} = f_1\left(\frac{d}{l_m}, \frac{r_o}{l_m}, \Theta\right) \qquad (41)$$

The above function can be presented by an expression of potential functions. The approximations yield the expression

$$\frac{s_{1o}}{l_m} = 5{,}7\cdot10^{-4}\left(\frac{r_o}{d}\right)^{0,25}\left(\frac{r_o}{l_m}\right)^2\left(\frac{l_m}{x}\right)^{1,6}\Theta^{-0,67} \qquad (42)$$

$$p_o = 0,\ t = 1h,$$

in which

$$x = \frac{G}{2{,}73\cdot10^5\,\gamma_s\,\Theta}$$

The expression of the total settlements s_1 by one potential equation gare unsatisfying results. Therefore, the differences s_{1p} of the settlements are determined. The approximation of these numerical results gives:

$$\frac{s_{1p}}{l_m} = 2500\cdot\left(\frac{r_o}{d}\right)^{0,25}\left(\frac{r_o}{l_m}\right)^2\frac{p_o}{G}\,\Theta^{1/3}, \quad t=1h \qquad (43)$$

The time dependent settlements may be presented by an equation of the form:

$$\frac{s_{1t}}{l_m} = \left(\frac{s_{1o}}{l_m} + \frac{s_{1p}}{l_m}\right)t^m, \quad t\ [h] \qquad (44)$$

where the exponent m is a function of the distinct parameters listed in Eq. (40). In accord with numerical evidence, the parameter m becomes:

$$m = 9\cdot10^{-6}\,\frac{r_o\,l_m}{\Theta\,d^2}\left(1+5{,}8\cdot10^6\,\frac{p_o\,\Theta}{G}\right) \qquad (45)$$

Even after 800 hours the computation results show no significant increase of the creep rate.

Equations (42) until (47) express the settlements of point 1 in Fig. 9 for a wide class of tunnels with frozen shells. To simplify the expression for s_{2t} as function of s_1 a separation into different parts is not performed. Therefore, numerical results are best fitted by the expression

$$s_{2t} = s_{1t}\exp\left(0{,}3\,\frac{p_o}{\gamma_s\,l_m} - 16{,}3\,\frac{r_o}{l_m}\left(\frac{d}{l_m}\right)^{0,5}\right) \qquad (46)$$

A comparison of numerical values and s_{2t} of Eq. (46) shows deviations up to 10%

Finally, the settlement curves may be presented by both settlements s_{1t} and s_{2t}. Therefore, a similar expression is introduced as Kany (1974) used in this model to describe the bearing pressure beneath a base plate. Consequently, the settlements of a surface point i at a distance x from the center line may be truly expressed by

$$s_{it} = \frac{s_{2t}\,s_{1t}}{s_{2t} + (s_{1t}-s_{2t})\left(\frac{x}{3r_o}\right)^2} \qquad (47)$$

Applying Eq. (47) the settlements at the two points 1 and 2 are s_{1t} and s_{2t} respectively. The application of Eqs. (42) until (47) is shown through an example: For a tunnel project the following parameters are given:

$2\cdot r_o$ = 8 m, d = 1,5 m, l_m = 28 m, p_o = 218,4 KPa, γ_s = 26,6 KN/m³, T_m = - 10°C, loading time t = 800 h.

The settlement curve of the ground surface is predicted as follows:

Shear modulus, Eq. (37), $II_s^{1/2}=0$, $T=T_m$:

$G = 2000(20 + 11{,}35\,|T|) = 267\cdot10^3$ KPa

Surface point 1 above the crown of the tunnel, t = 1 h, Eqs. (42) and (43):

$s_1 = s_{1o} + s_{1p} = 7{,}56$ cm

Settlement of point 1 after 800 hours, Eq. (44)

s_{1t} = 11,83 cm

Consequently, the settlement by creep deformations of the frozen soil yields

$s_{1K} = s_{1t} - s_1$ = 4,27 cm

We then find for the surface point 2 at a distance of 3 r_o from point 1; Eq. (46)

s_{2t} = 4,40 cm

In accordance with Eq. (47) the predicted settlement curve is presented in Fig. 10 .

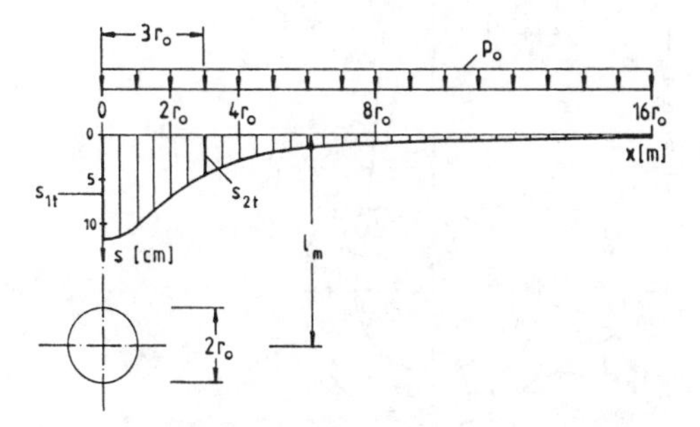

Fig. 10 Settlement curve after t = 800 h.

4 EFFECTS OF DISTINCT PARAMETERS

Ground movements may be controlled by both the thickness d of the frost shell and the freezing temperature T_m during the application of the freezing method in tunnel construction. The wall thickness again depends on both the freezing temperature and the freezing time until excavation. In order to gain some insight into the effectiveness of the two parameters, the settlements are plotted versus the relative values of d and T_m in Figures 11 and 12 respectively.

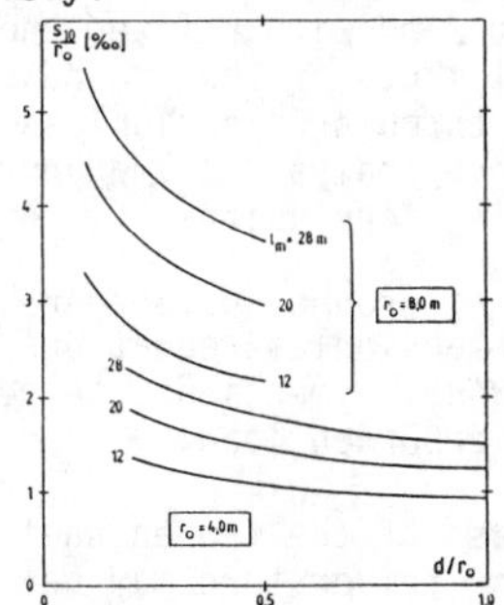

Fig. 11 Settlements vs wall thickness, Eq (42), p_o = 0, T =- 10°C, t= 1h

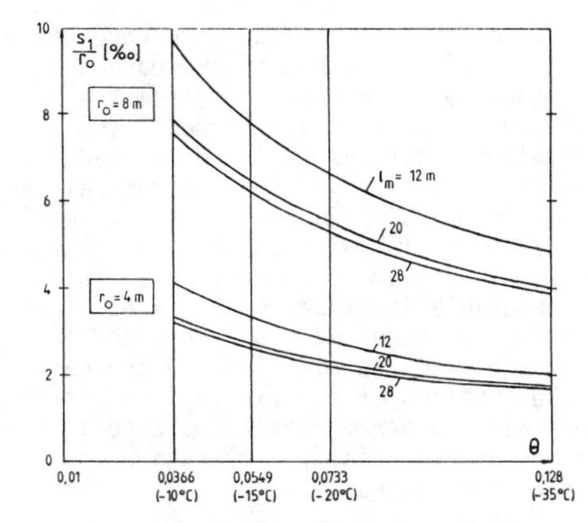

Fig. 12 Settlements vs. temperature, Eq. (43), p_o = 218,4 KPa, d =1,5m t = 1h

For clarity, graphs are shown for certain tunnel radii and values of l_m. At low values of d/r_o and Θ, it is clear that the rate of settlement is remarkably increasing. However, this tendency is more significant for higher values of r_o. Furthermore, it is interesting to note that, when applying different overburden loads, significant differences of the settlement rate are obtained for distinct values of l_m. Figure 13 depicts the settlements due to relative overburden load.

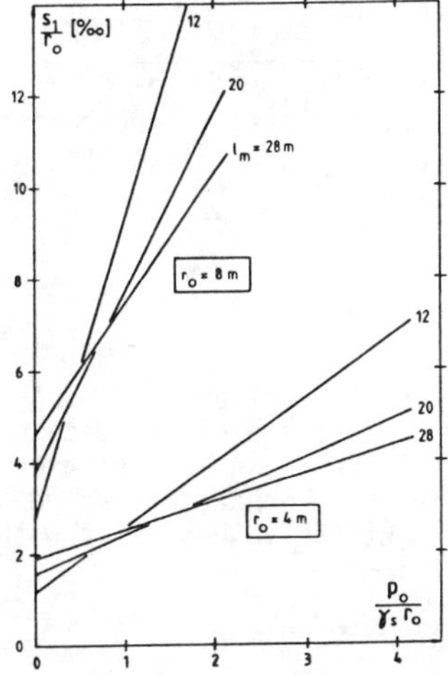

Fig. 13 Settlements vs. overburden load, Eq. (43), d = 1,5 m, T = 10°C, t = 1 h

As a matter of fact the rate of settlement is remarkably greater for lower values of l_m than for higher ones. It is well understood that the bearing capacity of the unfrozen soil substantially increases with the increase of the overburden height above the tunnel. Let us consider the case where the rates of settlement by increasing l_m are equal to those resulting from the increase of comparable overburden loads. Derivating equations (42) and (43) and solving for t_m yielded the theoretical value of $l_m \approx 230$ m. For values of l_m less than 230 m a fill of thickness $\Delta z = P_o/\gamma_s$ results in higher settlement rates than in comparable increases of $\Delta l_m = \Delta z$. Finally, the magnitudes of creep settlements are given in Fig. 14. The increase in the duration of the excavation process leads also to a remarkable increase in the settlement, especially with regard to large diameter values.

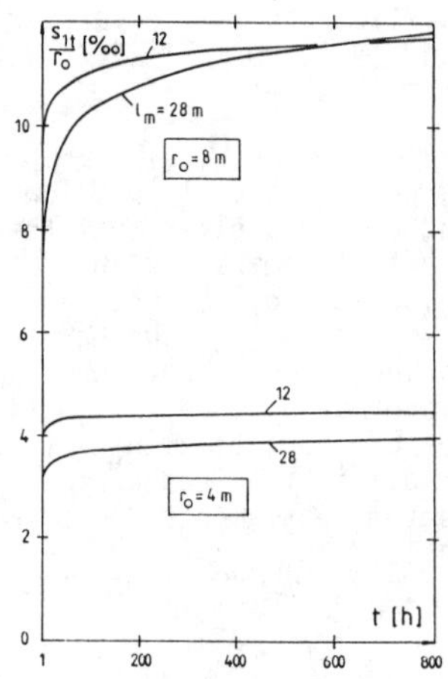

Fig. 14 Settlements vs. creep time, Eq. (44), p_o=218,4 KPa, d = 1,5 m t =- 10°C

5 MODEL TESTS

The check on the reliability of the numerical model is based on the data generated from several tests on instrumented tunnel models. The test processes are reported elsewhere but the values dealt with were as follows:

$0,45 \text{ m} \leq l_m \leq 0,60 \text{ m}$

$r_o = 0,15$ m

$d = 0,10$ m

$T_m \approx -2°C$ to $-3°C$

$0 \leq P_o \leq 250$ KPa

It is obvious that test results depend at least on the accuracy with which both parameters d and T_m were taken since they remain constant during the test process. Figur 15 illustrates the soil movements above the tunnel crown after the application of the finite element method as a tool for the verification of test conditions. Empirical and numerical results as well as tunnel convergences are in good agreement. Thus, in special and well defined conditions, the numerical model is a reliable one to describe ground movements caused by tunnel driving by means of frozen shell protection.

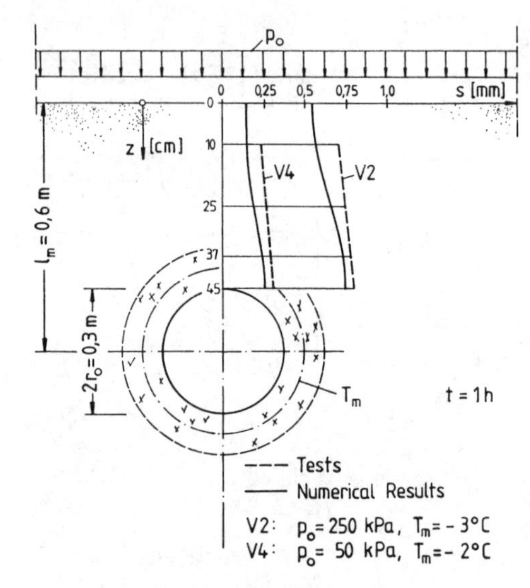

Fig. 15 Displacements from model tests and numerical analysis respectively.

6. CONCLUSION

The results of a study on ground movements due to tunnel driving by means of frozen shell protection are presented herein. The underground consists of a saturated medium sand with a fixed known density. Settlements of the ground surface are described through dimensionless relations. The above analysis allows us to draw up the following conclusions:

1. Settlement of ground surface are predictable by dimensionless equations depending on: tunnel dimensions, temperature, time and overburden load.

2. The thickness of the frozen shell and the freezing temperature may be optimized for certain boundary conditions of a tunnel project without affecting the restrictions on the settlements.

3. In some cases numerical analysis revealed a rupture of the tunnel. Significant creep strains arise from tunnels with large diameters and high loads acting on the ground surface.

4. Parametric studies will be carried on for different sand densities.

AKNOWLEDGEMENT

The investigations presented in this paper were supported by the Deutsche Forschungsgemeinschaft (DFG), Bonn, which is gratefully appreciated. Computations are performed both on the Univac 1100 of the University of Karlsruhe and the Siemens 2000 of the University of Kaiserslautern, West Germany.

REFERENCES

Goldscheider, M., 1976. Grenzbedingung und Fließregel von Sand. Mech.Res.com.Vol.3, Pergamon Press.

Kany, M., 1974, Berechnung von Flächengründungen, Berlin: Ernst & Sohn

Meißner, H., 1983, Tragverhalten achsial und horizontal belasteter Bohrpfähle in körnigen Böden. Essen: DGEG, Geotechnik 1 u. 3, S.1- 13 u. S 115 - 124.

Meißner, H., 1985, Bearing behaviour of frost shells in the construction of tunnels. Sapporo: Fourth Int. Symp. o. Ground Freezing, P. 37 - 45.

Orth, W., 1986. Gefrorener Sand als Werkstoff. Inst. f. Bodenmechanik und Felsmechanik, Universität Karlsruhe, Heft 100.

Stutz, P. 1972. Comportement elasto-plastique des milieus granulaires. Found.of. Plast. Int. Symp. Warschau

Ting, B.J.M., 1983. Tertiary Creep Model for Frozen Sands. J.of.Geotechn. Eng. ASCE, Vol. 109.

5 Rock hydraulics

Numerical Methods in Geomechanics (Innsbruck 1988), Swoboda (ed.)
© 1988 Balkema, Rotterdam. ISBN 90 6191 809 X

Finite difference and finite element modelling of an aquifer in Cretaceous chalk

P.Y.Bolly, A.G.Dassargues & A.Monjoie
Laboratory of Engineering Geology, Hydrogeology and Geophysical Prospecting, University of Liège, Belgium

ABSTRACT : The studied aquifer is located in Cretaceous chalk near Liège (Belgium). It is recharged by infiltration through the overlying loess and conglomerate. Wells and pumping adits produce a daily flow of 60000 m^3. Mathematical models have been worked out to foresee the evolutions of the water table level and to get some additional informations about the drainance axis. The finite difference model NEWSAM has been used for the steady state study of the aquifer. Some values of transmissivity obtained by the calibration of this model have been further introduced into the finite element models. Two- and three- dimensional finite element models have been developped using an adapted LAGATHER thermic conductivity program in order to study unsteady conditions of the aquifer. These two models are useful and helpful to enable good location of the permeability variations due to geological features, and good estimations of hydrogeological balances. Conclusions are drawn from the comparison and the complementarity of the two models for this kind of study.

1 INTRODUCTION

Two kinds of models are considered. The finite difference model NEWSAM is applied in quasi 3D conditions and steady state. The finite element model, using some developments of the LAGAMINE code has been adapted for modelling a water table aquifer in transient conditions with a fixed mesh structure. This new computer code is tested on uniaxial and two-dimensional sections, and then on a 2D cross-section of the Hesbaye aquifer. The 3D modelling of the whole aquifer is completed with 2670 brick finite elements.

Hydrogeological basin of Hesbaye covers an area of 350 km^2 in the NW of Liège in Belgium (figure 1). This water table aquifer is mainly composed of Cretaceous chalks which are intensively fractured in some places.

The alimentation of Liège and its suburbs is provided by the 60000 m^3/day collected in the aquifer by 45 km of adits. The hydrographic network is sparse and we can distinguish several dry valleys due to the immediate infiltration of the water in the fractured chalks. In the other places and especially on the hills, infiltration occurs through the overlying loess and conglomerates.

More than 500 wells, boreholes and piezometers have given a lot of informations about the piezometry and the geological data.

The portion of the lithostratigraphic sequence of interest consists of :

1. Primary rocks composed of Visean limestones in contact with Silurian shales, siltstones and sandstones by a fault shift.
2. Secondary layers with a 20 to 100 m thickness essentially composed by Cretaceous layers
   - the "smectite de Herve" (hardened calcareous clay) which is considered as the bottom of the aquifer
   - the compact solid white chalk, also called lower chalk, in which the main water circulations are present in the zones where the fracturation is more important
   - the Hard-Ground which is a thin layer (less than 1 m thick) of hardened chalk
   - the grey chalk or upper chalk which are exposed to deconsolidation, alteration and karst phenomena
   - the residual conglomerate which is the result of very large and superficial alteration phenomena.
3. Tertiary and Quaternary formations.
   - the loess with a thickness of 2 to 20 m

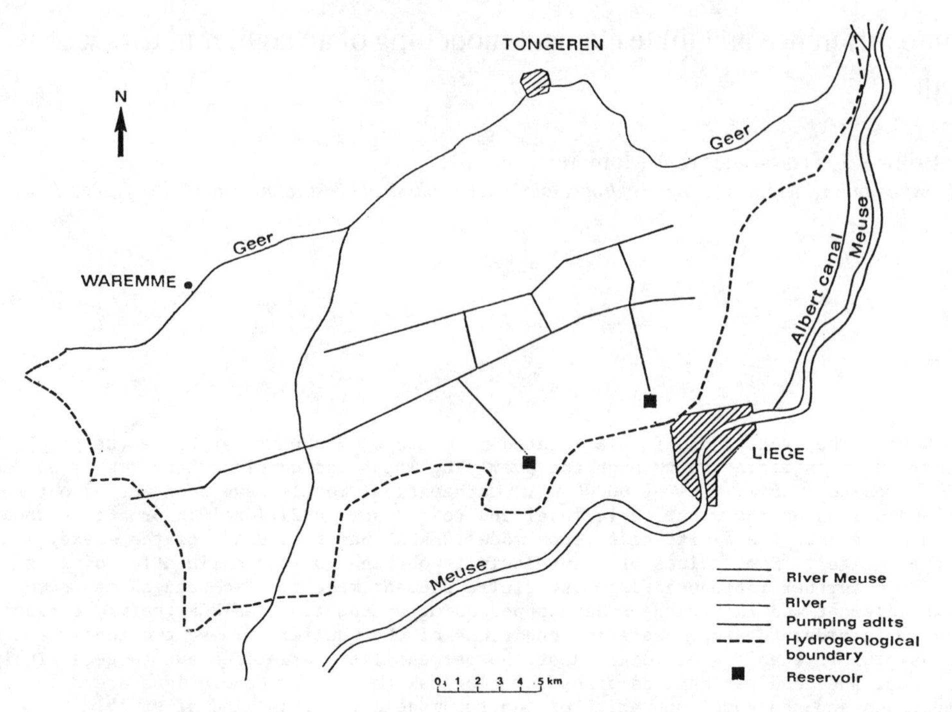

Figure 1 : Location map of Hesbaye.

- the recent alluvial and colluvial deposits especially in the valley of the River Geer

Hydrogeological parameters of the different geological units have been provided by 150 pumping tests. Between 1951 and 1984, piezometric maps have been drawn every year and piezometerhead evolution in function of the time has been recorded at different "check points".

2 FINITE DIFFERENCE MODEL

The finite difference computations are realized with an adapted NEWSAM program, solving the second-order partial differential equation of the flow in porous media. A quasi-three-dimensional idealization is obtained by the superimposition of 2 layers separated by an aquitard implicitely represented by vertical drainance coefficients.

Dirichlet conditions along the aquifer boundaries are settled to take into account the flow of the River Geer (in the North) and the lateral exchanges with the phreatic aquifer of the River Meuse (in the South-East). Initially, the pumping adits were also simulated by Dirichlet or Fourier conditions neglecting possible overpressure conditions in the adits. Later on, an explicit representation using an additional layer is prefered. The range of permeability coefficients of this special layer is from 1 to 20 m/sec, in order to compound with gradient problems (numerical stability, convergence) and with the reduction of the losses of energy.

The calibration procedure of the model is realized with a total of 4647 cells, for both years 1951 (low water level) and 1984 (high water level), assuming steady state and water table conditions.

The results appear satisfactory, seeing that :

- the main features of the water table are correctly reproduced (figure 2)
- the flows collected by the different sections of the pumping adits are correctly withdrawn, by way of reduced exchanges on the western, southern and eastern boundaries.
- the simulated global water-balances are in agreement with those calculated.

The main outputs of the model consist in :

- global groundwater balances, with quantification of the flows exchanged through the aquifer boundaries
- local groundwater balances and quantification of the laterally and

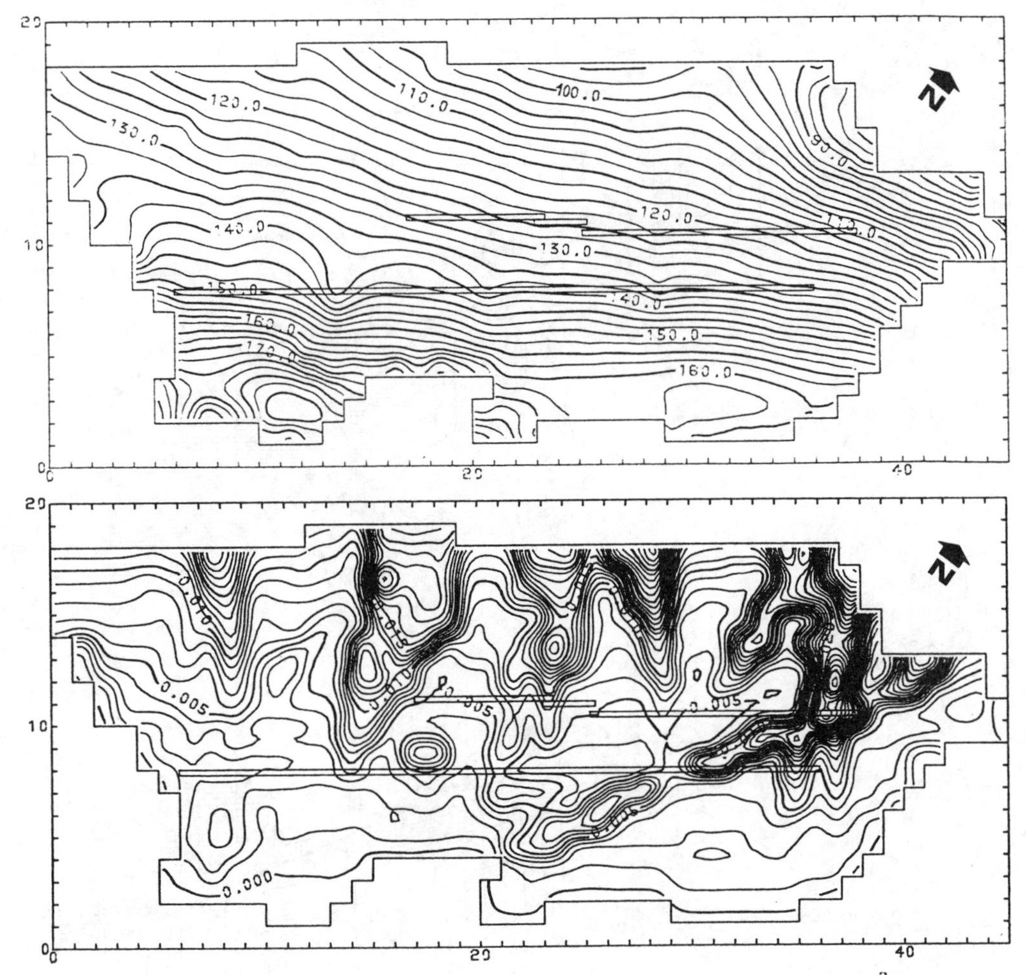

Figure 2 : Piezometerhead map and transmissivity field (equidistance 1.10^{-3} m²/sec).

vertically discharges drained by the wells and the pumping adits

- complete fields of permeability and vertical drainance coefficients; these fields, in agreement with the results of pumping tests, show obviously the drainage produced by a dry valley system and the "screen-effect" locally induced by the presence of the Hard-Ground.

For example, the figure 2 shows the transmissivity field of the upper chalk, with its particular dry valley system.

Transmissivity values are from 8.10^{-4} to $3.7\ 10^{-2}$ m²/sec for the upper chalk, from $1.1\ 10^{-4}$ to $6.6\ 10^{-3}$ m²/sec for the lower chalk and the vertical drainance coefficient is from 9.10^{-6} to $1.6\ 10^{-4}$ sec^{-1} for the Hard-Ground.

This NEWSAM model is also used in unsteady state to interpret multilayer pumping tests and the results are compared with those of the BRUGGEMAN method.

3 FINITE ELEMENT MODELS

3.1 Modelling 2D cross-section

Two different discretizations have been tested for modelling the cross-section (figure 3). Calibration of this section was really uneasy because of the lack of precision in the estimation of the withdrawal data to be considered in the adit for a 1 meter thick section. Lateral exchanges are neglected. The main constatation made from the comparison between

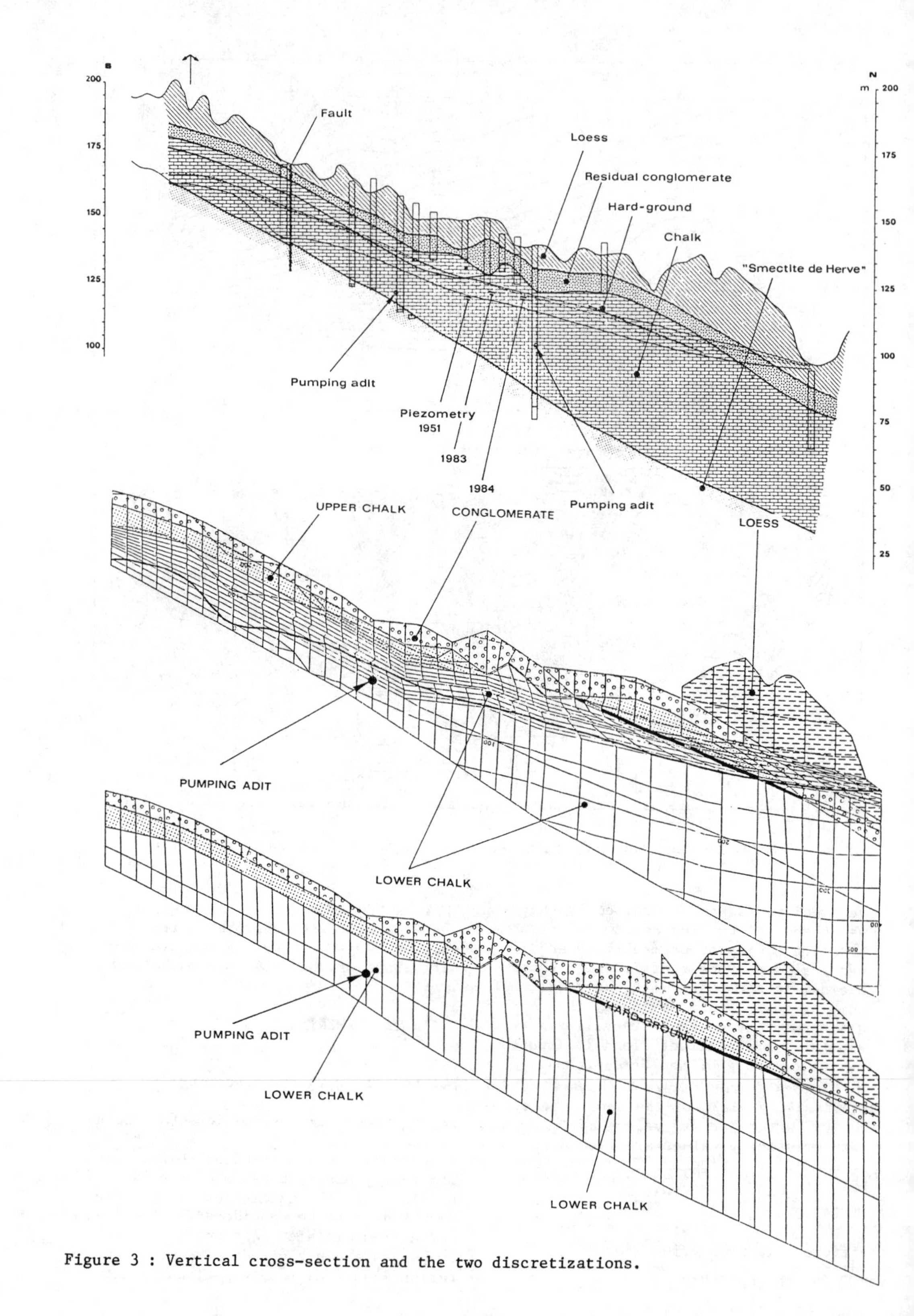

Figure 3 : Vertical cross-section and the two discretizations.

results of these two models (of the same cross-section) is that the small benefit in accuracy gained with fine mesh doesn't justify the use of so fine mesh in the structure for the 3D modelling.

3.2 3D Modelling

Finite elements used in this model are 8-nodes isoparametric bricks. Their edges are line segments and interpolating functions are linear and written in function of the three main coordinates. The change of variable is made using interpolation functions themselves which are satisfying the displacement compatibility and the geometrical compatibility of all the elements.

3D structure have to take into account a lot of geometrical data to obtain the final discretisation : pumping wells and adits, main geological faults, limits of geological units, hydrogeological basin limits ...

Five main strata of 534 elements have been distinguished : 6 levels of nodes. From one level to the next one, only Z coordinate changes. By this way, the running numbers of each node and of each element of the discretized structure can be computed from one station to the next one by simple addition of an increment. Complexity of the whole discretized structure is very high (figure 4).

Modelling of the pumping adits is realized with uniaxial "pipe elements" with infinite permeability in regard to the neighbouring elements. In the proximity of these pipe elements, the dimension of the mesh has been decreased in order to get accurate informations in this area from where a lot of water is withdrawn.

Prescribed heads (Dirichlet condition) are imposed on the northern boundary at the River Geer. The other boundaries are assumed to be impermeable (Neuman condition). As this aquifer is clearly a water table aquifer, the two hydrogeological parameters are the permeability and the effective porosity. Different "materials" have been defined corresponding essentialy to the different geological units.

Material 1 :
lower chalk $2.10^{-5} \leqslant K \leqslant 5.10^{-4}$ m/sec
$0.075 \leqslant S \leqslant 0.12$

Material 2 :
dry valley $2.10^{-4} \leqslant K \leqslant 2.10^{-3}$ m/sec
$0.075 \leqslant S \leqslant 0.20$

Material 3 :
Hard-Ground $5.10^{-6} \leqslant K \leqslant 2.10^{-4}$ m/sec
$0.075 \leqslant S \leqslant 0.10$

Material 4 :
upper chalk $2.10^{-4} \leqslant K \leqslant 1.10^{-4}$
$S \leqslant 0.15$

Figure 4 : The 4th layer of the discretized structure showing the complexity of the geometrical problem.

Material 5 :
alluvial deposits and/or residual conglomerate

$$3.10^{-5} \leqslant K \leqslant 1.10^{-3}$$
$$0.075 \leqslant S \leqslant 0.20$$

Material 6 :
conglomerate and loess

$$1.10^{-6} \leqslant K \leqslant 1.10^{-5}$$
$$0.075 \leqslant S \leqslant 0.10$$

Material 7 :
fault material

$$1.10^{-4} \leqslant K \leqslant 1.10^{-2}$$
$$0.075 \leqslant S \leqslant 0.15$$

The time step for the calibration procedure has been chosen to one year. The initial conditions are the piezometer-heads of the year 1951. The model has performed until 1984. Additional data are

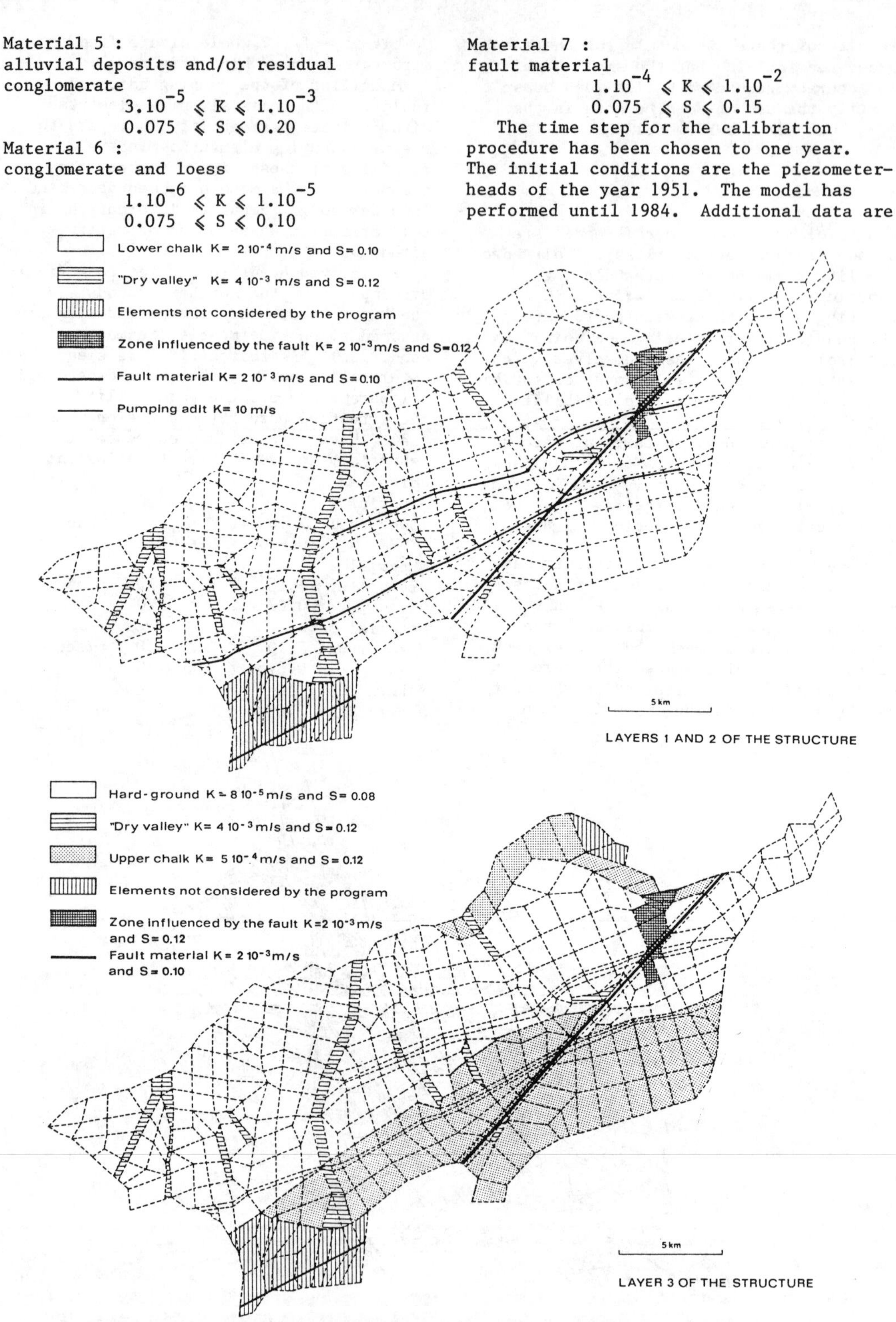

Figure 5 : Final repartition of the different materials in the structure for layers 1, 2 and 3.

the annual pumping rates in the adits and the values of effective infiltration during all these years.

The calibration procedure has been realized mainly in comparing measured piezometerheads and computed ones. This comparison is made with the piezometry maps, some cross-sections chosen in the structure and the "check points". During this procedure, values of the permeabilities and effective porosities have been setted up and some zones have been changed of "material". All these changes are made in the brackets of logical values of the parameters. At the end of the calibration, the computed flow of the River Geer is consistent with the reality and the computed piezometerheads are very close to the measurements. At this step, the final values of the materials parameters are deduced (figures 5 and 6). Figure 7 give an example of computed piezometry and flow maps of the aquifer in 1966.

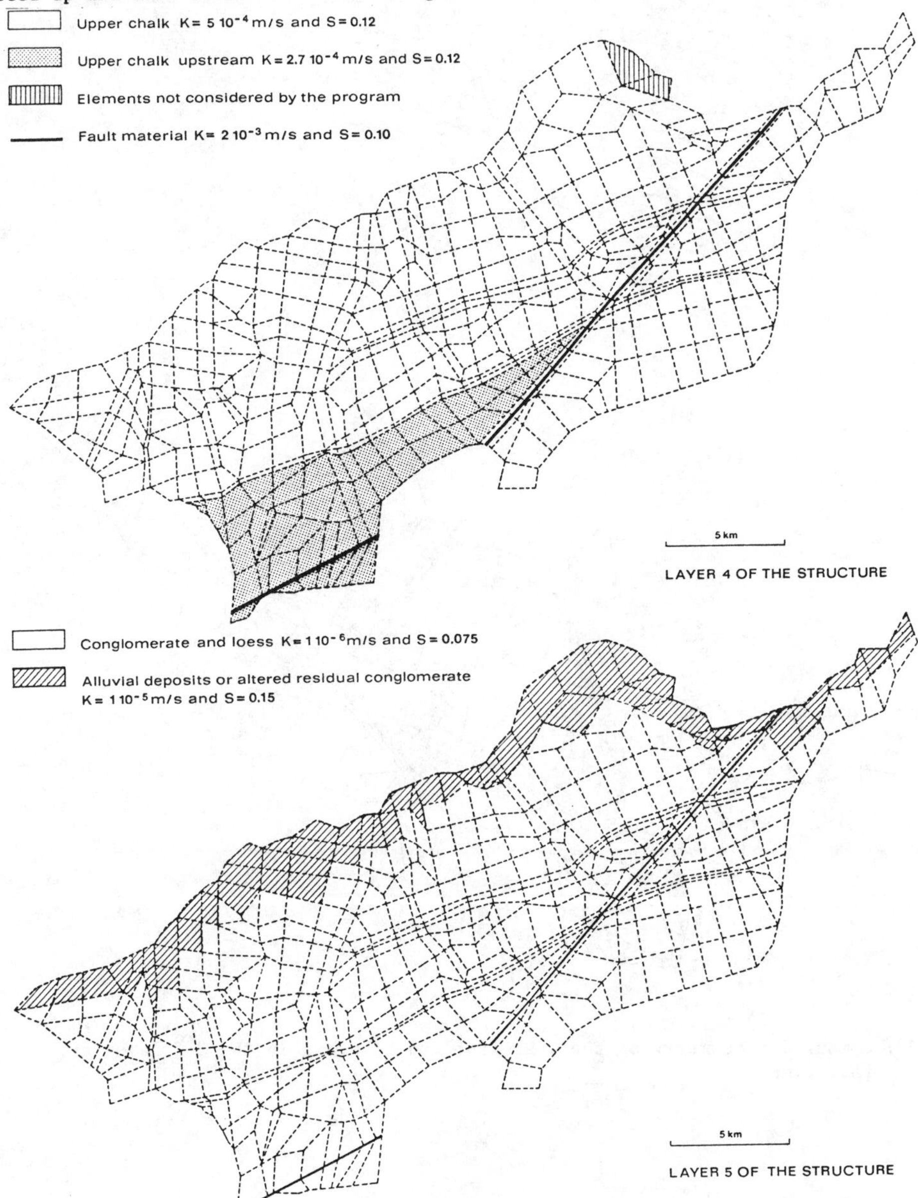

Figure 6 : Final repartition of the different materials for layers 4 and 5.

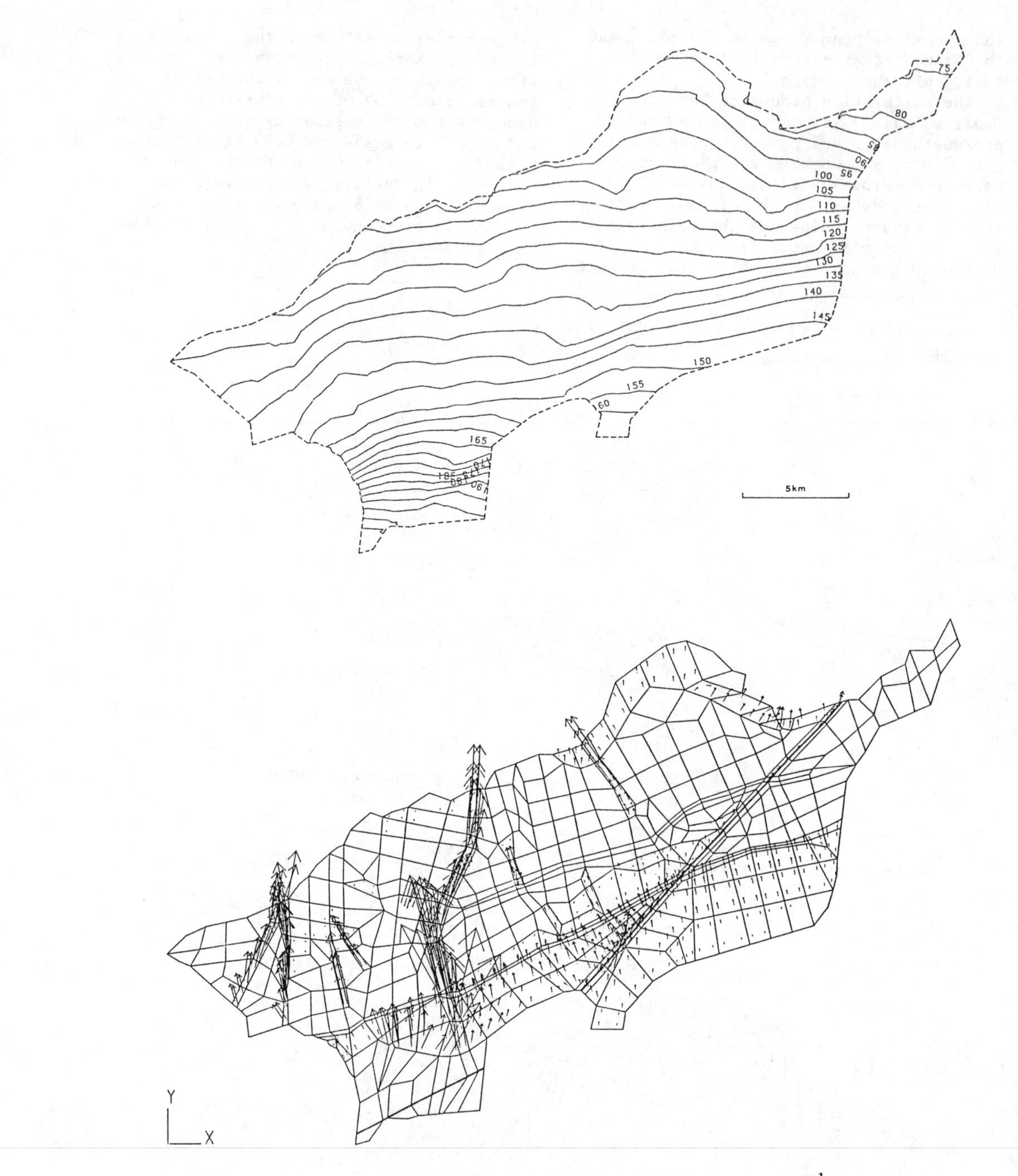

Figure 7 : Computed piezometry of the year 1966 and flow map in the 3rd layer at this time.

4 CONCLUSIONS AND FUTURE PROSPECTS

These two models allow us to start a "dynamic management" of this important aquifer. It's not our purpose to compare the approximations made in each model. In our case, we can emphasize on their good complementarity. The finite difference model in spite of big approximations about the geometry gives a good answer especially in the analysis of the transmissivity field in steady-state conditions. The finite element 3D model, more accurate in its conception gives very good results in transient conditions. The outputs of this LAGAMINE code are numerous and useful, giving piezometry maps, flow maps, cross-sections in the structure, evolution of the piezometerheads in function of the time at different "check points" ... The main problem of this model is the computation time; for a simulation from 1951 to 1984, it takes about 46 CPU hours on a Micro VAX II.

5 ACKNOWLEDGEMENTS

This work was carried out under a contract with the Water Office of the Walloon Region of Belgium in cooperation with Mr DERIJCKE of the Belgian Geological Survey who provided the NEWSAM code. The authors are grateful to R. CHARLIER, S. CESCOTTO and J.P. RADU of the University of Liège who have provided the numerical code LAGAMINE and have developped it for modelling the water-table aquifer in transient conditions.

6 REFERENCES

H. AHMED et D.K. SUNEDA, Non linear flow in porous media. Proc. A.S.C.E., Hydraulics Div., 95, HY6, 1847-59 - 1969.

G. ASSENS, Eléments d'hydrogéologie mathématique. Centre d'Informatique Géologique. E.N.S.M.P. 59 p - 1974.

K.J. BATHE et M.R. KHOSHGOFTAAR, Finite element free surface seepage analysis without mesh iteration. Int. J. Num. Anal. Meth. Geomech, 3, 13-22 - 1979.

S.I.E. BLOK, A. LEIJNSE, Report of the T.N.O. Commission for Hydrological Research. Groundwater Models and Numerical Computer Software. The Hague. 72 p. - 1978.

C.A. BREBBIA, S.Y. WANG, Finite Elements in Water Resources. Proceedings of the 5th International Conference Burlington, Vermont, U.S.A. 823 p. - 1984.

A. BRIXKO, L'alimentation en eau de l'agglomération liégeoise. in La technique de l'eau et de l'assainissement n° 402-403, p. 87-95. 1980.

E. BRUCH, Résolution par éléments frontières des écoulements permanents en milieu poreux, à surface libre éventuellement indéterminée. Travail de fin d'étude, F.S.A., Université de Liège. - 1984-1985.

G.A. BRUGGEMAN, Analyse van de bodemconstanten in een grondpakket, bestaande uit twee of meer watervoerende lagen gescheiden dor semi-permeabele lagen, Rapport Scientifique non publié, 1966.

L. CALEMBERT, Le Crétacé supérieur de la Hesbaye et du Brabant. Ann. Soc. Géol. de Belgique. Tome LXXX. pp. 129-165. - 1956.

R. CHARLIER, J.P. RADU, S. CESCOTTO, Simulation numérique des écoulements transitoires à surface libre en milieu poreux, Actes du premier congrès national belge de Mécanique Théorique et Appliquée, Bruxelles, Mai 1987, Editeur M. DEVILLE (U.C.L.).

G. GOMINI, S. DE GUIDICE, R.W. LEWIS, O.C. ZIENKIEWICS, Finite element solution of non-linear heat conduction problems with special reference to phase change. Int. J. For Num. Meth. in Eng., Vol. 8, pp. 613-624.- 1974.

G. de MARSILI, Hydrogéologie quantitative. Masson, Paris, pp. 1-215.-1981.

F. DERYCKE, Modèle mathématique déterministe de la nappe des "craies" du bassin de Mons.- 1978.

C.S. DESAI, Finite element residual scheme for unconfined flow. Int. J. Num. Meth. Eng., 10, pp. 1415-1418.- 1976.

C.S. DESAI et G.C. LI, A residual flow procedure and application for free surface flow in porous media. Int. J. Advances in Water Resources, 6, 27-35.- 1983.

Y. EMSELLEM, Construction de modèles mathématiques en hydrogéologie. Centre d'Informatique Géologique E.N.S.M.P. 125 p.- 1971.

G. FONDER, Application des ordinateurs au calcul des structures. Notes de cours. U.Lg.- 1985.

P. GHIJSEL, Crétacé de Hesbaye. Rapport final M.R.W.- C.I.L.E.- 1985.

J. GHABOUSSI et E.L. WILSON, Flow of compressible fluid in porous elastic media. Int. J. Num. Meth. Eng., 5, 419-442.- 1973.

P. JUNGELS, Sondages en Hesbaye. Professionnal Paper n° 15. 1968.

E. LEDOUX, Programme Newsam : principe et notice d'emploi. Centre d'Informatique Géologique. E.N.S.M.P. 55 p.- 1978.

R. LEGRAND, Précision sur le rejet de la Faille Bordière. Professional Paper n° 146.-1977.

E. LEROUX, J. RICOUR, G. WATERLOT, La surface piézométrique de la nappe de la craie du Nord de la France. Société Géologique du Nord. Annales. Tome LXXX. pp. 234-240.- 1960.

A. MONJOIE, Observations nouvelles sur la nappe aquifère de la craie en Hesbaye.- 1966.

A. MONJOIE, Hydrogéologie. Notes de cours U.Lg.- 1984.

A. MONJOIE, Compléments de Géologie de l'Ingénieur et d'hydrogéologie. Notes de cours. Inédit.- 1985.

S.P. NEUMAN, Saturated-unsaturated seepage by finite elements. ASCE, Hydraulics Div., 99, HY12, 2233-2250.- 1973.

J.C.J. NIHOUL, R. WOLLAST, Hydrodynamic and Dispersion Models. Boundary Fluxes and Boundary Conditions. pp. 11-198.- 1983.

J. PEL, Observations géologiques et hydrogéologiques sur le territoire de la commune de Vottem. Ann. Soc. Géol. de Belgique. Tome LXXXIII., pp. 345-350.- 1960.

Ch. PIETTE, Application des éléments finis à la détermination de la surface piézométrique d'une nappe d'eau souterraine. Travail de fin d'étude, F.S.A., Université de Liège.- 1976.

Ch. PIETTE et S. CESCOTTO, Application des éléments finis à la détermination de la surface piézométrique d'une nappe d'eau souterraine. Journée d'études. La méthode des éléments finis appliquée.

R.S. SANDHU et E.L. WILSON, Finite element analysis of seepage in elastic media. Proc. A.S.C.E., Engineering Mechanics Div., 95, EM3, 641-651.- 1969.

R.S. SANDHU, H. LUI et K.J. SINGH, Numerical performance of some finite element schemes for analysis of seepage in porous elastic media. Int. J. Num. Anal. Meth. Geomech., 1, n° 3, pp. 177-194.- 1977.

R.L. TAYLOR et C.B. BROWN, Darcy flow solution with a free surface. A.S.C.E., Soil Mechanics and Foundation Div., 98, SM11, pp. 1143-1162.- 1972.

J. THOREZ et A. MONJOIE, Lithologie et assemblages argileux de la smectite de Herve et des craies campaniennes et Maastrichtiennes dans le Nord-Est de la Belgique. Ann. Soc. Géol. de Belgique. Tome 96, pp. 651-670.- 1981.

R.E. VOLKER, Non linear flow in porous media by finite elements. Proc. A.S.C.E., Hydraulics Div., 95, HY6, 2093-114.- 1969.

Numerical Methods in Geomechanics (Innsbruck 1988), Swoboda (ed.)
© 1988 Balkema, Rotterdam. ISBN 90 6191 809 X

Free-surface tracking through non-saturated models

D.Aubry
Ecole Centrale de Paris, Chatenay Malabry, France
O.Ozanam
Coyne et Bellier, Paris, France

ABSTRACT : The first filling and the rapid drawdown of the reservoir of an earthdam contribute largely to the stability of it. A coupled mechanical-hydraulic analysis must take into account the effects of partially saturated parts of the dam. To deal with such an analysis three developments implemented in the computer code GEFDYN will be presented in the paper : hydraulic model for the non saturated soil ; an initial hydraulic condition formulated with respect to the initial degree of saturation ; a numerical formulation for the hydraulic boundary conditions on the potential seepage surface. The transient free surface may then be determined automatically. Some applications are presented.

Introduction

The first filling and drawdown of the reservoir of an earthdam induce numerical difficulties due to the transient free surface. To determine its position the properties of the partially saturated part of the dam (above the free surface) must be taken into account and introduced in the coupled mechanical-hydraulic modelling of the saturated soils. Then the free surface tracking is involved in the analysis : the variation of the degree of saturation is continous through the free surface, which is defined as the contour of the zero pore pressure.

Numerically the partial saturation introduces a nonlinear hydraulic behaviour in the non saturated part of the model in addition to the non linear mechanical behaviour. To deal with the movements of the reservoir, the potential seepage surface must be modelized properly : a numerical formulation for these particular hydraulic boundary conditions has been developped by introducing hydraulic interface elements [8].

The whole model is presented in the next sections with some emphasis on the hydraulic interface and its implementation in the finite element computer code GEFDYN. Some applications have been performed to study the numerical influence of this interface element in case of nondeformable and deformable matrix.

The following short and hand notation will be used in the remainder of this paper : for any vector field **v** and any stress field **s**, we shall write :

$$(\mathbf{v},\mathbf{v})_\Omega = \int_\Omega \mathbf{v}.\mathbf{v}\, d\Omega \,; \quad (\mathbf{s},\mathbf{s})_\Omega = \Sigma_{ij} \int_\Omega \mathbf{s}_{ij}\mathbf{s}_{ij}\, d\Omega$$

$$<\mathbf{v},\mathbf{v}>_\Gamma = \int_\Gamma \mathbf{v}.\mathbf{v}\, d\Gamma$$

1. BASIC EQUATIONS OF THE NON SATURATED MODEL

A brief review of the necessary equations is presented in this section. A more detailed description of the coupled mechanical-hydraulic modelling of partially saturated soils is given in Aubry et Al.[1]. While the porous media consists of three phases its behaviour depends on the balance equation and constitutive law of each constituant. However the pore air is assumed to flow at the same velocity as the pore water and its mass is negligible, so that no equation concerning the gazeous phase is written.

First the balance equation of the fluid is written [2] [4] :

$$(1) \quad c(\mathbf{u},p)\ \partial_t p + S(p)\ \mathrm{div}\,(\partial_t \mathbf{u}) - \mathrm{div}\,(\mathbf{K}(\mathbf{u},p)\,.\,\mathbf{grad}\,(p + \rho_w g.\mathbf{x})) = 0$$

in which the unknowns are the matrix displacement **u** and the pore pressure p.

The rate of storage c(**u**,p) is defined by :

$$c\,(\mathbf{u},p) = n\,(\mathbf{u})\,(\,S(p)\,\beta + S'(p)\,)$$

and is influenced by the compressibility β of the fluid [10], the rate of the degree of saturation S and the rate of porosity n. The displacement of the solid matrix **u** influences the fluid flow through the volumetric strain, which modifies the available pore space.

The filtration velocity **V** is related to the gradient of pore pressure by the generalized Darcy's law [6] :

(2) $\mathbf{V} = -\mathbf{K}.\mathbf{grad}\ (p + \rho_w\, g.\, \mathbf{x})$

The permeability tensor **K** is defined by :

$$\mathbf{K} = k_r(S)\, \mathbf{K'}(\mathbf{u}) / \mu$$

where the relative permeability k_r is an irreversible function of the degree of saturation, and the intrinsic permeability **K'** depends only on the geometrical distribution of the pores and thus of the porosity.

Moreover the overall mechanical equilibrium is considered in the porous medium. To deal with it, the extension of the principle of effective stresses to partially saturated soils has been chosen as proposed by Bishop [2] [3] :

(3) $\sigma = {}'\sigma' - S(p)\, p\, \delta$

where σ, σ', δ designate respectively the total and effective stress fields and the Kronecker symbol.

The total equilibrium equation becomes :

(4) $\mathbf{DIV}\, \sigma' - \mathbf{grad}\, (S(p)\, p) + \rho(\mathbf{u},p)\, \mathbf{g} = 0$

where the bulk volumetric mass ρ is defined by :

$$\rho(\mathbf{u},p) = n(\mathbf{u})\, S(p)\, \rho_w + (1 - n(\mathbf{u}))\, \rho_s$$

Lastly the constitutive soil behaviour is given by an elastoplastic theory, which relates the rate of effective stress to the rate of strain. It is not the purpose here to discuss the detailed specification of the used constitutive equations [5].

2. HYDRODYNAMICAL CHARACTERISTICS

The fonctions $S(p)$ and $k_r(S)$ are determined experimentally for each material. These relations are assumed to be biunivocal, so that the hysteretic behaviour of the degree of saturation during drainage and imbibition is neglected. A lot of analytical approximations have been proposed in the literature. The Van Genuchten model [11] [9] has been chosen, which is satisfactory enough for several porous media, and which is given in a simplified form by :

(5) $S(p) = S_r + (1 - S_r) / [\, 1 + (\alpha\, p / \rho_w\, g_z)^2\,]^{1/2}$
and
$k_r(S) = (S - S_r)^3 / (1 - S_r)^3$

S_r is the residual degree of saturation and α is a mathematical parameter, which influences the curvature of the function $S(p)$. It infers from these relations the storage function $c(p)$. These functions are represented on the figure 1. While the saturation S and the permeability K are monotonous functions of the pressure, the storage coefficient $c(p)$ presents a maximum value at a negative pressure p_M, i.e. at a degree of saturation S_M less than one. This special feature induces numerical difficulties in the choice of the auxiliary matrix as shown in section 4.

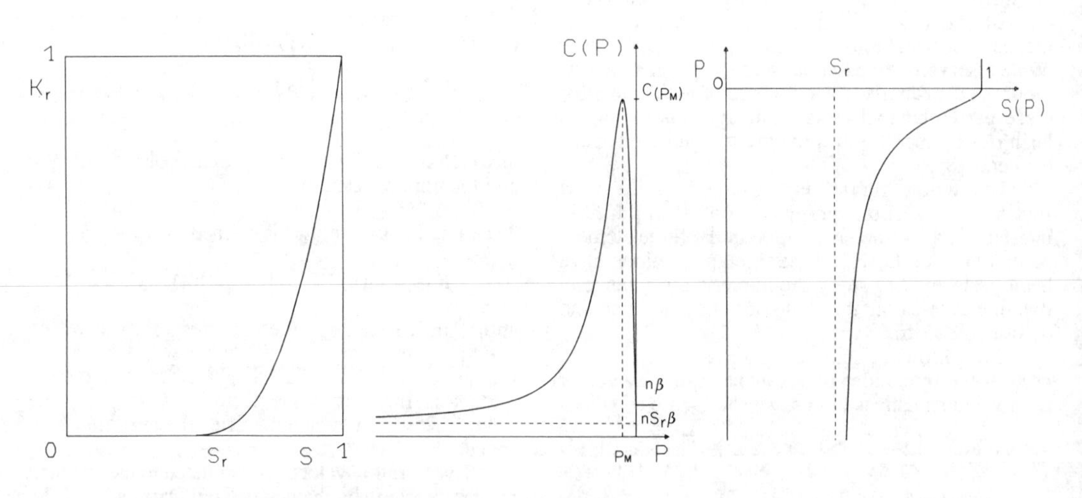

Fig.1 : Shape of the hydrodynamical characteristics $S(p)$, $k_r(S)$, $c(p)$

3. MODELLING OF THE SEEPAGE SURFACE

The hydraulic boundary conditions are usually of two types : imposed pressures on Γ_p or imposed fluxes on Γ_ϕ, which are complementary parts of the boundary of the domain. When dealing with partial saturation an additionnal and particular boundary condition must be introduced on the potential seepage surface Γ_s [7], the position of which is unknown and which may move. Along this surface, the boundary condition is either a pressure imposed or a flux imposed condition, depending on the hydraulic state at the interface (inside and outside). Such boundary conditions take place usually on the downstream face of dams, but also upstream if filling or drawdown of the reservoir occur. First the modelling of a seepage surface will be described and then will be generalized to the other cases.

If the external pressure is equal to zero, i.e. on the seepage surface, when the water flows its pressure must be in equilibrium with the atmospheric pressure. Hence as long as the pore pressure is negative the flux vanishes and as soon as the pore pressure becomes positive it must be equal to the ambiant atmospheric pressure. This leads to the following conditions at the interface Γ_s, represented in bold line on the figure 2 :

$$p \leq 0 \quad \text{and} \quad \phi \geq 0 \quad \text{and} \quad p \,.\, \phi = 0 \quad \text{on } \Gamma_s$$

It means, with our choice of hydrodynamical characteristics, that :

$$p = p_{atm} = 0 \qquad \text{if } S(x,t) = 1 \quad \text{on } \Gamma_s$$
$$\phi = 0 \qquad \text{if } S(x,t) < 1 \quad \text{on } \Gamma_s$$

The switch between Dirichlet boundary condition and Neumann boundary condition depending on the value of the saturation is the main difficulty because the flux boundary condition is automatically satisfied by the weak formulation. On the contrary the Dirichlet boundary condition must be imposed directly on the degrees of freedom. To deal with this difficulty a penalization technique is employed : the ϕ-p curve, represented on the figure 2, is penalized by introducing a coefficient k_v, called fictive permeability, such that :

$$(6) \quad \phi = 0 \qquad \text{if } p \leq 0 \text{ and } p_{ext} = 0$$
$$\phi = k_v\, p \qquad \text{if } p > 0 \text{ and } p_{ext} = 0$$

On the other hand if the external pressure is due to the hydrostatic water pressure and no more to the air pressure, the boundary condition on pressure may be imposed in the same way, by using the fictive permeability k_v and defining a flux through the surface with :

$$(7) \quad \phi = k_v\,(p - p_{ext}) \qquad \text{if } p_{ext} > 0 \text{ and } \forall p \in R$$

The term introduced in the weak formulation to take into account the seepage or the change of external pressure is written :

$$(8) \quad < \phi \,,\, q >_{\Gamma s} = < k_v\,(p - p_{ext})\,,\, q >_{\Gamma s^*}$$

where q is a virtual pressure and Γ_s* is the part of the surface Γ_s where seepage occurs (p_{ext}=0 and p>0) or where the external pressure is strictly positive (p_{ext}>0). If k_v is sufficiently large, the other terms in the weak formulation are negligible, so that $\phi \approx 0$, i.e. $p \approx p_{ext}$, on Γ_s*. The wanted boundary condition is satisfied.

The numerical modelling of this seepage condition will be developped in section 5. The applications presented in the section 7 will demonstrate that the fictive permeability must be understood as a numerical parameter.

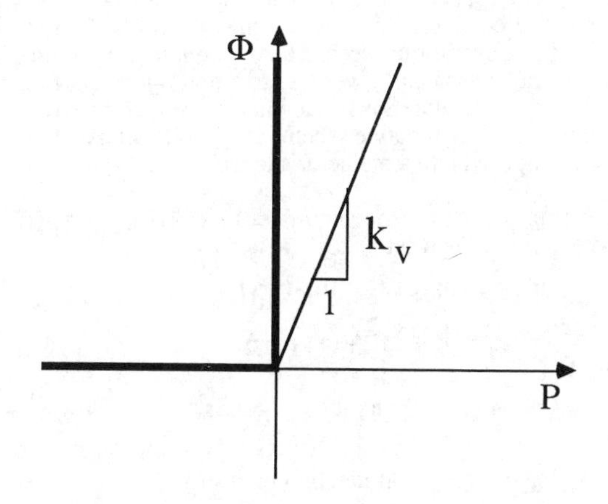

Fig.2 : Definition of the boundary conditions on the potential seepage surface
Description of the penalization method

4. WEAK FORMULATION AND TIME DISCRETIZATION

Let **V** be the space of admissible displacement **v** and let Q be the space of admissible pressure q the weak formulation of the continuous problem (eq. (1) and (4)) reads :

$$(9) \quad (\sigma', \varepsilon(\mathbf{v}))_\Omega - (S(p)p\,,\, \text{div}\,\mathbf{v})_\Omega = (\rho\,(\mathbf{u},p)\,\mathbf{g}\,,\,\mathbf{v})_\Omega \qquad ,\forall\, \mathbf{v} \in \mathbf{V}$$

$$(10)\ (S(p)\ \text{div}\,(\partial_t \mathbf{u})\,,\, q)_\Omega + (c(\mathbf{u},p)\,\partial_t p\,,\, q)_\Omega$$

$$+ (\mathbf{K}(p)\ \mathbf{grad}\ (p+\rho_w \mathbf{g}\cdot\mathbf{x}) , \mathbf{grad}\ q)_\Omega$$

$$+ <k_v (p-p_{ext}) , q>_{\Gamma s^*} = < \varphi , q>_{\Gamma\phi}$$

$$,\forall q \in Q$$

The couple ($\mathbf{u}$, p) must be a solution of the above weak form and furthermore satisfy the mechanical constitutive equation and the initial conditions. The displacements and pore pressures are expanded with the finite element basis functions w_{Ii} and r_I where I refers to a node number and i to a space direction so that :

$$\mathbf{u}_h(t) = \Sigma_{Ii}\, u_{Ii}(t)\, \mathbf{w}_{Ii} \ ; \ p_h(t) = \Sigma_I\, p_I(t)\, r_I$$

Full time discretization

The time variable plays an important role in the hydraulic equation. The total time interval [0, T] is cut into several time steps Δt. An *implicit* integration scheme is used and gives rise to a non linear system of algebraic equations that must be solved at each time step. The iterative scheme is the following and is decomposed into a predictor and corrector stages :

Predictor : At time step number n, let $\{\mathbf{u}_n , p_n, \sigma'_n\}$ be known. Then set :

$$\mathbf{u}^o_{n+1} = \mathbf{u}_n + \Delta t_{n+1} (\mathbf{u}_n - \mathbf{u}_{n-1})$$

$$p^o_{n+1} = p_n + \Delta t_{n+1} (p_n - p_{n-1})$$

$$\sigma'^o_{n+1} = \sigma'_n$$

Global corrector : At iteration number k,

let $\{ \mathbf{u}^k_{n+1} , p^k_{n+1} , \sigma'^k_{n+1} \}$ be known,

then search for $\{\Delta\mathbf{u}, \Delta p\}$ such that :

$$(\mathbf{D}_n^* \,\varepsilon(\Delta\mathbf{u}),\varepsilon(\mathbf{v}))_\Omega - (S_n^*\, \Delta p, \mathrm{div}\ \mathbf{v})_\Omega = F_M$$

$$,\forall \mathbf{v} \in \mathbf{V}_h$$

$$(S_n^* \,\mathrm{div}\,(\Delta\mathbf{u}) , q)_\Omega + (c_n^* \Delta p , q)_\Omega$$

$$+ \Delta t_{n+1} (\mathbf{K}_n^*\ \mathbf{grad}\ \Delta p, \mathbf{grad}\ q)_\Omega$$

$$+ \Delta t_{n+1} <k_v \Delta p, q>_{\Gamma s} = F_H$$

$$,\forall q \in Q_h$$

where $\Delta\mathbf{u} = \mathbf{u}^{k+1}_{n+1} - \mathbf{u}^k_{n+1}$

$$\Delta p = p^{k+1}_{n+1} - p^k_{n+1}$$

$$F_M = (\rho (\mathbf{u}^k_{n+1}, p^k_{n+1})\, \mathbf{g} , \mathbf{v})_\Omega - (\sigma'^k_{n+1}, \varepsilon(\mathbf{v}))_\Omega + (S(p^k_{n+1})\, p^k_{n+1} , \mathrm{div}\ \mathbf{v})_\Omega$$

$$F_H = \Delta t_{n+1} < \varphi_{n+1} , q>_{\Gamma\phi} - (S(p^k_{n+1})\, \mathrm{div}\, (\mathbf{u}^k_{n+1} - \mathbf{u}_n) , q)_\Omega - (c(\mathbf{u}^k_{n+1}, p^k_{n+1})\, (p^k_{n+1} - p_n), q)_\Omega - \Delta t_{n+1} (\mathbf{K}(p^k_{n+1}) . \mathbf{grad}\ (p^k_{n+1} + \rho_w \mathbf{g}\cdot\mathbf{x}), \mathbf{grad}\ q)_\Omega$$

$\mathbf{D}_n^*$ is an auxiliary elasticity matrix,

S_n^*, c_n^* and $\mathbf{K}_n^*$ are respectively auxiliary degree of saturation, storage capacity and permeability coefficients.

The modified Newton method is used to solve the nonlinear system. Two different choices of these auxiliary coefficients have been employed.

Firstly they are independant on the time t and equal to their maximum value :

$$S_n^* = 1 \ ; \ K_n^* = K_s = K'/\mu \ ; \ c_n^* = c(p_M).$$

In addition to the slow convergence of this iteration scheme, the three values cannot correspond to a real state of the porous medium : the storage capacity reaches never its maximum in a saturated soil.

The other possibility is to use the coefficients equal to the physical parameters at the beginning of each time step :

$$S_n^* = S(p_n) \ ; \ K_n^* = K(p_n) \ ; \ c_n^* = c(p_n).$$

The matrix must then be rebuilt at each time step or at least frequently. In certain cases, e.g., during drawdown, the convergence is faster than the first method.

5. HYDRAULIC INTERFACE FINITE ELEMENT

To compute the flux term due to seepage on Γ_s, a particular interface element has been implemented for 2D or 3D problems. This element is like a slip element without thickness and with two faces, which are useful for distinguishing between the inside and the outside. The pressure may vary either linearly or quadratically. On the outer face, the degrees of freedom are imposed equal to zero and the external pressure is computed as an hydrostatic pressure by knowing the height of the water level outside. On the inner face, the pressure is computed by :

$$p = \Sigma_I\, p_I\, r_I$$

where I refers to an inside node number and r_I are the basis functions of a line element in 2D and of a surface element in 3D. The contribution of this element in the global matrix is like :

$$\int_{\Gamma s} k_v\, r_I\, r_J\, d\Gamma_s = \Sigma_{e,l}\ k_v\, r_I(\xi_l)\, r_J(\xi_l)\ \pi_l\, J(\xi_l)$$

where J is the jacobian of the element,

ξ_l is the abscissae of the Gauss points,

π_l is the weight coefficient for Gaussian integration,

e refers to an element number.

The flux computed on the right hand side of the system is :

$$\int_{\Gamma s} k_v (p-p_{ext}) r_J d\Gamma_s = \sum_{e,l} k_v (p(\xi_l)-p_{ext}) r_J(\xi_l)\, \pi_l\, J(\xi_l)$$

To avoid a poorly conditionned matrix, the value of the parameter k_v must be not too large compared to the true permeability k_s : it is usually chosen from 5 to 100 times the value of the saturated permeability.

6. INITIAL HYDRAULIC CONDITION

Numerically the initial hydraulic conditions must be formulated with respect to the nodal pore pressure. But, practically in situ, the initial water content or degree of saturation is better controlled by the geotechnical engineer than the pore pressure. The initial degree of saturation is a characteristic of each material and then is given as a characteristic of each element. The pore pressure in each element is deduced from the initial saturation S_o and with a local smoothing the initial nodal pore pressure is determined. Because of the uniform value of S_o on the element, the smoothing is reduced to the means of the pore pressure on all the elements which contain each node.

7.APPLICATIONS

The first example is concerned with the drawdown of a 2 meters high and 3 meters large saturated column : the downstream water level goes down from the top to the bottom of the dam at the rate of 0.036 m/h. This model is purely hydraulic and hydraulic interface elements have been placed on the downstream face. Three different values of the fictive permeability have been used :

$$k_v = 10^{-6}\,s^{-1} \quad \text{or} \quad 10^{-5}\,s^{-1} \quad \text{or} \quad 10^{-4}\,s^{-1}$$

The characteristics of the clay which constitutes the dam are :

$$k_s = 8.5\ 10^{-7} m/s \quad S_r = 0.4236$$

$$\alpha = 2.2\ m^{-1} \quad n = 0.495$$

The figure 3 shows the contours of the pressures 0.(that gives the saturation front), -1000. and -2000. Pa at the time 55.6 hours. We remark that a too small fictive permeability impounds the water inside and that a too large value implies numerical errors that produce an apparent inwards flux, which really vanishes in the right hand side. A more satisfactory choice is $10^{-5}\,s^{-1}$.

The same model has been used to simulate the filling on the downstream face at the same rate as the drawdown. The column is initially desaturated and the contour of the zero pressure is at the bottom. The figure 4 shows the contours of the pressure 0., -1000. and -2000. Pa at the time 27.8 hours fo two different values of k_v. We remark that the zero pressure contour-line is in the two cases undeneath the external water level. The convergence of the iterative scheme is more difficult during the filling than during the drawdown.

The second example is the core of the Mont Cenis earthdam. This purely hydraulic model contains interface elements on the downstream and on the upstream faces. The initial state is the stationnary state when the reservoir is at the higher level. The drawdown is simulated with a time step equal to three days. The figure 5 shows the contours of the pore pressure at 210 days. We see that seepage occurs on the upstream face and also on the downstream. The characteristics of the clay are :

$$K_{sh} = 4.\ 10^{-3}\ m/s \quad \text{and} \quad K_{sv} = 10^{-3} m/s$$

$$S_r = 0.125 \text{ and } \alpha = 1.\ m^{-1}$$

$$\beta = 4.5\ 10^{-7}\ Pa^{-1}$$

and the fictiv permeability has been chosen to be $8.10^{-3}\ s^{-1}$. A computation made with $8.\ 10^{-2}\ s^{-1}$ gives about the same results.

The last example is a coupled modelling of an earthdam with a central core. The outer shells have been assumed perfectly drained so that the pore pressure was unknown only inside the core of the dam and inside the drains. The initial degree of saturation in the core is 0.9. The permeabilities of the clay and of the drains are respectively 2.10^{-8} and 4.10^{-5} m/s horizontally with a ratio of 4 between the horizontal and vertical premeabilities. The fictive permeability is $4.10^{-4}\ s^{-1}$. The construction of the dam, the filling of the reservoir and the drawdown have been simulated. The figure 6 shows the contours of the pore pressure at the end of the drawdown in the core and drains. The hydrostatic equilibrium is reached in the saturated part of the drains, but seepage occurs on the upward part of them. The figure 7 shows the deformed mesh of the entire dam at the end of the filling of the reservoir. The filling and the drawdown create supplementary settlements.

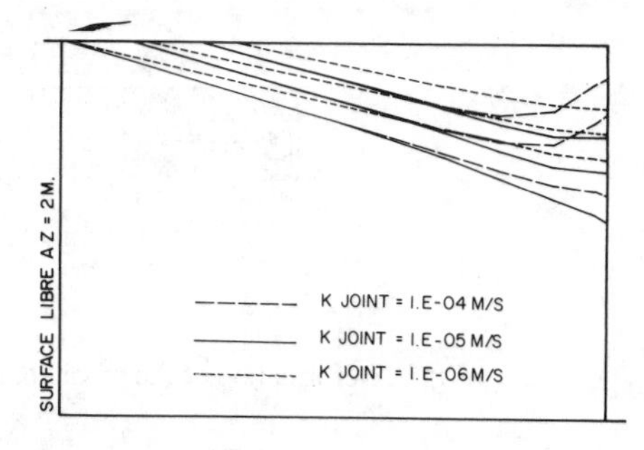

ig.3 : Drawdown : contours of the pore pressures (0., -1000., -2000 Pa) at 55.6 hours with differents fictive permeabilities

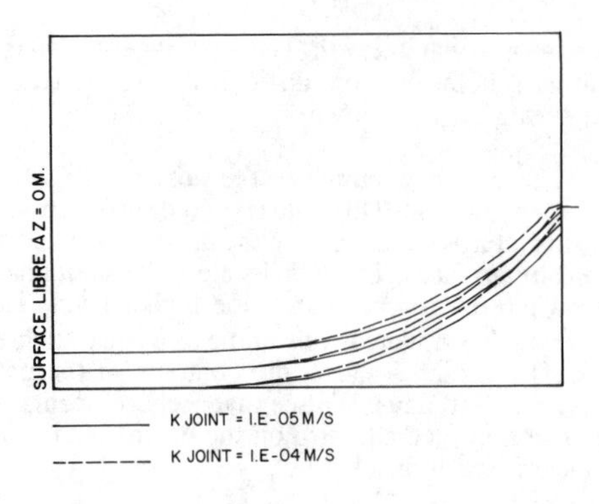

Fig.4 : Filling : contours of the pore pressures (0., -1000.,-2000. Pa) at 27.8hours with different fictive permeabilities

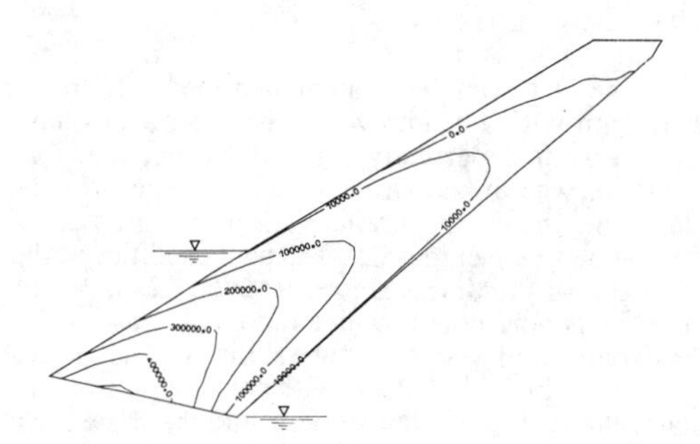

Fig.5 : Mont Cenis's core : contours of the pore pressures after drawdown

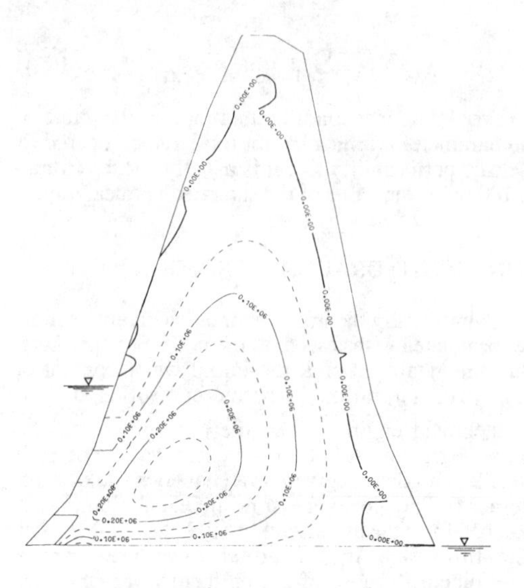

Fig.6 : Contours of the pore pressures in the core and the drains of the model dam at the end of the drawdown

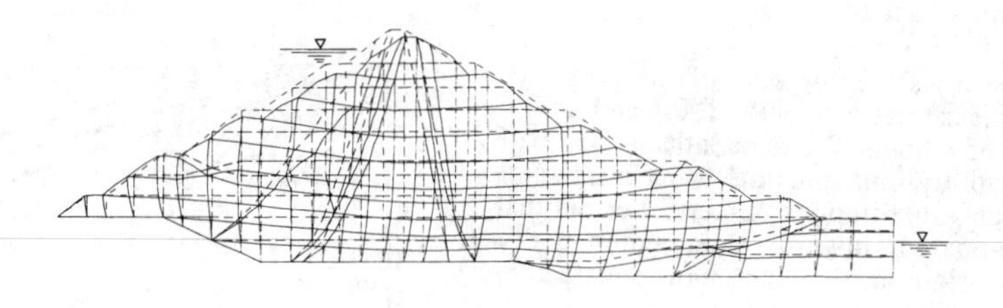

Fig.7 : Deformation of the model dam at the end of the filling of the reservoir

ACKNOWLEDGEMENTS

The analysis presented in this paper has been developped as a joint work between Ecole Centrale de Paris, Coyne et Bellier Consulting Engineers and Electricité de France. The support of Electricité de France Region de Chambéry through several research contracts is kindly acknowledged. Several fruitful discussions with M. Aufaure and M. Poupard from Electricité de France and with M. Person from Coyne et Bellier were very helpful. A. Modaressi and H. Modaressi have also taken their part in the overall implementation of the computer code *GEFDYN*.

REFERENCES

1. Aubry D., Chouvet D., Ozanam O., Person J.P. (1986) Coupled mechanical-hydraulic behaviour of earth dams with partial saturation, *Eur. Conf. Num. Meth. Geomech.*, Stuttgart, Vol.2
2. Bear J., Pinder G.F. (1981) Porous medium deformation in multiphase flow, *J. Geotech. Eng.*,ASCE, 109,N°5,734-737
3. Bishop A.W., Blight G.E. (1963) Some aspects of effective stress in saturated and unsaturated soils. *Geotechnique*, Vol. 13 177-197.
4. Ghaboussi J., Kim K.J. (1984) Quasistatic and dynamic analysis of saturated and partially saturated soils, Mechanics of Engineering Materials,(Ed. C.S. Desai and R.H. Gallagher)., Wiley.
5. Hujeux, J.C. (1985) Une loi de comportement pour le chargement cyclique des sols, Génie parasismique (Ed. V. Davidovici), Presses ENPC, 287-302.
6. Lewis R.W., Roberts P.M. (1984) The finite element method in porous media flow, 805-896,in Fundamentals of transport phenomena in porous media, (Ed. Bear and Corapcioglu), Martinus Nijhoff.
7. Neuman, S.P. (1973) Saturated-unsaturated seepage by finite elements, ASCE, *J. Hydr. Div.*, 99, 2233-2250.
8. Ozanam O. (1986) Modélisation numérique des sols élastoplastiques non saturés, Application aux barrages en remblais, Thèse de Doctorat (en préparation), Ecole Centrale de Paris.
9. Person J.P. (1981) Caractérisation des propriétés thermohydrauliques d'un sol non saturé, Thèse de Docteur-Ingénieur, Université de Grenoble.
10. Schuurman I.E. (1966) The compressibility of an air/water mixture and a theoretical relation between the air and water pressures, *Géotechnique*, Vol. 16, No 4, 269-281.
11. Van Genuchten (1980) A closed form equation for predicting the hydraulic conductivity of unsaturated soils, *Soil Sciences Am. Soc*, 44, 892-898.

Numerical Methods in Geomechanics (Innsbruck 1988), Swoboda (ed.)
© 1988 Balkema, Rotterdam. ISBN 90 6191 809 X

Finite element double-porosity model for deformable saturated-unsaturated fractured rock mass

Yuzo Ohnishi & Takuo Shiota
Kyoto University, Japan
Akira Kobayashi
Hazama-gumi, Tokyo, Japan

ABSTRACT: An idealized double porosity model is developed for the purpose of studying the coupled effects of flow in porous blocks and fractures as well as solid displacement in saturated-unsaturated medium. Subsequently, Galerkin formulation is used for the finite element method to develop a new technique to investigate coupled hydraulic-mechanical behavior in the double porosity reservoir. The verification are performed in comparison with an analytical solution of one-dimensional consolidation problem and experimental results of unsteady flow in the sand box. Finally, a secondary compression is examined with this model, the environment of rock mass in Lugeon test is simulated, and two-dimensional consolidation problem in a saturated and saturated-unsaturated media are investigated in comparison with the single porosity model.

1 INTRODUCTION

It is a common procedure to take account of the effects of seepage water on the design and construction of structure in the field of soil mechanics and rock mechanics. The way to totally examine the behavior of soil structure, involving the coupled effects between water pressure and deformation has been developed with the progress of the high speed digital computer.

The study on numerical method for analysis of coupled hydraulic and mechanical behavior of saturated porous media has been performed for a long time and notable among those are those due to Christian and Boehmer [1] and Sandhu and Willson [2]. Recently, the model considering elasto-plasticity or visco-plasticity of soils has been developed.

Unlike soils, fractures and their effects on deformation and seepage flow went unnoticed until recently. It is seemed that there is not authorized model, yet.

Flow of the ground water through rock mass may be strongly influenced by the fractures in it. Currently there are two theoretical approach to flow through fractured porous medium system. One is the discrete fracture approach which regards the fractured medium as a continuum in which there exist discrete discontinuities. (Noorishad et al.[3] and Ohnishi and Ohtsu[4]) The other is the equivalent continuum porous modeling which models the fractured system by an equivalent permeability tensor.(Long et al.[5] and Oda and Hatsuyama[6])

Moreover there is a hybrid method between two approaches mentioned above, which uses joint element for the large and clear fracture and uses equivalent porous modeling for the rock blocks which has small cracks.(Ohnishi et al.[7])

The similar methods are used for the analysis of the mechanical behavior of rock mass. In addition, the limit analysis like RBSM or DEM are studied by many investigators.

In this paper, we apply the double porosity modeling to the flow through jointed porous media. This model was introduced by Barenblatt et al.[8] and uses the concepts of statistical averaging, volume averaging, or the theory of mixtures. This type of flow model is expounded by Huyakorn et al.[9] and Warren and Root[10]. Sato et al.[11] represented the effect of permeability of rock matrix on the entire permeability of rock mass becomes strong with the decrees of fracture apertures by experimental approach using the realized double porosity model.

An extension of double porosity model that includes the coupling of flow and deformation was presented by Dugid and Lee [12] and Willson and Aifantis [13] Dugid and Lee derived equilibrium equation on

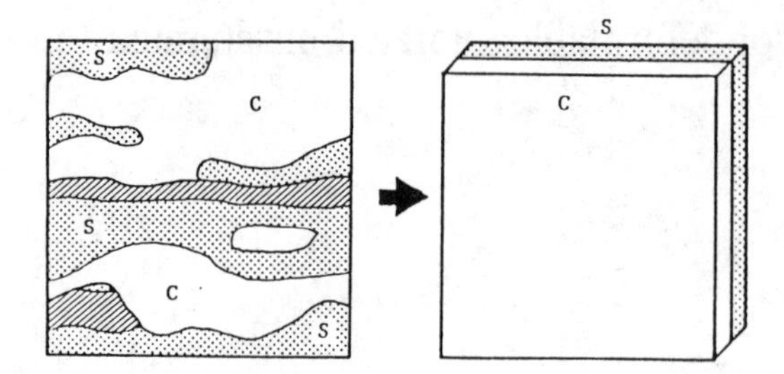

Fig.1. Double porosity modeling.

one porosity media of a pair. Willson and Aifantis derived equilibrium equation based on both porosity media and applied the definition of effective stress proposed by Nor and Byerlee [14] to the coupled term between flow and deformation in the equilibrium equation. These two approaches were developed for the saturated jointed media.

In this paper, we present the double porosity model for deformable saturated-unsaturated fractured rock mass.

Although the double porosity model has been developed in the field of oil engineering and the media for which that model is considered is rock mass, this analogy can also be applied to clay which consists of different types of soil structure and soil foundation which has many layers. For example, we can use this model like Fig.1 for stratified media in spite of making a very fine finite element mesh to express the complicated inhomogeneity.

We examine a few behaviors of various media with this newly developed model.

2 CONCEPT OF DOUBLE POROSITY MODEL

The double porosity model literally considers a medium by superposing two deferent porous media. Namely, two deferent water pressures are assumed to be at the same point in the medium and two continuous equations are made for both porous media, respectively.

Let p_1 be the average pressure of the fluid in one porous medium (we call it the primary porosity medium) and p_2 be the average pressure of the fluid in the other porous medium (the secondary porosity medium). For example of fractured porous media, p_1 represents the pressure in the rock matrix block of a given point and p_2 represents the one in the fractures of the given point.(see Fig.2)

Because that the permeabilities of both media are different, flow from one porous medium to the other medium is occurred due to the deference between p_1 and p_2 at the same point.

As mentioned above, two coupled equations are necessary to double porosity modeling. It is obvious that the analysis with this model has a meaning for the transient state, because inhomogeneity of the ground is expressed by coupled terms in two continuous equations.

There are two different water pressures at the same point in this model. Such a situation, however, is not occurred in the real ground. Thus, this model is regarded as not physical modeling but mathematical modeling and has a weak point that it is difficult to compare calculated values by this model with measured values.

But there exist cases that the measured values are not continuous. The pressure measured in the field has to be understood carefully.

To apply this model to the real field, we need consider the representative elementary volume(REV). Such a REV is a volume including two different porous media, which is large enough to exclude the microscopic effects and small enough to keep the element homogeneous.(see Fig.3) The REV also is necessary to analysis of the continuous porous media, but this is specially important for analysis of fractured rock mass with this model.

3 GOVERNING EQUATIONS

The assumptions used to derive the governing equations are following:

1) The medium is isotropic and poro-elastic for the mechanical behavior.
2) Darcy's law is valid for flow of water

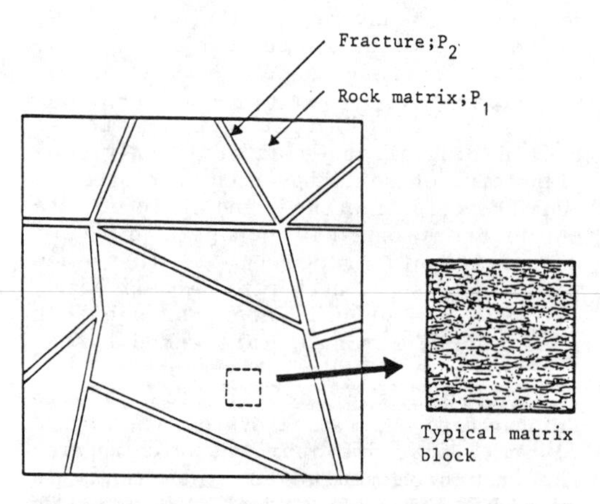

Fig.2. Double porosity model of fractured rock mass.

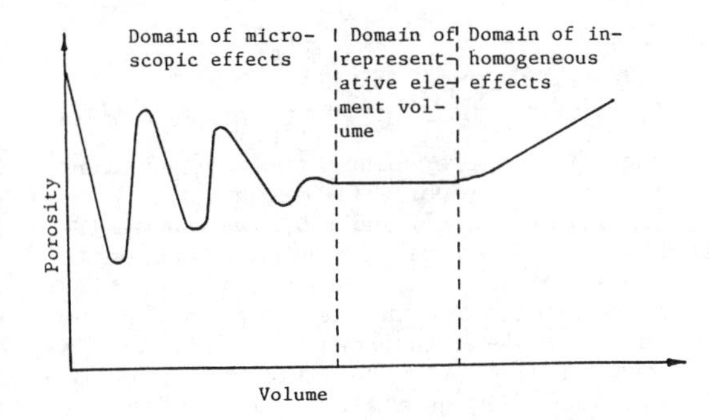

Fig.3. Representative elementary volume.

in the saturated-unsaturated fractured media.
3) The fluid is slightly compressible and the solid grains (solid mass) are incompressible.
4) The deformation is considered in the field including both porous media.
The water pressure concerned with the effective stress is assumed to be expressed with volumetric averaging ones of two media.
5) The unsaturated flow in the primary porosity medium (for example, rock block matrix) is calculated by using water retention curve. The one in the secondary porosity medium (for example, fractures) is neglected.

Using the assumptions mentioned above and extending the governing equations derived by Huyakorn and Pinder[15] to account for the saturated-unsaturated porous media, we present the system of basic equations used in our newly developed computer code.

3.1 Equilibrium equation

The total stress consists of the effective stress and the fluid pressures in the both porous media. After all, the governing deformation equation for elastic and isotropic fractured medium are almost identical to those for the conventional porous media. This equation can be written as follows by assuming that extensive stresses are taken as positive, and strains are infinitesimal.

$$\frac{\partial \sigma^{\circ}_{ij}}{\partial x_j} + F_i + (\lambda+G)\frac{\partial^2 u_j}{\partial x_j \partial x_i} + G\frac{\partial^2 u_i}{\partial x_j \partial x_j} - \frac{\partial p}{\partial x_i} = 0 \quad (1)$$

where σ°_{ij} is initial effective stress tensor, u_i is deformation vector, F is body force vector per unit volume of the medium and λ and G are Lamé's constants.
p is replaced by the weighted average of p_1 and p_2, i.e.,

$$p=(S_{r1}n_1p_1+S_{r2}n_2p_2)/(n_1+n_2) \quad (2)$$

where S_r and n are the degree of saturation and porosity. The subscript 1,2 represent the primary porosity medium and the secondary porosity medium, respectively.
There is the other definition of p in Eq.(1) proposed by Willson and Aifantis[13]. The values of water pressure calculated by their definition are mostly same as the one calculated by ours, however.
We let n_1 be the volume of the porosity in the primary porosity medium divided by the total volume including the both media. In the same way, n_2 is the volume of the porosity in the secondary porosity medium divided by total volume.

3.2 Continuity equation

The continuity equation for solid part is given as,

$$-\frac{\partial}{\partial x_i}\{(1-n_1-n_2)\rho_s v_{is}\} = \frac{\partial}{\partial t}\{(1-n_1-n_2)\rho_s\} \quad (3)$$

where ρ_s is the density of the solid mass, v_{is} is the deformation vector of the solid.
Using the substantial time differential derivative

$$D/Dt = \partial/\partial t + v_{is}\partial/\partial x_i \quad (4)$$

and assuming that the solid mass is incompressive, we can rewrite Eq.(3) as

$$Dn_1/Dt + Dn_2/Dt = (1-n_1-n_2)\partial v_{is}/\partial x_i \quad (5)$$

Now consider the continuity of flow in the primary pores. Let v_{i1} be the Darcy velocity through the primary porous medium. Because that Darcy's law is valid for the soil ground at rest, we can define the relative bulk velocity of the fluid as

$$v_{i1} = n_1S_{r1}(v_{if1}-v_{is})$$

in which v_{if1} is the real velocity of the fluid in the primary pores. The continuity equation of flow in the primary medium can be written in the form

$$-\frac{\partial}{\partial x_i}(n_1S_{r1}v_{if1}\rho_f) - \Gamma = \frac{\partial}{\partial t}(n_1S_{r1}\rho_f) \quad (6)$$

where Γ is the rate of fluid mass transferred from the primary porosity medium to the secondary one. Introducing the substantial derivative and the Darcy velocity v_i, Eq.(6) can be reduced to

$$-\frac{\partial}{\partial x_i}(\rho_f v_{i1}) = n_1S_{r1}\frac{D\rho_f}{Dt} + \rho_f n_1\frac{DS_{r1}}{Dt} +$$

$$\rho_f S_{r1}\frac{Dn_1}{Dt} + \rho_f n_1 S_{r1}\frac{\partial v_{is}}{\partial x_i} + \Gamma \quad (7)$$

Combining Eqs.(5) and (7) and using the fluid compressibility relation, $d\rho_f = \rho_{f0}\beta dp$ for the first term of right hand side of Eq.(7), we obtain

$$-\frac{\partial}{\partial x_i}(\rho_f v_{i1}) = n_1 S_{r1}\rho_{f0}\beta\frac{Dp_1}{Dt} + \rho_f n_1\frac{DS_{r1}}{Dt}$$

$$+\rho_f S_{r1}(1-n_2)\frac{\partial v_{is}}{\partial x_i} - \rho_f S_{r1}\frac{Dn_2}{Dt} + \Gamma \quad (8)$$

where ρ_{f0} is the reference density of the fluid, ρ_f is the density of the fluid and β is the compressibility of water.
According to the definition, n_2 is written as

$$n_2 = V_2/V \quad (9)$$

where V_2 is the volume of porosity of the secondary medium, V is the total volume. Differentiation of Eq.(9) with respect to time gives

$$Dn_2/Dt = (DV_2/Dt - n_2 DV/Dt)/V \quad (10)$$

By definition,

$$V = V_1 + V_2 + V_s \quad (11)$$

where V_1 and V_s denote the volume of porosity in the secondary porosity medium and the bulk volume of the solid mass, respectively. By assuming the solid mass is incompressible, we get

$$Dn_2/Dt = DV_1/Dt + DV_2/Dt \quad (12)$$

Substituting Eq.(12) into Eq.(10), we obtain

$$Dn_2/Dt = \{(1-n_2)DV_2/Dt - n_2 DV_1/Dt\}/V \quad (13)$$

Substituting the definition of compressibility of fluid

$$-(\beta n_1 S_{r1}V)Dp_1/Dt = DV_1/Dt,$$

$$-(\beta n_2 S_{r2}V)Dp_2/Dt = DV_2/Dt \quad (14)$$

into Eq.(13), we obtain

$$Dn_2/Dt = -(1-n_2)\beta n_2 S_{r2} Dp_2/Dt + n_2\beta n_1 S_{r1} Dp_2/Dt \quad (15)$$

Combination Eqs.(15) and (8) yields

$$-\frac{\partial v_{i1}}{\partial x_i} = n_1 S_{r1}\beta(1-n_2 S_{r1})\frac{\partial p_1}{\partial t} +$$

$$n_2(1-n_2)S_{r1}S_{r2}\beta\frac{\partial p_2}{\partial t} + n_1\frac{\partial S_{r1}}{\partial t} + (1-n_2)S_{r1}\frac{\partial v_{is}}{\partial x_i}$$

$$+\frac{\Gamma}{\rho_f} \quad (16)$$

where it is assumed that the fluid is slightly compressible and that $Dp_1/Dt \cong \partial p_1/\partial t$, $Dp_2/Dt \cong \partial p_2/\partial t$, $\rho_{f0} = \rho_f$ and $DS_{r1}/Dt \cong \partial S_{r1}/\partial t$
Now we let the following equation be valid.

$$n_1\frac{\partial S_{r1}}{\partial t} \cong \frac{\partial S_{r1}n_1}{\partial t} = \frac{\partial\theta_1}{\partial t} = \frac{\partial\theta_1}{\partial p_1}\frac{\partial p_1}{\partial t} \quad (17)$$

where θ_1 is the volumetric water content of the primary porosity medium.
Combining Eqs.(17) and (16), we obtain the flow equation of the primary porosity medium taking into account the coupled effects of flow and deformation in the saturated-unsaturated medium. Its counterpart is the continuity equation for the secondary porosity medium. This is derived using a similar procedure to that given. The result is given by

$$-\frac{\partial v_{i2}}{\partial x_i} = n_1(1-n_1)S_{r1}S_{r2}\beta\frac{\partial p_1}{\partial t} +$$

$$n_2 S_{r2}\beta(1-n_1 S_{r2})\frac{\partial p_2}{\partial t} + \frac{\partial\theta_1}{\partial p_1}\frac{\partial p_1}{\partial t} +$$

$$(1-n_1)S_{r2}\frac{\partial v_{is}}{\partial x_i} - \frac{\Gamma}{\rho_f} \quad (18)$$

To obtain the final form of the two equations, we use Darcy's law:

$$v_{i\ell} = -\frac{k_{\ell ij}}{\mu}\left(\frac{\partial p_\ell}{\partial x_j} + \rho_f g_j\right) \quad \ell=1,2 \quad (19)$$

where k_{1ij} and k_{2ij} are the permeability tensor of the primary porosity medium and the one of the secondary medium, respectively. μ is the kinematic viscosity and g is the ith component of gravitational acceleration.
The required flow equaitons take the form

$$\frac{\partial}{\partial x_i}\left\{\frac{k_{\ell ij}}{\mu}\left(\frac{\partial p_\ell}{\partial x_j} + \rho_f g_j\right)\right\} = n_\ell S_{r\ell}\beta(1-n_{\bar\ell}S_{r\ell})\frac{\partial p_\ell}{\partial t}$$

$$+ n_{\bar\ell}(1-n_{\bar\ell})S_{r\ell}S_{r\bar\ell}\beta\frac{\partial p_{\bar\ell}}{\partial t} + \frac{\partial\theta_\ell}{\partial p_\ell} + (1-n_{\bar\ell})S_{r\bar\ell}\frac{\partial v_{is}}{\partial x_i}$$

$$+\frac{\Gamma}{\rho_s} \quad (\bar\ell=2, \text{if } \ell=1.\ \bar\ell=1, \text{if } \ell=2) \quad (20)$$

3.3 Leakage term Γ

The formula describing fluid transfer between the primary porous medium and the secondary one is derived using dimensional analysis in Barenblatt's model. Such a derivation is not well grounded and the formula is based on the assumption of quasi-steady state leakage. As an alternative to the approach of Barenblatt et al.[8], a more elaborate approach which estimates transient fluid transfer rates has recently been developed by Huyakorn et al.[9].
In their model, the primary porosity medium (rock blocks) is idealized as a

series of equal size spheres or prismatic matrix blocks and the flow through the rock blocks into the fractures (the secondary porosity model) is solved using the convolution integral of the analytical solution or finite differences.
Dugid and Lee [12] presented the approach in which the flow rate between the primary porosity medium and the secondary one is obtained by the analytical solution of one dimensional transient heat transfer equation.
Comparing the result using the analytical solution of the transient equation with the one using the analytical solution of steady equation, they concluded that the approach with the analytical solution of the steady heat conduction equation has a enough accuracy because of little difference between them.
The rate of flow transfer between two different porous media at the steady state is presented as

$$\Gamma = \frac{k_1}{\mu} \frac{4n_2\rho_f}{\pi Cl}(p_1 - p_2) \tag{21}$$

where l is the matrix characteristics length (for example, the half length of a crack in the primary porosity media) and c is the void aperture.
We use this equation for the leakage term Γ.

3.4 Initial and boundary condition

There are initial and boundary conditions that should be considered. these can be expressed as
Initial conditions:

$$\begin{aligned} u_i(x_1,x_2,t) &= u_i^\circ(x_1,x_2,0) \\ p_\ell(x_1,x_2,t) &= p_\ell^\circ(x_1,x_2,0) \quad \ell=1,2 \end{aligned} \tag{22}$$

boundary conditions:

$$\begin{aligned} u_i(x_1,x_2,t) &= \hat{u}_i(x_1,x_2,t) \\ \tau_{ij}(x_1,x_2,t)n_j &= \hat{S}(x_1,x_2,t) \\ p_\ell(x_1,x_2,t) &= \hat{p}_\ell(x_1,x_2,t) \\ \{-\frac{k_{\ell ij}}{\mu}(\frac{\partial p_\ell}{\partial x_j} + \rho_f g_j)\}n_j &= -\hat{Q}_\ell(x_1,x_2,t) \\ \ell &= 1,2 \end{aligned} \tag{23}$$

4 ·Finite element discretization

We used here the Galerkin method to formulate the finite element discretization. Linear quadrilateral isoparametric element is employed in the code. The system of algebraic equations derived from finite element approximation is nonlinear due to the dependency of the permeability in the unsaturated primary porous medium on suction. Consequently, it is necessary to employ iterative method to obtain a solution.

5 NUMERICAL ANALYSES

The numerical procedures described in the foregoing sections have been implemented into a computer code capable of simulating coupled flow and deformation behavior in the fractured porous media. Now in order to demonstrate the function and the utility of this code, some example problems are solved by this code.

5.1 Comparison of numerical results with analytical and experimental ones.

To verify the numerical solution algorithms and asses their accuracy, the code was used to simulate two problems involving a seepage problem in a saturated-unsaturated porous media and one dimensional consolidation problem. The result of the consolidation analysis is compared with Terzaghi's theoretical solution. The comparison in the degree of consolidation is shown in Fig.4. It can be seen that there is good agreement between the analytical and numerical solution.
To calculate this problem, the water in the secondary porosity medium and the water moving between both media are setted not to flow. The overburden pressure is therefore taken by the water pressure of the primary porosity medium immediately after loading. The effective stress increases due to the decrease of the water pressure in the primary porosity medium during consolidation.
The saturated-unsaturated flow problem is analyzed by setting the deformation and

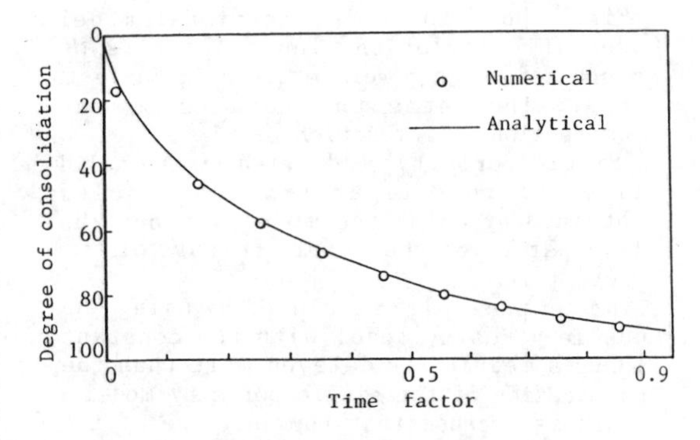

Fig.4. Comparison of analytical solution with numerical one.

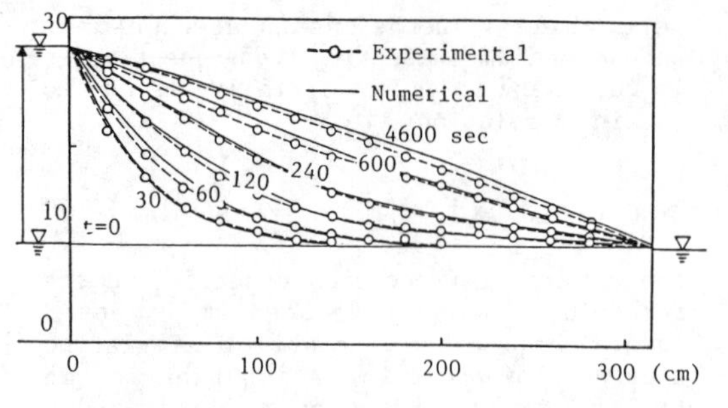

Fig.5. Comparison of experimental results with numerical ones.

the water pressure of the secondary porosity medium to be zero.

Comparing the numerical results with the experimental ones obtained by Akai and Uno [16], we get good agreements between two results.(Fig.5)

From the results mentioned above, the basic functions of this code are seemed to be valid.

5.2 Secondary compression problem

The analysis of the secondary compression problem is mostly performed by using the visco-elastic model.

Now we try to analyze this problem with the double porosity model by using the theory that the primary consolidation is due to a change in effective stress brought by the dissipation of excess pore water pressure of the macro pore, and the secondary compression is due to the dissipation of the water pressure of the micro pore.(see Fig.6)

Let the primary porosity be the micro pore, and the secondary porosity be the macro pore. The permeability of the primary porosity is setted to be different from that of the secondary one.

Fig.7 shows the one-dimensional model used in calculation. The properties of macro and micro pore are given in Table 1.

Fig.8 indicates the results of the compression as a function of time.

'Single porosity' indicated by the solid line in this figure is the results obtained by using the model without the dissipation of the water pressure of the micro pore.

The compression calculated by using the double porosity model with the constant Young's modulus is delayed more than the compression of the single porosity model.

This is because that the increase of the effective stress is delayed by the slow dissipation of the water pressure of the

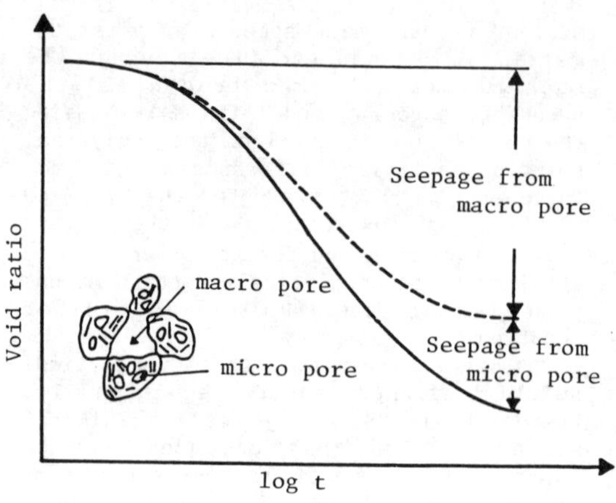

Fig.6. Schematic compression-time curve.

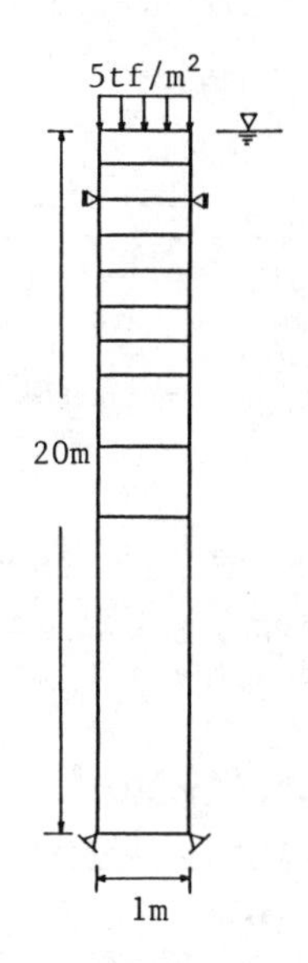

Fig.7. Finite element model.

Table 1. Data used in secondary compression analysis.

Properties	Values
Permeability of micro pore	1.0×10^{-12}m/s
Porosity of micro pore	0.2
Permeability of macro pore	1.0×10^{-9} m/s
Porosity of macro pore	0.5
Young's modulus	100 tf/m^2
Poisson's ratio	0.3
Void apeture	0.025 m
Matrix characteristic length	0.015 m

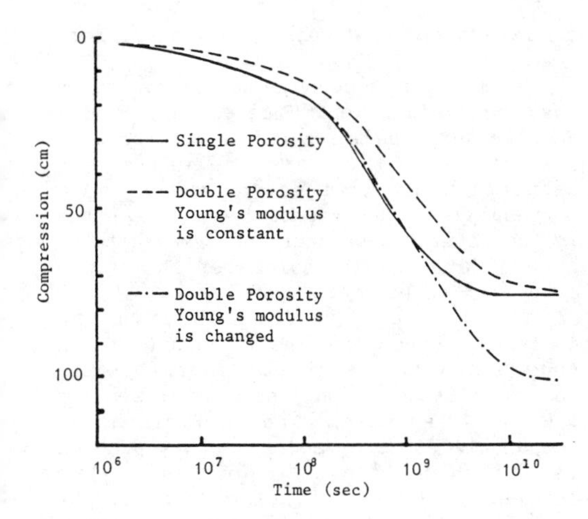

Fig.8. Compression-time curve.

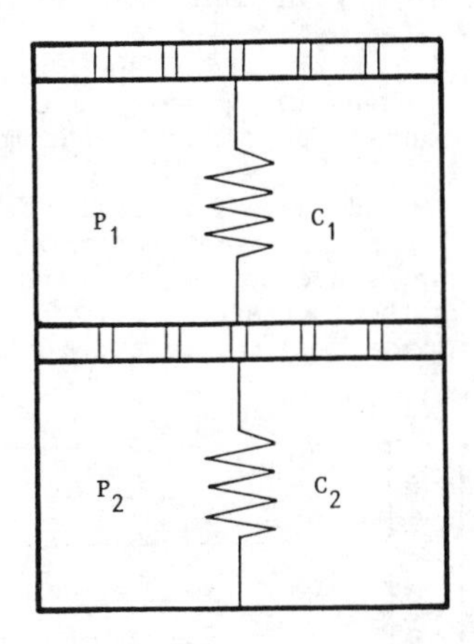

Fig.9 Mechanical analogue.

micro pore. The settlements after consolidation of the both model are same because that Young's modulus of the single porosity model is not different from that of the double porosity one.

We have to consider the change of Young's modulus of the ground due to the compression of the micro pore.

By using the mechanical analogue shown in Fig.9, the following equilibrium equation is introduced,

$$(C_1^{-1}+C_2^{-1})^{-1}\varepsilon - (C_1^{-1}+C_2^{-1})^{-1}(C_1^{-1}p_1+C_2^{-1}p_2)=\sigma \tag{24}$$

where C_1 is the elastic constant of the primary porosity medium(micro pore), and C_2 is that of the secondary porosity medium(macro pore).

C_2 is the parameter of the skeleton structure of the soil, and is assumed to be constant, as Terzaghi's theory is.

Because that the micro pore medium is expected to become hard by the drainage, Young's modulus of the primary medium is assumed to increase according to the following exponential function,

$$E_1 = E_{10}\exp(\Delta p_1^\circ - \Delta p_1) \tag{25}$$

where Δp_1° is the increment of the excess water pressure of the primary porosity medium immediately after loading. In this case, Δp_1° equals to 5 tf/m^2. Δp_1 is the excess water pressure of the primary porosity medium during consolidation.

With Eq.(24), Young's modulus of the soil involving both medium is obtained by $E_1=E_1E_2/(E_1+E_2)$.

The results calculated by setting initial Young's modulus of both medium to be 100 tf/m^2 are given by the dot-dash line in Fig.8. The analysis using the double porosity model with the assumptions mentioned above can represent the tendency of the secondary compression.

After all, to represent the secondary compression by using the theory that the secondary compression is occurred due to the drainage from micro pore, it is necessary to consider the mechanical properties taking account of the deformation of the micro pore medium.

Although we use the nonlinear function Eq.(25) in this paper, it is the future problem to identify such a mechanical parameter.

5.3 Simulation of Lugeon test

To examine the transmission of the water pressure in the fractured rock mass, the Lugeon test is simulated with the double porosity model.

The finite element mesh and boundary condition used in the problem is shown in Fig.10. Table 2 gives the data used for the analysis.

Fig.11 indicates the distribution of the water pressure head as a function of time.

Table 2. Data used for analysis of Lugeon test.

Properties	Values
Permeability of rock matrix	1.0×10^{-11} m/s
Porosity of rock matrix	0.2
Permeability of fracture	1.0×10^{-4} m/s
Porosity of fracture	0.05
Young's modulus	1.0×10^{4} tf/m^2
Poisson's ratio	0.3
Fracture aperture	0.001 m
Matrix characteristic length	0.5 m

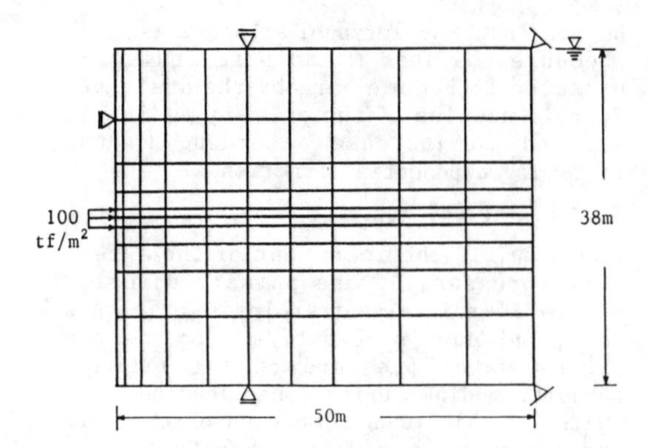

Fig.10. Finite element model.

Because that the permeability of the rock blocks is larger than that of the fractures, the transmission of the water pressure in the rock block is much slower than that in the fractures.

The double porosity model can represent the discontinuous situation between rock blocks and fractures shown in Fig.11 which can not be simulated by the conventional continuous method.

The real phenomena is seemed to be same as these results because that the inletted water flows mainly through the fractures.

5.4 Two-dimensional consolidation problem

Two-dimensional consolidation problem is analyzed using the finite element model and boundary condition shown in Fig.12.

We examine two cases. One is the saturated case that ground water table corresponds to the ground level. The other is the saturated-unsaturated case that the ground water table is located at a depth of 5 m from the ground level.

Table 3 gives the data used in the analysis. Young's modulus is assumed to be constant in this analysis. Therefore the secondary compression presented in Section 5.1 is not appeared. The analysis using the double porosity model in this problem is different from the one using the conventional consolidation model at the point of the change of the effective stress brought due to the flow in the primary porosity medium and the flow transfer between both media.

In this analysis, let the primary porosity be the micro pore of the clay, and the secondary porosity be the macro pore.

The upper boundary is setted to be the zero hydraulic pressure condition, and the other boundaries are setted to be no flow condition in the saturated case.

The all boundaries are setted to be no

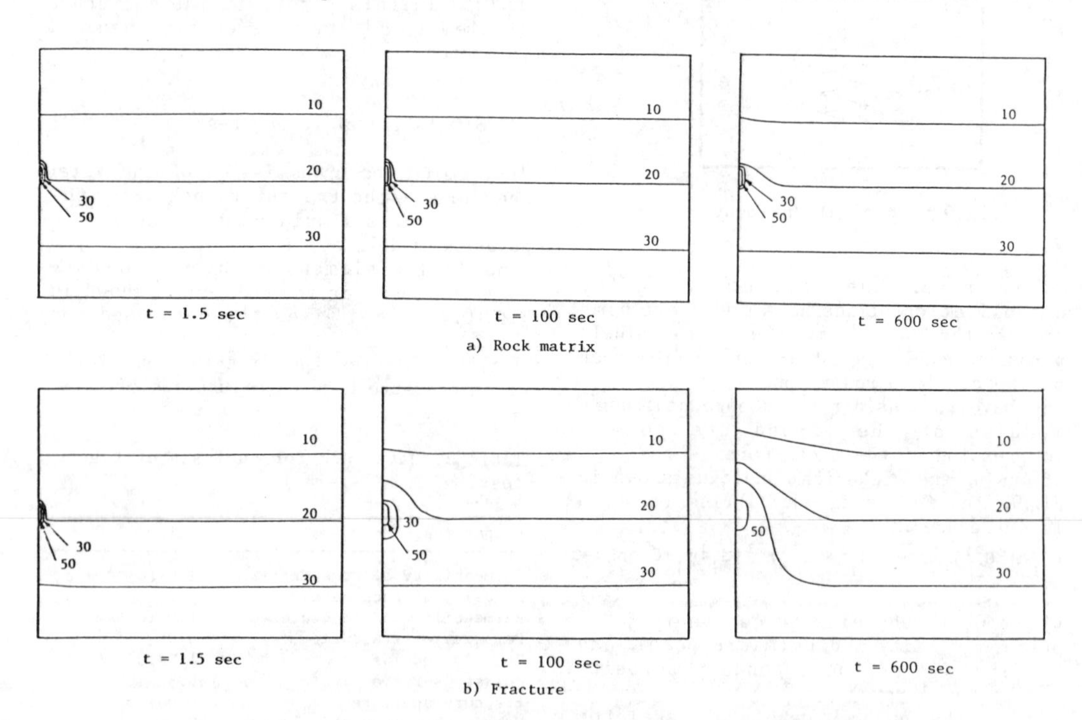

Fig.11. Distribution of pressure head as a function of time (tf/m²)

Table 3. Data used for two-dimensional consolidation problem.

Properties	Values
Permeability of primary porosity	1.0×10^{-12} m/s
Porosity of primary porosity	0.2
Permeability of secondary porosity	1.0×10^{-9} m/s
Porosity of secondary porosity	0.5
Young's modulus	100 tf/m^2
Poisson's ratio	0.33
Void apeture	0.02 m
Matrix characteristic length	0.05 m

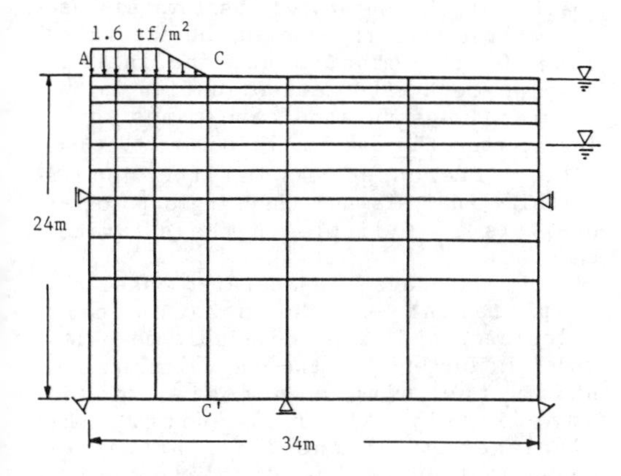

Fig.12. Finite element model.

flow condition in the saturated-unsaturated case.

To compare with the conventional method, the results calculated by using the Sandhu type consolidation model (single porosity model) which uses the linear quadrilateral isoparametric element for the water pressure and the quadratic quadrilateral element for the displacement are given with the results by our double porosity model.

Fig.13 shows the settlement as a function of time at point A.

The compression calculated by the double porosity model is delayed more than the one calculated by the single porosity model because of the slow increase of the effective stress due to the detain of the water in the primary porosity.

The compression at the early stage after loading in the saturated-unsaturated case is larger than the one in the saturated case, and the end of the consolidation in the saturated-unsaturated case is earlier than the one in the saturated case.

The settlement after consolidation in the saturated-unsaturated case is smaller than the one in the saturated case. These are probably because that the elastic compression immediate after loading becomes large, and the settlement by consolidation becomes small, as the load is burdened at the unsaturated part.

Fig.14 indicates the ground water tables as a function of time in the saturated-unsaturated case.

The change of the ground water table obtained by the single porosity model analysis is slight. On the other hand, the one by the double porosity model analysis is complicated. The water pressure under the loading part of the primary porosity in which the permeability

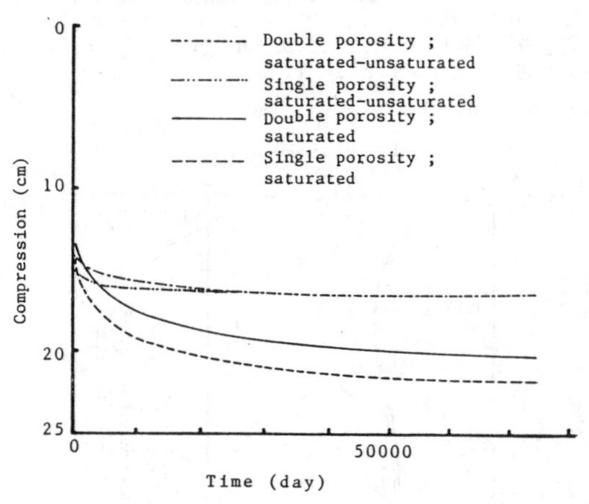

Fig.13. Compression-time curve at point A.

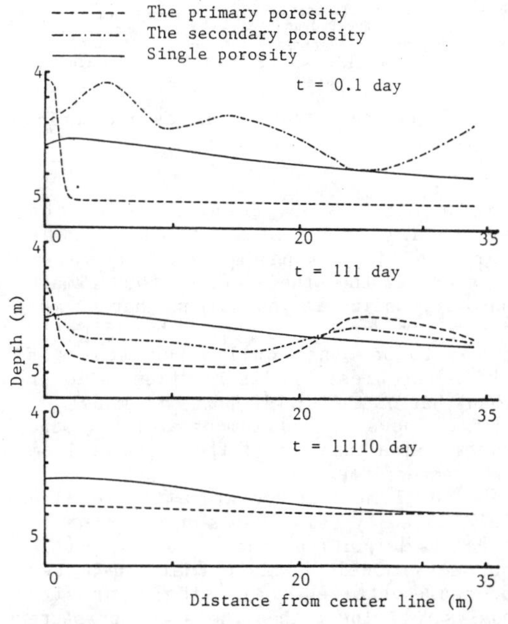

Fig.14 Ground water table as a function of time.

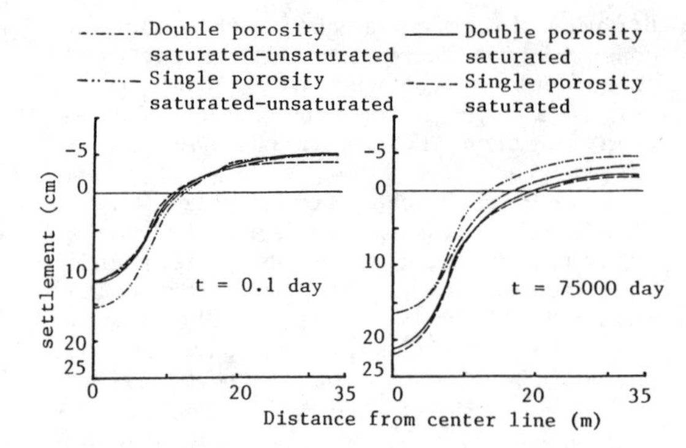

Fig.15. Settlement as a function of the distance from center line.

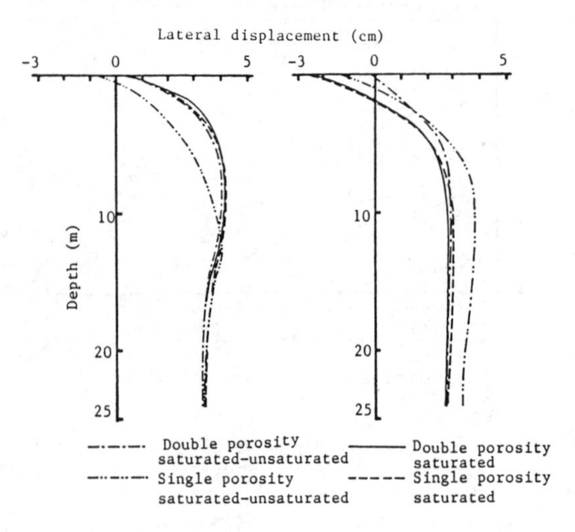

Fig.16. Lateral displacement as a function of depth.

is low is larger than that of the secondary porosity in which the permeability is high, and the water pressure at the other part of the primary porosity medium is lower than that of the secondary one. The water table finally comes to be the equilibrium state of which the pressure distribution is lower than that of the single porosity model.

Fig.15 shows the settlement at the ground surface as a function of the distance from the center line.

The settlement immediate after loading calculated by using the single porosity model is larger than the one by the double porosity model, because that the water pressure obtained by the single porosity analysis is lower than the water pressure of the secondary porosity by the double porosity model analysis. The pressure of the secondary porosity has more efficient on the deformation than that of the primary one.

The settlement after consolidation obtained by the single porosity analysis is, however, smaller than the one by the double porosity analysis, because that the water pressure distribution by the double porosity model analysis is lower than that by the single one.

Fig.16 shows the distribution of the lateral displacement at c-c'section.

The lateral displacement immediate after loading in the saturated-unsaturated case is smaller than the one in the saturated case. On the other hand, the lateral displacement after consolidation in the saturated-unsaturated case comes to be larger than the one in the saturated case.

The difference between the double porosity analysis and the single porosity analysis is a little in the saturated case.

In the saturated-unsaturated case, the displacement at the shallow part calculated by using the double porosity model is larger than the one calculated by using the single porosity model immediately after loading. The displacement at the deep part after consolidation by the double porosity analysis is smaller than the one by the single porosity analysis.

As mentioned above, the double porosity analysis can represent the behavior that the load burdened at the ground surface is not transferred to the deep part due to the existence of the upper unsaturated part.

6 CONCLUSION

The work presented here provides a new technique for investigation of coupled mechanical-hydraulic behavior of saturated-unsaturated fractured rock mass. The work is based on the double porosity modeling proposed by Barenblatt et al.[8] and the saturated-unsaturated seepage analysis method developed by Neuman[17].

In this presentation we apply this new method to the analysis of the secondary compression problem, the Lugeon test simulation and the two-dimensional consolidation problem with a free surface.

From our study the following conclusions can be drawn:

1.The coefficient matrix of the finite element equation derived to describe the behavior of the saturated-unsaturated deformable rock mass with the double porosity modeling is not symmetric.

2.The basic function of this newly

developed code is recognized to be sure in comparison of numerical results with the analytical and experimental results.

3.To represent the secondary compression with the double porosity model it is necessary to consider the elastic constant which takes account of the deformation due to the dissipation of the water from the micro pore.

By using such a method in addition to the visco-elastic analysis it is expected to get a clue to understand the secondary compression.

4.This method can be applied to the discontinuous phenomena like the flow through the fractured medium which is not explained by the conventional continuous method.

5.This method can reflect more dramatically the effect of the unsaturated part on the the total ground deformation than the conventional method.

REFERENCES

[1]J.T.Christian and J.W.Boehmer: Plane strain consolidation by finite elements, J. Soil Mech. and Foundation Div. ASCE, SM 4, 1435-1457 (1970)

[2]R.S.Sandu and E.L.Wilson: Finite element analysis of seepage in elastic media, J. Eng. Mech. Div. ASCE, EM 3, 641-652 (1969)

[3] T.N.Noorishad, P.A.Witherspoon and T.L.Brekke: A method for coupled stress and flow analysis of fractured rock masses, Geotechnical Engineering publication No.71-6, University of California, Berkeley. (1971)

[4] Y.Ohnishi and H.Ohtsu,: Coupled stress flow analysis of discontinuous media by finite elements, Transaction of the JSCE No.322, 111-120 (1982)

[5] J.C.S.Long, J.S.Remer, C.R.Wilson and P.A.Witherspoon:Porous media equivalents for networks of discontinuous fractures, Water Resouces Research 18, No.3, 645-658 (1982)

[6] M.Oda and M.Hatsuyama: Permeability tensor for jointed rock masses, Proc. of the Int. Symp.on Fundamentals of Rock Joints, 303-312, Bjorkiden (1985)

[7] Y.Ohnishi, H.Kagimoto and K.Nishino:Estimation of discontinuity characteristics and its application to rock hydraulics analysis, Transaction of JSCE, 241-248 (1986)

[8]G.I.Barenblatt, I.P.Zheltov and I.N.Kochika: Basic concepts in the thory of seepage of homogeneous liquids in fissured rocks (strata), PMM,Vol.24,852, 1286-1303 (1960)

[9] P.S.Huyakorn, B.H.Lester and C.R.Faust:Finite element techniques for modeling groundwater flow in fractured aquifers, Water Resources Research, Vol.19. No.4, 1019-1035 (1983)

[10] J.E.Warren and P.J.Root: The behavior of naturally fractured reservoirs, Soc. of Petroleum Engineers Journal, 245-255 (1963)

[11]K.Sato, T.Shimizu and Y.Itou: Fundamental study on permeability and dispersion in double porosity rock masses, Fifth International Conference on Numerical Methods in Geomechanis,Nogoya,657-664 (1985)

[12] J.O.Duguid and P.C.Y.Lee: Flow in fractured porous media, Research Report No.73-WR-1, Princeton University,(1973)

[13] R.K.Wilson,E.C.Ainfantis: On The Theory of Consolidation with Double Porosity, Int.J.Engng.Sci.Vol.20, No.9, 1009-1035 (1982)

[14] A.Nur and J.D.Byerlee: An exact effective stress law for elastic deformation of rock with fluids, J.G.R., Vol.76, No.26, 6414-6419 (1971)

[15] P.S.Huyakorn and G.F.Pinder: Computational method in subsurface flow,Academic Press, (1983)

[16] K.Akai and T.Uno:Study on the quasi-one-dimensional, non-steady seepage flow through soil, Transaction of JSCE, No.125, 14-22 (1966)

[17] S.P.Neuman:Saturated-unsaturated seepage by finite elements,Proc.,ASCE HY,Vol.99,No.12, 2233- 2250 (1973)

Numerical Methods in Geomechanics (Innsbruck 1988), Swoboda (ed.)
© 1988 Balkema, Rotterdam. ISBN 90 6191 809 X

An integral equation technique for the propagation of a vertical hydraulic fracture in a poroelastic formation

Emmanuel Detournay & Luc Vandamme
Dowell Schlumberger, Tulsa, Okla., USA
Alexandre H.-D.Cheng
University of Delaware, USA

ABSTRACT: This paper reports the progress in the development of a two-dimensional hydraulic fracturing model based on the Biot theory of poroelasticity. The first part of the paper describes an integral formulation for the pressure distribution in the fracture in terms of fracture opening and fluid loss into the formation. Using this formulation, a numerical algorithm is presented for the problem of a nonpropagating fracture with a given injection rate at the well. The paper concludes with some results on pore pressure distributions ahead of an opening edge dislocation moving at a constant speed.

1 INTRODUCTION

Mathematical models for hydraulic fracturing (a technique to enhance hydrocarbon production from underground reservoirs) must be designed to take the following processes into account:

1. fracture opening, caused by the pressure distribution inside the fracture, also referred to as fracture compliance;
2. viscous fluid flow inside the fracture, controlled by the fracture opening profile;
3. fluid leak-off into the formation; and
4. fracture propagation, caused by injection of fluid from the well.

The conventional assumptions, on which hydraulic fracture models are generally based, are that

1. the rock is a linear elastic material;
2. the fluid flow inside the fracture is laminar;
3. the fluid loss into the formation is evaluated using a one-dimensional diffusion solution (the "Carter model" [1]); and
4. the fracture propagation criterion is based on a critical value of the stress intensity factor, a parameter that characterizes the singular stress field ahead of the fracture tip [2].

Hydraulic fracturing models that rely on assumptions 1 and 3 treat the deformation of the rock and the diffusion of fluid in the formation as two independent processes. However, there is strong evidence that the mechanical response of porous rocks saturated with fluid is dominated by coupled diffusion-deformation effects:

- the deformation of the rock is controlled by an "effective" stress, dependent on the pore pressure in the formation;
- a body force, proportional in magnitude to the gradient of the pore pressure, has to be taken into account in expres- sing the equations of equilibrium for this effective stress;
- under undrained conditions (rapid loading), a change of the mean stress induces a variation of the pore pressure (the Skempton effect); and

- the volumetric deformation of the rock is sensitive to the rate of isotropic loading, i.e., the rock appears to be stiffer under fast than under slow loading rate.

All these coupled diffusion-deformation effects are accounted for, in a consistent manner, by the Biot theory of poroelasticity [3-5].

This paper reports the progress in the development of a two-dimensional model for hydraulic fractures, that includes poroelastic effects. As will be shown, a formulation based on the theory of poroelasticity enables one to treat the dependence of (i) fracture opening and (ii) fluid loss on the pressure distribution in the fracture in a consistent manner.

The paper is organized as follows. First the Biot theory of poroelasticity is briefly reviewed. Next, an integral formulation is presented, which relates any field variable (e.g., stress, pore pressure, displacement, ...) to the history of fracture opening and fluid leak-off. The numerical technique required to solve the coupled problem of a nonpropagating hydraulic fracture with a given injection rate at the well is described next. Finally, the problem of a steadily moving opening edge dislocation is analyzed using a simplified numerical algorithm.

2 THEORY OF POROELASTICITY

The theory of poroelasticity was first introduced by Biot in 1941 [3] to model the mechanics of fluid-infiltrated solids. The theory consists of (i) a constitutive equation, which rigorously couples the solid and fluid stress and strain parameters in a linear manner; (ii) an equilibrium equation for the effective stress, in which the fluid pressure gradient acts as a body force; (iii) Darcy's law for porous media flow; and (iv) a continuity equation for mass conservation (see [6] for a more detailed discussion).

The resulting model is fully defined by five material constant: G and ν are the conventional elasticity constants known as the shear modulus and Poisson ratio (here referred to as the "drained" Poisson ratio); κ is the permeability coefficient for porous media flow; ν_u, the "undrained" Poisson ratio, characterizes the bulk material Poisson effect when fluid is entrapped in the pores; and B, the Skempton pore pressure coefficient, describes the ratio of pore pressure rise over the variation of confining stress under undrained conditions. Using these constants, the set of coupled governing equations reveals a variety of physical phenomena [7,8] which cannot be predicted by the simpler, uncoupled theories.

3 INTEGRAL TECHNIQUE

The poroelastic hydraulic fracturing model is based on an indirect boundary integral method that evaluates the response of an infinite poroelastic medium to fracture opening and fluid leak-off, using variables that are defined on the fracture locus.

In the following, a two-dimensional, plane-strain, straight fracture propagating in tensile mode perpendicular to the minimum in situ stress, σ^o, is examined. The formulation for the more general case of a curved fracture is described in [9]. The model discussed here is consistent with the kinematic constraints (wedge-shaped fracture and constant height) of the so-called KGD model [10,11] used in the design of hydraulic fractures.

3.1 Integral formulation

A hydraulic fracture in a poroelastic medium is a surface across which the solid displacements and the normal fluid fluxes are discontinuous. Such a fracture can mathematically be simulated by a distribution, over time and over the fracture locus, of impulse point displacement discontinuities (DD) and impulse fluid sources. The sign convention for the discontinuities is defined in Figure 1. Knowing the corresponding singular

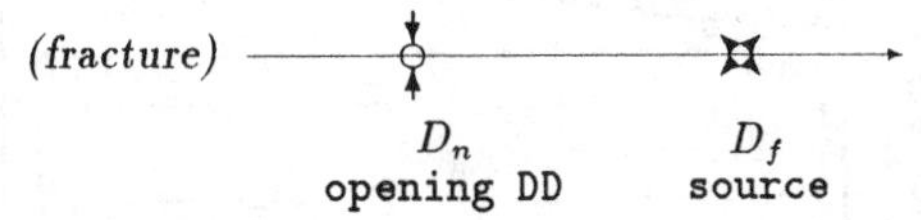

(fracture) D_n opening DD D_f source

Figure 1: Positive displacement and source discontinuities.

solutions [6], an integral representation of field parameters, such as displacement, flux, stress, and pore pressure, can be derived using the principle of superposition. This technique constitutes a generalization of the Displacement Discontinuity method [12], which has already been used to model fracture propagation in elastic media [13].

The displacement and flux discontinuities, D_n and D_f, are usually not known a priori, and must be determined from the pressure distribution, p_f, in the fracture. In the hydraulic fracturing problem, this pressure distribution is itself unknown and must be solved concurrently, using the additional equations discussed in Section 4.

Determination of the displacement and flux discontinuities in terms of p_f requires the solution of the following set of singular integral equations, which expresses the dependence of the induced normal stress and pore pressure along the fracture on the history of D_n and D_f:

$$\left\{ \begin{array}{c} p_f(x,t) - \sigma^o \\ p_f(x,t) - p^o \end{array} \right\} = \int_0^t \int_{L(t)} \begin{bmatrix} \sigma^*_{nn}(x-\chi, t-\tau) & \sigma^*_{nf}(x-\chi, t-\tau) \\ \sigma^*_{pn}(x-\chi, t-\tau) & \sigma^*_{pf}(x-\chi, t-\tau) \end{bmatrix} \left\{ \begin{array}{c} D_n(\chi,\tau) \\ D_f(\chi,\tau) \end{array} \right\} dL(\chi) d\tau \qquad (1)$$

where p^o is the far-field pore pressure and the quantities marked with an asterisk are influence functions: σ^*_{nn} and σ^*_{pn} are the normal stress and pressure, respectively, due to an impulse point DD; σ^*_{nf} and σ^*_{pf} are the normal stress and pore pressure, respectively, induced by an impulse point source. It is worth mentioning that the influence functions for the DD singularity are composed of two parts [6], an instantaneous component, which corresponds to the solution of an elastic DD singularity with undrained elastic constants, and a time-dependent component.

3.2 Numerical procedure

The numerical solution technique is based on the discretization of Equations (1), their integration in both time and spatial coordinates, and the collocation of the pressure p_f at discrete points along the crack surface. These operations result in a linear system of algebraic equations, which need to be solved for the displacement and flux discontinuities at each time step.

In this implementation, the fracture is divided into segments and the distribution density of displacement and flux discontinuities is assumed to be constant on each segment. The influence functions are then analytically integrated over each fracture segment. The time axis is also discretized and all discontinuities are assumed to vary linearly on each time segment. The convolutional integral is then solved using numerical quadrature rules.

Collocating for the normal stress and pressure at the midpoint of the fracture segments results in a linear system of equations which can be written in matrix notation, as follows:

$$(\boldsymbol{A}^o + \boldsymbol{\Delta A})\, \boldsymbol{D}^h = \boldsymbol{\sigma}^h + \sum_{r=1}^{h-1} \boldsymbol{B}^r \boldsymbol{D}^r \qquad (2)$$

where the superscripts r and h represent the current and previous time steps, respectively; σ is a column matrix containing normal stress, and fluid pressure; $\boldsymbol{D}^h$ is a column matrix containing the unknown displacement and flux discontinuities at time h; the matrices $\boldsymbol{D}^r$ are known from previous solutions (Note that the order of the matrices is $2N$, where N is the number of elements used). Once the time history of the DD's becomes available, the solution for stress, pressure, displacement and flux at any point in space and time can be found using the Duhamel principle. The details of this procedure are described in [9].

The influence-coefficient matrix, $\boldsymbol{A}^o$, is that due to the instantaneous response, and is therefore independent of time. If a constant time step is used, the matrix $\boldsymbol{\Delta A}$ remains constant and is computed only once. The matrices $\boldsymbol{B}^r$ are defined in a recurrence relationship, and only one of them is evaluated at each time step. As time increases, the tail of the convolutional integral can be approximated using only a few time steps such that the summation on the right hand side of Equation (2) contains a fixed number of terms.

For solutions at very large times, a variable time step scheme is more efficient. In that case, the matrices $\boldsymbol{\Delta A}$ and $\boldsymbol{B}^r$ need to be reevaluated whenever a change of time step takes place. The discontinuities are then interpolated along the time axis.

3.3 Pressurized Fracture

Consider now the problem of a linear crack of length $2L$ in an infinite poroelastic medium. At time $t = 0$, the inside of the crack is suddenly subject to a constant fluid pressure, p_f. This problem is solved by decomposing the loading into two fundamental modes: (i) a normal stress loading, and (ii) a pore pressure loading, characterized by the following boundary conditions:

$$\left\{ \begin{matrix} \sigma_n \\ p \end{matrix} \right\} = \left\{ \begin{matrix} -p_f \\ 0 \end{matrix} \right\} \text{ mode 1; } \left\{ \begin{matrix} 0 \\ p_f \end{matrix} \right\} \text{ mode 2;} \quad (3)$$

By superposition of the two loading modes, one obtains the solution for the general case where there is a far-field stress normal to the crack, σ^o, and a far-field pore pressure, p^o.

The width variation at the center of the fracture ($x = 0$) with respect to the dimensionless time, $\tau = L^2t/c$, is plotted in Figure 2. It can be seen on this figure, that mode 1 causes the crack to open while mode 2 induces closure.

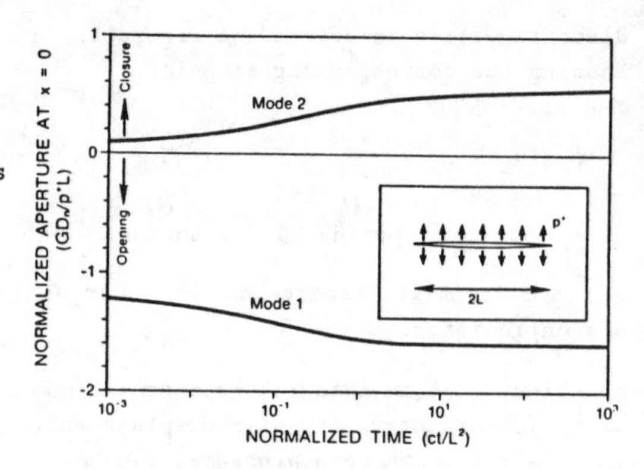

Figure 2: Variation with time of the displacement discontinuity at the crack mid-point for the two loading modes ($\nu = 0.2$, $\nu_u = 0.4$, $B = 0.8$).

4 MODELING HYDRAULIC FRACTURES

We now present the algorithm required to solve the problem of a non-propagating hydraulic fracture connected to a well at which an arbitrary discharge is prescribed. Considering only one fracture wing of length L, we aim at predicting the pressure, $p_f(x,t)$, in the fracture, its opening, $D_n(x,t)$, and fluid leak-off, $D_f(x,t)$, given the volumetric discharge $q_o(t)$ at the fracture inlet ($x = 0$). The other "boundary" conditions to be considered are (i) the in situ stress, σ^o, perpendicular to the fracture, (ii) the far-field pore pressure, p_o, and (iii) the no-flow condition at the fracture tip ($x = L$).

To solve this problem, the integral equations (1) relating pressure in the fracture, crack opening, and fluid loss have to be complemented by the equation governing the flow of fluid in the fracture. Resorting to the usual approximations of the lubrication theory with an incompressible Newtonian fluid, the mass and momentum balance equations can be written as:

$$\frac{\partial q}{\partial x} - \frac{\partial D_n}{\partial t} + D_f = 0 \quad (4)$$

$$q = -\frac{D_n^3}{12\mu}\frac{\partial p_f}{\partial x} \quad (5)$$

where μ is the fluid viscosity and q is

the fluid discharge along the fracture, defined as a volumetric rate per unit length. These two equations can be combined into the following:

$$\frac{D_n^3}{12\mu}\frac{\partial^2 p_f}{\partial x^2} + \frac{D_n^2}{4\mu}\frac{\partial D_n}{\partial x}\frac{\partial p_f}{\partial x} = -\frac{\partial D_n}{\partial t} + D_f \quad (6)$$

This equation, together with the integral equations (1), the boundary conditions $q(0,t) = q_o(t)$ and $q(L,t) = 0$, and the initial conditions, completely defines the problem under consideration.

Equation (6) is discretized using a finite-difference (FD) scheme which has the following features:

1. the formulation is implicit, i.e., Equation (6) is assumed to hold at the new time step, h;
2. Equation (6) is linearized by evaluating the terms D_n and $\frac{\partial D_n}{\partial x}$, in the left hand-side of Equation (6), at the previous time step, $h-1$ (iterative refinements are however possible);
3. the spatial nodes of the FD are located at the mid-point of the boundary elements used in the discretization of the integral equations (1);
4. the spatial derivatives are evaluated using the central difference operator; and
5. the boundary conditions for q at $x = 0$ and $x = L$ are directly be incorporated in the equations for the first and last nodes (since the two end nodes are not located at the extremities of the fracture proper account of the leak-off and storage between the fracture limits and the end FD node has to be given).

The finite-difference discretization of the fracture-flow equation (6) leads therefore to the formulation of N linear equations. Combined with Equation (2), these equations constitutes a linear system of $3N$ equations for the three unknowns D_n, D_f, and p_f, at time step h, that have defined at the N nodes on the fracture.

5 STEADILY MOVING DISLOCATION

5.1 Introduction

Among the simplest problems involving propagating cracks (dislocations) in a poroelastic medium is that of semi-infinite, quasistatic cracks moving at a steady-state speed, V. Indeed, the assumption of a steady-state speed, V, implies that there is no apparent change for an observer located at the moving crack tip. If one assumes that the crack is propagating along the x-axis in the positive direction, the dependence of any field quantity, F (e.g., pore pressure), on position and time is of the form $F(x - Vt, y)$.

This class of problems has been studied within the context of Earthquake Mechanics, using shear cracks or slip dislocations (see [7] for a review). The moving tensile crack, more relevant to hydraulic fracturing applications, has been analyzed by Cleary [14] and Ruina [15] in the case of impermeable fracture walls, and by Huang and Russell [16,17] for permeable fracture walls. In the following sections, we first discuss the solution of a moving tensile edge dislocation with impermeable walls, derived by Cleary [14], then recreate the result by distributing point DD's along a straight line segment whose length grows at a constant velocity.

5.2 Mathematical modeling for a moving dislocation

The exact solution for a moving normal dislocation is easily obtained from the solution of an impulse normal dislocation with origin at x', and created at time t' (Figure 3). In the space-time domain $[x, y, t]$, this solution corresponds to the introduction of a DD of the following form:

$$u_y(x, 0^-, t) - u_y(x, 0^+, t) = D_n H(x' - x)\delta(t - t') \quad (7)$$

where $H(x)$ is the Heaviside step function, $\delta(t)$ is the Dirac delta function, u_y are the displacements parallel to the y-axis,

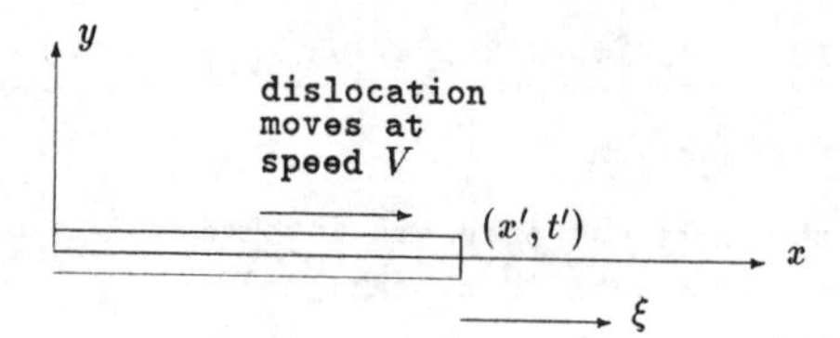

Figure 3: Coordinates of a moving dislocation.

and D_n is the magnitude of the DD (negative for an opening displacement). The solution, at a distance ξ ahead of the origin of a moving dislocation, is then determined by replacing x' by Vt' and x by $\xi + Vt$, and by integrating the transformed impulse solution from $t' = -\infty$ to $t' = t$. The pair (ξ, y) can be understood as a moving coordinate system at the origin of the dislocation (Figure 3).

For instance, from the known fundamental solution for the pore pressure field $p(\xi, y)$ generated ahead of a static dislocation [6], one can deduce the following expression for a moving dislocation:

$$p(\xi, y) = \frac{B(1+\nu_u)}{3\pi(1-\nu_u)} G D_n \int_{-\infty}^{t} \frac{[\xi + V(t-t')]\delta(t-t')}{[\xi + V(t-t')]^2 + y^2} - \frac{\xi + V(t-t')}{4c(t-t')} \exp\left\{-\frac{[\xi + V(t-t')]^2 + y^2}{4c(t-t')}\right\} dt' \quad (8)$$

where the first term under the integral sign represents the instantaneous elastic response to the impulse DD at time t and the second term represents the delayed responses to the impulse DD's at times $t' < t$.

Along the positive x-axis, Equation (8) can be simplified as follows [14]:

$$p(\xi, 0) = \frac{B(1+\nu_u)}{3\pi(1-\nu_u)} \frac{G D_n}{\xi} [1 - g(\frac{V\xi}{c})] \quad (9)$$

where

$$g(\frac{V\xi}{c}) = \frac{V|\xi|}{4c} \int_0^{\infty} \frac{1+x}{x^2} \exp\left\{-\frac{V|\xi|}{4c} \frac{(1+x)^2}{x}\right\} dx \quad (10)$$

5.3 Moving tip element

In the case of moving-crack problems, a special tip element can be implemented to allow the crack to extend at a speed V between two time steps. For the sake of simplicity, it will be assumed that the crack-tip profile remains self-similar during propagation.

Consider the general case where the opening shape of the tensile crack, in the neighborhood of the tip, is of the following form:

$$D_n(\xi) = A(-\xi)^{\alpha}, \quad \xi \le 0 \quad \alpha \ge 0 \quad (11)$$

where $\alpha = 0.5$ for static cracks in linear elastic media.

If the special tip element allows the crack to propagate from time t_{h-1} to time t_h, its contribution ΔF to any field quantity F can be expressed as follows:

$$\Delta F(\xi, y, t_h) = \int_{t_{h-1}}^{t_h} \int_{L_{h-1}}^{L_{h-1}+V(t'-t_{h-1})} \mathcal{F}_n(L_h + \xi - \chi, t_h - t') D_n(\chi, t')\, d\chi\, dt' \quad (12)$$

where

$$D_n = A\left[L_{h-1} + V(t' - t_{h-1}) - \chi\right]^{\alpha}, \quad (13)$$

L_{h-1} is the length of the fracture at time t_{h-1}, and $\mathcal{F}_n$ is the influence function for an impulse point normal DD.

In the case of a moving dislocation, $\alpha = 0$ and Equation (12) becomes simpler as the magnitude of the displacement discontinuity reduces to the constant A.

5.4 Numerical implementation

Using a poroelastic DD program, a steadily moving dislocation was approximated by a finite dislocation of length L, grown at a constant speed, V, from an initial zero length. This approximation is accurate, provided that the coordinate, ξ, of the field point is small compared to the fracture length, i.e., if $|\xi|/L << 1$.

In the numerical algorithm, the spatial integrals are performed exactly and the time integrals are solved numerically, using an adaptive Romberg-interpolation scheme that yields an arbitrarily small error. Hence, the number of boundary elements used to discretize the finite dislocation becomes unimportant. The results shown below were obtained with only one element: the special tip element.

Figure 4 shows a comparison between the pressure profile along the ξ-axis obtained using Equation (9) and those obtained using the poroelastic DD algorithm for dimensionless dislocation lengths LV/c equal to 20, 50, 100 and 200. First, it

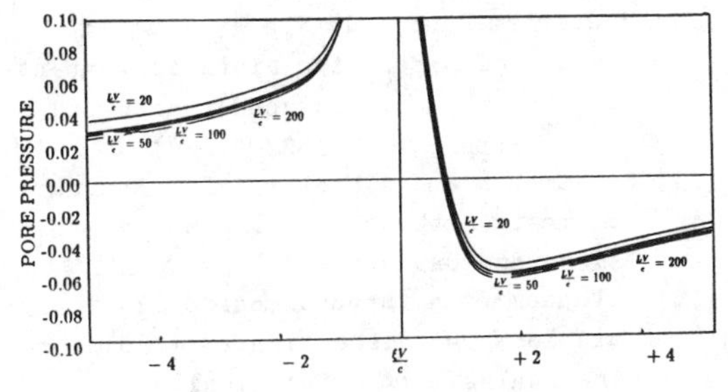

Figure 4: Pore pressure profile on the axis of a moving dislocation.

can be seen that the numerical output for a finite dislocation of length $LV/c = 200$ cannot be distinguished, at the precision of the graph, from the theoretical results of Equation (9). One should also notice that for the range of ξ under consideration, the pore pressure calculated for $LV/c = 20$, ahead of the dislocation, does not differ by more than 15% from the values predicted for the moving semi-infinite dislocation.

Figure 4 also indicates that the minimum in the pore pressure induced by the moving dislocation occurs at about $\xi V/c = 1.9$. Moving from this point toward the origin of the dislocation, the induced pore pressure progressively increases to infinity, (the pressure becomes positive at about $\xi V/c = 0.5$). It can be deduced from this plot that, in the limiting case of zero velocity, the pore pressure evolves toward the drained elastic solution for a static, continuous edge dislocation (i.e., $p = 0$ everywhere). In the limiting case of infinite velocity, the pore pressure evolves toward the corresponding undrained elastic solution. This latter solution is characterized by a jump of the pressure from $-\infty$ at $\xi = 0^+$ to $+\infty$ at $\xi = 0^-$.

It should also be noted that reading the plot of Figure 4 from right to left gives the history of the pore pressure at a fixed point on the x-axis as the dislocation is approaching and then passing by.

Finally, Figure 5 gives contour plots of the induced pore pressures in the vicinity of the origin of the dislocation.

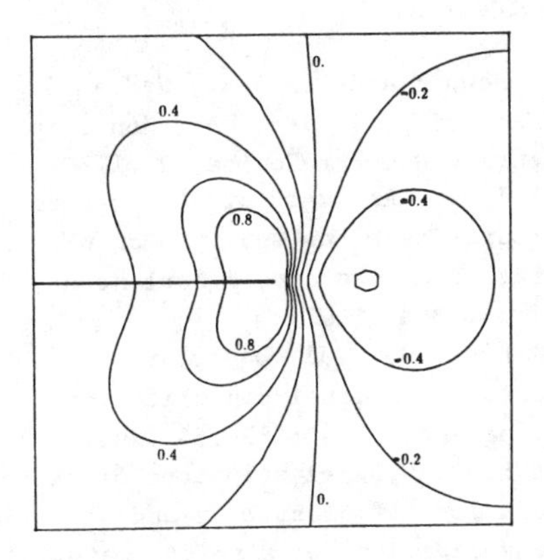

Figure 5: Pore pressure contours around the origin of a moving dislocation (side of contour grid is $10c/V$).

6 CONCLUDING REMARKS

The Biot theory of poroelasticity provides a rigorous framework for formulating a mathematical model for hydraulic fractures in porous formations. Not only are the coupled diffusion-deformation effects directly accounted for, but the response of the rock mass to the fracture opening and fluid loss are treated in a symmetric and consistent manner.

The numerical algorithm to model nonpropagating hydraulic fractures can directly be extended to the case of growing fractures. However, a consistent propagation criterion for tensile cracks in poroelastic material, which accounts for the pore pressure change ahead of the tip, needs first to be developed.

7 ACKNOWLEDGMENTS

The authors wish to acknowledge Dowell Schlumberger for allowing publication of this paper and Dr. J.-C. Roegiers for supporting this project. A.H.-D. Cheng wish to acknowledge the Gas Research Institute for financial support and Dr. I. Salehi for his interest in this research.

8 BIBLIOGRAPHY

[1] G.C.Howard and C.R.Fast, Hydraulic fracturing, Soc.Pet.Engr.Monograph, Henry L.Doherty Series, 2 (1970)

[2] G.R.Irwin, Analyses of stresses and strains near the end of a crack traversing a plate, J.Appl.Mech., 24:361-364 (1957)

[3] M.A.Biot, General theory of three-dimensional consolidation, J.Appl.Phys. 12:155-164 (1941)

[4] M.A.Biot, General solutions of the equations of elasticity and consolidation for a porous material, J.Appl.Mech., Trans. ASME, 78:91-96 (1956)

[5] J.R.Rice and M.P.Cleary, Some basic stress-diffusion solutions for fluid saturated elastic porous media with compressible constituents, Rev.Geophys.Space Phys., 14:227-241 (1976)

[6] E.Detournay and A.H.-D.Cheng, Poroelastic solution of a plane strain point displacement discontinuity, J.Appl.Mech., in print (1987)

[7] J.Rudnicki, Effects of pore fluid diffusion on deformation and failure of rock, Chap.15 in "Mechanics of Geomaterials," Proc. IUTAM William Prager Symp. on Mechanics of Geomaterials: Rocks, Concrete, Soils, (ed.) Z.P.Bazant, Wiley, 315-347 (1985)

[8] E.Detournay, A.H.-D.Cheng and J.-C.Roegiers, Some new examples of poroelastic effects in rock mechanics, Proc. 28th Rock Mech.Symp., 575-584 (1987)

[9] L.Vandamme, E.Detournay and A.H.-D.Cheng, A 2-D poroelastic displacement discontinuity method for hydraulic fracture simulation, subm. to Int.J.Num.Anal.Meth.Geomech. (1987)

[10] S.A.Khristianovich and Y.P.Zheltov, Formation of vertical fractures by means of highly viscous fluid, Proc. 4th World Pet.Cong. - Sect.II / T.O.P., 579-586 (1955)

[11] J.Geertsma and F.de Klerk, A rapid method of predicting width and extent of hydraulically induced fractures, J.Pet.Tech., 1572-1581 (1969)

[12] S.L.Crouch and A.M.Starfield, Boundary elements methods in solid mechanics, Allen and Unwin (1983)

[13] L.Vandamme, A three-dimensional displacement discontinuity model for the analysis of hydraulically propagated fractures, PhD thesis, U.Toronto, 1986

[14] M.P.Cleary, Moving singularities in elasto-diffusive solids with applications to fracture propagation, Int.J.Sol.Struct. 14:81-97 (1978)

[15] A.Ruina, Influence of coupled deformation-diffusion effects on the retardation of hydraulic fracture, Proc. 19th U.S.Rock Mech.Symp., 274-282 (1978)

[16] N.C.Huang and S.G.Russell, Hydraulic fracturing of a saturated porous medium-I: general theory, Theor.Appl.Fract.Mech., 4:201-213 (1985)

[17] N.C.Huang and S.G.Russell, Hydraulic fracturing of a saturated porous medium-II: special cases, Theor.Appl.Fract.Mech., 4:215-222 (1985)

Numerical Methods in Geomechanics (Innsbruck 1988), Swoboda (ed.)
© 1988 Balkema, Rotterdam. ISBN 90 6191 809 X

An analysis of hydraulic fracturing by the finite element method

A.Polillo Filho & A.Maia da Costa
PETROBRÁS/CENPES, Rio de Janeiro, Brazil
Nelson F.Favilla Ebecken
COPPE/UFRJ, Federal University of Rio de Janeiro, Brazil

ABSTRACT:

This paper discuss a finite element strategy to simulate the hydraulic fracturing technique for oil well stimulation. After a review of the analytical solution procedures, an elasto plastic finite element formulation is adopted. Infinite finite elements are also implemented. Some numerical comparisons are presented.

1 - INTRODUCTION

The hydraulic fracturing technique for oil well stimulation has been worldwide used and requires practical experience and geotechnical knowledgement.

The first step in the the work is the stress distribution analysis to stablish the hydraulic pressure to be applied inside the well which will provide the required fracture system.

This paper covers the application of finite element method for stress analysis considering adequate boundary conditions for simulation of underground excavations in an infinite medium, by the application of special elements. The stress distribution analysis is actually the simulation of the effect of well boring taking into account the fluid pressure inside the well and the fluid pressure inside the rock mass when a penetrating fluid is considered. The material behavior is considered to be elasto-plastic according to the Mohr-Coulomb yielding criterium.

At the elastic region some comparisons are made against the available analytical solutions.

2 - REVIEW OF THE ANALYTICAL ELASTIC SOLUTION FOR BREAKDOWN PRESSURE

The basic relations that describe the elastic behaviour of porous media can be applyed to the borehole breakdown pressure problem whether the borehole fluid penetrates or not into the formation. There are many works reporting to this subject in the technical literature, but in this paper the solution proposed by Haimson & Fairhurst /1/ has been chosen to be compared with the numerical one, and is also presented in reference /2/.

In this solution the formation is assumed to be linear elastic, porous, isotropic and homogeneous. There is also no interaction between the fracture fluid inside the wellbore and the formation fluid. Two kinds of stresses are defined: total and effective. The effective is the part of the total which acts upon the rock sKeleton. It is expressed by the equation below.

$$\sigma e = \sigma - P_p \qquad (1)$$

The stresses involved in the hydraulic fracturing problem are three principal ones. The vertical σve and the two horizontals σhe_2 and σhel . Assuming that no horizontal strain is present. Those stresses can be related by Hooke's law.

$$\sigma he2 = \nu(\sigma he1 + \sigma ve)$$
$$\sigma he1 = \nu(\sigma he2 + \sigma ve) \qquad (2)$$

In the absence of tectonisms equations (2) can be rearranged to:

$$\sigma he1 = \sigma he2 = \frac{\nu}{1 - \nu}(\sigma v - P_p) \qquad (3)$$

The stress redistribution development around the wellbore can be obtained by using the principal of superposition. Three stress fields are considered and are consequence of the following items:

(i) the redistribution of initial stresses by wellboring.

(ii) the difference between the fluid pressure in the wellbore and the formation fluid pressure.

(iii) the radial fluid flow through porous medium from the wellbore to the formation as a consequence of this pressure differential.

Each problem is a linear elastic one that can be solved independentely.

The first one is obtained from the theory of stress concentrations in an infinite plate with a small circular hole with the initial stresses acting on its edges.

The solution for the problem caused by the borehole inner pressure arises from Lame's solution for the thick-wall cylinder of porous material. The outer radius of cylinder is very large with respect to the inner radius. It is assumed plane strain state.

At last, the stresses induced by the radial flow of fluids through porous medium can be predicted by simply solving the poroelasticity equations developed by Biot (1941) using a thermoelastic-poroelastic analogy which has first been shown by Lubinski (1954). This analogy consists in keeping the same solution replacing the dilatant coefficient α by $(1 - \beta)(1 - 2\nu)/E$ and $\Delta T(r)$ by $\Delta P(r)$, where $\Delta P(r) = Pp(r) - Pe$ (4)

The resultant stresses are determined, as already mentioned, by the superposition of the three analysed fields.

In the pratical applications the effective stresses are the used ones and are given by

$$\sigma re = pw - pe \qquad (5)$$

$$\sigma\theta e = 3\sigma he_2 - \sigma he_1 - pw + 2pe + (1 - \beta)\frac{(1 - 2\nu)}{(1 - \nu)}(P_p - P_e) - P_p \qquad (6)$$

$$\sigma ze = \sigma re + 2\nu(\sigma he_2 - \sigma he_1) + (1 - \beta)\frac{1 - 2\nu}{1 - \nu}(P_p\ P_e) + Pe - Pp \qquad (7)$$

The fracture will be horizontal if $|\sigma e| \geqq |RT|$ or vertical if $|\sigma\theta e| \geqq |RT|$, where RT is the rock tensile strength.

The fluid effect is also considered in the analysis. This can be done considering the pore pressure Pp = pe for the non penetrating fuild or Pp = p(r) to the penetrating one. In the wellbore wall Pp = pe for the first one and pp = pw for the other.

Thus the equations for determining the wellbore inner pressure necessary to breakdown the formation are respectevely given by equations (08), for penetrating fluids and (09) for non penetrating fluids.

$$pw = \frac{3\sigma he_2 - \sigma he_1 - RT + pe}{2 - \frac{(1 - \beta)(1 - 2\nu)}{1 - \nu}} \qquad (8)$$

$$pw = 3\,\sigma he_2 - \sigma he_1 + pe - RT \qquad (9)$$

Both are valid for vertical fractures.

From $|\sigma ze| \geqq |RT|$, equation (10) is obtained

$$pw = \frac{\sigma ve - RT + pe}{1-(1-\beta)\frac{(1-2\nu)}{1 - \nu}} \qquad (10)$$

for predicting the wellbore inner pressure to cause a horizontal fracture, for penetrating fluids. Non-penetrating fluids cannot create a horizontal fracture since the vertical effective stress is not a function of pw.

3 - IMPLEMENTATION OF THE ELASTO-PLASTIC MODEL

It is adopted the yielding criterium of Mohr-Coulomb.

In the case of a non-strain hardening material. The yielding equation is a function only of the stress tensor.

$$F(\underset{\sim}{\sigma}) = \sigma m \, sen\phi + \tilde{\sigma}(\cos\theta - 1/3 \, sen\theta \, sen\phi) - c \cos\phi = 0 \qquad (11)$$

$$\tilde{\sigma} = 0.50(Sx^2 + Sy^2 + Sz^2 + \tau xy^2 + \tau yz^2 + \tau xz^2 \qquad (12)$$

$$\theta = 1/3 arc\ sen \frac{(-3\sqrt{3}}{2} \frac{\tau 3)}{\tilde{\sigma}^3} ; -\frac{\pi}{6} \leq \theta \leq \frac{\pi}{6} \qquad (13)$$

$$\tau_3 = Sx\ Sy\ Sz + 2\tau xy\ \tau yz\ \tau xz - \tau x^2 z\ Sy - Sx\ \tau y^2 z - Sz\ \tau x^2 y \qquad (14)$$

$$\sigma m = \frac{\sigma x + \sigma y + \sigma z}{3} \qquad (15)$$

The incremental plastic strain is calculates assuming the associative law.

$$d\underset{\sim}{\varepsilon}^{P} = d\lambda \cdot \frac{\partial F}{\partial \underset{\sim}{\sigma}} \qquad (16)$$

Calling $a = \frac{\partial F}{\partial a}$

The elasto-plastic matrix is given by.

$$D_{ep} = \frac{(\underset{\sim}{D} - \underset{\sim}{D}\ \underset{\sim}{a} \cdot \underset{\sim}{a}^T \cdot \underset{\sim}{D})}{\underset{\sim}{a}^T \cdot \underset{\sim}{D} \cdot \underset{\sim}{a}} \qquad (17)$$

$$d\underset{\sim}{\sigma} = d_{\underset{\sim}{ep}}\ d\underset{\sim}{\varepsilon} \qquad (18)$$

4 - FINITE ELEMENT FORMULATION

In an underground cave or a well boring, the initial gravitational stresses will not participate as nodal equivalent internal forces at the equilibrium equation. These stresses will be important for cheking the yielding condition of the rock and also in the determination of the plastic strain increment by the constitutive law of the material.

The initial gravitational stresses will be considered at the global equilibrium equation as equivalent nodal excavation forces at the perimeter of the cave, in order to vanish the stresses at this surface and the internal induced stresses will be added to the initial stresses inside the rock mass, fig. 1

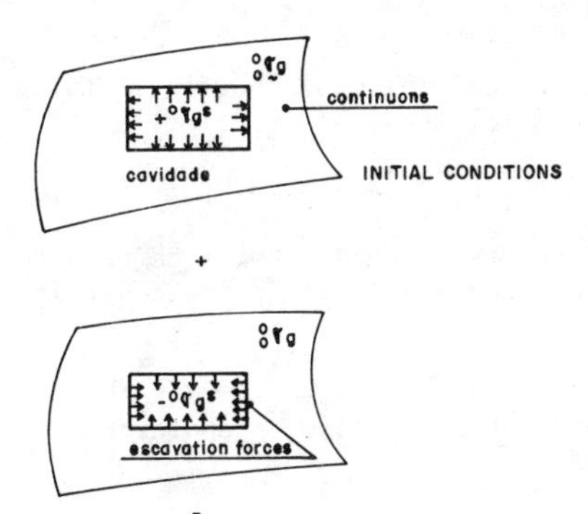

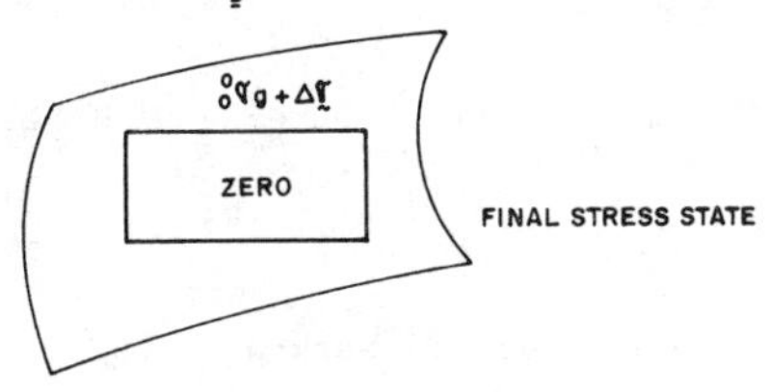

FIGURE 1

At the first step it will be applied the excavation forces at the perimeter of the cave. The equilibrium equation is obtained by the virtual work principle:

$$\int \delta\underset{\sim}{\varepsilon}^T \cdot \underset{\sim}{\sigma} dv - \int \delta \underset{\sim}{u}^T\ f\underset{\sim}{v}\ dv - \int \delta u^T \cdot f\underset{\sim}{s} ds = 0 \qquad (19)$$

$$\sigma = \underset{\sim}{D} \cdot \underset{\sim}{\varepsilon} + \underset{\sim}{\sigma r} + \sigma_T \qquad (20)$$

σr - residual stresses

$f\underset{\sim}{v}$ - volumetric forces

fs - surface pressure equal to the initial gravitacional stresses at the perimeter

σ_T = Thermal stresses

$$\underset{\sim}{\varepsilon} = \underset{\sim}{B} \cdot \underset{\sim}{U} \quad \delta\underset{\sim}{\varepsilon} = \underset{\sim}{B} \cdot \sigma\underset{\sim}{U} \qquad (21)$$

$$\sigma \underset{\sim}{U} = \underset{\sim}{N} \cdot \delta\underset{\sim}{U} \qquad (22)$$

Taking (21) and (20) in (19)

$$\int_V \underset{\sim}{B}^T . \underset{\sim}{D} . \underset{\sim}{B} . dV . \underset{\sim}{U} + \int_V \underset{\sim}{B}^T . {}^0\underset{\sim}{\sigma}r dV + \int_V \underset{\sim}{B}^T . {}^0\underset{\sim}{\sigma}T dV - \int_V \underset{\sim}{N}^T . \underset{\sim}{f}v . dV - \int_S \underset{\sim}{N}^T . {}^0\underset{\sim}{\sigma}g dS = 0 \quad (23)$$

Where ${}^0\sigma g$ initial gravitational stresses.

After solving (20) the final total stresses will be

$$\underset{\sim}{\sigma} = {}^0\underset{\sim}{\sigma}g + {}^0\underset{\sim}{\sigma}r + {}^0\underset{\sim}{\sigma}^T + \underset{\sim}{D}.\underset{\sim}{B}.\underset{\sim}{U} \quad (24)$$

The total stress will be used to calculate the non-linear plastic strain

$$\underset{\sim}{\varepsilon}^P = f(\underset{\sim}{\sigma}) \quad (25)$$

When the yielding condition is reached

$$\varepsilon^P \neq 0 \quad (26)$$

The final total stress will be given by

$$\underset{\sim}{\sigma} = \underset{\sim}{D}(\underset{\sim}{\varepsilon} - \underset{\sim}{\varepsilon}^P) + {}^0\underset{\sim}{\sigma}g + {}^0\underset{\sim}{\sigma}r + {}^0\sigma T \quad (27)$$

Using the virtual work principle:

$$\int \underset{\sim}{B}^T . \underset{\sim}{D}(\underset{\sim}{\varepsilon} - \underset{\sim}{\varepsilon}^P) + {}^0\underset{\sim}{\sigma} + {}^0\underset{\sim}{\sigma}r + {}^0\underset{\sim}{\sigma}T) \, dV = Rext + Fex \quad (28)$$

The forces $\int \underset{\sim}{B}^T \sigma g \, dV$ should be eliminated from the equilibrium equation, by just adding, those forces at the right side of the equation.

5 - DISCRETIZATION OF INFINITE AND SEMI-INFINITE MEDIUM

The application of numerical methods for simulation of infinite and semi-infinite medium, when submitted to external forces, requires a special procedure for the continuum modelling.

Presently, it has been given a great emphasis to the application of the Finite Element Method, by employing special elements designated as "infinite" elements.

The implemented infinite element is similar in its formulation to the isoparametric finite element, as it employs the usual function "serendipity" for the generation of interpolation functions. Figure 2 illustrates the type of element and the corresponding "maping" in natural coordinates.

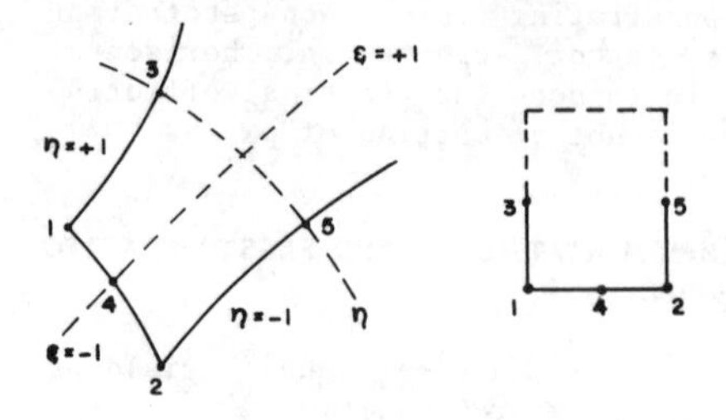

FIGURE 2

The geometric description of the element is conducted by association of the usual serenditipy functions multiplied by one term with singularity for $\xi = +1$

$$\phi = (a_1 + a_2\xi + a_3\eta + a_4\xi\eta + a_5\eta^2) \frac{1}{1-\xi} \quad (29)$$

The interpolation functions for displacements are obtained by the serendipity functions without the singular term, multiplied by a decay function of the type 1/r.

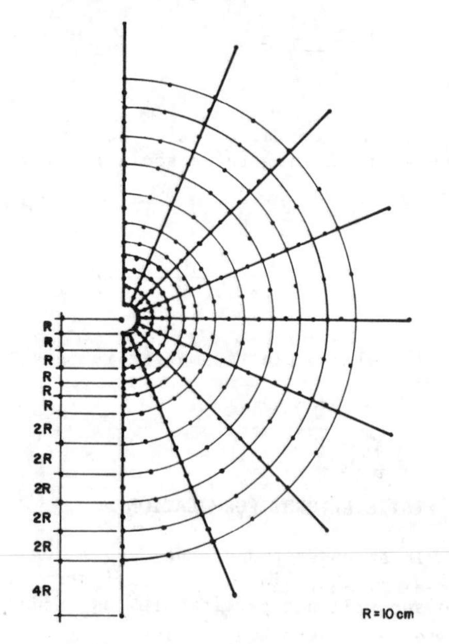

Radial mesh with 88 elements and 286 nodes

FIGURE 3

TABLE 1

Comparison between Finite Element and Analytical Solution (linear elastic approach)

Non Penetrant Fluid

RADIUS(cm)	σθe(MPa)	σθe(MPa)
	FEM	ANAL
11,06	21.3135	21.2915
22,10	14.5920	14.5811
32,08	13.4007	13.4045
42,11	12.9587	12.9603
64,22	12.6063	12.6087
315,05	12.3545	12.3536

Penetrant Fluid

RADIUS(cm)	σθe(MPa)	σθe(MPa)
	FEM	ANAL
11,06	2.0301	2.0492
22,10	9.0248	8.9912
32,08	10.5968	10.5549
42,11	10.7577	10.7140
64,22	11.1571	11.1123
315,05	12.0026	11.5149

The rigidity matrix of the elements is obtained with the usual formulation of the Finite Element Method /3/.

6 - RESULTS

A radial finite element mesh has been used to generate results. Figure 3 illustrates a general view of the mesh with its characteristics.

Two examples are presented. The first one makes a comparison between the results obtained by the finite element method and those obtained by the analytical solution presented in section 2; for the circunferencial, radial and vertical stresses. Table 1 describes the results obtained for the non - penetrant and penetrant fluid consideration.

The following data were used in example 1.

r_w = 10.0 cm
r_e = 30.0 Km
D = 3000m
p_e = 32.1855 M Pa
p_w = 33.5919 M Pa
E = 20,000. M Pa
ν = 0.25
$(1 - \beta)$ = 0.6
σ_{ve} = 37.0306 M Pa
$\sigma_{he1} = \sigma_{he2}$ = 12.3435 M Pa

Example 2 shows the stress distribution behaviour in a integration point very close to the wellbore wall for a penetrant fracture fluid.

The rock reservoir and were germetric parameters are presented below.

r_w = 10.0 cm
r_e = 30.0 Km
D = 3000m
p_e = 32.1855 M Pa
$(1- \beta)$ = 0.6
E = 26,346.4 M Pa
ϕ = 37.2
C = 8.1549 M Pa
RT = 4.16746 M Pa

Table 2 shows the values of circunferencial and radial stresses until rock failure as the difference of pressure between the wellbore and the formation fluid pressure increases.

Two different assumptions were made. The fisrt one considering a linear elastic medium and another treating it as an elasto-plastic one.

For the elasto plastic approach the breakdown pressure would be approximately (30.335 + 32.1855 M Pa) = pw otherwise in the elastic approach it would be near (23.649 + 32.1855) M Pa.

TABLE 2
Comparison between elastic and Elastoplastic Approaches

(pw - pe) M Pa	ELASTO PLASTIC		ELASTIC	
	σθe	σre	σθe	σre
0.0	31.73	3.127	31.73	3.127
100.0	16.58	1.706	16.58	1.706
200.0	1.428	0.2863	1.428	0.2863
250.0	-3.017	0.1056	-6.146	-0.4238
303.35	-4.188	0.1835	-14.227	-1.1815

7 - CONCLUSIONS

From the results obtained in the examples one conclude that the numerical analysis approaches the available analytical solution very precisely.

If one is able to obtain the simple couple of rock parameters as internal friction angle and cohesion, hydraulic fracture pressure can be predicted even if the behaviour is not linear elastic, but elasto-plastic. This enables the use of heavier "muds" to kill the well, during drilling work, avoiding the necessity of casing it, if no other alternative is present.

In a stimulation job, the producing engineer can predict the real wellbore inner pressure to cause fracture and optimize his design with safety, since where the rock behaviour is elast plastic, bigger values of pressure are obtained.

Numerical methods are the only available to analyze the problem in a reservoir with an elasto-plastic behavior since for those the analytical solution cannot be applied.

The numerical solution is the only avaiable approach for inclined and horizontal wells.

8 - REFERENCES

1 - Haimson, B. & Fairhurst, C. Initiation and Extension of Hydraulic Fractures in Rocks. Society of Petroleum Engineers Journal, 310-318, Sep. 1967.

2 - Campos, J.C.B., Stresses at the wall of a circular well according the Theory of Elasticity, Technical Petrobras Repport, Rio de Janeiro, 26(3): 209-216, Jul./Set. 1983.

3 - Costa, A.M.,"An application of Computational Methods and Rock Mechanics Principles in the Design and Analysis of Underground Excavations", Rio de Janeiro, D.Sc. Thesis - COPPE/UFRJ, 1984.

Numerical Methods in Geomechanics (Innsbruck 1988), Swoboda (ed.)
© 1988 Balkema, Rotterdam. ISBN 90 6191 809 X

Coupled models and free-surface seepage analysis without mesh iteration

M.Dysli & J.Rybisar
Soil Mechanics Laboratory, Swiss Federal Institute of Technology, Lausanne

ABSTRACT : In a finite element model with ground water table, the classical method consists of using repeated iterations to adjust the water table to the geometry of the mesh while modifying the form of the elements. This method can be advantageously replaced by one which does not modify the mesh geometry but which defines the water table by means of the relationship : Hydraulic conductivity = f (pore water pressure).

This method allows the detailed and precise characteristics of the unsaturated zone above the water table to be taken into account and can be used in a model coupling the groundwater flow equation with the displacement equilibrium equation, hence introducing the effect of variations in permeability as a function of void ratio.

1 INTRODUCTION

The finite element method is a powerful tool widely used today for seepage analysis. Compared with analytical or analogical methods, it has the advantage of being universally applicable, from the simplest to the most complex case, and of accepting any boundary conditions. What is more, it easily accomodates a study of that most important yet imprecise and highly variable parameter - hydraulic conductivity.

This is perhaps the determining factor in its use. Of course it requires a computer for its application; however, this is no longer a problem since finite element codes are available for personal computers and the use of high-performance processors, such as vectorial machines, no longer appears expensive when compared with the time necessary to prepare and interpret results.

The solution of the ground water flow equation - a Laplace equation with partial derivatives which defines, based on certain hypotheses, a velocity potential flow - using the finite element method presents a difficulty which has not yet been correctly overcome up to the present time. In the case of free surface seepage, the surface is a limit of integration and as such one of the limits governing the finite element discretization. However, this limit is not known a priori and the mesh must be adjusted step by step, each of the iterations consisting of the solution of a new system of equations based on new boundary conditions.

There are numerous techniques to carry out the iterative process. These techniques must avoid creating highly distorted elements in the neighbourhood of the water table which often necessitates the re-creation, at each step, of the entire mesh. In what follows, we shall refer to the method which adjusts the water table surface to the overall mesh geometry as the classical method.

However, a method does exist which requires no modification to the initial finite element discretization and which in addition allows the characteristics of the unsaturated zone above the free water surface to be taken into account. The principles of this method have been known for some time and were explained by, amongst others, K.J. Bathe and M.R. Khoshgoftaar (1979), but it does not seem to have given rise to many echoes in geotechnical circles. It requires a slightly more elaborate (non-linear) code than those used in practice today, but such codes exist and their use presents no major difficulties.

Since water table movements modify stress tensors and hence void ratios and hydraulic conductivity, it can prove interesting to couple the ground water flow equations with displacement equilibrium equations so as to take account of this variation in hydraulic conductivity and make concurrent calculations of stresses, strains and pore water pressures under either steady or transient state conditions.

2 SEEPAGE ANALYSIS

One of the classical forms of the ground water flow equation is

$$\frac{\delta}{\delta x}\left(k_x \frac{\delta h}{\delta x}\right) + \frac{\delta}{\delta y}\left(k_y \frac{\delta h}{\delta y}\right) + \frac{\delta}{\delta z}\left(k_z \frac{\delta h}{\delta z}\right) = M$$

steady state conditions $M = 0$

transient state conditions $M = m_v \gamma_w \left(\frac{\delta h}{\delta t}\right)$

where k_x, k_y and k_z are the hydraulic conductivities in directions x, y and z, h is the potential ($h = z+u/\gamma_w$, z being the elevation and u the pore pressure), m_v is the coefficient of volume change, γ_w is the density of water and t is the time.

In the traditional example of a well as shown in Figure 1, the boundary conditions are defined thus : $h = h_1$ on the face adjacent to the ground (L1), $h = h_2$ beneath the surface of the water in the well (L2), $\delta h/\delta n = 0$ on the lower impermeable face (L3) and $h = z$ along the seepage face (L4). On the water table surface (L5) $h = z$ and $\delta h/\delta n = 0$ where n is the normal vector at the limit L_n.

All this is well known and merely acts as a reminder.

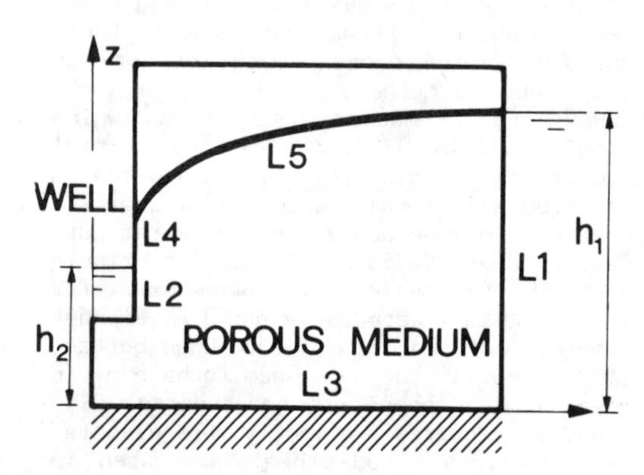

Fig.1 The well

3 WATER TABLE VARIATIONS $k = f(u)$

The limiting conditions which define the limit of the geometrical mesh of finite elements correspond well to the limit L5. However, if the numerical code allows the introduction of hydraulic conductivity as a function of pore pressure - thus becoming a nonlinear code - this conditions with limits ($h = z$, where $u = 0$ and $\partial h/\partial n = 0$) can be replaced by a relationship which virtually cancels out the hydraulic conductivity when pore pressure becomes negative (relationship I in Figure 2). Hence the geometrical model no longer needs to be adjusted little by little to the ground water table. This method nevertheless leads to the solution of a system of equations containing nonlinear partial derivatives also requiring an iterative process (Newton-Raphton, for example), which does not, however, have the same complexity as a mesh generator, particularly when dealing with three-dimensional problems, and does not present the danger of creating elements so distorted that the final result cannot be relied upon.

The classical method does not describe the material above the water table. With the relationship $k = f(u)$ of type I (Figure 2), this material is assumed to be impermeable, or almost so. This simplified hypothesis leads to exact correspondance between the method proposed here and the classical method.

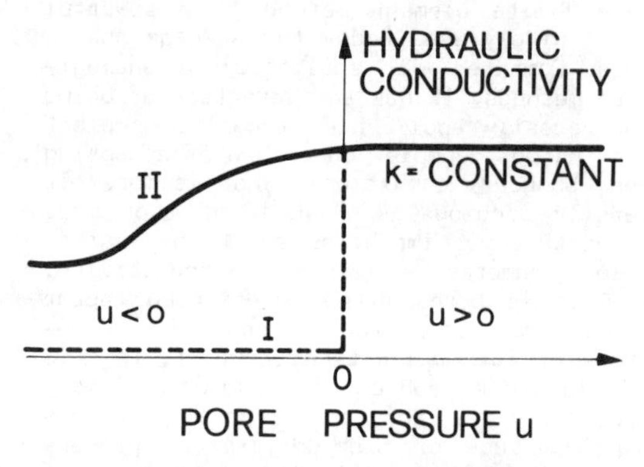

I Theoretical relationship corresponding to a calculation with adjustment of the finite element mesh to the ground water table

II Real relationship of the unsaturated medium above the water table.

Fig.2 Relationships between hydraulic conductivity and pore pressure

In reality, the material above the water table is not impermeable in the majority of cases. In this zone an unsaturated state develops, the characteristics of which can influence the flow in the saturated zone beneath the water table. From the hydraulic point of view, the unsaturated zone is characterized by a relation ship - hydraulic conductivity = f (suction) - of the type given as an example in Figure 3.

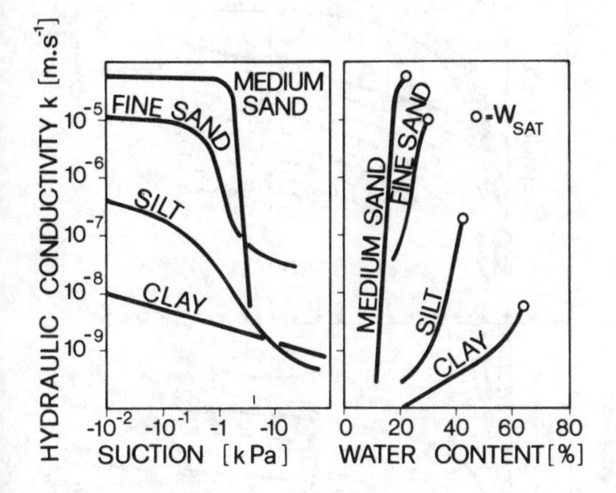

Fig.3 Example of the relationship : hydraulic conductivity - suction - water content

The very simple method proposed here can thus be applied to any hydraulic conductivity - pore pressure relationship in the negative pressure (suction) domain which characterizes the unsaturated medium above the water table (relationship II in Figure 2), and this is one of its great advantages.

4 APPLICATIONS

Since this paper aims above all at presenting a simple demonstration of the possibilities of the nonlinear method, let us reconsider the traditional example of the well.

The soil under consideration is a silt with an isotropic conductivity k of $5 \cdot 10^{-6}$ $m \cdot s^{-1}$, a coefficient of volume change m_v of $1 \cdot 10^{-4}$ kPa^{-1} (E_{oed} = 10'000 kPa) corresponding to a void ratio e of 0.60.

The finite element mesh used, together with its dimensions, are shown in Figure 4a. All the elements are eight-node isoparametric elements and the approach to the problem is by axial symmetry.

The well is represented by highly permeable elements to avoid defining limiting conditions against the wall in the unsaturated zone. The calculations are based on transient state conditions.

The first case studied (Figure 4b) corresponds to the relationship k = f(u) - I in Figure 2. The second (Figure 4c) corresponds to relationship II with k = f(u) specific to a silt. The third case is similar to the second except that continuous heavy rain at the soil surface is simulated (Figure 4d).

All calculations were carried out using the ADINAT program.

From Figures 4b and 4c it can be seen that although the introduction of a relationship k = f(u) corresponding to the chosen soil has little influence on the position of the water table (u = 0), it greatly modifies pore pressure distribution in the unsaturated zone. In the case of sand, the position of the water table undergoes much greater modification with the introduction of a real relationship k = f(u) into the unsaturated zone. On the other hand, surface rainfall rapidly modifies the position of the water table.

5 COUPLING

The speed of today's computers and the modest cost of using them allows the coupling of a numerical groundwater seepage model with a stress-strain calculation model using continuum mechanics linked to linear and nonlinear constitutive laws. Since the lowering of the water table around the well in our small example will cause an increase in the stresses found in that zone, and hence a reduction in the voids ratio, we shall try to demonstrate - by coupling a numerical groundwater seepage model with a "stress-strain" model - the effect that variations in hydraulic conductivity as a function of the voids ratio have on flow in the region of a well.

For this demonstration, the ADINAT program has been coupled to the ADINA program by means of an interprocessor.

An interprocessor achieves the coupling of two finite element codes by solving the displacement equilibrium equations and the groundwater flow equations.

Between each time step the interprocessor carries out a coupling function such as, for instance, a relationship of the type $\Delta p = f(\Delta\sigma')$ and restarts the two finite element codes, one after the other with new initial conditions. These two codes can be linear or nonlinear and, in the latter case, can have their own internal loop cycles. Figure 5 illustrates the principle of an interprocessor.

It can be seen that the interprocessor consists in fact of two codes :

- A GENU code to generate the initial conditions of the finite element code

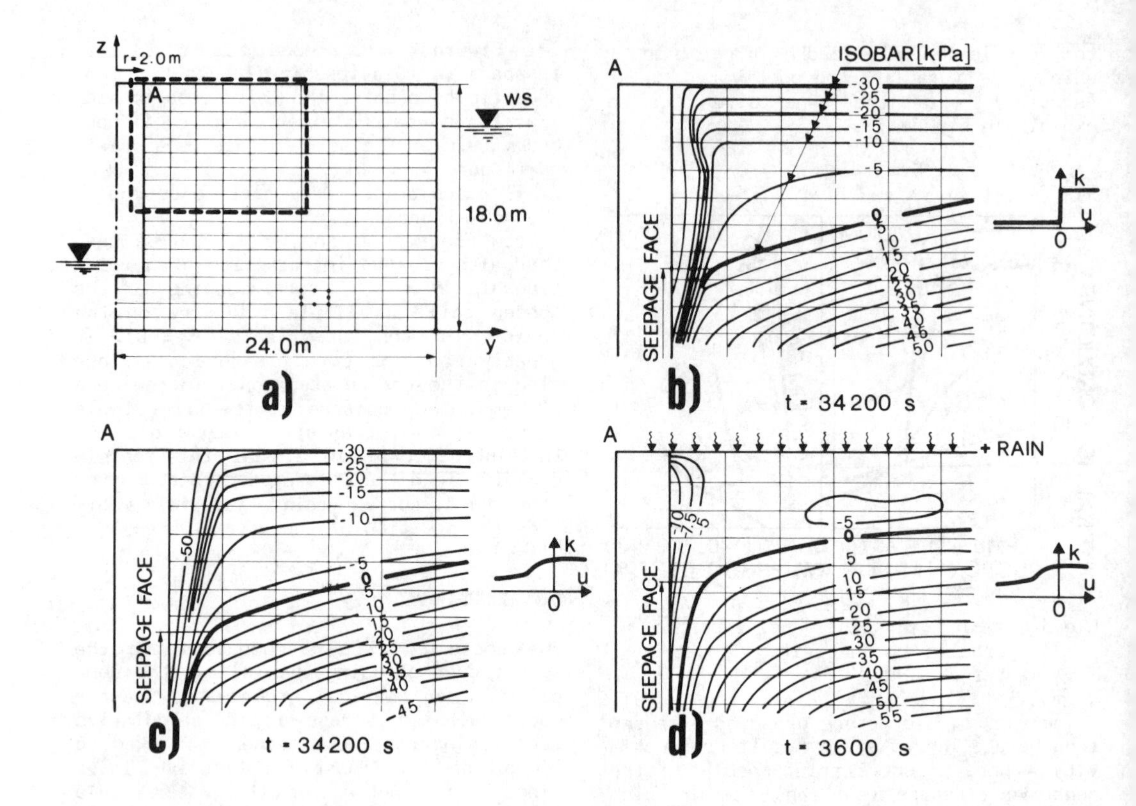

Fig.4 A few possibilities of the nonlinear method
a) Geometry of the model
b) Case corresponding to relationship I in Figure 2
c) Case corresponding to relationship II in Figure 2
d) As for c) but with rainfall at the surface

solving the groundwater flow equation (FEP) on the basis of stresses or strains calculated during the previous step by the finite element code solving the displacement equilibrium equation (FEC).

- A GENP code to generate element surface pressures or nodal loading for the finite element code solving the displacement equilibrium equation (FEC) on the basis of pore pressures calculated by the FEP code.

These two codes, constituting the interprocessor, are normally controlled by a command procedure file and employ scratch files to store information required for use in the text time step solution.

In addition to variations in pore pressure, the GENU code can calculate for each step new values of hydraulic conductivity, which are a function of the strain at that particular time.

At first sight coupling would appear to be a simple process. In practice, however, several problems must be solved. In particular :

- Generation of the initial conditions before the beginning of the actual response solution. The soil must in fact have a certain consolidation ratio before loads can be applied. If the constitutive law of the "stress-strain" finite element code (FEC) is nonlinear, it is impossible to introduce a priori initial stresses (principle of superposition not valid). The soil history must therefore be reconstructed prior to applying loads other than gravity. This can be done by a series of preparatory time steps.
- In nonlinear finite element codes, excavations are simulated by the "death" of elements at a given instant, and backfills for instance, by the "birth" of

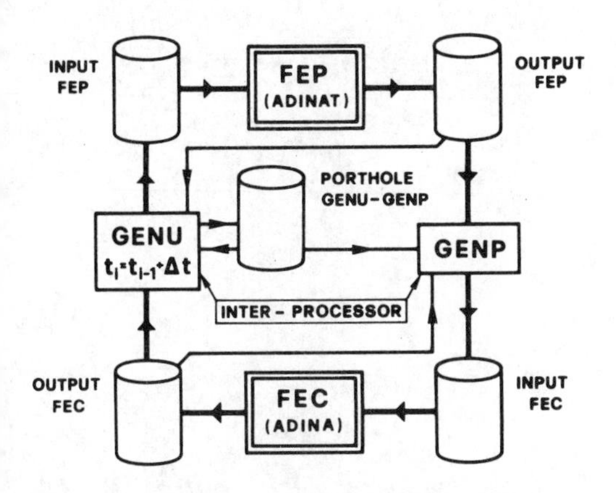

FEC = finite element code solving a displacement equilibrium equation
FEP = finite element code solving a groundwater flow equation
GENU = code generating initial conditions for the FEP ($\Delta p = f(\Delta\sigma)$ or $= f(\Delta\varepsilon)$)
GENP = code generating element surface pressures ($P = f(p)$)

Fig.5 Principle of an interprocessor

elements. In the groundwater flow finite element code (FEP) the birth and death of elements can create difficulties, which must be resolved by the GENU code of the interprocessor.

- Variations in water surface in the groundwater flow finite element code (FEP) can lead to serious convergence problems and a relatively small value for the time step Δt must be chosen.

The interprocessor dealt with in this paper has been developed for use with the ADINA (1981) and ADINAT (1981) codes. These two codes allow both static and dynamic analyses (steady-state and transient analyses with ADINAT) with various nonlinear constitutive laws and an extensive library of elements for one-, two- and three-dimensial analyses.

The law linking hydraulic conductivity k with void ratio is as follows :

$$k = k_0\, e^{\left(\frac{e-e_0}{D}\right)}$$

where k_0 = initial hydraulic conductivity
e_0 = voids ratio corresponding to k_0
D = parameter determined, for instance, by a consolidation test

In our example, D has been assumed to be 0.08. As regards the relationship between hydraulic conductivity and pore pressure, we have adopted that relating to Fig. 4c.

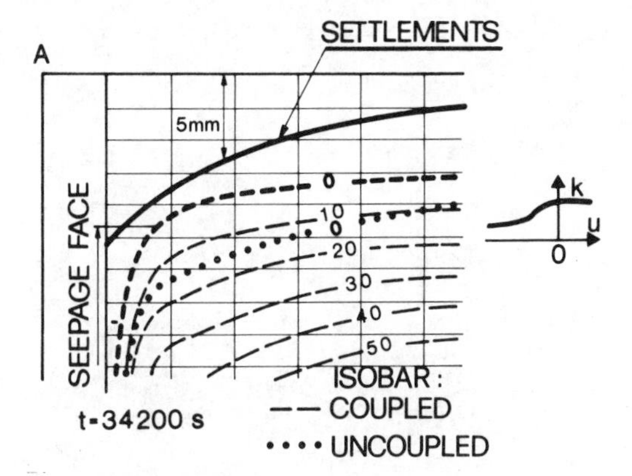

Fig.6 Effect of the variation in hydraulic conductivity as a funtion of void ratio which decreases in the region of the well (coupled model)

Figure 6 compares the result of the coupled calculation with that of the uncoupled calculation. The result is plausible.

The water surface is "retained" by the zone against the well where hydraulic conductivity has decreased, and the settlement obtained is correct. However, we freely admit to having come up against enormous problems of numerical stability in the unsaturated zone during these calculations. The very rapid decrease in hydraulic stability as a function of suction is the cause of this instability. This snag in the coupled method applied to unsaturated conditions has not yet been removed.

REFERENCES

ADINA 1981. A finite element program for automatic dynamic incremental nonlinear analysis. Adina Engineering, Inc., Watertown MA USA.

ADINAT 1981. A finite element program for automatic dynamic incremental nonlinear analysis of temperature. Adina Engineering, Inc., Watertown MA USA.

Bathe, K.J. & M.R. Khoshgoftaar 1979. Finite element free surface seepage analysis without iteration. Int. J. Num. and Anal. Meth. in Geomechanics Vol. 3 pp 13-22.

Dysli, M. 1985. Usage pratique de modèles couplés. C.R. 11ème congrès int. de méc. des sols et travaux de fondations, San Francisco.

Dysli, M. 1985. The practical use of coupled models in soil mechanics. Computer & Structures Vol. 21 No 1/2.

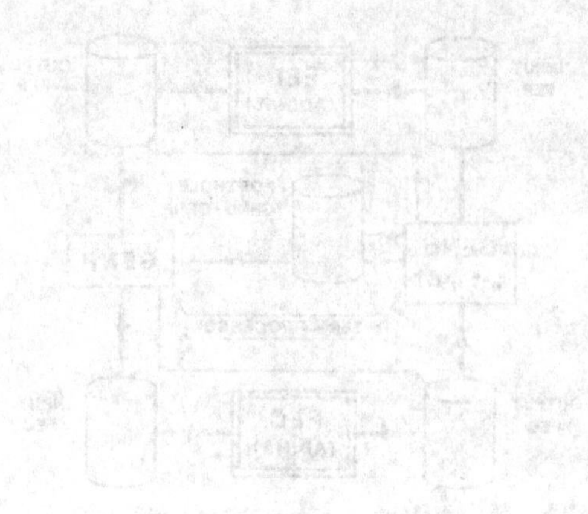

Numerical Methods in Geomechanics (Innsbruck 1988), Swoboda (ed.)
© 1988 Balkema, Rotterdam. ISBN 90 6191 809 X

Analysis of the growth and interaction of multiple-plane hydraulic fractures

V.M.Narendran
Singapore Science Council
K.Y.Lam
National University of Singapore

ABSTRACT: A general-purpose numerical formulation is presented for analysis of the quasi-static stable growth and interaction of fluid driven plane-strain fractures in inhomogeneous reservoir structures. Multiple fractures emanating from multiple well-bores are allowed to evolve along curved trajectories as dictated by non-uniform stress fields arising from interaction of fractures with one another and with material and stress variations.

1 INTRODUCTION

The generation of underground fractures with suitable extent, orientation and distribution is a problem that is currently receiving much attention in the resource extraction industry. The objective for these fractures is to increase the reservoir accessibility and the resource recovery rate by creating highly conductive paths (compared to reservoir permeability) some distance away from the wellbore into the resource bearing formation.

This study presents a general purpose analysis to describe both the asymmetric (or directionally biased) growth of fracture cross-sections and also the efficient automatic tracing of the evolution of one or more fractures in representative formations. Applications include the mechanical interaction of two or more fractures growing in a reservoir, and simulations tracing growth near reservoir inhomogeneities and under stress variations. Although apparently not seriously considered to date by the industry, this (curving) evolution of (multiple) fractures has important significance; for instance, in near-wellbore growth (including competing wings in radial fracture growth) and in efforts to achieve horizontal fractures where the vertical orientation is preferred. Other important motivations include the design of multiple fractures for selective stimulation of each zone in an open-hole completion or of each perforated interval (to obtain maximum post-treatment well productivity and to deplete each zone uniformly as the well is produced), and for stimulating naturally fractured reservoirs such as Devonian shales where the hydrafracs are expected to effectively intersect existing fractures and connect them to the wellbore; fracture intersection schemes may also be designed to kill blow-outs (Cleary, 1978 & 1980).

A comprehensive computer program has been developed, for which the formulations and numerical implementation are described. This model, called MULTIFRAC to emphasize the capability to analyze multiple fractures (Narendran, 83), not only provides a framework to solve the vast array of associated elasticity problems, but it also allows these fracture geometries to evolve quasi-statically in typical inhomogeneous reservoir structures, driven by fluid flow from any number of wellbores.

2 GENERAL EQUATIONS GOVERNING HYDRAULIC FRACTURING

There are seven separate identifiable phenomena which either contribute to or influence the hydrafrac process. These equations include:

i) Mass Conservation

ii) Crack Opening Relationship
iii) Frac-fluid rheology
iv) Frac-Reservoir Fluid Exchange and Induced Backstresses
v) Frac-Reservoir Heat Exchange
vi) Proppant Transport and Deposition
vii) Fracture Growth and Interaction

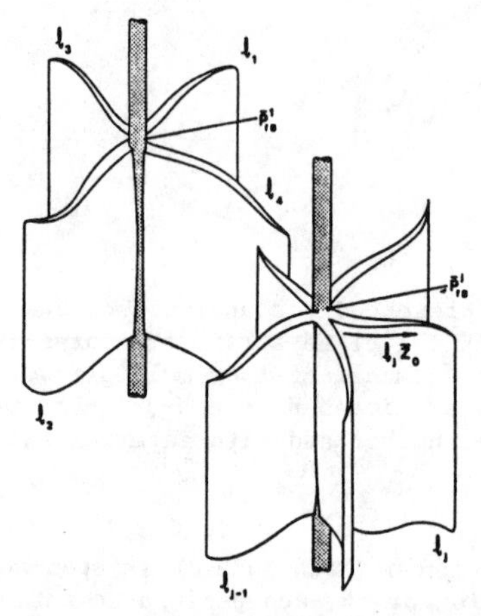

Fig. 1 Schematic of multiple arbitrarily shaped fractures emanating from multiple wellbores.

For any number of arbitrarily shaped fracture wings emanating from any number of well-bores (Fig. 1), the equations can be compacted and are as summarized.

The dominant result is the expression for the actual pressure variation in the j-th segment of fracture (Fig. 1):

$$\bar{p}_f^j(\bar{z}_o) = \bar{\sigma}_{cB}^j + \bar{\sigma}_j^B\,(1 \pm \lambda_j^m\,\psi_o^j(\bar{z}_o/\ell_j)), \tag{1a}$$

$$\bar{\sigma}_j^B = \bar{p}_{fB}^j - \bar{\sigma}_{cB}^j$$

where $\bar{p}_{fB}^j$ is the fracturing pressure driving each segment (i.e. B denotes the main "body" or "borehole" for each fracture, Fig. 1), and $\bar{\sigma}_{cB}^j$ is the confining stress acting at the junction of segments (e.g., in the reservoir pay zone). The fracture growth speed of the j-th wing is represented in the (eigenvalue) parameter λ_j (determined by satisfying the near-tip critical conditions):

$$\dot{\ell}_j = \lambda_j \ell_j / \tau_c^j \tag{1b}$$

in which there appears naturally a characteristic time

$$(\tau_c^j)^m \equiv \left[\left(\frac{\bar{E}_j}{\sigma_j^B}\right)^{2n+2-m}\frac{\mu_B}{\bar{E}_h}\right]\ell_j^{2m-2n} \tag{1c}$$

where the plane-strain crack-opening modulus $\bar{E}$ has the dimensions of stress (being $G/2(1-\nu)$ for an isotropic homogeneous medium).

The variation in pressure, due to channel flow of a power-law fluid, is described by the integral.

$$\psi_o^j(z_o^j) = \int_0^{z_o^j} \pm \frac{n\,dz}{\delta^{2n+1}}\,(\pm q\,\delta/\lambda_j)^m\,, \quad z_o^j \leqslant 1-\omega_j \tag{2a}$$

$$= \psi_o^j(1-\omega_j) + \sigma_\omega(z_o^j),\quad z_o^j \geqslant 1-\omega_j \tag{2b}$$

which is dominated by the dimensionless normalized fluid flow rate,

$$q\,\delta/\lambda_j = (2+\Lambda^j)\int_z^{1-\omega_j} ds\ \rho\delta(s)/\rho(z) + z\,\delta(z) + q_L^j(z) \tag{2c}$$

determined by the integral of mass conservation,

$$\frac{\partial}{\partial z}(\rho q\,\delta) + \frac{\partial}{\partial t}(\rho\delta) = \rho\bar{q}_L \tag{2d}$$

by employing a weak assumption of instantaneous self-similarity in successive crack shapes δ (which is found to adequately capture the fluid storage due to crack-opening, without allowable iteration to the "exact" solution)

$$\tau_c\,\frac{\partial}{\partial t}\,(\rho\delta) = -\lambda z\,\frac{\partial}{\partial z}\,(\rho\delta) \tag{2e}$$

The fluid loss distribution $q_L^j(z)$ describes in a dimensionless way the mass rate $w_L S_L(z)$ of fluid being lost between z and $1-\omega_j$ over the segment of lenght l_j:

$$q_L^j(z) = S_L^j(z)\dot{w}_L\bar{E}_j/\rho_B\,\bar{\sigma}_j^B\,\ell_j\,\dot{\ell}_j \tag{2f}$$

where the loss shape function $S_L^j(z)$ may take any desired form for the j-th segment. The pressure variation parameter Λ is defined by

$$\Lambda^j = (\dot{\rho}_B^j/\rho_B^j + \dot{\sigma}_j^B/\sigma_j^B)\, \ell_j/\dot{\ell}_j \tag{2g}$$

Note that the position of any point on a segment is denoted by the non-dimensional magnitude $z_o^j = |\bar{z}_o|/l_j$ of its curvilinear length $\bar{z}_o$ from the junction of segments. Also, a positive/negative $q\delta/\lambda_j$ requires use of the $\pm$ option in eqn. (2d), thereby allowing $p_f^{\prime} = dp_f/ds$ to be positive (e.g., for reverse flow), in the expression for dimensionless fluid speed, q, which is positive outward toward the tip, namely

$$\pm\, q\delta \equiv (\mp\, \delta^{2n+1}\, p_f^{\prime}/\eta)^{1/m} = \pm\, \dot{w}\tau_c \bar{E}/\bar{\sigma}_B \bar{\rho}_B \ell^2 \; ; \; \eta \equiv \bar{\eta}/\bar{\eta}_B \tag{3}$$

The elastic crack opening δ_E is determined, as described in (Narendran, 1985) by solution of the singular integral equations:

$$\bar{p}_f - \bar{\sigma}_c^B - \bar{\sigma}^v\Big|_{\bar{z}_o^i} = \sum_{j=1}^{M} \int_0^{\ell_j(t)} d\bar{z}\; \bar{\Gamma}(\bar{z}_o^j, \bar{z}^j)\; \bar{\delta}_E^{\prime}(\bar{z}^j, t) \tag{4a}$$

in which $\bar{\sigma}^v$ is the variation in confining stress σ_c (e.g., due to tectonic conditions, backstresses, etc.). We note that δ_E will be the total crack-opening displacement if $\bar{\sigma}^v$ includes the stresses generated by interfacing ot other schemes (e.g., hybridisation with finite elements) to capture all other effects (nonlinearity, inelasticity, inhomogeneity, etc.) The appropriate nondimensionalisations are the following:

$$\delta_E^j = \bar{E}_j \bar{\delta}_E^j / \bar{\sigma}_j^B \ell_j \qquad p_f = \bar{\sigma} / \bar{\sigma}^B$$

$$\bar{\sigma} = \bar{p}_f - \bar{\sigma}_c^B \qquad \sigma^v = \bar{\sigma}_c^v / \bar{\sigma}^B \tag{4b}$$

which lead to a dimensionless version of eqn. 4(a),

$$p_f - \sigma^v\Big|_{z_o^i} = \sum_{j=1}^{M} \int_0^1 ds\; \Gamma(z_o^i, z^j)\; \delta_E^{\prime}(z^j, t), \tag{4c}$$

in which the dimensionless singular kernel can be most simply written as

$$\Gamma(z_o^i, z^j) = \bar{\Gamma}(\ell_i z_o^i/\ell_j, z^j)/\bar{E}_j \tag{4d}$$

Here the influence function $\bar{\Gamma}$ should be chosen as that nearest the tip of the j-th wing, in order to correctly capture the singularity there. Having solve eqn. (4c) for $\delta_E(z^j, t)$, the opening displacement at any point on any segment, as required in eqns. (2a-c), may be obtained by integration

$$\delta_E^j = \delta_E(z^j, t) = \delta_{EB} + \int_0^z ds\; \delta^{\prime}(s^j, t); \tag{4e}$$

$$\delta_{EB} \equiv \bar{E}_j \bar{\delta}_{EB}^j / \bar{\sigma}_j^B \ell_j$$

Although the physical $\bar{\delta}_{EB}$ is the "central" opening displacement, independent of the segment under consideration, the output δ_{EB} of the program may look discontinuous between segments with different $\bar{E}_j$, because of the different nondimensionalisation factors for each; there may also be a "net entrapped dislocation", due to disconnected slippage or plasticity (as against material removal from fracture walls, e.g., in response to acidisation, which merely adds to the overall δ),

$$\bar{\delta}_{ED}^j = \bar{\delta}_{EB}^j + (\bar{\sigma}^B \ell_j/\bar{E}_j) \int_0^1 ds\; \delta_E^{\prime}(s^j, t) \tag{4f}$$

The solution for δ can also be related to a fracture criterion around each of the crack tips and it is this requirement that finally determines the values of λ_j which are needed to calculated the propagation rates for each segment. For example (Lam, 1984), we identify the stress-intensity factors $\bar{K}^j$ as the amplitudes of the square-root singularities (in dislocation density δ' and hence stress ahead of the tip),

$$\delta_E' \equiv F(z)/\sqrt{(1-z^2)}, \quad \bar{K}^j = (\bar{E}_j^T/\bar{E}_j)F^j\bar{\sigma}_j^B\sqrt{\pi\ell_j}, \tag{5a}$$

$$F^j = F(z^j=1)$$

Here we have distinguished between the overall modulus $\bar{E}_j$ governing opening of the j-th segment, eqn. (4b), and the local modulus $\bar{E}_j^T$ near the crack tip; high/low $\bar{E}_j/\bar{E}_j^T$ will serve as a shield/focus for energy transmission to fracture material in that vicinity. The pressure distribution in each segment will produce a distinct F^j at each tip, as indicated by the following insertion of eqn. (1a) into eqns. (4b,c):

$$p_f^i - \sigma_i^v = 1 - \sigma_i^v \mp \lambda_i^m \psi_o^i(z^i) \tag{5b}$$

$$\rightarrow F^j = F_i^j \mp \lambda_i^m F_i^j$$

By imposing the condition that the combination of all the pressure distributions produce the required value of $\bar{K}^j = \bar{K}_c^j$, $F_i^j = F_c^j$ at each tip,

$$\sum_{i=1}^{M} [F_i^j] \{\lambda_i^m\} = F_c^j - \sum_{j=1}^{M} F_i^j \equiv F_c^j - F_u^j \tag{5c}$$

$$F_c^j = \bar{K}_c^j \bar{E}_j / \bar{E}_j^T \bar{\sigma}_j^B \sqrt{\pi\ell_j} \tag{5d}$$

We obtain the desired result for propagation rates:

$$\lambda_j^m = [F_j^i]^{-1} \{F_c^i - F_u^i\}, \; i,j=1,\ldots,M \tag{5e}$$

3. NUMERICAL SOLUTION PROCEDURE

The foregoing equations need to be solved at each incremental growth step of the fracturing simulation. A very terse listing of the steps and equations is therefore adequate, allowing corrections and some new discussion:

1) Solve eqn. (4c) for $\delta_E' = F(s)/\sqrt{(1 - s^2)}$, using the latest pressure

2) Integrate the solution for F(s) to get the elastic crack opening, eqn. (4e)

3) Use eqns. (2a,b,c) to compute ψ_o^i; a simple quadrature formula (trapezoidal or Simpson's rule) is adequate for this, and δ_i is assumed known as input or from using any appropriate empirically derived function for the initial estimate.

4) Compute λ_i^m from eqn. (5e). Use it to compute new pressure and displacement curves

$$p_f^i = 1 \mp \lambda_i^m \psi_o^i(s_i) \tag{6a}$$

$$\delta_E = \delta_{Eu} - \sum_{i=1}^{M} \lambda_i^m \delta_{Ei}(s) \tag{6b}$$

5) Check convergence of λ^m by comparison with value computed on last step of iteration. Check also if the corresponding value of σ_c is close enough to that required in the physical situation; the latter is easy to achieve for the higher range of σ_c, so a fixed small ω may be adequate there.

6) If necessary, change ω_j and return to step (3) for further iteration.

7) Once satisfactory λ_j and σ_c^j are achieved, the relevant output parameters are printed and a check is provided on their physical consistency,

Further details can be obtained from Narendran, 86.

4. CRACK GROWTH SCHEME

While the iterative procedure described above solves for the crack tip propagation rates at each incremental position of the fracture trajectory, it does not provide any information as to the direction of growth in the succeeding step.

The growth of hydraulic fractures is an inherently quasi-static process, stabilised by pore-fluids and confining stresses on the exterior. Fracture paths in typical reservoir structures are associated with gradual changes in loading and these non-uniform stress fields give rise to paths that are generally curved. The smoothly turning paths of hydraulic fractures imply that $K_{II} = 0$ as the crack extends. This criterion allows us to determine the path taken by a growing fracture.

5. RESULTS AND DISCUSSION

The hydrafrac simulator has very wide applicability. Typical configurations which can be analyzed include : fracture realignment in a biased stress field, fracture interaction with bimaterial interfaces and a cylindrical inclusion, and multiple fractures propagating from adjacent boreholes. These configurations have in fact been analyzed and the results obtained compared favourably with laboratory experiments where available.

5.1 Fracture Realignment in a Biased Stress Field

Hydraulic fractures often initiate at an odd angle to the borehole and to the confining stress field. This stress field is often biased, in that the components of the tectonic, structural, residual and even fracturing induced stresses are often different in the three principal directions. As the fracture propagates, these biased stresses cause the wings to turn and to eventually become oriented in a direction normal to the minimum principal confining stress, as this direction offers the least opening resistance. The rate of turning into the preferred direction is determined by the shear stresses acting on the wings, which in turn depend upon the relative magnitudes of the confining stresses at the tip and the fracturing pressure at the borehole.

The numerical results in Fig. 2, for the growth of a fracture initially inclined at 45° to the maximum closure stress direction, in an isotropic homogeneous medium with no fluid-loss or backstress effects, confirm this predicted realignment. These trajectories also show an increasing rate of turning (into the preferred direction) with bias stress.

5.2 Fracture Interaction with Bimaterial Interfaces

The study of fracture growth near material inhomogeneities is one of considerable importance for evaluating complex reservoir production, and ranges from understanding the problem of fracture containment in the payzone to determining if a fracture intersects an inclusion (e.g., sand lens); the latter is critical for allowing economic production from tight gas sands (which have lenticular formations).

The direction and rate of crack propagation near an interface can be influenced at some distance from the interface by the differences in material properties. Variations in the shear modulus G and/or Poisson's ratio ν across an interface are conveniently expressed as differences in the crack opening modulus $\bar{E}(\equiv G/2(1-\nu)$ for plane strain). The effect of a single interface upon an initially inclined fracture is shown in Fig. 3. As it grows outward, the fracture is attracted by a softer adjacent stratum (i.e., it grows into the interface), but diverts away from a stiffer stratum. This growth behavior may be explained by considering the stress intensity factors K_1 and K_2 of an uniformly pressurized fracture, as shown in Figs. 4 for modulus contrasts. The variation of the opening mode stress intensity factor with distance from the interface also contributes to increasing (decreasing) crack-tip growth speed as the fracture approaches a softer (stiffer) adjacent stratum, as plotted in Fig. 5.

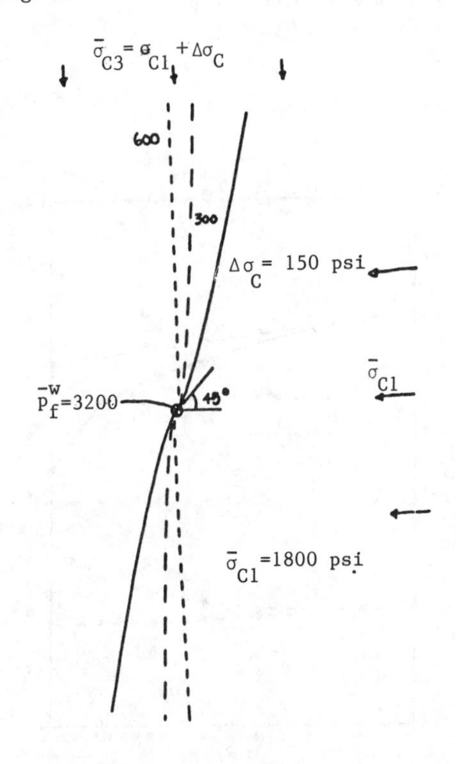

Fig. 2 MULTIFRAC generated trajectories for an initially inclined fracture, showing realignment in a biased stress field.

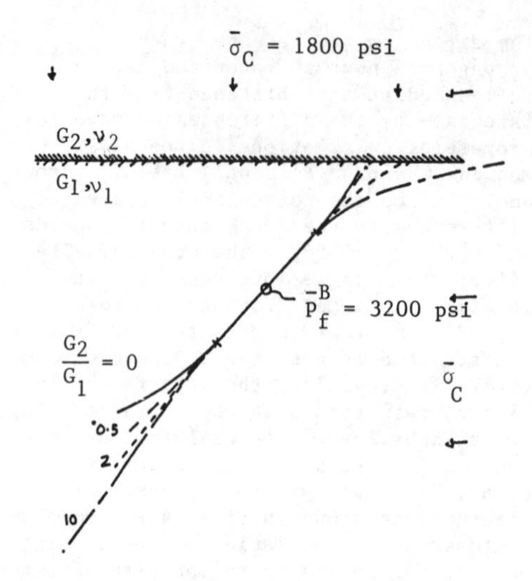

Fig. 3 Fracture trajectories near a single bimaterial interface, for different modulus contrast ratios between the fractured zone and the adjacent stratum.

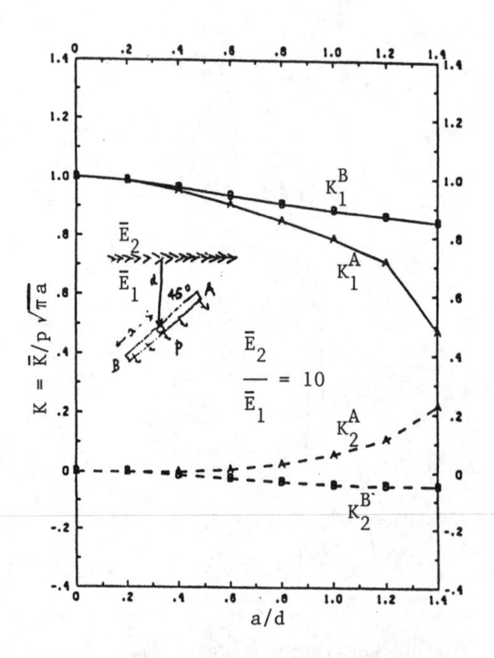

Fig. 4 Stress intensity factors for an inclined fracture near a bimaterial interface $\bar{E}_2/\bar{E}_1$ = 10.

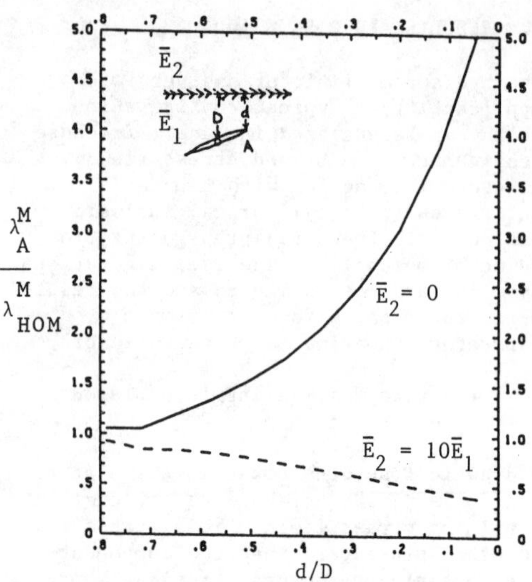

Fig. 5 Growth behavior of an initially inclined fracture tip near a bimaterial interface (a) propagation rate.

6. CONCLUSION

The formulations and numerical implementation of the framework for a multiple hydrafrac simulator are presented. Some capabilities of the simulator are discussed. These include fracture realignment in a biased stress field and interaction with bimaterial interfaces. Results are also available for more complex crack interaction.

REFERENCES

Cleary, M.P., Primary factors governing hydraulic fractures in heterogeneous stratified porous formations, Paper No. 78-Pet-47, ASME, Nov. 1978.

Cleary, M.P., Analysis of mechanisms and procedures for producing favourable shapes of hydraulic fractures, Paper No. SPE 9260, 1980.

Lam, K.Y. and M.P. Cleary, Slippage and Re-initiation of (Hydraulic) Fractures at Frictional Interfaces, I.J. Num. and Anal. Meth. in Geomech., Vol. 8, 589-604, 1984.

Narendran, V.M. and M.P.Cleary, Elastostatics interaction of multiple arbitrarily shaped cracks in plane inhomogeneous regions, Eng. Frac. Mech., Vol. 19, 481-506, 1984.

Narendran, V.M., Analysis of the Growth and Interaction of Multiple Plane Hydraulic Fractures, PhD thesis, M.I.T., 1986.

Numerical Methods in Geomechanics (Innsbruck 1988), Swoboda (ed.)
© 1988 Balkema, Rotterdam. ISBN 90 6191 809 X

Analysis of stress, water flow and heat transfer in fractured rocks

R.Hamajima
Saitama University, Japan

T.Kawai
Science University of Tokyo, Japan

M.Watanabe
Toshiba Ltd, Tokyo, Japan

H.Koide
Geological Survey of Japan, Ibaragi

M.Kusabuka
Hzama-gumi Ltd, Saitama, Japan

T.Yamada
Institute of Nonlinear Dynamics Co, Tokyo, Japan

Abstract: To analyze the coupled field problem of stress, groundwater flow and heat transfer, the governing equations are formulated besed in the integral form on the conservation law and them directly solved through discritization. It should be noted that direct discritization of the conservation law of various physical quantities using the "Rigid Bodies-Spring Model" makes the analysis not only simpler but also more flexible. Two coupled field problem of heat and groundwater flow as well as heat conduction and stresses are analyzed in this paper.

1 INTRODUCTION

Recently, the nuclear waste geological disposal, etc. have produced the problems of heat transfer and groundwater flow in fractured rocks. The analysis of heat and groundwater flow requires a evaluation of the effect of stress, and therefore it is necessary for this purpose to consider discontinity of the rock mass in these analysis. Various approaches have been proposed to the stress analysis of fractured rocks, and the effect of discontinuity was simulated by using of the Cundall model(Cundall 1974), the joint element (Goodman et al.1968), the rigid body spring model (Kawai.1980), etc. Further Kyoya et al.1985 and Oda et al. 1985 conducted an elaborate analysis in which discontinuity effect was considerated in the constitutive equation by using damage tensor and fabric tensor. In the evaluation of the stability of the underground rock caverns, tanks of storing LPG, cracking always occur around the caverns due to contracive strain, and therefore a coupled analysis of heat transfer, groundwater flow and stress in cracked rocks is actually necessary.

In this paper direct discritization of the conservation law is proposed using the RBSM to deal with these problems. As a matter of fact the RBSM can be considered a discrete model to fulfil the conservation law in the stress analysis. Under such consideration, extended use of the RBSM is attempted in these coupled field proflems of fractured rock mass.

2 ANALYSIS OF THE COUPLED FIELD OF HEAT TRANSFER AND GROUNDWATER FLOW

2.1 Formulation

The governing partial differential equations of unsteady coupled field proflem of heat transfer and groundwater flow can be given as follows(Utsugida 1985):

$$\rho c \frac{\partial T}{\partial t} + ((\rho c)_f v_i T)_{,i} = (\lambda T_{,i})_{,i} + Q \quad (1)$$

$$[\frac{k}{\mu}(P_{,i} + (1-\beta(T-T_o))\rho g \delta_{i3})]_{,i} = \frac{Ss}{\rho_f g}\frac{\partial P}{\partial t} \quad (2)$$

$$v_i = \frac{-k}{\mu}(P_{,i} + (1-\beta(T-T_o))\rho g \delta_{i3}) \quad (3)$$

Integrating them with respect to a region V and introducing Fourier's law;

$$q_i = -\lambda T_{,i} \quad (4)$$

and then rearranged using the Gauss' theorem, the equations (1) and (2) can be rearranged as follows;

$$\iiint \rho c \frac{\partial T}{\partial t} dv = -\iint (\rho c)_f v_i n_i T ds + \iiint Q dv - \iint q_i n_i ds \quad (5)$$

$$\iint \frac{k}{\mu}(P_{,i}+\rho_f g\delta_{i3})n_i ds \tag{6}$$

$$-\iint \frac{k}{\mu}\{\rho_f g\beta(T-T_o)\delta_{i3}\} n_i ds=\iiint \frac{Ss}{\rho_f g}\frac{\partial P}{\partial t} dv$$

where ρc: Equivalent heat capacity of rock masses, $(\rho c)_f$: Heat capacity of fluid, T_o: Initial temperature, β: Thermal expansion coefficient of the fluid, v_i: Components of seepage velocity vector, λ: Thermal conductivity, Q: Heat generation rate, q_i: Components of heat flux, P: Pore water pressure, K: Permeability coefficient, μ: Coefficient of fluid viscosity, g: Gravitaional acceleration, δ_{i3}(1 when i=3), Ss: Specific storage, ρ_f: Fluid density, n_i: Directional cosine of the surface normal.

The equations (3),(5) and (6) are discretized and are to be solved simultaneously.

The following assumptins are made before discretization of the equation (5).

1) The temperature within elements is constant.
2) The heat flux on the interelement boundary surfaces is constant.
3) The seepage velocity on the interelement boundary surface is also constant.

Now lets assume that the volume of element i is V_i, the contact area with the adjacent element j is S_{ij}, and the element temperature is T_i. The temperature at the boundary surface is an average of temperatures at adjacent elements.

$$T_{ij}=(T_i+T_j)/2 \tag{7}$$

Then the equation (5) is discretized as follows.

$$\rho_i c_i \dot{T}_i V_i=-\sum_j(\rho c)_f v_{nij}\left(\frac{T_i+T_j}{2}\right)S_{ij}+Q_i V_i-\sum_j q_{ij}s_{ij} \tag{8}$$

where V_{ij} and q_{ij} are the flow velosity and the heat flux at the boundary surface of elements in the direction of the outward normal, respectively. However the suffixes i and j in the equation (8) simply show the elements i and j, unlike the equations (1) through(6) which show a tensor expression. $\sum$ complies summation of the adjacent elements. The heat flux q_{ij} is expressed by using the heat pass coefficient k_{ij}as follows.

$$q_{ij}=-k_{ij}(T_j-T_i) \tag{9}$$

Here the definition of k_{ij} may be different balance according to the method of heat to/from the adjacent element, as follows.

1) In the case of heat conduction:
$$k_{ij}=\lambda/\ell_{ij} \tag{10}$$
(λ=Thermal conductivity, ℓ_{ij}=Distance between element gravitationnal centers)
2) In the case of convective heat transfer: $k_{ij}=\alpha$ (11)
(α:Coefficent of heat transfer)
3) In the case of thermal radiation:
$$k_{ij}=\sigma\varepsilon F(T_i+T_j)(T_i^2+T_j^2) \tag{12}$$
(σ : Stephan-Boltsman constant,
ε : Emissivity,
F : Geometrical factor,
T_i, T_j : Temperature of the elements i and j)

Introducing the mean load approximation to the time integration using a factor θ, the equation(8)becomes as shown below.

$$\left[\rho_i c_i V_i+\theta\Delta t\sum_j\left(\frac{1}{2}(\rho c)_f v_{nij}^t+k_{ij}^t\right)S_{ij}\right] T_i^{t+\Delta t}$$
$$+\left[\theta\Delta t\sum_j\left\{\left(\frac{1}{2}(\rho c)_f v_{nij}^t-k_{ij}^t\right)S_{ij}T_j^{t+\Delta t}\right\}\right] \tag{13}$$
$$=\left[\rho_i c_i V_i-(1-\theta)\Delta t\sum_j\left(\frac{1}{2}(\rho c)_f v_{nij}^t+k_{ij}^t\right)S_{ij}\right] T_i^t$$
$$-\left[(1-\theta)\Delta t\sum_j\left\{\left(\frac{1}{2}(\rho c)_f v_{nij}^t-k_{ij}^t\right)S_{ij}^t T_j^t\right\}\right]+\Delta t Q_i^t V_i$$

If θ= 0 in the equation, it will become the Euler method, which is unsuitable for the unsteady analysis to determine the temperature distribution over many hours as the equation will be divergent if a long time step is taken. Therefore the Crank-Nikorson method (θ= 0.5) is adopted in the present analysis.

Similarly the equation (6) for unsteady groundwater flow is like the equation (5). The following assumptions are taken prior to the discretization.

1) The pour water pressure within elements is constant.
2) The pressure gradient on the interelement boundary surface is constant.

If the pressure gradient at the interelement boundary surface is equal to the quotient of the difference of the pour water pressure between the adjacent elements divided by the distance between their gravitational centers, the equaiton (6) become as follows.

$$\left[A_i V_i+\theta\Delta t\sum_j \frac{k_{ij}}{\mu\ell_{ij}} S_{ij}\right] P_i^{t+\Delta t}$$
$$+\left[\theta\Delta t\sum_j\left(-\frac{k_{ij}}{\mu\ell_{ij}}\right)S_{ij}P_j^{t+\Delta t}\right] \tag{14}$$

$$= [A_i V_i - (1-\theta)\Delta t \sum_j \frac{k_{ij}}{\mu \ell_{ij}} S_{ij}] P_i^t$$

$$- [(1-\theta)\Delta t \sum_j (- \frac{k_{ij}}{\mu \ell_{ij}}) S_{ij} P_j^t] + \Delta t \sum_j \frac{k_{ij}}{\mu} \rho_f g n_{zij} S_{ij}$$

$$- \Delta t \sum_j \frac{k_{ij}}{\mu} \rho_f g \beta (\frac{T_i^t + T_j^t}{2} - T_o) n_{zij} S_{ij}$$

where n_{zii} is the vertical element of the outward unit normal vector. Next, the equation (3) for the flow velocity V_i obtained from the Darcy's law is discretized into the following equation.

$$v_{nij}^t = - \frac{k_{ij}}{\mu} [\frac{P_j^t - P_i^t}{\ell_{ij}} + \rho_f g n_{zij} - \rho_f g \beta (\frac{T_i^t + T_j^t}{2} - T_o) n_{zij}] \quad (15)$$

2.2 Numerical Results

Fig.1 shows an example of the analysis of groundwater flow in fractured rocks. This is the numerical example of a finite element analysis made by Kawamoto et al. 1977, indicating idealized lattice joint systems. The joint interval is 20m and the joint gap is 3mm. The permeabillity of each joint system can be obtained from $k_f = g(2b)^2/(12v) = 7 \times 10^{-2}$cm/s where (2b) is the gap between parallel model plates,

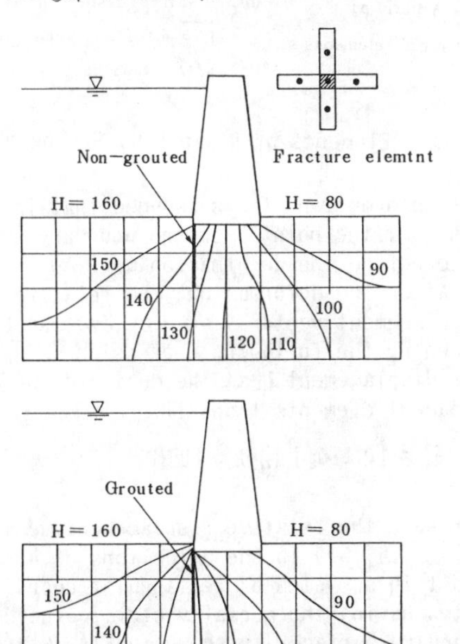

Fig.1 Analysis of groundwater flow in the dam foundation rock with joints

and is the kinetic viscosity coefficient. The permeability coefficient kg of the grouted joint is kg = 8×10^{-5}cm/s , if the opening of the grouted joint is assumed to be 0.1mm wide. The results of analysis are shown as the equipotential lines in the dam foundaiton rocks and are fairly good agreement with the Kawamoto et al. s results. Fig.2 shows an example of the coupled field analysis of heat transfer and groundwater flow in a simple two dimensional rock model. The analysis model is a 100 meter square and 10m thick model as shown in Fig.2a.with all the boundaries being impermeable, and the upper and lower surfaces made to be adiavatic boundayies, the model observes the occurrence and growth of the natural convection, and changes in the temperature distribution under initial rock temperature of 50°C with the left and the right boundary temperatures fixed at 50°C and 100°C respectively. Further, the static water pressure distribution is assumed at the initial condition. The material properties are the rock masses heat capacity: $\rho c = 2.44 \times 10^6$ J/m³°C, the fluid heat capacity: $(\rho c)_f = 4.19 \times 10^6$ J/m³°C, the rock thermal conductivity: λ = 2.9W/m°C, the rock specific storage: Ss = 3×10^{-7}/m, the rock permeability: K = 10^{-5}m/s, and the fluid thermal expansion coefficient: β = 0.00021/°C.

In a fractured rock masses model,cracks were treated in such that no water flow within the joints was considered and the heat conductivity in the direction perpendicular to the joints was calculated as 1/100. It can be seen that the convection flow occurs and the temperature distribution changes 15 years later. When the vertical cracks initiate, the temperature distribution is considerably affected by them but no appreciable difference is observed in the calculation of the flow velocity vector.

In the present analysis, the basic equation for heat transfer and groubdwater flow are expressed in the integral form, and discretized directly. This has enabled us to handle the problems in a similar way as in the analysis of the rigid body spring model, and find a clue to the coupled field analysis of heat transfer and the fluid flow under influeuce of crack initiation and growth.

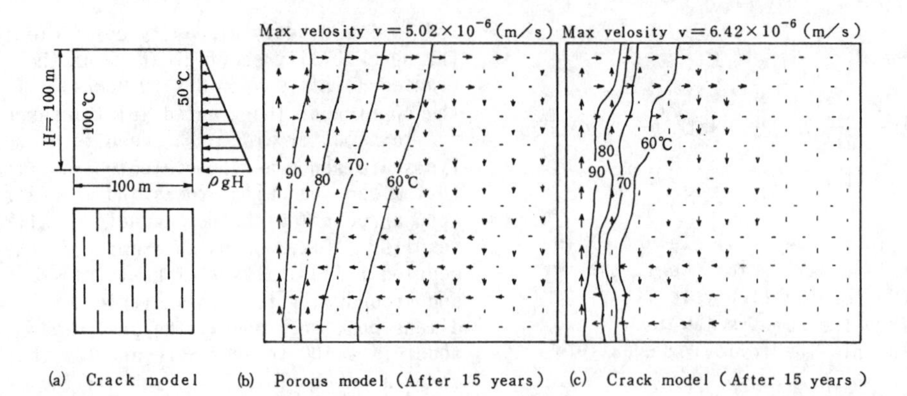

Fig.2 Analysis of heat convection flow in fractured rock masses

3 COUPLED FIELD ANALYSIS FOR HEAT CONDUCTION AND STRESSES

3.1 Formulation of Elasto-Plastic Thermal Stress Analysis of Three-Dimensional Fractured Rocks

The elasto-plastic thermal stress analysis is here formulated by using the new discrete three-dimensional element proposed by Kawai 1980. In this new discrete element, the element mode is only considared at the centroid and arbitrary elements shape can be used. For example elements shown in Fig. 3 are considerered. Each of them is assumed to be rigid, and their centroids of I and II are shown as points G_1 and G_2. After deformation, the displacement $\mathbf{u}$ of any arbitrary points within the rigid bodies I and II is shown by the following expressions.

$$\mathbf{u}=\begin{bmatrix} Q_1 & 0 \\ 0 & Q_2 \end{bmatrix}\begin{Bmatrix} U_1 \\ U_2 \end{Bmatrix}=[Q]\{U\} \tag{16}$$

$$\mathbf{U}_i=(u_i, v_i, w_i, \theta_i, \phi_i, \chi_i) \tag{17}$$

$$\mathbf{Q}_i=\begin{bmatrix} 1 & 0 & 0 & 0 & (z-z_i) & -(y-y_i) \\ 0 & 1 & 0 & -(z-z_i) & 0 & (x-x_i) \\ 0 & 0 & 1 & (y-y_i) & -(x-x_i) & 0 \end{bmatrix} \tag{18}$$

where $\mathbf{U}_i$ is the displacement vector at the centroid of the ith element, and $\mathbf{Q}i$ is the (3 x 6) matrix which depends on the coordinates components (x_i, y_i, z_i) at the centroid of the element and those (x,y,z) at any arbitrary point. The relative displacement of two adjacent elements occur at their interelement boundary surface after the deformation. It is assumed that in Fig. 3b, the point P on the boundary surface before the deformation move to P' and P'' after the deformation. The relative displacement vector $\boldsymbol{\delta}=(\delta x, \delta y, \delta z)$ can be shown by the following expression using the displacement $\mathbf{U}$ at the centroids of the adjacent elements I and II.

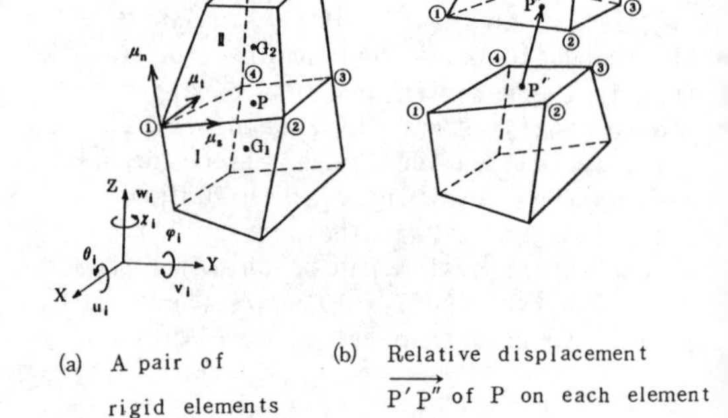

(a) A pair of rigid elements (b) Relative displacement $\overrightarrow{P'P''}$ of P on each element after loading

Fig.3 Elements of Rigid Body Spring Model

$$\{\delta\} = [Q_1, -Q_2]\begin{Bmatrix} U_1 \\ U_2 \end{Bmatrix} = [B]\{U\} \tag{19}$$

Further, the relative displacement vectots $\{\bar{\delta}\}=(\delta_n, \delta_s, \delta_t)$ in the directions of n, s and t in a system of rectangular coordinates having the normal vector to the boundary surface, as shown in Fig.3a, are given by the expression below using the coordinate transportation matrix $\mathbf{T}$.

$$\{\bar{\delta}\}=[T]\{\delta\}=[T][B]\{U\}=[A]\{U\} \tag{20}$$

Then, the strain vector $\boldsymbol{\varepsilon}=(\varepsilon_n, \gamma_s, \gamma_t)$ can be approximated by the expression below dividing the relative displacement vector $\{\delta\}$ by the distance h between the centroids of the elements I and II.

$$\{\varepsilon\}=\{\delta\}/h = [A]\{U\}/h \tag{21}$$

The stress vector is:

$$\{\sigma\} = \begin{Bmatrix}\sigma_n\\ \tau_s\\ \tau_t\end{Bmatrix} = \begin{bmatrix}k_n & & 0\\ & k_s & \\ 0 & & k_t\end{bmatrix}\begin{Bmatrix}\varepsilon_n\\ \gamma_s\\ \gamma_t\end{Bmatrix} = [D^e]\{\varepsilon\} \tag{22}$$

where k_n and k_s and k_t are the normal stiffness and the shear stiffness, which are separately defined with regard to the continuous media and the discontinuous media (Hamajima et al. 1985).

Next, the formulation of the elasto-plastic thermal stress analysis is considered.

The incremental stages are assumed to be $P^{(o)}, P^{(1)}, \ldots, P^{(n)}$. When the increment $\Delta P^{(n)}$ is considered with the value of $P^{(n)}$ stage assumed to be known, the virtual work is given by the following expression.

$$\iiint \delta\{\Delta\varepsilon\}^T(\{\sigma\}+\{\Delta\sigma\})dV - \iiint \delta\{\Delta u\}^T(\{\bar{F}\}+\{\Delta\bar{F}\})dV - \iint \delta\{\Delta u\}^T(\{\bar{q}\}+\{\Delta\bar{q}\})ds = 0 \tag{23}$$

where $\bar{F}$ and $\bar{q}$ represent the body force and the surface load. $\{\varepsilon\}$ and $\{U\}$ are given by the expressions (21) and (16), and the incremental stress vector $\{\Delta\sigma\}$ are giben by the following expression.

$$\{\Delta\sigma\}=[D^{ep}](\{\Delta\varepsilon\}-\{\Delta\varepsilon^T\}) \tag{24}$$

$[D^{eP}]$ is the elasto-plastic matrix which can easily be calcualted if the yield function is given. $\Delta\varepsilon^T=(\beta(T-T_o),0,0)$ is the thermal strain vector, β is the thermal expansion coefficient to rock masses, and To is the initial temperature. Thus the equation (23) becomes as shown below.

$$\{\Delta fs\}+\{\Delta f_v\}+\{\Delta f_T\}+\{\Delta r\}=[k^{eP}]\{\Delta u\}$$

$$\{\Delta f_T\}=\iint [A]^T[D^{eP}]\{\Delta\varepsilon^T\}dV$$

$$\{\Delta r\}=\iint [Q]^T\{\bar{q}\}ds \iiint [Q]^T\{\bar{F}\}dv - \iint [A]^T\{\sigma\}ds \tag{25}$$

$$[k^{eP}]= \frac{1}{h}\iint [A]^T[D^{eP}][A]ds$$

where $\Delta\mathbf{f}s$ and $\Delta\mathbf{f}_v$ represent the surface loads and the body forces, and the balance $\Delta\mathbf{r}$ is assumed to be $\Delta\mathbf{r} = 0$. Further, the surface integral to $\{\Delta f_T\}$ and $[k^{eP}]$ is calculated by means of numerical integration.

3.2 Coupled Analysis of Heat Conduction and Stress

The following expression is used as the governing equation for heat conduction.

$$\iiint \rho c \frac{\partial T}{\partial t}\, dV = \iiint Q dV - \iint q_i n_i ds \tag{26}$$

Here, a coupled field analysis is made for unsteady heat conduction and stress.As the first step, the unsteady analysis of the governing equation for heat conducion is made, and basd on the coupled temperature , an analysis is made on the thermal stress certain hours later. The Mohr-Coulomb's yield criterion for the constitutive equation of material was used.

$$f = \tau+\sigma\tan\phi-c=0 \;,\; \sigma<\sigma_t \tag{27}$$

where ϕ and C are the internal friction angle and the cohesion ,and σ_t is the allowable tensile stress. equation (27) is constitutive equation with tension cut, which does not consider the influence of the temperature stress.

As an example, an analysis was made on the rock cavern storing LPG to which the finite element analysis by Ishizuka et al. 1985. Dry rock masses were assumed and their physical properties were:
the internal friction angle: $\phi = 45°$,
Poisson's rasio: $\nu = 0.3$,
Young's modulus: $E = 6.8 \times 10^4 kg/cm^2$,
the cohesion: $C = 56kg/cm^2$,
the allowable tensile stress: $\sigma_t = 28kg/cm^2$
density: $\rho = 2.6g/cm^3$,
specific heat: $Cp = 0.753\ KJ/kg°C$,
the thermal conductivity: $\lambda = 3.13W/m°C$,
and the coefficient of thermal expansion: $\beta = 5 \times 10^{-6}/°C$
All of them were assumed to be constant irrespective of temperature.

The analysis results are as shown in Fig.4, which almost coincide with the Ishizuka et al. in respect to the thermal distribution characteristics. With regard to the tensile fracture region, however, the two results are not always the same as the tensile strength was assumed to be constant in this analysis although development of the tensile region are observed

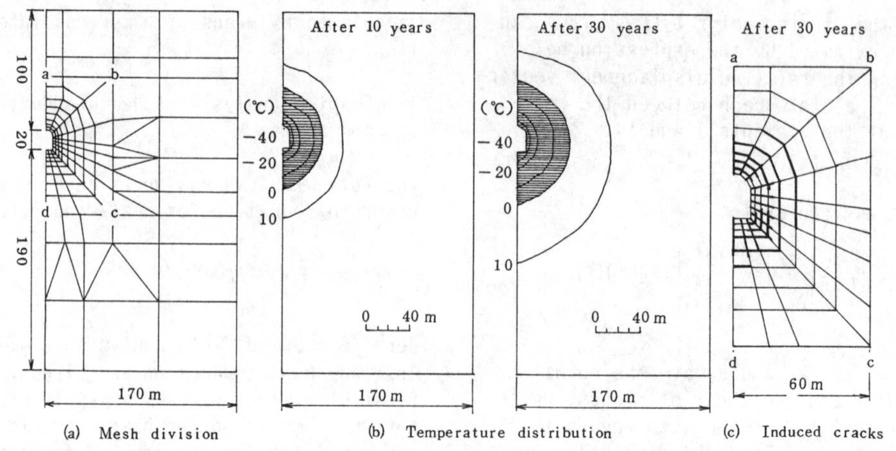

(a) Mesh division (b) Temperature distribution (c) Induced cracks

Fig.4 Temperature distribution and induced cracks in the rock cavern storing lPG

to be similar in both reults. This analysis can be made changing the heat transfer mode within the cracks from conduction into convective or thermal radiation, at the same time tracing the development of cracks, and furthermore the three-dimensional analysis may be possible. It is therefore considered to be effectively applicable to LPG and LNG, in future.

4 CONCLUSION

The RBSM can be applied successfully to analysis of crack as well as slip growth. The present authors examined possibility of application of this element to coupled field analysis of stresses and heat as well as water flow and they have drawn the following conclusions.

1) Concepts of the RBSM element can be equally applied to the heat transfer as well as the seepage flow analysis. Naturally element of any shape can be used and coupling effect of heat transfer and crack growth may be successfully studied.

2) Consideration of flow in the crack as well as along to vertical direction of the crack can be made in the heat transfer and seepage flow analysis.

3) Coupled field analysis of heat transfer and seepage flow may be possible under the influence of crack initiation and growth in discontinuous rock masses.

4) Coupled analysis of stresses and heat transfer may be possible for the 3D discontinuous rock foundation, similarily coupled analysis of stresses and seepage may be expected in near future.

REFERENCES

Cundall, P.A. 1971. A compater model for simulating progressive large-scale movements in blocky rock system, symp. on rock fracture, Nancy, Sec.2-B.

Goodmann, R.E., Tayler, R.I. and Brekke, T.L. 1968. Amodel for che mechanics of jointed rock, J. of soil mech.and found. Proc. A.S.C.E. SM3, P.637-659.

Hamajima, R., Kawai, T., Yamashita, K. and Kusabuka, M. 1985. Numerical analysis of cracked and jointed rock mass. Fifth int. conf. on num. meth. in geomecha. P.207-214.

Ishizuka, Y., Kinoshita, N., and Okuno, T. 1985. Stability analysis of a rock cavern with LPG storage under thermal stresses. Fifth int. conf. on num. metho. in geomech. P.1233-1240.

Kawai T. 1980. some considerations on the Finite Element Method. Int.J.for num. meth in engr. vol.16, P.81-120.

Kawamoto, T. and Kadota, S. 1977. Analysis of seepage flow in jointed rock masses. Hydro Electric Power, No.147.P.7-16.

Kyoya, T., Ichikawa, Y. and Kawamoto, T. 1985. A damage mechanics theory for discontinuous rock mass. Fifth int. conf. on num. meth. in geomecha. P.469-480.

Oda, M. and Maeshibu, T. 1985. Characterization of jointed rock masses by fabric tensor. Fifth int.con.on num. meth. in geomech. P.481-488.

Utsugida, Y. 1985. Coupled analysis of flow and heat around a high-level nuclear waste reposistory. Fifth int.conf.on num. meth. in geomecha. P.711-716.

Numerical Methods in Geomechanics (Innsbruck 1988), Swoboda (ed.)
© 1988 Balkema, Rotterdam. ISBN 90 6191 809 X

The coupled relationships between mechanical parameters and hydraulic ones of a structural surface and their applications

Sun Guangzhong, Liu Jishan & Guo Zhi
Institute of Geology, Chinese Academy of Sciences, Beijing

Abstract: Two coupled formulae between mechanical parameters and hydraulic ones of a structural surface were presented in this paper. The hydromechanics of a structural surface was studied by means of modelling test. The characteristics of seepage flux change in Gezhouba Sluice Foundation on the Second River was analysed according to these two formulae. The research results shown that they were the basis to study and predict the hydraulic characteristics of rock mass, and to carry out rock mass hydraulics analysis.

1. Introduction

The statistical result shown that 90% the failure of rock slopes was related to ground water, 60% the accidents in coal mines were related to ground water, 30-40% the accidents of dam were caused by permeability deformation. These facts shown that the move laws of ground water in rock mass and its effects on the deformation and the failure of rock mass has become an important problem urgent to be solved.

We must establish and develop rock mass hydraulics to solve these geotechnical problems.Rock mass hydraulics should combine rock mass mechanics with hydromechanics, and study the laws of re-deformation and re-failure of rock mass when existed the coupled action between rock mass and ground water, and study their applications in engineering practise.

We can conclude that these problems as follows must be solved when carried out rock mass hydraulics analysis.

(1) the hydraulic characteristics of rock mass and their determination
(2) the softening effect of water on rocks
(3) the coupled constitutive laws between rock mass and ground water flow
(4) the coupled relationships between mechanical parameters and hydraulic ones of rock mass etc.

The preliminary research result on the last problem was presented in this paper.

2. The theoretical derivation of coupled formulae between mechanical parameters and hydraulic ones of a structural surface

The coupled relationships betwwen mechanical parameters and bydraulic ones of rock mass were a tie to combine rock mass mechanics with hydromechanics,and the basis to carry out rock mass hydraulics analysis. The hydraulic characteristics of rock mass can be changed distinctly due to the change of loading conditions. Therefore, to establish the coupled relationships between mechanical parameters and hydraulic ones of rock mass was very neccesary for predicting the hydraulic characteristic change of orck mass, and evaluating the deformation and failure of rock mass.

Many researchers (Walsh,1981;Witherspoon et al, 1980) studied the coupled relations between mechanical parameters and hydraulic ones of a structural surface.But so far,the satisfying research result has not been seen.

The authors combined cubic law with mechanical parameters of a structural surface, and presented two coupled formulae (Liu Jishan, 1987) by means of the close deformation law of a structural surface(Sun Guangzhong, 1983).

2.1 Cubic Law

The cubic law is the relation between flow rate and aperture derived from solution of the Navier Stokes Eq.for laminar flow between two parallel plates which are not in contact.

In a shmplified form the cubic law may be written as:

$$Qf = (\gamma_w / 12\mu)(2b)^3 i \quad (1)$$

and the permeability coefficient may be written as:

$$Kf = (\gamma_w / 12\mu)(2b)^2 \quad (2)$$

Where 2b is the effective fracture aperture, i is the hydraulic gradient, μ is the dynamic viscosity of fluid, Qf is the flux, Kf is the permeability coefficient, γ_w is specific gravity.

2.2 The close deformation law of a structural surface

The close deformation curve of a structural surface was shown in fig.1. Ujo is the max. compressive deformation, Kn was the close stiffness, U was the close deformation, σ was normal stress.

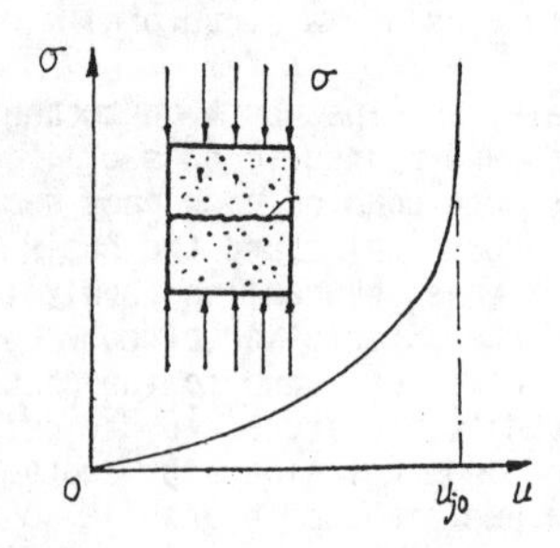

Fig.1 The close deformation curve of a fracture

$$U = Ujo\,(1 - e^{\sigma/Kn}) \quad (3)$$

The aperture of a structural surface may be written as:

$$2b = Ujo\ e^{-\sigma/Kn} \quad (4)$$

2.3 The theoretical derivation of coupled relationships

σ was replaced by σe(effective stress) in the Eq.(4). And then bring it into Eq.(1) and (2). Two coupled formulae may be derived as follows:

$$Kf = (\gamma_w / 12\mu)Ujo^2\ e^{-2\sigma_e/Kn} \quad (5)$$

$$Qf = -(\gamma_w / 12\mu)Ujo^3\ e^{-3\sigma_e/Kn} \partial H/\partial X \quad (6)$$

Where H was water head, x was the ordinate in the direction of water flow.

3. The modelling test research on hydromechanics in a structure suface

The authors studied the hydromechanics in a structural surface by modelling test in order to study the coupled relations between mechanical parameters and hydraulic ones of a structural surface.

3.1 Modelling test set-up and process

The modelling test set-up was shown in fig.2. The modelling test was completed on the compressive machine.

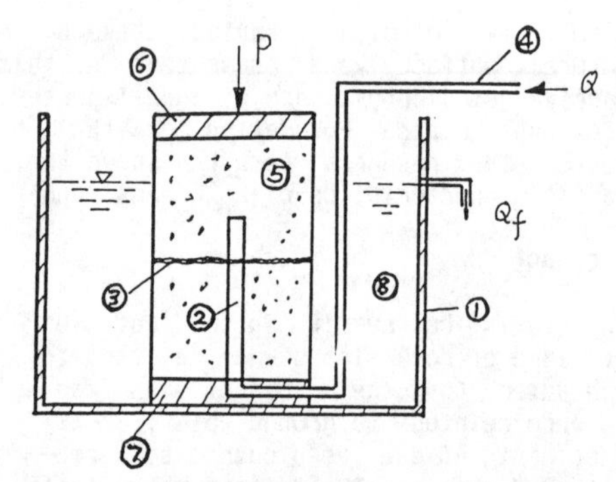

Fig.2 The diagram of modelling test set-up (1) Water basin (2)Pumping water hole (3)Fracture (4) Pumping pipe (5)Up steel plate (7) Down steel plate (8)water

First, the test specimen was satruated and then placed on the compressive machine. The close deformation, flux, and normal stress were measured at the same time.

3.2 Experimental results and their analysis

Three curves compaired experimental results with theoretical results were presented in fig.3.

The experimental results were analysed according to the cubic law and the coupled formula.

(1) The relationship between flux Qf and aperture 2b of a structure surface

$$Qf = (\pi \gamma_w Ho / 6\mu \ln R/r_o)(2b)^3 \quad (7)$$

Where R was the specimen radial, r_o was the radial of pumping water hole.

(2) The relationship between flux Qf and Normal stress σ

$$Qf = (\pi \gamma_w Ho Ujo^3 / 6\mu \ln R/r_o)\ e^{-3\sigma/Kn} \quad (8)$$

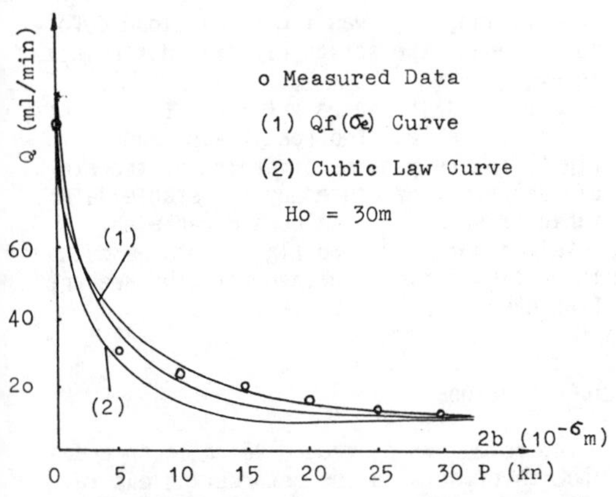

Fig.3 The comparison between the measured flux curve and two theoretical flux curves

In the process of analysis, Kn was determined by Eq.(9) or (10).

$$Kn=\partial\sigma/\partial u\,(U_{jo}-U) \quad (9)$$

or

$$Kn=\partial\sigma/\partial u\,(2b) \quad (10)$$

Actually, we considered Kn as a constant which was equal to the slope of the close deformation curve times the average aperture. In the process of analysis, we all considered the porous medium seepage flux or the matrix seepage flux.We can coclude that the measured flux curve coincided with two theoretical flux curves very well. This shown that the cubic law and the coupled formulae are valid.

4. The analysis on the characteristics of seepage flux change in Gezhouba Sluice Foundation on the Second River

The coupled effects betwwen stress field and seepage field of rock mass must be considered to solve rock mass hydraulics problems. This was proved by the characteristics of seepage flux change with seasons in Gezhouba Sluice Foundation on the Second River.

4.1 The characteristics of seepage flux change

The change process curves of up-stream and down-stream water tables, seepage flux of sluice foundation, and the settlement of sluice foundation plate, were presented in fig.4. We can conclude that the change characteristics of seepage flux with seasons were caused by the settlement and rebound

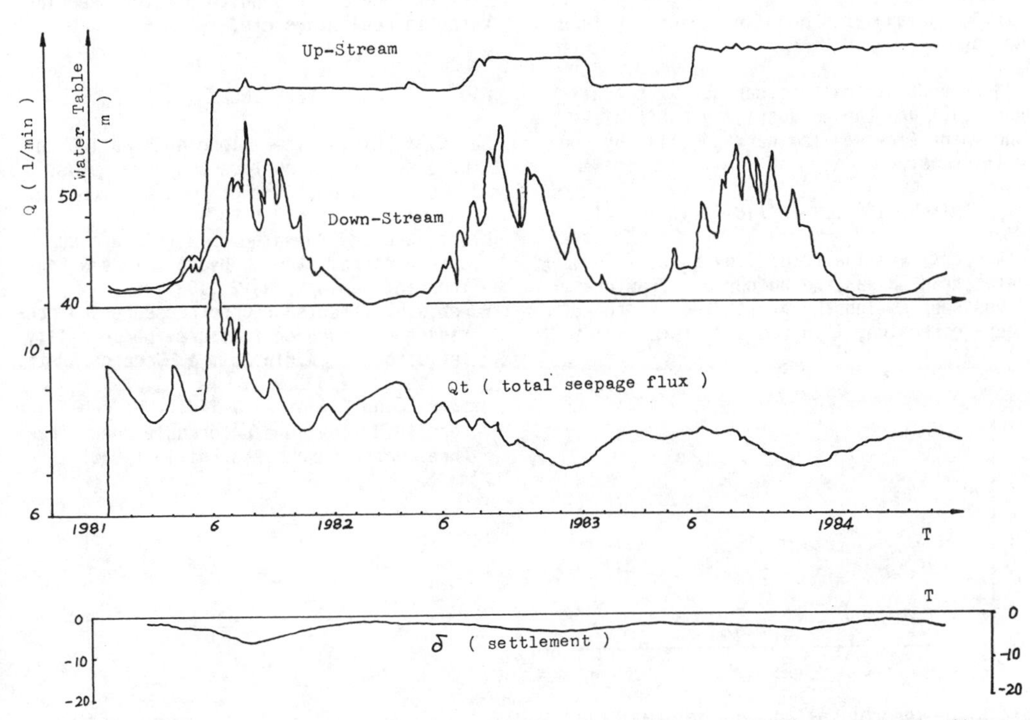

Fig.4 The change process curves of up-stream and down-stream water tables, seepage flux, and the settlement of sluice foundation

of sluice foundation due to the change of down-stream water table.
The foundation rock was the interlayer of muddy siltstone and sandstone, and almost horizontal. Seepage happened mainly along layer surfaces.

4.2 The theoretical analysis

A set of parallel fractures (layer surface) seepage model can be abstracted as shown in fig.5 according to the characteristics of rock mass structure and anti-permeability structure.

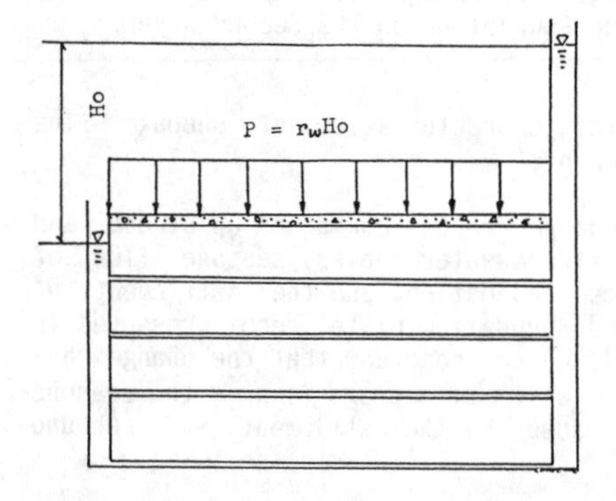

Fig.5 The analysis model of layer surface seepage

Surpose that fracture surface was continuous, Kf was the permeability coefficient, and water flow was laminar. Eq.(11) may be written as:

$$Q=(\gamma_w H o N / 12 \mu l) A [(U jo - U jt)/N]^3 \quad (11)$$

Where Q was the total flow rate, Ho was water head, N was the number of fractures, l was seepage length, A was the width of seepage section, Ujo was max. total close deformation, Ujt was the total close deformation under the action of some determined load.

Ujo and Ujt can be got from fig.4, let N=10. The calculated result was shown in fig.6. In the process of analysis, the stable seepage flux caused by the stable water table of up-stream was considered also.

We can conclude from fig.6 that the calculated flux curve coincided with the measured flux curve very well.

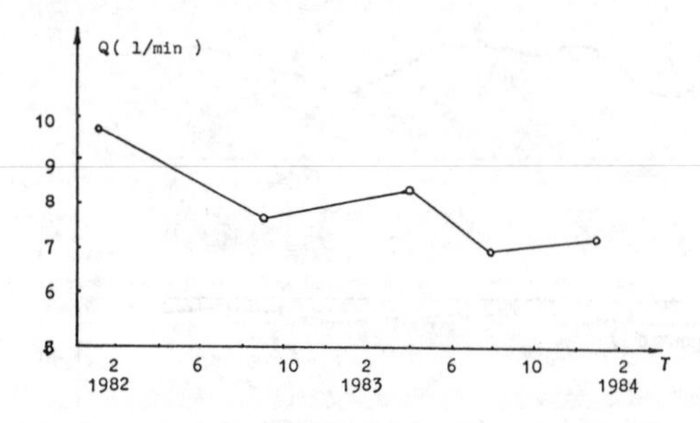

Fig.6 The theoretical curve between the seepage flux of sluice foundation and time

5. Conclusions

The move laws of ground water in rock mass and its effects on the deformation and failure of rock mass have become an important problem urgent to be solved. This research is one of the bases to solve it.

These two coupled formulae and hydromechanics modelling test are all based on the simplest model of rock mass hydraulics. But, they can be used in many fields. The formula between the seepage flux and deformation of a set of fractures can be used to solve some rock mass hydraulics problems under some determined conditions.

In addition, these two coupled formulae can be used to carry out jointed rock mass hydraulics analysis, to analyse the stability of rock slope, and to predict sweeping water in coal mines etc.

REFERENCES

Sun,G. & Lin,W. The close deformation law of a structural surface and the elastic deformation eqations of rock mass, Geological Sciences, No.2 (1983)

Liu,J. Seepage formulae of a fracture subject to normal stress, Hydrogeology & Engineering Geology, No.2 (1987)

Walsh,J.B. Effects of pore pressure and confining pressure on fracture permeability, Int.J.Rock Mech.Min.Sci. & Geomech.Abstr. Vol.18 429-435 (1981)

Witherspoon, P.A.et al, Validity of cubic law for fluid flow in a deformable rock fracture, Water Resour.Res.Vol.16 (1980)

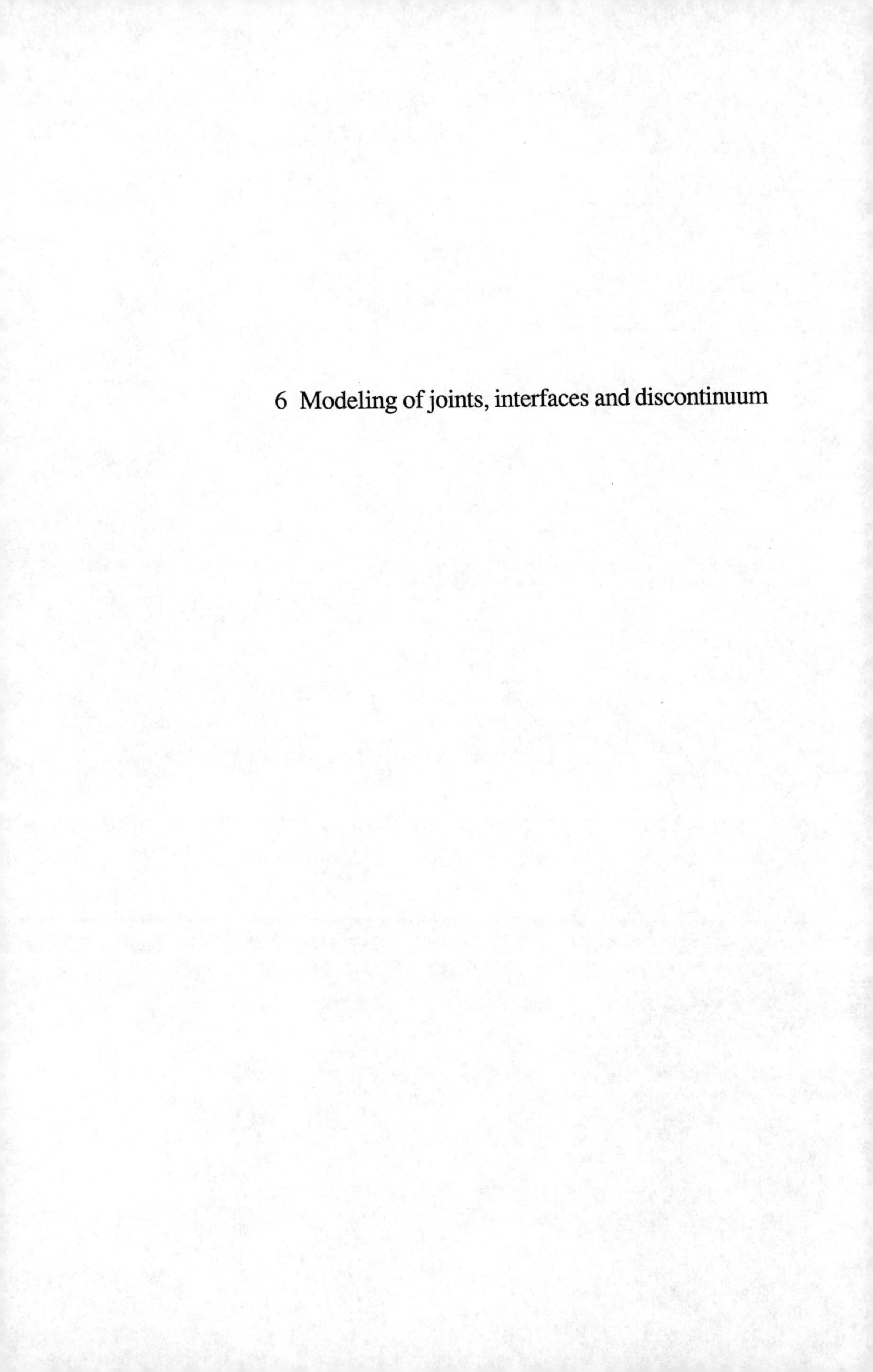

6 Modeling of joints, interfaces and discontinuum

Numerical Methods in Geomechanics (Innsbruck 1988), Swoboda (ed.)
© 1988 Balkema, Rotterdam. ISBN 90 6191 809 X

Elastic moduli of well-jointed rock masses

W.G.Pariseau & H.Moon
University of Utah, USA

ABSTRACT: A theoretical approach to the description of the overall elastic properties of well-jointed rock masses is outlined. The approach is based on a sequential application of two phase composite theory of linearly elastic media. The finite element method is used in numerical experiments to obtain data from which the true overall properties are calculated. Example results involving jointed and faulted rock masses show excellent agreement between theory and (numerical) experiment. The technique is confirmed by two entirely different programming and testing efforts that give the same results for equal rock and joint Poisson's ratios.

1 INTRODUCTION

Rock masses generally contain structural features that are not present in laboratory size samples. The most common structural features in rock masses are, of course, the discontinuities such as bedding planes, faults and joints. Although the properties of intact rock between the discontinuities ("joints" for brevity) and the properties of the joints themselves can be determined in the laboratory, the properties of the rock mass are not amenable to direct physical measurement. Obviously, rock mass properties can only be determined indirectly. A theory is needed for this purpose. The establishment of such a theory is a fundamental problem in rock mechanics. Such a theory must be testable in some independent way. It should also lead to useful engineering analyses of rock mass safety and stability.

In our approach, intact rock between the joints is idealized as a homogeneous, linearly elastic material that need not be isotropic. Isotropy of the overall properties is neither implied nor assumed. Joints are idealized as thin layers of material, also homogeneous and linearly elastic, but not necessarily isotropic. No inelastic slip on the joint or in the joint zone is allowed. These assumptions are entirely consistent with the concept of an elastic material. The common practice of including frictional slip on joints in the formulation of a secant shear modulus is avoided as is the inclusion of other nonlinearities that give rise to path dependent properties that are difficult to interpret as elastic parameters.

Of course, the actual behavior of jointed rock masses may be more complicated than that of a linearly elastic, anisotropic medium just as the behavior of any real material may be more complicated than some idealized material model used in its stead in an engineering analysis. However, the usefulness of the elastic idealization in rock mechanics is not the issue here, although, as a practical matter, elastic behavior is generally considered "normal" behavior and inelasticity a departure from the norm. In any event, the objective here is to arrive at a reasonable description of the overall elastic properties of an arbitrary but idealized sample volume of a well jointed rock mass.

In this regard, we also adopt the view that the concept of a representative sample volume is inappropriate in rock mechanics and, indeed, has been quite detrimental to progress. The reason for this is simple and due to the contradictory requirements of a sample volume that is large relative to the scale of the discontinuities but small relative to the scale of excavation. The two scales are very often of the same order of magnitude, so that meaningful analysis of stress is precluded even if the properties of the sample could be reliably determined.

Because the sample volume is arbitrary, the sample volume may be so small as to exclude joints. In this way, the sample volume can be as small as necessary for meaningful stress analysis. It also means that each elemental volume may have different properties. The heterogeneity of the rock mass caused by the discontinuities at one scale is therefore averaged up or redistributed to a scale that can be handled with current numerical models and computer technology. The approach is quite natural to the finite element technique that allows each (volume) element to be assigned different properties. This capability was incorporated into our two and three dimensional finite element codes UTAH2 and UTAH3 a number of years ago during a study of rock masses having random properties (within geologically defined units). Both programs are used in the present study to obtain directly the stress and strain fields needed to determine the overall properties of jointed rock mass samples of arbitrary volume.

2 THEORY

By definition, the rock mass elastic properties are the constants in the generalized Hooke's law that relates the average stresses to the average strains in the considered volume. Thus,

$$\{e^*\} = [S^*]\{s^*\} \quad \text{or} \quad \{s^*\} = [C^*]\{e^*\} \qquad (1)$$

where $\{e^*\}$ is a 6x1 vector of overall average strains, $\{s^*\}$ is a 6x1 vector of overall average of stresses, $[S^*]$ is a 6x6 matrix of the rock mass compliances and $[C^*]$ is a 6x6 matrix of rock mass moduli. The matrices of the elastic moduli $[C^*]$ and compliances $[S^*]$ are mutual inverses, thus

$$[C^*] = [S^*]^{-1}, \quad [S^*] = [C^*]^{-1}, \text{ and} \qquad (2)$$

$$\{e^*\}^t\{s^*\} = \{e^*\}^t[C^*]\{e^*\}$$
$$= \{s^*\}^t[S^*]\{s\} \qquad (3)$$

where $\{\}^t$ means transpose.

The considered volume may be thought of as an enormous test specimen taken from the parent rock mass to a gigantic laboratory for testing in the usual way. Strain measurements under known uniaxial loads or controlled displacements supply all the experimental data needed to calculate the elastic properties in $[C^*]$ or $[S^*]$.

If V is the volume of the rock mass sample, then the overall volume averages of stress and strain are

$$\{e^*\} = \int \{e\}dV, \qquad \{s^*\} = \int \{s\}dV \qquad (4)$$

The overall elastic properties are not simply the volume averages of the different rock mass constituent properties because

$$\{e^*\} = \int [S]\{s\}dV = ([S]\{s\})^* \neq [S]^*\{s^*\} \qquad (5)$$

If the intact rock and joint material are viewed as a rock matrix containing joint material inclusions each occupying 50 % of the sample, then volume averaging gives the same overall properties when the joint material is considered to be the matrix. This is physically unrealistic, especially with respect to shear properties that are governed largely by the matrix material. Simple volume averaging, therefore, cannot be expected to give very useful results.

The view of rock masses as aggregates of two constituents, intact rock and joints, suggests that two phase composite theory of elastic media may be applicable. Hill [2] in an important contribution describes a general, exact approach to the elastic properties of two isotropic materials firmly bonded together to form a mixture of any concentration. The overall properties of the sample are then determined from the properties of the two constituents, their volume fractions and influence matrices of stress and strain that relate global to local averages, thus

$$[S^*] = f_1[S_1][B_1] + f_2[S_2][B_2]$$
$$\text{and} \qquad (6)$$
$$[C^*] = f_1[C_1][A_1] + f_2[C_2][A_2]$$

where the f's are volume fractions, and the [A]'s and [B]'s are influence matrices,

$$f_1 = (V_1/V), \; f_2 = (V_2/V), \; V = V_1+V_2 \qquad (7)$$

$$\{\bar{e}_1\} = [A_1]\{e^*\}, \quad \{\bar{e}_2\} = [A_2]\{e^*\}$$
$$\{\bar{s}_1\} = [B_1]\{s^*\}, \quad \{\bar{s}_2\} = [B_2]\{s^*\} \qquad (8)$$

The formulas (6) show that the problem of determining the overall properties of a two phase composite is equivalent to the problem of determining the stress or strain influence matrices. If desired, the local averages of stress and strain can be found once the overall properties are determined.

However, the stress and strain influence matrices are generally not known beforehand, so the original problem remains. There are still far more unknowns than equations, so that additional information is needed. To see this clearly, one may consider local averaging of the stress strain law $\{e\} = [S]\{s\}$ which after averaging may be written as

$$\{e^*\} + \{\Delta\bar{e}\} = [S](\{s^*\} + \{\Delta\bar{s}\}) \qquad (9)$$

Equations (9) apply to any of the rock mass constituents and can be viewed as a system of 6 equations in the 12 unknown local averages of the global deviations. Six additional independent relationships are needed. If these were known, then the relationship between the overall average stress and strain could be found and the problem would be solved.

There are two assumptions that allow the analysis to proceed. The assumptions relate to the stress and strain deviation terms in (9) and are motivated by physical considerations.

First, there is the requirement of equilibrium across the joint. At the joint, the tractions must be continuous. We therefore assume that the stress deviation terms associated with the direction perpendicular to the joint vanish. There are three such terms.

Second, we assume that the local averages of the global strain deviations associated with a direction parallel to the joint plane also vanish. The joint layer is thin, relatively soft compared with the adjacent and much thicker intact rock and behaves elastically without slip. It seems reasonable under these conditions to assume the average strains parallel to the joint are very nearly equal to those in the intact rock. There are three strain deviation terms that are assumed to vanish.

When more than one joint is present in the sample volume, the analysis proceeds sequentially. Overall properties of the intact rock and first joint are found. These properties then replace the original intact rock properties and are combined with the second joint. The overall properties of this second combination are then combined with the properties of the third joint. The process continues until all joints in the sample volume have been taken into account.

The analysis involves a number of multiplications, partitions, additions, inversions, compositions and integrations of the matrices in (9). Space limitations prevent the full development and expression of the resulting formulas here, some of which are several pages in length. However, details for the compliances can be found in the dissertation by Moon [4]. A full discussion of the problem including a critical review of the literature and a comprehensive exposition of theory awaits publication elsewhere.

Of more immediate interest is the goodness of the approximation obtained. The nature of the problem precludes full scale physical testing of theory, of course. However, numerical experiments on a finite element model of a test volume from the idealized rock mass provide all the information needed to test the theory. There are considerable advantages to numerical testing of idealized models compared with physical testing, say, of artificial laboratory scale models of jointed rock masses. The numerical model corresponds precisely to the assumptions made. Any lack of agreement between theory and (numerical) experiment cannot therefore be explained away by the departure of the material from assumed behavior, instrument accuracy, and so forth. Numerical experiments when properly conducted provide a rigorous test of theory.

3 RESULTS

A sequence of increasingly complicated two and three dimensional results are contained in [4]. Only two sets of analysis can be presented here.

The first set compares theory and (numerical) experiment with respect to the overall properties of a two dimensional sample volume containing an inclined joint offset by a fault. The fault is a normal fault in some cases and a reverse fault in others. The fact that knowledge of the overall properties is equivalent to a knowledge of the local averages of stresses (and strains) within the intact rock, joint and fault is also demonstrated in this example set.

The second set deals with some of the purely numerical aspects of the finite element experimental calculations. A two dimensional slab containing three intersecting, non-orthogonal joints are contained in the sample volume.

Figure 1 shows seven sample volumes containing a joint offset by a fault. The seven samples are two dimensional rock mass slabs. The angle alpha in Figure 1 is the angle of rotation of the fault with respect to its zero position. Strike of the faults and joints is parallel to the z-axis (perpendicular to the plane of the page).

Intact rock, joints and faults are isotropic. The properties of the intact rock are:

Shear modulus: $G = 6.895\ GN/m^2$ (1.0 million psi)

Young's modulus: $E = 16.548\ GN/m^2$ (2.4 million psi)

Poisson's ratio: $v = 0.2$

Poisson's ratio is also 0.2 for the faults and joints. The moduli of the faults and joints are then specified by a modulus contrast number (MC) that is simply the ratio of rock modulus to fault or joint modulus. For example, an MC of 10 for a joint means that the joint shear modulus is $0.6895\ GN/m^2$; the same joint Young's modulus is then $1.6548\ GN/m^2$. In this regard, joint stiffnesses, joint thickness and moduli are related by expressions of the form

(modulus) = (thickness)(stiffness)

Physical tests for stiffness and a prudent, but somewhat arbitrary choice of thickness, allows the specification of the joint layer moduli. A scan of the literature (e.g. [1] and [3]) reveals a large range of joint normal and shear stiffnesses. For joints 2.54 cm thick (1 in.), a realistic range of experimental values of stiffnesses is included by a modulus contrast range of 5 to 1000 which is the range used in [4].

Determination of the overall properties of the slabs in Figure 1 requires uniaxial normal loading in the x and y directions and a xy shear test. Accordingly, uniform tractions of $68.95\ MN/m^2$ (1000 psi) are sequentially applied to the surface of each test volume. Members of the overall compliance matrix are then determined by dividing the overall average strains by the overall average stress which is also the surface average stress or applied traction. For example,

$$S_{12} = e_2/s_1, \quad \text{while} \quad S_{21} = e_1/s_2 \qquad (10)$$

where the first subscript refers to the loading and the second to the resulting strain., i.e., 1 refers to uniaxial loading in the x direction; 2 refers to uniaxial loading in the y direction, and 3 refers to loading in pure shear. The calculated volume average stresses are always checked against the applied tractions. Agreement is always better than one hundredth of one percent. The diagonal members of the compliance matrix [S] are the reciprocals of the moduli, thus

$$E_x = 1/S_{11}, \ E_y = 1/S_{22}, \ G_{xy} = 1/S_{33} \qquad (11)$$

One also expects that the matrix of overall properties should be symmetric, that is

$$[S] = [S]^t, \qquad (12)$$

and in particular $(v_{xy}/E_x) = (v_{yx}/E_y)$.

Figure 2 compares the results of theory with numerical experiment and shows the variation of normalized Young's moduli with the parameter alpha. The normalized moduli are simply the overall moduli divided by the original intact rock modulus. The normalized overall Young's moduli of the rock mass test volume are simply fractions of the original (isotropic) intact rock Young's modulus. The results are shown for three values of modulus contrast (MC). The response at low modulus contrast (5 and 10) is similar in shape but shifted downwards. At high MC (100), the response is different. The overall Young's moduli range from 9% to 72% of the intact rock Young's modulus. The prediction is within 6% of the (numerical) experimental results over the entire range of alpha and MC.

Figure 3 shows the variation of normalized shear modulus with alpha. The normalized shear moduli are the overall rock mass shear moduli divided by the original intact rock shear modulus. The overall rock mass shear moduli range from 9% to 68% of intact rock shear modulus. The prediction is within 5% of the (numerical) experimental results.

Figure 4 shows the variation of overall Poisson's ratio with alpha. The rock mass test volumes are obviously not isotropic; the test results and the predictions show this to be the case. Two Poisson's ratios are obtained. They may be as low as 0.04 and as high as 0.46 depending on the modulus contrast and fault and joint geometries. A high modulus contrast (MC=100) shows the greatest sensitivity to fault and joint geometry. The prediction is within 8% of the (numerical) experimental results.

Figure 5 shows example results concerning the local averages of stress. Instead of determining the overall properties first, it is possible to determine influence matrices that give the local averages of stress and strain. Some insight regarding the heterogeneity of stress and strain within the test volume can be gained in this way without doing a finite element calculation. The latter do provide at test of theory, however. One can also do this after determining the overall properties.

The stress components in the inclined faults (not the offset joints) are shown in Figure 5 for a modulus contrast of 10 and a uniaxial normal load in the vertical x direction. The prediction is within 6% of the (numerical) experimental results.

Because the experimental data are numerical, it is of interest to examine some of the purely numerical questions associated with the work. Two questions are touched upon here that involve mesh refinement and equation solvers. Enhanced versions of the UTAH2 and UTAH3 codes offer an equation solver option.

Figure 6 shows a test volume containing three intersecting joints that is subdivided into a coarse and a refined mesh containing constant strain triangles and quadrilaterals composed of four such triangles. Table 1 shows the results obtained with the two meshes using a Gaussian elimination equation solver and a Gauss-Seidel iterative equation solver (with over relaxation). The differences in numerical results due to the equation solvers are scarcely noticeable at a modulus contrast of 10. Although not shown here, the differences become even less at 100 and vanish to four decimal points at a modulus contrast of 1000. The differences in numerical results due to mesh refinement are also slight at a modulus contrast of 10 and become even less at higher modulus contrast values. The improved performance at the higher modulus contrast is due to the fact that the slender elements in the joint regions behave better. (The product of shear modulus and aspect ratio dominates slender element behavior.) Finally, the differences between theory and numerical results in the case of the three joint test volume are less than 3% for the moduli and less than 10% for Poisson's ratios in the worst case.

4 CONCLUSION

An approximate description for the overall elastic properties of well jointed rock masses that treats the elastic response of joints as thin layers of elastic material has been proposed. The concept and approach is valid for test volumes of arbitrary size and joint numbers. The sample volume need not be and is probably not representative of the parent rock mass. Neither isotropy nor orthogonal joint angle intersections are assumed. The example results presented here show excellent agreement between theory and experiment. Agreement is within a few percent using realistic rock and joint properties. These results are indicative of our experience to date. Although the concept of treating a joint as a thin layer is certainly open to discussion, we are encouraged by the results in the elastic regime and are hopeful of extending the approach beyond the elastic domain towards the description of rock mass strength.

5 ACKNOWLEDGEMENTS

Scholarship support for H. Moon from the W. C. Browning Fund, the Graduate Research Fellowship Fund of the University of Utah, and the Mining and Minerals Resources Research Institute is gratefully acknowledged. Computational funds granted by the University Computer Center are deeply appreciated.

Table 1. Overall moduli and Poisson's ratios for three intersecting joints and a modulus contrast of 10 using coarse and refined meshes and elimination and iterative equations solvers. Experiment from finite element analyses. Prediction from PM theory.

Overall Moduli	Prediction	Experiment			
		Coarse		Refined	
		Elimination	Iteration	Elimination	Iteration
E_x^*/E_r	0.5353	0.5390	0.5392	0.5369	0.5370
E_y^*/E_r	0.5498	0.5404	0.5404	0.5376	0.5375
G_{xy}^*/G_r	0.5197	0.5227	0.5227	0.5206	0.5203
v_{xy}^* (a)	0.2112	0.2212	0.2212	0.2216	0.2216
v_{yx}^* (a)	0.2169	0.2218	0.2218	0.2219	0.2219

*(a) $v_r = v_j = 0.2$

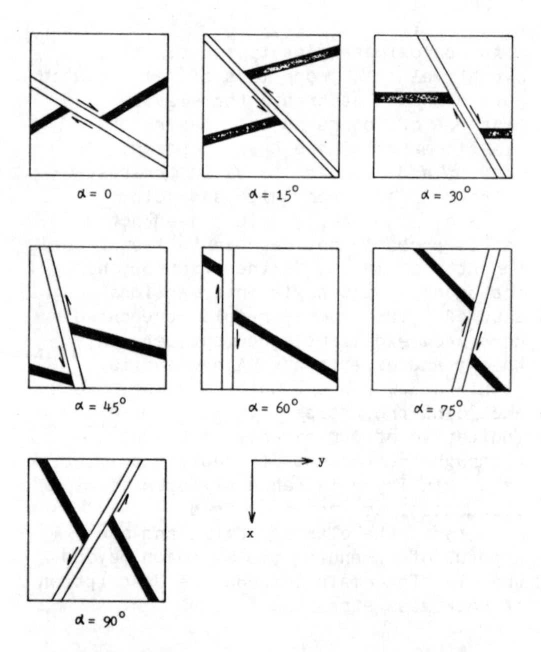

Figure 1. Seven jointed and faulted test volumes, reference axes and orientation.

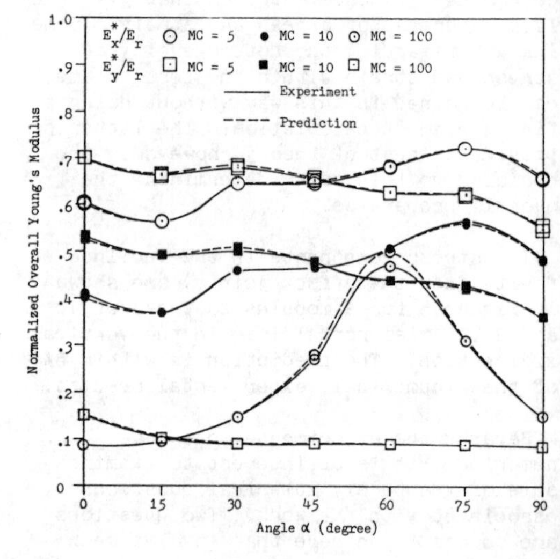

Figure 2. Overall Young's modulus for seven joint and fault geometries and three values of modulus contrast.

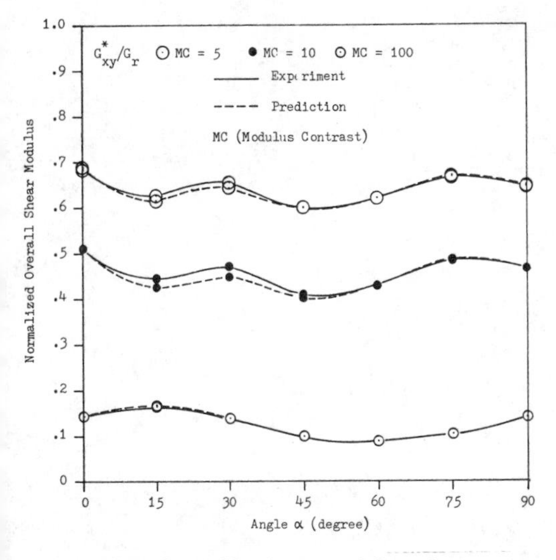

Figure 3. Overall shear modulus for seven joint and fault geometries and three values of modulus contrast.

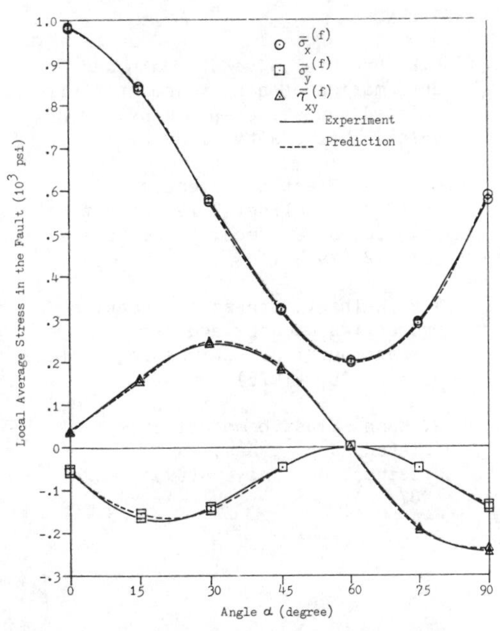

Figure 5. Localized average stresses for seven joint and fault geometries and three values of modulus contrast. Test volume is under a unit vertical (x-direction) load.

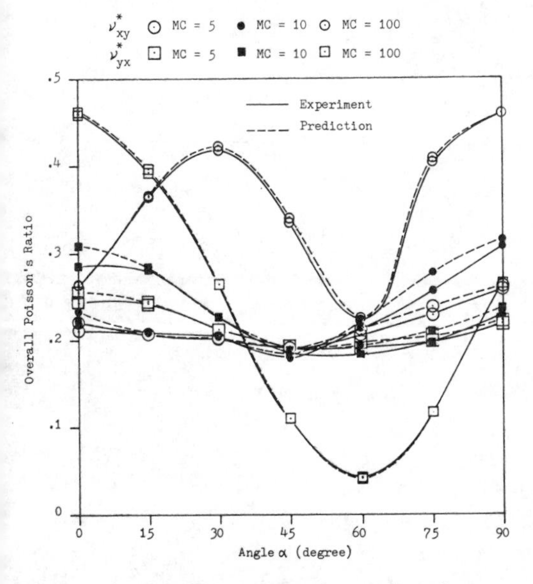

Figure 4. Overall Poisson's ratios for seven joint and fault geometries and three values of modulus contrast.

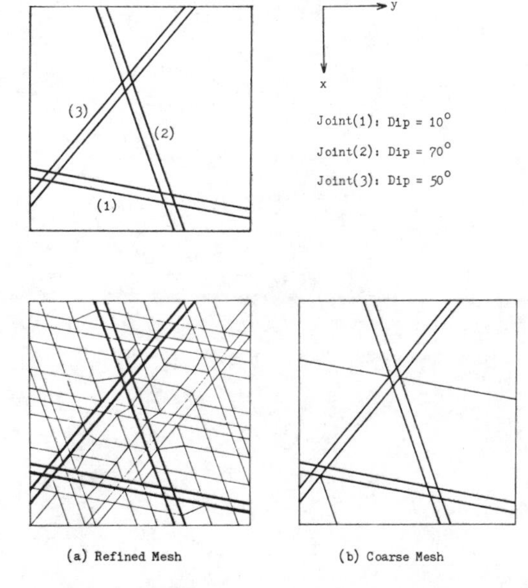

Figure 6. Test volume containing three intersecting joints in a coarse and a refined mesh.

6 REFERENCES

[1] C.E. Brechtel, The strength and deformation of singly and multiply jointed sandstone specimens. M.S thesis, University of Utah, 1978.

[2] R. Hill, Elastic properties of reinforced solids: Some theoretical principles. J. Mech. Phys. Solids. 11, 357-372 (1963).

[3] F.H. Kulhawy. Stress deformation properties of rock and rock discontinuities. Engineering Geology. 9, 327-350 (1975).

[4] H. Moon. Elastic moduli of well jointed rock masses. Ph.D dissertation, University of Utah, 1987.

Numerical Methods in Geomechanics (Innsbruck 1988), Swoboda (ed.)
© 1988 Balkema, Rotterdam. ISBN 90 6191 809 X

KEM in geomechanics

Peter Gussmann
University of Stuttgart, FR Germany

Abstract: The Kinematical Element Method (KEM) was presented first by the author in 1982. The method was designed as a general procedure for any kind of limit-state problem of Coulomb material. This paper will show a summary of the theory together with relevant applications.

1. Geometry

The geometry of an arbitrary straight-line bounded kinematical element f will be described with reference to its nodal coordinates, Fig. 1. Its area can be calculated by the formula of Gauss to

$$A = 1/2\ \Sigma i\ (x_i z_{i+1} - x_{i+1} z_i)\ ;\ P_{m+1} = P_1 \quad (1)$$

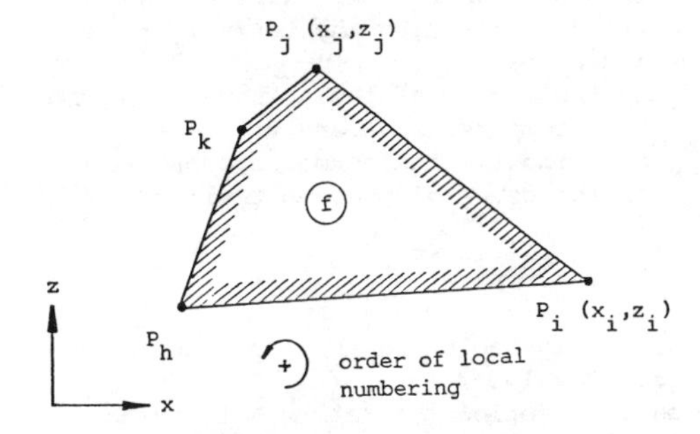

Figure 1 Nodal points

The boundary s, which separates either
- two adjacent elements c and f
- rigid area (element 0) and element f
- movable boundary element $\hat{f}$ and f
- free space and surface element f

is defined only by its outward normal **e** according to Fig. 2. Referring to element f, the following notations and relations can be verified:

$$e_x = l_{c/f} = \cos\alpha_{c/f} = z_{i,j}/d_{i,j} = -l_{f/c}$$

$$e_z = n_{c/f} = \cos\gamma_{c/f} = -x_{i,j}/d_{i,j} = -n_{f/c} \quad (2)$$

$$l^2 + n^2 = 1\ ;$$

$$\mathbf{e}_{c/f} = \{l_{c/f}; n_{c/f}\} = -\mathbf{e}_{f/c}$$

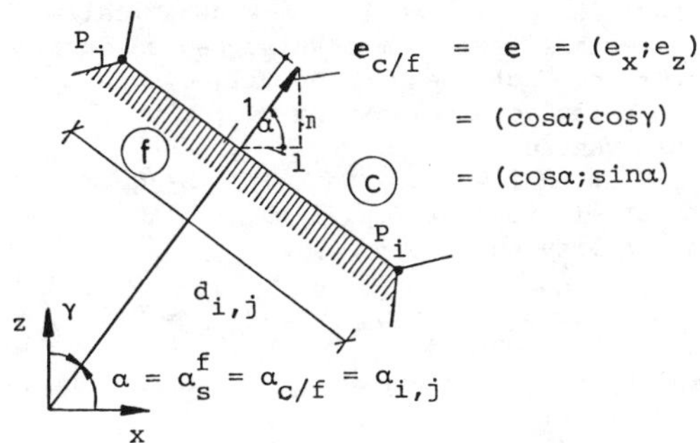

Figure 2 Boundary

with the abbreviations

$$\alpha_{c/f} = \alpha_s^f = \alpha_{i,j}\ ;\quad \gamma_{c/f} = \gamma_s^f = \gamma_{i,j}$$

$$x_{i,j} = x_j - x_i = -x_{j,i}$$

$$z_{i,j} = z_j - z_i = -z_{j,i} \quad (3)$$

$$d_{i,j} = (x_{i,j}^2 + z_{i,j}^2)^{0.5} = d_{j,i}$$

This notation is different to that used mainly in the previous papers, Gussmann (1982), Gussmann (1986a),but it has the great advantage, that it can be extended to 3D-problems, Gussmann (1986b).

2. Failure mechanism and Kinematics

The idealisation of failure mechanisms in real earth structures consists of failure lines and plastic zones on one hand and rigid zones and elasto-plastic zones on the other. The KEM even simplifies these facts to just rigid parts, separated by failure lines; the latter are assumed to be straight lines. The kinematical behaviour

within the failure lines is assumed to just sliding (in this paper). An idealized failure mechanism therefore consists of straight-line bounded rigid elements, which can perform translatoric movements, but no rotations.
The simplification to straight lines instead of circles for instance is not a question of the kinematics -this could be assumed also, see Goldscheider/Gudehus (1974)- but rather a question of the definiteness of the statics.
These simplifications seem to be very crude, but nevertheless they allow the modelling of even highly sophisticated failure mechanisms with plastic zones, elastic zones and curved failure lines, as it can be seen from Fig. 8, which shows an idealized failure mechanism for a bearing capacity problem.
The compatibility for two displaced adjacent elements c and f of a kinematically admissible failure mechanism can be derived from the fact that the normal component of the relative displacement vector $\mathbf{v}_{c/f}$ must vanish.
This is easily expressed with the geometric relations of Fig. 3 to give the compatibility equation

$$v_{c/f,x} \cos\alpha_{c/f} + v_{c/f,z} \cos\gamma_{c/f} = 0 \quad (4)$$

which leads together with the abbreviations

$$\mathbf{v}_{c/f} = \mathbf{v}^f - \mathbf{v}^c ;$$

$$v_{c/f,x} = v_x^f - v_x^c \ ; \quad v_{c/f,z} = v_z^f - v_z^c \qquad (5)$$

and eq. (2) to

$$l_{c/f} v_x^f + n_{c/f} v_z^f + l_{f/c} v_x^c + n_{f/c} v_z^c = 0 \quad (6)^1$$

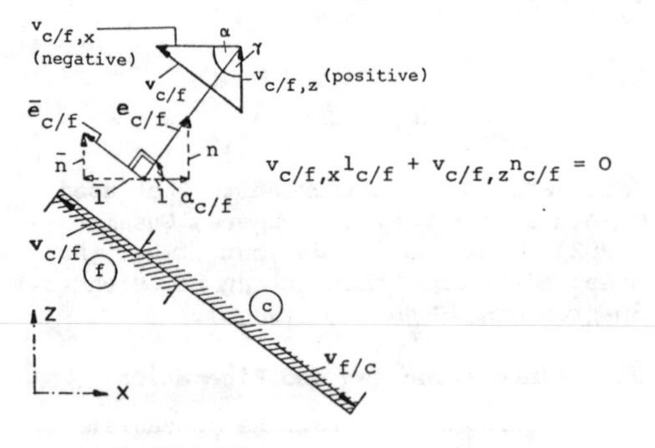

Figure 3 Compatibility in case of sliding

The position of any 2D-element f with the introduced properties is uniquely defined by exactly two compatibility equations for two boundaries c/f and e/f, Fig. 4 - if the slopes are not parallel - and therefore we derive the second compatibility equation for the element f to be

$$l_{e/f} v_x^f + n_{e/f} v_z^f + l_{f/e} v_x^e + n_{f/e} v_z^e = 0 \quad (6)^2$$

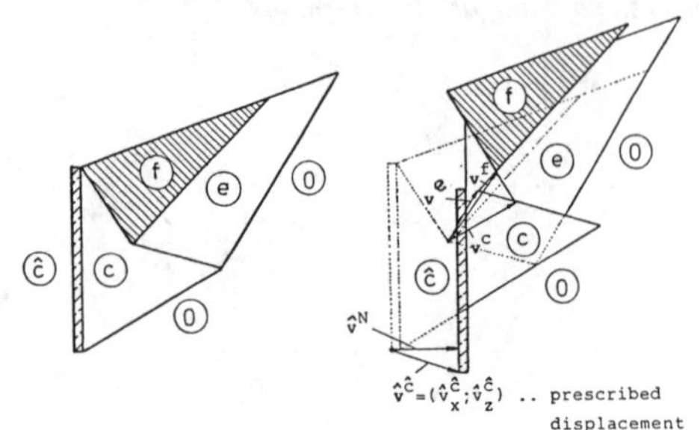

Figure 4 Kinematics

If a kinematically admissible failure mechanism consists of m real elements, we can derive just 2m compatibility equations for the 2m unknown displacement components v_x, v_z.
Together with the notations $\hat{\mathbf{v}}^m$ for the movable boundary and $\hat{\mathbf{v}}^0$ for the rigid boundary the kinematics of the whole structure can be given by a set of linear equations

$$\mathbf{K}\,\mathbf{v} + \hat{\mathbf{v}}_N = \mathbf{0} \qquad (7)$$

where

K is a non-symmetrical "kinematics-matrix", containing the terms $l_s = \cos\alpha_s$ and $n_s = \cos\gamma_s$,
v is the vector of the unknown displacement components v_x and v_z,
$\hat{\mathbf{v}}_N$ contains all the normal components of the given displacement via :

$$\hat{v}_N^{\hat{f}} = l_{f/\hat{f}} \hat{v}_x^{\hat{f}} + n_{f/\hat{f}} \hat{v}_z^{\hat{f}} \ ; \quad \hat{v}_N^{o} = 0$$

for the moveable and the rigid boundary respectively.
How in practice the matrix **K** is established to represent a particular failure mechanism will be shown in chapter 4.
The solution of eq. (7) can be found by the elimination method of Gauss together with pivoting for example. Then, the direction cosines of the relative displacement vectors can be calculated together with eq. (5) by

$$\bar{e}_x = \bar{l}_{c/f} = \cos\bar{\alpha}_{c/f} = v_{c/f,x} / |v_{c/f}|$$

$$\bar{e}_z = \bar{n}_{c/f} = \cos\bar{\gamma}_{c/f} = v_{c/f,z} / |v_{c/f}| \qquad (8)$$

$$|v_{c/f}| = (v_{c/f,x}^2 + v_{c/f,z}^2)^{0.5}$$

As the unit vectors $\mathbf{e}_{c/f}$ and $\bar{\mathbf{e}}_{c/f}$ are perpendicular (Fig. 3) to each other, we can introduce the alternate notation

$$\bar{l}_{c/f} = -\delta_{c/f}\, n_{c/f}$$
$$\bar{n}_{c/f} = \delta_{c/f}\, l_{c/f}$$
$$\delta_{c/f} = \text{sign}(\mathbf{v}_{c/f}) = \mathbf{v}_{c/f}/|\mathbf{v}_{c/f}| \qquad (9)$$

with the sign definition: a relative displacement vector $\mathbf{v}_{c/f}=\mathbf{v}^f_s$ is defined positive, if it turns anti-clock wise within the element f, Fig. 5. Therefore $\mathbf{v}_{f/c}=\mathbf{v}^c_s$ will also be positive.

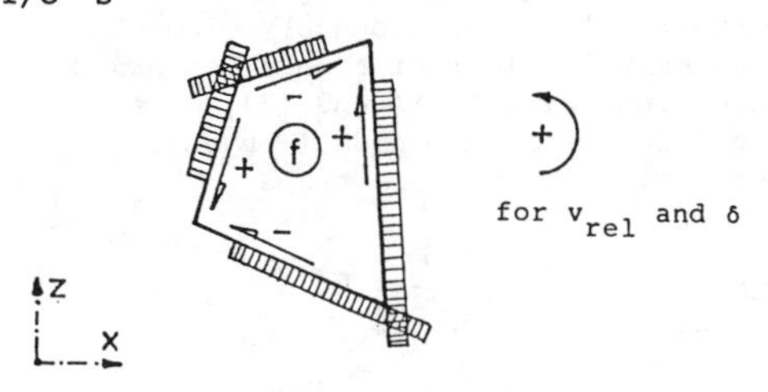

Figure 5 Sign convention

Comparing eq. (8) and (9) yields

$$\delta_{c/f} = -\frac{v_{c/f,x}}{|v_{c/f}|\ l_{c/f}} = +\frac{v_{c/f,z}}{|v_{c/f}|\ l_{c/f}} \qquad (10)$$

For numerical reasons it is better to avoid division if possible and to determine the sign by

$$\delta_{c/f} = \text{sign}(-n_{c/f}\, v_{c/f,x} + l_{c/f}\, v_{c/f,z}) \qquad (11)$$

instead of using eq. (9.3) or (10).

3. Statics

The force **S** which acts at the boundary s onto the element f can be splitted into a normal component N and a tangential component T. These components are directed just opposite to the normal vectors **e** and **e** respectively. Together with Fig. 6 we derive the cartesion components

$$N_x = -l\,N \;;\; T_x = -\bar{l}\,T \;;\; S_x = N_x + T_x$$
$$N_z = -n\,N \;;\; T_z = -\bar{n}\,T \;;\; S_z = N_z + T_z \qquad (12)$$

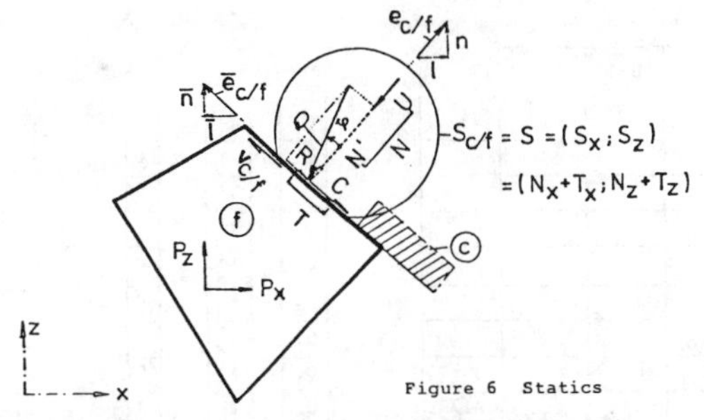

Figure 6 Statics

The total normal force N consists of the effective normal force N' and the neutral pore pressure force U, the tangential force consists of friction force R and cohesion force C in case of a coulomb material via

$$N = N' + U \;;\; T = R + C$$
$$N' = Q\cos\varphi \;;\; R = Q\sin\varphi \qquad (13)$$
$$C = c\,d \;;\; U = u\,d\,.$$

Q is the resultant of N' and R, φ is the friction angle, c the cohesion and u is a mean porepressure over the boundary length d. Thus the components of a resultant force $\mathbf{S}_{c/f}$ which act onto the element f within the boundary s=c/f can be expressed as

$$S_{c/f,x} = -\{l_{c/f}\cos\varphi_{c/f} + \bar{l}_{c/f}\sin\varphi_{c/f}\}Q_{c/f} + \bar{l}_{c/f}C_{c/f} + l_{c/f}U_{c/f}\}$$
$$(14)$$
$$S_{c/f,z} = -\{n_{c/f}\cos\varphi_{c/f} + \bar{n}_{c/f}\sin\varphi_{c/f}\}Q_{c/f} + \bar{n}_{c/f}C_{c/f} + n_{c/f}U_{c/f}\}$$

By use of the sign definitions of eq. (9) and (11) we can write alternatively

$$S_{c/f,x} = -\{(l_{c/f}\cos\varphi_{c/f} - \delta_{c/f}n_{c/f}\sin\varphi_{c/f}) * Q_{c/f} - \delta_{c/f}n_{c/f}C_{c/f} + l_{c/f}U_{c/f}\}$$
$$(14)^1$$
$$S_{c/f,z} = -\{(n_{c/f}\cos\varphi_{c/f} - \delta_{c/f}l_{c/f}\sin\varphi_{c/f}) * Q_{c/f} + \delta_{c/f}l_{c/f}C_{c/f} + n_{c/f}U_{c/f}\}$$

Equation (14) has the advantage of being easily extended to 3D-problems. But for 2D-problems eq. $(14)^1$ seems to be more appropriate, as it needs less computation effort during the variation of the geometry, see chapter 6. With respect to the global structure we can introduce the abbreviations

$$\tilde{k}^f_{s,x} = -(l^f_s\cos\varphi_s - \delta_s n^f_s\sin\varphi_s) \equiv -\tilde{l}^f_s$$
$$\tilde{k}^f_{s,z} = -(n^f_s\cos\varphi_s + \delta_s l^f_s\sin\varphi_s) \equiv -\tilde{n}^f_s$$
$$(15)$$
$$F^f_{s,x} = \delta_s\, n^f_s\, C^f_s - l^f_s\, U^f_s = C_x + U_x$$
$$F^f_{s,z} = -\delta_s\, l^f_s\, C^f_s - n^f_s\, U^f_s = C_z + U_z$$

and get the shorter formulation

$$S^f_{s,x} = \tilde{k}^f_{s,x}Q_s + F^f_{s,x}$$
$$(14)^2$$
$$S^f_{s,z} = \tilde{k}^f_{s,z}Q_s + F^f_{s,z}$$

Together with the body forces **P** the equilibrium equations for one element can now be expressed as

$$P^f_x + \Sigma\, S^f_{s,x} = 0$$
$$P^f_z + \Sigma\, S^f_{s,z} = 0 \qquad (16)$$

Introducing eq. (14)[1] into (16) and summing up all m elements of the structure yields to the system of linear equations

$$\tilde{K} * Q + F = 0 \qquad (17)$$

where

$\tilde{K}$ is the "friction matrix" containing the terms $\tilde{k}_{sx}$ and $\tilde{k}_{sz}$

Q is the vector of unknown static forces Q_s

F is the load vector containing body forces P^f and cohesion and water pressure forces F_s.

which can be solved by standard procedures.

4. Global structure

In principle the numbering order of nodes, boundaries and elements is arbitrary. But from a numerical point of view it seems to be preferable to have always nonzero terms on the diagonal of the matraces K and $\tilde{K}$ respectively.
How this can be achieved will be shown with the following example, Fig.7, with 5 real elements, 1 movable boundary element (=element 6), 10 real boundaries and 2 surface boundaries (= 11 and 12), 8 real nodes and 1 supplementary node (= node 9) for describing the surface slope.

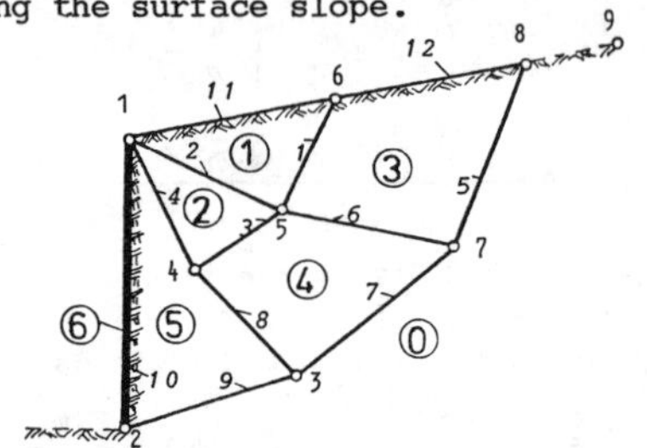

Figure 7 Global structure

Elements

Boundaries	1 x	1 z	2 x	2 z	3 x	3 z	4 x	4 z	5 x	5 z
1	$l_{3/1}$	$n_{3/1}$			$-l_{3/1}$	$n_{3/1}$				
2	$l_{2/1}$	$n_{2/1}$	$-l_{2/1}$	$-n_{2/1}$						
3			$l_{4/2}$	$n_{4/2}$			$-l_{4/2}$	$-n_{4/2}$		
4			$l_{5/2}$	$n_{5/2}$					$-l_{5/2}$	$-n_{5/2}$
5					$l_{0/3}$	$n_{0/3}$				
6					$l_{4/3}$	$n_{4/3}$	$-l_{4/3}$	$-n_{4/3}$		
7							$l_{0/4}$	$n_{0/4}$		
8							$l_{5/4}$	$n_{5/4}$	$-l_{5/4}$	$-n_{5/4}$
9									$l_{0/5}$	$n_{0/5}$
10									$l_{6/5}$	$n_{6/5}$

$$\hat{v}^6_N = l_{5/6}\, \tilde{v}^6_x + n_{5/6}\, \hat{v}^6_z$$

v		$\hat{v}_N$		0
v^1_x				0
v^1_z				0
v^2_x				0
v^2_z				0
* v^3_x	+	0	=	0
v^3_z				0
v^4_x		0		0
v^4_z				0
v^5_x		0		0
v^5_z		$\hat{v}^6_N$		0

Kinematics

$$K * v + \hat{v}_N = 0$$

Boundaries

Elements		1	2	3	4	5	6	7	8	9	10
1	x	$-\tilde{l}_{3/1}$	$-\tilde{l}_{2/1}$								
	z	$-\tilde{n}_{3/1}$	$-\tilde{n}_{2/1}$								
2	x		$\tilde{l}_{2/1}$	$-\tilde{l}_{4/2}$	$-\tilde{l}_{5/2}$						
	z		$\tilde{n}_{2/1}$	$-\tilde{n}_{4/2}$	$-\tilde{n}_{5/2}$						
3	x	$\tilde{l}_{3/1}$				$-\tilde{l}_{0/3}$	$-\tilde{l}_{4/3}$				
	z	$\tilde{n}_{3/1}$				$-\tilde{n}_{0/3}$	$-\tilde{n}_{4/3}$				
4	x			$\tilde{l}_{4/2}$			$\tilde{l}_{4/3}$	$-\tilde{l}_{0/4}$	$-\tilde{l}_{5/4}$		
	z			$\tilde{n}_{4/2}$			$\tilde{n}_{4/3}$	$-\tilde{n}_{0/4}$	$-\tilde{n}_{5/4}$		
5	x				$\tilde{l}_{5/2}$				$\tilde{l}_{5/4}$	$-\tilde{l}_{0/5}$	$-\tilde{l}_{6/5}$
	z				$\tilde{n}_{5/2}$				$\tilde{n}_{5/4}$	$-\tilde{n}_{0/5}$	$-\tilde{n}_{6/5}$

Q		F		0
Q_1		F^1_x		0
Q_2		F^1_z		0
Q_3		F^2_x		0
Q_4		F^2_z		0
* Q_5	+	F^3_x	=	0
Q_6		F^3_z		0
Q_7		F^4_x		0
Q_8		F^4_z		0
Q_9		F^5_x		0
Q_{10}		F^5_z		0

Statics

$$\tilde{K} * Q + F = 0$$

The key for a "good numbering" can be stated by two rules:
- try to number the elements beginning with the one with just 2 unknown forces and carry on to the last element with the movable boundary
- number just 2 boundaries with unknown forces in the element i with the numbers (2i-1) and 2i; so the free surface boundaries get numbers > 2m.

The structure of **K** corresponds to $\tilde{\mathbf{K}}^t$. So both matraces **K** and $\tilde{\mathbf{K}}$ can be easily established by just computing the 4m terms within the marked diagonal band and introducing in the negative terms outside of it by knowing their position.

5. Objective function

The external work $\hat{E}$ of the inducing displacements $\hat{\mathbf{v}}$ of the movable boundary and the adjacent boundary forces **S** can be expressed as

$$\hat{E} = \hat{\mathbf{v}}^t \mathbf{S} \qquad (18)$$

An alternate formulation reads

$$\hat{E} = D-E = \mathbf{v}_s^t \mathbf{T}_s - \mathbf{v}^t \mathbf{P} \qquad (19)$$

with D = dissipation in the boundaries
E = kinetical work

The decisive geometry is that which minimizes $\hat{E}$. If **X** is a vector containing n variable nodal coordinates x_i or z_i respectively we can write the optimization problem

$$f = \hat{E} \rightarrow f_{min} = \hat{E}_{min} (\mathbf{X}^*) \qquad (20)$$

together with the restrictions

$$L_i \geq 0 \qquad (21)$$

$$A_j > 0 \; ; \; j=1 \; (1) \; m \qquad (22)$$

$$Q_j \geq 0 \; ; \; j=1 \; (1) \; 2m \qquad (23)$$

where L_i is a linear restriction (to describe a concrete boundary for example), A_j is the area of an element, which has to remain positive and Q_j has to be positive as a pressure force.

In case of a movable boundary, which can be defined by constant nodal point coordinates, the energy terms of the cohesion and the pore pressure of eq. (18) are constant. So we can choose the following objective function instead of eq. (18)

$$f = \tilde{\mathbf{v}}^t \mathbf{Q} \qquad (18)^1$$

with

$$\tilde{v}_{f/\hat{f}} = \tilde{l}_{f/\hat{f}} \hat{v}_x^{\hat{f}} + \tilde{n}_{f/\hat{f}} \hat{v}_z^{\hat{f}} \; ; \; \tilde{v}_{f/0} = 0 \qquad (24)$$

which is minimized with the same solution vector $\mathbf{X}^*$ as eq. (18). This has some computational advantage as can be seen later on.

The problem of slope stability requires a different objective function, if there is no real footing on the slope. The author recommends to define the factor of safety via Fellenius rule to be the objective function, Gussmann (1986 a). This problem can only be solved by a certain iteration technique which is not discussed within this paper.

6. Optimization

The optimization is the main computational problem of the KEM. The author recommands 4 different optimization programmes:
- direct search with simplex, see Nelder/-Mead (1965))
- direct search with complex, see Box (1965)
- evolution theory of Rechenberg (1973), see Schwefel (1982)
- Quasi-Newton-Methods, see Davidon/-Nazareth (1977a),(1977b)

All Gradient-Methods and all Quasi-Newton--Methods require the determination of the gradient vector **g**

$$\mathbf{g}^t = (g_1, .. g_i, .. g_n); \; g_i = \partial f / \partial x_i \qquad (25)$$

which should be zero at a smooth minimum of f. Considering the eq. (17), (18)¹ and (24) we can differentiate and get

$$g_i = \tilde{\mathbf{v}}^t \frac{\partial \mathbf{Q}}{\partial x_i} = -\tilde{\mathbf{v}}^t \tilde{\mathbf{K}}^{-1} \left(\frac{\partial \mathbf{F}}{\partial x_i} + \frac{\partial \tilde{\mathbf{K}}}{\partial x_i} \mathbf{Q}\right)$$

or (26)

$$g_i = -\mathbf{w}^t \left(\frac{\partial \mathbf{F}}{\partial x_i} + \frac{\partial \tilde{\mathbf{K}}}{\partial x_i} \mathbf{Q}\right) \; ; \; \mathbf{w}^t = \tilde{\mathbf{v}}^t \tilde{\mathbf{K}}^{-1}$$

The last formulation is very efficient as the vector **w** is build up only once for all gradients **X** and as a change of just one coordinate affects only the elements around the considered point. the derivatives can be found by analytical differentiation according to the terms shown in the following table for homogenous elements.

In case of nonhomogenius problems the gradients may alternatively computed by

$$\frac{\partial f}{\partial x} \approx \frac{\Delta f}{\Delta x} = \frac{f(x + \Delta x) - f(x)}{\Delta x} \qquad (28)$$

or by using eq.(26) together with the approximations

$$\frac{\partial \mathbf{F}}{\partial x} \approx \frac{\Delta \mathbf{F}}{\Delta x} \quad \text{and} \quad \frac{\partial \mathbf{K}}{\partial x} \approx \frac{\Delta \mathbf{K}}{\Delta x}$$

7. Application

Fig.8 shows the well-known bearing capacity problem for a rough and a smooth footing with two different failure mechanisms. The adjacent bearing loads differ by a factor of 2. Experimental results with a Schneebeli-model can be seen from Fig.9, which

term t	formula	$\partial t/\partial x_i$	$\partial t/\partial z_i$	$\partial t/\partial x_j$	$\partial t/\partial z_j$
A_Δ	$0.5(x_{i,j}z_{i,k}-x_{i,k}z_{i,j})$	—	—	$\frac{1}{2}z_{i,k}$	$-\frac{1}{2}x_{i,k}$
d	$(x^2_{i,j}+z^2_{i,j})^{0.5}$	n	-l	-n	l
l	$z_{i,j}/d$	$-\frac{l\,n}{d}$	$-\frac{n^2}{d}$	$\frac{l\,n}{d}$	$\frac{n^2}{d}$
n	$-x_{i,j}/d$	$\frac{l^2}{d}$	$\frac{l\,n}{n}$	$-\frac{l^2}{d}$	$-\frac{l\,n}{d}$
U_x	$-d\,l\,u_m$	0	u_m	0	$-u_m$
U_z	$-d\,n\,u_m$	$-u_m$	$\frac{1}{2}\gamma_w\,d\,l$	u_m	$\frac{1}{2}\gamma_w\,d\,l$
C_x	$\delta\,d\,n\,c_m$	$\delta\,c_m$	0	$-\delta\,c_m$	0
C_z	$-\delta\,d\,l\,c_m$	0	$\delta\,c_m$	0	$-\delta\,c_m$
$\tilde{l}$	$l\cos\varphi-\delta\,n\sin\varphi$	$-\frac{l\,\tilde{n}}{d}$	$-\frac{n\,\tilde{n}}{d}$	$\frac{l\,\tilde{n}}{d}$	$\frac{n\,\tilde{n}}{d}$
$\tilde{n}$	$n\cos\varphi+\delta\,l\sin\varphi$	$\frac{l\,\tilde{l}}{d}$	$\frac{n\,\tilde{l}}{d}$	$-\frac{l\,\tilde{l}}{d}$	$-\frac{n\,\tilde{l}}{d}$

assumptions: (27)

$x_{i,j}=x_j-x_i$; $z_{j,i}=z_j-z_i$

$u_m = \frac{1}{2}(u_i+u_j)$; $u = \gamma_w(h-z)$; h=head=const. ; c=const. ; φ=const.

$\frac{dt}{dx} = \frac{\partial t}{\partial x} - \frac{l_p}{n_p}\frac{\partial t}{\partial z}$; $e(l_p,n_p)$ = normal of a restricting polygon, where point i or j has to lie on

$\frac{dt}{dz} = \frac{\partial t}{\partial z} - \frac{n_p}{l_p}\frac{\partial t}{\partial x}$

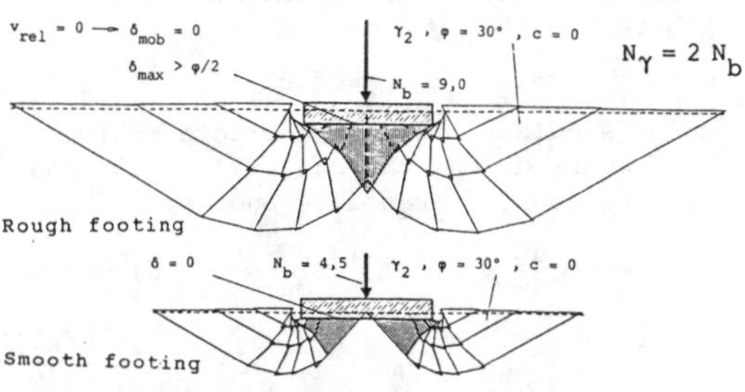

Figure 8 Bearing capacity for rough and smooth footing

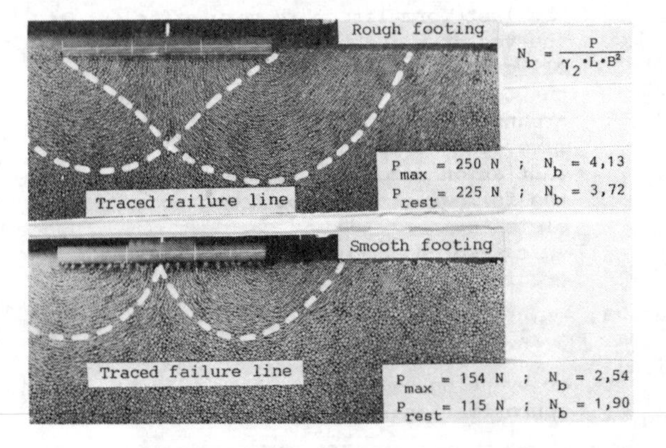

Figure 9 Model tests for rough and smooth footing

$\gamma_2 = 14,7\ kN/m^2$; $\Phi = 24°$ to $25°$; L = 4,7 cm ; B = 30 cm

$N_{b,DIN\ 4017} = 4,15$.

prove the different kinematics and statics as well. Figures 10 and 11 show the influence of different element-meshes for an active and a passive earth-pressure problem respectively.

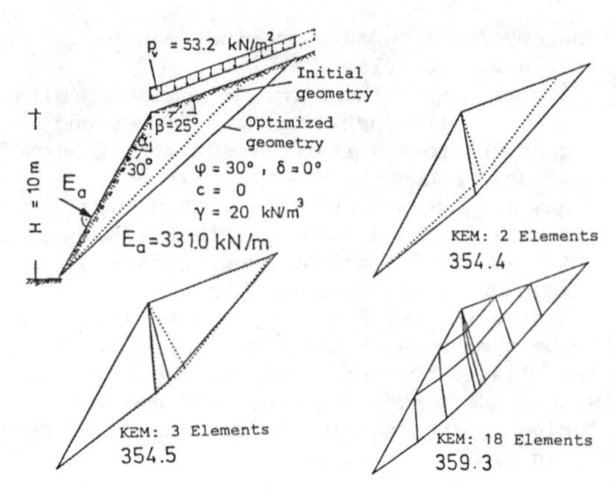

Figure 10 Active earth pressure

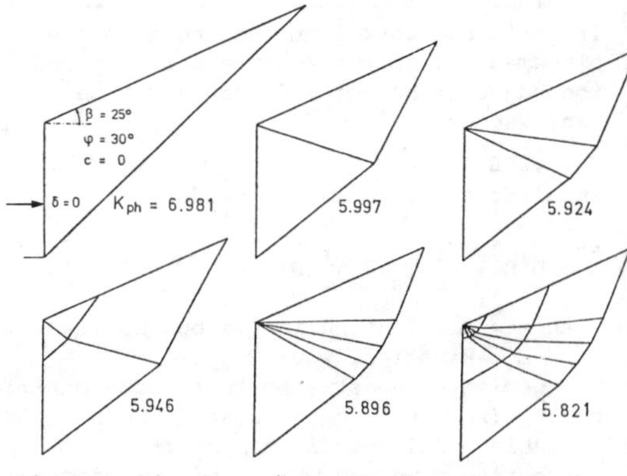

Figure 11 Passive earth pressure

References

Box, M.J. 1965 : A new method of constrained optimization and a comparision with other methods, Comp.J.8, 42-52

Davidon, W.C./ Nazareth, L. 1977 : DRVOCRA FORTRAN Implementation of Davidon's Optimally conditioned Method, Argonne National Lab., Ill. (USA)

Davidon, W.C./ Nazareth, L. 1977 : OCOPTRA Derivative Free Implementation of Davidon's Optimally Cond. Method, Argonne National Lab., Ill. (USA)

Goldscheider, M./Gudehus, G. 1974 : Verbesserte Standsicherheitsnachweise, Vorträge Baugrundtagung, Frankfurt, S. 99-127

Gussmann, P. 1982 : Kinematical Elements for Soil and Rocks, Proc. 4.Int. Conf. on Num. Meth. in Geomech., Edmonton, S. 47-52

Gussmann, P. 1986a : Die Methode der Kinematischen Elemente, Mitteilung 25 des Baugrundinstituts Stuttgart

Gussmann, P. 1986b : Kinematical Element Method for 3-D-Problems in Geomechanics, Proc. ECONMIG 86, Stuttgart, Vol.2

Nelder, J.A. and Mead, R. 1965 : A Simplex method for function minimization, The Computer Journal 7, 308-313

Rechenberg, I. 1973 : Evolutionsstrategie: Optimierung technischer Systeme nach Prinzipien der biologischen Evolution, Fromann-Holzboog Verlag, Stuttgart

Schwefel, H.P. 1982 : Numerical optimization of computer models, John Wiley and Sons, New York

Numerical Methods in Geomechanics (Innsbruck 1988), Swoboda (ed.)
© 1988 Balkema, Rotterdam. ISBN 90 6191 809 X

A macro joint element for nonlinear arch dam analysis

J.-M.Hohberg & H.Bachmann
Swiss Federal Institute of Technology (ETH), Zurich

ABSTRACT: The family of isoparametric joint elements with zero thickness is reviewed as standing between thin-layer and spring-box models. A threedimensional version, compatible with conventional 20-node solid elements, has been refined to capture in coarse discretization the full range of joint nonlinearity in arch dams and their abutment rock in a seismic event. This paper concentrates on the opening/closing performance and discusses remedies to the notorious oscillatory tendency in the family.

1 THE ARCH DAM PROBLEM

Arch dams are not constructed monolithic but endowed with a variety of discontinuities: vertical block and horizontal construction joints, a weak concrete-rock interface, sometimes a perimetral joint, Fig. 1a. If no tensile strength were attributed to these interfaces, the dam structure would resemble a discontinuum hold together by compressive forces (selfweight, water pressure), similar to the tectonically damaged rock against which it abuts.

Alterations to the carefully designed balance of loads - due to low temperature, rock deformation, or earthquake - may bring about tensile stresses which cannot be reconciled in homogeneous-elastic analysis. Joints are then assumed to beneficially relieve tension, and to provide wellcome damping in seismic response. However, the redistribution of stresses with possible secondary shear and compression concentrations, and the effect of hydraulic joint pressure need to be followed. A recent survey explores the state of nonlinear joint analysis for dams and in related disciplines of precast building construction and geotechnical engineering [1].

Simulating the seismic behaviour of arch dams beyond the linear range requires refined modelling of the adjacent media, Fig. 1b, as the misrepresentation of their contribution (resonance, energy radiation) would offset any gain in predictability of structural nonlinearity. This aspect is investigated in a concurrent part of the ETH project.

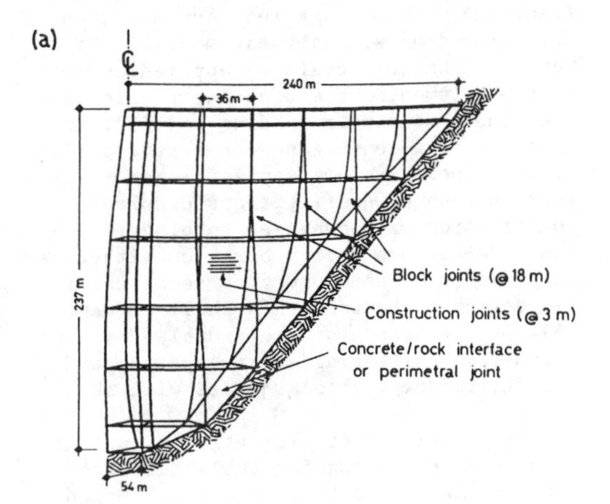

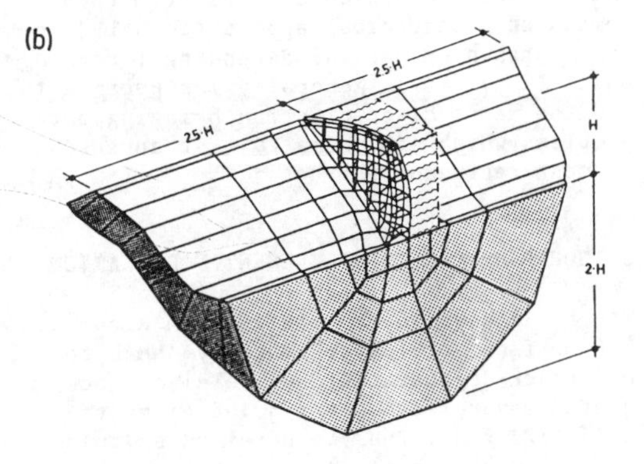

Figure 1. FE mesh generated by a commercial arch dam program (water not discretized)

2 MACRO DISCRETIZATION

Arch dams are highly hyperstatic structures, the safety of which is not determined by local material strength but by overall deformation, shift of stress resultants, etc. This explains the success of various design methods with poor local resolution of the 3-D stress state (e.g. arch-cantilever methods, finite shell elements). Finite solid("brick") elements of higher order are the most versatile in transition to adjacent media and were very early shown [2] to give competitive results even in much coarser discretization than that in Fig. 1, where the vertical element boundaries are made to coincide with every second block joint. To include (rock) joints in such a macro model has been suggested as early as in 1970 [3].

The seismic response is inertia-dominated and hardly influenced by local impact of closing joints. The spatial and temporal discretization may safely disregard high-frequency wave propogation. And although not persued here, nonlinear modal or Ritz vector techniques could be applied.

Joint elements were used for tension relief under monotonic loading [4] [5], exhibiting slow convergence or status oscillation. Known dynamic applications - apart from an attempt of fictitious displacement modifications at unchanged topology [6] - use node-to-node spring boxes on either dam face [7] or moment-curvature relations for discontinuity between thin shell elements [8]. Both approaches were troubled by high-frequency "chattering" of the joints and are restricted to block joints with shear keys.

Isoparametric joint elements, in contrast, also include sliding for small displacements (compared to the element dimension). The present discussion differs from shear-dominated geotechnical applications [9] in the emphasis on partial debonding ("rocking mode" [10]). It is generally instructive to see the macro discretization bringing out problems which are normally just appeased by mesh refinement.

3 ISOPARAMETRIC JOINT ELEMENT FORMULATION

The basic coordinate relations on a curved 3-D surface were established by AHMAD for his brick-degenerated shell element and quoted several times for joint elements [11] [12] [13]. Thus, a brief recapitulation may suffice.

Geometry and global displacements are interpolated from the values at the coinciding double-nodes by the same hierarchical shape functions (n=4..9):

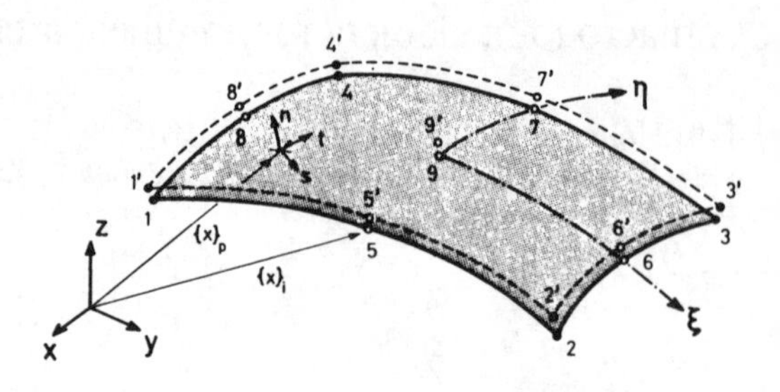

Figure 2. Geometry of 3-D curved interface element with 4 to 9 double-nodes

$$\{x\} = \sum_{i=1}^{n} h_i(\xi,\eta)\ \{\hat{x}\}_i \tag{1}$$

$$\{u_x\} = \sum_{i=1}^{n} h_i(\xi,\eta)\ \{\hat{u}_x\}_i$$

The Lagrangian double-node 9-9' would be desirable but had to be constrained in use with 20-node Serendipity brick elements.

In the stress points p, local tangential coordinate systems $\{n,s,t\}_p$ are defined through derivates $\partial\{x_p\}/\partial\xi$, $\partial\{x_p\}/\partial\eta$ and their vector product. Rotated to local coordinates, the displacement differences between top nodes p' and bottom nodes p are multiplied with the contact modulus matrix $[D]_p$ to yield local contact stress components which are rotated back to global components and numerically integrated (over p=1..nip quadrature points) to equivalent global forces. The global joint element stiffness matrix reflects these steps by

$$[K_x] = \int_{A^e} [\bar{B}]^T [D]\ [\bar{B}]\ dA^e \tag{2}$$

$$= \sum_{p=1}^{nip} \begin{bmatrix}-H\\+H\end{bmatrix} \begin{bmatrix}\hat{n}\hat{s}\hat{t}\\ \check{}\check{}\check{}\end{bmatrix} \begin{bmatrix}k_{nn} & & \\ & \cdot & \\ & & k_{tt}\end{bmatrix} \begin{bmatrix}\langle n\rangle\\ \langle s\rangle\\ \langle t\rangle\end{bmatrix} [-H,+H]\ \alpha\cdot dA$$

where all factors (shape function matrix $[H_i(\xi_p,\eta_p)]$, local direction cosines, contact moduli k, integration weight α and contributing area dA) may vary from one quadrature point p to the next.

In coding, the displacement differences are evaluated before interpolation, force vector and stiffness matrix computed only for the bottom nodes and finally expanded back to the top nodes (sign reversed):

$$\begin{Bmatrix} f_{bot} \\ f_{top} \end{Bmatrix} = \begin{bmatrix} \hat{K} & -\hat{K} \\ -\hat{K} & \hat{K} \end{bmatrix} \begin{Bmatrix} u_{bot} \\ u_{top} \end{Bmatrix} \tag{3}$$

Only the quadrant matrix $\hat{K}$ needs to be computed.

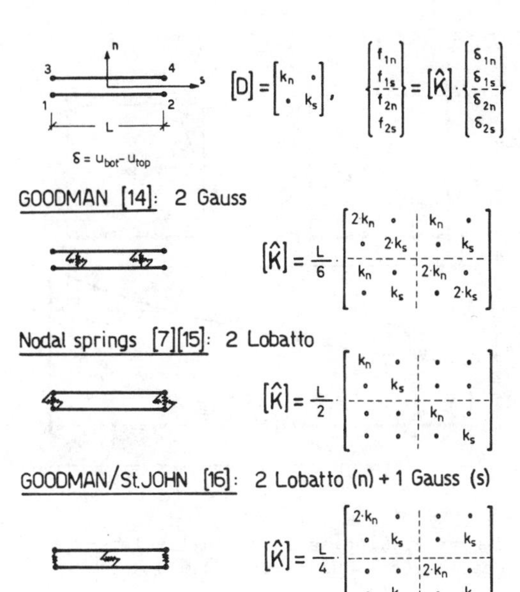

Figure 3. Known 2-D linear joint elements

Collapsing the element in one direction results in a contact line and a straight joint element, when the midside double-node is omitted. From that, all familiar 2-D joint elements can be recovered just by selecting the appropriate quadrature rule, equivalent to springs at the integration points [1], Fig. 3.

For the same 2-D geometry, the elastic performance of a thin solid and a joint element are compared in Fig. 4: With diminishing thickness $t \to 0$, the main diagonal terms start to dominate. The coupling of nodes through the element increases (to the ratio 2:1 for exact integration), while the spatial components decouple (for $\nu=0$), approaching the joint behaviour with $k_n=E/t$ and $k_s=G/t$. This was already stated in [3].

Hence, using a thin solid or a joint element makes no difference with view to the magnitude of penalty. In fact, expressing k_n in equivalent thickness provides a fair guideline for its selection [10]. One major advantage of the joint element, however, is the dispense with through-thickness integration. This eliminates spurious coupling between the two faces [17] and facilitates partial debonding.

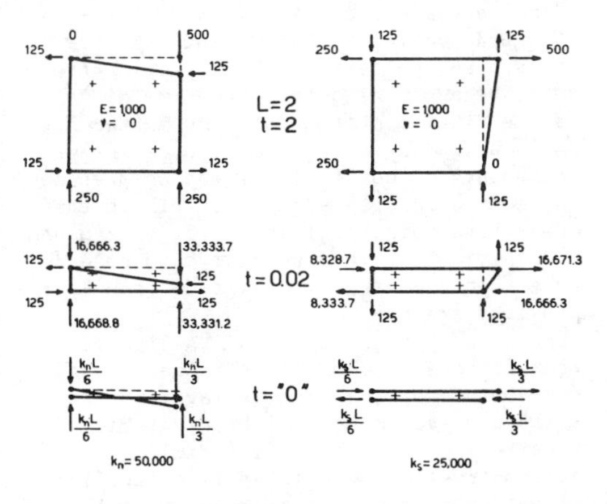

Figure 4. Limiting case of a thin continuum

4 QUADRATURE OF PARTIAL CONTACT AREA

In partial opening, the magnitude and position of contact stress resultants must not change abruptly whenever the neutral axis crosses a quadrature point. On the other hand, the integration of closed joints should not be burdened by heavy overintegration such as 10x10 rectangular rules used for reinforced concrete column sections. Thus, an idea of "floating" quadrature points was taken up, which seems to have been first mentioned for uniaxial bending by ÅLDSTEDT and BERGAN in 1978 but was later abandoned [18]. Only recently, it found new application in thermal phase change problems [19].

The idea here is to approximate the curve of zero (or a specified) gap width in the parent element by a straight line and to find the intersection of this "neutral axis" with the element edges. Along every edge, the existance of a compression zone is checked from the quadratic displacement interpolation, L_2 in Fig. 5, then the corners of the edge zones are connected by straight lines ($\overline{C_3C_8}$ and $\overline{C_6C_7}$) giving an idealized polygonal compression area into which the 2x2 or 3x3 Gauss integration points are contracted. The shape functions, integration weights and areas and - for a curved joint - the local direction cosines have to be reevaluated at the new stations.

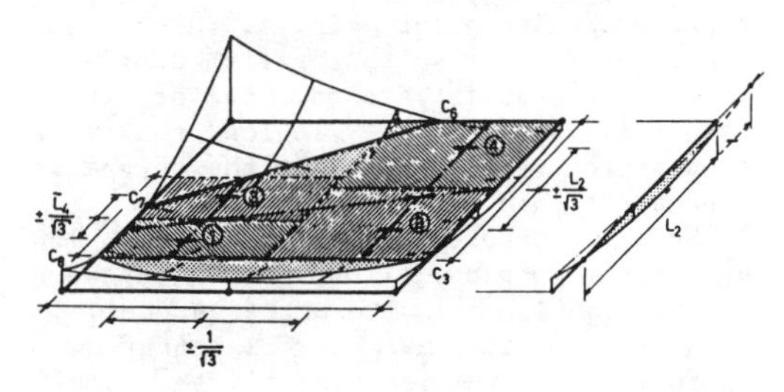

Figure 5. Quadrature point contraction

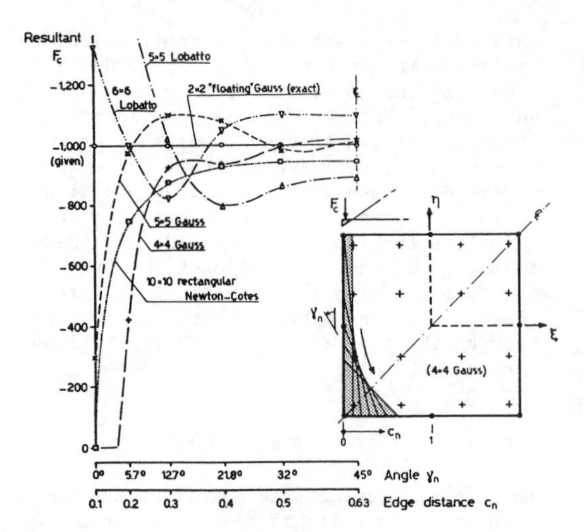

Figure 6. Stress resultant of a rotating compression zone as depending on quadrature

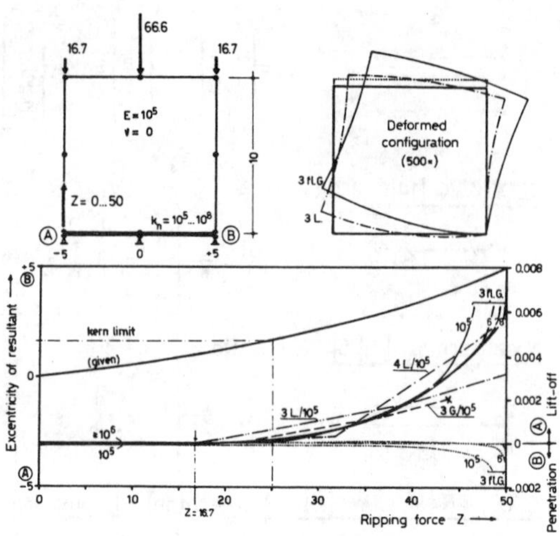

Figure 7. Ripping up a single-element joint (parameters: quadrature rule / k_n)

Figure 6 demonstrates the ability to track a 5% compression zone rotating around an element corner, where fixed-point quadrature schemes oscillate or fail if too coarse. (Note also the permanent underestimation by rectangular Newton-Cotes [18]).

The floating quadrature points are proposed for hyperelastic contact. The scheme looses attractiveness for path-dependent constitutive laws (friction) requiring memory interpolation, and whenever the debonded area is to be integrated as well, e.g. for joint water pressure or shear transmission by interlock.

The following paragraphs discuss typical problems in joint element application. They arise in 2-D and 3-D alike, so that 2-D examples are chosen for simplicity and familiarity.

5 STATIC PERFORMANCE

The first example is a tensile version of the GOODMAN/St.JOHN single-element test [15]: An increasing tension force Z rips up a precompressed joint, Fig. 7. The remaining compressive resultant falls outside the joint when Z=50 is reached. However, the joint fails numerically in load control as soon as the resultant passes the outermost quadrature point (see 3-pt. Gauss).

The joint becomes more flexible with lower contact modulus, allowing larger penetration in (B), and with higher number of quadrature points. Large overintegration approximates piecewise linearly the opening curve of "floating" quadrature.

The joint behaviour does not differ very much for $k_n=10^7 \div 10^8$, which corresponds to $t/L=0.01 \div 0.001$. Incidentally, this is about the illconditioning limit in 36-bit single-precision arithmetic. A higher penalty k_n aggrevates the overshoot by large unbalanced forces and causes (if not underrelaxed) delayed and oscillating convergence in full Newton-Raphson iteration.

Note the special case of a decoupled stiffness matrix, here for 3-pt. Lobatto. This "lumping" to nodal spring boxes results in premature opening as soon as the local force Z exceeds the consistently distributed precompression. Thereafter Z(A) remains zero, whereas it turns tensile for any other quadrature rule even before the resultant leaves the kern.

The second example is the one of restrained opening by TAYLOR and KUO [6]. It poses a severe test on the iteration procedure, because the joint opens immediately to the final contact state, no matter how small a load step is chosen. Underintegrated joints do not fail, as the compression force excentricity adjusts itself to the quadrature point position. The joint mouth opening is also less sensitive to quadrature rule and penalty magnitude than in the previous example.

However, the joint draws its flexibility from different "mechanisms": Coarse quadrature point spacing permits large unpenalized overlap to occur inbetween, Fig. 8a, frustrating the penetration limit. If, alternatively, the overlap is controlled by "floating" quadrature or high overintegra-

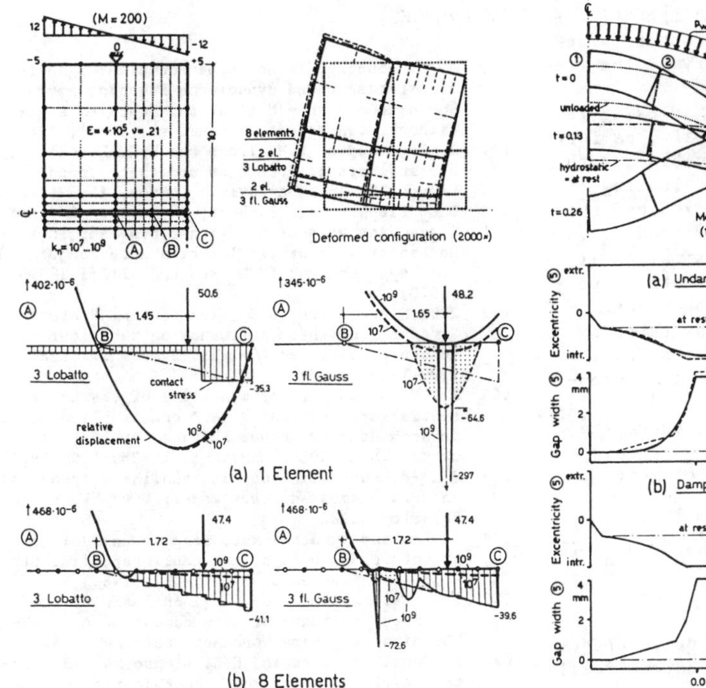

Figure 8. Restrained opening problem with rel. displacements and contact stresses

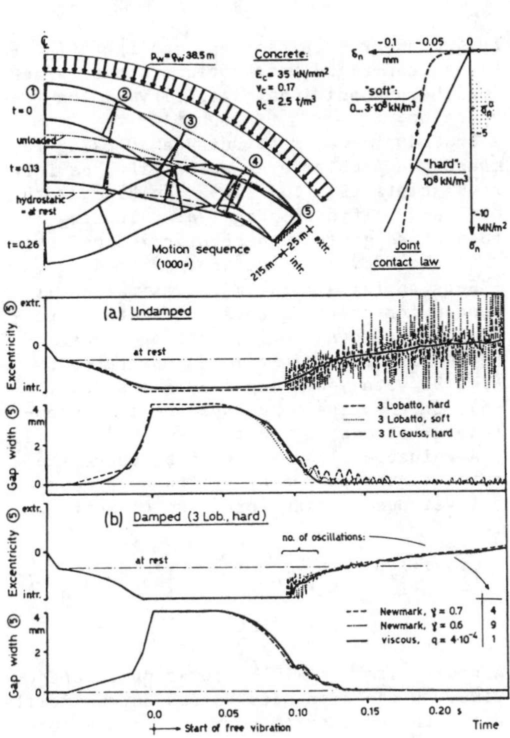

Figure 9. Arch dam ring with 5 quadratic joint elements, 1st eigenfrequency 2.05 Hz

tion, the contact area can be determined more accurately but turns out to depend on the contact stiffness, Fig. 8b: As the quadratic displacement function attempts to model the discontinuity, the contact area localizes and nodes separate which would be in compression at finer discretization.

If stresses are of interest in macro joint elements, careful tuning of the contact modulus is needed. A plastic compression limit would help although costly and somewhat in misuse. For smoothing is underintegration far more efficient.

The discontinuity is better captured with finer discretization, but inconsistent node displacements still occur at steep gradients, producing undulated overlap of the two faces [20]. Lumping the contact matrix by 3-pt. Lobatto quadrature averages the displacements of adjacent joint elements and gives the fastest convergence [15].

6 DYNAMIC PERFORMANCE

The "chattering" of joints is investigated next. As test problem, a horizontal slice through an actual arch dam under water pressure (similar to [6]) is deplaced until all joints are open and then allowed to swing freely about its at-rest position.

Abrupt changes of the stiffness matrix introduce spurious vibrations, the frequency of which depends on contact penalty and nodal mass terms and is usually too high to permit propagation through the large solid elements. Apart from possible excitation of low-energy modes they remain unaffected, reflecting the noise and locking it in the joint forever, if the time integrator is non-dissipative. This is the case for the Newmark scheme with $\gamma=\frac{1}{2}$ and $\beta=\frac{1}{4}(\gamma+\frac{1}{2})^2$ in:

$$[\tilde{K}]_n \{\Delta u\}^{i+1} = \{\tilde{p}\}_{n+1} \tag{4}$$

$$[\tilde{K}]_n = \frac{1}{\beta\Delta t^2}[M] + \frac{\gamma}{\beta\Delta t}[C] + [K]_n \tag{5}$$

$$\{\tilde{p}\}_{n+1} = \{p\}_{n+1} - \{r\}^{i+1}_{n+1} + (\frac{1}{\beta\Delta t}\{\dot{u}\}_n + (\frac{1}{2\beta}-1)\{\ddot{u}\}_n)[M] + ((\frac{\gamma}{\beta}-1)\{\dot{u}\}_n + (\frac{\gamma}{2\beta}-1)\Delta t\{\ddot{u}\}_n)[C] \tag{6}$$

The noise enforces step reduction for convergence, but errors in joint status evaluation may still become so large (possibly triggered by step reexpansion) that fictitious near-rigid body motions occur, wrecking the unconditional stability of the time operator.

Introducing a "soft surface" closing law makes the problem even worse, if – as for rock joints [9] [15] [16] – combined with a high end stiffness, Fig. 9a. "Floating" quadrature gives good regularization, even when the midside joint were omitted.

Substantial algorithmic dissipation is needed for coarse quadrature, with a spectral radius $\rho(A)<\sim 0.7$ to achieve decay reasonably fast. This is demonstrated by Fig. 9b with $\gamma=0.6$ and 0.7 in eqs. (5), (6). In practise, one uses the more accurate α-methods [8] or the ρ-method of [21].

A valuable alternative is to curbe the noise at the source by stiffness-proportional viscous damping forces on element level:

$$\{f_d\}_{n+1}^{i+1} = \left(-\frac{\gamma}{\beta\Delta t}\{\Delta u\}^{i+1} + \left(\frac{\gamma}{\beta}-1\right)\{\dot{u}\}_n + \left(\frac{\gamma}{2\beta}-1\right)\Delta t\,\{\ddot{u}\}_n\right) q\,[K]_n \quad (7)$$

A coefficient $q=4\cdot 10^{-4}$ proves quite effective, as the time steps drops automatically below the chattering period to reach convergence (here: Δt = 10→0.2 ms for ~350 Hz noise). Unlike algorithmic dissipation, this method can also be used for explicit (central difference) time integration.

7 CONCLUSIONS

Partial opening and closing is well captured by contact penalties equivalent to a solid aspect ratio of $10^{-2}..10^{-3}$. The lowest possible stiffness should be chosen with regard to the effect on stresses in the solids, or on system eigenfrequencies, and compared to the discretization error.

Overintegration and "floating" quadrature points regularize the opening/closing problem, but entail contact localization. Underintegration and lumped stiffness are more efficient as long as several joint elements in a row stay in contact. Thus, the quadrature rule may well become part of the solution strategy, if it can be changed during computation [21].

A regularized joint with moderate penalty is virtually free of chattering. Numerical damping, if necessary, can be provided by viscous joint forces on element level.

ACKNOWLEDGEMENTS

The continuing support of the project by the Swiss Federal Office of Water Resources, Berne, is gratefully acknowledged.

REFERENCES

[1] J.-M. Hohberg, Trennflächenformulierungen für die statische und dynamische Berechnung von Bogenstaumauern – Motivation, Modelle, Algorithmen, IBK report, Birkhäuser, Basel, 1987

[2] J. Ergatoudis, B.M. Irons and O.C. Zienkiewicz, 3-D analysis of arch dams and their foundations, ICE Arch Dam Symp., London, II/4 (1968)

[3] O.C. Zienkiewicz, B. Best, C. Dullage and K.G. Stagg, Analysis of non-linear problems in rock mechanics with particular reference to jointed rock systems, 2nd ISRM, Beograd, III:501-509 (1970)

[4] R.E. Ricketts and O.C. Zienkiewicz, Preformed "cracks" and their influence on behaviour of concrete dams, Dam Analysis Symp., Swansea, 1129-1147 (1975)

[5] J.P.F. O'Connor, The modelling of cracks, potential crack surfaces, and construction joints in arch dams by curved surface interface elements, 15th ICOLD, Lausanne, II:389-406 (1985)

[6] J.S.-H. Kuo, Joint opening nonlinear mechanism: Interface smeared crack model, UCB/EERC-82/10, Berkeley, 1982

[7] D. Row and V. Schricker, Seismic analysis of structures with localized nonlinearities, 8th WCEE, San Francisco, IV:475-482 (1984)

[8] J.F. Hall and M.J. Dowling, Analysis of the nonlinear response of arch dams to earthquakes, US-China Arch Dams Workshop, Beijing, 1987

[9] I. Carol, A. Gens and E.E. Alonso, A 3-D elasto-plastic joint element, Rock Joints Symp., Björkliden, 441-451 (1985)

[10] C.S. Desai, M.M. Zaman, J.G. Lightner and H.J. Siriwardane, Thin-layer element for interfaces and joints, Int.J.Num.Anal.Meth.Geomech. 8, 19-43 (1984)

[11] B. Tardieu and P. Pouyet, Proposition d'un modèle de joint tridimensionelle courbé, 3rd ISRM, Denver, IIB:833-836 (1974)

[12] D.N. Buragohain and V.S. Shah, Curved isoparametric interface surface element, J.Struct.Div. ASCE 104, 205-209 (1978)

[13] G. Beer, An isoparametric joint/interface element for FE analysis, Int.J.Num.Meth.Eng. 21, 585-600 (1985)

[14] R.E. Goodman, R.L. Taylor and A.M. Brekke, A model for the mechanics of jointed rock, J. Soil.Mech.Div. ASCE 94, 637-659 (1968)

[15] H.M. Hilber and R.L. Taylor, A finite element model of fluid flow in systems of deformable fractured rock, UCB/SESM-76/5, Berkeley, 1976

[16] R.E. Goodman and C.St. John, FE analysis for discontinuous rock. Numerical Methods in Geotechnical Engineering (Desai, Christian eds.), McGraw-Hill, New York, 1977

[17] D.V. Griffiths, Numerical modelling of interfaces using conventional finite elements, 5th ICONMIG, Nagoya, II:837-844 (1985)

[18] P.G. Bergan, Some aspects of interpolation and integration in nonlinear FE analysis of reinforced concrete structures, Concrete Structures Analysis Conf., Split, I:301-316 (1984)

[19] M. Storti, L.A. Crivelli and S.R. Idelsohn, Numerical implementation of a discontinuous FE algorithm for phase-change problems, Adv.Eng. Software 9, 66-73 (1987)

[20] L.R. Herrmann, FE analysis of contact problems, J.Eng.Mech.Div. ASCE 104, 1043-1057 (1978)

[21] E. Anderheggen et al., FLOWERS user's manual, 3rd ed., Inst. of Informatics, ETH Zürich, 1985

Numerical Methods in Geomechanics (Innsbruck 1988), Swoboda (ed.)
© 1988 Balkema, Rotterdam. ISBN 90 6191 809 X

Shear element techniques for tunnel design

G.Rome
Melbourne and Metropolitan Board of Works, Australia
W.E.Bamford
Melbourne University, Australia

ABSTRACT: The use of a simple shear element and associated techniques for the estimation of rock loads, ground reaction curves and failure modes in blocky rock where rock strength is dominated by weak joints and sheared zones is described. The application of these techniques is extended to estimation of the primary support requirement and performance assessment for a major rock chamber constructed in Victoria, Australia.

1 INTRODUCTION

The analysis presented demonstrates the versatility of a simple shear element which was published by Rome (1982), after it had been introduced in connection with rock slope stability analysis techniques carried out at Melbourne University under the supervision of Bamford. Further development of these techniques has been published by Rome (1986). This paper provides a further application in the analysis of rock support requirements in heavily sheared and jointed rock masses where low strength joints dominate the rock behaviour. Somewhat similar approaches have been introduced by Goodman (1976) and Desai et.al (1984) but to the Authors' knowledge have not been utilised in the manner presented.

2 SHEAR ELEMENT ANALYSIS METHOD

Briefly the analysis approach, involving the shear element, is to concentrate almost exclusively on the behaviour of joint intersection zones. The term joint is deemed to cover all rock defects such as sheared zones etc. which are concentrated on definable planes or surfaces.

The assembly of shear and block elements for the analysis and assessment of rock loading for the design of primary support systems is illustrated in Figure 1. A segment of a model based on the geological investigations carried out for a 14m diameter valve chamber for the Thomson Water Supply Scheme for the City of Melbourne in Australia is illustrated together with a detail showing the generated joint system at a typical joint intersection zone. Joint thicknesses range from 10mm to 150mm but have

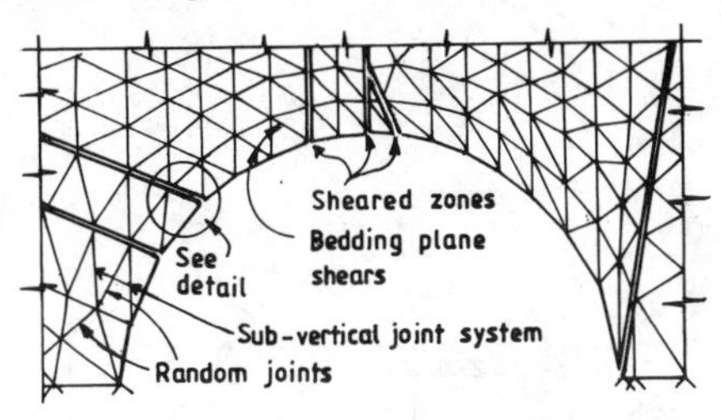

(a) Typical Model Segment

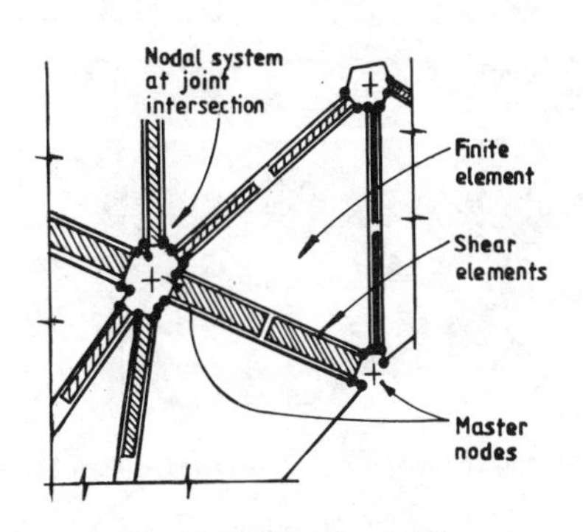

(b) Typical Joint Intersection Detail

Fig 1 Typical Arrangement of Elements

been enlarged in the detail for illustration purposes.

Full details of the shear element adopted have been provided in the reference given, however the main properties will be summarised as a background to the analysis results provided.

The concept of the shear element is illustrated in Figure 2(a). It is orthogonal in behaviour with shear displacement on the joint face and shear stress being controlled exclusively by a

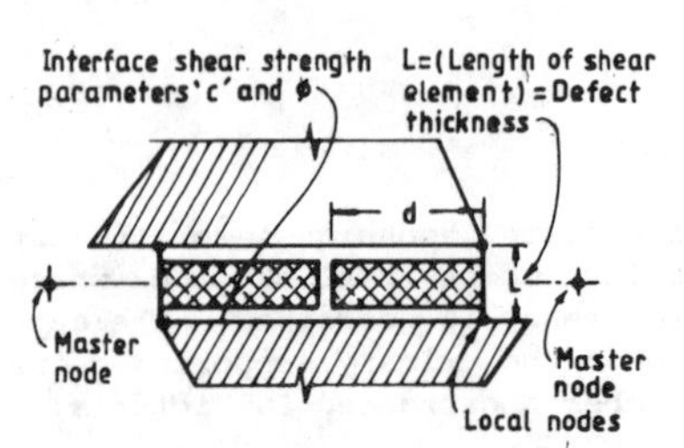

L(Length of shear element) = Defect thickness
d(Depth of shear element = 1/2 Distance between master nodes
b(Width) = Unity for given case

(a) Pair of Shear Elements

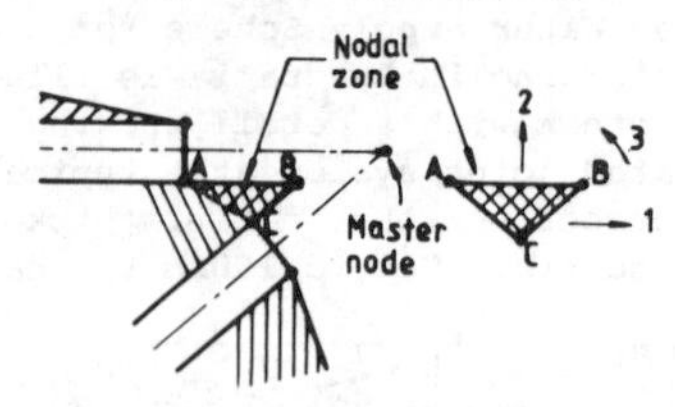

(b) Nodal Zone

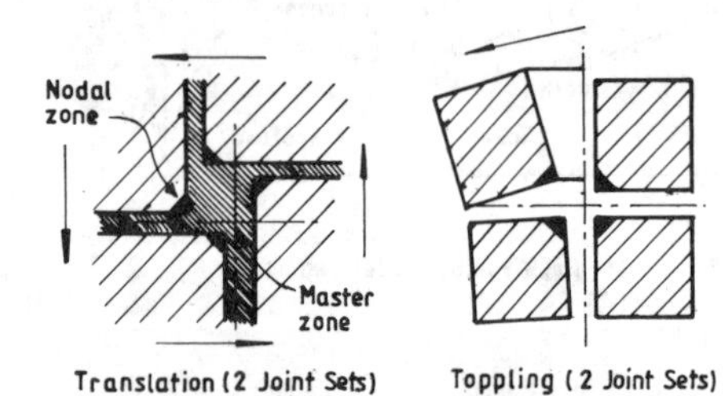

Translation (2 Joint Sets) Toppling (2 Joint Sets)

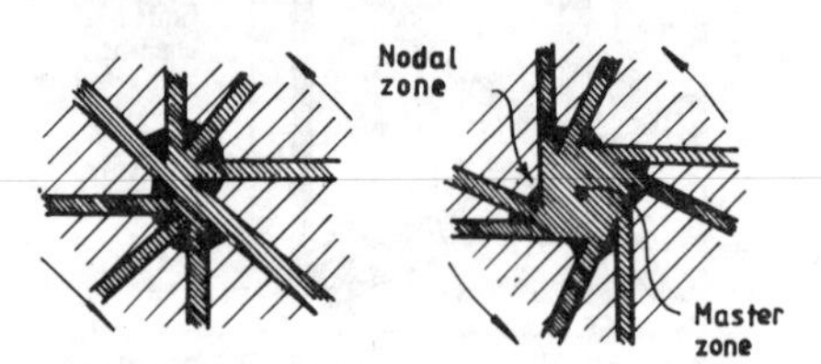

Translation (4 Joint Sets) Rotation (4 Joint Sets)

(c) Typical Displacement Modes

Fig 2 The Shear Element

stress/strain dependent shear modulus. Axial deformation ie. joint opening and closing, is similarly controlled by stress or strain dependent deformation moduli which may be elastic. Axial loading or stress is independent of shear displacement and shear forces are independent of the axial displacement although joint strength may be put to zero in tension.

Mohr-Columb strength assessments may thus be readily applied. Where overstress is identified, redistribution is attained by modifying shear modulus and so on.

Shear elements are generated in pairs adjacent to the edges of each face of rock blocks which are in practice defined by the joint system. Generation is achieved by the adoption of a master node system defining joint centre-lines.

To remove the possibility of "hang up" all shear elements are generated perpendicular to block faces to double precision accuracy (14 digits) hence very large shear displacements may be accomodated with negligible artificial restraint. A nodal zone concept, essential to this approach, is illustrated in Figure 2(b) whereby adjacent nodes of different co-ordinates are analysed as a zone of material having one set of freedoms only. The relative displacement between element nodes within a nodal zone is in effect considered to be zero. Displacement modes have been indicated in Figure 2(c). This approach is extended to shotcrete and rockbolt modelling.

3 GROUND REACTION CURVE DEVELOPMENT

The philosophy adopted with respect to assessing loading from the "ground reaction curve" computation for blocky jointed rock is first that the fully elastic deformation, without joint slip, would indicate a stable rock mass. The support system is therefore required to compensate for the overstress which must be transferred to it by joint relaxation. This support reaction is shown to be deformation dependent and by plotting the free elastic deflection on this curve an estimate of the maximum support requirement is obtained.

By adopting a logarithmic plot procedure the well defined "collapse" deflection and minimum support requirement is estimated.

The 'ground reaction curve' is estimated by inserting into the modelled cavity a set of radial elements. For any given stiffness of these elements, a relationship between deformation and load may be obtained at any point on the cavity circumference after all over-stress in the rock system has been released. This relationship is shown in Figure 3 to follow well defined paths.

This use of shear elements to develop ground reaction curves for a number of field stress conditions on the valve chamber illustrated has been given by Rome (1986). The ground reaction curve for the crown zone only is reproduced in Figure 3. An empirical evaulation of rock load is included for reference.

Lateral field stress is modelled by pre-stressing the rock blocks against rigid lateral supports at a suitable distance from the cavity. Vertical field stress is modelled by similarly applying a vertical prestress but it is considered preferable to provide freedom to the upper limit of the rock model and apply additional loading equivalent to the vertical field stress. All elements are given their gravitational forces. The shear elements are not prestressed and providing the volume of these elements is small they experience the field stress applied.

Pore water or hydrostatic pressure is "injected" into the shear elements where applicable, hence the stresses recorded by shear elements are the "effective" stresses between rock block faces.

The rockbolt design assumptions for the valve chamber are indicated in Figure 3 and are simplistic in nature. Visual observation of the excavation as it proceeded, indicated that the shorter 1.8 rockbolts and 75mm shotcrete might have provided adequate support without the introduction when space permitted, of the longer 4m bolts. However no adequate computational approach was available at that time and knowing that some form of full crown support was probably essential and not an insurance measure, the additional bolting was allowed to proceed.

Excavation was by road header. Crown deflections were substantially less than predicted.

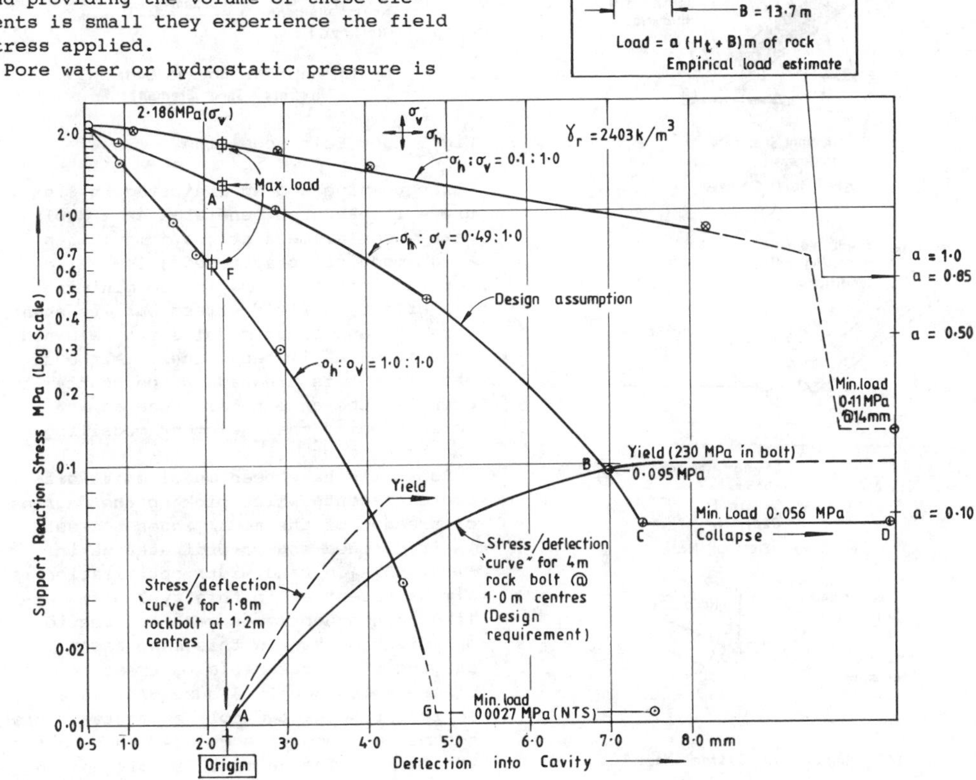

Fig 3 Ground reaction curves for crown of valve chamber

A retrospective assessment of this decision using improved techniques is now being investigated.

The major aspects now introduced into the analysis are the direct modelling of the rockbolts and the shotcrete. Field stress, joint modelling etc. are similar in form to those previously described and published.

4 SHOTCRETE AND ROCKBOLT MODELLING

To model the shotcrete, further use has been made of the shear element. A typical situation is indicated in Figure 4. For the rock type being considered the shotcrete applied to the block face will in general be weaker and more flexible than the intact rock to which it adheres. Its contribution to support in relatively undisturbed rock is consequently associated with joint

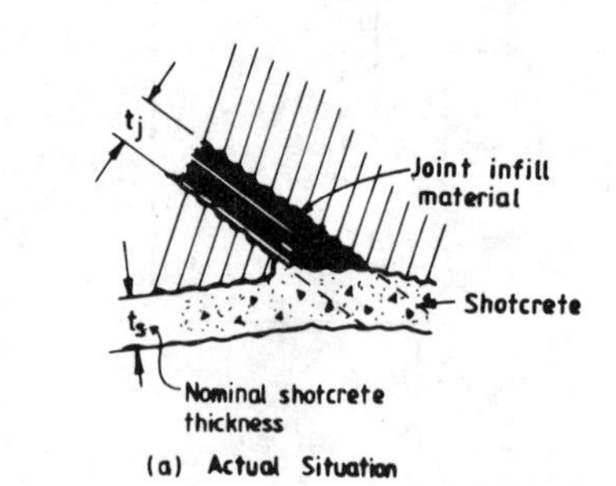

(a) Actual Situation

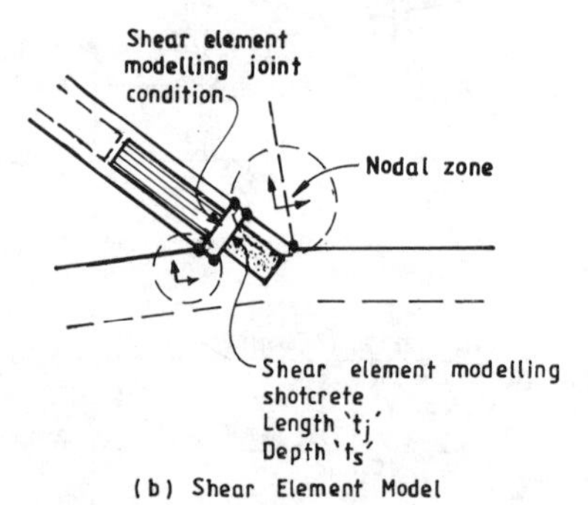

(b) Shear Element Model

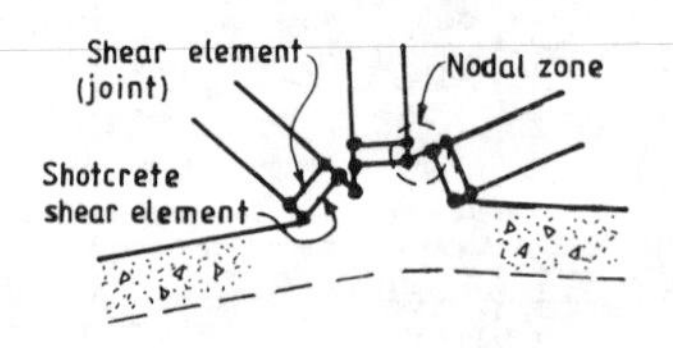

(c) Complex Joint System Model

Fig 4 Shotcrete Modelling

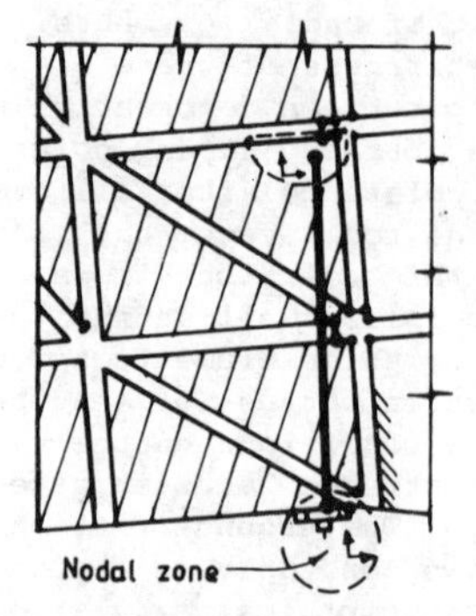

(a) Simple Anchored Rockbolt (Discrete Element)

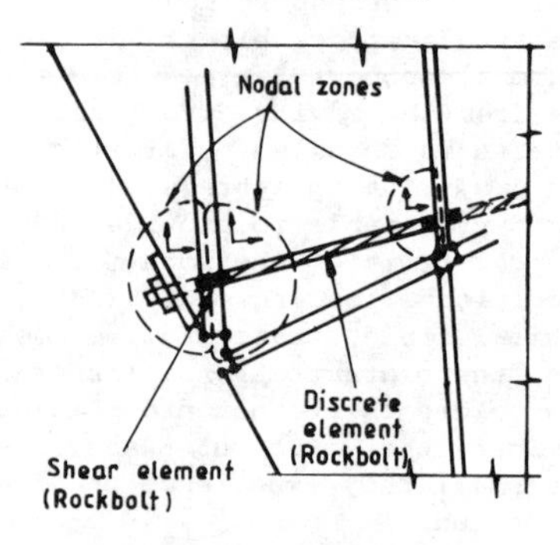

(b) Continuous Grouted Bolt (Discrete + Shear Element)

Fig 5 Rock Bolt Modelling

strengthening and as indicated in Figure 4 it has been generated as a duplicate shear element at the joint. Its axial modulus is assigned a low value for the initial iteration to minimise its effect on field stress but its shear strength and shear modulus are retained at test or estimated values. Although the element is treated independently it occupies the same nodal zones as its corresponding shear element modelling the rock joint.

Rockbolts have been modelled as discrete elements which pick up the degrees of freedom of the nodal zones nearest their defined end co-ordinates as in Figure 5(a). Band width optimisation is almost essential in this type of modelling, consequently as vertical section optimisation has in this case been adopted it is desirable to treat inclined bolts, as in Figure 5(b) as a series of connected bolt segments. This becomes analogous to grouted rock bolt treatment. The segment of bolt on the joint may be treated as a shear element but unlike the shotcrete element its shear modulus in this instance is kept small. Axial properties are retained. The assumption is that the high theoretical shear stresses which might be introduced in the bolt could not be maintained by the rock fabric at the joint face and hence should be excluded from the analysis. The rockbolt both sustains tensile forces and contributes to the intrinsic rock joint strength by maintaining a state of compression.

Failure of shotcrete is treated in the same way as clay gouge in joints. It is provided with "peak" and "residual" friction and cohesion parameters. Given that at any stage in the analysis the shear modulus 'G' and shear stress 't_s' is estimated by the previous analysis iteration, the shear displacement '$\triangle$' is given by:

$$\triangle = \frac{t_s L}{G} \qquad (1)$$

where 'L' is the real or arbitrary joint thickness. The shear strength parameters are made displacement dependent with respect to overstress and joint slip is modelled by proportional reduction in shear modulus. The simple relationship adopted is indicated in Figure 6.

Rockbolt modelling is carried out using strain-dependent moduli techniques. For convenience the rock bolt is inserted in the initial model but by providing a low modulus for the 1st iteration it experiences strain with minimal stress. After the 1st iteration it is activated by introducing its true stiffness. Adopting the treatment given by Rome (1976) for stress-dependent moduli used for tunnel lining design in shattered rock, a stress vs. strain diagram is prepared as shown in Figure 7. From this a strain dependent

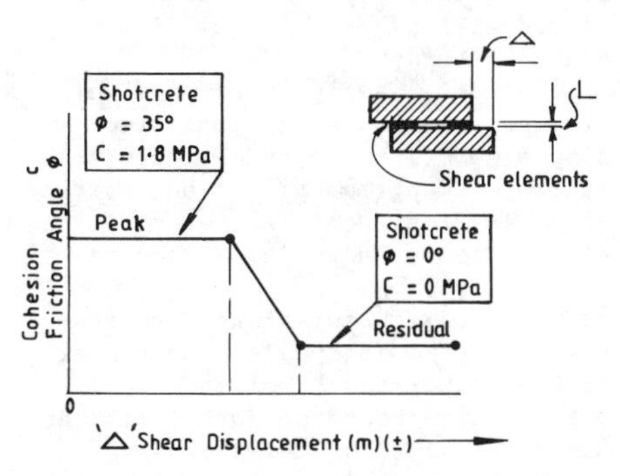

Fig 6 shotcrete - Shear Element Strength Relationship

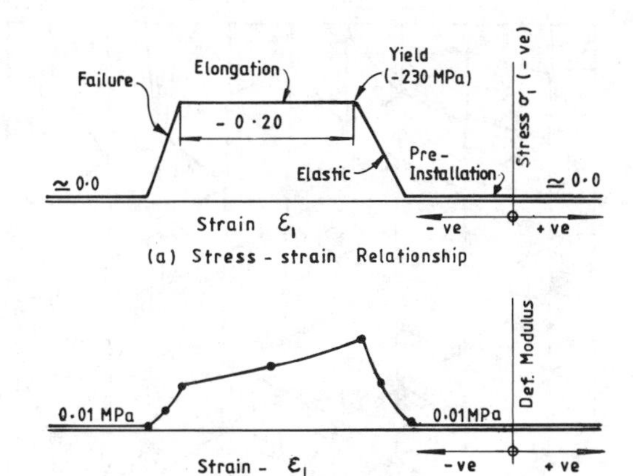

Fig 7 Strain dependent modulus for rockbolt

modulus relationship is prepared. In practice it is convenient to compute this relationship within the computer program using a preliminary (additional) iteration. Damping on rate of moduli change is necessary for both elements.

5 PRELIMINARY ANALYSIS RESULTS

The valve chamber model, as originally used for the ground reaction curve previously introduced, was refined to permit the inclusion of rockbolts as indicated in Figures Nos 1 and 8. The number of rock mass defects included was increased but apart from the major sheared zones present, defect spacing etc. in the model still greatly exceeded that of the actual rock condition.

The effect of the initially introduced 1.8m 25mm dia. rockbolts which had been installed on a 1.2m square grid was first examined. Within the limitations of the model, this rockbolt pattern was incorporated as shown in Figure 8. This Figure also provides some general data on joint strength etc. For the fully excavated cavern rapid failure was indicated, the failure mode being as illustrated. It should be noted that no attempt was made to scientifically locate rockbolts on key blocks etc. as this was not, in general feasible or reliable in practice.

Shotcrete, 75mm thickness, was then introducted into the model but stabilisation of the cavity was not achieved until a shotcrete shear strength of 3.0

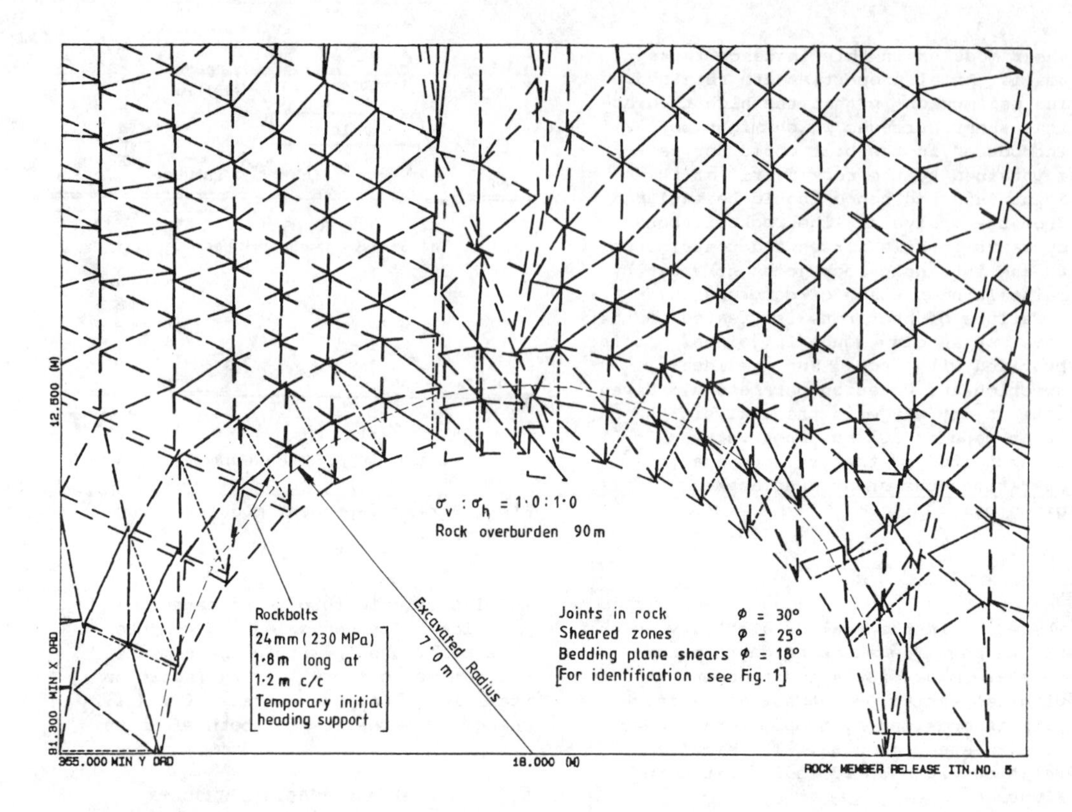

Fig 8 Possible failure mode for crown 1.8m rockbolts @ 1.2m c/c only

MPa approx. and a tensile strength of 1.5 MPa was provided. Although ultimate shotcrete strengths of this order were probably achieved, these would not have been accepted for design purposes.

Water inflow, apart from some damp patches was restricted to the invert, being largely controlled by shaft excavations and access tunnels. However water pressure was introduced into some analyses.

The design requirement called for 4m rockbolts at 1.0m centres placed as soon as space permitted after the initial shotcrete application. Although some reduction in the number of these longer rockbolts was accepted the decision to retain this general requirement appears in retrospect to have been prudent. The study is still proceeding.

REFERENCES

Desai C. S. et.al (1984). Thin layer element for interfaces and joints, Int. Jnl. for Numerical and analytical Methods in Geomechanics. Vol. 8, 19-43

Goodman R.E. (1976). Methods of geological engineering in discontinuous rocks, West Publishing Co., St. Paul.

Proctor R.V. and White T.L. (1968). Rock tunnelling with steel supports, Commercial Shearing and Stamping Co., Youngstown, Ohio.

Rome G. (1976). Equivalent rock column design system for tunnel liners and underground structures incorporating rock restaint. 2nd Australian Conf. on Tunnelling, Aust. Inst. of Mining and Metallurgy.

Rome G. (1982). Analysis of massive blocky rock slopes incorporating simple generated joint elements, Finite element methods in engineering, Proc. 4th Int. Conf. in Australia on Finite Element Methods, Melbourne University.

Rome G. (1986). Equivalent Parameter Method for rockslope dam and tunnel design, Published: Melbourne and Metropolitan Board of Works, Melbourne Australia. ISBN 0 7241 4065 4.

Numerical Methods in Geomechanics (Innsbruck 1988), Swoboda (ed.)
© 1988 Balkema, Rotterdam. ISBN 90 6191 809 X

A comprehensive numerical model of rock joints and its implementation in a finite element model

H.Werner & J.Bellmann
Technische Universität München, FR Germany
S.Niedermeyer, H.Bauer & W.Rahn
Ingenieur-Geologisches Institut Dipl.-Ing. S.Niedermeyer, Westheim, FR Germany

ABSTRACT: The numerical model of rock joints comprises stress-, path- and size-dependent coupling of shear stresses, shear displacements, dilation, normal stresses, joint closure and tension failure. The efficiency of the numerical model and the algorithms are demonstrated using practice relevant examples.

1. Introduction

Deformability and strength of a rock mass are strongly influenced by the presence of discontinuities such as joints and bedding planes. These frequently lead to anisotropic effects with respect to strength and the elastic behaviour of the rock system.

The small or even zero tensile strength, the relatively small shear strength and the surface roughness of these discontinuities are of great significance in assessing the performance and stability of rock structures. The shear strength of smooth dis- continuites is determined by pure friction between the adjacent rock components. The shear strength of rough discontiniuities is moreover influenced by the roughness of the discontinuity surfaces and the strength of the rock material.

In this paper a comprehensive numerical model of rough rock joints is described. It comprises stress- and path-dependent coupling of shear stresses, shear displacements, dilation, normal stresses, joint closure and tension failure and takes the mobilisation or reduction of joint roughness during shearing into account.

The incorporation of the model into a non-linear finite-element program and special numerical techniques for obtaining numerical stability are described. The efficiency of the numerical model and the algorithms are demonstrated by examples.

2. Mechanical behaviour of rock discontinuities

Fig. 1 shows a rock mass element containing a discontinuity loaded by normal and shear stresses. Moreover the local coordinate system used in this paper is defined in fig. 1.

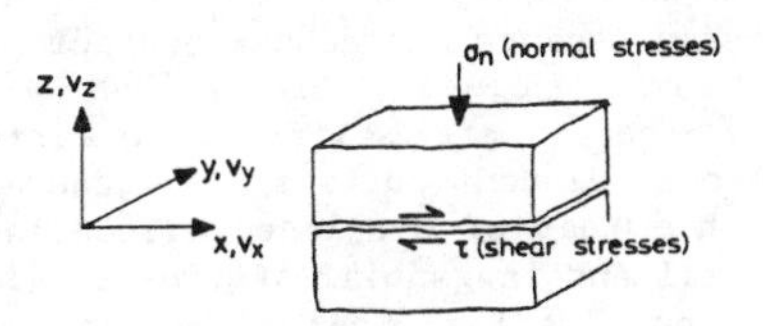

fig. 1: Coordinate system and stresses related to a discontinuity plane

The mechanical behaviour of rough rock discontinuities is schematically explained by fig. 2.

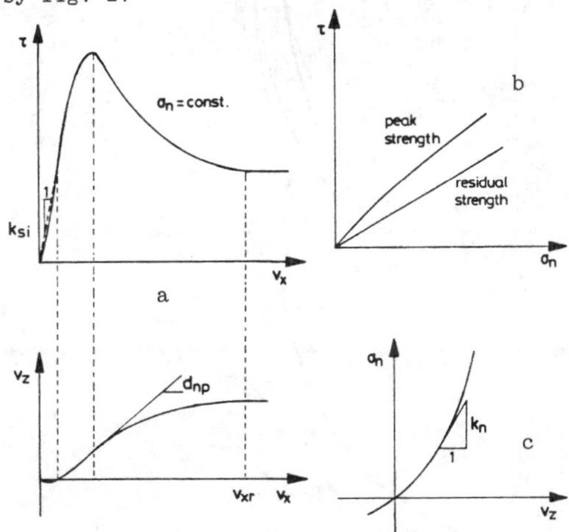

Fig. 2: Mechanical behaviour of rough rock discontinuities
a) Shear strength and dilation in direct shear tests
b) Failure criteria
c) Stress dependency of the normal stiffness k_n

Small compaction and beginning dilation in the first section of the dilation curve of fig. 2a represent the interlocking of the two opposite rock probes. In this stage of shear process a constant (initial) shear stiffness k_{si} can be assumed.

The angle of dilation d_{np} has a maximum at peak shear resistence τ_p. After the residual shear strength of the rock discontinuity is reached, no further dilation occurs ($d_{np}=0$), because the asperities of the discontinuity have been sheared off. The peak shear strength is a function of the frictional properties of the rock and the active strength of the rock, the roughness of the discontinuity, the strength of the rock and the active normal stresses. The residual shear strength, however, depends only on the frictional properties of the rock. Therefore the peak shear strength is described by a non-linear function and the residual shear strength by a linear one (see fig. 2b).

The normal stiffness (see fig. 2c) increases non-linearly with increasing normal stresses because the actual contact area of rough rock joints increases as well. Obviously the normal stiffness of interlocked rock discontinuities is generally greater than that of displaced discontinuities. Normal and tangential stiffness values vary during loading, deloading and reloading, and they change with the number of loading cycles, as it is shown in figs. 3 and 4.

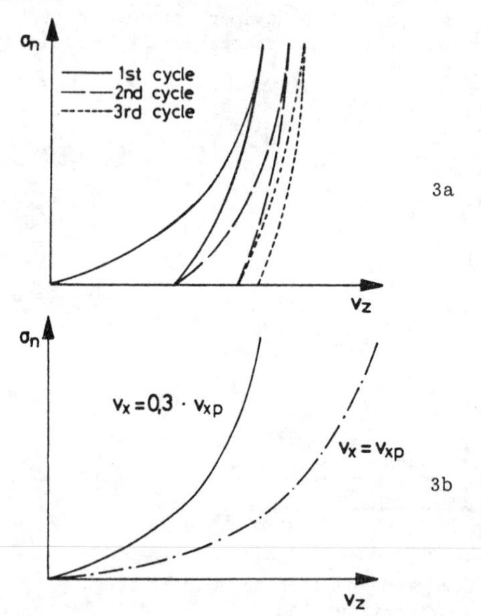

Fig. 3: Variation of normal stiffness (schematic)
a) as a function of loading cycles
b) as a function of shear displacements

Since the mechanical behaviour of rough rock discontinuities is influenced by the geometry (roughness) of the discontinuity plane and the strength of the rock material scale effects should be considered. With an increasing size of a discontinuity the peak shear strength decreases, but the displacement v_{xp} which characterizes the peak shear resistance increases, so that the tangential stiffness decreases.

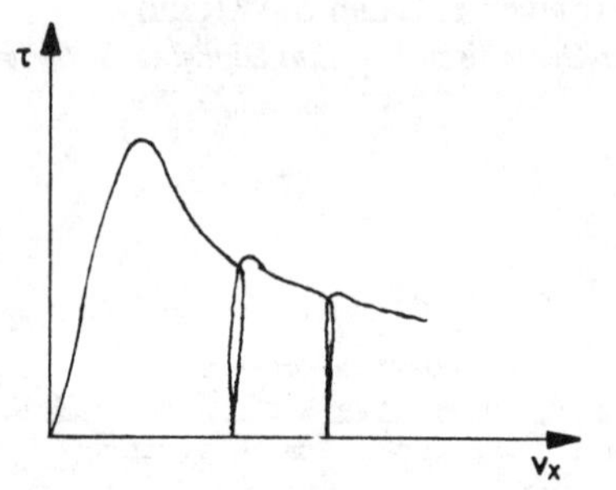

Fig. 4: Variation of tangential stiffness (schematic)

3. Numerical model for the description of the mechanical behaviour of rock discontinuities

3.1 Mathematical formulation of the model

The mathematical formulation of the shear behaviour of rock discontinuities was chosen according to Barton's /1/ formulations. Additionally the normal deformability is described by a hyperbolic normal stress - normal displacement correlation, and a tensile strength of rock discontinuities is allowed.

The modified model of Barton according to /2/ takes the stress dependency of the shear strength into account. The stress and displacement history of a rock discontinuity are considered by using the JRC_m-concept (JRC_m = Mobilized Joint Roughness Coefficient). The failure condition for shear failure is given by equ. (1).

$$f = |\tau| + \sigma_n * \tan(JRC_m * \lg(JCS/|\sigma_n|) + \phi_r) \quad (1)$$

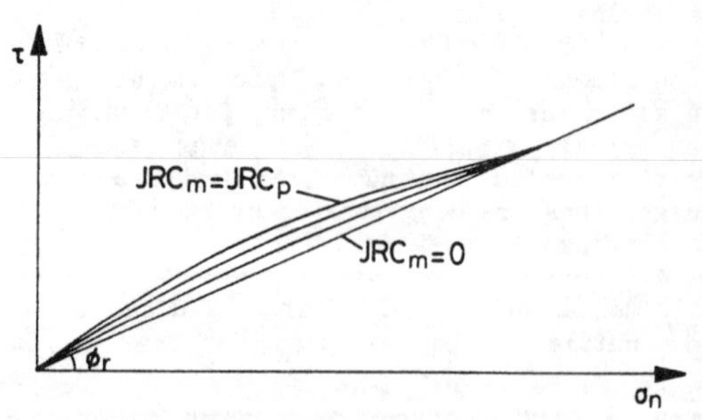

Fig. 5: Failure condition (yield function) for shear failure

Fig. 5 demonstrates the shape of the failure criterion for different values of JRC_m.

The actual mobilized Joint Roughness Coefficient JRC_m can be evaluated from the diagram in fig. 6 using the input parameters JRC_p (which is a "material constant" if the normal stress dependency of JRC_p is neglect) and v_{xp}.

In Fig. 6 the normalized JRC_m/JRC_p, the shear resistance τ and the dilation v_z are schematically shown as functions of the normalized shear displacement v_x/v_{xp}.

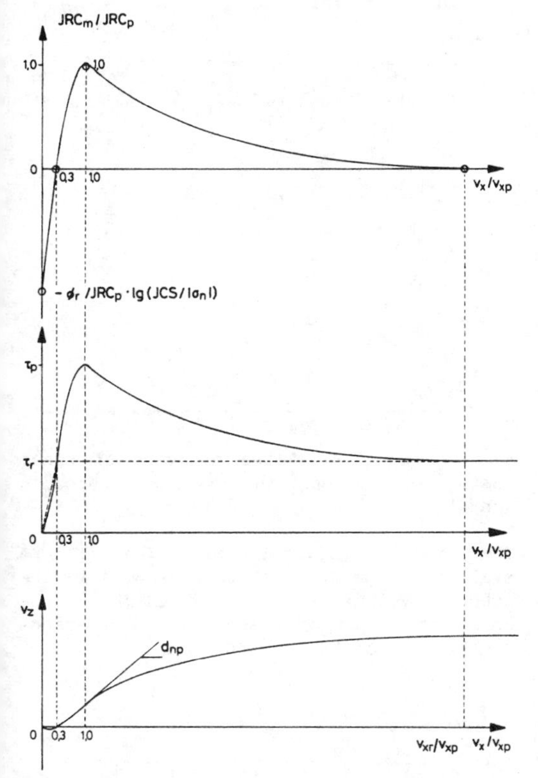

Fig. 6: Barton's model

Barton has found from numerous shear tests of rock joints of hard rocks like granite, basalt, gneiss etc. that in a direct shear test dilation begins with a shear displacement which is approximately 0.3 * v_{xp}. Additional analyses of shear tests of rock joints of relatively weak rocks (mesozoic sedimentary rocks of Germany) done by the author show, that the above described relationship is also valid for discontinuities of weak rocks and, moreover, that the shape of the JRC_m/JRC_p-curve is independent of the rock material and the discontinuity roughness in the pre-peak-stage . On the other hand the course of the JRC_m/JRC_p- curve and the dilation curve vary in the post-peak-stage with different rocks and discontinuity roughnesses. This is contrary to Barton's results.

The irreversible portion of shear deformation is described by using a plastic potential and a non-associated flow rule (utilizing the angle of dilation d_n) for the complete stress-displacement history of the rock discontinuity:

$$g = |\tau| + \sigma_n * \tan d_n \quad (2)$$

$$d_n = A * JRC_m * \lg (JCS / |\sigma_n|) \quad (3)$$

$$dv_{z,pl} = (\delta g / \delta \sigma_n) * d\lambda \quad (4a)$$

$$dv_{x,pl} = (\delta g / \delta \tau) * d\lambda \quad (4b)$$

The parameter A is dependent of the rock type and the discontinuity geometry. Moreover, the authors' investigations have yielded a stress dependency of the parameter A which may be neglected for practical purposes.

The elastic and plastic behaviour of a rock discontinuity having a (small) tensile strength and being loaded with normal stresses is demonstrated by fig. 7.

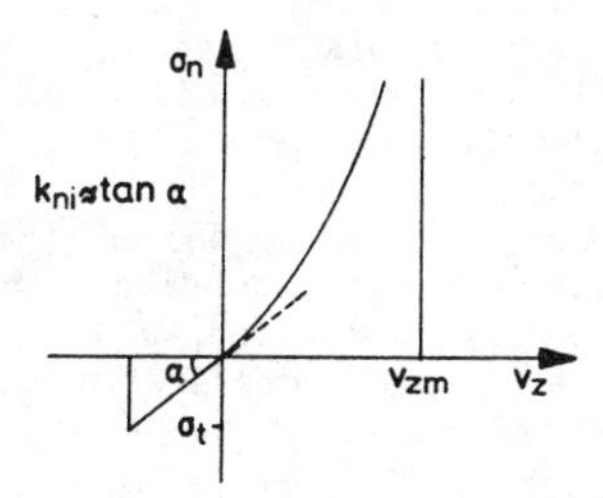

Fig. 7: Normal stress - normal displacement behaviour of rock discontinuity applied for the material model.

After tension failure has occured the tensile strength σ_t is set to zero.

The initial yield function f for tensile stresses normal to the discontinuity plane is given by equ. (5a)

$$f = \sigma_n - \sigma_t \leq 0 \quad (5a)$$

The post-failure yield function reduces to equ. (5b)

$$f = \sigma_n \leq 0 \quad (5b)$$

For tensile stresses the elastic behaviour of the discontinuity is described by the initial normal stiffness k_{ni} whereas a hyperbolic function (equ. (6)) is assumed in a compressive stresses state.

The normal stiffness is a function of the actual vertical displacement v_z, of the initial normal stiffness and of the initial maximum possible closure of the discontinuity v_{zm}.

$$\sigma_n = k_{ni} * v_z / (1-v_z/v_{zm}) \quad (6)$$

$$k_n = k_{ni} / (1-v_z/v_{zm})^2 \quad (7)$$

It is assumed for practical purposes that the parameters of the hyperbolic function k_{ni} and v_{zm} are constant for σ_n=const.

The mathematical model also assumes that the tangential stiffness is independent of the actual shear displacement, independent of the direction of shear displacement and independent of normal stresses, which seems to be an acceptable simplification for practical purposes. The value of the tangential stiffness k_{si} is determined as a secant of the elastic portion of the $k_{ni} - v_x$ - curve between $v_x = 0$ and $v_x = 0.3\ v_{xp}$ (see fig. 6):

$$k_{si} = \sigma_n \tan \phi_r / (0.3\ v_{xp})$$
$$= \tau_r / (0.3\ v_{xp}) \quad (8)$$

3.2 Range of applicability and input data for the discontinuity model

Analyses of direct shear tests of discontinuities (bedding planes and joints) of mesozoic sedimentary rocks and granites of Germany proved the applicability of Barton´s model to these types of rock discontinuities. But they also showed that the application of the model to discontinuities of rocks of very low strength (e.g. silstones with a unaxial compressive strength (σ_c = 1.5 M Pa) and of extremely rough discontinuities of hard rocks (due to high values JRC and JCS the calculation of the over-all angle of friction ϕ_m may give values of ϕ_m = 90° or even greater) is problematic.

Representative results of the analyses are listed in table 1 (the JCS-values used for the analyses were estimated from the results of uniaxial compressive tests on rock samples). The scatter of the listed values for the rock types is due to different surface geometries of the tested discontinuities.

The shear strength function, calculated by inserting the evaluated model parameters into equ. (1) and the results of the individual shear tests are plotted together into the τ-σ -diagram of fig. 8.

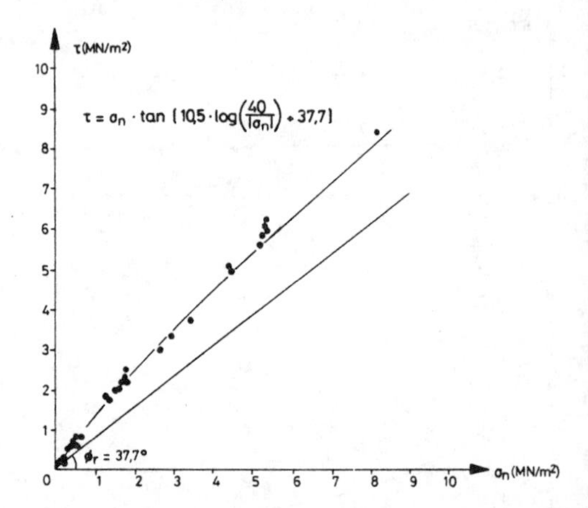

Fig. 8: Results of the analysis of shear tests on discontinuities of mesozoic sandstones

The parameter values listed in table 1 were evaluated from shear tests performed in the laboratory. As mentioned in chapt. 2 the shear behaviour of discontinuities depends on the size of the discontinuity.

Table 1: Input parameters for mesozoic sedimentary rocks and granite

Rock-material	Number of tests	JRC_p	A	v_{xp}	v_{xr}/v_{xp}	k_{ni}	v_{zm}
-	-	-	-	in % of L	-	GPa/m	mm
Sandstone	31	9.0...12.1	0.52...0.96	0.12...0.20	45.2...86.4	12.4...16.9	0.24...0.29
Granite	8	4.3...6.6	0.58...1.02	-	18.8...26.6	19.2...23.6	0.07...0.09
Wellenkalk* (mm 3)	6	6.2...13.4	0.42...0.56	0.71...1.33	7.6...10.6	12.1...23.7	0.19...0.32
Wellenkalk* (mm 1)	6	7.2...13.2	0.17...0.37	-	17.7...26.1	8.7...17.6	0.28...0.42
Graukalk* (mm 1)	5	7.5...15.3	0.18...0.66	-	13.6...19.4	9.1...20.9	0.29...0.46
Limestone* (mm 1)	3	3.0...6.2	0.94...1.20	0.35...0.61	13.2...16.2	5.4...7.2	0.15...0.25

*) different types of limestone of Middle Triassic (Muschelkalk, mu)

Barton et. al. /2/ have found the following empirical correlations:

$$v_{xp} = L/500 * (JRC/L)^{0.33} \quad (9)$$

$$JRC = JRC_0 * (L/L_0)^{-0.02\,JRC_0} \quad (10)$$

$$JCS = JCS_0 * (L/L_0)^{-0.03\,JRC_0} \quad (11)$$

(the index $_0$ characterizes the parameter value determined from laboratory scale tests).

These correlations given by equa. (9), (10), (11) could not be verified by the authors because of lack of data. Therfore in the numerical model equations (9), (10), (11) are not yet activated and the input parameter values for large size discontinuities have to be estimated or taken directly from laboratory scale tests.

Barton et.al. /2/ have found empirical relationships for k_{ni} and v_{zm} for interlocked discontinuities loaded by normal compressive stresses $\sigma_n \leq 1$ kN/m: (this is equivalent to the dead weight of the upper rock sample in a laboratory shear test).

$$k_{ni} = -7.15 + 1.75*JRC_0 + 0.02*(JCS_0/a_j) \quad (12)$$

$$v_{zm} = C * (JCS_0/a_j)^D \quad (13)$$

$$\text{with } a_j = JRC_0/5 * (0.2*\sigma_c/JCS_0 - 0.1) \quad (14)$$

The values for k_{ni} and v_{zm} listed in table 1 were evaluated by using equations (12), (13), (14) and with C = 8.57 and D = -0.68.

These parameter values vary with increasing shear displacements. Our investigations have shown, that a linear reduction of k_{ni} in the interval $0 \leq v_x \leq v_{xp}$ is acceptabe for σ_n=const. The initial normal stiffness $k_{ni,p}$ and the maximum possible joint closure $v_{zm,p}$ in the state of reaching the peak shear strength ($v_x = v_{xp}$) can be estimated by equs. (15) and (16).

$$k_{ni,p} = (1 - 0.05 * JRC_p) * k_{ni} \quad (15)$$

$$v_{zm,p} = (1 + 0.002 * JCS * JRC_p)*v_{zm} \quad (16)$$

The parameter values used in the numerical analyses discribed in chap. 5 were evaluated by using these relation-ships.

4. Implementation of the joint model in the Finite-Element-Program-Chain SET

4.1 The program chain SET

The joint model described above was implemented in the program chain SET /3/ (Structural Engineering and Tunneling) at the Technical University of Munich. SET has been developed for computer aided design in structural engineering and in tunneling. On a finite element basis stresses and strains are computed for two- and three-dimensional structures which may change in progress of the analysis and may have linear or non-linear material properties. Plain stress or plane strain elements, 3D continuum elements as well as shells, beams, spring and boundary elements are available. The sequence of the simulation modules is controlled by the engineer. The GENSET module is a tool for the generation of the input data, NONSET is used for both linear and nonlinear computations, and PLOTSET displays the results graphically. These programs are (among some 15 others) connected by a common data base where the progress of the simulation is updated.

4.2 The joint element

The joint model has been implemented into a joint element. This element is used as a separate boundary element lying between two-dimensional continuum elements which model the rock. It joins the nodes of the continuum on one side with associated nodes on the opposite side (fig. 9). If it is used as a bedding-element, the associated nodes serve as supports. Elements with two pairs (linear displacement) or three pairs of nodes (quadratic displacement) are available.

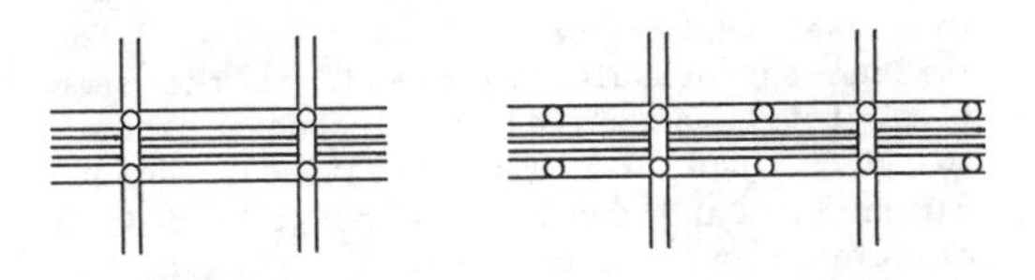

a: linear b: quadratic

Fig.9 Joint element

4.3 The implemented joint model

The joint element is based on a model (extended Barton model) with the following properties:

- stress- and shear-dependent normal and shear stiffness
- non-associative yield law with dilation

4.3.1 Normal stiffness

The normal stiffness (elastic resistance against compression) grows hyperbolically with the compression v_z (equa. 7), starting with stiffness k_{ni}.

At maximal joint closure v_{zm} the elastic problem becomes a contact problem. In order to avoid numerical difficulties, for $v_{zel} > 0.9\ v_{zm}$ we use the value of k_n at $v_{zel} = 0.9\ v_{zm}$ also for higher compression.

4.3.2 Shear stiffness

The shear stiffness changes nonlinearly with shear displacement v_x and normal stress σ_n. Therefore the stiffness $k_s = \tau/v_x$ as a secant modulus is also dependent on the final state. The momentary k_s value is computed by an iteration starting from the state of the previous step. The iteration starts with k_{si} calculated by equa. 8, where, as an approximation, σ_n is estimated by the value corresponding to $v_{zel} = 0.5\ v_{zm}$.

4.3.3 Joint roughness

With given normal stress σ_n and tangential displacement v_x, the activated joint roughness can be computed from the input parameters JRC_p and v_{xp}, using the JRC_m concept. The curve shown in fig. 6 has been implemented as a standard. It represents medium values valid with joints in limestone. However, alternative curves for the related joint roughness may be defined as well.

4.3.4 Plastification

In case that shear stress exceeds the maximum supportable value and, at the same time, total tangential displacement amounts to more than 0.3 v_{xp}, additional deformations Δv_x and Δv_z must be considered in a plastic computation. This effect is described by the plastic potential. An elastoplastic matrix DEP relates the incremental displacements to the incremental change of stress during plastification.

$$(\Delta\sigma , \Delta\tau) = DEP * (\Delta v_z , \Delta v_x) \qquad (17)$$

The normal deformation of the joint element is then a combination of the elastic compression Δv_{zel} and the plastic part Δv_{zpl} resulting from dilation:

$$\Delta v_z = \Delta v_{zel} + \Delta v_{zpl} \qquad (18)$$

4.3.5 Unloading and re-loading

Even after a plastification, the starting stiffness k_{si} is kept if the shear force in the joint decreases. Fig. 10 shows an unloading process, normal stiffness staying constant throughout loading and unloading.

Reloading of the joint in the same direction as first loading results in a deformation following the path of unloading (branch 2 in fig. 10) at first, until plastification starts again (branch 3).

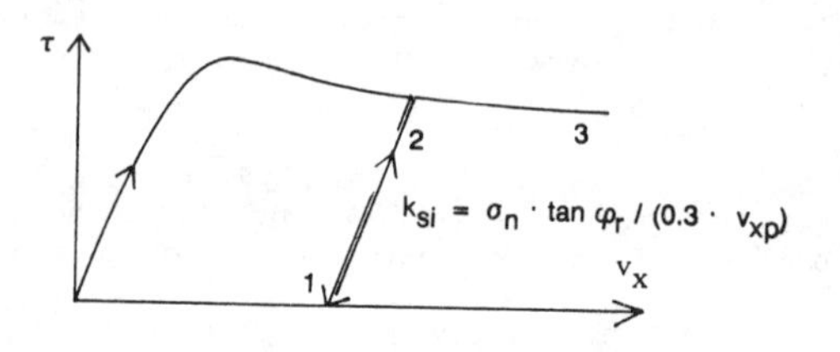

Fig.10 Unloading and re-loading

4.4 Input of parameters

A practical computation of displacements and stresses in rocks with joints requires the following input data:

1. topology and elastic constants of the joint elements
2. normalized JRC_m-curve unless standard curve is to be used
3. joint parameters k_{ni}, v_{zm}, v_{xp}, JRC_p, JCS and ϕ_r

4.5 Algorithm

For the nonlinear computation the load is imposed in several load steps. A computation with the starting values yields the displacements of the structure. Using the constant starting stiffnesses and considering possible primary stresses a first approximation of the stresses is obtained. From the relative displacements of the joint nonlinear stresses can be computed for every node of the joint. The difference of linear and nonlinear stresses yields residual stresses which are integrated over the joint element, thus giving residual forces at the nodes. These residual forces have to be eliminated in an iterative computation, correcting the displacements of the structure in correspondence to the residual forces.

After convergence of the iteration has been established the next load step is performed. Using the momentary stress state a new stiffness can be calculated.

The strong influence of the tangential displacements onto the normal displacements (dilation effect) often causes convergence problems. Therefore some special considerations have to be taken into account in the nonlinear computation.

If the computation is performed in at least two load steps, a significant stabilization of the iteration can be obtained neglecting the influence of dilation in the first load step. The normal stress in the joint converges well and a good estimation for the maximal shear force can be obtained, along with a good starting value for the shear modulus for the following load step iteration.

In further load steps both normal and shear forces are iterated simultaneously.

5. Examples

5.1 Shear experiment

As a first model a simplified shear experiment shall be simulated. A joint element is bedded between two plain strain elements, joined at the two node pairs (fig. 11). The additional spring prevents the slipping away of the upper element when the maximal shear resistance is exceeded.

In order to come close to the actual conditions of the experiment and to demonstrate the behavior of the joint element only vertical loads are used in the first NONSET-run. Thus a constant compression stress in the joint element is obtained. The load is imposed in 20 steps. Fig. 12 shows the nonlinear elastic behavior on the compression range and the linear assumption above 0.9 v_{zm}.

In a second NONSET-run the horizontal load is imposed and thus shear stresses in the joint element are produced.

Fig. 13 shows the shear stress in the joint for the 20 load steps, fig. 14 shows the dilation starting at 0.3 v_{xp}.

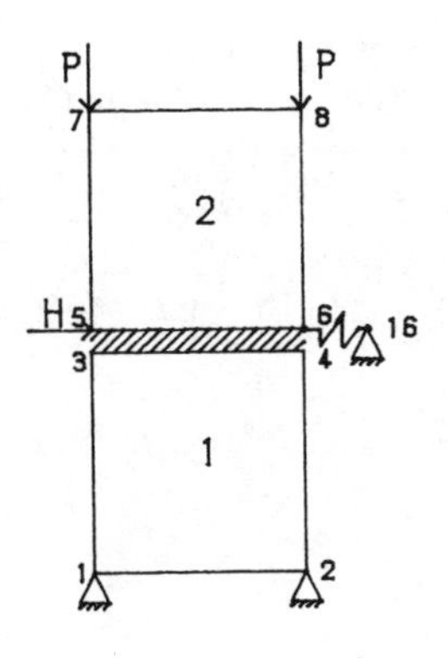

Fig.11 Shear experiment

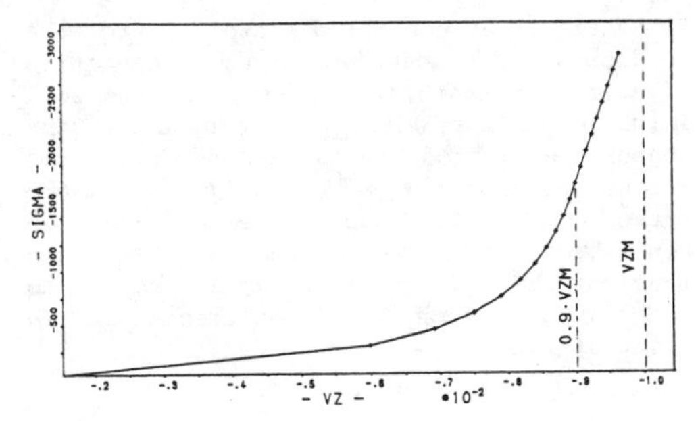

Fig.12 Normal stress (first run)

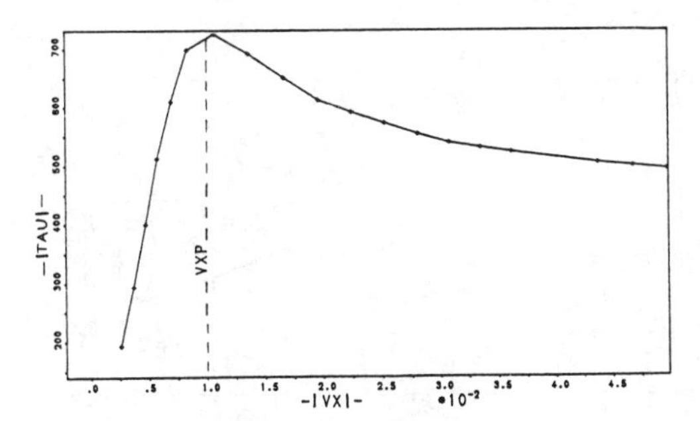

Fig.13 Shear stress

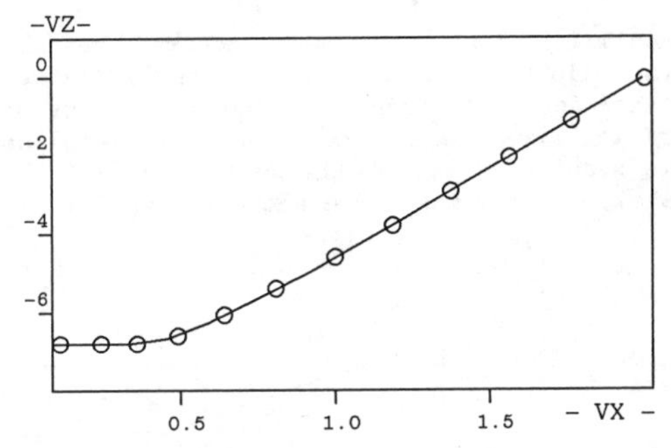

Fig.14 Dilation

5.2 Slope protection

For the construction of a tunnel along a slope by a cut-and-fill method the ground had to be protected by two walls of drilled piles. The rock masses consist of sandstone and siltstone of the Bunter Sandstone (lower triassic). Since the bedding planes are inclined parallel to the surface (slope angle 20 degree), sliding of the slope had to be investigated with special care. The following joint parameters were used: k_{ni}=7000 MN/m^3, v_{zm}=0.01 m, v_{xp}=0.05 m, JRC_p=5.0, JCS=5.0 MN/m^2, ϕ_r=15°. The walls of drilled piles were secured by 8 rows of anchors.

In the finite element model three discrete

discontinuities oriented parallel to the surface were defined. After the computation of a primary state in a first step the soil in front of the walls was removed and the anchors activated in a second step. It turned out that the wall displacements were large (fig. 15, point A-A´) but an equilibrum state was found. The soil in front of the pile was strongly lifted due to the plastification of the elements (B) in front of it.

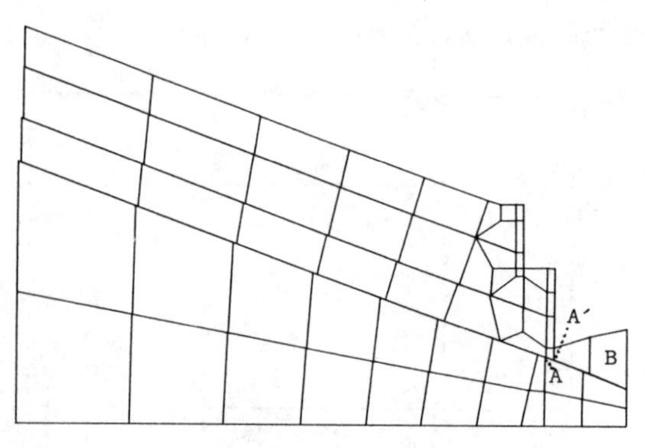

Fig.15 Slope displacements

Maximal shear deformation occured beneath the pile where already a post-failure stage was reached. It turns out that the expansion of the rock masses above the sliding plane is stabilized especially by the peak of the shear-deformation-curve above the wall (fig. 16).

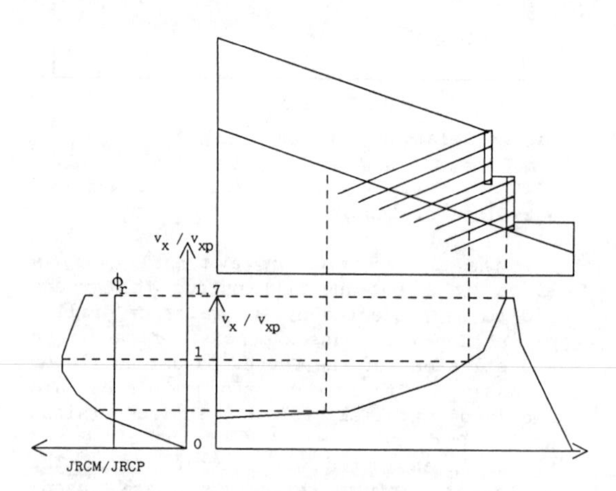

Fig.16 Mobilized joint roughness

REFERÈNCES

/1/ Barton, N. 1976. The shear strength of rock and rock joints. Int. J. Rock. Mech. Min. Sci & Geomech. Abstr., 13, 255 - 279

/2/ Barton, N., Bandis, S. & Bakhtar, K. 1985.Strength, deformation and conductivity. Int. J. Rock. Mech. Min. Sci & Geomech. Abstr. 22, 121-140

/3/ Werner, H. 1985. SET, Technische Entwurfsberechnungen, Fachgebiet elektronisches Rechnen im konstruktiven Ingenieurbau, Techn. Univ. München

Numerical Methods in Geomechanics (Innsbruck 1988), Swoboda (ed.)
© 1988 Balkema, Rotterdam. ISBN 90 6191 809 X

Adaptive creation of kinematical discontinuities

G.Adjedj & D.Aubry
Ecole Centrale de Paris, GRECO Géomatériaux, France

ABSTRACT : In order to capture the discontinuity surfaces where rupture occurs in situ inside geomaterials we need to further the computations until kinematical discontinuities appear. However because the displacement finite element method implies numerical stress vector discontinuities we first need to rub them out by refining the mesh according to the results given by an estimator. Then using an adaptive data structure we can dynamically generate interface elements where the slip criterion is no more available

INTRODUCTION

It is now well accepted that one of the major drawbacks of the classical implementation of the finite element method is the lack of a posteriori estimated error which is the key to a reliable numerical technique. Obviously if we want to capture the discontinuity surfaces where rupture occurs inside materials we need to further the computations until kinematical discontinuities appear. However because the displacement finite element method implies numerical stress discontinuities we first need to rub them out by refining the mesh according to the results given by an estimator. This computation of the error must lead to a dynamical process of refinement so that the user may reach the level of accuracy he is willing to pay through a somewhat optimal path. Unfortunately the current data structure in FEM software is essentially locked at the beginning of the analysis and so is not well adapted to the above process.

The aim of this paper is to discuss the steps involved in the process of a dynamical creation of discontinuities:
- a numerical refinement which must get rid of the stress discontinuities implied by the use of a displacement field method.
- a rheological refinement which must allow the creation of new elements once the kinematic inside the element has been relaxed.

We will present the general setting for the linear case as well as for the non-linear case.

1. THE OBTENTION OF AN OPTIMAL MESH

The optimal mesh we want to get is one which minimizes the numerical discontinuities involved in the displacement field FEM. Therefore we need to construct a reliable estimator of the discretization error which must provide good informations about the areas to refine.

1.1 Error estimation in FEM

In the framework of the elasticity theory we present here a method to compute an approximate a posteriori error which would be called an estimation of the truncation error in the field of ordinary differential equations [2].

Let Ω be the domain occupied by the elastic material. Let $\mathbf{u}$, $\varepsilon(\mathbf{u})$, $\sigma(\mathbf{u})$, $\mathbf{t}(\mathbf{u})$ be respectively the real displacement field, strain and stress fields on Ω and stress vector field on Γ_σ where Γ_σ is the part of the boundary where tractions are imposed, $\mathbf{u}$ satisfies exactly the elasticity equations :

$$\mathbf{Div}\ \sigma(\mathbf{u}) + \mathbf{f} = 0 \qquad , x \in \Omega$$
$$\mathbf{t}(\mathbf{u}) = \mathbf{g} \qquad , x \in \Gamma_\sigma. \tag{1}$$

where **Div** is the classical divergence operator of the equilibrium equations, **f** the given body force vector and **g** is the given surface traction vector on Γ_σ.

Let **V** be the vector field of kinematically admissible displacements i.e. statisfying the following equation on Γ_u the complementary part of the boundary Γ_σ where displacements are imposed :

$V = \{ v / (\sigma(v), \varepsilon(v)) < \infty, v = 0 \text{ on } \Gamma_u \}$,

where $(\sigma(v), \varepsilon(v))$ denotes the dot product for the two tensor fields σ and ε, that is :

$$(\sigma, \varepsilon) = \Sigma_{i,j} \int_\Omega \sigma_{ij}(x) . \varepsilon_{ij}(x) \, dV , \quad i,j=1,3$$

The weak form of the equations of elasticity applied to the domain Ω reads classically as :
Find $\mathbf{u} \in \mathbf{V}$ such that :

$$(\sigma(\mathbf{u}), \varepsilon(\mathbf{v})) = (\mathbf{f}, \mathbf{v}) + \langle \mathbf{g}, \mathbf{v} \rangle \quad \forall \mathbf{v} \in \mathbf{V} \tag{2}$$

where the following notations of the dot product for the vector fields hold :

$$(\mathbf{f}, \mathbf{v}) = \int_\Omega \mathbf{f}(x) . \mathbf{v}(x) \, dV$$

and

$$\langle \mathbf{g}, \mathbf{v} \rangle = \int_\Gamma \mathbf{g}(x) . \mathbf{v}(x) \, dS.$$

The conforming finite element method basically consists in designing a finite dimensional subspace $\mathbf{V_h}$ of the vector space $\mathbf{V}$ such that $\mathbf{V_h} \subset \mathbf{V}$ and to set the weak form on it. Then the approximate solution $\mathbf{u_h}$ solves the following problem :
Find $\mathbf{u_h} \in \mathbf{V_h}$ such that :

$$(\sigma(\mathbf{u_h}), \varepsilon(\mathbf{v_h})) = (\mathbf{f}, \mathbf{v_h}) + \langle \mathbf{g}, \mathbf{v_h} \rangle \quad \forall \mathbf{v_h} \in \mathbf{V_h} \tag{3}$$

Let $\mathbf{e_h}$ be the discretization error defined by :

$$\mathbf{e_h} = \mathbf{u} - \mathbf{u_h} \tag{4}$$

A classical result is the orthogonality of the error $\mathbf{e_h}$ to $\mathbf{V_h}$ with respect to the weak forms directly from Eq. (2) and (3) :

$$(\sigma(\mathbf{e_h}), \varepsilon(\mathbf{v_h})) = 0 \quad \forall \mathbf{v_h} \in \mathbf{V_h} \tag{5}$$

From the fact that $\mathbf{u_h}$ does not verify exactly the Eq (1) it is possible to define a residual force consisting of two parts $(\mathbf{f_h}, \mathbf{t_h})$:

$$\begin{aligned} \mathbf{Div}\, \sigma(\mathbf{u_h}) + \mathbf{f} &= \mathbf{f_h}, & x &\in e_i \\ \mathbf{t}(\mathbf{u_h}) - \mathbf{g} &= \mathbf{t_h}, & x &\in \Gamma_\sigma \cap \partial e_i \\ \mathbf{t}_i(\mathbf{u_h}) + \mathbf{t}_j(\mathbf{u_h}) &= \mathbf{t_h} & x &\in \partial e_i \cap \partial e_j \end{aligned} \tag{6}$$

where e_i is the element number of the triangulation and ∂e_i is the boundary of this element. Thus we get :

$$(\sigma(\mathbf{e_h}), \varepsilon(\mathbf{v})) = -\Sigma_i (\mathbf{f_h}, \mathbf{v})_{ei} - \Sigma_i \langle \mathbf{t_h}, \mathbf{v} \rangle_{\partial e} \quad \forall \mathbf{v} \in \mathbf{V} \tag{7}$$

From (6) and (7) we deduce that :

$$\Sigma_i (\mathbf{f_h}, \mathbf{v_h})_{ei} + \Sigma_i \langle \mathbf{t_h}, \mathbf{v_h} \rangle_{\partial ei} = 0 \quad \forall \mathbf{v_h} \in \mathbf{V_h} \tag{8}$$

On one hand what is interesting if we compare (7) and (8) is that the residual forces do no work over the field $\mathbf{v_h}$ itself but in the virtual work sense a richer field like $\mathbf{v}$ is able to detect the local equilibrium defect of $\mathbf{v_h}$ [1].
On the other hand we can see that the computational discretization error is the sum of a quantity corresponding to the residual inside an element and of the sum of the stress vector jumps (normal and tangential) at the interface between two elements [7] [8]. Numerically the first term is often negligible with respect to the second one the sake of brievety we only consider for a computational point of view the jump of stress vector.

This error analysis had two aims :
(i) To estimate a computable error which will indicate whether some further refinement is necessary; this can be done using the estimator of the error given by the equation (8);
(ii) To locate the regions where to refine the mesh; the error indicator will be merely the average of the stress jumps across each element's edges.
So the error estimator will be simply the sum of the element error indicators.
Let η_i denote the error indicator in the element e_i and η_{max} be the maximum value of η_i, then an element e_i will be refined if η_i is greater than $C\eta_{max}$. At this stage C has been chosen to be 0.25 and this has shown to be very efficient in getting a uniform distribution of the error.

1.2 The adaptive mesh refinement with hierarchical basis functions

The chosen technique for refinement is the "h-version" method that carries out refinement by subdivision of elements which moreover are hierarchical : the shape fucntions of the "root" degrees of freedom remain unchanged through the refinement process [1][7][8]. We will give in this section a summary of all the computational steps involved for a full serie of successively embedded meshes.

1.2.1 The expansion of the Hierarchical Finite Element Method (HFEM).

First in the refinement process the generated finite dimensional vector spaces $\mathbf{V}_{h\alpha}$ where α stands for the level of discretizationwill all realize an internal approximation of the vector space $\mathbf{V}$ and the following inclusions will hold :

$$\mathbf{V}_{h1} \subset \mathbf{V}_{h2} \subset \ldots \subset \mathbf{V}_{h\alpha} \subset \mathbf{V}_{h\alpha+1} \subset \ldots \subset \mathbf{V}_{h\nu} \subset \mathbf{V}.$$

In the HFEM if $\mathbf{w}_{\alpha Ii}$ is the basis function created at the level α for the node I in the spatial direction i (i=1 to 3), the subsequent inclusion holds :

$$\mathbf{w}_{\alpha Ii} \in \mathbf{V}_{h\alpha}.$$

Then the approximate displacement field at the level ν, $\mathbf{u}_{h\nu}$ is expanded as follows :

$$\mathbf{u}_{h\nu} = \Sigma_{\alpha Ii}\, u_{\nu\alpha Ii}\, \mathbf{w}_{\alpha Ii}$$

with a natural notation for $u_{\alpha Ii}$ the components of $\mathbf{u}_{h\nu}$ with respect to the basis.

It is clear that in the HFEM the previously created basis functions are still active and are computed only when they are created. The $u_{\alpha Ii}$ appear thus as correction terms to the preceeding values. This is in opposition to the standard FEM where the support of the basis functions intersect only over the elements whose one local node corresponds to the global node of the basis function. In the latter the computations inside one element are completely localized and limited to the local basis functions of this given element nodes.

1.2.2 The HFEM linear system

At a given ν-level the linear system to be solved may be written:

Find $u_{\nu\alpha Ii}$ such that :

$$\Sigma_{\alpha Ii}\, u_{\nu\alpha Ii}\ (\sigma(\mathbf{w}_{\alpha Ii}), \varepsilon(\mathbf{w}_{\beta Jj})) = (\mathbf{f}, \mathbf{w}_{\beta Jj}) + \langle \mathbf{g}, \mathbf{w}_{\beta Jj} \rangle$$

$$\alpha,\beta \in (1 \ldots \nu),\ I \in (1 \ldots N_\alpha),\ J \in (1 \ldots N_\beta),\ i,j \in (1 \ldots 3) \qquad (9)$$

where N_β is the number of nodes at the β-level. Of course the different integrals appearing in the above system are classically splitt-up over each element E_γ at the highest level of refinement where it occurs so that a typical term is the following :

$$\Sigma_{E\gamma}\, (\sigma(\mathbf{w}_{\alpha Ii}), \varepsilon(\mathbf{w}_{\beta Jj}))_{E\gamma}$$

where E_γ indicates that this element has been created at level γ.

In the HFEM due to the above-mentionned element coupling two loops are necessary. The first loop through the elements is the same as in the standard FEM : it corresponds to the decomposition of the integral over each element in the finest mesh as these elements realize a partition of the domain. Then if $\alpha<\nu$ and $\beta<\nu$ the coefficient has already been computed at a lesser refined level. Else this coefficient must be computed so that this first loop is actually limited to newly created elements.
The global stiffness matrix has a bordered block-diagonal structure and at each new level we add to it a new diagonal matrix corresponding to the standard stiffness coefficients of the newly created elements and a border matrix corresponding to the interactions between the newly created elements and those issued from previous refinements. This border matrix can be splitt-up in several blocks as shown below.

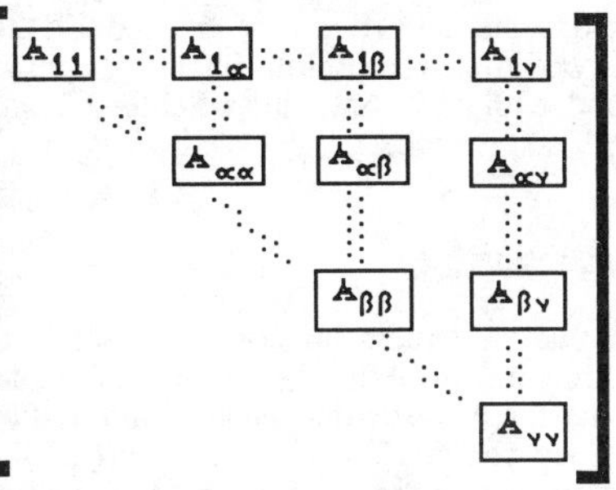

$$\begin{bmatrix} A_{11} & \cdots & A_{1\alpha} & \cdots & A_{1\beta} & \cdots & A_{1\nu} \\ & \ddots & \vdots & & \vdots & & \vdots \\ & & A_{\alpha\alpha} & & A_{\alpha\beta} & & A_{\alpha\nu} \\ & & & \ddots & \vdots & & \vdots \\ & & & & A_{\beta\beta} & & A_{\beta\nu} \\ & & & & & \ddots & \vdots \\ & & & & & & A_{\nu\nu} \end{bmatrix}$$

Figure 1 : block structured matrix

The linear system (9) to be solved should certainly not be solved by a direct method as the whole stiffness matrix would have to be triangularized for each level. An iterative method very closed to a multi-grid algorithm is preferred and is proposed in the next section which will be aesily generalized to the nonlinear case [1][6].

As explained before the implementation of this algorithm implies the computation of the interaction terms between "senior" and "junior" shape functions when their support contains the same element. This can be done efficiently only by using an adequate data structure, that is a data tree structure , which is able to give for any node of the tree the relation of inclusion in the sense of element area with the other nodes situated above or below on the tree [1][3][4].

1.2.3 Iterative solution of the HFEM linear system

The block structure of the HFEM matrix suggests naturally an iterative algorithm which would take benefit of such a structure.

For the sake of simplicity the imposed traction vector **g** will be assumed now to vanish.

Let consequently $\mathbf{u}_v^n$ be defined as an approximate field for $\mathbf{u}_{hv}$ by the following iterative scheme at iteration number n:

a. Assume $\mathbf{u}_v^{n-1}$ to be known.

b. Compute $\mathbf{u}_v^n$ such that :

$$(\sigma(\mathbf{u}_{v\beta}^n), \varepsilon(\mathbf{w}_{\beta Jj})) = (\mathbf{f}, \mathbf{w}_{\beta Jj}) - \sum_{\alpha<\beta} (\sigma(\mathbf{u}_{v\alpha}^n), \varepsilon(\mathbf{w}_{\beta Jj})) - \sum_{\alpha>\beta} (\sigma(\mathbf{u}_{v\alpha}^{n-1}), \varepsilon(\mathbf{w}_{\beta Jj}))$$

$\alpha,\beta \in (1 \dots v)$, $I \in (1 \dots N_\alpha)$, $J \in (1 \dots N_\beta)$,

$i,j \in (1 \dots 3)$ (10)

The whole system will not be solved this way because it would be too prohibitive but a multigrid-like method with an initial Gauss-Seidel relaxation scheme will be prefered.

1.2.4 Some results

We present below some results on a surface load on a semi infinite massif. After the initial mesh input by the user the three following meshes automatically computed are shown. Then the isovalues for the vertical stress is sketched in for the second and the fourth grid.

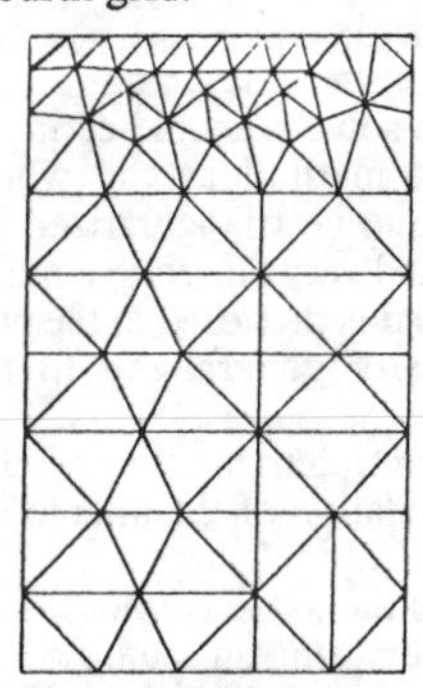

Figure 2: initial mesh
138 dof

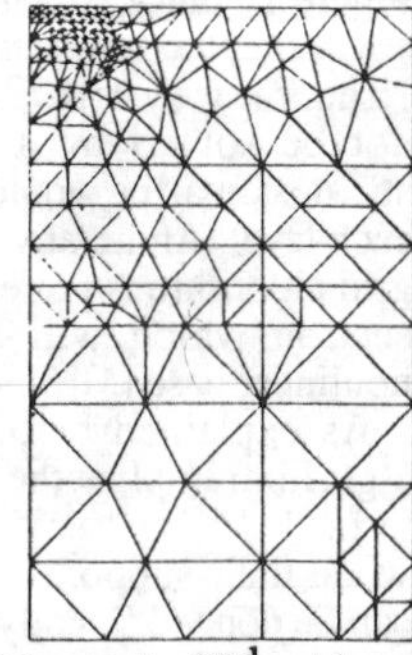

Figure 3 : 2nd grid
346 dof

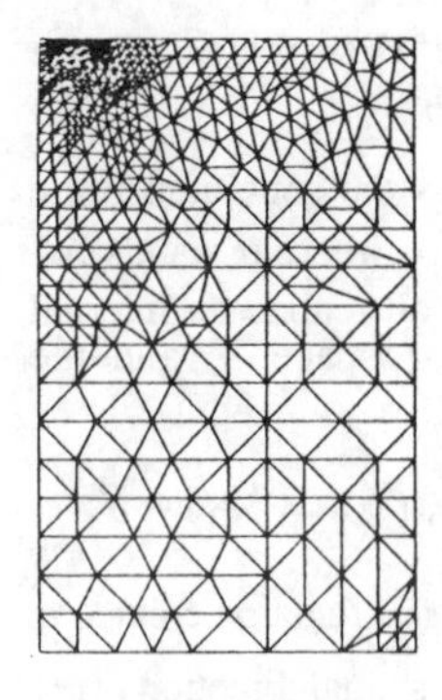

Figure 4: 3rd grid
938 dof

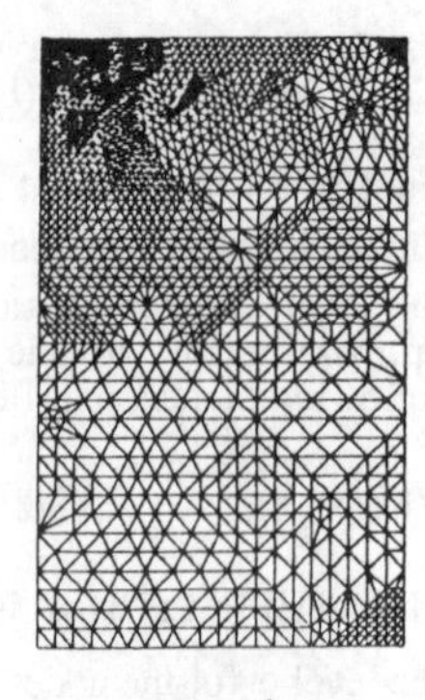

Figure 5 : 4th grid
4316 dof

Figure 6 : vertical stress isovalues

a) for the 2nd grid b) for the 4th grid

2.ALGORITHM IN THE CASE OF NON-LINEAR PROBLEM

This section is concerned with the use of the previous numerical methods in the case of non-linear problems. We still assume that the domain Ω is a polygon of $\mathbf{R}^3$ and for the sake of simplicity that there are no traction applied on the boundary Γ of the domain.

To fix ideas we describe the procedure in the case of plasticity :

$$\mathbf{Div}\ \sigma + \mathbf{f} = 0$$

$$d\sigma/dt = \mathbf{C}^{ep}(\sigma, d\varepsilon/dt) \cdot d\varepsilon/dt$$

where $\mathbf{C}^{ep}$ stands for the elasto-plastic matrix.

The system should be solved by an iterative method on the time interval [0,T]. At each time step t_n, $\mathbf{u}_n$

and σ_n are known. At the time step t_{n+1} using a weak formulation of the system of equations the problem is :

Find $\mathbf{u}_{n+1}, \sigma_{n+1}$ such that :

$$0 = (\mathbf{f}_{n+1}, \mathbf{v}) - (\sigma_{n+1}, \varepsilon(\mathbf{v}))$$
$$\sigma_{n+1} - \sigma_n = \mathbf{C}^{ep}(\sigma_{n+1})(\varepsilon(\mathbf{u}_{n+1}) - \varepsilon(\mathbf{u}_n)) \qquad (13)$$

The solution will be found using an iterative algorithm similar to the one presented before. Let the superscript m denote the number of the iteration step, $\mathbf{u}^m{}_{n+1}$ and $\sigma^m{}_{n+1}$ are known , the problem is to find $\mathbf{u}^{m+1}{}_{n+1}$. The operator of the iterative scheme will be the elastic part of the matrix $\mathbf{C}^{ep}$ named the auxiliary operator $\mathbf{C}^*$. The weak formulation (14) can still be written :

$$(\mathbf{C}^* \varepsilon(\mathbf{u}_{n+1} - \mathbf{u}_{n+1}), \varepsilon(\mathbf{v})) = (\mathbf{f}_{n+1}, \mathbf{v}) - (\sigma_{n+1}, \varepsilon(\mathbf{v})) \qquad (14)$$

i.e at the iteration step m :

Find $\mathbf{u}^{m+1}{}_{n+1}$:

$$(\mathbf{C}^* \varepsilon(\mathbf{u}^{m+1}{}_{n+1} - \mathbf{u}^m{}_{n+1}), \varepsilon(\mathbf{v})) = (\mathbf{f}^m{}_{n+1}, \mathbf{v}) - (\sigma^m{}_{n+1}, \varepsilon(\mathbf{v})) \qquad (15)$$

Once $\mathbf{u}^{m+1}{}_{n+1}$ is computed we just need to get $\sigma^{m+1}{}_{n+1}$through the constitutive equations:

$$\sigma^{m+1}{}_{n+1} = \sigma_n + \mathbf{C}^{ep}(\sigma^{m+1}{}_{n+1})(\varepsilon(\mathbf{u}^{m+1}{}_{n+1}) - \varepsilon(\mathbf{u}_n)) \qquad (16)$$

Obviously this algorithm is very closed to the one presented in the previous section in Eq (10) which is well adapted because it is able to handle a right hand side with an energy term. The difference stands in the fact that the unknown of the iterative scheme is the increment of displacement $\Delta\mathbf{u}$ and that we need to update the stress tensor at each step.

3.ADAPTIVE CREATION OF DISCONTINUITIES

3.1 Physical motivations

A lot of progresses have been made in the area of constitutive equations along a priori given kinematic discontinuity surfaces. In most of the cases in Geomechanics analysis people use interface elements where the constitutive equations are applied.

In some of existing softwares it is possible either at the beginning of the localization or when discontinuity surfaces have been developped to start such a process. However because the discontinuity creation process is mainly dynamic it needs a non-fixed mesh, so that only an adaptive algorithm which is able to create and eventually remove elements can fit with this aim.

An important theoritical research has been done which allows a detailed investigation of the volume model/ surface model relations together from the development of the strain speed jump till the displacement jump.

Basically the idea is to find out in each element of the mesh whether a slip criterion is still available and then, if it is necessary, to create an interface element with its own behaviour equations and modelling the slipping phenomenon. The technique is not to represent some a priori discontinuity surfaces but to track a discontinuity while it is propagating.

3.2 Interface element creation

The basic idea is to loop over the whole mesh to check out each element individually and to compute if necessary the direction of the normal to the discontinuity surface. Then an interface element is created if it is required along the indicated direction and for the sake of simplicity starting at the midpoint of one of the edge of the element as shown on Figure 7 :

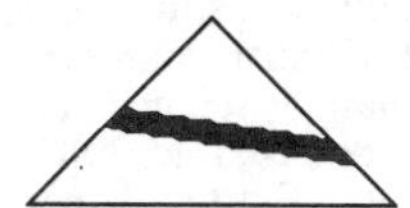

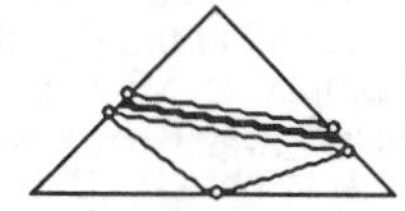

Figure 7: creation of an interface element.

Because we need to keep on working on triangular elements the part of the "macro"-triangle which results to quadrilater is to be subdivided. Of course the design of the new elements must be controlled by a quality measure of the triangle shape. This algorithm will be the one used for the mesh generator. An interface element has its own constitutive equations and is defined as a lineic finite element by two nodes at each end which have different numbers but the same coordinates.

In order to still be able to handle the continuity required by a classical finite element method we will not interpret the interface elements as stand alone elements but as a contribution to a weakening of the stiffness of the element where a slipping has been pointed out. For that purpose we must write the weak formulation of each macro-element having an interface element and take into account the virtual work on this interface that will be named Σ_e. So that we get at time step t_{n+1} :

$$0 = (\mathbf{f}_{n+1}, \mathbf{v})_e - (\sigma_{n+1}, \varepsilon(\mathbf{v}))_e - <\tau_{n+1}, [\mathbf{v}] >_{\Sigma e} \quad (17)$$

where [**v**] denotes the displacement jump and where the constittutive equation of the interface element is given by :

$$d\tau/dt = \mathbf{C}^{ep}{}_{\Sigma e} \; [d\mathbf{u} /dt] \quad (18)$$

The whole solution of the problem with discontinuity may be solved by the general adaptive and hierarchical software in a similar fashion as the one described before.

4 CONCLUSIONS

The techniques presented for the measure of the discretization error and adaptive refinement have shown good results for the linear case. A theoritical study has been performed for the non-linear case which will be soon implemented in the existing software. Then the simulation of kinematical discontinuities which is under progress should be completed as an application of the non-linear case.

BIBLIOGRAPHY

[1] Adjedj G., Aubry D., Development of a hierarchical and adaptive finite element software, to be published at the Eighth International Conference on Computing Methods in Applied Sciences and Engineering, INRIA, Versailles, France, Dec. 1987.
[2] Axelsson O., Barker V.A., Finite element solution of boundary problems, Academic Press (1984) .
[3] Bank R., Sherman, Weiser A., Refinement algorithms and data structures for regular local mesh refinement, Scient. Comp, (1983) .
[4] Bank R., PLTMG User's guide, Technical report, University of California, San Diego, (1981) .
[5] Demkowicz L., Oden J.T., Strouboulis T., Adaptive finite elements for flow problems with moving boundaries, part1, Comp. Meth. Appl. Mech. Eng., 46, 217-251, (1984) .
[6] Stuben K., Trottenberg U., Muligrid Methods: Fundamental Algorithms, Model Problem Analysis and Applications, Multigrid Methods, Ed by Hackbusch and Trottenberg, (1981) .
[7] Zienkiewicz O.C., Kelly D.W., The hierarchical concept in finite element analysis, Comp. Struct., 16., (1983) .
[8] Zienkiewicz O.C., Kelly D.W., de S.R. Gago J.P., Babuska I. , 'Hierarchical' finite element approaches, error estimates and adaptive refinement,MAFELAP 1981,pp 313-346.

Numerical Methods in Geomechanics (Innsbruck 1988), Swoboda (ed.)
© 1988 Balkema, Rotterdam. ISBN 90 6191 809 X

Modelling of contaminant movement through fractured or jointed media with parallel fractures

R.K.Rowe
Geotechnical Research Centre, Faculty of Engineering Science, University of Western Ontario, London, Canada
J.R.Booker
School of Civil and Mining Engineering, University of Sydney, Australia

ABSTRACT: A solution for advective-dispersive contaminant movement through a system of parallel cracks is developed for the case where the mass of contaminant available for transport into the rock is assumed to be finite and, as a consequence, the concentration of contaminant in the source decreases with time. The solution is developed for assumed one-dimensional advective-dispersive transport along the fractures and one-dimensional diffusive transport into the rock matrix perpendicular to the fractures. The analytical technique involves taking the Laplace transform of the governing equations, boundary conditions and initial condition, finding a closed form solution in transform space and then numerically inverting the transform. The numerical inversion can be readily performed on a micro-computer.

The application of the technique is illustrated and it is shown that the mass of contaminant can have a significant effect on the attenuation of contaminant. It is also shown that the rate of contaminant transport may be very much slower than that deduced from simple consideration of the seepage velocity along the fracture.

1 INTRODUCTION

Waste disposal sites have been, and are being, constructed in fractured rock formations. The evaluation of potential impact of these facilities on the ground water system involves consideration of the mechanisms of predominantly advective-dispersive transport along the fractures and diffusive transport into the matrix of the rock between the fractures. Various investigators have considered this problem. For example, Neretnieks (1980) and Tang et al. (1981) developed a solution for 1D contaminant transport along a single fracture together with 1D diffusion of contaminant into the matrix of the rock adjacent to the fracture. Sudicky and Frind (1982) extended this solution for the case of multiple parallel fractures. Again, they provide an analytic expression for the concentration of contaminant at any point along the fracture, however evaluation of this expression involved the numerical evaluation of two infinite oscillating integrals which requires considerable care and a substantial amount of computer time. Barker (1982) also presented results for contaminant migration along an infinite series of parallel cracks. A primary difference between the work by Sudicky and Frind and Barker was the fact that in the former case, evaluation of the solution involved a double numerical integration while in the latter case, evaluation involved numerical inversion of the Laplace Transform. In each case it was assumed that:

1. The width of each fracture is much smaller than its length.
2. Transverse diffusion and dispersion within each fracture assures complete mixing across its width at all times.
3. The permeability of the intervening porous matrix is low and transport within the matrix is mainly by molecular diffusion.
4. Transport along each fracture is much faster than transport within the matrix.
5. The source concentration remains constant for all time (i.e. the source of contaminant is effectively infinite).

In parallel with the development of the analytic solution discussed above, others have developed numerical solutions for contaminant transport in fractured specimens. For example, Grisak and Pickens (1980) obtained a finite element solution for the same problem but without the restriction of assuming constant concentration across the width of the crack or one-dimensional migration in the crack and matrix system. Tang et al. (1981) compared the results obtained from their analytic solution with Grisak and Pickens's (1980) Finite Element results and found excellent agreement between the two solutions even for the case where there was large diffusive loss into the matrix (and consequently a short penetration distance along the fissure). They concluded that this good agreement was strongly supportive of the validity of the assumptions underlying the general analytical solution. The advantage of the analytical solution being that it does not suffer from numerical difficulties associated with high advective velocities along the fracture (numerical dispersion) or with high concentration gradients into the matrix of the rock.

From the foregoing, it may be concluded that considerable progress has been made in the development of analytical solutions for the case where the source of contaminant is effectively infinite.

In their previous work, the present authors have developed a series of theoretical solutions for contaminant migration in a continuous (but layered) porous medium for 1D, 2D and 3D conditions (eg. Rowe and Booker, 1985a, b; 1986, 1987). This approach has involved the use of Laplace and, where appropriate, Fourier transforms to obtain an analytic solution in transformed space. It then remains to numerically invert the Laplace (and Fourier) transforms. In this paper, the authors extend this work to consider the migration of contaminant in a parallel system of cracks. The work parallels that of earlier investigators (eg. Tang et al., 1981; Barker, 1982; Sudicky and Frind, 1982) insofar as it is assumed that the contaminant transport can be regarded as one dimensional along the cracks and one dimensional into the matrix perpendicular to the fractures. However, in this paper, it will not be assumed that concentration of contaminant in the source remains constant, but rather, the more realistic assumption of finite mass of contaminant will be adopted. An analytic solution will then be obtained in transform space and it simply remains to numerically invert the Laplace Transform. Using a modern algorithm for inverting the Laplace Transform, it turns out that this approach is both less restrictive and numerically far more efficient than the Sudicky and Frind solution where they analytically invert the Laplace Transform and then numerically evaluate two integrals (assuming constant source concentrations). The implication of considering the finite mass of contaminant will be examined.

2 THEORY

Consider the situation shown schematically in Fig. 1. Initially, a fixed mass of contaminant is placed in a landfill, the material adjacent to the landfill consists of a matrix of horizontal layers, of depth 2H, interspersed by a series of equally spaced horizontal fissure planes with thickness 2b.

It will be assumed that the fissures are parallel to the xy plane and that the fluid in the fissure moves with a velocity v_f in the x direction. Neglecting possible sorption on the face of the cracks, the equation for advective-dispersive transport of the contaminant along a typical fissure is:

$$n_f D_f \frac{\partial^2 c_f}{\partial x^2} - n_f v_f \frac{\partial c_f}{\partial x} = n_f \frac{\partial c_f}{\partial t} + \frac{F}{b} \qquad (1)$$

where c_f is the concentration of the contaminant within the fissure;
n_f is the porosity of the fissure material if the fracture is "open" (i.e. free of material impeding flow along the fracture) then $n_f = 1$;
v_f is the average linearized groundwater velocity along the fracture;
D_f is coefficient of hydrodynamic dispersion (incorporating the effects of molecular diffusion and mechanical dispersion) in the fissure;
F is the flux leaving the fissure and moving into the matrix.

It will be assumed that there is no advection in the matrix and that, to sufficient approximation, all diffusion occurs in the vertical (z) direction. It then follows that

$$n_m D_m \frac{\partial^2 c_m}{\partial z^2} = n_m \frac{\partial c_m}{\partial t} + \rho_m K_m \frac{\partial c_m}{\partial t} \qquad (2)$$

where c_m is the concentration of contaminant in the matrix
n_m is the porosity of the matrix
D_m is the coefficient of molecular diffusion in the matrix
ρ_m is the dry density of the rock matrix
K_m is a distribution or partitioning coefficient for removal of contaminant from solution due to

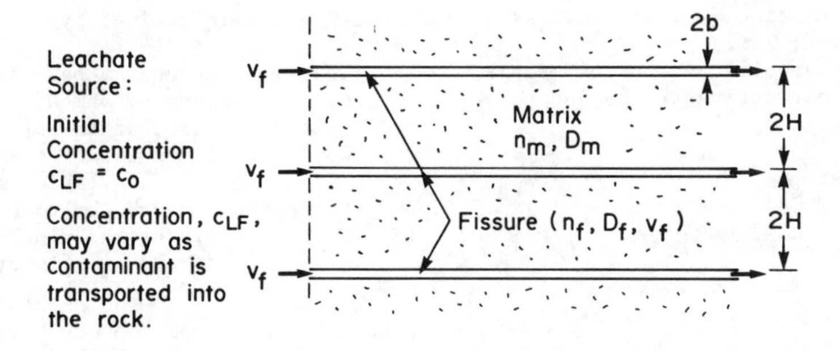

Figure 1. Problem considered - migration along parallel cracks

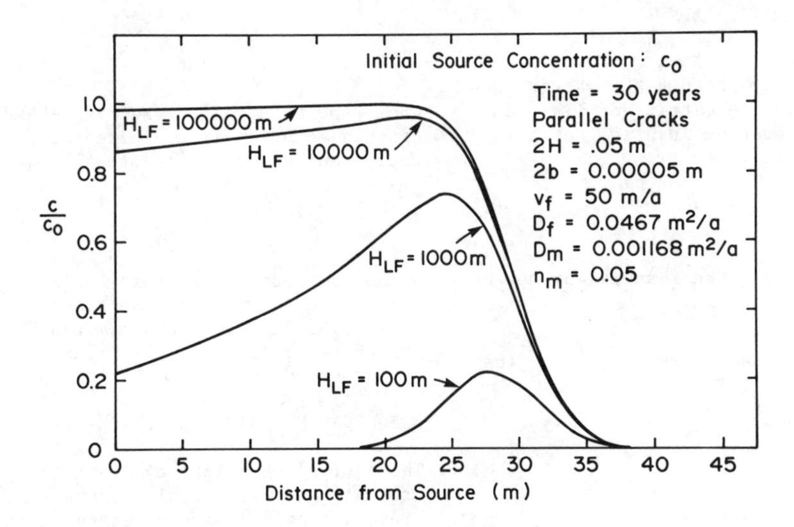

Figure 2. Contaminant plume after 30 years for different assumed values of H_{LF}

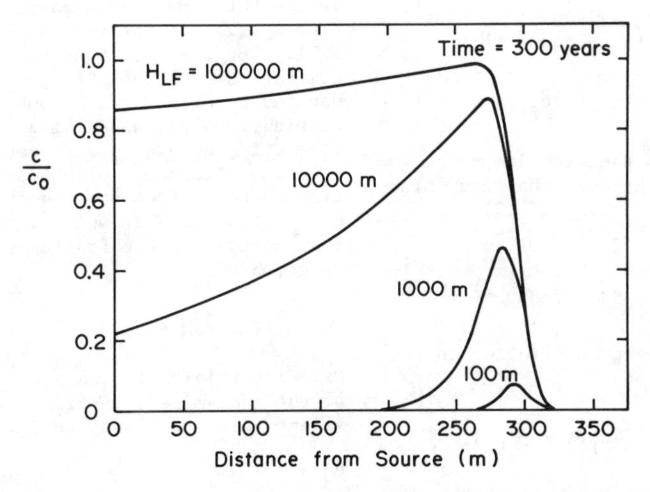

Figure 3. Contaminant plume after 300 years for different assumed values of H_{LF}

linear sorption (eg. see Sudicky and Frind, 1982).

Initially, the concentration in the matrix is zero and so introducing the Laplace Transform

$$\bar{c}_m = \int_0^\infty c_m(t)e^{-st}\,dt \qquad (3)$$

it is found that

$$\frac{\partial^2 \bar{c}_m}{\partial z^2} = \mu^2 \bar{c}_m \qquad (4)$$

where

$$\mu^2 = (1+\rho_m K_m/n_m)s/D_m.$$

It is convenient to locate the origin of coordinates midway through the matrix then recognizing that the matrix and fissure concentrations must be identical at their interface it follows that

$$\bar{c}_m = \bar{c}_f \frac{\cosh \mu z}{\cosh \mu H} \qquad (5)$$

The flux χ entering the fissure may now be calculated and

$$\bar{F} = -n_m D_m \frac{\partial \bar{c}_m}{\partial z} \quad \text{when } z = H \qquad (6a)$$

so that

$$\bar{F} = -\chi \bar{c}_m \qquad (6b)$$

where $\chi = n_m D_m \mu \tanh \mu H$.

Initially, the concentration in the fissure is zero and thus, taking the Laplace Transform of equation (1) and combining it with equation (6b), it is found that

$$D_f \frac{\partial^2 \bar{c}_f}{\partial x^2} - v_f \frac{\partial \bar{c}_f}{\partial x} = (s + \frac{\chi}{n_f b})\bar{c}_f \qquad (7)$$

It will be assumed that the length of each fissure is far greater than its thickness and thus that solution of (7) must vanish as $x \to \infty$ so that

$$\bar{c}_f = \bar{c}_{LF}\, e^{-\lambda x} \qquad (8)$$

where c_{LF} denotes the concentration in the landfill and

$$\lambda = \frac{1}{2D_f}\{- v_f + (v_f + 4D_f(s + \frac{\chi}{n_f b}))^{1/2}\}$$

It remains to compute the concentration in the landfill. To do this, we introduce the equivalent height/fissure (H_{LF}) of the landfill which is defined by

$$H_{LF} = \frac{\text{Volume of leachate, per unit width of fissure, which, at an initial concentration } c_o\text{, would contain the entire mass of contaminant available for transport through the rock}}{\text{Number of fissures x width of fissures}} \qquad (9)$$

It then follows from conservation of mass that

$$H_{LF}[c_{LF}(o)-c_{LF}(t)] = -\int_o^t n_f(D_f \frac{\partial c_f}{\partial x} - v_f\, c_f)\,dt \qquad (10)$$

where the integral on the right-hand side of this equation is evaluated at x=0, so that upon Laplace transformation we find

$$\bar{c}_{LF} = \frac{c_{LF}(o)}{s+n_f(D_f\lambda+v_f)/H_{LF}} \qquad (11)$$

The complete solution in Laplace Transform space is given by equations (8) and (11). The Laplace Transform can be inverted using Talbot's algorithm (Talbot, 1979).

3 EQUIVALENT HEIGHT OF LEACHATE

The equivalent height of leachate H_{LF} represents a mathematically convenient way of representing the mass of contaminant available for transport into the rock. To illustrate the concept, suppose that M_{TC} is the total mass of a particular contaminant species of interest within the landfill. Suppose also that the volume of infiltration into the site, per unit time, is denoted by q_i and the volume of leachate passing into the underlying rock is q_a (the difference q_i-q_a corresponds to the volume of leachate collected by the collection system), then the proportion of the mass of contaminant which is available for transport along the fractures in the rock is given by

$$M_o = M_{TC} \cdot q_a/q_i \qquad (12)$$

Dividing this by the maximum concentration of the contaminant species gives the equivalent volume of leachate V_o, i.e.,

$$V_o = M_o/c_o \qquad (13)$$

This volume corresponds to the volume of leachate which would be required if all the mass M_o was initially in solution at the maximum concentration. (It should be noted that this volume of leachate need not correspond to the actual volume of leachate; rather, it is a mathematical convenience based on the conservative assumption that the available mass of contaminant can be quickly leached from the solid waste). It is also convenient to express the volume V_o in terms of an "equivalent height of leachate" per fracture H_{LF}, where H_{LF} is defined as the volume of leachate divided by the net cross-sectional area through which the contaminant may pass. Assuming that the gross area perpendicular to contaminant flow is A_f (typically the width of the landfill multiplied by the thickness of the fractured rock) and the bulk porosity is n_b (n_b = 2b/2H), then the net area of fracture through which the contaminant may pass is $A_f \cdot n_b$ and hence

$$H_{LF} = V_o/(A_f n_b) \qquad (13a)$$

or

$$H_{LF} = \left(\frac{M_{TC}}{c_o A_f}\right) \cdot (q_a/q_i) \cdot (1/n_b) \qquad (13b)$$

This representation can be thought of as a column of leachate of width 2b which extends back a distance H_{LF} from the face of the fracture. Since the fracture width may be very small even modest volumes of leachate V_o may give rise to values of H_{LF} of 1000 m or more.

4 RESULTS

To illustrate the application of the theory outlined in the previous sections, consider the migration of a conservative species of contaminant (K_m=0) from a hypothetical waste disposal site located in a fractured rock mass which has horizontal parallel fractures with spacing (2H) of 0.05 m and an average thickness (2b) of 50 μm (and hence a bulk porosity n_b= (2b)/(2H)=0.001). Assuming that the diffusion coefficient in the fractures for the contaminant species of interest is 0.0467 m^2/a (14.8×10^{-6} cm^2/s), the coefficient of hydrodynamic dispersion in the fractures is given by

$$D_f = 0.0467 + \alpha v_f \ m^2/a$$

where α is the dispersivity (in m) and v_f is the seepage velocity in the fracture. A velocity v_f = 50 m/a has been adopted for this example. Three values of α were considered, namely no dispersion as the basic case (i.e., α=0), α=.3 m and α=30 m. A matrix porosity n_m of 0.05 was adopted for the rock. Finally, the tortuosity for diffusion in the rock matrix was taken to be 0.025 giving a matrix diffusion coefficient D_m = 0.025 x 0.0467 = 0.01168 m^2/a.

Analyses were performed using the theory described in the preceding section to determine the variation in contaminant concentration with distance from the contaminant source at three times (30, 300 and 3000 years) for an equivalent height of leachate per fracture, H_{LF}, ranging from 100 m to 100 000 m. The results from these analyses are shown in Figures 2, 3 and 4.

The situation where the source concentration is assumed to be constant (eg. as assumed by Sudicky and Frind, 1982 and Barker, 1982) is equivalent to assuming that the mass of contaminant is infinite (i.e., $H_{LF} = \infty$). For this case, the process of diffusion of contaminant from the fracture into the rock matrix would serve to slow the rate of advance of the contaminant front, but eventually the concentration of contaminant at each point along the fracture would approach the source value of c_o.

If consideration is given to the more realistic situation where the mass of contaminant is assumed to be finite (and expressed here in terms of H_{LF}), then diffusion into the rock matrix not only slows down the advance of the contaminant front but also gives rise to a predicted decrease in concentration with time within the leachate and significant attenuation of the contaminant plume as it moves away from the source. For example, examining Figures 2, 3 and 4, it can be seen that for all finite values of H_{LF} there is a decrease in the predicted leachate concentrations with time, the effect being most pronounced for values of H_{LF} of 1000 m or less. It can also be seen that for each value of H_{LF} and time, the concentration increases with distance away from the source until a peak concentration c_p is reached at some distance x_p, and then decreases with further distance from the source. The magnitude of this peak depends on the mass of contaminant (i.e., H_{LF}), but in all cases the peak is seen to decrease with increasing time and distance from the source. This clearly has important implications with respect to the potential environmental impact of the landfill upon water resources at some distance from the waste disposal site.

The rate of movement of the contaminant plume warrants some comment. The average linearized groundwater velocity in the fissure was assumed to be 50 m/a. Based on

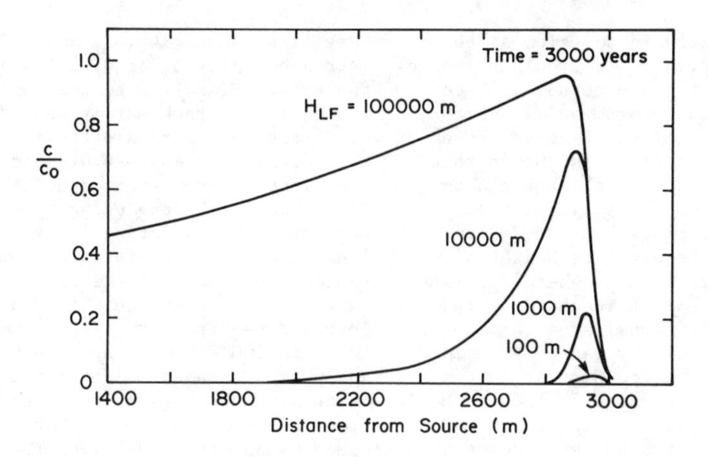

Figure 4. Contaminant plume after 3000 years for different assumed values of H_{LF}

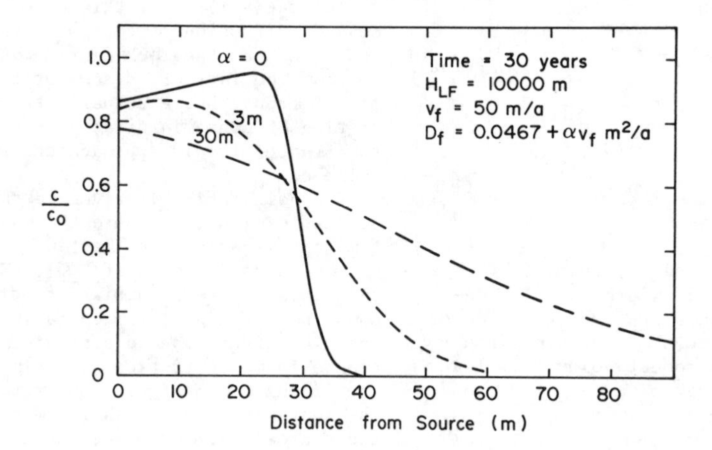

Figure 5. Effect of dispersivity α on contaminant plume after 30 years

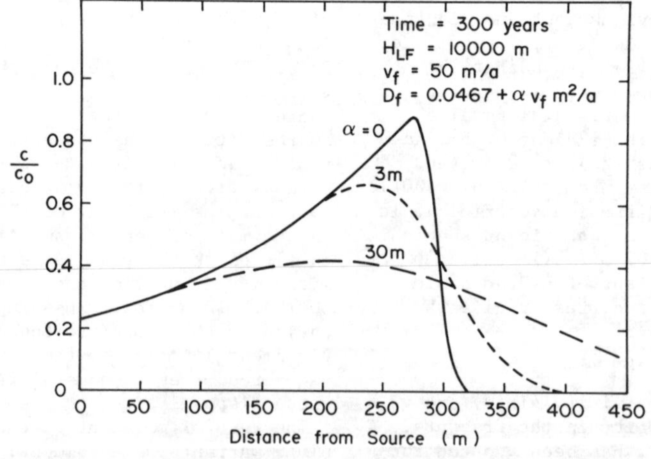

Figure 6. Effect of dispersivity α on contaminant plume after 300 years

simple considerations of advective transport, this would suggest that the contaminant front should have advanced 1500 m in 30 years and 150 km in 3000 years! With this in mind, inspection of Figures 2-5 shows the very significant effect of diffusion into the rock matrix on the rate of advance of the contaminant plume. Considering the advance of the point where the concentration is 50% of the original source concentration c_o as being an indication of the location of the plume, it can be seen that in 30 years the plume has moved approximately 29.5 m rather than 150 m and that the velocity of plume movement is a little less than 1 m/a. Thus, after 300 years, the plume has only moved about 295 m rather than the 15 km that would be predicted based on simple advective calculations.

An obvious, but nevertheless important, practical observation that can also be drawn from Fig. 2 is that for the assumed conditions, monitoring wells located 40 m or more from a waste disposal site would not detect any contaminant within a 30 year monitoring period after the commencement of leachate migration; however, this does not imply that there is no contaminant plume.

The foregoing results and discussion have been for the case where it is assumed that there is no mechanical dispersion as contaminant moves along the fissures. The phenomenon of mechanical dispersion in fractured systems is not well understood. Limited empirical evidence would suggest that the dispersivity α increases with the distance of travel. To illustrate possible impact of considering mechanical dispersion, analyses were performed for a dispersivity α of 3 m and 30 m, for the case where H_{LF} = 10 000 m. The calculated concentration plumes are shown in Figs. 5 and 6 at times of 30 years and 300 years respectively.

Dispersion serves to increase the spread of the contaminant which implies that the rate of advance of the front of the contaminant plume is increased and the concentration gradients within the fracture are decreased. For example, considering the point where the concentration is 10% of the initial value ($c/c_o = 0.1$), as an indication of the position of the contaminant front, it is seen that dispersivities of 0, 3 and 30 m give travel distances of approximately 34 m, 49 m and 94 m respectively after 30 years and 315 m, 350 m and 450 m respectively after 300 years.

Dispersion also serves to decrease the magnitude of the peak concentration and actually slows the advance of the peak (i.e., decreases the distance x_p to the peak at a given time). Again referring to Figs. 5 and 6, it is seen that dispersivities of 0, 3 and 30 m give peak concentrations at approximately 22 m, 8 m and 0 m after 30 years and 280 m, 237 m and 200 m after 300 years. It can also be seen that the decrease in the magnitude of the peak concentration due to dispersion increases with time. Thus at a point 300 m from the source, inspection of Fig. 6 would suggest that the peak concentration for α = 30 m would be less than half the peak value that would be obtained for α = 0. Analyses indicate that 3000 m from the source, the peak for α = 30 m is less than one quarter (25%) of that obtained for no dispersion (α = 0). Clearly then, while it may be conservative to consider high dispersivities when calculating times for contaminant to reach a given point, it is not conservative to consider high dispersivities when assessing the potential peak degradation of groundwater quality.

The results shown in Figs. 5 and 6 were obtained for H_{LF} = 10 000 m. It is noted in passing that the effects of dispersion discussed above do depend on H_{LF}, and that the effects of dispersivity may be even more pronounced for lower values of H_{LF}.

5 CONCLUSION

This paper extends previous research into advective-dispersive contaminant movement through a system of parallel cracks by presenting a solution for the case where the mass of contaminant available for transport into the rock is assumed to be finite and, as a consequence, the concentration of contaminant in the source decreases with time. The solution is in a form that can be readily implemented on a micro-computer. Examination of a typical situation involving migration in fractured rock indicates that the mass of contaminant can have a significant effect on the attenuation of contaminant and should be considered if a realistic assessment is to be made of the potential impact on nearby water resources. It has also been shown that the rate of contaminant transport may be very much slower than that deduced from simple consideration of the seepage velocity along the fracture.

6 ACKNOWLEDGEMENT

The work described in this paper forms part of a general programme of research into contaminant migration in soil and rock supported by grant No. A1007 from the Natural Science and Engineering Research Council of Canada. Additional funds for senior author's travel to Sydney have been provided by the Civil Engineering Foundation at The University of Sydney.

7 REFERENCES

[1] J.A. Barker, Laplace transform solutions for solute transport in fissured aquifers, Advanced Water Resear. 5(2), 98-104 (1982).

[2] G.E. Grisak and J.F. Pickens, Solute transport through fractured media, 1, Water Resource Research 16, 731 (1980).

[3] I. Neretnieks, Diffusion in the rock matrix: An important factor in radio nuclide retardation, J. Geophys. Res. 85 (B8), 4379-4397, (1980).

[4] R.K. Rowe and J.R. Booker, 1-D pollutant migration in soils of finite depth, J. Geot. Engrg, ASCE, 111, GT4, 479-499 (1985a).

[5] R.K. Rowe and J.R. Booker, 2D pollutant migration in soils of finite depth, Canadian Geotech. J., 22, 4, 429-436, (1985b).

[6] R.K. Rowe and J.R. Booker, A finite layer technique for calculating three-dimensional pollutant migration in soil, Geotechnique, 36, 2, 205-214, (1986).

[7] R.K. Rowe and J.R. Booker, An efficient analysis of pollutant migration through soil, Chapt. 2 in Numerical methods in transient and coupled systems, Lewis, Hinton, Bettess & Schrefler (Eds.), John Wiley & Sons, 13-42, (1987).

[8] E.A. Sudicky and E.O. Frind, Contaminant transport in fractured porous media: Analytical solution for a system of parallel fractures, Water Resources Research, 18, 1634-1642, (1982).

[9] D.H. Tang, E.O. Frind and E.A. Sudicky, Contaminant transported in fractured porous media: Analytical solution for a single fracture, Water Resource Research, 17, 555-564, (1981).

[10] A. Talbot, The accurate numerical integration of Laplace transforms, J. Inst. Maths. Applics., 23, 97-120, (1979).

Numerical Methods in Geomechanics (Innsbruck 1988), Swoboda (ed.)
© 1988 Balkema, Rotterdam. ISBN 90 6191 809 X

Numerical interpretation of borehole jack test in consideration of joint stiffness

Chikaosa Tanimoto & Shojiro Hata
Department of Civil Engineering, Kyoto University, Sakyo, Japan
Akira Nishiwo
Hazamagumi Co., Tokyo, Japan
Kozo Miki & Jun'ichi Miyagawa
Kawasaki Geo-Consulting Co., Osaka, Japan

ABSTRACT: It is well known that the mechanical behaviour of fractured rock is strongly influenced by the state of joints. Joint orientation, contact pressure at joints and physical properties such as fracture (joint) frequency, aperture, roughness, alteration and moisture content are predominant parameters in the mechanical behaviour of rock mass. The authors have investigated the mechanical behaviour of jointed rocks through the plate bearing test, the rock shear test, and the borehole loading test in association with the laboratory test. However, it has been realized that joint behaviour is very hard to be analyzed in the laboratory test and the employment of limited number of specimens give rather unreliable solution apart from reality. It is concluded that the physical properties of joints must be determined in situ in undisturbed condition with a reasonable interpretation. Then the authors have tried to solve it by introducing the concept of joint stiffness. The joint stiffnesses for normal and shear directions are considered as the overall indexes which cover roughness, alteration and aperture at joints at the same time. The authors' objective is to clarify the relationship between physical properties of joints and joint stiffnesses through the execution of the borehole jack loading test.

1 INTRODUCTION

In order to determine the stress distribution and intensity in composite load-bearing structures it is necessary to know the load-deformation characteristics of the component materials. Most materials are more or less continuous and homogeneous such that the behaviour of large masses can reasonably be predicted from experiments on small laboratory samples.

When dealing with rock masses the engineer is faced with an unusual set of problems. The material at any site is unique and its properties cannot be deduced on the basis of experience gained elsewhere, except in general terms. Rock masses are heterogeneous and usually discontinuous assemblages of material with the result that the scale of an experiment to some extent determines the result of the experiment. Comparisons of the results of in situ field tests on rock masses with the results of laboratory tests on the same rock show that laboratory tests invariably lead to an overestimate of the stiffness of the rock mass.

The principal reason for this discrepancy is the presence of discontinuities such as jointing and bedding, microcracks, faults, localized altered rock zones etc.[Stagg,1968] Through the experience with rock problems in Japan it has been well known that the mechanical behaviour of fractured rock is strongly influenced by the state of joints contained in them. Predominant parameters for the mechanical behaviour of rock are joint orientation, contact pressure at joints and physical properties such as fracture frequency (or joint spacing in an other term), aperture, roughness, alteration and moisture content. Specially among the several predominant parameters, 'aperture', 'roughness', and 'alteration' at joints are very difficult to be characterized quantitatively.

Then, the authors have tried to solve this problem by employing the concept of joint stiffness which Goodman [1968;1976] and Barton [1972] have proposed. The joint stiffness is conceptional and consists of normal and shear joint stiffnesses corresponding with the directions normal to and parallel with joint. They are described in the quantitative manner, and are considered as overall indexes which express roughness, alteration and aperture at joints altogether by two parameters. The authors' concern lies on the relationship between physical properties of joints and joint stiffnesses through the execution of the borehole loading test. It is because this test is remarkably simple, economical, and time-saving in comparison with the plate bearing test, the rock shear test, and other conventional ones.

As the fundamental study, several simplified models are discussed with following parameters: joint spacing, orientation, normal joint stiffness and shear joint stiffness, and elastic constants of intact rock. Next, extending this concept to several field tests, the mechanical behaviour of joints posing in an arbitrary three dimensional state has been discussed by the numerical models in association with the experimental study in which the 40 cm cubic specimens have been loaded in the biaxial state.

Based on the results obtained by the numerical and experimental studies, the relationships between the deformability of jointed rock, orientation, loading direction and joint spacing were discussed in this paper.

2 ANALYTICAL PROCEDURE OF JOINT STIFFNESS

The total deformation of rock mass is the sum of the deformation of intact rock and that of joint. Let us consider of a three dimensional parallel joint system as shown in Fig. 1. Joints are regularly distributed. Then, trying to express a grobal deformation coefficient $D(\theta)$ in the direction with an angle θ counter-clockwise from E(East) direction when a partial uniform load acts onto the borehole wall, $D(\theta)$ can be shown by Eq.(1) as the function of θ, E, N, i, h_i, k_{ni}, k_{si}, α_i, and β_i.

$$D(\theta) = \left[\frac{1}{E} + \sum_{i=1}^{N} \frac{A_3^{(i)}}{h_i}\left\{\frac{1}{k_{si}}\left(A_1^{(i)}+A_2^{(i)}\right) + \frac{1}{k_{ni}} A_3^{(i)}\right\}\right]^{-1} \quad (1)$$

where

$$A_j^{(i)} = \left(a_{j1}^{(i)}\cdot\cos\theta + a_{j2}^{(i)}\cdot\sin\theta\right)^2 \quad (j = 1, 2, 3)$$

$$\left[a_{jk}^{(i)}\right] = \begin{bmatrix} \cos\alpha_i\cdot\cos\beta_i & -\sin\alpha_i\cdot\cos\beta_i & -\sin\beta_i \\ \sin\alpha_i & \cos\alpha_i & 0 \\ \cos\alpha_i\cdot\sin\beta_i & \sin\alpha_i\cdot\sin\beta_i & \cos\beta_i \end{bmatrix}$$

where θ : counter-clockwise angle from E-direction, E: Young's modulus of intact rock [kN/m^2], N: number of predominant directions of joints, i: number of joint groups, h_i: average interval (spacing) of i-th joint group [m], k_{ni} & k_{si}: normal and shear stiffnesses of i-th joint group (kN/m^3), and α_i & β_i: strike α_i (Nα_iE) and dip β_i (β_iSE) of i-th joint group.

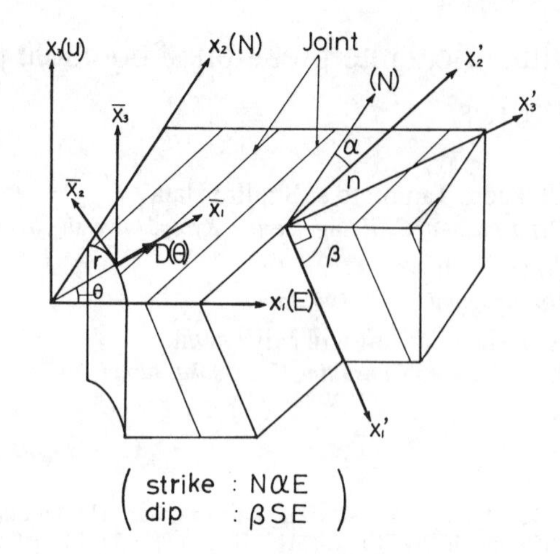

Fig. 1 Jointed Rock Model

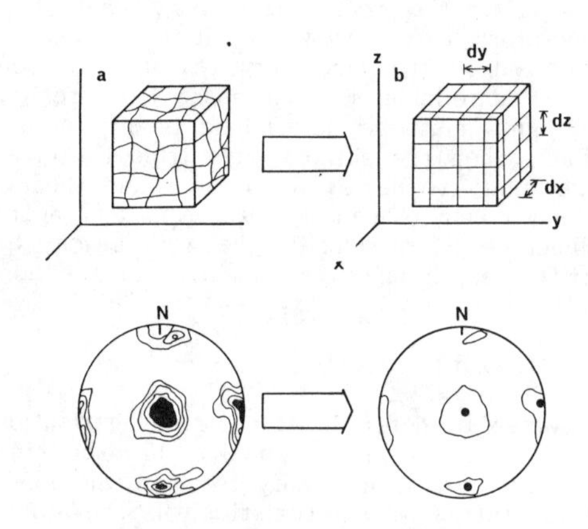

Fig. 2 Simplified model by the stereo-graphic projection

Also, when the borehole jack test is applied to the vertical borehole as shown in Fig. 1, the deformation coefficient $Dj(\theta)$ which is obtained by loading in the direction of θ is the complex consisting of the deformations of intact (solid) rock and discontinuity, and it is considered to correspond with $D(\theta)$ in Eq.(1).

Then, assuming that $Dj(\theta)$ is proportional to $D(\theta)$, $Dj(\theta)$ is expressed as follows.

$$Dj(\theta) = K \cdot D(\theta) \quad (2)$$

where K: constant.

When deformation coefficient Dj(θ), spacing and orientation of joint system and Young's modulus of intact rock can be known by the borehole jack test, the observation with the borehole TV or scope, and the uniaxial loading test on rock specimens respectively, Eq.(2) gives the function with respect to joint stiffnesses. And besides, it is not so difficult to apply jack-loading in various directions (θ) corresponding with a number of unknowns. Therefore, stiffnesses of kn and ks can be obtained by solving simultaneous equations in the same manner as Eq.(2).

In general, as it is common that discontinuities in rock masses show several prominent directions and form specific joint systems, the statistical interpretation of joint systems by means of the stereographic projection provides a simplified model as easily as shown in Fig. 2.

Fig. 3 illustrates the relationship between D(θ) in various directions and stiffnesses kn and ks of joint systems which were obtained by applying Eq.(1) to the modelled rock with the regularly spacing layers. As you see, the deformation coefficient D(90) in the direction of strike is not affected by the joint system.

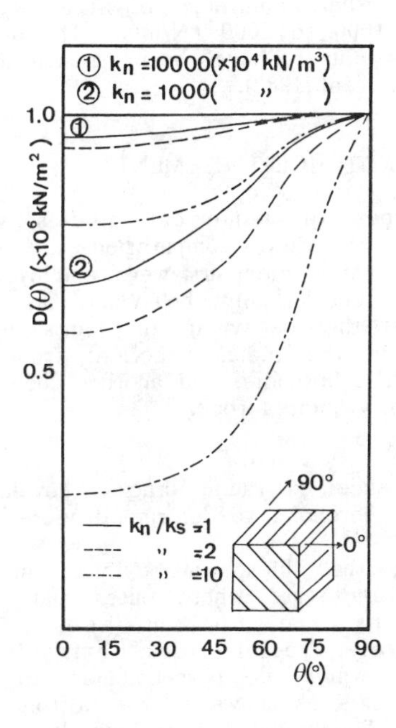

Fig.3 Deformability obtained by jointed rock model

3 LABORATORY TEST ON JOINT MODELS

In order to investigate the relationship between characteristics of joints and deformation coefficients, the biaxial loading cell with the steel cylinder of 60 cm in dia., 40 cm high and 2.5 cm thick was prepared. After setting the joint model assemblage in the cell, four pieces of flat jacks with the size of 40 cm square were installed onto the four sides, and the space between the assemblage and the cell was filled tightly with the stiff cement mortar as shown in Fig. 4. The joint model was formed by assembling many cubes or rectangular prisms made of plaster. The mixing ratio is 3 to 2 for plaster to water, and physical properties of a plaster block are 5.27×10^6 kN/m^2 and 0.20 for Young's modulus E and Poisson's ratio ν, respectively. These values were constant during the testing period.

Model No.1 was prepared for the continuous rock without discontinuity, and Model No.2 to No.5 for jointed rock masses were designed so as to correspond with the joint frequencies n′[in number of joints per meter] of 7.5, 15, 25, and 35, respectively. Joint frequency n can be defined by dividing the total length [in meter] of joints with respect to a horizontal plane by the objective area [in square meters]. This has an equal unit to 'number of joints per meter' along an objective line.

Arbitrary combinations of confining pressures (or contact pressures at joints) can be applied to the models by controlling surrounding flat jacks. The loading pattern which was employed in the laboratory test was a stepwisely repeated loading, and the peak load in each loading process was divided into four and increased step by step with the increment of a quarter of the peak load. The magnitude of peak load was 1.0 to 4.0×10^3 kN/m^2 corresponding with the same level of confining pressure of 1.0 to 4.0×10^3 kN/m^2 respectively. The loading directions were as shown in Fig. 5.

Deformation coefficient Dj is obtained from the gradient of straight portion of the envelope line to the 'load(p)-displacement(u)' curve by means of the equation given by Goodman(1968). The values obtained in such manner were needed to compensate for the actual loading area, for the loading area of the jack does not agree with the actual loading area due to the unevenness of the borehole wall when an applied load is not so high in comparison of the hardness of rock.

Fig. 6 illustrates the relationship between the loading direction θ , and deformation

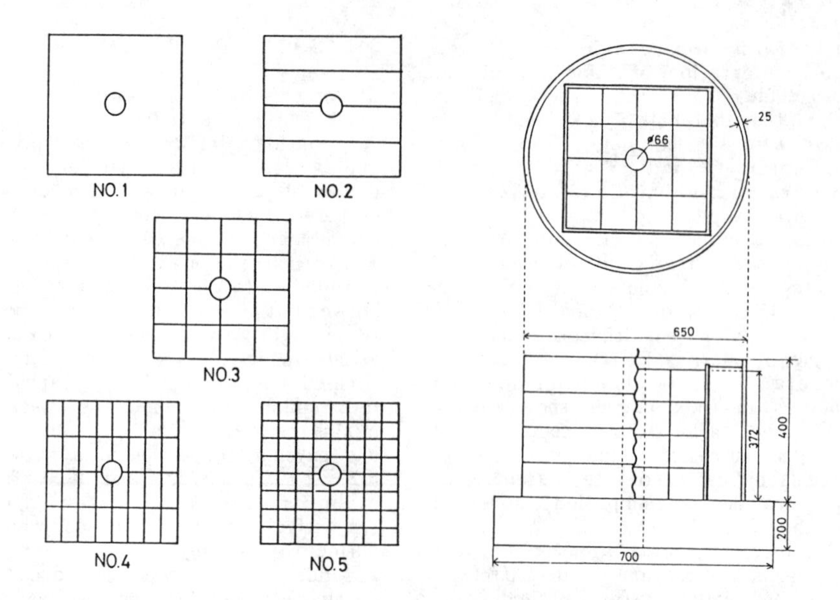

Fig.4 Modelled rock mass and loading cell

coefficient Dj for Model No.2. The change of Dj due to different θ shows the anisotropic behaviour of jointed rock. And, the change of Dj becomes small in association with the increase of confining pressure and the behaviour similar to an isotropic body has been observed.

The effect of joint frequency to deformability is another concern of ours. Fig. 7 illustrates the relationship between joint frequency and 'deformability ratio', namely the ratio of average $\bar{D}j$ for jointed rock versus that of solid rock with no joint $\bar{D}jo$. It is easily recognized that deformation coefficient is highly affected by the degree of joint frequency. And, the higher 'joint frequency' is, the more remarkable the reduction of deformation coefficient becomes under lower confining pressure. It suggests that the shear resistance (strength) of jointed rock highly depends on the magnitude of confining pressure when confining pressure is low such as 1000 to 2000 kN/m^2. The details on this matter have been presented by Tanimoto,et al.(1983).

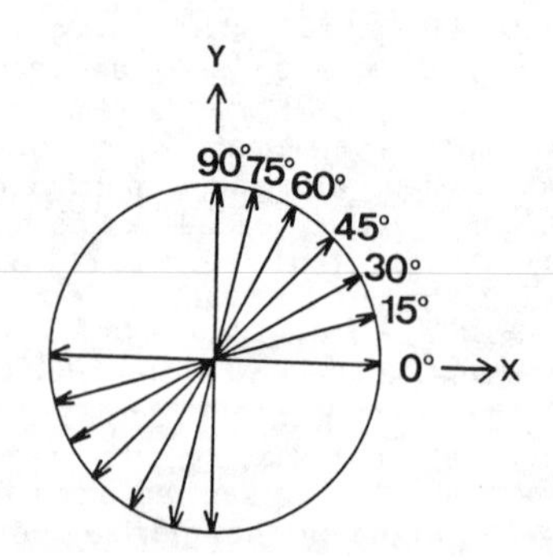

Fig.5 Loading direction in borehole jack test

4 FEM WITH JOINT ELEMENTS

Two types of numerical models were analyzed by FEM computation in which additional joint elements were employed in order to consider joint behaviour. One type simulates the behaviour of joints which develop in the radial directions from the wall of the borehole, and another does that of regularly jointed rock.

4.1 Influence of radial cracks, developing from the borehole, to deformability

Assuming the thick wall cylinder in the plane strain state whose outer and inner diameters are 60 cm and 6 cm respectively, radial cracks exist in x-direction which is perpendicular to the borehole axis as shown at the bottom half of Fig.9. The model points along the cracks include joint elements. Under arbitrary

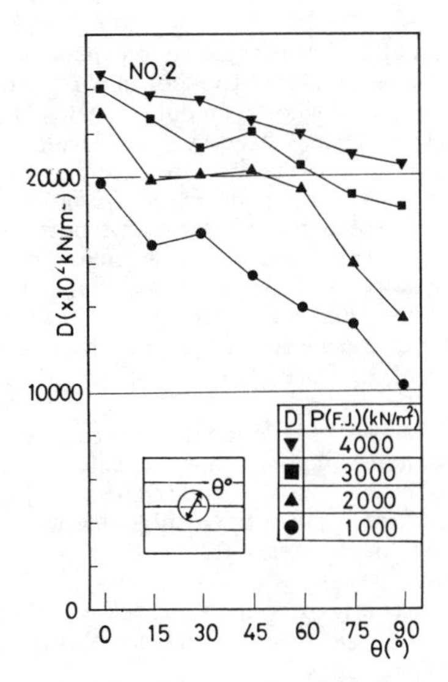

Fig.6 Relationship between deformation coefficient and loading direction

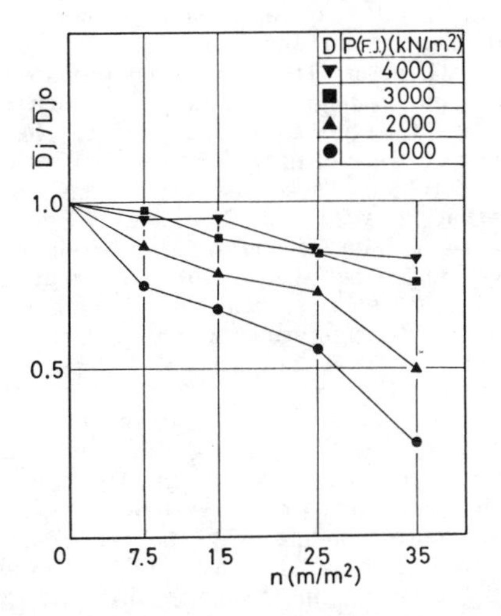

Fig.7 Influence of joint frequency to deformability

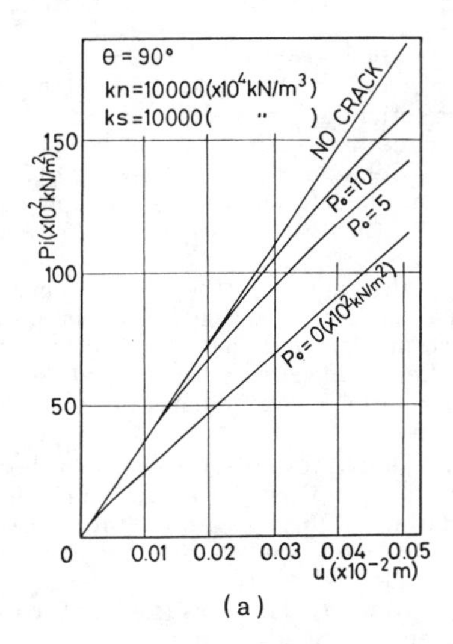

(a)

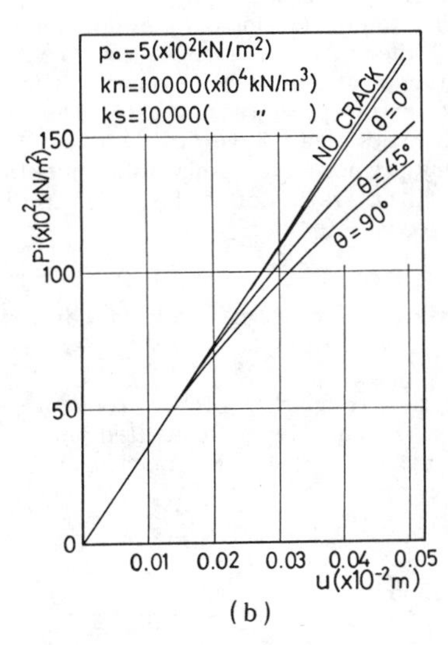

(b)

Fig.8 Load-displacement curves

pressure p_o which acts on the outer boundary, the magnitude of jack pressure (p_i) was calculated by applying uniform displacement (u) to the area (B) of 3 cm wide subject to the jack action. Input data given to the computation are as follows; $E=1.0 \times 10^6$ [kN/m^2], $\nu=0.2$, $kn=1.0 \times 10^7$ & 1.0×10^8 [kN/m^3], and $ks=1.0 \times 10^7$ & 1.0×10^8 [kN/m^3]. The yielding condition of joint elements is

$\tau_y = 0.8\sigma_n + 5.0$ [x 10^2kN/m^2] for $\sigma_n \geq 0$ (closed crack)

or $\tau_y = 0.0$ for $\sigma_n < 0$ (open crack).

The loads were applied in the various directions as shown in Fig.5.

Fig.8 illustrates the summary of the some results for the load-displacement (p_i-u curves). Fig.8(a) corresponds with the case that confining pressure changes under the condition of θ = 90 degree, and Fig.8(b) does with the one that a direction of applied load (θ) changes under the constant confining pressure. From Fig.8(a) it can be seen that the lower 'confining pressure' is, the smaller 'deformation coefficient' becomes in association with the easier opening even if p_i is low. In Fig.8(b), When the direction of applied load is $\theta = 0$ degree, the model appears to be almost in elastic behaviour, but the noticeable reduction of deformation coefficient can be seen for the cases of $\theta = 45$ and 90 degree.

From the tangent of the p_i-u curve the value of deformation coefficient is obtained by Goodman's equation (Goodman, et al; 1968). Let us denote it $Dj(\theta)$ and let make a deformation coefficient for continuous body Djo.

Fig.9 illustrates the relationship between the ratio of $Dj(\theta)$ to Djo and loading direction. Closed and open circles correspond with the cases that cracks are closed and open respectively. It is known that while θ increases, deformation coefficient decreases easily.

Then, when the jack face does not meet cracks, the ratio of p_i/p_o shows remarkably wide range of scattering, namely the bigger the ratio of p_i versus p_o is, the smaller the ratio of $Dj(\theta)$ versus Djo is. On the contrary, when the jack face meets with cracks, the ratio of $Dj(\theta)$ versus Djo is nearly constant.

4.2 Behaviour of the regularly jointed rock model

Next, the regularly jointed rock as shown in Fig.10 has been modelled with joint elements. For the comparison the orthogonally anisotropic (continuous) model, which is equivalent to the model with joint elements and whose modulus ratio D_1/D_2 is 2.0, has been discussed as well. Fig.10 illustrates the results in which the solid curves, the dotted curves, and closed circles correspond with the theoretical ones obtained from Eq.(1), the ones from the orthogonally anisotropic model, and the ones from the discontinuous model with joint elements, respectively. Input data given to this analysis are the same as stated in Item 4.1.

As shown in Fig.10, there are some discrepancies among those results. This fact suggests that stress distribution given by the partial uniform displacement (in the borehole jack test) is rather complicate, and that tension in rock easily accelerates the progress of open crack and causes remarkable variation in deformability.

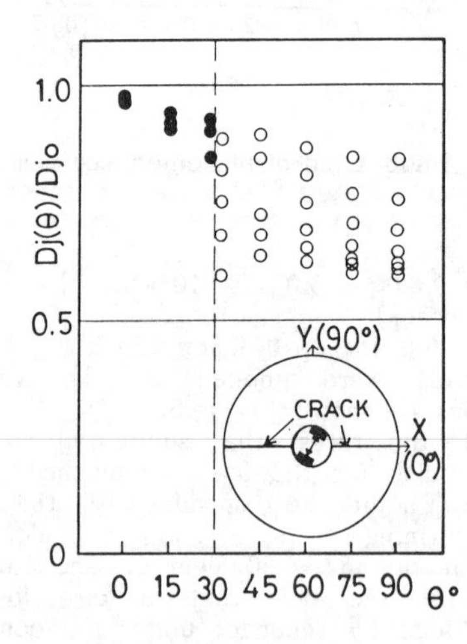

Fig.9 Relationship between deformation coefficient and loading direction in the model with cracks

5 APPLICATION TO THE FIELD TEST

Fortunately the authors had been given the opportunity to evaluate deformation coefficient from the point of view on the joint stiffness. The site was prepared for the dam foundation and it consisted of fractured schalstein and chert. Several boreholes were drilled for the application of the borehole jack test and the precise observation with the borehole TV camera was carried out ahead of the borehole jack test. Based on the careful observation of jointing systems, several parts which showed clear joint orientation with regularity were chosen for the purpose of obtaining the fundamental data.

Fig.11 illustrates the example of the results. The values of deformation coefficient were calculated from the straight portions of the load-displacement curves corresponding with the range of p_i = 3.0 to 5.0×10^3 kN/m^2. It clealy shows that variation in loading direction gives remarkably different deformability. It could be considered that the effective loading area was constant in spite of different loading direction because of the smoothness of the borehole wall.

It is concluded that the variation in deformation coefficient was attributed to the physical properties and orientations of discontinuities in this field test. Then, the obtained data had been analyzed in terms of Eq.(1). It must be noted that the applied load to the field test was not high, but rather at low level such as 5×10^3 kN/m^2 at maximum, therefore, it can be accepted to assume that the relation of load-displacement is proportional in both

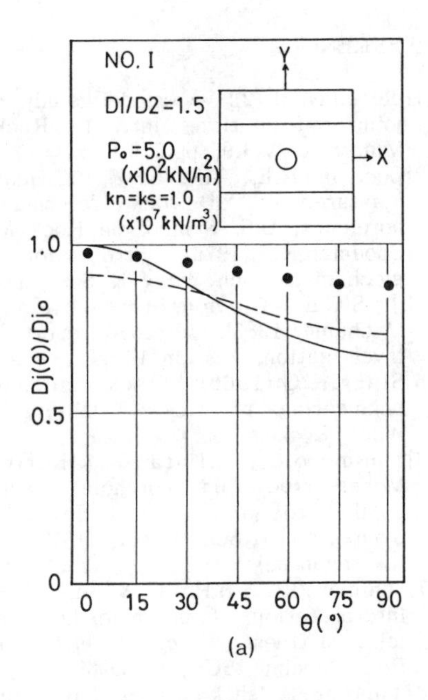

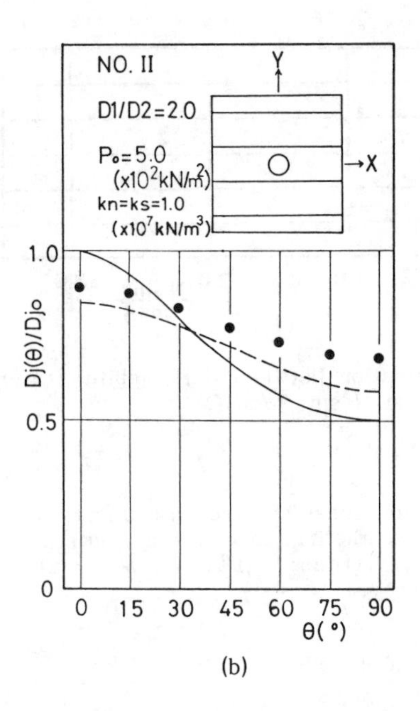

Fig.10 The results from theoretical and numerical analyses, experiment for jointed rocks

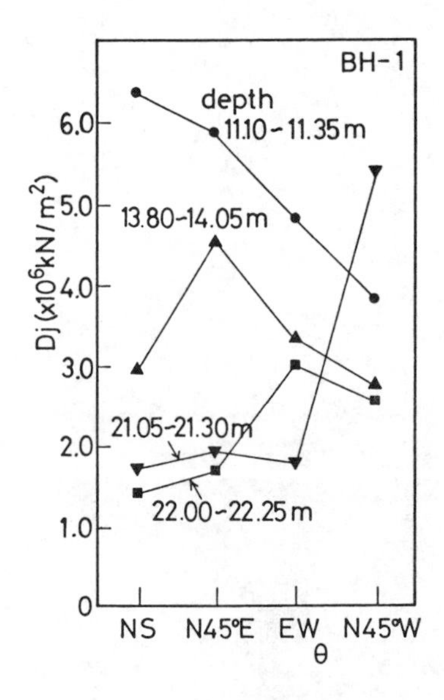

Fig.11 Results of the borehole jack test in the in-situ field test

directions of normal and parallell to the discontinuity and stiffness of kn and ks take constant values. As Yamabe, et al.(1983) reported that the initial shear stiffness was observed to be almost equal to the normal stiffness in their experiment concerning the joint behaviour of tuff, we assume that kn is equal to ks for simplification.

As it is clear in the Fig.3, deformation coefficient becomes maximum and minimum in the directions of strike and dip, and the ratio of maxmum value to minimum value is given as follows:

$$\frac{D_{max}}{D_{min}} = 1 + \frac{E \cdot \sin^2\bar{\beta}}{h \cdot kn} = 1 + \frac{E \cdot}{0.25kn}\left(\frac{N \cdot \sin^2\bar{\beta}}{\cos\bar{\beta}}\right) \qquad (3)$$

where $h = \cos\bar{\beta}/n = 0.25\cos\bar{\beta}/N$

N : number of continuities observed in the loading area of 0.25 m long.

n : joint frequency [number of joints per m],

and, $\bar{\beta}$: average dip angle of discontinuities

Fig.12 illustrates the relationship between the ratio of maximum value to minimum one of Dj (Djmax/Djmin) and the parameter of N $\sin^2\bar{\beta}/\cos\bar{\beta}$ in Eq.(3). There are some scatterings. It suggests the

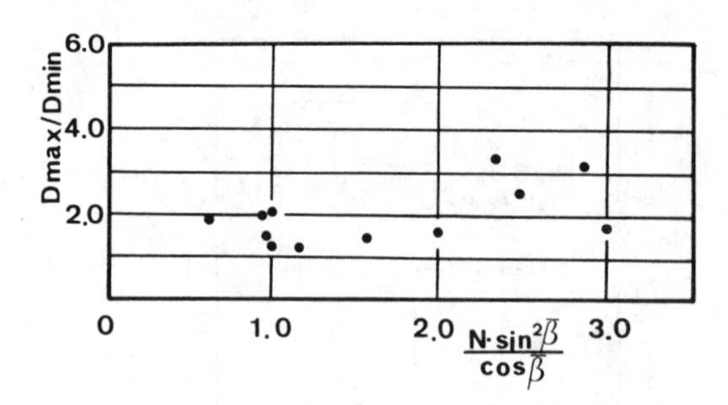

Fig.12 Relation between deformability ratio and N $\sin^2\bar{\beta}/\cos\bar{\beta}$

necessity to consider the influences of roughness, weathering, and filling materials to joint stiffness qualitatively and quantitatively.

6 CONCLUSION

The authors proposed the concrete application of joint stiffness to discontinuous rock mass through the theoretical, numerical and experimental approaches and field test. More fundamental discussions and practical data are necessary, and it must be emphasized that reliable information about joint orientation and characteristics of joints are essential in this study. The application of the borehole TV system newly developed by the authors was quite helpful and recommendable to many readers who are interested in joint survey. This issue is still in consideration, and further comments will appear in the near future.

ACKNOWLEDGEMENT

Messrs. Ohyabu, Yamaguchi and Hara - Water Resources Development Authority of Japan - offered us the opportunity to initiate this study. Messrs A.Iizuka, A.Moriki and T.Inada were very helpful in typing this manuscript and in drawing figures. The authors wish to express thier cordial thanks to these people for their cooperation.

REFERENCES

[1]Barton,N.(1972): A model study on rock joint deformation, Int. J. Rock Mech. Min. Sci., Vol.9, pp.579-602

[2]Goodman,R.E., T.K.Van & F.E.Heuze(1968): Measurement of rock deformability in boreholes, U.S. Sympo. on Rock Mech.

[3]Goodman,R.E.(1976): Introduction to rock mechanics, John Wiley & Sons, pp.186-193

[4]JSCE Committee on Rock Mechanics(1983): Internal report on rock investigation, Session III

[5]Stagg,K.G.(1968): Rock mechanics in engineering practice, Chapter 5, John Wiley & Sons, pp.125-156

[6]Tanimoto,C., S.Hata & K.Kariya(1982): Model study on borehole loading test appiled to jointed rock, Proc. of 14th Sympo. on Rock Mech., JSCE, pp.10-14 (in Japanese)

[7]Tanimoto,C., S.Hata & A.Nishiwo(1985): Interpretation of borehole jack test with joint stiffness, Proc. of 17th Sympo. on Rock Mech., JSCE, pp.85-89 (in Japanese)

[8]Yamabe,T.,Ishikawa & Hoshino(1983): Experimental study on the estimation of deformability of jointed rocks, Proc. 38th Annual Meeting of JSCE (in Japanese)

Numerical Methods in Geomechanics (Innsbruck 1988), Swoboda (ed.)
© 1988 Balkema, Rotterdam. ISBN 90 6191 809 X

A correction for sampling bias on joint orientation for finite size joints intersecting finite size exposures

P.H.S.W.Kulatilake
University of Arizona, Tucson, USA

ABSTRACT: In joint surveys, biases are introduced in sampling for orientation. Terzaghi (1965) developed a correction for orientation bias which depends only on the orientation of the joint with respect to the sampling domain. This bias is applicable only for joints of infinite size intersecting infinite size exposures. In reality, both the joint sizes and the size of the sampling domain are finite. The paper provides a methodology to derive corrections for orientation bias in the case of finite size joints intersecting finite size exposures. The derivation of the correction for a thin circular joint intersecting a finite size rectangular domain is given in the paper. The obtained expression for the correction clearly shows that the bias depends on the sizes and shapes of both the joint and the sampling domain in addition to the orientation of the joint with respect to the sampling domain. For joints of infinite size intersecting infinite size domains, the developed corrections get reduced to the Terzaghi's (1965) correction. The ways one can apply this derived correction in practice are discussed.

1 INTRODUCTION

An accurate representation of joint geometry is extremely important in the study of hydraulic and mechanical properties of jointed rock masses. One of the key parameters used in modeling of joint geometry is the joint orientation. Joint orientation is usually determined in the field by sampling along a line, as in a borehole, or over an area, as on an outcrop surface. However, the observed frequency differs from actual frequency, because the probability of a joint being sampled is proportional to the probability of that joint intersecting the sampling domain. This is called sampling bias in orientation. The observed frequency for each observed orientation is one. The corrected frequency may be obtained by assigning a weight for each observed joint through a weighting function, which is proportional to the reciprocal of the probability of the joint being sampled. Terzaghi (1965) has suggested a correction for sampling bias in orientation. However, this correction is applicable only for joints of infinite size intersecting infinite size exposures. In reality both the joint sizes and the size of the sampling domain are finite. This paper presents a methodology to derive corrections for sampling bias in orientation for joints of finite size intersecting vertical rectangular exposures.

2 CORRECTION OF SAMPLING BIAS

Unfortunately, at the present time, the experimental evidence available on joint shape is inconclusive. Robertson (1970) after analyzing nearly 9000 joint trace lengths from the De Beer mine concluded that the strike and dip trace lengths have about the same distribution, possibly implying joints to be equidimensional. Two other studies on joint shape (Bridges, 1975; Einstein at al., 1979) indicated that joints are non-equidimensional. In joint modeling, joints have been represented as thin circular discs (Baecher et al., 1977; Barton, 1978; among others) and as Poisson planes (Veneziano, 1978, Dershowitz and Einstein, 1987). In this paper the joints are considered as thin circular discs.

The probability of intersection of a

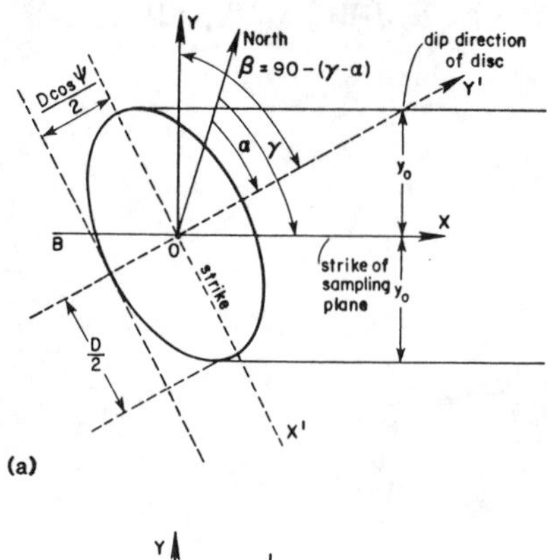

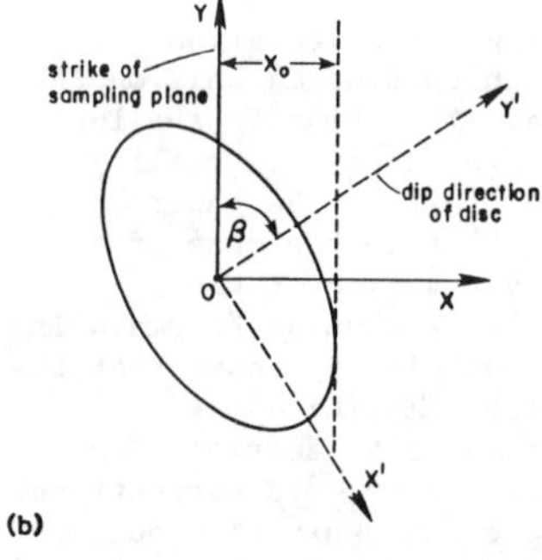

Fig.1 Possible Plan views for the intersection between the disc and the sampling plane.

joint with a vertical rectangular sampling area (a window of width, w, and height h) can be considered as (a) the probability of intersection with the window fixed and the joint randomly located in space, or (b) the probability of intersection with the joint fixed and the window randomly located in space. For case (a), the probability of intersection is proportional to the volume within which the center of the joint should lie in order to intersect the window. For case (b), the probability of intersection is proportional to the volume within which the center of the window should lie in order to intersect the joint. Case (a) was solved in an earlier paper (Kulatilake and Wu, 1984) in which the joints were represented as circular discs. For practical applications, it will be desirable to extend the solution to joints with non-circular shapes. In this endeavor, it is easier to approach through case (b) than case (a). The derivation for case (b) is presented in this paper. The joints are assumed to be circular discs. Derivations for other joint shapes such as rectangles, parallelograms and triangles will be given in a future paper.

We first identify the volume prescribed by the extreme positions that the window can assume while still intersecting the disc. Figure 1(a) shows a plan view of a disc of diameter D intersecting a vertical sampling plane (window). Strike and dipping direction of the disc are ox' and oy' respectively. Its dip angle and dip direction angle are ψ and α respectively. OX is the strike of the window. The angle of the sampling plane from north is γ.

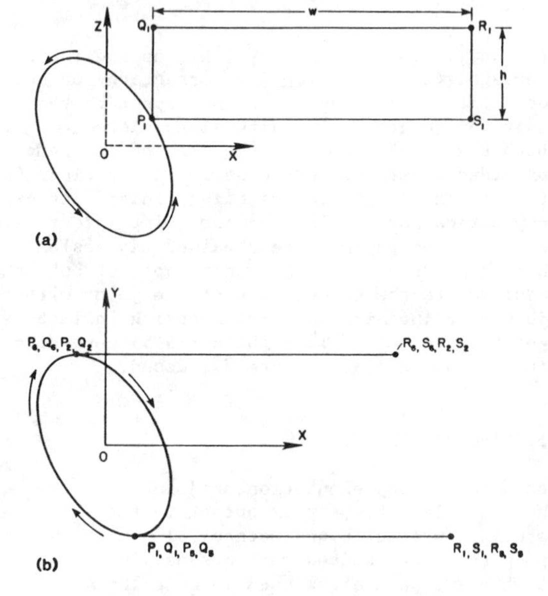

Fig.2 Generation of the volume for the intersection between the circular disc and the window. (a) Elevation view (b) plan view

Consider the extreme positions of a window, PQRS, that will allow it to intersect the disc. Let the lower corner just touch the edge of the disc as shown in Fig. 2(a). This position is identified as $P_1Q_1R_1S_1$. A volume is generated by moving the corner P all the way around the edge of the disc as indicated by the arrows in Fig. 2. The volume generated is shown in Fig. 3(a). This volume consits of: V_1, the rectangular prismoid (Fig. 3(b)) generated by the window as it is moved from $P_1Q_1R_1S_1$ to $P_2Q_2R_2S_2$ (also see Fig. 2(b)). V_2, the oblique cylinder generated by the line PS as P_1 is moved

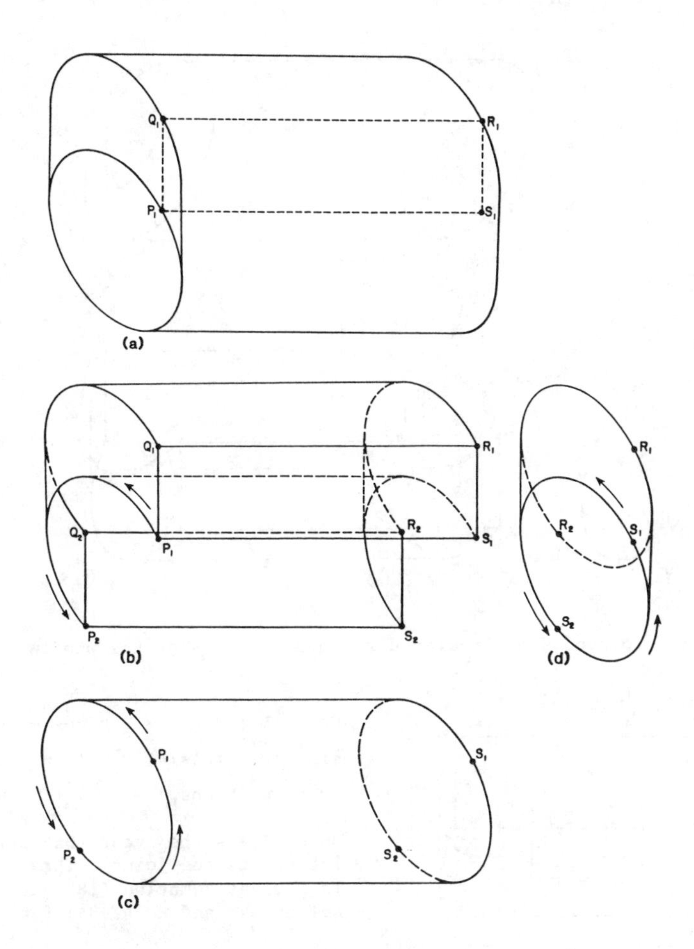

Fig.3 Generated volumes for circular disc when corner P of the window touches the disc.

from P_1 to P_2 to P_1 (Fig. 3(c)); and V_3, the oblique cylinder generated by the line RS when S is moved from S_1 to S_2 to S_1 (Fig. 3(d)). Any window with corner P touching the surface of the disc lies inside the volume shown in Fig. 3a. Thus, for this type of intersection the center of the window should lie within a volume equal to $V_1/4 + V_2/2 + V_3/2$.

At the other extreme, consider a window with its upper right corner, R, touching the disc as shown in Fig. 4 (a). This is position $P_3Q_3R_3S_3$. If the corner R is moved all the way around the edge of the disc, the volume generated is shown in Fig. 4(a), (b) and (c) as V_1, V_2, and V_3, respectively. For this type of intersection the center of the window also should lie within a volume equal to $V_1/4 + V_2/2 + V_3/2$. Two other extreme positions are obtained by putting corners Q and S on the edge of the disc. These are denoted by $P_5Q_5R_5S_5$ (Fig. 5(a)) and $P_7Q_7R_7S_7$ (Fig. 5(b)) respectively. The volume, V_1, is generated, by the movement of the corner Q along the edge of the disc as shown in Fig. 5(a). Plan view of $P_5Q_5R_5S_5$ and $P_6Q_6R_6S_6$ are shown in (Fig. 2(b). A similar volume is generated as shown in Fig. 5(b) when the corner S is moved along the edge of the disc. These four positions include all possible intersections between the window and the disc. Therefore, in order for the disc to intersect the window, the center of the window should lie within a volume

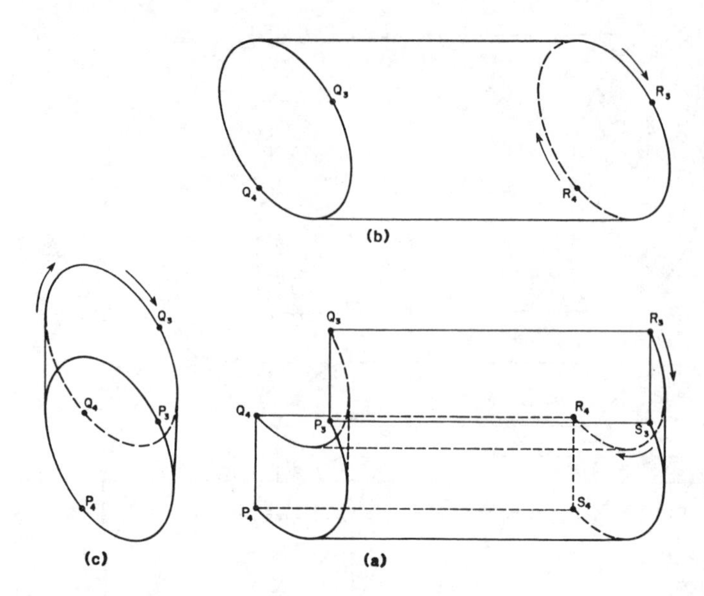

Fig.4 Generated volumes for circular disc when corner R of the window touches the disc.

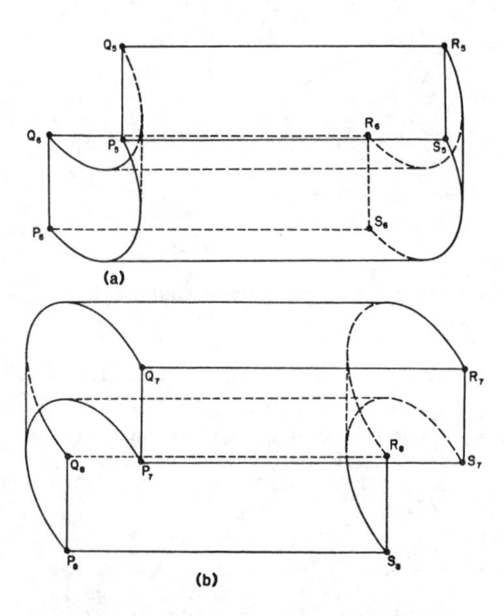

Fig.5 Generated volumes for circular disc when (a) corner Q (b) corner S of the window touches the disc.

$$V_T = V_1 + V_2 + V_3 \qquad (1)$$

The derivations for the volumes are not given in the paper due to space limitation. The probability of intersection is proportional to V_T and is

$$Pr(Is|D,\psi,\beta,h,w) \propto [Dwh(\cos^2\psi \cos^2\beta + \sin^2\beta)^{1/2} + (\pi/4)\, D^2 w\, |\sin\beta|\, \sin\psi + (\pi/4)\, D^2\, h \cos\psi] \qquad (2)$$

Where Is is the event that the window intersects the joint. This expression is same as Equation (18) given in Kulatilake and Wu (1984) for case (a).

$$Pr(Is|D,\psi,\beta,\ h/D \rightarrow \infty,\ w/D \rightarrow \infty) \propto D(\cos^2\psi \cos^2\beta + \sin^2\beta)^{1/2} \qquad (3)$$

We note that equations (2) and (3) were derived for the orientation of the disc shown in Fig. 1(a). The other possibility is shown in Fig. 1(b) in which the vertical sampling plane lies on the other side of the dip direction, OY'. For this case, the probabilities of intersection are given by:

$$Pr(Is|D,\psi,\beta,h,w) \propto [Dwh\ (\cos^2\psi \sin^2\beta + \cos^2\beta)^{1/2} + (\pi/4)\, D^2\, w\, |\cos\beta|\, \sin\psi + (\pi/4)\, D^2 h\cos\psi] \qquad (4)$$

and

$$Pr\ (Is|D,\psi,\beta,\ h/D \rightarrow \infty, w/D \rightarrow \infty) \propto D(\cos^2\psi \sin^2\beta + \cos^2\beta)^{1/2} \qquad (5)$$

The weighting function, W, which is used to correct the observed frequencies of orientation, should be inversely proportional to probability of intersection. Therefore,

$$W(\psi,\beta,D,H,w) = [D(\cos^2\psi \cos^2\beta + \sin^2\beta)^{1/2} + (\pi D^2/4h) |\sin\beta| \sin\psi + (\pi D^2/4W) \cos \psi]^{-1} \quad (6)$$

$$W(\psi,\beta,D,h/D \to\infty, w/D\to\infty) = (\cos^2\psi\cos^2\beta + \sin^2\beta)^{-1/2} (1/D) \quad (7)$$

for the condition in Fig. 1(a), and

$$W(\psi,\beta,D,h,w) = [D(\cos^2\psi\sin^2\beta + \cos^2\beta)^{1/2} + (\pi D^2/4h) |\cos\beta| \sin\psi + (\pi D^2/4w) \cos\psi]^{-1} \quad (8)$$

$$W(\psi,\beta,D,h/D\to\infty,w/D\to\infty) = (\cos^2\psi\sin^2\beta + \cos^2 \beta)^{-1/2} (1/D) \quad (9)$$

for the condition in Fig. 1(b).

3 DISCUSSION AND CONCLUSIONS

The weighting functions can be derived by considering the probability of intersection of either (a) a given window with a randomly located joint or (b) a given joint with a randomly located window. The case (a) was solved in Kulatilake and Wu (1984) for a circular joint. However, in order to derive the weighting functions for other joint shapes, it seems the approach in case (b) is easier than in case (a). Therefore, in this paper the weighting functions were derived for case (b) for a circular joint and it was shown that the weighting functions are same as for case (a). Because of the space limitation, the weighting functions for other shapes will be given in a future paper.

The weighting functions obtained for the sampling bias correction are functions of the size of the joint, size of the sampling domain and the orientation of the joint with respect to the the sampling plane. The weighting functions given in this paper reduce to the Terzaghi's (1965) weighting function for joints of infinite size intersecting infinite size domains.

All the weighting functions (Equation (6) through (9)) contain the joint diameter D as a variable. The values of D for a joint set are unknown. However, it is possible to apply these equations in practice by assuming the same value for D, for all the joints in the cluster. In such a case, effect of different D values will not be reflected on the correction. A reasonable value for D may be the mean diameter, $\bar{D}$. A procedure to estimate the mean diameter from the trace length measurements is given in Kulatilake and Wu (1986). Effect of $\bar{D}$ on the correction may be studied separately by choosing different values of $\bar{D}$ for the whole joint set.

ACKNOWLEDGEMENTS

The writer is grateful to the Arizona Mining and Mineral Research Institute, for support of this study.

REFERENCES

Baecher, G.B., Lanney, N.A. and Einstein, H.H. 1977. Statistical description of rock properties and sampling. 18th U.S. Symposium on Rock Mechanics, Colorado, 5C1-8.

Barton, C.M. 1978. Analysis of joint traces. 19th U.S. Symposium on Rock Mechanics, Vol. 1, 38-41.

Bridges, M.C. 1975. Presentation of fracture data for rock mechanics. 2nd Australia - New Zealand Conference on Geomechanics, Brisbane, 144-148.

Dershowitz, W.S. and Einstein, H.H. 1987. Three dimensional flow modeling in jointed rock masses. 6th International Congress on Rock Mechanics, Montreal, Canada, 87-92.

Einstein, H.H. et al. 1979. Risk analysis for rock slopes in open pit mines, Part I and IV, contract J0275015 MIT.

Kulatilake, P.H.S.W. and Wu, T.H. 1984. Sampling bias on orientation of discontinuities. Rock Mechanics and Rock Engineering, Vo. 17, 243-253.

Kulatilake, P.H.S.W. and Wu, T.H. 1986. Relation Between Discontinuity Size and Trace Length. 27th U.S. Symposium on Rock Mechanics, Tuscaloosa, Alabama, pp. 130-133.

Robertson, A. 1970. The interpretation of geologic factors for use in slope theory. Proceedings of the Symposium on the Theoretical Background to the Planning of Open Pit Mines, Johannesburg, South Africa, 55-71.

Terzaghi, R. 1965. Sources of error in joint surveys, Geotechnique, Vol. 15, 287-304.

Veneziano, D. 1978. Probabilistic model of joints in rock. Internal report, MIT, CAmbridge, MA.

Numerical Methods in Geomechanics (Innsbruck 1988), Swoboda (ed.)
© 1988 Balkema, Rotterdam. ISBN 90 6191 809 X

Mathematical model of probabilistic analysis for a jointed rock slope

Zhang Xing
Beijing University of Iron and Steel Technology, People's Republic of China

ABSTRACT: This paper is concerned with the failure probability analysis for rock slopes with multiple planar slip surfaces or with multiple wedges. The geomechanical parameters on any slip surface of a rock slope may be corrected with those on the other slip surfaces. A Monte-Carlo simulation technique, based on the random sampling of vectors with a jointly n-dimensional normal distribution associated with the geomechanical parameters on different planar slip surfaces or on slip surfaces of different wedges, is used to simulate the failures of the slopes. The failure probabilities of rock slopes which have different slope lengths, different numbers of planar slip surfaces or different numbers of wedges are calculated. Moreover, the effect of the correlation on the failure probabilities of the rock slopes is also discussed.

1 INTRODUCTION

Conventional studies of rock slope stability, based on deterministic principles, are increasingly being supplemented by risk and reliability analysis. The uncertainty associated with significant parameters can be reduced to some extent by proper site investigation and testing but it can never be eliminated [1].

The use of some probabilistic approachs for rock slopes was presented by McMahon [2], Piteau and Martin [3], Marek and Savely [4], Moriss and Stotter [5], Coates [6]. In all these studies, attention has been concentrated on the calculation of failure probability of individual slip surfaces. The failures of slip surfaces are regarded as independent events so the correlation between the failures can not be taken into account. Moreover, their methods can calculate only the entire failure probability of a rock slope rather than the actual failure probability at each slope heights where the slope face is intersected by slip surfaces.

Chowdhury [7] recently presented a risk model for the occurence of succesive failures based on a bivariate normal probability distribution of the safety margins associated with a pair of slip surfaces. In his method techniques for numerical integration was used to calculate the failure probability of the slope. His method is not appropriate for calcuating the failure probability of a slope whose slip surfaces are more than two.

In the mathematcal model to be presented in this paper, a Monte-Carlo simulation technique, based on the random sampling of vectors with a jointly n-D normal distribution, are used to simulate the failures of rock slopes with multiple planar slip surfaces or with multiple wedges.

It is unnecessary to assume that the failures along different slip surfaces are independent. The correlation between the failures may be taken into account in this model. Through theoretical demostration, the entire failure probability of a rock slope may be considered as the sum of the actual failure probabilities at different slope heights. Consequently, this method can calculate not only the entire failure probabilities of rock slopes but also the actual failure probabilities along any slip surfaces and the failure probabilities in different areas for the slopes with multiple wedges.

A subsidary objective of the paper is to provide a means for investigating the effect of correlation on failure probability of rock slopes. It should be pointed out that this method can also be used to study the correlation between different random parameters on the same slip surfaces of a rock slope. However, for the sake of simplity in introducing this model, the random parameters on the same slip surfaces are regarded as independent variables with a normal distribution.

2 RISK MODEL FOR A ROCK SLOPE WITH MULTIPLE PLANAR SLIP SURFACES

Consider a simple rock slope with three potential planar slip surfaces a, b and c only (see Fig. 1). The shear strength parameters c, ϕ and the geometric parameter,

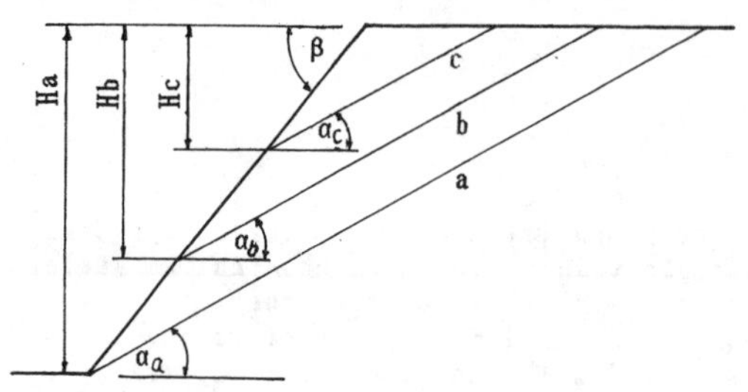

Fig.1 Slope with three planar slip surfaces

dip α, of each slip surface are considered as random parameters. Let A, B and C denote the events that failures occur on a, b and c respectively. The slope will fail when one or more of A, B and C occur. Thus there are $2^3-1=7$ possible cases in the slope such as; ABC, $AB\bar{C}$, $A\bar{B}C$, $A\bar{B}\bar{C}$, $\bar{A}BC$, $\bar{A}B\bar{C}$ and $\bar{A}\bar{B}C$. Here, ABC means that failures occur simultaneously along A, B and C; $AB\bar{C}$ means that failures occur simultaneously along A and B, and so on. The event that the slope fails, Tp, is the union of the seven events. Through the union and intersection oprations of the events, Tp may be written as;

$$\begin{aligned} Tp&=ABC\cup AB\bar{C}\cup A\bar{B}C\cup A\bar{B}\bar{C}\cup \bar{A}BC\cup \bar{A}B\bar{C}\cup \bar{A}\bar{B}C\\ &=AB\cup \bar{A}B\cup \bar{A}\bar{B}C\\ &=A\cup B\bar{A}\cup C\bar{A}\bar{B} \end{aligned} \quad (1)$$

moreover,

$$A\cap B\bar{A}=\phi \quad (2a)$$
$$A\cap C\bar{A}\bar{B}=\phi \quad (2b)$$
$$B\bar{A}\cap C\bar{A}\bar{B}=\phi \quad (2c)$$

where ϕ is an impossible event. Clearly, the events A, $B\bar{A}$ and $C\bar{A}\bar{B}$ are mutually exclusive events so the entire failure probability of the slope P(Tp) may be written as;

$$P(Tp)=P(A)+P(B\bar{A})+P(C\bar{A}\bar{B}) \quad (3)$$

Now the conclusion can be extended to a rock slope with n potential planar slip surfaces, $s_1, s_2, \cdots, s_n$. Let S_i denote the event that failure occurs along the ith slip surface (i=1, 2, ⋯, n), while S_1 denote that failure occurs along the lowest slip surface. Repeating the derivations in equation (1), one can write the event, Tp, that a slope with n potential planar slip surface fails, as;

$$Tp=S_1\cup S_2\bar{S}_1\cup\cdots\cup S_n\bar{S}_{n-1}\cdots\bar{S}_1 \quad (4)$$

where $\bar{S}_i$ is the complementary event of event S_i. Also there are following relations

$$(S_i\bar{S}_{i-1}\cdots\bar{S}_1)\cap(S_j\bar{S}_{j-1}\cdots\bar{S}_1)=\phi \quad (5)$$
$$(i=1, 2, \cdots, n;\ j=1, 2, \cdots, n,\ i\neq j)$$

It is evident that the n events in equation (4) are mutually exclusive events. Thus, the entire failure probability P(Tp) should be;

$$P(Tp)=P(S_1)+P(S_2\bar{S}_1)+\cdots+P(S_n\bar{S}_{n-1}\cdots\bar{S}_1) \quad (6)$$

The actual failure probability along ith slip surface, $P(P_i)$, is

$$P(P_i)=P(S_i\bar{S}_{i-1}\cdots\bar{S}_2\bar{S}_1) \quad (i=1, 2, \cdots, n) \quad (7)$$

Hence, equation (6) may be written as equation (8)

$$P(Tp)=\sum_{i=1}^{n}P(P_i) \quad (8)$$

It should be noted that if failures simultaneously occur along several slip surfaces this should be considered as the actual failure of the lowest one among the several slip surfaces. In other words, when an actual failure occurs along the ith slip surface, no failure occurs along all i-1 slip surfaces below the ith slip surface.

3 FAILURE PROBABILITY FOR A ROCK SLOPE WITH MULTIPLE PLANAR SLIP SURFACES

Consider the potential planar slip surfaces s_i. The safety margin of this slip surface may be expressed as;

$$SP_i=C_i H_i/\sin\alpha_i+N_i\,tg\,\phi_i-T_i \quad (9)$$

where N_i and T_i are normal and tangential forces along the slip surface respectively.

$$N_i=0.5\frac{tg\beta-tg\alpha_i}{tg\beta\,tg\alpha_i}H_i^2\,\gamma\cos\alpha_i \quad (10)$$

$$T_i=0.5\frac{tg\beta-tg\alpha_i}{tg\beta\,tg\alpha_i}H_i^2\,\gamma\sin\alpha_i \quad (11)$$

where the shear strength parameters of slip surfaces C_i, ϕ_i and the dip of slip surfaces, α_i, are regarded as random variables. The points intersected by slip surfaces with slope face H_i and the unit weight of rock mass γ are regarded as constants.

If each failure along the n slip surfaces is considered as independent event, the actual failure probability along the ith slip surface $P(P_i)$ in equation (7) may be written as

$$P(P_i)=P(S_i)P(\overline{S}_{i-1})\cdots P(\overline{S}_2)P(\overline{S}_1)$$
$$=P(SP_i \leqslant 0)P(SP_{i-1}>0)\cdots P(SP_2>0)P(SP_1>0) \quad (12)$$

in which each random parameter is given by

$$V=X+RD \quad (13)$$

where V is a random variable used in calculation; X is the mean value of random parameters; D is the standard deviation of random parameters; R is independent random variable with a standard normal distribution.

However, in fact, there is no reason to assume that the failures are independent each other, i.e. the random parameters between different slip surfaces are uncor related.

It is important to consider why there should be a correlation between random parameters of different slip surfaces. They should be correlated because not only they belong to the same slope but also the random parameters of different slip surfaces have the same statistical distribution. By considering the correlation, the actual failure probability along the ith slip surface in equation (7) should be written as a joint probability:

$$P(P_i)=P(S_i\overline{S}_{i-1}\cdots\overline{S}_2\overline{S}_1)$$
$$=P(SP_i\leqslant 0 \text{ and } S_{i-1}>0 \text{ and}\cdots\text{and } SP_2>0 \text{ and } SP_1>0)$$
$$(i=1,2,\cdots,n) \quad (14)$$

where the random parameters used in calculation should be dependent variables with a normal distribution or $n\times 1$ vectors with a jointly n-D normal distribution. The vectors can be obtained through a linear transformation to be introduced.

4 RANDOM SAMPLING OF VECTORS WITH A JOINTLY N-D NORMAL DISTRIBUTION

Through a linear transformation one can obtain a $n\times 1$ vector of random variables V having a jointly n-D normal distribution and a given $n\times n$ covariance matrix M. Here, use the prime symbol ′ to indicate the tran spose of a vector. Thus $V'=(v_1,v_2,\cdots,v_n)$. The detailed steps for obtaining the vector V are as follows:

1. Construct a vector of independent random variables having a standard normal distribution, $R'=(R_1,R_2,\cdots,R_n)$. The covariance of vector R is

$$E(R_K R_L)=\begin{cases}1 & ,k=l\\ 0 & ,k\neq l\end{cases} \quad (15)$$

2. Construct a middle vector of random variables $Y'=(y_1,y_2,\cdots,y_n)$ from equation (16). Then choose a transformation matrix A so that Y has a given $n\times n$ covariance matrix $M=(m_{KL})$.

$$Y=AR \quad (16)$$

or

$$\left.\begin{array}{l}Y_1=a_{11}R_1\\ Y_2=a_{21}R_1+a_{22}R_2\\ \vdots\\ Y_n=a_{n1}R_1+a_{n2}R_2+\cdots+a_{nn}R_n\end{array}\right\} \quad (17)$$

where A is in the following form

$$A=\begin{pmatrix}a_{11} & & & \\ a_{21} & a_{22} & & 0\\ \vdots & \vdots & \ddots & \\ \vdots & \vdots & & \ddots\\ a_{n1} & a_{n2} & \cdots & a_{nn}\end{pmatrix} \quad (18)$$

from equation (15) through (18), the covariance of the vector Y may be written as

$$E(Y_K Y_L)=E\left[\left(\sum_{i=1}^{K}a_{Ki}R_i\right)\left(\sum_{j=1}^{L}a_{Lj}R_j\right)\right]$$
$$=\sum_{j=1}^{K}a_{Kj}a_{Lj}$$
$$=(AA')_{KL} \quad (19)$$

moreover

$$E(Y_K Y_L)=m_{KL} \quad (20)$$

Comparing equations (19) with (20), one can obtain the expression of elements of matrix A from a column to another one.

$$a_{KL}=\frac{m_{KL}-\sum_{j=1}^{L-1}a_{Kj}a_{Lj}}{\sqrt{\left(m_{KL}-\sum_{j=1}^{L-1}a_{Lj}^{2}\right)}} \quad , \quad \sum_{j=1}^{0}a_{Kj}a_{Lj}=0$$
$$(1\leqslant l\leqslant k\leqslant n) \quad (21)$$

3. The vector V may be obtained from following equations:

$$\begin{array}{l}V_1=X_1+Y_1\\ V_2=X_2+Y_2\\ \vdots\\ V_n=X_n+Y_n\end{array} \quad (22)$$

Here it can be demonstrated that vector V has a vector of mean values X and a given covariance matrix M.

$$E(V_K)=E(X_K+Y_K)$$
$$=E(X_K)+E(Y_K)$$
$$=X_K \quad (1\leqslant k\leqslant n) \quad (23)$$

$$E\,[(V_K - E(V_K))(V_L - E(V_L)] = E\,[(V_K - X_K)(V_L - X_L)] = E(Y_K Y_L) = m_{KL} \quad (1 \leqslant l \leqslant k \leqslant n) \qquad (24)$$

Through the linear transformation, one can respectively obtain these vectors Vc, Vϕ, and Vα for random parameters C, ϕ and α. Then the failure probability of a rock slope with multiple planar slip surfaces, associated with the correlation between different slip surfaces, can be calculated.

5 RISK MODEL FOR A ROCK SLOPE WITH MULTIPLE WEDGES

Consider a rock slope with n wedges w_1, w_2, ..., w_n cut by two joint sets A, B, slope face and slope surface. The vertical distance from slope crest to the line of intersection is H for the wedges from w_1 to w_i, and H_h for the rest wedges (Fig. 2).

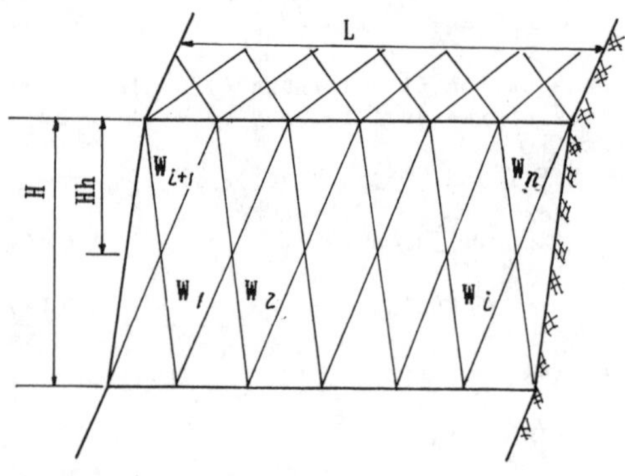

Fig. 2 Rock slope with multiple wedges

Whether these wedges fail or not depends on eight random parameters of joint set A and B, including the strength parameters Ca, C_b, ϕa, ϕ_b, geometric parameters, dips α_a, α_b, direction θa, θ_b. Let W_i denote the event that the ith wedges w_i fails (i=1, 2, ... n).

If any one or more of the n wesges fail, we define that the slope fails. In other words, if none of the n wedges fails, the slope will not fail, and let $P(\overline{W}_1 \overline{W}_2 \cdots \overline{W}_n)$ denote the probability that the slope will not fail. Thus, the entire failure probability of the slope is

$$P(Tw) = 1 - P(\overline{W}_1 \overline{W}_2 \cdots \overline{W}_n) \qquad (25)$$

From the equation presented by Hoek and Bray [8], one can calculate the safety margins of these wedges, SW_i (i=1, 2, ..., n). By considering the correlation between wedges, P(Tw) should also be written as a joint probability.

$$P(Tw) = 1 - P(SW_1 > 0 \text{ and } SW_2 > 0 \text{ and } \cdots \text{ and } SW_n > 0) \qquad (26)$$

Thus eight n×1 vectors Vca, Vcb, Vϕa, Vϕ Vαa, Vαb, Vθa, Vθb for random parameter Ca, Cb, ϕa, ϕb, αa, αb, θa and θb in equation (26) can be obtained by the random sampling introduced in the previously section.

Also we can calculate the failure probabilities of different areas of a rock slope with multiple wedges. Let P(Lw) denote the failure probability of lower area of the slope, including these wedges from w_1 to w_i. Let P(Uw) denote the failure probabilit of upper area of the slope, including the rest wedges. P(Lw) may be obtained by following equation

$$P(Lw) = 1 - P(\overline{W}_1 \overline{W}_2 \cdots \overline{W}_i) = 1 - P(SW_1 > 0 \text{ and } SW_2 > 0 \text{ and } \cdots \text{ and } SW_i > 0) \qquad (27)$$

moreover

$$P(Tw) = P(Lw) + P(Uw) \qquad (28)$$

From equations (27) and (28), P(Uw) is

$$P(Uw) = P(SW_1 > 0 \text{ and } SW_2 > 0 \text{ and } \cdots \text{ and } SW_i) - P(SW_1 > 0 \text{ and } SW_2 > 0 \text{ and } \cdots \text{ and } SW_n) \qquad (29)$$

6 RESULTS FOR SOME EXAMPLE PROBLOMS

A computer program incorporating the probabilistic models presented by this paper was developed. This program may be used to analyse rock slopes with multiple planar slip surfaces or with multiple wedges. The characters of the program are that the number of planar slip surfaces or the number of wedges may be changed from one to arbitary number; the correlation coefficients betwee random parameters of different slip surface ρ_{KL} (k=1, 2, ..., n; l=1, 2, ..., n; k≠l) may be changed from -1 to +1; the variation coefficients of random parameters VA may be changed arbitarily.

The first example chosen was a rock slope with four potential planar slip surfaces and with an angle of slope face β=45°. The unit weight of the slope material was assumed to be constant at valu γ=22 KN/M. The correlation coefficients ρ_{CKL}, $\rho_{\phi KL}$, $\rho_{\alpha KL}$ for C, ϕ and α between the four slip surfaces were simultaneously changed from zero to 0.75. The variance coefficient of random parameters VAc, VAϕ, and VAα for C, ϕ and α were 0.25. The numbers of slip surfaces were respec-

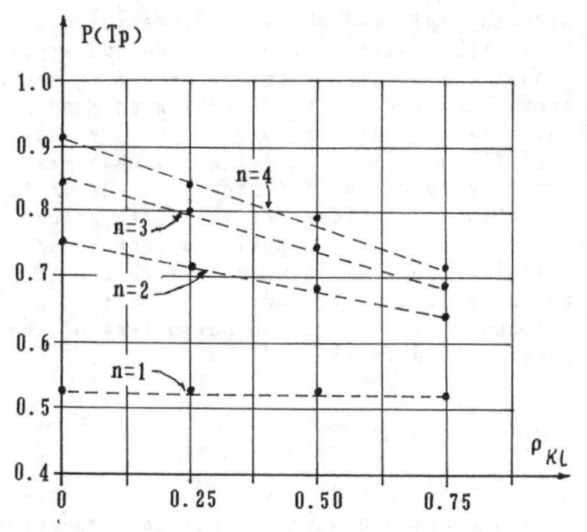

Fig.3 Relations between ρ_{KL} and P(Tp) with different numbers of planar slip surfaces, n

Table 1. Parameters used and results obtained for the first example

Parameters used:

$\beta=45°$ $\gamma=22KN/M$

$H_1=35$ M $H_2=30$ M $H_3=25$ M $H_4=20$ M

$C_1=C_2=C_3=C_4=20KN/M$ $\phi_1=\phi_2=\phi_3=\phi_4=30°$

$\alpha_1=\alpha_2=\alpha_3=\alpha_4=35°$ $VAc=VA\phi=VA\alpha=0.25$

$\rho_{cKL}=\rho_{\phi KL}=\rho_{\alpha KL}$

Numbers of slip surfaces (n)	1 (S_1)	2 (S_1,S_2)	3 (S_1,S_2,S_3)	4 (S_1,S_2,S_3,S_4)
Entire failure probabilities P(Tp):				
$\rho_{KL}=0$	0.5302	0.7627	0.8385	0.9048
$\rho_{KL}=0.25$	0.5302	0.7302	0.8186	0.8453
$\rho_{KL}=0.50$	0.5302	0.6929	0.7608	0.7934
$\rho_{KL}=0.75$	0.5302	0.6596	0.6965	0.7128
$P(Tp)=1-\prod_{i=1}^{n}{}'(S_i)$ *	0.5302	0.7498	0.8579	0.9133

Actual failure probabilities of each slip surfaces for the slope with 4 slip surfaces:

slip surface	S_1 $P(P_1)$	S_2 $P(P_2)$	S_3 $P(P_3)$	S_4 $P(P_4)$
$\rho_{KL}=0$	0.5302	0.2325	0.0956	0.0465
$\rho_{KL}=0.25$	0.5302	0.2000	0.0884	0.0267
$\rho_{KL}=0.50$	0.5302	0.1624	0.0682	0.0326
$\rho_{KL}=0.75$	0.5302	0.1294	0.0369	0.0163

* The equation was presented by Coates [6]

tively considered from one to four in this calculation. The points, from H_1 to H_4, where the slope face was intersected by slip surfaces S_1, S_2, S_3 and S_4 were 35M, 30M, 25M and 20M respectively. The parameters used in the calculation and the results obtained are shown in Table 1. The relations between ρ_{KL} and P(Tp), when the numbers of slip surfaces are different, are shown in Fig. 3.

The results show that the more the planar slip surfaces are, the greater the entire failure probability P(Tp) is. If a plus correlation cofficient is considered in calculation, the P(Tp) and the actual failure probability along each slip surface $P(P_i)$ will decrease. This is because the various slip surfaces will have either favorable conditions or unfavorable conditions, in obedience to some probability, simultaneously rather than randomly, when there is a plus correlation coefficient between random parameters of different slip surfaces. In other words, failures wil occur or not along different slip surfaces, in obedience to some probability, simultaneously rather than randomly. It should be noted that the results calculated by using this method, when $\rho_{KL}=0$, are almost the same with the results from the equation presented by Coates [6]. This is because the failures, in both cases, are regarded as independent.

The second example was a rock slope in which there was two joint sets A and B (see Fig.4).

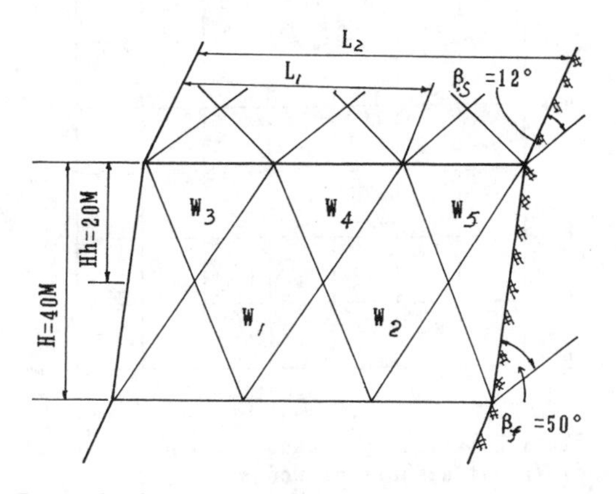

Fig.4 Rock slope with five wedges

The dip of the slope face β_f was 50° and the dip direction of the slope face λ_f was 185°. The dip of the slope surface βs was 12° and the dip direction of the slope surface λs was 195°. The slope height, H, where the intersection lines of two wedges, W_1 and W_2, intersect the slope face, was 40M, while the slope height, Hh, where the intersection lines of the other three wedges, W_3, W_4, W_5 intersect the slope face was 20M. The unit weight of rock mass γ was 22KN/M. The rest parameters used in the calculation and the results obtained are shown in Table 2. The relations between ρ_{KL} and P(Tw) with different numbers of wedges is shown in Fig.5.

The results calculated indicate that

Table 2. Parameters used and results obtained for the second example problem

Parameters used:
Ca=20KN/M Cb=25KN/M ϕ_a=25° ϕ_b=30°
α_a=45° α_b=70° θ a=105° θ b=235°
VAc=VAϕ=VAα=VAθ=0.25(for both joint sets)
$\rho_{cKL}=\rho_{\phi KL}=\rho_{\alpha KL}=\rho_{\theta KL}$ (for both joint sets)

Number of wedges (n)		ρ_{KL} =0	ρ_{KL} =0.25	ρ_{KL} =0.50	ρ_{KL} =0.75
3	P(Uw)	0.2020	0.1791	0.1586	0.1272
	P(Lw)	0.5414	0.5414	0.5414	0.5414
	P(Tw)	0.7436	0.7205	0.7000	0.6686
5	P(Uw)	0.1665	0.1386	0.1138	0.0800
	P(Lw)	0.7975	0.7475	0.7250	0.7075
	P(Tw)	0.9240	0.8861	0.8388	0.7875

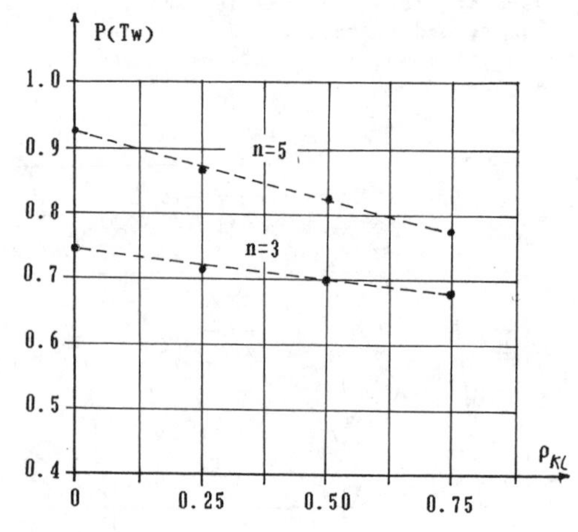

Fig.5 Relations between ρ_{KL} and P(Tw) with different numbers of wedges, n

the more the wedges in the slope are, the greater the failure probability of the slope is. Also, if a plus correlation between various wedges is considered, the failure probabilities of the slope will decrease.

7 CONCLUSION

In this paper, it has been shown that the risk of rock slopes with multiple planar slip surfaces or with multiple wedges may simply be analyzed by using this mthod. This method can calculate not only the entire failure probability of rock slopes but also the actual failure probability along each planar slip surface and the failure probability in different areas for rock slopes with multiple wedges. The entire failure probability will increase as the number of planar slip surfaces or the number of wedges increases. The correlation between random parameters of different slip surfaces has a significant effect on failure probability. The results calculated show that the failure probability decreases as the correlation coefficient which is positive increases.

It is important to note that this method may be extended to study the correlation between different random parameters of the same slip surfaces.

ACKNOWLEDGEMENT

The author would like to acknowledge Professor Liao Guohua for his encouragement, assistance and guidance. Many of the ideas presented in this paper originated from dicussions with him. The author also acknowledge Mr.Qu Si-je for revising this manuscript.

REFERENCES

[1] R.V.Whitman, Evaluating calculated risk in geotechnical engineering, J.Geotechnical engineering ASCE, 110, 145-188 (1984)

[2] B.K.McMahon, Probability of failure and expected volume of failure in high rock slopes, Proc. 2nd Australia-New Zealand Conf.on Geomechanics, 308-314, Brisbane, (1975)

[3] D.R.Piteau and D.C.Martin, Slope stability analysis and design based on probability techniques at Cassiar mine, Can.Min.Metall.J., 1-12 (1977)

[4] J.M.Marek and J.P.savely, Probabilistic analysis of plane shear failure mode, 19th U.S. Symp.on Rock Mechanics,40-44 (1978)

[5] P.Morris and H.J.Stotter, Open-cut design using probabilistic methods, Proc. 5th Int. Congr. on Rock Mechanics Vol.1, 107-113 (1983)

[6] I.F.Coates, Pit slope manual charpter 5 Design, Minister of supply and Canada Oteawa, 1977

[7] R.N.Chowdnury, Geomachanics risk model for multiple failures along rock discontinuities, Int.J.Rock Mech.Sci.& Geomech.Abstr. 23, 337-346 (1980)

[8] E.Hoek and J.Bradgn, Rock slope engineering, Institution of Mining and Metallurgy, London, 1974

Numerical Methods in Geomechanics (Innsbruck 1988), Swoboda (ed.)
© 1988 Balkema, Rotterdam. ISBN 90 6191 809 X

Analysis of fault initiation and fault spacing in elasto-plastic sedimentary materials

H.W.M.Witlox
Shell Research B.V., Rijswijk, Netherlands

ABSTRACT: In elasto-plastic sedimentary materials a fault appears as a thin, plastic softening zone (shear band) that is bounded by broader elastic zones consisting of material points that have unloaded. Associated with a certain prescribed incremental strain an incremental "elasto-plastic" stress (in the case of plastic yielding) and an incremental "elastic" stress (after unloading) can be determined. The ratio of these two stresses is identified to be the quantity that controls the initiation and spacing of the faults. Faults form when this ratio is sufficiently large and the fault spacing increases with this ratio. The results obtained are verified numerically for a Drucker-Prager type yield condition.

1. INTRODUCTION

The elasto-plastic deformation of sedimentary rock involves pressure-sensitive yielding, a non-associated flow rule and plastic hardening/softening [1,2]. The strain may localise along thin plastic softening zones called shear-bands or faults [3].

Mandel [4], Rudnicki and Rice [5], Vardoulakis [6] and Vermeer [7] studied the inception of localisation of deformation as a bifurcation from an homogeneous deformation. For the Drucker-Prager and Mohr-Coulomb yield criteria they formulated a bifurcation condition that needs to be satisfied for the initiation of a fault (shear-band formation). They derived an expression for a critical value of the hardening modulus and a formula for the shear-band orientation.

To trigger shear-band formation in finite-element simulations the original (perfect) material is slightly modified with the help of "imperfections". Imperfection analysis of strain localisation has been applied to metals (with associated plasticity) by Tvergaard et al. [8], Needleman and Tvergaard [9], and Triantafyllidis et al. [10], to over-consolidated clay (with a simple von Mises criterion) by Bardet and Mortazavi [11], and to rock (with non-associated plasticity) by Walters and Thomas [12].

The author [13] performed a finite-element simulation of faulting in a horizontal rock layer. A Drucker-Prager yield criterion with a non-associated flow rule was adopted, and eight-noded finite elements with reduced integration were utilised. A series of faults progressing away from an imperfection were generated. The fault spacing was found to increase with increasing layer thickness, increasing rate of softening and decreasing shear modulus.

In this paper a mathematical small-strain analysis of the initiation and spacing of faults is described. Thus the rigorous localisation theory (initiation of faults) of Rudnicki and Rice [5] is confirmed in a somewhat intuitive manner, and extended to the spacing of faults. The mathematical analysis is verified by the finite-element simulation of faulting in rock described above.

2. MATERIAL MODEL

An elasto-plastic material model is adopted. The vector notation is used for stresses and strains (see, for example, Table 4.2 in Reference 14), and the stress/strain relations are expressed in matrix form (see, for example, Section 7.3 in Reference 15). The model is characterised in the usual manner by a yield function F and a plastic potential function G. The functions $F = F(\underline{\sigma},\kappa)$ and $G = G(\underline{\sigma},\kappa)$ depend on the stress vector $\underline{\sigma}$ and the hardening parameter κ; κ is the integration of the equivalent incremental plastic strain $d\bar{\epsilon}^p$,

$$\kappa = \int d\bar{\epsilon}^p \ , \quad d\bar{\epsilon}^p = (2\, de^p_{ij}\, de^p_{ij})^{1/2} , \quad (1)$$

where de^p_{ij} (i,j=1,2,3) are the components of the incremental deviatory plastic strain tensor and the summation convention is adopted.

The constitutive law relates the incremental stress $\underline{d\sigma}$ and incremental hardening parameter $d\kappa$ to the incremental total strain $\underline{d\epsilon}$. In the interior F<0 of the yield surface Hooke's law applies,

$$\underline{d\sigma} = \mathbf{D}^e \underline{d\epsilon} \ , \ d\kappa = 0 \quad (2a)$$

and on the yield surface F=0,

$$\underline{d\sigma} = \mathbf{D}^{ep}(\underline{\sigma},\kappa)\ \underline{d\epsilon} \ , \ d\kappa = \underline{c}^T(\underline{\sigma},\kappa)\ \underline{d\epsilon} \ . \quad (2b)$$

Here $\mathbf{D}^e$ is the usual matrix of elastic constants (see, for example, Table 4.3 in Reference 14) and the superscript T denotes a vector transpose; the elasto-plastic matrix $\mathbf{D}^{ep}$ and the vector $\underline{c}$ are defined by

$$\mathbf{D}^{ep} = \mathbf{D}^e - \mathbf{D}^e \underline{b}\ \underline{c}^T ,$$
$$\underline{c} = (A + \underline{a}^T \mathbf{D}^e \underline{b})^{-1}\ \mathbf{D}^e \underline{a} \ . \quad (3)$$

Here $\underline{a}$, $\underline{b}$ are the stress gradients of F and G, and A is the hardening modulus. For a three-dimensional model $\underline{a}$, $\underline{b}$ are given by

$$\underline{a} = [\frac{\partial F}{\partial\sigma_{11}},\frac{\partial F}{\partial\sigma_{22}},\frac{\partial F}{\partial\sigma_{33}},\frac{\partial F}{\partial\sigma_{23}},\frac{\partial F}{\partial\sigma_{31}},\frac{\partial F}{\partial\sigma_{12}}]^T ,$$
$$\underline{b} = [\frac{\partial G}{\partial\sigma_{11}},\frac{\partial G}{\partial\sigma_{22}},\frac{\partial G}{\partial\sigma_{33}},\frac{\partial G}{\partial\sigma_{23}},\frac{\partial G}{\partial\sigma_{31}},\frac{\partial G}{\partial\sigma_{12}}]^T , \quad (4)$$

with $\underline{b}$ assumed to be normalised such that $b_{ij} b_{ij} = 0.5$ applies for its associated tensor components. The hardening modulus A defines the rate of expansion of the yield surface

$$A = - \frac{\partial F}{\partial\kappa} \ . \quad (5)$$

It is assumed that initially, for small κ, the yield surface expands (plastic hardening: A>0) and subsequently contracts (plastic softening: A <0). With the consistency condition dF = 0 and equations (4),(5), A can be written as a ratio of elastic strain $\underline{d\epsilon}^e$ to plastic strain $\underline{d\epsilon}^p$

$$A = (\underline{a}^T \underline{d\sigma})/d\kappa = (\underline{a}^T \underline{d\epsilon}^e)/\ d\bar{\epsilon}^p \ . \quad (6)$$

The phases in the elasto-plastic deformation of a material point are schematised in Figure 1. A material point will first deform elastically according to (2a) and its stress vector $\underline{\sigma}$ will move towards the yield surface F=0 (phase I). Subsequently $\underline{\sigma}$ will move along the yield surface, in the course of which the yield surface first expands and then contracts (phases II, III). Points on a fault will continue to soften (phase IVa), while points adjacent to a fault unload and return back to the elastic state (phase IVb). In the next section a parameter is derived that expresses the tendency to unload and therefore increases with the width of the elastic unloading zones, i.e. the fault spacing.

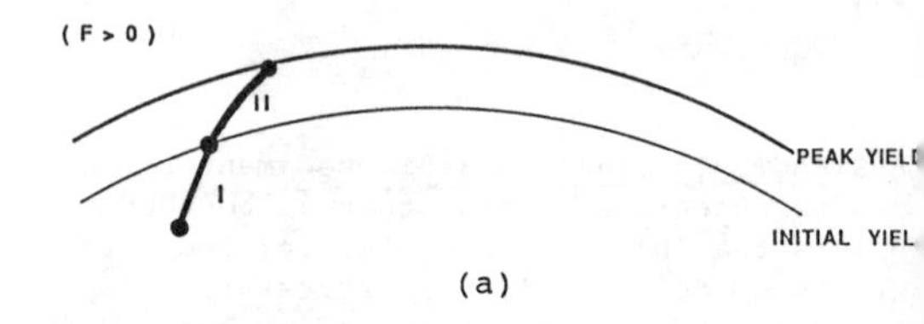

(a)

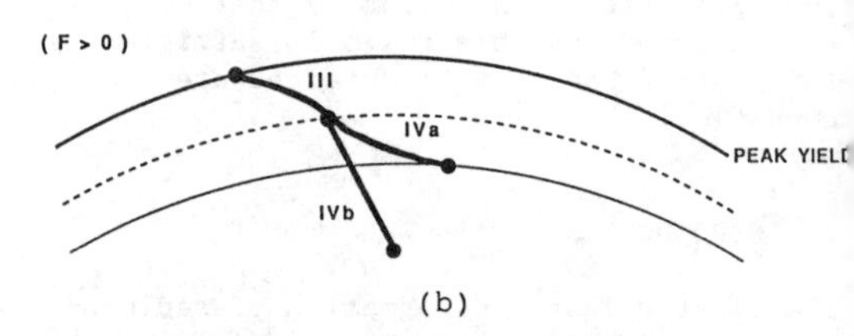

(b)

Figure 1. Phases in motion of stress vector $\underline{\sigma}$ relative to the yield surface $F(\underline{\sigma},\kappa)=0$
(a) elasticity (I), hardening (II)
(b) softening (III,IVa), unloading (IVb)

3. TENDENCY TO UNLOAD

To study the tendency to unload of a material point $\underline{x}$, we consider its stress vector $\underline{\sigma} = \underline{\sigma}(\underline{x})$ and apply a strain increment $\underline{d\epsilon}$ at $\underline{x}$.

For $\underline{\sigma}$ located on the yield surface, we define $d\sigma^{ep}$ to be the component along the inward normal to the yield surface of the elasto-plastically determined incremental stress $\underline{d\sigma} = \mathbf{D}^{ep}\underline{d\epsilon}$. This component pushes $\underline{\sigma}$ towards the interior of the yield surface in the case of plastic softening (see Figure 2), and increases with the rate of softening (i.e. the softening modulus -A). From equations (5), (6) and (2b) we find

$$(\underline{a}^T\underline{a})^{1/2} d\sigma^{ep} = - \underline{a}^T \mathbf{D}^{ep} \underline{d\epsilon} = - A\ \underline{c}^T d\epsilon \ . \quad (7)$$

For $\underline{\sigma}$ located just within the yield surface, we define $d\sigma^e$ to be the component along the outward normal to the yield surface of the elastically determined incremental stress $\underline{d\sigma} = \mathbf{D}^e\underline{d\epsilon}$. This component pushes $\underline{\sigma}$ towards the yield surface (see Figure 2), and increases with the elastic shear modulus G. From equations (3) and (2b) we find

$$(\underline{a}^T\underline{a})^{1/2} d\sigma^e = - \underline{a}^T \mathbf{D}^e \underline{d\epsilon} = [A + \underline{a}^T \mathbf{D}^e \underline{b}]\ \underline{c}^T \underline{d\epsilon}\ . \tag{8}$$

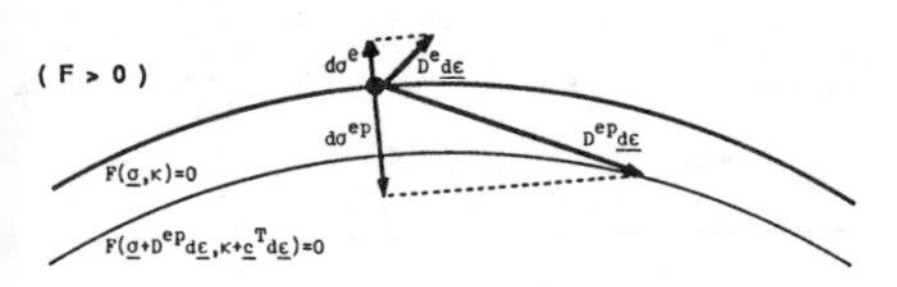

Figure 2. Counteracting stresses affecting the tendency to unload from the yield surface

The stresses $d\sigma^{ep}$ and $d\sigma^e$ counteract. The tendency to unload is expected to increase if the relative magnitude s of these stresses increases. From equations (7) and (8) we find

$$s = \frac{d\sigma^{ep}}{d\sigma^e} = [1 + A/(\underline{a}^T \mathbf{D}^e \underline{b})]^{-1} - 1\ . \tag{9}$$

For the usual material models $\underline{a}^T \mathbf{D}^e \underline{b}$ is proportional to the elastic shear modulus G, and hence s increases with increasing ratio r = -A/G of the plastic softening modulus -A and G.

The value of s must be sufficiently large for fault initiation. This rather intuitive observation is in agreement with the rigorous shear-band localisation theory of Rudnicki and Rice who derived an upper limit of r (see equation (20) in Reference 5). In addition, the fault spacing, i.e. the width of the elastic unloading zones, is expected to increase with the unload parameter s. These theoretical predictions are verified below with a numerical example.

4. NUMERICAL EXAMPLE

4.1. Mathematical model

A horizontal rock layer with initial gravitational stresses is extended by a uniform strain acting along its entire base. Figure 3 depicts the layer geometry and the imposed boundary conditions. The maximum basal displacement u_0 is built up in a sequence of small increments. An incision at the top of the middle of the rock layer is applied to trigger faulting.

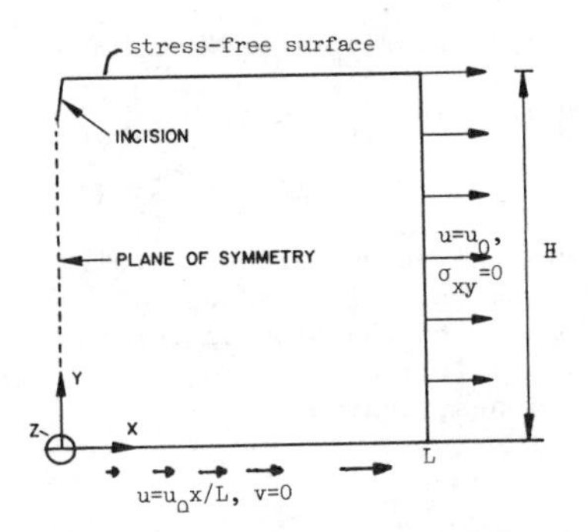

Figure 3. Geometry of rock layer and imposed boundary conditions

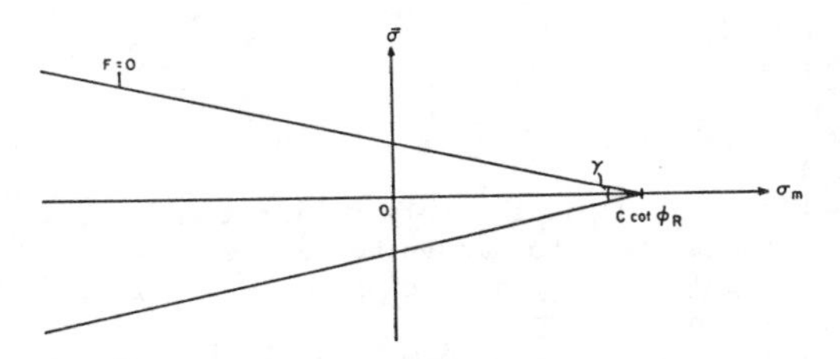

Figure 4. Drucker-Prager yield surface $F(\underline{\sigma},\kappa)$ in σ_m, $\bar{\sigma}$ - plane; $\gamma = 2\ \arctan(3a)$

The rock is assumed to deform according to an extended Drucker-Prager criterion. The yield function F and the plastic potential function G are defined by (see Figure 4)

$$F(\underline{\sigma},\kappa) = \bar{\sigma} - 3\ a(\kappa)\ (\sigma_m - C(\kappa)\ \cot\ \phi_r), \tag{10}$$
$$G(\underline{\sigma},\kappa) = \bar{\sigma} - 3\ \beta(\kappa)\ \sigma_m\ ,$$

with the mean stress σ_m and the equivalent stress $\bar{\sigma}$ given by

$$\sigma_m = \frac{1}{3}\ \sigma_{kk}\ ,\ \bar{\sigma} = [\frac{1}{2}(\sigma_{ij}-\sigma_m)(\sigma_{ij}-\sigma_m)]^{1/2}\ . \tag{11}$$

The friction angle ϕ and the dilation angle ψ are related through a and β,

$$a = \hat{a}(\phi) = (9 + 3\ \sin^2\phi)^{-1/2}\ \sin\ \phi\ , \tag{12}$$
$$\beta = \hat{\beta}(\psi) = (9 + 3\ \sin^2\psi)^{-1/2}\ \sin\ \psi\ ,$$

to the hardening parameter κ by

$$\alpha(\kappa) = \alpha_R + 4(\alpha_M - \alpha_R)\, B_2\, \kappa\, (B_2+\kappa)^{-2},$$
$$\beta(\kappa) = \beta_R + 4(\beta_M - \beta_R)\, B_2\, \kappa\, (B_2+\kappa)^{-2}, \quad (13)$$

with

$$\alpha_R=\hat{\alpha}(\phi_R),\ \alpha_M=\hat{\alpha}(\phi_M),\ \beta_R=\hat{\beta}(\psi_R),\ \beta_M=\hat{\beta}(\psi_M). \quad (14)$$

The cohesion C is related to κ by

$$C = C_M + 4(C_R - C_M)\, B_1\, (B_1+\kappa)\, (2B_1+\kappa)^{-2}. \quad (15)$$

Figure 5 depicts the above hardening rule; the material point initially undergoes friction hardening and cohesion softening, and subsequently friction softening and continued cohesion softening.

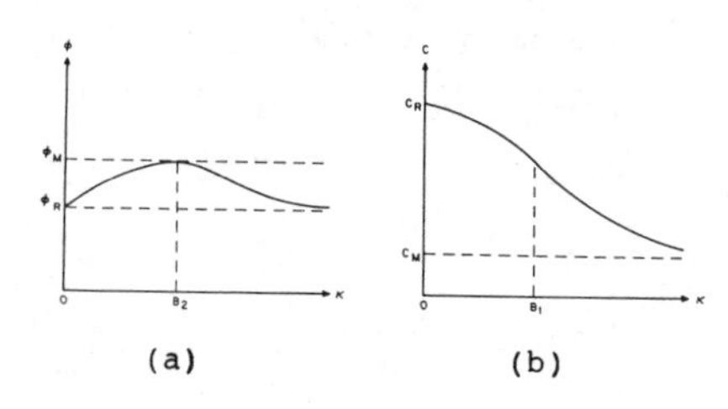

(a) (b)

Figure 5. Hardening rule relating shape of yield surface to hardening parameter κ
(a) relationship between friction angle ϕ and κ
(b) relationship between cohesion C and κ

With equations (4-6) and (8) the following expressions are easily derived for the hardening modulus A and the unload parameter s:

$$A = (d\bar{\sigma} + 3\alpha\, d\sigma_m)/\, d\kappa$$
$$= 3\, \frac{\partial\alpha}{\partial\kappa}(C \cot\phi_R - \sigma_m) + 3\alpha\, \frac{\partial C}{\partial\kappa} \cot\phi_R\ , \quad (16)$$
$$s = [1 - r(1 + 3\alpha\beta K/G)]^{-1} - 1\ .$$

with K being the elastic bulk modulus. For realistically chosen parameters $3\alpha\beta K/G << 1$ usually applies, and hence s increases with increasing $r = -A/G$.

For further details of the mathematical model the reader is referred to Reference 13. The mathematical model is solved numerically by the finite element method. Eight-noded two-dimensional isoparametric elements are used with reduced integration [16]. The solution method utilises a sub-incremental initial-stress approach with a modified acceleration procedure [17].

4.2. Numerical results

The finite element was refined near the incision. The location of the faults was found to be independent of the mesh chosen.

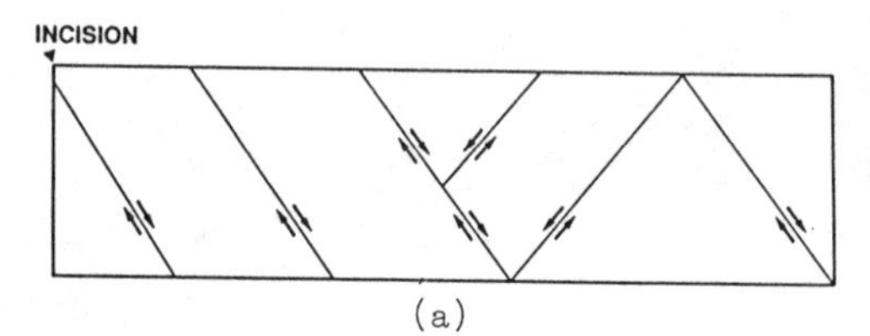

(a)

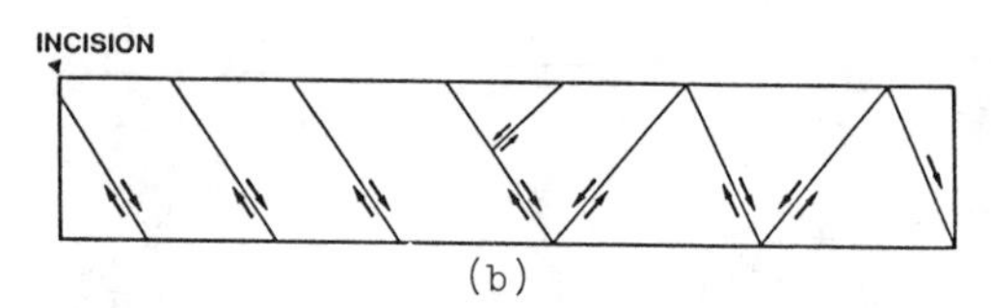

(b)

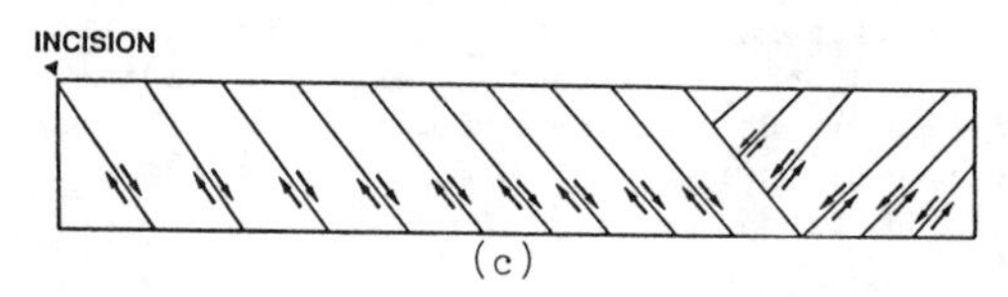

(c)

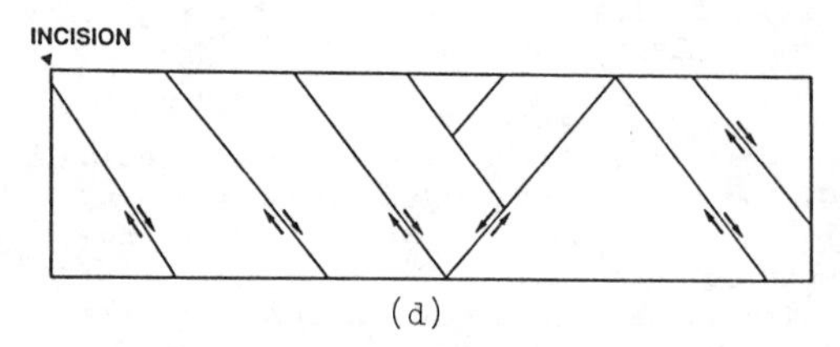

(d)

Figure 6. Dependence of fault spacing on layer thickness H and material properties
(a) H=2000 m, $G=1.6\times10^{10}\ N/m^2$, $B_1=B_2=5.0\times10^{-4}$
(b) H=1200 m, $G=1.6\times10^{10}\ N/m^2$, $B_1=B_2=5.0\times10^{-4}$
(c) H=1200 m, $G=1.6\times10^{10}\ N/m^2$, $B_1=B_2=1.0\times10^{-3}$
(d) H=2000 m, $G=2.4\times10^{10}\ N/m^2$, $B_1=B_2=5.0\times10^{-4}$

Figure 6 includes the fault patterns obtained for the right-hand half of the rock layer, for which the material properties $\phi_R=30^o$, $\phi_M=45^o$, $\psi_R=\psi_M=0^o$, $C_R = 5\times10^6\ N/m^2$ and $C_M = 0\ N/m^2$ were adopted. Faulting occurs propagating away from the triggering incision. Figure 6 clearly demonstrates the increase of fault spacing with the layer thickness (compare Figures 6a and 6b), the softening modulus -A (compare Figures 6b and 6c in which the higher softening modulus is reflected by a lower value for B_1 and B_2), and the shear modulus G (compare Figures 6c and 6d). Variation of other parameters did not lead to a significant change in the fault spacing. The reader is referred to Reference 13 for further details.

These numerical results are in complete agreement with the theoretical observation in Section 2 that the fault spacing increases with the dimensionless ratio $r = -A/G$. In the dimensionless formulation of the problem it is easily shown (see Section 2 of Reference 13) that r, and hence the dimensionless fault spacing, is independent of the layer thickness. Thus the dimensional fault spacing should indeed vary linearly with the layer thickness.

In the case of a Drucker-Prager yield criterion, Rudnicki and Rice derived the bifurcation condition and calculated the angle between the shear band and the direction of maximum principal compressive stress. The corresponding formulas (see equations (19) and (20) in Reference 5) are expressed in terms of α, β and the principal stresses. In our numerical analysis the values for these were calculated at the plastic integration points and subsequently inserted in these formulas. Figure 7 includes the shear-band directions thus obtained. Note that the fault is indeed parallel to these directions, and therefore our numerical results are in agreement with the theory of Rudnicki and Rice.

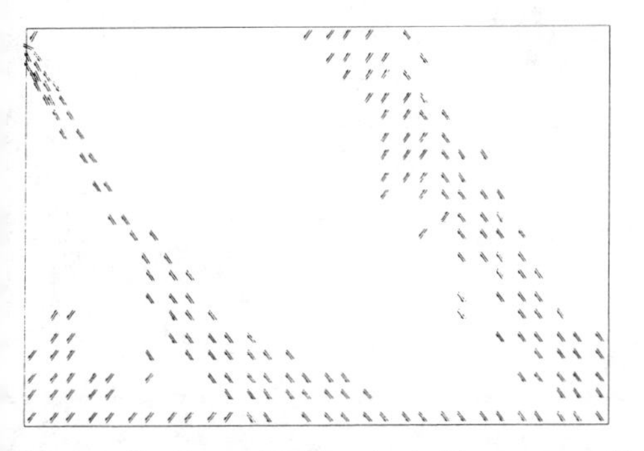

Figure 7. Shear-band orientations according to Rudnicki and Rice at softening integration points in rock layer; the faults correspond to the left-most faults in Figure 6a

4.3. Comparison with experimental results

The linear variation of fault spacing with layer thickness was also confirmed experimentally by Vendenville [18]. However, the dependence of fault spacing on the material properties does not seem to have been studied experimentally. A verification of the theory presented in this report requires a precise knowledge of the hardening rule in order to determine the hardening modulus A. A method of evaluation of the yield function $F = F(\underline{\sigma},\kappa)$ as a function of the hardening parameter κ has been indicated by Tisa and Kovari [19].

REFERENCES

[1] P.A. Vermeer and R. De Borst, Non-associative plasticity for soils, concrete and rock, Heron 29 (1984).

[2] J.W. Rudnicki, A class of elastic-plastic constitutive laws for brittle rock, Journal of Rheology 28, 759-778 (1984).

[3] S. Sture and H.Y. Ko, Strain softening of brittle geological materials, Int. J. Num. Anal. Meth. Geom. 2, 237-253 (1978).

[4] J. Mandel, Conditions de stabilite et postulat de Drucker, Proc. IUTAM Symp. on Rheology and Soil Mechanics, Grenoble, 331-341, Springer Verlag, Berlin (1966).

[5] J.W. Rudnicki and J.R. Rice, Conditions for the localization of deformation in pressure-sensitive dilatant materials, J. Mech. Phys. Sol. 23, 371-394 (1975)

[6] I. Vardoulakis, Shear band inclination and shear modulus of sand in biaxial tests, Int. J. Num. Anal. Meth. Geom. 4, 103-119 (1980).

[7] P.A. Vermeer, A simple shear-band analysis using compliances, Proc. IUTAM Conf. on deformation and failure of granular materials, Delft, 31 Aug. - 3 Sept. 1982, 493-499, Balkema, Rotterdam.

[8] V. Tvergaard, A. Needleman and K.K. Lo, Flow localization in the plane strain tensile test, J. Mech. Phys. Sol. 29, 115-142 (1981).

[9] A. Needleman and V. Tvergaard, Finite element analysis of localization in plasticity, Finite elements - Special Problems in Solids Mechanics 5, 94-157, Prentice Hall.

[10] N. Triantafyllidis, A. Needleman, and V. Tvergaard, On the development of shear bands in pure bending, Int. J. Sol. Str. 18, 121-138 (1982).

[11] J.P. Bardet and S.M. Mortazavi, Simulation of shear band formation in overconsolidated clay, Proc. of the Second Int. Conf. on Constitutive Laws for Engineering Materials: Theory and Applications, 5-8 January 1987, Tucson, 805-812.

[12] J.V. Walters and J.N. Thomas, Shear zone development in granular materials, Proc. of the Fourth Int. Conf. on Num. Meth. in Geomechanics, Edmonton, 263-274 (1982).
[13] H.W.M. Witlox, Finite element simulation of basal extensional faulting within a sedimentary overburden, Eur. Conf. Num. Meth. Geom., Stuttgart, September 16-18 (1986).
[14] K. J. Bathe, Finite element procedures in engineering analysis, Prentice Hall, New Jersey (1982).
[15] D.R.J. Owen and E. Hinton, Finite elements in plasticity: theory and practice, Pineridge Press, Swansea (1980).
[16] O.C. Zienkiewicz, The finite element method, Third edition, McGraw Hill, London (1977).
[17] J.N. Thomas, An improved accelerated initial-stress procedure for elasto-plastic finite element analysis, Int. J. Num. Anal. Meth. Geom. 8, 359-379 (1984).
[18] B. Vendeville, Champs de failles et tectonique en extension: modelisation experimentale, Thèse presentee devant l'Universite de Rennes I (1987).
[19] A. Tisa and K. Kovari, Continuous failure state direct shear tests, Rock Mech. and Rock Eng. 17, 83-95 (1984).

Numerical Methods in Geomechanics (Innsbruck 1988), Swoboda (ed.)
© 1988 Balkema, Rotterdam. ISBN 90 6191 809 X

Application of a joint model to concrete-sandstone interfaces

J.P.Carter & L.H.Ooi
University of Sydney, Australia

ABSTRACT: An elastoplastic shear hardening and softening model for the behaviour of bonded concrete-sandstone interfaces is presented. The model incorporates the gradual degradation of the cohesive component of shear strength (i.e. softening) with shear displacement, as well as the mobilisation of the frictional component of strength (i.e. hardening) as the shearing creates a rupture surface. The model is capable of matching well the behaviour of bonded interfaces observed in laboratory direct shear tests.

1. INTRODUCTION

Understanding and being able to predict the mechanical behaviour of interfaces between structural elements and rock masses are important in foundation engineering. Typically in the stronger rocks, this behaviour is highly non-linear and involves significant coupling between the shear and dilational modes of behaviour. Relative or shear displacement of the two sides of the interface is very often accompanied by some normal separation of the interface (dilation).

Predicting the mechanical behaviour of natural rock joints has received much attention in the literature (e.g. Patton, 1966; Jaeger, 1971; and many others). In many of these works, models have been developed to describe and predict the behaviour of perfectly mated and perhaps rough, naturally formed interfaces in rock. In most cases no actual bonding or cementation across the interface has been assumed. In these cases the shear strength of the interface is derived from the combined effects of friction, interlocking and the tendency for the rough interface to dilate.

In contrast to natural joints, the man-made interfaces usually involve a degree of bonding between the two surfaces as, for example, when a concrete foundation is cast into a rock socket. In this case the cementation or bonding may make a significant contribution to the shear strength of the interface and will combine, usually in some complex way, with the other factors like surface roughness, friction and dilation to determine the complete shear behaviour of the interface.

In this paper a set of constitutive relations are presented to describe the coupled shear and dilational behaviour of bonded interfaces between concrete and Hawkesbury sandstone. The model derived here is based on the theory of shear hardening/softening elastoplasticity.

2. TYPICAL BEHAVIOUR

The constitutive model presented in this paper was derived after consideration of experimental data from direct shear tests on bonded concrete-sandstone interfaces (Ooi and Carter, 1987; Ooi, 1988). The specimens tested were manufactured from a length of core of Hawkesbury sandstone, 76 mm in diameter, and from concrete. The unconfined compressive strengths of the sandstone and the concrete were found to be 19 MPa and 59 MPa, respectively.

The interface was formed across approximately a diametral plane of the core, which had been roughened with specially prepared asperities. These asperities were cut into the sandstone using a carborundum file guided by a template and producing a saw-tooth pattern in cross-section. Each asperity had a total base width of 10 mm and a height of 2.5 mm. The concrete was cast onto the roughened surface of the sandstone and allowed to cure for at least 28 days prior to shear testing.

The direct shear tests were carried out under conditions of constant normal stiffness, using the device described by Ooi and Carter (1987), and typical data (from tests A and B) are given in Figs 1 to 3. It can be

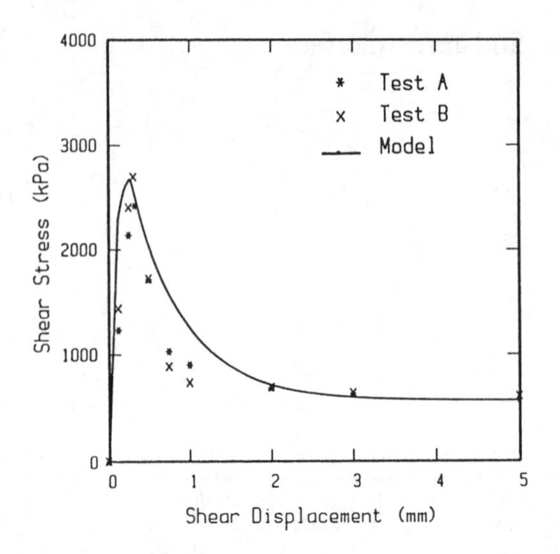

Figure 1. Shear stress versus shear displacement

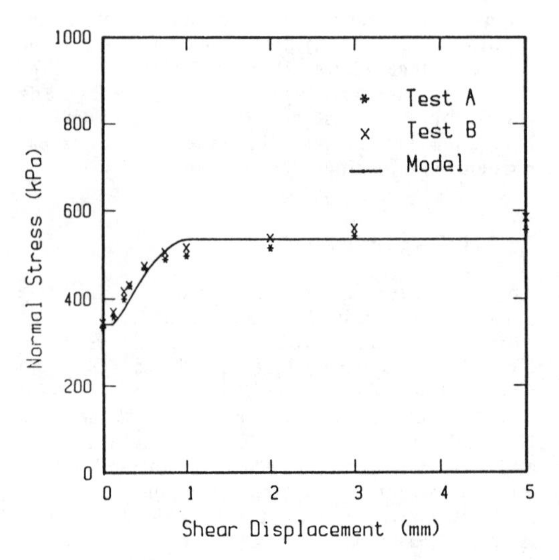

Figure 2. Normal stress versus shear displacement

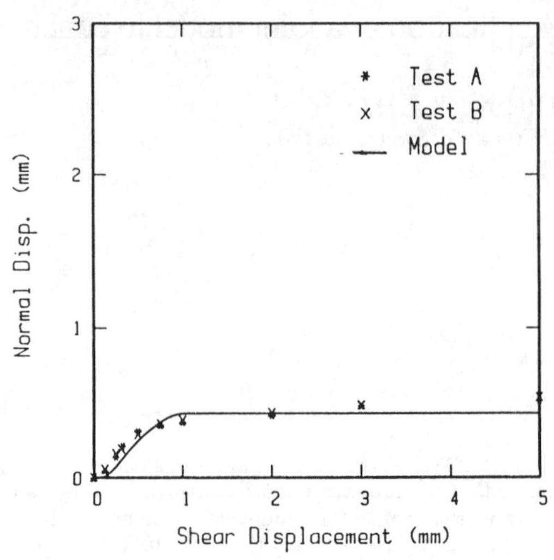

Figure 3. Normal displacement versus shear displacement

observed that with this kind of apparatus any dilation (normal displacement) that accompanies the shearing of the interface results in an increase in normal stress - see Figs. 2 and 3. The rates of dilation and normal stress increase are high at first but eventually the dilation abates.

Of course, an increase in normal stress will tend to enhance the frictional component of shear resistance. However, Fig. 1 shows that the shear stress rises to a peak at low values of shear displacement and then subsequently weakens with further shearing. This suggest that some damage to the bonding of the interface results from continued shearing and that this form of weakening tends eventually to dominate the strengthening effects of dilation.

3. PROPOSED CONSTITUTIVE MODEL

From the above discussions it should be apparent that any reasonable model for the bonded interface ought to include some form of shear strength weakening. The slight non-linearity of the pre-peak response shown in Fig. 1 indicates that perhaps some hardening may also be included in the model.

It is assumed that the bonded interface will initially behave in a linear elastic manner. The constitutive equations for elastic behaviour, which relate the increases in shear and normal stresses, τ and σ, to the shear and normal displacements, u and v, may be written in incremental form as

$$\begin{Bmatrix} d\tau \\ d\sigma \end{Bmatrix} = \begin{bmatrix} k_s & 0 \\ 0 & -k_n \end{bmatrix} \begin{Bmatrix} du \\ dv \end{Bmatrix} \tag{1}$$

in which k_s and k_n are the elastic shear and normal stiffness of the interface respectively. Compressive normal stress and dilative normal displacement have been assumed positive.

Once the interface yields plastically (slips), the displacements are assumed to be composed of elastic and plastic parts, i.e.

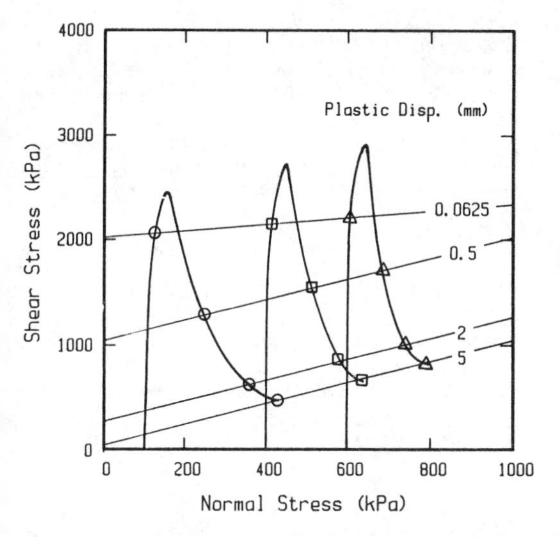

Figure 4. Procedure for determining strength parameters

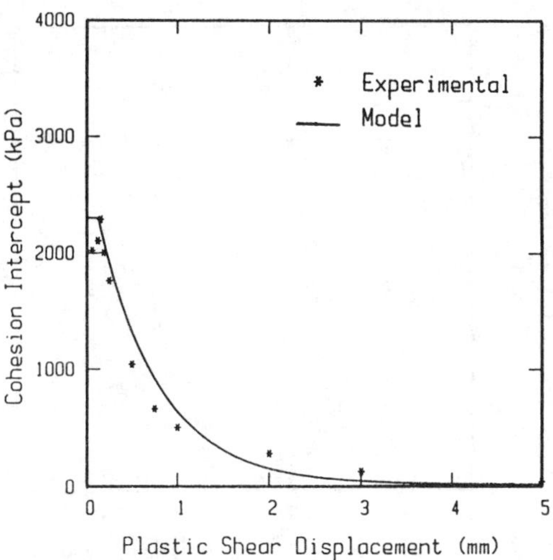

Figure 5. Cohesion intercept versus plastic shear displacement

$$du = du^e + du^p \quad (2a)$$

$$dv = dv^e + dv^p \quad (2b)$$

In this model it is assumed that plastic yielding of the interface will occur whenever the following criterion is satisfied:

$$\tau = c + \sigma \tan\varphi \quad (3)$$

in which c and φ are the interface strength parameters. Equation (3) is of course the well known Mohr-Coulomb criterion, but when used in this model it will be further assumed that the parameters c an φ may vary with the plastic shear displacement, u^p.

The precise manner in which the strength parameters vary during shearing has been determined from the results of over 20 direct shear tests, covering a range of initial normal stress from 50 to 800 kPa and imposed external normal stiffness from 390 kPa/mm to 1250 kPa/mm. Stress paths for each test were plotted and superimposed on these paths were contours of plastic shear displacement. This procedure is illustrated schematically in Fig. 4. These contours were found to be reasonably linear and in fact they represent yield surfaces at various stages of the deformation. In this way it was possible to determine the progress of plastic yielding in a large number of tests and the dependence of the yield surface on plastic shear displacement. In particular, it is possible to determine the variation of the cohesion intercept (on the shear stress axis) and the slope of the yield surface, with plastic shear displacement. The variations determined for the bonded interfaces are shown in Figs 5 and 6.

The data in Fig. 5 demonstrate that the softening of the interface is due to a reduction in the cohesive component of strength (destruction of bonding or cementation) and that it is possible to relate this to the plastic shear displacement, u^p, in the following form

$$c = c_0, \qquad u^p \leqslant \delta \quad (4a)$$

$$c = c_0 \exp[-k_1(u^p - \delta)/u_b], \qquad u^p > \delta \quad (4b)$$

where c_0 = the maximum (initial) cohesion intercept and δ is the threshold value of plastic shear displacement at which the damage commences. The data in Fig. 5 indicate that this threshold value is small, typically 0.13 mm. Beyond this plastic displacement the cohesive strength is degraded exponentially at a rate determined by the dimensionless parameter k_1. The constant u_b is unit displacement (usually 1 mm) and has been introduced to normalise the plastic shear displacement. The data in Fig. 5 also indicate that k_1 has a value of about 1.5.

Fig. 6 indicates that the frictional strength of the interface is not mobilised immediately after the commencement of plastic shearing. Instead, the frictional strength is mobilised over a finite range of plastic displacement, albeit at a relatively rapid rate. This form

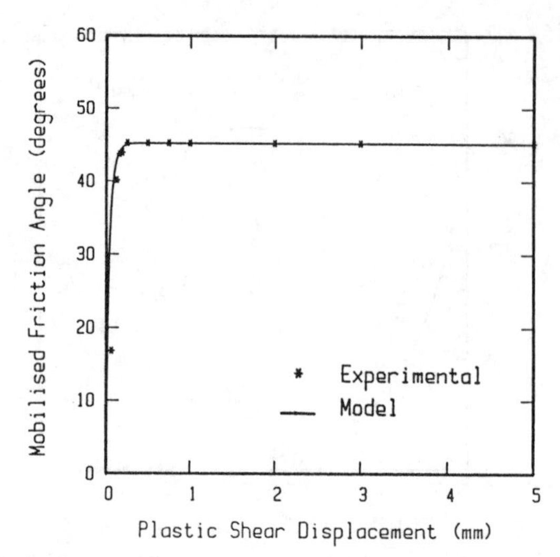

Figure 6. Mobilised friction angle versus plastic shear displacement

of mobilisation can be described by the following equation

$$\tan\varphi = \tan\varphi_m(1 - \exp[-k_2\ u^p/u_b]) \quad (5)$$

in which k_2 describes the rate of mobilisation and φ_m is the maximum friction angle for plastic sliding. The data appear to suggest that 20 and 45° are reasonable values for k_2 and φ_m, respectively.

To complete the description of the elastoplastic model a plastic flow rule is required. The data in Fig. 3 indicate that dilation is rapid in the early stages of plastic shearing of these bonded interfaces, i.e. as progressive rupture of the bonding takes place. However, the dilation appears to cease after about 1 mm of total shear displacement. One form of flow rule that can adequately describe this behaviour is given by the following equations:

$$\frac{dv^p}{du^p} = \tan\psi_0 \exp[-k_3\sigma/((1 - u^p/\lambda)q_u)] \qquad 0 \leqslant u^p \leqslant \lambda \quad (6a)$$

$$= 0, \quad u^p > \lambda \quad (6b)$$

in which ψ_0 is the initial dilation angle for the interface, k_3 is an empirical parameter introduced to define the rate at which dilation is suppressed, λ is the magnitude of the plastic shear displacement at which dilation ceases, and q_u is the unconfined compressive strength of the sandstone, introduced only to normalise the normal stress and its influence on dilation. Fig. 3 shows how well this flow rule fits the observed behaviour in tests A and B when $\psi_0 = 40°$, $k_3 = 10$, $\lambda = 1.1$ mm, and $q_u = 19$ MPa.

For this ideal interface the incremental constitutive relations describing the elastoplastic behaviour can be written as

$$\begin{bmatrix} d\tau \\ d\sigma \end{bmatrix} = \begin{bmatrix} k_{11} & k_{12} \\ k_{21} & k_{22} \end{bmatrix} \begin{bmatrix} du \\ dv \end{bmatrix} \quad (7)$$

where

$$k_{11} = k_s(\alpha k_n \tan\varphi + c^* + \beta)/f$$

$$k_{12} = -k_s k_n \tan\varphi/f$$

$$k_{21} = \alpha\ k_s\ k_n/f$$

$$k_{22} = -k_n(k_s + c^* + \beta)/f$$

and

$$c^* = 0, \qquad u^p \leqslant \delta$$

$$c^* = -k_1\ c_0 \exp[-k_1(u^p - \delta)/u_b]/u_b, \qquad u^p > \delta$$

$$f = k_s + \alpha\ k_n \tan\varphi + c^* + \beta$$

$$\beta = k_2\ \sigma \tan\varphi \exp[-k_2\ u^p/u_b]$$

$$\alpha = \tan\psi_0 \exp[-k_3\sigma/((1 - u^p/\lambda)q_u)]$$

4. PREDICTIONS OF SHEAR STRENGTH

It was demonstrated in Figs 1 to 3 that the model can be made to fit the shear and dilational behaviour of a number of bonded interfaces through a large range of monotonically increasing shear displacement. It will now be demonstrated that the model is also capable of adequately predicting the peak and residual strengths of these interfaces over wider ranges of normal stiffness and initial normal stress than those employed in determining the basic model parameters.

Predictions have been made of the behaviour of bonded concrete-sandstone interfaces for the following values of initial normal stress, σ_0 = 100, 500 and 1000 kPa and external normal stiffness, k = 100, 500 and 1000 kPa/mm. The stress paths followed in these numerical 'tests' have been plotted in Fig. 7 using the axes τ/σ versus σ.

Also superimposed on this plot are experimental data for the peak and residual strength envelopes determined by Ooi and Carter (1987). The experimental data come from a large number of tests on bonded

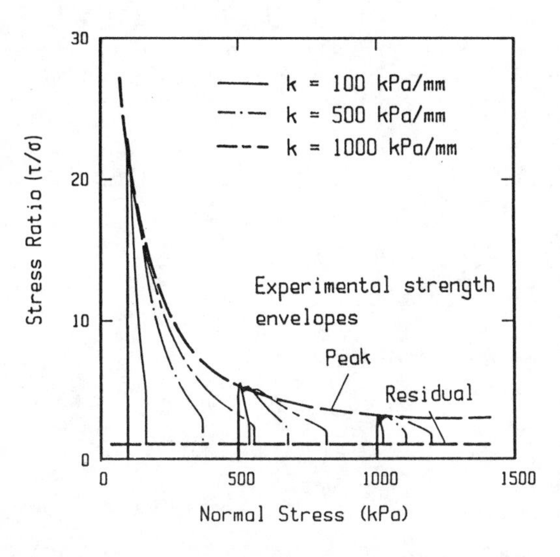

Figure 7. Comparison of predicted stress paths and measured strength envelopes

interfaces and include a number of different sets of data for different asperity roughness and normal stiffness. The experimental data indicate that the peak and residual strength envelopes, when plotted in this form, appear to be independent of the stress path to failure. This finding was in agreement with earlier work by Leichnitz (1985), who tested artificially prepared sandstone interfaces.

Fig. 7 indicates that very good general agreement is obtained between the strength predictions of the model and the experimental data, thus providing further confidence in the use of the model in practice.

5. CONCLUSIONS

A constitutive model has been presented for the coupled shear and dilational behaviour of rough, bonded concrete-sandstone interfaces. The model is based on the theory of work hardening/softening elastoplasticity. Its predictive capabilities were demonstrated and it was found to match quite well many of the essential features observed in laboratory direct shear tests. In particular it captured the coupling of shear and dilational behaviour, some pre-peak hardening and post-peak softening. It also successfully predicted the shear behaviour when subject to the additional constraint of deformation under a constant (external) normal stiffness, such as might occur at the interface between foundation concrete and a surrounding rock formation. This was achieved by allowing the cohesive component of shear strength to degrade exponentially with shear displacement, and the frictional component of strength to be mobilised over a small range of shear displacement, as observed in the experiments. The gradual destruction of the bonding between the concrete and sandstone as a rupture surface develops forms the physical basis for this approach.

At present the model has been applied only to cases of monotonically increasing shear displacement. The extension of the model to cases where the shear displacement is repeatedly reversed, i.e. cyclic loading, is currently under investigation.

ACKNOWLEDGEMENTS

Support for this work from a University of Sydney Research Grant, which provided a post-graduate scholarship for the second author, is gratefully acknowledged.

REFERENCES

Jaeger, J.C. (1971), Friction of Rocks and the Stability of Rock Slopes, Geotechnique, Vol. 21, No. 2, pp 97-134.

Leichnitz, W. (1985), Mechanical Properties of Rock Joints, Int. J. Rock Mechanics and Mining Sciences, Vol. 22, No. 5, pp 313-321.

Ooi, L.H. (1988), Forthcoming Ph.D. Thesis, University of Sydney.

Ooi, L.H. and Carter, J.P. (1987), A Constant Normal Stiffness, Direct Shear Device for Static and Cyclic Loading, Geotechnical Testing Journal, ASTM, Vol. 10, No. 1, pp. 3-12.

Patton, F.D. (1966), Multiple Modes of Shear Failure in Rock, Proc. 1st Congress, Int. Soc. Rock Mech., Lisbon, Vol. 1, pp 509-513.

Numerical Methods in Geomechanics (Innsbruck 1988), Swoboda (ed.)
© 1988 Balkema, Rotterdam. ISBN 90 6191 809 X

Numerical analysis of discontinuous block systems

J.Cañizal & C.Sagaseta
University of Cantabria, Santander, Spain

ABSTRACT: A numerical model is presented for the analysis of discontinuous systems formed by parallelepiped blocks. It can be of application to such problems as rubblemound breakwaters or severily jointed rock masses. The unit blocks are considered as rigid, with concentrated deformations at their contacts. The block-by-block placement process is simulated. Friction and no-traction limits at the contacts are considered leading to global and local instabilities. The analysis is carried out in two dimensions.

1 INTRODUCTION

A large number of engineering materials are formed by elements which can be easily separated, being known as "discontinuous systems" or "particulate media". Soils and rock masses fall into this category of materials.

In the analysis of behaviour of soils under external loads, the great number of particles involved makes unpractical their consideration as a discontinuous material, and hence the Theoretical Soil Mechanics has been developped based on the continuum approach. The mean particle size is small enough to justify this assumption. Only a few works exist which consider the discontinuum character of soils (Rodríguez-Ortiz 1974, Cundall & Strack 1979, Scott & Craig 1980). On the other hand, in rock masses, the block size is of the same order of the problem scale, and hence discontinuous models have been used.

Rockfill masses or rubblemound breakwaters constitute a range of problems somewhat midway between the two described cases. The "particle" size is usually very large (up to several meters in breakwaters) and, on the order hand, the contacts between them are mainly of the point type (corner-to-edge or corner-to-face), which makes it similar to the case of soils, in contrast with rock masses where the contacts are of face-to face type (plane discontinuities).

In this paper, a model is presented for the analysis of this kind of problems.

It has been conceived for breakwaters formed by parallelepiped concrete blocks. The system response to the complex external actions depends on its structure,which in turn is governed by the block shape and dimensions, process of placement, etc. The influence of these parameters is analyzed by the model.

2 GENERAL ASSUMPTIONS

Due to the great number of blocks and contacts involved, the problem is analyzed in two dimensions (plane strain), considering the blocks as infinite prisms of rectangular cross-section (Figure 1).

The position of each individual block is defined by the coordinates (x_G, y_G) of its centroid and its orientation, α.

The contacts between blocks are usually of corner-to-face type. The contact angle, θ, is defined as the smaller of the angles formed by the edges of the corner with the side in the contact with it. When a face-to-face contact is found, it is simulated by two corner-to-face contacts at its ends, with zero contact angle (θ=0). (Figure 1-b)

3 MECHANICAL BEHAVIOUR OF THE CONTACTS

The blocks are assumed to be rigid. The system deformations are mainly due to relative movements (separation, rotation or sliding) between the blocks. At each contact, there exist only a normal (N) and a shear force (T).

The contact forces are limited by the two conditions of no-traction and of friction resistance:

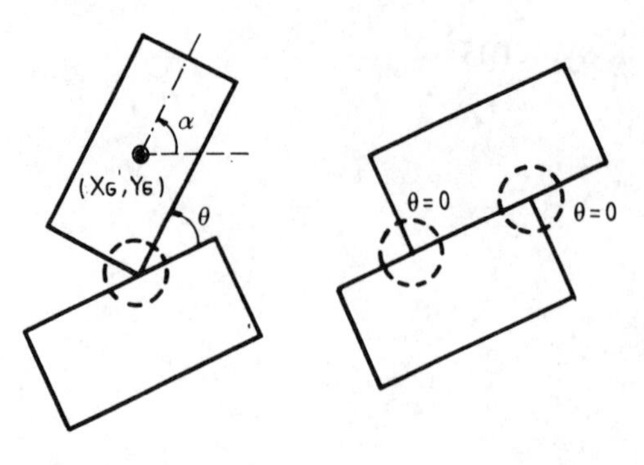

a) Block definition. Point contact.

b) Simulation of a plane contact.

Fig. 1 Block and contact definition

$$N \geq 0$$
$$|T| \leq \mu . N \tag{1}$$

where μ is the coefficient of friction between blocks, which is assumed to be constant.

Within the above limits, the system deformations are small and practically negligible. However, some finite stiffness is needed in order to solve the problem. It is assumed that the normal and shear forces are linearly dependent of the corresponding displacement components, with no coupling between them (no dilatancy effects), i.e.:

$$N = k_N \, . \, \Delta_N$$
$$T = k_T \, . \, \Delta_T \tag{2}$$

Where Δ_N and Δ_T are the normal and shear relative displacements at the contact point.

The parameters k_N and k_T in eq. (2) are the normal and shear sitffness of the contact. Their absolute values are not meaningful. However, the ratio between them (k_T/k_N) has a great influence on the contact force distribution. Similarly, the variation of both k_N and k_T from contact to contact due to the different contact angles θ is also important.

For the derivation of the normal and shear stiffness parameters, k_N and k_T, in terms of the contact angle θ, it is considered that at the contact point, some yield is produced in the corner, so that the normal stress does not exceed the uniaxial compressive strength of the material. Beyond this plastic zone, an elastic region is considered in which the solution given by Michell (1900) for the stress and strain fields in an elastic wedge is used. As a result, the following expressions are found for the stiffnesss:

$$k_N = \frac{4.\sigma_c}{a_N + b_N.\sin 2\theta}$$
$$k_T = \frac{4.G_c}{a_T + b_T.\sin 2\theta} \tag{3}$$

where a_N, b_N, a_T and b_T are material constants, and G_c and σ_c are the material shear modulus and uniaxial compressive strength, respectively. A complete description of the derivation has been given by Cañizal (1987).

Equations (2) are valid for the elastic range. When a contact is at one of the limit situations (eqs. 1), equations (2) must be replaced by the following relationsships:

- For the separation case:

$$N = 0$$
$$T = 0 \tag{4}$$

- For the sliding case:

$$N = k_N \, . \, \Delta_N$$
$$T = \mu \, . \, k_N \, . \, \Delta_N \tag{5}$$

4 NUMERICAL MODEL

A contact between two blocks can be considered as a joint beam element with two nodes (Figure 2).

The stiffness equations of the element can be written in matrix form as:

$$\{X'_e\} = |K'_e| . \{\delta'_e\} \tag{6}$$

where $|K'_e|$ is the elementary stiffness matrix (eqs. 2, 4 or 5), and $\{\delta'_e\}$ and $\{X'_e\}$ are the vectors of nodal displacements and forces.

The assembly of the elementary matrix into the general system has some important differences with the usual procedure in conventional finite element analyses, because in the present model, the elements are not directly in contact between them through the nodes, but instead they are linked to the rigid blocks. The compatibility conditions must provide that the displacements of the elements adjacent to a given block are compatible with its rigid movement. Similarly, the equilibrium condition at each block implies that both the reseultant force and moment are zero.

The most efficient way of introducing these conditions is to refer all the variables (forces and displacements) to the

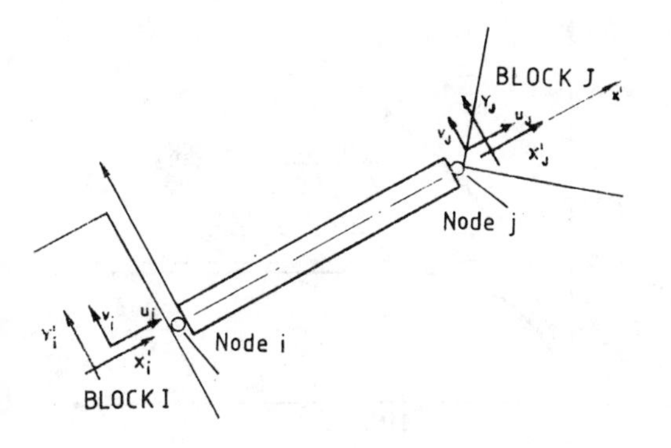

Fig. 2 Joint element

centroids of the blocks. Each block is then considered as a "macronode", with three degrees of freedom: the two components of the displacement of its centroid and the block rotation about it. Similarly, the right-hand terms of the stiffness equations are the external forces and moments applied at the centroids. This procedure has been followed also by Dowding et al. (1983) for the analysis of jointed rock masses. These transformations lead to:

$$\{F_e\} = |K_e| \cdot \{\Delta_e\} \qquad (7)$$

where:

$$\{F_e\} = |A| \cdot |R| \cdot \{X'_e\}$$

$$\{\Delta_e\} = |A| \cdot |R| \cdot \{\delta'_e\}$$

$$|K_e| = |A| \cdot |R| \cdot |K'_e| \cdot |R|^T \cdot |A|^T$$

and where $\{F_e\}$ and $\{\Delta_e\}$ are the generalized actions and movements, respectively, at the block centroid. $|R|$ is the transformation matrix from local to global coordinates and $|A|$ is the transformation matrix from vectors at the contact points to the block centroid.

The matrices of equation (7) can now be assembled in the usual manner to form the global matrix equation for the whole structure.

The solution of the above system gives the block movements and from them the contact forces can be obtained. However, the multiplicity of stiffness matrices (eqs. 2, 4 and 5) results in a non-linear problem. Several procedures can be applied to solve it. Iterative relaxation algorithms have proved to be of poor efficiency, due to two main reasons: first, there are two possible situations of inadmissibility (separation or sliding), and second, the number of inadmissible contacts is very high. This makes the solution highly dependent on the order in which these contacts are relaxed.

The conclusion is that an incremental procedure is needed, following as closely as possible the actual loading history, i.e., the block-by-block placement process.

The placement of an individual block is carried out in two phases:

- An initial position is found for the block, leaning on the previously placed ones. This is carried out by a routine called "simple placement".
- The whole system is structurally analized. The weight of the placed block is applied in steps small enough so that at each step no more than one contact can reach one of the limit conditions.

Initially, each block is horizontal and ideally suspended from a cable attached to its centroid. The starting abscissa of the centroid can be imposed or randomly generated. Then, it is gradually lowered until it touches one of the previously placed blocks or the boundary (Figure 3-a). From here on, the block movement is simulated from kinematical and static considerations, assuming that the rest of the blocks are fixed and unaffected. While there is only one contact, the three acting forces (block weigth, cable force and contact reaction) must be in equilibrium. This gives enough restrictions for definition of the block trajectory. When a second contact is reached (Figure 3-b), it is checked that the abscissa of the centroid lies between the two contacts and, in the case of the two contacts belonging to the same block side, that the slope of this side is less than the friction angle. If one of these conditions does not hold, the block continues tilting or sliding, respectively, until a new contact is generated.

Once a potentially stable position is reached, the action of the block weight is considered, and the complete structure is analyzed.

At first, all the existing contacts are considered in the elastic range (even if they have reached previosly the sliding condition) and the whole block weigth is applied. The resulting incremental forces are then accumulated to the previous ones. Then all the contacts are checked and the load is reduced so that only one of them reaches the limit condition. The stiffness of this contact is changed to the appropiate one (eqs. 4 or 5) and the rest of the load is applied. The process is repeated until the total weigth of the block is applied and a new block can be placed.

At any loading step, it may happen that the system matrix becomes singular. This means that some block or blocks are unstable. A special algorithm is needed to identify the unstable blocks.

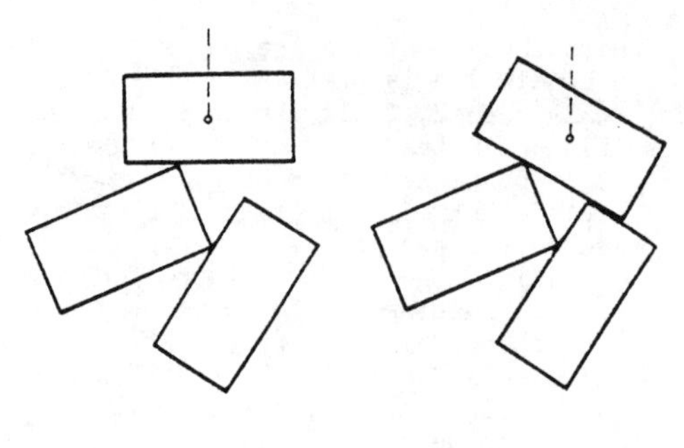

Fig. 3 Simple placement

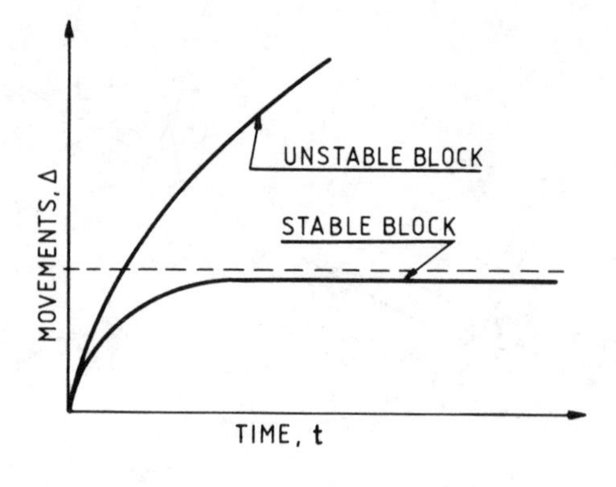

Fig. 4 Identification of unstable blocks

A viscous relaxation procedure has been selected by which the current load increment is applied again assuming some viscous damping which prevents infinite movements to take place. Equation (7) is replaced by:

$$\{F_e\} = |K_e| \cdot \{\Delta_e\} + |C_e| \cdot \{\frac{\partial \Delta_e}{\partial t}\} \qquad (8)$$

where $|C_e|$ is a diagonal matrix, defined by an arbitrarily chosen damping factor, C. Equation (8) is solved by implicit finite differences. After a number of time increments, the movements of stable blocks tend to asymptotic values, while those of the unstable ones increase infinitely. This provides an easy identification of the unstable blocks (Figure 4).

The unstable blocks are then allowed to move to a new more stable position with a routine similar to the "simple placement" one, the movement trends resulting from the viscous analysis being used as a guide. In the case of a group of blocks, such as in Fig. 5, the contact forces are used for identifying the "key block". Then the upper blocks are removed and the key block moves to the new position. The removed blocks are then replaced in the usual way.

5 APPLICATION

In order to assess the model validity, it has been applied to some simple cases of known solution. One of these cases is shown in Figure 5. It consists of three blocks with tilting instability, placed in the order shown in the figure. The key block is successfully identified as No. 2. The final situation is shown in Fig. 6 for different friction values between blocks.

As a practical application, a breakwater formed by 25 blocks of dimensions ratio of 1.4:1 has been analyzed. Figure 7 shows the final structure and the contact forces for, $K_T/K_N = 0.5$ and $\mu = 0.84$.

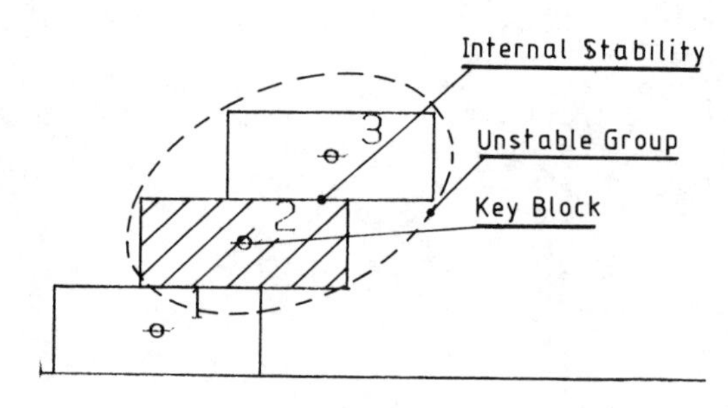

Fig. 5 Group instability. Key block

The situation of each contact can be defined by a load ratio between the existing shear force and the maximun allowable one, $T/\mu.N$. The average of the load ratio of all the contacts is a measure of the structure stability, and is called "instability factor".

Fig. 8 shows the instability factor against the stiffness ratio, K_T/K_N, for a friction coefficent of 0.84. As it can be expected, it increases with the stiffness ratio, but for a value of about 1.0 it shows an abrupt decrease, due to the fact that a local collapse is produced, and the blocks rearrange in a more stable position. The same factor is plotted in Fig. 9 against the friction coefficient for stiffness ratio of 0.5. Here, the instability factor shows a continuous decrease with the friction.

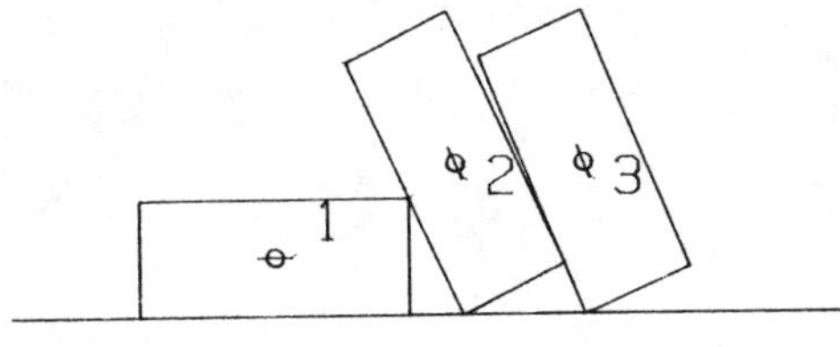

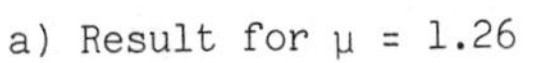

a) Result for μ = 1.26

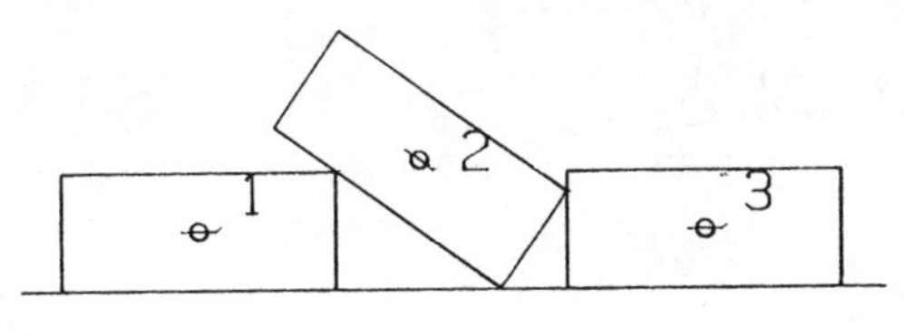

b) Result for μ = 0.15

Fig. 6 Final structure for the case of figure 5

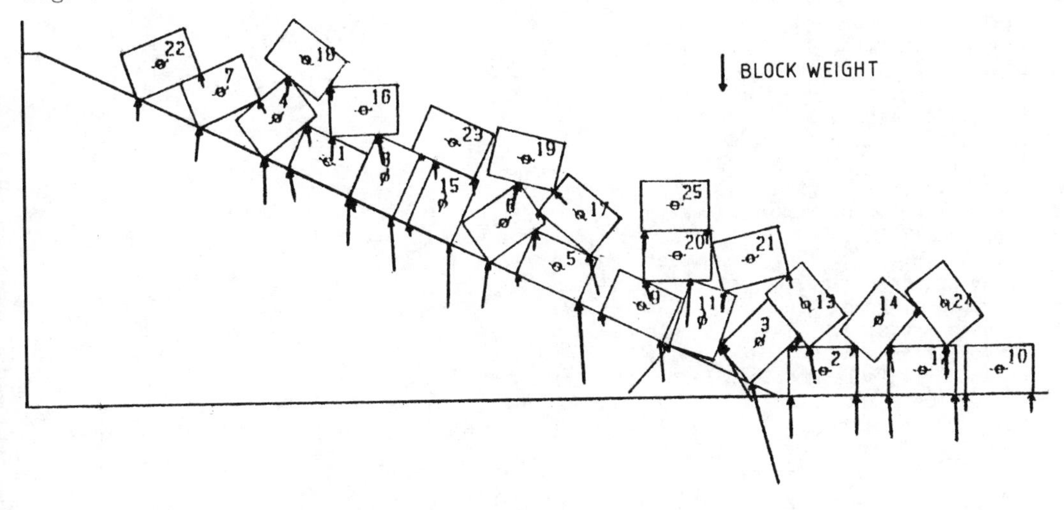

Fig. 7 Breakwater. Final structure and contact forces for $k_T/k_N = 0.5$ and $\mu = 0.84$

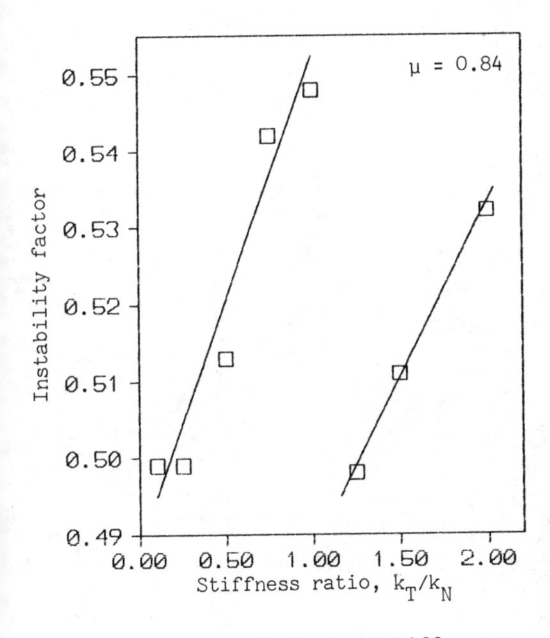

Fig. 8 Instability factor vs. stiffness ratio, k_T/k_N

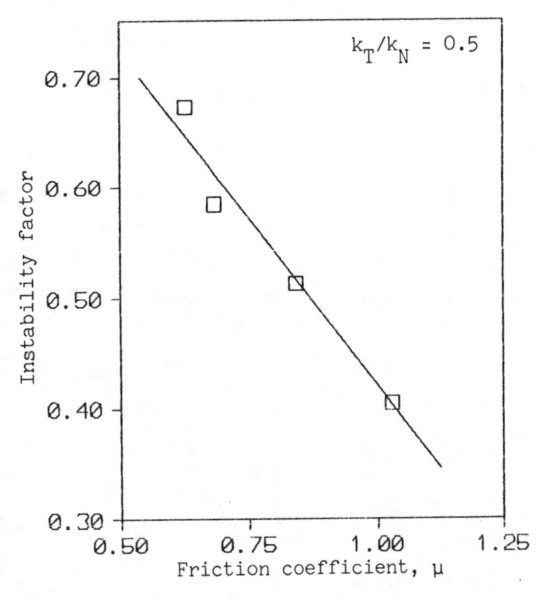

Fig. 9 Instability factor vs. friction coefficient, μ

6 ACKNOWLEDGEMENTS

The present work has been supported by the CAICYT (Project No. 369-C02-02) of the spanish Ministerio de Educación y Ciencia.

REFERENCES

Cañizal, J. 1987. Análisis del comportamiento mecánico de un sistema de bloques paralelepipédicos con consideración especial de su proceso de colocación. Doctor Dissert. Univ. of Cantabria.

Cundall, P.A. & Strack, O.D.L. 1979. A discrete numerical model for granular assemblies. Géotechnique 29, 1:47-65.

Dowding, C.H., Belytschko, T.B. & Yen, H.J. 1983. A coupled finite element -rigid block method for transient analysis of rock caverns. Int. Jour. Num. Anal. Meth. Geomech. 7:117-127.

Michell, J.M. 1900. The stress distribution in an aelotropic solid with infinite boundary plane. Proc. London Math. Soc. 1,32.

Rodríguez Ortiz, J.M. 1974. Estudio del comportamiento de medios granulares heterogéneos mediante modelos discontinuos analógicos y matemáticos. Doctor Dissert. Univ. Polit. Madrid.

Scott, R.F. & Craig, J. 1980. Computer modelling of clay structure and mechanics. Journ. Geot. Eng. Div. ASCE 106, GT1:17-34.

Numerical Methods in Geomechanics (Innsbruck 1988), Swoboda (ed.)
© 1988 Balkema, Rotterdam. ISBN 90 6191 809 X

Landslide effects to buried pipelines

Yaw-Huei Yeh
National Taiwan Institute of Technology

ABSTRACT: During earthquakes, a buried pipeline may experience significant loading as a result of large relative displacements of the ground along its length. Large ground movements can be caused by faulting, liquefaction, lateral spreading, landslides, and slope fąilures.

Landslides are mass movement of the ground that can be triggered by seismic shaking. The failure pattern of a buried pipeline is usually displaced by large landslide movement which involves slope instibility.

The objective of this research is to present a precise analysis for buried peipelines under effect of large landslide movement. The proposed analysis model includes the concepts of the beam on elastic foundation and the ultimate passive soil pressure for presenting horizontal soil resistance to movement.

1. Failure modes of buried pipelines

A pipeline transmission system is a linear system which traverses a large geographical area, and thus, may encounter a wild variety of seismic hazards and soil conditions. The major seismic hazards which can significantly affect a pipeline system are

1) Differential fault movement and ground rupture
2) Ground shaking
3) Liquefaction
4) Landslides
5) Tsunamis or seiches

The considerations for failure patterns of pipeline systems subjected to large soil movement would be herein discussed.

Land-slides are mass movements of the ground that can be triggered by seismic shaking. A great many landslide patterns are possible, and a typical landslide pattern associated with a displaced buried pipeline is shown in Fig.1.

A primary concern for buried pipelines is their ability to accommodate abrupt ground distortions or differential displacements due to land-slides. An analysis model for pipeline buried through sliding zone, which includes the concepts of the beam on elastic foundation and the ultimate passive soil pressure for presenting horizoital soil resistance to movement, is thus proposed and also illustated in Fig.2.

Fig.2 shows a soil movement perpendicular to the axis of buried pipeline resulting a down-ward displacement. This movement will require the pipe to elongate within sliding region and will result in axial tensile strain in the pipe.

Recently, Audibert and Nyman (1) have performed experiments on the soil resistance against horizontal motion of pipe. They found that the soil pressure-displacement relationship was non-linear, showing a greater increase of displacement (y) at larger earth pressure (p). They also pointed out that in most cases the p-y curves reach a well defined plateau at a large lateral displacement.

As can be seen, the soil reaction force mobilized on the pipe would be almost kept constant under the effect of mass movement.

There are two possible sliding patterns affecting the behavior of buried pipeline. First of all, if the length of sliding region is relatively small, the pipe segment buried within sliding zone can be considered as a beam segment, subjected to the lateral earth pressure distributed along the entire segment, resting on two elastic soil foundation at both ends. A proposed analysis model is shown in Fig.3. Secondly, if the region of landslide is extended tremendously, the configuration of the displaced pipeline could be taken as two independent failure modes as shown in Fig.4.

2. Analyses for two failure mechanisms

2.1 Small movement region

From Fig.5, the resulting moment, "M_x", along the central pipe segment is expressed as

$$M_x = -\frac{q}{2}x^2 + Ny + M_o \quad (1)$$

The governing differential equation is thus derived

$$y'' - \frac{N}{EI}y = -\frac{q}{2}\frac{x^2}{EI} + \frac{M_o}{EI} \quad (2)$$

where q = the ultimate load bearing capacity of the soil
$= (\nu\ Z\ N_q)D$

ν = unit weight of soil
Z = buried depth to center of pipe
D = diameter of pipe
M_o = resulting moment at mid-point of pipe segment within sliding sone

Based on the derivation of Hetenyi (2), the rotation at the interface is given below

$$\theta = \frac{1}{EI}\frac{1}{3\alpha^2-\beta^2}\left[\frac{q\ell}{2} - 2\alpha M_f\right] = y'|_{x=\frac{\ell}{2}} \quad (3)$$

where

$\alpha = \sqrt{\lambda^2 + N/4EI}$

$\beta = \sqrt{\lambda^2 - N/4EI}$

$\lambda^4 = K_s/EI$

K_s= spring constant of soil

M_f= resulting moment at the interface
$= \left(\frac{-qEI}{N} + M_o\right)\cosh\frac{c\ell}{2} + \frac{qEI}{N}$

$c = \sqrt{N/EI}$

Thus

$$M_o = \frac{a}{b} \quad (4)$$

where $a = m\left(\frac{q\ell}{2} + \frac{2\alpha q}{\ell^2}\cosh\frac{c\ell}{2} - \frac{2\alpha q}{c^2}\right) + \frac{q}{Nc}\sinh\frac{c\ell}{2} - \frac{q\ell}{2N}$

$b = 2\alpha m\cosh\frac{c\ell}{2} + \frac{c}{N}\sinh\frac{c\ell}{2}$

and the expression for displacement is obtained

$$y = \left(-\frac{qEI}{N^2} + \frac{M_o}{N}\right)\cosh cx + \frac{q}{2N}x^2 + \frac{qEI}{N^2} - \frac{M_o}{N} \quad (5)$$

where

$m = 1/(\sqrt{K_s EI} + N)$

For a typical case of ℓ=50^m, the relationship between M_o and N is also shown in Fig.6.

2.2 Large movement region

For general consideration, the angle of buried pipeline crossing sliding region may not be 90°. A geometric configuration and the force distribution of buried pipeliae are shown in Fig.7 and Fig.8 respectively.

The expression for displacement "y" can be obtained by satisfying all the boundary conditions

$$y = \frac{-d_B\theta_B+d_B}{2d_A^3}x^3 + \frac{3(d_A\theta_B-d_B)}{2d_A^2}x^2 - \theta_B x + d_B \quad (6)$$

The equilibrium equations for the curved pipe are derived as below.

$$\Sigma F_x = 0$$

$$F_B\cos\theta_B + V_A\sin\theta_A - F_A\cos\theta_A = p_p d_B - V_B\sin\theta_B - f_p d_A \quad (7)$$

$$\Sigma F_y = 0$$

$$F_B\sin\theta_B - V_A\cos\theta_A - F_A\sin\theta_A = V_B\cos\theta_B - p_p d_A - f_p d_B \quad (8)$$

$$\Sigma M_o = 0$$

$$F_B d_B\cos\theta_B - V_A d_A\cos\theta_A - F_A d_A\sin\theta_A = -M_B - 2f_p A_o - d_B V_B\sin\theta_B - 0.5p_p(d_a^2 - d_B^2) \quad (9)$$

where A_o = total area under the curve of deformed pipe between points "A" and "B".

$$V_B = -\lambda M_B = -\lambda(2\lambda\theta EI) \quad (10)$$

Based upon Cramer's rule, the unknowns of "F_B", "F_A" and "V_B" can be obtained by solving the given eqns. (7)-(9).

The resulting stresses along the pipe within transition zone can be acquired in the similar manner by cutting a pipe segment as a free-body element at the location as desired.

As mentioned previously, the parameters θ_B, M_B and d_A are correlated to each other, once one of the parameters is determined, the rest of them can be easily calculated out using eqns. (7) and (11). If the value of "d_A", for example, is assigned appropriately, the deformed curve "y" within transition zone is thus determined by employing eqn.(7); and the acquisition of "θ_B", "M_B" will be subsequently achieved without problem.

As shown in Fig.7, the considerable length difference between the segments A'B and AB is known as "Geometric elongation", Δ_G.

Note that the $\overline{A'B}$ is the initial length of pipe segment sliding zone. An expression for "Δ_G" (or A*A) is thus introduced as follows;

$$\Delta_G = \int_0^{d_A} y(x)\,dx - \overline{A'B} \qquad (11)$$

The total elongation obtained from integrating axial strain along the semi-infinite pipe must be compatible with the geometric elongation. For convenience, the integrated axial strain is designated as permissible elongation "Δ_p".

To satisfy the condition of compatibility, the following relation must be held

$$\Delta_G = \Delta_P \qquad (12)$$

3. Results and conelusions

A series of parametric analyses has been undertaken to study the influence of soil movement effects on the design of buried pipline. The numerical data of those parameters for the analyses are listed as follows

1) soil movement (Δ_f).
   210, 240, 270*, 300, 360, and 450 cm.
2) Angle crossing a sliding zone (β) = 45, 55, 60, 65, 70*, 75, and 80°.
3) Soil/pipe friction angle (ϕ_p) = 5, 10, 15, 20*, 25, 30, 45, and 50°.
4) Buried depth from top of the pipe = 30, 60, 90*, 120, 150, and 180 cm.

only one of the parameters is investigated to find its effects on the buried pipeline; the rest (of reference values) are kept constant. The symbol of (*) shown above indicates the reference values of parameters for the series of analyses.

After those investigations, the critical value for each parameter can be obtained by using the computer program of the iterative analysis, they are

1) Soil movement $\leq$ 270 cm (8.9ft)
2) Crossing angle $\geq$ 70°
3) Soil/pipe frictional angle $\leq$ 20°
4) Buried depth $\leq$ 91 cm (3ft)

From other investigation, it is also concluded that the combined effects of axial pipe force and bending moment occuring along the pipe segment near the sliding region are significant, and the negligence of geometric effect which caused an occurrence of resulting maximum moment within the transition zone, may lead to a non-conservative design of pipeline buried through a fault zone.

ACKNOWLEDGMENT

This research was funded by the National Science Council under grant number NSC 75-0410-P011-09. Much appreciation is also expressed to National Taiwan Institute of Technology for the encouragement and supports.

REFERENCES

1. J.M.E. Audibert, and K.J. Nyman, "Soil resistant Against Horizontal Motion of Pipe", Journal of The Geotechnical Engineering Division, ASCE, Vol.103, No.GT10, Oct. 1977.
2. M. Hetenyi, "Beams on Elastic Foundation", University of Michigan Press, 1946.

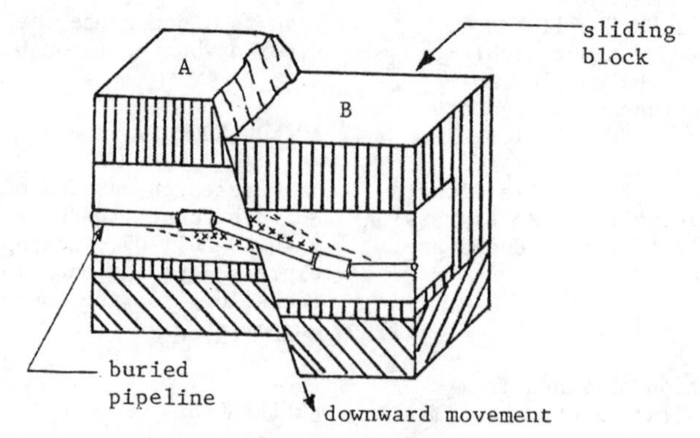

Fig.1 Configuration of buried pipeline under the effect of landslide.

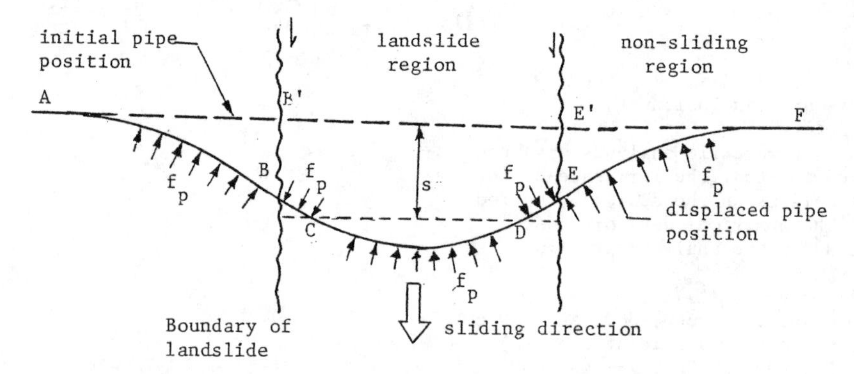

Fig.2 Proposed analysis model for buried pipeline subjected to landslide movement.

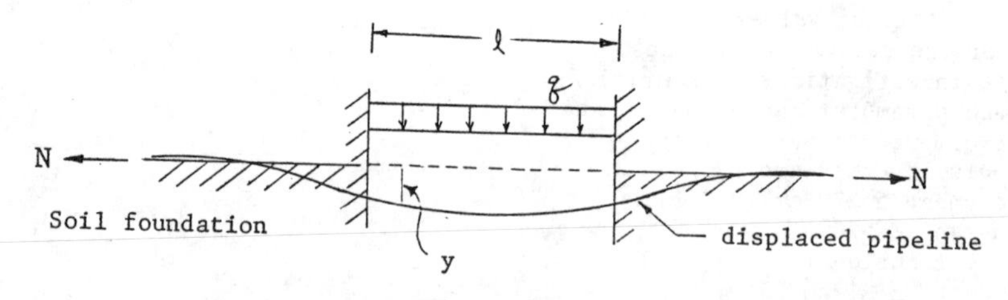

Fig.3 Analytical model for buried pipeline subjected to small scale landslide movement.

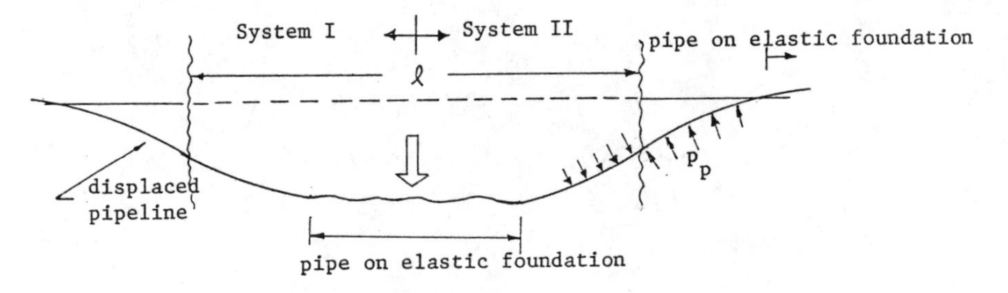

Fig.4 Analytical model of buried pipeline subjected to large landslide.

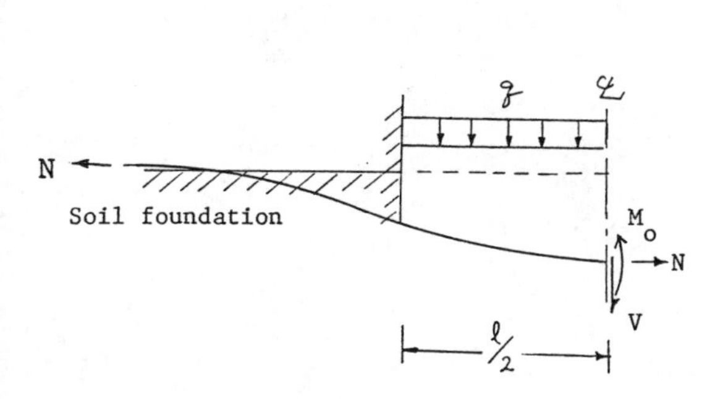

Fig.5 Force distrbution along left portion of pipeline under the effect of small amount of landslide movement.

Axial forces (N) 0 1000 2000(tons)

ℓ = 50^m

moment

500 1000 1500 (t-m)

Fig.6 $N - M_o$ relationship

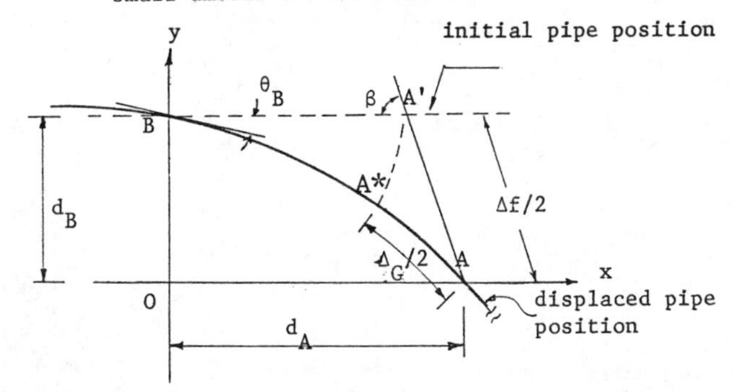

Fig.7 Geometric configura-
Geometric configuration of buried pipeline under the effect of large landslide movement.

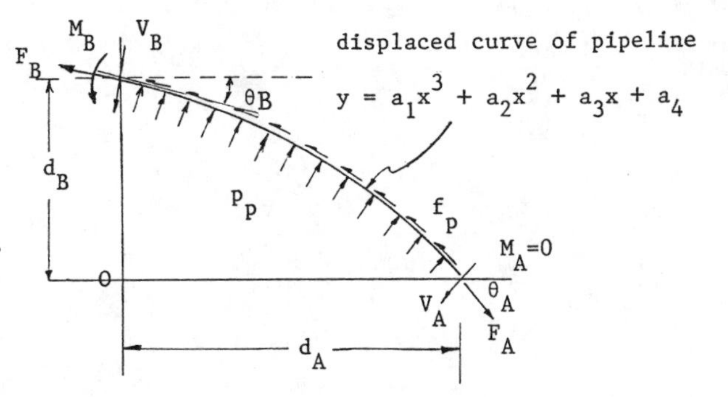

Fig.8 Force distribution along left portion of pipeline for the case of large landslide movement.

Numerical Methods in Geomechanics (Innsbruck 1988), Swoboda (ed.)
© 1988 Balkema, Rotterdam. ISBN 90 6191 809 X

Application of the thin-layer interface element to geotechnical problems

H.F.Schweiger
Institute for Soil Mechanics, Rock Mechanics and Foundation Engineering, Technical University of Graz, Austria
W.Haas
TDV (Technische Datenverarbeitung), Graz, Austria

ABSTRACT: Many geotechnical problems involve interaction between structure and soil or rock. If numerical methods are applied to obtain solutions to this type of problems, proper modelling of the contact zone strucure-soil may be important. Many formulations of joint and interface elements have been proposed by a number of researchers but an easy to handle, robust element for routine anylyses has not yet been presented. In this paper the application of a thin-layer element which is a combination of the thin-layer element proposed by Desai and the Multilaminate model introduced by Pande is discussed.

The element is implemented into a commercial finite element program and a viscoplastic algorithm is used to solve the nonlinear equations. Investigations concerning optimum time step lenghts are briefly presented. Examples of a buried pipe, a diaphragm wall and an avalanche protection tunnel demonstrate the applicability of the element.

1 INTRODUCTION

Because of the importance of proper modelling of interfaces extensive research in this subject has lead to various formulations capable of modelling interaction phenomena. Some successful applications have been reported using conventional solid elements /1/,/2/ but in general these elements cannot be used because of numerical problems due to very high aspect ratios. Additionally meaningful stresses - especially shear stresses - in the contact zone are difficult to obtain.

A commonly used interface element is the one proposed by Goodman, Taylor and Brekke /3/ with relative displacements being the nodal unknowns. The thickness is often assumed to be zero and to avoid overlapping and penetrating a very high normal stiffness of the interface, which is not in line with the actual behaviour, has to be introduced.

Ghaboussi, Wilson and Isenberg /4/ developed an element considering relative motions within the interface element as independent degrees of freedom. A similar element, based on an isoparametric, "shell-type" - formulation has been proposed by Beer /5/ for two- and threedimensional applications.

Modified spring elements have also been used and although the desired effect may be obtained, data preparation, mesh generation and interpretation of results turn out to be very involved /6/.

Desai /7/ proposed a thin-layer element based on an isoparametric formulation. The main features are the introduction of an independent shear modulus in the elastic range and the possibility of taking into account stick, slip, debonding and rebonding modes. Although some applications have been reported /7/,/8/ the performance of this element is not fully investigated.

The approach used in this paper combines the idea of the thin-layer element with the Multilaminate model introduced by Pande /9/ for the analysis of jointed rocks, the main advantage being an easier implemention into existing finite element codes.However, if a viscoplastic algorithm is used, the critical time step length can be determined only empirically. Some investigations to find out optimum time step lengths are presented. As shown in one of the examples 3-point Gauss integration is necessary to obtain consistent stresses in the interface elements. Solutions to the problem of a buried pipe, a diaphragm wall and an avalanche protection tunnel applying the proposed element are briefly presented.

2 BASIC ASSUMPTIONS OF THE PROPOSED INTERFACE ELEMENT

The use of an element based on an isoparametric formulation has the advantage of being easily implemented into existing finite element codes. Furthermore , data preparation and mesh generation can be done in a routine manner. However, if conventional solid elements are used, numerical ill-conditioning is likely to occur and stresses obtained in the interface are not very meaningful. The element used in this study is based on the formulation presented by Desai in the elastic range and on the Multilaminate model developed by Pande , i.e. the interface behaviour is modelled by a constitutive law rather than by introducing relative displacements, zero thickness etc.

The constitutive relation for the interface behaviour is given in the elastic range by

$$\{\sigma\} = [D]\{\varepsilon\}$$

where D, the constitutive matrix, is given by

$$D = \begin{bmatrix} Dnn & Dns \\ Dsn & Dss \end{bmatrix}$$

where Dnn is the normal component, Dss is the shear component and Dns, Dsn represent coupling terms which are neglected at present.

Experimental determination of Dnn and Dss is not easy and much more work has to be done to provide realistic parameters for the numerical analyses. In this study the normal component has been assumed to be the same or slightly less than in the surrounding soil and the shear component was related to the thickness of the element (approximately an order of magnitude smaller than the dependent elastic isotropic shear modulus). This formulation allows large relative displacements in the elastic range without causing numerical difficulties.

The nonlinear deformation characteristics of the interface are described applying the Multilaminate model. Extensive coverage of the theretical background is given elsewhere /9/ and only a brief summary is given here for continuity.

The model assumes an uniformly and homogeneously distributed influence of discontinuities over the region under consideration i.e. the properties of joints or discontinuitied are smeared over the element. It assumes that stresses in joints and intact material are the same and that total strains are the summation of its individual components. 3-D stress components are transformed to normal and shear components on the interface and the Mohr-Coulomb criterion is applied to give viscoplastic strains.Opening and closing of the interface is monitored by observing the normal component of the viscoplastic strain vector.

The basic steps are outlined in the following:

1. Calculate D - matrix for thin-layer elements with independent shea modulus and assemble into the overall stiffness matrix (note that 3-point Gauss integration is essential for calculating stiffnesses of inter face elements).

2. Transform stresses to obtain normal and shear stresses in interface elements.

3. Check yielding in interface and resolve till convergence is achieved to specified tolerance.

3 NOTE ON TIME STEP LENGTH

The time step length applied in a viscoplastic algorithm cannot be chosen arbitrarily but it must be smaller than a critical value to ensure convergence. For this type of element there is no theoretical time step limit and empirical rules must be utilized. In the first instance the critical time step length for the Multilaminate model, also determined empirically, has been used. However, convergence was found to be not satisfactory in some cases

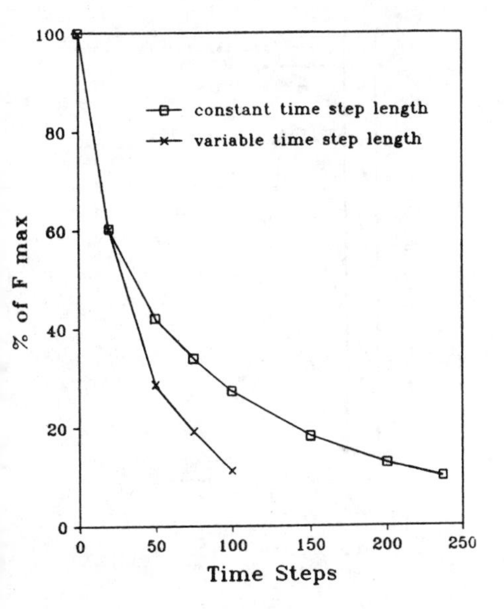

Figure 1. Convergence of iteration.

This was the motivation to investigate the use of larger time steps to speed up convergence. It was found that time steps cannot be increased from the starting point of the iterations (the problems diverged immediately) but after performing a number of steps (in the order of 2Ø) with the theoretical time step limit.Time steps 2 to 3 times larger lead to a reduction of total steps necessary to obtain convergence by a factor of 2 (Figure 1). However, monitoring viscoplastic strains is essential to discover any oscillations which may occur and may lead to wrong results. In addition it has been found, that high normal stiffness of the interface - somehow unrealistic anyway - causes serious problems as far as convergence is concerned. On the other hand problems were encountered for the case of linear elastic materials on both sides of the interface when the interface itself was assumed to behave elasto-plastic.

4 EXAMPLES

4.1 Buried pipe

In the first example a buried pipe is analyzed for the loadcase "backfilling". Thin-layer elements were used around the pipe and on the contact zone of backfilling and in situ soil. The analysis showed that the influence of the interface elements on internal stresses of the pipe is not very significant but displacements are slightly affected. However, the example serves to demonstrate the importance of the 3-point Gauss integration. If Figures 2 and 3 are compared it is obvious that consistent stresses in the interface only can be obtained using higher order integration rules.

Figure 2. Principal stresses 2x2 integration

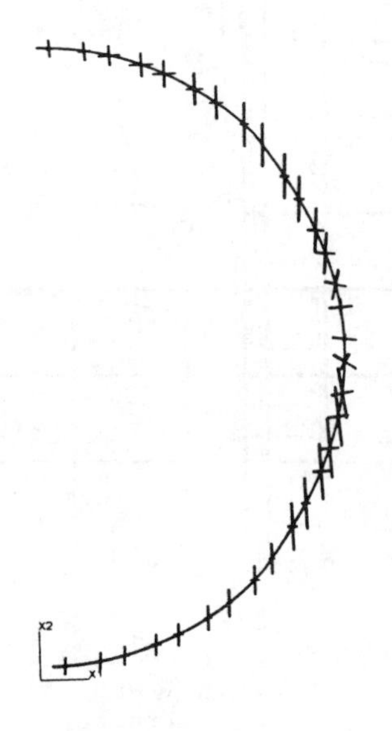

Figure 3. Principal stresses 3x3 integration.

4.2 Diaphragm wall

Diaphragm walls, often used as retaining walls for deep excavations, are designed in most cases applying the subgrade reaction method. Alternatively the finite element method may be used with th possible advantage of easier handling of the initial stress state and a more realistic simulation of the excavation process. However, in the analysis presented here the excavation (1Ø m deep) was simulated in one step by specifying the initial stress state and applying equivalent nodal forces on the free surfaces. The Young's modulus and the cohesion for the soil were assumed to increase with depth and the vertical modulus was assumed to be twice the horizontal modulus. A detailed description of the parameters is omitted here.

Three types of analyses have been performed:

1. Thin layer elements as described in section 2 of this paper (independent shear modulus) to simulate the contact behaviour soil-wall.

2. Conventional solid elements (dependent shear modulus) with Multilaminate material law.

3. No interface elements.

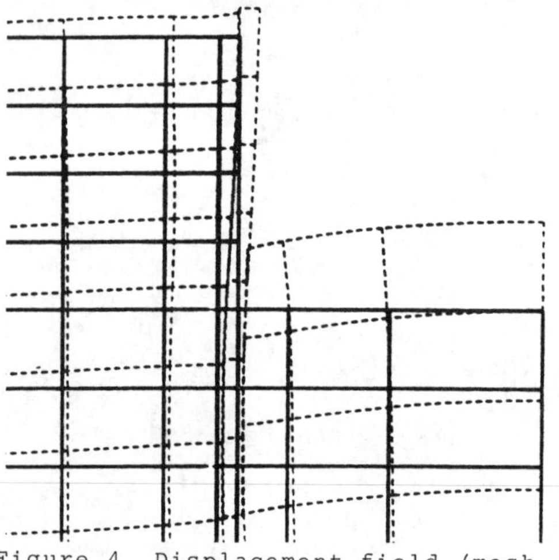

Figure 4. Displacement field (mesh detail)

Figure 4 shows the displacement field for analysis 1 and it can be seen that a large slip can be modelled (approx. 1 cm compared to a maximum heave of approx. 3 cm). Analysis 2 yielded similar displacements but analysis 3 showed a very small horizontal movement of the wall. Figures 5 and 6 compare horizontal and shear stresses in the interface and it can be seen that stresses are smoother if the independent shear modulus is used. In Figure 7 shear stresses in adjacent soil elements are plotted and the influence of the interface elements is obvious for this case. Figures 8 to 1Ø show internal forces in the wall and the difference in the bending moments between analysis 1 and 2 and in the normal forces between analysis 1 and 3 is remarkable.

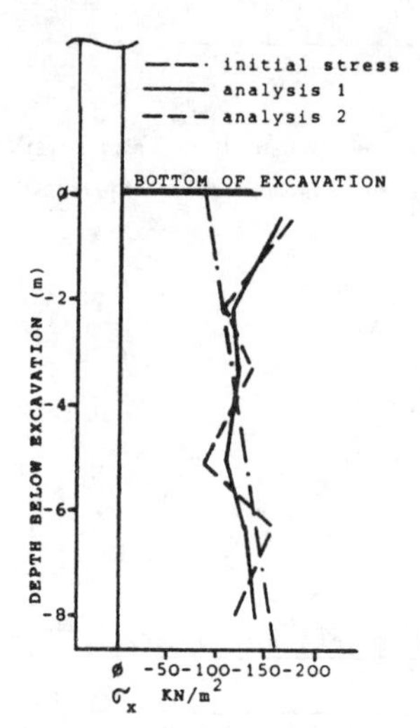

Figure 5. Horizontal stress distribution in interface.

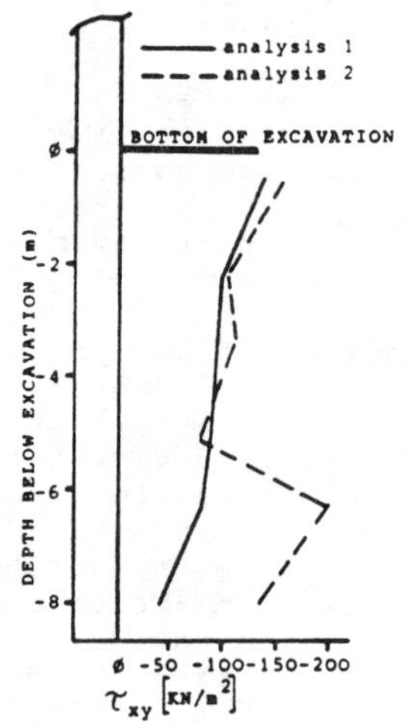

Figure 6. Shear stress distribution in interface.

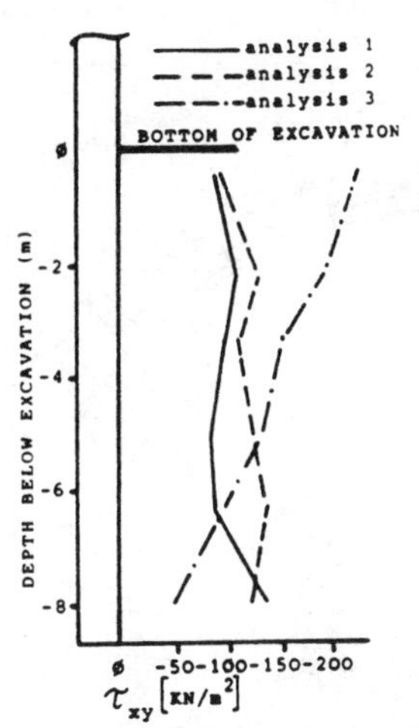

Figure 7. Shear stress distribution in adjacent soil.

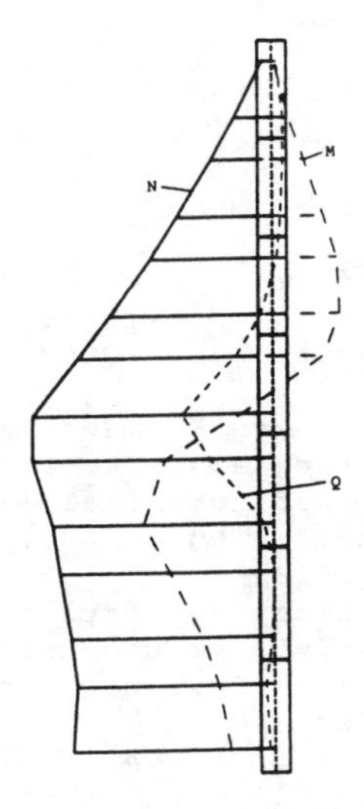

Figure 8. Internal forces, analysis 1.

4.3 Avalanche Protection tunnel

The examples presented obove served mainly to investigate the behaviour of the suggested formulation of the interface element. Following these studies the element has been applied to the solution of a practical problem namely an avalanche protection tunnel. Because of space limitations only the results obtained for the loadcase "backfilling + avalanche" are shown in Figure 11. Internal forces are plotted together with the contact stresses (σ_n and τ) in the interface.

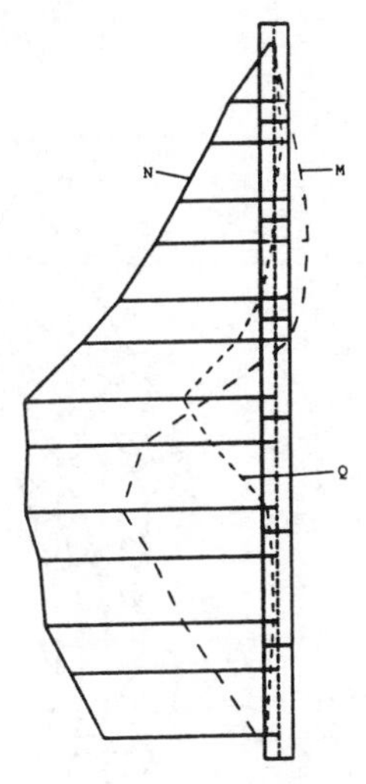

Figure 9. Internal forces, analysis 2.

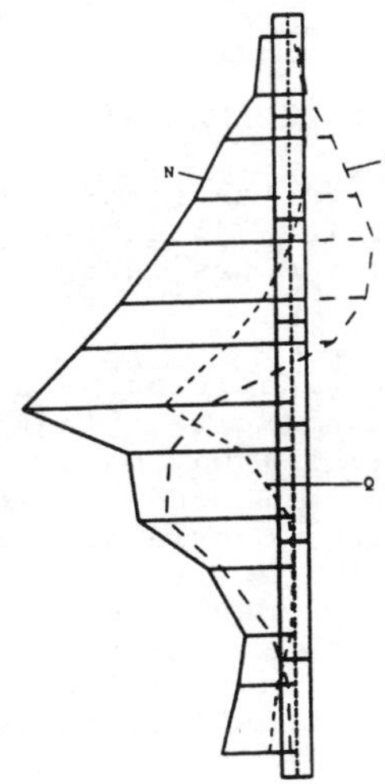

Figure 1Ø. Internal forces, analysis 3.

5 CONCLUSION

The application of an interface element based on an isoparametric formulation has been presented. The advantages of the proposed element are:

1. Because of the independent shear modulus large relative displacements are possible even in the elastic range without causing numerical difficulties.

2. At the same time aspect ratios of approx. 1:1ØØ are admissible.

3. A realistic stress distribution

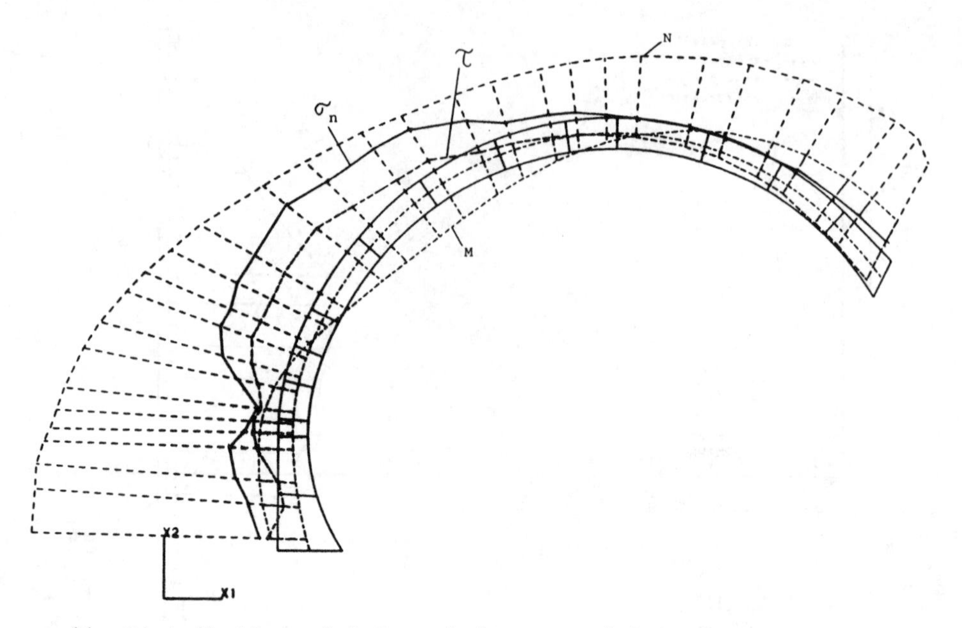

Figure 11. Distribution of internal forces and interface stresses.

is obtained in the contact elements rather than in adjacent elements provided 3-point Gauss integration is used.

4. In the plastic range standard material laws (Multilaminate model, tension cut off) may be activated without sophisticated routines to check opening, overlapping etc.

5. Easy implementation into existing FE-codes (standard mesh generation, no complications are introduced for relative displacements as unknowns).

The results presented indicate that modelling the contact behaviour between structure and soil or rock can be significant in some cases but for "standard analyses" the additional computational effort may not be justified in many cases.

Empirical improvements regarding the critical time step length in the viscoplastic algorithm have been investigated.

The applicability of the element has been demonstrated by a number of examples.

However, a lot of work has to be done on the experimental side to provide realistic data for the numerical modelling of contact zones.

6 REFERENCES

/1/ D.V.Griffiths, Numerical modelling of interfaces using conventiona finite elements, 5th ICONMIG, Nagoya 1985

/2/ G.N.Pande and K.G.Sharma, On joint/interface elements and associated problems of numerical ill-conditioning, Int.Jou.Num.a.Analyt.Meth.i Geomech., 3, 1979

/3/ R.E.Goodman, R.L.Taylor and T.L. Brekke, A model for the mechanics of jointed rock, J.Soil Mech.Found.Div. ASCE,94, 1968

/4/ J.Ghaboussi, E.L.Wilson and V.Isberg, Finite element for rock and joint interfaces, J.Soil Mech.Found. Div.ASCE,99, 1973

/5/ G.Beer, An isoparametric joint/ interface element for finite element analysis, Int.Jou.Num.Meth.Eng.,21, 585-600, 1985

/6/ R.Frank et al., Numerical analysis of contacts in geomech., 4th ICOMIG, Edmonton, 1982

/7/ C.S.Desai, M.M.Zaman, J.G.Lightn and H.J.Siriwardane, Thinlayer eleme for interfaces and joints,Int.Jou.Nu a.Analyt.Meth.i.Geomech.,8, 1984

/8/ M.M.Zaman, Evaluation of thin-la er element and modelling of interfac behaviour in soil-structure interaction, 5th ICONMIG, Nagoya, 1985

/9/ G.N.Pande and W.Xiong, An improved multilaminate model of jointed rock masses,2nd NUMOG, Zürich, 1982

Numerical Methods in Geomechanics (Innsbruck 1988), Swoboda (ed.)
© 1988 Balkema, Rotterdam. ISBN 90 6191 809 X

Generating a 3-D block pattern: A key to blocky system study

J.Arcamone
Centre d'Etudes et de Recherches de Charbonnage de France, Verneuil
D.Héliot
Laboratoire de Mécanique des Terrains, ENSMIM, Nancy, France

ABSTRACT: Since more than ten years, several methods have been developed for studying the mechanical behaviour of heavily jointed rock masses. Among those methods, the simplest ones deal with single block stability analysis, whereas the most sophisticated ones provide a means for studying large block systems. In both cases an initial block pattern is needed and is an important parameter. Hence we followed the idea of developing a computer program for rebuilding 3D blocky systems. To this end, a probabilistic approach has been used so as to make the most of the often-sparse available data. The paper describes the Block Generator and presents a way to use such a tool in a coal mine whose roof is partly instable.

1 INTRODUCTION

Since more than ten years, several methods have been developed for studying the mechanical behaviour of heavily jointed rock masses. Among those methods, the simplest ones deal with single block stability analysis [1, 2], whereas the most sophisticated ones -especially the Distinct Element Method [3, 4, 5]- provide a means for studying large block systems. In both cases displacements in the structure are driven mainly by the initial block pattern, the geometry of which is obviously three-dimensional.

It is why has been followed the basic idea of developing a computer program for rebuilding 3D blocky systems.

Since data on joints are generally sparse and statistically known, the block generator must be a stochastic model. According to JOURNEL [6], a stochastic model can be seen as a probabilistic algorithm that allows to generate equiprobable different numerical outputs. Here all such solutions provide different possible images of the real block pattern, honoring the same input conditions, especially orientation and spacing distributions.

This paper aims at presenting both the general structure and the main concepts of the Block Generator command [7]. This command generates a block database intended to serve as input in several other codes (Cf figure 1). One of these codes is featured: it permits to image the numerical 3D block pattern and to display statistical informations on it. Then a case study is presented, showing one among the possible use of such a computer tool and providing evidences for its interest in rock mechanics engineering.

2 THE BLOCK GENERATOR

2.1 Block generation concepts

The rock mass is represented by an assemblage of polyedra.

The spatial organisation of joints can be seen as the result of a long and sometimes complex structural history: several successive tectonic episodes, followed by the excavation, have built the present pattern. Each of these tectonic episode is characterized by its own stress field, and produces one or several sets of fractures. Because these fractures affect an already jointed media, they do not locate anywhere.

Here comes the idea of simulating joints set by set, from the older to the younger, taking into account so the preexisting joint structure at each step.

To this end, only a basic procedure for splitting a father block into two son blocks is needed. The very problem lies in specifying at each step what blocks must be splat by what joints. To avoid a tedious data input, a specific computer language has been developed: the Block Generation Language (**BGL**). Using it, any user has only to describe the structural history of the rock mass in a file called "scenario file". The Block Generator command will then interprete this information to generate a "block database file". This block database file stores the numerical blocky system to serve as input in other programs.

Notice that different block databases can be generated from the same scenario file, according to their simulation number. Each simulation number is associated with a different pseudo-random series of number, and then with a different image of the rock mass described in the scenario file. So, the command syntax is:

bg <scenario file> <simulation number>

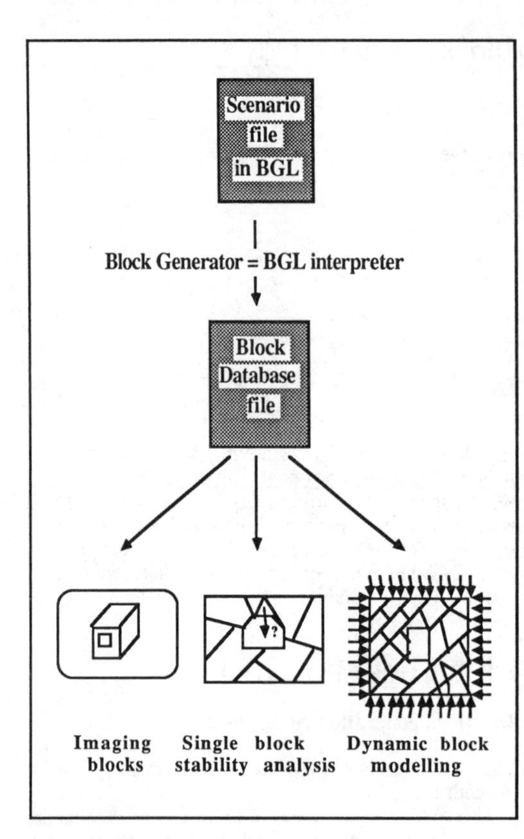

Figure 1. Possible uses of the block database generated by the Block Generator.

2.2 The BGL: Block generation language

The scenario file written according to the BGL grammar describes a way to rebuild episode after episode the block pattern. The scenario file includes at first a description of the zone of interest and then a succession of Sequence, Excavation, Partition, Select and Save statements.

a -The Zone of interest statement

The zone of interest is a parallelepiped whose orientation is specified by a First Orientation (**FO**) and a Second Orientation (**SO**) standing respectively for the X-axis and the Z-axis. The Y-axis is chosen normal both to the X-axis and the Z-axis, defining so the Global Reference. The zone of interest dimensions are then specified with regard to those axis:

```
zone_of_interest (
        FO = {8, 90}; SO = {90, 90};
        X = 20,30; Y= -10,10; Z = -5,20;
        )
```

These orientations are specified in the Georeference (X-axis:West->East, Y-axis:South->North and Z-axis: -gravity). For instance {8,90} means N 90E, 8.

b -The Partition and Select statements

The Partition statement makes possible the definition of a bipartition among blocks. By superimposing such bi-partitions, any partition of the zone of interest can be defined.

The Select statement allows to specify the set of blocks to be taken into account by the following Sequence, Excavation or Partition statements. It uses the subzones previously defined by Partition statements. Several operators (OR, AND, NOT, MINUS) permit to select the relevant set of blocks:

```
select  ( East_of_Fault OR
        ( West_of_Fault AND Layer_b) )
```

c -The Sequence and Excavation statements

The Sequence statement groups parameters characterizing a set of joints splitting the selected blocks.

We can distinguish at first the well known joint planes, that are specified by their orientation in the georeference or in the global reference: **{<dip> , <dip direction>}** or **[<x> , <y> , <z>]**, and then by a point of the plane: **(<x> , <y> , <z>).**

```
sequence Fault1 (
        plane = {83, 123}, (8., 5., 5.);
        ...
        )
```

For the statistically characterized sets of joints, orientation and spacing distributions are defined by one or several kernels, each specified by:

<w> : <m> (<s or d>)

...where **<w>** is the weigth of the kernel, **<m>** the mean of the kernel, **< s >** the standard deviation (in the case of linear data) and **< d >** the dispersion parameter (in the case of directional data). So doing permits to fit any distribution, approximating it with the sum of several small normal ditributions [7, 8, 9, 10, 11].

Joints are assumed to abut always one onto another. So, their extent is specified whether **SET** if all the selected blocks are splat together, or **BLOCK** if blocks are splat independently one from another.

```
sequence Joint1 (
        extent = SET;
        pole = 1.: {45, 123} (100.);
        spacing = .8: 2.2 (1.); .2: 3.5 (1.);
        )
```

Each Excavation statement describes a convex polyhedral subzone to be removed from the rock mass. It is specified by a list of planes bounding it and a point lying inside it.

Besides, a Save statement permits to save in a file postfixed ".bd" the current block database between two Episode or Excavation statements. This file is intended to serve as input in other codes.

2.3 Command using the block database

The **bd** (Block display) command allows to image the block database generated by the "bg" command and to display informations about it. Its syntax is:

bd < block database file >

It groups a set of subcommands permiting to display and image the block database in several ways: **"z"** displays general data about the zone of interest and the block database. **"c"** makes a cross section through the zone of interest. **"s"** images the block pattern in 3D. **"e"** displays an "eclaté" view of the zone of interest. **"b"** images individually one block. ...etc.

As shown in the foregoing case study, only imaging the 3D block pattern, can be worthwhile.

Other commands exist, or are in development (Cf figure 1): **sbsa** permits to analyze single block stability using WARBURTON's method [1]. **tofb** allows to select cross-sections through the 3D block pattern and to prepare the geometrical input for FEBLK, an hybrid Finite Element-Rigid Block code developed by ACIRL [12]. A similar interface is intended to be developed with UDEC, the Distinct Element code developed in Minneapolis [3, 4, 5].
All these commands are grouped in a package named RESOBLOK.

3 AUMANCE MINE, FRANCE: A CASE STUDY

Aumance Mine is a coal mine in the center of France, near Moulins. The coal is stephano-autunian (primary era) with a roof composed mainly of shales and friable sandstones. Eight tectonic episodes have influenced the deposit with compression and extension stages [13,14].

One 4-meters thick, 400 meters deep seam is extracted with room and pillar method [15]. Before removing coal in a 50x100 meters room, roadways are driven around it. The aim of the study was to predict the blocky system in the roof of the room with a view to foresee any great danger and improve the planning from both safety and productivity points of view. This prediction must be done mainly with the geological informations gathered in the roadways.

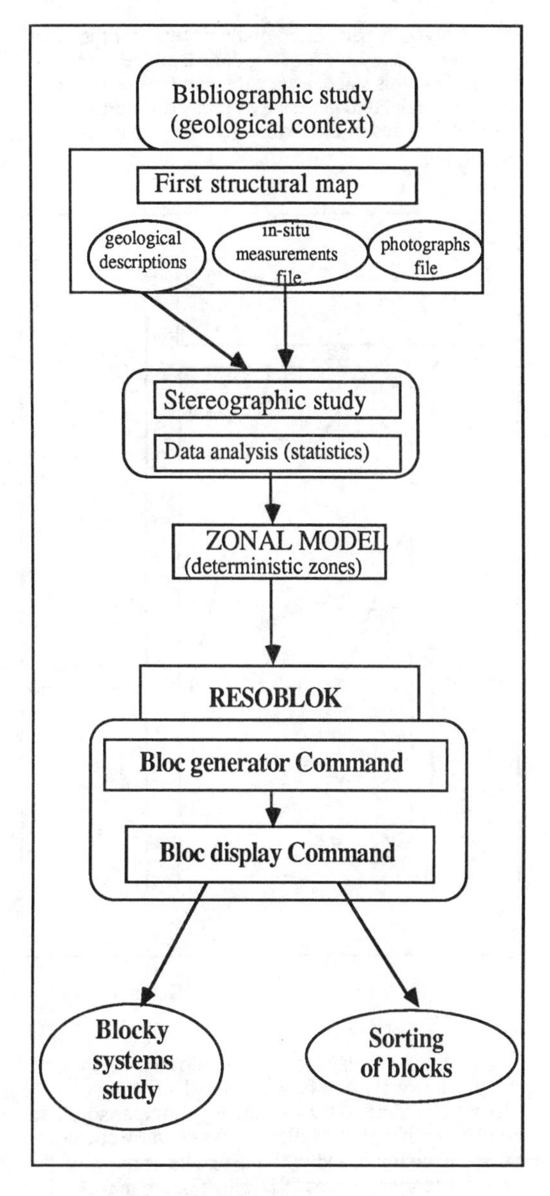

Figure 2: Basic methodology for RESOBLOK.

The basic methodology which has been followed is given on figure 2 [16]. After a bibliographic study of both the geology [13, 14] and the mine geotechnical history [15, 17], a first structural map is drawn according to the informations gathered during the in-situ investigations: geological descriptions, measurement recording sheets and photographs. The zonal model can be drawn representing the main deterministic zones (Cf figure 3). Then a stereographic study permits to determine the main discontinuity sets. A data analysis is performed on these sets. All the information necessary for running the Block Generator is available at this stage: a scenario can thus be written as aforexplained to generate as many different possible images of the hidden roof blocky geometry as wanted.

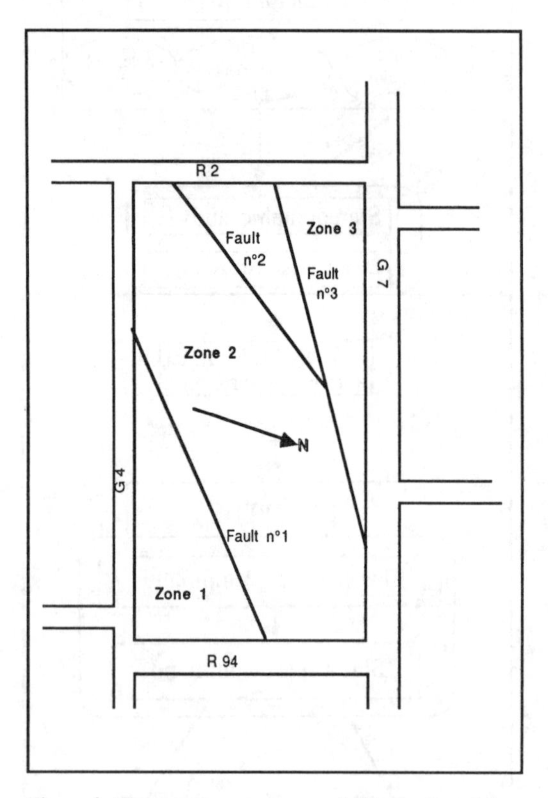

Figure 3: The Aumance mine panel VI zonal model.

The Block Display Command allows a viewing of the blocky systems (Cf figure 4) and a sorting of blocks according to their geometry. Comparisons can be done both with the fallen blocks actually encountered in the mine, and with the results of the Key Block Theory approach [2]. This feature makes possible to plan soundly the coal removing and to warn the staff in the most dangerous zones. In this respect, the graphic capabilities of RESOBLOK provide a very educational tool.

4 CONCLUSION

The RESOBLOK package, and especially the Block Generator command developed at CERCHAR have been outlined. One very simple among its possible applications has been presented on a first case study. Further tests, involving Distinct Element codes, have been undertaken and will be presented

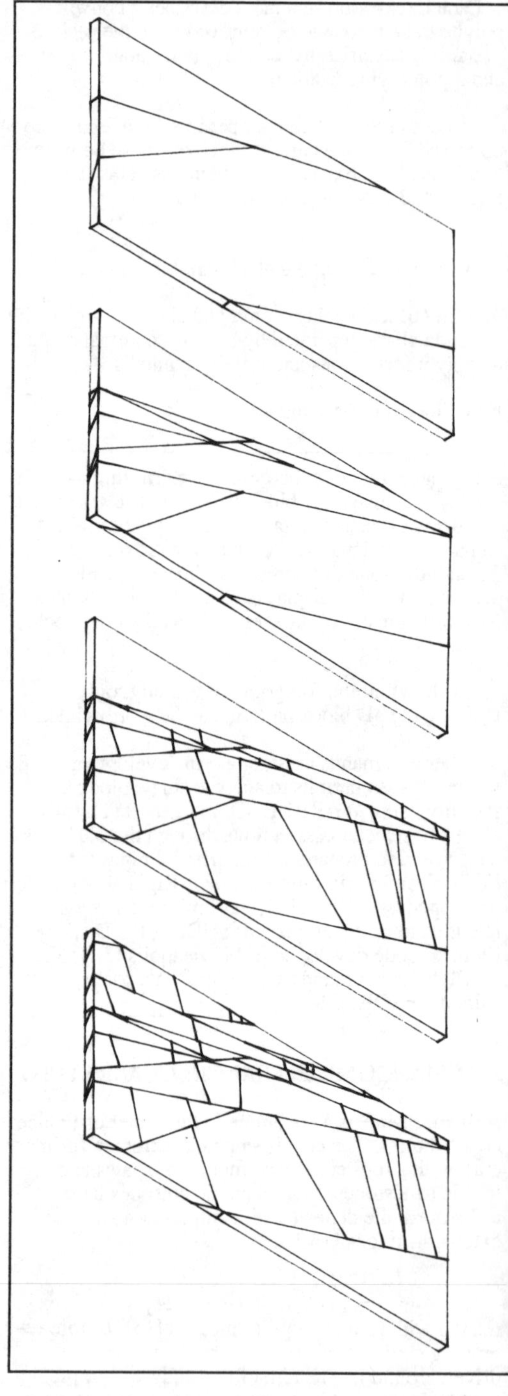

Figure 4: Successive steps in the rebuilding by RESOBLOK of the hidden block pattern of panel VI roof (Aumance mine).

elsewhere. The authors think that this kind of tools, that make the most not only of graphic and interactive computer capabilities, but also of C.A.D. (Computer Aided Design) techniques is about to renew the modeling practice in Rock Mechanics Engineering.

ACKNOWLEDGEMENTS

This work was conducted through the cooperation of the LIAD. The authors wish to thank Pr. J.L. Mallet from ENSG for many useful discussions and suggestions. They are indebted towards Aumance coalmine staff and especially its Director M. FRANCO for providing facilities and time. M. GONZE must also be acknowledged for carrying out most of the Aumance case study presented here.

BIBLIOGRAPHY

[1] P.M. Warburton, Vector Stability of an Arbitrary Polyedral Rock Block with any Number of Free Faces. Int. J. Rock Mech Minning Sci & Geomech. Abstr., Vol 18, 415-427 (1981).
[2] R.E. Goodman , G.H. Shi, Block Theory and its Application to Rock Engineering. Prentice Hall, Englewood Cliffs,N.J., 1985
[3] P.A. Cundall, A computer model for simulating progressive large-scale movements in blocky rock systems. Symposium of the I.S.R.M., Nancy. II-8 (1971)
[4] P.A. Cundall, O.D.L. Strack, A discrete Numerical Model for Granular Assemblies. Geotechnique, Vol 29, 47-65. (1979)
[5] J.V. Lemos, R.D. Hart, P.A. Cundall, A generalized distinct element program for modelling jointed rock mass. Proc. Int. Symp. on Fundamentals of Rock Joints, Bjorkliden, Sweden, 335-343. (1985)
[6] A.G. Journel, C.J. Huijbregts, Mining Geostatistics. Academic Press. New York. 1978
[7] D. Heliot, Generating a blocky rock mass. Submitted to the Int. J. Rock Mech Minning Sci & Geomech. Abstr. (1979)
[8] M. Rozenblatt, Remarks on some non parametric estimates of a density function. Annals of math. stat. 27, 832-835. (1956)
[9] E. Parzen, On estimation of a probability density function and mode. Annals of math. stat. 33, 1065-1076 (1962)
[10] J.L. Mallet, J.M. Epitalon, F. de Beaucourt, Discrimination non linéaire par indicatrices floues: Application à la reconnaissance des formes. Proc. of the Fourth Int. Symp. Data Analysis and Informatics. Versailles. (1985)
[11] P.J. Diggle, N.I. Fisher, Sphere: a contouring program for spherical data. Computer & Geosciences. 11. 725-766. (1985)
[12] G. A. Puckett, P.A. Mikula, Development of Geotechnical Mathematical Modelling Techniques for Coal Mine Design. ACIRL Published Report 84-5. (1984)
[13] S. Bonnion, Structuration du bassin houiller de l'Aumance (Allier) - Analyse structurale des dépôts de charbon et stériles. Thèse. Université de Dijon. 1983 [14] Y. Paquette, Le Bassin Autunien de l'Aumance (Allier) - Sédimentologie, Tectonique syndiagénétique. Thèse. Université de Dijon. 1980
[15] J. Arcamone, Etude de synthèse des îlots réduits à la mine de l'Aumance. Doc. Int. CERCHAR. (1984)
[16] M. Gonze, Caractérisation, analyse et simulation des systèmes de blocs. Application au cas de la mine de l'Aumance. Rapport de stage ENSGN. 1987
[17] H. Gonard, Boulonnage d'un toit très diaclasé. Industrie technique et Minérale, Janvier 1978. 27-30. (1978)

Numerical Methods in Geomechanics (Innsbruck 1988), Swoboda (ed.)
© 1988 Balkema, Rotterdam. ISBN 90 6191 809 X

The constitutive relation of joints and boundary surfaces in geomechanics

Fan Wen-ching
Division of Mathematics, Beijing College of Commerce, People's Republic of China
Yin You-quan
Department of Mechanics, Beijing University, People's Republic of China

ABSTRACT: Considering the joints and weak surfaces and other displacement discontinuous surfaces as the limit cases of a thin mass layer with zero thickness, we deduce its constitutive relation from the constitutive law of a continuous medium. Various features, such as dilation and softness (strain soften and permeation soften), are properly described in this constitutive relation. The joint elements based on this constitutive relation are used successfuly in some rock engineering analysis.

1 INTRODUCTION

Great progress in non-linear FEM analysis in geomechanics has been made since R.E. Goodman et al. presented the joint element. Using the joint element to simulate the joints, faults and other displacement discontinuous surfaces has become the most officient method in numerical analysis in geomechanics. As we know, joint surface has the properties of dilation and softening, these properties have very important influences on the stabilities in rock engineerings. In order to consider these properties, we must establish a rational constitutive relation in the joint surfaces. Only based on these constitutive relation, can we make analysis and decision in rock engineering more consistent with practices.

2 THE CONSTITUTIVE RELATIONS IN JOINT SURFACES IN CONSIDERATION OF THE DILATION AND THE STRAIN SOFTENING

In order to establish the constitutive relation of joint surfaces, we consider the joint surface not only as a displacement discontinuous surface geometrically, but also as a mass surface with mass property. It is just like the concept of a mass particle in dynamics, where the particle is a geometric point and a mass point in the same time. Across the joint-surface, discontinuity in displacement will take place. These discontinuity can be represented by a three-dimensional vector as;

$$\langle u \rangle = [u^+ - u^-, v^+ - v^-, w^+ - w^-]^t \quad (1)$$

where the "+" and "-" represents the values of the displacements in each side of the surface respectively. The stress vector, conjugating with the displacement-discontinuity vector in energy, is the stress components in the discontinuous surface:

$$\bar{\sigma} = [\tau_{xz}, \tau_{yz}, \sigma_z]^t \quad (2)$$

where the x,y,z is the local coordinate system, z is in the direction of normal to the joint-surface. The constitutive relation presented in this paper is the relation between the displacement-discontinuity vector $\langle u \rangle$ and the stress vector $\bar{\sigma}$.

If take the irreversible deformation in rock and soils as a plastic deformation in macroscopic view (it is substantially different with the metalic material in microscopic view), so we can obtain the incremental constitutive relation of the rock medium

$$d\sigma = (D-D_p)\ d\varepsilon \qquad (3)$$

where σ and ε denote a six-dimensional stress and strain vector respectively, D is the elastic matrix.
In equ. (3),

$$D_p = \frac{1}{A} D \frac{\partial g}{\partial \sigma} \left(\frac{\partial f}{\partial \sigma} \right)^t D \qquad (4)$$

$$A = \left(\frac{\partial f}{\partial \sigma} \right)^t D \left(\frac{\partial g}{\partial \sigma} \right) - \frac{\partial f}{\partial k} h \qquad (5)$$

$$h = \begin{cases} \sigma^t \dfrac{\partial g}{\partial \sigma} & \text{if } k = w^p \\ e^t \dfrac{\partial g}{\partial \sigma} & \text{if } k = \theta^p \\ \left[\left(\dfrac{\partial g}{\partial \sigma} \right)^t \dfrac{\partial g}{\partial \sigma} \right]^{\frac{1}{2}} & \text{if } k = \bar{\varepsilon}^p \end{cases} \qquad (6)$$

$$e^t = (1\ 1\ 1\ 0\ 0\ 0) \qquad (7)$$

where f and g are the yield function and the plastic potential respectively.

Now we consider a thin layer with finite thickness, and take the medium in the layer as an elastoplastic medium. It is strongly anisotropic in strength. Yield and break only present in shear crack along the layer surface and tensile break normal to the surface. The yield function and plastic potential can be written as:

$$f = (\tau_{xz}^2 + \tau_{yz}^2)^{\frac{1}{2}} + \mu \sigma_z - C \qquad (8)$$

$$g = (\tau_{xz}^2 + \tau_{yz}^2)^{\frac{1}{2}} + \bar{\mu} \sigma_z \qquad (9)$$

where c and μ are the cohesion force and frictional coefficient in the layer surface, both of them are functions of internal variable K.

If $\left(\frac{\partial c}{\partial k} - \sigma_z \frac{\partial \mu}{\partial k} \right)$ takes a positive, zero and negative values respectively, it will denote to the strain hardening, the ideal plasticity and the strain softening correspondingly, here $\bar{\mu}$ is a parameter representing the dilatation

$$0 \leq \bar{\mu} \leq \mu \qquad (10)$$

The medium will have no dilatation effect when $\bar{\mu} = 0$. The associated flow situation will take place $\left(\frac{\partial f}{\partial \sigma} = \frac{\partial g}{\partial \sigma} \right)$ when $\bar{\mu} = \mu$, the responding ratios of shear to dilatation will be larger than that of the experimental value. So the values of $\bar{\mu}$ in the range between 0 and μ should be determined according the experimental data. Instituting equations (8) and (9) into (3) and (4), the constitutive relations of a layered medium material will be obtained.

Then, let the thickness approaches zero, the constitutive relation of joint surface can be obtained from that of the layered medium. (ref. fig. 1). For the joint surface, the thickness b is much smaller than the characteristic length L, i.e.

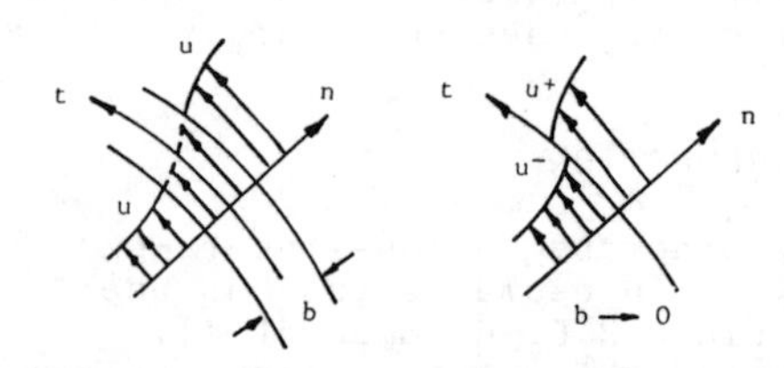

Fig. 1. The discontinuity surface looked upon as the limit of a material layer.

$b \ll L$, then we can assume that the stress and strain are distributed uniformly along the thickness, so,

$$\Delta u = br_{zx},\ \Delta v = br_{zy},\ \Delta w = b\ \ _z \qquad (11)$$

Furthermore, if we denote the order of the displacement by u, then, we have

$$\varepsilon_x \sim u/L \quad \varepsilon_y \sim u/L,\ \varepsilon_z \sim u/b$$
$$\gamma_{xz} \sim u/b \quad \gamma_{yz} \sim u/b,\ \gamma_{xy} \sim u/L \qquad (12)$$

Then we can neglect ε_x, ε_y, γ_{xy} in comparing with ε_z, γ_{zx}, γ_{zy}, so the constitutive relation of joint surface can be deduced from the limit case of that of a layered material as following (Yin, et al., 1981):

$$d\bar{\sigma} = (\bar{D} - \bar{D}_p)\ d\langle u \rangle \qquad (13)$$

$$\bar{D} = \begin{bmatrix} K_t & 0 & 0 \\ 0 & K_t & 0 \\ 0 & 0 & K_t \end{bmatrix} \tag{14}$$

$$\bar{D}_p = \begin{cases} \frac{1}{a} p r^t & \text{if } r^t d \langle u \rangle > 0 \\ 0 & \text{if } r^2 d \langle u \rangle \leq 0 \end{cases} \tag{15}$$

$$P = \left[\frac{K_t \tau_{xz}}{\sqrt{\tau_{xz}^2 + \tau_{yz}^2}} \quad \frac{K_t \tau_{yz}}{\sqrt{\tau_{xz}^2 + \tau_{yz}^2}} \quad \bar{\mu} k_n \right]^t \tag{16}$$

$$\gamma = \left[\frac{K_t \tau_{xz}}{\sqrt{\tau_{xz}^2 + \tau_{yz}^2}} \quad \frac{K_t \tau_{yz}}{\sqrt{\tau_{xz}^2 + \tau_{yz}^2}} \quad \mu K_n \right]^t \tag{17}$$

$$a = \mu \bar{\mu} k_n + K_t + h \left(\frac{\partial c}{\partial \kappa} - \frac{\partial \mu}{\partial \kappa} \sigma_z \right) \tag{18}$$

$$h = \begin{cases} c + (\bar{\mu} - \mu) \sigma_z & \text{if } \kappa = w^p \\ \bar{\mu} & \text{if } \kappa = \theta^p \\ (1 + \bar{\mu}^2)^{1/2} & \text{if } \kappa = \bar{\varepsilon}^p \end{cases} \tag{19}$$

where k_t and k_n are the tangential and the normal elastic stiffness of the joint surface respectively. If we know the elastic constants and the thickness b of the joint surface we can calculate them from the following equations:

$$k_t = G/b \qquad k_n = (k + \tfrac{4}{3} G)/b \tag{20}$$

We also can determine K_t and K_n directly by experiment using the method presented by Goodman to measure the joint-stiffness of the jointed rock samples. While parameter $\bar{\mu}$ reflects the delatation property of the joint surface, so $\frac{\partial c}{\partial \kappa} - \frac{\partial \mu}{\partial \kappa} \sigma_z < 0$ coresponding to the softening stage of the joints.

3 JOINT-SURFACE CONSTITUTIVE RELATION CONSIDERING THE PERMEATION SOFTENING

In order to investigate the influences of the underground water to the stability problem in rock engineerings, we must consider the permeation softening properties of the joint surfaces. In that case, the cohesion force c and frictional coeficient μ depend not only on the plastic internal variables κ, but also on the permeation parameter η,

$$c = c(\kappa, \eta) \qquad \mu = \mu(\kappa, \eta) \tag{21}$$

The concrete form of the above function and its parameters of permeation are selected according the components of the filling minerals in the joints and the property of the permeation (e.g. offinity of it). In the present, we can take the water content as parameter η, where the total stress $\bar{\sigma}$ in the joint surfaces are composed of effective stress $\bar{\sigma}'$ and permeation pressure q, written as:

$$\bar{\sigma} = \bar{\sigma}' + \bar{e} q \tag{22}$$

and

$$\bar{\sigma}' = [\tau_{xz}', \tau_{yz}', \sigma_z']^t \tag{23}$$

$$e = [0, 0, 1]^t \tag{24}$$

the corresponding constitutive relations are:

$$d\bar{\sigma} = (\bar{D} - \bar{D}_p) d\langle u \rangle + \bar{D}_\eta d\eta + \bar{e} dq \tag{25}$$

where

$$D_p = \begin{cases} \frac{1}{a} p \gamma^t & \text{if } \gamma^t d\langle u \rangle + Q_\eta d\eta > 0 \\ 0 & \text{if } \gamma^t d\langle u \rangle + Q_\eta d\eta \leq 0 \end{cases} \tag{26}$$

$$\bar{D}_\eta = \begin{cases} -\frac{1}{a} Q_\eta P & \text{if } \gamma^t d\langle u \rangle + Q_\eta d\eta > 0 \\ 0 & \text{if } \gamma^t d\langle u \rangle + Q_\eta d\eta \leq 0 \end{cases} \tag{27}$$

$$Q_\eta = -\frac{\partial c}{\partial \eta} + \sigma_z \frac{\partial \mu}{\partial \eta} \tag{28}$$

The definitions of other parameters can be found in equs (13)-(19). From (27) and (28) we can see that the increases of plastic internal variable κ and permeation parameter η can lead results of plastic loading and the softening of the joint surfaces. Vector $\bar{D}_\eta$ repesents the coupling actions of the plastic property with the water.

4 A PRACTICAL EXAMPLE

In non-linear program NOLM83, which is designed to analyze the stabilities of structures in rock engineerings, we used the joint-element with the constitutive relations mentioned above in this paper. With this program we examined the mechanism of collapsing of a mountain body of a phosporon mine. The main sliding profile of the mountain body is shown in fig. 2. Considering the permeation process of rain water from the back boundary to the front

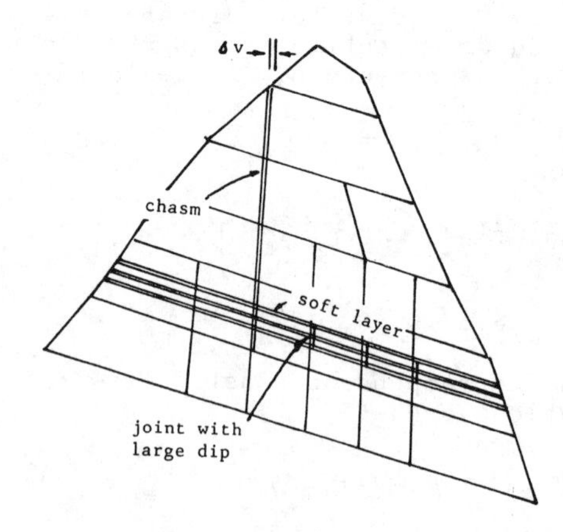

Figure 2. The main slide section of the mountain and its finite element net.

part caused by a heavy rain, we divided the permeation load into 7 load increase steps, each cores-ponding to a certain asigned level of the rain-water permeation. In the 5th step the plastic region of slip joint element in the back boundary sliding forward, is connected with the plastic region, which is appeared earlier in the front part. In the 6th sept, the stress greatly reduced, and step-wise crack surfaces appeared, the mountain bodies lost their stabilities. The numerical results represent that the permeation pressure and the softening function of the filling silt in the joints have a great influence to the collapse of the mountain bodies, they are main exciting factors of the collapse of the mountain. We also found that immediately before the collapse there is a accelerate tension cracking process in the back boundary faults. The theortical study and situ measurement about this phenomenon may have very important practical meanings.

REFERENCES

Goodman, R.E. 1976. Methods of geological engineering in discontinuous rocks. West Publishing Co., St. paul, Minn.

Hu Haitao, Jia Xuelang & Xu Kaixiang, The mechanism of mountain block slide-avalanche and its movement ways at Yan-chi-he river, Proceedings of International Symposium on Engineering Geolodical Environment in Mountainous Areas, 383-394, Beijing, China, 1987.

Wang Ren & Yin Youquan 1981. On the elasto-plastic constitutive equation of rock-like engineering materials. Acta Mechanic Sinica. 13:317-325.

Yin Youquan & Zhang Hong 1981. The one-dimensional joint element analysis of rock mass engineering. Mechanics and Practice. 3:52-54.

Yin Youquan & Zhang Hong 1985. The finite element program NOLM for elasto-plastic stress-strain analysis and stability analysis of the rock-soil system. In "The Records of Geological Research", 48-56, Peking University Press.

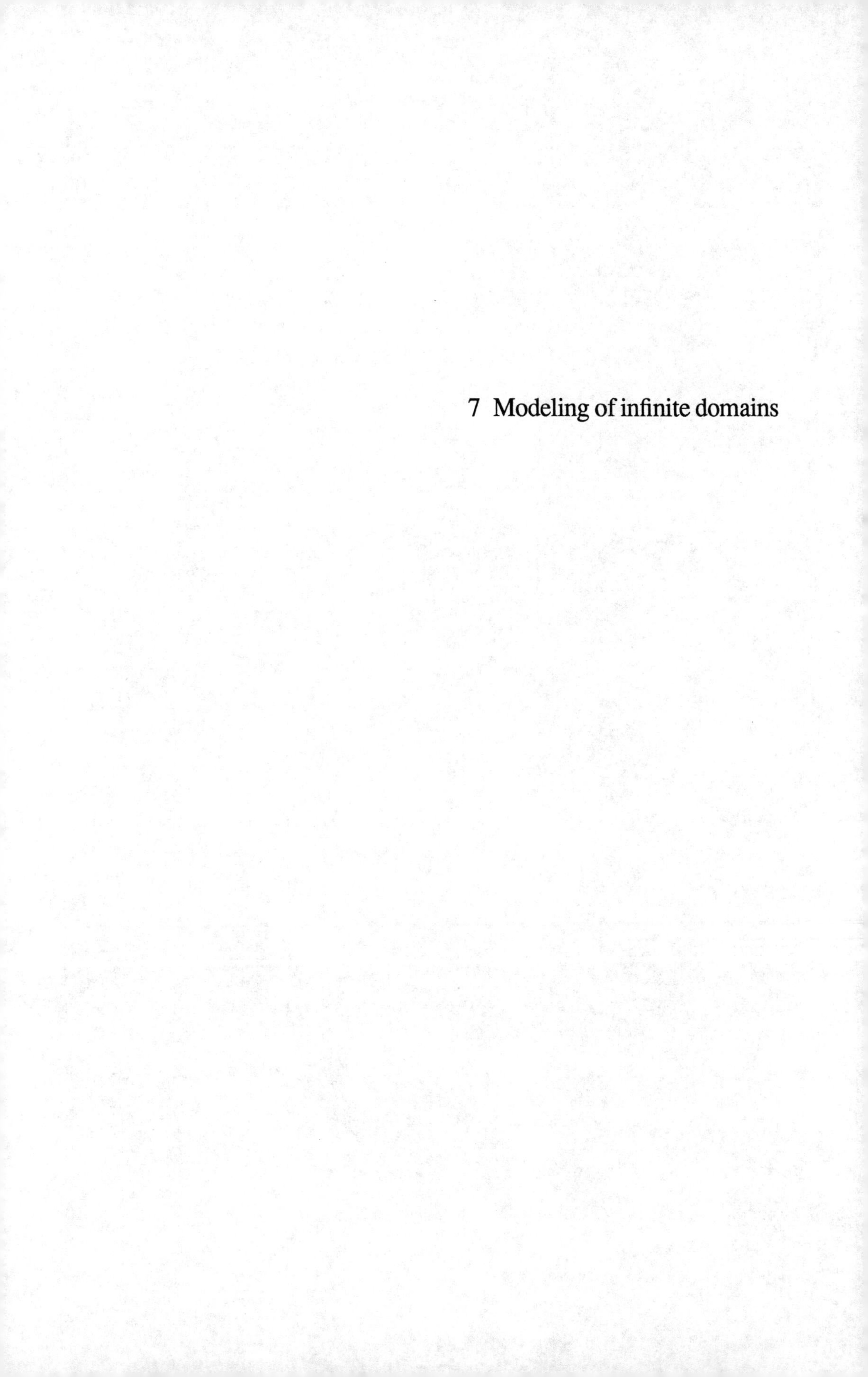

7 Modeling of infinite domains

Numerical Methods in Geomechanics (Innsbruck 1988), Swoboda (ed.)
© 1988 Balkema, Rotterdam. ISBN 90 6191 809 X

Elastic analysis of layered soils by boundary elements: Comparative remarks or various approaches

Giulio Maier & Giorgio Novati
Department of Structural Engineering, Technical University, Milan, Italy

ABSTRACT: The linear elastostatic analysis of layered soils often required for parametric studies in preliminary design of foundations, is tackled here by various boundary integral equation approaches apt to exploit the chain pattern of the system topology. These are: (a) a boundary element, transfer matrix method proposed by Banerjee and Butterfield; (b) a successive stiffness, boundary element procedure; (c) a successive compliance, boundary element method using half-space fundamental solutions and applicable to horizontally layered systems; (d) a procedure resting on Fourier series expansions and on the use of suitable Green's functions. The seemingly new method (c) is dealt with in some detail, whereas procedures (b) and (d) expounded elsewhere by the writers are merely outlined. Numerical performances are discussed in general, "a priori" terms related to the underlying mechanics.

1 INTRODUCTION

Boundary elements (BE) methods (i.e. methods resting on the discretization of boundary integral equations) are competing with the traditional finite element methods in certain restricted but practically important areas of mechanics. The main reason is that BE linear analyses deal with unknowns defined only on the boundary and, for piecewise homogeneous domains, on interfaces between subdomains. This fact reduces by one the dimensionality of the problem and alleviates both the data preparation and the size of the final algebraic system to solve (but its coefficient matrix turns out to be dense and non-symmetric). Unknowns in internal points are determined subsequently by "quadratures" only, selectively where needed. Unboundedness and three-dimensionality of domain, frequent features of geomechanical problems, represent other factors in favour of BE analysis, with respect to domain approaches, in linear-elastic and potential problems.

Therefore, BE approaches are well suited for the elastic analysis of layered soils. These either consist of physically distinct, individually homogeneous layers, or simulate by a sequence of such layers a "Gibson soil" model. A linear elastostatic analysis may be regarded as sufficiently realistic in parametric studies for preliminary design of shallow foundations (e.g. for a gravity offshore platform). It is quite natural and promising to exploit the peculiar topological features or "chain-pattern" of the system under consideration, schematically illustrated in fig.1. To such purposes, besides "ad hoc" domain methods (see e.g. [1][2]), a number of "ad hoc" BE methods have been devised by various authors (including the present writers) and are briefly outlined and critically compared in what follows. Details can be found in referenced papers.

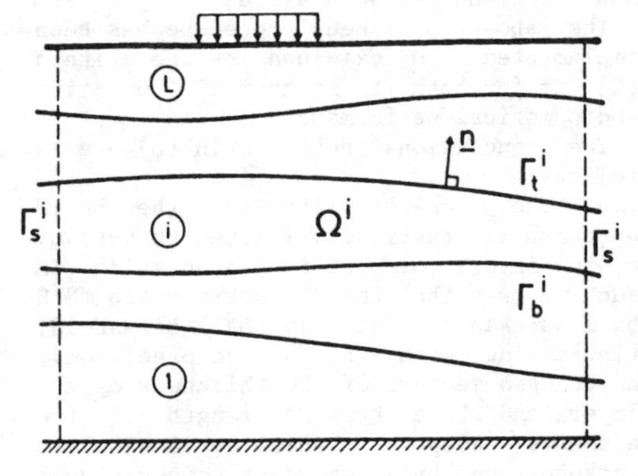

Fig.1 Schematic representation of a typical stratified soil on rigid rock.

2 TRANSFER MATRIX METHOD

An elegant procedure has been proposed in [3][4][5] which combines the BE method with the notion of "transfer matrices" (TM). Specifically this BE-TM procedure can be outlined as follows:

(1) For each layer the traditional Somigliana integral equation is taken to the boundary, discretized by the BE and "collocated" in all boundary nodes so that linear equations are obtained containing the nodal displacements and tractions on the boundary.

(2) Through a trivial manipulation of the equations generated in (1), for each layer i the transfer matrix $\underset{\sim}{T}_i$ is generated such that the nodal variables at the top interface (with layer i+1) are transformed by it into the nodal variables of bottom interface (with layer i-1).

(3) Using the matching conditions at interfaces, i.e. expressing displacement compatibility and traction equilibrium, by successive substitutions one relates the variables at the very top surface to the variables at the very bottom surface, through an overall transfer matrix which is simply the product of the transfer matrices of all layers.

(4) Imposing the top and bottom boundary conditions (usually: given loads at the top, zero displacements at the bottom), a linear system has to be solved which contains only the nodal unknowns pertaining to the boundary of the layer set, all the other unknowns having been eliminated (both at the interfaces and inside the layers).

(5) After the solution of the above reduced linear system, which contains useful information such as the displacements beneath the foundation, other unknowns, wherever desirable, are determined by straightforward back-substitutions and quadratures.

The above outlined procedure has been implemented and examined by the writers [6] as for both its mechanical foundation and numerical performance.

The conclusions achieved in [6] are as follows:

(a) The applicability of the BE-TM approach is drastically limited to sets of a few layers, each of them discretized in such a way that the "geometry ratio" H/D be less than a relatively small threshold, (increasing with the machine precision), H being a measure of the thickness of the layer and D a typical length of the elements adopted. This conclusion is corroborated by an abundant computational evidence, e.g. by the plots of fig.2.

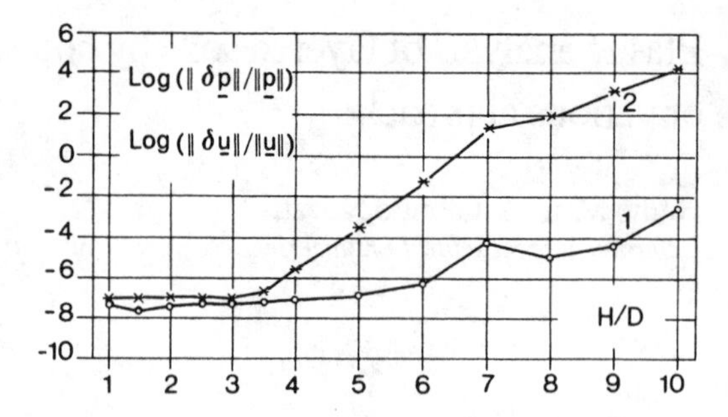

Fig.2 Analysis of a single, two-dimensional layer, with given loads on upper boundary, by the BE-TM method for various geometry ratios H/D: logarithms of "errors" with respect to standard BE solutions for top displacements (1) and bottom tractions (2).

These graphs refer to a simple layer in plane strain with constrained lower boundary and loaded upper boundary, both boundaries discretized by "constant" BE. The errors affecting the resulting top displacements and bottom reactions, are plotted versus the geometry ratio showing a dramatic, somewhat paradoxical degradation of accuracy as H/D increases (i.e., as the boundary element mesh is refined for a given situation). A standard BE solution has been taken as a reference; the norm of the difference between its results and the corresponding ones supplied by the TM solution, has been used to define the "errors" plotted in fig.2.

(b) The mechanical and mathematical reasons of the above accuracy deterioration and its links with the machine precision adopted have been elucidated in [6]. Indeed the ill-posedness of the elliptic b.v. problem underlying the concept of transfer matrix is reflected, for discretized continua, by possible ill-conditioning of certain submatrices to be inverted. The relevant "condition numbers" have been shown, by mechanical arguments, to necessarily increase for increasing geometry ratio, thus eventually leading to significant error amplification (earlier in the resulting reactions and later in the top displacements).

3 SUCCESSIVE STIFFNESS METHOD

The above conclusions have stimulated the search for alternative procedures capable of avoiding the limitations affecting the otherwise attractive BE-TM method, but still apt to exploit the chain-like pattern of layered soils. A procedure which attains these goals has been developed and tested in [7] according to the following strategy ("successive stiffness method").

(1) For each layer the same BE equations are constructed as in the earlier BE-TM method.

(2) Considering the lowest layer, simple algebraic passages, starting from the relevant BE equation, lead to the "stiffness matrix" which transforms nodal displacements on its top surface into the nodal tractions on the same surface.

(3) The matching conditions at the interface with the subsequent (second) layer, combined with a re-arrangement of the relevant BE equation, lead to a stiffness matrix, in the same sense as before, for the upper surface of the second layer. A recursive scheme can be established at this stage: it generates the stiffness matrix at the i-th interface for the subset of the i layers under it. When i equals the total number L of layers, the relevant stiffness matrix becomes the coefficient matrix of the linear equations to be solved as soon as the top boundary conditions (normally given loads) are introduced.

(4) After the solution of the equations in the top nodal unknowns alone, easy back-substitutions permit to define the variables at the interfaces and, if needed, within some layer.

The following circumstances are worth noting (and have been pointed out in [7]:

(a) All matrix manipulations implied in this procedure, in contrast to some of those required by the BE-TM method, can be mechanically interpreted as solution processes for well-posed b.v. problems and the condition numbers of the matrices involved are generally small and unaffected by the geometry ratio of the layer; hence, convergence is expected in any event as the BE mesh is refined. Fig.3 presents the errors (defined as earlier) affecting the results achieved by the present method for the same two-dimensional example: these graphs corroborate the theoretically expected intensitivity of the accuracy with respect to the geometry ratio H/D and, hence, the unrestricted applicability of the method in terms of geometric features of the discretized layers.

(b) In order to compare alternative solution procedures from the computational standpoint, a fairly meaningful and reliable index is represented by the number of arithmetic operations required

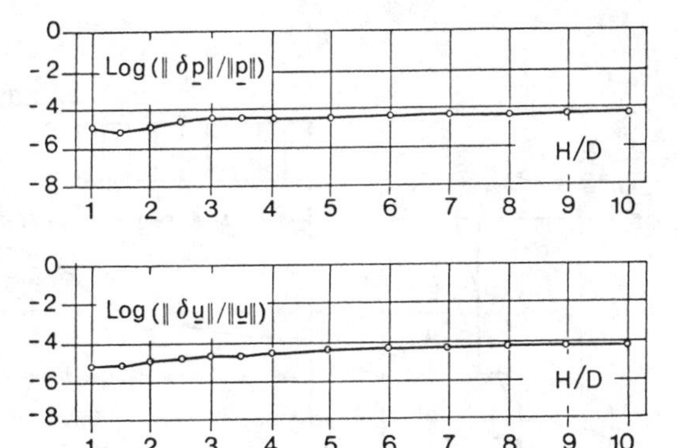

Fig.3 "Errors" in the same sense and for the same layer as in figure 2, but obtained by means of the successive stiffness method.

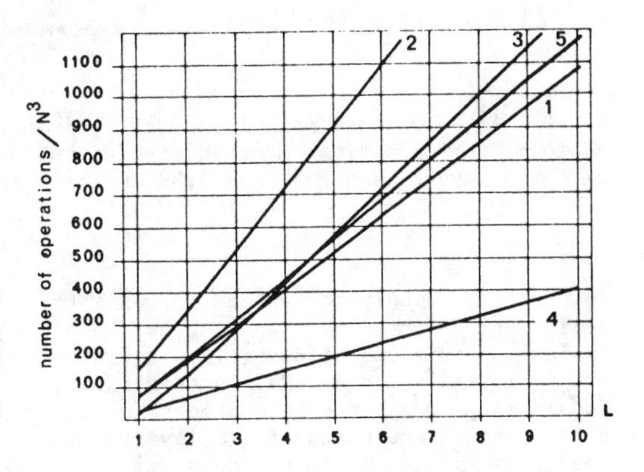

Fig.4 Number of operations divided by N^3 (N being the number of nodes at each interface) required by different BE procedures, versus layer number L: successive stiffness 1, transfer matrix 2; "zoned" BEM 3; successive compliance 4 and transfer matrix 5 with half-space fundamental solution.

for the determination of unknown top displacements and bottom tractions for a single set of boundary data. In the present context such count can be made explicitly in terms of the number N of the nodes assumed on the top surface, provided some simplifying hypotheses are adopted as specified in [6][7]. The main simplifications are as follows: N is large enough to make N^2 negligible with respect to N^3; the node number is the same on all

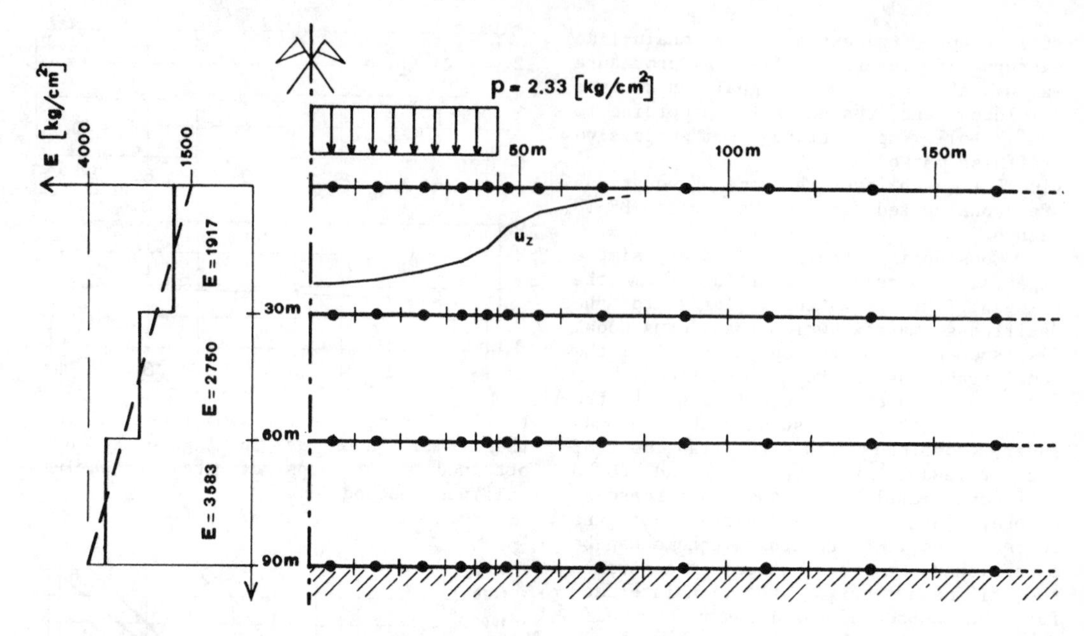

Fig.5 A two-dimensional Gibson soil simulated by individually homogeneous layers under uniform strip loading: adopted discretization and vertical top displacements (max: 5.93 cm for 3 layers as shown; max: 5.99 cm for 5 layers).

interfaces; widely accepted approximate estimates define the computing burden of typical matrix operations (e.g. inversion of a matrix of order N). On this basis, considering only the operations needed after the BE equations of all layers have been generated, fig.4 compares the successive stiffness method (plot 1) to the transfer matrix method (plot 2) and to the standard BE method "by zones" (plot 3). The last method involves also interface variables, but exploits the banded structure of the matrix (of order 4NL) in the final equations. These comparisons concern two-dimensional situations but may be easily extended to three-dimensional problems. The computational savings implied by the first method are thus quantified and found to increase with the layer number.

An illustrative application of the successive stiffness method is presented here (fig.5), additional to the examples contained in ref.[7]. A Gibson soil with elastic modulus varying from 1500 kg/cm^2 to 4000 kg/cm^2 and constant Poisson ratio $\nu = 0.3$ has been simulated in plane strain by three and, subsequently, five fictitious homogeneous layers, with increasing E. The results, part of which are depicted in fig.5, confirmed the theoretical prediction about the performance of the method and, in particular, showed that a linear increase with depth of the elastic modulus can be replaced by a step-wise one, in a few steps, with practically negligible effects on the main results (vertical displacements u_z and stresses).

4 SUCCESSIVE COMPLIANCE METHOD

We propose in this Section a noteworthy variant of the preceding method, with reference to unbounded layers with plane parallel interfaces. This variant will turn out to be particularly suitable for such special configurations. In these cases it is quite natural and advantageous to use a half-space fundamental solution (Melan's and Midlin's for two- and three-dimensional problems, respectively, e.g. [8]), its traction-free surface coinciding with the top boundary Γ_t^i of each layer i. As a consequence of this choice, and of the adoption of infinite BE (instead of fictitious side boundaries Γ_s^i), the BE equations formulated for layer i on the basis of the customary "direct" approach, in the absence of body forces for simplicity, acquire the following peculiar structured form:

$$\begin{bmatrix} \underset{\sim}{I} & \underset{\sim}{H}^i_{tb} \\ \underset{\sim}{0} & \underset{\sim}{H}^i_{bb} \end{bmatrix} \begin{Bmatrix} \underset{\sim}{u}^i_t \\ \underset{\sim}{u}^i_b \end{Bmatrix} = \begin{bmatrix} \underset{\sim}{G}^i_{tt} & \underset{\sim}{G}^i_{tt} \\ \underset{\sim}{G}^i_{bt} & \underset{\sim}{G}^i_{bb} \end{bmatrix} \begin{Bmatrix} \underset{\sim}{p}^i_t \\ \underset{\sim}{p}^i_b \end{Bmatrix} \quad (1)$$

Here and henceforth quantities pertaining to the upper (top) and lower (bottom) boundaries of each separate layer are denoted by the subscript t and b, respectively. Matrices (and column vectors) are indicated by underscored symbols; identity and null matrices by $\underset{\sim}{I}$ and $\underset{\sim}{0}$, respectively. Vectors $\underset{\sim}{u}^i$ and $\underset{\sim}{p}^i$ contain as components nodal displacements and tractions; these and the matrices of relevant coefficients, $\underset{\sim}{H}^i$ and $\underset{\sim}{G}^i$, are partitioned according to the above boundary categorization.

The circumstance

$$\underset{\sim}{H}^i_{tt} = \underset{\sim}{I} \quad , \quad \underset{\sim}{H}^i_{bt} = \underset{\sim}{0} \qquad (2)$$

is due to the vanishing of fundamental tractions on Γ^i_t: in fact, then the integrals over Γ^i_t containing unknown displacements are zero.

Suppose now that a compliance matrix $\underset{\sim}{C}^{i-1}$ is known, which relates the nodal tractions to the nodal displacements on the interface i-1, conceived as free top surface of the set of the first i-1 layers:

$$\underset{\sim}{u}^{i-1} = \underset{\sim}{C}^{i-1} \underset{\sim}{p}^{i-1}_t \qquad (3)$$

At every node of each interface we will assume a single unit normal vector, specifically directed outwards for the lower (i-th) and inwards for the upper (i+1) layer. As a consequence, interface compatibility and equilibrium at nodes are enforced by the following conditions:

$$\underset{\sim}{u}^i_t = \underset{\sim}{u}^{i+1}_b \quad , \quad \underset{\sim}{p}^i_t = \underset{\sim}{p}^{i+1}_b \qquad (4)$$

Through matching equations (4), relation (3) can be transferred to the kinematic and static nodal variables of the i-th stratum at its bottom face:

$$\underset{\sim}{u}^i_b = \underset{\sim}{C}^{i-1}_b \underset{\sim}{p}^i_b \qquad (5)$$

By substituting eq.(5) into the latter of eqs.(1), we obtain:

$$\underset{\sim}{p}^i_b = (\underset{\sim}{H}^i_{bb} \underset{\sim}{C}^{i-1} - \underset{\sim}{G}^i_{bb})^{-1} \underset{\sim}{G}^i_{bt} \underset{\sim}{p}^i_t \qquad (6)$$

Subsequent substitutions of eqs.(5) and (6) into the former of eqs.(1) lead to:

$$\underset{\sim}{u}^i_t = [\underset{\sim}{G}^i_{tt} - (\underset{\sim}{H}^i_{tb} \underset{\sim}{C}^{i-1} - \underset{\sim}{G}^i_{tb}) \cdot (\underset{\sim}{H}^i_{bb} \underset{\sim}{C}^{i-1} - \underset{\sim}{G}^i_{bb})^{-1} \underset{\sim}{G}^i_{bt}] \underset{\sim}{p}^i_t \equiv \underset{\sim}{C}^i \underset{\sim}{p}^i_t \qquad (7)$$

This is clearly a recursive equation, which permits to generate the compliance matrix $\underset{\sim}{C}^i$ at interface i for all i underlying strata, using the analogous matrix $\underset{\sim}{C}^{i-1}$ at interface i-1 and matrices concerning only the additional i-th stratum.

For the first layer (i = 1), the fixity of its bottom surface $\underset{\sim}{u}^1_b = \underset{\sim}{0}$ (fig.1) can be expressed by $\underset{\sim}{C}^o = \underset{\sim}{0}$. Setting $\underset{\sim}{C}^o = \underset{\sim}{0}$ in eq.(7), this equation provides the expression apt to calculate $\underset{\sim}{C}^1$. After repeated, recursive use of eq.(7) for i = 2,...L, one arrives at:

$$\underset{\sim}{u}^L_t = \underset{\sim}{C}^L \underset{\sim}{p}^L_t \qquad (8)$$

Since $\underset{\sim}{p}^L_t$ equals the given nodal tractions $\bar{\underset{\sim}{p}}$, eq.(8) defines the unknown top displacements.

The solution will be accomplished by straightforward backsubstitutions. In fact, the nodal tractions $\underset{\sim}{p}^i_b$ at each interface can be evaluated in descending sequence (i = L-1,....,1) by means of eq.(6) combined with interface equilibrium, eq.(4b). Subsequently, the interface displacements are directly determined, wherever needed, by means of eq.(5).

The present solution procedure is being implemented (and expounded elsewhere) and, hence, is not numerically tested here. However, when applicable, it turns out to be an advantageous alternative to all other BE solutions considered, in the light of a comparative "a priori" assessment of computational merits based on the same simplifying assumptions mentioned at the end of the preceding Section. On this basis , for each one of the L layers, the implementation of the recursive formula (7), impliying four products and one inversion of 2N-order matrices, approximately requires $4(2N)^3 + 4/3(2N)^3$ multiplications. For the lowest layer resting on rigid rock, the specialization of eq.(7), i.e. $\underset{\sim}{C}^o = \underset{\sim}{0}$, reduces the above operation count to $2(2N)^3 + 4/3(2N)^3$. Summing up, the "total computing cost" (besides the "cost" of the preliminary generation of matrix $\underset{\sim}{H}^i$ and $\underset{\sim}{G}^i$) amounts to: $(42.6L - 16)N^3$.

This estimation leads to the linear plot 4 in fig.4, which evidences that the procedure put forward in this Section computationally compares very well with all the preceding procedures since the "total cost" in the above sense is drastically reduced for any numbers L of layers. However, a trade-off arises here between the a priori vanishing of entries in matrix $\underset{\sim}{H}^i$ according to eq.(2), and the extra cost implied in the generation of $\underset{\sim}{H}^i$ and $\underset{\sim}{G}^i$ by the additional term of half-plane fundamental solution with respect to Kelvin's. It may be reasonably assumed that these conflicting factors roughly compensate each other.

It may be interesting to use the same criteria for a computational assessment of the transfer matrix procedure (a) when applied to the present case of horizontally layered soils and half-plane fundamental solution. This estimation, see [6], leads to line 5 in the graph of fig.4.

5 FOURIER SERIES, GREEN'S FUNCTION METHOD

The analysis of horizontally layered systems can be performed using Fourier series (or Fourier transform) in this way the original two- or three-dimensional problem can be reduced to a sequence of one-dimensional problems and easily solved using micro-computers. We outline in this section a method of solution, expounded in [9], which adopts Fourier series expansions in the horizontal direction and a Green's function approach to determine the Fourier's coefficient distribution across the thickness of each layer. At difference from the finite layer method [1], no discretization is required. The only approximations which need be made concern the loading distribution on the top surface which is assumed to be spacially periodic along the horizontal direction and synthesized by a truncated series expansion. However any isolated strip loading can be represented with the desired accuracy provided the spacial period and the number of harmonic terms be large enough. Of course the search for the response to each harmonic loading of frequency $\beta_n = n\pi/L$ decouples into a symmetric and a skew-symmetric problem (symmetry being meant with respect to the z axis). Referring for definitness to a symmetric problem in the generic frequency, making use of suitably defined Green's functions for the half-plane, for each layer it is possible to generate four algebraic equations linking the eight relevant Fourier coefficients pertaining to its upper and lower surfaces. In matrix form such relationships are formally identical to eq.(1) provided $\underset{\sim}{u}$ and $\underset{\sim}{p}$ be now interpreted as the displacement coefficient vector $\{u^s_{nx} \quad u^c_{nz}\}^T$ and the traction coefficient vector $\{p^s_{nx} \quad p^c_{nz}\}^T$ respectively, where subscript n is the harmonic index while superscripts s or c denote association with sine- or cosine-harmonics. Thus the successive compliance procedure described in the previous section can be applied formally unchanged to the present context leading to the determination of the displacement coefficients at the top followed by the evaluation of the coefficients at all intermediate interfaces in descending order. Successively, displacement or stress coefficients at any level within each layer can be easily evaluated in terms of those at the layer boundaries, this throughly completing the solution to the single-frequency problem.

Adding together the solutions for the harmonic loadings of the various frequencies (i.e. using the principle of superposition), the response to any surface loading can be obtained.

Fig.6a shows the three-layer system used to study the convergence of the proposed method. The layers have different elastic

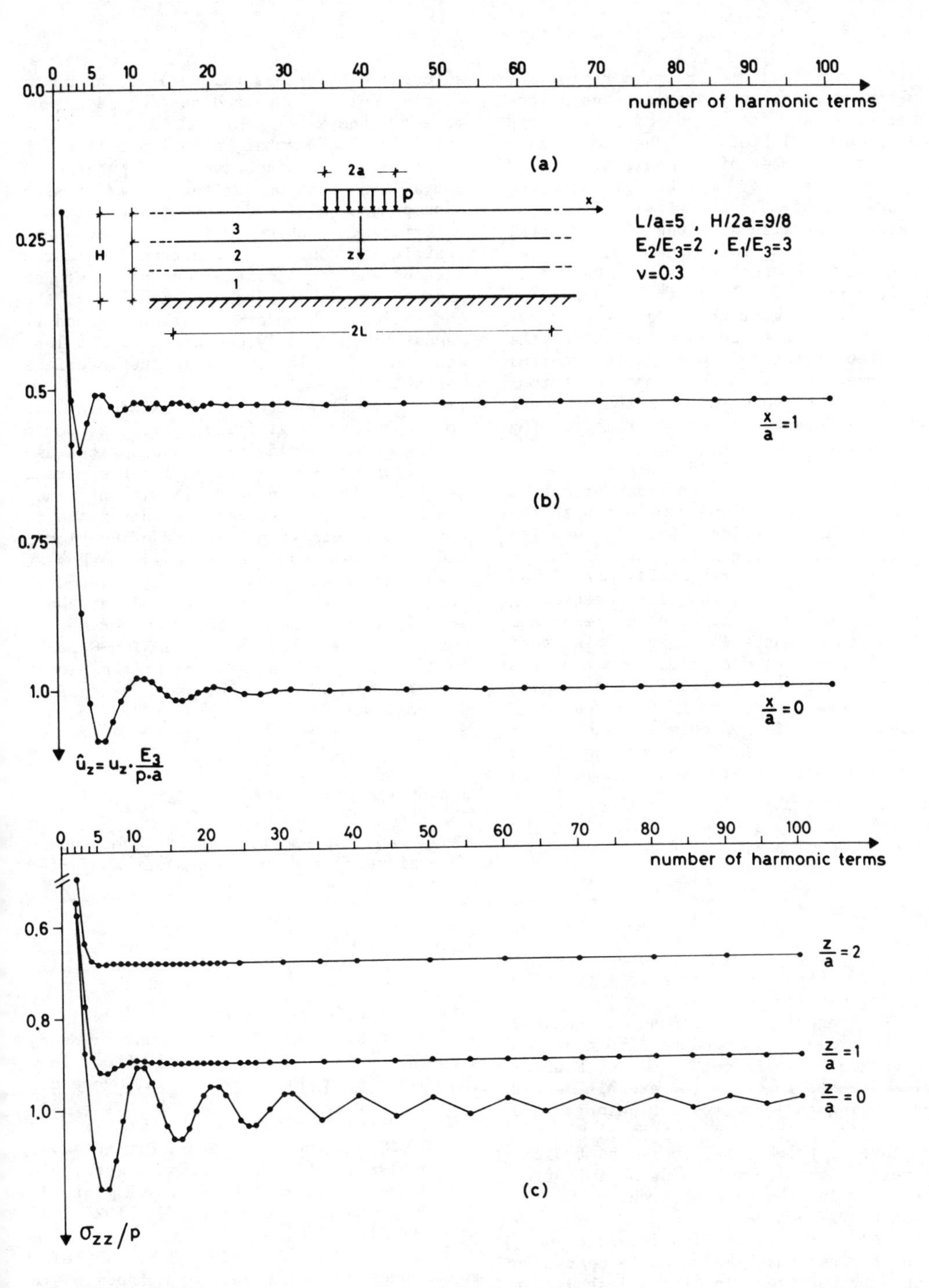

Fig.6 The three-layer illustrative example (a); convergence of vertical top displacements (b) and normal vertical stresses along central axis (c) in relation to the number of harmonic terms.

moduli but the same Poisson ratio. The resulting vertical top displacement at the center and at the edge of the strip loading are plotted in the same figure versus the number of harmonic terms used in the analysis; fig.6c depicts the convergence of the vertical stress at different locations along the central axis. The non-dimensional quantities represented in the plots only depend on the four non-dimensional ratios appearing in fig.6a. In fig.6c the lowest plot shows that an accurate representation of the given top traction even at the central point requires a high number of terms. However convergence of all the response quantities becomes very fast at increasing values of z/a.

As an alternative to this successive compliance procedure, the transfer notion can be reconsidered and combined with the present Fourier series - Green's function approach, thus originating a different version of the transfer matrix method. As one can expect the ill-conditioning intrinsic to such a methodology reappears in this context as well. This is discussed in [10] where the condition number of the transfer matrices (here of the fourth order because relevant to a single-frequency transfer problem) is shown to increase unboundedly at increasing values of the non-dimensional parameter $\beta_n H$ (H is the layer thickness), which hence plays the same role as the ratio H/D in the method of Section 2.

6 CONCLUSIONS

The foregoing considerations can be summarized and supplemented by the following closing remarks.

(i) The "successive stiffness" method (b) and the related procedures resting on "successive compliances" (c) and Fourier series-Green's function (d) exhibit matrix conditioning and accuracy not deteriorated by increasing ratio of a typical measure of layer thickness over a typical discretization length. Thus a limitation intrinsic in the elegant transfer matrix method (a) is avoided.

(ii) As for computational efficiency, rough but reasonably acceptable operation counts (visualized in fig.4) show that method (c) applied to horizontally layered soils (with Melan's fundamental solutions) implies substantial savings, whereas method (a) turns out to be not especially economical even in such peculiar situations (plot 5 in fig.4). A parallel comparative conclusion applies to successive stiffness (plot 1) vs. transfer matrix (plot 2) procedures for layers of more general configurations, using Kelvin's fundamental solutions. It worth stressing that the above computational comparisons rest on various rather drastic assumptions (e.g. node number equal at all interfaces, which is a necessary restriction only for the transfer matrix method) and for a single loading case. For parametric studies varying the loads, the conventional boundary element method by zones would clearly compare less and less well, as it involves also the interface variables.

(iii) Finally, from the standpoint of application flexibility, it is worth noting that rigid block foundations and inhomogenuities by zones inside the strata (such as piles) can be dealt with by minor changes both in the transfer matrix and in the two recursive procedures of Section 3 and 4. Cavities and coarser meshes at lower interfaces can be easily allowed in the successive stiffness method (as shown in [7]), not so in the transfer-matrix method. The Fourier expansion-Green's function approach exhibits the least versability as for the above extensions and variants.

ACKNOWLEDGEMENT

This paper presents results achieved in the frame of a research project supported by MPI.

BIBLIOGRAPHY

[1] Y.K.Cheung and S.C.Fan, Analysis of pavements and layered foundations by the finite layer method, Proc. Int. Conf. on Num. Meth. in Geomech., Aachen, Vol.3, Balkema, Rotterdam, 1129-1135 (1979).

[2] J.C.Small and J.R.Booker, Finite layer analysis of layered elastic materials using a flexibility approach, Int. J. Num. Meth. Engrg., 20, 1025-1037 (1984).

[3] P.K.Banerjee and R.Butterfield, Boundary element methods in geomechanics, in Finite Elements in Geomechanics, Ed. G.Gudehus, Wiley, New York, 529-570 (1977).

[4] R.Butterfield, BEM in geomechanics, in Boundary Element Techniques in Computer-Aided Engineering, Ed. C.A.Brebbia, NATO-ASI Series, 1984.

[5] M.Ohga, T.Shigematsu and T.Hara, A combined finite element - transfer matrix method, J. Eng. Mech. Div. ASCE, 110, 9, 1335-1349 (1984).
[6] G.Maier and G.Novati, On boundary element-transfer matrix analysis of layered elastic systems, Engineering Analysis, 3, 208-216 (1986).
[7] G.Maier and G.Novati, Boundary element elastic analysis of layered soils by a successive stiffness method, Int. J. Num. Anal. Meth. Geomech., to appear (1987).
[8] C.A.Brebbia, J.C.F.Telles and L.C.Wrobel, Boundary Element Techniques, Springer-Verlag, Berlin (1984).
[9] G.Novati and A.De Crescenzo, A method for the elastic analysis of horizontally layered systems by Fourier series and Green's functions, to appear.
[10]G.Novati, On the analysis of elastic layers by a Fourier series, Green's function appraoch, Rendiconti dell'Accademia dei Lincei, Classe Sci., to appear (1987).

Numerical Methods in Geomechanics (Innsbruck 1988), Swoboda (ed.)
© 1988 Balkema, Rotterdam. ISBN 90 6191 809 X

Applications of the boundary element method for the analysis of discontinuities

F.S.Rocha, W.A.Nimir & W.S.Venturini
São Carlos School of Engineering, University of São Paulo, Brazil

ABSTRACT: A direct boundary element formulation to model discontinuities is proposed. The solution technique follows previous schemes developed to model other nonlinear behaviours. The quadrupole effects are now considered to remove stresses along discontinuities. The applicability of the numerical algorithm developed is demonstrated by practical engineering analyses.

1 INTRODUCTION

The boundary element formulation for the analysis of two and three-dimensional continuum mechanics problems was introduced in the end of the sixties by Rizzo [1] and Cruze [2]. In 1976, Hocking [3] has presented a work on the application of boundary techniques to rock mechanics analysis by the first time. Other works on geomechanics have been published two years later by Brady [4] and Wardle [5] in which the assumption of continuity between layers and other structural elements has been made.

No-tension, plastic and viscoplastic behaviours were introduced into the boundary methods few years later employing initial stress schemes with constant matrices [6,7].

The inclusion of slip and separation along discontinuities shown by rock material has been made by Hocking [8] in 1978. In this formulation, the author has adopted the type of singularities called quadrupoles to deal with discontinuities. This kind of nonlinearities can also be modelled by introducing non boundary integral equations based on special joint element relations [9].

In the following section the nonlinear boundary element formulation with initial stress terms is changed to model discontinuity effects. An iteractive scheme similar to that used for no-tension, plastic and viscoplastic analyses is also achieved. Simple examples are shown in order to illustrate the accuracy the algorithm gives. A dam foundation with discontinuities is analysed to show the numerical efficiency of the proposed process to solve practical problems.

2 BASIC IDEAS

Discontinuity effects in an elastic domain is modelled in this work by a convenient distribution of ficticious loads along lines where slips or separations are expected. The domain is initially assumed without any discontinuity and can be formed by several subregions. After computing the elastic values, displacements and stresses, using the usual direct boundary element formulation, at any necessary boundary or internal points, one can verify the lines over which discontinuities appear according to a failure criterion previously adopted. Similarly to the no-tension case [6] initial stress values can be computed along those lines. By applying these computed stress values back into the domain an iterative scheme is achieved to enforce the failure criterion assumptions.

Although any failure criterion could be modelled along discontinuities, the classical Mohr-Coulomb envelope is adopted here (see Figure 1). The constitutive equation, in this case, is given as a function of material cohesion, c, and the internal friction angle, ϕ, as follows,

$$|\tau| = -\sigma \tan \phi + c \tag{1}$$

For any point, P_1, inside the envelope (Figure 1), the criterion is not reached and no discontinuity is observed. Stress state represented by P_2 or P_3 means crite-

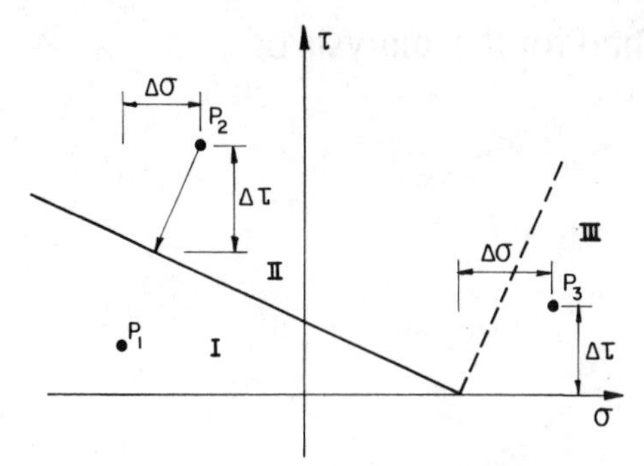

Fig.1 Mohr-Coulomb envelope

rion not obeyed and initial stress values, $\Delta\tau$ and $\Delta\sigma$, have to be applied. It is important to observe that the removal of $\Delta\tau$ and $\Delta\sigma$, in any case, is also governed by the aperture of the discontinuity.

3 BOUNDARY ELEMENT FORMULATION

In order to model discontinuities, tractions along both surfaces have to be corrected according to the criterion adopted, equation (1). This correction can be made using pairs of ficticious forces called dipoles, or to avoid usual stress perturbation in the normal stress in the discontinuity direction, one can also use quadrupole fundamental solutions [8].

Displacement and stress integral representations for a point without discontinuity, in an elastic domain with internal or boundary lines of quadrupoles, can be easily written in the following form,

$$C_{ik}(P)u_k(P) = -\int_\Gamma u_k(S)p^*_{ik}(P,S)d\Gamma(S) + \int_\Gamma p_k(S)u^*_{ik}(P,S)d\Gamma(S) + \int_\Omega b_k(s)u^*_{ik}(P,s)d\Omega(s) + \int_{\Gamma_d} q_k(S)u^{**}_{ik}(P,S)d\Gamma(S) \quad (2)$$

$$\sigma_{ijk}(p) = -\int_\Gamma u_k(S)S_{ijk}(p,S)d\Gamma(S) + \int_\Gamma P_k(S)D_{ijk}(p,S)d\Gamma(S) + \int_\Omega b_k(s)D_{ijk}(p,s)d\Omega(s) + \int_{\Gamma_d} q_k(S)\sigma^{**}_{ijk}(p,S)d\Gamma(S) \quad (3)$$

where $q_k(S)$ is the intensity of the ficticious forces applied along the discontinuity, Γ_d; u^{**}_{ik} and σ^{**}_{ijk} are quadrupole fundamental solutions. Equation (2) is valid for boundary or internal points out of discontinuity lines. For points on a discontinuity line, the singularity of the last integral term must be analysed to give aperture values.

For stress determination, equation (3) can be avoided for nodes along boundaries. In this case, finite difference schemes can be used for simplicity. No singularity arises when equation (3) is applied for nodes on discontinuities. As for boundary nodes, this equation can not be used to compute stresses at corner nodes.

After discretizing the boundary and internal discontinuity lines into elements the displacement representation is transformed into a matrix equation as follows,

$$\underset{\sim}{H}\underset{\sim}{U} = \underset{\sim}{G}\underset{\sim}{P} + \underset{\sim}{D}\underset{\sim}{B} + \underset{\sim}{K}\underset{\sim}{Q} \quad (4)$$

where $\underset{\sim}{K}$ is a matrix obtained by integrating the quadrupole fundamental solution over discontinuities; $\underset{\sim}{Q}$ is ficticious force values to be determinated in the nonlinear process, and $\underset{\sim}{D}\underset{\sim}{B}$ stands for the body force influences.

Special attention must be paid for the discontinuity line discretization. Nodes at element ends must be avoided to make possible integration over singular elements.

Equation (3) can also be written in its matrix form,

$$\underset{\sim}{\sigma} = -\underset{\sim}{H}'\underset{\sim}{U} + \underset{\sim}{G}'\underset{\sim}{P} + \underset{\sim}{D}'\underset{\sim}{B} + \underset{\sim}{K}'\underset{\sim}{Q} \quad (5)$$

By interchanging rows between matrices $\underset{\sim}{H}$ and $\underset{\sim}{G}$ in equation (4) according to the boundary prescribed values, the system to be solved becomes

$$\underset{\sim}{A}\underset{\sim}{X} = \underset{\sim}{F} + \underset{\sim}{K}\underset{\sim}{Q} \quad (6)$$

Where $\underset{\sim}{X}$ contains boundary unknowns and $\underset{\sim}{F}$ stands for all prescribed load or displacement influences.

Notice that in equation (6) only algebraic equations related to boundary displacement representations are included.

Equation (5) can be modified in a similar way to give

$$\underset{\sim}{\sigma} = \underset{\sim}{A}'\underset{\sim}{X} + \underset{\sim}{F}' + \underset{\sim}{K}'\underset{\sim}{Q} \quad (7)$$

From equation (6) one can compute boundary unknowns as follows,

$$\underset{\sim}{X} = \underset{\sim}{A}^{-1}\underset{\sim}{F} + \underset{\sim}{A}^{-1}\underset{\sim}{K}\underset{\sim}{Q} \quad (8)$$

Substituting equation (8) into equation (7) gives

$$\underset{\sim}{\sigma} = (\underset{\sim}{A}'\underset{\sim}{A}^{-1}\underset{\sim}{F}+\underset{\sim}{F}') + (\underset{\sim}{A}'\underset{\sim}{A}^{-1}\underset{\sim}{K}+\underset{\sim}{K}')\underset{\sim}{Q} \quad (9)$$

Equations (8) and (9) can be conveniently used to model discontinuities. The first terms of (8) and (9) represent elastic solution and terms with $\tilde{Q}$ give the influence of load removals along discontinuities.

4 SOLUTION TECHNIQUE

As it has already been mentioned in previous sections an iterative process can be idealized to model discontinuities. The problem is highly nonlinear, therefore loads must be applied in increments. Within each increment, iteractions are necessary to eliminate the amount of stresses which exceeds the assumed failure values.

The complete scheme to obtain the final solution used to model the examples shown in subsequent sections is given by the following steps:

a) Compute elastic solution after applying a load increment. Fist terms of equations (8) and (9)

$$\tilde{X}^e = \tilde{A}^{-1}\tilde{F} \text{ and } \tilde{\sigma}^e = \tilde{A}'\tilde{A}^{-1}\tilde{F} + \tilde{F}' \qquad (10)$$

b) Avaliation of the exceeding stress values $(\Delta\sigma, \Delta\tau)$ according to the criterion adopted.

c) Find ficticious forces to be applied to each node using stress values determinated in "b".

d) Compute new stress distribution given by the application of internal pairs of fictious forces distributed along discontinuities using equation (9)

$$\tilde{\sigma} + (\tilde{A}'\tilde{A}^{-1}\tilde{K}+\tilde{K}')\tilde{Q} \rightarrow \tilde{\sigma} \qquad (11)$$

e) If convengence has not been achieved according to a tolerance previously adopted go back to step "b".

f) Compute total boundary values and stress distributions at non discontinuity points using accumulated values of ficticious forces, $\tilde{Q}^a$, using (8) and (9)

$$\tilde{X} + \tilde{A}^{-1}\tilde{K}\tilde{Q}^a \rightarrow \tilde{X}$$
$$\tilde{\sigma} + (\tilde{A}'\tilde{A}^{-1}\tilde{K}+\tilde{K}')\tilde{Q}^a \rightarrow \tilde{\sigma} \qquad (12)$$

g) go to step "a" for a new load increment

5 EXAMPLES

Two examples are presented in this section to illustrate both accuracy and applicability of the algorithm developed. First, a simple example was run to emphasize the accuracy one can achieve with the technique. The second example solved is a dam case practical analysis where changes in stress distributions due to the discontinuity is observed.

5.1 Square domain analysis

This example consists of analysing a square domain with a possible discontinuity located at the middle section. The geometric characteristics, and boundary and internal discretizations are given in Figure 2. Plane strain behaviour has been assumed with Poisson's ratio and the elastic modulus taken equal to 0.3 and 1000kN/cm^2 respectively. Shear stresses along discontinuous lines are governed by a Coulomb law characterized by cohesion, c, taken equal to 20N/cm^2 and friction angle, ϕ, equal to 0.0^o.

Precribing horizontal displacements along the top line equal to $6\text{x}10^{-4}\text{cm}$ and all other values, along horizontal sides, equal to zero, displacements along a vertical line (Figure 3) indicate that discontinuities arise. Slip occurs along all discontinuity points and the final shear stress values, 20N/cm^2, are precisely computed.

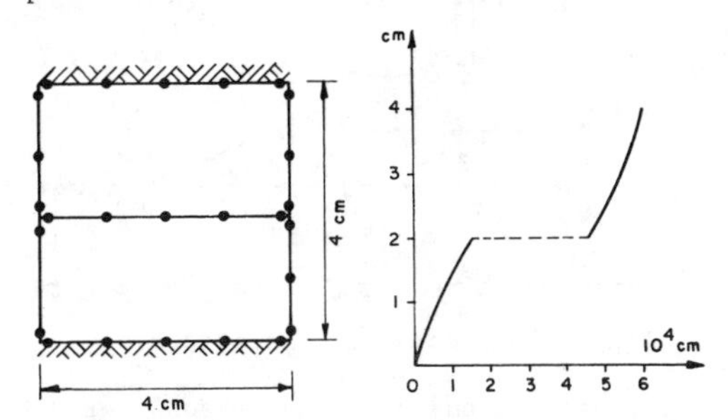

Fig.2 Square domain. Geometry and discretization

Fig.3 Displacements along a vertical side

The example has been run several times to verify discontinuities along boundaries, interfaces and for other situations. In all cases, final stress distributions computed were equal to the corresponding exact solution.

5.2 Dam foundation analysis

In order to show the applicability of the proposed process a practical example is solved. A simplified domain of a concrete dam is analysed including possible slip and separation along basalt layers. This case has already been analysed by Souza Lima [10] where complete data for the dam design are described and all parameter adopted are extensively discussed. Herein, only the section through the spillway dam is studied to show how the boundary algorithm works. The simplified section adopted in this analysis is shown in Figure 4 together with the discretization used. Finite boundaries have been adopted to avoid too many changes in the original example. Subregions have been considered to model different materials. The foundation is modelled by two sobregions for which young modulus values equal to 7 GPa and 10 GPa have been adopted while Poisson's ratios have been assumed 0.3. For concrete region, the elastic modulus is taken equal to 12 GPa and Poisson's ratio 0.2. Along the interface between subregions 2 and 3 constituted by two different basalts, a joint is assumed. The joint behaviour is considered without any cohesion and with internal friction angle, ϕ, equal to 28°.

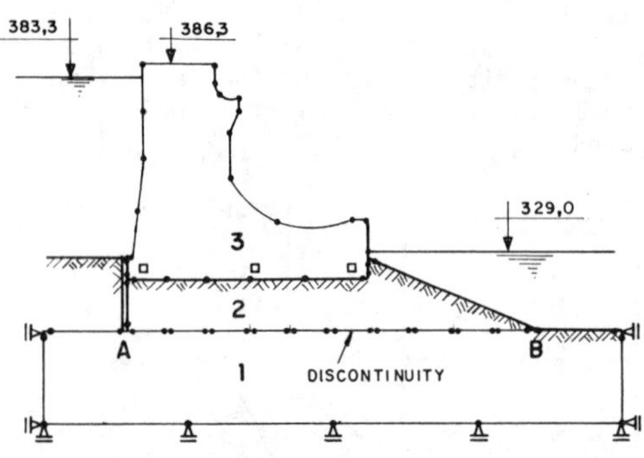

Fig.4 Dam section. Discretization

Subregions 2 and 3 have been considered with original stresses defined by material specific weight equal to 22MN/m^3 and earth pressure coefficient at-rest, K_o, 1.0. Body forces, computed with the specific weight equal to 25MN/m^3, have been considered for concrete region only. The water pressures used for the example are given by the water level (Fig.4). Uplift pressures on the discontinuity line have been computed according to Bureau of Reclamation [11].

The problem was first run considering continuous interaction between subregions. The elastic results obtained indicated clearly a possible joint formation between subregions 2 and 3 according to the failure criterion adopted. Then, the final solution has been modelled by applying the total load in increments. Discontinuities arise on the right quarter of the idealised joint. Elastic and final stress distributions are shown in Figure 5 together with ultimate shear stresses given by the failure criterion.

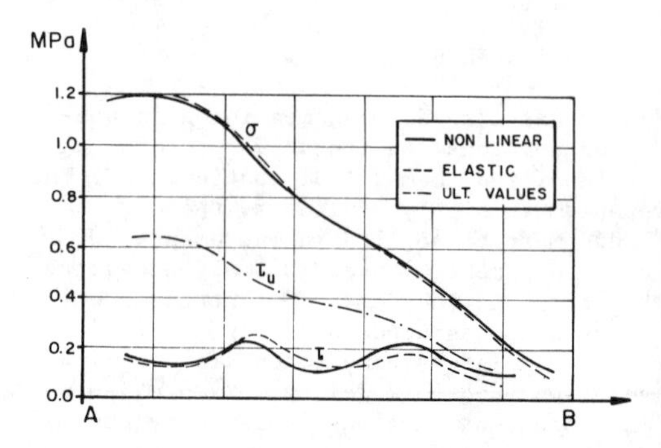

Fig.5 Elastic and nonliner stresses along interface between subregion 2 and 3

6 CONCLUSION

A direct boundary element formulation for the analysis of discontinuities was presented. Since the technique adopted is similar to those used for other nonlinearities, complex domains including several types of material behaviour will be easily analysed. In the same iteractive step for the algorithm proposed one can model, for example, plastic and discontinuity effects. The technique avoids the system resolution at each iteraction as necessary in discontinuity formulation with special joint element relations. The applicability of the technique for the analysis of practical engineering problem is pointed out with the examples shown.

7 REFERENCES

1 F.J.Rizzo, An integral equation approach to boundary value problems of classical elastostatics, Quart. Appl.Math., 25, 83-95 (1967)
2 T.A.Cruse, Numerical solutions in three-dimensional elastostatics, Int. J. Solids Struct., 5, 1259-1274 (1969)
3 G.Hocking, Three dimensional elastic stress distribution around the flat end of a cylindrical cavity, Int. J. Rock Mech. Min. Sci., 13, 331-337(1976)
4 B.H.G.Brady and J.W.Bray, The Boundary element method for determining stress and displacements around long openings in a triaxial stress field, Int. J. Rock Mech. Min. Sci., 15, 21-28 (1978)
5 L.J.Wardle and J.M.Crotty, Two-dimen-

sional boundary integral equation analysis for non-homogeneous mining applications, In recent advances in boundary element methods, ed. by C.A.Brebbia, Pentech Press, 1978

6 W.S.Venturini and C.A.Brebbia, The boundary element method for the solution of no-tension materials, In boundary element methods, ed. by C.A.Brebbia, Springer-Verlag, 1981

7 W.S.Venturini, Boundary element method in geomechanics, Lecture notes in engineering, Springer-Verlag, 1984

8 G.Hocking, Stress analysis of underground exacavations incorporating slip and separation along discontinuities, In recent advances in boundary element methods, ed. by C.A.Brebbia, Pentech Press, 1978

9 W.S.Venturini and C.A.Brebbia, Formulation of the boundary element method for discontinuous domains, Proceeding of the IV Latin-American congress on numerical methods in engineering, Santiago, Chile (in portuguese), 1983

10 V.M.Souza Lima, R.A.Abrahão, R.Pinheiro and J.C.Degaspare, Rock foundations with marked discontinuities criteria and assumptions for stability analysis, Quatorzième congrès des grands barrages, Rio de Janeiro, 1982

11 Bureau of reclamation, Design of gravity dams, Denver, 1976

Numerical Methods in Geomechanics (Innsbruck 1988), Swoboda (ed.)
© 1988 Balkema, Rotterdam. ISBN 90 6191 809 X

Elasto-plastic analysis of tunnel excavation in layered rock medium by coupled FEBEM

R.B.Singh
Central Soil and Materials Research Station, New Delhi, India
K.G.Sharma & A.Varadarajan
Indian Institute of Technology, New Delhi

ABSTRACT: The elasto-plastic analysis of a circular tunnel excavation in layered rock medium using finite element method(FEM), the coupled finite element and boundary element method(FEBEM) is presented. The coupling procedure, the elasto-plastic formulation and application to the layered medium are discussed. The results of FEBEM analysis are compared with those obtained from FEM. The computation time and the number of iterations as required by FEBEM and FEM are compared. The effect of variation of modulus of elasticity on deformation, yielding and stresses is presented.

1 INTRODUCTION

Underground structures range from temporary mine openings to permanent nuclear power stations and are increasingly adopted in the recent times. The design, construction and performance evaluation of the underground openings often require the determination of displacements and stresses in the surrounding rock due to excavation. An analysis procedure, known as coupled FEBEM, developed by coupling Finite Element Method, FEM, with Boundary Element Method, BEM is ideally suited for studying various aspects of the underground structures.

Zienkiewicz et al. (1977), among other researchers, proposed the coupling of FEM with BEM. Beer and Meek(1981) adopted this method for the analysis of underground openings. Varadarajan et al.(1985) investigated various aspects affecting the coupled FEBEM in the analysis of underground openings. Varadarajan et al.(1987) extended the above study to elasto-plastic analysis. Swoboda et al. (1987) adopted coupled FEBEM in the viscoplastic analysis of underground structures. Beer and Swoboda(1986) proposed the coupled FEBEM to investigate the underground structures in layered medium near the ground surface. Singh et al.(1986) gave an elastic coupled FEBEM for analysing the deep underground openings in layered media.

Herein, an elasto-plastic analysis of the underground structures in layered media by coupled FEBEM is presented. Included in it are a two layered medium and the effect of variation of the modulus of deformation. The displacements, stresses, yieldings, computation time and the number of iterations are presented. These results are also compared with those obtained from FEM.

2 FORMULATION OF FEBEM

The finite elements are coupled with boundary elements using direct formulation of BEM. The detailed procedure of coupling has been presented by Varadarajan et.al. (1985). It is briefly discussed here.

The equilibrium equation of the system is given by

$$[K]\ \{u\} = \{Q\} \tag{1}$$

where $\{u\}$ is the nodal displacement vector, $\{Q\}$ is the nodal load vector and $[K]$ is the total stiffness matrix given by

$$[K] = [K_f] + [K_b] \tag{2}$$

in which $[K_f]$ and $[K_b]$ are the stiffness matrices of finite element and boundary element regions, respectively, and are defined as

$$[K_f] = \sum_{i=1}^{n} \int_V [B]^T\ [D]\ [B]\ dV \tag{3}$$

$$[K_b]=\frac{1}{2}([C][U]^{-1}[T]+([C][U]^{-1}[T])^T) \quad (4)$$

where n is the number of finite elements, [B] is the strain-displacement matrix, [D] is the elasticity matrix, [C] contains the collocated area for all the boundary elements, and [U] and [T] are the matrices of the Kernel shape functions which are obtained by the procedure given by Watson (1979).

3 ELASTO-VISCOPLASTICITY

Elasto - viscoplasticity has been used as an artifice to obtain elasto-plastic solutions(Zienkiewicz and Cormeau(1974)). The viscoplastic strain rates are given by the flow rule

$$\{\dot{\varepsilon}^{vp}\} = \mu < F > \frac{\partial F}{\partial\{\sigma\}} \quad (5)$$

in which $\{\dot{\varepsilon}^{vp}\}$ is the viscoplastic strain rate vector, μ is the fluidity parameter (taken as 1 for elastoplastic analysis), F is the yield function, $\{\sigma\}$ is the stress vector, and $< >$ is used to indicate that if $F \leqq 0$, $<F> = 0$ and $F > 0$, $<F> = F$.

The Mohr-Coulomb yield criterion has been adopted for the present study. The criterion in terms of the principal stresses σ_1 and σ_3 can be written as

$$F=(\sigma_1-\sigma_3)-(\sigma_1+\sigma_3)\sin\phi - 2c\cos\phi = 0 \quad (6)$$

where c is the cohesion and ϕ is the angle of internal friction of rock.

For $F \leqq 0$, the rock mass is linear elastic and for $F > 0$, it starts yielding and the flow rule given by equation (5) is used to calculate the plastic strains.

The critical time stepping scheme as suggested by Zienkiewicz and Cormeau(1974), has been used in the iterations to obtain elasto-plastic solution.

4 THE PROBLEM

A circular tunnel of 5m diameter,found in biotite gneiss has been analysed with the Young's modulus E1 of lower layer taken equal to 4.48×10^6 t/m^2. For the upper layer, three values of Young's modulus(E2) have been considered with the ratio E1/E2 =1,2 and 3. Thus the lower layer is harder than the upper layer. Poisson's ratio of 0.18 is assumed for both the layers.

Based on the unit weight of the rock equal to 2.94 t/m^3 and a height of overburden of 300m, insitu vertical stress is calculated as 882 t/m^2. An insitu stress ratio K_o equal to 1.0 has been chosen for the present study. The value of cohesion, c equal to 160 t/m^2 and angle of internal friction, $\phi = 41°$ have been adopted.

5 ANALYSIS

A two-dimensional analysis with plane strain condition has been carried out. The excavation of the opening is simulated in single step. The discretizations of the circular opening for FEBEM and FEM are shown in Figs.1 and 2, respectively. In FEBEM, the boundary between finite element and boundary element regions has been taken at a distance of four times the radius of the opening from the centre of the opening(Varadarajan et.al.(1985)). The boundary has been fixed at eight times the radius of the opening in FEM. Although half the section is adequate for the analysis with a horizontal layered medium, the full section of the opening has been discretized to facilitate the use of the same mesh for inclined layered rock media also(Singh et al.(1986)).

The stiffness matrix $[K_f]$ for the finite elements is developed by using the different material properties for the elements in two layers. For developing the boundary element stiffness matrix(BESM), the steps used for the horizontal layered rock medium are as follows(Singh(1985)).

5.1 Horizontal layering

The interface for the horizontal layering is shown by H-H in Fig.1.

1. The stiffness matrix $[K_1]$ is generated for half the part with unit Young's modulus following the condensation procedure given by Sharma et al.(1985).
2. The matrix $[K_1]$ is multiplied by E1 and E2 to obtain matrices $[K_{E1}]$ and $[K_{E2}]$ for the lower and upper layers, respectively.
3. The matrices $[K_{E1}]$ and $[K_{E2}]$ are assembled to get the required BESM, $[K_b]$, i.e.

$$[K_b] = [K_{E1}] + [K_{E2}] \quad (7)$$

This stiffness matrix $[K_b]$ is assembled with the finite element stiffness matrix,

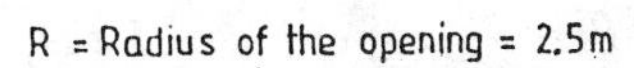

FE Region

Elements = 84
Nodes =276

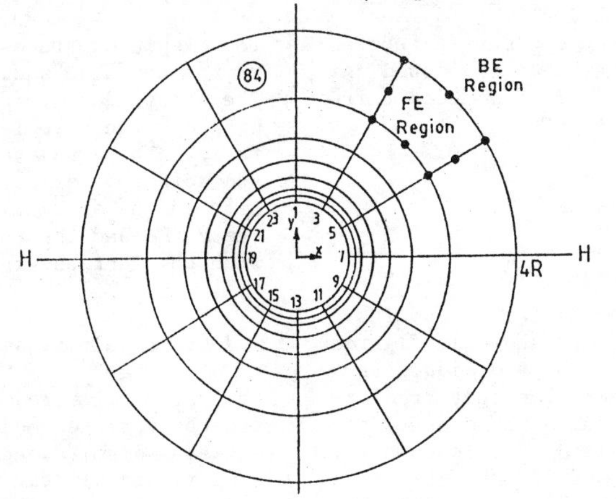

BE Region

Elements = 12
Nodes = 24

Fig.1 FEBEM discretization

Finite elements = 108
Nodal points = 348
R = Radius of the opening = 2.5 m

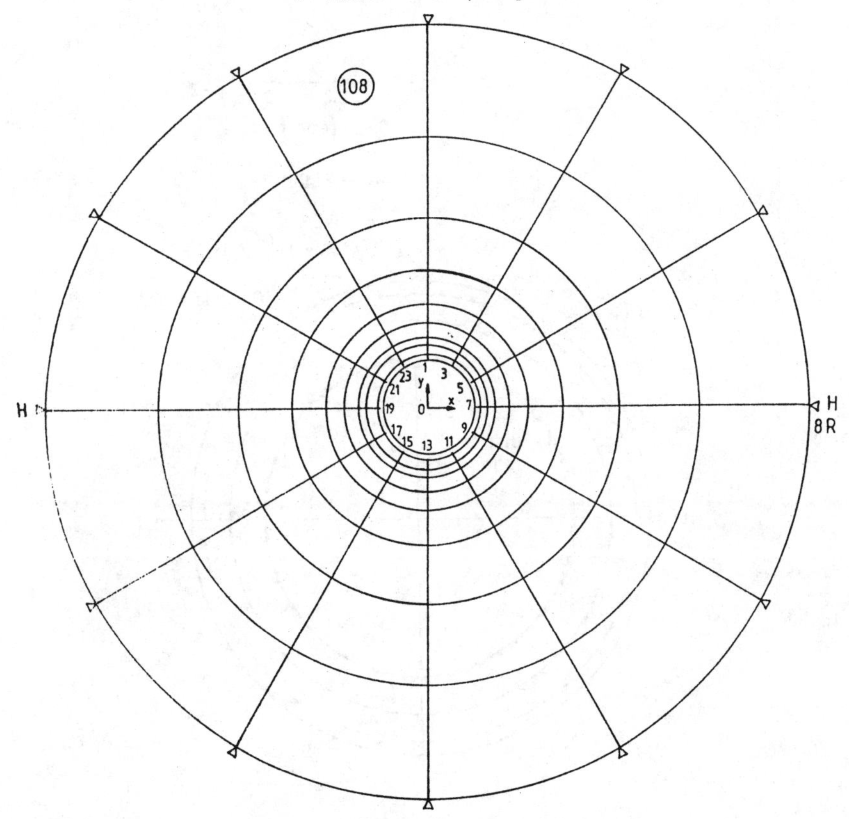

Fig.2 FEM discretization

$[K_f]$ to obtain the coupled stiffness matrix [K] as in equation 2.

Based on the procedure discussed above, a computer program has been developed on ICL2960 and the analyses have been carried out using FEBEM and FEM.

6 RESULTS AND DISCUSSION

6.1 Yielded zone

The yielded zones for different modulus ratios of 1, 2 and 3 are shown in figures 3 to 5 respectively. The yielding given by the FEBEM is higher than that predicted by the FEM for modulus ratio equal to 1. However, the yielded zone is same for modulus ratios of 2 and 3. The stiffer rock shows larger yielding. The highest amount of yielding is found near the contact line H-H at section OA. The yielding is slightly higher for the modulus ratio of 3(figure 5) as compared to the modulus ratio of 2(figure 4).

6.2 Displaced shape

The displaced shapes for different modulus ratios are also shown in figures 3 to 5. The magnitudes of displacements increase with the increase in the value of modulus ratio. The displacements are higher in the softer material. The difference between the displacements obtained from the FEBEM and the FEM analyses increases with the increase in modulus ratio.

6.3 Principal stresses

The principal stresses at sections OA (in the stronger medium) and OB (in the weaker medium) along the radial line (see figure 5) for modulus ratios of 1,2 and 3 are shown in figures 6 to 8 respectively. In the case of elasto-plastic analysis, the highest magnitude of stresses for modulus ratio of 1 are found to be at a distance of about 2.97m from the centre of the opening(figure 6).

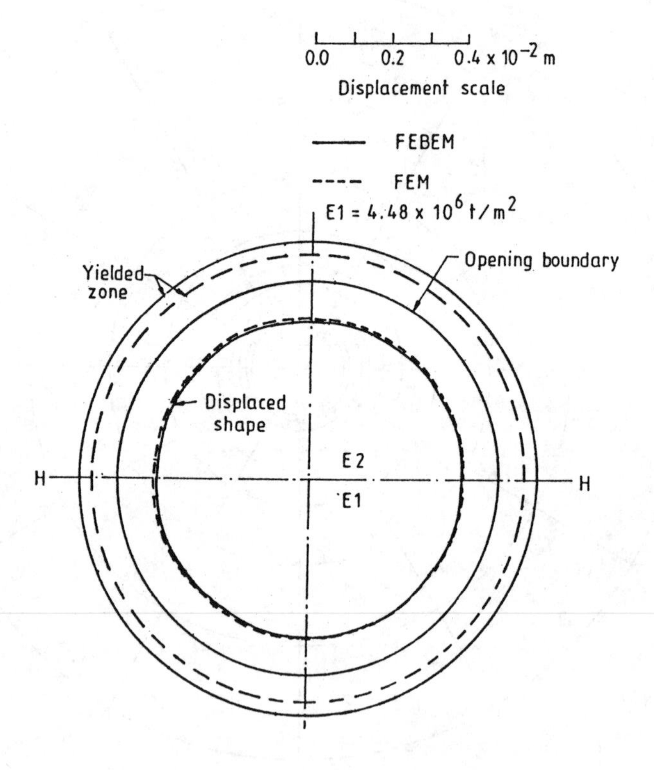

Fig.3 Yielded zone and displaced shape of opening for K_o=1.0 and E1/E2=1

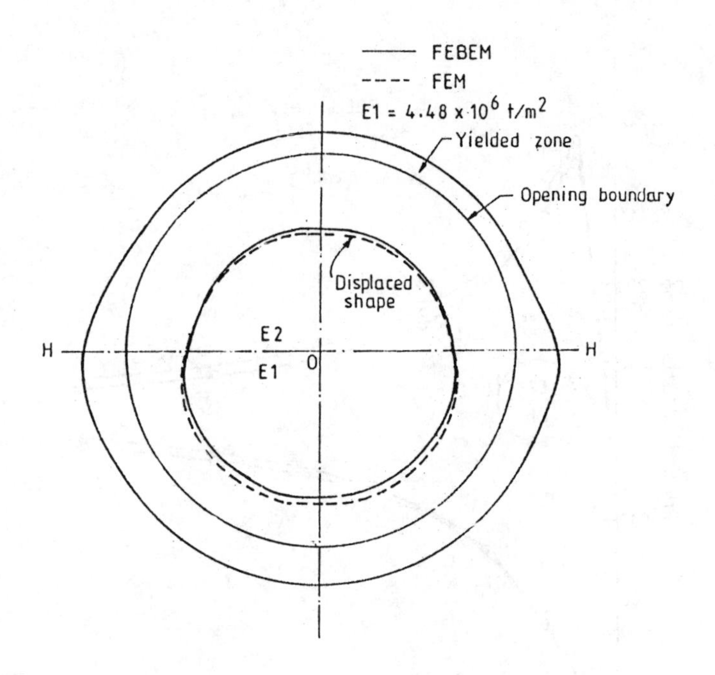

Fig.4 Yielded zone and displaced shape of opening for K_o=1.0 and E1/E2=2

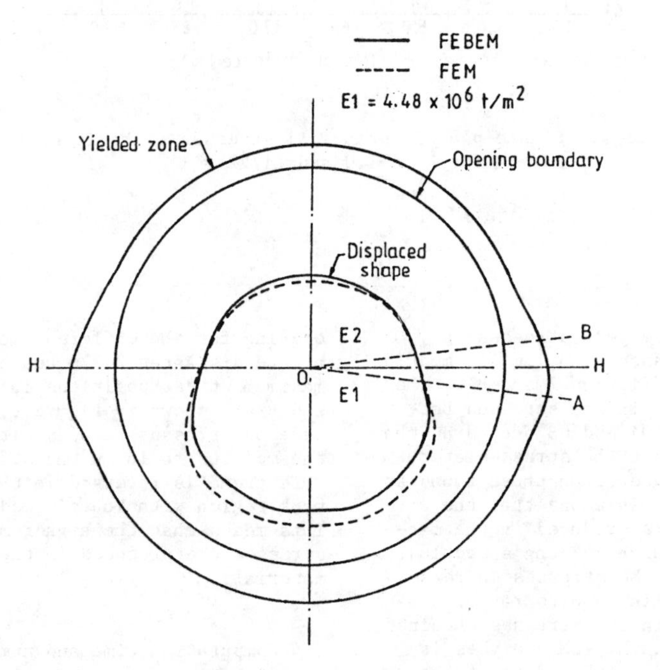

Fig.5 Yielded zone and displaced shape of opening for K_o = 1.0 and E1/E2=3

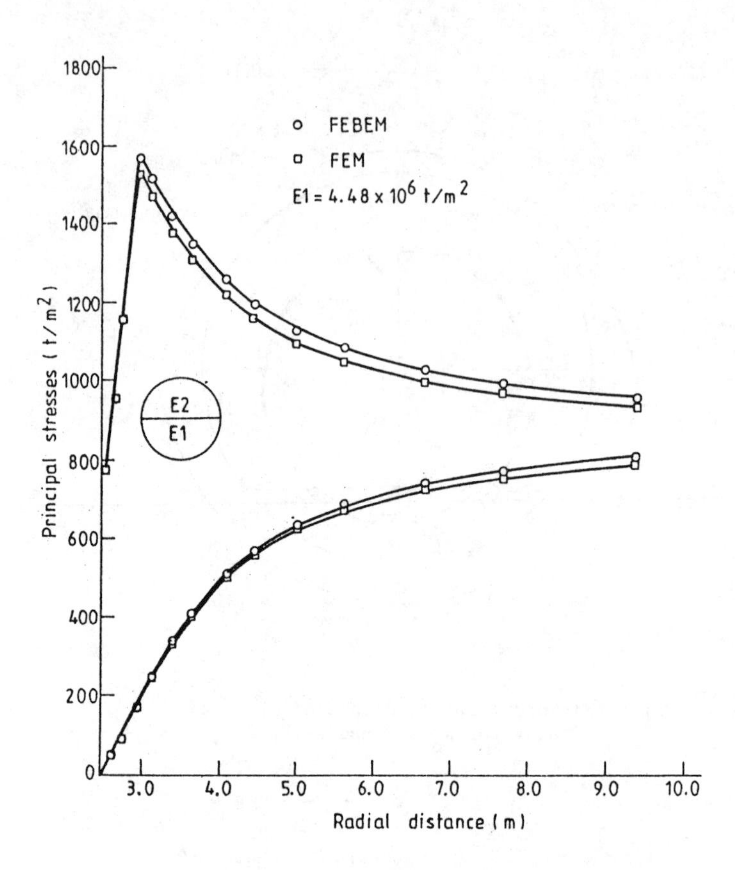

Fig.6 Elasto-plastic principal stresses of horizontal layers for K_o = 1.0 and E1/E2 = 1

The position of the peak stresses is shifted to a distance of 3.15m for modulus ratios of 2 and 3(figures 7 and 8). The maximum difference in the stresses between the sections OA and OB occurs at the peak stress points. The stresses at the section OA are higher than those found at the section OB. This means that the higher stresses are produced in the material having the higher Young's modulus. The magnitudes of the stresses increase with the increase in modulus ratio.

The difference in the stresses obtained from the FEBEM and the FEM analyses is insignificant near the boundary of the opening, but it increases as one moves away from the boundary i.e. in the elastic region.

The major principal stresses around the opening for the different modulus values at the distances 2.97m and 3.15m (the maximum stress positions in figures 6,7 and 8 are shown in figure 9. The magnitude of stresses reduces with increase in the modulus ratio in the softer rock. This trend is reversed in the stiffer rock region with Young's modulus of E1. This means that the higher magnitudes of stresses are produced in the stiffer material.

6.4 Computation time and number of iterations

For the elasto-plastic analysis, a comparison between FEBEM and FEM is shown in table 1 in terms of the computation time

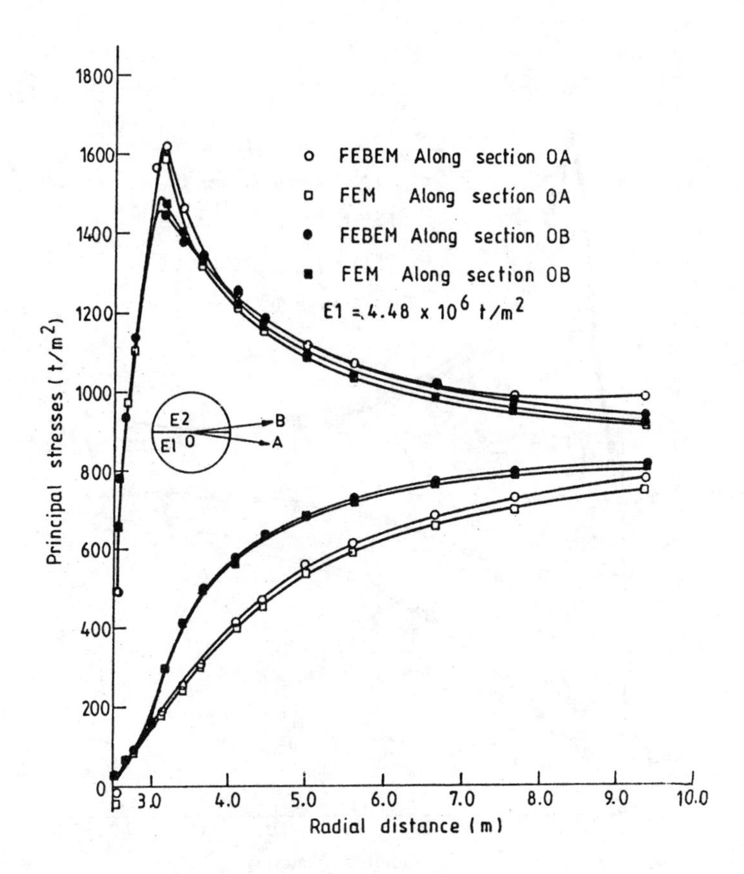

Fig.7 Elasto-plastic principal stresses of horizontal layers for K_o=1.0 and E1/E2=2

Table 1. Computation time and number of iterations taken by FEBEM and FEM for different modulus ratios.

Modulus Ratio E1/E2	FEBEM		FEM		% age increase in computation time
	CPU time Seconds	Number of iterations	CPU time Seconds	Number of iterations	
1	1049	103	1266	99	20.7
2	1890	207	2359	203	24.8
3	2616	292	3248	290	24.2

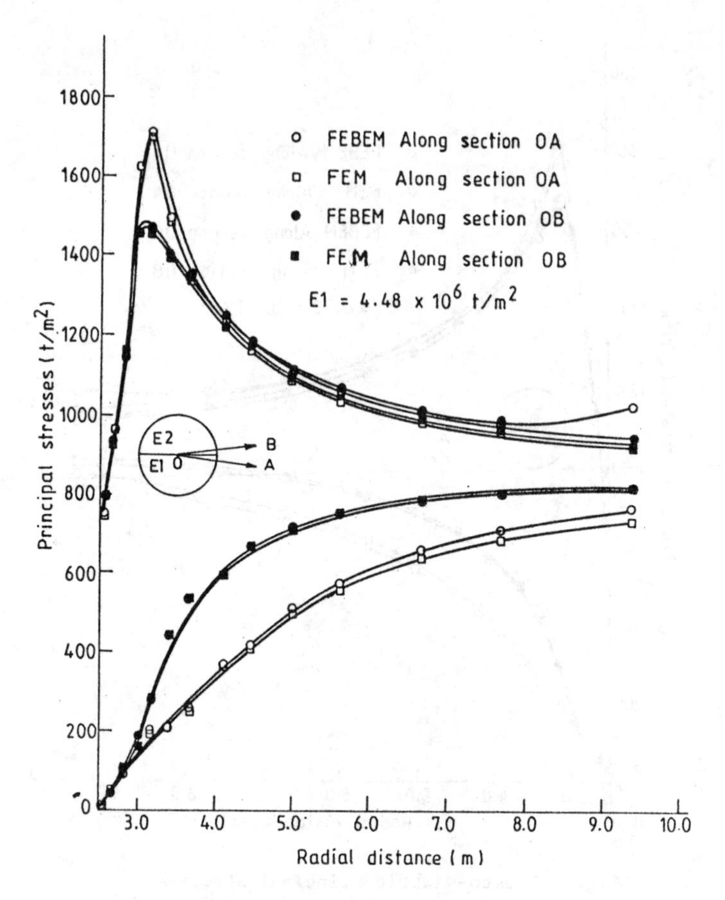

Fig.8 Elasto-plastic principal stresses of horizontal layers for K_o=1.0 and E1/E2=3

and the number of iterations taken for the convergence.

The number of iterations and hence the computation time increase as the modulus ratio increases. The number of iterations (103 to 292) taken by the FEBEM is slightly more than that taken by the FEM (99 to 290) but the computation time is much less for the FEBEM(1049 to 2616 seconds) than that for the FEM(1266 to 3248 seconds). This is due to the larger number of equations involved in the FEM, 552 in the FEBEM as compared to 696 in the FEM(see figures 1 and 2).

The difference in computer time for the FEBEM and the FEM for the three modulus ratios is also presented in Table 1. For E1/E2=1 i.e., homogeneous medium, the FEM shows 20.7 per cent increase in the computer time. The modulus ratios 2 and 3 show higher percentage difference (24.8 % and 24.2 %). It is interesting to note that higher yielding (E1/E2=3) has not resulted in higher increase in computer time in the FEM.

7 CONCLUSIONS

From the elasto-plastic analyses of a circular underground opening excavated in a rock medium with horizontal layering, it is found that the displacements and stresses increase with the increase in the E1/E2 value. The maximum displacements occur in the rock having smaller modulus value and the maximum stresses are observed in the rock having higher modulus value.

The highest stresses are found in the

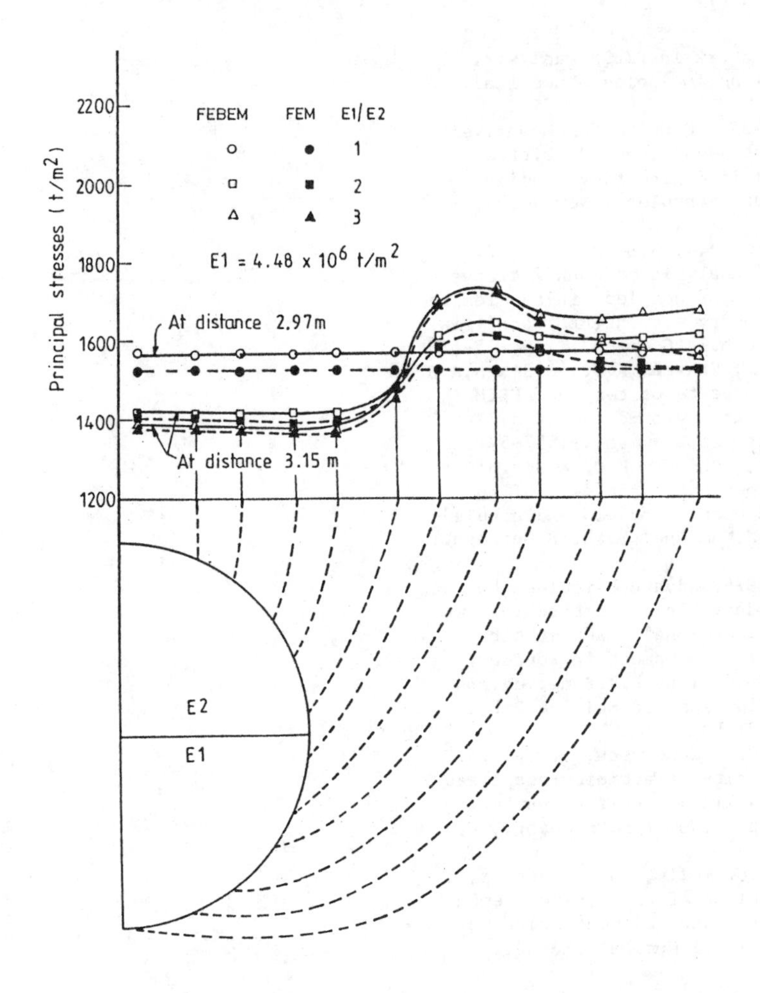

Fig.9 Elasto-plastic principal stresses of horizontal layer at the distances of 2.97m and 3.15m around the opening

stiffer rock near the contact line of the two-rock media. The yielding is more in the stiffer rock than in the softer rock. The number of iterations taken are more for FEBEM, but the computation time is much less than for FEM. However, with the increase in the E1/E2 value, the percentage increase in computer time for FEM remains almost same.

The results of the FEBEM and the FEM are having the same trend for all the analyses and the values predicted in general are higher in FEBEM particularly at the point of stress concentration near the contact line of the two rocks. Both the methods show similar results. However in general, the FEBEM shows higher magnitudes of stresses and displacements.

REFERENCES

Beer,G & Meek,J.L. 1981, The coupling of boundary and finite element methods for infinite problems in elasto-plasticity. Proceeding 3rd Int. Seminar,Irvine, California, C.A. Brebbia(Ed.), p.575-591.

Beer,G. & Swoboda, G. 1986. On the efficient analysis of shallow tunnels. Computers and Geotechnics 1, p.15-31.

Sharma, K.G., Varadarajan,A & Singh,R.B. 1984. Condensation of boundary element stiffness matrix in FEBEM analysis. Communications in Applied Numerical Methods.

Singh,R.B. 1985. Coupled FEBEM analysis of underground openings. Ph.D.thesis, Dept. of Civil Engineering, Indian Institute of Technology, New Delhi, India.

Swoboda,G., Mertz,W. & Beer, G. 1987. Rheological analysis of tunnel excavations by means of coupled finite element (FEM) - boundary element(BEM) analysis. Int.J.Num.Meth.in Geomech.11, p.115-129.

Varadarajan,A., Sharma, K.G., & Singh,R.B. 1985. Some aspects of coupled FEBEM analysis of underground openings, Int.J. Num.Anal.Meth.in Geomech.,9:557-571.

Varadarajan,A., Sharma, K.G.,& Singh,R.B. 1987. Elasto-plastic analysis of an underground opening by FEM and coupled FEBEM. Int.J.Num.Anal.Meth.in Geomech., in print.

Watson,J.O. 1979. Advanced implementation of the boundary element method for two and three dimensional elastostatics. Chapter 3 of Development in Boundary Element Method -1 by P.K.Banerjee and R.Butterfield. Applied Science Publications Ltd. 31-63.

Zienkiewics, O.C. & Cormeau, I.C. 1974. Visco-plasticity, plasticity and creep in elastic solids: a unified numberical approach. Int.J.Numer method eng., 8, p.821-845.

Zienkiewics,O.C. Kelly,D.W. & Bettess,P. 1977.The coupling of the finite element method and boundary element solution procedures. Int.J.Num.Meth.Engng. 11: 355-375.

Numerical Methods in Geomechanics (Innsbruck 1988), Swoboda (ed.)
© 1988 Balkema, Rotterdam. ISBN 90 6191 809 X

The spline indirect boundary element method and its application in analyzing stress and pressure around underground opening

Lin Yu Liang
Guangxi University, Nanning, People's Republic of China

ABSTRACT: This paper puts forward a new calculation method named the spline indirect boundary element method.And with this method,the author of this paper has solved the problems in analyzing and calculating the displacement,stress and rock pressure around underground opening.It turns out that this method has such advantages as high precision,simplicity and readiness and that its total matrix is a symmetrical and belt-like one.So this method can be used more conveniently than the Boundary Element Method and the Spline Direct Boundary Element Method.

1 THE FUNDAMENTAL PRINCIPLES OF THE SPLINE INDIRECT BOUNDARY ELEMENT METHOD

On the basis of BEM,Qin Rong[1] has advanced a new method of the spline direct boundary element method(SDBEM) and he has ever used his new method to analyze and calculate the problem of thin plate and been satisfied with good results.Also the author of this paper[2] has ever applied SDBEM to the problems related to the displacement,stress and rock pressure around underground opening.Indeed SDBEM turned out to be a good one to solve the above peoblems.This paper now puts forward a newer method the Spline Indirect Boundary Element Method.It is better than SDBEM,the author thinks.

In the Indirect Boundary Element Methords (IBEM),the boundary quantities are divided into two kinds:the real quantities(e.g. the real boundary displacement) and the fictitious quantities(e.g. the fictitious boundary displacement).And it is considered that the real quantities are induced by the united function of the fictitious quantities.If the fictitious quantities are fictitious stresses,this IBEM is called the Fictitious Stress Method;if the fictitious quantities are fictitious displacements.it is called the Fictitious Displacement Method.

We repesent R in Fig.1 as a two-dimensional plane.The region of R is bounded by a contour C.Along C,Y_s refers to the real sear component,Y_n to the real normal component,X_s to the fictitious shear compoment

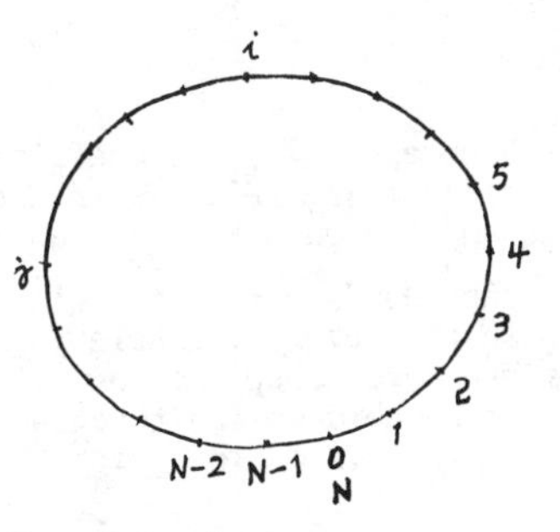

Fig.1 Boundary elements

and X_n to the fictitious normal component.

We divide C into a number of equal elements.Let n be this number.

$$s_o < s_1 < s_2 < \cdots < s_n$$

$$s_i = s_o + 2ia$$

$$a = l/2n$$

Where

l—the length of the total boundary
s_i—the joint point
2a—the length of the segment

Thus generally in IBEM,it is considered that the fictitious variables in every element are constants and we can obtain the following relations:

$$Y_s^i = \sum_{j=1}^{n} C_{ss}^{ij} X_s^j + \sum_{j=1}^{n} C_{sn}^{ij} X_n^j \qquad i=1 \text{ to } n \quad (1.1)$$

$$Y_n^i = \sum_{j=1}^{n} C_{ns}^{ij} X_s^j + \sum_{j=1}^{n} C_{nn}^{ij} X_n^j$$

Where $C_{ss}^{ij}, C_{sn}^{ij}, C_{ns}^{ij}, C_{nn}^{ij}$ etc. are the boundary influence coefficients.

If we consider that the fictitious quantities are variables in each element,i.e. they are the functions of the co-ordinates: $X_s(s)$ and $X_n(s)$,then for every point P in the boundary,the relations between the real variables $Y_s(P)$ and $Y_n(P)$ and the fictitious variables can be obtained as follows:

$$Y_s(P)=\oint_c [C_{ss}(P,s)X_s(s)+C_{sn}(P,s)X_n(s)]ds$$

$$Y_n(P)=\oint_c [C_{ns}(P,s)X_s(s)+C_{nn}(P,s)X_n(s)]ds \qquad (1.2)$$

Where the boundary influence coefficients become those that are relaed with the co-ordinate s and the poit P. $\oint_c$ is an integral along the boundary of the region.

We suppose that every variable is represented by nth power B spline function (in this paper the cubic spline function is used) and suppose $P=S_l$,i.e. the point P is on the point S_l.

$$Y_s(s_l)=\sum_{i=0}^{n} b_i^s \phi_i(s_l) \qquad (1.3)$$

$$Y_n(s_l)=\sum_{i=0}^{n} b_i^n \phi_i(s_l) \qquad (1.4)$$

$$X_s(s)=\sum_{i=0}^{n} a_i^s \phi_i(s) \qquad (1.5)$$

$$X_n(s)=\sum_{i=0}^{n} a_i^n \phi_i(s) \qquad (1.6)$$

$$C_{ss}(s_l,s)=\sum_{i=0}^{n}\sum_{j=0}^{n} g_{ij}^{ss} \phi_i(s_l)\phi_j(s) \qquad (1.7)$$

$$C_{sn}(s_l,s)=\sum_{i=1}^{n}\sum_{j=0}^{n} g_{ij}^{sn} \phi_i(s_l)\phi_j(s) \qquad (1.8)$$

$$C_{ns}(s_l,s)=\sum_{i=0}^{n}\sum_{j=0}^{n} g_{ij}^{ns} \phi_i(s_l)\phi_j(s) \qquad (1.9)$$

$$C_{nn}(s_l,s)=\sum_{i=0}^{n}\sum_{j=0}^{n} g_{ij}^{nn} \phi_i(s_l)\phi_j(s) \qquad (1.10)$$

$$l=0,1.2.\cdots n.$$

Where $b_i^s, b_i^n, g_{ij}^{ss}, g_{ij}^{sn}, g_{ij}^{ns}, g_{ij}^{nn}$ are coefficients they can be determined by the spline interpolation method. a_i^s and a_i^n are coefficients awaiting solution. $\phi_i(s)$ and $\phi_j(s)$ are n-th power B spline function.i.e.

$$\phi_i(s)= \phi_n\left(\frac{s-s_i}{2a}\right) \qquad (1.11)$$

Using $P=s_l$ in (1.2) we can get:

$$Y_s(s_l)=\oint_c [C_{ss}(s_l,s)X_s(s)+C_{sn}(s_l,s)X_n(s)]ds$$

$$Y_n(s_l)=\oint_c [C_{ns}(s_l,s)X_s(s)+C_{nn}(s_l,s)X_n(s)]ds$$

$$l=0,1,2,\cdots n. \qquad (1.12)$$

If we suppose

$$\{Y(s_l)\}=\begin{Bmatrix} Y_s(s_l) \\ Y_n(s_l) \end{Bmatrix}=[Y_s(s_0)\ Y_s(s_1)\cdots Y_s(s_n)\ Y_n(s_0)\ Y_n(s_1)\cdots Y_n(s_n)]^T \qquad (1.13)$$

$$\{X(s)\}=\begin{Bmatrix} X_s(s) \\ X_n(s) \end{Bmatrix} \qquad (1.14)$$

$$[C]=\begin{bmatrix} C_{ss}(s_l,s) & C_{sn}(s_l,s) \\ C_{ns}(s_l,s) & C_{nn}(s_l,s) \end{bmatrix} \qquad (1.15)$$

then (1.12) turns into

$$\{Y(s_l)\}= \oint_c [C]\{X(s)\}ds \qquad (1.16)$$

When we rearrange the matrix elements of (1.13) and combine (1.13) with (1.3) and (1.4) ,we can get:

$$Y(s_1)=\sum_{i=0}^{2n+1} b_i\phi_i(s_1) \quad (1.17)$$

In the same way,we can get:

$$X(s)=\sum_{i=0}^{2n+1} a_i\phi_i(s) \quad (1.18)$$

$$C(s_1,s)=\sum_{i=0}^{2n+1}\sum_{j=0}^{2n+1} g_{ij}\phi_i(s_1)\phi_j(s) \quad (1.19)$$

Using (1.17),(1.18) and (1.19) in (1.16) ,we obtain:

$$\sum_{i=0}^{2n+1} b_i\phi_i(s_1)=\oint\sum_{i=0}^{2n+1}\sum_{j=0}^{2n+1} g_{ij}\phi_i(s_1)\phi_j(s)\cdot \sum_{i=0}^{2n+1} a_i\phi_i(s)ds. \quad s_1=0,1,\cdots 2n+1 \quad (1.20)$$

We rewrite the above equation into:

$$\sum_{i=0}^{2n+1} b_i\phi_i(s_1)=\sum_{i=0}^{2n+1}\sum_{j=0}^{2n+1}(\phi_i(s_1)g_{ij}\oint\phi_i(s)\phi_j(s)ds\cdot a_i) \quad s_1=0,1,\cdots 2n+1 \quad (1.21)$$

and write this set of equations into the following matrix form:

$$[B]\{b\}=[B][K]\{a\} \quad (1.22)$$

where

$$[B]=[\phi_i(s_1)] \quad (1.23)$$

$$[K]=[K_{ij}] \quad (1.24)$$

$$K_{ij}=g_{ij}\oint\phi_i(s)\phi_j(s)ds \approx g_{ij}\sum_{l=0}^{2n+1}\phi_i(s_l)\phi_j(s_l)\cdot 2a \quad (1.25)$$

$$\{b\}=[b_0\ b_1\cdots b_{2n+1}]^T \quad (1.26)$$

$$\{a\}=[a_0\ a_1\cdots a_{2n+1}]^T \quad (1.27)$$

Eliminating [B] from both sides of (1.22), we can obtain:

$$\{b\}=[K]\{a\} \quad (1.28)$$

That is the fundamental equation of SIBEM .After solving this fundamental equation, we can get coefficients a_i.Using (1.18) we can obtain the fictitious variables X(s) .After knowing the fictitious variables, we can solve the unknown quantities on the boundary,the stresses and the displacements in the region R by IBEM.

From (1.25) and the characteristics of the spline function,we can see that the [K] matrix is a symmetrical and belt-like matrix,and that the belt width is very narrow,therefore it is more convenient to solve the equation set (1.28).

2 SOLVING THE DISPLACEMENT STRESS AND ROCK PRESSURE AROUND UNDERGROUND OPENING BY SIBEM

2.1 The displacement and stress around the underground opening which is of linear elasticity and without support

In order to solve the problems of underground rock opening,the stress around the underground opening is generally resolved into the field stress and the induced stress i.e.

$$(\sigma_{ij})_t=(\sigma_{ij})_f+(\sigma_{ij})_i \quad (2.1)$$

Where

$(\sigma_{ij})_f$—the field stress

$(\sigma_{ij})_i$—the induced stress

$(\sigma_{ij})_t$—the total stress around opening

Generally speaking,we take three steps to analyze stresses they are as follows:

1. Determining the field stresses;
2. determining and solving the boundary value problem of induced stresses;
3. Adding the induced stresses to the field stresses to make the stresses around the opening.

The field stresses can be determined according to the results of measuring or calculating.In this paper we suppose that they have been known.Here we only do the determination and calculation of the boundary values of induced stressed.

The boundary values of the induced stress

can be calculated by the following formula:

$$t_i=(t_i)_t-(t_i)_f \tag{2.2}$$

Where
t_i—the boundary values of induced stresses
$(t_i)_t$—the total stresses around opening
$(t_i)_f$—the field stresses

If the opening is unsupported, then $(t_i)_t=0$, then

$$t_i=-(t_i)_f \tag{2.3}$$

Take the fictitious stress method for example to explain how to use the SIBEM to solve the problems of stress, displacement and rock pressure around the underground opening. As is well known, in the problems of unsupported opening, the ordinary boundary conditions are the stress conditions:

$$Y_s(s_1)=t_s(s_1)=(t_s(s_1))_f\cdot(-1) \tag{2.4}$$

$$Y_n(s_1)=t_n(s_1)=-(t_n(s_1))_f \tag{2.5}$$

The influence coefficients of the boundary stresses are

$$C_{ij}^{ss}=A_{ij}^{ss}=-2(1-\mu)(\sin2\gamma\cdot\overline{F}_2-\cos2\gamma\cdot\overline{F}_3)-\overline{y}(\sin2\gamma\cdot\overline{F}_4+\cos2\gamma\cdot\overline{F}_5) \tag{2.6}$$

$$C_{ij}^{sn}=A_{ij}^{sn}=(1-2\mu)(\cos2\gamma\cdot\overline{F}_2+\sin2\gamma\cdot\overline{F}_3)-\overline{y}(\cos2\gamma\cdot\overline{F}_4-\sin2\gamma\cdot\overline{F}_5) \tag{2.7}$$

$$C_{ij}^{ns}=A_{ij}^{ns}=\overline{F}_2-2(1-\mu)(\cos2\gamma\cdot\overline{F}_2+\sin2\gamma\cdot\overline{F}_3)-\overline{y}(\cos2\gamma\cdot\overline{F}_4-\sin2\gamma\cdot\overline{F}_5) \tag{2.8}$$

$$C_{ij}^{nn}=A_{ij}^{nn}=\overline{F}_3-(1-2\mu)(\sin2\gamma\cdot\overline{F}_2-\cos2\gamma\cdot\overline{F}_3)+\overline{y}(\sin2\gamma\cdot\overline{F}_4+\cos2\gamma\cdot\overline{F}_5) \tag{2.9}$$

Where
μ ---the Poisson's ratio
γ ---the angle of the two local co-ordinate axises of the joint points i and j

$$\overline{F}_2=\frac{1}{4\pi(1-\mu)}\left[\ln\sqrt{(\overline{x}-a)^2+\overline{y}^2}-\ln\sqrt{(\overline{x}+a)^2+\overline{y}^2}\right] \tag{2.10}$$

$$\overline{F}_3=\frac{-1}{4\pi(1-\mu)}\left[\arctan\frac{\overline{y}}{\overline{x}-a}-\arctan\frac{\overline{y}}{\overline{x}+a}\right] \tag{2.11}$$

$$\overline{F}_4=\frac{1}{4\pi(1-\mu)}\left[\frac{\overline{y}}{(\overline{x}-a)^2+\overline{y}^2}-\frac{\overline{y}}{(\overline{x}+a)^2+\overline{y}^2}\right] \tag{2.12}$$

$$\overline{F}_5=\frac{1}{4\pi(1-\mu)}\left[\frac{\overline{x}-a}{(\overline{x}-a)^2+\overline{y}^2}-\frac{\overline{x}+a}{(\overline{x}+a)^2+\overline{y}^2}\right] \tag{2.13}$$

$$\overline{x}=(x^i-x^j)\cos\beta^j+(y^i-y^j)\sin\beta^j \tag{2.14}$$

$$\overline{y}=-(x^i-x^j)\sin\beta^j+(y^i-y^j)\cos\beta^j \tag{2.15}$$

x^i, y^i—the co-ordinates of the joint point i
x^j, y^j—the co-ordinates of the joint point j
β^j—the angle of the local and the total co-ordinate axises in the joints point j

In this paper the cubic B spline function is made use of

$$\phi\left(\frac{s-s_i}{2a}\right)=\frac{1}{6(2a)^3}\begin{cases}0 & s\leq s_{i-2}\\(s-s_{i-2})^3 & s\in[s_{i-2},s_{i-1}]\\(s-s_{i-2})^3-4(s-s_{i-1})^3 & s\in[s_{i-1},s_i]\\(s_{i+2}-s)^3-4(s_{i+1}-s)^3 & s\in[s_i,s_{i+1}]\\(s_{i+2}-s)^3 & s\in[s_{i+1},s_{i+2}]\\0 & s\geq s_{i+2}\end{cases} \tag{2.16}$$

The graph is shown in Fig.2. From the figure, we can see that, for every joint point s_j, the formula (2.16) at most has three terms that are not equal to zero.

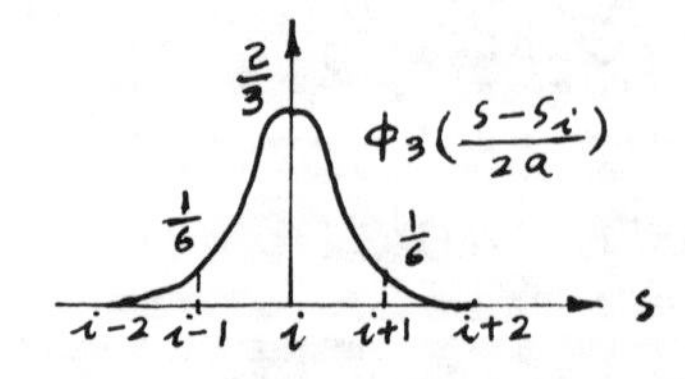

Fig.2 Cubic B spline function

$$u(s_j)=a_{i-1}\phi_{i-1}(s_j)+a_i\phi_i(s_j)+a_{i+1}\phi_{i+1}(s_j) \quad (2.17)$$

Therefore having determined $Y(s_1)$ and $C(s_1,s)$, we can use (1.17) and (1.19) to determine the coefficients b_i and g_{ij} with the interpolation method. And we can use (1.25) to determine the elements K_{ij} of matrix [K]. Then we can obtain the fundamental equation of SIBEM.

2.2 Rock pressure problems of the opening

The calculation model is shown in Fig.3.

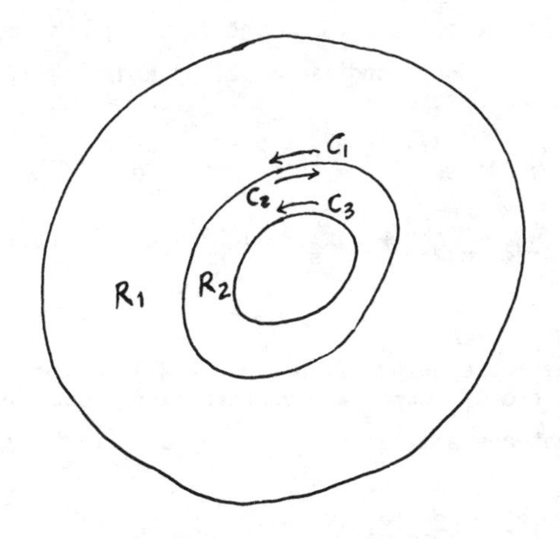

Fig.3 Calculation model of country rock pressure

Suppose that P_0 indicates the field stress; μ_1, E_1 and G_1 the elastic constants of country rock; μ_2, E_2 and G_2 the elastic constants of supporting material; C_1 the contour of the interior interface of the country rock R_1; C_2 the contour of the exterior interface of the supporting material R_2; C_3 the contour of the interior interface of the supporting material R_2. Let P be the normal pressure of the country rock acting on the supporting material. Then

$$P=P_0+t_n^{c_1} \quad (2.18)$$

Where $t_n^{c_1}$ is the induced stress that is normal to the contour C_1, so long as the induced stress $t_n^{c_1}$ is worked out, then using (2-18), we can arrive at the rock pressure P by calculation.

Because the total surface stress is equal to zero, the induced surface stress of the contour C_3 can be shown as follows:

$$t_n^{c_3}=-P_o$$
$$t_n^{c_3}=0 \quad (2.19)$$

Let $t_n^{c_2}$ represent the induced stress which is normal to C_2; $t_s^{c_2}$ the induced stress which is in the tangent direction of C_2.

So the problem of solving the induced stress turns into another one, i.g. the region of interest consists of an annulus R_2 with elastic constants μ_2 and G_2 inside a circular hole in a large plate with elastic constants μ_1 and G_1. The inside wall of the annulus is subjected to a normal stress $\sigma_r=-P_o$, and the plate is unstressed at infinity.

So for the contact point Q the stress continuity conditions are

$$\sigma_{s(Q)}^{[1]}=\sigma_{s(Q)}^{[2]}$$
$$\sigma_{n(Q)}^{[1]}=\sigma_{n(Q)}^{[2]} \quad (2.20)$$

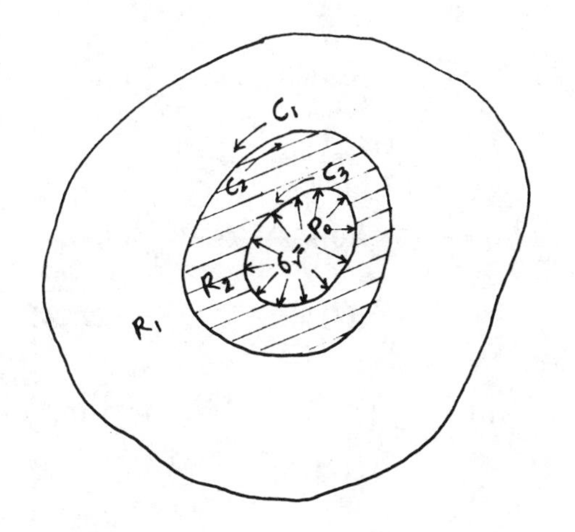

Fig.4 Field of induced stresses

The displacement continuity conditions can be written as:

$$u_{s(Q)}^{[1]} = -u_{s(Q)}^{[2]}$$
$$u_{n(Q)}^{[1]} = -u_{n(Q)}^{[2]} \quad (2.21)$$

For the country rock region R_1, equations can be written in the following forms:

$$Y_{s(P)}^{(1)} = \oint_{c_1} [C_{ss}^{(1)}(P,s)X_s^{(1)}(s) + C_{sn}^{(1)}(P,s)X_n^{(1)}(s)] \cdot ds$$

$$Y_n^{(1)}(P) = \oint_{c_1} [C_{ns}^{(1)}(P,s)X_s^{(1)}(s) + C_{nn}^{(1)}(P,s)X_n^{(1)}(s)] \cdot ds \quad (2.22)$$

For the supporting material region R_2 equations can be written as follows:

$$Y_s^{(2)}(P) = \oint_{c_2+c_3} [C_{ss}^{(2)}(P,s)X_s^{(2)}(s) + C_{sn}^{(2)}(P,s) X_n^{(2)}(s)]ds$$

$$Y_n^{(2)}(P) = \oint_{c_2+c_3} [C_{ns}^{(2)}(P,s)X_s^{(2)}(s) + C_{nn}^{(2)}(P,s) X_n^{(2)}(s)]ds \quad (2.23)$$

If the point P is on the contact line of the two regions, for example, on the point Q of it, then the above two sets of equations can be transformed into the following ones respectirely.

$$Y_s^{(1)}(Q) = \oint_{c_1} [C_{ss}^{(1)}(Q,s)X_s^{(1)}(s) + C_{sn}^{(1)}(Q,s) X_n^{(1)}(s)]ds$$

$$Y_n^{(1)}(Q) = \oint_{c_1} [C_{ns}^{(1)}(Q,s)X_s^{(1)}(s) + C_{nn}^{(1)}(Q,s) X_n^{(1)}(s)]ds \quad (2.24)$$

and

$$Y_s^{(2)}(Q) = \oint_{c_2+c_3} [C_{ss}^{(2)}(Q,s)X_s^{(2)}(s) + C_{sn}^{(2)}(Q,s) X_n^{(2)}(s)]ds$$

$$Y_n^{(2)}(Q) = \oint_{c_2+c_3} [C_{ns}^{(2)}(Q,s)X_s^{(2)}(s) + C_{nn}^{(2)}(Q,s) X_n^{(2)}(s)]ds \quad (2.25)$$

Now let's take the fictitious stress method for example to explain how to deal with these equations. From the stress contact conditions(2.20) we arrive at: $Y_s^{(1)}(Q) = Y_s^{(2)}(Q)$ and $Y_n^{(1)}(Q) = Y_n^{(2)}(Q)$. Combining(2.24) with (2.25):

$$0 = \oint_{c_1} [C_{ss}^{(1)}(Q,s)X_s^{(1)}(s) + C_{sn}^{(1)}(Q,s)X_n^{(1)}(s)]ds - \oint_{c_2+c_3} [C_{ss}^{(2)}(Q,s)X_s^{(2)}(s) + C_{sn}^{(2)}(Q,s)X_n^{(2)}(s)]ds$$

$$0 = \oint_{c_1} [C_{ns}^{(1)}(Q,s)X_s^{(1)}(s) + C_{nn}^{(1)}(Q,s)X_n^{(1)}(s)]ds - \oint_{c_2+c_3} [C_{ns}^{(2)}(Q,s)X_s^{(2)}(s) + C_{nn}^{(2)}(Q,s)X_n^{(2)}(s)]ds \quad (2.26)$$

If the point P is on the line c_3, then from the boundary conditions (2.19) and (2.20), we can obtain:

$$0 = \oint_{c_2+c_3} [C_{ss}^{(2)}(P,s)X_s^{(2)}(s) + C_{sn}^{(2)}(P,s)X_n^{(2)}(s)]ds$$

$$-P_0 = \oint_{c_2+c_3} [C_{ns}^{(2)}(P,s)X_s^{(2)}(s) + C_{nn}^{(2)}(P,s)X_n^{(2)}(s)]ds \quad (2.27)$$

Therefore using (2.26) and (2.27), and supposing $c = c_1 + c_2 + c_3$, we can establish equations that are similar to the equations (1.2).

$$Y_s(P) = \oint_c [C_{ss}(P,s)X_s(s) + C_{sn}(P,s)X_n(s)]ds$$

$$Y_n(P) = \oint_c [C_{ns}(P,s)X_s(s) + C_{nn}(P,s)X_n(s)]ds \quad (2.28)$$

Where

$$Y_s(P) = 0$$

$$Y_n(s) = \begin{cases} 0 & P \in c_1 + c_2 \\ -P_0 & P \in c_3 \end{cases}$$

$$C_{ss}(P,s) = \begin{cases} C_{ss}^{(1)}(P,s) & P \in c_1 + c_2,\ s \in c_1 \\ -C_{ss}^{(2)}(P,s) & P \in c_1 + c_2,\ s \in c_2 \\ 0 & P \in c_3,\ s \in c_1 \\ C_{ss}^{(2)}(P,s) & P \in c_3,\ s \in c_2 + c_3 \end{cases}$$

$$C_{sn}(P,s)=\begin{cases} C_{sn}^{(1)}(P,s) & P\in C_1+C_2, s\in C_1 \\ -C_{sn}^{(2)}(P,s) & P\in C_1+C_2, s\in C_2+C_3 \\ 0 & P\in C_3, \quad s\in C_1 \\ C_{sn}^{(2)}(P,s) & P\in C_3, \quad s\in C_2+C_3 \end{cases}$$

$$C_{ns}(P,s)=\begin{cases} C_{ns}^{(1)}(P,s) & P\in C_1+C_2, \ s\in C_1 \\ -C_{ns}^{(2)}(P,s) & P\in C_1+C_2, s\in C_2+C_3 \\ 0 & P\in C_3, \quad s\in C_1 \\ C_{ns}^{(2)}(P,s) & P\in C_3, \quad s\in C_2+C_3 \end{cases}$$

$$C_{nn}(P,s)=\begin{cases} C_{nn}^{(1)}(P,s) & P\in C_1+C_2, \ s\in C_1 \\ -C_{nn}^{(2)}(P,s) & P\in C_1+C_2, s\in C_2+C_3 \\ 0 & P\in C_3, \quad s\in C_1 \\ C_{nn}^{(2)}(P,s) & P\in C_3, \quad s\in C_2+C_3 \end{cases}$$

$$X_s(s)=\begin{cases} X_{s(s)}^{(1)} & s\in C_1 \\ X_{s(s)}^{(2)} & s\in C_2+C_3 \end{cases}$$

$$X_n(s)=\begin{cases} X_{N(s)}^{(1)} & s\in C_1 \\ X_{n(s)}^{(2)} & s\in C_2+C_3 \end{cases}$$

Consequently, if we start with (2.28) and adopt the general SIBEM, we can evaluate the fictitious stressed $X_s(s)$ and $X_n(s)$. After that using the general IBEM, we can work out the induced stresses of contact lines $Y_s(P)$ and $Y_n(P)$. Finally, using (2.18) we can eualuate the rock pressure P.

2.3 The problems of cracks

In order to deal with the influence of cracks in IBEM, the joint elements are introduced. The components of displacement discontinuity D_n' and D_s' are used for the fictitious variable of the joint element. suppose that the common element numbers equal m, the total element numbers equal n. If in the formula (1.28), $i\leqslant 2m+1$, then the meaning of every symbol is just the same as the above ones respectirely; if $2m+2\leqslant i\leqslant 2n+1$, then X(s) are represened by the components of the displacement, discontinuity D_s' and D_n'. The real variable Y(P) and the influence coefficients C_{ss} etc. must be expanded to all elements. When evaluating the matrix elements g_{ij} and K_{ij}, we can do it in the same way. Eventually we can use the SIBEM to solve crack problems.

3 CONCLUSIONS

1. The SIBEM is a new numerial method. it is a more convenient method the BEM and SDBEM. Meanwhile, this method has such advantages as high precision, simplicity and readiness and that its total matrix is a symmetrical and belt-like one.

2. Using SIBEM in the problems of rock mechanics, it is also simple, convenient and practicable. So it will be widely applied.

This paper is transtated by Wang Da Peng and the auther of this paper. Edited by Prof. Wei Shu Ying.

REFERENCES

[1] Qin Rong. The principle and applying of spline boundary element method. Pro. of 5th international conference on boundary element method. 1983.

[2] Y.L.Lin(Lin Yu Liang). The calculation for displacement, stress and rock pressure around underground opening by spline direct boundary element method. Pro. of first chinese calculation mechanics in geomechanics conference. 1987.

[3] S.L.Crouch, A.M.Stavfield. Boundary element methods in solid mechanics. George Allen and Unwin, 1982.

[4] Li Yue Sheng, Qi Dong Xu. The method of spline function. The science press. China. 1979.

Numerical Methods in Geomechanics (Innsbruck 1988), Swoboda (ed.)
© 1988 Balkema, Rotterdam. ISBN 90 6191 809 X

Underground stress analysis using mapped infinite elements

P.Marović
University of Split, FGZ, Yugoslavia
F.B.Damjanić
University of Ljubljana, FAGG, Yugoslavia

ABSTRACT: In the large spectrum of practical engineering geomechanical problems, there is always a question how to treat unbounded continua in the most rational way to get accurate results. This paper considers the use of mapped infinite elements in conjuction with finite elements in both linear and non-linear stress analysis sumarising previous research on the use of mapped infinite elements in the linear elastic and non-linear stress analysis. Material non-linear behaviour is described by Perzyna's elasto-visco-plastic model in widely accepted standard finite element formulation. Additional attention is given to gravity loading and its common action with other loadings. The gravity loading can be treated in two different ways: (a) by evaluating element nodal forces from the actual values of gravity stresses in Causs points or element volume weight; (b) by incorporating gravity stress from Gauss points directly in the total stress vector as initial ones. In the linear and non-linear analysis when the problem is discretized only with finite elements both strategies give the same results. Discretizing the half-space under consideration with infinite elements too, only the second strategy gives the correct results. The applicability of the suggested technique is examined and the benefits of employing proposed finite/infinite element approach in both linear and non-linear analysis is then illustrated by the solution of few unbounded geomechanical problems.

1 INTRODUCTION

In many problems of engineering interest, the boundaries of a real continua are clearly defined. However, some classes of practical engineering problems involve unbounded continua Specific examples can be found in fields such as heat and mass transfer, fluid and solid mechanics, acoustic, electrics and electromagnetics, etc. Probably the most common situation in civil engineering in which unbounded domains arise is in geomechanical applications where modelling the far field behaviour can often present considerable numerical difficulties. Analytical solutions neither in close form /1/ nor tabulated /2/ are very often unsufficient and usualy treat only simple problems. In order to overcome this problem and get the best results, different techniques and numerical procedures are developed which include analytical matching of the far field solution, mapping the exterior domain onto an interior finite one, finite element method using truncation techniques, boundary integral methods, infinite elements, etc. A very valuable reviews on this topic, of the different techniques for unbounded domain suitable for use in the numerical analysis are given in Ref. /3-6/.

In numerical analysis a usual engineering approach is simple truncation in which the infinite domain is terminated at a finite distance and the appropriate infinite boundary condition imposed. However the location of the finite boundary is essential for accurate solution and the selection of its minimum distancefrom the region of interest is often a question of experience and intuition.

A recent development is the mapped infinite element /3,7-9/. In this paper the mapped infinite elements /4-6, 10-14/ are coupled with conventional finite isoparametric elements to solve undefined field problems in non-linear stress analysis.

The application of this mapped infinite elements is more simplier then previously developed types, and as their formulation follows the standard finite element procedure /15,16/, they can be very easily and efficiently used in conjuction with conventional finite elements.

In the solution procedure of loaded half-space which arises in the field of geomechanics, stress state in the half-space is usually determined only for external loads, although there is a stress state due to gravity loading. If the material behaviour is described by linear and elastic model, total stress state can be achieved simply by adding these two stresss states. If the material behaviour is described by some non-linear model the before mentioned principle of superposition cannot be used. Namely, in the elasto-visco-plastic analysis the global stress state is defined by the total stress vector according to which visco-plastic strain and stress rates are evaluated if yielding occurs. Because of that, stress state due to gravity loading cannot be added at the end of the analysis, i.e. the gravity loading has to be incorporated at the begining of the analysis. Generally the gravity loading can be taken into account in two different ways: (a) by evaluating equivalent nodal forces on element due to element body forces or gravity stresses in Gauss points; (b) by incorporating gravity stresses from Gauss points directly into the total stress vector as initial stresses. Both strategies hold true for finite element idealization /17/, and for finite/infinite element discretisation extensive investigations have been made to find the right treatment of gravity loading /6/.

In this analysis, a material non-linear behaviour is described by Perzyna's elasto-visco-plastic model /18/ in widely accepted standard finite element formulation /4,6,13,16/.

The practical applicability of this finite/infinite element approach in solving the undefined field problems is shown by the solution of few usual geomechanical problems: loaded linear elastic half-space, linear elastic half-space with excavation, loaded non-linear half-space and loaded non-linear half-space with excavation, all with and without influence of gravity loading.

2 BRIEF CONCEPT OF MAPPED INFINITE ELEMENTS

The concept of mapped infinite elements is based upon the idea of mapping an infinite region onto a finite one. Appropriate element shape functions are developed for such geometry mapping, which for instance for two-dimensional infinite element extending to infinity in one direction are evaluated as a product of special mapping functions for infinite direction and standard shape functions for finite direction.

However, the field variables are interpolated using standard finite interpolation functions. The standard procedure of the parametric finite element formulation /15,16/ may then be applied.

Marques /5,13/ have worked out and tabulated the mapping functions for a large range of commonly used elements in a more simplier and systematic way then has been given by Zienkiewicz et al. /3/. The historical development and extensive treatment of infinite elements can be found in /19/.

Some care should be exercised in positioning the poles of infinite elements as they are origin of the geometric and the solution polynomial expansion.Although arbitrary, the positions of poles are very important in order to ensure unique solution, continuity of the solutions between elements and the most accurate solutions. They must be external to the infinite element, so that the element sides extending towards infinity can be and must be parallel or divergent to avoid the overlapping of elements and preserve mapping uniqueness. To ensure continuity of the solutions between infinite and finite elements or adjacent infinite elements, across their common sides, the numbers and positions of the connecting nodes must coincide.

3 IMPLEMENTATION OF MAPPED INFINITE ELEMENTS IN STRESS ANALYSIS

The standard matrix formulation of equilibrium equaition for discretised domain can be written as

$$\underline{K}\ \underline{\delta} + \underline{f} = 0 \tag{1}$$

where $\underline{K}$ stiffness matrix, $\underline{\delta}$ displacement matrix and $\underline{f}$ external load matrix.

Element stiffness matrix, given as usual for finite element as

$$K_{ij}^{e} = \int_{V^e} \underline{B}_i^T\ \underline{D}\ \underline{B}_j\ dv^e \tag{2}$$

is obtained for an infinite element integrating expression (2) numerically using Gaussian quadrature and looping over each integration point:

- * Evaluate the Gauss abscissae and weights, W_i.
- * Call the mapping function subroutine which evaluates mapping functions and its first partial derivates over natural coordinates.
- * Compute the coordinate Jacobian matrix $\underline{J}$ using first partial derivates of mapping functions, which defines geometric mapping.
- * Compute the determinant, det $\underline{J}$, and the inverse Jacobian matrix, $\underline{J}^{-1}$.
- * Call the standard shape function subroutine which evaluates standard shape functions and its first partial derivatives over natural coordinates.
- * Compute the Cartesian derivates in standard way.
- * Evaluate the strain matrix, $\underline{B}$, using Cartesian derivatives.
- * Evaluate the elasticity matrix, $\underline{D}$.
- * Evaluate $\underline{B}_i^T\ \underline{D}\ \underline{B}_j\ w_i w_j$ det $\underline{J}$ and sum into K_{ij}^{e}.

Thus, the geometric mapping functions are used to relate the local and global coordinates in the computation of the Jacobian matrix, $\underline{J}$, which inverted is used to obtain the

Cartesian shape function derivatives needed to express the strain matrix, $\underline{B}$. Further solution process of equation (1) follows standard finite element procedure /15, 16/.

Although the possibility of forming a mesh using mapped infinite elements only, exists, experience shows that solutions are generally better when the unbounded domains are discretised using a combination of finite and infinite elements /4, 6, 11, 12/. Further, the mapped infinite elements exhibit more advantages and practical benefits where far fields are not of interest, and are only used to model effects of infinity, bounding some finite region.

The modelling of domains far from the region of interest using infinite elements results in a saving in computational costs. Consequently, the mesh in the regions of interest may be refined resulting in greater accuracy.

4 SHORT DESCRIPTION OF ELASTO-VISCO-PLASTIC MATERIAL MODELLING

A more detailed introduction to the elasto-visco-plastic finite element analysis can be found in /16/, where the adopted Perzyna's elasto-visco-plastic model /18/ is incorporated in finite element formulation for small strains and small deformatins given by Zienkiewicz and Cormeau /20/. Obtained finite element formulation /16/ is widelly accepted and particurarly is used in /4-6, 10-14/. Especially, only the essential expressions for description of the elasto-visco-plastic material behaviour which were used in this paper can be found in /21/.

The basic assumption of the adopted Perzyna's elasto-visco-plastic model is decomposition of the total strain rate into its elastic and visco-plastic parts. The visco-plastic deformations, which are the function of the current stress state, occur only in the case when equivalent stress state according to adopted yield criterion (Tresca, Von Mises, Mohr-Coulomb, Drucker-Prager) overflows the uniaxial or effective yield stress. In this analysis the associated visco-plasticity is used.

5 TREATMENT OF GRAVITY LOADING

Detailed description of stress state in the subsoil due to different kinds of loading and especially of gravity loading can be found in /22/. In investigations carried out in this work it is taken that in any point of the soil the vertical and horizontal component of stress due to gravity loading is present according to

$$\sigma_v = \gamma h \tag{3}$$

$$\sigma_h = \frac{\upsilon}{1-\upsilon} \gamma h = k_o \gamma h \tag{4}$$

where γ is the specific weight of the soil material, h is depth of observed point below surface, υ is Poisson's ratio and k_o is coefficient of pressure at rest.

More detailed information considering gravity loading and determination of stress state in soil with finite element method can be found in /17/. Consequently, two basic approaches are: (a) after discretisation of finite domain with finite elements, gravity loading is treated as volumetric force on element or gravity stress in Gauss points from which equivalent element nodal forces are evaluated using shape functions and element volume; (b) initial stresses due to gravity loading, whose components are shown with eqs. (3) and (4), are added in the first load increment directly into the total stress vector. These approaches are shematically shown in Fig. 1.

The first approach is more convenient for problems with complex boundaries and/or excavations, but with this approach very often unwanted settlements due to gravity loading are obtained. To eliminate this settlements additional analysis due to gravity loading only has to be carried out and two sets of obtained results have to be superposed. The second approach is generaly more convenient and especially when the non-linear analysis is performed.

When we want to idealize any problem of half-space using combination of finite and infinite elements and with gravity loading included in the analysis, then we can solely use the second approach /6/. Namely, during the calculation of the element nodal forces from element volumetric force for the infinite element one obtain almost infinite volume due to which large nodal forces are evaluated and which cause unequal nodal displacements (see Fig. 2) and disturbed stress state. With better finite/infinite element idealization and adequate infinite element pole positioning these effects can be diminished but still with great solution errors.

So, for discretisation of undefined domain with finite and infinite elements gravity loading can be treated only using concept of initial stresses without evaluation of equivalent nodal forces /6/. This holds true for the linear and non-linear stress analysis. Adopted concept is applicable for the analysis of loaded half-spaces with or without natural cavities. For the analysis of the problems with some excavations (underground storages, tunels, etc.), the influence of the excavated material have to be taken into account with additionaly applied stress cancelling tractions (normal and tangentional) to the boundaries of the excavations /17/.

6 NUMERICAL EXAMPLES

The described mapped infinite elements and

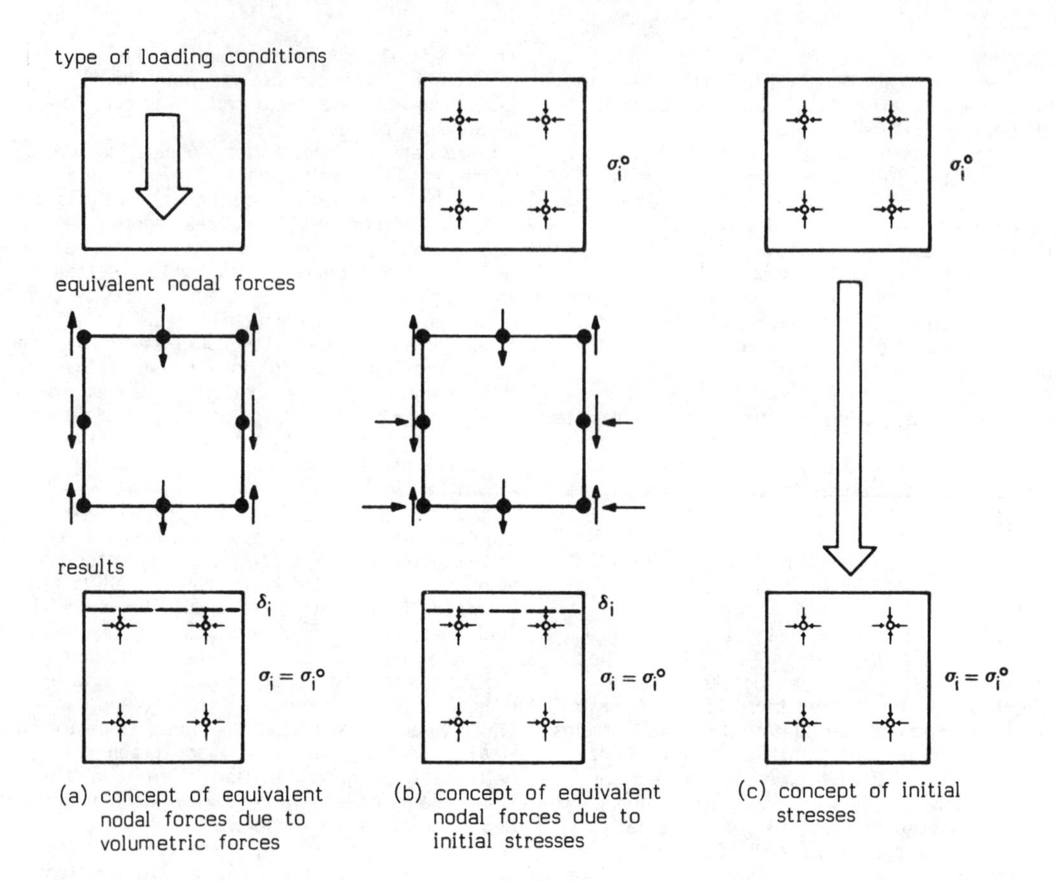

Fig. 1 Modelling of gravity loading

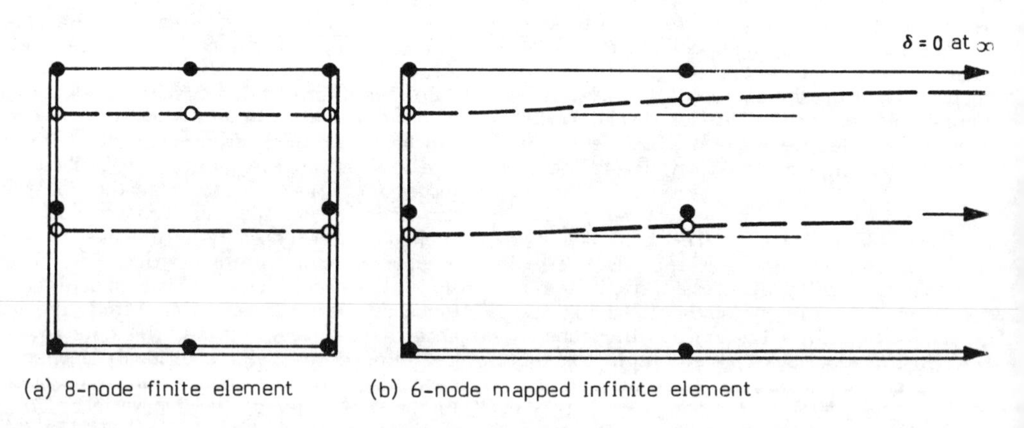

Fig. 2 Element nodal displacements due to equivalent nodal forces

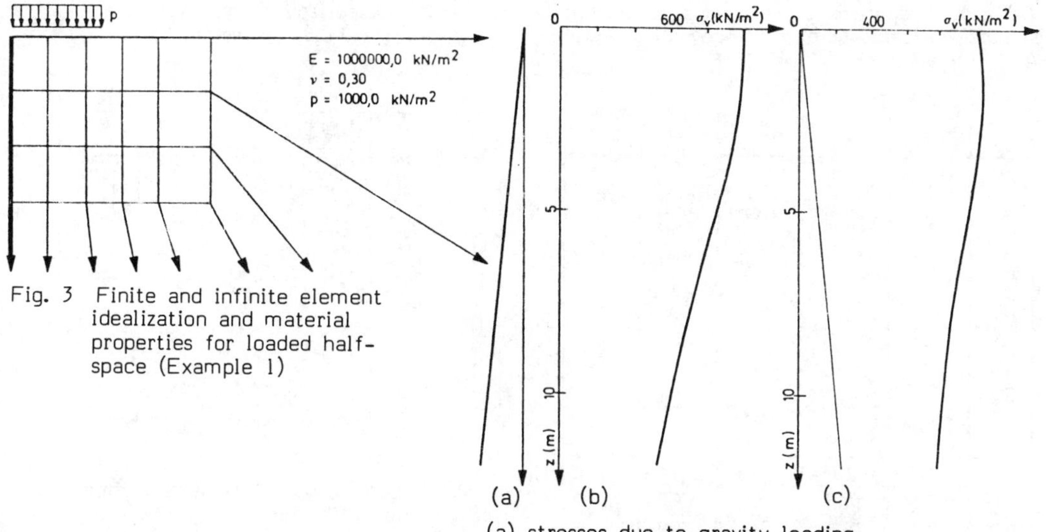

Fig. 3 Finite and infinite element idealization and material properties for loaded half-space (Example 1)

(a) stresses due to gravity loading
(b) stresses due to external loading
(c) stresses due to common action of gravity and external loadings

Fig. 4 Distribution of vertical stresses in section x=0.42 m (Example 1)

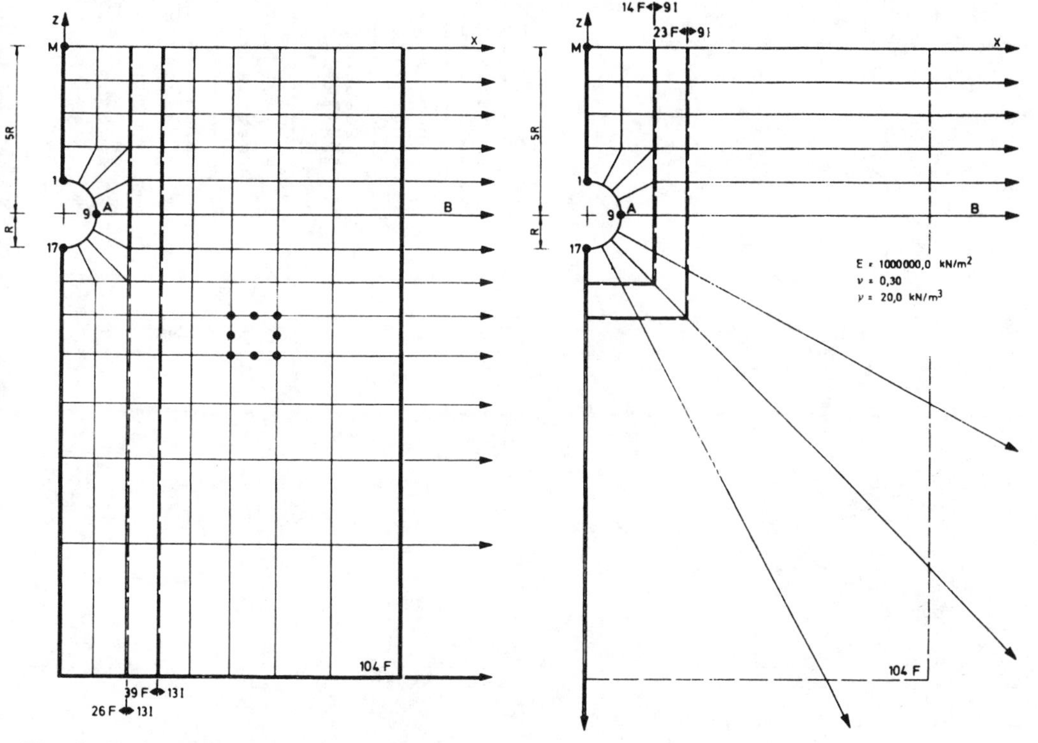

Fig. 5 Finite and infinite element meshes employed in the solution of the half-space with excavation (Example 2)

Table 1. Comparison of vertical and horizontal stresses in Gauss-point line above the section A-B (Example 2)

x (m)	104F	39F131	26F131	23F91	14F91
(a)	VERTICAL STRESSES				σ_v (kN/m²)
6.04	960.93	956.81	945.81	958.78	953.73
8.94	625.11	623.29	616.88	623.97	620.54
11.06	568.45	566.64		566.83	
12.68			542.82		543.40
13.94	529.14	527.92		527.83	
16.06	515.65				
18.94	501.45				
19.02		504.08		503.01	
21.27	494.60				
24.73	487.18				
27.52	483.45				
31.68	479.51				
34.98	477.70				
39.86	476.02				
43.83	475.38				
47.32			480.42		480.53
49.81	474.93				
70.98		477.68		478.11	
(b)	HORIZONTAL STRESSES				σ_H (kN/m²)
6.04	137.26	137.23	136.44	136.78	137.28
8.94	238.87	240.60	242.73	240.41	242.14
11.06	229.56	232.96		230.51	
12.68			232.50		230.51
13.94	224.20	228.99		226.99	
16.06	221.74				
18.94	217.57				
19.02		221.87		218.76	
21.27	215.66				
24.73	213.27				
27.52	211.82				
31.68	209.99				
34.98	208.86				
39.86	207.55				
43.83	206.81				
47.32			205.49		206.18
49.81	206.25				
70.98		204.50		205.29	

Table 2. Comparison of relative surface settlement, z=0.0 m (Example 2)

x (m)	104F	39F131	26F131	23F91	14F91
0.0	0.000	0.000	0.000	0.000	0.000
2.5	0.003	0.003	0.003	0.003	0.003
5.0	0.012	0.011	0.011	0.011	0.011
7.5	0.025	0.024	0.024	0.023	0.022
10.0	0.040	0.038	0.039	0.037	0.036
12.5	0.055	0.053	-	0.051	-
15.0	0.069	0.064	-	0.061	-
17.5	0.082	-	-	-	-
20.0	0.092	-	0.088	-	0.080
23.0	0.101	-	-	-	-
26.0	0.108	-	-	-	-
29.2	0.113	-	-	-	-
30.0	-	0.098	-	0.087	-
33.2	0.116	-	-	-	-
37.42	0.118	-	-	-	-
41.64	0.118	-	-	-	-
46.82	0.118	-	-	-	-
52.0	0.118	-	-	-	-

Table 3. Comparison of relative vertical displacements of the nodes on the excavation line (Example 2)

NODE	104F	39F131	26F131	23F91	14F91
1	0.371	0.368	0.362	0.370	0.366
2	0.369	0.366	0.360	0.367	0.364
3	0.336	0.332	0.327	0.334	0.331
4	0.301	0.298	0.293	0.299	0.296
5	0.260	0.258	0.253	0.259	0.256
6	0.222	0.220	0.216	0.221	0.219
7	0.175	0.173	0.170	0.174	0.171
8	0.092	0.092	0.090	0.092	0.091
9 *	0.000	0.000	0.000	0.000	0.000
10	0.096	0.095	0.093	0.096	0.095
11	0.185	0.183	0.180	0.185	0.184
12	0.240	0.239	0.235	0.239	0.238
13	0.286	0.284	0.280	0.284	0.283
14	0.337	0.335	0.330	0.335	0.333
15	0.382	0.380	0.376	0.382	0.380
16	0.427	0.425	0.420	0.427	0.424
17	0.430	0.428	0.424	0.431	0.429

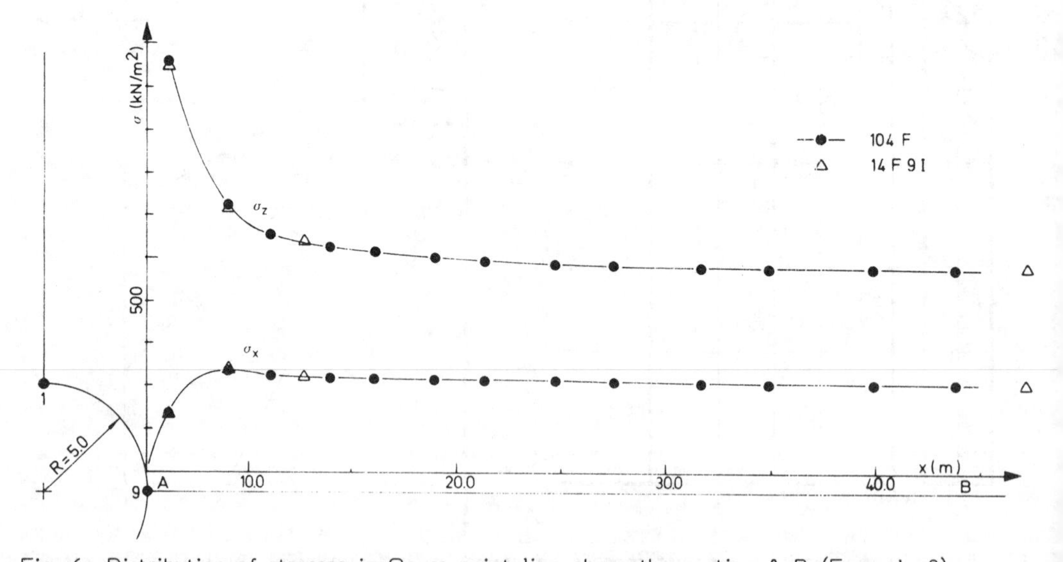

Fig. 6 Distribution of stresses in Gauss-point line above the section A-B (Example 2)

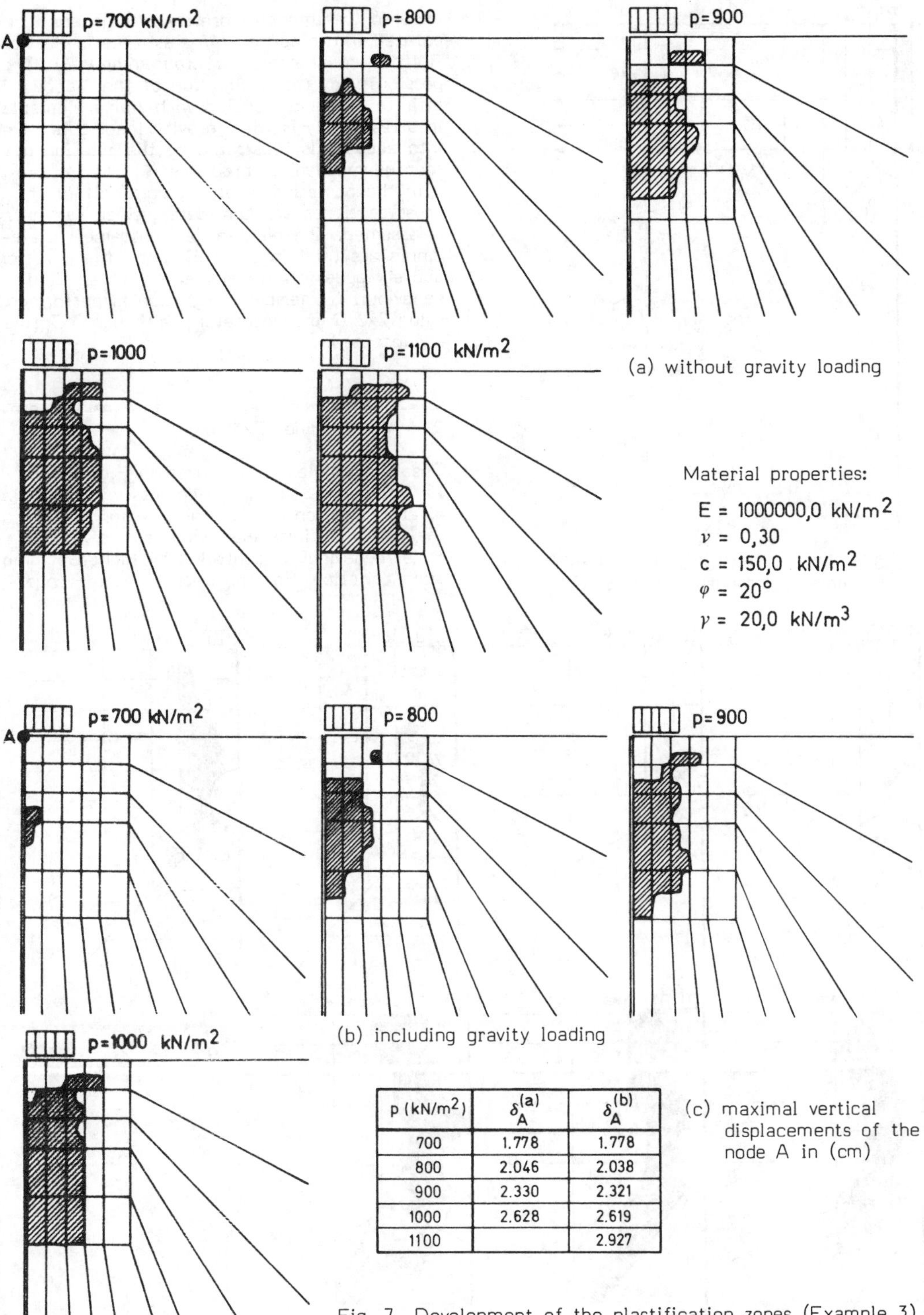

p (kN/m²)	$\delta_A^{(a)}$	$\delta_A^{(b)}$
700	1.778	1.778
800	2.046	2.038
900	2.330	2.321
1000	2.628	2.619
1100		2.927

(c) maximal vertical displacements of the node A in (cm)

Fig. 7 Development of the plastification zones (Example 3)

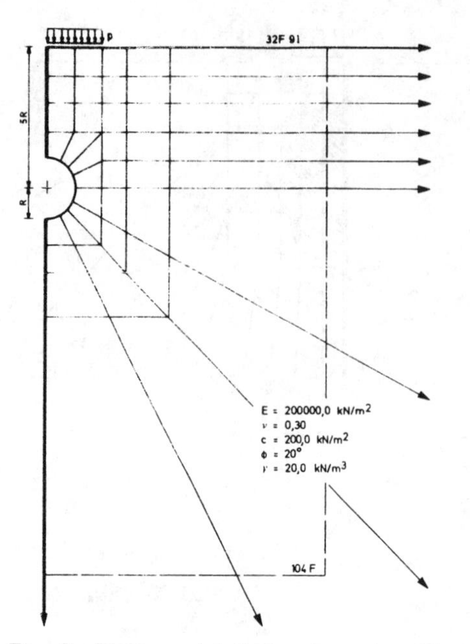

Fig. 8 Finite and infinite element mesh with material properties for Example 4

adopted treatment of gravity loading are incorporated into modified /4/ elasto-visco-plastic finite element computer programme /16/. The possibility of the application of the mapped infinite elements coupled with finite elements in stress analysis when gravity loading is taken into account is illustrated by the solution of four geomechanical problems /6, 21, 23/.

In the numerical analysis, only half of the cross sectiones are considered, since symmetry is assumed. The eight-node isoparametric elements are used for modelling the finite regions and six-node infinite elements for far field behaviour. Numerical integration is performed using 2x2 Gauss quadrature, although 3x3 rule can be used.

6.1 Loaded linear elastic half-space and gravity loading (Example 1)

Observed problem is idealized in Fig. 3 with 15 finite and 8 infinite elements with 11 poles after truncation of the finite region /6, 24/ of the basic 32 finite element example /25/. Material behaviour is controled by Mohr-Coulomb yield criterion. For the sake of better under-

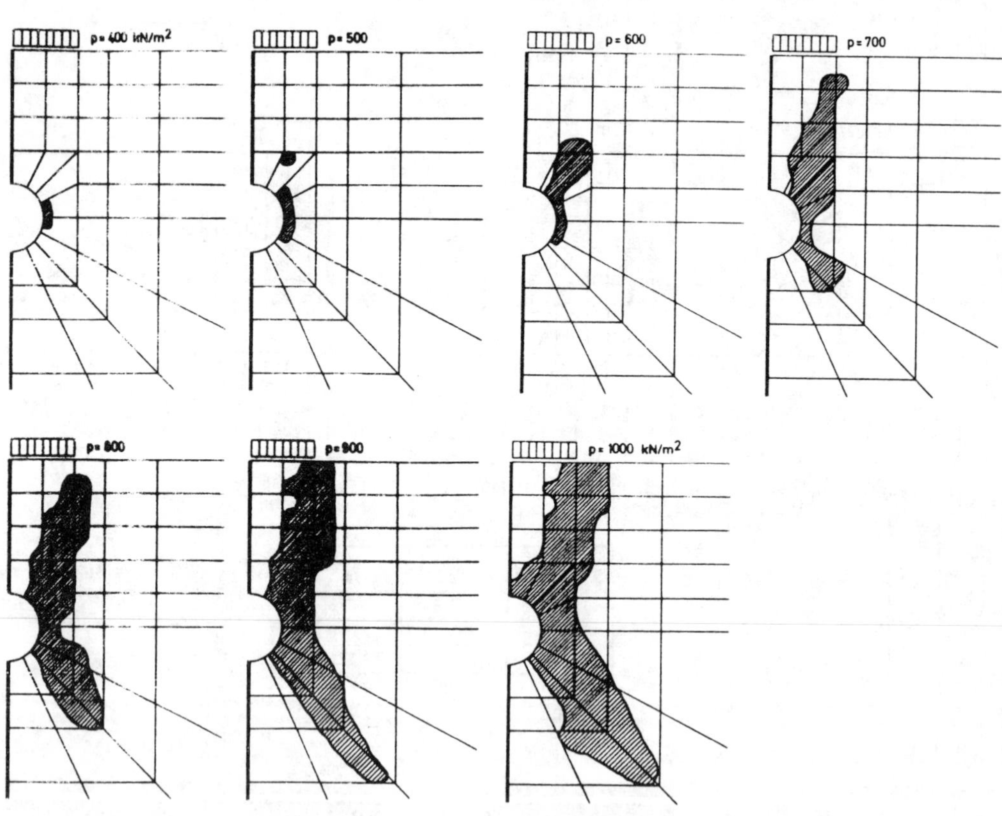

Fig. 9 Development of the plastification zone (Example 4)

standing of gravity loading treatment and analysis of obtained results, Fig. 4 shows the vertical stress distribution for different loadings in elastic range. Similar diagrams for the elasto-visco-plastic analysis can not be evaluated because strain and stress rates are function of the total stress state when the yielding occurs. The obtained results are displacements due to external loading and stresses due to common action of gravity and external loading.

6.2 Linear elastic half-space with excavation and gravity loading (Example 2)

The extensive investigations have been performed /6/ to find the most adequate combination of finite/infinite element mesh (see Fig. 5) with respect to the most accurate results. Because of the lack of vertical gravity forces and presence of additional stress-cancelling tractions over the boundary of the excavation, meshes expand upwards. The stress states (Table 1. and Fig. 6) and the relative displacements of the nodes on the surface (Table 2.) and on the excavation line (Table 3.) are very similar for different meshes. But, one have to have on his mind that the results obtained with the use of infinite elements too, are calculated with smaller total number of elements and with less computational time and costs.

6.3 Loaded non-linear half-space (Example 3)

The same problem as in the Example 1 is analysed here, but with 25 finite and 10 infinite elements. The external load is varied in increments of 100 kN/m² until the plastic flow invade the space of the first infinite element. Material behaviour is again controled by Mohr-Coulomb yield criterion and the numerical integration is performed using Euler explicit time scheme with time step $\Delta t=0.1$ sec. Comparing the development of the plastification zones in Fig. 7, one can see the influence of including the gravity loading on the stress state. The differences in spreading directions of the plastification zones can be partly explained by the characteristics of the stress states given by the diagrams (b) and (c) in Fig. 4.

6.4 Loaded non-linear half-space with excavation (Example 4)

To illustrate the possibilities of the developed numerical algorithm one complicated geotechnical problem is chosen. The problem of loaded half-space with excavation is discretised with 32 finite and 9 infinite elements as is shown in Fig. 8. The external load is varied in increments of 100 kN/m² and material behaviour is controled by Mohr-Coulomb yield criterion. Numerical integration is performed using Euler explicit time scheme with time step $\Delta t=1.0$ sec. In Fig. 9 the development of the plastification zone is shown until the plastic flow invade the space of the first infinite element. This example can be a solid base for further investigations in the analysis of different stress and deformation states around shallow and deep tunnels.

7 CONCLUSIONS

The paper has shown the use of two-dimensional mapped infinite elements coupled with conventional finite isoparametric elements to solve undefined field problems in linear and non-linear stress analysis when gravity loading is taken into account. In that case the gravity loading has to be treated as gravity stresses in Gauss points which are then added to the total stress vector as initial stresses. The obtained results are displacements due to some external loading and stresses due to some external and gravity loading. When the problem under consideration is half-space with some excavation then stress-cancelling tractions have to be applied additionaly to the boundaries of the excavation.

Experience to date indicates that a coupled finite/infinite element approach to unbounded domain problems results in substantial savings in computational time, storage and costs.

Furthermore, performed investigations show that a coupled finite/infinite element approach provides more accurate solution then a simple finite element modelling using truncation approach.

Since the formulation of mapped infinite elements follows the standard finite element procedure very closely, they can be readily implemented into existing finite element programms.

REFERENCES

/ 1/ E.Nonveiller, Soil mechanics and foundation design, (in croatian), Školska knjiga, Zagreb, 1979

/ 2/ H.G.Poulos and E.H.Davis, Elastic solutions for soil and rock mechanics, John Wiley & Sons, New York, 1974

/ 3/ O.C.Zienkiewicz, P.Bettess, T.C.Chiam and C.Emson: Numerical methods for unbounded field problems and a new infinite element formulation, Winter annual meetting of the A.S.M.E., Washington D.C., also: J.appl.mech.div., ASCE, 115-148 (1981)

/ 4/ F.B.Damjanić, Reinforced concrete failure prediction under both static and transient

conditions, Ph.D.Thesis, C/Ph/71/83, University of Wales, Swansea, 1983
/ 5/ J.M.M.C.Marques and D.R.J.Owen: Infinite elements in quasi-static materially non-linear problems, Comp. & Struct. 18, 739-751 (1984)
/ 6/ P.Marović, Contribution to the continuum problems solution using finite and infinite elements, (in croatian), Ph.D.Thesis, University of Zagreb, 1987
/ 7/ P.Bettess: Infinite elements, Int.j.num. meth.engng. 11, 53-64 (1977)
/ 8/ P.Bettess: More on infinite elements, Int. j.num.meth.engng. 15, 1613-1626 (1980)
/ 9/ O.C.Zienkiewicz, C.Emson and P.Bettess: A novel boundary infinite element, Int.j. num.meth.engng. 19, 393-404 (1983)
/ 10/ F.B.Damjanić, V.Šimić and P.Marović, Numerical analysis of undefined field problems using infinite elements - I part, (in croatian), Report 15C/83, Faculty of civil engineering, Split, 1983
/ 11/ F.B.Damjanić, V.Šimić and P.Marović, Numerical analysis of undefined field problems using infinite elements - II part, (in croatian), Report 24/84, Faculty of civil engineering, Split, 1984
/ 12/ P.Marović and F.B.Damjanić: Infinite elements in stress analysis, (in croatian), Proc. VI Int.conf. PPPR, Zagreb, Ref. B1.02, 179-184 (1984)
/ 13/ J.M.M.C.Marques, Finite and infinite elements in static and dynamic structural analysis, Ph.D.Thesis, C/Ph/78/84, University of Wales, Swansea, 1984
/ 14/ P.Marović and F.B.Damjanić: Mapped infinite elements in linear elastic stress analysis, work presented at Congress GAMM-85, Dubrovnik, (1985)
/ 15/ E.Hinton and D.R.J.Owen, Finite element programming, Academic Press, London, 1977
/ 16/ D.R.J.Owen and E.Hinton, Finite elements in plasticity: theory and practice, Pineridge Press Ltd., Swansea, 1980
/ 17/ D.J.Naylor, G.N.Pande, R.Simpson and R.Tabb, Finite elements in geotechnical engineering, Pineridge Press Ltd., Swansea 1981
/ 18/ P.Perzyna: Fundamental problems in viscoplasticity, in: Recent advances in applied mechanics, Academic Press, New York, 9, 243-377 (1966)
/ 19/ P.Bettess and J.A.Bettess: Infinite elements for static problems, Eng.compt. 1, 4-16 (1984)
/ 20/ O.C.Zienkiewicz and I.C.Cormeau: Viscoplasticity, plasticity and creep in elastic solids - a unified numerical solution approach, Int.j.num.meth.engng. 8, 821-845 1974
/ 21/ P.Marović and F.B.Damjanić: Mapped infinite elements in non-linear stress analysis with special respect to gravity loading, Proc. 3rd Int. Conf. Num. Meth. for Non-linear Problems, Dubrovnik, Eds. C. Taylor, D.R.J.Owen, E.Hinton and F.B. Damjanić, Pineridge Press Ltd., Swansea, Vol. 3, 1263-1275 (1986)
/ 22/ J.Feda, Stresses in subsoil and methods of final settlement calculation, Elsevier Sci.Publ.Comp., Amsterdam, 1978
/ 23/ P.Marović: Numerical analysis of continuum using finite and infinite elements with special respect to gravity loading, (in croatian), Proc VIII Int.conf. PPPR, Zagreb, Ref. B1.06, 189-194 (1986)
/ 24/ P.Marović and F.B.Damjanić: Mapped infinite elements in the analysis of geomechanical problems, Proc. Int. Conf. Num. Meth. Engng.: Theory and applications, Swansea, Eds. J.Middleton and G.N.Pande, A.A.Balkema Publ., Rotterdam, Part 2, 749-753 (1985)
/ 25/ O.C.Zienkiewicz, C.Humpheson and R.W. Lewis: Associated and non-associated visco-plasticity in soil mechanics, Geotechnique 25, 671-687 (1975)

Numerical Methods in Geomechanics (Innsbruck 1988), Swoboda (ed.)
© 1988 Balkema, Rotterdam. ISBN 90 6191 809 X

Application of the common boundary element software COMVAR to the analysis of soils

Bulent A.Ovunc
University of Southwestern Louisiana, Lafayette, USA

ABSTRACT: The soil is assumed to be a semi infinite plate made of anisotropic materials, having its upper boundary described by the surface of the ground and extending to infinity in the other directions. For the elastic analysis of anisotropic plates a new displacement function has been developed in function of the location independent variable rather than material characteristic related independent variable. The stress and displacement boundary conditions of the plates as well as the displacements, stresses and strains at any arbitrary point of the plate can be easily evaluated through the displacement function without requiring any additional transformation. The new displacement function is introduced into the software COMVAR for the common boundary element method.

INTRODUCTION

Various approximate methods have been applied to determine the stress distributions in soils. The approximate analytical solutions can be divided into the limit equilibrium approach (Sarma 1979) and the stress-strain approach (Carothers 1920). The Scharz-Christoffel transformation has been suggested in order to map the semi infinite half plane with protrusion into a half plane without protrusion. However, the the suggested procedure has not been carried on (Finn 1960). The stress distribution within and under an elastic embankment has been obtained by transforming the boundary of the soil including the embankment into a semi-infinite half plane (Perloff, et all 1967) and the influence diagrams for the stresses have been given (Perloff 1975). Similar problems have been also investigated by applying a numerical scheme to the differential equation of the conformal mapping function (Silvestri and Tabib 1983). The nonlinear behavior of the soil has been modeled by cubic (Desai 1972) and bicubic (Lakmazaheri 1983) spline functions and the isoparametric formulation of the finite element method is used for the analysis of soil systems. Although the finite element method is a powerful method, the infinite elements which may be used to satisfy the boundary conditions of the soil at the infinity may have some defficiencies (Beer 1984). In the boundary element method, the boundary conditions are satisfied as the integral part of the analysis (Venturini 1983). Another advatage of the boundary element method is that the algorithm of subregions takes automatically the continuity of both pore pressure and mass flux between the adjacent layers into the consideration by means of two parameters only: permeability and thickness (Aramaki et all 1982). The soil has been considered as viscoelastic material and analyzed by combining the finite element and boundary element methods (Kaneko et all 1985). In the common boundary element method, only the common boundaries adjacent to two elements or between an element and the external support are divided into discrete points. Since the accuracy improves by the use of the large size of elements, the region occupied by the body must be divided into minimum number of elements (Ovunc and Owokoniran 1985). A solution has been derived for stresses in an orthotropic quarter plane subjected to various type of loadings and the results may find their application in reinforced-earth, unsupported excavations of anisotropic soils, etc, (Constantinou and Greenleaf 1987).

Herein, the analysis of anisotropic plates by new stress function (Ovunc 1987, 1987) is extended to the analysis by new displacement function. The displacement function appearing in the differential equation is determined in terms of two independent complex functions and their derivatives. They are expressed by means of a single, location related independent variable, rather than a material characteristic independent variable (Lekhnitskii 1981, Constantinou and Gazetas 1986, Constantinou and Greenleaf 1987). Thus, the displacement and stress boundary conditions and the stress and strain states at any arbitrary point of the plates can be easily obtained in a similar way as in the case of isotropic plates (Muskhelishvili 1963). A semi infinite half plane of anisotropic material is considered as an example. The characteristic matrices and the load vectors to be used in the common boundary element method are generated and introduced in the corresponding software COMVAR (Ovunc 1982, 1987).

PROCEDURE OF ANALYSIS

For the anisotropic plates, subjected to in-plane forces, the displacement function is obtained by the integration of the related differential equation without the introduction of an extra parameter.

For an anisotropic material, the stress strain relationships can be written as,

$$\begin{aligned}\sigma_x &= b_{11}\varepsilon_x + b_{12}\varepsilon_y + b_{16}\gamma_{xy}\\ \sigma_y &= b_{21}\varepsilon_x + b_{22}\varepsilon_y + b_{26}\gamma_{xy}\\ \tau_{xy} &= b_{61}\varepsilon_x + b_{62}\varepsilon_y + b_{66}\gamma_{xy}\end{aligned} \qquad (1)$$

where, the coefficients b_{ij} are related to the characteristics of the anisotropic material.

In the complex domain, the displacement function can be written as,

$$D(z,\bar{z}) = u(x,y) + iv(x,y) \qquad (2)$$

where, u and v are the components of the displacement function D, along x and y directions, respectively.

If the strain-displacement relationships,

$$\begin{aligned}\varepsilon_x &= \frac{\partial u}{\partial x} = \tfrac{1}{2}\left[\frac{\partial D}{\partial z} + \frac{\partial D}{\partial \bar{z}} + \frac{\partial \bar{D}}{\partial z} + \frac{\partial \bar{D}}{\partial \bar{z}}\right]\\ \varepsilon_y &= \frac{\partial v}{\partial y} = \tfrac{1}{2}\left[\frac{\partial D}{\partial z} - \frac{\partial D}{\partial \bar{z}} - \frac{\partial \bar{D}}{\partial z} + \frac{\partial \bar{D}}{\partial \bar{z}}\right]\\ \gamma_{xy} &= \frac{\partial u}{\partial y} + \frac{\partial v}{\partial x} = -i\left[\frac{\partial D}{\partial \bar{z}} - \frac{\partial \bar{D}}{\partial z}\right]\end{aligned} \qquad (3)$$

are substituted into the stress-strain relationships (Eq. 1), one has,

$$\begin{aligned}\sigma_x &= \tfrac{1}{2}(b_{11}+b_{12})\left[\frac{\partial D}{\partial z}+\frac{\partial \bar{D}}{\partial \bar{z}}\right]+\tfrac{1}{2}(b_{11}-b_{12})\left[\frac{\partial D}{\partial \bar{z}}+\frac{\partial \bar{D}}{\partial z}\right] -ib_{16}\left[\frac{\partial D}{\partial \bar{z}}-\frac{\partial \bar{D}}{\partial z}\right]\\ \sigma_y &= \tfrac{1}{2}(b_{22}+b_{21})\left[\frac{\partial D}{\partial z}+\frac{\partial \bar{D}}{\partial \bar{z}}\right]-\tfrac{1}{2}(b_{22}-b_{21})\left[\frac{\partial D}{\partial \bar{z}}+\frac{\partial \bar{D}}{\partial z}\right] -ib_{26}\left[\frac{\partial D}{\partial \bar{z}}-\frac{\partial \bar{D}}{\partial z}\right]\\ \tau_{xy} &= (b_{61}+b_{62})\left[\frac{\partial D}{\partial z}+\frac{\partial \bar{D}}{\partial \bar{z}}\right]-(b_{61}-b_{62})\left[\frac{\partial D}{\partial \bar{z}}+\frac{\partial \bar{D}}{\partial z}\right] -ib_{66}\left[\frac{\partial D}{\partial \bar{z}}-\frac{\partial \bar{D}}{\partial z}\right]\end{aligned} \qquad (4)$$

The substitution of the above stress-displacement relationships (Eq. 4) into the equilibrium equations of an infinitesimal element

$$\begin{aligned}\frac{\partial \sigma_x}{\partial x} + \frac{\partial \tau_{xy}}{\partial y} &= 0\\ \frac{\partial \tau_{xy}}{\partial x} + \frac{\partial \sigma_y}{\partial y} &= 0\end{aligned} \qquad (5)$$

yields,

$$b_{11}F_1+(b_{12}+b_{66})F_2-b_{66}F_3+i[(2b_{61}-b_{16})F_4 +2b_{62}F_5+b_{16}F_6] = 0 \qquad (6)$$

$$b_{22}F_5+(b_{21}+b_{66})F_6-b_{66}F_6-i[(2b_{62}-b_{26})F_2 +2b_{61}F_1+b_{26}F_3] = 0 \qquad (7)$$

where,

$$\begin{aligned}F_1 &= \left(\frac{\partial^2}{\partial z^2} + 2\frac{\partial^2}{\partial z\partial \bar{z}} + \frac{\partial^2}{\partial \bar{z}^2}\right)(D + \bar{D})\\ F_2 &= \left(\frac{\partial^2}{\partial z^2} - \frac{\partial^2}{\partial \bar{z}^2}\right)(D - \bar{D})\\ F_3 &= \left(\frac{\partial^2}{\partial z^2} - 2\frac{\partial^2}{\partial z\partial \bar{z}} + \frac{\partial^2}{\partial \bar{z}^2}\right)(D + \bar{D})\\ F_4 &= \left(\frac{\partial^2}{\partial z^2} - \frac{\partial^2}{\partial \bar{z}^2}\right)(D + \bar{D})\\ F_5 &= \left(\frac{\partial^2}{\partial z^2} - 2\frac{\partial^2}{\partial z\partial \bar{z}} + \frac{\partial^2}{\partial \bar{z}^2}\right)(D - \bar{D})\\ F_6 &= \left(\frac{\partial^2}{\partial z^2} + 2\frac{\partial^2}{\partial z\partial \bar{z}} + \frac{\partial^2}{\partial \bar{z}^2}\right)(D - \bar{D})\end{aligned}$$

The above two equations (Eqs. 6 and 7) can be satisfied by selecting the displacement function $D(z,\bar{z})$, as follows,

$$D(z,\bar{z})=\kappa\phi(z)+z\overline{\phi_1(z)}+z^2\overline{\phi_2(z)}+\ .\ .\ +z^k\overline{\phi_k(z)}+ -\overline{\psi(z)}+\bar{z}^3\psi_3(z)+.\ .\ +\bar{z}^k\psi_k(z)+ \qquad (8)$$

The functions ϕ_1, ϕ_2, ϕ_k and ψ_3, ψ_k which appear in the expression of the displacement $D(z,\bar{z})$ (Eq. 8), can be determined in terms of the two main components $\phi(z)$ and $\psi(z)$, by substituting the displacement D, into the stress equilibrium equations (Eq. 5 and 7), and by setting the coefficients of the independent variable z^k to zero.

Thus, one has,

$$\phi_1 = -\phi' + \frac{(\alpha_1'\gamma_1 - \bar{\alpha}_1\gamma_1')}{2(\bar{\alpha}_1'\beta_1 + \alpha_1\bar{\beta}_1')}\psi' \qquad (9)$$

$$\phi_2 = \frac{(\bar{\alpha}_1'\beta_1 - \alpha_1\beta_1')}{(\alpha_1\alpha_1' + \bar{\alpha}_1\bar{\alpha}_1')}\phi'' + \frac{(\beta_1\gamma_1' + \beta_1'\gamma_1)}{2(\alpha_1'\beta_1 + \bar{\alpha}_1\bar{\beta}_1')}\psi'' \qquad (10)$$

and for k, $(k=3,4,\ldots,n)$

$$\phi_{k+2} = -\frac{\bar{\alpha}_1 B_{k+2} + \alpha_1' A_{k+2}}{(k+1)(k+2)(\bar{\alpha}_1\bar{\gamma}_1' + \alpha_1'\bar{\gamma}_1)} \qquad (11)$$

$$\psi_{k+2} = \frac{\gamma_1 B_{k+2} - \gamma_1' A_{k+2}}{(k+1)(k+2)(\bar{\alpha}_1\bar{\gamma}_1' + \alpha_1'\bar{\gamma}_1)} \qquad (12)$$

where,

$$A_{k+2} = \gamma_1\phi_k'' + \alpha_1\psi_k'' + 2(k+1)(\beta_1\phi_{k+1}' + \bar{\beta}_1\psi_{k+1}')$$

$$B_{k+2} = \gamma_1'\phi_k'' + \bar{\alpha}_1'\psi_k'' - 2(k+1)(\bar{\beta}_1'\phi_{k+1}' - \beta_1'\psi_{k+1}')$$

and α_1, β_1, γ_1, α_1', β_1', γ_1' are the parameters related to the characteristics of the anisoptropic materials.

Once the displacement function $D(z,\bar{z})$ is determined, the strains and the stresses at any arbitrary point of the body can be obtained from the strain-displacement and stress-displacement relationships (Eqs. 6, 7).

If some type of soils are assumed to be orthotropic, the characteristics of the orthotropic materials can be determined from those of the anisotropic ones by setting the following characteristics of the anisotropic materials to zero,

$$b_{16} = b_{26} = b_{61} = b_{62} = 0$$

leading to the imaginary parts of the previously defined parameters be also zero,

$$\alpha_1 = \alpha = b_{11} + b_{12} \quad , \quad \alpha_1' = \alpha' = b_{22} + b_{21}$$

$$\beta_1 = \beta = b_{11} + b_{66} \quad , \quad \beta_1' = \beta' = b_{22} + b_{66}$$

$$\gamma_1 = \gamma = b_{11} - (b_{12} + 2b_{66}), \quad \gamma_1' = \gamma' = b_{22} - (b_{21} + 2b_{66})$$

$$\kappa = (\alpha\beta' + \alpha'\beta)/\alpha\alpha''$$

The components $\phi_k(z)$ and $\psi_k(z)$ of the displacement function $D(z,\bar{z})$ (Eq. 8), are obtained from their expressions (Eqs. 9,10 11 and 12), by substituting the characteristics of anisotropic material by those of the orthotropic materials.

IMPLEMENTATION TO CODE COMVAR

The main two components $\phi(z)$ and $\psi(z)$ of the displacement function $D(z,\bar{z})$ can be uniquely determined either from the displacement or stress boundary condition or from both. The determination of these two main components depends on the geometry of the boundary of the body. The bodies whose boundaries are formed by piecewise smooth curves are transformed into bodies with smooth boundaries. The bodies whose boundaries can not be easily transformed into smooth ones are divided into elements.

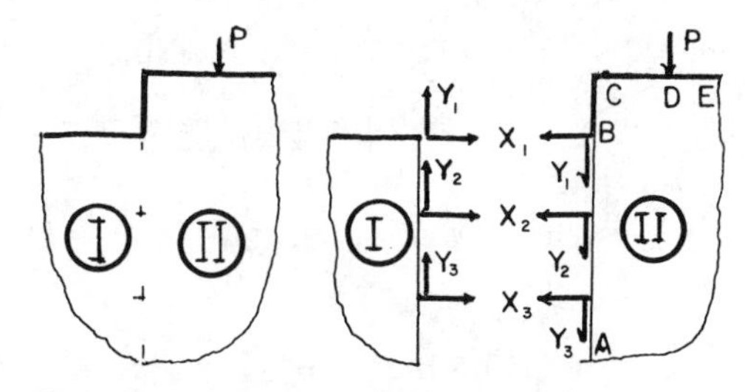

Fig. 1- Tractions at discrete points.

Only the common boundaries adjecent to two elements or between an element and the external support are discretized (Fig. 1). In the discretized elements, the unknown tractions are shown by Z_n,

$$Z_n = X_n + iY_n \qquad (13)$$

Each element is mapped into a unit circle by a transformation function,

$$z = \omega(\zeta) \qquad (14)$$

In order to determine the two main components $\phi(z)$ and $\psi(z)$ of the stress function $U(z,\bar{z})$, the stress boundary condition for the element can be written as, ($\zeta = \sigma$ on Γ)

$$\phi(\sigma) + \omega(\sigma)\frac{\overline{\phi'(\sigma)}}{\overline{\omega'(\sigma)}} + \overline{\psi(\sigma)} + \sum\Big[\frac{(\omega(\sigma))^k}{(\omega'(\sigma))^{k+2}}(a_k\bar{\psi}_1^{(k-2)} + b_k\bar{\phi}_1^{(k-2)} + k\frac{(\overline{\omega(\sigma)})^k}{(\omega'(\sigma))^{k-1}}(a\,\psi_1^{(k-1)} + b_k\phi_1^{(k-1)}))\Big] = F(\sigma) \qquad (15)$$

where the functions $\phi_1^{(k)}$ and $\psi_1^{(k)}$ are defined as the k'th differentials of the funtions $\phi(\zeta)$, $\psi(\zeta)$, and $F(\sigma)$ is the summation of the unknown boundary tractions Z_n.

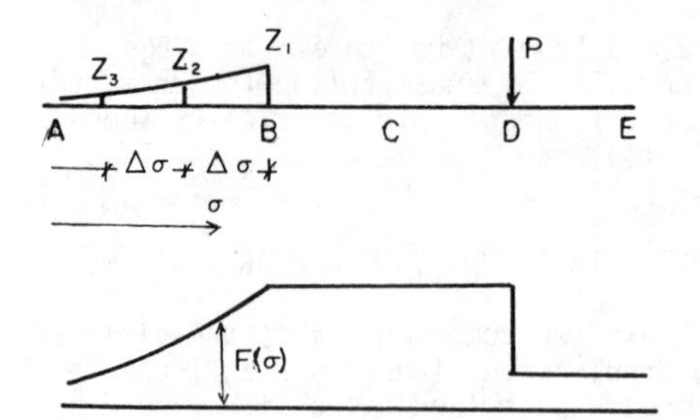

Fig. 2 - Tractions and summation of tractions

For instance, by considering the element II (Fig.1), the value of the function $F(\sigma)$ at a discrete point j along the common boundary, can be written as,

$$F_j=F_{j-1}+0.5(Z_j+Z_{j+1})\Delta\sigma + \text{a function of P}$$

and at an intermediate point between j and j+1,

$$F(\sigma)=F_j+0.5(\sigma-j\Delta\sigma)(Z_{j+1}(\frac{\sigma}{\Delta\sigma}-j)+Z_j(2-\frac{\sigma}{\Delta\sigma}+j))$$
$$+ \text{a function of P}$$

An expansion of the function $F(\sigma)$ into the Fourier series yields,

$$F(\sigma) = \sum(A_n + B_n)\sigma^n$$

where,

$$A_n = \{g\}^T\{Z\}$$

and B_n is a known coefficient in terms of the external loads.

The two main components $\phi(\zeta)$ and $\psi(\zeta)$ are assumed in series form. The unkown coefficients of these two functions are evaluated by substituting them into the stress boundary condition. The substitution yields to,

$$\phi(\zeta) = \{\alpha(\zeta)\}^T([L]\{Z\}+\{D\}) + \phi_0(\zeta) \quad (16)$$

$$\psi(\zeta) = \{\alpha(\zeta)\}^T([N]\{Z\}+\{E\}) + \psi_0(\zeta) \quad (17)$$

where the only unknown vector is {Z} which is the vector of the unknown tractions at the discrete points along the common boundary. $\phi_0(\zeta)$ and $\psi_0(\zeta)$ are the holomorphic functions within the domain.

The displacement boundary condition provides the characteristic matrices and the load vectors for the elements. By writing the displament boundary condition in the mapped region, one has,

$$D\Big|_{\zeta=\sigma}=\frac{\omega'(\sigma)}{\sigma}[a\xi(\sigma)+\frac{\omega(\sigma)}{\sigma}\overline{\xi'(\sigma)}+. \ .] =g(\sigma) \quad (18)$$

where, $g(\sigma)$ is the function of the unknown displacements at the discrete points along the common boundaries. By substituting the coordinates σ_j of the discrete points j in the equation of the displacements along the boundary (Eq. 18), one has,

$$\{g\} = [\Gamma]\{Z\} + \{\Gamma_0\} \quad (20)$$

or,

$$\{Z\} = [K]\{g\} + \{Z_0\} \quad (21)$$

where for the element,

$[K]=[\Gamma]^{-1}$ is the characteristic or the stiffness matrix,

$\{Z_0\}=[\Gamma]^{-1}\{\Gamma_0\}$ is the load vector,

{g} is the vector of the displacements at the discrete points,

{Z} is the vector of tractions at the discrete points.

It can be easily noticed that the characteristic matrices of the elements depend on the length of the of the common boundary of the element compared to its total perimeter.

The characteristic matrix and the load vector for the system are obtained from those of the elements (Eq. 21), by considering the equality of the displacements at the discrete points along the common bounaries. Once the unknown tractions all along the common boundaries are determined from the equilibrium of the system, their corresponding parts for the elements {Z}, are substituted in the two main components of the displacement function for the individual elements. Thus the displacement function for each element is obtained.

If the expressions of the two main components are written in simpler form,

$$\phi_k = a_k\phi^{(k)} + b_k\psi^{(k)} \quad (22)$$

$$\psi_k = c_k\phi^{(k)} + d_k\psi^{(k)} \quad (23)$$

where the exponent on a function $(\ )^{(k)}$, represents k'th differenciation of the function ( ) with respect to z,

the displacement function $D(z,\bar{z})$ becomes,

$$)=\kappa\phi - z(\bar{\phi}' - b_1\bar{\psi}') + \sum_2 z^k (a_k \bar{\phi}^{(k)} + b_k \bar{\psi}^{(k)}) - \bar{\psi} + \sum_2 \bar{z}^k (c_k \phi^{(k)} + d_k \psi^{(k)}) \quad (24)$$

All the coefficients a_k, b_k, c_k, d_k which appear in the expressions of the displacement, stresses are all related to the characteristics of the anisotropic or orthotropic materials and independent of the geometry and the loading of the system. Therefore, for a given material they can be calculated once and they can be stored in a database for their future uses. Moreover, when the characteristics of the anisotropic materials tend to those of the isotropic materials, the displacement function of anisotropic materials tends to that of isotropic materials.

For the bodies with piece-wise smooth boundaries, like: quarter infinite plane, infinite and semi-infinite strips, rectangle, triangles, etc., their boundaries are conformally mapped into a semi-infinite plane by Schwarz-Christoffel transformation then into a circle by a bilinear transformation.

REFERENCES

Aramaki, G., Kuroki, T. and Onishi, K. 1982. Consolidation analysis by boundary element method, Boundary Element Methods in Engineering, Springer Verlag.

Beer, G., 1984. Infinite elements. Int. J. Engineering Computations, 1,3:290-292.

Carothers, S.D., 1920. Direct determination of stresses, Proc. Royal Soc. London Series A. 49:100.

Constantinou, M.C. and Gazetas, G. 1986. Loading of anisotropic quarter plane, ASCE, 112, EM10:1021-1040.

Constantinou, M.C. and Greenleaf, R.S.1987. Stress systems in orthotropic quarter plane, ASCE, 113, EM11:1720-1738.

Desai, S.C. 1972. Nonlinear analysis using spline function, ASCE, 98, SM9:967.

Finn, W.D.L. 1960. Stresses in soil masses under various boundary conditions, Ph. D. thesis, Univ. of Washington.

Kaneko, N., Shinokawa, T., Yoshida, N. and Kawahara, M. 1985. Numerical analysis of viscoelasticity using combined finite and boundary element methods, 5'th Int. Conf. Numerical Methods in Geomechanics, 1049-1054.

Lakmazaheri, S. 1983. Nonlinear analysis of anisotropic materials by finite element method, M. S. thesis, Univ. Southwestern Louisiana.

Muskhelishvili, N.I. 1963. Some basic problems of the mathematical theory of elasticity, P. Noordhoff, Groningen.

Ovunc, P.A. 1982. Application of complex variable theory to the boundary element method, Boundary Element Methods in Engineering, Springer Verlag, 500-514.

Ovunc, B.A. Owokoniran, O. 1985.Analysis of plates by common boundary element method, Abst., SIAM Fall Meeting.

Ovunc, B.A.1987. Anisotropic infinite plane with circular hole, Abst., ASCE/EMD Specialty Conf., State Univ. of New York at Buffalo: 242.

Ovunc, B.A.1987. Analysis of plates subjected to in-plane or transversal loads, Boundary Elements XI, Computational Mechanics Publications, Springer Verlag, 2: 81-96.

Perloff, W.H., Baladi, G.K. and Harr, M.E. 1967. Stress distribution within and under long embankments, Highway Research Record, 181:12-40.

Perloff, W.H. 1975. Pressure distributions and settlements, in Foundation Engineering Handbook, Edited by H.F. Winterkorn and H.Y. Fang, Van Nostrand Reynold Comp., : 751.

Sarma, S.K. 1979. Stability analysis of embankments and slops, ASCE, 105, GT12: 1511-1524.

Silvestri, V. and Tabib, C. 1983. Exact determination of gravity stress in finite elastic slops, Canadian Geotech. J., 20:47-60.

Venturini, W.S. 1983. Boundary Element Methods in Geomechanics, Springer Verlag.

Numerical Methods in Geomechanics (Innsbruck 1988), Swoboda (ed.)
© 1988 Balkema, Rotterdam. ISBN 90 6191 809 X

BE-BEM and IFE-BEM: Two models for unbounded problems in geomechanics

R.Pöttler
ILF Consulting Engineers, Innsbruck, Austria
W.Haas
TDV Pircher, Graz, Austria

ABSTRACT: By coupling infinite elements and beam elements a new model for the analysis of underground openings is presented. It is compared to the coupled beam-boundary element model. The field of application of these models are unbounded problems in geomechanics, where the rock behavior can be described by the theory of elasticity.

1 INTRODUCTION

By coupling boundary elements and beam elements an efficient computational model (BE-BEM) was created requiring neither sophisticated computer installations nor excessive computation time in order to determine the behavior of the rock mass and of the support measures in a realistic way. /1/ /2/ /3/ The only shortcoming of BE-BEM is the fully populated stiffness matrix. Therefore the solving of equations, in particular for non-linear problems, is time consuming. To avoid this problem beam elements are coupled with infinite elements (IFE-BEM). In the present paper the results of this type of analysis are compared to the results of the boundary elements. A practical example for a tunnel project was finally calculated with both methods: BE-BEM and IFE-BEM.

2 THEORY

2.1 Beam element - boundary element model

The theory of the BE-BEM is extensively described in /1/. For coupling the boundary elements (BE) with the beam elements linear 2-node-BEs were used. The stiffness matrix of the BE region was symmetrized for computational convenience and efficiency and led to the well known matrix

$$[K]^{BE} = \frac{1}{2}\left[\left[[M][C]\right]^T + [M][C]\right] \qquad (1)$$

The difference between the results obtained by using the asymmetric matrix for the BE- region and the symmetric matrix is negligible for problems in the engineering practice. /4/

A 2-node beam element with 6 degrees of freedom representing the shotcrete lining was implemented. The transverse displacement of the beam is defined in terms of the usual third degree Hermitian polynomials associated with transverse displacements and rotations at the nodes. The longitudinal displacement of the beam is defined by a linear shape function. In the stiffness matrix of the beam element

$$[K]^{BEM} = \int_V [B]^T [D] [B] \, dV \qquad (2)$$

the non-linear material behavior is considered in the matrix $[D]$, which is a function of the actual stress-strain level of the cross section.

Coupling of the beam elements with the boundary element region is done by means of two-node constraint elements. /5/ The constraint element takes the highly non-linear behavior of the interaction between shotcrete and rock mass into account.

2.2 Infinite element - beam element model

From their first publication /6/, infinite elements have attracted researchers and engineers interested in solving problems of "unbounded" domains. /7/ /8/ /9//10/ Different investigators have concentrated their attention on questions of convergence, accuracy, and on various applications from static to dynamic or radiation problems, thus making also literature on infinite elements almost an "unbounded" domain. Also the question of coupling infinite elements versus boundary elements to "normal" finite elements has created an enormous and often emotional and amusing discussion. /11/ /12/

The common characteristic of infinite elements is that special assumptions for the variation of the unknown variable in the infinite direction are made. If the solution is approximated by

$$\phi = \sum N_i \ \phi_i \tag{3}$$

then the N_i is composed of the standard shape function (e.g. isoparametric quadratic) multiplied by some decay function (e.g. proportional to $1/r^n$).

In this study, so-called "mapped" infinite elements have been used. /13/ In this element the infinite domain is mapped onto a local system $-1 \leq \xi \leq +1$. The mapping for the local coordinate is

$$\xi = 1 - \frac{2a}{r} \tag{4}$$

i.e. proportional to $1/r$.

This is a good approximation for the far-field analytic solution of many physical problems.

The element has been used frequently and successfully in the context of finite element analyses. /14/

For the IFE-BEM the same beam elements and constraint elements as mentioned above are used.

3 TEST OF ACCURACY

The coupling of BE resp. IFE and beam elements was done by non-conforming shape functions (see Table 1). The influence on the results in the BE-BEM was shown by /1//2/ /15/. The influence of the non-conformity of the shape functions of the IFE and the beam element on the results of IFE-BEM can be calculated with the same formulas as presented by /1/ /3/. This problem is not dealt with in the present paper.

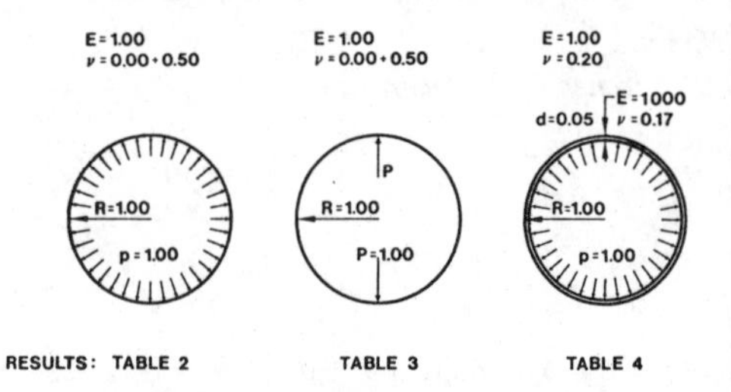

Fig. 1. Test of accuracy: problems studied

Three problems were studied: an unlined circular opening under circumferential uniform pressure and under two single loads as well as a lined opening under circumferential uniform pressure (see Fig. 1). Once the problem is solved by using BE, once by using IFE. The results are shown in Table 2 (uniform load), Table 3 (nodal loads) and Table 4 (lined opening). From Table 2 it becomes obvious that the results (radial displacements) of the IFE are generally more accurate than those of the BE. There is also considerable influence of the Poisson's ratio ν on the results by using BE, which does not happen by using IFE. Upon subdividing the tunnel cross section into 16 elements, the BE's provide satisfactory results for practical engineering purposes, if the Poisson's ratio is in a realistic range of up to $\nu \leq 0.45$.

The results of the nodal loading in Table 3 show that neither the BE nor the IFE can handle this problem in a proper way, up to 32 subdivisions of the opening. The reason seems to be that the used linear shape function of the BE and IFE cannot cope with the singularity of the nodal loads (Fig. 2). The "accuracy" of the result is influenced by the subdivision and also by the Poisson's ratio using BE and IFE. The "exact" solution was obtained by using a FE model with 8-noded isoparametric elements.

Table 4 shows that the accuracy of BE-BEM and of IFE-BEM is in the same range. The results of BE-BEM are slightly more accurate.

Table 1. Shape functions for the Beam Element, Infinite Element and Boundary Element

SHAPEFUNCTION	BEAM-ELEMENT		BE/IFE-ELEMENT		
	NODE 1	NODE 2	NODE 1	NODE 2	
N_1, N_4	$1-\xi$	ξ	$1-\xi$	ξ	$u=1$ CONFORM
N_2, N_5	$1-3\xi^2+2\xi^3$	$3\xi^2-2\xi^3$	$1-\xi$	ξ	$w=1$ NONCONFORM
N_3, N_6	$L(\xi-2\xi^2+\xi^3)$	$L(-\xi^2+\xi^3)$	—	—	$\varphi=1$

Table 2. Test of accuracy: circular openings under internal pressure. Radial displacement Δr, error %.

ν	n	4 Δr	4 %	8 Δr	8 %	16 Δr	16 %	32 Δr	32 %	EXACT
0.000	IFE	0.750	25.0	0.946	5.4	0.987	1.3	0.997	0.3	1.000
	BE	0.704	29.6	0.906	9.4	0.975	2.5	0.994	0.6	
0.200	IFE	0.900	25.0	1.135	5.4	1.184	1.3	1.196	0.3	1.200
	BE	0.788	34.3	1.055	12.1	1.160	3.3	1.190	0.8	
0.400	IFE	1.050	25.0	1.324	5.4	1.382	1.3	1.395	0.4	1.400
	BE	0.742	47.0	1.081	22.8	1.300	7.1	1.374	1.9	
0.450	IFE	1.088	25.0	1.371	5.4	1.431	1.3	1.445	0.3	1.450
	BE	0.661	54.4	0.962	33.7	1.272	12.3	1.401	3.4	
0.490	IFE	1.118	25.0	1.409	5.4	1.471	1.3	1.485	0.3	1.490
	BE	0.539	63.8	0.581	61.0	0.928	37.7	1.282	14.0	
0.499	IFE	1.124	25.0	1.417	5.5	1.480	1.3	1.494	0.3	1.499
	BE	0.500	66.6	0.353	76.5	0.338	77.5	0.612	59.2	

Table 3. Test of accuracy: circular opening under two single loads. Radial displacement at load point Δr, error %

ν	n	4 Δr	4 %	8 Δr	8 %	16 Δr	16 %	32 Δr	32 %	"EXACT"
0.000	IFE	1.125	70.0	1.547	58.8	1.685	55.1	1.722	54.2	3.756
	BE	1.199	68.1	1.768	52.9	2.266	39.7	2.273	39.5	
0.200	IFE	1.125	70.5	1.535	59.7	1.685	55.8	1.727	54.7	3.809
	BE	1.289	66.2	1.826	52.1	2.289	39.9	2.378	37.6	
0.400	IFE	0.875	77.3	1.222	68.3	1.480	61.6	1.581	59.0	3.854
	BE	1.218	68.4	1.683	56.3	2.094	45.7	2.294	40.5	
0.450	IFE	0.741	80.0	0.994	74.3	1.313	66.0	1.489	61.5	3.865
	BE	1.144	70.4	1.570	59.4	1.985	48.6	2.238	42.1	
0.490	IFE	0.602	84.5	0.643	83.4	0.874	77.4	1.206	68.9	3.874
	BE	1.047	73.0	1.367	64.7	1.778	54.1	2.175	43.9	
0.499	IFE	0.567	85.4	0.517	86.7	0.542	86.0	0.675	82.6	3.875
	BE	1.018	73.7	1.269	67.3	1.562	59.7	1.970	49.2	

Table 4: Lined circular opening under circumferential pressure. Radial displacement Δr; exact Δr = 0.200.

n	4	8	16	32
BE-BEM	0.141	0.184	0.196	0.199
IFE-BEM	0.139	0.181	0.193	0.196

n .... number of elements

4 APPLICATION OF BE-BEM AND IFE-BEM

By means of the following example of a stability analysis for the heading with an invert arch installed immediately after excavation, the results of a BE-BEM calculation are compared with those of an IFE-BEM calculation. The assumed load configuration is verified by measurements which have been performed during tunnel construction in the course of an express railway line in Germany. /16/ In Fig. 3 the

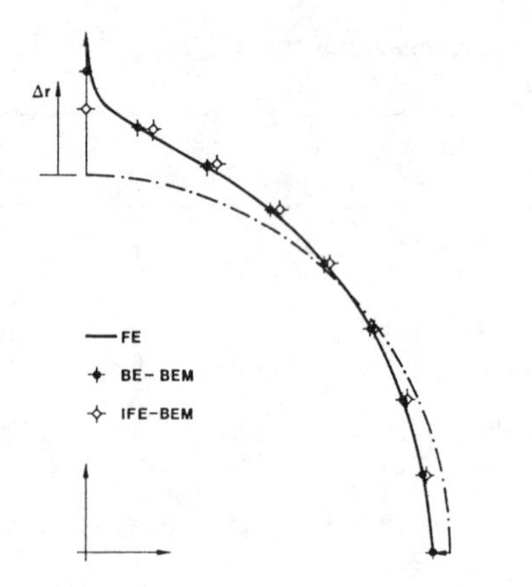

Fig. 2. Test of accuracy: singularity of load point

load configuration and the dimensions of the system are shown.

The results can be seen in Fig. 4. The results of the BE-BEM calculation in the first line of fig. 4 represent the exact solution. This was proved by refining the mesh. The results of the IFE-BEM model depend very much on the placements of the poles. Some of the results obtained are completely unrealistic (see Fig. 4, lines 2 - 5).

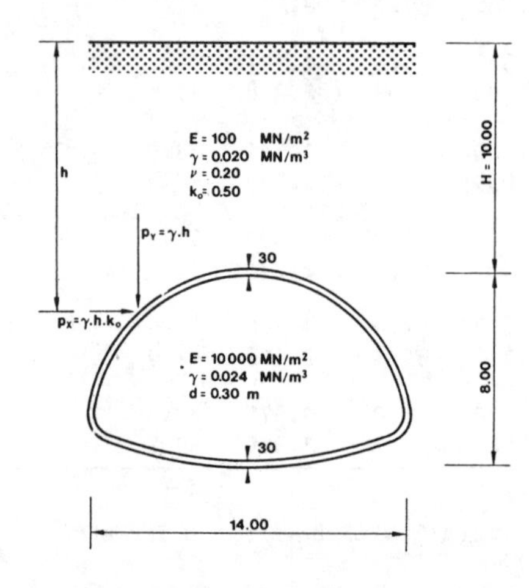

Fig. 3. Application of BE-BEM and IFE-BEM: geometry and load configurations of the system

5 COMPARISON IFE-BEM TO BE-BEM

IFE-BEM and BE-BEM are both applicable to unbounded problems in geomechanics with some restrictions. Both models cannot handle concentrated loads. Relatively better results are obtained by using BE-BEM for this load configuration, whereas for uniform loads and simple circular geometries better results are obtained by using IFE-BEM. The results of the IFE-BEM are the better, the closer the stress-strain distribution is proportional to the analytic solution 1/r.

Using 2-noded-IFEs for simulating the rock mass and coupling them directly with the beam elements (without the interaction of a constraint element), there is one important advantage in respect to using BE: the band width of the equation system does not increase. The IFE-BEM is therefore a real alternative to the coefficient of subgrade reaction method (CSRM). The IFE-BEM neither possesses the shortcomings of the CSRM /15/ nor the shortcoming of the BE-BEM with the fully populated stiffness matrix which results in a time-consuming solution of the equation system.

In case of a symmetric system, the advantages of the IFE-BEM even increase: the computation time for calculating and assembling the stiffness matrices of a symmetric system is only half of an unsymmetric one. Using BE-BEM, the computation time is not reduced to 50 %: the influence of part I on part II must be calculated. The total matrix is

$$[K]=[K_{II}]-[K_{III}]\ [K_{IIII}]^{-1}[K_{III}] \tag{5}$$

By using IFE, only the first part of the matrix of equation (5) is to be calculated. For complex geometric situations, however, the IFE-BEM does not give any accurate answers. As a main drawback, the results depend on the placement of poles for the decay function, which can only be determined by empirical rules.

6 CONCLUSION

The IFE-BEM represents a model to describe unbounded problems in geomechanics, generally similar to BE-BEM. If there are concentrated loads, e.g. a group of prestressed anchors, or complex geometries, it can be assumed that the BE-BEM calculation will lead to more accurate results. It remains to be investigated in a further step if the behavior of the IFE-BEM in these cases can be improved by introducing higher order polynomial terms in the decay function of the infinite element.

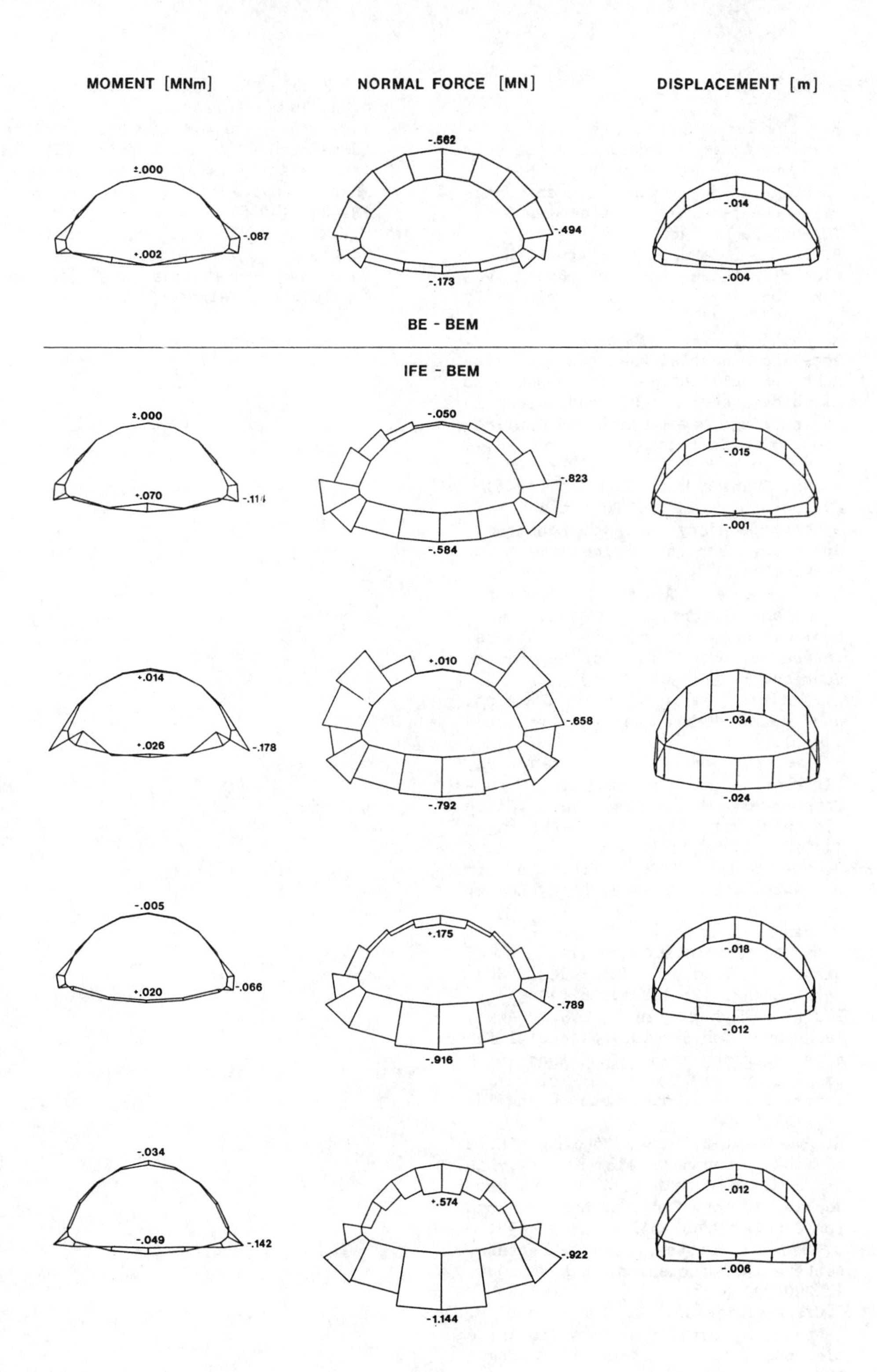

Fig. 4. Application of BE-BEM and IFE- BEM: results of calculation

REFERENCES

/1/ R. Pöttler, Gekoppeltes Stabwerk-Boundary-Element-Modell zur Berechnung von Tunnelbauten unter Berücksichtigung des nichtlinearen Materialverhaltens von Stahlbeton, Diss. TU Innsbruck, 1986

/2/ R. Pöttler, Gekoppelte Stabwerk-Randelement-Methode zur Berechnung von Tunnelbauten, Bauingenieur 62, 109-116 (1987)

/3/ R. Pöttler, G.A. Swoboda and G. Beer, Non-linear coupled beam-boundary element (BE-BEM) analysis for tunnels on microcomputers, In: Microcomputers in Engineering: Development and Application of Software, Ed. by B.A. Schrefler and R.W. Lemis, Prineridge Press, Swansea U.K.: 161-176 (1986)

/4/ C.A. Brebbia, I.C.F. Telles and L.C. Wrobel, Boundary element techniques, Springer: Berlin, Heidelberg, New York, 1984

/5/ M.G. Katona, A simple contact-friction interface element with applications to buried culverts, Intern. J. Numer. Analyt. Methods in Geomech. 7: 371-304 (1984)

/6/ P. Bettess, Infinite elements, Int. Jou. Num. Meth. Engng 11: 53-64 (1977).

/7/ P. Bettess and O.C. Zienkiewicz, Diffraction and refraction of surface waves using finite and infinite elements, Int. Jou. Num. Meth. Engng 11: 1271-1290 (1977)

/8/ G. Beer and L. Meek, Infinite domain elements, Int. Jou. Num. Meth. Engng: 43-52 (1981)

/9/ F. Medina and R.L. Taylor, Finite element techniques for problems of unbounded domains, Int. Jou. Num. Meth. Engng. 19: 1209-1226 (1983)

/10/ T.J.R. Hughes and I.E. Akin, Techniques for developing special finite element shape functions with particular reference to singularities, Int. Jou. Num. Meth. Engng 15: 733-751 (1980)

/11/ G. Beer and L. Meek, Coupled finite element - boundary element analysis of infinite domain problems in geomechanics, Int. Conf. on Num. Methods for Coupled Problems, Swansea, 1981

/12/ G. Beer, P. Bettess, and J. Bettess, Letters and Discussion, Eng. Comput. 1: 290-292 (1984)

/13/ O.C. Zienkiewicz, C. Emson and P. Bettess, A novel boundary infinite element, Int. Jou. Num. Meth. Engng 19: 393-404 (1983)

/14/ W. Haas, MISES3 User Manual, Rev. 8.06, 1987

/15/ R. Pöttler and G.A. Swoboda, Coupled Beam-Boundary Element Model (BE-BEM) for the Analysis of Underground Openings, Computers and Geotechnics 2: 239-256 (1986)

/16/ M. John, R. Pöttler and G. Heissel, Vergleich verschiedener Rechenmethoden untereinander und mit Meßergebnissen, Felsbau 3: 21-26 (1985)

Numerical Methods in Geomechanics (Innsbruck 1988), Swoboda (ed.)
© 1988 Balkema, Rotterdam. ISBN 90 6191 809 X

Back analysis method of elastoplastic BEM in strain space

Zhang De-cheng & Gao Xiao-wei
Ningxia University, Yinchuan, People's Republic of China
Zheng Yingren
Air Force College of Engineering, Xian, People's Republic of China

ABSTRACT: In this paper the back analysis of elastoplastic BEM is presented on the basis of the plastic theory in strain space. The method can be applied to back analysis of underground cavity. And the initial stresses in rock mass and Young's modulus can be calculated from the displacements measured along the side of cavity. In the paper the approximate middle-point stiffness method to solve non-linear equations is proposed, with which convergence can be obtained more quickly.

1 INTRODUCTION

In the computation of underground cavity the initial stresses in rock mass and Young's modulus as the basic parameters, are commonly difficult to be determined accurately. The measured displacements of cavity side in situ is liable to be obtained, therefore, in present calculation the displacements of cavity side are increasingly regarded as the calculating parameters instead of ground stresses and material constants of cavity, but elastoplastic BEM is rarely used at present for back analysis from displacement. The authors presented the back analysis of elastoplastic BEM based on the plastic theory in stress space in reference (Zheng Yingren, etc. 1986). In this paper the back analysis of elastoplastic BEM in strain space is proposed, and the calculating results are showing that the back analysis of displacement in strain space has the advantage over that in stress space.

2 ELASTOPLASTIC CONSTITUTIVE EQUATIONS FORMULATED IN STRAIN SPACE

Stress increment $\dot{\sigma}_{ij}$ and elastic strain increment $\dot{\varepsilon}^{e}_{ij}$ can be expressed as follows

$$\dot{\sigma}_{ij} = C_{ijkl}\dot{\varepsilon}^{e}_{kl} = C_{ijkl}(\dot{\varepsilon}_{kl} - \dot{\varepsilon}^{p}_{kl})$$
$$= C_{ijkl}\dot{\varepsilon}_{kl} - \mathring{\sigma}^{p}_{ij} \quad (2\text{-}1)$$

where the over notation e represents elastic components, p, plastic components, and repeat index represents summation.

$$C_{ijkl} = \frac{E}{1+\nu}\left[\delta_{ik}\delta_{jl} + \frac{\nu}{1-2\nu}\delta_{ij}\delta_{kl}\right] \quad (2\text{-}2)$$

$$\mathring{\sigma}^{p}_{ij} = C_{ijkl}\dot{\varepsilon}^{p}_{kl} \quad (2\text{-}3)$$

Let the yield function formulated in strain space be

$$Q(\varepsilon_{ij}, \varepsilon^{p}_{ij}, w^{p}) = g(J_{1e}, \sqrt{J'_{2e}}) - K(w^{p}) = 0 \quad (2\text{-}4)$$

where J_{1e} and J_{2e}' are respectively the first invariant of elastic strain and the second invariant of elastic deviation strain.

The normal flow law in strain space is

$$\mathring{\sigma}^{p}_{ij} = \dot{\lambda}\frac{\partial Q}{\partial \varepsilon_{ij}} \quad (2\text{-}5)$$

According to the above formula, the following generalized expression of elastoplastic constitutive equations can be derived

$$\dot{\sigma}_{ij} = \left(C_{ijkl} - \frac{\alpha(\hat{g})}{A}\frac{\partial Q}{\partial \varepsilon_{ij}}\frac{\partial Q}{\partial \varepsilon_{kl}}\right)\dot{\varepsilon}_{kl} \quad (2\text{-}6)$$

where $\alpha(\hat{g})$ is loading or unloading

factor

$$\alpha(\hat{g}) = \begin{cases} 1 & \text{when } g>0 \text{ (loading)} \\ 0 & \text{when } g<0 \text{ (unloading)} \end{cases} \quad (2\text{-}7)$$

$$\hat{g} = \frac{\partial g}{\partial \varepsilon_{ij}} \dot{\varepsilon}_{ij} \quad (2\text{-}8)$$

$$A = \frac{1}{E}\left[3\beta_1^2 (1-2\nu) + \frac{1+\nu}{2}\beta_2^2\right] + H \quad (2\text{-}9)$$

$$\beta_1 = \frac{\partial Q}{\partial J_{1e}}, \quad \beta_2 = \frac{\partial Q}{\partial \sqrt{J'_{2e}}} \quad (2\text{-}10)$$

$$H = \frac{dK(w^p)}{dw^p}(\beta_1 J_{1e} + \beta_2 \sqrt{J'_{2e}}) \quad (2\text{-}11)$$

where $dK(w^p)/dw^p$ is determined through experiments.

$$\frac{\partial Q}{\partial \varepsilon_{ij}} \frac{\partial Q}{\partial \varepsilon_{kl}} = \beta_1^2 \delta_{ij}\delta_{kl} + \frac{\beta_1 \beta_2}{2\sqrt{J'_{2e}}}(e^e_{ij}\delta_{kl} + e^e_{kl}\delta_{ij}) + \frac{\beta_2^2}{4J'_{2e}} e^e_{ij} e^e_{kl} \quad (2\text{-}12)$$

where e^e_{ij} is deviator of elastic strain. (2-6) can be expressed in matrix form

$$[\dot{\sigma}] = ([D] - [D_p])[\dot{\varepsilon}] = [D_{ep}][\dot{\varepsilon}] \quad (2\text{-}13)$$

$[D]$, $[D_p]$ and $[D_{ep}]$ are elastic, plastic and elastoplastic matrixes. The elements in the matrixes are obtained by taking respectively the indexes ij and kl in C_{ijkl} and $\frac{\alpha(\hat{g})}{A}\cdot\frac{\partial Q}{\partial \varepsilon_{ij}}\frac{\partial Q}{\partial \varepsilon_{kl}}$ as the positions of "row" and "column", and their values can be found according to the order of 11, 22, 33, 23, 31, 12.

We can obtain elastoplastic matrix in the problem of plane strain

$$[D_{ep}] = \frac{E}{(1+\nu)(1-2\nu)}\begin{bmatrix} 1-\nu & \nu & 0 \\ \nu & 1-\nu & 0 \\ 0 & 0 & 1-2\nu \\ \nu & \nu & 0 \end{bmatrix} - \frac{\alpha(\hat{g})}{A}\begin{bmatrix} S_1^2 & S_1 S_2 & S_1 S_3 \\ S_2 S_1 & S_2^2 & S_2 S_3 \\ S_3 S_1 & S_3 S_2 & S_3^2 \\ S_4 S_1 & S_4 S_2 & S_4 S_3 \end{bmatrix}$$

$$S_1 = \beta_1 + \frac{\beta_2 e^e_{11}}{2\sqrt{J'_{2e}}}, \quad S_2 = \beta_1 + \frac{\beta_2 e^e_{22}}{2\sqrt{J'_{2e}}}$$

$$S_3 = \frac{\beta_2 e^e_{12}}{2\sqrt{J'_{2e}}}, \quad S_4 = \beta_1 + \frac{\beta_2 e^e_{33}}{2\sqrt{J'_{2e}}}$$

The yield function (2-4) formulated in strain space can be seen in Reference (Zheng Yingren, Chu Jian, 1986)

3 METHOD OF APPROXIMATE MIDDLE-POINT STIFFNESS

The relation between stress increment and strain increment in elastoplastic problem is

$$[\dot{\sigma}] = [D_{ep}][\dot{\varepsilon}] \quad (3\text{-}1)$$

where $[D_{ep}]$ is the function of strain ε_{ij}. For $[\dot{\varepsilon}]$ of the (i+1)th load increment if ε_{ij} at a certain point between point i and point (i+1) can make $[D_{ep}]$ equal to the slope of line connecting i with i+1 (Fig.1),

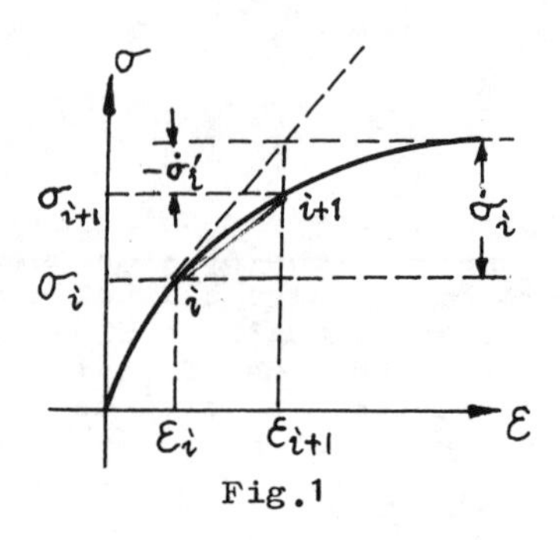

Fig.1

then $[\dot{\sigma}]$ calculated from (3-1) would be accurate. But the position of point i is unknown, it can only be replaced by the known approximate point. Obviously, it is better to replace it by using middle point $i+\frac{1}{2}$. In this paper, the middle-point stiffness will be calculated with the help of Taylor's series.

Expanding $[D_{ep}]$ into Taylor's series at point i and taking the linear term as the approximate one, we obtain

$$[D_{ep}] = [D_{ep}]_i + \frac{\partial [D_{ep}]_i}{\partial \varepsilon}[\delta\varepsilon] = [D_{ep}]_i + [D_{ep}]'_i \quad (3\text{-}2)$$

where $[D_{ep}]'_i = \frac{\partial [D_{ep}]_i}{\partial \varepsilon}[\delta\varepsilon]$

$$[\delta\varepsilon] = \frac{1}{2}[\dot{\varepsilon}] \quad (3\text{-}3)$$

substitution of (3-2) into (3-1) gives

$$[\dot{\sigma}] = [\dot{\sigma}]_i + [\dot{\sigma}]'_i \quad (3\text{-}4)$$

where $[\dot{\sigma}]_i = [D_{ep}]_i[\dot{\varepsilon}]$ is stress increment calculated bv stiffness at point i, and $[\dot{\sigma}]'_i = [D_{ep}]'_i[\dot{\varepsilon}]$ is the correction value of stress, which can be written as the following component form

$$\dot{\sigma}'_{ij} = \frac{1}{2}\frac{\partial D^{ep}_{ijkl}}{\partial \varepsilon_{rs}} \dot{\varepsilon}_{rs}\dot{\varepsilon}_{kl} \quad (3\text{-}5)$$

where

$$D^{ep}_{ijkl}=C_{ijkl}-\frac{\alpha(\hat{g})}{A}\frac{\partial Q}{\partial \varepsilon_{ij}}\frac{\partial Q}{\partial \varepsilon_{kl}} \quad (3\text{-}6)$$

It can be seen that as soon as the yield function Q is known the correction value of stress can be obtained. In general differentiating (3-5) is very complicated. But it is easy when Q is the linear function of J_{1e} and $\sqrt{J'_{2e}}$. In that case, from (2-2) and (2-12) we can get

$$\frac{\partial D^{ep}_{ijkl}}{\partial \varepsilon_{rs}}=-\frac{\beta_1\beta_2}{2A\sqrt{J'_{2e}}}(\delta_{kl}\delta_{ir}\delta_{js}+\delta_{ij}\cdot\delta_{rk}\delta_{sl}-\frac{2}{3}\delta_{ij}\delta_{kl}\delta_{rs})+\frac{\beta_1\beta_2 e_{rs}}{4A(J'_{2e})^{3/2}}\cdot(e_{ij}\delta_{kl}+e_{kl}\delta_{ij})+$$

$$\frac{\beta_2^2 e_{ij}e_{kl}e_{rs}}{4A(J'_{2e})^2}-\frac{\beta_2^2}{4AJ'_{2e}}\left[\left(\delta_{ir}\delta_{js}-\frac{1}{3}\delta_{ij}\delta_{rs}\right)\cdot e_{kl}+\left(\delta_{rk}\delta_{sl}-\frac{1}{3}\delta_{rs}\delta_{kl}\right)e_{ij}\right] \quad (3\text{-}7)$$

Substitution of (3-7) into (3-5) gives

$$\sigma'_{ij}=\frac{\alpha(\hat{g})G\beta_2}{4A}\left[\frac{\beta_2\dot{w}^2}{I_2'^{\,2}}\tau_{ij}+\frac{\beta_1\dot{w}}{(I_2')^{3/2}}\cdot\left(\dot{w}\delta_{ij}+\dot{J}_1\tau_{ij}\right)-\frac{\beta_2}{I_2'}(\dot{w}\dot{c}_{ij}+2\dot{J}'_2\tau_{ij})-\frac{2\beta_1}{\sqrt{I_2'}}(2\dot{J}'_2\delta_{ij}+\dot{J}_1\dot{c}_{ij})\right] \quad (3\text{-}8)$$

where $\dot{J}_1=\dot{\varepsilon}_{ii}$, $\dot{J}'_2=\frac{1}{2}\dot{c}_{rs}\dot{e}_{rs}$, $I'_2=(2G)^2\cdot J'_{2e}$

$\dot{w}=\tau_{ij}\dot{c}_{ij}=\tau_{ij}\dot{e}_{ij}$, $\dot{c}_{ij}=\dot{\varepsilon}_{ij}-\frac{1}{3}\dot{\varepsilon}_{rr}\delta_{ij}$

After gaining (3-8) the computation of increment is carried out by (3-4).

4 BACK ANALYSIS OF ELASTOPLASTIC BEM

The basis formulations of elasto-plastic boundary element method expressed in matrix form is

$$\left.\begin{aligned}&[H][\dot{U}]=[G][\dot{T}]+[D][\dot{\sigma}^p]\\&[\dot{UT}]=[GT][\dot{T}]+[HT][\dot{U}]\\&\quad+[DT][\dot{\sigma}^p]\end{aligned}\right\} \quad (4\text{-}1)$$

where $[\dot{U}]$ and $[\dot{T}]$ stand for the increasements of displacement and traction at boundary nodes respectively, $[\dot{UT}]$ for displacement increasement at nodes in internal cells, $[\dot{\sigma}^p]$ for plastic stress in sidecells, [H], [G], [GT] and [HT] for influence coefficients as known, [D] and [DT] for nonlinear coefficients.

Before excavation of underground opening, rock mass is in static balance. The displacements U'_i will be produced at points in rock mass due to excavation. And they are regarded as what is caused by releasing stresses σ'_{ij}. After excavation of cavity re-distributing of the stresses in rock mass takes place. And when the new balance is achieved the stresses and displacements in rock mass are as follows

$$\left.\begin{aligned}\sigma_{ij}&=\sigma^0_{ij}+\sigma'_{ij}\\U_i&=U^0_i+U'_i\end{aligned}\right\} \quad (4\text{-}2)$$

where U^0_i are initial displacements in rock mass, which needn't be considered in computation. The principle of back analysis from displacement is that the ground stresses σ^0_{ij} and Young's modulus E are calculated by the releasing displacements U'_i measured in site along cavity side.

In underground cavity supported by anchor-shotcrete the releasing distribution forces are written as

$$[T]=[F]-[P] \quad (4\text{-}3)$$

where the distribution forces [F] of shotcrete acted upon rock mass are computed by the following formula

$$-[MT][F]=[KT][UB] \quad (4\text{-}4)$$

where [MT] is transformation matrix between distribution force and equivalent node force, [KT] is stiffness matrix of shotcrete, [UB] is displacement on interface between rock mass and shotcrete. [P] in the above formula is boundary traction corresponding to ground stress σ^0_{ij}, its matrix form is as follows

$$[P]=[M][\sigma^0] \quad (4\text{-}5)$$

$$[M]=[[n(q^1)]\cdots[n(q^N)]]^T \quad (4\text{-}6)$$

where N stands for the number of boundary nodes, q^i for position of boundary nodes, n for direction cosine, and $[\sigma^0]$ for ground stress.

supposing that the displacement produced along cavity side before shotcreting is expressed in [UO], the total displacement may be expressed in increasement as follows

$$[\dot{U}]=[\dot{UB}]+[\dot{UO}] \quad (4\text{-}7)$$

From (4-5), (4-4), (4-3) and (4-1), and by solving the gained simultaneous equations, we can obtain

$$\begin{aligned}[\dot{P}]&=[X][\dot{\sigma}^P]+[\dot{Y}]\\ [\dot{UT}]&=[XT][\dot{\sigma}^P]+[\dot{YT}]\end{aligned} \quad (4\text{-}8)$$

where

$$\left.\begin{aligned}&[X]=[G]^{-1}[D]\\ &[XT]=[DT]-[GT][G]^{-1}[D]\\ &[\dot{Y}]=-[MT]^{-1}[KT][\dot{UB}]\\ &\qquad -[G]^{-1}[H][\dot{U}]\\ &[\dot{YT}]=([HT]+[GT][G]^{-1}\\ &\qquad \cdot[H])[\dot{U}]\end{aligned}\right\} \quad (4\text{-}9)$$

Noticing that [G] , [D] and [DT] all include 1/E, the coefficients in (4-8) can be generalized and rewritten as

$$\left.\begin{aligned}[\dot{P}^*]&=[X][\dot{\sigma}^{*P}]+[\dot{Y}^*]\\ [\dot{UT}]&=[XT^*][\dot{\sigma}^{*P}]+[\dot{YT}]\end{aligned}\right\} \quad (4\text{-}10)_1$$

where $[\dot{P}^*]=\frac{1}{E}[\dot{P}]$, $[\dot{Y}^*]=\frac{1}{E}[\dot{Y}]$

$[XP^*]=E[XT]$, $[\dot{\sigma}^*]=\frac{1}{E}[\dot{\sigma}^P]$

Correspondingly, (4-5) can be written as

$$[\dot{P}^*]=[M][\dot{\sigma}^{*0}],\ [\dot{\sigma}^{*0}]=\frac{1}{E}[\dot{\sigma}^0] \quad (4\text{-}10)_2$$

5 OPTIMIZATION COMPUTATION AND INCREASEMENT ITERATION

From (4-9), it can be seen that if [UO] and [UB] is known, $[\dot{Y}]$ and $[\dot{YT}]$ can be calculated by $[\dot{UO}]$ and $[\dot{UB}]$ of each grade load. From $(4\text{-}10)_1$ [P*] and [UT] can be calculated through iteration procedure, then from $(4\text{-}10)_2$ $[\sigma^{*0}]$ will be found. But the number of elements in [P*] is commonly more than that of ground stress, so the optimization procedure must be carried out. The authors have derived the following formula from the extreme value principle of function, i.e.

$$[\sigma^{*0}]=([M]^T[M])^{-1}([M]^T[P^*]) \quad (5\text{-}1)$$

If E is known the ground stress $[\sigma^0]$ will be solved from (4-10). Conversely, supposing that the vertical component σ_{22} of ground stress is equal to the rock gravity, that is, $\sigma_{22}=\gamma H$ (γ — unit weight of rock mass; H — depth situated under ground surface), then E can be found from $(4\text{-}10)_2$

$$E=\gamma H/\sigma_{22}^{*0} \quad (5\text{-}2)$$

Once E is found the ground stresses in other directions will be obtained from $(4\text{-}10)_2$.

The strain increasements in cells in iteration can be found from $[\dot{UT}]$ at various nodes by means of finite element formula:

$$[\dot{\varepsilon}]=[B][\dot{UT}] \quad (5\text{-}3)$$

where [B] stands for strain matrix of element.

The plastic stress increasements can be found from the following formula

$$\dot{\sigma}_{ij}^P=\dot{\sigma}_{ij}^e-\dot{\sigma}_{ij} \quad (5\text{-}4)$$

where $\dot{\sigma}_{ij}^e$ and $\dot{\sigma}_{ij}$ are calculated by elastic and elastoplastic constitutive equations respectively with the same $\dot{\varepsilon}_{ij}$. The detail procedures of iteration on each grade load are as follows:

1. At first iteration let $[\dot{\sigma}^{*P}]=0$ or let $[\dot{Y}]=[\dot{YT}]=0$. The initial ground stress increasement $[\dot{\sigma}^0]$ and displacement $[\dot{UT}]$ at various nodes are calculated from (4-10) and (5-1).

2. Element strain increasement $\dot{\varepsilon}_{ij}$ is calculated with $[\dot{UT}]$ from (5-3), then, elastic stress increasement $\dot{\sigma}_{ij}^e$ is calculated with the help of the generalized Hook's law and the shorthand is $\dot{\sigma}^e$.

3. The real stress increasements $\dot{\sigma}_{ij}$ is calculated with $\dot{\varepsilon}_{ij}$. In (2-4) let the value of $K(w^P)$ in current grade load be K_0 and the stress before applying this grade load be σ, and do such a computation that $\sigma'=\sigma+\dot{\sigma}^e$. Let $f_1=g(\sigma)$, $f_2=g(\sigma')$, and discuss the three cases as fo-

llows

a. If $f_1 < K_0$, $f_2 < K_0$, then $\dot{\sigma}_{ij} = \dot{\sigma}^e_{ij}$;
b. If $f_1 \geqslant K_0$, $f_2 \geqslant K_0$, then $\dot{\sigma}_{ij}$ is calculated by σ from (3-4);
c. If $f_1 < K_0$, $f_2 \geqslant K_0$, then this element is the crossing element, the following computation is needed

$$\Delta f = f(\dot{\sigma}^e),\quad m = |(f_2 - K_0)/\Delta f|,$$
$$\sigma^{ep} = \sigma + (1 - m)\dot{\sigma}^e$$

$\dot{\sigma}^{ep}$ is calculated by σ^{ep} from (3-4), so

$$\dot{\sigma}_{ij} = (1 - m)\,\dot{\sigma}^e_{ij} + m\,\dot{\sigma}^{ep}_{ij}$$

4. Calculate the initial stress increasements $\dot{\sigma}^p_{ij}$ from (5-4), then examine whether they are convergent.

5. Ground stresses, real stress and displacement are computed

$$\sigma^0_{ij} = \sigma^0_{ij} + \dot{\sigma}^0_{ij} \qquad \sigma_{ij} = \sigma_{ij} + \dot{\sigma}_{ij}$$

$$UT_i = UT_i + \dot{UT}_i$$

6. The computation should be continued for the next cell beginning from the second procedure until the computations of all cells are accomplished.

7. Make a new iteration after returning to the first procedure. The iteration computation should be carried out till the results of all cells are convergent, then a new load is applied, and the above mentioned seven procedures are repeated until all loads are added.

6 EXAMPLE

Now, we take a certain underground opening as an example, its geometric dimension is shown in Fig.2

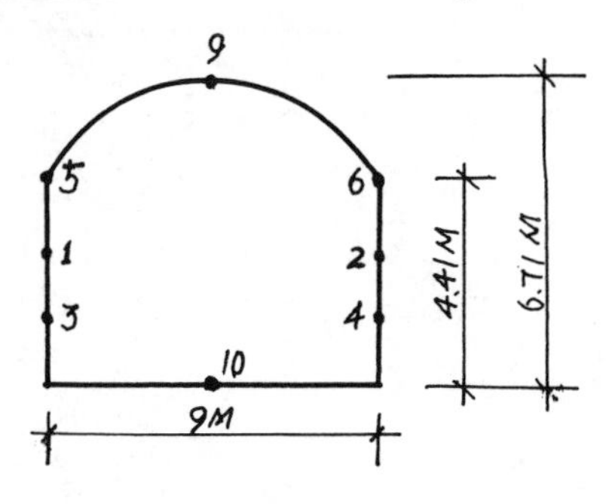

Fig.2

$C = 0.104\text{MPa}$, $\varphi = 31°$, $\nu = 0.34$
$\gamma = 0.0027\text{kg/m}^3$, $H = 42.5\text{m}$

For checking the correctness of the process elaborated in the paper, we have calculated displacements with following initial stresses and Young's modulus firstly

$$\sigma^0_{xx} = -0.95\text{MPa},\quad \sigma^0_{yy} = -1.15\text{MPa}$$
$$\sigma^0_{xy} = 0,\quad E = 980\text{MPa}$$

Results Calculated

Points in Fig.2	1-2	3-4	5-6	9-10
Relative Displacements (mm)	18.5	16.5	17.0	30.0

Then the calculation of back analysis is made according to the displacements in above table.

1. Having known E = 980MPa, we have calculated ground stresses as follows

σ^0_{xx}	σ^0_{yy}	σ^0_{xy}
-0.9501208	-1.1501050	0.0000205

2. If $\sigma^0_{yy} = -1.15$, and initial Young's modulus E_0=1000MPa, we have calculated E and other ground stresses

E Iteration Time	E	σ^0_{xx}	σ^0_{xy}
1	979.958	-0.950020	0.00001
2	979.959	-0.950019	0.00001

3. If $\sigma^0_{yy} = -1.15$, and E_0=900MPa, the results computed are as follows

E Iteration Time	E	σ^0_{xx}	σ^0_{xy}
1	979.948	-0.950020	0.000011
2	979.95	-0.950019	0.000011

Thus it can be seen that the accuracy of back analysis is higher, and no matter how the initial value E_0 is taken, the gained results all converge to an identical value. It shows that the calculating results are stable.

REFERENCES

Zheng Yingren, Chang Decheng, Gao Xiaowei, 1986, BEM of elastoplastic back analysis, proceeding of the third yearly session of Civil Engineering Association of China

Zheng Yingren, Chu Jian, Xu Zhenhong, 1986, strain space formulation of the elastoplastic theory and its Finite Element Implementation, Computers and Geotechnics 2, 373-388

N. Shimizll and S. Sakurai, 1983, Application of Boundary Element Method for back analysis associated with tunnelling problems, Boundary Element, Hiroshima, Japen

Telles, 1983, The Boundary Element Method applied to inelastic problems, Springer

Numerical Methods in Geomechanics (Innsbruck 1988), Swoboda (ed.)
© 1988 Balkema, Rotterdam. ISBN 90 6191 809 X

Analysis of stress and slide of underground openings by BEM with nonlinear slip elements

Shen Jiaying, Dai Miaoling & Zhang Yang
Hohai University, Nanjing, People's Republic of China

ABSTRACT: Boundary element method (BEM) for linear elasticity is unfasible to the analyses of stress and slide of underground openings with faults and clay courses nearby, because of their notable nonlinearity. The difficulty is overcome by introducing nonlinear slip elements in BEM for linear elasticity. The analyses of the underground openings in Laxiwa engineering in China is put forward and some interesting results have been obtained.

1 INTRODUCTION

In rock engineering, in the foundation of rock bodies there often exist weak structural faces like clay courses and joints, which may play a controlling function in the strength and stability of structures. With the finite element method, the deformation features are modelled with joint elements, Considering the advantage of the boundary element method that the elements need only be divided on the boundary, some people have tried to use it on weak structural faces. Crouch has used discontinuous displacement method to model the deformation features of structural faces. There are still some people who have applied direct boundary element method to calculate the plane strain for weak structural face. In this paper, the 2-dimensional and 3-dimensional direct boundary element methods are considered together for the weak structural face and the discussion of the weak structural face is divided into elastic contact, relative slide, and pulling-open. Computer programs have been compiled and some examples have been introduced.

2 CRITERION OF FAILURE FOR WEAK STRUCTURAL FACES

According to the characteristics of the weak structural face, its failure can be divided into two types: the pulling-open in the normal direction and the relative slide in the tangential direction. At the same time, two yield function are used:

$$F_n = t_n - R_T \quad \text{for the normal direction} \qquad (1)$$

$$F_s = |t_s| - (C - f t_n) \quad \text{for the tangential direction} \qquad (2)$$

where R_T is the tensile strength of the structural face, t_n is the compression strength for the normal direction, t_s is the shear stress, C and f are cohesion and friction factor respectively.

After slide and pulling-open of the structural face:

$$F_n = t_n \qquad (3)$$

$$F_s = |t_s| - (C_r - f t_n) \qquad (4)$$

where C_r is the cohesion for shear failure.

The working states of the structural are divided into 3 kind:

1 $F_n < 0$ and $F_s < 0$ for elastic contact

2 $F_n < 0$ and $F_s \geq 0$ for shear failure

3 $F_n \geq 0$ for tensile failure

3 DIRECT BOUNDARY ELEMENT METHOD WITH WEAK ROCK FORMATION ELEMENT

From the physical equation for deformation it can be known that the boundary integral equation for static linear elastic problems has the same form in the whole and local rectangular coordinate systems, i.e.

$$\bar{C}_{ki}(P)\,\bar{u}_i(P) = \int_S \bar{t}_i(Q)\,\bar{u}^*_{ki}(P,Q)\,dS(Q) - \int_S \bar{t}^*_{ki}(P,Q)\,\bar{u}_i(Q)\,dS(Q) \qquad (5)$$

where $\bar{u}_i$ and $\bar{t}_i$ are the displacement and

planar force of the boundary element respectively, for 2-dimensional problems, i=1 indicating tangential direction, i=2 indicating normal direction; for 3-dimensional problems, i=1, 2 indicating tangential direction, i=3 indicating normal direction.

From Eq. (5) we can obtain

$$\sum_{Q=n_1^r}^{n_2^r} \overline{H}_{ki}^{PQ} \, \overline{u}_i^{(Q)} = \sum_{Q=n_1^r}^{n_2^r} \overline{G}_{ki}^{PQ} \, \overline{t}_i^{(Q)} \tag{6}$$

where n_1^r and n_2^r are the numbers of the starting and ending nodes in the region r.

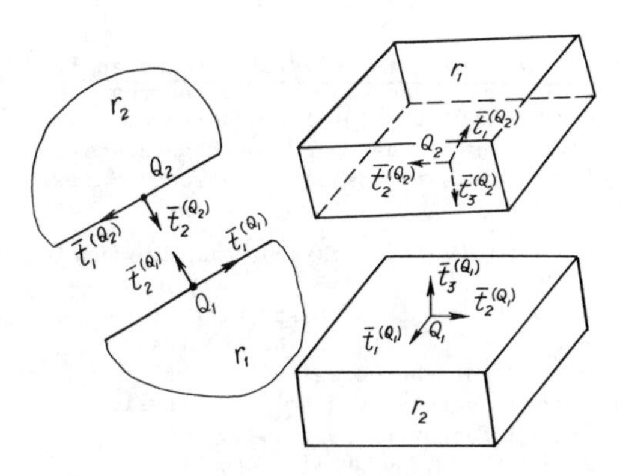

Fig. 1

Now consider the 2 regions r_1 and r_2 shown in Fig. 1, between which there is a weak structural face. Q_1 and Q_2 are two opposite points. In the following, the displacements of Q_1 and Q_2 and the relation between the displacements and the face will be discussed.

1. For $F_n < 0$ and $F_s < 0$, and elastic contact between Q_1 and Q_2 ; we have

$$\overline{t}_i^{(Q_1)} = - K_i \, (\overline{u}_i^{(Q_1)} + \overline{u}_i^{(Q_2)}) \tag{7}$$

$$\overline{t}_i^{(Q_2)} = \overline{t}_i^{(Q_1)} \tag{8}$$

where K_i is the coefficient of rigidity.

2. For $F_n < 0$ but $F_s \geqslant 0$, shear failure will occur between Q_1 and Q_2 .

For 2-dimensional porblems:

The tangential planar forces at Q_1 and Q_2 obey the friction law, i.e.

$$|\overline{t}_1^{(Q_1)}| = C_r - f_r \, \overline{t}_2^{(Q_1)} \tag{9}$$

$$|\overline{t}_2^{(Q_2)}| = C_r - f_r \, \overline{t}_2^{(Q_2)} \tag{10}$$

The coordinative condition for the normal planar forces and displacements at Q_1 and Q_2 are

$$\overline{t}_2^{(Q_1)} = \overline{t}_2^{(Q_2)} \tag{11}$$

$$\overline{u}_2^{(Q_1)} = - \overline{u}_2^{(Q_2)} \tag{12}$$

For 3-dimensional problems:

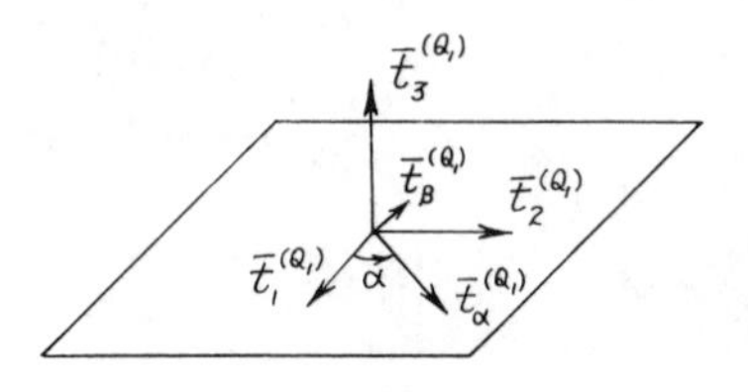

Fig. 2

The direction of $\overline{t}_\alpha^{(Q_1)}$ in Fig.2 is the direction of the maximum shear stress obtained from the former trial calculation, i.e.the direction of the resultant force of $\overline{t}_1^{(Q_1)}$and $\overline{t}_2^{(Q_1)}$, which can be taken as the direction of slide at Q_1 for this trial calculation. According to the friction law, we have the following approximate expression:

$$\overline{t}_\alpha^{(Q_1)} = C_r - f_r \, \overline{t}_3^{(Q_1)} \tag{13}$$

Similarly, for Q_2 we have

$$\overline{t}_\alpha^{(Q_2)} = C_r - f_r \, \overline{t}_3^{(Q_2)} \tag{14}$$

Besides, the coordinative conditions for the planar forces and displacements normal to the slide direction at Q_1 and Q_2 are

$$\overline{t}_\beta^{(Q_1)} = \overline{t}_\beta^{(Q_2)} \tag{15}$$

$$\overline{u}_\beta^{(Q_1)} = - \overline{u}_\beta^{(Q_2)} \tag{16}$$

The normal planar forces at Q_1 and Q_2 are the same as in the case of elastic contact, satisfying

$$\overline{t}_3^{(Q_1)} = - K_3 \, (\overline{u}_3^{(Q_1)} + \overline{u}_3^{(Q_2)}) \tag{17}$$

$$\overline{t}_3^{(Q_2)} = - K_3 \, (\overline{u}_3^{(Q_1)} + \overline{u}_3^{(Q_2)}) \tag{18}$$

if the difference between the direction of the maximum shear stress from this trial calculation and that from the former trial calculation is smaller than a present value, the direction of the maximum shear stress from this trial calculation can be taken as the direction of slide, otherwise, trial calculation should be carried on.

3. When $F_n \geqslant 0$, Q_1 and Q_2 are pulled open, and the planar force conditions for the interface are

$$\bar{t}_i^{(Q_1)} = 0 \quad (19)$$

$$\bar{t}_i^{(Q_2)} = 0 \quad (20)$$

For J-dimensional problems, there are J unknown displacements and J unknown planar forces at Q_1 and Q_2 respectively. However, when the boundary element method is used to find the solutions, there can only be J unknowns, therefore, the J additional unknowns have to be expressed by means of the connection conditions at Q_1 and Q_2. Take 3-dimensional slide for example, the unknowns at Q_1 can be taken as $\bar{U}_\alpha^{(Q_1)}$, $\bar{U}_\beta^{(Q_1)}$, and $\bar{U}_3^{(Q_1)}$, the unknowns at Q_2 as $\bar{U}_\alpha^{(Q_2)}$, $\bar{t}_\beta^{(Q_2)}$, and $\bar{U}_3^{(Q_2)}$, while $\bar{t}_\alpha^{(Q_1)}$, $\bar{t}_\beta^{(Q_1)}$, and $\bar{t}_3^{(Q_1)}$ at Q_1 can be solved from Eqs. (13), (15) and (17) respectively, and $\bar{t}_\alpha^{(Q_2)}$, $\bar{U}_\beta^{(Q_2)}$, and $\bar{t}_3^{(Q_2)}$ can be found from Eqs. (14), (16), and (18) respectively, other cases can be treated the same way. In order to have a unified form, let the unknown at Q_1 be $X_m^{Q_1}$, and the unknown at Q_2 be $X_m^{Q_2}$, then, the displacement and planar force at Q_1 and Q_2 under various connection conditions can be expressed as

$$\bar{u}_i^{(Q_1)} = C_{im}^{r_1 Q_1} x_m^{Q_1} + C_{im}^{r_2 Q_2} x_m^{Q_2} + \delta\, \bar{u}_i^{(Q_1)} \quad (21)$$

$$\bar{t}_i^{(Q_1)} = d_{im}^{r_1 Q_1} x_m^{Q_1} + d_{im}^{r_1 Q_2} x_m^{Q_2} + \delta\, \bar{t}_i^{(Q_1)} \quad (22)$$

$$\bar{u}_i^{(Q_2)} = C_{im}^{r_2 Q_1} x_m^{Q_1} + C_{im}^{r_2 Q_2} x_m^{Q_2} + \delta\, \bar{u}_1^{(Q_2)} \quad (23)$$

$$\bar{t}_i^{(Q_2)} = d_{im}^{r_2 Q_1} x_m^{Q_1} + d_{im}^{r_2 Q_2} x_m^{Q_2} + \delta\, \bar{t}_i^{(Q_2)} \quad (24)$$

where $C_{im}^{r_1 Q_1}$, $d_{im}^{r_1 Q_1}$, $\delta\, \bar{U}_i^{(Q_1)}$ etc. are coefficients related with the connection conditions at Q_1 and Q_2, and can be derived from Eqs. (7)-(20) according to corresponding connection conditions.

For any point P_1 in the region r_1, from Eq. (8) we have

$$\text{------} + \bar{H}_{ki}^{P_1 Q_1}(C_{im}^{r_1 Q_1} x_m^{Q_1} + C_{im}^{r_1 Q_2} x_m^{Q_2} + \delta\, \bar{u}_i^{(Q_1)}) + \cdots\cdots$$

$$= \text{------} + \bar{G}_{ki}^{P_1 Q_1}(d_{im}^{r_1 Q_1} x_m^{Q_1} + d_{im}^{r_1 Q_2} x_m^{Q_2} + \delta\, \bar{t}_i^{(Q_1)}) + \text{------} \quad (25)$$

let

$$\left.\begin{aligned} \bar{A}_{km}^{P_1 Q_1} &= \bar{H}_{ki}^{P_1 Q_1} C_{im}^{r_1 Q_1} - \bar{G}_{ki}^{P_1 Q_1} d_{im}^{r_1 Q_1} \\ \bar{A}_{km}^{P_1 Q_2} &= \bar{H}_{ki}^{P_1 Q_1} C_{im}^{r_1 Q_2} - \bar{G}_{ki}^{P_1 Q_1} d_{im}^{r_1 Q_2} \\ \bar{b}_k^{P_1 Q_1} &= \bar{G}_{ki}^{P_1 Q_1} \delta\, \bar{t}_i^{(Q_1)} - \bar{H}_{ki}^{P_1 Q_1} \delta\, \bar{u}_i^{(Q_1)} \end{aligned}\right\} \quad (26)$$

Eq. (25) can be written as

$$\cdots + \bar{A}_{km}^{P_1 Q_1} x_m^{Q_1} + \cdots + \bar{A}_{km}^{P_1 Q_2} x_m^{Q_2} + \cdots = \cdots + b_k^{P_1 Q_1} + \cdots \quad (27)$$

After the same treatment for Q we can obtain

$$\sum_{Q=1}^{n_p} \bar{A}_{km}^{PQ} x_m^{Q} = \sum_{Q=1}^{n_p} b_k^{PQ}, \quad P = 1 \text{------} n_p \quad (28)$$

where n_p is the total number of the boundary nodes.

After X_m^Q is obtained, the planar forces and displacements at Q_1 and Q_2 can found from Eqs (21)-(24).

4 EXAMPLE

4.1 Original data of the underground openings in Laxiwa engineering

The underground openings in Laxiwa engineering consist of the left opening (the main power-house), the middle opening (the main transformer room), and the right opening (tail race regulation opening), as shown in Fig.3(a). The surrounding rock is granite, the elastic modulus being 320,000 kg/cm^2, and the Poisson's ratio 0.21. On the top of the openings there is a weak rock formation Hf-8.

In calculation, the problem was considered as a plane strain problem, the initial stress field in the section normal to the opening axis was given by the designing institute: $\sigma_x = -129.098$ kg/cm^2, $\sigma_y = -152.274$ kg/cm^2, $\tau_{xy} = -17.132$ kg/cm^2.

4.2 The influence of the variation of the elastic modulus of the rock formation on the stress and displacement of the surrounding rock

Since the weak rock is located on the top of the openings, whether the weak rock should be treated or not and to which extent the treatment should be depend mainly on the influence of the elastic modulus of the weak rock on the stress and displacement of the surrounding rock. Calculations were performed for 4 cases: the elastic modulus for the weak rock being taken as 40000 kg/cm^2, 120000 kg/cm^2 and 200000 kg/cm^2 the Poisson's ratio as 0.27, and both the elastic modulus and the Poisson's ratio being taken as the same as those for granite (320000kg/cm^2, 0.21 respectively).

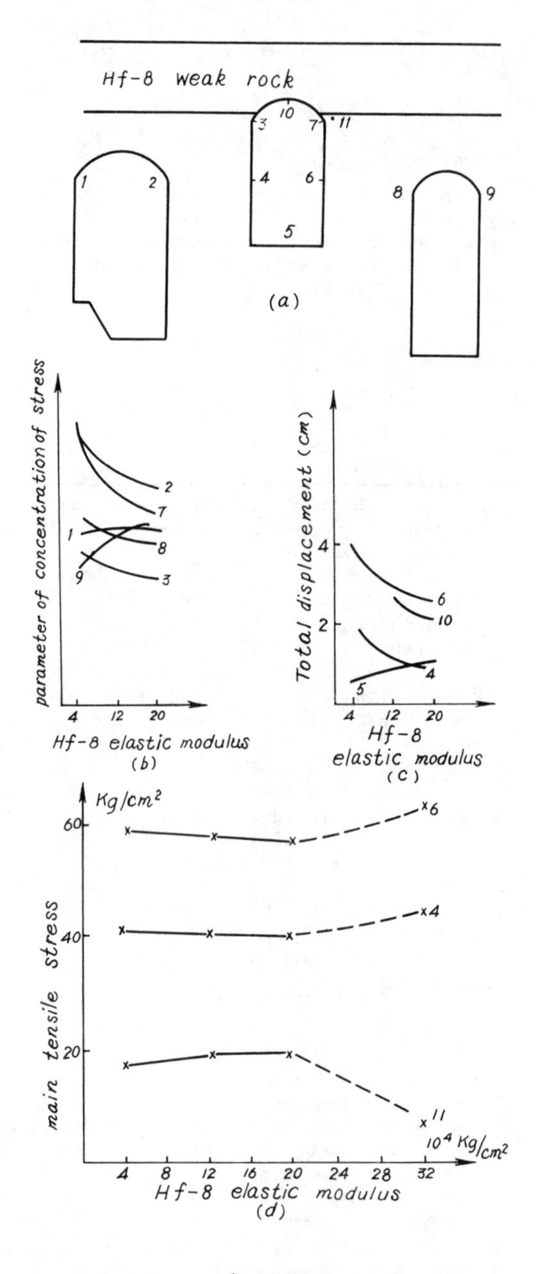

Fig. 3

Calculations show that the smaller the elastic modulus of the weak rock the more serious the concentration of the stress at the abutment and the angular flange (see Fig.3(b) with the exception of the case for the right abutment of the right opening. The right abutment of the middle opening is the most sensitive to the change of the elastic modulus of the weak rock: when the elastic modulus is 40000 kg/cm², the circumferential stress at that location is 1.36 times that without the weak rock, but when the elastic modulus is 120000 kg/cm², only 1.14 times. The smaller the elastic modulus of the weak rock, the larger the displacement of the arch crown and the middle part of the side wall (see Fig.3(c)). The arch crown of the middle opening is most seriously affected by the elastic modulus of the weak rock: when the elastic modulus is 40000 kg/cm² the displacement at that place is 3.2 times that without the weak rock, but when the elastic modulus is 120000kg/cm² only 1.58 times. The change of the elastic modulus of the weak rock does not have much effect on the tensile stress (see Fig.3(d)). When the Poisson's ratio is taken as 0.27 the smaller the elastic modulus, the larger the tensile stress. It can also be seen from the figure that the Poisson's ratio of the weak rock has some effect on the tensile stress, the maximum tensile stress occurs when the ratio is small (0.21) and the elastic modulus is large (320000 kg/cm²)

From the above analyses we can reach the conclusion: if the elastic modulus of the weak rock reaches 120000 kg/cm², the weak rock will not have much influence on the stress and displacement of the surrounding rock.

5 CONCLUSION

Calculations show that the boundary element method with nonlinear slip elements is suitable for the calculation and analysis of the stability and stress of the underground engineering with weak rock. Compared with the finite element method, this method can save a large amount of work for data preparation and result processing.

REFERENCES

Crouch, S. L. and Stanfield, A. M., Boundary Element Methods, London, George Allen and Unwin, 1983.

Crotty, J. M. and Wardle, L. J., Boundary Integral Analysis of Piecewise Homogeneous Media with Structural Discontinuities, Int. J. Rock Mech. Min. Sci. and Geomech. Abstr. Vol. 22, No.6, Printed in Great Britain, 1985.

Numerical Methods in Geomechanics (Innsbruck 1988), Swoboda (ed.)
 ISBN 90 6191 809 X

Elastoplastic boundary element analysis of well-uniform self-supporting structure

Yonghe Li
Xi'an Mining Institute, Shaanxi, People's Republic of China

ABSTRACT: This paper first deals with the supporting idea of the well-uniform self supporting structure (W.U.S.S.S.). Then, the application of the elastoplastic boundary element method is presented for the solution of such problem by means of general pseudo-body method. Here, the numerical method involves the coupled problem of three different yield materials, i.e. borehole, rock bolt and rock. Finally, several examples are given to outline the efficiency and applicability of this numerical method and this supporting structure.

1 INTRODUCTION

What the form, technique and theory of support are used will have been given much attention and research by scholars at home and abroad because more plastic deformation, expansion and ground swelling take place in the excavation. In order to solve this complex problem, in the paper author, according to the idea of " self supporting rock ring " given by Roest [5], discusses a more new type of the supporting structure. This structure is formed by coupling of two supporting techniques such as borehole and rock bolt so that this structure is named "W.U.S.S.S. as shown in Fig. 4. Although the elastoplastic analysis of this problem is very difficult, it is easy to be solved by the elastoplastic boundary element method given in this paper.

2 NUMERICAL METHOD OF B.E.M.

Basic Equation and General Body Force

According to elastoplastic theory, the extended form of Navier equation without body force in terms of displacement rate can be expressed as

$$\dot{U}_{j,ll} + \frac{V}{1-2V}\dot{U}_{l,lj} = -\dot{f}^p_j / G \quad (1)$$

where

$$\dot{f}^p_j = -2G\left(\dot{\varepsilon}^p_{ij,i} + \frac{V}{1-2V}\dot{\varepsilon}^p_{kkj}\right)$$

From eqn. (1), it can be seen that the contribution of plastic strain rate $\dot{\varepsilon}^p_{ij}$ to the displacement is similar to the application of the elastic body force. Thus, $\dot{f}^p_j$ is defined as a general pseudo-body corresponding to $\dot{\varepsilon}^p_{ij}$. Here, "General" means three different yield materials of borehole, rock bolt and rock.

For eqn. (1), the solution of the total displacement rate may be composed of two parts such as

$$\dot{U}_i = \dot{U}^e_i + \dot{U}^p_i \quad (2)$$

Where $\dot{U}^e_i$ corresponding to elastic displacements caused by external load, and $\dot{U}^p_i$ is plastic displacement caused by plastic strain $\dot{\varepsilon}^p_{ij}$.

Similarly, for total stress rate, yield

$$\dot{\sigma}_{ij} = \dot{\sigma}^e_{ij} + \dot{\sigma}^p_{ij} \quad (3)$$

Where $\dot{\sigma}^e_{ij}$ is elastic stress caused by external load, and $\dot{\sigma}^p_{ij}$ is plastic stress and it can generally be written in the following form

$$\dot{\sigma}^p_{ij}(x) = \int_{\Omega^p} \phi_{ijkl}(x-x')\,\dot{\varepsilon}^p_{kl}(x')\,d\Omega^p \qquad x' \in \Omega^p \quad (4)$$

in which Ω^p denotes plastic zone, ϕ is called intergrand influence coefficient function of $\dot{\varepsilon}^p_{ij}$.

The discreted form for eqn.(4) can be given by

$$\dot{\sigma}^p = \phi \dot{\varepsilon}^p \tag{5}$$

Where ϕ is influence coefficient matrix of $\dot{\varepsilon}^p$ and it can directly obtained by boundary element equation.

The above equations are valid for plane stress if V is replaced by $\bar{V} = V/(1+V)$.

2.1 Boundary Element Formulation

From Betti reciprocal theorem or weighted residual method, the boundary intergal equations for two dimensional elastoplastic problem without body force are presented as follows. The range of subscript indices in all the following equation is 1, 2.

$$C_{ij}\dot{u}_j = \int_\Gamma U_{ij}\dot{t}_j d\Gamma - \int_\Gamma T_{ij}\dot{u}_j d\Gamma + \int_\Omega \Sigma_{ijk} \dot{\varepsilon}^p_{jk} d\Omega \tag{6}$$

$$\dot{\sigma}_{ij} = \int_\Gamma D_{ijk}\dot{t}_k d\Gamma - \int_\Gamma S_{ijk}\dot{u}_k d\Gamma + \int_\Omega \Sigma'_{ijkl} \dot{\varepsilon}^p_{kl} d\Omega - \frac{G}{4(1-V)} (2\dot{\varepsilon}^p_{ij} + \dot{\varepsilon}^p_{ll}\delta_{ij}) \tag{7}$$

Where $C_{ij} = \delta_{ij}/2$ on smooth surface and $C_{ij} = \delta_{ij}$ for interior points. The kernels $U_{ij}, T_{ij}, \Sigma_{ijk}, D_{ijk}, S_{ijk}$ and Σ'_{ijkl} are known sigular solution and are available in many references such as [3].

Assuming L boundary elements and Z internal cells, the discretized forms of eqns. (6) and (7) for a boundary node S_i is given by

$$\underset{\sim}{C}(S_i)\dot{\underset{\sim}{u}}(S_i) = \sum_{j=1}^{l} \left(\int_{\Gamma_j} \underset{\sim}{U}\underset{\sim}{N} d\Gamma\right) \dot{\underset{\sim}{t}}^{(n)} - \sum_{j=1}^{l} \left(\int_{\Gamma_j} \underset{\sim}{T}\underset{\sim}{N} d\Gamma\right) \dot{\underset{\sim}{u}}^{(n)} + \sum_{j=1}^{z} \left(\int_{\Omega_j} \underset{\sim}{\Sigma}\underset{\sim}{N} d\Omega\right) \dot{\underset{\sim}{\varepsilon}}^{p(n)} \tag{8}$$

$$\dot{\underset{\sim}{\sigma}}(S_i) = \sum_{j=1}^{l} \left(\int_{\Gamma_j} \underset{\sim}{D}\underset{\sim}{N} d\Gamma\right) \dot{\underset{\sim}{t}}^{(n)} - \sum_{j=1}^{l} \left(\int_{\Gamma_j} \underset{\sim}{S}\underset{\sim}{N} d\Gamma\right) \dot{\underset{\sim}{u}}^{(n)} + \sum_{j=1}^{z} \left(\int_{\Omega_j} \underset{\sim}{\Sigma}'\underset{\sim}{N} d\Omega\right) \dot{\underset{\sim}{\varepsilon}}^{p(n)}_{ij} + \underset{\sim}{C}(S_i)\dot{\underset{\sim}{\varepsilon}}^p_i(S_i) \tag{9}$$

Where $Z = Z_1 + Z_2 + Z_3$, and Z_1, Z_2 and Z_3 are internal cells of borehole, rock bolt, and rock, resprectively. l is boundary elements of tunnel. $\underset{\sim}{N}$ stands for interpolated functions of linear elements.

Then, equations(8) and (9) can be written in matrix forms:

$$[H]\{\dot{u}\} = [G]\{\dot{t}\} + [\Sigma]\{\dot{\varepsilon}^p\} \tag{10}$$

$$\{\dot{\sigma}\} = [G']\{\dot{t}\} - [H]\{\dot{u}\} + [D + C]\{\dot{\varepsilon}^p\} \tag{11}$$

Applying boundary condition to eqns.(10) and (11), we have

$$[A]\{\dot{x}\} = \{\dot{f}\} + [\Sigma]\{\dot{\varepsilon}^p\} \tag{12}$$

$$\{\dot{\sigma}\} = -[A']\{\dot{x}\} + \{\dot{f}'\} + [D' + C']\{\dot{\varepsilon}^p\} \tag{13}$$

If we solve equation(12), we obtain

$$\{\dot{x}\} = [A]^{-1}\{\dot{f}\} + [A]^{-1}[\Sigma]\{\varepsilon^p\} \tag{14}$$

By substituting equn.(14) into (13), yields

$$\{\dot{\sigma}\} = \{\dot{\sigma}^e\} + \{\dot{\sigma}^p\} \tag{15}$$

where $\{\dot{\sigma}^e\} = \{\dot{f}'\} - [A'][A]^{-1}\{\dot{f}\}$

is the fictitious elastic stress rate corresponding to external loads; and

$$\{\dot{\sigma}^p\} = ([D'+C'] - [A'][A]^{-1}[\Sigma])\{\dot{\varepsilon}^p\} = \phi\dot{\varepsilon}^p$$

represents the corrective stress rate caused by plastic strain $\dot{\varepsilon}^p$. ϕ is just influence coefficient matrix in eqn.(5).

2.2 Yield Criteria

It is known from the numerical equations (14) and (15) that these formulations involve the elastoplastic stress-strain relations of three different yield criteria of borehole, rock bolt and rock. Here, we notice that the internal cells of borehole is similar to these of rock bolt. Then, the borehole can be looked the special problem of rock bolt without pseudobody force. Therefore, only the problems of rock bolt and rock will be considered here.

According to the shear sliding experiment of rock bolt element, we obtain the stress-deformation curve shown in Fig.1, a,b.From

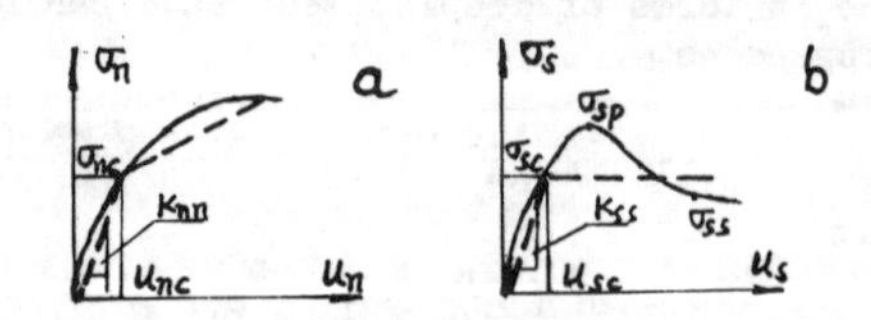

Fig. 1 Stress-deformation curve

this curve, we know that the main damage of rock bolt boundary depends on shear sliding behaviour, and its stear stain represents hardening and softenting features after the material becomes plastic yield. For simplicity, we only discuss the problem of ideal plastic shown dot line in Fig. 1. a.b. Thus, yield criteria of rock bolt can be given by Mohr - Coulomb Criteria:

$$F = |\sigma_s| - C + \sigma_n \tan\phi = 0 \quad (16)$$

Where ϕ and C are the internal friction angle and cohesion of rock bolt cells, respectively.

Similarly, the yield criteria for rock can directly be represented by

$$F = \beta I_1 + J^{\frac{1}{2}} + K = 0 \quad (17)$$

Where for different yield criteria such as Mohr - Coulomb, Drucker - Prager and so on, the coefficients β and K can be given by [4].

Constitutive Relation

According to the theory of plastity, the stress-strain relation for post yield behaviour can be written in follows :

$$d\sigma = (D_e - D_p)\, d\varepsilon = D_{ep} d\varepsilon \quad (18)$$

Where D_{ep} is elastic-plastic matrix, D_e is elastic matrix known by the theory of elastity and D_p is plastic matrix written in the following form :

$$D_p = D_e \left\{\frac{\partial G}{\partial \sigma}\right\}\left\{\frac{\partial F}{\partial \sigma}\right\} / \left(A + \left\{\frac{\partial F}{\partial \sigma}\right\} D_e \left\{\frac{\partial G}{\partial \sigma}\right\}\right) \quad (19)$$

in which F is yield function; A hardening parameters; and for relative flow law, F = G.

Taking A = 0, F = G and (16) into (19) and (18), the constitutive relation for rock bolt element can be obtained by

$$\begin{Bmatrix} \dot{\sigma}_{11} \\ \dot{\sigma}_{12} \end{Bmatrix} = \frac{-1}{S_o} \begin{bmatrix} K_{11}^2 - K_{11} S_o & K_{11} S_1 \\ K_{11} S_1 & S_1^2 - K_{22} S_o \end{bmatrix} \begin{Bmatrix} \dot{\varepsilon}_{11} \\ \dot{\varepsilon}_{12} \end{Bmatrix} \quad (20)$$

Where $S_o = K_{11} + K_{22} \tan^2\phi$, $S_1 = K_{22} \cdot \tan\phi$; Here, $K_{11} = K_s$ and $K_{22} = K_n$ are stiffness of the rock bolt shown in Fig. 1.

Similarly, the constitutive relation of rock elements can be obtained from (17),(18) and (19). Here, the plastic matrix is written in follows.

$$D_p = \frac{1}{S_o} \begin{bmatrix} S_1 & & \\ S_1 S_2 & S_2^2 & \\ S_1 S_3 & S_2 S_3 & S_3^2 \end{bmatrix}$$

where

$$S_o = S_1 \bar{\sigma}_{11} + S_2 \bar{\sigma}_{22} + S_3 \bar{\sigma}_{12}$$

$$S_3 = G \bar{\sigma}_{12}, \quad S_i = D_{i1} \bar{\sigma}_{11} + D_{i2} \bar{\sigma}_{22}$$

$$\bar{\sigma}_{ii} = \frac{\partial F}{\partial \sigma_{ii}} = \beta + (2\sigma_{ii} - \sigma_{ij})/6J_2^{1/2}$$

$$\bar{\sigma}_{12} = \sigma_{12}/J_2^{1/2} \qquad (i=1,2, j=2,1)$$

3 NUMERICAL EXAMPLES

By means of the above numerical method, the following examples are computed by using microcomputer (IBM). For the comparision, the materials of rock bolt and rock in the examples are all considered to be perfectly plastic as shown in the following table.

Table 1 The elastoplastic parameters in the following examples

BOLT	M - C	ROCK	D - P	UNIT
E_b	1200	E	500	kpa
V_b	0.32	V	0.2	
C_b	0.6	C	0.25	kpa
ϕ_b	30°	ϕ	25.8°	

In the table, M - C and D - P mean Mohr - Coulomb and Drucker - Prager criterias, respectively.

Ex. 1, As shown in Fig. 2a, a circular tunnel with 12 boreholes is subjected to the initial uniform stress field P. Here, boreholes is long $\frac{3}{4}$ r and diameter 0.06 meter. Then, the results of this example are shown in Fig. 2.

Ex. 2 A circular tunnel with 12 rock bolts are also subjected to the initial uniform stress field P. Here, suppose that the rock bolts is the same size as the borehores. The results of this problem is shown in Fig. 3.

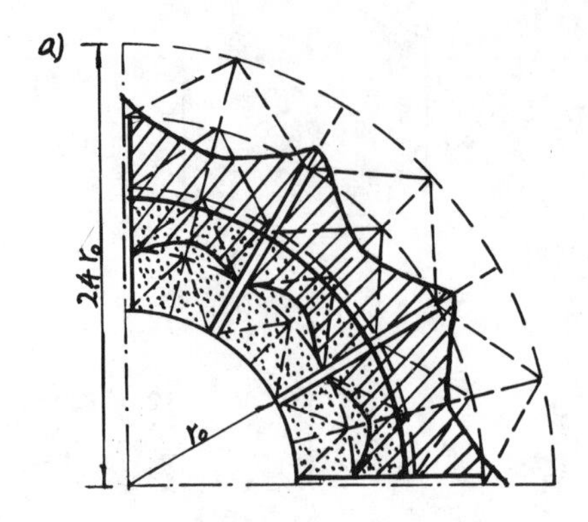

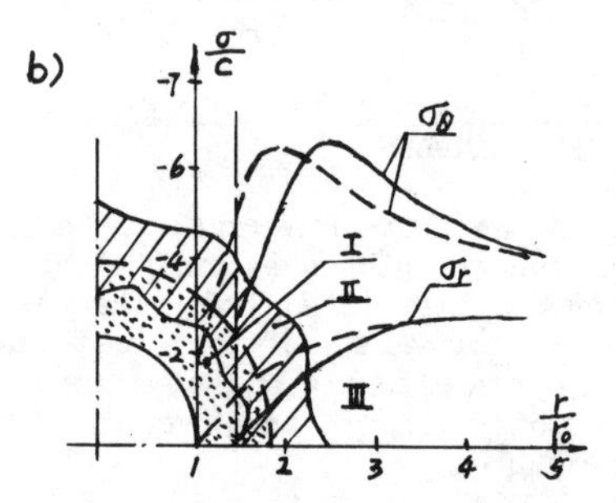

Fig. 2 The destressing problem of boreholes for the circular tunnel
a) The different spread of plastic zone; b) The stress distribution along the horizontal section through the medium. In the Fig. 2, " ▨ " and " —— " express the plastic zone and the stress curve for the problem with boreholes; But, " ⁙ " and " - - - -" express for the problem without boreholes.

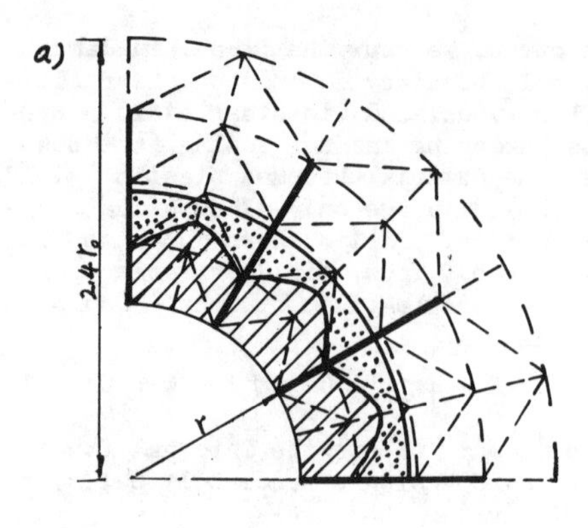

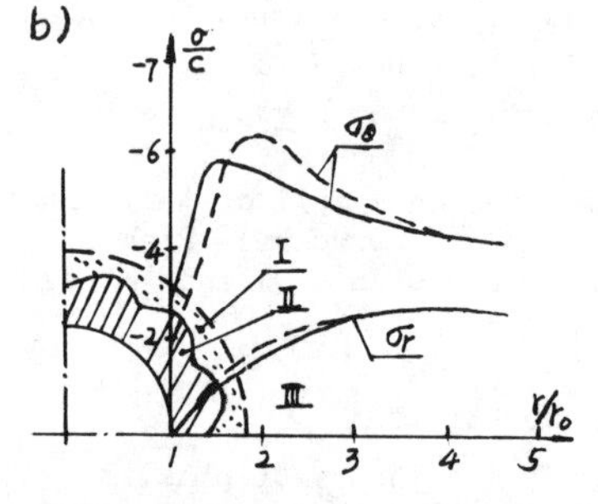

Fig. 3 The stiffening problem of rock bolts
a) The spread of plastic zone; b) The stress distribution along the horizontal section through the medium. In the Fig.3, "▨" and " ——" are the plastic zone and the stress curve with rock bolt. But, " ⁙ " and " - - - " mean without rock bolt.

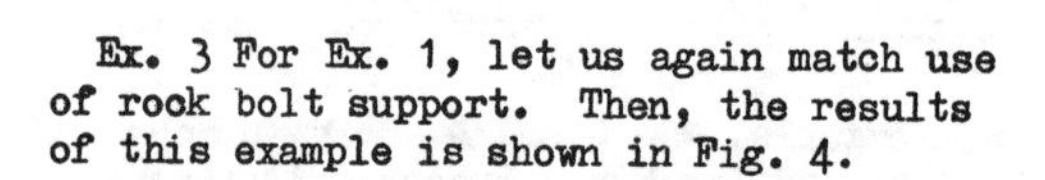

Ex. 3 For Ex. 1, let us again match use of rock bolt support. Then, the results of this example is shown in Fig. 4.

4. CONCLUSION

A. The destressing problem of boreholes

It is known from Ex. 1 that the destressing of boreholes can makes the stress peak or plastic zone (I) away from the gallery walls into the internal surrounding rock, i.e zones (II) and (III). Thus, zone(I) becomes a destressed zone; zones (II) and (III) are changed into a plastic zone and elastic zone. Because this supporting system is composed of the above three rock ring zones, it is named "self supporting structure" as shown in Fig. 2. The working behaviour of this supporting structure is that the strength of the internal surrounding rock of the plastic zone (II)and elastic (III) is used as more as possible, and the destressed zone (I) itself is formed a self supporting rock ring. These results are in closing agreement with the experiment results given by Roest [5] .

B. The stiffening problem of rock bolts

As shown in Fig. 3 from Ex. 3, by means of the stiffening of rock bolt the plastic zone(I) is reduced to zone (II), and the stress peak in the zone (II) is lower than zone (I). These results show that the stiffening behaviour of rock bolts not only can increase the strenth of the plastic zone, bult also can confine the size and deformation of the plastic zone; and these results are in good agreement with the experimental results in [6] .

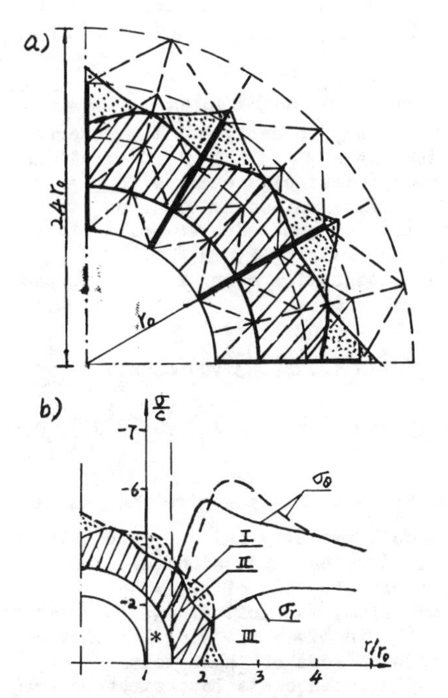

Fig. 4 W.U.S.S.S. In the Fig. 4, "[dotted]" and " ——— " are the plastic zone and the stress curve for the destressing problem of boreholes; "[hatched]" and "—" are for the stiffening of rock bolts. * is the destressed zone self supporting rock ring.

C. W.U.S.S.S.

According to the supporting properties of Ex.1 and Ex.2, a new supporting structure shown in Ex. 4 is discussed. This supporting structure is formed by coupling of two supporting techniques as follows.

First, the destressing technique of boreholes makes the stress peak (plastic zone) from the gallery walls into the internal rock so that the self supporting structure can be formed as shown in Fig. 2.

Then, the stiffening technique of rock bolts is again used to increase the strength of the plastic zone and reduce the size of plastic zone so that the self supporting structure can be changed into a more ideal supporting system, i.e. W.U. S.S.S. as shown in Fig. 4.

Letting us compare the numerical results in Fig. 4 with these in Fig. 2 and in Fig. 3, it can be seen that the elastoplastic behaviours for M.U.S.S.S. have more advantages over these for other two supporting structures. Therefore, it shows that the application of the working mechanisms between the destressing of boreholes and the stiffening of rock bolts is significant for the solution of the supporting problem in the excavation.

D. Numerical method

From the results of the above examples, it can be seen that these results given in the paper are in good agreement with the experimental results by Roest [5] and in [6]. Thus, it also shows that this numerical method is an extremely versatile and powerful tool for solving the elastoplastic problem for the borehole - rock bolt support and other complex structure in the underground engineering.

REFERENCES

[1] P.K. Banerjee and R. Butterfield. Boundary Element Method in Engineering Science, McGraw-Hill, London, 1981.

[2] P.K. Banerjee and D.N. Cathie, A Direct Formulation and Numerical Implementation of the Boundary Element Method for Two-Dimensional problem of Elastoplastity. Int. J. Mech. Sci., 22, 233-245, 1980.

[3] C.A. Brebbia, Boundary Element Method, Springer- Verlag Berlin Heidelberg New York, 1981.

[4] J.C.F. Telles, The Boundary Element Method Applied to Inelastic Problem, Spring - Verlag Berlin Heidelberg New York Tokyo, 1983.

[5] J.P.A. Roest, The self supporting Rock Ring - Port, Proceeding of Mine construction Session, China Institute of Mining and Technology, Cd4, China, 1985.

[6] X.F. Yu, Y.R. Zheng, H.H. Liu and I.C. Fang, The steady Analysis in Underground Engineering, Coal Publications, China, 1983.

[7] C.A. Brebbia and J.J. Connor, Fundamentals of Finite Element Techniques for Structural Engineers, Butterworths, 1973.

[8] S.G. Mikhlin, The Problem of the Minimum of a Quadratic Functional, Translated by A. Feinstein, Holden-day Inc, 1965.

APPENDIX A LINEAR INTERPOLATION FUNCTION AND SINGULAR INTEGRAL IN THE DOMAIN

The domain intergal of pseudobody (initial strain) in Eqns.(8) and (9) can be carried out numerically. In what follows the procedure to obtain the initial strain integral is discussed.

For example, the initial strain intergral over any single cell (j) in Eqn.(8) can be written as

$$I = \int_{\Omega_j} (\Sigma_{ijk} \underset{\sim}{N} d\Omega) \dot{\varepsilon}^p \qquad (A.1)$$

For linear triangular cells of the rock the interpolation functions are expressed in terms of a homogeneous coordinate system (ξ_1 ξ_2), i.e.

$$N = [\xi^1 \ \xi^2 \ \xi^3] \qquad (A.2)$$

where

$$\xi^\alpha = \frac{1}{2A} [2A^0_\alpha + b_\alpha X + a_\alpha y] \qquad (A.3)$$

in which A is the area of cell and

$$a_i = x_1^k - x_1^j$$
$$b_i = x_2^j - x_2^k \qquad (A.4)$$
$$2A^0_\alpha = x_1^j x_2^k - x_1^k x_2^j$$

with $i = 1,2$, $j = 2,3$ and $k = 3,1$

This interpolation function can also be written in a polar coordinate, i.e.

$$\xi^\alpha = \xi^\alpha(p) + \frac{1}{2A}(b_\alpha \cos\theta + a_\alpha \sin\theta)\, r \qquad (A.5)$$

where p is the load point and

$$\xi^\alpha(p) = \frac{1}{2A} [2A^0_\alpha + b_\alpha x_1(p) + a_\alpha x_2(p)] \qquad (A.6)$$

The fundamental solution Σ_{ijk} can be written as,

$$\Sigma_{ijk}(p, \theta, r) = \Sigma^*_{ijk}(p, \theta) \frac{1}{r} \qquad (A,7)$$

Substituting the equations (A.5) and(A.7) into (A.1) each cell integral can be becomes,

$$I = \int_\Omega \Sigma^*_{ijk}(p, \theta) [\xi^\alpha(p) + \frac{1}{2A}(b_\alpha \cos\theta + a_\alpha \sin\theta)\, r]\, dr\, d\theta \qquad (A.8)$$

This equation can be integrated over r remaining only to carry out the integral over the angle θ. The θ integral is performed numerically using a four or six Gauss integration rule. Once these integrals are computed over cells one obtains the matrix [Σ] in Eqn.(10).

After integration over r the expression (A.8) can becomes

$$I = \int_\theta \Sigma^*_{ijk}(p, \theta) [\xi^\alpha(p)(R_2 - R_1) + \frac{1}{4A}(b_\alpha \cos\theta + a_\alpha \sin\theta)(R_2^2 - R_1^2)]\, d\theta \qquad (A.9)$$

where R_1 and R_2 are given in function and cell geometry and can be easily calculated using expression (A.6) over an appropriate side of cell.

In addition, the initial strain integral in Eqn.(9) involves the singular characteristics of some of these terms.

According to Hooke's law, stresses are related to displacement derivatives as follows

$$\sigma_{ij}(p) = \frac{2G}{1-2V} \delta_{ij} \frac{\partial u_1(p)}{\partial x(p)} + G\left[\frac{\partial u_j(p)}{\partial x_j(p)} + \frac{\partial u(p)}{\partial x_j(p)}\right] \qquad (A.10)$$

Substituting above equations(A.1),(A.5), (A.7) into (A.10), the derivatives of the singular integrals arise.The singularity is due to the first term in Eqn.(A.5) ,i.e.

$$D = \frac{\partial}{\partial x_j(p)} \int_\Omega \Sigma^*_{ijk}(p, \theta) \xi^\alpha(p)/r \, d\Omega \qquad (A.11)$$

One can treat this derivatives by means of Mikhlin [8], in which the derivatives can be introduced into the integral and another term is added, so that the equation is written as

$$D = \int_\Omega \frac{\partial}{\partial x_j(p)} [\Sigma^*_{ijk}(p, \theta)/r]\, \xi^\alpha(p) d\Omega - \xi^\alpha(p) \int_0^{2\pi} \Sigma^*_{ijk}(p, \theta) \cos\theta_j \, d\theta \qquad (A.12)$$

where $\cos\theta_j$ is $\cos\theta$ or $\cos(\theta - \frac{\pi}{2})$ for j = 1 or 2, respectively.

It can be known that the kernels function Σ^*_{ijkl} and so on in Eqn.(7) can be solved by means of above method. Similarly, for the linear bolt cells of rock bolt we can also solved.

Numerical Methods in Geomechanics (Innsbruck 1988), Swoboda (ed.)
© 1988 Balkema, Rotterdam. ISBN 90 6191 809 X

The progressive failure of the rock mass surrounding an opening by boundary element method

Chen Jin & Yuan Wenbo
China Institute of Mining Technology, Beijing

ABSTRACT: When an underground opening is excavated, the rock mass surrounding the opening will be progressively damaged due to ground pressure. This is one of important fields to be investigated in geomechanics. In the present paper, boundary element method is developed to simulate and estimate the progressive failure (post-failure) behaviour of rock. To guarantee the stability and convergence of numerical method, an optimum method is used to control the load and deformation of rock. Some examples are solved to illustrate the efficiency of boundary element method and its application to underground opening.

1 INTRODUCTION

Rock mass, as a natural and heterogeneous material, is characterized by microcracks and microstrustures and their evolution and interaction govern microscopic mechanical response of rock. When rock mass is stressed beyond its strength, it shows progressive failure, i.e. weakening or softening of bearing ability and dilation in volume. This progressive failure has an strong effect on the stability of underground rock structure such as mine opening, tunnel et. al.. Study of the failure process of rock mass would be helpful to the construction and maintenance of underground engineering. Theory of progressive failure was studied by Dougill (1976), Preost and Heog (1975), Bazant(1984) et. al.. Brown (1983), Yuan and Chen (1986) has used strain softening theory to analyse a circular opening in simple conditions and gave some useful results. And analysis of rock failure by finite element method was performed by Nayak and Zienkiewicz(1972), Pietruszczuk (1982), Lee (1985) et. al.. But as pointed by Bazant (1984), Crisfield(1984) et. al., in the analysis of progressive failure there remain some problems to be solved, such as instability in theory, size effect and convergence of numerical analysis.

Boundary element method has been a wide field of application in engineering. The technique is of great advantage to geomechanics problems as only the boundary of the domain has to be discretized, except where the boundary goes to infinity, in such case no discretization is required. Boundary element method requires a smaller number of data to run the problem. Until now, only a few authors applied the technique to progressive failure of rock mass. Perice (1983) gave a paper about the analysis of a square opening by boundary element method and his model of rock is simple.

The purpose of the present paper is to extend the boundary element method to progressive failure of rock mass such as those present in underground opening. In this paper, a generalized Drucker-Prager function which contains internal variables is used to progressive failure of rock mass. These internal variables are used to represent total effect of progressive damage in rock mass. The constitutive relationship is developed on the basis of the plastic theory with internal variables. And an optimum process is used to overcome size effect and instability of numerical analysis. Finally, several examples of mine openings are solved to illustrate the failure mechanism of underground opening. And some suggestions for improvement of maintenance and support underground opening are given.

2 COSTITUTIVE RELATIONSHIP

Progressive failure is a decline of strength of rock at increase strain and a dilation in volume, but the decline rate decreases with increasing of confining stess. Fig. 1 and Fig. 2 show the progressive failure of soft rock.

From microscopic point of view, rock mass contains a lot of microcracks and microvoides and their evolution and interaction

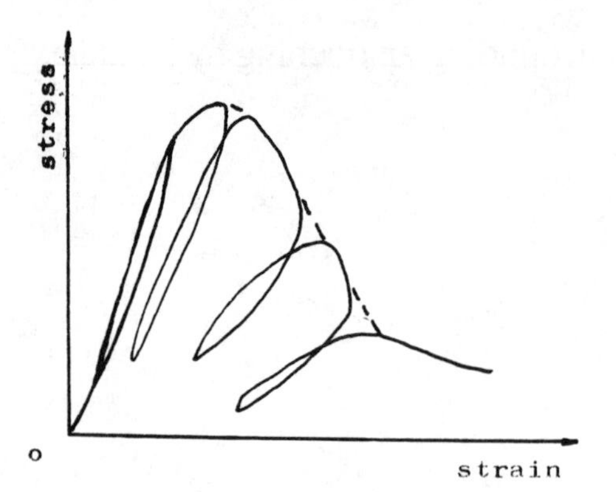

Fig. 1

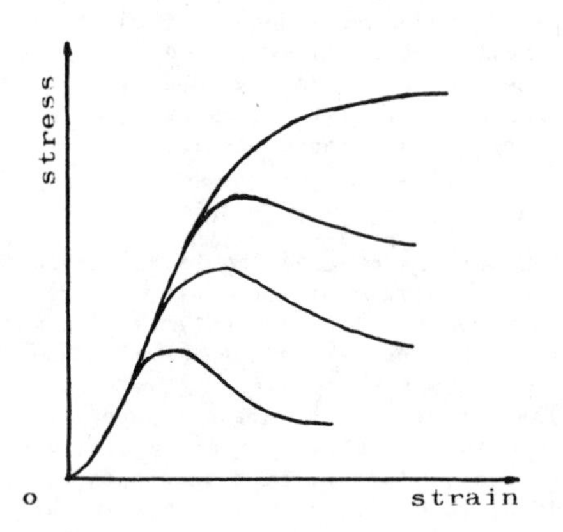

Fig. 2

govern macroscopic mechanical response of rock. Failure mechanism of rock mass consists of progressive distributed damage. The total response of the damage can be represented as internal variables such as plastic work or effective plastic strain et. al.. As a homogenization technique, the constitutive relationship of progressive failure can be developed on the basis of plastic theory with internal variables.

In general the failure function is given below;

$$F=f(\sigma_{ij}\ , q^{k}_{ij}\ ) \tag{1}$$

where σ_{ij} is stress tensor; q^{k}_{ij} are internal variables.

The yield criterion commonly used in rock is Drucker-Prager or Mohr-Coulomb criterion as follows;

$$\alpha I_1+\sqrt{J_2}+K=0 \tag{2}$$

where α and K are constants.

This criterion can be extended as a failure function to indicate progressive failure of rock. That is

$$\alpha(q^K)I_1+\sqrt{J_2}+K(q^K)=0 \tag{3}$$

where α (q^K) and $K(q^K)$ are any form functions of q^k respectively based on the simplified model and experimental results of rock.

Plastic flow in rock is non-associated because there is a dilation in volume. Non associated flow rule requires to difine a potential function. The general form of poternal function is

$$G=g(\sigma_{ij}\ , q^K) \tag{4}$$

For progressive failure, Drucker's stability postulate in stress space is violated, Ilinshins' postulate in strain space can be applied to flow process. Considering the exchange formulae in stress space with that in strain space, the flow rule is represented as follows;

$$\dot{\varepsilon}_{ij}=-\dot{\lambda}\frac{\partial q}{\partial \sigma_{ij}} \tag{5}$$

Consistency condition is

$$dF=\frac{\partial F}{\partial \sigma_{ij}}d\sigma_{ij}+\frac{\partial F}{\partial q^K}dq^K=0 \tag{6}$$

From above equations, stress increment equations are given below;

$$\dot{\sigma}_{ij}=D^{eP}_{ijKl}\ \dot{\varepsilon}_{Kl} \tag{7}$$

where

$$D^{eP}_{ijKl}=D^{e}_{ijKl}-\frac{D^{e}_{ijtu}\left(\frac{\partial F}{\partial \sigma_{rs}}\right)\left(\frac{\partial G}{\partial \sigma_{tu}}\right)D}{h+\left(\frac{\partial F}{\partial \sigma_{mn}}\right)D_{mnpq}\left(\frac{\partial G}{\partial \sigma_{pq}}\right)}$$

$$h=-\frac{\partial F}{\partial q^K}\frac{\partial q}{\partial \varepsilon^{P}_{ij}}\frac{\partial G}{\partial \sigma_{ij}}$$

D^{e}_{ijKl} is material stiffness tensor.

Rock is considered as no-tension material because its tension ability is small. Failure criterion in tension range has the following form;

$$F_1=\sigma_1\leqslant 0 \tag{8}$$

where σ_1 is the maximum principal stress.

3 BOUNDARY INTEGRAL FORMULE

An isotropic linear elasoplastic problem is completely defined by the following set of governing equations:

Displacement-strain relationships

$$\dot{\varepsilon}_{ij} = \frac{1}{2} (\dot{u}_{i,j} + \dot{u}_{j,i}) \quad (9)$$

where $\dot{u}_{ij}$ are the displacement rate derivatives.

Equilibrium equations

$$\dot{\sigma}_{ij,j} + \dot{b}_i = 0 \quad \text{in} \quad V \quad (10)$$

where b is body force rate components and V is the domain.

Boundary conditions

$$\dot{t}_i - \dot{\sigma}_{ij} n_j = 0 \quad \text{on } L \quad (11)$$

where t_i is the prescribed components of boundary L. n_j is direction cosines of outword normal to the boundary of the body.

Stresses-strains relationship

$$\dot{\sigma}_{ij} = 2G(\dot{\varepsilon}_{ij} - \dot{\varepsilon}^P_{ij}) + \frac{2G\nu}{1-2\nu} (\dot{\varepsilon}_{ll} - \dot{\varepsilon}^P_{kk}) \delta_{ij} \quad (12)$$

where G and ν are the shear modulus and poisson's ratio respectively.

Using Betti's theorem or the weighted residual technique, boundary integral equations can be obtained, i.e.

$$C_{ij} \dot{U}_j + \int_L P^*_{ij} \dot{U}_j dL = \int_L U^*_{ij} \dot{P}_j dL + \int_V \varepsilon^*_{ijk} \dot{\sigma}^P_{jk} dV \quad (13)$$

$$\dot{\sigma}_{ij} = \int_L U^*_{ijk} \dot{P}_k dL - \int_L P^*_{ijk} \dot{U}_k dL + \int_V \varepsilon^*_{ijkl} \dot{\sigma}^P_{kl} dV - \frac{1}{8(1-\nu)} (2\dot{\sigma}^P_{ij} + (1-4\nu) \dot{\sigma}^P_{kk} \delta_{ij}) \quad (14)$$

for plane strain problem in initial stress.

where C_{ij} are constants; P^*_{ij} and P^*_{ijk}, U^*_{ij} and U^*_{ijk}, ε^*_{ijk} and ε^*_{ijkl} are fundamental solutions for tractions, displacements and strains respectively.

The no-tension solution was achieved by Venturini (1981). In the solution, the equilibrium of body is maintained by an iterative process in which a series of initial stresses to compensate the tension stresses.

4 NUMERICAL IMPLEMENTATION

4.1 Spatial discretization

For numerical analysis, it is required to make the spatial discretization of the integral equations. In this case, the boundary is discretized into a series of elements over which dicplacements and tractions are piecewise interpolated between the element nodal points. The part of the domain in which non-zero inelastic strain is expected is discretized using linear internal cells for integration. Applying this process to equation (13) and (14), the following matrix relationships are obtained.

$$\underset{\sim}{H} \dot{\underset{\sim}{U}} = \underset{\sim}{G} \dot{\underset{\sim}{P}} + \underset{\sim}{D} \dot{\underset{\sim}{\sigma}}^P \quad (15)$$

$$\dot{\underset{\sim}{\sigma}} = \underset{\sim}{G'} \dot{\underset{\sim}{P}} - \underset{\sim}{H'} \dot{\underset{\sim}{U}} + \underset{\sim}{E} \dot{\underset{\sim}{\sigma}}^P \quad (16)$$

where matrixes H, G, D and E correspond to the previous integral equations.

4.2 Loading control

A loading control is required to overcome the numerical instability in progressive failure analysis. The principle of this control is to maintain an equilibrium state of system energy in every loading step, that is

$$\frac{dW}{dS} = 0 \quad (17)$$

where W is the total potential energy of the system; S is loading step length.

The optimum method such as search process is perfect for the loading control. Usually the linear search is enough. However when the failure function is simplized to linear, the loading control is unnecessary.

4.3 Numerical procedures

The algorithm in progressive failure requires incremental loading and complete knowledge of the inelastic stress distribution within the failure region. This algorithm can be described as follows:

- Computing elastic coefficients
- Obtaining elastic solution
- Scaling the elastic solution such that the highly streses node at failure
- Controlling loading steps
- Determining failure region and inelastic cofficients
- Determing the stresses and strain within failure and tension region
- Computing internal stresses and displacements

4.4 Accuracy and convergence

In order to investigate the accuracy and

convergence of the numerical analysis, the progressive failure of a circular rock opening was studied. And the results obtained are compared with analytical solution in [6] . Table 1 shows the summary below:

Table 1 (cm)

METHOD ELEMENTS	FAILURE REGION	MAXIMUM DISPLACEMENT
ANALYSIS [6]	341.50	3.547
BEM 4	274.43	2.291
BEM 8	322.62	3.166
BEM 12	340.18	3.520

The above results show that the boundary element method presented is stable and effective and the size effect and unconvergence et.al. that occurred in finite element method can be overcome.

5 APPLICATION TO MINE OPENING

Some practical examples of progressive failure and no-tension behaviour of rock mass are given below to illustrate the failure mechanism of the rock mass surrounding a mine opening.

Example 1

The example of a circular in Fig.3 shows the discretization used for boundary element results and total spread of plastic zone and broken zone.

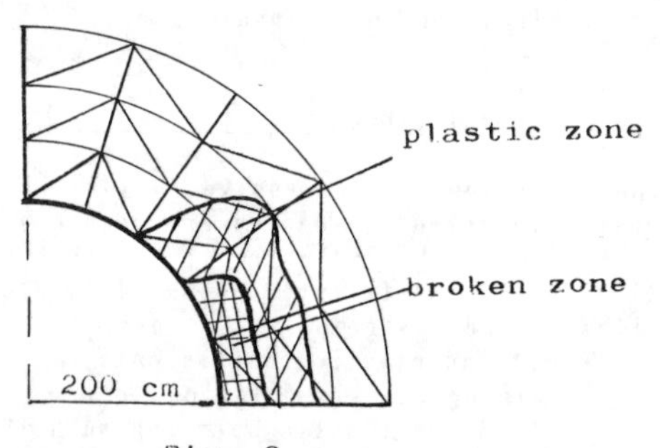

Fig. 3

Example 2

This is a common mine opening in China. Figure 4 shows the tatal spread of the failure region.

Comparing these results with that of ideal elastoplastic analysis, the progressive failure and no-tension behaviour of rock mass have a considerable effect on the stability of underground opening. Considering

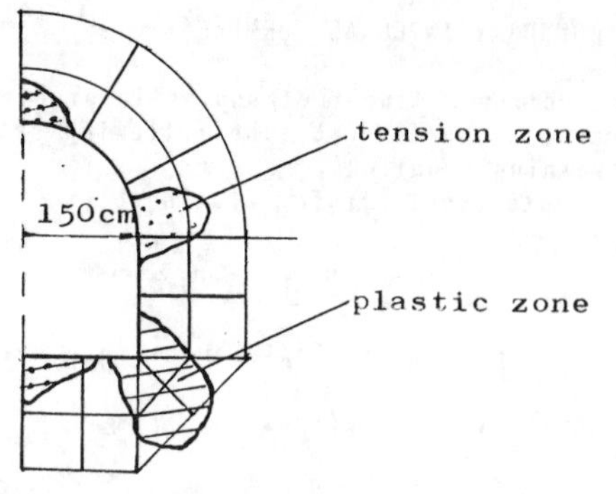

Fig. 4

this point, it is worthy of paying more attention to the internal reinforcement of rock (such as rock bolts, grouting) for controlling the failure of rock and increasing its strength.

6 CONCLUSION

(1) Rock mass has progressive failure and no-tension characteristic. But in the numerical analysis of progressive failure there remain some problems to be solved such as size effect, instability et. al..

(2) Boundary element method is effective fo the problem in geomechanics. In the present work, this technique is used to simulate an estimate the progressive failure and no-tension of rock mass in which the progressive failure is modelled by a Drucker-Prage function with internal variables and the load is controlled by the optimum method.

(3) Progressive failure and no-tension beha viour of rock mass have a harmful effect on underground engineering. It is necessary to develop some methods of internal reinforcement of rock.

REFERENCES

[1] W.B. Yuan, J. Chen: Effects of strain-softening behaviour of rock mass on stresses around a circular opening, Trans. Tech. Publications, Clausthal-Zellerfeld, 1987, 706-713

[2] J.W.Dougill: On stable progressively fracturing solids, J.of apply.math.& phys. 27,423-437 (1976)

[3] J.H.Prevost and K.Hoeg: Soil mechanics and plasticity analysis of strain softening, Geotechnique, 25, 279-297 (1975)

[4] Z.P. Bazant, B. Belytschko and T.P. Chang: Continuum theory for strain-softening, J.of engng. mech. ASCE, 110,

1662-1692 (1984)
[5] E.T. Brown et.al.; Ground response curves for rock tunnels, J.of geote. engng. ASCE, 109, 16-31 (1983)
[6] W.B. Yuan and J. Chen; Analysis of plastic zone and loose zone around an opening in softening rock mass, J. of China coal society, 77-86 (1986)
[7] G.C. Nayak and O.C. Zienkiewicz; Elastic-plastic stresses analysis; a generalization for various contitutive relation including strain-softening, Int.J.numer. meth. in. engng. 5,113-135 (1972)
[8] S. Pietruszczak and Z. Morz; Finite elemant analysis of deformation of stain-softening materials, Int.J. numer. meth. in engng. 17, 327-334 (1981)
[9] W.C. Pariseaa;A finite element approach to strain softening and size effects in rock mechanics, Proc. 3rd Int. conf. on numer. meth. in gromachanics, 1979
[10] T. Matsumoto and H.Y. Ko; Finite element analysis of strain-softening soils, 4th Int. conf. on numer. meth. in geomachanics, 1982
[11] H.K. Lee and H.S. Yang; An elastoplastic anslysis for post-failure behaviour of strain-softening rocks, 5th Int. conf. on numer. meth. in geomachanics, 1985 165-171
[12] A.P. Peirce and J.A. Ryder; Extended boundary element method in the modelling of brittle rock behaviour, 5th Int. conf. rock mech., 1983 F159-166
[13] M.A. Crisfield; Overcoming lilmit point with material softening and strain localization, Numer. meth. for non-linear problem 1984 244-277
[14] J.C.F. Telles; The boundary element method applied to inelasitc problem, Spring-Verlag, Berlin, 1983
[15] W.S. Venturini and C. Brebbia; The boundary element method for the solution of no-tension materials, boundary element method, Spring-Verlag, Berlin, 1981
[16] J. Chen; Progressively failure theory of the underground rock structure, Ph.D. thesis to be submitted to the China Institute of Mining Technology, 1987

Numerical Methods in Geomechanics (Innsbruck 1988), Swoboda (ed.)
© 1988 Balkema, Rotterdam. ISBN 90 6191 809 X

Infinite element and its application to geotechnical problems

Liu Huaiheng & Yang Gengshe
Xian Mining Institute, People's Republic of China

ABSTRACT The present paper deals with the infinite element modelling the infinite or semi-infinite domain and the application of it,and proposes the computation formulas and the calculating examples of infinite element.On the basis of the examples the problems asscoiated with the infinite element are discussed,such as the influence of the decay function,the mapped infinite element,the advantage of infinite element in the calculating accuracy and the economic effect and so on.In the end of this paper some computing examples are introduced.An utility FEM program with the infinite element on micro- or personal computer has been developed by the authers.

1 INTRODUCTION

The infinite or semi-infinite domain are quite often found when analysing the problems of geotechnics using the numerical method. But in the finite element method(FEM) a finite region must be truncated from the infinite domain ,and the approximation boundary conditions are assumed with, which will introduce the calculating errors of the numerical solution.The same problems are also encountered in the boundary element method (BEM) for some special case. So that the infinite elements are very useful and effective for soving the above-mentioned problems.

Infinite element is a special element simulating the infinite or semi-infinite domain,and its idea is first proposed by R.L.Ungless(1977), O.C.Zienkiewicz(1983), G.Beer and Meek(1981,1983),etc. A lot of papers has discussed the fundment concept of infinite element.However,the application of it are not many. In this paper, we introduced the two dimentional infinite element and the joint infinite element, and its application to geoechanical problems. The concept and the principle are also used to three-dimentional problems.

Practice indicated that, using the infinite element is very effective. So We not only can conveniently solve the difficulty of boundary conditions, but may also obtain satisfactory results applying a few elements, and the conputer time can be shortened greatly.

2 THE CONCEPT AND FUNDMENT FORMULA

The characterization of infinite element is that the geometric size are tended to infinity at one direction. Thus in oder to form the element charactor matrix, we must intergrate on the infinite domain. It can be finished by two ways.That is

a) using the mapping infinite element.

b) taking the general integration on the infinite domain and the decay function of displacement.

For the sake of simplicity,in this study the two dimensional formulation is discussed and presented,the prenciple and method can be appied to three-dimensional problems.

2.1 Infinite element funded by mapping method

Usually, the 4- or 8-nodal points isoparametric elemnet are used for two dimensional perobleems, so that corresponding infinite element as showing in Fig.1 are used. The transformation relationshap beteen the Cartesian coordinate system and the - local coordinate system may be determined by

$$X=\sum_{i=1}^{n} N_i X_i \qquad Y=\sum_{i=1}^{n} N_i Y_i \qquad (1)$$

where n is the number of nodel points of element, N_i is the shape function .The infinite element in Fig.1a can be mapped into a square element as showing in Fig. 1b, and the mapping coordinate system are taken as

$$\xi = 2\xi'(1-\xi'), \qquad \eta = \eta'$$
$$\text{or} \quad \xi' = \xi/(2+\xi), \qquad \eta' = \eta \tag{2}$$

so when the mapping coordinate $\xi = 1$ the local and the general coordinate will tends to infinity on the correspending direction. Thus, integration of element over the infinite domain may be changed into the finite domain over the ξ' and η'. There we have the shape function of infinite elemnt

$$N_i = -N_i(\eta)\xi, \qquad (i=1,2,3)$$
$$N_{i+3} = N_i(\eta)(1+\xi) \tag{3}$$

and the intervanient function of displacement

$$N_i' = N_i(\eta')(\xi'-1)\xi', \qquad (i=1,2,3)$$
$$N_{i+3}' = N_i(\eta')(1-\xi'^2) \tag{4}$$

Where the $N_i(\eta)$ or $N_i(\eta')$ is the standardizing Lagrage polynomial for η (or η'). Thus we can write out the displacement functions

$$u = \sum_{i=1}^{n} N_i' u_i, \qquad v = \sum_{i=1}^{n} N_i v_i \tag{5}$$

The shape function (3) mapping into the system $\xi'-\eta'$ we can finished the integration over the infinite domain. The intervanient function of displacements are immediateiy founded in the $\xi-\eta$ system, which the displacement conditions at infinity is satisfied.The following differential formula are useful:
where

$$\begin{Bmatrix} \dfrac{\partial N_i}{\partial \xi'} \\ \dfrac{\partial N_i}{\partial \eta'} \end{Bmatrix} = \begin{bmatrix} \dfrac{\partial \xi}{\partial \xi'} & \dfrac{\partial \eta}{\partial \xi'} \\ \dfrac{\partial \xi}{\partial \eta'} & \dfrac{\partial \eta}{\partial \eta'} \end{bmatrix} \begin{Bmatrix} \dfrac{\partial N_i}{\partial \xi} \\ \dfrac{\partial N_i}{\partial \eta} \end{Bmatrix} = [J_2] \begin{Bmatrix} \dfrac{\partial N_i}{\partial \xi} \\ \dfrac{\partial N_i}{\partial \eta} \end{Bmatrix}$$

wher

$$[J_2] = \begin{bmatrix} \dfrac{2}{(1-\xi')} & 0 \\ 0 & 1 \end{bmatrix} \tag{6}$$

$$\text{so that} \quad \det|J| = \frac{2}{(1-\xi')^2} \tag{7}$$

thus the element stiffness matrix may be obtained from the following integration formula

$$[K] = \iint [B]^T [D] [B] \, dA =$$
$$= \int_{-1}^{1}\int_{-1}^{1} [B]^T [D] [B] \det|J_1| \, d\xi \, d\eta$$
$$= \int_{-1}^{1}\int_{-1}^{1} [B]^T [D] [B] \det|J_1| \det|J_2| \, d\xi' \, d\eta' \tag{8}$$

It should be noted that the $\det|J_1|$ is computed using the shape function (3), and the matrix [B] is computed by intervanient function (4).In this case the decay function of displacements is not needful.

2.2 Infinite element using the decay function

An another way for funding the infinite element is choice acorrect shape function and a decay function of displacemwent,and they should satisfy the geometric and the displacement conditions at infinity.For an example, we given out the shape function of the Fig.2b as follows

when $\xi \le 1$

$$N_i = -\tfrac{1}{2}\eta_i\eta(1+\eta_i\eta)\xi, \qquad (i=1,4)$$
$$N_i = \tfrac{1}{2}(1+\eta_i\eta)(1+\xi), \quad (i=2,3) \tag{9}$$
$$N_5 = -(1-\eta^2)\xi,$$

When $\xi > 1$

$$N_i = -\tfrac{1}{2}(1+\eta_i\eta)\xi/(1-\xi), \qquad (i-1,4)$$
$$N_i = \tfrac{1}{2}(1+\eta_i\eta) + \tfrac{1}{2}(1+\eta_i\eta)\xi/(1-\xi) \qquad (i=2,3)$$
$$N_5 = 0 \tag{10}$$

The above shape function satiafy the charactor of the intervenient function. e.t.

when $\xi \le 0$ $\quad N = 1$

when $\xi > 0$ and $\xi \to 1$, $N \to \infty$

The displacement function can be written as follows

$$u = \sum_{i=1}^{n} M_i u_i \qquad v = \sum_{i=1}^{n} M_i v_i \tag{11}$$

$$M = N \, . f(\xi, \eta) \tag{12}$$

The N may be choice the same form as the shape function, and the decay function $f(\xi,\eta)$ should satisfy with the zero displacement at infinity.

Here the stiffness matrix can be integrated by the general formula:

$$[K] = \int_{-1}^{1}\int_{-1}^{1} [B]^T [D] [B] \det|J_1| \, d\xi \, d\eta \tag{13}$$

However the matrix [B] should be computed from the displacement function.it is

$$[B] =$$

where $[B] = [B_1, B_2, \cdots\cdots, B_n]$

$$\begin{Bmatrix} \frac{\partial M_i}{\partial x} \\ \frac{\partial M_i}{\partial y} \end{Bmatrix} = [J]^{-1} \begin{Bmatrix} \frac{\partial M_i}{\partial \xi} \\ \frac{\partial M_i}{\partial \eta} \end{Bmatrix}$$

$$\frac{\partial M_i}{\partial \xi} = \frac{\partial N_i'}{\partial \xi} f(\xi,\eta) + N_i' \frac{\partial f(\xi,\eta)}{\partial \xi}$$

$$\frac{\partial M_i}{\partial \eta} = \frac{\partial N_i'}{\partial \eta} f(\xi,\eta) + N_i' \frac{\partial f(\xi,\eta)}{\partial \eta}$$

The decay function should represent of displacement conditions and its varying law on infinite domain.One of the proper method is considering the theory solution of the same type problem and choice the corresponding form.The following forms are usually used for geomechancis problems

$$f(\xi,\eta) = (r_i/r^b)^a$$

or $$f(\xi,\eta) = \exp(1-r/r_i) \qquad (14)$$

Where r and r_i is the distance from the computing point and the nodal points of infinite element to the decay centre,respectively, which are also the function of ξ and η, the a and b are constants.

3. JOINT INFINITE ELEMENT

The joint infinite element is shown in Fig.3. If the mapping method is used to form the infinite element (see Fig.3a.) the shape function can be taken as.

$$N_i = -\xi \;, \quad (i=1,2) \qquad (15)$$
$$N_i = 1+\xi \;, \quad (i=3,4)$$

and the intervenient function of displacement is written as

$$N_i = \tfrac{1}{2}\xi'(\xi'-1) \qquad (i=1,2) \qquad (16)$$
$$N_i = (1-\xi') \qquad (i=3,4)$$

In the second way, the shape function (Fig.3b.) is

$\xi \le 0$

$$N_i = \xi(2\xi+1) \;, \qquad (i=1,6)$$
$$N_i = 4\xi(\xi+1) \;, \qquad (i=2,5) \qquad (17a)$$
$$N_i = (2\xi+1)(\xi+1) \;, \qquad (i=3,4)$$

$\xi > 0$

$$N_i = (1-4\xi)/(1-\xi) \;, \qquad (i=1,6)$$
$$N_i = 2\xi/(1-\xi) \;, \qquad (i=2,5) \qquad (17b)$$
$$N_i = \xi(2\xi-1)/(1-\xi) \;, \qquad (i=3,4)$$

,

and the decay function may choose the same form as the formula (14).In considering of the joint is defined by the normal stiffness Kn,and the tangential stiffness Ks and the corresponding stress and , the stiffness matrix can be obtained by

$$[Ke] = \iint_A [B]^T[D][B]\,dA$$
$$= \iint_A [B]_j^T[D]_j[B]_j\,dA \qquad (18)$$

Where the matrix [B] and [D] correspond to local coordinate of joint.The integration of [Ke] over the infinite element is simular to the formula (8) or (13) and dependent on the type of joint infinite.

4. CONSIDERING THE VOLUME FORCE

Considering the volume force in the infinite element has not difficult if it is constant in a element. Because of the above mentioned shape function Ni is satisfactory to the geometric conditions at infinity of the problem.So the formula of the equivanlent nodal forces of the volume force can be written as foiiows

$$\{Fe\} = \int_{-1}^{1}\int_{-1}^{1}[N]^T\begin{Bmatrix} Q_x \\ Q_y \end{Bmatrix} \det|J_1|\det|J_2|\,d\xi'\,d\eta'\,t$$
$$(19b)$$
$$\{Fe\} = \int_{-1}^{1}\int_{-1}^{1}[N]^T\begin{Bmatrix} Q_x \\ Q_y \end{Bmatrix} \det|J|\,d\xi\,d\eta$$

Similarly, the mass matrix [m] and the damping matrix [Ce] of infinite elementin the dynamic analysis can be obtained by

$$[m_e] = \rho t \int_{-1}^{1}\int_{-1}^{1}[N]^T[N]\det|J_1|\det|J_2|\,d\xi'\,d\eta' \qquad (20a)$$

or $$[m_e] = \rho t\int_{-1}^{1}\int_{-1}^{1}[N]^T[N]\det|J|\,d\xi\,d\eta \qquad (20b)$$

so $$[Ce] = \alpha[m_e] + \beta[Ke] \qquad (21)$$

Where the (19a) and (20a) are corresponding the mapping element, and the formulas (19b) and (20b) are used to the infinite element with the decay function. The shape function Ni is taking from the corresponding formula as the mentioned above.

5. RESEARCH OF APPLICATION OF INFINITE ELEMENT

In this section we discussed the application of infinite element and researched some problems of application in practice.

Example 1. A circular tunnel is loaded with the axialsymmetry. We choise this example so as to compare the numerical solution with the theory one.

The initial stress in the ground is $\sigma_{ox} = \sigma_{oy} = 10$ mpa, and parameters of the rock mass are : E = 5X10 mpa and $\nu = 0.21$, C =1.1 mpa, $\varphi = 31^\circ$. The FEM meshes are shown in Fig.4. We had carried out the elastic and the elastic-plastic solutions.They indicated that. using the infinite element to the FEM the computing results and comparision of them are shown in Fig.5 and Fig.6.

All solutions either the mapping infinit element or with the various decay function are agrement with theoretic results.

In order to research the influence of various decay functions, we chose three forms, that is $f_1=ri/r$, $f_2=\exp(1-r/ri)$ and $f_3=(ri/r)^2$. The computing results are listed in table 1, and shown in Fig.7.

Example 2. It is the same tunnel as example 1. Now we research the economical effect and the computing time of infinite element. Therefore, the meshes showing in Fig.4.a. and 4b are used to compute, and the resluts are listed in table 2. It is shown the computing time using of infinite element is short about 40-60% than common FEM, and the volume of prer-and post-precessing of FEM can be shortened greatly.

Example 3. A mining opening is shown in Fig.8. The initial stresses is $\sigma_{ox}=10$ mpa, $\sigma_{oy}=6$ mpa, and the parameters of rock mass and joint is:

rock mass	E=10 mpa,	$\nu=0.35$
	C=10 mpa ,	$\varphi=40°$,
joint	Kn= 10 mpa ,	Ks=15 mpa
	Cj=1.0 mpa ,	$\varphi_j=28°$

The calculating results are shown in Fig.9 and the comparision of computing time are listed in table 4.

It is shown that the joint infinite element proposed in this paper are also useful and the computing occuracy are satisfactory.

6. CONCLUSION

Introduction of infinite element into the FEM is a useful approach for solving the infinite or semi-infinite domain problems. It can overcome the difficulties of determination the calculating region and the boundary conditions in aassumed regionand may increase the computing precesion.

The other adevantage of infinite element are decreasing the computing region, shorting the computing time and request a little storage capacity. Therefore, it is a useful appreach for solving the larg topic by micro- or personal-computer. All the examples in this paper are finished by a personal computer IBM-PCXT

The dacay function is important factor influence the computing results but the solution of different dacay functions are similar in the near region, and it is the most interesting region for engineering practice. So that, to useing which form of decay function is not important, if the convergance over the infinite domain can be ensured, then their computing precision in the near region will be satisfied. The mapping infinite element may be the better approach.

Using the infinite element to solving the dynamic problems is also effect. The present paper we could not more deadlly discussed with it.

REFRANCE

1 Petter Bettes, Infinite element. Int. J. Num. meth. in Engn. Val.5, 1977.
2 Petter Beetess, More on infinite element Int.J. Num.meth.Engn.Val.15,. 1980.
3. Zienkiewicz, O.C., Emoson, C., Better, P., A novel boundary infinite element. Int. J. Num. Meth. Engn. Val.19, 194-404, 1983.
4. Beer, G. Infinite element in finite element analysis of underground excavations. Int. J. Num. and Analysis Meth. ENgn. Val. 7, 1983.

Table 1. The influence of differert decay functoin to the computing solutions (diametral displacement)

Destance (m)	Theory solution	Decay function $f_1=r_i/r$	$f_2=\exp(r/r_i)$	$f=(r_i/r)^2$
2.0	-0.1823	-0.1853	-0.1823	-0.1778
2.5	-0.1403	-0.1453	-0.1448	-0.1407
3.5	-0.1053	-0.1028	-0.1018	-0.0973
4.5	-0.0704	-0.0772	-0.0772	-0.0789
6.0	-0.0504	-0.0502	-0.0551	-0.0563
7.5	-0.0400	-0.0391	-0.0413	-0.0409
9.5	-0.0263	-0.0210	-0.0261	-0.0295
11.5	-0.0152	-0.0133	-0.0197	-0.0243
13.5	-0.0122	-0.0103	-0.0148	0.0196

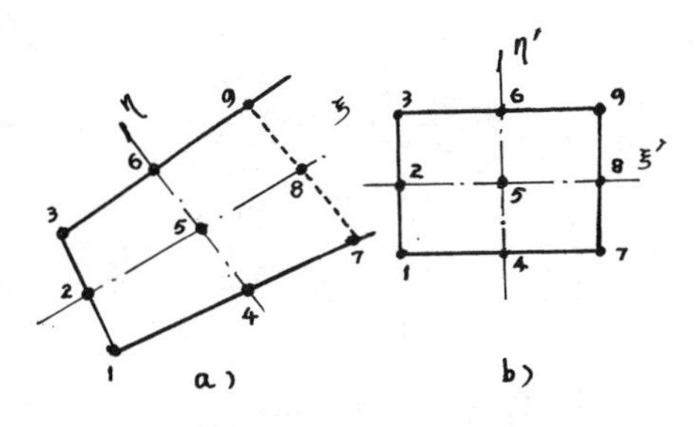

Fig.1 Infinite element and its mapping coordinate
a) local coordinate;
b) mapping coordinat

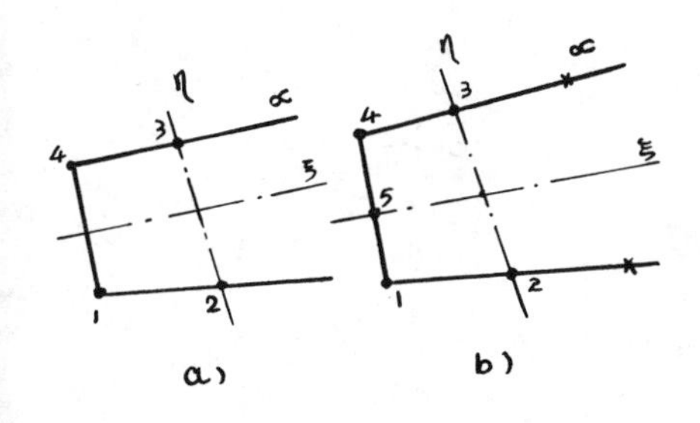

Fig.2 The 4-or 5-nodal points infinite element

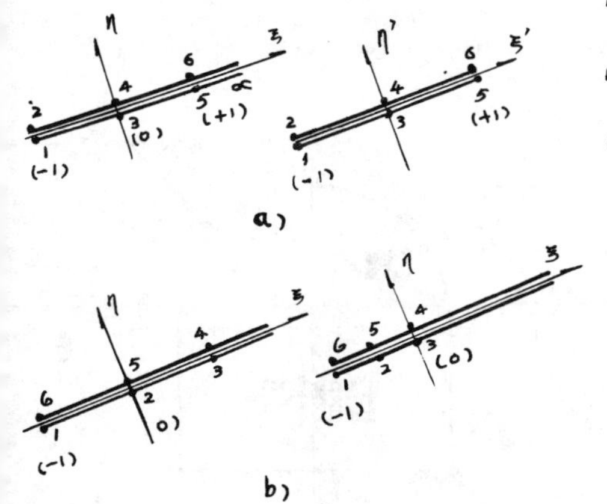

Fig.3 The joint infinite element
a) The mapping element
b) The element using the decay function

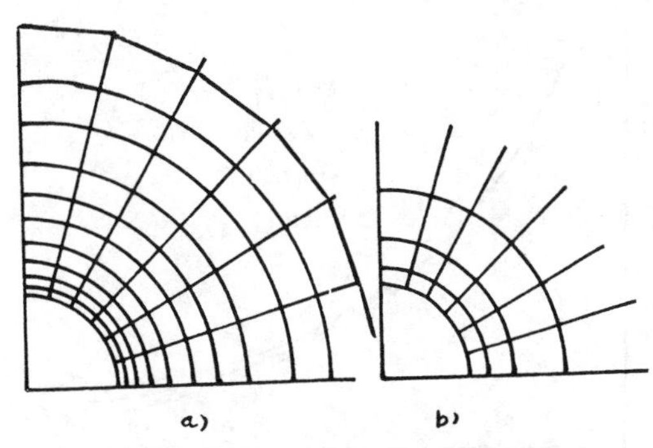

Fig.4 The meshes of finite elements and infinite elements.

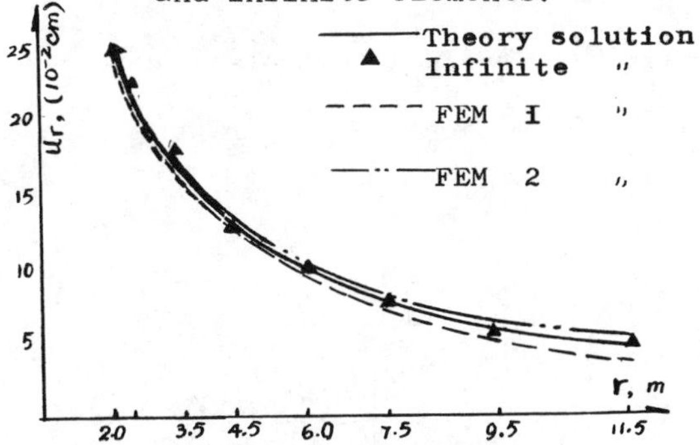

Fig.5 The elaseo-plastic displacement of analytical and numerical solutions

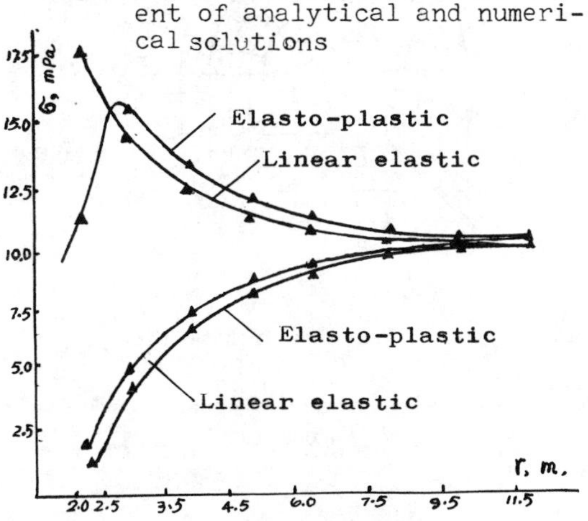

Fig.6 The stress of infinite element and analytical solutions

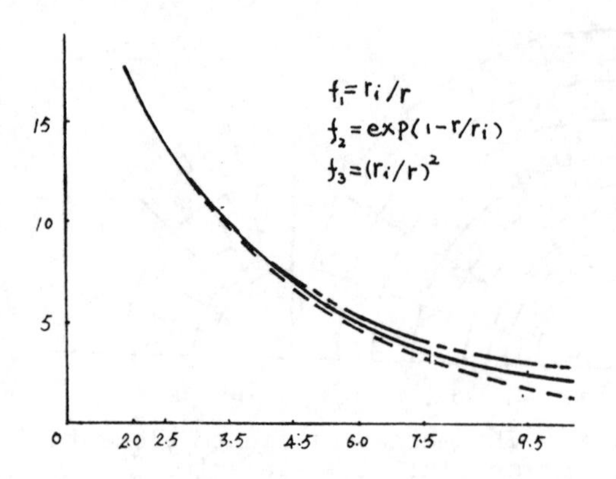

Fig.7 The infiunce to displacement of different decay functions.

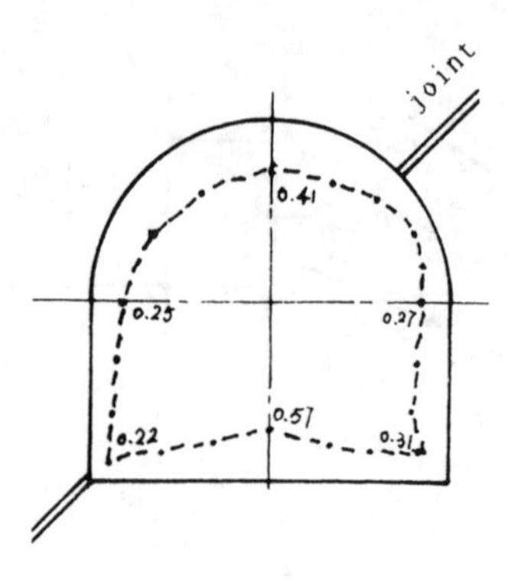

a) Displacement of the rock mass (elastic)

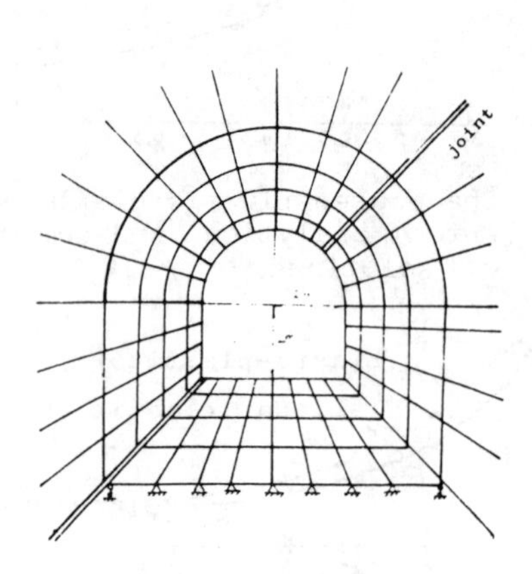

Fig.8 The meshes of a arch secton opening

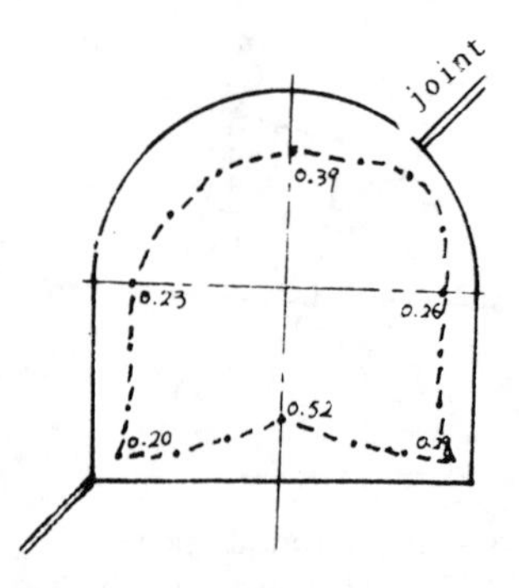

b) Displacement of the rock mass (plastic)

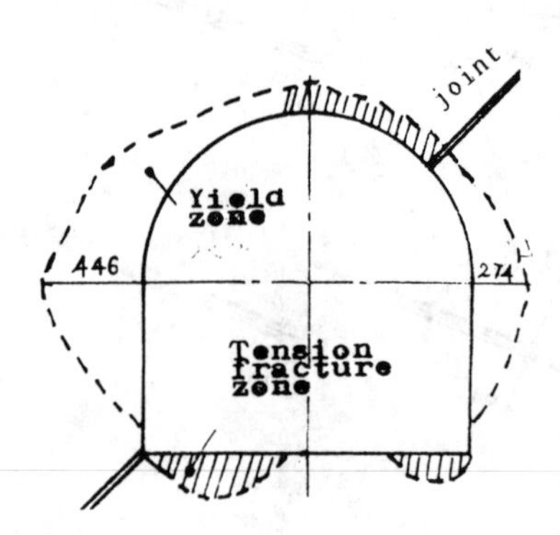

c) Yied and tension fracture in the rock mass

Fig.9 The computing results of a arch opening

Table 2. The computer time and resuits of elasto-plastic analysis by FEM, inFEM and theory

Diametral destance	Solving method					
	Theoretic		inFEM		FEM	
	u_r,(cm)	σ_r, mpa	u_r,(cm)	σ_r, mpa	u_r,(cm)	σ_r, mpa
2.5	-0.2482	-10.40	-0.2556	-10.38	-0.2381	-10.21
2.5	-0.1990	-15.60	-0.2070	-15.53	-0.1879	-15.15
3.5	-0.1423	-13.42	-0.1480	-13.38	-0.1319	-13.08
4.5	-0.1135	-11.99	-0.1175	-11.93	-0.0993	-11.65
6.0	-0.0860	-11.17	-0.0907	-11.01	-0.0711	-10.90
7.5	-0.0723	-10.75	-0.0784	-10.67	-0.0529	-10.48
Computer time (sec)			2101		3410	

The FEM computing boundary is from the centre to the 5 diametral destance of the opening.

Table 3. The computing plane and time of the arch sectional opening

Computing method	Number of element				computer time (min)	
	piane elements	infinite elements	joint elemen	jiont infinite elements	linear	non-linear
FEM	226	0	14	0	16.3	46.6
EM with inFEM	112	8	20	2	7.7	28.1

Numerical Methods in Geomechanics (Innsbruck 1988), Swoboda (ed.)
© 1988 Balkema, Rotterdam. ISBN 90 6191 809 X

The application of the boundary element method to the stress analysis of anisotropic and heterogeneous solid medium

Ge Xiurun & Yang Wei
Institute of Rock & Soil Mechanics, Academia Sinica, Wuhan, People's Republic of China

ABSTRACT: A method considering "amended fictitious body forces" is presented to deal with the problems of stress analysis of heterogeneous and anisotropic solid medium in boundary element method. An example solved with this method shows that the numerical method is quite effective and accurate.

INTRODUCTION

This paper presents a method to deal with the problems of stress analysis of heterogeneous and anisotropic solid medium in boundary element method.

The need for considering complicated anisotropic and heterogeneous madia often appear in geomechanical analysis, and for these problems the so called fundamental solutions are difficult to be found. For solving this kind of problems, if the coupled b.e.-f.e. method is used, the set of linear algebraic equations would be very large. In this paper, the influence of anisotropy and heterogeneity is replaced by so called "amended fictitious body forces" based on the reference medium. The actual stress and displacement field is approximated by means of the iterative process. The process in this method is similar to the modified Newton-Raphson method, which is commonly used in finite element analysis (Zienkiewicz 1977). Some concerned parts of the interior domain must be discreted and in this paper four-node quadrilateral elements are used in the subdivision.

A typical example of computation for heterogeneous problem is presented bellow and we can see that the results of numerical calculation is quite coincident with the theoretical solution.

Furthermore, the method presented here can also easily be used in the computation of the heterogeneous and/or anisotropic viscoelastic problems with a few improvements, which will not cause any difficulty.

BOUNDARY INTEGRAL EQUATIONS

Just as in discussion of elasto-plastic problems suggested by Telles (1983), we have the following incremental integral equations

$$c_{ij}(s)\dot{u}_j(s)=\int_S U^*_{ij}(s,Q)\dot{p}_j(Q)dS(Q) -\int_S P^*_{ij}(s,Q)\dot{u}_j(Q)dS(Q)+\int_\Omega U^*_{ij}(s,q)\cdot \dot{b}_j(q)d\Omega(q)+\int_\Omega \varepsilon^*_{jki}(s,q)\dot{\sigma}^a_{jk}(q)d\Omega(q) \quad (1)$$

$$\dot{\sigma}_{ij}=\int_S U^*_{ijk}\dot{p}_k dS-\int_S P^*_{ijk}\dot{u}_k dS+\int_\Omega U^*_{ijk}\dot{b}_k d\Omega+\int_\Omega \varepsilon^*_{ijkl}\dot{\sigma}^a_{kl}d\Omega+g_{ij}(\dot{\sigma}^a_{kl}) \quad (2)$$

where u_j is the actual boundary displacement and σ_{ij} is the actual stress distribution in region Ω. U^*_{ij}, P^*_{ij}, U^*_{ijk}, ε^*_{jki}, ε^*_{ijkl} and g_{ij} are the same as those used by Telles (1983). When s is on the boundary, c_{ij} are functions of boundary geometry, and within region Ω, $c_{ij}=\delta_{ij}$.

Integrating Eqs. (1) and (2) by the component of time and assuming that in some region $\sigma^a_{ij}=0$, we have

$$c_{ij}u_j=\int_S U^*_{ij}p_j dS-\int_S P^*_{ij}u_j dS+\int_\Omega U^*_{ij}b_j d\Omega$$

$$+\int_{\Omega 1}\varepsilon^*_{jki}\sigma^a_{jk}d\Omega \qquad (3)$$

$$\sigma_{ij}=\int_S U^*_{ijk}p_k dS-\int_S P^*_{ijk}u_k dS+\int_\Omega U^*_{ijk}\cdot$$

$$b_k d\Omega+\int_{\Omega 1}\varepsilon^*_{ijkl}\sigma^a_{kl}d\Omega+g_{ij}(\sigma^a_{kl}) \qquad (4)$$

Ω_1 is subregion of Ω, where $\sigma^a_{ij}\neq 0$.

Eqs. (3) and (4) can be used as fundamental equations in stress analysis in anisotropic and/or heterogeneous solid media.

If stress σ^a_{ij} is prescribed, we can easily obtain displacement u_j and stress σ_{ij} from Eqs. (3),(4). Discretizing the whole boundary S and interior Ω_1 and by means of Gaussian numerical integration on each boundary and interior region element, we can obtain a group of algebraic equations:

$$Ax=f+D\sigma^a \qquad (5)$$

where A is determined by the geometry of the boundary and D by the geometry of both the boundary and subregion Ω_1; f is a vector made up of the known terms of displacements and/or tractions on boundary S and the distributing body forces in Ω; σ^a is a vector made up of σ^a_{ij} distributing in Ω_1. When considering that the whole region Ω is made up of one kind of material C_{ijkl} and σ^a_{ij} is the initial stress (see the following paragraph), we can obtain the internal displacements from Eq. (3) by taking $c_{ij}=\delta_{ij}$, and furthermore

$$\sigma=Bx+Ff+H\sigma^a \qquad (6)$$

BASIC EQUATIONS AND ITERATIVE PROCESS

Supposing there is a solid medium Ω and it is naturally divided into two subregions Ω_o and Ω_1, as shown in Fig. 1. In Ω_o, we suppose the rela-

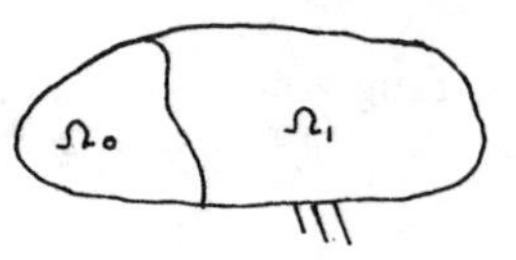

Fig. 1 A solid medium made up of two kinds of materials

tion between stress and strain is isotropic and can be written in the following form:

$$\sigma_{ij}=C_{ijkl}\varepsilon_{kl} \qquad (7)$$

where C_{ijkl} is the 4th-order isotropic tensor of elastic constants

$$C_{ijkl}=\frac{2G\nu}{1-2\nu}\delta_{ij}\delta_{kl}+G(\delta_{ik}\delta_{jl}+\delta_{il}\delta_{jk}) \qquad (8)$$

In subregion Ω_1, the relation between stress and strain can be written as

$$\sigma_{ij}=D_{ijkl}\varepsilon_{kl} \qquad (9)$$

in which D_{ijkl} can be any 4th-order tensor of elastic constants to describe the material character in subregion Ω_1.

In Eqs. (7) and (9), stress σ_{ij} and strain ε_{kl} are actual ones in subregions Ω_o and Ω_1 respectively. Supposing region Ω is made up of one kind of material, which is called "the reference material", its elastic constants are C_{ijkl}, and in both Ω_o and Ω_1, its deformation is correctly represented by actual strain ε_{kl}. Then the true stress in Ω_1 noted by σ^t_{ij} can be written as

$$\sigma^t_{ij}=D_{ijkl}\varepsilon_{kl}=C_{ijkl}\varepsilon_{kl}-\sigma^a_{ij} \quad \text{in } \Omega_1 \qquad (10)$$

where

$$\sigma^a_{ij}=(C_{ijkl}-D_{ijkl})\varepsilon_{kl} \quad \text{in } \Omega_1 \qquad (11)$$

Under a specific boundary condition, if either σ^a_{ij} or ε_{ij} is prescribed, we can have the whole solution of this problem. But both the tensors are not prescribed and an iterative method under the fundamental solution of the so called "reference medium" is presented here to approach the actual solution of the problem. We should seek σ^a_{ij} above all by use of iterative process. In the further discussion, we consider that the influence of non-zero stresses σ^a_{ij}, which are taken as "amended fictitious body forces" for its effect when the iterative process is made, is just similar to that of body forces. If a pure anisotropic problem appears, we should have $\Omega_1=\Omega$, and Ω_o becomes the only fictitious reference medium.

For deriving the iterative process, we express the problem as follows

$$\left.\begin{array}{ll}\sigma_{ij,j}+f_i=0 & \\ \varepsilon_{ij}=(u_{i,j}+u_{j,i})/2 & \\ \sigma_{ij}=D_{ijkl}\varepsilon_{kl} & X\in\Omega_1 \\ \sigma_{ij}=C_{ijkl}\varepsilon_{kl} & X\in\Omega_o \\ u_i=\bar{u}_i \quad X\in S_u, \quad t_i=\bar{t}_i & X\in S_t\end{array}\right\}\quad(12)$$

The actual problem (12) can be considered as the equivalent obtained by means of superposition of the following two problems

$$\left.\begin{array}{ll}\sigma^{(1)}_{ij,j}+f_i=0 & \\ \varepsilon^{(1)}_{ij}=(u^{(1)}_{i,j}+u^{(1)}_{j,i})/2 & \\ \sigma^{(1)}_{ij}=C_{ijkl}\varepsilon^{(1)}_{kl} & X\in\Omega_1 \\ \sigma^{(1)}_{ij}=C_{ijkl}\varepsilon^{(1)}_{kl} & X\in\Omega_o \\ u^{(1)}_i=\bar{u}_i \quad X\in S_u, \quad t^{(1)}_i=\bar{t}_i & X\in S_t \\ & \\ \sigma^{(2)}_{ij,j}=0 & \\ \varepsilon^{(2)}_{ij}=(u^{(2)}_{i,j}+u^{(2)}_{j,i})/2 & \\ \sigma^{(2)}_{ij}=A^1_{ij}(\varepsilon^{(2)}_{kl}) & X\in\Omega_1 \\ \sigma^{(2)}_{ij}=A^o_{ij}(\varepsilon^{(2)}_{kl}) & X\in\Omega_o \\ u^{(2)}_i=0 \quad X\in S_u, \quad t^{(2)}_i=0 & X\in S_t\end{array}\right\}\quad(13)$$

where

$$\left.\begin{array}{ll}\sigma^{(1)}_{ij}+\sigma^{(2)}_{ij}=\sigma_{ij} & \text{in } \Omega_o U\Omega_1 \\ \varepsilon^{(1)}_{ij}+\varepsilon^{(2)}_{ij}=\varepsilon_{ij} & \text{in } \Omega_o U\Omega_1\end{array}\right.\quad(14)$$

Then we can easily obtain

$$\left.\begin{array}{l}A^o_{ij}(\varepsilon^{(2)}_{kl})=C_{ijkl}\varepsilon^{(2)}_{kl} \\ A^1_{ij}(\varepsilon^{(2)}_{kl})=D_{ijkl}\varepsilon^{(2)}_{kl}-E_{ijkl}\varepsilon^{(1)}_{kl}\end{array}\right\}\quad(15)$$

where $E_{ijkl}=C_{ijkl}-D_{ijkl}$

According to the same principle $\sigma^{(2)}_{ij}$ can be also written as

$$\sigma^{(2)}_{ij}=\sigma^{(3)}_{ij}+\sigma^{(4)}_{ij}\quad(16)$$

where $\sigma^{(3)}_{ij}$ and $\sigma^{(4)}_{ij}$ satisfy the following partial differential equations respectively

$$\left.\begin{array}{ll}\sigma^{(3)}_{ij,j}=0 & \\ \varepsilon^{(3)}_{ij}=(u^{(3)}_{i,j}+u^{(3)}_{j,i})/2 & \\ \sigma^{(3)}_{ij}=C_{ijkl}\varepsilon^{(3)}_{kl}-E_{ijkl}\varepsilon^{(2)}_{kl} & \Omega_1 \\ \sigma^{(3)}_{ij}=C_{ijkl}\varepsilon^{(3)}_{kl} & \Omega_0 \\ u^{(3)}_i=0 \quad X\in S_u, \quad t^{(3)}_i=0 \quad X\in S_t & \end{array}\right\}\quad(17,a)$$

$$\left.\begin{array}{ll}\sigma^{(4)}_{ij,j}=0 & \\ \varepsilon^{(4)}_{ij}=(u^{(4)}_{i,j}+u^{(4)}_{j,i})/2 & \\ \sigma^{(4)}_{ij}=D_{ijkl}\varepsilon^{(4)}_{kl}-E_{ijkl}\varepsilon^{(3)}_{kl} & \Omega_1 \\ \sigma^{(4)}_{ij}=C_{ijkl}\varepsilon^{(4)}_{kl} & \Omega_o \\ u^{(4)}_i=0 \quad X\in S_u, \quad t^{(4)}_i=0 \quad X\in S_t & \end{array}\right\}\quad(17,b)$$

Finally, the problem (12) is equivalent to superposition of the following series of problems

$$\left.\begin{array}{l}\sigma^o_{ij,j}+f_i=0 \\ \varepsilon^o_{ij}=(u^o_{i,j}+u^o_{j,i})/2 \\ \sigma^o_{ij}=C_{ijlm}\varepsilon^o_{lm} \quad X\in\Omega_1 U\Omega_o \\ u^o_i=\bar{u}_i \quad X\in S_u, \quad t^o_i=\bar{t}_i \quad X\in S_t\end{array}\right\}$$

$$\left.\begin{array}{l}
\sigma^1_{ij,j}=0 \\
\varepsilon^1_{ij}=(u^1_{i,j}+u^1_{j,i})/2 \\
\sigma^1_{ij}=C_{ijlm}\varepsilon^1_{lm}-E_{ijlm}\varepsilon^o_{lm} \quad X\in\Omega_1 \\
\sigma^1_{ij}=C_{ijlm}\varepsilon^1_{lm} \quad X\in\Omega_o \\
u^1_i=0 \; X\in S_u, \; t^1_i=0 \quad X\in S_t
\end{array}\right\}$$

$$\cdots \quad \cdots$$

$$\left.\begin{array}{l}
\sigma^{(n-1)}_{ij,j}=0 \\
\varepsilon^{(n-1)}_{ij}=(u^{(n-1)}_{i,j}+u^{(n-1)}_{j,i})/2 \\
\sigma^{(n-1)}_{ij}=C_{ijlm}\varepsilon^{(n-1)}_{lm}-E_{ijlm}\varepsilon^{(n-2)}_{lm} \quad \Omega_1 \\
\sigma^{(n-1)}_{ij}=C_{ijlm}\varepsilon^{(n-1)}_{lm} \quad \Omega_o \\
u^{(n-1)}_i=0 \; X\in S_u, \; t^{(n-1)}_i=0 \quad X\in S_t
\end{array}\right\}$$

$$\left.\begin{array}{l}
\sigma^{(n)}_{ij,j}=0 \\
\varepsilon^{(n)}_{ij}=(u^{(n)}_{i,j}+u^{(n)}_{j,i})/2 \\
\sigma^{(n)}_{ij}=D_{ijlm}\varepsilon^{(n)}_{lm}-E_{ijlm}\varepsilon^{(n-1)}_{lm} \quad \Omega_1 \\
\sigma^{(n)}_{ij}=C_{ijlm}\varepsilon^{(n)}_{lm} \quad \Omega_o \\
u^{(n)}_i=0 \; X\in S_u, \; t^{(n)}_i=0 \; X\in S_t
\end{array}\right\} \tag{18}$$

By analysis of above equations, we know that each group of equations (except n-th group) should be solved with the isotropic medium of material constants C_{ijkl} taken into consideration, and the known terms at the lefthand side of constitutive equations are so called "amended fictitious body forces", which can be easily determined by strains ε_{lm} obtained from previous computation of iterative process. And we have

$$\left.\begin{array}{l}
\sigma_{ij}=\sum\limits_{\lambda=0}^{n} C_{ijkl}\varepsilon^{(\lambda)}_{kl} \quad X\in\Omega_o \\
\sigma_{ij}=\sum\limits_{\lambda=0}^{n-1} C_{ijkl}\varepsilon^{(\lambda)}_{kl}-\sum\limits_{\lambda=0}^{n-1} E_{ijkl}\varepsilon^{(\lambda)}_{kl} \quad X\in\Omega_1
\end{array}\right\} \tag{19}$$

If $\varepsilon^{(n)}_{ij}\to 0$ and $\sigma_{ij}\to\infty$ for $X\in\Omega_1$ when $n\to\infty$, we have a method to solve Eqs. (5) and (6) by the following process

(i) for $\sigma^a=\{0\}$ (20)

we have $x=A^{-1}f$ (21)

and from Eq. (12)

$$\sigma=(BA^{-1}+F)f \tag{22}$$

$$\varepsilon_{ij}=(C_{ijkl})^{-1}\sigma_{kl} \tag{23}$$

$$\sigma^t_{ij}(1)=\sigma_{ij} \tag{24}$$

$$\sigma^a_{ij}(1)=(C_{ijkl}-D_{ijkl})\varepsilon_{kl} \tag{25}$$

(ii) $\sigma^t_{ij}=\sigma^t_{ij}(1)-\sigma^a_{ij}(1)$ (26)

for $f=\{0\}$ (27)

we obtain

$$x=A^{-1}D\sigma^a(1) \tag{28}$$

and

$$\sigma=(BA^{-1}D+H)\sigma^a(1) \tag{29}$$

$$\varepsilon_{ij}=(C_{ijkl})^{-1}\sigma_{kl} \tag{30}$$

$$\sigma^a_{ij}(2)=(C_{ijkl}-D_{ijkl})\varepsilon_{kl} \tag{31}$$

$$\sigma^t_{ij}(2)=\sigma^t_{ij}+\sigma_{ij} \tag{32}$$

(iii) If the value of

$$\sigma^t_{ij}(n)-\sigma^t_{ij}(n-1)/\sigma^t_{ij}(n-1) \tag{33}$$

is small enough, the iterative process is finished.

Finally we obtain

$$\varepsilon^t_{ij}=(D_{ijkl})^{-1}\sigma^t_{kl}(n) \tag{34}$$

and furthermore, in subregion Ω_1:

$$\sigma^a_{ij}=(C_{ijkl}-D_{ijkl})\varepsilon_{kl} \tag{35}$$

The principle of the iterative process mentioned above is shown in Fig. 2.

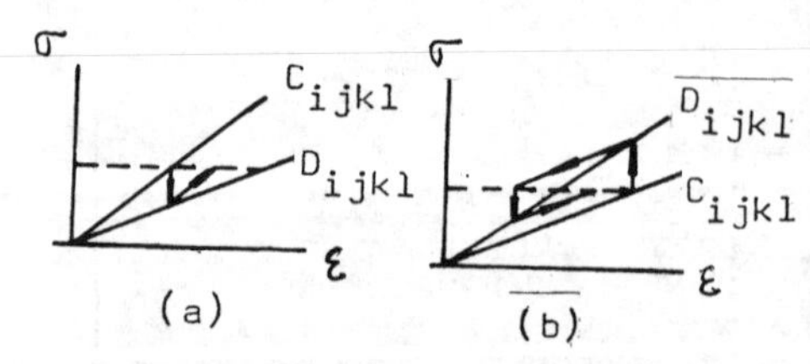

Fig. 2 The principle of the iterative process in the present paper

Under the condition displayed in Fig. 2a, convergency of iterative process can always be reached, but in Fig. 2b, convergency is not always satisfied and should be more deeply discussed.

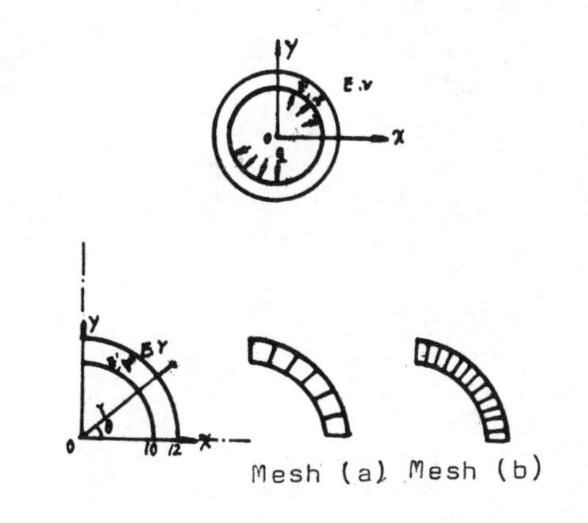

EXAMPLE

The elastic moduli for the bedrocks of the annulus around a circular cavity and the outer region are E' and E, respectively. Poisson's ratio ν for both is 0.2. The cavity is subjected to internal hydraulic pressure q=100 KPa. For the two cases listed in Table 1 b.e. calculation for heterogeneous problems is performed using the method proposed in the present paper, and for the purpose of comparison these problems are also solved theoretically. The distribution of **radial** and tangential stresses is given in Fig. 3 and displacements are listed in Table 1. These results display a rather good coincidence of the numerical and theoretical solutions, which provides support for the feasibility of the principle of the method presented here to deal with heterogeneity and anisotropy by means of BEM on the basis of the concept of "reference material".

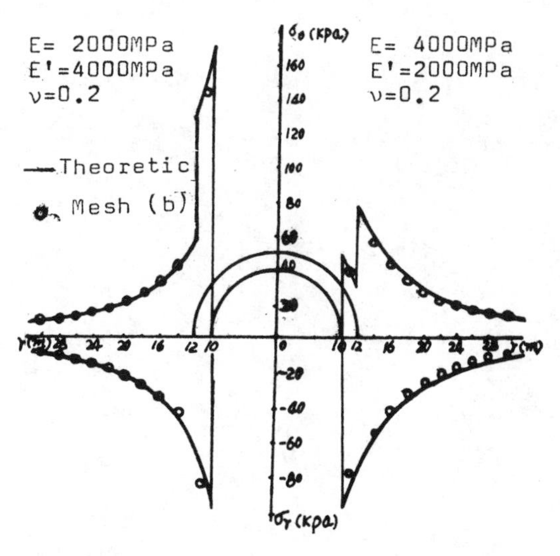

Fig. 3 Comparison of stress distribution surrounding the cavity by theoretical and numerical methods

Table 1. Comparison of results of boundary displacements

Material	Mesh	u_r (mm)	Theoretc u_r (mm)	Error (%)
E=4000 E'=2000 (MPa)	a	0.3584	0.3703	-3.2
	b	0.3695	0.3703	-0.9
E=2000 E'=4000 (MPa)	a	0.4650	0.4749	-2.1
	b	0.4709	0.4749	-0.8

REFERENCES

Telles, J.C.F. 1983. The boundary element method applied to inelastic problems, Springer-Verlag (Lecture Notes In Engineering, ed. by Brebbia, C.A. et al)

Zienkiewicz, O.C. 1977. The finite element method, (the third, expanded and revised edition of The Finite Element Method In Engineering Science)

Numerical Methods in Geomechanics (Innsbruck 1988), Swoboda (ed.)
© 1988 Balkema, Rotterdam. ISBN 90 6191 809 X

Boundary element analysis of linear quasi-static coupled transient thermoelasticity

D.W.Smith & J.R.Booker
University of Sydney, Australia

ABSTRACT: Traditionally, transient thermoelastic problems have been solved by assuming that the determination of temperature can be uncoupled from the determination of strain. Thus it is postulated that a change in temperature causes the material to strain, but that straining causes no temperature change. This semi- coupled approach usually provides an adequate approximation, however, a more theoretically sound treatment recognises the coupling of strain and temperature fields. In this paper, a boundary integral equation is derived and a numerical procedure based on this equation is implemented. A comparison of analytic and numerical solutions provides a means of assessing the accuracy of numerical solutions.

1. INTRODUCTION

In conventional thermal analyses, it is usual to assume material straining induces no temperature change. This semi-coupled theory is often implemented numerically using finite elements. More recently, boundary element methods have been used. The most satisfactory boundary integral equation for uncoupled thermoelasticity has been presented by Sladek and Sladek (1984). This integral equation has been used as the basis of a numerical procedure by Sharp and Crouch (1986), but the procedure involves time stepping which is computationally expensive, and domain integrals which limit its usefulness, especially in those applications with infinite domains. However, by using a Laplace transformation both these difficulties are removed, and this approach has been successfully implemented by Smith and Booker (1987).

There are few numerical studies of coupled thermoelasticity and fewer still analytic solutions. Sladek and Sladek (1983) have presented a boundary integral equation and fundamental solutions for coupled thermoelastic problems, but do not proceed to a numerical implementation. In this paper we present a practical numerical procedure that uses the Laplace transform technique, and thereby has the advantages of avoiding time stepping and domain integrals as in the semi-coupled analysis.

2. GOVERNING EQUATIONS

The process of coupled thermoelastic deformation is governed by the following set of equations.

(a) Equations of Internal Equilibrium

$$\frac{\partial \sigma_{ij}}{\partial x_j} + F_i = 0 \qquad (1)$$

where σ_{ij} = the components of the incremental stress tensor

F_i = the components of the incremental body force vector.

(b) Stress Strain Relations

$$\sigma_{ij} = -\beta\theta\delta_{ij} + \lambda\epsilon_v\delta_{ij} + 2G\epsilon_{ij} \qquad (2)$$

where ϵ_{ij} = components of incremental strain tensor

ϵ_v = ϵ_{kk} is the incremental volume strain

λ = Lame modulus

G = shear modulus

α = coefficient of thermal expansion

β	=	$\alpha(3\lambda + 2G)$ is the thermal modulus

(c) Strain Displacement Relation

$$\epsilon_{ij} = \tfrac{1}{2}\left(\frac{\partial u_i}{\partial x_j} + \frac{\partial u_j}{\partial x_i}\right) \quad (3)$$

where u_i = components of the displacement vector (measured from the initial state)

(d) Constitutive Relation for Heat Flow

$$h_i = -k\frac{\partial \theta}{\partial x_i} \quad (4)$$

where

h_i	=	component of heat flux vector
θ	=	increment in temperature from initial isothermal state
k	=	thermal conductivity

(e) Conservation of Energy

$$\rho c\ \theta + T_c\ \beta\ \epsilon_v = + \int_0^t \left(q - \frac{\partial h}{\partial x_i}\right) dt \quad (5)$$

where

ρ	=	density
c	=	specific heat (at constant volume)
T_c	=	current absolute temperature
t	=	time
q	=	rate of generation per unit volume.

The term $T_c\beta\epsilon_v$ accounts for the temperature changes induced by material straining and is ignored in the semi-coupled analysis. Equation (5) may be linearised by assuming T_c is equal to the initial absolute temperature T_0. If θ is small compared to T_c, the error introduced by this assumption is also small.

Equations (1-5) must be satisfied throughout the region under consideration, together with appropriate boundary conditions.

(f) Equations of Boundary Equilibrium

$$\sigma_{ij}\ n_j = T_i \quad (6)$$

where

n_j	=	components of the outward normal to the surface of the body
T_i	=	components of traction.

(g) Displacement Boundary Condition

$$u_i = d_i \quad (7)$$

where d_i = component of prescribed boundary displacement.

Compatible tractions and displacements may be specified over the surface S.

(h) Thermal Boundary Conditions

These boundary conditions may be expressed

$$Mh_n + N\theta_s = W \quad (8)$$

where

h_n	=	outward normal component of flux on the surface
θ_s	=	surface temperature
M,N	=	known functions of position
W	=	known function of position and time.

By suitable specialisation of the functions M, N and W, most of the commonly occurring boundary conditions may be specified. For example if $M = 0$, $N = 1$ a boundary temperature is prescribed, and if $M = 1$, $N = 0$ a boundary flux is prescribed. If $M = 1$, $N = H$, $W = H\theta_a$ where H is the surface heat transfer coefficient and θ_a is the temperature of the surroundings, then a convection boundary condition is prescribed. Mixed boundary data may be also be prescribed over the surface S.

If we introduce the Laplace transform of some function f(t) as,

$$f(s) = \int_0^\infty e^{-st}\ f(t)\ dt \quad (9)$$

we may use this to temporarily remove the time dependence from equations (1) to (5) and boundary conditions (6) to (8). If we introduce equation (4) into equation (5) it is convenient to express the result for a homogeneous isotropic material as,

$$(A - M)\bar{\epsilon}_v = \beta\left[\frac{\kappa}{s}\ \nabla^2\ \bar{\theta} - \bar{\theta} + \frac{\bar{q}}{s\rho c}\right] \quad (10)$$

where M = isothermal confined modulus

= $(\lambda + 2G)$

A = adiabatic confined modulus

= $M + \dfrac{\beta^2 T_0}{\rho c}$

3. BOUNDARY INTEGRAL EQUATION

Equations (1) to (3) are equivalent to an equation of virtual work. Using Betti's reciprocal theorem and the Duhamel-Neuman body force analogy, Nowacki (1962), we find

$$\int_S (\overline{T}_i^* \overline{u}_i - \overline{T}_i \overline{u}_i^*) dS =$$

$$\int_V (\beta \overline{\theta} \overline{\epsilon}_V^* - \beta \overline{\theta}^* \overline{\epsilon}_V) dV - \int_V \overline{F}_i \overline{u}_i dV \tag{11}$$

where $\overline{F}_i^*$ = components of body force vector in starred system of temperatures and displacements.

Then introducing equation (10) into (11), (with no heat sources present in the unstarred system) assuming constant material properties over the domain and using the divergence theorem leads to

$$\int_S (\overline{T}_i^* \overline{u}_i - \overline{T}_i \overline{u}_i^*) dS =$$

$$\frac{\beta^2}{(A - M)} \frac{\kappa}{s} \int_S (\frac{\partial \overline{\theta}^*}{\partial n} \overline{\theta} - \frac{\partial \overline{\theta}}{\partial n} \overline{\theta}^*) dS$$

$$+ \frac{\beta^2}{(A - M)} \frac{1}{s} \int_V \frac{\overline{\theta q}^*}{\rho c} dV - \int_V \overline{F}_i^* \overline{u}_i dV \tag{12}$$

We may now choose the starred system of temperatures and displacement as due to a line heat source (or point heat source in 3D), that is,

$$\overline{q}^* = \rho c \delta (r - r_0) \tag{13}$$

and $\overline{F}_i^*$ = 0

where δ = dirac delta function

r_0 = position vector of source point

r = position vector of field point.

Then using the property

$$\int_V f \delta (r - r_0) dV = f(r_0) \tag{14}$$

where f is an arbitary differentiable function we find

$$\frac{\epsilon (r_0) \beta^2 \overline{\theta}(r_0)}{(A - M) s} = \int_S (\overline{T}_i^* \overline{u}_i - \overline{T}_i \overline{u}_i^*) dS$$

$$+ \frac{\beta^2}{(A - M)} \frac{\kappa}{s} \int_S (\overline{\theta}^* \frac{\partial \overline{\theta}}{\partial n} - \overline{\theta} \frac{\partial \overline{\theta}^*}{\partial n}) dS \tag{15}$$

where $\epsilon(r_0)$ = 1 inside the domain

= ½ on a smooth boundary

= 0 outside the domain

Alternatively, we may choose the starred system of temperatures and displacements as due to a line load (or a point load in 3D), that is,

$$\overline{F}_i^* = \delta (r - r_0) \tag{16}$$

$\overline{q}^* = 0$ to find

$$\epsilon (r_0) \overline{u}_i (r_0) = \int_S (\overline{T}_i^* \overline{u}_i - \overline{T}_i \overline{u}_i^*) dS -$$

$$\frac{\beta^2}{(A - M)} \frac{\kappa}{s} \int_S (\overline{\theta}^* \frac{\partial \overline{\theta}}{\partial n} - \overline{\theta} \frac{\partial \overline{\theta}^*}{\partial n}) dS \tag{17}$$

Thus, providing there are no body forces are heat sources present, equation (15) expresses the temperature at the field point r_0, and equation (17) the displacement at field point r_0 in terms of boundary integrals only. If we choose the starred systems to be due to the derivative of the dirac delta function in a given direction, then we may use the property

$$\int_V f \frac{\partial \delta (r - r_0)}{\partial x_i} dV = - \frac{\partial f(r_0)}{\partial x_i} \tag{18}$$

to find the flux and strains at an internal point. If there are body forces and heat sources present, and the distribution of both satisfy Laplace's equation, then the volume integrals resulting from the quantities may also be expressed as boundary integrals, though details are not given here.

4. PRINCIPAL SOLUTIONS

The effect of a line load or a line source acting in an infinite coupled thermoelastic medium have been given by Nowacki (1964) and Sladek and Sladek (1983). These principal solutions have also been presented by Cleary (1977) in the analogous poroelastic problem. It is found that the displacements due to a line load may be conveniently regarded as the sum of two solutions.

$$\overline{U}^* = \overline{U}^*_a + \overline{U}^*_b \tag{19}$$

where U^*_a is the well known integrated Kelvin solution, Flugge (1965), with the isothermal moduli replaced by adiabatic moduli. It can be shown that,

$$\nu_a = \frac{\lambda_a}{2(\lambda_a + G)} \tag{20}$$

where ν_a = adiabatic Poisson's ratio

λ_a = $\lambda + (A - M)$

U^*_b is a transient component that ensures that U^* approaches the integrated Kelvin solution with isothermal moduli in the long term. For a line load acting in the x direction, the transient component is given by

$$\begin{bmatrix} \overline{U}^*_x \\ \\ \overline{U}^*_y \end{bmatrix}_b = \frac{-(A - M)}{MA\mu^2} \left\{ \begin{bmatrix} \chi_1 - \Omega_1 \\ 0 \end{bmatrix} + x(\chi_2 - \Omega_2) \begin{bmatrix} x \\ y \end{bmatrix} \right\} \tag{21}$$

and where χ_1, Ω_1, χ_2, Ω_2 are given by setting m = 1, 2 in

$\chi_m = \frac{(-1)^m}{2\pi} \frac{2^{m-1}}{r^{2m}} (m - 1)!$

$\Omega_m = \frac{(-1)^m}{2\pi} \frac{\mu^m}{r^m} K_m(\mu r)$

K_m = modified Bessel function of order m.

$\mu = \left[\frac{A}{M} \frac{s}{\kappa}\right]^{\frac{1}{2}}$

= root with positive real component.

$r = |\mathbf{r} - \mathbf{r}_0|$

$\mathbf{r}$ = position vector of field point

$\mathbf{r}_0$ = position vector of source point

For a heat source of unit strength in two dimensions the temperature distribution can be shown to be

$$\overline{\theta}^* = \frac{1}{2\pi\kappa} K_0(\mu r) \tag{22}$$

where

$$\kappa = \frac{k}{\rho c}$$

5. NUMERICAL PROCEDURE

Let us assume that the boundary of a region may be represented by N straight line segments, and that the continuous functions of temperature, flux, displacement and traction on the boundary may be approximated by piecewise constant functions, being constant over each element. Higher order approximations both to the boundary geometry and functions may be used, but the simple approximations made here are sufficient to establish the procedure. Then equation (15) may be approximated as

$$\frac{\epsilon(\mathbf{r}_0)\beta^2\,\overline{\theta}(\mathbf{r}_0)}{(A - M)\,s} = \sum_{k=1}^{N} \Bigg[\overline{u}_i \int \overline{T}^*_i\, dS_k - \overline{T}_i \int \overline{u}_i\, dS_k + \frac{\beta^2}{(A - M)} \frac{\kappa}{s} \times \left(\frac{\partial\overline{\theta}}{\partial n} \int \overline{\theta}^*\, dS_k - \overline{\theta} \int \frac{\partial\overline{\theta}^*}{\partial n} dS_k \right) \Bigg] \tag{19}$$

and equation (17) by

$$\epsilon(\mathbf{r}_0)\,\overline{u}_i(\mathbf{r}_0) = \sum_{k=1}^{N} \Bigg[\overline{u}_i \int \overline{T}^*_i\, dS_k - \overline{T}_i \int \overline{u}^*_i\, dS_k - \frac{\beta^2}{(A - M)} \frac{\kappa}{s} \times \left(\frac{\partial\overline{\theta}}{\partial n} \int \overline{\theta}^*\, dS_k - \overline{\theta} \int \frac{\partial\overline{\theta}^*}{\partial n}\, dS_k \right) \Bigg] \tag{20}$$

where

$\int f\, dS_k$ = integral of f over kth element

and the starred system in equation (19) is due to a point heat source and in equation (20) is due to a point load acting in the ith direction.

For a well posed boundary value problem, either the temperature or flux (or a linear combination of these), and either the displacement or traction (or a compatible combination of these) is prescribed for each boundary element. Hence at each boundary element there are 3 unknown quantities for two dimensional problems. To use equations

(19) and (20) to determine the temperature or displacement at an internal point we must first determine the unknown boundary quantities.

To do this, we choose the point r_0 to be at the midpoint of a boundary element, thereby involving boundary quantities only in equations (19) and (20). It can be shown analytically that $\epsilon(r_0)$ is equal to one half, and the integral over the element containing the singularity is evaluated in the Cauchy principal value sense, the singularity point being excluded from the integration. The starred system is chosen to be due to a line heat source and line loads in the x_1 and x_2 directions, thereby generating 3 linearly independent equations from equations (19) and (20). The procedure is repeated at each of the N elements, and the resultant system of 3N equations may be written as

$$[G]\ (Q) = [H]\ (P) \qquad (21)$$

where

$$(Q) = \begin{bmatrix} \frac{\partial \overline{\theta}_k}{\partial n} \\ \overline{T}_{k\ell} \end{bmatrix} \qquad (P) = \begin{bmatrix} \overline{\theta}_k \\ \overline{u}_{k\ell} \end{bmatrix}$$

where

$$\frac{\beta^2}{(A - M)} \quad \frac{\kappa}{s} \int \overline{\theta}^* \, dS_k, \int \overline{u}_i^* \, dS_k$$

= coefficients in G matrix

$$\frac{\beta^2}{(A - M)} \quad \frac{\kappa}{s} \int \frac{\partial \overline{\theta}^*}{\partial n} \, dS_k, \int \overline{T}_i^* \, dS_k$$

= coefficients in H matrix

$\overline{T}_{k\ell}$ = Traction in ℓth direction over kth element.

$\overline{u}_{k\ell}$ = Displacement in ℓth direction over kth element.

k = 1, 2, ..., N

ℓ = 1, 2.

After solution all the boundary quantities are determined, and hence equations (19) may be used to find an estimate of the temperature at an internal point, and equation (20) to find an estimate of the displacement at an internal point. By approximating the exact boundary integral equation for the strain at a point in the same way as for displacement at a point, and using the approximation for temperature at an internal point, the stress at the point is found. Numerical inversion of the transform solution was done using an efficient algorithm due to Talbot (1979).

6. RESULTS

A ratio of adiabatic to isothermal confined moduli (A/M) provides a measure of the degree of coupling between the strain and temperature fields. In the following examples, an A/M ratio of 2 is chosen when comparing numerical and analytic solutions. In most materials this ratio is much less. For each of the examples, 64 boundary elements are used to approximate each surface. Symmetry is used and tension is considered positive. Analytic solutions were obtained by numerical inversion of the Laplace transform solution.

Example 1

A slab, infinite in extent but of finite thickness is initially at a uniform temperature θ_b, and both surfaces traction free. From time t = 0 the upper surface is subjected to a constant temperature θ_a, and the lower surface is maintained at θ_b. Stresses induced in the horizontal plane by transient thermal gradients are shown below.

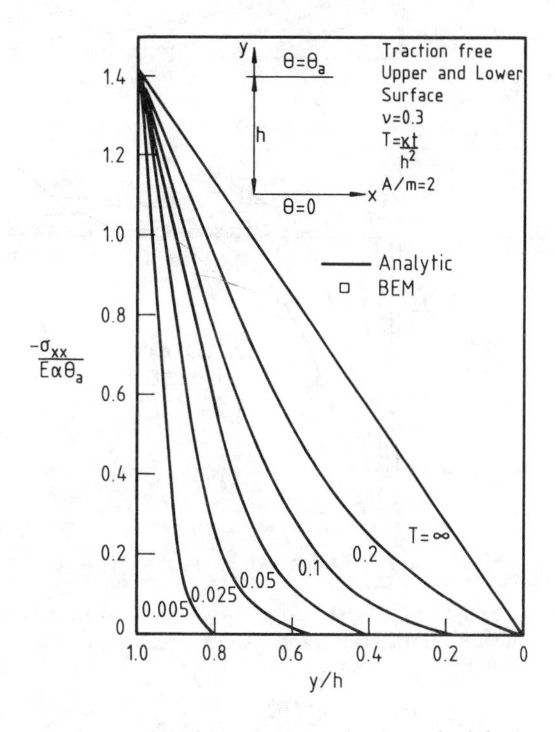

Figure 1. Transient stresses in heated slab.

Example 2

An infinite solid containing a traction free circular hole initially at a uniform temperature is subjected to a constant boundary temperature θ_a from time t = 0. Radial and tangential stresses induced by the transient thermal gradients are shown.

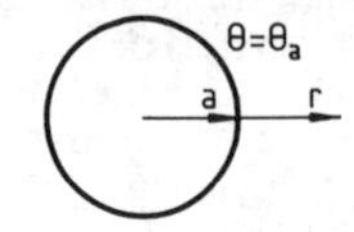

Figure 2(a). Hole in infinite solid

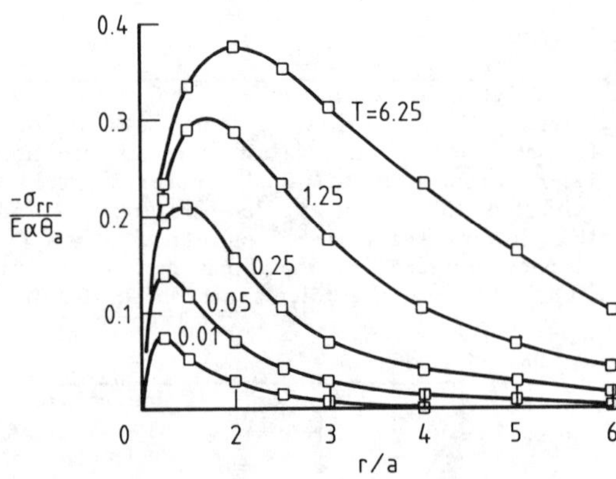

Figure 2(b). Radial stress

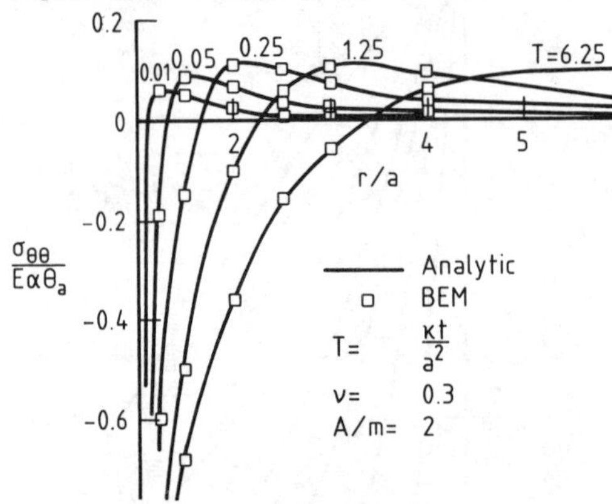

Figure 2(c). Tangential stress

For each of the problems extremely accurate thermal and stress estimates were obtained. Generally

$$\left| \frac{\sigma_{ij}(A) - \sigma_{ij}(B)}{E\ \alpha\ \theta_a} \right| < \frac{1}{400}$$

where $\sigma_{ij}(A)$ = analytic stress

$\sigma_{ij}(B)$ = boundary element stress

σ_a = prescribed boundary temperature

E = Young's Modulus

α = coefficient of thermal expansion.

7. CONCLUSIONS

By using a Laplace transformation to temporarily remove the time dependence of the governing equations the evaluation of domain integrals and time marching is avoided. Though a relatively simple interpretation of the exact boundary integral equation was programmed, very accurate thermal and stress data were computed using few elements. Similar accuracy is expected for more complicated boundary conditions when higher order interpolating functions are used.

8. REFERENCES

Flugge, S. (editor) 1965. Encylopedia of Physics, Mechanics of Solids, V4, pp. 73.

Cleary, M.P. 1977. Fundamental solutions for a fluid saturated porous solid, Vol. 8, pp. 413-418.

Nowacki, W. 1964. Green functions for thermoelastic medium. I and II, Bull. Acad. Polon. Sci., Ser. Sci. Tech., Vol. 12, pp. 465-472.

Nowacki, W. 1962. Thermoelasticity, Pergamon Press, London.

Sharp, S. and Crouch, S.L. 1986. Boundary integral methods for thermoelastic problems. Journal of Applied Mechanics, Vol. 53, pp. 298-302.

Sladek, V. and Sladek, J. 1983. Boundary integral equation method in thermoelasticity part I: general analysis. Applied Mathematical Modelling, Vol. 7, pp. 241-253.

Sladek, V. and Sladek, J. 1984. Boundary integral equation method in thermoelasticity part III: uncoupled thermoelasticity. Applied Mathematical Modelling, Vol. 8, pp. 413-418.

Smith, D.W. and Booker, J.R. 1987. Boundary integral analysis of transient thermoelasticity. Fifth International conference on Finite Element Methods in Australia. Paragon Printers ACT Aust.

Talbot, A. 1979. The accurate numerical inversion of Laplace transforms. Jl. Inst. Math. Applics Vol. 23 pp. 97-120.

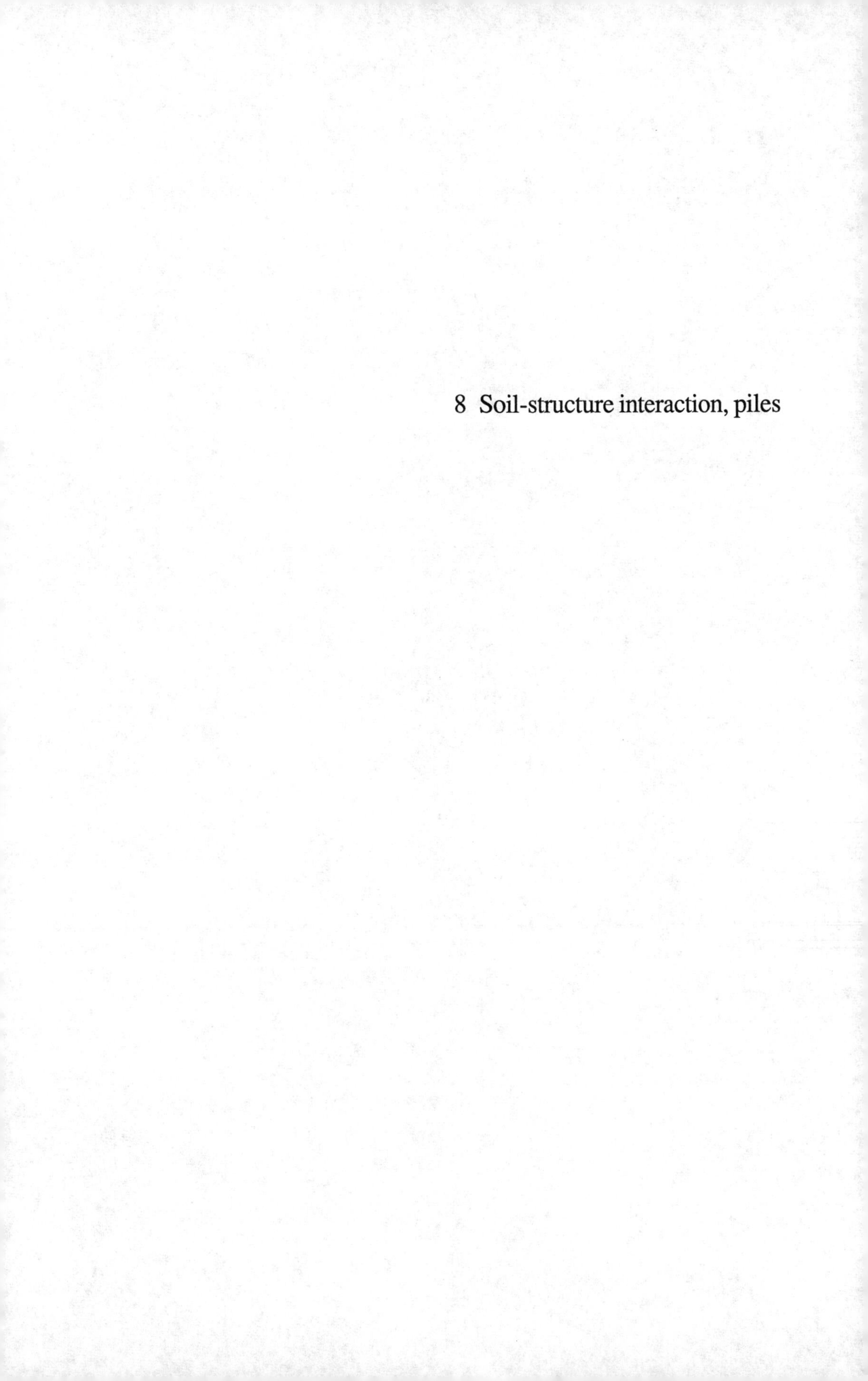

8 Soil-structure interaction, piles

Numerical Methods in Geomechanics (Innsbruck 1988), Swoboda (ed.)
© 1988 Balkema, Rotterdam. ISBN 90 6191 809 X

Computation of the ultimate pressure of a laterally loaded circular pile in frictional soil

P.A.Lane & D.V.Griffiths
Manchester University, UK

ABSTRACT: The paper presents numerical solutions for the ultimate pressure to cause failure of a laterally loaded circular pile in frictional soil. The analysis considers a slice of pile at sufficient depth below ground level such that approximately plane strain conditions apply. The analysis is performed using a 'non-axisymmetric' approach involving Fourier expansions in the tangential directions and this allows a rather simple transformation from rough to smooth conditions at the pile/soil interface. The pile is displaced incrementally into the elasto-plastic soil, and the reactions back-figured from the converged stresses. The effects of tension behind the pile are fully accounted for, and the influence of volumetric changes in the soil during shearing demonstrated. The ultimate pressure is presented in the form of dimensionless 'bearing capacity factors', and comparisons with closed-form solutions made where available.

1 INTRODUCTION

The failure of a laterally loaded pile is governed by the soil resistance and the pile strength [1,2]. As an initial step to the numerical analysis of the full-depth pile, the behaviour of the pile at depth may be considered as a 'slice' of pile and soil under plane strain conditions.

The limiting lateral pressure on the pile disc has been assesed as between 8.3 and 11.4 Cu for a purely cohesive soil [2]. A characteristic solution by Randolph and Houlsby [3] refined this further. Their results of $(4\sqrt{2} + 2\pi)$Cu and $(6 + \pi)$Cu for perfectly rough and perfectly smooth discs respectively, are used here for comparison.

Several features are considered as a basis for future work on a full-depth model, to include pile/soil detaching and soil with both cohesion and internal friction. As no suitable analytical solution was available for this problem a parallel analysis using a conventional plane strain, finite element mesh, technique was conducted wherever possible and good agreement obtained.

2 ANALYSIS METHOD AND TECHNIQUES

The behaviour of a rigid pile at depth was modelled as a disc displaced laterally into a slice of elasto-visco-plastic material, under plane strain conditions. A rigid boundary was placed at 10 pile diameters distant, sufficient to minimize boundary effects for most cases. The pile/soil interface was considered as perfectly rough or smooth by the use of decoupled freedoms in the mesh [4].

A Mohr-Coulomb yield criterion was applied to the soil and the effects of reducing tension in the soil behind the pile were examined by the imposition of a 'no tension' criterion.

$$\sigma_3 < \text{TVAL} \qquad (1)$$

where TVAL may vary from zero (no tension) to some permitted limit.

Within the pile and soil slice tangential stresses, strains and displacements were modelled using Fourier expansions [5]. This technique for modelling axisymmetric structures subjected to non-axisymmetric loading has been shown to be an efficient and flexible method with potential savings in storage and computational costs compared to the conventional plane strain analysis [6].

For the cases considered here NHAR harmonics were used with NANG angular sampling points between 0° and 180°. The relationship between NANG and NHAR to obtain the 'best' solution varies for different soil properties. For example, a purely cohesive soil, allowing tension to develop, needs only odd harmonics (1,3,5 etc) and where the number of harmonics (NHAR) is given by:

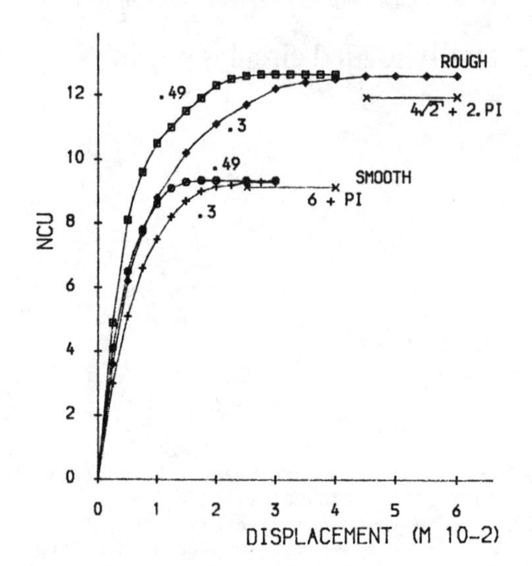

Figure 2.1 Load/displacement response for a rigid pile in a purely cohesive soil permitting tension,Cu=100,E=10E5.

where

$$NCu = \frac{P}{Cu.D} \qquad (5)$$

and P is the load per unit length of pile (kN/m) with diameter D (m).

$$NHAR = \frac{NANG - 1}{2} \qquad (2)$$

When no tension is permitted:

$$NHAR = \frac{NANG + 1}{2} \qquad (3)$$

with all harmonics (odd, even and zero) necessary for a good solution [7].

The Fourier amplitudes from each expansion are conveniently summed using the repeated Trapezium rule e.g.

$$a_i = \frac{2}{\pi} \int_0^{\pi} f(\Theta).\cos i\Theta.d\Theta \qquad (4)$$

The problem was first solved for a purely cohesive soil, permitting tension, and excellent agreement obtained with the closed-form solution [3,8], see Fig 2.1.

3 'NO TENSION' CRITERION

The introduction of a criterion for permissable tension in the soil is necessary to account for the gapping that occurs behind a laterally loaded pile. It was considered here to examine the reduction in stiffness and ultimate pressure that results. The elastic case shown in Fig 3.1 was first solved with full tension permitted, giving exact agreement with the closed-form solution of Baguelin et al [9]. The no tension case has a lower stiffness.

When a similar plastic analysis was performed on a cohesive soil the reduced stiffness and ultimate pressure was also seen, although the maximum is not well defined.

When plastic displacement vectors are plotted for the soil with and without ten-

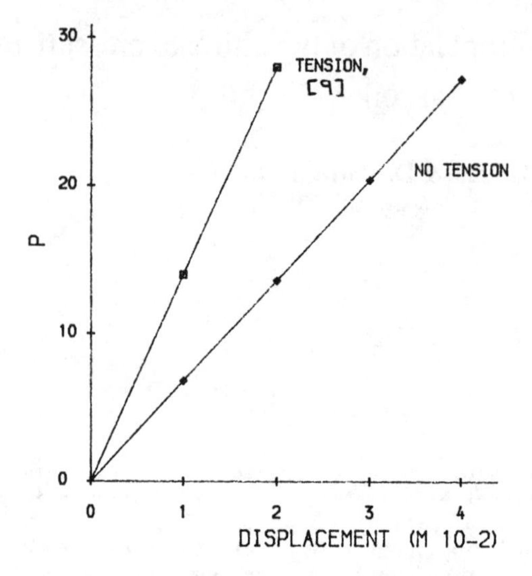

Figure 3.1 Elastic response of a pile in soil with and without tension.

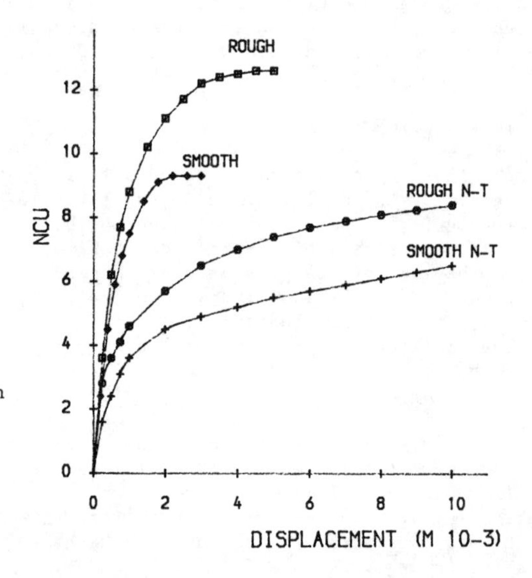

Figure 3.2 Effect of permissable soil tension on the load/displacement response of a pile in cohesive soil.

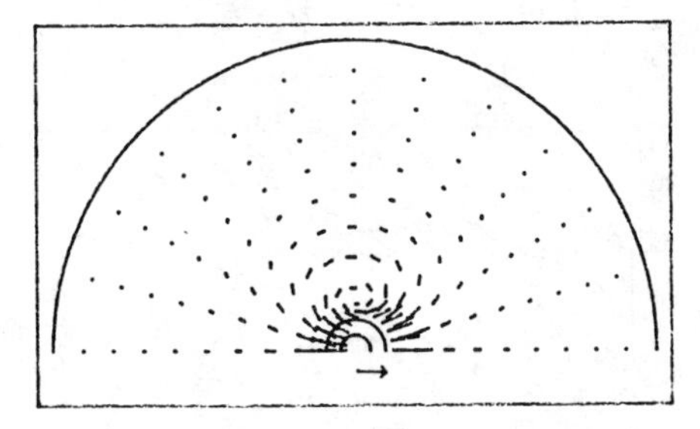

Figure 3.3 Plastic displacement vectors for cohesive soil with tension (NU=0.3)

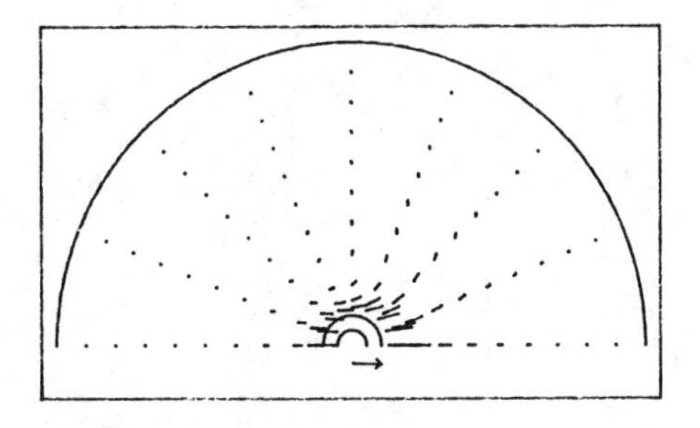

Figure 3.4 Plastic displacement vectors for cohesive soil without tension (0.3)

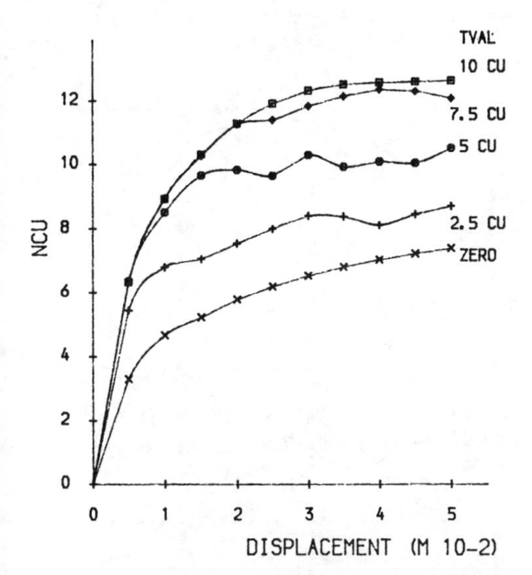

Figure 3.5 Effect of permissable tension on the load/displacement response of a rough pile in cohesive soil (NU=0.3).

sion, the effect of the criterion is seen reducing soil movement behind the pile and changing the overall mechanism, Figs 3.3 and 3.4, (some movement still occurs behind the pile to maintain mesh continuity). The change in the nature of the displacements illustrates the change in the load/displacement response in Fig3.2.

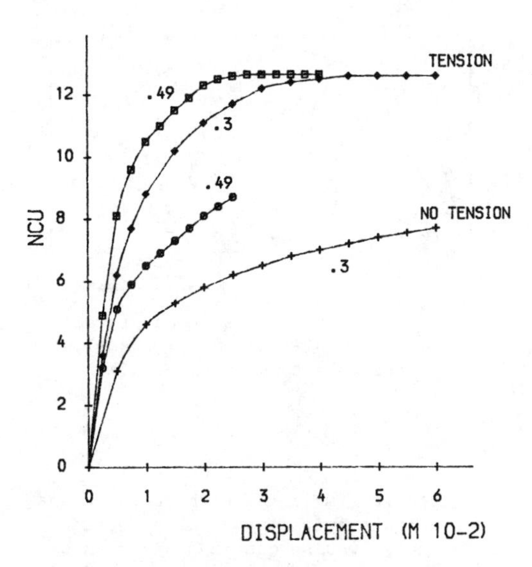

Figure 3.6 Effect of Poisson's Ratio on the load/displacement response of a rough pile in cohesive soil with and without tension.

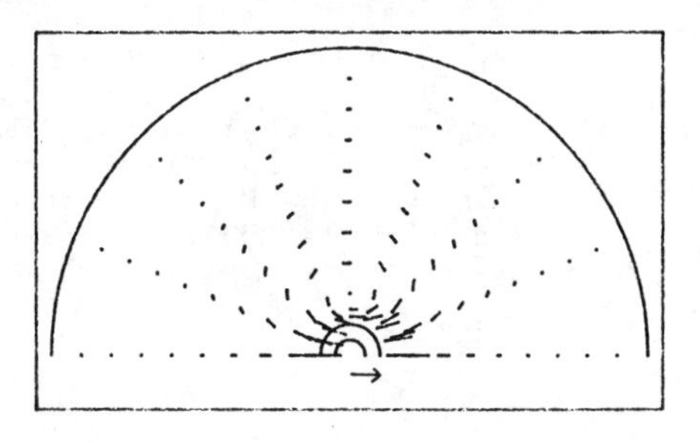

Figure 3.7 Displacements for a rough pile in cohesive soil with no tension (NU=0.49)

The variation in stiffness and ultimate load with permissable tension is shown in Fig 3.5. In practice TVAL $\ngtr$0.5 Cu approximates to no tension and TVAL >10 Cu to full tension.

With a higher Poisson's Ratio (NU) the stiffness of a 'no tension' case increases but it is still lower than with full tension Fig 3.6. The effect of Poisson's Ratio on the plastic displacements can be seen from Fig 3.7 (NU = 0.49) compared to Fig 3.4 (NU = 0.3).

The reduced soil disturbance for a smooth pile can be seen from comparing Fig 3.8 to Fig 3.7. The smaller disturbance pattern of the smooth pile accounts for the reduced stiffness and ultimate pressure compared to the rough (Fig 3.2).

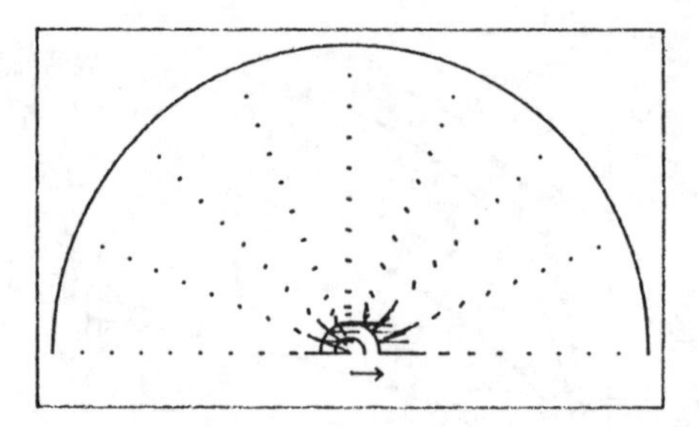

Figure 3.8 Plastic displacement vectors for a smooth pile in cohesive soil with no permissable tension (NU=0.49).

4 FRICTIONAL SOIL

A weightless material with cohesion and interparticle friction was also considered As the model was so confined dilation was reduced to zero with a non-associated flow rule (PSI=0) in the following examples to prevent locking of the mesh.

The effect of the internal angle of friction (PHI) is shown in Fig 4.1, where

$$Nc = \frac{P}{c.D} \qquad (6)$$

There is an increase in strength with PHI although the ultimate pressure is not well defined for high PHI. The displacement required to reach 'ultimate' pressure (Δult) also increases with PHI (Fig 4.1).

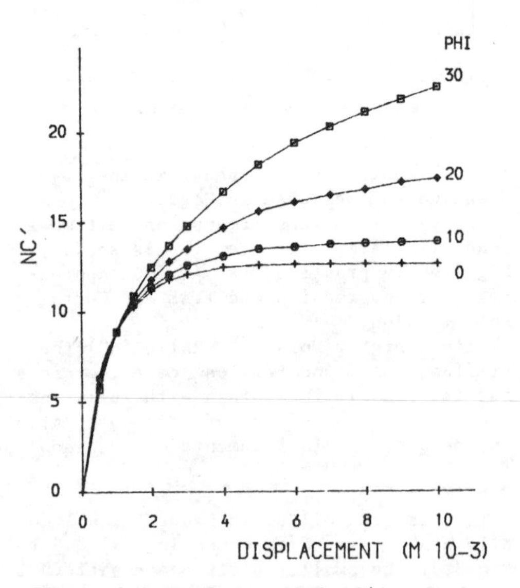

Figure 4.1 Variation in load/displacement response with PHI for a rough pile in soil with tension (c=10, NU=0.3).

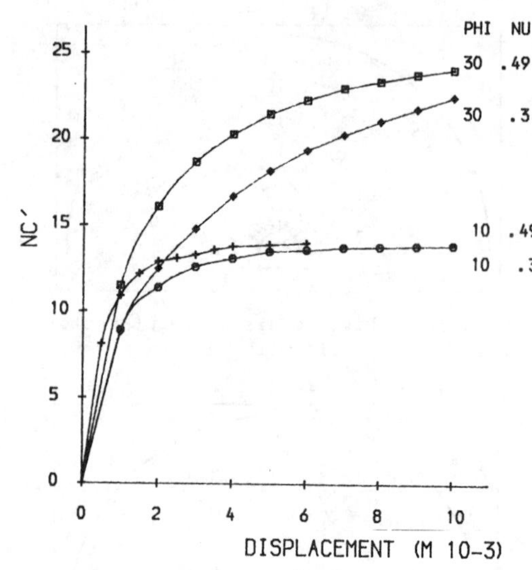

Figure 4.2 Effect of Poisson's Ratio on the load/displacement response of a rough pile in frictional soil with tension.

With a higher Poisson's Ratio the increased initial stiffness, reduced Δult and similar ultimate pressure seen in Fig 2.1 is again demonstrated, Fig 4.2.

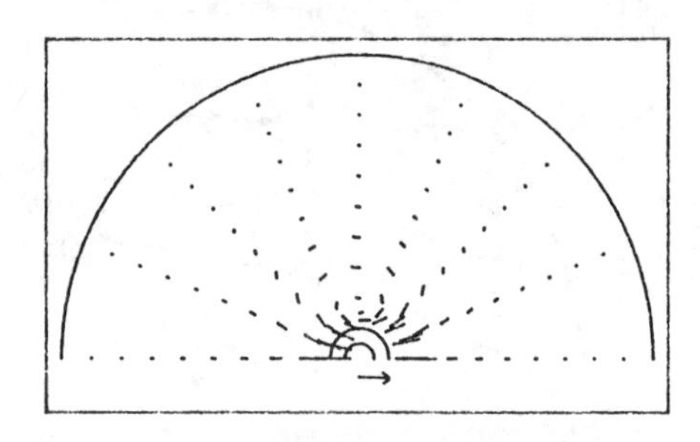

Figure 4.3 Plastic displacement vectors for a rough pile in frictional soil with tension (PHI = 10°)

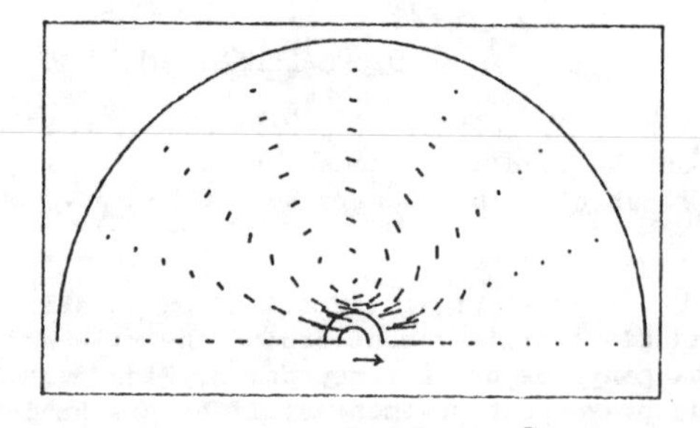

Figure 4.4 Plastic displacement vectors for a rough pile (PHI =30°).

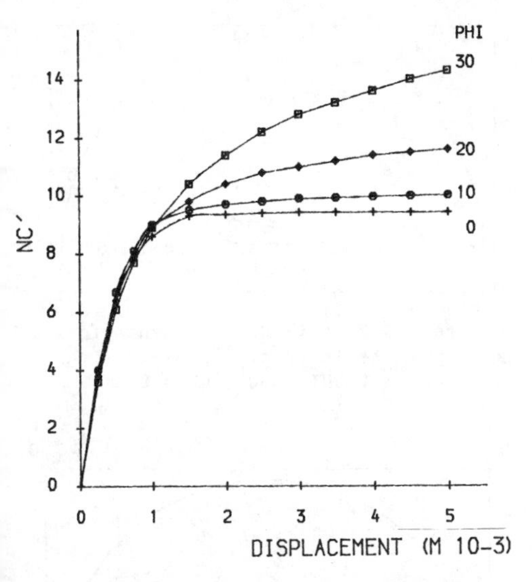

Figure 4.5 Effect of PHI on the load/displacement response for a smooth pile in frictional soil with tension (NU=0.49)

The change in the mechanism of the plastic displacements of the soil with PHI is shown in Figs 4.3 and 4.4.

Figure 4.5 illustrates the reduced stiffness and ultimate pressure for a smooth pile with varying PHI compared to a rough pile (Fig 4.2).

Figure 4.6 summarizes the variation in Nc with PHI and NU at the given Δ.

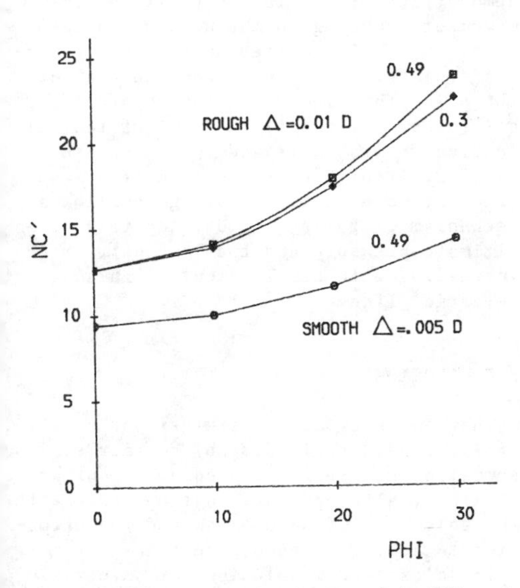

Figure 4.6 Variation in Nc with PHI and NU in frictional soil with tension.

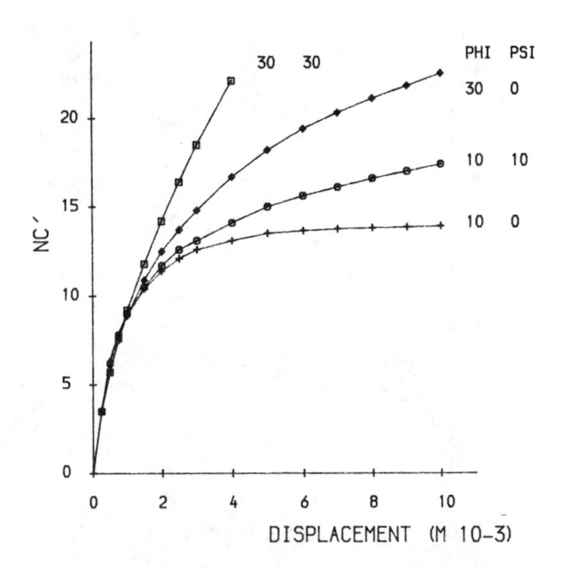

Figure 4.7 Effect of associated and non-associated flow rules on the response of a rough pile in frictional soil (NU=0.3)

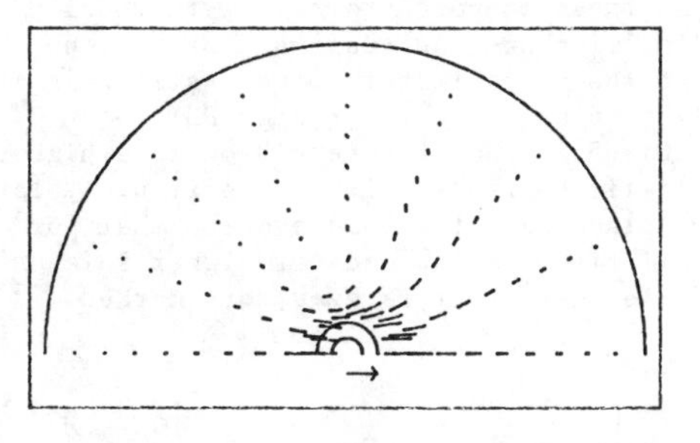

Figure 4.8 Plastic displacement vectors for associated flow for a rough pile in frictional soil (PHI=PSI=30°).

To demonstrate the problems expected with dilation in this confined model an associated flow rule was used, see Figs 4.7 and 4.8. Compared with Fig 4.4 the mechanism of plastic flow is no longer predominately rotational, with the soil in front moving towards the boundary increasing the apparent strength (Fig 4.7).

5 FRICTIONAL SOIL WITH NO TENSION

When the Mohr-Coulomb and No Tension yield criteria were implemented together for a frictional soil the pattern of results was rather different to those obtanied with the purely cohesive soil (Section 2). For low angles of friction the reduced stiffness and ultimate pressure and increased Δult were observed. For higher PHI an apparent increase occured in the 'plastic'

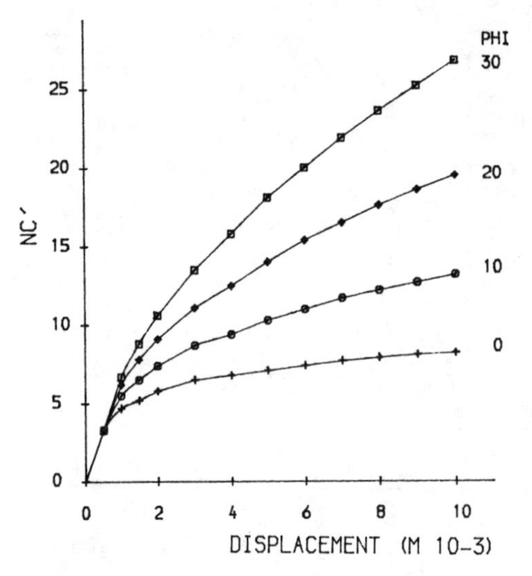

Figure 5.1 Variation in the response with PHI for a rough pile in frictional soil with no tension (NU = 0.3).

tiffness compared to soil with tension. Fig 5.1 shows the response for various PHI and the approximately linear stiffness at large displacement for high PHI.

Fig 5.2 shows the result of this higher plastic stiffness with no tension. At large displacements the load exceeds that for soil with tension and for higher Poisson's Ratio the effect is even more marked.

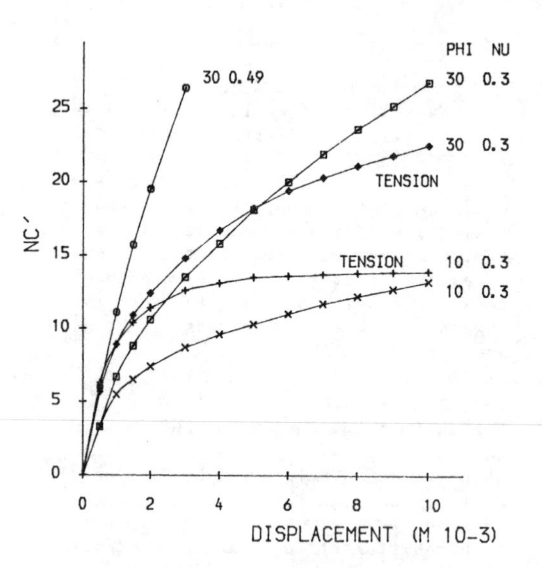

Figure 5.2 Comparison of load/displacement response for a rough pile in frictional soil with and without tension.

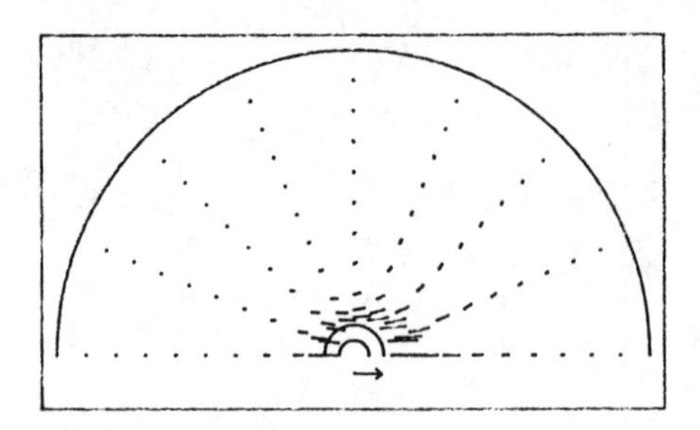

Figure 5.3 Plastic displacements for a rough pile in frictional soil with no tension (PHI = 30°, NU = 0.3).

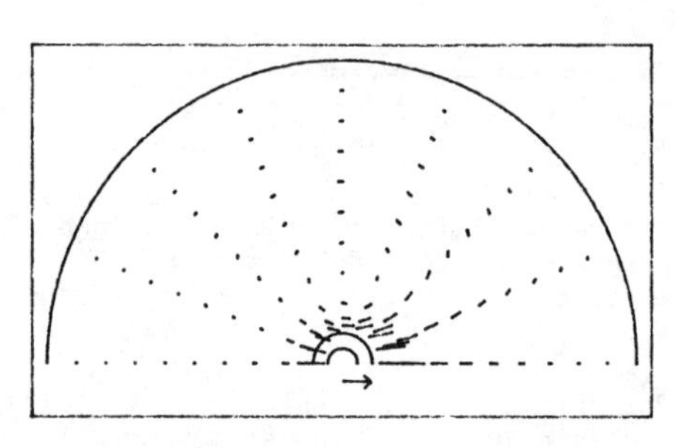

Figure 5.4 Plastic displacements for a rough pile in frictional soil with no tension (PHI = 30°, NU = 0.49).

An examination of the plastic displacement vectors shown in Figs 5.3 (NU = 0.3) and 5.4 (NU = 0.49) compared to Fig 4.4 demonstrates the relatively little soil movement occuring in the no tension case.

The no tension criterion is causing stress redistribution away from the back of the pile, this increases the apparent stiffness of the soil in front of the pile leading to its increased load response, Fig 5.2. At high angles of friction insufficient movement is occuring to allow a mechanism of failure to develop to give an ultimate pressure and the load carries on increasing with displacement in an apparent 'elastic' line.

6 CONCLUSIONS

The non-axisymmetric analysis of a laterally loaded rigid disc by Fourier series summation has so far proved efficient and flexible, although care must be taken with the selection of NANG, NHAR and the harmonics to achieve a good result.

The closed-form solution for an elastic material was reproduced. Good agreement was obtained with a closed-form solution for both rough and smooth piles in a purely cohesive soil. The ultimate pressure on a laterally loaded rigid disc in a cohesive plastic soil is not dependent on Poisson's Ratio for the soil.

When no soil tension is permitted there is a reduction in stiffness and ultimate pressure for a laterally loaded rigid disc in cohesive soil. The ultimate pressure of such a soil is not always well defined but 65% of the full-tension Nc value may be obtained at increased Δult. The effective range of permissable tension is 0.5Cu to 10Cu.

The ultimate pressure on a laterally loaded disc is dependent on the internal angle of friction of the soil with both friction and cohesion. In this confined model non-associated flow rules are used as dilation causes 'locking' of the mesh and an over-stiff, over-strong response at high angles of friction.

The reduction of permissable soil tension in a frictional soil reduces the initial stiffness of the load/displacement response. The ultimate pressure is not defined for high PHI in such a material due to the 'elastic' response at large displacements. The ultimate pressure for a 'no tension', frictional soil is not independent of the soil's Poisson's Ratio.

The features of various soil properties (c, PHI, NU), the pile/soil interface condition (rough or smooth) and the maximum permissable soil tension are all factors in the load/displacement behaviour and ultimate pressure of the disc. They have been considered in this 'at depth' model to illustrate their individual and combined effects on the response of a rigid laterally loaded pile.

7 REFERENCES

[1] B.B.Broms, Lateral resistance of piles in cohesionless soils, ASCE(SM3),123-156 (1964)

[2] B.B.Broms, Lateral resistance of piles in cohesive soils, ASCE(SM3), 27-63 (1964)

[3] M.F.Randolph and G.T.Houlsby, The limiting pressure on a circular pile loaded laterally in cohesive soil, Geotechnique 34, 613-623 (1984)

[4] D.V.Griffiths, Simple modelling of smooth interfaces using finite elements Internal Report, Simon Eng. Labs., Manchester University, U.K. (1985)

[5] L.A.Winnicki and O.C.Zienkiewicz, Plastic behaviour of axisymmetric bodies subjected to non-axisymmetric loading, IJNME 14, 1399-1412 (1979)

[6] D.V.Griffiths, HARMONY- A program for predicting the response of axisymmetric bodies subjected to non-axisymmetric loading, Report to Fugro Geotechnical Engineers B.V. (1985)

[7] P.A.Lane, Numerical analysis of laterally loaded pile failure mechanisms, Internal Report, Simon Eng. Labs., Manchester University, U.K. (1986)

[8] D.V.Griffiths and P.A.Lane, The influence of interface roughness on problems of axisymmetric soil/structure interaction, Proc. 2nd Int. Conf. Constitutive Laws for Eng. Materials : Theory and Applications, Tuscon, U.S.A. , 1051-1058 (1987)

[9] F.Baguelin, R.Frank and Y.H.Said, Theoretical study of the lateral reaction mechanism of piles, Geotechnique 27, 405-434 (1977)

8 ACKNOWLEDGEMENTS

Developement of the HARMONY program was supported in part by Fugro Geotechnical Engineers B.V., P.O.Box 63, 2260 AB Leidschendam, Netherlands. This support is gratefully acknowledged.

Numerical Methods in Geomechanics (Innsbruck 1988), Swoboda (ed.)
© 1988 Balkema, Rotterdam. ISBN 90 6191 809 X

Elastoplastic analysis of laterally loaded piles

A.P.Kooijman & P.A.Vermeer
Faculty of Civil Engineering, University of Technology, Delft, Netherlands

ABSTRACT

The numerical analysis of laterally loaded piles becomes unwieldy when a fully three-dimensional discretization of the surrounding soil is used. Moreover, a full three-dimensional analysis is unnecessary, as the vertical displacement components can be disregarded with respect to the horizontal components. A quasi three-dimensional analysis is obtained by neglecting vertical displacements, and the results of such analyses appear to agree extremely well with full three-dimensional solutions for elastic soil behaviour.

The second part of the paper, is concerned with elastoplastic soil behaviour and with the interface behaviour at the pile-soil contact. Both the non-linear interface behaviour and the elastoplastic soil behaviour seem to necessitate the use of powerful mainframe computers. In the present analysis, however, this is avoided by means of a sub-structuring technique in which the soil is divided into interacting horizontal layers. First, attention is focused on the analysis of such an individual layer using the finite element method. Next, this layer-model is incorporated in the main program for laterally loaded piles, and a complete elastoplastic continuum analysis of a laterally loaded pile is made. Finally the method is evaluated and a judgment is made on the applicability of the method to practical problems.

1 INTRODUCTION

Offshore foundation piles are indirectly loaded by the jacket structure they support. In order to be able to examine the role the foundation piles play in the overall structural behaviour, the jacket is disconnected and replaced by boundary conditions on the pile heads. These boundary conditions can be resolved into axial and lateral components. Strictly speaking, the response of the soil cannot be examined separately for these components. For reasons of simplicity, however, an integrated analysis is avoided in practice, and in fact the distinction between axial and lateral pile response computations forms one of the basic elements in offshore pile foundation design. A practical justification of this distinction is the fact that, for most circumstances encountered in offshore engineering, the lateral load transfer to the soil is concentrated in the upper layers of the soil, while the axial load transfer is concentrated in the lower soil layers. With the presently available computers an integrated analysis of axial and lateral loads based on a continuum approach can only be made at the cost of accepting other restrictions. For example, an axially symmetric geometry is demanded. The method presented here does not pursue an integrated approach. On the contrary, it exploits the customary distinction between axial and lateral loading explicitly. In this way a model can be developed that offers promising prospects for the application of continuum models for engineering purposes. The basic ideas of the substructuring technique is outlined in the next section. Then results are presented for elastic soil behaviour (see also Verruijt & Kooijman, 1987). Finally, attention is focused on elastoplastic soil behaviour.

2 BASIC IDEAS OF THE PILE-SOIL MODEL

The basic idea of the model is a substructuring of the pile-soil system at two levels. At the first level, the pile-soil system is separated into two subsystems, representing the pile and the soil respectively. In the analysis the two systems are coupled by satisfying compatibility and quilibrium conditions. At the second level, the soil is subdivided into a number of interacting layers. As a consequence, three major components can be distinguished in the model: a model for the pile, a model for a soil layer, and a coupling routine.

The pile is analysed by means of a finite difference discretization of the equations for a beam on elastic foundation (Hetenyi, 1946). Let z be the coordinate along the pile, M the bending moment and u the lateral displacement, then we have

$$\frac{\partial^2 M}{\partial z^2} = K u - f \qquad (1)$$

$$\frac{\partial^2 u}{\partial z^2} = - \frac{M}{EI} \qquad (2)$$

where f is a given lateral load distribution, K is a spring constant and EI the flexural rigidity of the pile. The actual values of these parameters will be obtained from the analysis of the deformation of the soil.

The soil is considered to be a layered continuum, with a cylindrical hole, in which lateral forces are acting which represent the interaction of the soil and the pile. In order to avoid the gigantic system of numerical equations that would result from a complete three-dimensional modelling of this continuum, the soil is divided into a number of interacting layers. For this purpose some simplifications in the description of the deformation of the soil continuum are necessary. It is assumed that the vertical displacement component w is much smaller than the displacements u and v in a horizontal plane. For each layer of the soil the basic equations can now be obtained by averaging the equations of horizontal equilibrium, and by disregarding all terms involving the vertical displacement w. This leads to the equations of equilibrium for a plate,

$$\frac{\partial \sigma_{xx}}{\partial x} + \frac{\partial \sigma_{yx}}{\partial y} + Q_x = 0 \qquad (3)$$

$$\frac{\partial \sigma_{xy}}{\partial x} + \frac{\partial \sigma_{yy}}{\partial y} + Q_y = 0 \qquad (4)$$

where σ_{xx}, σ_{xy} and σ_{yy} are the average stresses in the horizontal plane, and wher Q_x and Q_y represent the forces transmitted to the layer by shear stresses from the layers above and below it, i. e.

$$Q_x = (\sigma_{zx}^{+} - \sigma_{zx}^{-})/H \qquad (5)$$

$$Q_y = (\sigma_{zy}^{+} - \sigma_{zy}^{-})/H \qquad (6)$$

where H is the thickness of the layer, and where the superscripts + and - indicate the values at the bottom and top surfaces of a layer, respectively.

The shear stresses σ_{zx} and σ_{zy} at the top and bottom of the i-th layer can be expressed in the shear strains ϵ_{zx} and ϵ_{zy} by Hooke's law. Disregarding the vertical displacement component w, Hooke's law gives

$$\sigma_{zx} = G \left[\frac{\partial u}{\partial z} + \frac{\partial w}{\partial x} \right] \approx G \frac{\partial u}{\partial z} \qquad (7)$$

$$\sigma_{zy} = G \left[\frac{\partial v}{\partial z} + \frac{\partial w}{\partial y} \right] \approx G \frac{\partial v}{\partial z} \qquad (8)$$

By using a finite difference approximation of the derivatives in the vertical direction, the forces Q_x and Q_y, which act as body forces in the system of equations for the horizontal layer, eqs. (3) and (4), can be determined for each estimate for the displacements in the various layers. The result of this procedure is that for each layer the system of equations is reduced to the familiar equations for plane stress deformations, with given body forces representing the interaction between the layers. The analysis of stresses and strains in each layer is performed by using the finite element method.

The coupling of the pile and soil subsystems is achieved in the following way. The response of the soil layers to the forces transmitted to them by the pile can be represented formally as

$$F_i = \sum_{j=1}^{N} K_{ij} u_j \qquad (9)$$

where u_j is the displacement of layer j at the pile-soil interface. Quite formally, equation (9) can also be written as

$$F_i = k_i u_i + f_i \qquad (10)$$

where f_i represents the contribution of all layers except layer i. The value of the spring constants k_i can be determined from the soil model, by running the program with the boundary conditions that $u_i = 1$ and that the displacements of all other layers are zero. The value of the force at layer i following from the finite element analysis of the soil system then gives the value of the spring constant K_i. The interaction of the soil and the pile is now represented by an equation of the form (10), which is just what is needed for the analysis of the pile, see eq. (1). Thus the complete system can be analyzed by an iterative procedure, in which the pile and soil subsystems are analyzed separately in successive steps. The analysis of the pile system results in values for the displacements at the pile-soil interface in each layer. On the basis of these data the soil subsystem is analyzed, which results in values for the forces F_i at each level. This leads to an updated value for the part f_i in the response function (10). Then the pile analysis can be repeated, using the same values for the spring constants and updated values for the coupling terms f_i.

3 ELASTIC SOLUTIONS

The model has been validated by comparison of the results with those obtained by other (numerical) solutions based on a continuum approach. In order to give an idea of the consequences of the above-mentioned simplifications, some results of the validation are reproduced here.

The finite element grid used for the validation of the model is shown in Fig. 1. The outer boundary is considered as having zero displacements. The infinite lateral extent of the soil mass is simulated by reducing the modulus of elasticity in the outer ring of elements by a factor 2. When the radial dimension of these elements is one half of the diameter of the entire mesh, such a reduction is in agreement with a decrease in the lateral stresses which is inversely proportional to the square of the distance in an infinite soil. The diameter of the outer boundary of the grid has been taken as 30 times the pile diameter. The lower layer of the layered system is assumed to be rigid, thus simulating a fixed boundary condition at a certain depth. The depth of that boundary is taken at one-quarter of the pile length below the tip of the pile.

The soil is modelled by a system of 25 layers, or, for very flexible piles, 50 layers. In each layer the same finite element grid of 88 eight-noded isoparametric elements is used. Because of symmetry only one-quarter of the elements has been taken into account in the actual computations. In the analysis a nine-point Gaussian integration scheme is employed (Bathe & Wilson, 1976). In the finite difference approximation of the pile, two elements are used for each layer of soil, which means that the pile is subdivided into either 40 or 80 elements.

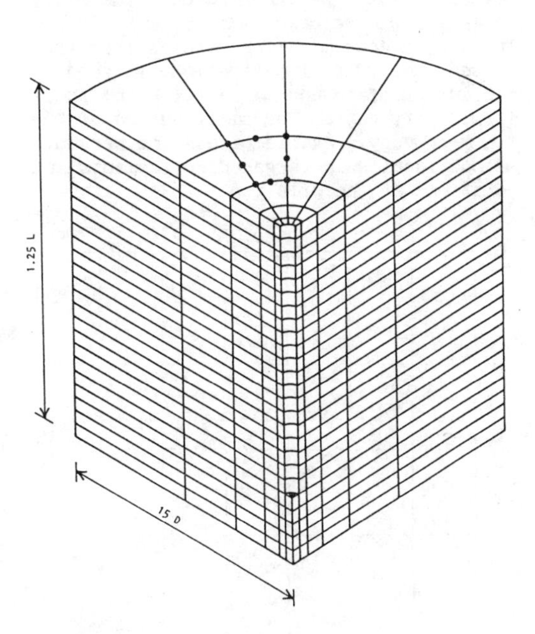

fig. 1 Finite element mesh

For the case of a vertical pile of constant flexural rigidity in a homogeneous elastic half-space, with modulus of elasticity E_s and Poisson's ratio ν_s, a comparison is made with the solutions obtained by Poulos (1971). In this analysis the solutions are expressed in terms of the ratio of pile length to pile diameter (L/D), and a dimensionless factor K_r, defined by

$$K_r = \frac{(EI)_p}{E_s L^4} \tag{11}$$

which characterizes the stiffness ratio of the pile and the soil. When the pile is loaded at its top by a lateral force H, the displacement ρ and the rotation θ of the top of the pile are expressed in terms of the influence factors

$$I_{\rho H} = \rho E_s L/H \tag{12}$$

and

$$I_{\theta H} = \theta E_s L^2/H \tag{13}$$

The maximum moment in the pile is represented by the dimensionless factor M/HL.

For the case of a pile having a length 25 times its diameter (L/D = 25) the results of the numerical calculations are shown in Figs. 2 - 4 as a function of the flexibility ratio K_r. The value of Poisson's ratio is taken as 0.5. The fully drawn lines have been taken from Poulos. The single dots mark the results from the present analysis. The agreement is good for intermediate and large values of the flexibility ratio. For small values of K_r, i.e. for very flexible piles, the present method results in larger displacements and rotations at the pile top.

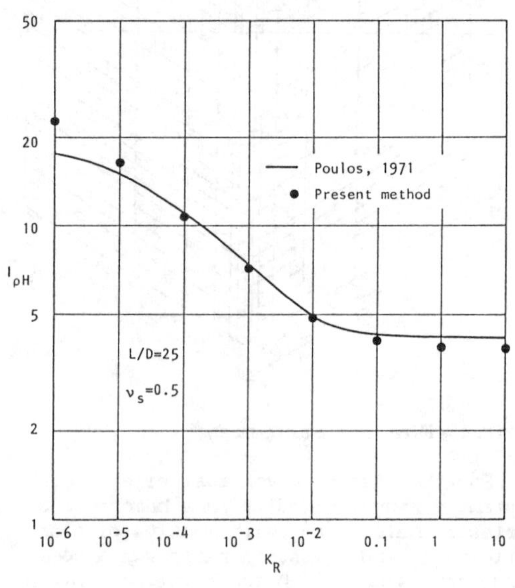

fig. 2 Displacement influence factor

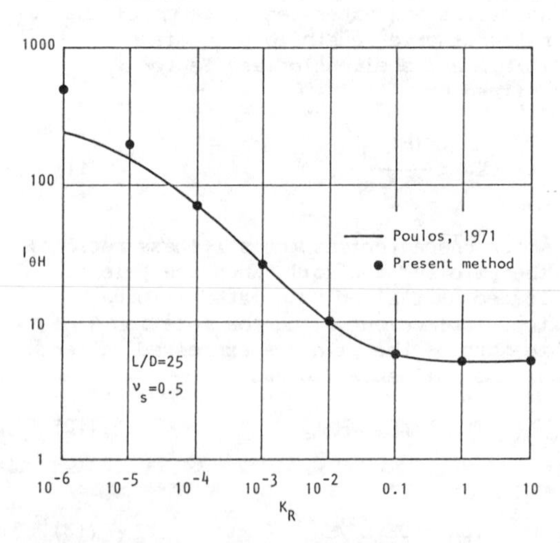

fig. 3 Rotation influence factor

This is in agreement with the findings of Evangelista & Viggiani (1976), who reported that the accuracy of Poulos' original analysis is strongly dependent upon the length of the elements near the top of the pile, at least for piles of large flexibility. A subdivision into a larger number of elements than the 21 elements used by Poulos gives a considerable increase of the displacement at the top of the pile. For the category of medium-stiff to very stiff piles the number of elements used by Poulos is amply sufficient, and in this region the results of the present analysis agree fairly well with those of Poulos.

The maximum bending moment occurring in the pile is shown in Fig. 4. Although in general the values are somewhat larger than those obtained by Poulos, the agreement is good over the entire range of flexibility factor.

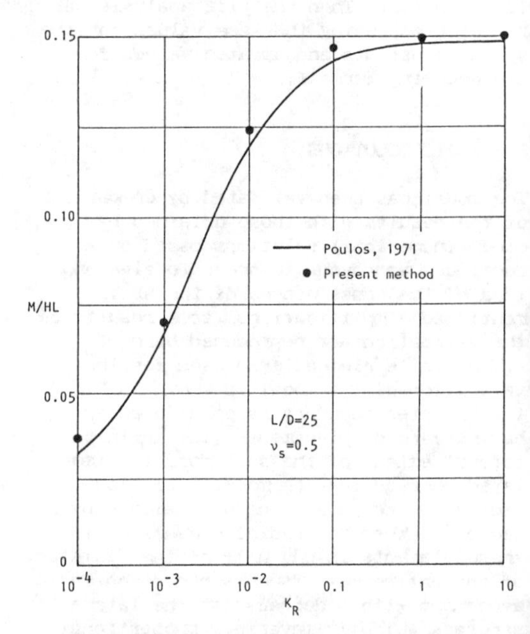

fig. 4 Maximum bending moment

In addition to the case reported above, many other cases have been investigated and been compared with results from the literature. The differences between the results obtained from the other continuum models are small and can be attributed to the various approximations that are made in each method for computational reasons.

4 ELASTOPLASTIC RESPONSE OF A SINGLE LAYER

As has been stated before, one of the basic components of the numerical model presented here for a laterally loaded pile is the finite element representation of a soil layer. Before extending the description of the soil behaviour to elastoplasticity in the main program, it is useful to examine the behaviour of a single isolated soil layer.

The analysis is restricted to cohesive soil behaviour, i.e. undrained clay, with perfect plasticity according to the Tresca yield criterion

$$f = \tfrac{1}{2}|\sigma_3 - \sigma_1| - c_u \leq 0 \qquad (14)$$

A change of stress invokes only elastic strain increments as long as $f<0$, whereas both elastic and plastic strain increments may occur for $f=0$. A tension cut-off is not implemented. The numerical implementation of this constitutive model in the finite element method by means of an initial-stress type method is described by, for example, Vermeer (1980).

Before discussing the application of the numerical model described above, it is useful to recollect the closed-form solution for the limit load on a laterally loaded circular pile-segment in a cohesive soil under plane strain conditions, obtained by Randolph & Houlsby (1984). Starting from a perfectly plastic material model, they managed to find identical upper and lower bound solutions for the limit load, depending on the roughness of the pile, expressed in the ratio of adhesion and undrained shear strength:

$$\alpha = c_a / c_u \qquad (15)$$

The limit P load varies from $(6+\pi)c_uD$ for a smooth pile ($\alpha=0$) to $(4\sqrt{2}+2\pi)c_uD$ for a rough pile ($\alpha=1$). The general equation is

$$P = [\pi + 2\Theta + 2\cos\Theta + 4(\cos\tfrac{1}{2}\Theta + \sin\tfrac{1}{2}\Theta)] * c_uD \qquad (16)$$

where $\Theta = \arcsin \alpha$. This closed-form solution will be used for the validation of the numerical analysis presented here.

The elastoplastic finite element calculations are carried out for a mesh representing the area within a radius of 5 pile diameters. A 2*2 Gaussian integration scheme is used. A local convergence criterion for each integration point, as described by de Borst & Vermeer (1984), is employed. Within each iteration a statically admissible stress field and a stress field which complies with the yield criterion are calculated. The iteration process is interrupted if the statically admissible stress field does not deviate from the yield surface by more than the specified norm ϵ, and the maximum difference between the two stress fields is smaller than $4\epsilon^2$. The calculations presented here are performed with an accuracy tolerance of $\epsilon=0.05$.

In Fig. 5 the load-displacement curve is shown for a rough pile ($c_a=c_u$). The curve remains almost linear up to about half the limit load; then a strongly non-linear response is obtained. The analytical limit

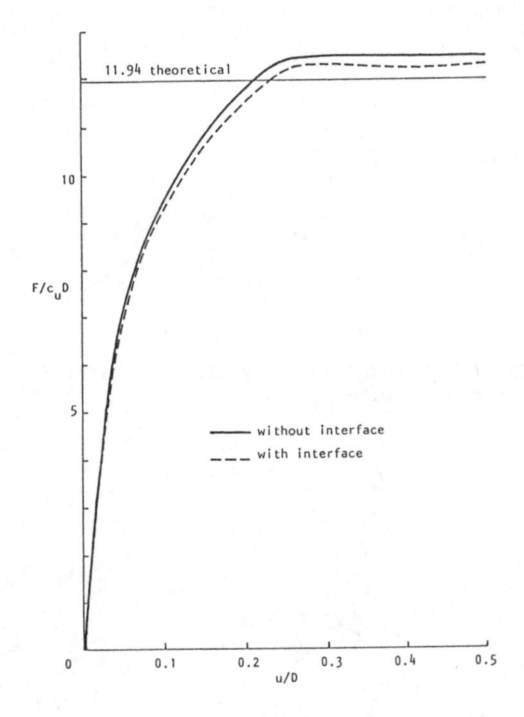

fig. 5 Computed load-displacement curves for a single layer with a rough pile

load of 11.94 c_uD is overestimated by about 5%. These results are obtained by using the finite element grid of Fig. 6 with 96 eight-noded elements and a total

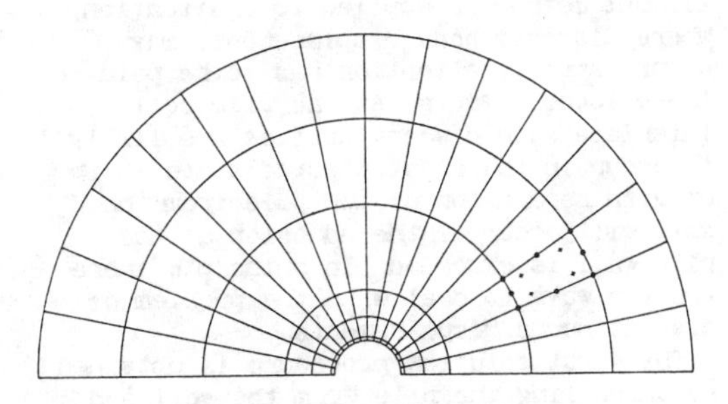

fig. 6 Finite element grid without interface elements

of 333 nodes. Fig. 7a shows the computed velocity field at failure. Fig. 7b does not show the true velocity field, but the velocities relative to the pile; it shows a very realistic flow pattern. Although the quadratic interpolations in the elements dominate the velocity field near the pile, the inner ring of elements, which has a thickness of D/20, is capable of giving a reasonable approximation of the-actually-discontinuous, pile-soil transition.

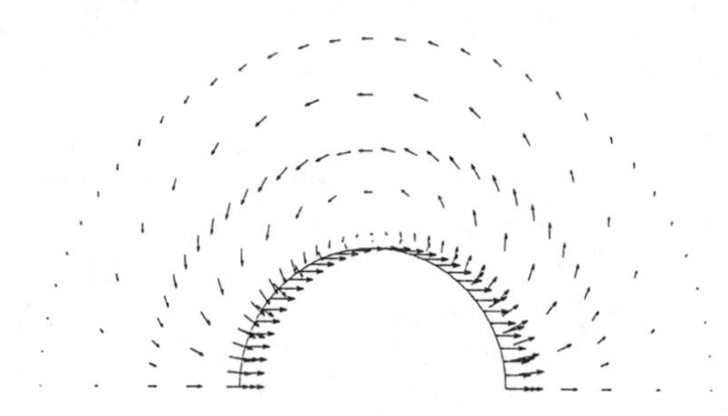

fig. 7a Computed velocity field at failure for a rough pile

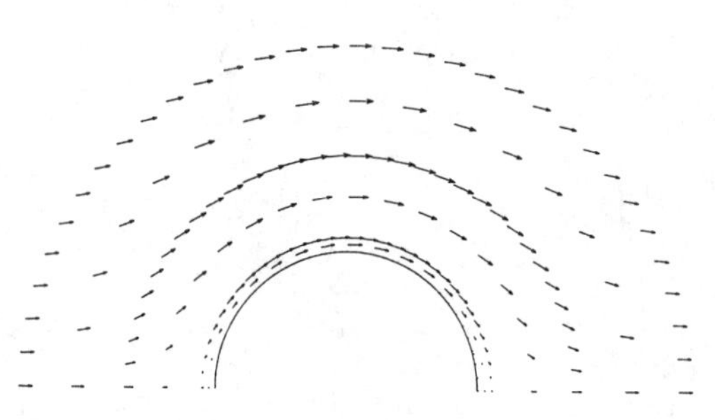

fig. 7b Flow pattern when plotting velocities relative to the pile

5 THE USE OF INTERFACE ELEMENTS

When a continuum model like the finite element method is applied to a situation where discontinuous displacements may occur, special attention has to be paid to these local effects. At the pile-soil interface such discontinuities are likely to occur in the elastoplastic state. Slip or even separation of the pile from the soil will occur if the adhesion at the pile wall is exceeded. In principle there are two ways to deal with the problem of discontinuous displacements.

The first solution procedure is obtained by uncoupling the pile from the soil and adjusting the mutual boundary conditions if necessary. This approach requires an updating of the stiffness matrix for every change of mode in a grid node. When computation time plays an important role in the practical applicability of a program, this solution is not very attractive.

The second solution procedure consists in a smoothing of the discontinuities by representing them by continuous displacements in a thin zone, with very large displacement gradients. This approach has the advantage that the initial-stress technique which is used for the non-linear soil behaviour can also be employed for this thin zone. No updating of the stiffness matrix will be necessary. On the other hand, the strain localization in a very thin zone around the pile will considerably increase the number of iterations.

For reasons of computer time optimization, the second approach is adopted here. However, in order to minimize the number of iterations, an option has been installed to use special interface elements. Besides, these elements offer a simple tool to reduce the pile-soil adhesion to a value lower than the undrainded shear strength of the soil. The type of interface element that is used is the six-noded isoparametric element with an adjusted interpolation in the direction perpendicular to the interface, and a simple constitutive relation, expressed in terms of a normal stiffness K_n and a tangential stiffness K_s (Sharma et al., 1986). The geometrical interface thickness is zero, and therefore the normal and the tangential stiffness have a finite value only for computational reasons.

The flow criterion for the interface is derived from Tresca's flow criterion, and is adapted to the interface material behaviour, as follows:

$$f_w = \tau * \mathrm{sign}(\tau) - c_a \qquad (17)$$

where τ is the shear stress and c_a the adhesion at the pile-soil interface. This yield function is also used as a potential function for deriving the pseudo-strains in the interface elements. The implementation of the six-noded interface element is straightforward.

The calculations mentioned in the previous section are also performed with the element grid shown in Fig. 8. In comparison with the mesh used earlier, the inner ring of eight-noded elements is omitted and interface elements are added. The interface stiffness is obtained by assuming a virtual interface thickness t_v of D/20, i. e. :

$$K_n = E_s / t_v \qquad (18)$$

and

$$K_s = G_s / t_v \qquad (19)$$

In Fig. 5 the displacement curves for both the first mesh and the second mesh with interface elements are shown. In the elastic range the differences are small. In the elastoplastic range, however, the mesh with interface elements shows a more flexible behaviour. The analytical limit load is overestimated by less than 4%,

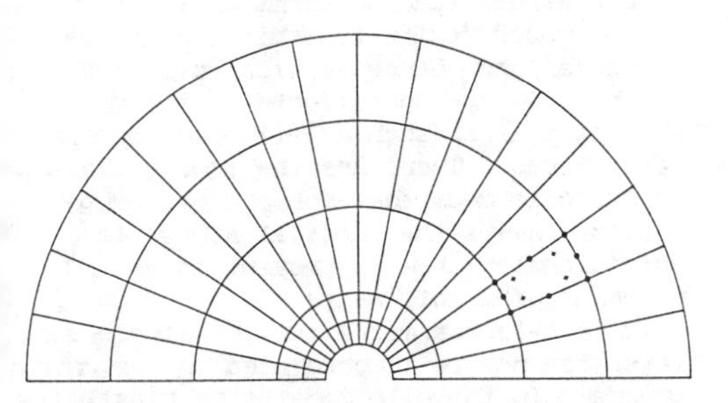

fig. 8 Finite element grid with interface elements along the pile

but the most important advantage is that the computer run time has decreased by a factor of 2.3 compared with the previous solution. The influence of the finite interface stiffness on the load-displacement curve appears to be small. Of course, the influence can be further reduced by increasing the interface stiffness. However, the computer run time and, to a lesser extent, the condition of the system of equations, make opposite demands.

6 ANALYSIS FOR REDUCED ADHESION

The advantage of the interface elements is not only the short computer run times, but also the possibility of simulating any adhesion at the pile-soil interface. Computational results are to be presented for a smooth pile (α=0), a realistic pile (α=0.5) and a rough pile (α=1.0).

In contrast to previous computations we now use a very fine grid of elements as plotted in Fig. 9. Here the number of eight-noded quadrilaterals is 112 and there are 16 six-noded interface elements. The computed load-displacement curves are shown in Fig. 10. Because of the very fine grid and the additional interface elements, very accurate predictions for the limit loads are obtained. Indeed, the numerical

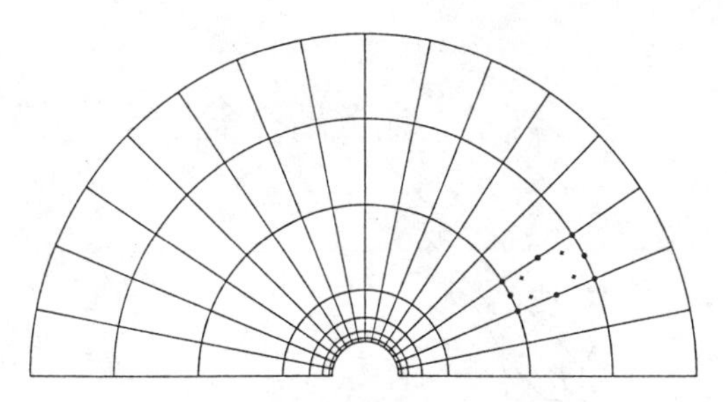

fig. 9 Finite element grid for computations with reduced adhesion

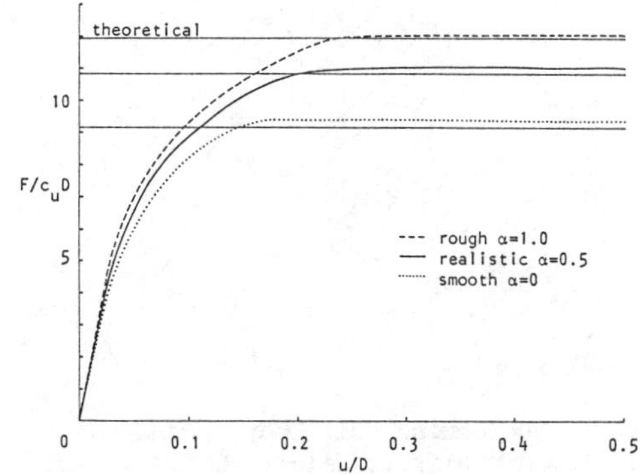

fig. 10 Computed load-displacement curves for reduced adhesion

values deviate less than 2.5 percent from the theoretical solutions by Randolph & Houlsby (1984). Similarly the final velocity fields correspond well with the theoretical slip-line fields.

An increase of the roughness not only gives a higher limit load, but also a much larger zone of deforming soil around the pile. This is indicated in Figs. 11a, 11b and 11c.

From the examples presented above, it can be concluded that the elastoplastic model is capable of making an accurate analysis of a soil layer loaded by a segment of a pile. However, it should be noted that the region near the pile, especially the pile-soil connection, needs special attention when the mesh grid is designed. The use of interface elements is advantageous from a cpu-time point of view, and in addition, provides means to adjust the pile-soil adhesion in an elegant manner. This model, including the six-noded interface elements, will now be incorporated in the main computer program for an entire pile.

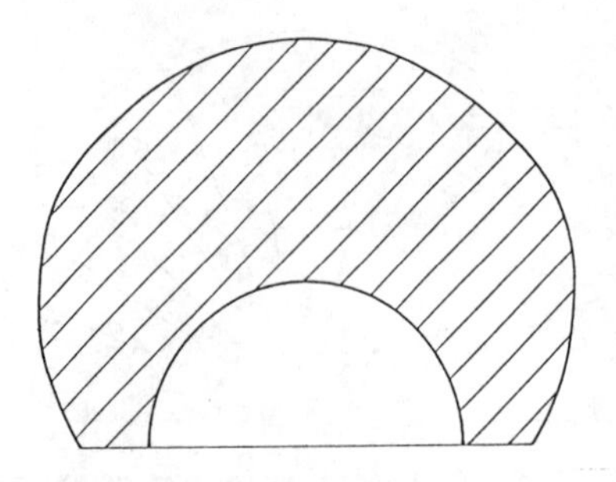

fig. 11a Deforming soil body around the pile for a smooth pile (α=0)

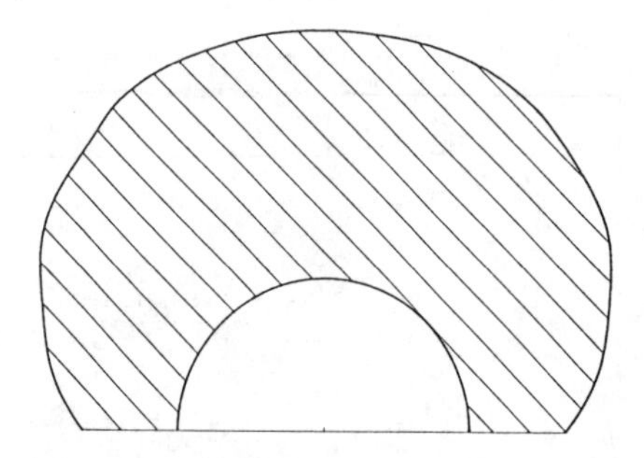

fig. 11b Deforming soil body around the pile for a realistic pile (α=0.5)

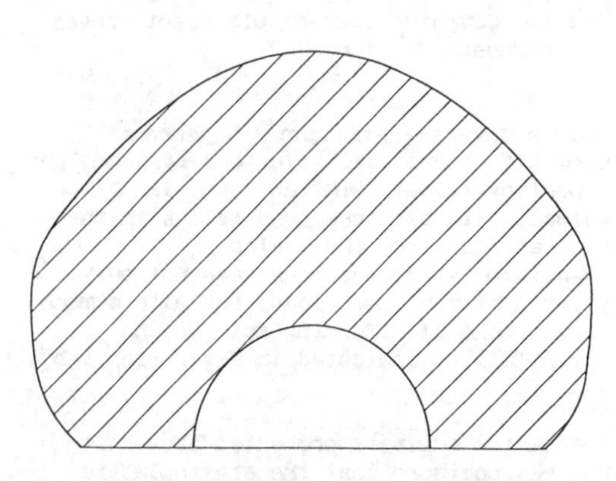

fig. 11c Deforming soil body around the pile for a rough pile (α=1.0)

7 EXTENSION OF THE PILE-SOIL MODEL TO ELASTOPLASTICITY

When the elastic model for a laterally loaded pile is extended to elastoplastic material behaviour by incorporating the previously described single layer model in the main program, an extra assumption has to be made. Apart from the internal strains of a layer, there are two other strain components to be considered in the multi-layer soil-model, namely, γ_{zx} and γ_{zy}. Since these shear strain components are not included in the plastic potential function, the restriction is imposed that they remain purely elastic, i.e.

$$\Delta\gamma_{zx}{}^{p} \text{ and } \Delta\gamma_{zy}{}^{p} = 0 \qquad (20)$$

Furthermore, a remark has to be made concerning the overburden of the soil close to the surface. If the vertical stress σ_{zz} is not sufficiently large, a different failure mechanism may occur. The soil will move upwards, and, in terms of the p-y curve procedures, will exhibit a wedge--type failure. Since vertical displacements of the soil are not allowed in the model presented here, such a failure mechanism cannot occur. Therefore the implicit assumption is made that for all pile-supporting layers the vertical stress is sufficiently large to prevent an upward movement of a soil wedge.

The non-linear model for the single soil layer can now be incorporated in the main program. In the pile model, the plasticity can be introduced as follows:

$$F_i = K_i\ (u_i - u_i{}^{p}) + f_i \qquad (21)$$

where $u_i{}^{p}$ is the non-linear displacement component. The subscript i indicates the number of an individual layer. This equation can also be written as:

$$F_i = K_i\ u_i + h_i, \qquad (22)$$

where

$$h_i = f_i - K_i\ u_i{}^{p} \qquad (23)$$

This means that the coupling procedure used in the elastic model, can be applied to the elastoplastic model without drastic changes. In order to prevent a cumulative error in the pile-soil equilibrium when eq. (22) is used with incremental displacements, Δh_i has to be related to the total interaction force F_i:

$$\Delta h_i = F_i - K_i\ (u_i + \Delta u_i) - h_i \qquad (24)$$

The analysis can now be performed in a way similar to the elastic solution.

8 ELASTOPLASTIC SOLUTIONS

Rather as an illustration of the application of the model presented here than as validation, two examples will be considered. Relatively simple soil profiles are selected for this purpose. These two cases concern a relatively flexible and a relatively stiff rough pile, respectively; that is, according to the elastic flexibility ratio K_r. Both piles are loaded at the top by a horizontal force.

The finite element grid used is the same as described earlier for the elastic solutions, except that the elastoplastic version of the program analyses one half, instead of one quarter, of the cylindrical mesh. Furthermore, interface elements having a virtual thickness t_v=0.1 m are used on the pile wall. The soil is represented by 25 layers. In the finite difference approximation of the pile, one element is used for each layer of soil; this means that the pile is subdivided into 20 elements. A 2*2 Gaussian integration scheme is employed.

The flexible pile has an outer diameter of 1 m and a wall thickness of 0.05 m. The pile length is 50 m. The soil is characterized by a modulus of elasticity E_s=1 MN/m^2, a Poisson's ratio of 0.5, and an undrained shear strength of 10 kN/m^2 (K_r=5.67e-4). The load-deflection curve is computed, including one unload-reload cycle (Fig. 12). Both the non-linear behaviour and the effect of cyclic mobility can be observed in that figure.

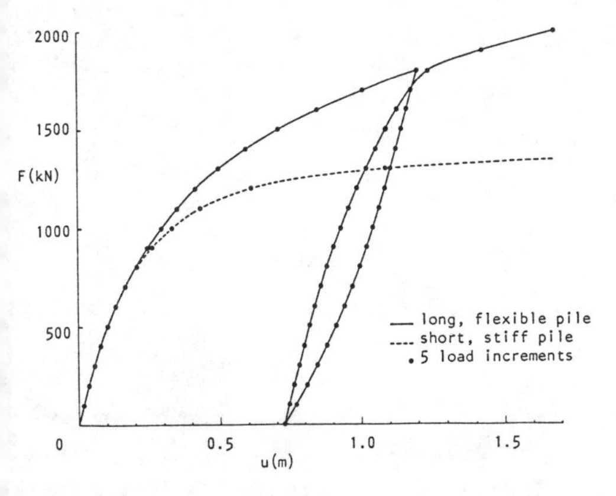

fig. 12 Elastoplastic responses for piles in homogeneous soil layers

The changes that occur in the load transfer to the soil, when the lateral load increases, are visualized in a series of pile displacement profiles for increasing load levels (Fig. 13). The point of zero deflection moves downward in the successive profiles, illustrating the concentration of the load transfer in the lower soil layers, when the upper layers reach the elastoplastic state. This mechanism is not possible for stiff piles, because the lower soil layers are already mobilized at low load levels. Therefore, it can be expected that a stiff pile exhibits a more brittle behaviour. The stiff pile

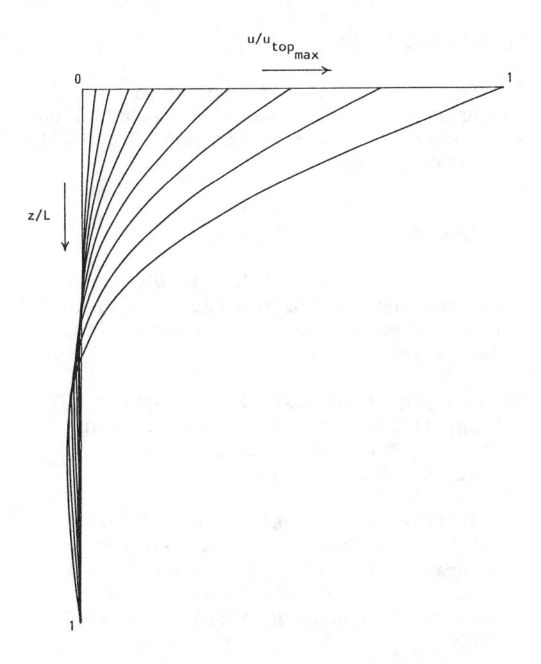

fig. 13 Successive displacement profiles for the flexible pile

examined here is obtained from the flexible pile by reducing its length to 25 m and increasing its wall thickness to 0.10 m. The remaining data are kept unchanged. These measures increase the flexibility ratio K_r by a factor of 1000. The load-displacement curve is shown in Fig. 12. It can be seen that the contradictory measures of increasing the wall thickness and reducing the length of the pile result in a very small increase in pile stiffness in the elastic stage. In the elastoplastic stage, however, the curves diverge considerably.

9 CONCLUSIONS

The model presented here offers an alternative and an extension of the p-y curve method for the design, analysis, and assessment of a single laterally loaded pile in a layered elastoplastic continuum. The numerical calculations can be performed on a personal computer. Both points contribute to the practical applicability of the model. As a limitation of the present version of the model, the relatively simple stress-strain relation can be mentioned. A very promising feature of the method is the possibility to extend the model to the analysis of pile groups.

ACKNOWLEDGEMENTS

The Delft University of Technology research project on laterally loaded piles was sponsored by the Netherlands Technology Foundation (STW).

REFERENCES

Bathe, K.J. & Wilson, E.L. (1976). Numerical methods in finite element analysis. Prentice Hall, Inc., Englewood Cliffs, New Jersey.

de Borst, R. & Vermeer, P.A. (1984). Possibilities and limitations of finite elements for limit analysis. Géotechnique 34, No. 2, 199-210.

Evangelista, A. & Viggiani, C. (1976). Accuracy of numerical solutions for laterally loaded piles in elastic half-space. Proc. 2nd Int. Conf. Num. Meth. Geomech., Blacksburg, Virginia 3, 1367-1370.

Hetenyi, M. (1946). Beams on elastic foundation. University of Michigan Press, Ann Arbor.

Poulos, H.G. (1971a). Behaviour of laterally loaded piles : I-Single piles, J. Soil Mech. Found. Div., ASCE 97, no. SM5, 711-731.

Randolph, M.F. & Houlsby, G.T. (1984). The limiting pressure on a circular pile loaded laterally in cohesive soil. Géotechnique 34, No. 4, 613-623.

Sharma, K.G., Varadarajan, A. & Chinnaswamy, C. (1986). Finite element analysis of dam foundations with seams. Proc. 2nd Int. Conf. Num. Mod. Geomech., Ghent, 473-480.

Vermeer, P.A. (1980). Formulation and analysis of sand deformation problems. Dissertation, Delft University of Technology, Delft.

Verruijt, A. & Kooijman, A.P. (1987). Laterally loaded piles in a layered elastic medium. Submitted for publication.

Numerical Methods in Geomechanics (Innsbruck 1988), Swoboda (ed.)
© 1988 Balkema, Rotterdam. ISBN 90 6191 809 X

A simplified analysis of pile penetration

Annamaria Cividini
Technical University of Milan, Italy
Giancarlo Gioda
Technical University of Milan and University of Udine, Italy

ABSTRACT: Some preliminary results are discussed of a study concerning the finite element analysis of pile penetration in quasi static conditions. The shear interaction between the pile and the surrounding soil is taken into account through axisymmetric interface elements. For the elasto-plastic soil both drained and undrained conditions are considered. Some step-by-step procedures for the numerical simulation of the penetration process are outlined and the advantages of one of them are shown.

1. INTRODUCTION

The numerical analysis of the stress and strain evolution produced in the ground by the driving process of a pile, or by a penetration test, has a relevant interest in geotechnical engineering. In fact, an analysis of this type could help the designer of deep foundations in predicting the vertical force required during the pile advancing, or to evaluate the bearing capacity of the pile and its resistance to tensile loads [1,2]. In addition, such a numerical tool can be used in the interpretation of the data recorded during in situ penetration tests [3] or during laboratory model tests [4].

A variety of numerical procedures have been proposed and applied to the analysis of penetration. Some of them are based on relatively simple (one dimensional) geometrical schemes, like the so called Cavity Expansion Methods [5] that reduces the driving process to the uniform expansion of a cylindrical cavity.

Other, more sophisticated, solution methods operate in two dimensional conditions and consider large strain effects [6,7]. Also in this case some simplifying assumptions can be introduced in order to reduce the computational cost, like e.g. when the shear interaction between pile and soil is neglected.

Another possible approach consists in treating the penetration as a two dimensional limit analysis, or plastic flow, problem, see e.g. [8-11]. It has to be considered, however, that this solution method aims in particular at determining the force required for penetration, and does not provide in general a complete description of the stress and strain fields in the soil.

Among the mentioned techniques, the approaches based on the finite element method, see e.g. [2,12,13], are perhaps those that permit to take into account the various factors influencing the problem (e.g. the non linear behaviour of soil and interfaces; the development of large strains in the zone close to the pile tip; the two phase nature of saturated soils; etc.) in the most integrated and consistent manner.

The finite element analysis of pile penetration here discussed tries to reach an "engineering" compromise between the conflicting requirements of limited computational cost, from the one hand, and of realistic simulation of the penetration process, from the other hand. The analysis is carried out in quasi static conditions, thus neglecting the dynamics effects. Since the study has not been completed yet, only some preliminary results will be presented.

The discussion is focused on the main characteristics of the computer code developed for the purpose of this study. They concern in particular the interface elements (governing the interaction between the lateral surface of the pile and the ground) and the procedure for simulating the advancing process. Other important features which are not yet operative in the finite element code, like e.g. the part concerning the consolidation [14] induced by the penetration in saturated soils, will be discussed separately.

2. INTERFACE ELEMENTS

Interface, or joint, elements represent an essential ingredient in the finite element analysis of axially loaded piles [15]. In fact, particular elements are required in order to model properly the highly disturbed zone between the pile wall and the soil, due to its small thickness and to the high stress gradient within it.

The most commonly used interface elements can be grouped into two main classes. Those belonging to the first group, which are referred to as "zero thickness" elements [16-18], are characterized by the fact that their thickness can be equal to zero (i.e. the corresponding nodes on the two sides of the element can have the same coordinates) without introducing any singularity in their formulation. The actual thickness of the joint is only used when computing the element properties.

A second group of interface elements are the so called "thin layer" elements [19]. In this case the behaviour of the interface is described through suitable two-dimensional elements for which a small (but different from zero) thickness is assumed. In order to avoid numerical problems, the ratio between their thickness and the size of adjacent elements should be between 0.1 and 0.01. Particular constitutive relationships are used for the thin layer elements that allow for the various deformation modes of the joint.

Two interface elements of the first category, with 4 and 6 nodes (cf. fig.1), were considered in this study. They are used, respectively, in conjunction with 4 and 8 node quadrilateral isoparametric elements discretizing the soil. Details on the formulation of interface elements were presented in the above mentioned papers. Here only the main characteristics of the adopted joints are recalled.

The displacement distribution along the side I of the element in the local reference system is expressed through its components $u'_I(s)$ and $v'_I(s)$ in the s and n directions. These are related to the displacements $\underline{u}_I$ (in the global axisymmetric system r-z) of the nodes of side I by the following relationship,

$$\begin{Bmatrix} u'_I(s) \\ v'_I(s) \end{Bmatrix} = \underline{N}_I(s)\ \underline{T}_I\ \underline{u}_I \tag{1}$$

where $\underline{N}$ is the matrix of the interpolation functions and $\underline{T}$ is a transfer matrix between local and global reference systems.

After writing an equation analogous to the above one for the displacements of side II, the relationship between nodal $\underline{u}$ and relative $\underline{\delta}$ displacements is easily obtained.

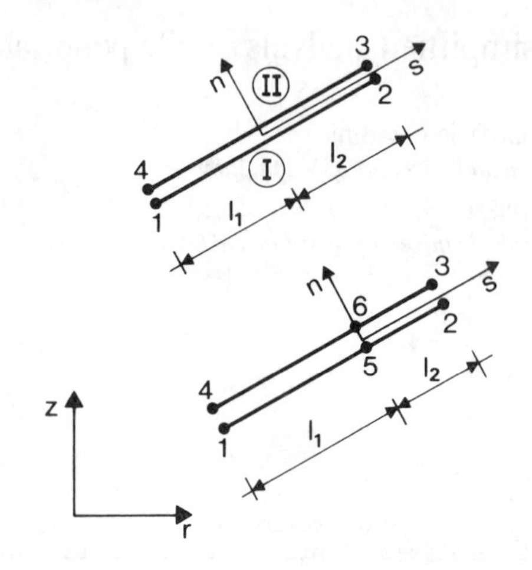

Fig. 1 Interface elements

$$\underline{\delta}(s)= \begin{Bmatrix} u'_{II}(s)-u'_I(s) \\ v'_{II}(s)-v'_I(s) \end{Bmatrix} = \underline{N}(s)\ \underline{T}\ \underline{u} \tag{2}$$

As to the constitutive relationship, in the linear range it is expressed as,

$$\underline{\sigma}(s) = \underline{D}\ \underline{\delta}(s) \tag{3}$$

where,

$$\underline{\sigma}(s) = \begin{Bmatrix} \tau(s) \\ \sigma(s) \end{Bmatrix} \quad ; \quad \underline{D} = \begin{bmatrix} k_t & 0 \\ 0 & k_n \end{bmatrix} \tag{4a,b}$$

and k_t and k_n are, respectively, the shear and normal stiffness of the joint.

The stiffness matrix $\underline{K}$ of the interface is expressed by the following equation.

$$\underline{K} = \underline{T}^T \{\int \underline{N}(s)^T\ \underline{D}\ \underline{N}(s)\ r\ ds\}\ \underline{T} \tag{5}$$

The non linear behaviour of the interface is here defined through a simple elasto-perfectly plastic model with tension cut-off and no dilatancy. This model is governed by the relationships between shear and normal stresses at failure, and between these stresses and the corresponding relative displacements δ_s and δ_n shown in fig.2. Of course, more complex models can be adopted, but their use did not seem necessary in this preliminary study.

3. SIMULATION OF THE ADVANCING PROCESS

Some approximated procedures were considered for simulating the quasi-static advancing of the pile. They reduce the continuous process to a finite number of increments, each involving a new geometry of the mesh and leading to the stress and

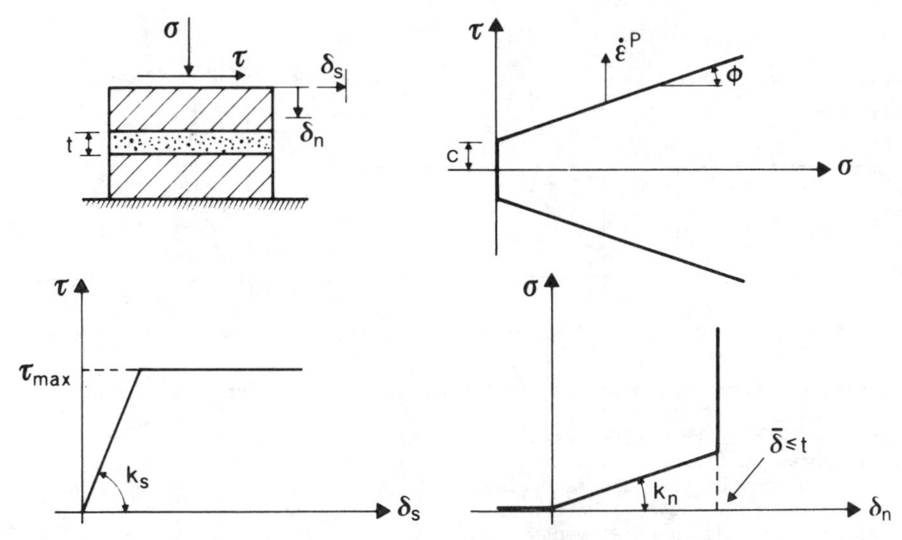

Fig. 2 Schematic representation of the interface behaviour

strain states in the ground for a chosen depth of penetration.

A first attempt was based on the scheme shown in fig.3, that does not require the discretization of the pile. Interface elements are introduced around the pile, and between the first soil element below the pile tip and the mesh axis. Dummy joint elements (with negligible shear strength) are placed on the remaining part of the symmetry axis. Note that for graphical reasons the thickness of these elements in the figure has been magnified.

All the nodes situated on the pile surface and on the symmetry axis are constrained, but the node representing the pile tip. This node is displaced horizontally by small increments. A non linear stress analysis is carried out for each of them by adopting, for instance, an updated Lagrangian approach (see e.g. [6,20]) and modifying the nodal coordinates following each increment. In these analyses the shear stress along the joints is kept at the limit value corresponding to the actual normal stress on the pile.

When the horizontal displacement of the node is completed, a new interface element is placed below the pile tip and an additional step of analysis is carried out.

This simple procedure, that leads to results quite similar to those of the so called cavity expansion method, does not properly estimate the bearing capacity of the pile. In fact, the summation of the vertical forces at the constrained nodes basically corresponds to the skin resistance of the pile, while its end resistance is grossly underestimated.

To overcome this problem a second pro-

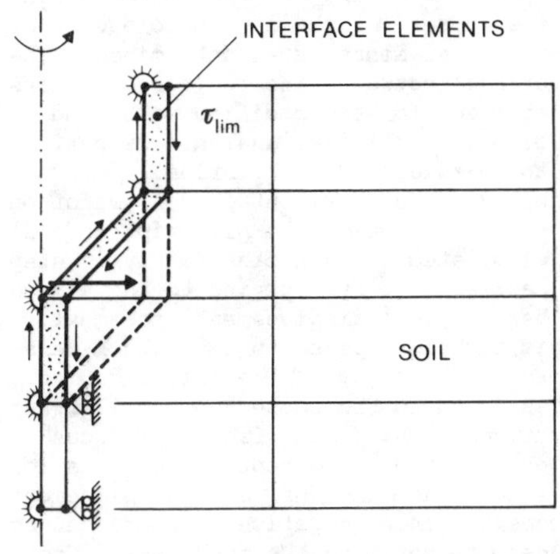

Fig. 3 Simulation of the penetration process as non uniform expansion of a cavity

cedure was considered in which the actual movement of the pile with respect to the soil is simulated. In this case both pile and soil are modeled through a grid of finite elements (cf.fig.4a). Interface elements are introduced around the pile and dummy joint elements are placed on the mesh axis below the pile tip. The horizontal displacements of the nodes belonging to both dummy joint and soil elements are constrained (nodes 1 and 2), but for the node (3) corresponding to the pile tip.

A first vertical displacement increment δ_1 is imposed to the pile and is subdivided into sub-increments. Each of them requires a non linear stress analysis and a modification of the nodal coordinates. A qualitative deformed shape of the discrete

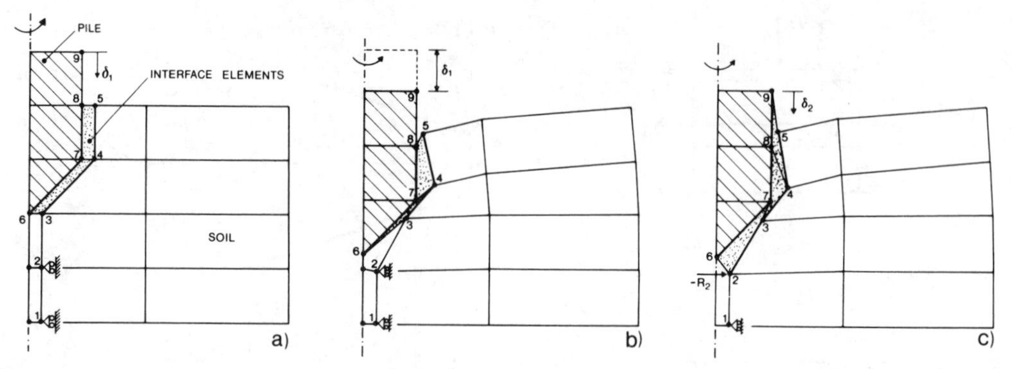

Fig. 4 An alternative procedure for simulating the advancing process of the pile

model at the end of this first stage of analysis is shown in fig.4b.

Now the incidences of the interface elements are updated (cf.fig.4c). The stress-strain states in the new elements are evaluated on the basis of the results of the previous step, and the forces acting on the nodes of the new and old sets of interface elements are determined. The difference between the two force vectors is applied to the modified mesh, and an additional non linear analysis is performed to re-establish equilibrium.

Finally, the horizontal constraint on the first node below the pile tip (node 2) is eliminated, and another analysis step is carried out by applying to the pile a further vertical displacement increment.

This technique persents a drawback related to the updating of the interface incidences, and to the subsequent equilibrium iterations. This, in fact, produces a "jump" in the load-depth of penetration curve at the end of each displacement increment. This negative effect can be reduced by refining the mesh, at the cost of a marked increase of computation time.

Since the above schemes did not yield satisfactory results, a third procedure was attempted (cf.fig.5) in which the pile is represented by a rough, rigid "rail" moving downward by small increments. The rail consists of a series of straight segments to which the soil is constrained by means of rollers that cannot support tensile normal forces. The shear interaction between pile and soil is taken into account by applying to each roller on the pile wall a tangential force evaluated through the previously mentioned constitutive model of the interface.

At the beginning of the calculations all rollers are located on the axis of symmetry and no tangential forces are applied to them (cf.fig.5a). The pile is moved downward by small increments (fig.5b); new positions of the rollers are determined by

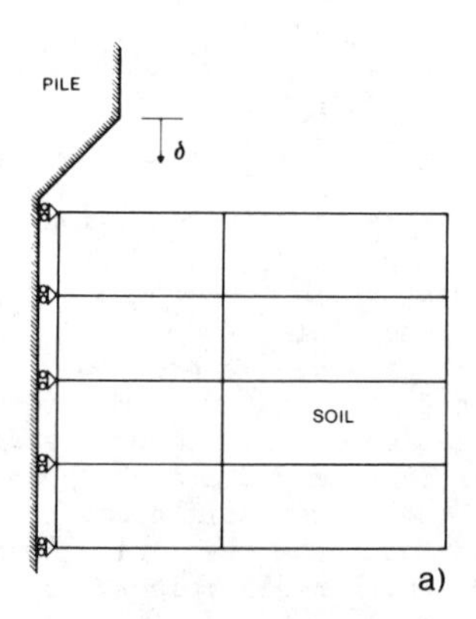

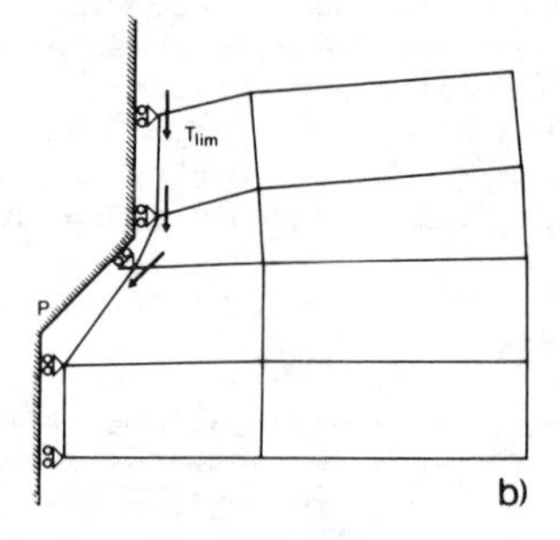

Fig. 5 Simulation of the penetration process through the advancing of a rough, rigid "rail"

means of non linear analyses at the end of which the nodal coordinates are updated.

Some particular provisions has to be introduced when a roller reaches the corner between two adjacent segments of the rigid rail (e.g. point P in fig.5b). It could happen, in fact, that the roller remains "trapped" in the corner for some subsequent steps of analysis, during which behaves as a "hinge" subjected to known vertical displacements. The details of the technique adopted for handling this prob-

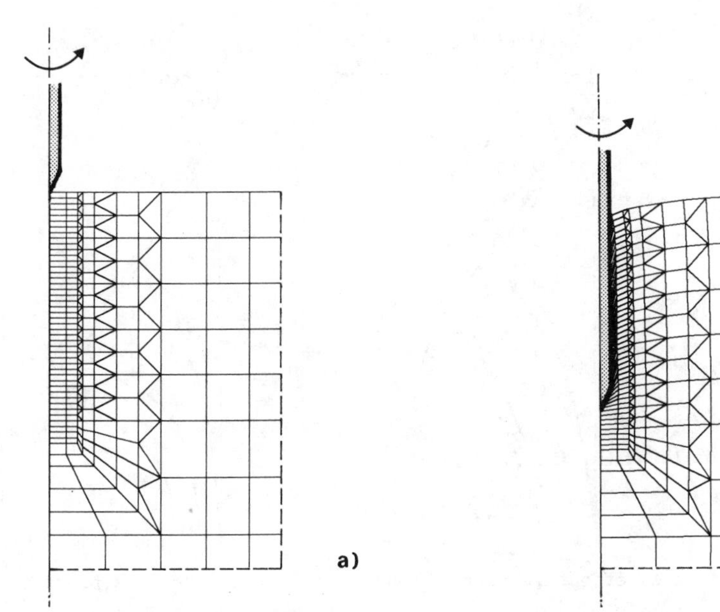

Fig. 6 Undeformed (a) and deformed (b) finite element meshes.

lem are not described here for briefness.

Among the three mentioned procedures, the last one turns out to be the most suitable for simulating the effects of penetration and it was used for the example discussed in the next section.

4. AN ILLUSTRATIVE EXAMPLE

A detail, close to the symmetry axis, of the axisymmetric finite element mesh adopted in the analyses (the complete mesh consists of 341 elements) is shown in figs.6a and 6b. They refer respectively to the undeformed geometry and to a deformed state during the advancing process.

An elastic-perfectly plastic behaviour, with Drucker-Prager yield criterion and associated flow rule, was chosen for the soil adopting, in terms of effective stresses, a unit value of cohesion c' and a friction angle ϕ' of 30°. Various values of the effective friction angle ϕ'_a between pile and soil were considered: $\phi'_a/\phi'=0$ (smooth pile), 0.5 and 1.0.

The penetration process was analysed in drained and undrained conditions. The so called coupled approach was used in the undrained case, considering as free variables the pore pressures and the nodal displacements and assuming the bulk modulus of water equal to 100 times that of the soil.

The cone representing the pile tip had a hight equal to its diameter D. The penetration process was studied up to a depth

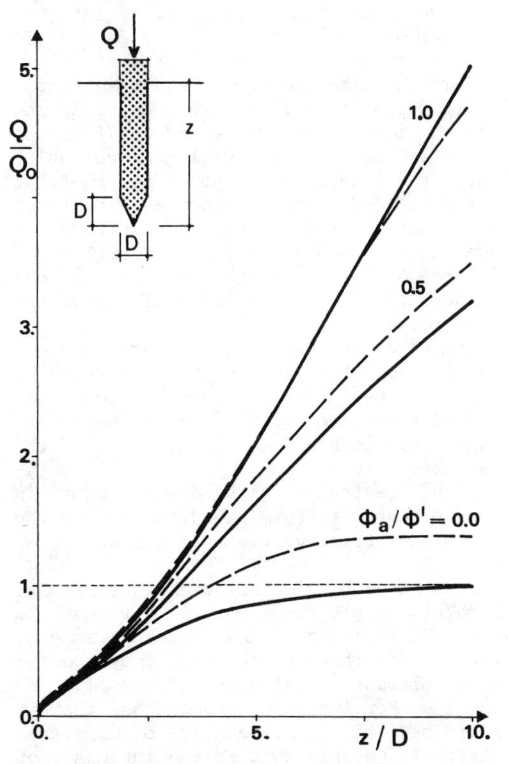

Fig. 7 Variation of the total load Q on the pile with the depth of penetration z (the results of drained and undrained analyses are represented, respectively, by solid and dashed lines).

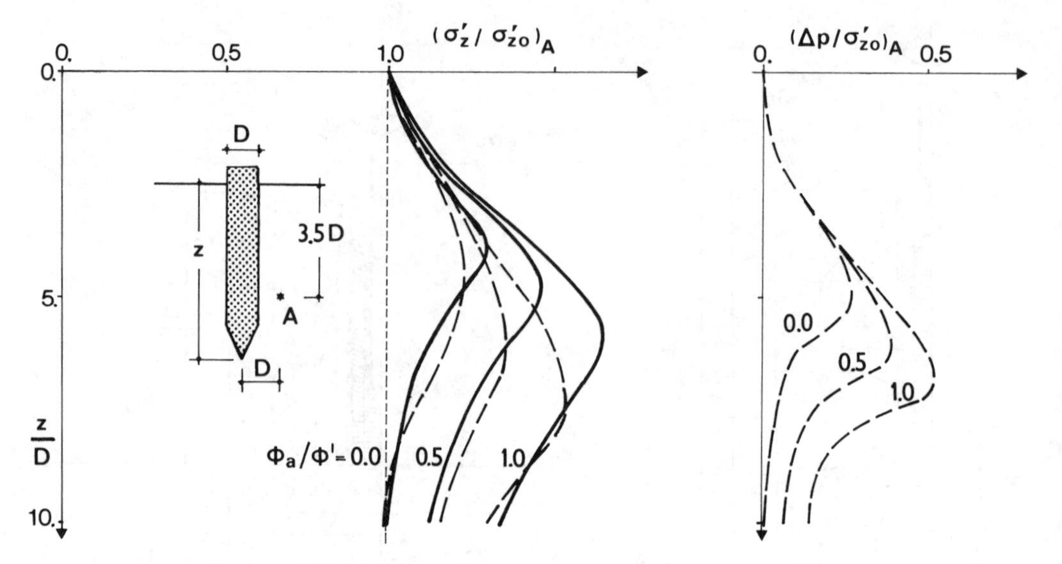

Fig. 8 Variation of the effective vertical stress σ'_z and of the pore pressure Δp at point A during penetration.

of 10D and it was subdivided into 160 steps for the drained analyses and 320 steps for the undrained analyses.

Some of the numerical results are summarized in figs.7 and 8. The first one shows, for various values of the friction angle between pile and soil, the variation with the penetration depth z of the total vertical load Q applied to the pile. Solid and dashed curves refer, respectively, to drained and undrained conditions. In this figure Q_o represents the maximum load obtained from the drained analysis of the penetration of a smooth pile.

Fig.8 shows the variation during penetration of the excess pore pressure Δp (with respect to its hydrostatic value) and of the effective vertical stress σ'_z at point A, located in the vicinity of the symmetry axis at a depth of 3.5D. Effective and neutral stresses are presented in non dimensional form dividing them by the in situ effective vertical stress σ'_{zo} at point A.

Even though this test example refers to a rather simple situation, with respect to those met in practice, it shows some of the effects that can be studied by the finite element simulation of the pile advancing. For instance, among those that it would be difficult to analyse by more traditional calculations is the influence of the roughness of the pile wall on the pore pressure developed during an undrained penetration.

5. CONCLUSIONS

A finite element technique has been summarized that, among other possible procedures, seems the most suitable for studying various aspects of the driving of piles through elastic-plastic soils. Some preliminary results show that this numerical tool permits to define quantitatively some important effects of penetration, like the variation of the neutral and effective stresses in the ground, and the increase of the total load applied to the pile.

This indicates that such a type of analysis has a potential for practical applications in foundation engineering, and for the interpretation of the results of in situ penetration tests. On these bases it seems advisable to continue the study toward the use of constitutive models more sophisticated than the simple elastic-ideally plastic law here adopted, taking also into account the effects of consolidation during the advancing process.

ACKNOWLEDGEMENTS

This study is supported by the National Research Council (CNR) and by the Ministry of Education (MPI) of the Italian Government. The authors wish to thank F. Genna and A. Pandolfi for stimulating discussions and computational help.

REFERENCES

[1] K.T. Law, "Numerical analysis of pile loading and pulling tests", Proc.4th. Int.Conf.on Numerical Methods in Geomechanics, Edmonton, pp.825-833, 1982

[2] R. De Borst, P.A. Vermeer, "Finite element analysis of static penetration tests", Proc.2nd European Symp. on Penetration Testing, Amsterdam, pp.457-462, 1982.

[3] M.M. Baligh, V. Vivatrat, C.C. Ladd, "Cone penetration in soil profiling" J.of the Geotechnical Engin.Div., ASCE, 106, No.GT4, pp.447-461 (1980)

[4] P.K. Banerjee, T.G. Davies, R.C. Fathallah, "Behaviour of axially loaded driven piles in saturated clay from model studies", in Development in Soil Mechanics and Foundation Engineering, Applied Science Publ. LTD, pp.169-195, 1983.

[5] M.M. Baligh, "Undrained deep penetration", Geotechnique, Vol.36, No.4, pp.471-501 (1986)

[6] P.K. Banerjee, R.C. Fathallah, "An Eulerian formulation of the finite element method for predicting the stresses and pore pressures around a driven pile", Proc.3rd Int.Conf.on Numerical Methods in Geomech, Aachen, pp.1053-1060, 1979

[7] P.S. Selvadurai, "Large strain and dilatancy effects in pressuremeter", J.of Geotechnical Engin.Div., ASCE, Vol.110, No.3, pp.421-436 (1984)

[8] G.T. Houlsby, C.P. Wroth, "Determination of undrained strength by cone penetration tests", Proc.2nd European Symposium on Penetration Testing, Amsterdam, pp.585-590, 1982.

[9] G.T. Houlsby, A.A. Wheeler, J. Norbury, "Analysis of undrained cone penetration as a steady flow problem" Proc.5th Int.Conf. on Numerical Methods in Geomechanics, Nagoya, pp.1767-1773, 1985

[10] T.T. Mehemet, B.A. Yalcin, H.C.Murat, R. Narayanan, "Flow field around cones in steady penetration", J.of Geotechnical Eng. Div., ASCE, 111, No.2, pp.193-204 (1985)

[11] P. De Simone, G. Golia, "Theoretical analysis of the CPT in sands", Proc. Int.Symp.on Penetration Testing, Orlando, 1988

[12] C.S. Desai, "Effects of driving and subsequent consolidation on behaviour of driven piles", Int.J.Num.Anal. Methods in Geomechanics, Vol.2, pp.283-301 (1978)

[13] P.D. Kiousis, G.Z. Voyiadjis, M.T. Tumay, "Large strain elastoplastic analysis of the cone penetration test", 1st World Congress on Computational Mechanics, Austin, 1986

[14] R.S. Sandhu, E.L. Wilson, "Finite element analysis of seeage in elastic media", J.Engineering Mechanics Division, ASCE, Vol.95, No.EM3, pp.641-652 (1969)

[15] Y.K. Chow, I.M. Smith, "Static-dinamic analysis of axially loaded piles" Proc. 4th.Int.Conf.on Numer. Methods in Geomechanics, Edmonton, pp.819-824, 1982

[16] R.E. Goodman, R.L. Taylor, T.L. Brekke, "A model for the mechanics of jointed rock", J.Soil Mech.and Found. Div, ASCE, Vol.94, No.SM3 (1968)

[17] O.C. Zienkiewicz, B. Best, C.Dullage, K.G. Stagg, "Analysis of non linear problems in Rock mechanics with particular reference to jointed rock systems", Proc.2nd Congress Int. Society of Rock Mechanics, Beograd, Vol.3, 1970.

[18] J. Ghaboussi, E.L. Wilson, J. Isemberg, "Finite element for rock joints and interfaces", J.Soil Mech.and Found.Div., ASCE, 99, No.SM10 (1973)

[19] C.S. Desai, M.M. Zaman, J.C.Lightner, H.J. Siriwardane, "Thin-layer element for interfaces and joints", Int.J. Num.Anal. Methods Geomech., Vol.8, pp.19-43 (1984)

[20] J.P. Carter, J.R. Booker, E.H. Davis, "Finite deformation of an elasto-plastic soil", Int.J.Num.Anal.Methods Geomech., Vol.1, pp.25-43 (1977)

Numerical Methods in Geomechanics (Innsbruck 1988), Swoboda (ed.)
© 1988 Balkema, Rotterdam. ISBN 90 6191 809 X

Analysis of laterally loaded piles by quasi-three-dimensional finite element method

T.Shibata & A.Yashima
Disaster Prevention Research Institute of Kyoto University, Japan
M.Kimura
Kyoto University, Japan
H.Fukada
Fudo Construction Co. Ltd, Japan

ABSTRACT: Analyses of the behaviour of pile groups subjected to lateral loads are carried out by the quasi-three-dimensional finite element method (GPILE : Ground Pile Interaction of Laterally Loading Effect). In this method, the elastic ground is modeled by stratified panels connected with shear springs, and the piles are modeled by beam elements. Thus, the interaction effect between the piles and the ground can be considered. The results of lateral loading tests on model free-headed pile groups are compared with those obtained from this method, and it is found that the measured and predicted values are generally in good agreement. Some non-tension analyses are also made using a nonlinear-elastic model.

1 INTRODUCTION

Almost all piled foundations are subjected to lateral loading; for example, unsymmetrical pressure or seismic force. With this type of lateral loading, the bearing capacity of the pile groups is reduced. Thus, the question of how to evaluate the reduction factor of lateral resistance becomes the most important point when designing piled foundations.

Until now extensive studies have been conducted on this problem. That is, the empirical equation for group efficiency has been estimated from model tests[1], and theoretical studies have been carried out using the theory of elasticity[2],[3]. Numerical analyses have also been made with the Finite Element Method (FEM)[4].

When a two-dimensional model has been adopted for FEM, the piles are treated as a sheet pile foundation. We cannot, therefore, correctly consider the interaction between the piles and the ground by this two-dimensional model. Furthermore, since the three-dimensional model[5] requires much CPU time and memory, it is not practical.

In this study, analyses of the behaviour of pile groups subjected to lateral loads are carried out by quasi-three-dimensional FEM (GPILE). In order to check the validity of this analysis we used a soil chamber for the model test[6]. The condition of the experimental pile top was free-headed, and the parameters for the test model were the arrangement of piles, the spacing between pile center lines and the number of embedded piles. The conditions for the analytical studies correspond to those of the model tests.

Since only elastic analysis is treated here, the necessary ground parameters are Young's modulus E_S and Poisson's ratio ν of the soil. When assuming ν to be 0.33 for Young's modulus of the ground, we supposed two different types of ground. One is the C-type (cohesive soil) for homogeneous soil, and the other is the S-type (sandy soil) for stiffness proportional to depth. These values are calculated (using the back analysis technique) from the top pile displacement taken from the model tests on single piles. The analyses for the behaviour of group piles are conducted using the above soil parameters.

Since the lateral displacement, stress, shear force and bending moment of each pile are calculated using this analysis, these values are then compared with the experimental results. These predicted values are also compared with Randolph's equation (PIGLET)[4], which assumes the soil as an elastic material.

In addition to the above, assuming the soil is non-resistant to tension, non-tension analyses are also made using a non-linear elastic model. In this way we will attempt to establish a more practical analytical approach.

2 OUTLINE OF ANALYSIS

2.1 Quasi-three-dimensional FEM model

The Quasi-Three-Dimensional FEM model used in this study is explained, and the validity of this analytical technique is represented. Firstly, as shown in Fig.1, both the ground and the structure are modeled as laterally stratified panels, and each panel is assumed to be a two-dimensional plane strain model. Between each panel, shear springs exist for x and y directions. Consequently, the forces can be transmitted from the upper panels to the lower ones.

Secondly, the way to determine the values of the shear springs between each panel is related. If we suppose the shear deformation for a soil block as shown in Fig.2, the values of the shear springs for

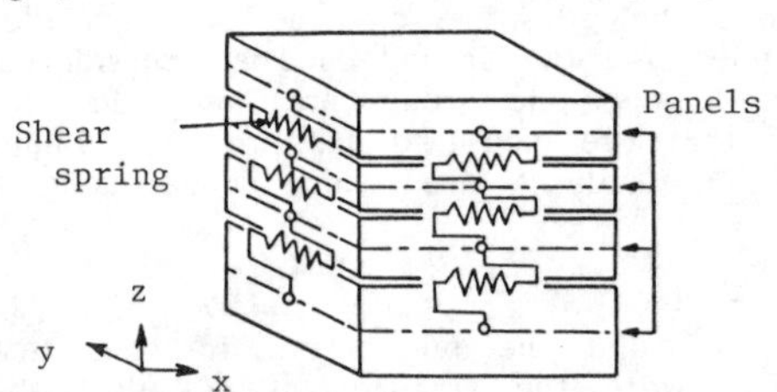

Fig.1 Modeling of ground

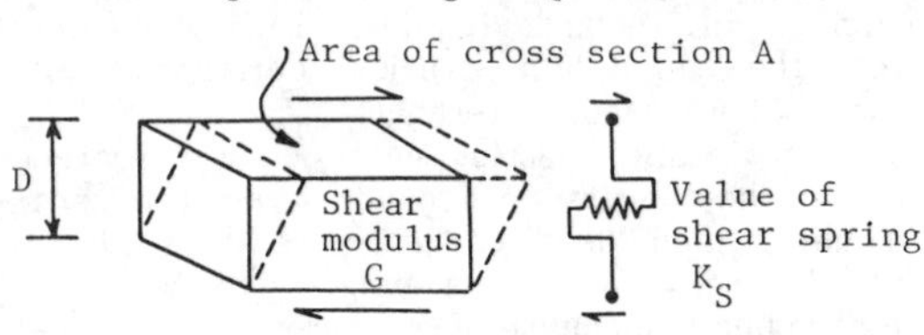

Fig.2 Shear deformation of soil block

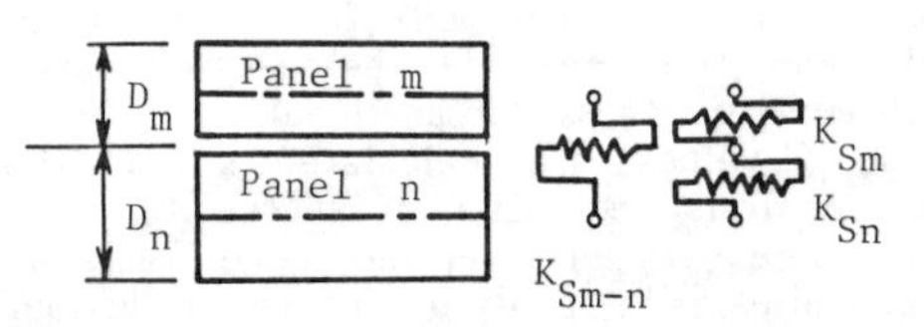

Fig.3 Shear spring between two panels

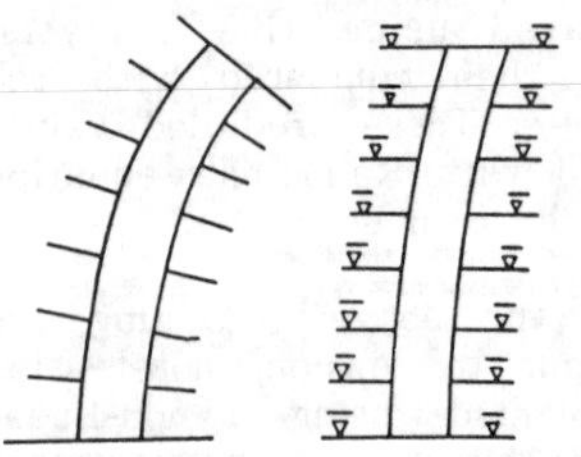

Fig.4 Bending and shear deformation of piles

each block can be represented by the following equation.

$$K_S = G \cdot A / D \qquad (1)$$

where, K_S : value of shear spring
G : shear modulus
A : area of cross-section of block
D : thickness of block

The deformation of each panel is represented by the middle of the panel thickness. And each shear spring is connected in series as shown in Fig.3. Consequently, the values of the shear springs connecting the panels are represented in the following equation.

$$K_{Sm-n} = 1/(1/K_{Sm} + K_{Sn}) \qquad (2)$$

where, K_{Sm-n} : value of shear spring between m-th and n-th panels; K_{Sm} , K_{Sn}: value of shear spring for m-th and n-th panels

$$K_{Sm} = G_m \cdot A_m/(D_m/2) = 2G_m \cdot A_m/D_m \qquad (3)$$
$$K_{Sn} = 2G_n \cdot A_n/D_n \qquad \text{(see, Fig.3)}$$

Actually, the shear spring values are calculated for each element and divided equivalently into nodal points which make up the element.

The panels are divided in the same way as the piles, however, as a pile does not have freedom in the vertical direction, only shearing deformation, can occur and not bending deformation as shown in Fig.4. Since piles resist external forces by bending rigidity, vertical beam elements are introduced in this method. Also, plane elements corresponding to the piles do not include shear springs.

2.2 Model test and analytical model

The object of this analytical study is to explain the lateral loading tests using

Table 1 Properties of material and Young's modulus

Properties of material		Ground constant
		(Lateral load 8kgf, δ_s=1.98mm)
D_{50}= 0.60 mm	e = 0.76	S-type : $\Delta E_S/\Delta z$=0.8 kgf/cm³, E_S=0
U_c = 30	e_{max}= 0.82	C-type : E_S=5.0 kgf/cm², $\Delta E_S/\Delta z$=0
γ_t = 1.74 gf/cm³	e_{min}= 0.52	

Table 2 Arrangement of piles for model tests

Arrangement	Number of piles	s/d	
Parallel	2,3,4	2.0,2.5,5.0	Parallel Box
Series			Series
Box	4,9,16		

model piles embedded in the indoor soil chamber. Details of these model tests are shown in reference[6].

The lateral loading tests were carried out using sand in the cylindrical chamber (height 105cm, diameter 165cm). The model piles were made of aluminum pipes, 80cm in length, 20mm in outside diameter, 1.6mm in thickness and 2.48×10^5 kgf·cm^2 in bending rigidity. The composition of the model ground was done through use of the boiling technique, and the ground was made to be as homogeneous as possible.

The properties of the sandy material for the model ground are represented in Table 1. The arrangement of model piles, as shown in Table 2, consisted of three different types, namely, the Parallel, Series and Box arrangement piles from the loading direction. The number of piles (in each arrangement) for the Parallel and Series arrangement piles was 2,3,4 and for the Box arrangement pile was 4,9,16. The spacing between pile centers (s) was 2.0d, 2.5d and 5.0d (d represents the pile diameter), and thus by changing the conditions, 27 different kinds of model tests were conducted. The items of measurement were lateral displacement and a lateral load, and also bending moment and shear force at a depth of 8cm from the pile top. In this analysis too, calculations were made with the same number of tests as for the FEM model. The characteristics of the test ground are divided into the following two types. One is the C-type (cohesive soil) for homogeneous soil, and the other is the S-type (sandy soil) with stiffness proportional to depth. Thus, a total of 54 analyses were carried out.

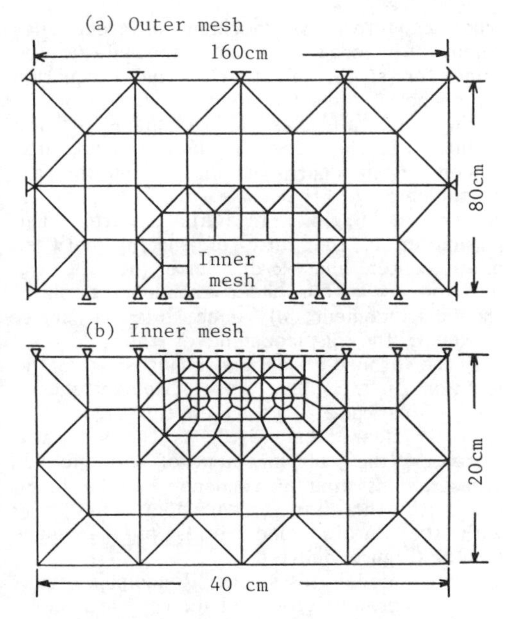

Fig.5 FEM mesh for Box arrangement pile

The typical FEM mesh is shown in Fig.5. This mesh represents the experimental soil chamber, and because of its symmetrical shape, only half of the area was calculated. The size of the mesh is 160cm x 80cm. Though aluminum pipes were used for the piles in the model tests, we assumed them to be dense octagonal piles at the time of modeling. This FEM mesh has so many nodal points and elements that it would be very difficult to represent it with only one figure. Consequently, as shown in Fig.5, we use an outer mesh and an inner to express it. Accordingly, the inner mesh actually enters into the center of the rectangular area of the outer mesh.

Each in-depth panel is divided into eight layers as shown in Fig.6, with the ground being represented by the third and all lower layers from the top. To simulate the model ground, the first and second layers represent the area above the ground surface, because the loading plate is attached at the pile top. The elements surrounding the piles above the ground surface are assumed as dummy elements, and reduce Young's modulus to a great extent.

2.3 Determination of ground parameters

In order to simulate the model tests, it is necessary to determine the ground parameters of the sandy soil in the soil chamber. In this GPILE analysis, we assume that the ground consists of linear-elastic material. Thus, the necessary ground parameters are Young's modulus of the ground E_S and Poisson's ratio ν. However, since there are various problems in determining

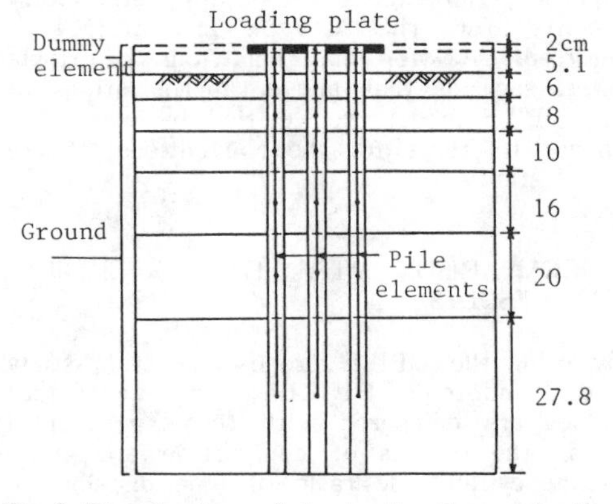

Fig.6 Distribution of in-depth piles and pile elements

the exact values, we conducted back analysis with GPILE using the top pile displacement obtained from the lateral loading tests on single piles.

Since the model ground consists of sandy soil, Young's modulus E_S is assumed to increase proportionally to the depth. However, since the thickness of the model ground is only 1m at most, E_S is thought to be nearly constant in depth. We, therefore, carried out analyses on group piles assuming the following two types of ground.

(1) S-type ; Young's modulus E_S is ground of sandy soil which increases proportionally to the depth.

(2) C-type ; Young's modulus E_S which is constant to the depth.

According to lateral loading tests for single piles, the mean value of lateral displacement (δ_S) of the pile top is δ_S= 1.98mm when the free standing length h= 6.1cm and lateral load H=8kgf. From both of these above values and the material constants of the model piles, the results of back analyses by GPILE are summed up as in Table 1. In other words, when assuming the C-type ground, Young's modulus E_S = 5.0kgf/cm^2, and for the S-type ground, $\Delta E_S/\Delta z$=0.8kgf/cm^3. Thus, each value is used in the following analyses.

Fig.7 gives an example of the distribution of in-depth bending moment for a single pile. According to this figure, we obtain a good agreement between this GPILE back analysis and Randolph's equation which assumes the ground as a linear-elastic material. Since the bending moment is measured at only one point (about 8cm below the pile top), the distribution in depth cannot be discussed, but the actual measurement at that one point corresponds to the analytical result. Fig.7 shows the results of the S-type ground. Comparing this analysis with Randolph's or Chang's equation, however, we obtain satisfactory results for the C-type ground.(Figure omitted.) As for the behaviour of single piles, such as that above, appropriations of the GPILE model are verified by comparing them with those for the conventional theory of elasticity.

3 COMPARISON WITH THE EXPERIMENTAL RESULTS

By using the GPILE proposed in this study and Randolph's PIGLET, the analytical values are compared with the experimental ones. The objects of comparison are group efficiency and the ratio of load distribution for the piles.

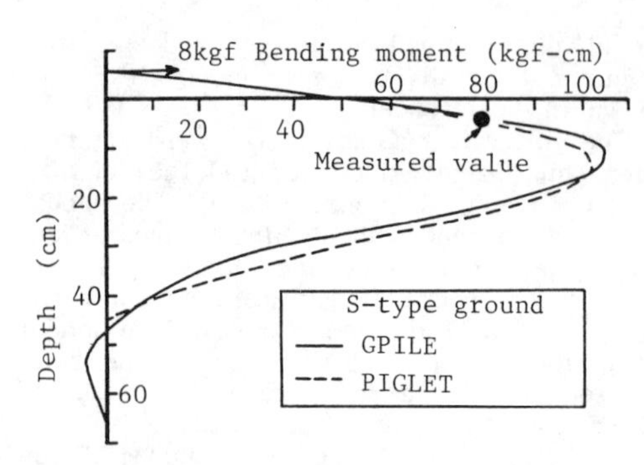

Fig.7 Bending moment for single pile

3.1 Group efficiency

The bearing capacity of group piles(Q_g) is not equal to the product of the single pile's bearing capacity(Q_s) and the number of embedded piles in the groups(N). The factor of reduction is generally called group efficiency (e), and represents the following equation.

$$e = Q_g / N \cdot Q_s \tag{4}$$

Group piles have more displacement than a single pile even when the average load of each pile is same. Thus, the group efficiency (e) can also be expressed in the following way.

$$e = \delta_s/\delta_g \tag{5}$$

Where δ_g=lateral displacement of group piles when the average load of each pile is the same for single pile. In this study, equation (5) concerning displacement is adopted.

First of all, as an example of the results of the GPILE distributional displacement in depth in piles, the Box arrangement, in the case of 3x3 piles is shown in Fig.8. In this figure, the parameter is the ratio of spacing between pile center lines (s/d). Since the pile top is fixed by a pin under the loading plate, the displacement of each pile is same. However, the displacement of the piles under the ground's surface changes with the location of the piles. The displacement of the corner pile is shown in this figure.

As is clear from Fig.8, as the s/d ratio increases, the displacement of the pile decreases. This kind of tendency can be found not with the Box arrangement, but also with the Parallel and Series pile arrangements. (Figure omitted.)

Next, model tests and Randolph's equation concerning group efficiency are compared with GPILE for the Box arrangement

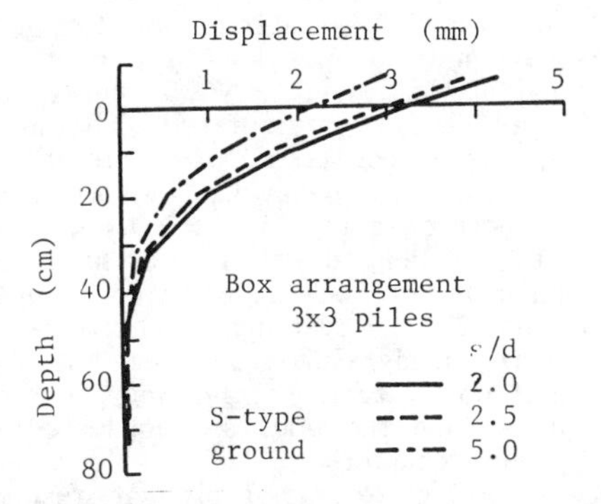

Fig.8 Lateral displacement of group piles

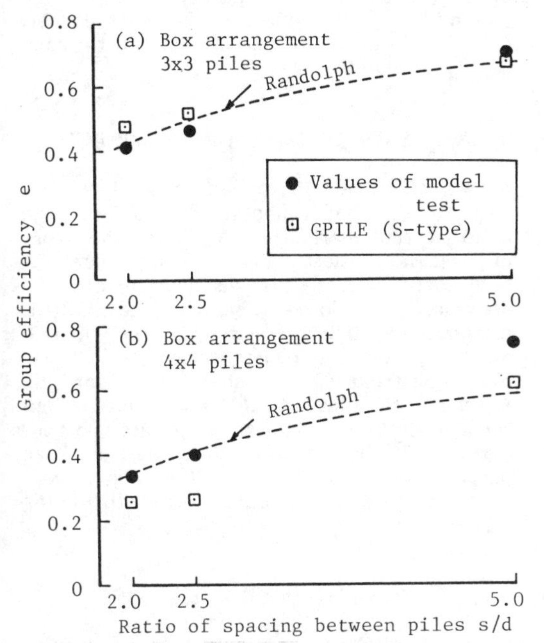

Fig.9 Group efficiency and ratio of spacing between pile center lines

piles (3x3 piles and 4x4 piles). Fig.9 shows these results. The lateral load per pile is 8kgf for any case.

According to the Fig.9, the GPILE values represent the tendency that along with an increase in the spacing between pile centers, the group efficiency also increases. In particular, one can see Fig.9(a), that the GPILE results are in quite good agreement with both Randolph's and the experimental results, however, when there are many piles as in Fig.9(b), the spacing between the piles is narrow, and the GPILE values are lower in comparison to the test results. Furthermore, in comparing Fig.9(a) with Fig.9(b), it is clear that a decrease in group efficiency can be seen in proportion to an increasing number of embedded piles, if the ratio of spacing between the pile centers s/d is the same. This tendency is notable with the smaller s/d ratio area.

3.2 Ratio of load distribution

The distribution of load is different for each pile depending on the location of the pile subjected to the lateral load. From the results of the model tests, the distribution of load for the front row (to the direction for load) is larger than that for the back row. With this analytical model and Randolph's equation, assuming the ground is an elastic material and resists tension, the results of the model tests (see Fig.11) cannot, therefore, be completely explained. However, the distribution of bending moment and shear force are different for the front and second row piles. We can express this difference in distribution by the location of the piles. Thus, according to the position of the piles, we can represent the differences in shear force and, at the same time, compare the top head shear force for the ratio of the load distribution. Next, we shall consider the

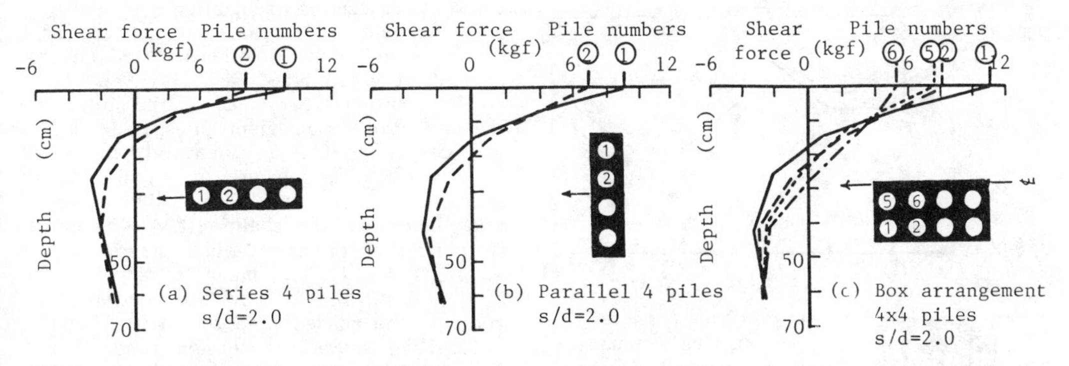

Fig.10 Distribution of shear force for group piles (C-type ground)

influence of the position of the piles and the spacing between the piles.

The distribution of shearing force for each pile (Parallel, Series (4 piles) and Box arrangement(4x4 piles)) is shown in Fig.10 as an example of the results of GPILE. Here we suppose that the ground is of the C-type and the spacing between the piles is s/d=2.0d. As for the Parallel or Series piles, it is clear that the shearing force of the pile top of pile No.1 is larger than that of pile No.2. A large difference cannot be found, however, for the shear force distribution of piles in the ground. The shear forces of the pile top for each pile in the Box arrangement, that of the No.1 pile is the largest, and the shear forces decrease in the following order; pile No.2, pile No.5, pile No.6. The pile top shear forces of both pile Nos.2 and 5 are similar. The shear force of pile No.6, however, is extremely small; about 60% of that for pile No.1.

For the ratio of load distribution, these GPILE analyses are compared to Randolph's equation and the experimental results. Fig.11(a) and (b) show the results of the Box arrangement pile (4x4 piles, s/d= 2.5,5.0). As the shear force of the pile top in pile No.1 sets the standard, the ratio of load distribution for other piles is calculated. Also, the analytical ground is supposed to be of the S-type. The following items are pointed out in Fig.11.

(1) Since both GPILE and Randolph's equation assume the ground as an elastic material, the soil can resist tension, and

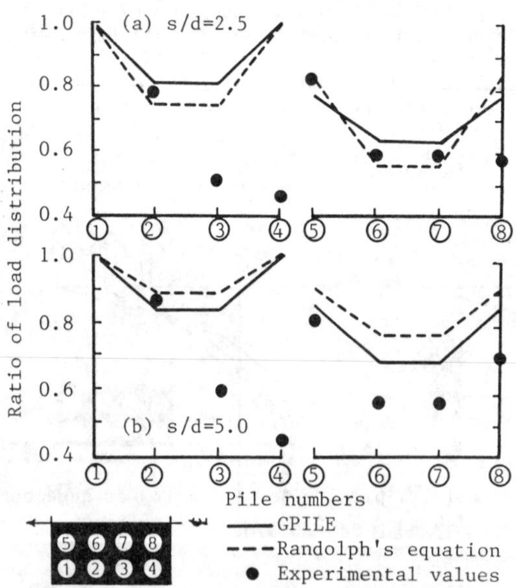

Fig.11 Comparison of ratio of load disribution for group piles (S-type ground)

thus, the ratio of load distribution for piles in the symmetrical positions (for example; Nos.1 and 4, Nos.2 and 3, Nos.5 and 8, and Nos.6 and 7) are calculated equally, however, the model tests do not represent this kind symmetric behaviour. In particular, the piles positioned in the back (Nos.3 and 4) for the loading direction are subjected to a much lesser load than the front piles (Nos.1 and 2). Regarding this difference between the analytical results and the experimental ones, studies considering the tension zone behind the piles are conducted in the following chapter.

(2) According to the analytical results, when comparing s/d=2.5 to s/d=5.0, a tendency exists for the ratio of space between pile center lines to becomes larger, and the lateral load is distributed to each pile equivalently.

4 EXTENSION OF ANALYTICAL MODEL

As we mentioned in the above section, concerning the experimental results for the ratio of the distribution of load, the front pile is the largest, and generally speaking, twards the back, the distribution of load decreases. However, with the GPILE method, assuming the ground is an elastic material, since it resists tension, the ratio of the distribution of load at the front pile is equal to that at the back pile. Consequently, these analytical results do not agree with the experimental ones. For the purpose of improving on the above differences, we conducted the following studies.

4.1 Tension zone behind piles

First of all, it is necessary to recognize the stress conditions of the ground around the group piles. Here, the area around these piles shows subjection to quite a large amount of tension stress. We shall discuss this next together with the behaviour of group piles.

The tension zone around the piles for the 4x4 Box arrangement piles (s/d= 2.5,5.0 and C-type ground) is illustrated in Fig.12. In this analysis, the ground is modeled by stratified panels, but the panels shown in this figure are the shallowest. As long as the criterion of the tension stress is $\sigma_m < -0.01 kgf/cm^2$ (σ_m ; mean principal stress $\sigma_m=(\sigma_1+\sigma_2+\sigma_3)/3$), we show the tension stress zone by the shaded area shown in Fig.12.

Fig.12(a) shows the tension zone of s/d= 2.5 for the narrow spacing between pile centers. According to this figure, there exist tension zones in six out of eight piles in the ground. In the two center piles of front area, however, tension zones do not appear. The reason for this is that the ground in the front pile area is compressed by the back piles. Also, the tension zone occurring behind the group piles is larger than that of Fig.12(b), as it extends over about twice as large an area as the loading plate.

When s/d in Fig.12(b) is somewhat larger than that in Fig.12(a), tension zones exist behind each pile. If we compare this pattern with that shown in Fig.13, we find that the behaviour of each pile is similar to that of a single pile. Also, the tension zone behind the group piles is comparatively smaller than that in the case of s/d=2.5, and the group piles have little effect on the surrounding ground. Though omitting the figure, there exists a tension zone behind the group piles as with the Parallel and Series piles, and depending on the arrangement of the piles and the space between the pile centers, its shape is influenced.

According to the above, to investigate the behaviour of group piles subjected to lateral loads, it is necessary to conduct non-tension analysis for the occurrence of stress in the ground. As a first step to this non-linear analysis, the non-tension analyses are conducted in the following section.

4.2 Nonlinear behaviour for non-tension analysis

As shown in the preceding section, there exists quite a large tension zone in the surrounding ground of group piles subjected to lateral force. Young's modulus of elements exceeding the criterion of tension stress are reduced to a value of nearly zero. Thus, we apply a non-tension analysis so as to avoid an exceeding tension stress [7].

In the linear analyses using GPILE, the lateral load per pile is 8kgf. In the non-linear analyses, the 8kgf load is divided into 4 steps, and thus, the increment of (each) subjected load becomes 2kgf. Also, as a standard for the tension stress (as in above section) the limitation for tension stress is σ_m=-0.01kgfcm^2. If the tension stress of the element exceeds the above criterion, Young's modulus E_S changes to the E_d=0.01kgf/cm^2 for the rigidity of the dummy elements. Two models, namely, the single pile and the Box arrangement 3x3 piles (s/d=2.0) in the S-type ground are used for calculating here.

Firstly, in each step the tension occurring in the single pile (as shown in Fig.13), and the relation between the single pile displacements in the linear analysis and non-tension analysis are shown in Fig.14. According to Fig.13, the tension zone does not occurs in the first and second steps, however, in the third and fourth steps, the tension zone is extensive. Thus, it is found that the load vs. displacement curve separates from that of linear elastic

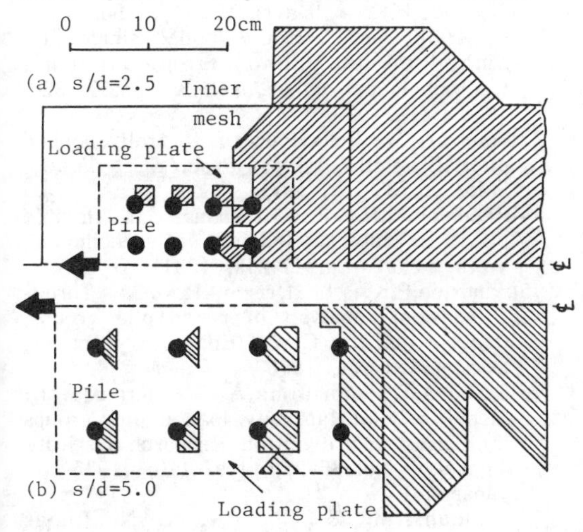

Fig.12 Tension zone around piles (C-type ground)

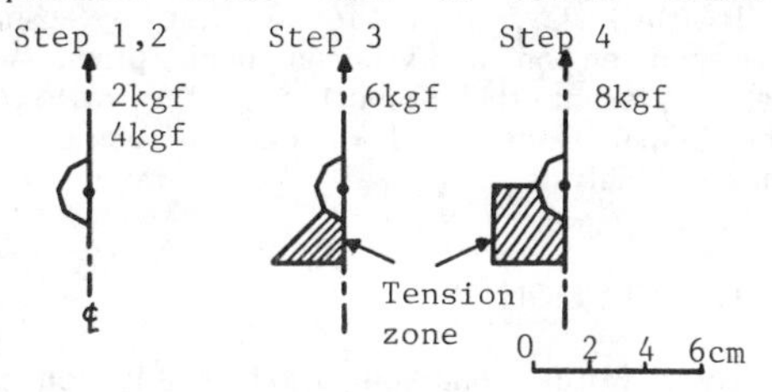

Fig.13 Tension zone behind Single pile

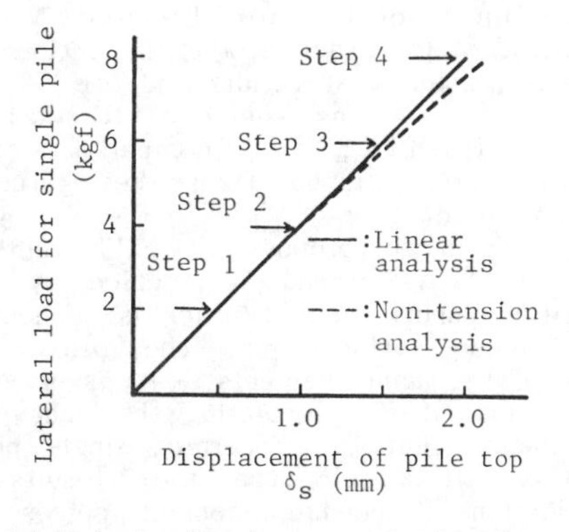

Fig.14 Relationship between load and displacement for Single pile

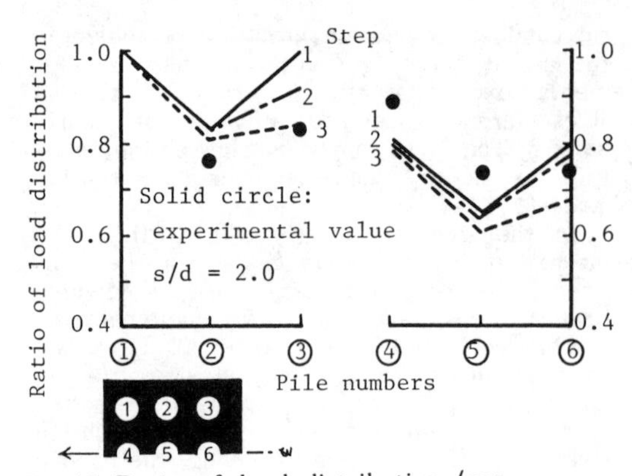

Fig.15 Ratio of load distribution (non-tension analysis)

analysis.

Secondly, the change in the ratio of the distribution of load for Box arrangement 3x3 piles is represented as Fig.15. In the first step, the tension stress has not yet reached the limitation of σ_m<-0.01kgf/cm^2. Thus, the ratio of the distribution of load at the symmetrical positions (Nos.1 and 3 and Nos.4 and 6 piles) is the same. However, as the steps increase, the ratio of the distributional load for pile Nos.3 and 6 gradually decrease. This tendency is in relatively good agreement with the results of the model tests(dark, solid circles).

In this way, it is found that by using the non-tension analysis on back piles, we can represent the reduction in the ratio of the distributional load instead of using the elastic analysis.

5. CONCLUSION

To clarify the behaviour of laterally loaded group piles, many numerical analyses are conducted with the newly developed Quasi-Three-Dimensional Finite Element Model (GPILE). To examine the formation of modeling, model test results and one of the analytical solutions, namely, Randolph's equation (PIGLET), are compared. The main results obtained from these studies are as follows;

(1) The newly proposed GPILE method —— the elastic ground is modeled by the stratified panels each of which is connected with shear springs, and the piles are modeled by beam elements —— is a very useful method for calculating the behaviour of piles. That is, concerning single piles subjected to lateral force, the results of distribution of bending moment correspond well to Randolph's and Chang's equations. In addition, there is almost no discrepancy with Randolph's equation in regards to group pile efficiency and the ratio of load distribution. Nevertheless, as GPILE is based on the theory of elasticity (similar to the conventional method), there are limitations to the applicability of these model tests.

(2) We considered the tension zone behind the group piles by means of GPILE. As the result, we see that the larger the spacing of piles becomes, the smaller the tension zone becomes. And, the behavior of each pile is similar to that of single pile, after all making the distributional load uniform.

(3) As mentioned above, the tension stress occurs behind the group piles. However, from the point of view that soil cannot resist the tension stress, non-tension analyses are performed to cut the tension stress. As a result, it is impossible to clearly explain the exact values of the ratio of distributional load by the conventional theory of elasticity. Therefore, according to non-tension analysis, it is found that this GPILE method is quite valuable for analyzing the behaviour of group piles.

Moreover, since this method can understand the stress condition of the ground around piles, not only elastic analyses, but also the method including non-linearity of the ground, shows much potential for future development.

REFERENCES

[1]Tamaki,O.,Mihashi,K. & Imai,T.: Horizontal resistance of a pilegroup subjected to lateral load, Proc.JSCE, No.192, pp.79-89, 1971 (in Japanese)

[2]Banerjee,P.K. & Davis,T.G.: The behaviour of axially and laterally loaded single pile embedded in nonhomogeneous soils, Geotechnique, Vol.28, No.3, pp.309-326, 1978

[3]Poulos,H.G.: Behaviour of laterally loaded piles II - pile groups, Proc.ASCE, Vol.97, SM5, pp.733-751, 1971

[4]Randolph,M.F.: The response of flexible piles to lateral loading, Geotechnique, Vol.31, No.2, pp.247-259, 1981

[5]Banerjee,P.K. & Driscoll,R.M.C.: Three-Dimensional analysis of raked pile groups, Proc. Instn. Civ. Engrg., Part 2, Dec.,pp.653-671, 1976

[6]Shibata,T., Yashima,A. & Kimura,M.: Model test of laterally loaded pile groups (3), Disaster Prevention Research Institute Annuals, No.28B-2, pp.97-110, 1985 (in Japanese)

[7]Duncan,J.M. & Chang,C.Y.: Nonlinear analysis of stress and strain in soils, Proc.ASCE, Vol.96, SM5, pp.1629-1653, 1970

Numerical Methods in Geomechanics (Innsbruck 1988), Swoboda (ed.)
© 1988 Balkema, Rotterdam. ISBN 90 6191 809 X

The problems of a slab on an elastic solid foundation in the light of the finite element method

Anastasios M.Ioannides
University of Illinois, Urbana, USA

ABSTRACT: A state-of-the-art report on numerical methods of analysis for a slab on an elastic solid foundation is presented. Finite element results are used to develop predictive equations, similar to those by Westergaard for the dense liquid subgrade. The limitations of this approach are accounted for in the formulae derived. The use of lumped, non-dimensional ratios offers a sound method of accounting for the host of variables involved.

1 INTRODUCTION

This Paper presents an evaluation of recent progress accomplished toward bringing the solution of the problem of a slab on an elastic half-space "to the same stage of completion as the Westergaard solution" [21], applicable to the corresponding dense liquid case. The main tool in this investigation is ILLI-SLAB, a two-dimensional finite element program capable of accommodating several foundation idealizations [16]. This is based on classical medium-thick plate theory and employs the 4-noded, 12-degree-of-freedom plate bending element. Results are used to derive predictive equations for the response of a slab loaded by an edge or a corner load. Theoretical closed-form solutions have long been available only for interior loading [19]. A brief description of a three-dimensional investigation [18] is also presented.

2 STATE-OF-THE-ART IN NUMERICAL PLATE-ON-ELASTIC SOLID ANALYSIS

Finite element models: The first application of the finite element method to two-dimensional analysis of a plate resting on an elastic solid is due to Cheung and Zienkiewicz [4]. Their conclusion was that the "Winkler type spring approximation introduced to avoid mathematical difficulties need no longer be used where continuous foundations are presented." More than two decades later, however, the "spring approximation" is still by far the most widely used subgrade representation, mainly due to practical limitations encountered in the application of their approach.

A major difficulty is the large computer storage required, in view of the fact that the subgrade stiffness matrix is symmetric but not banded. An iterative scheme was proposed by Huang [10] to convert this matrix into a banded one and thus reduce the memory core required. This is achieved using a "pseudo-bandwidth." An alternative method was proposed by Cheung [3]. Modification of the foundation stiffness matrix is based on the assumption that the deflection at a point is affected significantly only by forces acting at nearby points.

Other numerical models: It is sometimes difficult to interpret finite element data, since it may not be clear whether they reflect real system responses or they are merely due to the numerical intricacies of the finite element method. It is, therefore, desirable to consider responses from a number of other tools, as well.

Analysis of axisymmetric slabs of finite extent supported by an elastic solid foundation may be performed using computer program CFES [14]. This employs the method of concordant deflections [2] and is much less taxing on computer resources than corresponding finite element programs. Limited to the interior loading condition, CFES can determine stresses and deflections at the slab-subgrade interface. The program has been particularly useful in investigating subgrade stress distributions [11; 14].

Maximum bending stress, σ_e, developing in a slab on an elastic solid under an edge load may be determined using H51ES [11], a computerized adaptation of the corresponding Pickett chart [21]. Its solution is, therefore, akin to a theoretical, closed-form one. It can accommodate both single- and multi-wheel gears, and requires a relatively small memory core.

An existing finite difference solution [20] was incorporated into the FIDIES computer program [15]. This permits an investigation of all three fundamental loading conditions, generally making more modest demands on computer resources than ILLI-SLAB.

3 GUIDELINES FOR THE INVESTIGATION OF THE THREE FUNDAMENTAL CASES OF LOADING

The slab-subgrade system response under any loading condition, as well as the convergence of numerical solutions, may be investigated in terms of several non-dimensional ratios [11; 14; 15]. The effect of slab size, for example, can be examined in terms of (L/ℓ_e), where L is the length of the (square) slab, and ℓ_e is the radius of relative stiffness of the slab-subgrade system, determined from the equation:

$$\ell_e = \sqrt[3]{[Eh^3 (1-\mu_s^2) / \{6(1-\mu^2)E_s\}]} \quad (1)$$

in which:

- E : Young's modulus for the slab;
- μ : Poisson ratio for the slab;
- h : slab thickness;
- E_s : Young's modulus for the foundation;
- μ_s : Poisson ratio for the foundation.

Minimum slab size requirements for the development of infinite-slab responses vary between 3 and 10 ℓ_e, depending on the loading condition (interior, edge or corner) and the response considered (deflection, subgrade stress, bending stress).

Similarly, the effect of the finite element mesh fineness can be expressed in terms of the ratio (2a/h), where 2a is the short side of an element. This is the single most important factor influencing the accuracy of the numerical solution. With respect to ILLI-SLAB, the value of 0.8 for this ratio was found to be satisfactory in most cases [11; 16].

Restrictions in the size of computer memory core often dictate using a non-uniform mesh, which will inevitably include non-square elements. This gives rise to two additional effects. The first is related to element aspect ratio, α, which is the quotient of the long side, 2b, of a (rectangular) element divided by its short side, 2a. This was found to be considerably more important in elastic solid computations [16], compared to the relatively low sensitivity of the dense liquid solution [11]. Since the Cheung and Zienkiewicz method is strictly applicable only to square elements [16], aspect ratios greater than 3.0 should be avoided.

In addition, a transition zone is necessary between the critical regions (which should be fine and consist of square elements) and the coarser portions further away. Gradation, χ, describes how this is accomplished. It is defined as the ratio of the distance over which mesh is fine and uniform, to the element size in the fine region, $2a_{min}$. Dense liquid results suggest that χ should be greater than 2 or 3 [11].

4 INTERIOR LOADING

This case is by far the best understood and most widely studied among the three fundamental loading conditions. Hogg [6; 7] and Holl [8; 9] performed numerical calculations for an axisymmetric slab of infinite extent resting on an elastic solid subgrade, but fell short of developing closed-form equations similar to those presented earlier by Westergaard [22] for the dense liquid foundation. This task was completed by Losberg [19].

An extensive finite element investigation of interior loading was presented in an earlier paper [16], in which guidelines for better accuracy were developed. Comparing ILLI-SLAB responses and the closed-form solutions indicates that even when the preceding guidelines are adhered to, the finite element solution may still exhibit a discrepancy of the order of 10 to 20%. This is particularly true for high load size ratios, (c/ℓ_e), in which c is the side length of a square load, and may be attributed to the way the finite element method converts applied distributed loads to concentrated nodal force components. Preliminary results using the CRAY X-MP/24 supercomputer indicate that this discrepancy is eliminated when the mesh is refined further in the region under the load [13]. The requirement for a more stringent mesh fineness criterion in the critical area under the load was also confirmed using the CFES program [14].

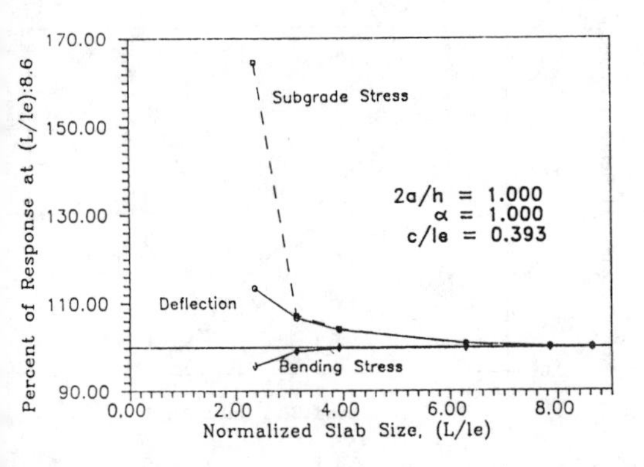

Figure 1. Effect of slab size on maximum edge loading responses.

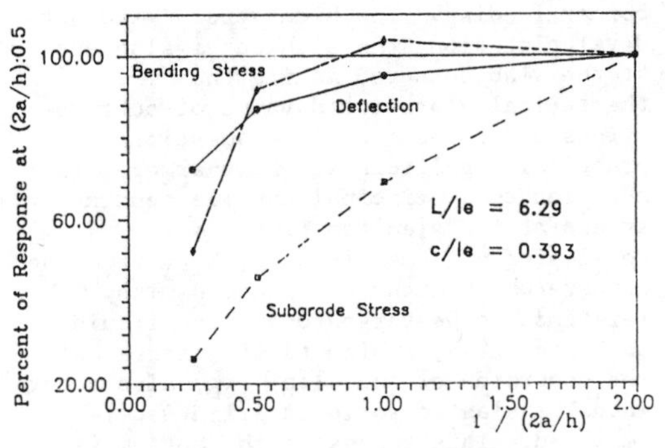

Figure 2. Effect of mesh fineness on maximum edge loading responses.

5 EDGE LOADING

Theoretical solutions for this case have been unavailable for a long time, with the exception of a graphical method for the determination of the maximum bending stress proposed by Pickett, et al. [21]. Using a computerized version of this chart (H51ES), a predictive equation for σ_e was derived [11]. The variation of the Load Placement Effect Factor, i.e. the ratio of σ_e to the maximum stress under interior loading, σ_i, has also been examined [12]. This was shown to be solely a function of the load size ratio, (a/ℓ_e), where a is the radius of the applied load (or some other measure of its size). A detailed investigation of edge loading using the finite difference method has been presented elsewhere [15]. A similar examination using ILLI-SLAB follows.

Effect of slab size: Convergence patterns obtained are best illustrated by the non-dimensional plot in Figure 1. Trends exhibited are very similar to those obtained during earlier studies [15; 16; 17]. Slab size requirements for the development of infinite-slab responses under edge loading range between 4 and 6 ℓ_e.

Effect of mesh fineness: Results are shown in a normalized plot in Figure 2, in which the abscissa is the inverse of the mesh fineness ratio. Convergence patterns are again similar to those obtained in previous investigations [11; 15; 16]. Much higher sensitivity is exhibited by the maximum subgrade stress, q_e, which appears to require a value of (2a/h) below 0.5. This trend can be explained with reference to the high stress concentrations developing at the slab edges. The presence of these gradients was identified under interior loading [5; 14], and was later confirmed under both edge and corner loading conditions, as well [15; 18]. The numerical solution for q_e is usually an underestimate of a much larger value predicted by theory at the slab edge. It is worth noting that for a perfectly rigid slab, the theoretical q_e value according to Boussinesq tends to infinity. Pursuing this value by additional refinement of the mesh may not be worthwhile, however, since the theoretical stress concentrations are limited to a very narrow zone along the slab edge (of the order of 0.1 ℓ_e in width) [14; 18]. Soil yielding will preclude their development in the field.

Effect of element aspect ratio and gradation: Results show that all three maximum responses decrease as α_{max} increases. When α_{max} exceeds 4.0, however, responses may fluctuate wildly (increasing or decreasing in a random fashion) rendering the solution unreliable. The detrimental effects of high α_{max} are more serious when χ is low. In view of these results, the recommendation for α_{max} not to exceed 2.0 or 3.0, whenever possible [16], is reiterated here. In addition, all three responses increase as χ increases, as long as α_{max} remains below 4.0. This indicates an improvement in the accuracy of the solution, similar to that observed upon increasing mesh fineness. It is recommended that values of χ in excess of 2.0 be used, whenever this is feasible.

6 CORNER LOADING

The only systematic and relatively complete examination of this loading condition to date is the one conducted using the finite difference computer program FIDIES [15]. This revealed several aspects of slab-subgrade system behavior that are unique to corner loading. The high stress gradients developing along the slab edges also appear at the slab corners, making the theoretically predicted value of subgrade stress under the corner of questionable practical significance. Furthermore, in addition to a principal tensile bending stress at the slab top fiber, σ_{ct}, developing at some distance, X_{1t}, from the corner (which might be expected with reference to Westergaard's dense liquid analysis [22]), a high tensile stress at the underside of the slab, σ_{cb}, also occurs when the elastic solid idealization is employed. This stress at the bottom is often of the same order of magnitude as the stress at the top, and occurs near the centroid of the applied load. It seems, therefore, that while in the dense liquid case the corner region of the slab acts like an unsupported cantilever, its behavior when supported by an elastic solid is more akin to that of a propped cantilever. These observations are confirmed by ILLI-SLAB results.

Effect of slab size: Convergence patterns obtained are similar to those in Figure 2, as well as those from previous studies [15; 17]. Bending stress, σ_{ct}, is particularly sensitive. Slab size requirements for the development of infinite-slab responses under corner loading range from 4 to 5 ℓ_e.

Effect of mesh fineness: The sensitivity of corner loading results to mesh fineness is striking. Using (2a/h) of 1.0, maximum subgrade stress, q_c, is only 22% of the value obtained with (2a/h)=0.2. The corresponding percentages for maximum bending stress, σ_{ct}, and deflection, δ_c, are 59% and 71%, respectively. The great sensitivity of q_c to mesh fineness is the result of the high gradients developing in a narrow zone along the edges and at the corners of the slab. To obtain an accurate estimate of the theoretical corner loading responses, therefore, a very fine mesh is necessary (2a/h = 0.2 to 0.3).

7 DEVELOPMENT OF PREDICTIVE EQUATIONS

The user guidelines developed above may be employed in designing a series of runs, to provide the database for the development of a set of formulae for the determination of maximum responses under edge and corner loading, which have been unavailable up to now. This may be achieved by investigating the relation between normalized responses and the load size ratio (c/ℓ_e). A similar set of predictive equations was derived earlier using results from FIDIES [15]. Those formulae were limited, however, to the case of a point load applied near a slab edge or corner.

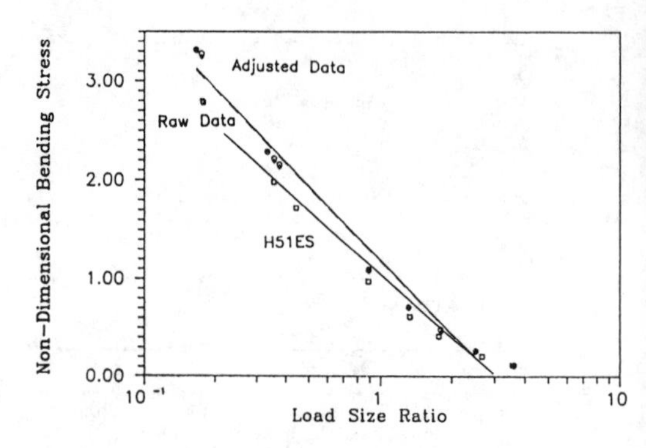

Figure 3. Development of predictive equation for maximum bending stress under edge loading.

Using results from ILLI-SLAB, a similar set of equations were obtained for small loaded areas of finite size. Original data obtained were adjusted to make up for deficiencies caused by inadequate mesh refinement. This adjustment was on the average an increase of 3% for δ_e, 30% for q_e, and 1% for σ_e. Figure 3 shows the non-dimensional lumped group for maximum edge stress, ($\sigma_e h^2/P$), plotted in a semi-log graph against the load size ratio, (c/ℓ_e). Similar plots were obtained for ($\delta_e D/P\ell_e^2$) and ($q_e\ell_e^2/P$). P denotes the total applied load, while D is the slab flexural stiffness [$=Eh^3/12(1-\mu^2)$]. The scatter in the data is fairly small, indicating the soundness of the non-dimensional lumped parameters approach. Only subgrade stresses exhibit a somewhat greater variability, in view of the high gradients existing along the edge. These would require a much finer mesh for an accurate examination.

The plot of non-dimensional σ_e values also shows a line obtained using H51ES [11]. Agreement between the two programs is fairly good except for very low values of (c/ℓ_e). It is considered that ILLI-SLAB results will converge to those from H51ES upon mesh refinement and slab expansion, since H51ES is akin to a closed-form solution for an infinite slab.
Equations describing the best-fit straight lines for the adjusted data are presented in Table 1, where coefficient of determination, R^2, values are also given. For the maximum edge bending stress, the equation for the H51ES data is used.

In view of the higher sensitivity of corner loading results to mesh fineness, a much more significant adjustment is

Table 1. Best-fit predictive equations based on adjusted ILLI-SLAB results.

General form of the equations:
Non-dimensional response = $A\log_{10}(c/\ell_e)+B$
except for: $(X_{1t}/\ell_e) = A\sqrt{(c/\ell_e)}+B$

N-D response	A	B	R^2
Edge Loading:			
$(\delta_e D/P\ell_e^2)$	-0.185	0.251	0.95
$(q_e \ell_e^2/P)$	-2.040	1.587	0.91
$(\sigma_e h^2/P)$*	-2.148	1.042	0.98
Corner Loading:			
$(\delta_c D/P\ell_e^2)$	-0.593	0.495	0.94
$(q_c \ell_e^2/P)$	-43.973	26.087	0.81
$(\sigma_{ct} h^2/P)$	-2.044	0.879	0.81
(X_{1t}/ℓ_e)	1.961	-0.412	0.86

*: Based on H51ES results.

required. Thus, δ_c was increased by an average of 30%, q_c by 75%, and σ_{ct} by 45%; X_{1t} was decreased by 35%. The magnitude of these adjustments suggests that the accuracy of the equations developed can be significantly improved if a larger machine (e.g. a supercomputer) is used. The scatter of the data is also considerably greater. Results, however, may still be fitted with straight lines, the equations for which can be used to approximate system response, within the limits of accuracy imposed by the computer used in this study. Equations for the adjusted data, expressed in terms of the lumped groups $(\delta_c D/P\ell_e^2)$, $(q_c \ell_e^2/P)$, $(\sigma_{ct} h^2/P)$ and (X_{1t}/ℓ_e) are also given in Table 1.

8 THREE-DIMENSIONAL ANALYSES

An existing three-dimensional finite element program (GEOSYS) was recently adapted to provide a more realistic representation of the slab-on-elastic solid problem, and to evaluate the accuracy of two-dimensional models routinely used in engineering practice. A detailed description of that investigation is presented elsewhere [18].

GEOSYS was developed in the early 1970's for analyzing rock-structure interaction [1]. The version used consisted of about half the original code and employed the linear, isoparametric brick element for both the slab and the foundation. Numerous convergence studies were performed to establish guidelines for setting up the finite element mesh. These included sensitivity analyses with respect to the subgrade lateral and vertical extents, the boundary conditions, as well as vertical and horizontal mesh fineness. The overriding importance of the latter was confirmed, as was the requirement for a more stringent mesh fineness criterion under the loaded area.

The most significant conclusion reached was that the three-dimensional analyses reinforce the validity and desirability of conventional two-dimensional studies. When these disagreed, probable causes were sought and were usually found to be due to the coarse three-dimensional mesh used. From a more practical viewpoint, results confirmed the existence of very high subgrade stress gradients along the edges and at the corners of the slab. These become even steeper under edge and corner loading conditions. Using an iterative scheme, local soil yielding was adequately modeled. Overstressed elements were found to constitute only about 15% of the total number of soil elements and to occur in the upper 5 to 7 ft (1.5 to 2.0 m) of the subgrade. Local yielding tends to move the response of the elastic solid foundation toward that of the dense liquid model, if only slightly. The latter seems to be closer to real (stress dependent) subgrades than is often assumed.

9 CONCLUSION

Investigations focusing on analytical methods for the problem of a slab on an elastic solid foundation have resulted in major advances toward bringing its solution to the same stage of completion as Westergaard did for the corresponding dense liquid case. At the heart of this effort is a systematic approach to the interpretation of numerical analysis data, based on the use of lumped, non-dimensional ratios. Conclusions reached during this study are reinforced and validated by comparison with analyses performed using several other slab-on-elastic solid models. The most ambitious of these is an adaptation of an existing three-dimensional finite element program, whose great potential can only be fully exploited on a supercomputer. The systematic approach presented will be indispensable in future studies, given the tremendous volume of data generated by these mammoth machines. Three-dimensional results to date reinforce the validity of conventional two-dimensional analysis. They indicate that notwithstanding difficulties inherent in the elastic solid model, such solutions are feasible, useful and enlightening.

ACKNOWLEDGMENTS

Investigations were supported by Grant No. AFOSR-82-0143 from the U.S. Air Force Office of Scientific Research, Air Force Systems Command, Bolling AFB, DC. The project manager was Lt. Col. L.D. Hokanson.
The author acknowledges the cooperation of Messrs J.P. Donnelly and M.W. Salmon, formerly Graduate Research Assistants, during the three-dimensional finite element investigation. Professors C. Tanimoto, E.J. Barenberg and M.R. Thompson reviewed the manuscript.

REFERENCES

[1] Agbabian Associates, Analytic modeling of rock-structure interaction, U.S. Department of Defense, Report No. AD-761-648 (1973)
[2] S.G.Bergström, E.Fromén and S. Linderholm, Investigation of wheel load stresses in concrete pavements, Swedish Cement and Concrete Research Institute, Proceedings No. 13 (1949)
[3] M.S.Cheung, A simplified finite element solution for the plates on elastic foundation, Computers and Structures 8, 139-145 (1978)
[4] Y.K.Cheung and O.C.Zienkiewicz, Plates and tanks on elastic foundations: An application of finite element method, Int. J. of Solids and Structures 1, 451-461 (1965)
[5] Y.T.Chou, Subgrade contact pressures under rigid pavements, J. of Transp. Eng. ASCE 109, 363-379 (1983)
[6] A.H.A.Hogg, Equilibrium of a thin plate, symmetrically loaded, resting on an elastic foundation of infinite depth, Philosophical Magazine Series 7 25, 576-582 (1938)
[7] A.H.A.Hogg, Pavement slabs on non-rigid foundations, Proc. 2nd Int. Conf. on Soil Mech. and Found. Eng. 3, 70-74 (1948)
[8] D.L.Holl, Equilibrium of an thin plate, symmetrically loaded, on a flexible subgrade, Iowa State College J. of Science 12, 455-459 (1938)
[9] D.L.Holl, Thin plates on elastic foundations, Proc. 5th Int. Congress on Applied Mechanics, 71-74 (1939)
[10] Y.H.Huang, Finite element analysis of slabs on elastic solids, Transp. Eng. J. ASCE 100, 403-416 (1974)
[11] A.M.Ioannides, Analysis of slabs-on-grade for a variety of loading and support conditions, Ph.D. thesis, U. of Illinois, Urbana, U.S.A. (1984)
[12] A.M.Ioannides, Discussion of Response and performance of alternate launch and recovery surfaces ALRS containing stabilized-material layers, by R.R. Costigan and M.R.Thompson, Transp. Res. Rec. 1095, 70-71 (1986)
[13] A.M.Ioannides, Insights from the use of supercomputers in analyzing rigid and flexible pavements, 66th Annual Meeting of the Transp. Res. Board, Washington, D.C., U.S.A. (1987)
[14] A.M.Ioannides, Axisymmetric slabs of finite extent on elastic solid, J. of Transp. Eng. ASCE 113, 277-290, (1987)
[15] A.M.Ioannides, A finite difference solution for plate on elastic solid, accepted for publication, J. of Transp. Eng. ASCE. (1987)
[16] A.M.Ioannides, M.R.Thompson and E.J. Barenberg, Finite element analysis of slabs-on-grade using a variety of support models, Proc. 3rd Int. Conf. on Concrete Pavement Design and Rehabilitation, Purdue University, Ind., U.S.A., 309-324 (1985)
[17] A.M.Ioannides, M.R.Thompson and E.J. Barenberg, Westergaard solutions reconsidered, Transp. Res. Rec. 1043, 13-23 (1985)
[18] A.M.Ioannides, J.P.Donnelly, M.R. Thompson and E.J.Barenberg, Three-dimensional finite element analysis of a slab on stress dependent elastic solid foundation, U.S. Air Force Office of Scientific Research, Report No. TR-86-0143 (1986)
[19] A.Losberg, Structurally reinforced concrete pavements, Doktorsavhandlingar Vid Chalmers Tekniska Högskola, Göteborg, Sweden (1960)
[20] G.Pickett, M.E.Raville, W.C.Janes and F.J.McCormick, Deflections, moments and reactive pressures for concrete pavements, Kansas State College Bulletin No. 65, U.S.A. (1951)
[21] G.Pickett, S.Badaruddin and S.C. Ganguli, Semi infinite pavement slab supported by an elastic solid subgrade, Proc. 1st Congress on Theor. and Appl. Mechanics, Indian Inst. of Technology, 51-60 (1955)
[22] H.M.Westergaard, Stresses in concrete pavements computed by theoretical analysis, Public Roads 7, 25-35 (1926)

Numerical Methods in Geomechanics (Innsbruck 1988), Swoboda (ed.)
© 1988 Balkema, Rotterdam. ISBN 90 6191 809 X

Numerical modelling of pile behaviour under low-strain integrity testing

H.Balthaus & H.Meseck
University of Braunschweig, FR Germany

INTRODUCTION

Sonic and impulse testing methods for pile integrity are in use for about a decade now. The established methods based on one-dimensional wave propagation theory usually render unambiguous results for pile length and distinct deviations from normal pile shape or properties.

However, in numerous cases e.g. when only slight deviations from normal pile response occur or when strong effects of soil embedment overshadow the pure pile response, parametric studies and a method of evaluating the different interrelated effects of shape and soil-pile-interaction helps in interpreting test results and improves the quality and reliability of judgement.

For the complexity of the involved wave propagation processes, closed form analytical methods cannot model pile behaviour with satisfactory exactness.

In this contribution a numerical computer model for pile integrity analysis will be introduced and shown in a number of applications.

PILE INTEGRITY TESTING AND COMPUTER ANALYSIS

Fundamentals of low-strain integrity testing

In the construction of bored piles it occurs repeatedly that zones of necking, bulging, low quality concrete or uncovered reinforcement form along the pile. Such deficiencies and deviations from required pile length can lead to serious reductions in pile bearing capacity.

To detect the hidden defects impulse testing of piles has been developed. The pile top is hit with a falling weight or a hammer and the generated force pulse of 1N to 20N in amplitude moves down the pile as a strain wave of low magnitude ("low strain testing"). Soil-pile interaction, cross-sectional changes along the pile and reflection characteristics at the pile tip influence the travelling wave. From the response at the pile top, that can suitably be measured with a hand-held acceleration transducer, information about pile shape, length, and integrity can be gained if the measurement results are interpreted in the light of one-dimensional wave propagation theory (Timoshenko & Goodier (1970), Graff (1975), Garbrecht 1976), Balthaus & Meseck (1984)). A typical setup for data acquisition and processing is shown in Fig. 1.

In a free pile, that is slender, straight, homogeneous, elastic with a Young's modulus E and of constant cross-sectional area A, the one-dimensional wave equation

$$\frac{\partial^2 u}{\partial t^2} = \frac{E}{\rho}\frac{\partial^2 u}{\partial x^2} = c^2\frac{\partial^2 u}{\partial x^2} \qquad (1)$$

describes the propagation of a mechanical disturbance (Timoshenko & Goodier, 1970). ρ is the specific density of the pile material, c the wave velocity, u the displacement, x the length coordinate and t the time.

The homogenous differential equation can be solved by d'Alembert's integration method. The general solution

$$u(x,t) = u_i(x-ct) + u_r(x+ct) \qquad (2)$$

is composed of an incident portion u_i moving downwards and a reflected portion u_r moving upwards, both with the velocity c. In the location-time diagram of Fig. 2 the propagation of a triangular disturbance is shown as movement along the so-called characteristics.

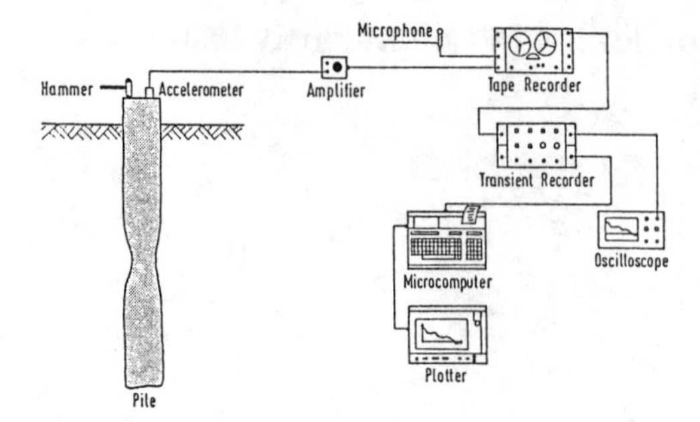

Fig. 1: Measuring system for digital data processing in integrity testing

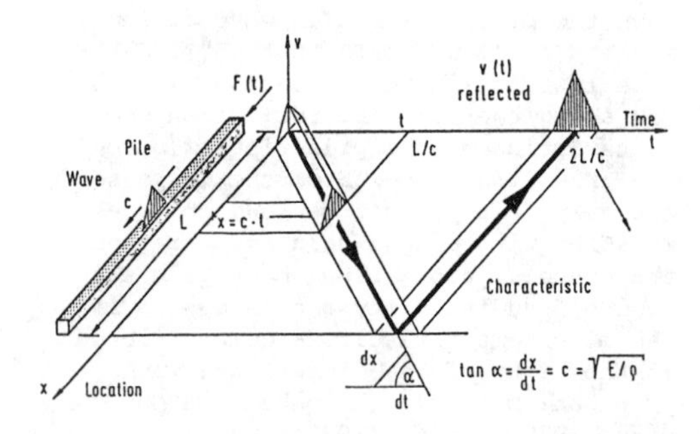

Fig. 2: Illustration of wave propagation as movement along chracteristics

When the strain wave front reaches a pile location, where the cross-sectional area or material properties changes, the wave will be reflected and transmitted to certain amounts depending on the change in mechanical impedance.

$$Z = EA/c = A \cdot \sqrt{E \cdot \rho} \tag{3}$$

The magnitudes of reflected and transmitted wave portions are related to the mechanical boundary and transmission conditions at the interface.

Detailed descriptions of the reflection laws can be found elsewhere (Graff (1975), Middendorp & van Brederode (1984) and Balthaus & Meseck (1984)). Examples will be discussed later in connection with results of numerical simulations.

Numerial treatment of pile-soil system

Due to the embedment of the pile, soil resistances are generated at the pile skin and tip during the passage of a stress wave.

All soil resistances generated during wave propagation are composed of a static portion depending on displacements u and a dynamic portion usually seen as velocity dependent.

The dynamic resistance can physically be explained by dissipation and radiation damping processes that occur both at the pile skin and tip.

Because cross-sectional changes along a pile usually occur gradually, the basic wave equation (1) has to be supplemented by a term that accounts for the changing cross-sectional area (Graff, 1975).

For a conical pile, bedded elastically with viscous damping along its skin a modified equation

$$\frac{\partial^2 u}{\partial t^2} + \frac{k'}{A\rho} u + \frac{b'}{A\rho} \cdot \frac{\partial u}{\partial t} = c^2 \quad \frac{\partial^2 u}{\partial x^2} + \frac{1}{A(x)} \frac{dA}{dx} \cdot \frac{\partial u}{\partial x} \tag{4}$$

with stiffness and damping coefficients k' and b' describes the wave propagation which is now accompanied by dispersion effects. To solve equ. (4) elaborate numerical procedures must be applied.

In the following a simple numerical approach will be shown that accounts for the additional terms in equ. (4) by appropriately introducing discrete soil resistances and stepwise cross-sectional changes and thereby modifying passing waves through the generation of reflected wave portions.

For the simulation of pile driving processes and data evaluation of dynamic pile test measurements a computer program has been developped that is based on the method of characteristics solution (2) of the basic one-dimensional wave equation (1). The pile is discretized in such a way that during a constant time step Δt wave portions propagate by a distance equal to the pile element length

$$\Delta x_i = c_i \cdot \Delta t \tag{5}$$

The additional terms for displacement and velocity dependent pile skin bedding in equation (4) are accounted for by introducing discretized bilinear elasto-plastic and linear viscous skin resistances as shown in Fig. 3. More sophisticated resistance laws were also used, but did not improve computed results and will therefore not be discussed here.

At the pile tip a resistance law like the one at the skin resistance points was chosen. There, however, it is possible to exclude tensile stress trans-

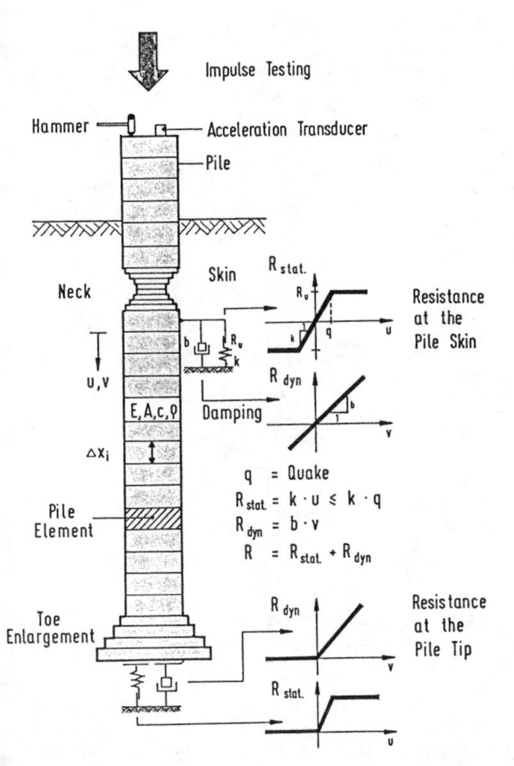

Fig. 3: Discretized bilinear elasto-plastic and linear viscous soil resistance model

fer to the soil.
Because computational algorithms with similar features have been described elsewehere (e.g. Jansz et al., 1976), a detailed discussion is omitted here and emphasis is instead placed on program reliability and applications.

The FORTRAN 77 computer program for microcomputers in which the wave propagation algorithm was incorporated is designed to handle both measured and synthetic load histories. Alternatively forces or velocities can be prescribed as loading boundary conditions at any arbitrary pile position. Vibrations can be imposed as well as impact loads.

A detailed description of the programm can be found in Balthaus (1986).

Suitability for conical pile shapes

The simple numerical model described before only allows stepwise cross-sectional changes. Such a model does not necessarily describe conical pile shapes with sufficient accuracy. To investigate the exactness of the numerical approach with respect to equ. (4) a conical pile was loaded at the top with a sinusoidal force function $F(t) = P_o \cdot \sin \omega t$. The pile response can be described by a Bessel differential equation the solution of which in terms of velocity can be expressed as:

$$v(a,t) = \frac{P_o}{Z_o} \cdot \frac{\omega^2 a^2}{c^2 + \omega^2 a^2} \cdot \left(\sin \omega t + \frac{c}{\omega a} \cdot \cos \omega t\right) \qquad (6)$$

Parameters are explained in Fig. 4. As this Fig. shows, agreement between numerical and analytical solution is excellent.

Integrity analysis procedure

A computer model to be used to support integrity test interpretation has to fulfil a number of requirements:

- accuracy in modelling the shape of the impulse loading
- in corporation of damping, tip and mantle resistance
- detailed representation of possible pile geometries
- true modelling of dispersion effects from conical pile shape and embedment.

How the described computer program can be used in integrity analysis of piles is illustrated by Fig. 5. It is essential to know that an iterative prodecure of changing pile geometry and soil properties has to be followed to obtain the required goodness of fit between measured and computed velocity response at the pile top.

So far this iterative process is based on a sequence of educated guesses and experience with pile integrity test interpretation. In an outlook a numerical search routine will be suggested that can help to automate the search for optimum fit.

APPLICATION AND EXAMPLES

General

How the computer model can be used in integrity analysis shall be shown with a number of examples and practical applications. Computer re-analysis of pile test data can help in the identification, distinction, and quantification of pile defects and the general interpretation of integrity test data.

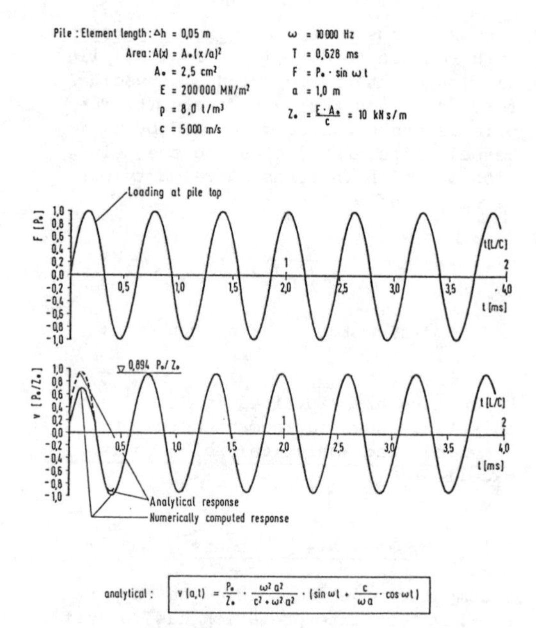

$$v(a,t) = \frac{P_o}{Z_o} \cdot \frac{\omega^2 a^2}{c^2 + \omega^2 a^2} \cdot \left(\sin \omega t + \frac{c}{\omega a} \cdot \cos \omega t\right)$$

Fig. 4: Conical pile under sinusoidal loading at pile top. Analytical and computed velocity response at position x = a = 1 m

Identification of defects

Computer simulation is a useful tool of integrity test data analysis.

Different cross-sectional changes and locally concentrated mantle resistance, e.g. in a hard layer, give distinctly different velocity responses at the pile top.

Fig. 6 illustrates the effects of three different features for the velocity plot at the pile top. On the basis of such model calculations pile defects can be identified with more certainty and accuracy than by pure experience.

Distinction between similar defects

In the interpretation of integrity test data it can be a most demanding task to distinguish between pile defects that give similar pile top response. Model analyses can support judgement in such cases.

An example is given in Fig. 7 where different gradual variations between necking and an open crack across the pile are shown in their effect on velocity response during low strain testing. As tension transfer across a crack depends on prestressing of the pile joint,

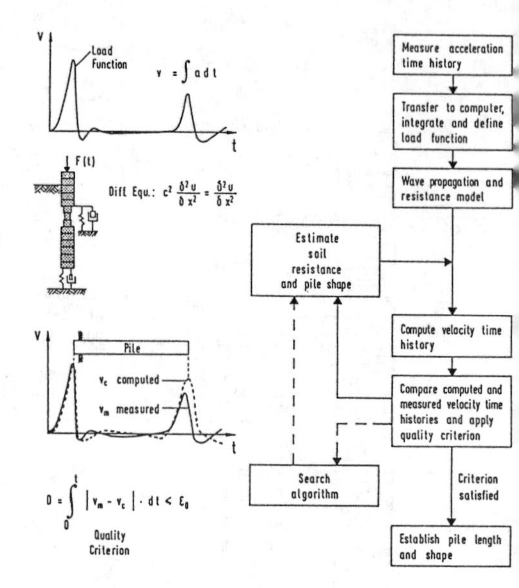

Fig. 5: Pile shape and integrity analysis procedure

or in other words on the prevailing stress conditions at the crack, the response of a broken pile with an opened joint (d, pure pressure transfer) may resemble closely that of a very long and sound pile.

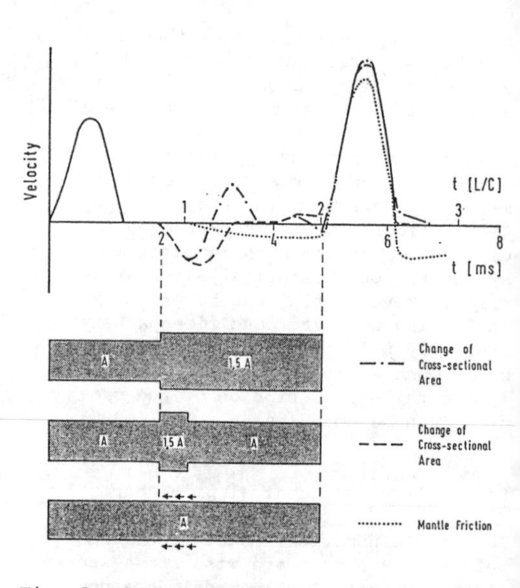

Fig. 6: Characteristic velocity plots for typical cross-sectional changes and local mantle friction

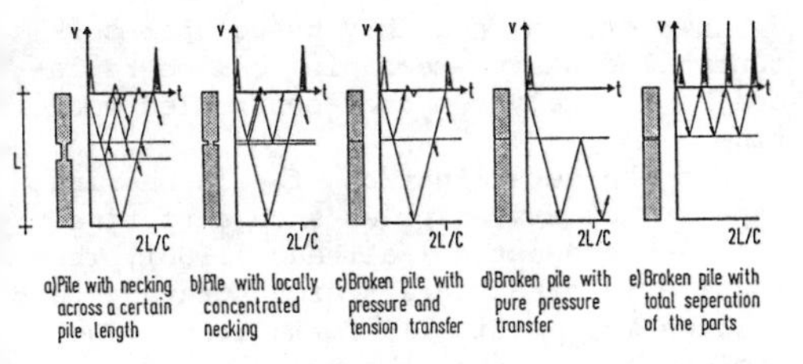

Fig. 7: Theoretical velocity plots for piles of same length with different deficiencies

Computer analysis can also proof, as shown in Fig. 8, that integrity testing can only reveal changes in mechanical impedance. As equ. 3 defines corresponding changes in elastic modulus or cross-sectional area can lead to the same characteristic reponse. Usually however the fact that a pile has deviations from the required quality is most important.

Quantification

Quite often for an obviously detected pile defect a measure for the amount of necking or the changes in impedance has to be given.

Then the parameter in question can be varied systematically to relate e.g. an amplitude ratio to this parameter. From the obtained curve the required quantity can be read if the measurement plot is evaluated appropriately. Fig. 9 illustrates the described method of parametric variation.

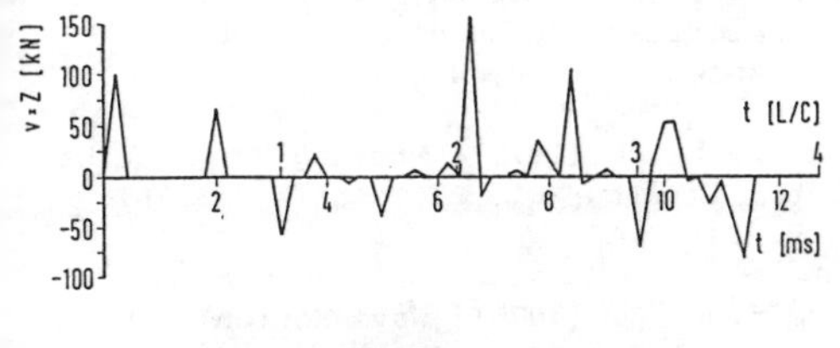

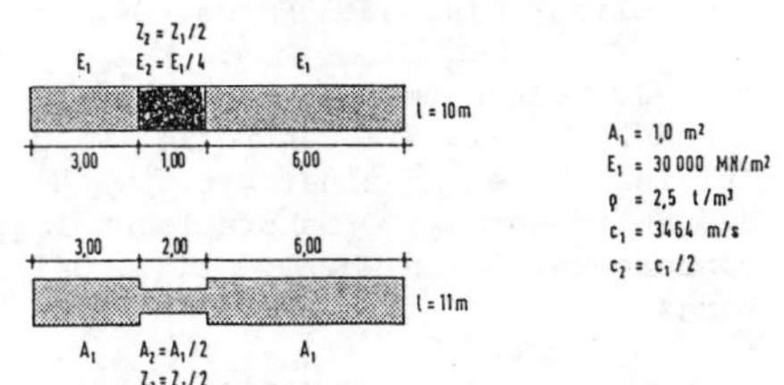

Fig. 8: Ambignity in integrity test results

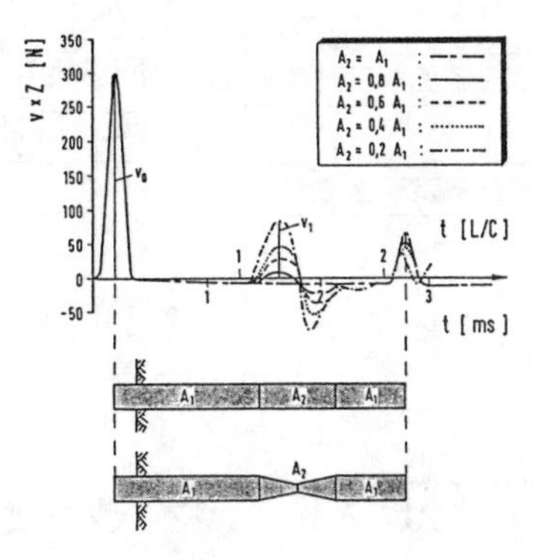

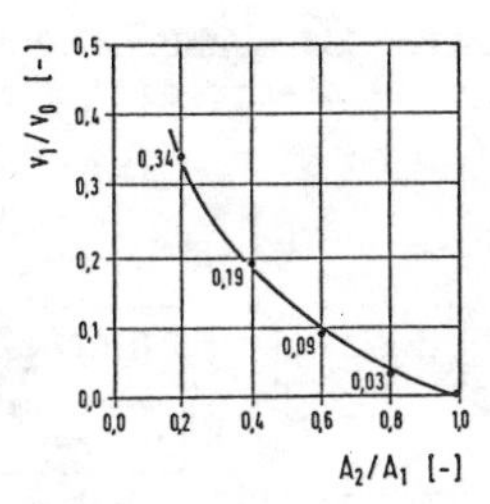

Fig. 9: Parametric study as a means for quantifying the amount of necking

Shape and integrity analysis by curve fitting

If doubts exist about a certain integrity test interpretation, a full computer analysis according to Fig. 5 can be used to give reliable support and an objective and independent judgement. For two different test piles of 13,00 m and 13,60 m in length (compare Ulrich & Stocker, 1983)

Computed and measured velocity response and computer-estimated and actual shape are compared in Figs. 10 and 11.

However the results also reveal that apart from generally good fit the amount of local necking or bulging can hardly be reproduced exactly. Consequently the ability to quantify such defects is only enhanced to a limited amount by computer analysis.

CRITICAL APPRAISAL AND POSSIBLE DEVELOPMENTS

The previous examples of computer-supported integrity analysis of piles have also pointed out some of the weaknesses of the applied method:

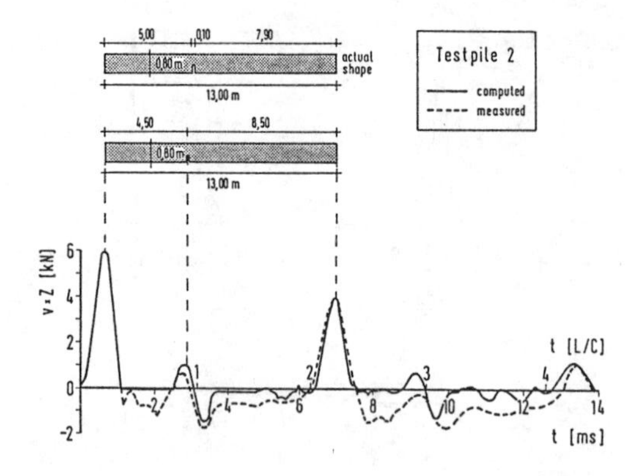

Fig. 10: Comparison of measured and computed time histories for testpile 2

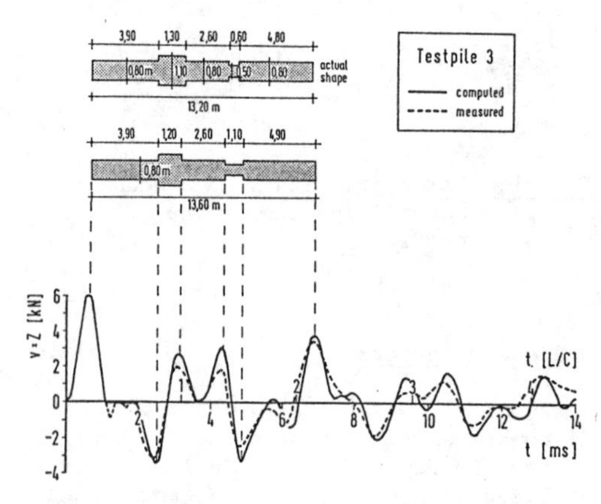

Fig. 11: Comparison of measured and computed time histories for testpile 3

- o The method of user-controlled parameter variation and curvefitting is time consuming and leads to operator-dependent results.
- o Different parameter combinations can lead to the same result (Fig. 8)
- o Due to unavoidable measurement errors and differences between reality and computer model usually a complete fit is not possible. Therefore a margin for engineering judgement remains.

Because the "optimum fit" of measured and computed pile integrity plots is the result of a sophisticated process of interactions within a wide set of geometrical and soil resistance parameters a certain fuzzyness will always prevent an unambiguous judgement of the test data.

However, the possibly based operator-controlled search mechanism can be replaced by a purely computer-controlled routine.

For the determination of pile bearing capacity of driven piles such a routine has been suggested (Balthaus, 1986). The routine is based on a search strategy that follows mathematically formulated principles of evolution theory (Schwefel, 1977). The search is terminated when a quality criterion (compare Fig. 5) is satisfied or in other words the target function has an absolute minimum. Details of the method can be found in Balthaus (1986).

If for each pile element a size or diameter parameter is introduced and included in the parameter variation, the strategy can be applied in integrity testing.

With the increasing availability of high speed computing facilities automatized pile shape analysis on the basis of integrity test results is an attractive alternative to the present tedious methods of data interpretation.

REFERENCES

Balthaus, H. (1986): Tragfähigkeitsbestimmung von Rammpfählen mit dynamischen Pfahlprüfmethoden; Doctoral Thesis, Techn. Univ. Braunschweig

Balthaus, H. & Kielbassa, S. (1986): Numerical Modelling of Pile Driving; 2. Int. Cont. on Numerical Models in Geomechanics Ghent, 1.-4. April

Balthaus, H. & Meseck, H. (1984): Integritätsprüfungen an Ortbetonpfählen; TIS, 8, 470 - 478

Garbrecht, D.(1976): Prüfung von Pfahlbeton in situ; Die Bautechnik 7/1976, S. 241-244

Graff, K. J. (1975): Wave Motion in Elastic Solids; Clarendon Press, Oxford

Jansz, J., van Hamme, V. Gerritse, A. & Bomer, H. (1976): Controlled Pile Driving above and under Water with a Hydraulic Hammer; Offshore Technology Conference, Paper No. OTC 2477, Dallas, Texas

Meseck, H. & Simons, H.(1984): Bericht über Integritätstests auf dem Bauhof der Firma Bauer; Institut für Grundbau und Bodenmechanik, TU Braunschweig

Middendorp, P. & Van Brederode, P.(1984): A field monitoring-technique for the integrity testing of foundation piles; 2nd Intern. Conf. on the Application of Stress-Wave Theory on Piles, Stockholm, May

Schwefel, H.-P. (1977): Numerische Optimierung von Computer-Modellen mittels der Evolutionsstrategie; Birkhäuser Verlag, Basel, 1. edition

Timoshenko, S. P. & Goodier, J. N. (1970): Theory of Elasticity; Mc Graw-Hill, New York, 3. edition

Ulrich, G. & Stocker, M. (1983): Integritätsuntersuchungen an präparierten Betonpfählen; Symp. Meßtechnik im Erd- und Grundbau, München, Nov. 1983, S. 243-247

Numerical Methods in Geomechanics (Innsbruck 1988), Swoboda (ed.)
© 1988 Balkema, Rotterdam. ISBN 90 6191 809 X

Evaluation of stiffness characteristics of Bombay High calcareous sediments for numerical methods of soil-foundations interactive analysis

V.S.Golait
Government College of Engineering, Karad, India
R.K.Katti
Indian Institute of Technology, Bombay

ABSTRACT: The subsurface deposits in the Bombay High offshore oilfield are calcareous in nature. It has been observed that their stress-strain and strength behaviour is governed by the carbonate content of the material. As such, the soil stiffness parameters, viz., initial Young's modulus, secant modulus, tangent modulus etc., which are required in the numerical methods of soil-structure interactive analysis of offshore foundations are evaluated in terms of the carbonate content dependent stress-strain model for these soils.

1 INTRODUCTION

In the numerical methods of soil-foundation interactive analyses of offshore structures the soil response in respect of its stress-strain and strength behaviour is required to be suitably modelled to generate various soil stiffness parameters such as modulus values and modulus ratios. The soil stiffness characteristics which is needed as input data in such analyses is governed by various conditions and factors, viz., mode of loading, strain level, confining pressure, void ratio etc. For the conventional soil systems of sands and clays a number of non-linear soil models are available (Duncan & Chang, 1970; Ramberg & Osgood, 1943; Desai, 1971; Hardin & Drnevich, 1972; Desai & Wu, 1976; Shibata, 1977).

The strcutures constructed on submarine soil deposits for oil exploitation are subjected to various types of loading, mainly the static loads coming from the superstructure, the cylic loading due to wave action and the dynamic loads of earthquake and machine induced vibrations. In many parts of the world the subsoil deposits of the continental shelves, where most of the present day offshore construction activities are concentrated, often consist of calcareous sediments. For the calcareous soils encountered in the Bombay High offshore region off the west coast of India, the stress strain and the strength behaviour under offshore loading conditions is observed to be related to the carbonate content of the material (Katti et. al, 1985; Golait & Katti 1987b). The data on the triaxial shear behaviour of 13 selected Bombay High samples has been analysed by the authors earlier and an idealized stress-strain relation is developed (Golait & Katti, 1987a). This soil model is now further utilized to evaluate the stiffness characteristics of calcareous sediments and to quantify the soil parameters required in the soil-foundation interactive analyses with specific relevance to foundations of Bombay High offshore structres. The paper outlines this study.

2 THE STRESS-STRAIN MODEL

An idealized constitutive law for Bombay High calcareous soils of different textural classes (calcareous sands, silts and clays) under various offshore loading conditions (conventional static loading 'CU-conv' and trapezoidal wave loading 'CU-TWL') is represented by an empirical hyperbolic model involving four non-dimensional carbonate content dependent parameters a, b, n_r and R_f (Golait & Katti, 1987a). Accordingly, the stress-strain nature and the strength under triaxial stress condition are represented as:

$$\sigma d = \frac{\sigma_3}{n_r} \cdot \frac{\epsilon}{a + b\epsilon} \quad (1)$$

$$\sigma d_f = R_f \cdot \frac{\sigma_3}{b.n_r} \quad (2)$$

where, σd = deviator stress,
σd_f = deviator stress at failure,
σ_3 = confining pressure,

ϵ = axial strain,
a and b = parameters defining the shape of hyperbolic stress-strain curve,
n_r = transform parameter for normalized deviator stress,
R_f = failure ratio.

The initial Young's modulus, E_i, and the asymptotic value of deviator stress, σd_{ult}, are related to a, b and n_r parameters as:

$$E_i = \frac{\sigma_3}{a.n_r} \quad (3)$$

$$\& \quad \sigma d_{ult} = \frac{\sigma_3}{b.n_r} \quad . \quad (4)$$

3 EVALUATION OF SOIL STIFFNESS PARAMETERS

In the non-linear soil response, the soil stiffness is characterized by tangent modulus, E_{tg}, and secant modulus, E_{sec}, alongwith the decrease in these moduli with stress or strain levels. These parameters are evaluated from the above mentioned soil model.

3.1 Tangent modulus

The tangent modulus is the slope of stress-strain curve at a point. Hence,

$$E_{tg} = \frac{\partial (\sigma d)}{\partial \epsilon} \quad .$$

Substituting for σd from Eq.1 and differntiating,

$$E_{tg} = \frac{\sigma_3}{n_r} \cdot \frac{a}{(a + b\epsilon)^2} \quad . \quad (5)$$

From Eq.1, $\quad \epsilon = \dfrac{n_r.a.\sigma d}{\sigma_3 - n_r.b.\sigma d} \quad .$

Substituting for ϵ in Eq.5 & simplifying,

$$E_{tg} = \frac{\sigma_3}{a.n_r} \cdot \left(1 - \frac{n_r.b.\sigma d}{\sigma_3} \right)^2 \quad . \quad (6)$$

Also, substituting the value for E_i, the tangent modulus ratio, E_{tg}/E_i, is expressed as:

$$\frac{E_{tg}}{E_i} = \left(1 - \frac{n_r.b.\sigma d}{\sigma_3} \right)^2 \quad . \quad (7)$$

This ratio can also be expressed in terms of deviator stress ratio, s (= $\sigma d/\sigma d_f$) as:

$$E_{tg}/E_i = (1 - R_f.s)^2 \quad . \quad (8)$$

3.2 Secant modulus

The secant modulus is the slope of the straight line joining the origin to a point on stress-strain curve. Thus,

$$E_{sec} = \frac{\sigma d}{\epsilon} \quad .$$

Substituting the value of σd and ϵ, and simplifying,

$$E_{sec} = \frac{\sigma_3}{a.n_r} \cdot \left(1 - \frac{n_r.b.\sigma d}{\sigma_3} \right) \quad . \quad (9)$$

The secant modulus ratio can be expressed as:

$$\frac{E_{sec}}{E_i} = \left(1 - \frac{n_r.b.\sigma d}{\sigma_3} \right) \quad (10)$$

$$\text{or } E_{sec}/E_i = (1 - R_f.s) \quad . \quad (11)$$

The equations 7 and 10 define the decrease in the modulus ratios with stress level. They may also be used to evaluate this decrease with strain level as the stresses and strains are related by Eq.1.

4 STIFFNESS CHARACTERISTICS OF SOIL

The soil stiffness parameters such as initial Young's modulus, secant modulus, tangent modulus etc. are evaluated for all the Bombay High calcareous soil samples under static and cyclic wave loading conditions using the expressions derived under section 3. The results are interpreted to bring out the effects of confining pressure, type of loading, strain level, void ratio etc. on the stiffness nature of the soils.

4.1 Initial Young's modulus and its variation with confining pressure and void ratio

The initial Young's modulus (or the low strain modulus) increases with increasing effective confining pressure. The void ratio of soil has also its effect on this parameter.

Janbu (1963) has correlated the initial Young's modulus with the confining pressure as:

$$E_i = K.p_a \cdot \left(\frac{\sigma_3}{p_a}\right)^n \quad (12)$$

where, p_a = atmospheric pressure,

K = coefficient,
n = an exponent.

Prakash (1981) suggested that the effect of void ratio, e, can be incorporated by a simple function $F(e) = 0.3 + 0.7e^2$. Hence, the combined effect of void ratio and confining pressure is proposed to be in the following form of equation for E_i of Bombay High soils:

$$E_i = \frac{K.p_a}{0.3+0.7e^2} \cdot \left(\frac{\sigma_3}{p_a}\right)^n \qquad (13)$$

Applicability of this relation to Bombay High calcareous soils is verified by plotting E_i values against σ_3/p_a on log-log scale. Typical variations are shown in Figure 1. A linear variation as is seen in the figure confirms the validity of Eq.13 for the Bombay High samples.

4.2 Variation of modulus ratio with strain level

It has been well-recognized that an increase in strain causes a decrease in deformation modulus of soils. The compilations of test data showing this trend were given by Hardin and Drnevich (1972). The data on Bombay High samples indicate certain trends regarding modulus variation with increasing strain.

The tangent modulus and the secant modulus are normalized by dividing the modulus value by the initial Young's modulus of the respective sample. This normalized modulus is termed 'modulus ratio'. The strain level during deformation process is denoted by ϵ/ϵ_f (where, ϵ_f is the strain at failure). The data for all the samples under both the loading modes was obtained on computer using the Eqs. 7 & 10. Typical variations of the modulus ratios with strain level for calcaresous sands are shown in Figure 2 for static loading and in Figure 3 for cyclic loading. It is seen that both the ratios decrease non-linearly with strain level with a comparatively rapid rate in the beginning. A similar trend is exhibited in both the cases of loading. This is consistent with the non-linear stress-strain behaviour of soil.

On further analysis of the data, it is observed that when E_{sec}/E_i is plotted against $\log(\epsilon/\epsilon_f)$ a linear variation is evident (Figure 4). This relationship can be expressed as:

$$\frac{E_{sec}}{E_i} - \frac{(E_{sec})_f}{E_i} = I_{md}.\log\left(\frac{1}{\epsilon/\epsilon_f}\right).. \qquad (14)$$

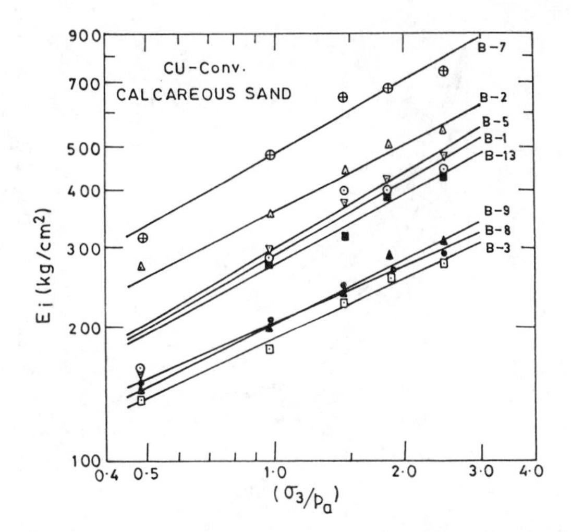

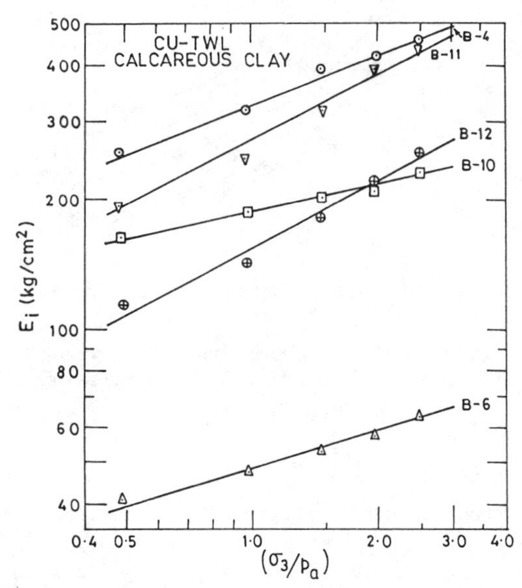

Fig.1 Typical log E_i vs. log (σ_3/p_a) plots.

where, $(E_{sec})_f$ = secant modulus at failure $= \sigma d_f/\epsilon_f$

I_{md} = slope of (E_{sec}/E_i) versus $\log(\epsilon/\epsilon_f)$ plot and it is designated as 'modulus decrease index'.

The modulus decrease index indicates the rate of decrease of secant modulus with strain. Its values for all the samples under both the loading modes are evaluated. It varies within a narrow range of 0.50 to 0.53. An average value of 0.515 can therefore be assigned to I_{md} in Eq.14. The secant modulus at any strain level ϵ with

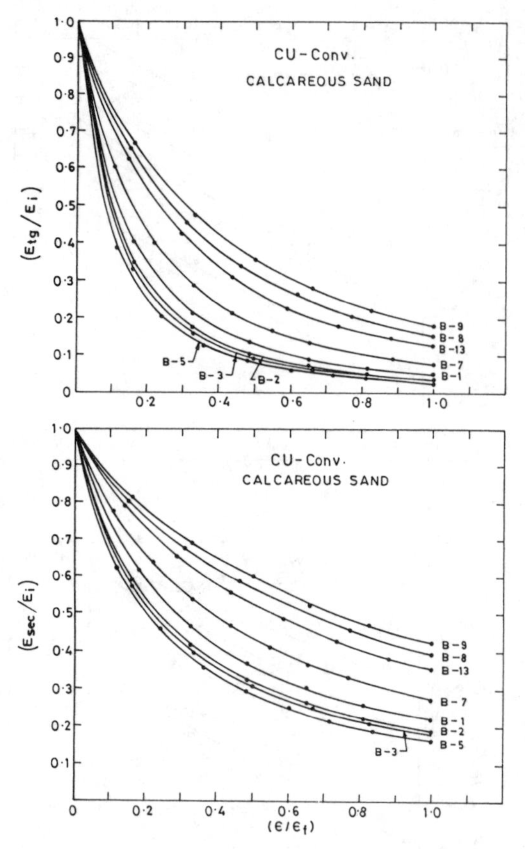

Fig.2 Variation of modulus ratios with strain level for calcareous sand samples under static loading

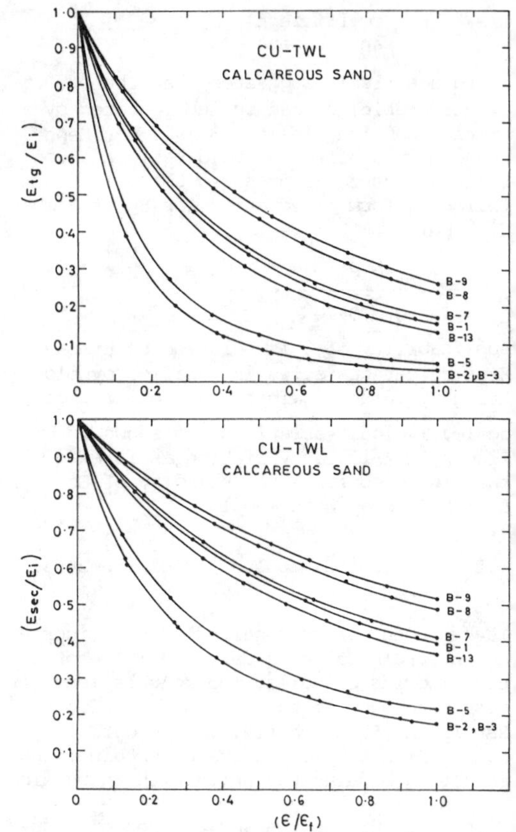

Fig.3 Variation of modulus ratios with strain level for calcareous sand samples under cyclic loading

respect to failure strain ϵ_f for the soils studied is thus proposed to be represented as :

$$E_{sec} = (E_{sec})_f + 0.515\ E_i . \log\left(\frac{1}{\epsilon/\epsilon_f}\right) \cdot \quad ..(15)$$

As regards tangent modulus ratio, its variation with $\log(\epsilon/\epsilon_f)$ is non-linear.

The modulus-strain relationship is generalized by representing the modulus ratios against log(b.ε/a). The points for all the cases of the base data are found to fall on singular lines for the two modulus ratios (Figure 5). The relationship for (E_{sec}/E_i) vs. log(b.ε/a) very nearly fits in a straight line. The slope of this line is 0.52 which value is about the same as the modulus decrease index defined earlier. The variation of tangent modulus ratio is however non-linear as seen in the figure.

5. SUMMARY AND CONCLUDING REMARKS

The soil stiffness characteristics in terms of various soil moduli and modulus ratios is evaluated for the Bombay High calcareous soil samples subjected to simulated offshore loadings. The effect of strain level, mode of loading and confining pressure on the nature of variation in these stiffness moduli is highlighted. A generalized relation, as shown in Figure 5, for the variation of soil stiffness with strain is found to hold good for various textural classes of Bombay High calcareous samples investigated under various offshore loading conditions. It is hoped that the expressions & the relations given in the paper will provide a simple aid in using appropriate input data for soil stiffness in the numerical methods of soil-foundation interactive analyses for foundations in calcareous soil with its specific relevance to Bombay High calcareous sediments. The accuracy of the

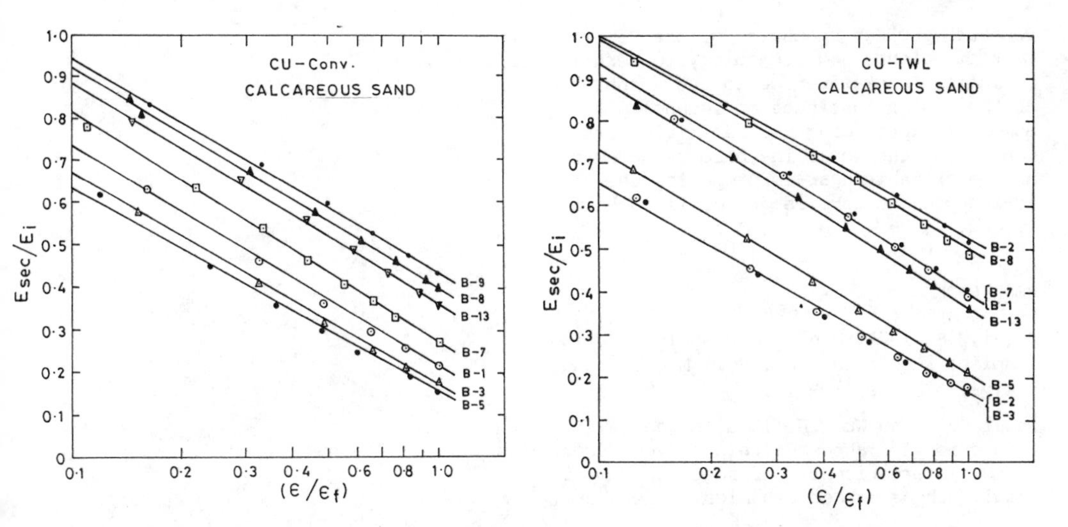

Fig.4 Secant modulus ratio versus log(strain level) for calcareous sand

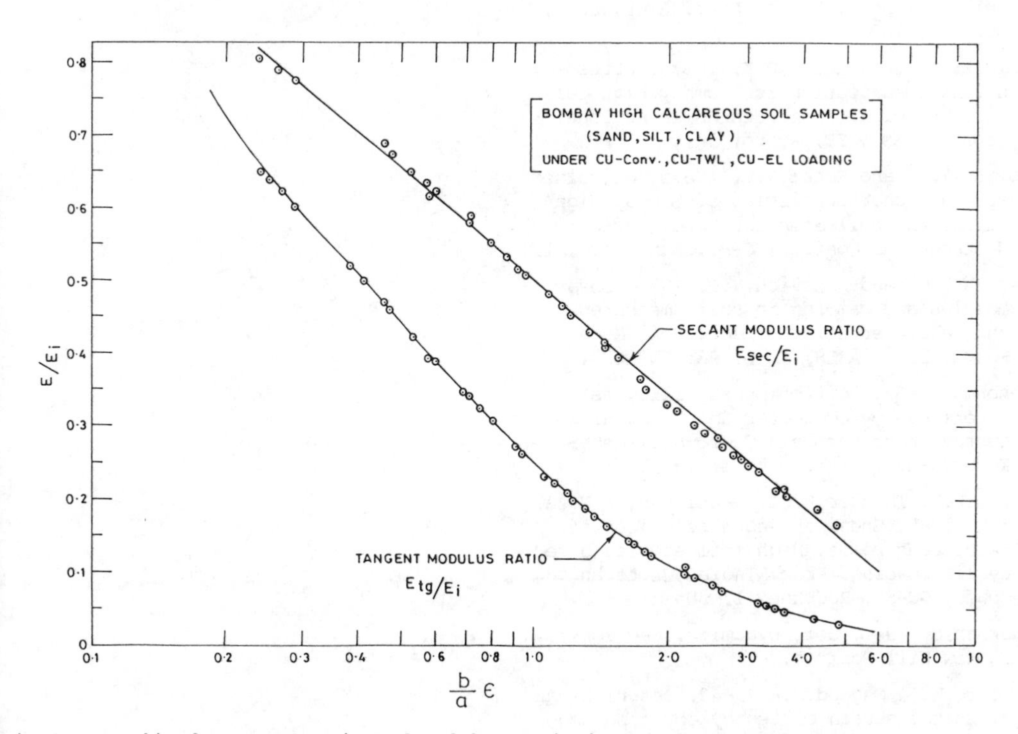

Fig.5 Generalized representation of modulus variation with strain

results will, however, depend on selecting the appropriate values of a, b, n_r and R_f parameters of hyperbolic stress-strain model for the given type of calcareous soil. For Bombay High samples these values are reported by the authors (Golait & Katti, 1987a).

ACKNOWLEDGEMENTS

The study formed further extension of the

research project sponsored by the Department of Science and Technology, Government of India. Thanks are due to the authorities of the Indian Institute of Technology, Bombay for providing the facilities for conduct of the work. The help rendered by Mr. Manoj Kshirsagar in preparing the paper in its present form is gratefully acknowledged.

REFERENCES

Desai,C.S. 1971. Non-linear analysis using spline functions. Jnl. Soil Mech. Found. Div., ASCE, 97, SM-10.

Desai,C.S. and Wu,T.H. 1976. A general function for stress-strain curves. Proc. 2nd. International Conference on Numerical Methods in Geomechanics, Blacksburg.

Duncan,J.M. and Chang,C.Y. 1970. Non-linear analysis of stress and strain in soils. Jnl. Soil Mech. Found. Div., ASCE, SM-5, 96: 1629-1653.

Golait,Y.S. and Katti,R.K. 1987a. Stress-strain idealization for Bombay High calcareous soils. Proc. 8th Asian Regional Conf. on SM & FE, Kyoto, July.

Golait,Y.S. and Katti,R.K. 1987b. Undrained strength characteristics of Bombay High calcareous soil samples. Proc. Indian Geotechnical Conf., IGC-87, Bangalore, 1.

Hardin,B.O. and Drnevich,V.P. 1972. Shear modulus and damping in soils-measurement and parameter effects. Jnl. Soil Mech. Found. Div., ASCE, SM-6, 98: 603-624.

Janbu,N. 1963. Soil compressibility as determined by odeometer and triaxial tests. Proc. European Conf. Soil Mech. Found. Eng., Wiesbaden, 1: 19-25.

Katti,R.K.,Thacker,K.C., Katti,D.R. & Moza K.K. 1985. Shear strength behaviour of calcareous Bombay High soil samples under cyclic loading. Proc. Indian Geotechnical Conf. IGC-85, Roorkee, 1: 305-310.

Prakash,S. 1981. Soil Dynamics. New York: McGraw Hill Book Co.

Ramberg,W. & Osgood,W.R. 1943. Description of stress-strain curves by three parameters. Tech. Note 902, Nat. Advisory Com. on Aeronautics, Washington, D.C.

Shibata,T. 1977. Panel discussion on State of the Art Report 'Soil dynamics and its application to foundation engineering', Proc. 9th Int. Conf. SM & FE, Tokyo, 3 : 439-441.

Numerical Methods in Geomechanics (Innsbruck 1988), Swoboda (ed.)
© 1988 Balkema, Rotterdam. ISBN 90 6191 809 X

Buckling of buried flexible structures of noncircular shape

I.D.Moore
University of Newcastle, NSW, Australia

ABSTRACT: The finite element method is used to examine the buckling strength of flexible metal tubes buried in elastic ground. The hoop thrust which elastically destabilises circular, elliptical, square and rectangular tubes is calculated. The numerical procedure is described, and the use of geometrical symmetries to obtain efficient linear buckling solutions is examined. A study is also made to determine the element sizes needed to obtain accurate estimates of buckling strength. The strength of various structures is reported, and the implications for metal culvert design are briefly discussed.

1. INTRODUCTION

When a metal culvert is buried to form a culvert under a railway or road embankment, large compressive thrusts can develop in the structure as a result of live and earth loads. However the flexural stiffness of these structures is generally low and there is a possibility of elastic buckling failure. A number of studies have been reported concerning the stability of circular tubes buried in elastic ground (e.g. Forrestal and Herrmann (1965), Duns and Butterfield (1971), Moore and Booker (1985), Moore (1987a)). Other structural shapes have received less attention and this work seeks to rectify that problem.

A finite element solution for predicting the elastic stability of buried structures of general shape is briefly described. Linear buckling analysis of circular, elliptical, square and rectangular tubes is reported. Finite element mesh selection and the use of geometrical symmetries to simplify the problem are investigated. Finally the results of a parametric study of tube shape effect are outlined.

2. PROBLEM DEFINITION

Figure 1 shows the four different tube geometries being considered. The geometry of the circle and square can be defined using one parameter, but the ellipse and rectangle require two. In each case the circumference C of the structure is scaled

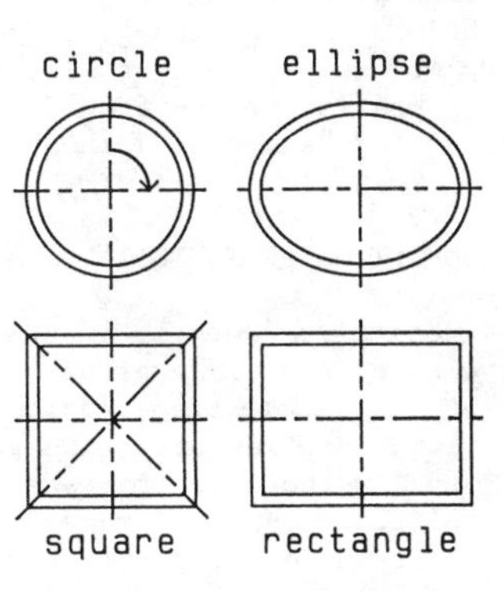

Figure 1: Geometry of buried tube

to give $r = C/2\pi$, the primary measure of tube size (r is the radius of a circle of equal circumference). The ellipse and rectangle are also defined using a/b, the ratio of "span" to "rise" of the tube.

The tube is assumed to have uniform flexural rigidity EI, and is considered under plane strain conditions, where there are no variations in geometry, material properties or loads along the tube axis. It is assumed that the structure is stressed by a uniform "hoop" thrust N.

The ground is modelled as a uniform isotropic elastic continuum, with shear modulus G_s and Poisson's ratio ν_s. A dimensionless flexural stiffness ratio $S_f = EI/G_s\, r^3$ is specified which will be used to characterise the flexural stiffness of the structure relative to the ground. In this study only the response of structures embedded within an infinite elastic solid are considered. This corresponds to a study of "deeply buried" tube buckling; the behaviour of tubes located close to the edge of an elastic half-space (i.e. "shallow buried") will be left for another study.

The critical hoop force distribution which elastically destabilises the flexible structure in its undeformed state is used to characterise the buckling strength. A recent study by Moore (1987b) reveals that linear solutions of this type are quite useful, and the linear elastic buckling solution for buried circular tubes has been calibrated for use in design using available experimental data. It was concluded then that discrepancies between theoretically predicted and experimentally measured values of critical hoop thrust are largely due to inelastic soil behaviour rather than prebuckling deflections, initial imperfections or mode change to a nonlinear buckling mechanism. It was also shown that continuum models for the ground are superior to Winkler (or "subgrade") theories.

3. DESCRIPTION OF NUMERICAL ANALYSIS

Moore (1985) gives details of the elastic finite element procedure being employed to solve the linear buckling problem for the buried flexible structure. The procedure divides the system into ground and structural components. Figure 2 shows one of the meshes used later in the work for

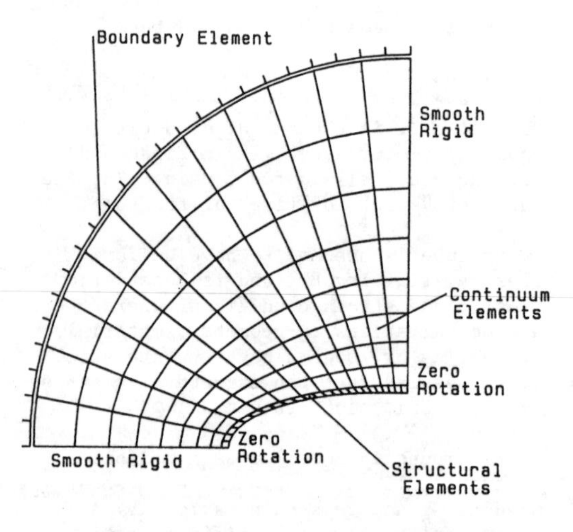

Figure 2: Example of Finite Element Mesh

the analysis of an elliptical culvert.

The elastic stiffness of the ground is calculated using the circular boundary element of Moore and Booker (1982) as well as eight noded isoparametric elements (e.g. Zienkiewicz (1977)). The equations are reduced so that the elastic stiffness of the ground at the soil-structure interface K_1 is obtained, viz.

$$K_1 \underset{\sim}{u}_1 = \underset{\sim}{f} \tag{1}$$

where $\underset{\sim}{u}_1$ are the ground deformations and $\underset{\sim}{f}$ are the interaction forces. Structural stiffness is a function of the magnitude of hoop thrust, and is calculated using straight beam-column elements similar to those of Yang (1973),

$$(K_s + N\, K_N)\, \underset{\sim}{u}_2 = -\underset{\sim}{f} \tag{2}$$

where K_s and K_N are the static and stability matrices of the structure and $\underset{\sim}{u}_2$ are the structural deformations at the interface. Compatability and equilibrium conditions at the interface are used to connect ground to structure, leading to a linear eigenvalue equation:

$$\det\,(K_1 + K_s + N\, K_N) = 0 \tag{3}$$

(this expression is for the case of bonded interaction where $\underset{\sim}{u}_1 = \underset{\sim}{u}_2$ - other interface conditions are treated using the initial stress soil-structure interaction procedure of Rowe, Booker and Balaam (1978)). A conventional eigenvalue routine is used to determine the lowest eigenvalue N_{cr}.

4. SOLUTIONS FOR A CIRCULAR TUBE

It is important to validate any finite element solution, and to develop experience with the discretisation necessary to obtain a solution of acceptable accuracy. That is accomplished in this Section through the examination of a circular tube embedded within an infinite elastic solid, since this problem has an analytical solution (e.g. Forrestal and Herrmann (1965), Moore and Booker (1985)).

Figure 3 shows finite element solutions for two different circular tube problems. Critical hoop force N_{cr} is normalised using N_{ref} defined by:

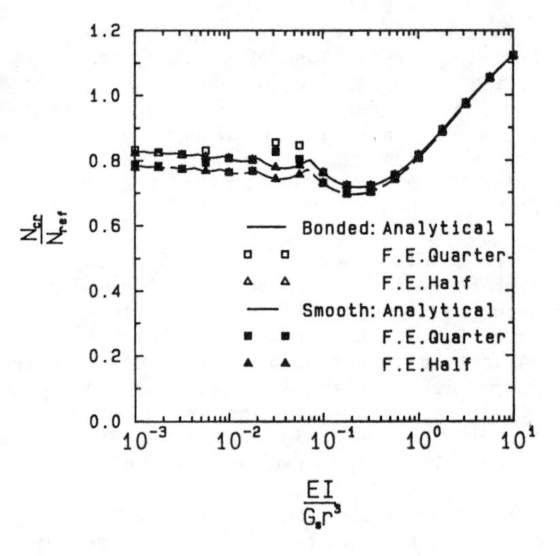

Figure 3: Influence of Symmetry Conditions for Buried Circular Tube

Figure 4: Influence of Finite Element Size

$$N_{ref} = \frac{3EI[(\frac{G_s r^3}{EI})^{2/3} + 1]}{r^2} \quad (4)$$

and results are presented for a range of structural flexibilities. Firstly a "rough" or bonded tube is considered, where there is full transfer of shear and normal stresses between ground and structure. Secondly the stability of a "smooth" structure is examined, where normal stresses are transferred between soil and structure but shear stresses are zero. In both cases $\nu_s = 0.3$.

The analytical solutions are given for both cases, as well as two finite element results. The numerical solutions are for the analysis of one quarter and one half of the system, with 24 and 48 structural elements respectively. The radial boundaries of the soil-structure system are assumed to be smooth and rigid, with zero rotation of the structure.

The results indicate that if one half of the system is modelled the numerical predictions match the analytical solutions closely. However there are some discrepancies if only one quarter of the problem is modelled. This occurs because the buckling deformations are harmonic in nature, and as the structure becomes more flexible the harmonic number increases from 2 (corresponding to elliptical deformations) to 3, 4 and so on. The boundary conditions for the analysis of one quadrant are such that only even numbered harmonics can be accommodated. Both odd and even harmonics are covered if one half is modelled. The discrepancies occur whenever the critical harmonic number is odd, and the single quadrant analysis predicts a critical hoop thrust which is too high.

Now consider the influence of finite element discretisation on solution accuracy. Figure 4 shows various finite element solutions for the case of bonded interface with $\nu_s = 0.3$. One half of the buried tube problem has been modelled with a variable number of structural elements (6, 12, 24 and 48). The analytical solution is also shown.

A reasonable numerical estimate of critical hoop force is obtained when the buckling deformations are adequately modelled in the analysis. For high values of structural stiffness $EI/G_s r^3$, the circular tube deforms in the elliptical mode and only small numbers of elements are needed in the analysis. However as the structure becomes more flexible the wavelength of the harmonic buckling deformations reduces, and more elements are needed to model the structure. Theoretical studies (Moore and Booker (1985)) indicate that the wavelength of the buckling deformation is approximately given by $2\pi(EI/G_s)^{1/3}$, so an examination of Figure 4 suggests that 10 elements are needed within each buckle to obtain a reasonable representation of the structural deformations. Even with 48 structural elements the numerical solution is unreliable for $EI/G_s r^3 < 0.001$.

Now the errors associated with a finite element model of only one quarter of the buried tube diminish and become negligible when $EI/G_s r^3$ is small (i.e. harmonic number is high). Therefore an efficient choice of finite element mesh will feature one half of the system modelled with a small number of elements when the structure is stiff (say 24 structural elements for $EI/G_s r^3 > 0.01$), but only one quarter of the problem with a larger number of elements when the tube is more flexible (e.g. 48 elements in one quadrant).

5. ELLIPTICAL TUBE ANALYSIS

Attention is now directed towards the analysis of elliptical tube stability. Elliptical culverts are often used in practice, and studies of static ellipse response are available (Abel, Mark and Richards (1973)). The buckling problem has not been solved analytically, although a numerical solution using the Winkler soil model has been reported (Kurt and Mark (1981)). Although elliptical culverts are normally located close to the ground surface and are nonuniformly stressed, the present study is restricted to a simple examination of "deeply" buried structures subjected to uniform hoop thrust.

To commence the study, Figure 5 shows numerical predictions of critical hoop thrust for a range of structural stiffnesses and three different treatments of structural symmetry. An ellipse with span to rise ratio of 3 (or 1/3) has been examined, and the bonded interface condition has been employed with $\nu_s = 0.3$.

The results shown in Figure 5 indicate that half of the structure must be modelled when the flexural stiffness lies in the range $0.003 < EI/G_s r^3 < 0.1$. This is a reflection of the occurrence of odd numbered "harmonic" buckling deformations of a type similar to those which develop around a circular tube. It does not appear to matter which half of the structure is modelled.

Figure 6 shows predictions of critical hoop thrust for elliptical structures of various shapes. The solutions for span to rise ratios a/b of 1, 3, 6 and infinity are given. There appears to be a slight strengthening associated with ellipse flattening. Very flexible tubes with large a/b appear to have the same buckling strength as circular structures, but the value of N_{cr}/N_{ref} for circular tubes increases gradually to this level whereas for elliptical structures N_{cr}/N_{ref} remains almost constant for $EI/G_s r^3 < 0.1$. Now this limiting value for flexible structures corresponds to the response of a straight structure supported by elastic ground. Curved structures approach this limit as structural flexibility increases, since the ratio of the buckle wavelength to the

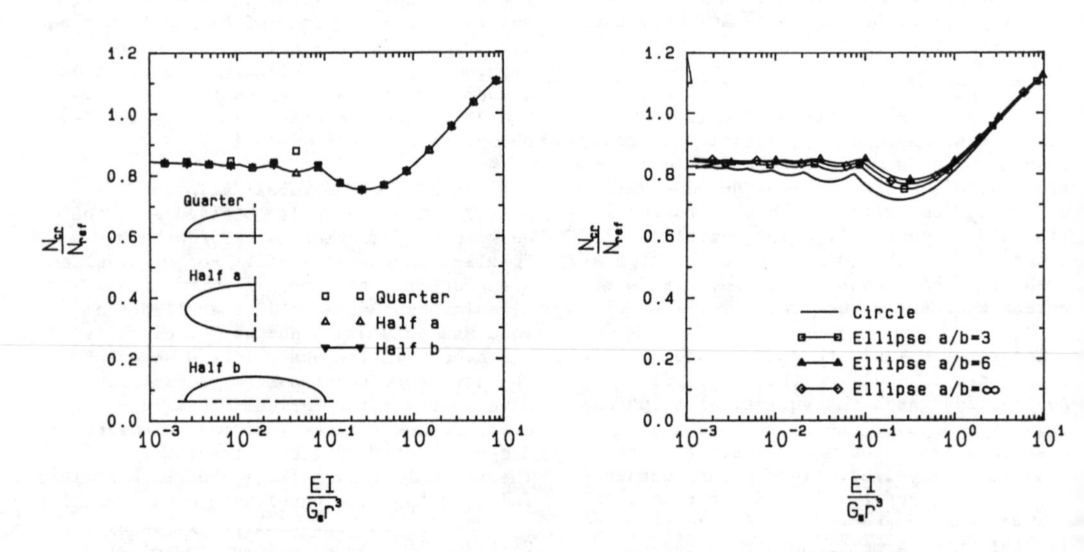

Figure 5: Influence of Symmetry for Elliptical Tube

Figure 6: Buckling Strength of Various Elliptical Tubes

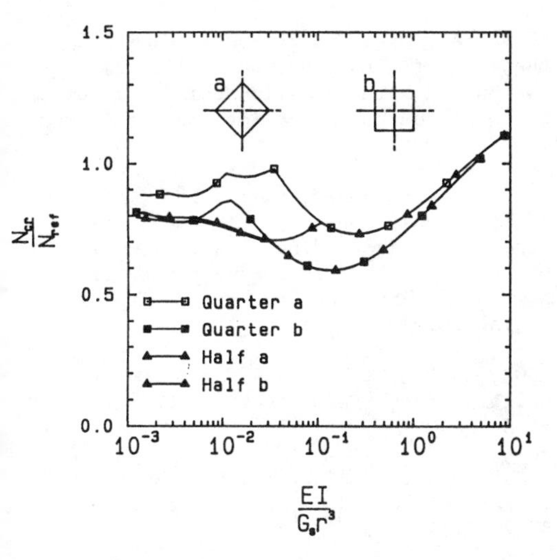

Figure 7: Influence of Symmetry for Square Tubes

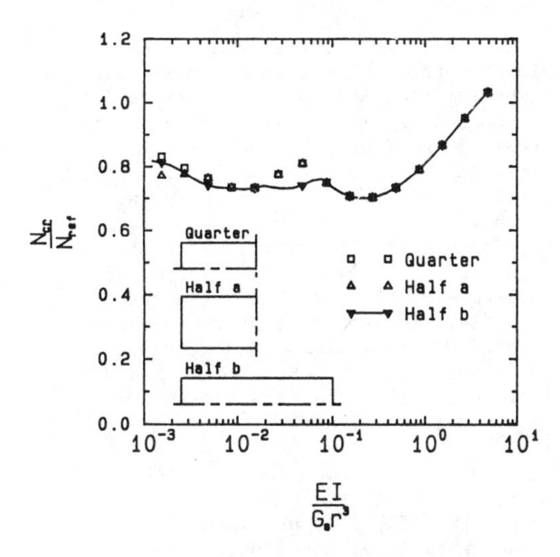

Figure 8: Influence of Symmetry for Rectangular Tubes

radius of curvature decreases towards zero. Ellipses, with large radius of curvature along most of their length, simply approach the "flat" tube limit more rapidly.

6. SQUARE TUBE RESPONSE

Metal box structures are also buried to form culverts, so the buckling strength of square tubes will be examined here as an introduction to rectangular structures of general shape.

The use of structural symmetry in the finite element analysis is considered in Figure 7. Four solutions are shown, corresponding to the analysis of one quarter and one half of the structure for lines of symmetry cutting through vertices (case a in Figure 7) and sides (case b). It is clear that the choice of symmetry conditions has a profound influence on the critical hoop thrust calculated for the structure.

Firstly, the critical thrust for a square tube restrained against rotation at its vertices is in general different to that for a tube restrained at midside points. Analysis of one quadrant only will lead to problems for tubes of intermediate stiffness as it did for circular and elliptical structures, although errors associated with this will again be of less significance once the flexural tube stiffness is low. An efficient mesh selection scheme can be selected through careful consultation with Figure 7, and it will involve the use of "case b" type symmetry conditions.

7. RECTANGULAR TUBES

A study of the symmetry conditions of rectangular tubes is made in Figure 8, and this again reveals that one half of the structure should be modelled for tubes of intermediate flexural stiffness. In this case it appears to be important to use the symmetry state labelled "half b" in the figure. This is probably associated with buckling modes which feature rotations at the midpoint of the long side of the rectangle, since these are prevented when the "half a" condition is adopted.

8. CONCLUSIONS

The finite element analysis of buried flexible structures has been described and its use in predicting the hoop thrusts which elastically destabilise circular, elliptical, square and rectangular tubes has been illustrated. Over the last few decades design practice for flexible metal culverts has been primarly based on field experience, but theoretical studies of the soil-structure interaction problem are now possible.

The use of geometrical symmetries to simplify the finite element analysis has been examined in some detail, as well as the effect of element size on numerical predictions of critical hoop force. Efficient solutions to the linear buckling problem can be obtained by analysing half the buried tube problem when the flexural stiffness of the structure is relatively high, and by analysing one quarter of the problem when the structure is very flexible. It appears that reasonable stability calculations result when about 10 or more of the structural elements employed in this analysis are used within each buckle wavelength.

In general the hoop thrust which elastically destabilises a tubular structure deeply buried within an elastic continuum is not greatly affected by the shape of the tube. This is quite different to the conclusions which arise if a "spring" (i.e. Winkler) model is employed for the soil. Deeply buried elliptical structures have higher critical hoop thrust than do circular structures with identical circumference. The results presented here indicate that the critical thrust for square structures is rather lower than that of circular tubes, although the buckling strength becomes more like that of circles and ellipses as the geometry becomes rectangular.

REFERENCES

ABEL, J.F., MARK, R. and RICHARDS, R., 1973, Stresses Around Flexible Elliptical Pipes, American Society of Civil Engineers, Journal of the Soil Mechanics and Foundations Division, 99: 509-526.

DUNS, C.S. and BUTTERFIELD, R., 1971, Flexible Buried Cylinders, Part 3 - Buckling Behaviour, International Journal of Rock Mechanics and Mineral Sciences, 8: 613-627.

FORRESTAL, M.J. and HERRMANN, G., 1965, Buckling of a Long Cylindrical Shell Surrounded by an Elastic Medium, International Journal of Solids and Structures, 1: 297-309.

KURT, C.E. and MARK, K.Y., 1981, Collapse of Noncircular Supported Thermoplastic Pipelines, Proceedings International Conference on Underground Plastic Pipe, New York, American Society of Civil Engineers.

MOORE, I.D., 1985, The Stability of Buried Tubes, Ph.D. Thesis, University of Sydney.

MOORE, I.D., 1987a, The Elastic Stability of Shallow Buried Tubes, Geotechnique, 37(2): 151-161.

MOORE, I.D., 1987b, Elastic Buckling of Buried Flexible Tubes - A Review of Theory and Experiment, Department of Civil Engineering and Surveying, University of Newcastle Research Report No: 024.09.1987.

MOORE, I.D. and BOOKER, J.R., 1985, Simplified Theory for the Behaviour of Buried Flexible Cylinders Under the Influence of Uniform Hoop Compression, International Journal of Solids and Structures, 21: 929-941.

ROWE, R.K., BOOKER, J.R. and BALAAM, N.P., 1978, Application of the Initial Stress Method to Soil-Structure Interaction, International Journal of Numerical Methods in Engineering, 13(5): 873-880.

YANG, T.Y., 1973, Matrix Displacement Solution to Elastica Problems of Beams and Frames, International Journal of Solids and Structures, 9: 829-842.

ZIENKIEWICZ, O.C., 1977, The Finite Element Method, New York, McGraw-Hill.

Numerical Methods in Geomechanics (Innsbruck 1988), Swoboda (ed.)
© 1988 Balkema, Rotterdam. ISBN 90 6191 809 X

Non-linear effects in soil-pipeline interaction in a ground subsidence zone

A.P.S.Selvadurai & S.Pang
Department of Civil Engineering, Carleton University, Ottawa, Ontario, Canada

ABSTRACT: This paper focusses on the interaction between a buried pipeline and the surrounding soil which is induced by a discontinuous ground movement. The pipeline is modelled as a shell structure and the soil medium is modelled as an ideal elastic plastic solid. A finite element technique is employed to study the flexural interaction between the pipeline and the soil. Numerical results presented in the paper illustrate the manner in which the flexural stresses in the pipeline are influenced by the relative stiffness of the soil-pipeline system and the yielding of the surrounding soil.

1 INTRODUCTION

Long distance buried pipelines are used quite extensively for the bulk transportation of oil, natural gas, slurried coal and other fluidized particulates (see e.g. Ariman et al., 1979; Pickell, 1983; Selvadurai and Lee, 1981, 1982; Selvadurai et al. 1983, 1985; Jeyapalan, 1985). In energy resource recovery endeavours in particular, long distance pipelines are used for the transportation of both on-shore and off-shore oil and natural gas. In contrast to above ground pipelines, the structural analysis, design and risk assessment should take into consideration the mutual interaction that exists between the flexible pipeline and the surrounding soil. Such interactions can be induced by the effect of service loadings such as expansion of the pipeline due to temperature and internal pressure, loadings of a geotechnical origin such as ground subsidence, frost heave, thaw settlement, and the action of external loadings such as berm construction, roadway traffic loads, landslides (both onshore and submarine), relief due to uplift and uplift due to soil inundation. To examine the soil-pipeline interaction problem related to ground subsidence in particular a number of facets of the problem need to be considered. These include (i) the modelling of the mechanical response of the soil surrounding the pipeline, (ii) the modelling of the mechanical behaviour of the pipeline (iii) the modelling of the mechanical response of the soil-pipeline interface (iv) the geometry or orientation of the pipeline in relation to the subsidence feature and (v) general features of the terrain in which the pipeline is located. With long distance buried pipelines in particular probable variations and uncertainties that are associated with spatial distributions in the properties of the soil strata need to be given careful consideration.

The objective of this paper is to consider a plausible elementary model of the soil-pipeline system with a view to establishing primarily the influences of soil non-linearity on the flexural interactive behaviour of the pipeline. The flexural interaction is induced by a discontinuous ground displacement at the base of a soil layer containing the pipeline (Figure 1). The numerical treatment of the soil-pipeline interaction is achieved via a non-linear finite element scheme which models the pipeline as a cylindrical shell and the surrounding soil mass is modelled as an ideal elastic-plastic solid with a Drucker-Prager type yield criterion and an associative flow rule. The interface between the pipeline shell and the surrounding soil is modelled by a perfectly adhesive contact. Several three dimensional soil-pipeline interaction problems are examined to generate results of engineering interest.

2 FINITE ELEMENT FORMULATION

A full three-dimensional finite element

technique needs to be used to examine the class of soil-pipeline problems associated with the discontinuous ground subsidence problem. In the ensuing, we shall briefly outline the salient features of the formulation; details of the developments will be presented elsewhere (Pang, 1988; Selvadurai and Pang, 1988).

Modelling of the Pipeline

The pipeline in the soil-pipe interaction problem is modelled via a superparametric Ahmad-type general shell element. One of the major reasons for choosing this element is that it can be easily adopted for use in conjunction with a conventional 20-node isoparametric brick element. The latter is employed to model the soil medium in the three-dimensional elastic and non-linear analysis. The shell element is essentially a 'degenerated' solid element (Figure 2). It has 8 nodes each of which has three translational and two rotational degrees-of-freedom which give rise to a total of 40 degrees-of-freedom. The parabolic interpolation function of the top and bottom element faces allow for the modelling of both regular and irregular curved surfaces. The sections across the element thickness are generated by straight lines representing a linear strain variation. The constraint of straight 'normals' is introduced to reduce computing time and the strain energy corresponding to stresses perpendicular to the mid-surface is ignored to improve numerical conditioning. Thus, the assumption in conventional shell theory that normals to the middle surface in the undeformed body remain normal to the middle surface after deformation is not enforced. This feature permits the shell to experience transverse shear deformations in addition to the flexural deformations. Since transverse shear is an important feature in thick shell situations, this superparametric shell element is preferred over the various shell elements with formulations based upon conventional shell theory. Also, the highly sophisticated semi-loof shell element is not considered here. The superparametric element discussed here has been proved to adequately meet the accuracy and versatility required for the analysis of the class of soil-pipeline interaction problems discussed in the previous section. The two major differences between the formulation of shell element and the solid brick element are as follows:

(i) the usual three dimensional stress-strain relationships are not invoked in the shell element formulation. The shell element can be visualized as a stack of isoparametric elements which are in a state of plane stress;
(ii) instead of having 20 translation degrees-of-freedom in each of the three principal orthogonal directions, the displacement field throughout the element is defined by three Cartesian components of the mid-surface nodal displacement and two rotational degrees-of-freedom about the two orthogonal directions normal to the shell thickness (see Figure 2).

The shell element is nonetheless a specialized version of the brick element. The geometry and interpolation concepts that are used in the isoparametric continuum elements can also be used in the evaluation of shell element matrices. In order to determine the properties of the shell element, the essential stresses and strains need to be defined; the relevant strain components are given by

$$[\underset{\sim}{\varepsilon}']^T = [\varepsilon_{x'}, \varepsilon_{y'}, \gamma_{x'y'}, \gamma_{x'z'}, \gamma_{y'z'}]$$

$$= [\frac{\partial u'}{\partial x'}, \frac{\partial v'}{\partial y'}, (\frac{\partial u'}{\partial y'} + \frac{\partial v'}{\partial x'}),$$

$$(\frac{\partial w'}{\partial x'} + \frac{\partial u'}{\partial z'}), (\frac{\partial v'}{\partial z'} + \frac{\partial w'}{\partial y'}) \qquad (1)$$

The strains in the z-direction are neglected in keeping with usual assumptions of a shell theory. In general, the coordinate system (x', y', z') will not coincide with those of the curvilinear coordinates (r,s,n). The strain-displacement relationship at a given point i is given by

$$[\underset{\sim}{B}_i] = \begin{bmatrix} a_i & 0 & 0 & -d_i\, l_{2i} & d_i\, l_{1i} \\ 0 & b_i & 0 & -e_i\, m_{2i} & e_i\, m_{1i} \\ 0 & 0 & c_i & -g_i\, n_{2i} & g_i\, n_{1i} \\ b_i & a_i & 0 & -e_i l_{2i} - d_i m_{2i} & e_i l_{1i} + d_i m_{1i} \\ 0 & c_i & b_i & -g_i m_{2i} - e_i n_{2i} & g_i m_{1i} + e_i n_{1i} \\ c_i & 0 & a_i & -d_i n_{2i} - g_i l_{2i} & d_i n_{1i} + g_i l_{1i} \end{bmatrix}$$

$(i = 1,2,3, \ldots, 8)$ (2)

In (2)

$$a_i = J^*_{11} \frac{\partial N_i}{\partial r} + J^*_{12} \frac{\partial N_i}{\partial s} ;$$

$$b_i = J^*_{21} \frac{\partial N_i}{\partial r} + J^*_{22} \frac{\partial N_i}{\partial s} \quad ;$$

$$c_i = J^*_{31} \frac{\partial N_i}{\partial r} + J^*_{32} \frac{\partial N_i}{\partial s} \quad ;$$

$$d_i = \frac{t_i}{2} (a_i n + J^*_{13} N_i)$$

$$e_i = \frac{t_i}{2} (b_i n + J^*_{23} N_i)$$

$$g_i = \frac{t_i}{2} (c_i n + J^*_{33} N_i) \qquad (3)$$

where l_{1i}, l_{2i}, ... etc. are the direction cosines of unit vectors at the point i with respect to the global axes x, y, z; $J^*_{11}, \ldots, J^*_{33}$ are components of the inverted Jacobian matrix; N_i is the equivalent shape function at point i; t_i is the shell thickness at point i. The stress-strain relationship for the elastic material composing the pipeline shell is the usual isotropic linear elastic stress-strain relationship

$$[\sigma'] = [S'][\varepsilon'] = \{\sigma_{x'}, \sigma_{y'}, \tau_{x'y'}, \tau_{y'z'}, \tau_{z'x'}\} \qquad (4)$$

where

$$[S'] = \frac{E}{(1-\nu^2)} \begin{bmatrix} 1 & \nu & 0 & 0 & 0 \\ & 1 & 0 & 0 & 0 \\ & & (\frac{1-\nu}{2}) & 0 & 0 \\ \text{sym} & & & (\frac{1-\nu}{2k}) & 0 \\ & & & & (\frac{1-\nu}{2k}) \end{bmatrix} \qquad (5)$$

This 5x5 material property matrix is defined with respect to the local axes and can be extended to incorporate any anisotropic elastic properties. From the definition of the displacement field within the element it may be noted that the shear stress distribution is linear, whereas in reality the shear stress distribution is approximately parabolic. Thus the factor k included in the stress strain relationship appropriate for $\tau_{y'z'}$ and $\tau_{z'x'}$ is taken as 1.2 in order to improve the shear displacement approximation. Since the matrix $[B]$ yields strains in the global directions further transformation must be applied to convert $[S']$ to $[S]$ in the global directions. Further details of the shell element formulation are given by Zienkiewicz (1979) and Bathe (1982).

Modelling of the Soil Medium

The solid brick element used to model the soil behaviour can be classified as a 20-node three dimensional element (Figure 3). One of the major advantages of this 20 node element is its capability to model curved and irregular boundaries due the presence of a midside variable node at each edge of the element. The isoparametric formulation employs a 'natural' coordinate system (r,s,n) which is defined by the local element geometry rather than by the element orientation in the global (x,y,z) coordinate system. This solid element has three translational degrees-of-freedom per node associated with the global (x,y,z) system giving a total of 60 degrees-of-freedom. The apparent flexibility of the element has allowed for its successful performance in the modelling of a variety of problems in solid mechanics including bending stress predictions. For a fully three dimensional analysis, the strain components are defined as

$$[\varepsilon]^T = [\varepsilon_x, \varepsilon_y, \varepsilon_z, \varepsilon_{xy}, \varepsilon_{yz}, \varepsilon_{zx}]$$
$$= [\frac{\partial u}{\partial x}, \frac{\partial v}{\partial y}, \frac{\partial w}{\partial z}, (\frac{\partial u}{\partial y} + \frac{\partial v}{\partial x}), (\frac{\partial v}{\partial z} + \frac{\partial w}{\partial y}), (\frac{\partial u}{\partial z} + \frac{\partial w}{\partial x})] \qquad (6)$$

After appropriate manipulations and matrix partitioning, the strain displacement relationship can be expressed as

$$[\varepsilon] = \{[B]_1, [B]_2, \ldots, [B]_{20}\} [a] \qquad (7)$$

where $[a]$ is the displacement vector. A typical sub-matrix $[B]_i$ is given as

$$[B]_i^T = \begin{bmatrix} \frac{\partial N_i}{\partial x} & 0 & 0 & \frac{\partial N_i}{\partial y} & 0 & \frac{\partial N_i}{\partial z} \\ 0 & \frac{\partial N_i}{\partial y} & 0 & \frac{\partial N_i}{\partial x} & \frac{\partial N_i}{\partial z} & 0 \\ 0 & 0 & \frac{\partial N_i}{\partial z} & 0 & \frac{\partial N_i}{\partial y} & \frac{\partial N_i}{\partial x} \end{bmatrix} \qquad (8)$$

where N_i are shape functions evaluated at point i.

Material Models for the Soil Medium.

The pre-yield behaviour of the soil medium is represented by an isotropic linear elastic response. The three-dimensional stress-strain matrix is given by

$$[S_s] = \frac{E(1-\nu)}{(1+\nu)(1-2\nu)} [S^*] \quad (9)$$

where

$$[S^*] = \begin{bmatrix} 1 & \frac{\nu}{1-\nu} & \frac{\nu}{1-\nu} & 0 & 0 & 0 \\ & 1 & \frac{\nu}{1-\nu} & 0 & 0 & 0 \\ & & 1 & 0 & 0 & 0 \\ & & & \frac{(1-2\nu)}{2(1-\nu)} & 0 & 0 \\ & \text{sym} & & & \frac{(1-2\nu)}{2(1-\nu)} & 0 \\ & & & & & \frac{(1-2\nu)}{2(1-\nu)} \end{bmatrix} \quad (10)$$

In order to account for the non-linearity of the soil medium, the elastic-plastic Drucker-Prager yield criterion is used in conjunction with an associative flow (normality) rule. The criterion can be written in the form

$$F = \sqrt{J_{2D}} - \alpha J_1 - k \quad (11)$$

where α and k are positive material parameters; J_1 is the first invariant for the stress tensor and J_{2D} is the second invariant of the deviatoric stress tensor.

The normality rule dictates that the incremental plastic strain vector, $d\varepsilon_{ij}^{p}$, is always perpendicular to the yield surface. The incremental plastic strain vector is given by

$$d\varepsilon_{ij}^{p} = d\lambda \frac{\partial Q}{\partial \sigma_{ij}} \quad (12)$$

where Q is the plastic potential function, $d\lambda$ is a positive scalar factor of proportionality. However, the Drucker-Prager (1952) yield criterion used in this context gives rise to a negative volumetric component in the expression for $d\varepsilon_{ij}^{p}$ which indicates dilatation (volume increase) at failure.

In the three dimensional principal stress space, this criterion represents a right circular cone where the base of the cone forms a circle on the π-plane. When α is equal to zero, the yield surface reduces to the von Mises criterion which is suitable for representing a purely cohesive material. The values of α and k are dependent on the angle of internal friction ϕ and the apparent cohesion c.

For the axi-symmetric case, α and k can be expressed as

$$\alpha = \frac{2\sin\phi}{\sqrt{3}\,(3-\sin\phi)}$$

$$k = \frac{6c\cos\phi}{\sqrt{3}\,(3-\sin\phi)} \quad (13)$$

The required input parameters for implementation of this model are the four basic material constants, E, ν, c and ϕ, all of which may be determined by standard geotechnical laboratory tests such as the triaxial test. Further details of relevant finite element implementation procedures may be found in Siriwardane and Desai (1984).

Coupling

As mentioned earlier, incompatibility exists between the shell element and the brick element. The method of 'static condensation' has been adopted to deal with this situation. This approach involves the elimination of the terms associated with the two rotational degrees-of-freedom at each of the eight nodes during the formation of the shell element stiffness matrix. Hence, the reduced element stiffness matrix has a size of 24 by 24. It should be noted that the effects of rotation are accounted for by the resulting translational degrees-of-freedom after all the matrix manipulations have been completed.

The shell element force-displacement relationship takes the form

$$[P_s^o] = [K_s^o]\,[U_s^o] \quad (14)$$

The stiffness matrix may be partitioned such that the above equation may be written as

$$\begin{bmatrix} [P_s] \\ [M_s] \end{bmatrix} = \begin{bmatrix} [K_{11}^o] & [K_{12}^o] \\ [K_{21}^o] & [K_{22}^o] \end{bmatrix} \begin{bmatrix} [\delta_s] \\ [\theta_s] \end{bmatrix} \quad (15)$$

As a result of the matrix manipulations associated with 'static condensation', the reduced force-displacement relationship is established with respect to translational degrees-of-freedom such that

$$[\bar{P}_s] = [\bar{K}_s]\,[\delta_s] \quad (16)$$

$[P_s^o]$ = original load vector, 40 x 1

$[\bar{P}_s]$ = reduced translational load vector, 24 x 1

$[K_s^o]$ = original shell element stiffness matrix, 40 x 40

$[\bar{K}_s]$ = reduced shell element stiffness matrix, 24 x 24

$[U_s^o]$ = original displacement vector, 40 x 1

$[\delta_s]$ = translational displacement vector, 24 x 1

3 THE SOIL-PIPELINE INTERACTION PROBLEM

The finite element procedure outlined in the previous section is used to examine the soil-structure interaction response of a pipeline which is located in a discontinuous shear zone (Figure 1). For purposes of illustration and in order to minimize computational resources, attention is restricted to the idealized soil-pipeline interaction problem associated with a finite region (Figure 4). The associated three dimensional finite element discretization is shown in Figure 5. There are 44 isoparametric solid elements each of which has 20 nodes or 60 translational degrees-of-freedom. Of those solid elements 32 assume soil material properties while the rest are 'rigid' elements used to induce abrupt ground displacement. There are 20 8-noded shell elements with 24 degrees-of-freedom each (after static condensation). In order to simulate the behaviour of a pipe with a 'rough' surface, the entire soil-pipe interface is assumed to be fully-bonded throughout the analyses. For a solid element, there are 27 Gaussian integration points, i.e. 3x3x3 with respect to the three local axes. A 3x3x2 configuration for a total of 18 points is used for the shell element with only 2 points in the 'thickness' direction.

Furthermore, a plane of asymmetry exists at section A-A in Figure 5. All the in-plane displacements with respect to the global x and y directions are prescribed to be zero. Owing to the asymmetric conditions associated with deformation, only half of the soil-pipeline system needs to be discretized. The dimension in both the x and the y directions are chosen such that the effects of imposed boundary conditions can be minimized. The dimension in the z-direction allows for the simulation of ground movement along a relatively long section of the pipeline. The abrupt upward displacement is induced by applying upward pressures on the rigid base. It should be noted that the load is not being directly applied to the pipe. This finite element mesh configuration is chosen after considering the trade-off between computational accuracy and reasonable computing (CPU) time within the capabilities of the available computing system (a Honeywell Level 66 Mainframe operating under the CP-6 system).

4 NUMERICAL RESULTS AND DISCUSSION

In this section, results of engineering interest regarding interaction between a buried pipeline and the surrounding soil are presented. This problem has several variable parameters; but the scope of our investigations concentrates on the evaluation of the flexural behaviour of the pipe. The relationship between the flexural response of the pipe and the relative stiffness of the soil-pipe system is emphasized. In order to present the numerical results in a concise manner, a relative stiffness parameter RS is being used. This parameter relates the bending stiffness of the pipeline to the elastic properties of the soil as follows

$$RS = \frac{E_p I}{E_s D^4} \qquad (17)$$

where E_p = Young's modulus of pipe
I = moment of interia of pipe cross-section
D = average diameter of pipe
E_s = Young's modulus of soil

In this paper, a typical value of 10^4 psi for E_s is being used as this is a typical value for many soils in their natural conditions. The thickness of the pipe is chosen to be equal to 0.03 in. which gives a D/t value of 200 (thin-walled pipe). As for the pipe itself, values of 1.5×10^7 psi, 5.0×10^6 psi and 5.0×10^5 psi are used. These values approximately correspond to typical values applicable for cast iron, concrete and PVC pipes respectively. The combination of the above parameters gives rise to RS vaues of 0.0408, 0.0137 and 0.0014. For the purpose of non-linear analyses, combinations of ν=0.48 and c=10.0 psi are used to simulate the purely cohesive soil. Figures 6 - 8 illustrate the manner in which the flexural moments along the length of the buried pipeline is influenced by the relative stiffness of the soil-pipeline system. These results indicate that the maximum flexural moment at the transition zone is altered by the yielding response of the cohesive soil and the relative stiffness RS of the soil-pipeline system. The results presented in this paper are of a preliminary nature; the accuracy of the various distributions can be improved by suitable refinement of the finite element discretization and the possible incorporation of infinite elements to model a pipeline of infinite extent. These investigations are relegated to future studies. The preliminary results presented in the paper indicates the degree to which soil yielding at a ground discontinuity zone can alter the flexural stresses that are generated in the buried pipeline.

REFERENCES

[1] Ariman, T., Liu, S.C. and Nickell, R.E., "Lifeline Earthquake Engineering - Buried Pipelines, Seismic Risk and Instrumentation", PVP34, American Society of Mechanical Engineers, 1979.

[2] Bathe, K.-J., "Finite Element Procedures in Engineering Analysis", Prentice-Hall Inc., 1982.

[3] Drucker, D.C. and Prager, W., "Soil Mechanics and Plastic Analysis of Limit Design", Quarterly J. of Applied Mathematics, Vol.10, No.2, pp.157-165, 1952.

[4] Jeyapalan, J.K. (ed.), Proceedings of "International Conference on Advances in Underground Pipeline Engineering", American Society of Civil Engineers, 1985.

[5] Pang, S., "Soil-Structure Interaction in Discontinuous Shear Zones", Ph.D. Thesis, Department of Civil Engineering, Carleton University, Ottawa, Ontario, Canada, 1988.

[6] Pickell, M.B. (ed.), "Pipelines in Adverse Environments II", Proc. of the ASCE Specialty Conf., San Diego, Ca., 1983.

[7] Selvadurai, A.P.S. and Lee, J.J., "Soil Resistance Models for the Stress Analysis of Buried Fuel Pipelines", Interim Report, Technical Research and Development Projects, Research and Engineering, Bechtel Group Inc., San Francisco, Ca., December, 1981.

[8] Selvadurai, A.P.S. and Lee, J.J., "Soil Resistance Models for the Stress Analysis of Buried Fuel Pipelines, Final Report, Technical Research and Development Project, Research and Engineering, Bechtel Group Inc., San Francisco, Ca., December 1982.

[9] Selvadurai, A.P.S., Lee, J.J., Todeschini, R.A.A. and Somes, N.F., "Lateral Soil Resistance in Soil-Pipe Interaction", Pipelines in Adverse Environments II, Proc. of the ASCE Specialty Conf., San Diego, Ca., 1983.

[10] Selvadurai, A.P.S., "Numerical Simulation of Soil-Pipeline Interaction in a Ground Subsidence Zone", Proc. of Inter. Conf. on Adv. in Underground Pipeline Engineering, American Society of Civil Engineers, 1985.

[11] Selvadurai, A.P.S. and Pang, S., "Soil-Pipeline Interaction Studies in Discontinuous Shear Zones", 1988 (in preparation).

[12] Siriwardane, H.J. and Desai, C.S., "Constitutive Laws for Engineering Materials (with Emphasis on Geologic Materials)", Prentice-Hall, 1984.

[13] Zienkiewicz, O.C., "The Finite Element Method", 3rd edition, McGraw-Hill, 1979.

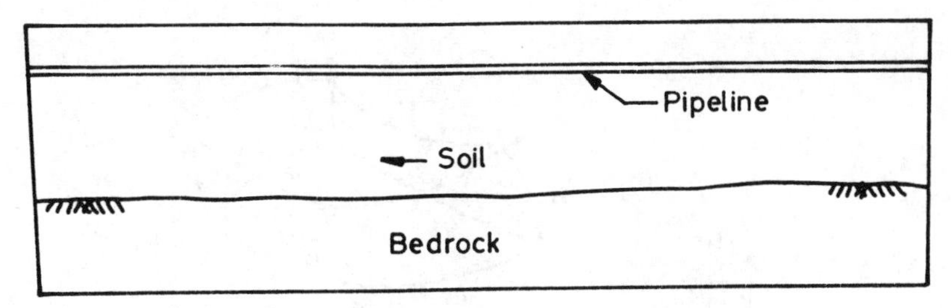

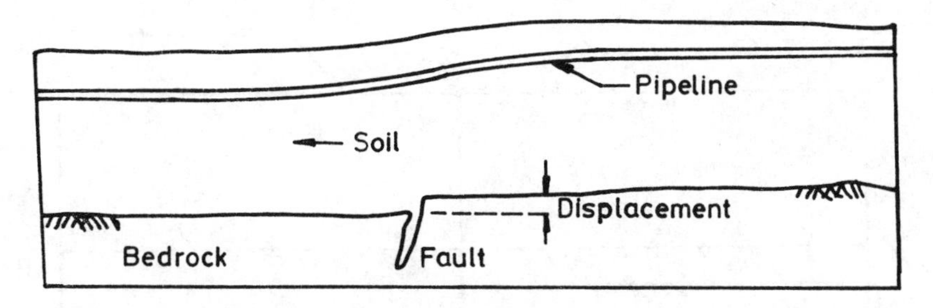

Figure 1: Soil-Pipeline Interaction in a Ground Subsidence Zone

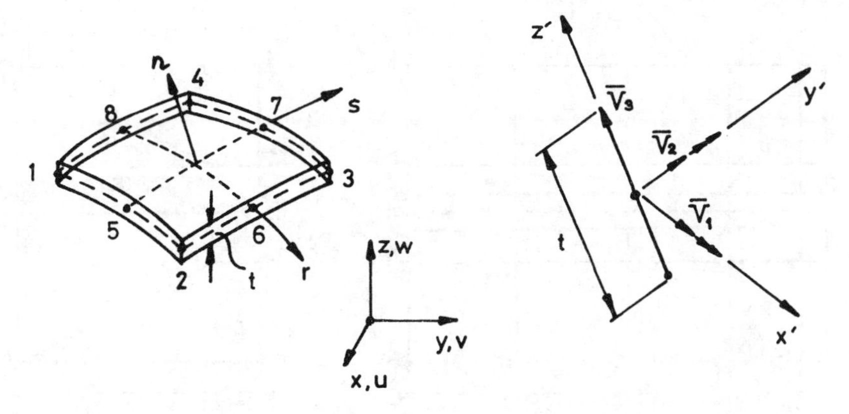

Figure 2: General Shell Element

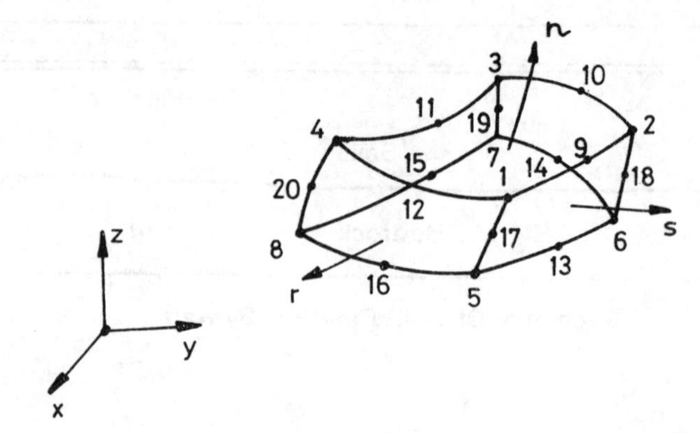

Figure 3: Solid 'Brick' Element

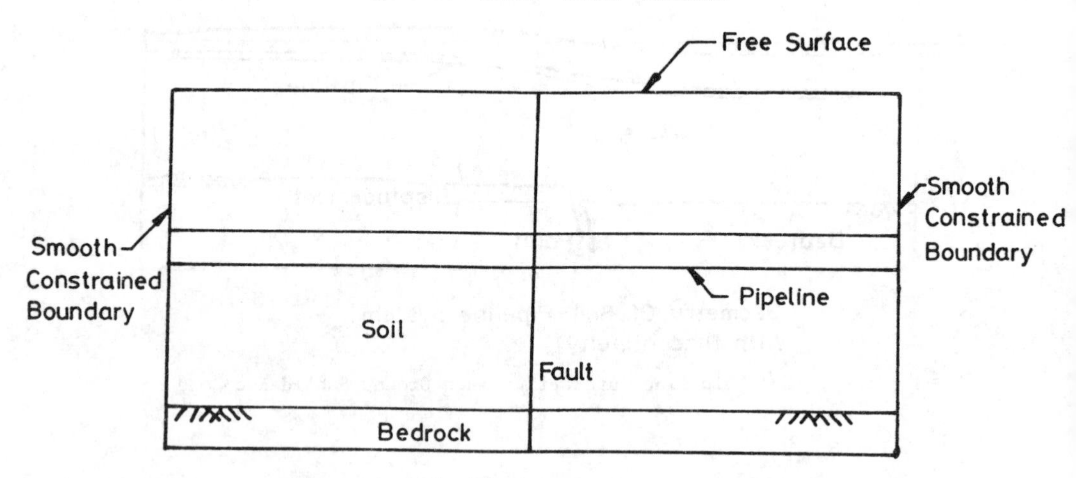

Figure 4: Idealized Soil-Pipeline Systems

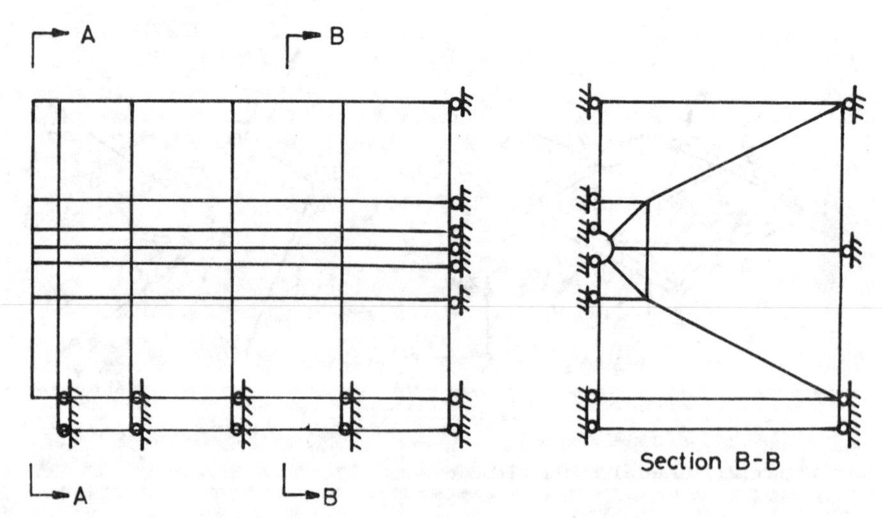

Figure 5: Finite Element Discretization of Idealized Soil-Pipeline System

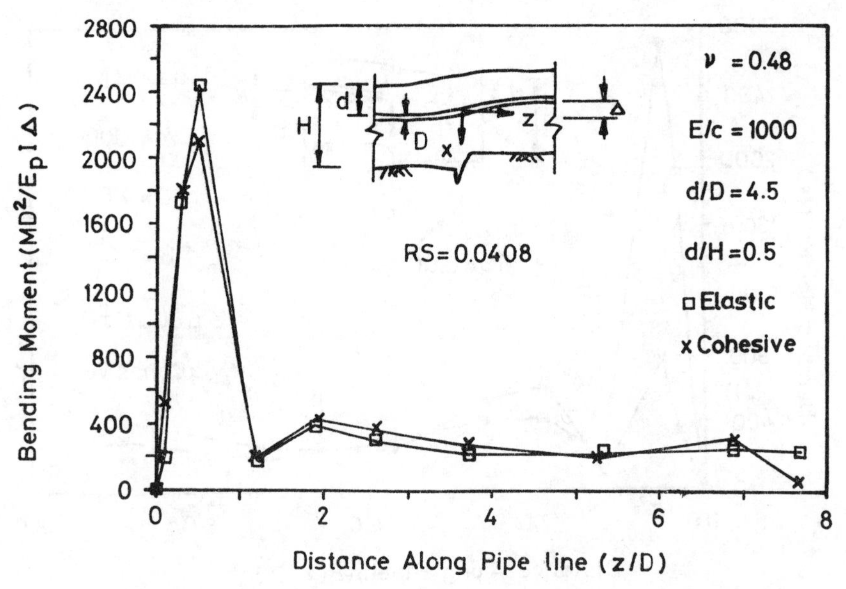

Figure 6: Bending Moment Distribution Along Pipeline (RS=0.0408)

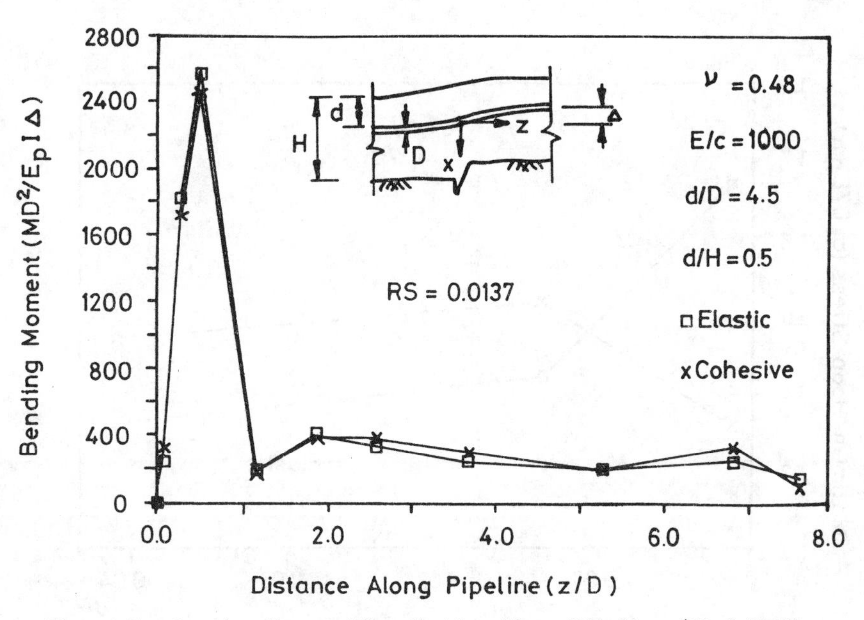

Figure 7: Bending Moment Distribution Along Pipeline (RS=0.0137)

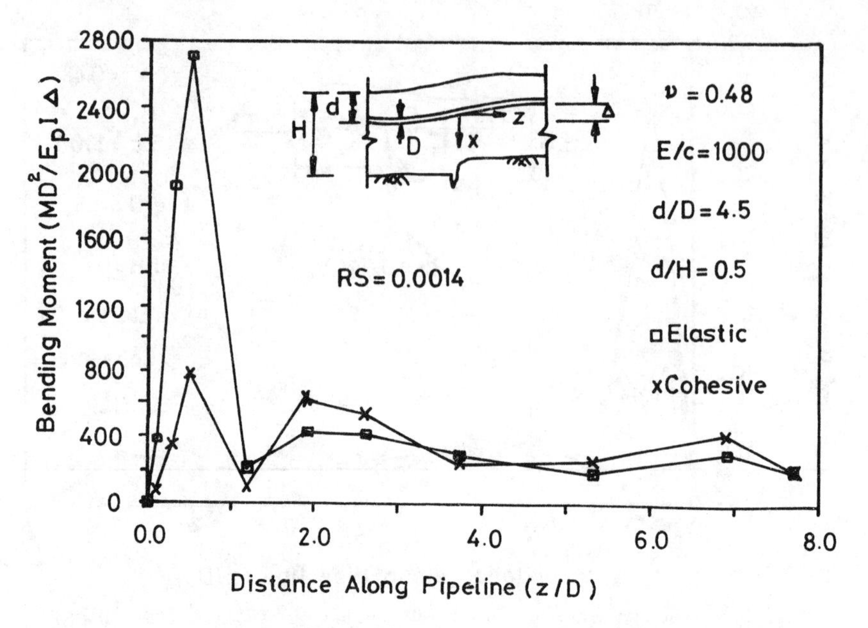

Figure 8: Bending Moment Distribution Along Pipeline (RS=0.0014)

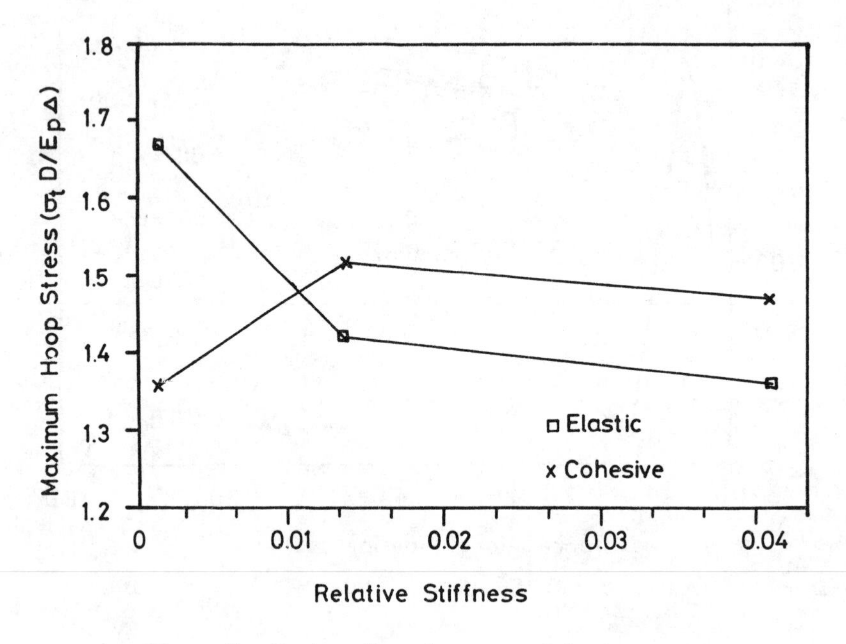

Figure 9: Maximum Hoop Stress vs Relative Stiffness

Numerical Methods in Geomechanics (Innsbruck 1988), Swoboda (ed.)
© 1988 Balkema, Rotterdam. ISBN 90 6191 809 X

Undrained strength from CPT and finite element computations

P. van den Berg
Delft Geotechnics, Netherlands

P.A. Vermeer
Delft University of Technology, Netherlands

ABSTRACT: The relationship between the cone resistance and the undrained shear strength of clay is considered. Having given evidence for a semi-theoretical solution, results of small-strain finite element computations are reviewed. As such analyses tend to underestimate the cone resistance, a large strain Eulerian formulation is used. Computational results are presented both for a soft and a stiff clay.

1 INTRODUCTION

In its original form the cone penetration test (CPT) found its main application in testing of competent soil layers for the determination of the bearing capacity of foundation piles. With its greater present sensitivity it is also applied for testing soft soils. For many years engineers have tried to correlate the cone resistance with shear strength and deformation moduli of the soil.

A powerful tool to analyse the problem of deep steady penetration is the finite element method. Obviously, cone penetration is a large deformation problem. The finite element analysis of such situations is much more difficult than for situations with small strains. Great care is required in the programming of large strain calculations to ensure that proper account is taken of all terms in the equilibrium equations. Therefore results should be carefully checked against simple solutions. The arrangements of the contexts of this paper are as follows. First theoretical evidence is given for a simple solution of the undrained clay problem. Then results of small strain finite element computations are reviewed. Finally the equilibrium equations for large strain problems are presented together with the results of a large strain analysis of cone penetration in soft and stiff clay.

2 EMPIRICAL AND THEORETICAL RELATIONS

To model the penetration process in clay empirical relations have been proposed of the following form :

$$q_c = N_c c_u + \sigma_{vo} \qquad (1)$$

where :

q_c : cone resistance
N_c : cone factor
c_u : undrained shear strength
σ_{vo} : total overburden pressure

Typical values for N_c are in the range from 9 to 17.

Theoretical methods can be subdivided into three approaches.
A first approach is the bearing capacity method based on the original work by Prandtl. The problem is considered as an incipient failure of a rigid plastic material.
An alternative approach refers to the theory of cavity expansion in an elasto-plastic material. In particular Vesic (1972) gives relationships between the penetration resistance q_c and the material parameters based on this method.
The third approach is based on the work done by Baligh (1975). He has pointed out that the deep steady penetration process is a basically strain controlled problem. Based hereon he proposed a "strain-path-method" to solve this kind of problems.

3 CONE PENETRATION MODEL

The most promising theoretical approach is to analyse the problem of the penetration process in clay as a steady state process. Based on the original work by Baligh (1975) the total cone resistance can be interpreted as the work done per unit area to push the cone a unit distance. Then according to Baligh the penetration process can be reasonably simplified by dividing the process into two contributions. The first contribution is the work required to give the tip of the cone a virtual vertical displacement. The second part is the work done to expand the soil around the shaft in the horizontal direction.
According to this theoretical model the cone resistance can be expressed in the following form :

$$q_c = N_p c_u + \sigma_{h0} + \Delta\sigma_h \qquad (2)$$

The first term in equation (2) is the pressure needed for a virtual vertical displacement. The partial cone factor N_p can be computed by means of the stress characteristics method. For a cone with a 60 degree sharp point this analysis results in values for N_p as presented in table 1. These values are taken from Luger et al. (1982) and Koumoto and Kaku (1982). The partial cone factor appears to vary as a function of the shaft adhesion α_s and the adhesion α at the pointed cone.

Table 1. Partial cone factor N_p for a smooth cone with 60 degree sharp point. Influence of adhesion to cohesion ratio on shaft (α_s/c_u) and cone tip (α/c_u).

α_s/c_u	0.0	0.5	1.0
α/c_u = 0.0	7.57		
α/c_u = 0.5	8.84	9.76	
α/c_u = 1.0	9.86	10.76	11.49

The second contribution to equation (2) is the increase of the horizontal stress at the shaft of the cone. The increase, $\Delta\sigma_h$, can be derived from the cylindrical cavity expansion theory (Gibson et al. (1961) and Vesic (1972)). The increase of the horizontal stress is equal to the expansion pressure required to produce a cylindrical cavity around the shaft of the cone.

$$\Delta\sigma_h = c_u + c_u \ln (G/c_u) \qquad (3)$$

Inserting equation (3) into equation (2) yields the semi-theoretical cone formula, relating the cone resistance to deformation and strength parameters for undrained behaviour of clay.

$$q_c = [N_p + 1 + \ln (G/c_u)]\, c_u + \sigma_{h0} \qquad (4)$$

4 RESULTS OF SMALL STRAIN ANALYSES

In principle, the finite element method is capable of solving the cone penetration problem.
One of the first computations, making use of a small strain finite element formulation, were performed by De Borst (1982). For a penetrometer with a smooth shaft it was found that the cone factor was about 10. In these calculations eight-noded quadrilateral elements were used. A similar analysis using the same type of elements was performed by Griffiths (1982). He arrived at a cone factor of about 9.5.
Good results were obtained for both a smooth and a rough tip of the cone by De Borst and Vermeer (1984) using special higher order elements. The fifteen-noded triangular elements were also capable of solving the problem of a rough shaft. Again a cone factor was found about equal to ten.

The studies outlined above are incomplete; they can not give results for the full cone factor. The factor calculated is independent of the stiffness of the soil. In the analyses the cone is introduced into a pre-bored hole, a small strain plasticity calculation is carried out and the resulting vertical pressure identified as the cone resistance. The build up of the horizontal pressure can not be modelled completely in a small strain analysis. To assess the full cone factor the stiffness-dependent contribution (equation (4)) is needed.

5 THEORY FOR LARGE PLASTIC DEFORMATION

In order to erase incorrectly estimated initial stress conditions a large deformation analysis is needed to simulate a displacement of at least one or two times the diameter of the cone. Then the stiffness dependent increase of the shaft pressure due to the effect of cavity expansion will develope completely to give the full cone factor. A realistic build up of the shaft pressure can not be modelled at all by a small strain calculation.

One of the first developments in the field of large strain analyses was published by Hibbit, Marcal and Rice (1970). Improvements were made by Mc. Meeking and Rice (1975) : the updated Lagrange approach was introduced.

In such an analysis the following relations are used to model large deformation of materials. First of all proper account is taken of the stress equilibrium equations. In short index notation the equilibrium can be written as follows :

$$\frac{\partial \sigma_{ij}}{\partial x_j} + \gamma_i = 0 \qquad (5)$$

where σ_{ij} is the Cauchy stress and γ_i are the components of the volume weight. For elasto-plastic material behaviour stresses and displacements are calculated incrementally. So, considering a current state with stresses σ_{ij} and coordinates x_i, the equations for continued equilibrium are :

$$\frac{\partial \Delta\Sigma_{ij}}{\partial x_j} = 0 \qquad (6)$$

where $\Delta\Sigma_{ij}$ is the increment of the first Piola-Kirchhoff stress.

$$\Delta\Sigma_{ij} = \Delta\sigma^c_{ij} + \omega_{ik}\sigma_{kj} - \sigma_{ik}\varepsilon_{kj} \qquad (7)$$

For $\Delta\sigma^c_{ij}$ usually the Jaumann stress increment is used. The relationship between this constitutive stress increment and the Cauchy stress increment is :

$$\Delta\sigma^c_{ij} = \Delta\sigma_{ij} - \omega_{ik}\sigma_{kj} + \sigma_{ik}\omega_{kj} \qquad (8)$$

The rotation tensor ω_{ij} is defined as follows :

$$\omega_{ij} = 0.5 \left(\frac{\partial \Delta u_i}{\partial x_j} - \frac{\partial \Delta u_j}{\partial x_i} \right)$$

$$\omega_{ij} = - \omega_{ji} \qquad (9)$$

Equations (7) and (8) express that the Cauchy stress σ_{ij} changes as a result of both deformation and rotation of the material. The special case of material rotation without additional deformation implies stress rotation. In the most simple small strain analysis these two rotation terms are omitted to obtain : $\Delta\sigma^c_{ij} = \Delta\sigma_{ij}$.

6 CONSTITUTIVE MODEL

For the constitutive behaviour of the corotational stress we have an incremental stress-strain relation of the following form :

$$\Delta\sigma^c_{ij} = S_{ijkl}\,\Delta\varepsilon_{kl} \qquad (10)$$

in which S_{ijkl} is the stiffness tensor containing the material properties. An elasto-plastic material model is used, whereas for the yield criterion the Von Mises plasticity model is used. This criterion can be written as follows :

$$s_{ij}\,s_{ij} = (2\,c_u)^2 \qquad (11)$$

where c_u is the undrained cohesion and s_{ij} the deviatoric stress, defined as :

$$s_{ij} = \sigma_{ij} - \frac{1}{3}(\sigma_{kk}\delta_{ij}) \qquad (12)$$

in which δ_{ij} is the Kronecker delta tensor.

Equation (10) can also be written as follows :

$$\Delta\sigma^c_{ij} = (\,S^e_{ijkl} - S^{pl}_{ijkl}\,)\,\Delta\varepsilon_{kl} \qquad (13)$$

where :

$$S^e_{ijkl} = G\left(\delta_{ik}\delta_{jl} + \delta_{il}\delta_{jk} + \frac{2\nu}{1-2\nu}\delta_{ij}\delta_{kl}\right) \qquad (14)$$

and :

$$S^{pl}_{ijkl} = \frac{G}{2c_u^2}(s_{ij}s_{kl}) \qquad (15)$$

7 EULERIAN APPROACH

A large strain analysis is usually characterized by the adaption of the finite element mesh during the incremental solution procedure. This method, the so-called updated Lagrange approach, implies that the nodal points follow the displacements of the material. In such an approach the elements retain neither their initial shape nor their original position. During continued deformation the shape of the elements changes continuously, which may lead to severe numerical problems. Indeed, for very large local deformations the elements are too much elongated or even turned inside out.

The above problems can be avoided if the material displacements are uncoupled from the nodal points. In that case we have a fixed element mesh and material flowing through the elements. For this so-called Eulerian approach, the position of the nodes is not adjusted, but now we adjust the state of the material by taking into account convection (Huetink, 1982). This approach is applied in the numerical analysis of the cone penetration problem.

8 RESULTS OF CONE PENETRATION ANALYSIS

The large strain analysis of the cone penetration problem is carried out using the finite element code DIEKA. For more detailed information about this program the reader is referred to Huetink (1986). As this program does not have the option of using fifteen-noded triangular elements, another element type has to be chosen.

As a first option eight-noded isoparametric quadrilateral axisymmetric elements were used. The analyses were restricted to a penetrometer with an entirely smooth shaft as the quadrilateral elements had failed for the rough problem (de Borst and Vermeer, 1984). To minimize the internal kinematic constraints both the shaft and the tip were taken smooth. Obviously many kinematic constraints remain, as there also is the internal constraint of incompressibility. However, using this element type with nine-point integration it proved to be impossible to reach a steady state solution.

A second type of elements used were four-noded constant dilatancy elements. To increase the ratio degrees of freedom - number of constraints a variational principle is used in which the dilatational strain increment and the displacement increments are present as independent variables. For convergence the dilatational strain increment has to be constant within an element. For more detailed information the reader is referred to Nagtegaal et al. (1974).

The finite element discretization consists of 75 elements with four Gaussian points per element (figure 1).

Two calculations were performed : the first one (run I) with a relatively soft clay and the other (run II) with a much stiffer clay. The material parameters were taken as follows :

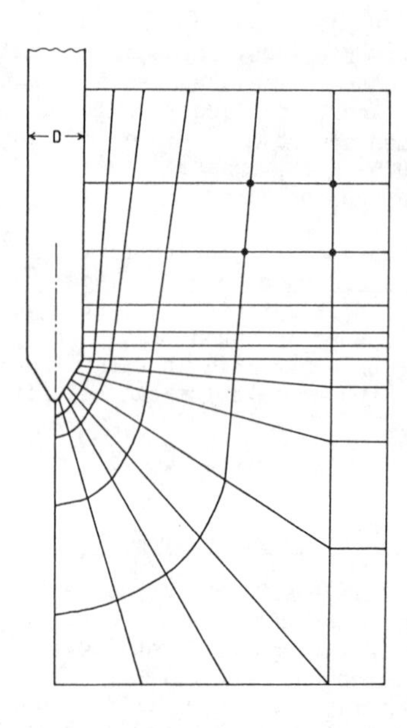

Fig. 1. Finite element discretization

Run I : $G = 50\ c_u$
$c_u = 20$ kPa
$\nu = 0.49$

Run II: $G = 300\ c_u$
$c_u = 20$ kPa
$\nu = 0.49$

The initial state of the material was assumed to be stress free.

In the Eulerian analysis carried out the material flows through the fixed mesh. At the bottom of the mesh prescribed material displacement increments of 0.5 mm were introduced. A modified Newton-Raphson iteration procedure was applied and about three iterations were necessary to reach a near-equilibrium state. The adopted convergence tolerance is 0.5 percent, i.e. the out-of-balance forces are demanded to be less than a half percent of the external forces.

The load-displacement curves computed for runs I and II are presented in figure 2. As is shown in figure 2 after a material displacement of about 2.5 - 3 times the diameter of the cone (i.e. about 100 mm) a steady state is nearly reached with respect to the reaction forces at the tip of the cone. The computed full cone factor is about 9.5 for the soft clay with $G=50c_u$, whereas this factor increases to about 12 for the stiff clay with $G=300c_u$.

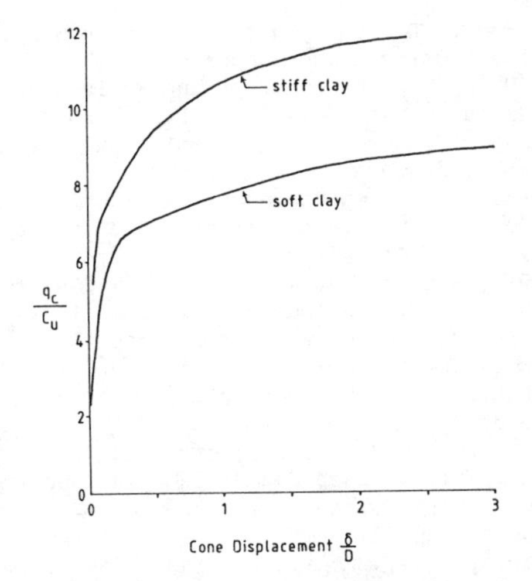

Fig. 2. Computed load-displacement curves

The results for the soft clay may be compared with previous results from a small strain analysis, as both computations were performed for $G=50c_u$. In a small strain analysis the limit load is reached after a displacement of 0.2D. For this displacement the large strain analysis shows by far not a steady state, but the curve in figure 2 does show a strong non-linearity and than a nearly constant slope. The further near-linear increase of the resistance from 0.2D to 3.0D is caused by the stiffness dependent increase of the horizontal pressure at the shaft due to the effect of cylindrical cavity expansion.

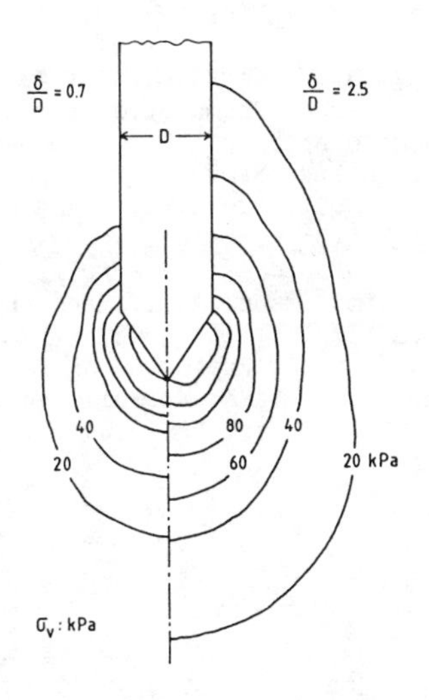

Fig. 3. Isolines for vertical stresses at a soil displacement of 0.7D(a) and 2.5D(b)

Figures 3a and 3b respectively show isolines representing the vertical stresses for a soil displacement of 0.7D and 2.5D. The stress distribution is fairly uniform indicating consistency of the results.

Contour lines for the horizontal stresses are presented in figures 4a and 4b. These figures clearly show the increase of the horizontal pressure at the shaft of the cone.

Finally figures 5a and 5b show the plastic zone and the isolines for the plastic strain at the two stages of the computation. Within the contoured zone the undrained shear strength is completely mobilized. As can be seen in these figures the plastic region moves upward along the shaft of the cone. Underneath the pointed cone the plastic strain distribution does not change anymore in a large measure between a displacement of 0.7D and 2.5D.

9 CONCLUSIONS

The results of the large strain analyses confirm the semi-theoretical equation (4) in the sens that the cone factor increases with the stiffness of the soil. According to the semi-theoretical formula the cone factor would increase from 12.5 to 14.3 when G/c_u increases from 50 to 300. A corresponding increment of about two and a half is obtained by the present finite element calculation.

However, it is unfortunate that the present computational results did not incorporate a genuine steady state performance. Indeed, the slopes of the load-displacement curves had not vanished for the applied displacement of two times the cone diameter. Therefore the

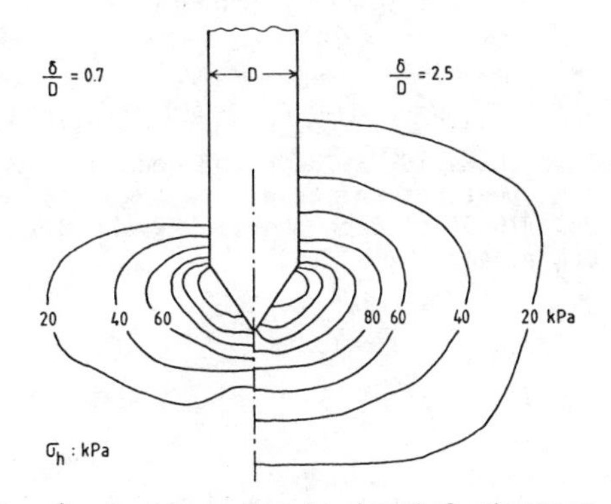

Fig. 4. Isolines for horizontal stresses at a soil displacement of 0.7D(a) and 2.5D(b)

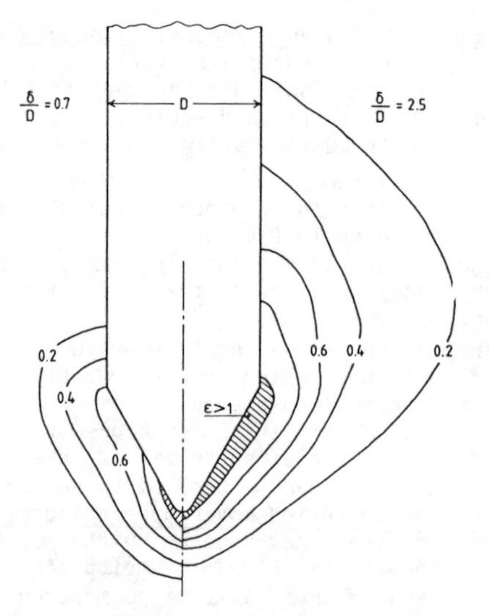

Fig. 5. Plastic region and isolines for plastic strains at a soil displacement of 0.7D(a) and 2.5D(b)

predictions of the semi-theoretical formula are not validated in an absolute sense.

The fact that a steady state solution was not yet obtained for large displacements of twice the cone diameter may have a numerical reason. It seems that the grid of four-noded elements behaves overly-flexible in the present axisymmetric problem. Here it should be realized that we have relaxed the incompressibility of the material by imposing the constant volume condition for the average deformation of the four-noded element rather than at all Gaussian integration points. This approach from Nagtegaal et al. (1974) was shown to be reliable for plane strain problems but as yet hardly applied to axisymmetric problems.

REFERENCES

Baligh, M.M. 1975. Theory of deep static cone penetration resistance. Publication No. R75-56 of Dept. Civil Engg of MIT, Cambridge.

De Borst, R. 1982. Calculation of collapse loads using higher order elements. Proc. IUTAM Symp. Deformation and Failure of Granular Materials (eds. P.A. Vermeer and H.J. Luger), pp. 503-513, Balkema, Rotterdam.

De Borst, R. and Vermeer, P.A. 1984. Possibilities and limitations of finite elements for limit analysis. Geotechnique 34, No. 2, pp. 199-210.

Gibson, R.E. and Anderson, W.F. 1961. In situ measurements of soil properties with the pressuremeter. Civ. Eng. and Public Works Review, Vol. 56, pp. 615-618.

Griffiths, D.V. 1982. Elasto-plastic analysis of deep foundations in cohesive soil. Int. J. Num. Meth. Geomech., No. 6, pp. 211-218.

Hibbit, H.D., Marcal, P.V. and Rice, J.C. 1970. A finite element formulation for large strain and large displacement. Int. J. Solids Struct. 6, pp. 1069-1086.

Huetink, H. 1982. Analysis of metal forming processes based on a combined Eulerian-Lagrangian finite element formulation. Proc. Int. Conf. Num. Meth. Industr. Forming Processes, Pineridge Press, Swansea, pp. 501-509.

Huetink, H. 1986. On the simulation of thermo-mechanical forming processes. Ph. D. Thesis, University of Enschede, The Netherlands.

Koumoto, T. and Kaku, K. 1982. Three-dimensional analysis of static cone penetration into clay. Proc. ESOPT II, 2, pp. 635-640, Amsterdam.

Luger, H.J., Lubking, P. and Nieuwenhuis, J.D. 1982. Aspects of penetrometer tests in clay. Proc. ESOPT II, 2, pp. 683-687, Amsterdam.

McMeeking, R.M. and Rice, J.C. 1975. Finite element formulation for problems of large elastic-plastic deformation. Int. J. Solids Struct., 11, pp. 611-616.

Nagtegaal, J.C., Parks, D.M. and Rice, J.C. 1974. On numerical accurate finite element solutions in the fully plastic range. Comp. Mech. Appl. Mech. Eng., 4, pp. 153-177.

Vesic, A.S. 1972. Expansion of cavities in infinite soil mass. J. Soil Mech. Found. Div. ASCE, 98, pp. 265-290.

Numerical Methods in Geomechanics (Innsbruck 1988), Swoboda (ed.)
© 1988 Balkema, Rotterdam. ISBN 90 6191 809 X

The estimation of design bending moments for retaining walls using finite elements

A.B.Fourie
University of Queensland, St. Lucia, Australia
D.M.Potts
Imperial College, London, UK

ABSTRACT: The finite element method is employed to investigate the influence of prop position and in-situ stresses on the maximum prop forces and bending moments in propped retaining walls. The soil is modelled as an elasto-plastic material with a Mohr-Coulomb yield criterion. It is shown that for retaining walls formed by excavation in soils with a high K_0 value, bending moments are grossly underpredicted by the limit equilibrium approach. Empirical factors are established which may be used with simple limit equilibrium calculations to arrive at design bending moments and prop forces.

1 INTRODUCTION

Present design methods for retaining walls are commonly based on the limit equilibrium techniques of analysis. These techniques provide no information about the magnitude and distribution of soil movements or their effect on earth pressure distributions and wall bending moments. Common practice is to design for overall stability, with the incorporation of a safety factor, which involves:-

1. The use of a simple limit equilibrium analysis in which the active and passive earth pressure distributions on the wall are assumed to be linear. Any modification to these pressures due to the restraining effect of props or anchors is ignored.
2. The inclusion of a lumped factor of safety which is intended to account for uncertainities of soil conditions, the method of stability analysis and the loading conditions as well as restricting soil movements to an acceptable level.
3. The calculation of the prop force from horizontal equilibrium. The wall depth obtained from the stability analysis is used, together with the assumed distributions of active and passive pressure and the value of F.
4. The calculation of wall bending moments from the assumed earth pressures and the calculated prop force.

The approximate method outlined above has a number of shortcomings, including:

1. Nothing is known about the distribution of earth pressures at working load conditions. To carry out the above calculations a distribution of earth pressure has to be assumed. This usually involves the modification of the classical linear distributions by the design value of F.
2. No account is taken of the initial state of stress in the ground. The same approach is prescribed for retaining walls constructed in stiff, overconsolidated clays as for those constructed in normally consolidated clay deposits.

2 PROBLEM GEOMETRY AND MATERIAL PROPERTIES

In this paper the finite element method has been used to check the validity of using the limit equilibrium method to evaluate retaining wall bending moments and prop forces. The finite element mesh chosen for this study is shown in Figure 1, and consists of 165 eight-noded isoparametric elements. The retaining wall is shown shaded and was assumed to be 20m long and 1m wide. Linear elastic properties of Young's Modulus E of 23 x 10^6 kN/m^2 and Poisson's Ratio $\mu = 0.15$ were used. The wall is thus typical of an uncracked reinforced concrete diaphragm retaining wall. The soil is assumed to be elastic-perfectly

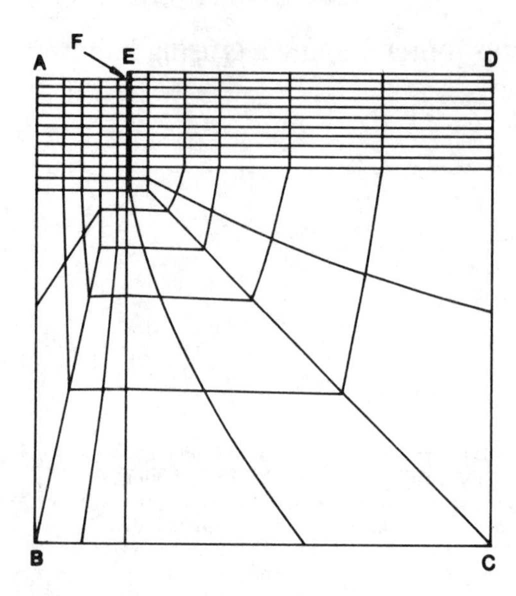

Figure 1 Finite element mesh

plastic with a Mohr Coulomb yield surface and strength parameters of c' = 0 and $\phi' = 25°$. Fully associated flow conditions were used to enable comparison with classical earth pressure solutions based on plasticity theory. Poisson's Ratio $\mu = 0.2$ and Young's Modulus $E = 6000\ Z\ (kN/m^2)$, where Z = depth below ground surface were used for all analyses.

The boundaries AB and CD were assumed free to move vertically but restrained horizontally, while boundary BC was restrained both horizontally and vertically. The geometry thus represents a 40m wide excavation with AB as the axis of symmetry. A number of analyses were carried out in which the action of a horizontal rigid prop was modelled by applying a horizontal restraint at the depth of propping. Separate analyses were carried out for prop depths of 0m, 1.5m, 3.25m, 5m, 7m and 9.26m below ground surface. Props were installed immediately the depth of excavation reached the required depth of propping. Initial stresses with $\sigma_V' = \gamma Z$, where $\gamma = 20\ kN/m^3$, and $\sigma_H' = K_o\ \sigma_V'$ were specified in the soil. K_o values of 0.5 and 2.0 were used to represent two possible extremes of in-situ stress conditions.

Excavation to depths of 13.26m and 15.26m below original ground level was simulated using a procedure outlined in some detail in a previous publication (Potts and Fourie [1]) and will not be repeated here. With increasing excavation depth it was found necessary to excavate a row of elements over a number of increments of the analysi To ensure overall equilibrium and satisfac tion of the soil constitutive law it was necessary for the analysis with $K_o = 2$ and a prop at ground level to use 75 increment to reach an excavation depth of 15.26m.

3 STABILITY ANALYSES

The results of conventional stability analyses using the limit equilibrium method with full wall friction are given in Table 1. In these analyses the earth pressures were modified by the value of F so as to ensure equilibrium. The prop forces and maximum bending moments predicted by the limit equilibrium approac are shown in Tables 2 and 3. These result indicate that lowering the position of the prop increases the magnitude of the predicted prop force for a particular dept of excavation, while the maximum bending moment value decreases.

Table 1. Dependence of factor safety Fr on depth of prop.

Depth of prop below ground level (m)	Depth of Excavation (m)		
	9.26	13.26	15.26
0.0	6.16	2.0	1.0
1.5	6.33	2.05	1.01
3.25	6.70	2.16	1.02
5.0	7.24	2.33	1.10
7.0	8.32	2.65	1.25
9.26	11.39	3.44	1.60

4 RESULTS OF FINITE ELEMENT ANALYSES

4.1 Retaining wall displacements and bending moments.

The horizontal wall displacements predicted by the finite element method for various prop positions are given in Figure 2 for an excavation depth of 15.26m and a K_o value of 2.0. (The predicted deflections for the prop at ground surface are not shown since an excavation depth of 15.26m corresponds to the limiting conditions for this case). For the prop located at, or above, 5m depth the displacement mode is essentially one of rotation about the prop, with large deflections into the excavation below the prop position. With the prop located further down the wall, movement is

Table 2. Variation of prop force (kN) with depth of prop below ground surface

Depth of Excavation (m)	K_o = 2.0		K_o = 0.5		LIMIT EQUILIBRIUM	
	13.26	15.26	13.26	15.26	13.26	15.26
Depth of prop below ground surface (m)						
0	1013	938	237	330	320	360
1.5	1267	1188	245	352	350	387
3.25	1185	1533	289	404	392	431
5.0	1874	1883	347	472	446	488
7.0	2045	2184	420	562	529	573
9.26	1953	2299	465	658	670	715

Table 3. Variation of maximum bending moment (kNm) with depth of prop below ground surface.

Depth of Excavation (m)	K_o = 2.0		K_o = 0.5		LIMIT EQUILIBRIUM	
	13.26	15.26	13.26	15.26	13.26	15.26
Depth of prop below ground surface (m)						
0	4400	4770	2220	2030	2135	2510
1.5	4100	4600	1124	1942	1903	2248
3.25	3450	4050	915	1565	1610	1924
5.0	2550	2900	672	1241	1270	1565
7.0	1400	-1700	-346	766	818	1086
9.26	-2050	-2900	-709	-590	260	483

more akin to that of a cantilever, with the maximum horizontal movement occurring at the top of the wall for a prop located at 9.26m depth. It should be noted however that this displacement has in fact reduced from a maximum value which occurred at the time the prop was installed.

The corresponding predicted bending moment distributions are shown in Figure 3. The effect of lowering the prop position is to steadily reduce the maximum value from that which occurs for a prop at the top of the wall, until eventually the sign of the maximum bending moment actually changes. Although the limit equilibrium predictions, as shown in Table 3, confirm the trend of a decreasing moment with increasing depth of prop position, the method does not predict a change in sign of the maximum bending moment, which is due to the unrealistic assumptions of earth pressure distribution inherent in the method.

4.2 Earth pressure distributions.

Horizontal effective stresses in front of and behind the wall at an excavation depth of 15.26m are given in Figure 4. Results for only three prop positions are included to avoid cluttering the figure.

Although an excavation depth of 15.26m corresponds to a low factor of safety for all prop positions considered (not greater than 1.6), the pressure behind the wall does not reach the active value at any point on the wall. In fact, for two of the prop positions shown the earth pressure actually exceeds the K_o value. On the active side of the retaining wall a bulge of earth pressure occurs for all prop locations, usually at a small depth below the prop level. This pressure bulge results in prop forces markedly in

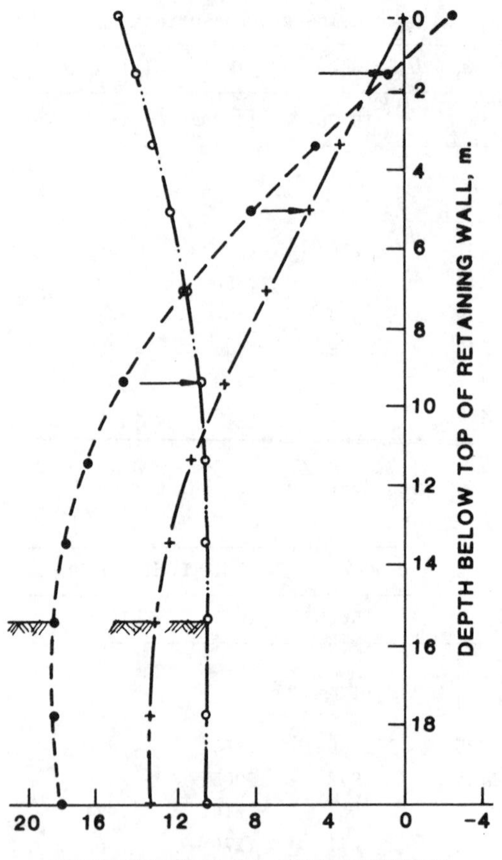

Figure 2. Horizontal wall displacements for prop depths of 1.5m, 5.0m and 9.26m. Excavation depth = 15.26m.

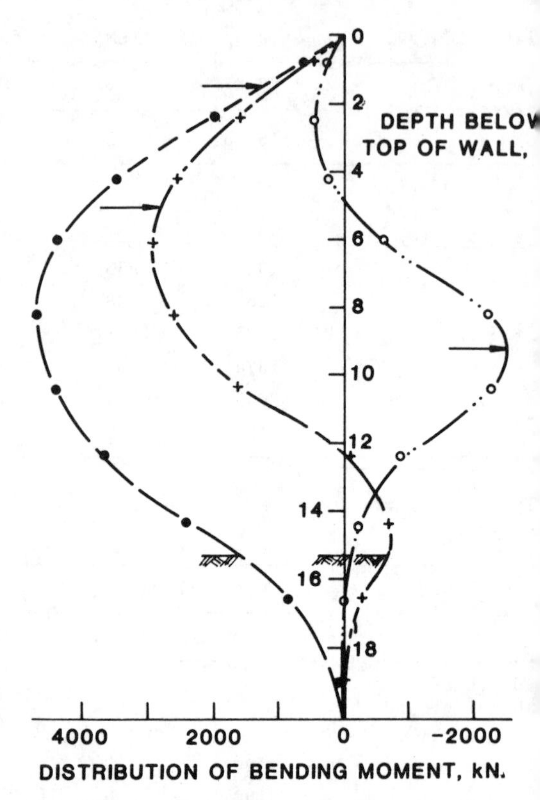

Figure 3. Bending moment distribution for prop depths of 1.5m, 5.0m and 9.26m, and excavation depth = 15.26m.

excess of those predicted by the limit equilibrium method, as discussed later. In front of the retaining wall the earth pressure exceeds the passive value over the entire embedded depth for all prop levels. This behaviour is consistent with results presented previously by the authors (Potts and Fourie, [2]) for a rigid retaining wall rotating about the top of the wall. In this study it was also found that on the passive side earth pressures significantly in excess of the classical value occurred as a factor of safety of unity was approached. Similar results were obtained experimentally by Bros ([3]) for laboratory tests on small scale retaining walls.

The effect of varying the prop location may be better appreciated by tracing the development of earth pressure distribution for one prop location as excavation proceeds to the final depth of 15.26m. The earth pressure distributions shown in Figure 5 are for a prop position of 7m below ground level. The corresponding wall displacements are shown in the inset to Figure 5. Until installation of the prop the wall displaces as a cantilever, and as shown in Figure 5(a) the pressure drops to near active conditions above the excavation level. Below the excavation level passive conditions are mobilised over almost half of the embedded depth in front of the wall. As excavation then proceeds below the prop position the wall can be seen to rotate about the prop in a clockwise direction. Above the prop the wall displaces into the retained soil, and vice versa below the prop. The earth pressure behind the wall undergoes corresponding changes. Above the prop the pressure increases, and even exceeds the original in-situ value for deep excavation levels, whereas below the prop the pressure continues to drop although

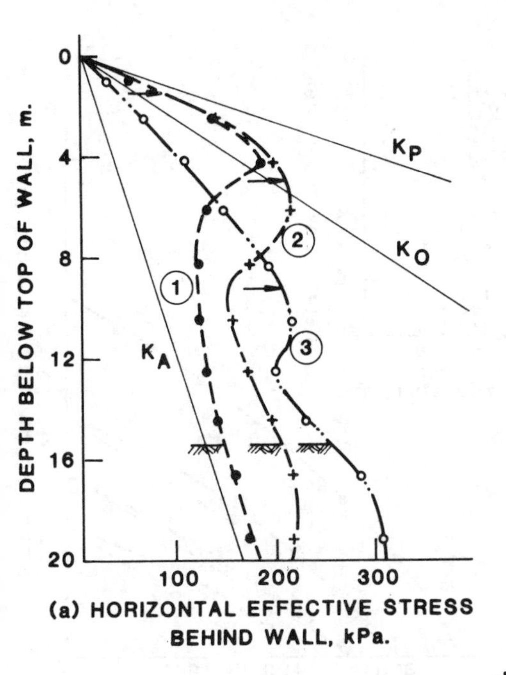

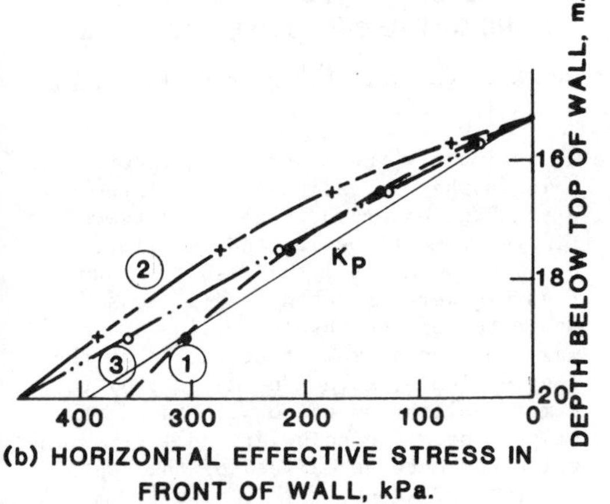

Figure 4. Earth pressure distributions for prop depths of 1.5m, 5.0m and 9.26m. Excavation depth = 15.26m.

fully active conditions are not achieved anywhere. The earth pressure diagram corresponding to the final excavation depth can now be seen to have a bulge at approximately the prop location.

5 COMPARISON OF FINITE ELEMENTS AND LIMIT EQUILIBRIUM

5.1 Prop forces.

The predicted prop force is shown in Table 2 for all the prop positions considered, and for excavation depths of 13.26m and 15.26m. Results are included for K_0 values of 0.5 and 2.0, together with the limit equilibrium values. For a given depth of excavation, all the analyses predict an increasing prop force with increasing depth of prop position. The limit equilibrium predictions agree exceptionally well with the finite element results for low in-situ stresses (K_0 = 0.5), particularly for an excavation depth of 15.26m, which corresponds to a factor of safety $F_r \leq 1.6$. The limit equilibrium value can be seen to always err on the conservative side, which is of course desirable. If the retaining wall is constructed in a high K_0 soil, however, the magnitude of the prop force is grossly underestimated. Referring to the results for an excavation depth of 13.26m, which corresponds to an F_r value of 2 - 3 (i.e. the "working stress" range), the limit equilibrium method underestimates the finite element value by a factor of approximately three.

5.2 Retaining wall bending moments.

The predicted maximum bending moment values are summarised in Table 3. In this case the results indicate a decrease in maximum value with increasing depth of prop position. A significant difference between the finite element and limit equilibrium results does however occur.

As the depth of prop position increases beyond 7m below ground level the finite element analyses predict a change in sign in the maximum bending moment, whereas no change in sign is predicted by the limit equilibrium method. This is once again attributable to the unrealistic assumptions of earth pressure distribution inherent in the limit equilibrium method, which neglects pressure concentrations at positions of fixity. The limit equilibrium method provides a conservative estimate of maximum moment for retaining walls in low K_0 soils, except when the depth of prop location increases beyond about 7m below ground surface. At working stress conditions the limit equilibrium method again provides an underestimate of the value for a retaining wall in a high K_0 soil. A difference of approximately 2 results, although for a prop at 9.26m depth the two methods predict bending moments of different sign.

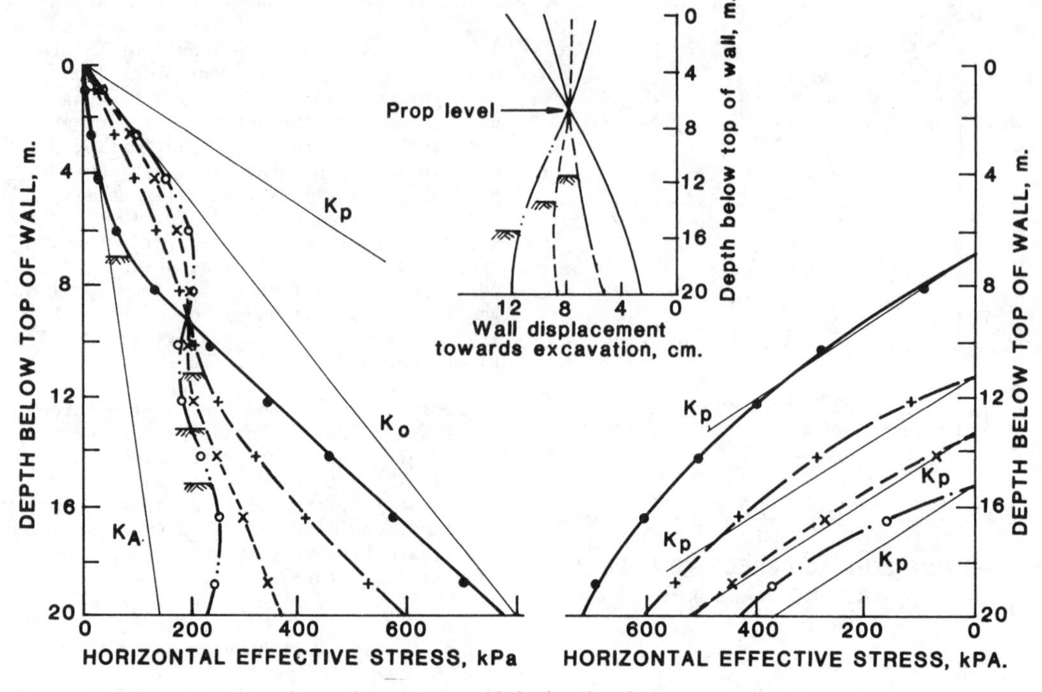

Figure 5. Development of earth pressure with depth of excavation. Prop depth = 7.0m.

6 PRACTICAL IMPLICATIONS

In certain field applications, e.g. deep basements, it may be necessary to provide some form of horizontal restraint to a retaining wall at some depth below the original ground surface. In the analyses described in the preceding sections the effect on prop force and wall bending moment of changing the depth of a single rigid prop was investigated. In general it was found that the conventional limit equilibrium method provided conservative and realistic estimates of the maximum prop force and bending moment for a retaining wall formed by the diaphragm wall or secant pile technique in a soil with low in-situ horizontal stresses. The finite element results for a wall formed in a high K_o soil such as an overconsolidated clay, however, markedly exceeded the limit equilibrium values. Although the data presented in this paper is not sufficient to draw definitive conclusions, it appears that for single-propped retaining walls installed in soils where K_o is large, it may be necessary to increase the prop force and maximum bending moment calculated from the limit equilibrium method by factors of approximately three and two respectively.

To further investigate the existence of a constant of proportionality between the finite element results and the limit equilibrium method, the results shown in Figure 6 were produced. This shows the variation of the absolute value of the maximum bending moment predicted by the finite element method normalised by the limit equilibrium value with the depth of prop position below ground level. This value has been normalised by the depth of excavation, which in this case is 13.26m ($F_r \geq 2.0$). It can be seen that up to a prop depth of approximately one-third the excavation depth a reasonably constant ratio between the finite element and limit equilibrium results exists. It is particularly notable that the constant of proportionality is remarkably similar to the value of K_o used in the analysis. At prop depths in excess of one-third the excavation depth, the ratio between the maximum bending moments varies erratically. This occurs because for deep props the limit equilibrium method assumes unrealistic earth pressure distributions.

The results presented in this paper indicate that for rigid retaining walls installed in high K_o soils, bending moments calculated using the limit

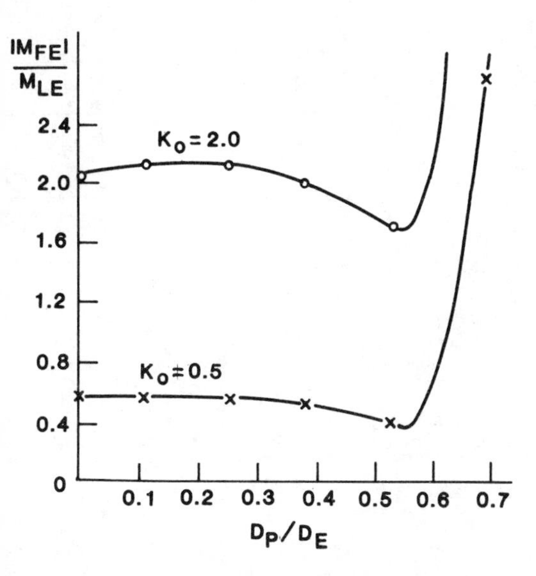

Figure 6. Variation of bending moment ratio with depth of excavation.

equilibrium method may severely underestimate the field value. To further investigate the validity of the approach used in this paper the authors studied the effects of wall flexibility on the behaviour of a propped embedded retaining wall (Potts and Fourie [4]). It was found that for flexible walls of the sheet pile type the predicted bending moments were in fact much lower than the limit equilibrium value. This result is in complete agreement with the model tests of Rowe ([5]), and was found to occur irrespective of the value of K_o.

7 CONCLUSIONS

The results presented in this paper indicate that the limit equilibrium method should be applied with caution when attempting to predict bending moments and prop forces for retaining walls installed in high K_o soils. The indications are that, irrespective of the depth of installation of a rigid prop, the limit equilibrium method severely underestimates the above quantities. A correction factor has been postulated which may be applied to bending moments calculated using the conventional approach, and this factor has been tentatively linked to the in-situ stress ratio K_o. This relationship is further corroborated by the finite element results for retaining walls installed in low K_o soils which indicate the limit equilibrium method overestimates bending moments and prop forces by a factor of two. These results therefore imply that the value of the initial stress ratio K_o cannot be ignored in the design of propped retaining walls. In particular, retaining walls installed in heavily overconsolidated soils may be significantly underdesigned if the importance of the K_o value is ignored.

REFERENCES

[1] D.M. Potts and A.B. Fourie. The behaviour of a propped retaining wall: results of a numerical experiment. Geotechnique 34 No. 3, 383-404 (1984).

[2] D.M. Potts and A.B. Fourie. A numerical study of the effects of wall deformation on earth pressures. Int. Jnl. for Num and Anal. Meth. in Geomech. 10. No. 3, 383-404 (1986).

[3] B. Bros. The influence of model retaining wall displacements on active and passive earth pressure in sand. Proc. 5th European Conf. on Soil Mech. and Foun. Eng., Madrid, 1, 241-249 (1972).

[4] D.M. Potts and A.B. Fourie. The effect of wall stiffness on the behaviour of a propped retaining wall. Geotechnique 35, No. 3, 347-352. 1985.

[5] P.W. Rowe. Anchored sheet-pile walls. Proc. Inst. Civ. Engrs. Part 1.1, 27-70. 1952.

Numerical Methods in Geomechanics (Innsbruck 1988), Swoboda (ed.)
© 1988 Balkema, Rotterdam. ISBN 90 6191 809 X

Finite element analysis of downdrag on piles

J.C.Small
School of Civil and Mining Engineering, University of Sydney, Australia

ABSTRACT: A method of analysis is presented which enables the downdrag forces acting on a single compressible pile to be computed. The tip of the pile is assumed to be resting on a perfectly rigid base, while the surrounding soil is treated as a poroelastic material which is consolidating under a surface loading. Slip is allowed between the pile and the soil according to an effective stress law.

Results are presented for the distribution of shear stresses with time along the pile shaft and for the shortening of the pile with time.

1. INTRODUCTION

Downdrag on piles occurs when the soil surrounding the pile is able to move downwards relative to the pile, thus causing a downward force through the skin friction developed between the pile and the soil. This situation often occurs when a pile is driven through soft material to rock or to a firmer bearing stratum, and subsequently embankments or other earthworks are constructed on the surface of the soft layer or if the soil consolidates due to a drawdown of the water table.

The forces developed during downdrag will cause additional settlement of the pile, partly due to compression of the pile shaft and partly due to penetration of the pile tip into the bearing stratum. They may even be large enough to cause failure of the pile through crushing. It is therefore desirable to be able to predict the extra compression of the pile as this is an important factor in the pile design.

Approximate methods of calculating the final downdrag force due to negative friction have been presented by Terzaghi and Peck (1948) and Zeevaert (1959). A summary of these methods is given by Locher (1965) who also quotes an empirical expression for the downdrag force given by Elmasry (1963). A theoretical analysis based on elastic theory has been presented by Salas and Belzunce (1965) however these authors took no account of pile compressibility, and assumed that the soil had the properties of a Boussinesq half space. Poulos and Mattes (1969) presented a method of computing the magnitude and distribution of downdrag forces on a pile which had a finite stiffness and which was driven to a firm base. Slip was allowed between the pile and the soil.

The time-dependent behaviour of downdrag forces was examined by Poulos and Davis (1972) who used simple one-dimensional consolidation theory to evaluate the excess pore water pressures. The pore water pressures were then used to compute the effective stresses acting along the pile shaft, and these were used to determine if slip had occurred between the pile and the surrounding soil. Pile crushing and the time at which the pile was installed after the placing of a surcharge were also considered. Poulos and Davis (1975) have compared the results of their analysis with measured values with reasonably close agreement being obtained. Walker and Darvall (1973) mention a finite element method of obtaining the downdrag forces on piles, in which a settlement profile is calculated at some distance away from the pile and then an iterative approach is used to compute a shear stress distribution throughout the soil mass.

In this paper a finite element method is used to compute the downdrag forces which develop as pore water pressures caused by the surface surcharge dissipate. The consolidating soil is treated as a poroelastic material which behaves according to Biot's (1941) theory. The pile is treated as being elastic and slip between the pile and soil is assumed to take place according to an effective stress law.

2. PROBLEM DEFINITION

Suppose that we consider the single circular elastic pile shown in Figure 1 which has a diameter d and length L. Suppose also that the pile is surrounded by an elastic layer of soil which is carrying a uniform surcharge q on its upper surface z = 0. The soil is assumed to be fully saturated and so consolidation of the clay layer under the applied load is governed by Biot's (1941) equations. While the soil layer is to be treated as homogeneous in this paper, it may be anisotropic with respect to elastic parameters and permeability, and such features are easily incorporated into finite element codes.

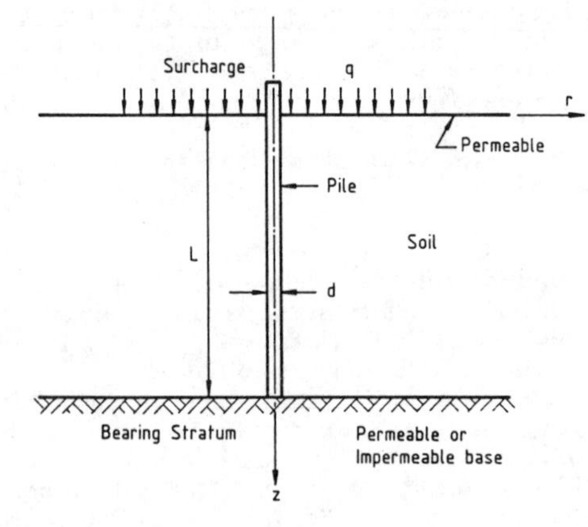

Fig. 1 End bearing pile in a compressible soil subject to a surcharge.

2.1 Finite Element Equations - Soil

If the soil is treated as a saturated porous elastic medium which behaves according to Biot's (1941) theory then it is well known (Sandhu and Wilson (1969)) that we may write the finite element equations which govern the consolidation process as follows:

$$\begin{bmatrix} K_E & -L^T \\ -L & -(1-\alpha)\Delta t\Phi \end{bmatrix} \begin{bmatrix} \Delta\delta \\ \Delta q \end{bmatrix} = \begin{bmatrix} \Delta m \\ \Phi\Delta t\, q_t \end{bmatrix} \tag{1}$$

where

K_E is the stiffness matrix for the soil
Φ is the flow matrix
L is the coupling matrix
δ is the vector of nodal deflections
q_t is the vector of nodal excess pore pressures at time t
t is time
m is the force vector
Δ signifies a change in a quantity between a time t and a time $t + \Delta t$
α is a parameter which must lie between 0 and 0.5 if unconditional stability is to be maintained (see Booker and Small (1975)).

By a forward marching process the solution at a time $t + \Delta t$ may be found from the solution at time t. This requires setting up the right hand side of equations (1) at each time step and solving. To initiate the process we may assume that the pore pressures are zero (i.e. $q_t = 0$). We may then compute the displacements and pore pressures at time $t + \Delta t$ from

$$\begin{aligned} \delta_{t+\Delta t} &= \delta_t + \Delta\delta \\ q_{t+\Delta t} &= q_t + \Delta q \end{aligned} \tag{2}$$

2.2 Finite Element Equations - Pile

For the pile which is considered to be elastic we may simply write a finite element relationship:

$$K_p \; \Delta\delta_p = \Delta f \tag{3}$$

where
K_p is the stiffness matrix for the pile
δ_p is the vector of nodal deflections for the pile
f is the force vector for the pile.

3. INTERACTION ANALYSIS

We can consider the pile and the soil separately as shown in Figure (2). The pile will be acted upon by interface forces due to skin friction which may be considered to act at the nodes of the finite element mesh. These are shown as forces N_i, T_i in Figure (2).

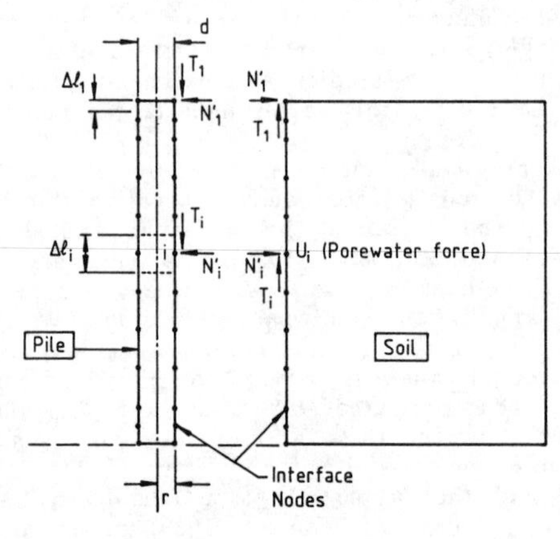

Fig. 2 Nodal forces at pile-soil interface.

The soil will be acted upon by the surface surcharge q as well as the forces N_i, T_i which will be equal and opposite to the forces on the pile.

In writing the interaction equations it is assumed that the displacement of the pile and soil due to any load applied to the top of the pile have taken place and that the forces shown in Figure (2) are all increases due to the application of the surcharge q. Any deflections computed will therefore be additional displacements caused by the downdrag on the pile.

We may rewrite equation (1) in the following way

$$\begin{bmatrix} K_E & -L^T \\ -L & -(1-\alpha)\Delta t\Phi \end{bmatrix} \begin{bmatrix} \Delta\delta \\ \Delta q \end{bmatrix} = \begin{bmatrix} \Delta f_p \\ 0 \end{bmatrix} + \begin{bmatrix} \Delta f_Q \\ 0 \end{bmatrix} + \begin{bmatrix} 0 \\ b \end{bmatrix} \qquad (4)$$

where

Δf_p is that part of the force vector due to forces at the interface nodes.

Δf_Q is that part of the force vector due to other forces applied to the soil (in this case the surcharge).

$b = \Phi\Delta t q_t$ is the vector which arises from the pore pressures existing at the previous time step.

Because the above system of equations (Equations 4) is linear we may write the solution as:

$$\Delta S = \Delta S_p + \Delta S_Q + \Delta S_b \qquad (5)$$

where $\Delta S = (\Delta\delta, \Delta q)^T$

and ΔS_p, ΔS_Q, ΔS_b are the solutions of equations (4) with the right hand side equal to $(\Delta f_p, 0)^T$, $(\Delta f_Q, 0)^T$ and $(0, b)^T$ respectively.

At any time step the vectors Δf_Q and b will be known, however the value of the forces acting at the interface nodes will not, and hence the vector Δf_p is not known. To determine the interface forces, we must make use of the fact that before slip at the pile-soil interface occurs, the deflection of the pile and the soil will be the same and that after slip, the shear forces T_i and the normal forces N_i are related by the failure criterion for the interface.

We can write the deflections ΔS_p (for the interface nodes only) as follows:

$$[I_{soil}]\, \Delta f_p^I = \Delta S_p^I \qquad (6)$$

where

$[I_{soil}]$ is a 3n x 2n influence matrix (n is the number of interface nodes), the columns of which are made up of solutions for the deflections and pore pressures at interface nodes due to unit loads being applied to each interface node of the soil in turn (in the horizontal and the vertical directions) and the superscript 'I' denotes that the vector refers to interface nodes only.

Hence we may write equation 5 (for the interface nodes only) as:

$$\Delta S^I = \Delta S_Q^I + \Delta S_b^I + [I_{soil}]\Delta f_p^I \qquad (7)$$

In a similar fashion we may write the deflections in the pile in terms of an influence matrix

$$[I_{pile}] \cdot -\Delta f_p^I = \Delta\delta_p^I \qquad (8)$$

where

$[I_{pile}]$ is a 2n x 2n influence matrix for the pile the columns of which are composed of the deflections at the interface nodes due to unit loads placed at each of the interface nodes in turn (both vertically and horizontally).

If we have conditions of no slip along the pile-soil interface the deflections computed for the soil and the deflections computed for the pile must be equal. Hence we may substitute Equation (8) into Equation (7) to obtain the final set of equations to be solved:

$$\begin{bmatrix} I_{soil} + I_{pile} & 0 \\ & -I \end{bmatrix} \begin{bmatrix} \Delta f_p^I \\ \Delta q^I \end{bmatrix} = -\Delta S_Q^I - \Delta S_b^I \qquad (9)$$

In the above equation 'I' is an n x n unit matrix and the addition of the influence matrices for the soil and pile is such that the elements of the 2n x 2n matrix for the pile are added to those of the 3n x 2n matrix for the soil in the appropriate positions (i.e. those corresponding to the displacements).

A solution of Equation (9) will yield the change in interface forces Δf_p^I and the change in pore pressures at the interface nodes Δq^I at any particular timestep.

4. CONDITIONS FOR SLIP

When the shear stress between the pile and

the soil becomes large enough slip will occur. The conditions for slip may be written according to the Mohr-Coulomb criterion in terms of effective stresses. For slip we must have

$$\tau = \sigma'_n \tan \delta + c'_a \quad (10)$$

where

τ is shear stress
σ'_n is the normal effective stress
c'_a is the adhesion between the pile and the soil
δ is the angle of friction between pile and soil.

If each side of Equation (10) is multiplied by the area of the annulus which is considered to contribute to the forces at a node, see Figure (2) we may write for node i:

$$T_i = (N_i + N'_{oi} - U_i) \tan\delta + C'_a \quad (11)$$

where

$C'_a = c'_a \times 2\pi r \Delta \ell_i$ is the cohesive force
$U_i = p_i \times 2\pi r \Delta \ell_i$ is the water force
N_i is the total force change from initial conditions.
N'_{oi} is the initial effective force due to the self weight of the soil and perhaps pile installation.

The above method may also be used for layered soils where material properties change from layer to layer, however for simplicity a uniform soil is assumed.

Once slip has taken place the relationship of Equation (11) must still hold. In terms of incremental quantities we may write:

$$\Delta T_i = (\Delta N_i - \Delta U_i) \tan\delta \quad (12)$$

Hence when a node has been detected to have slipped we need to modify equations (9) by inserting an equation like (12) in place of the equation for compatibility of displacement. For example if node 1 has slipped we replace the equation for vertical displacement compatibility of node 1 with

$$\Delta T_1 = (\Delta N_1 - \Delta p_1 . 2\pi r \Delta \ell_1) \tan \delta \quad (13)$$

One difficulty which does arise during the first slip at a node is that the nodal forces may go outside the failure criterion. This requires correction of the forces back to the yield surface which was done here by correcting ΔT keeping the increments of normal force ΔN and pore pressure force ΔU constant.

Once the forces at the interface node are bought onto the failure line they will remain there as all subsequent increments of T, N, U are computed from Equation 12.

5. EXAMPLES

As an example of the use of the proposed method, a solution was obtained to the problem of downdrag occurring on a pile which has a length to diameter ratio of L/d = 25. The finite element meshes used for the pile and the soil were such that 13 interface nodes were used. The surcharge was a uniform load of magnitude q applied to the entire upper surface of the mesh. Drainage was allowed only at the surface of the layer with the pile and the base of the layer being treated as impermeable.

Parameters chosen for the analysis were:

$K = E_p/E_s = 1000$ the ratio of the modulus of the pile to that of the soil.
$\nu'_s = 1/3$ the Poisson's ratio of the soil
$\nu'_p = 0.1$ the Poisson's ratio of the pile

Extra parameters need to be defined which control the slip which takes place according to the effective stress law of equation 10. The parameters used were:

$K'_s = 0.5$ $\qquad \gamma' L/q = 0.5$

$c'_a/q = 0.5$ $\qquad \delta = 28.7°$

K'_s is the coefficient of earth pressure which is used to determine the initial stresses normal to the pile
γ' is the submerged unit weight of the soil (assuming that the water table is at the surface).

Hence initial normal stresses at depth z are given by $\sigma'_n = K'_s \gamma' z$.

A plot of the shear stress computed to be acting along the shaft of the pile at various times is shown in Figure 3. The section of pile shaft over which slip has occurred is marked with an asterisk in the figure and it may be seen that as the time factor increases, the section of shaft which has reached this condition increases in length with the slip moving from the top of the pile to the bottom.

The finite element results are compared with a solution obtained by using the method of Poulos and Davis (1972). Although the plots are of similar shape in general, there are some marked differences. The finite element results show a much smaller shear stress at the top of the pile shaft, this being due to the low normal stresses caused by the way the soil deforms

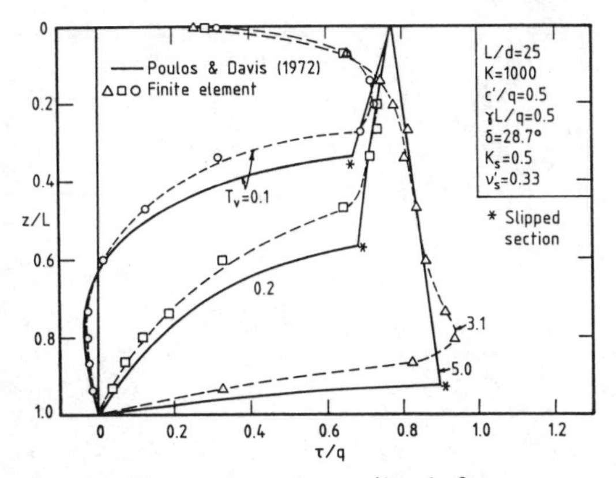

Fig. 3 Shear stress along pile shaft.

in this region. As well, the length of shaft along which slip has occurred is less at any time for the finite element solution.

Figure 4 shows the vertical displacement of the pile head with time which is caused by the shear stresses along the shaft. For the case where $\nu'_S = 1/3$ it may be seen that the solutions predicted by the finite element analysis and those predicted by the theory of Poulos and Davis (1972) are in reasonably close agreement even though the calculated distributions of shear stress differ as shown in Figure 3. However, if the Poisson's ratio of the soil is taken to be $\nu'_S = 0$, then the finite element and the Poulos and Davis solutions do not agree.

This is due to the fact that the normal stresses acting on the pile shaft (which control the maximum shear stress and slip) are computed in different ways in the two methods. For the finite element solution, the effective stress normal to the pile shaft σ'_n, is initially computed from the overburden stress (i.e. $\sigma'_n = K'_S \gamma' z$), but during consolidation of the soil under the surcharge, the normal stress changes depend upon the

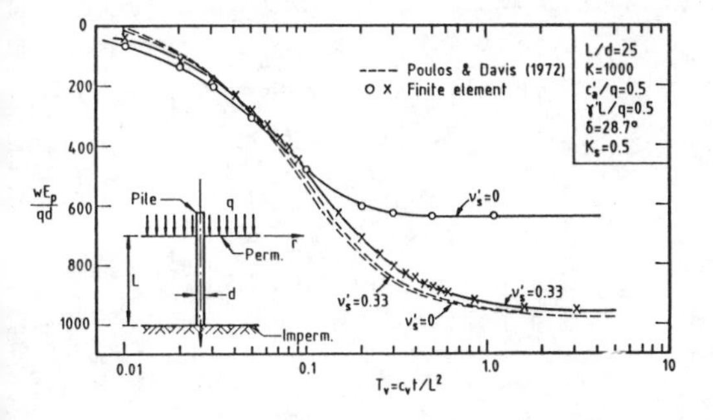

Fig. 4 Vertical deflection of pile head with time.

Poisson's ratio of the soil. Poulos and Davis make the assumption that the normal stress is always equal to $K'_S \sigma'_V$ and therefore their solution is not particularly sensitive to the Poisson's ratio chosen for the soil. Theoretically $K'_S = \nu'_S/(1 - \nu'_S)$ and so by choice of the appropriate Poisson's ratio any K'_S value may be simulated.

6. REFERENCES

Biot, M.A. (1941) "General Theory of Three Dimensional Consolidation", Jl. Appl. Phys., Vol. 12, pp. 155-164.

Booker, J.R. and Small, J.C. (1975) "An Investigation of the Stability of Numerical Solutions of Biot's Equations of Consolidations", Int. Jl. Solids Structs., Vol. 11, pp. 907-917.

Elmasry, M.A. (1963) "The Negative Skin Friction of Bearing Piles". Thesis Presented to the Swiss Federal Institute of Technology, Zurich.

Locher, H.G. (1965) "Combined Cast-in-Place and Precast Piles for the Reduction of Negative Friction Caused by Embankment Fill", Proc. 6th Int. Conf. Soil Mech. Found. Eng., Montreal, Vol. 2, pp. 290-294.

Poulos, H.G. and Davis, E.H. (1972) "The Development of Negative Friction with Time in End-Bearing Piles", Aust. Geomechs Jl., Vol. 62, No. 1, pp. 11-20.

Poulos, H.G. and Davis, E.H. (1975) "Prediction of Downdrag Forces in End-Bearing Piles", Jl. Geotech. Eng. Div., ASCE, Vol. 101, GT2, pp. 189-204.

Poulos, H.G. and Mattes, N.S. (1969) "The Analysis of Downdrag in End Bearing Piles", Proc. 7th Int. Conf. Soil Mech. Found. Eng., Mexico City, Vol. 2, pp. 203-209.

Salas, J.A.J. and Belzunce, J.A. (1965) "Resolution Theoretique de la Distribution des Forces dans des Pieux", Proc. 6th Int. Conf. Soil Mech. Found. Eng., Montreal, Vol. 2, pp. 309-313.

Sandhu, R.S. and Wilson, E.L. (1969) "Finite Element Analysis of Seepage in Elastic Media", Jl. Eng. Mech. Div., ASCE, Vol. 95, EM3, pp. 641-652.

Terzaghi, K. and Peck, R.B. (1948) "Soil Mechanics in Engineering Practice", New York, John Wiley and Sons.

Walker, L.K. and Darvall, P. Le P. (1973) "Downdrag on Coated and Uncoated Piles", Proc. 8th Int. Conf. Soil Mech. Found. Eng., Moscow, Vol. 2, pp. 257-262.

Zeevaert, L. (1959) "Reduction of Point Bearing Capacity of Piles because of the Negative Friction", First Pan-American Conference on Soil Mechanics and Foundation Engineering, Mexico, Vol. 3, pp. 1145-1152.

Numerical Methods in Geomechanics (Innsbruck 1988), Swoboda (ed.)
© 1988 Balkema, Rotterdam. ISBN 90 6191 809 X

Finite element analysis of deep foundations subjected to uplift loading

A.P.Ruffier
CEPEL, Electrical Energy Research Centre, Rio de Janeiro, Brazil
C.F.Mahler
EE & COPPE/UFRJ, Engineering School and Coordination of Postgraduate Programs in Engineering, Federal University of Rio de Janeiro, Brazil

ABSTRACT: This work deals with finite element simulation of the uplift of deep foundations. The program uses joint elements for the soil-structure interface and two-dimensional and axisymmetric elements with four nodes for the soil and the foundation. Several full scale tests performed in tropical residual soil near Rio de Janeiro are simulated. Both soil non-linearity and plastification are taken into account. The results obtained suggest that the proposed procedure can be used for the prediction of pull-out resistance of similar foundations.

1 INTRODUCTION

The study of foundations subjected to pull out forces has developed very much in the last years in Brazil. The necessity of self-supportting and guyed transmission towers, with greater dimensions, is basically a product of a possible future growing demand of electrical energy in the great urban centres. As this power, in Brazil, comes from far locations, as the Itaipu dam for example, a secure and economical design for each tower will avoid extra costs.

The usual procedures for estimating pull out resistance of deep foundation was developed using examples of sedimentary soils. These methods, which are usually used in design, only predict the failure load of the foundations. However, in Brazil, the foundation of the great majority of transmission towers are on residual soils,with quite different characteristics from sedimentary soils. Experience has shown that there is lack of knowledge about uplift loading in residual soils. In this paper, there is an attempt to advance a little more in the understanding of this problem.

Several cases of full-scale tests performed [1] in a residual soil near Rio de Janeiro are examined. It is the objective of this paper to determine the failure load taking into account the whole soil-structure system and the non-linear behaviour and plasticity of the soil. A finite element program was developed.

This program uses an incremental-iterative formulation for simulated pull-out load tests. Joint elements were used for the soil-structure interaction and axisymmetric elements for the soil and the pier foundation. The comparison of the results obtained with field measurements are presented.

2 METHOD OF ANALYSIS

2.1 Incremental-iterative approach

Consider a solid body surrounded by a non linear soil mass. The model idealized to represent this system consists of the soil, the structure and the interface between soil and structure.

When an axial load increment is applied to the structure the tendency of separation between soil and structure is observed. Joint elements were adopted as an alternative approach in order to accommodate the relative displacement involved.

Based on the fundamental energy equation and following the usual finite element procedure one arrives at the well known finite element equation:

$$\underset{\sim}{K}\underset{\sim}{u} - \underset{\sim}{R} = \underset{\sim}{0} \quad (1)$$

which represents the equilibrium equations for static boundary conditions where:

$\underset{\sim}{K}$ - the stiffness matrix,
$\underset{\sim}{u}$ - the displacement vector,
$\underset{\sim}{R}$ - the nodal forces vector.

The non-linear material effect is incorporated by adopting the incremental-iterative Newton-Raphson procedure. For each increment equation (1) may be transformed to:

$$^{m}\underset{\sim}{K}\ ^{m-1}\underset{\sim}{u} - {}^{m}\underset{\sim}{F} = \underset{\sim}{0} \qquad (2)$$

where:

$^{m}\underset{\sim}{K}$ - tangent stiffness increment,

$^{m}\underset{\sim}{u}$ - displacement increment vector,

$^{m}\underset{\sim}{F}$ - out-of-balance forces vector,

m - represents the m increment or general iteration.

The initial stress state necessary to initialize $^{m}\underset{\sim}{K}$ is evaluated using the "gravity-turn-on" process where the calculated displacements are neglected. As the pier foundation is surrounded by a cohesive over-consolidated soil, the escavation effect is not taken into account. Also the simulation of the pier foundation concreting is not considered.

Re-writing equation (2) in a more convenient form leads to:

$$^{m}\underset{\sim}{K}^{i-1}\ ^{m}\underset{\sim}{u}^{i-1} - {}^{m}\underset{\sim}{R} + {}^{m}\underset{\sim}{F}^{i-1} = \underset{\sim}{0} \qquad (3)$$

where:

m - m increment,
i - i iteration,
$^{m}\underset{\sim}{R}$ - m external load increment,

$^{m}\underset{\sim}{F}^{i-1}$ - i-1 balanced load for the m-increment

$$^{m}\underset{\sim}{F}^{i-1} = -\int \underset{\sim}{B}^{T}\ \Delta\underset{\sim}{\sigma}\ d\ \text{volume} \qquad (4)$$

and

$\underset{\sim}{B}^{T}$ - initial linear strain-displacement transformation matrix,
$\Delta\underset{\sim}{\sigma}$ - incremental stresses vector.

The difference $^{m}\underset{\sim}{R} - {}^{m}\underset{\sim}{F}^{i-1}$ represents the out-of-balance forces for the iteration i-1 at increment m.

2.2 Convergence criterion

The convergence criterion is defined by the following equation:

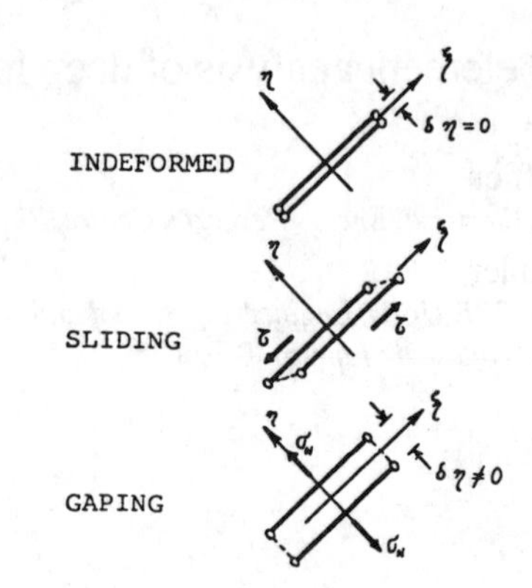

Figure 1. Joint element

$$\frac{1}{n}\left\| \frac{^{m}u^{i} - {}^{m}u^{i-1}}{^{m}u^{i}} \right\|_{2} < \varepsilon_{d}$$

where:

n - number of degrees of freedom,
ε_d - represent the tolerance and is defined by the program,
$^{m}\underset{\sim}{u}^{i}$ - is a already defined in this report.

In this report, the analysed problems (apresented in future sections) required an average five iterations to converge to the adopted tolerance ($\varepsilon_d = 10^{-4}$)

2.3 Adopted elements

The 4 - constant - strain - triangles element (4CST) is used to simulate the soil and the structure. Good performance of this element for axisymmetric problems was checked again. The join element [2], originally developed for analysis of jointed rock, was used for the interface. This element is indeformable for compression and permits two kinds of movements, sliding and gaping, for tangential and tensile stress respectively (Figure 1).

2.4 Material non-linearity

The non-linear, stress dependent stress-strain characteristics of the soil were considered. For this a hyperbolic stress-strain relationship [3] was adopted and used to obtain the tangent value of Young's modulus. The tangent value of Poisson's ratio was obtained through an exponencial formulation [4], although, is most cases, it was maintained constant.

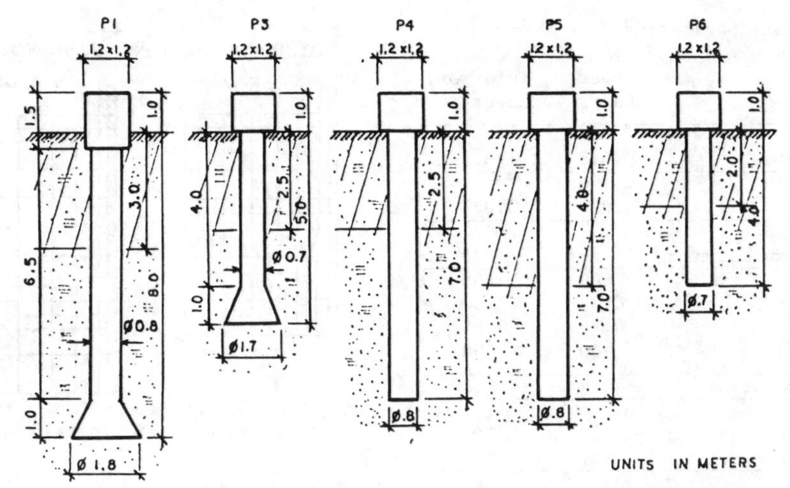

Figure 2. Pier foundation geometry and soil layers

For the interface, the value of k_n (normal stiffness of the joint) is maintaned constant, and the value of k_s (tangencial stiffness of the joint) changes with stress and strain in the following manner:

$$k_s = \frac{c + \sigma_n \tan \phi}{\varepsilon\eta} \qquad (6)$$

where:

c - cohesion or tangential stress for $\sigma_n = 0$,

ϕ - friction angle,

σ_n - normal stress,

$\varepsilon\eta$ - tangential strain

For both elements two failure plastification criteria were adopted:

a) Mohr-Coulomb,

b) Traction limits.

For the 4CST element, in both cases, once it fails, the program adopts a really low value for tangent- elasticity modulus. For the joint element k_s and k_n are considered equal to zero.

3 PARAMETERS DETERMINATION

The field tests were done on a plateau situated on a hill near Rio de Janeiro. The site consists of a layer of tropical/residual soil, mature, originated from a gneissic rock with something like 2.5m depth, superimposed to a less weathered very thick layer. The water table was not reached by sounding until 15m.

The in-situ tests consisted of a series of pull-out essays in pier foundations and footings [1]. For this study only five pier foundations are reported, two being with enlarged base (P1 and P3) and three without base (P4, P5 and P6). The least distance between the piers is about 8.5m. There dimensions and the depth of the first layer in each case are shown in Figure 2.

The adopted parameters were obtained from laboratory tests [1] and from backwards analysis of plate loading tests [5] and pull-out load tests. Their determination is well described in [6]. The parameters used here are presented in Tables 1, 2 and 3, for the interface, soil and pier foundation, respectively. The method of determination of the hyperbolic parameters can be seen in [3]. The great value adopted for k_n has the obvious finality of avoiding an interpenetration of two neighbouring elements.

Table 1 - Interface Parameters

	Upper Layer	Inferior Layer
k_n (MPa/m) (normal stiffness)	1000000	1000000
k_s (MPa/m) (tangential stiffness)	500	500
c (MPa) (cohesion)	0.044	0.035
ϕ (degree) (friction angle)	18.9	20.3
σ adm (MPa) (admissible tensile)	0.001	0.001

Table 2 - Soil parameters

		Upper Layer	Inferior Layer
γ (kN/m³) (unit weight)		16.5	18.0
c (MPa) (cohesion)		0.029	0.023
Ø(degree) (angle of internal friction)		27	29
Hyperbolic parameters	K	600	350
	n	0.52	0.50
	R_f	0.75	0.85
	K_{ur}	900	590
γ(Poisson's ratio)		0.4	0.4
E_R (MPa) (residual tangent elasticity's modulus)		0.1	0.1
σ adm (MPa) (tensile stress)		0.001	0.001

Table 3 - Concrete parameters

γ (kN/m³) (unit weight)	25
E (MPa) (elasticity's modulus)	2.1×10^4
γ (Poisson's ratio)	0.20

4 NUMERICAL EXAMPLES

Five cases of pier foundations subjected to uplift loading have been simulated by the above described method. Two meshes that have been used in the finite element analysis use are presented in Figure 3.

One difficulty in analysing the infinite boundary surface problem by the Finite Element Method (F.E.M.) is an adequate choice of the limit boundaries. Another question is the choice of the degree of meshes complexity. Studies of those factors have been made. The meshes presented are final satisfactory examples.

To simulate the uplift process, a 25kN increment load has been chosen.

The failure process initiates at the base of the pier foundation and propagates, element by element, along the shaft towards the surface.

For the pier foundation with an enlarged base plastification of the soil is also observed. It initiates at the extreme lateral joint of the base and propagates until a certain vertical portion on the soil mass. For both kinds of foundations the noticed failure surface is in agreement with the field observations. Examples of the evolution of soil-foundation interaction and soil plastification

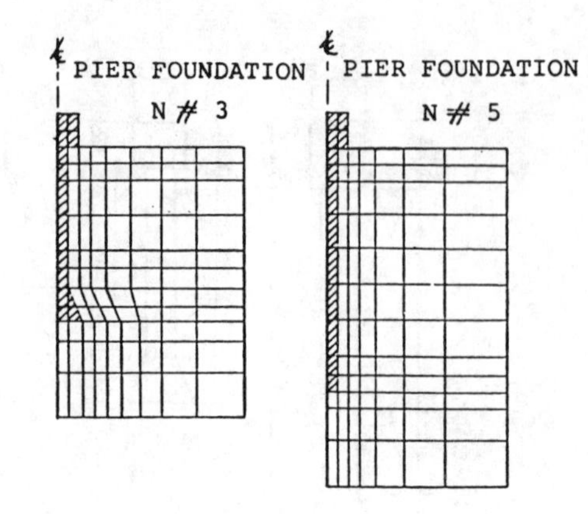

Figure 3. Examples of finite element meshes

for continuous loading are presented in Figure 4, and examples of pier's displacements in Figure 5. These figures where obtained using a post-processing plotter system, linked to the developed program.

The comparison between the predicted and observed ultimate resistance is displayed in Table 4. Satisfactory agreement can be noticed. For comparison purposes, results obtained using the method proposed by Martin [7], are also presented in Table 4. This method fusnished the best results between the traditional methods for determination of the failure load.

Examples of predicted and observed displacements versus uplift forces are presented in Figures 6 and 7. Again, reasonable agreement can be seen.

Table 4 - Pull-out resistance at diferent pier foundations

Pier	P1	P3	P4	P5	P6
Field Test	217.5	106.0	151.0	97.5	95.5
Martin's Method	209.5	102.5	85.7	85.8	32.3
F.E.M.	185.0	97.5	115.0	110.0	50.0

Units: $N \times 10^4$

5 CONCLUSIONS

The Finite Element Method has been applied to analyse the behaviour of pier foundations subjected to uplift forces. Two kinds of pier foundation, with and without enlarged base, have been considered in the investigation of the pier displacements,

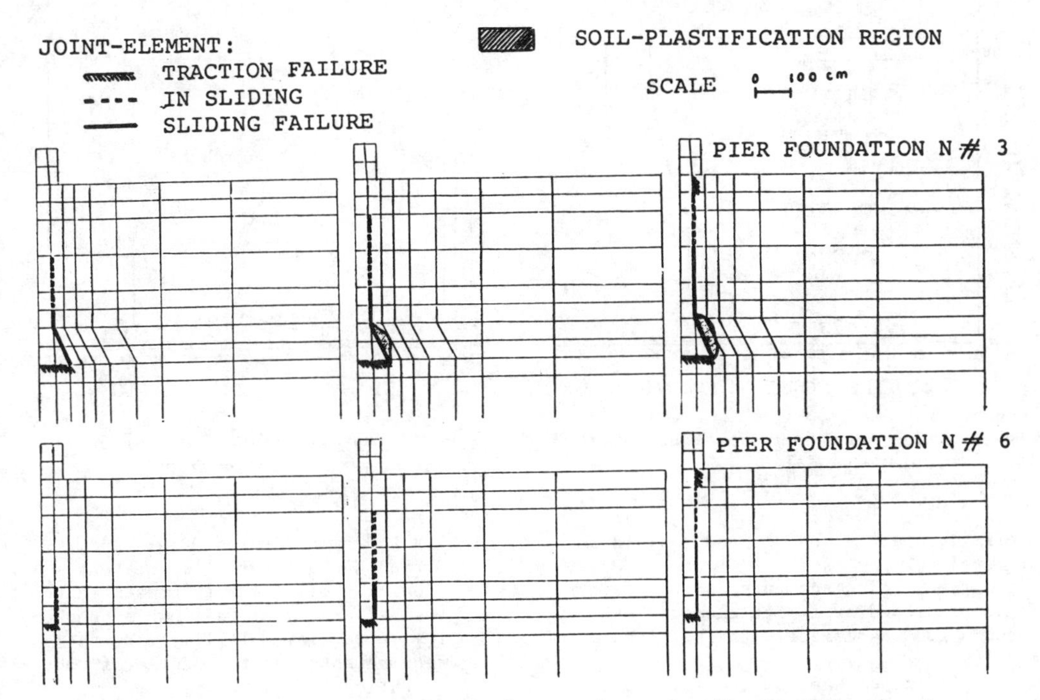

Figure 4. Examples of the soil-foundation interaction and soil plastification for continuous loading

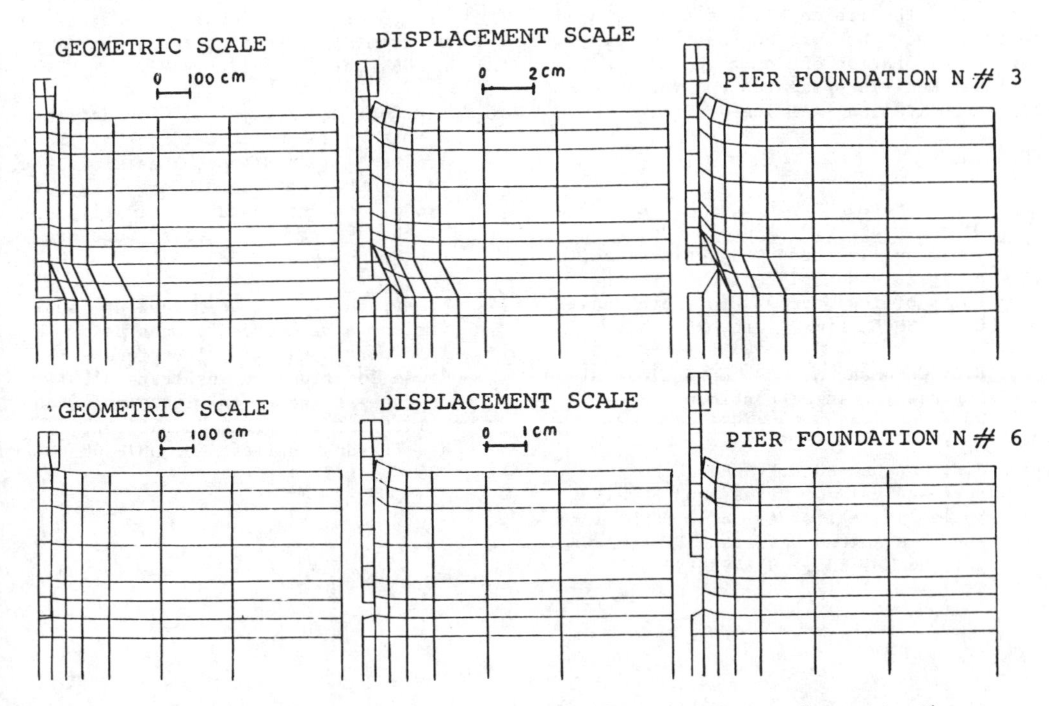

Figure 5. Examples of the evolution of pier displacements for continuous loading

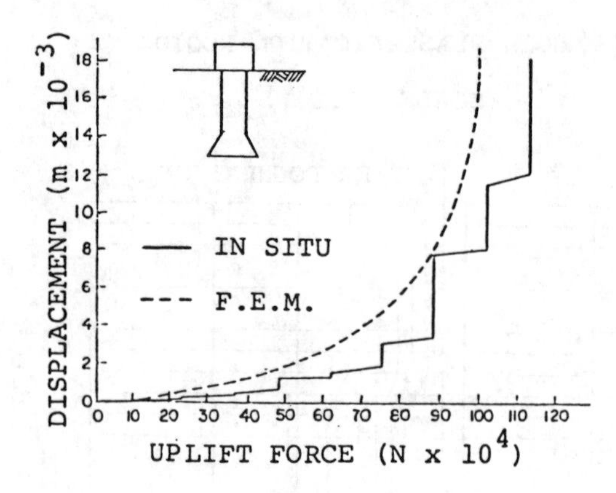

Figure 6. Top displacement versus uplift force for pier foundation N#3

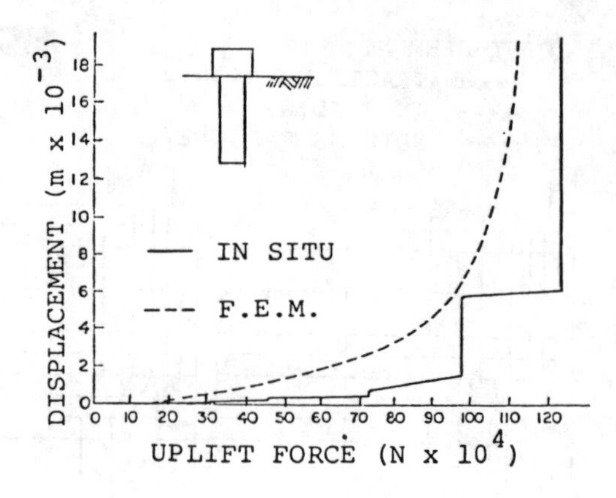

Figure 7. Top displacement versus uplift force for pier foundations N#5

the failure loads and the shape of failure surfaces.

The pier displacements and the failure loads predicted by finite element have shown a good agreement with the measured results.

The predicted shape of the failure surface clearly shows that the failure process starts around the pier foundation base and moves along the shaft towards the surface.

Finally, the finite element program used in this report has proved to be a very strong computation option to be used as an instrument for research and for secure and economical engineering design.

REFERENCES

[1] F.E. Barata, M.P. Pacheco, F.A.B. Danziger and C.P. Pinto, Foundations under pulling loads in residual soils Analysis and application of the results of load tests, Proc. 6th Panam. Conf. SMFE, Lima, Peru, 1979.

[2] R.E. Goodman, Methods of Geological Engineering in Discontinuous Rocks. West Publishing Co, 1976.

[3] J.M. Duncan and C.Y. Chang, Non-Linear Analysis of Stress and Strain in Soils. Journal of the Soil Mechanics and Foundations Division, Proc. ASCE Vol. 96 No. SM5, 495-498 (1970).

[4] P.V.Lade, The Stress-Strain and Strength Characteristics of Cohesionless Soils, Thesis presented to the University of California, Berkeley in partial fulfillment of the requirements for the degree of Doctor of Philosophy, 1972.

[5] M.L.G. Werneck, W.F.D. Jardim, and M.S.S. Almeida, Deformation Modulus of a Gneissic Residual Soil Dertermined from Plate Loading Tests, Solos e Rochas Vol. 2, 3-17 (1979).

[6] A.P.Ruffier, Analysis of Foundations submitted to pull-out forces by the Finite Element Method, Thesis presented to COPPE at Federal University of Rio de Janeiro in partial fulfillment of the degree of Master of Sciences (in Portuguese), 1985.

[7] D. Martin, Calcul des Pieux et foundations a Dalle des Pylones de Transport D'Énergie Électrique - Étude Théorique et Resultats d'Éssais en laboratoire et in-situ. Annales de L'Institut Technique du Batiment et des Travaux Publics, No. 307-308, 106-131 (1973).

Numerical Methods in Geomechanics (Innsbruck 1988), Swoboda (ed.)
© 1988 Balkema, Rotterdam. ISBN 90 6191 809 X

Static analysis of a space frame foundation structure

M.Srkoč, I.Jašarević, S.Sekulić & M.Vrkljan
Civil Engineering Institute, Zagreb, Yugoslavia

ABSTRACT: Static analysis of the space frame foundation structure of a hotel in Yugoslavia is presented. The space model consists of grid beams and piles of different lengths and diameters. Piles are modeled as space beam elements in horizontal layered linearly elastic medium with different mechanical characteristics in two layers and a verticaly elastic stratum at the end of piles. The use of space element on linearly elastic soil stiffness matrix enables the calculation of complex foundation model with personal computers.

1 INTRODUCTION

Hotel Epidaurus in Cavtat near Dubrovnik was designed und built under complex engineering geological and geotechnical conditions, in the immediate vicinity of the Adriatic sea, between two existing hotels that it is attached to. The building of the hotel is a space frame reinforced concrete structure covering a surface area of a 70•55 m base and the foundation is a combination of grid beams and piles (Fig. 1).

The specifics of the foundation of this building are due to the following reasons:

1 Poor surface layers
2 The bearing bed is at varied depths from 2 to 10 meters
3 Deep and shallow foundations are necessary
4 An area of high seismic activity

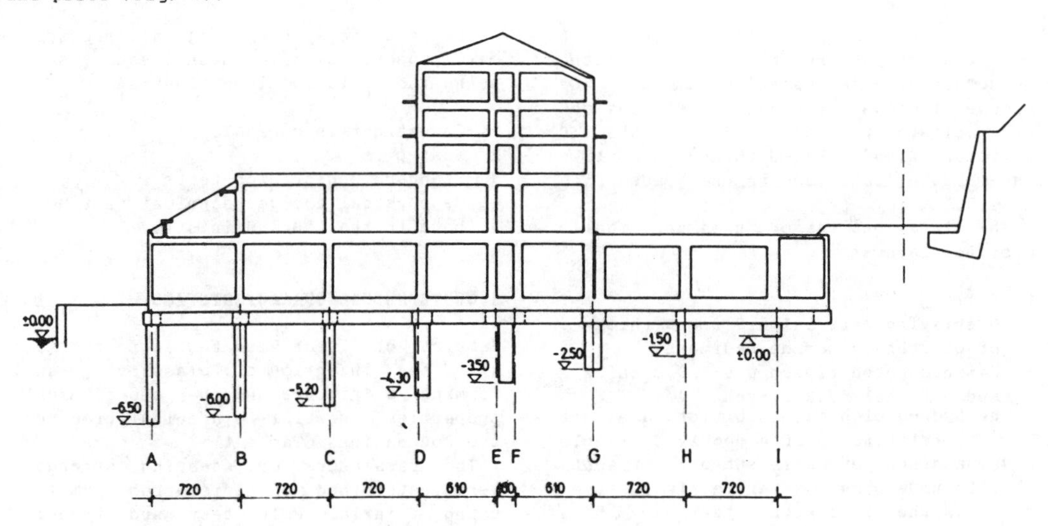

Fig. 1 Crossection of building and foundation at axis 5

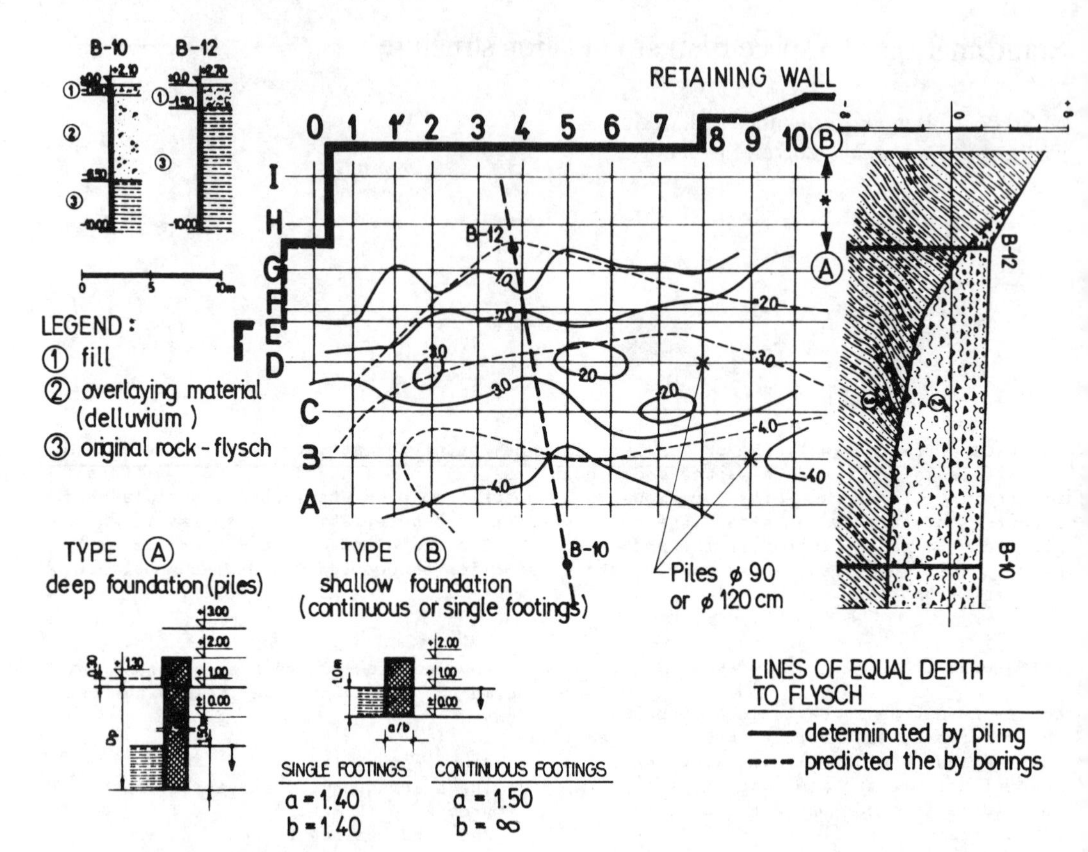

Fig. 2 Engineering geological situation with types of foundation

2 ENGINEERING GEOLOGICAL FEATURES

Through engineering geological research of the slopes in the area of the structure it was established that the Eocene-Oligocene litogenetic complex of fine clastc and coarse rock consists of siltstone with isolated thinner slab-like interbeds of sandstone on the grade (Fig. 2).

The slope consists of three genetically distinct layers :

1 Fill
2 Overlaying material, 0.5-7.0 m thick of Quartenary marine sediment
3 Desintegrated flysch, 0.5-5.0 m thick and original rock flysch

The hydrogeological conditions are set by the relatinship of a poorly permeable cover and non-permeable subsoil (flysch). Shallow underground water is always present in the slope with stationary flow towards the sea, usually along the contact between overlaying material and desintegrated flysch.

The parameters of physical and mechanical properties have been established using the following investigations :

1 Investigation boreholes
2 Test pits
3 Standard penetration tests
4 Geophysical and geotechnical testing both in the lab and in situ

3 GEOTECHNICAL CHARACTERISTICS

Samples of flysch were tested in the lab with the intention of classifying and forming a picture of the geotechnical properties essential to a calculation for the foundations (Table 1).

The parameters of shearing strength were determined on undisturbed samples using triaxial (CID) test and through direct shearing, while on disturbed samples in the Bishop's machine for ring shear (Fig. 3).

Table 1

ROCK KINDS	GRANULOMETRIC COMPOSITION [%]				CONTENT OF $CaCO_3$ [%]	MOISTURE CONTENT [%]	LIQUID LIMIT [%]	PLASTIC LIMIT [%]	PLASTICITY INDEX [%]	CONSISTENCY INDEX	NATURAL VOLUME WEIGHT [Mg/m^3]	SPECIFIC GRAVITY [Mg/m^3]	SOIL CLASSIFICATION AC	SEISMIC VELOCITY $\frac{v_p}{v_s}$ [ms^{-1}]	
	GRAVEL	SAND	SILT	CLAY <0.002 mm										DOWN HOLE	ULTRASONIC
CLAYEY - SANDY SILTSTONE	0-5	0-7	70-85	14-17	19-27	6,8-11,7	36-40	13-25	15-23	1,25-1,56	2,45	2,80	CI	2000-2700 / 1000-1200	1800-2500 / 350-570

The on site testing consisted of geophysical borehole measurement, shearing blocks of 50•50 cm (Fig. 3) and testing of the deformability by performing the plate-load test (Fig. 4).

Modulus of elasticity of flysch is :

$$E = (\Delta\sigma \cdot d \cdot I_W)/\epsilon_{el} = tg\ \alpha$$

$$E = (500 \cdot 0.60 \cdot 1)/0.0016 = 187500\ kN/m^2$$

d - diameter of the test plate
I_W- factor of the shape of the test plate

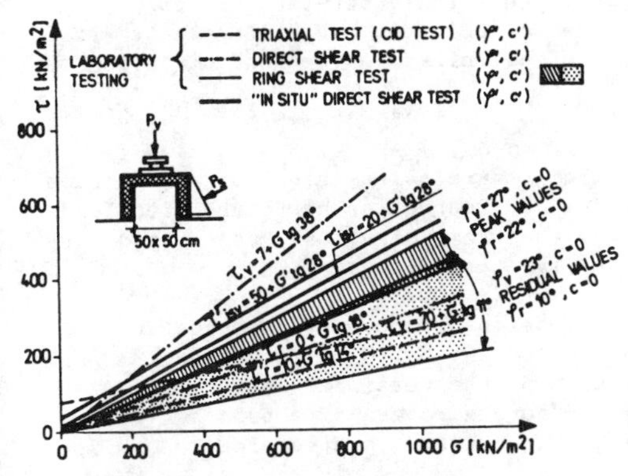

Fig. 3 Strength parameters for flysch shearing

4 GEOTECHNICAL MODEL

Through engineering geological, geophysical and geotechnical investigations three characteristic media are found :

1 Fill
2 Overlaying material (gravel, sand, silt, mud)
3 Desintegrated and firm flysch

with the following geotechnical parameters :

Medium		2	3
Angle of internal friction			
peak value	φ_p =	20°	28°
residual value	φ_r =	18°	28°
Cohesion			
peak value	c_p =	10	50 kN/m²
residual value	c_r =	5	20 kN/m²
Weight density	γ =	20	24 kN/m³
Submerged w. density	γ' =	10	14 kN/m³

The coefficient of horizontal soil reaction for the Winkler soil model that is used was determined according to work (3) for loose to medium relative density of the submerged sandy material:

$k = 5430 - 16280\ kN/m^3$ Medium 2

and according to work (5) for clayey materials for short-term static load on undrained strength :

$k = 80000\ kN/m^3$; $q = 169\ kN/m^2$ Medium 3

$k = 200000\ kN/m^3$; $q = 256\ kN/m^2$ Medium 3

q - unconfined compressible strength

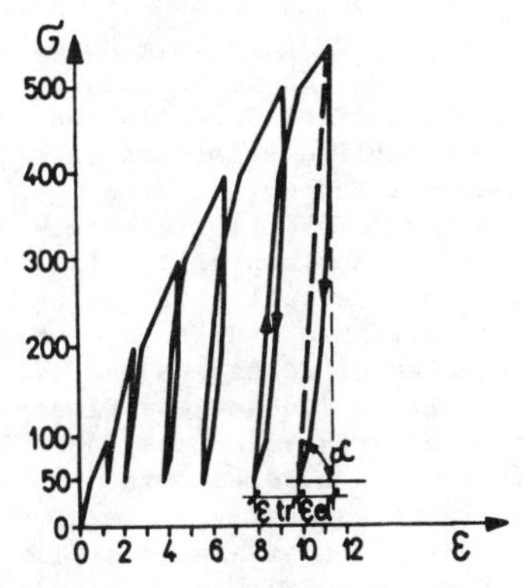

Fig. 4 Deformability of flysch determined by the plate-load test

The coefficient of the vertical soil reaction at the pile tip for the Winkler soil model for medium 3 according to work (1) is :

$$k = Es/(B*(1-\mu^2)*Iw) = 235\ 000\ kN/m^3$$

$Es = 187\ 000\ kN/m^2$
$\mu = 0.30$
$Iw = 0.88$
$B = 1.00\ m$

The optimal number and distribution of piles was determined according to the bearing capacity of piles, such that :

$N = 2193\text{-}3830\ kN$

depending on the length and diameter of the piles.

5 DESCRIPTION OF FOUNDATION STRUCTURE

The structure of Hotel Epidaurus in Cavtat is a space frame structure on bored piles interconnected by grid beams. The building is partially built into the hillside along its whole length, and retaining wall was necessary (Fig. 1, 2).
The foundation structure consists of 86 bored piles of various lengths and diameters (90 and 120 cm in diameter).

Due to high seismic activity the piles are made to stand 1.50 meters in firm flysch. The retaining wall is included in the structure of the building. The earth pressure is transferred through the ground floor slab and the coupled foundation structure onto the piles (Fig. 1,7).

6 A MODEL FOR THE SPACE FRAME STRUCTURE

The foundation structure is modeled as a space frame structure consisting of grid beams and piles (Fig. 6). In the space model the coupling beams are beam elements not soil-supported. They connect the pile heads. The piles are beam elements that are horizontally elastically supported on the soil with an elastic support at pile tip. Piles are modeled with two beam elements on an elastic foundation with different coefficients of soil reaction with which the horizontal soil stiffness is modeled (Fig. 5). The model contains a relatively small number of joints (518), beam elements (481) and degrees of freedom (2447).

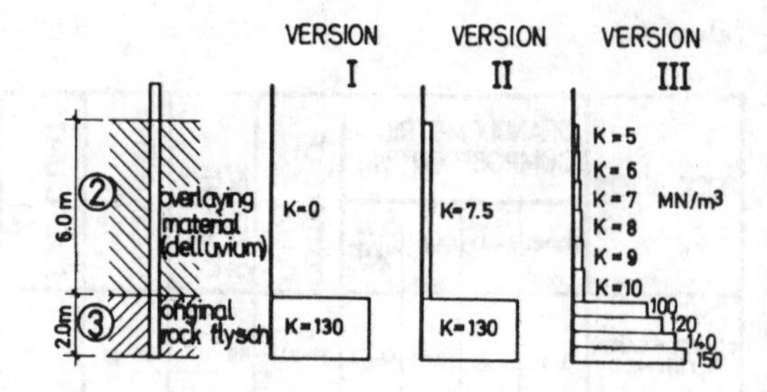

Fig. 5 Versions of pile model

Due to the unreliability of evaluation of real soil stiffness as well as the need for designing a rational structure, a parallel calculation for a plane model of foundation structure on axis 5 (Fig. 7) was also made for three versions of horizontal soil stiffness (Fig. 5).
Version I with an elastic support in firm flysch alone.
Version II with elastic support in both desintegrated flysch and firm flysch (a constant distribution).
Version III with elastic support in desintegrated flysch (linear distribution described with 10 constant values).
The coefficient of stiffness at pile tip of all piles in vertical direction is:

$$k = 250\ 000\ kN/m^3$$

The static calculation for the space model of foundation structure with loads from the frame structure was carried out for five basic loads:

1 The weight of foundation structure
2 Load of the building structure
3 Load of the retaining wall
4 Earthquake forces-direction X
5 Earthquake forces-direction Y

All elements were simultaniously designed using the same computer program for the least favorable combination of the above loads.

7 STIFFNESS MATRIX OF BEAM ELEMENT ON LINEARLY ELASTIC FOUNDATION

In the structural analysis using the finite element method, the soil elasticity most frequently perpendicular on beam element is modelled in terms of elasticity in the model with additional elastic springs with a stiffness corresponding to the stiffness of the soil.

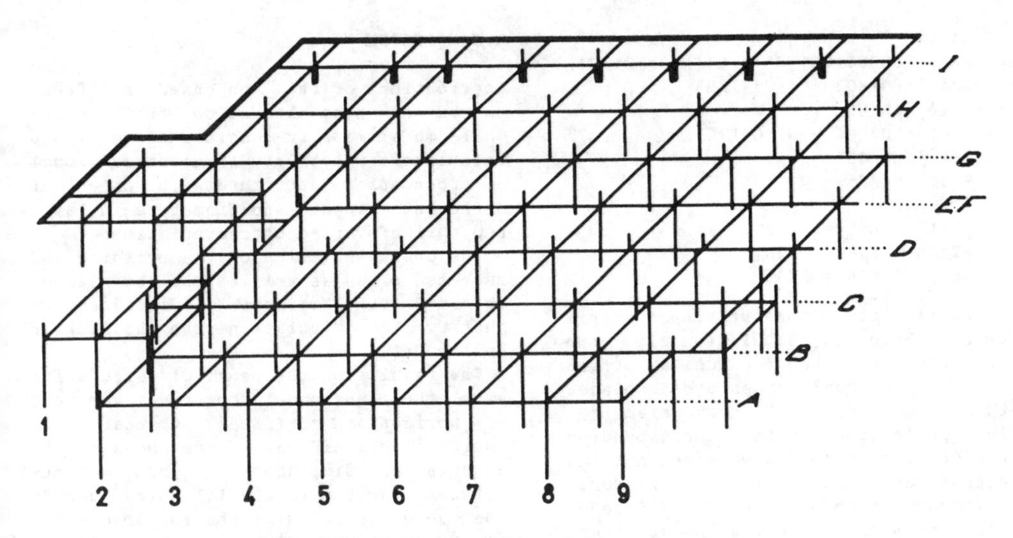

Fig. 6 Space model of foundation structure

In this way the number of degrees of freedom of the system is significantly increased, due to which the space model is impossible, sometimes, to resolve. The accuracy of the calculation depends on the number of replacing spring elements used.

Such an approach to work may avoid the use of the stiffness matrix of beam element on the Winkler elastic foundation (4). The stiffness matrix of plane beam element with the degrees of freedom according to Figure 8 is :

$$\delta = \sqrt[4]{k/4EJ}$$

$$K = \begin{bmatrix} K11 & 0 & 0 & K14 & 0 & 0 \\ & K22 & K23 & 0 & K25 & K26 \\ & & K33 & 0 & K35 & K36 \\ & & & K44 & 0 & 0 \\ & & & & K55 & K56 \\ & & & & & K66 \end{bmatrix}$$

$$G1 = ch\ \delta l \cdot \cos \delta l$$
$$G2 = (ch\ \delta l \cdot \sin \delta l + sh\ \delta l \cdot \cos \delta l)/2$$
$$G3 = 1/2 \cdot sh\ \delta l \cdot \sin \delta l$$
$$G4 = (ch\ \delta l \cdot \sin \delta l - sh\ \delta l \cdot \cos \delta l)/4$$
$$G5 = G3^2 - G2 \cdot G4$$

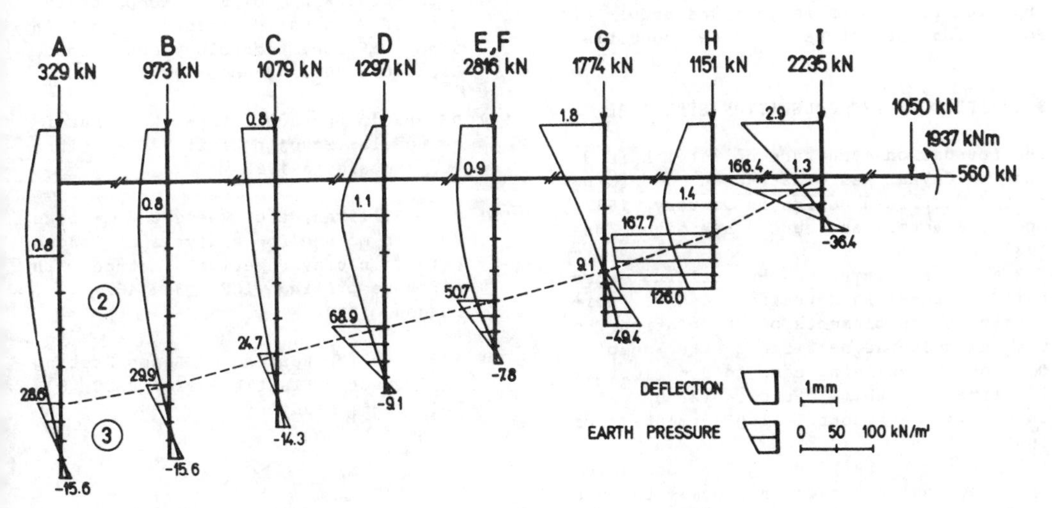

Fig. 7 Plane model of the foundation structure in axis 5

$$K11 = EA/l$$
$$K22 = k/4\delta \cdot (G3/G4 \cdot (G1 \cdot G3 + 4 \cdot G4^2)/G5 - G1/G4)$$
$$K23 = k/4\delta^2 \cdot (G1 \cdot G3 + 4 \cdot G4^2)/G5$$
$$K25 = -k/4\delta_3 \cdot G2/G5$$
$$K33 = k/4\delta^3 \cdot (G2 \cdot G3 - G1 \cdot G4)/G5$$
$$K35 = -k/4\delta^2 \cdot G3/G5$$
$$K36 = k/4\delta^3 \cdot G4/G5$$

$$K14 = -K11 \qquad K55 = K22$$
$$K26 = -K35 \qquad K56 = -K23$$
$$K44 = K11 \qquad K66 = K33$$

The application of the stiffness matrix described here in calculation with the finite element method produces a strict solution of internal forces and displacements at the ends of the beam element, while application of the corresponding number of elastic spring elements obtains an approximate solution with a considerably larger number of degrees of freedom in the system model. Using the stiffness matrix of the plane beam the stiffness matrix of the space beam can be deduced.

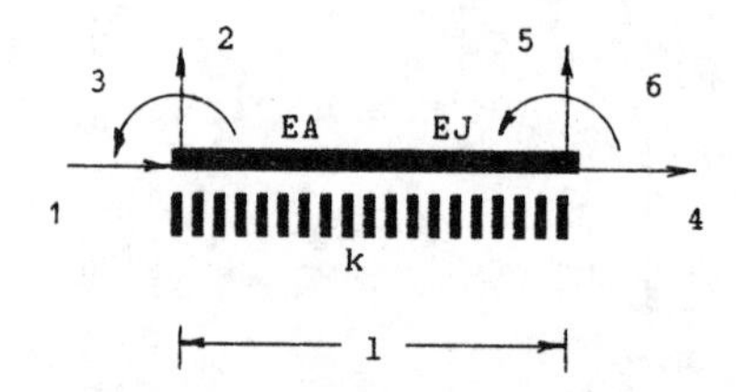

Fig. 8 Degrees of freedom of beam element on an elastic foundation

By programming the stiffness matrix of the space element on elastic foundation in the SPAN computer program for static and dynamic analysis of frame structures, the modelling of a large space model of foundation structure is made possible.

8. BUILDING OF THE FOUNDATION STRUCTURE

The foundation structure of Hotel Epidaurus in Cavtat was constructed according to the described design in January 1987, and the hotel has been in use since June 1987.

When the columns of the building were build, benchmarks were installed to facilitate the settlements of foundation. The maximum measured settlement after completion of the building came to 8 mm (piles in line H), while the calculated value was 10 mm. Horizontal displacements were insignificant.

During the building, the designers followed the construction process through design supervision.

9 CONCLUSION

Application of the condensed stiffness matrix of beam element on the Winkler soil model made it possible to model and calculate a relatively large space model of foundation structure with piles of different lenghts and diameters in layered soil of varied characteristics.

The characteristics of the building and selected piles as well as the low values obtained for horizontal displacements justify the calculation with the Winkler soil model.

The results of the parallel calculation with the plane model for three versions of horizontal stiffness of the soil indicate insignificant differences in the amounts of displacements and moments between version II and III which justify the selection of distribution of coefficients of the horizontal stiffness of the soil in the space model in two layers.

The entire calculation of the space model with automatic design of the piles and coupling beams was performed using an IBM/XT personal computer.

REFERENCES

(1) J.E. Bowles, Foundation analysis and design, Mc Graw Hill, New York 1982

(2) I. Jašarević, M. Srkoč, S. Sekulić, M. Vrkljan, Proračun prostorne temeljne konstrukcije hotela Epidaurus u Cavtatu, VII kongres SDGKJ, Knjiga K, Cavtat, Yugoslavia 1987

(3) L.C. Reese, W.R. Cox, F.D. Koop, Analysis of laterally loaded piles in sand, Offshore technology conference, Paper OTC 2080, Houston 1974

(4) M. Srkoč, Proračun štapne konstrukcije na elastičnoj podlozi, Građevinar 10, Yugoslavia 1983

(5) W.R. Sullivan, L.C. Reese, C.W. Fenske Unified method for analysis of loaded piles in clay, Numerical methods ih offshore pilling, ICE, 135-146, London, 1980

(6) A.S. Vesić, Bending of Beams Resting on Isotropic Elastic Solid, ASCE, 87, EM2, 35-51, 1961

Numerical Methods in Geomechanics (Innsbruck 1988), Swoboda (ed.)
© 1988 Balkema, Rotterdam. ISBN 90 6191 809 X

Effect of loading sequence and soil nonlinearity on the response of a pile group foundation using a three-dimensional finite element analysis

Y.M.Najjar & M.M.Zaman
University of Oklahoma, Norman, USA

ABSTRACT: Effects of loading sequence and soil nonlinearity on the deformation behavior of a pile group: in particular, the distribution of displacements, stresses, axial forces, shear forces and flexural moments in individual piles and in the pile cap, are investigated using a nonlinear three-dimensional finite element (FE) technique. Computer codes were written for post-processing of FE results to determine the bending moments in the piles and the cap, as well as to plot the contour of stress and displacement components of the pile-cap-soil system. An automated scheme implemented in the code enables generation of FE mesh. Emphasis is given to identify the manner in which loads are shared by individual piles in the group.

1 INTRODUCTION

In the past, many analytical/numerical methods for analysis of a pile group have employed simplified assumptions such as replacing the soil medium by Winkler springs, treating the soil medium as an elastic continuum, and neglecting the interaction between various components, namely, the pile-cap, piles, and soil medium [1]. It is certainly desirable, if not mandatory, to consider in the analysis the nonlinear behavior of soil medium and to treat all components of the pile group foundation together as a three-dimensional (3-D) system so that realistic response of the system can be obtained.

In this paper, the behavior of a pile group foundation consisting of four symmetrically located vertical piles of square cross-section, and subjected to arbitrary 3-D loading is analyzed by using an incremental-iterative finite element (FE) procedure. The nonlinearity of soil medium is represented by a path-dependent generalized plasticity model developed recently by Desai and his coworkers [3]. The major objective is to investigate the effects of loading sequence and soil nonlinearity on the overall behavior of the system, particularly on the distribution of forces in individual piles, on the displacements of the pile cap and individual piles, and on the distribution of bending moments and shear forces in each pile.

2 FINITE ELEMENT PROCEDURE

The finite element method is well documented in literature [2]. Hence, it is not detailed herein, only a brief description is presented.

In this study, a three-dimensional displacement FE procedure is employed. Using the principle of minimum potential energy, the following FE equation can be obtained:

$$[K_t] \{\Delta q\} = \{\Delta Q\} + \{\Delta R\} \tag{1}$$

where $[K_t]$ = tangent stiffness matrix, $\{\Delta q\}$ = vector of incremental nodal displacement, $\{\Delta Q\}$ = vector of incremental nodal loads, and $\{\Delta R\}$ = vector of incremental residual loads. An incremental-iterative technique is used for the solution of Eq. (1). Strain and stress increments are calculated as follows:

$$\{\Delta\varepsilon\} = [B] \{\Delta q\} \tag{2}$$

$$\{\Delta\sigma\} = [C]^{e-p}\{\Delta\varepsilon\} \tag{3}$$

where, $[B]$ = strain-displacement transformation matrix and $[C]^{e-p}$ = tangent elasto-plastic constitutive matrix.

The three-dimensional 8-noded hexahedral element [2], is used herein to discretize the domain of interest. The components of displacement, u, v, and w in the x, y, and z directions, respectively, are defined as

$$u = \Sigma N_i u_i, \quad v = \Sigma N_i v_i, \quad \text{and} \quad w = \Sigma N_i w_i \tag{4}$$

where u_i, v_i, and w_i (i=1,2,3...,8) are the nodal displacement components, and N_i (i=1,2,...,8) represents the shape functions.

3 CONSTITUTIVE MODEL

Different constitutive models have been reported in the literature [7,8]. In this study, behavior of pile cap and piles is assumed to be linearly elastic. Non-linearity of the soil medium is represented by the generalized single surface plasticity model, developed by Desai and co-workers [3] In this model, the yielding function is represented in the form [3]

$$F \equiv \sqrt{J_{2D}} + \alpha J_1 - \beta J_1 J_3^{1/3} - \gamma J_1 - k^2 = 0 \tag{5}$$

where J_{2D} = second invariant of the deviatoric stress tensor, J_1, J_3 = first and third invariants of the total stress tensor, respectively, k = measure of the cohesive strength of the material, and α, β, γ = material response functions.

The elasto-plastic constitutive matrix, $[C]^{e-p}$, is derived using the flow rule and the consistency condition. The final expression can be written in the form [3]

$$[C]^{e-p} = \frac{[C]\left\{\frac{\partial f}{\partial \sigma}\right\}\left\{\frac{\partial f}{\partial \sigma}\right\}^T [C]}{\left\{\frac{\partial F}{\partial \sigma}\right\}[C]\left\{\frac{\partial F}{\partial \sigma}\right\} - \gamma_F \cdot \frac{\partial F}{\partial \xi}} \tag{6}$$

where [C] = constitutive relation matrix for linearly elastic behavior, $\partial F/\partial \sigma$ = gradient of F with respect to stress tensor σ_{ij}, γ_F = length of the gradient vector in the stress space, and ξ is the trajectory of plastic strain.

4 PROBLEM DESCRIPTION

The problem under consideration is a pile group foundation, consisting of four vertical piles having uniform cross-section (0.45m x 0.45m) and equal length (14.0m), fully embedded in the soil medium. The pile cap is 3.0m x 3.0m in plan dimensions and 0.25m thick, the piles are arranged symmetrically in both X and Y directions. The cap is subjected to a vertical load (P) applied at the center of the cap, and a horizontal load (Q) applied along the centerline of the cap.

5 FINITE ELEMENT IDEALIZATION

Considering the symmetry of the problem with respect to x-axis, only half of the pile cap is analyzed here. The finite element (FE) mesh used for discretization of the cap and the underlying soil medium consists of 434 eight-noded hexahedral elements connected at 684 nodes. In FE idealization, a small gap between the pile cap and the soil is assumed to avoid direct transfer of any load from the cap to the underlying soil medium. The loads are thus transerred by means of piles only. Similar idealization has been used previously in many investigations [5].

In actual field conditions, the soil medium is of infinite extent in both the horizontal and vertical directions. In the FE idealization, the horizontal boundaries of the soil block in the x and y directions are placed from the pile cap origin at a distance of 10.0m, which is more than six times the half-length of the pile cap in that direction. Also, the depth (25.0m) of the soil block is assumed to be 11.0m deeper than the embedded length of the pile (14.0m) so that the appropriate boundary conditions can be imposed at the bottom of the soil block. A similar idealization was used by Muqtadir and Desai [5], however, the present study employs a more refined mesh.

6 MATERIAL PROPERTIES

The piles and the cap are assumed to behave as linearly elastic material during the loading process. The material properties used for the generalized single surface plasticity model to represent the nonlinear soil medium [5] are: Young's modulus (E_s) = 1.12 x 10^5 KPa, Poisson's ratio (ν_s) = 0.36, α = 25.28 KPa, k = 0.0 KPa, $\bar{\beta}_a$ = 0.005, and η = 0.74 where $\bar{\beta}_a$ and η are the material hardening constants. Elastic properties of piles and cap are Young's modulus (E_p) = 3.0 x 10^6 KPa, and Poisson's ratio (ν_s) = 0.17.

7 EFFECT OF LOADING SEQUENCE ON NONLINEAR RESPONSE

The following loading cases are considered herein to investigate the effects of loading sequence: case I - vertical load, P = 200 kN, and lateral load, Q = 200 kN; case II - vertical load, P = 200 kN, and lateral load, Q = 400 kN; and Case III - vertical load, P = 400 kN and lateral load, Q = 200 kN. The vertical and horizontal loads described in each case are applied incrementally in the following sequence: 1) vertical load (P) is applied in five increments, then lateral load (Q) is applied in five increments, 2) both horizontal and vertical loads are applied simultaneously in ten increments, and 3) lateral load (Q) is applied in five increments, then vertical load (P) is applied in five increments.

The calculated pile forces for the three loading cases (I, II, and III) and for different loading sequences considered herein are shown in Table 1. It is observed that for each loading case the axial force (F_z) in each pile is approximately the same regardless of the loading sequence used, while the lateral force (F_x) in each pile is significantly affected, when the lateral load (Q) is applied first. The lateral force (F_x) in pile #1 becomes maximum when the lateral load (Q) is applied first as in sequence #3.

Table 1 further shows that due to an increase in the lateral load, the axial force in pile #1 increases, although the applied vertical load remains unchanged (200 kN). An opposite behavior is observed for pile #2. Also, for all the loading cases and loading sequences studied herein, the shear forces in pile #1 are always larger than the shear forces in pile #2, because the x-coordinate of the applied lateral load is closer to the x-coordinates of pile #1 than that of pile #2. For instance, for P=200 kN and Q=400 kN, (F_x) in pile #1 and pile #2 are 145.7 kN and 54.3 kN, respectively, the ratio between these forces being approximately 2.7. Increasing P to 400 kN, while keeping Q constant, this ratio increases to about 5.9, because of the added 200 kN the vertical load (P) will increase the shear force in pile #1 and decrease it in pile #2. This is due to the interaction and the three-dimensional effects. This trend is the same for axial pile forces (i.e., axial forces are more evenly distributed between piles when Q is comparatively small, as expected). From design/analysis point of view these observations are extremely important because the classical methods often assume that the vertical and the lateral loads are shared equally between various piles in the group [1].

Table 1 Load distribution in piles for different loading sequences

Loading Sequence	Pile forces (kN)			
	Pile #1		Pile #2	
	F_x	F_z	F_x	F_z
Case I: P = 200 kN and Q = 200 kN				
1. P in 5 increments the Q in 5 increments	145.7	134.4	54.3	66.6
2. P and Q simultaneously in 10 increments	146.9	133.5	53.0	66.6
3. Q in 5 increments then P in 5 increments	162.7	132.8	37.4	67.3
Case II: P = 200 kN and Q = 400 kN				
1. P in 5 increments the Q in 5 increments	264.4	167.9	135.5	32.1
2. P and Q simultaneously in 10 increments	274.5	167.7	125.5	32.3
3. Q in 5 increments then P in 5 increments	275.8	168.5	124.2	31.6
Case III: P = 400 kN and Q = 200 kN				
1. P in 5 increments the Q in 5 increments	171.0	232.1	28.9	167.9
2. P and Q simultaneously in 10 increments	176.3	232.6	23.65	167.5
3. Q in 5 increments then P in 5 increments	199.4	232.4	0.6	167.7

The displacement fields for the two loading sequences in loading case II (P=200 kN and Q=400 kN) at longitudinal section (Fig. 2) are presented in Figs. 3 and 4. Same scale is used in these figures so that they can be superimposed directly for comparison. It is noticed that the nodal displacements are significantly different (both magnitude and orientation) for the two loading sequences. The rotation of the pile cap remains smaller when the vertical load is applied first or simultaneously with the lateral load than the case in which the lateral load is applied first. This is reasonable because application of the vertical load first makes the rotation of the pile cap more difficult. In terms of practical consideration, the third loading sequence is impractical because vertical loads such as those due to the self weight always precedes the lateral loads (e.g., wind). Note that Figs. 3 and 4 are drawn on much exaggerated scale for clarity. The actual displacements are relatively small, even for the loading sequence #3.

Distribution of axial forces along pile #1 for different loading sequences are shown in Fig. 6. It is observed that the axial force in pile #1 is maximum at the surface and minimum at the tip, as expected. A part of the load carried by the pile is transferred to the soil medium through skin friction. As a result, the axial force in the pile reduces with the depth. Although the loading sequence is noticed to have pronounced effects on stress distribution in soil medium [6], their effects are relatively insignificant for axial force, particularly in terms of the maximum axial force.

In summary, based on the numerical results, it is observed that the sequence of loading can be an important factor in nonlinear analysis. The pile group foundation under consideration are those used to support structures subjected to lateral forces, which means that the lateral force may occur when the structure is already there. The logical sequence of loading for this case is that the vertical load followed by the lateral load. This sequence of loading is used for all nonlinear analyses discussed subsequently.

8 NONLINEAR ANALYSIS OF A PILE GROUP FOUNDATION RESPONSE

The analysis is performed by considering the nonlinearity of soil only. The generalized single surface plasticity model is used, and the soil is treated as a non-tensional material.

8.1 Vertically loaded pile group

For this case, only a vertical load (P) of magnitudes 100, 200, 300 and 400 kN is applied at the center of the cap. No lateral force is applied. The calculated pile forces for the linear and nonlinear analyses for this case are shown in Table 2. The vertical force (F_z) predicted by the linear and nonlinear analyses are the same in both piles. The pile forces in the x and y directions predicted by the linear analysis is about 47% of the applied vertical load, while for the nonlinear analysis it is only about 43% of the applied vertical load, for P = 100 kN and decreases to 40% for P = 400 kN.

Table 2 Load distribution due to vertical loads

P	Pile head forces (kN)					
	Pile #1			Pile #2		
(kN)	F_x	F_y	F_z	F_x	F_y	F_z
(a) Linear Analysis						
100.0	23.4	23.4	50.0	-23.4	23.4	50.0
200.0	46.8	46.8	100.0	-46.8	46.8	100.0
300.0	70.2	70.2	150.0	-70.2	70.2	150.0
400.0	93.6	93.6	200.0	-93.6	93.6	200.0
(b) Nonlinear Analysis						
100.0	21.3	21.3	50.0	-21.3	21.3	50.0
200.0	40.7	40.7	100.0	-40.7	40.7	100.0
300.0	61.0	61.0	150.0	-61.0	61.0	150.0
400.0	81.0	81.0	200.0	-81.0	81.0	200.0

8.2 Combined vertical and lateral loading

For this case, the vertical load (P) is first applied at the center of the pile cap and then the lateral load (Q) is applied along the center line of the pile cap. The calculated pile forces for the linear and nonlinear analysis are shown in Table 3. It is noted that the linear analysis underpredicts the vertical force (F_z) in pile #1 for all the loading cases considered. The difference is minimum (about 4%) when the applied vertical load (P) is maximum (400 kN), and maximum (about 15%) when the applied lateral load (Q) is maximum (400 kN). Also, it can be observed that linear analysis overpredicts the magnitude of the vertical force (F_z) in pile #2. The difference is minimum (about 5%) when the applied vertical load (P) is maximum (400kN) and maximum (about 40%) when the applied lateral load (Q) is maximum (400 kN). An intermediate (about 14%) difference is noticed when both the vertical and the lateral loads are equal in magnitude. Note that, the linear analysis results are used as a basis in calculating these percentages.

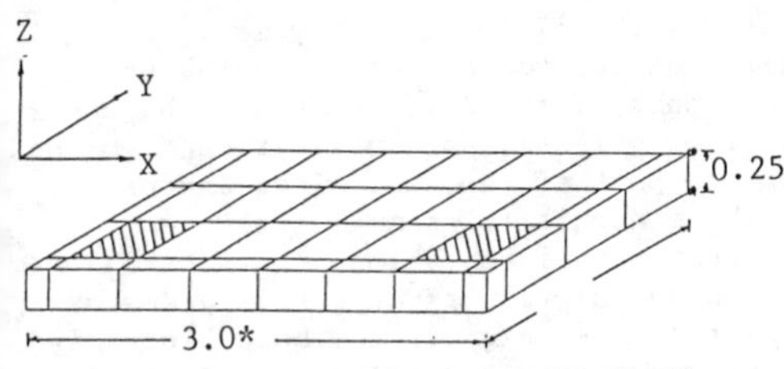

Figure 1. Finite element mesh of pile cap

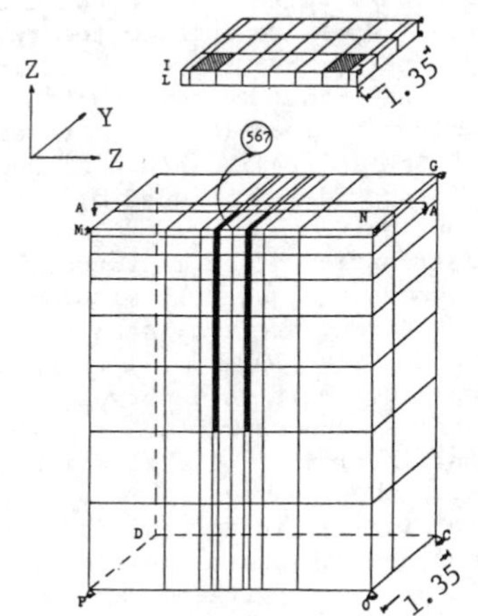

Figure 2. Longitudinal Sections in pile cap and soil block

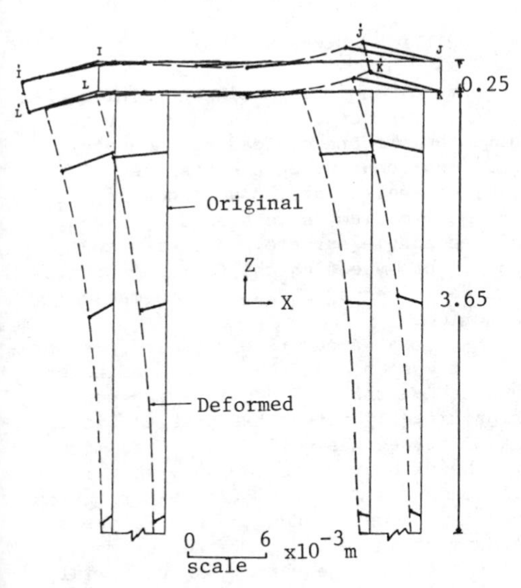

Figure 3. Displacement fields and deformed shape for loading sequence #1

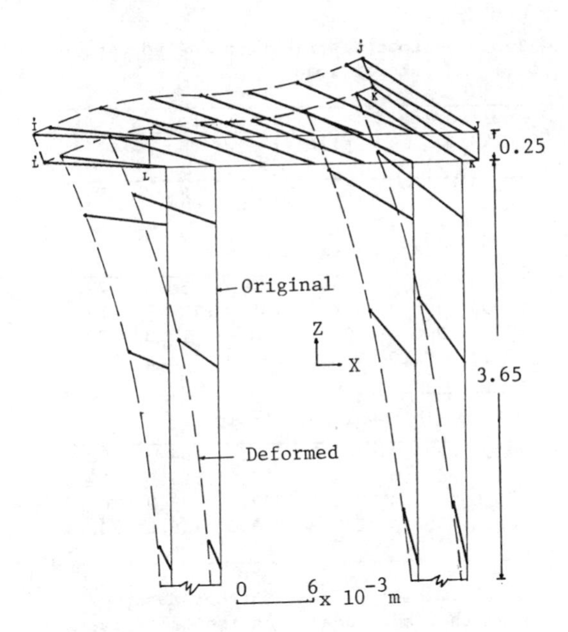

Figure 4. Displacement fields and deformed shape for loading sequence #3

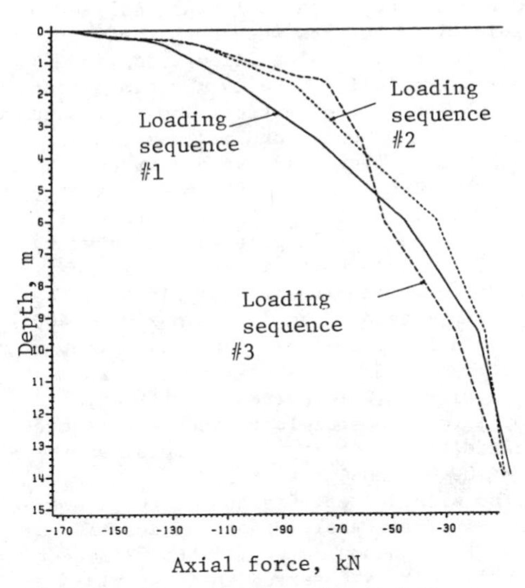

Figure 5. Distribution of axial force along pile #1

*All dimensions are in meters

Table 3 Load distribution due to vertical and lateral loads

Applied loads (kN)		Pile head forces (kN)			
		pile #1		pile #2	
P	Q	F_x	F_z	F_x	F_z
(a) Linear Analysis					
200.0	200.0	143.1	122.9	56.8	77.0
200.0	400.0	239.2	145.9	160.3	54.1
400.0	200.0	189.9	222.8	10.0	177.0
400.0	400.0	286.1	245.9	113.5	154.1
(b) Nonlinear Analysis					
200.0	200.0	145.7	133.4	54.3	66.6
200.0	400.0	264.1	167.9	135.5	32.1
400.0	200.0	171.1	232.1	28.9	167.9
400.0	400.0	281.3	268.0	118.6	132.1

From Table 3, it is further observed that the magnitude of horizontal force (F_x) in pile #1 as predicted by the nonlinear analysis is larger than that predicted by linear analysis when the applied lateral load (Q) is maximum (400 kN) and it is smaller when the applied vertical load (P) is maximum (400 kN). Both the linear and nonlinear analyses yield close results when both the vertical and the lateral applied loads are equal. The difference is minimum (about 2%) when both the vertical and the lateral loads are equal, and maximum (about 10%) when either the vertical or the lateral load is maximum (P = 400 kN or Q = 400 kN). Moreover, the magnitude of shear force (F_z) in pile #2 predicted by the linear analysis is larger than that predicted by nonlinear analysis when the applied lateral load (Q) is maximum (400 kN) and it is smaller when the applied vertical load (P) is maximum (400 kN).

In view of the increased interaction for larger lateral load, the results presented in the following are obtained for P = 200 kN and Q = 400 kN.

The displacement fields for the linear and nonlinear analyses at section MNOP in Fig. 2 are shown in Fig. 6 and 7, respectively. It is observed that the displacement field predicted by the nonlinear analysis for nodes located in the upper right part of the soil block and the upper part of pile #2 are significantly larger than the displacement field predicted by the linear analysis. Also, a change in the nodal displacement orientation is noticeable in this region. The effect of soil nonlinearity is less significant for nodes located in the upper left part of the soil block and the upper part of pile #1, and insignificant in the bottom part of the soil block and piles.

The contours for normal stress σ_{zz} evaluated at integration points on a longitudinal section through soil elements (section A-A, Fig. 2) for both linear and nonlinear analyses are presented in Figs. 8 and 9, respectively. The concentration of stresses (σ_{zz}) for the linear analysis appears to be in the upper middle part of the soil block, while the concentration of stresses (σ_{zz}) are shifted towards the upper left part, when the nonlinearity of soil is considered. This indicates that larger amount of load (F_z) is carried by the left pile (pile #1) when nonlinearity of soil is considered, caused by stress redistribution.

The distribution of axial forces, and flexural moments along pile #1 predicted by the linear and the nonlinear analyses are shown in Figs. 10 and 11, respectively. From Fig. 10 it is observed that the linear analysis underpredicts the axial force distribution in pile #1 which agrees with the findings in Figs. 8 and 9. The difference in values predicted by the linear and nonlinear analyses is minimum in the bottom part and maximum in the upper part of pile #1. Fig. 11 shows that the linear analysis underpredicts the positive and the negative flexural moments along the length of the pile.

9 CONCLUDING REMARKS

A three-dimensional nonlinear finite element (FE) procedure is used to investigate the effects of loading sequence and soil nonlinearity on the response of a pile-cap-soil system. Nonlinearity of the soil medium is idealized by a generalized single surface plasticity model in which both yielding and failure surfaces are described by a single mathematical function.

Based on the numerical results presented, it was observed that the loading sequence and the soil nonlinearity can significantly affect the distribution of axial forces, as well as flexural moments in the pile group, especially when lateral loads are involved. The distribution of displacements and stresses in the soil medium are more sensitive to the change in the loading sequence than the axial forces and flexural moments. Three-dimensional finite element analysis can be used to obtain an improved estimate of

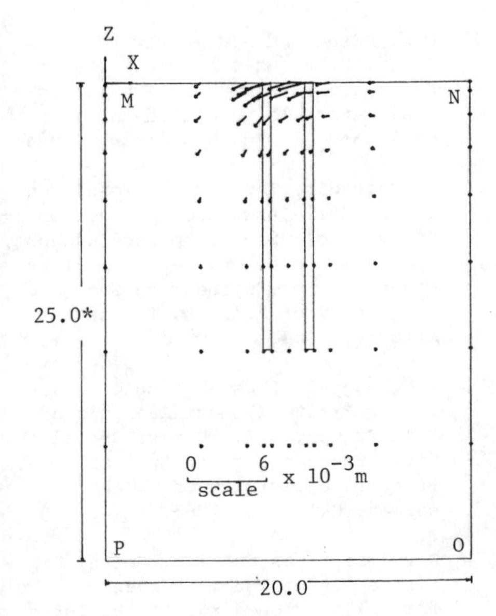

Figure 6. Displacement fields by linear analysis

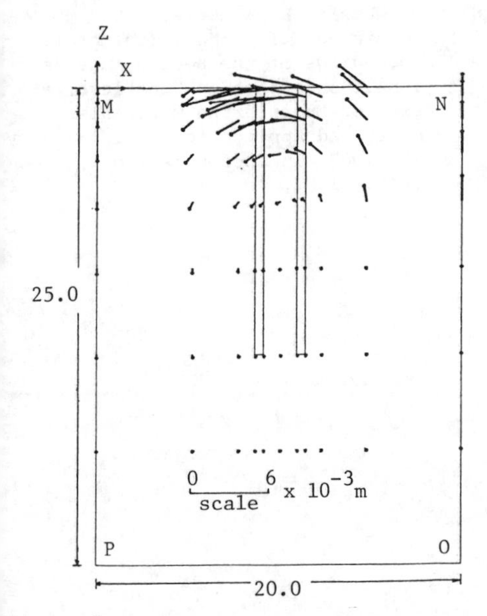

Figure 7. Displacement fields by non-linear analysis

* All dimensions are in meters

Figure 8. Distribution of σ_{zz} (KPa) in soil elements by linear analysis

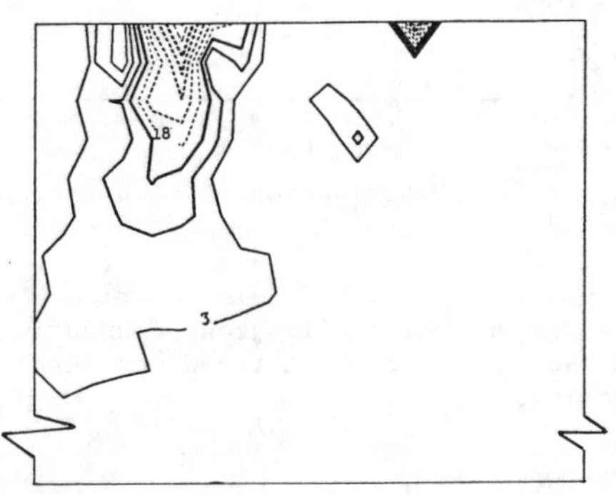

Figure 9. Distribution of σ_{zz} (KPa) in soil elements by nonlinear analysis

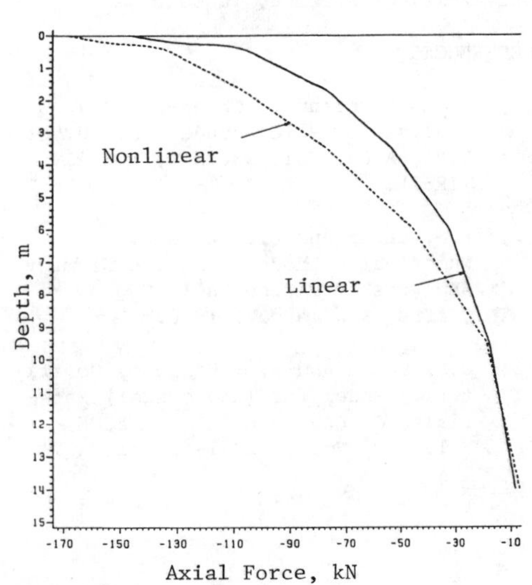

Figure 10. Distribution of axial force along pile #1

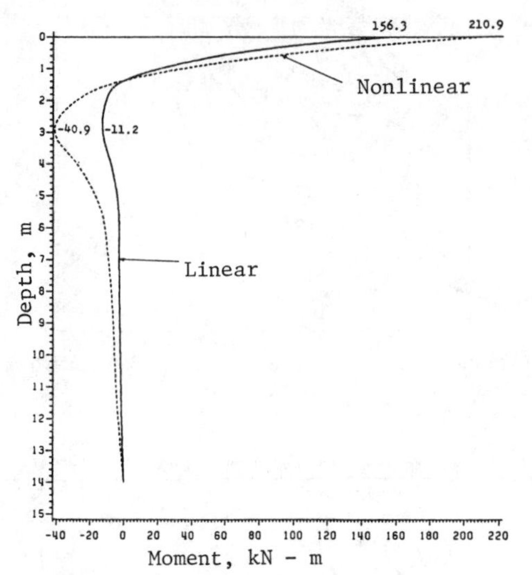

Figure 11. Distribution of moment along pile #1

the response of a pile group foundation subjected to a fully three-dimensional loading.

ACKNOWLEDGEMENT

The computer code used in this study was developed under the guidance of Professor C.S. Desai and was procured from the University of Arizona, Tucson.

REFERENCES

[1] R. Aschenbrenner, Three-Dimensional Analysis of Pile Foundation. J. St. Div., ASCE, Vol. 93, ST1, 201-219 (1967).

[2] K.J. Bathe and E.L. Wilson, Numerical Methods in Finite Element Analysis, Prentice-Hall, Englewood Cliffs, New Jersey, 1976.

[3] C.S. Desai and M.O. Faruque, Constitutive Model for (Geological) Materials, J. Eng. Mech. Div., ASCE, Vol. 110, No. 9, 1391-418 (1984).

[4] C.S. Desai and H.J. Siriwardane, Constitutive Laws for Engineering Materials with Emphasis on Geological Materials, Prentice-Hall, Englewood Cliffs, New Jersey, 1984.

[5] A. Muqtadir, Three-Dimensional Nonlinear Soil-Structure Interaction Analysis of Pile Groups and Anchors, Ph.D. Dissertation, Dept. Civil Engineering and Engineering Mechanics, University of Arizona, Tucson, Arizona, 1984.

[6] Y.M. Najjar, Three-Dimensional Nonlinear Finite Element Analysis of Pile Groups, M.S. Thesis, School of Civil Engineering and Environmental Science, University of Oklahoma, Norman, Oklahoma, 1986.

[7] D.V. Griffiths and I.M. Smith, Experience with a Double Hardening Model for Soil, Proceedings of the International Conference on Constitutive Laws for Engineering Materials: Theory and Application, held January 10-14, 1983, in Tucson, Arizona, U.S.A., 553-559 (1983).

[8] C.S. Desai, E. Kremple, P.D. Kiousis, and T. Kundu, (Editors), Proceedings of the Second International Conference on Constitutive Laws for Engineering Materials: Theory and Application, held January 5-8, 1987, in Tucson, Arizona, U.S.A., 1987.

Numerical Methods in Geomechanics (Innsbruck 1988), Swoboda (ed.)
© 1988 Balkema, Rotterdam. ISBN 90 6191 809 X

Soil-structure interaction – A possibility for elastic-plastic calculation of foundation slabs

R.Stark & J.Majer
Institut für Festigkeitslehre und Flächentragwerke, University of Innsbruck, Austria

ABSTRACT: An economical numerical procedure is described for the analysis of arbitrarily shaped raft foundations.The procedure uses available finite element programs, modified and adapted for the modulus of compressibility method. By the iteratively working method with variable moduli of subgrade reaction, the deformation behaviour of the subsoil can be simulated by various models and compatibility is reached in the plate-soil interface. Due to the supporting of the slab which is realized by discrete rods, connected with the plate on its nodal points, the stiffness matrix of the plate remains unaffected during the iteration process. Plastic limit states in the soil may be considered by a prior calculation of plastic soil pressures for regions of the plate which are to be determined in an iterative way.

1 INTRODUCTION

The finite element method nowadays is a standard procedure, successfully applied in nearly all domains of engineering practice and in geomechanics as well. However, for the design of raft foundations of arbitrary shape, a three-dimensional analysis of the subsoil requires excessive computer resources, as well as data preparation time, making the method uneconomical except for special projects. Therefore the soil-foundation interaction problem is usually simulated by simple models such as the Winkler- or Pasternak model. These models can easely be incorporated into finite element procedures for analysis of elastic plates. Although it is common opinion that the model of the isotropic elastic half-space is more convenient to describe the response of the subsoil, this model is used more rarely owing to its complexity.

Results obtained from analysis based on the the Winkler medium, may be reasonable for the design of the plate in special cases, but the surface deflection can not be described in a realistic manner. Improved methods for the calculation of vertical deformation can be found for instance in works of Chow [4], Brown and Gibson [3], Fraser and Wardle [5] and Selvadurai [10].

Because finite element programs are widely available for plate analyis, an attempt has been made to modify such pro-grams to permit realistic calculation of plate and surface deflection.

1.1 Description of the procedure

At the Institut für Bodenmechanik, Felsmechanik und Grundbau at the University of Innsbruck, a simple example was developed [11] to simulate the deformation behaviour of the linear elastic isotropic half-space by means of springs of variable stiffness in the soil-foundation interface. These springs correspond to a variable distribution of the modulus of subgrade reaction. Each spring was assigned to a surface element situated in a mesh at the soil-foundation interface (Fig.4). It was assumed that the distribution of pressure per element is constant. The stiffness distribution of the spring elements supporting the foundation was varied in an iteration process untill the deflection of the plate was identical with the surface deflection.

Working with springs which join the plate at the nodal points offers the advantage that the stiffness matrix of the plate does not change during the whole iteration process.

1.2 Calculation possibilities and assumptions

1. The finite element method can be applied to calculate plates of arbitrary shape and stiffness. Since the analytical solutions applied in the analysis of vertical deformation are valid for loaded areas of rectangular shape both plate and soil-foundation interface are discretized by rectangular elements. Because the subdivisions do not have to be regular, the grid may be refined, for instance near concentrated loadings. The generation of the meshes for the plate and the soil is automatically done by the program.

2. A simple coordinate transformation performed when processing the flexibility matrix of the subsoil enables the determination of the mutual interaction of arbitrarily placed structures.

3. The procedure allows evaluation of the response of both homogeneous, isotropic elastic half-space and layered elastic soil media.

4. This method permits not only the description of linear elastic response, but also, at least phenomenologically, consideration of the plastic behaviour or complete failure of the soil, for instance in marginal zones of the plate, according to the common models of soil mechanics. For the coupled system soil-foundation this offers a realistic possibility to determine the limit state when the subsoil approaches its bearing capacity, whereas the plate still responds predominantly elastic.

5. A 'relaxed' boundary condition is assumed at the interface of the foundation and the soil medium such that the tangential shear stresses are zero. Physically, this implies a perfectly smooth base for the foundation.

2 ELASTIC MODELS OF SOIL BEHAVIOUR

There are several models available for the determination of the soil pressure distribution under an elastic foundation. The most frequently used models will be described briefly.

2.1 Winkler model

The model proposed by Winkler [14] is used in the modulus of subgrade reaction method. The assumption of this model is that the deflection, s, of the soil medium at any point on the surface is directly proportional to the pressure, p, applied at that point and independent of pressures applied at other locations.

$$p(x,y) = ks(x,y) \tag{1}$$

where k is termed the modulus of subgrade reaction.

2.2 Pasternak model

The model of Pasternak assumes that similar to the Winkler model the soil medium consists of a system of spring elements. The spring elements however are no longer mutually independent but connected to a layer of incompressible vertical elements which deform in transverse shear only.

$$p(x,y) = ks(x,y) - G\nabla^2 s(x,y) \tag{2}$$

G denotes the shear modulus and ∇^2 is the Laplace differential operator in rectangular cartesian coordinates. The Winkler case can be recovered as a limiting case as G tends to zero. The Pasternak model already avoids the disadvantage of the Winkler model where surface deflections are limited to the loaded regions only. Wieghart's formulation [13] is consistent with the Pasternak model and was refined by Majer [7] in order to avoid discontinuities with respect to the pressures.

2.3 Elastic isotropic continuum model

The modulus of compressibility method is based on the the model of the isotropic elastic continuum. The deflections are not only limited to the loaded area, but they also will occur outside the loaded region, owing to the reciprocal deformation influence between all points of the continuum.

Employing the model of the elastic isotropic half-space to the analysis of vertical deformation all elements become completely mutually coupled. Hence the stiffness matrix of the entire system gets densed and no longer banded as in use of the modulus of subgrade reaction method. Working with influence coefficients for uniformly loaded elements, symmetry is lost if the soil-foundation interface domain is discretized by irregular subdivisions. This of course leads to a noticeable increase of computer resources.

The method described as follows, conforms to the modulus of subgrade reaction method with regard to the simplicity in the numerical mode of operation, and yields results which are identical to those obtained from an analysis with the modulus of compressibility method.

3 NUMERICAL PROCEDURE

3.1 Flexibility matrix of the subsoil

The fundamental solution given by Boussinesq [2] is an expression for the displacement field in an elastic isotropic medium caused by a concentrated load acting perpendicular to the traction-free surface.

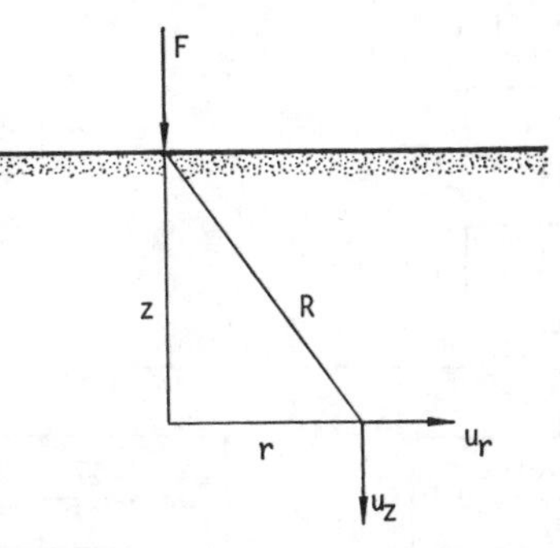

Figure 1. Local coordinate system for the Boussinesq solution.

Using the notation of Fig. 1 and terming E and μ the elastic constants of the soil the axisymmetric displacement field u may be written as

$$u_r = \frac{F(1+\mu)}{2\pi ER}\left\{\frac{rz}{R^2} - \frac{(1-2\mu)r}{R+z}\right\} \quad (3)$$

$$u_z = \frac{F(1+\mu)}{2\pi ER}\left\{2(1-\mu) + \frac{z^2}{R^2}\right\} \quad (4)$$

$$R = \sqrt{r^2+z^2}$$

Since we will work subsequently with loaded areas, Boussinesq's solution cannot directly be applied. The concentrated load of equation (4) must be substituted by an equivalent uniformly distributed load, p, acting on a rectangle with length, a, and breadth, b.

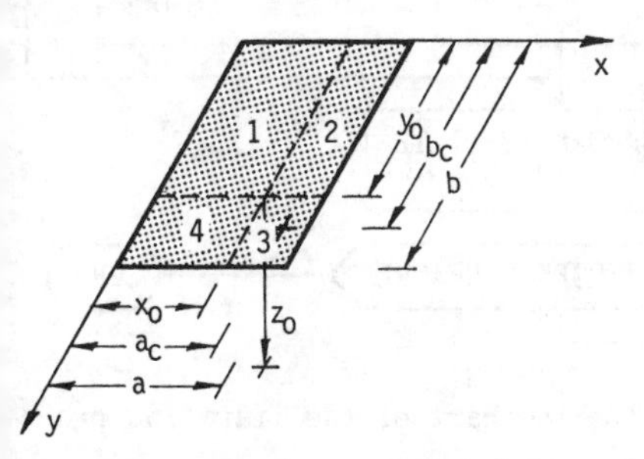

Figure 2. Considered Point O lies in the interior of the loaded rectangular area.

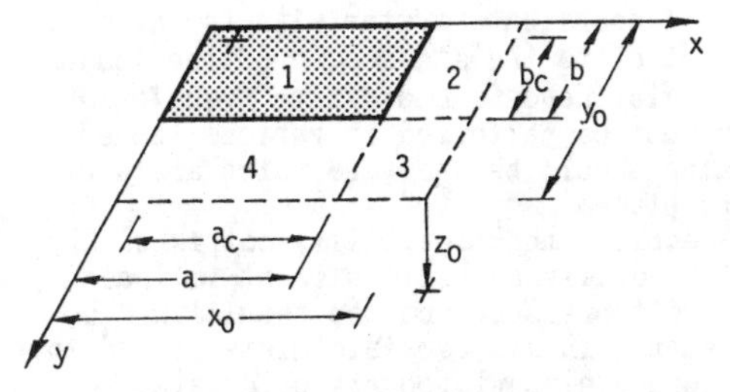

Figure 3. Considered Point O lies in the exterior of the loaded rectangular area.

Performing the integretion over the loaded domain, according to Steinbrenner [12] we find for the vertical displacement, s, of a point lying in the depth beneath the corner of a rectangular loaded area

$$s(z) = \frac{ap(1-\mu^2)}{2\pi E}\left\{\ln\frac{c+m}{c-m} + m\ln\frac{c+1}{c-1} - \frac{(1-2\mu)}{(1-\mu)}\left[n\tan^{-1}\frac{m}{nc}\right]\right\} \quad (5)$$

$$m = \frac{b}{a} \qquad n = \frac{z}{a}$$

$$c = \sqrt{1+m^2+n^2}$$

By recognizing the principle of superposition to be valid, the vertical displacement of any point within the mass can be obtained by means of a summation process. Rewriting equation (5)

$$s(z) = \frac{ap(1-\mu^2)}{2\pi E}\ \alpha \quad (6)$$

the displacement of any point in the interior or exterior of the loaded area (Fig.2,3) may be evaluated using equation (7) or (8) respectively.

$$s(z) = \frac{ap(1-\mu^2)}{2\pi E}\{\alpha_1+\alpha_2+\alpha_3+\alpha_4\} \quad (7)$$

$$s(z) = \frac{ap(1-\mu^2)}{2\pi E}\{\alpha_{1+2+3+4}-\alpha_{2+3}-\alpha_{3+4}+\alpha_3\}) \quad (8)$$

With respect to the numerical computation according to Poulos and Davis [9] the principle of superposition can be written as

$$s(z) = \frac{ap(1-\mu^2)}{2\pi E}\{\alpha(x_o,y_o)-\alpha(x_o-a,y_o)- \alpha(x_o,y_o-b)+\alpha(x_o-a,y_o-b)\} \quad (9)$$

Calculating the values for α it should be noticed that the loaded rectangular element and the considered point O whose displacement should be evaluated are lying in a local coordinate system with the axis x, y parallel to the sides a, b of the loaded rectangle. Hence a coordinate transformation must be performed if various loaded domains should be analysed which are arbitrary placed.

Assembling the flexibility matrix of the soil gradually one soil element is loaded by a unit pressure and the resulting displacements in all remainig elements and in the loaded element too are calculated (flexibility coefficients).

Because the formulae of Steinbrenner are valid for surface loadings, however, one single plate element is supposed to be rigid with respect to the subsoil, the flexibility coefficient f_{ij}, i=j, due to a load acting on the element itself is calculated by evaluating the displacement of the 'characteristic point'. For all remaining elements the flexibility coefficients f_{ij}, i#j, correspond to the deformation of element i due to a unit pressure applied at element j. It should be noticed that the point of element i for which the deformation is actually calculated coincides with the corresponding nodal point of the plate to which the spring element is connected. At this location, compatibility between plate displacement and surface deflection is required. Graßhoff [6] defined the 'characteristic point' to be that point of a surface area loaded by a uniformly distributed pressure, where the deflection due to that pressure is identical with the displacement of a rigid foundation of the same shape and loading. In the case of a rectangle, the characteristic point possesses the coordinates: $a_c=0.87a$, $b_c=0.87b$ (Fig.2,3).

By evaluating these flexibility coefficients with the real soil pressure and using the principle of superposition, the actual size and shape of the surface deflection of the foundation is obtained during the iteration process.

3.2 Algorithmus

In the flowchart the following notation is used:

$^{(i)}$ ...subscript, denoting iteration step i
$_j$ ...subscript, denoting element j
A ...area
F ...external load
c ...spring stiffness rate
p ...actual soil pressure
p^P ...soil pressure of an ultimate state
s ...ordinate of soil surface deformation
w ...nodal point displacement of the plate
ϵ ...small value for spring stiffness rate tending to zero

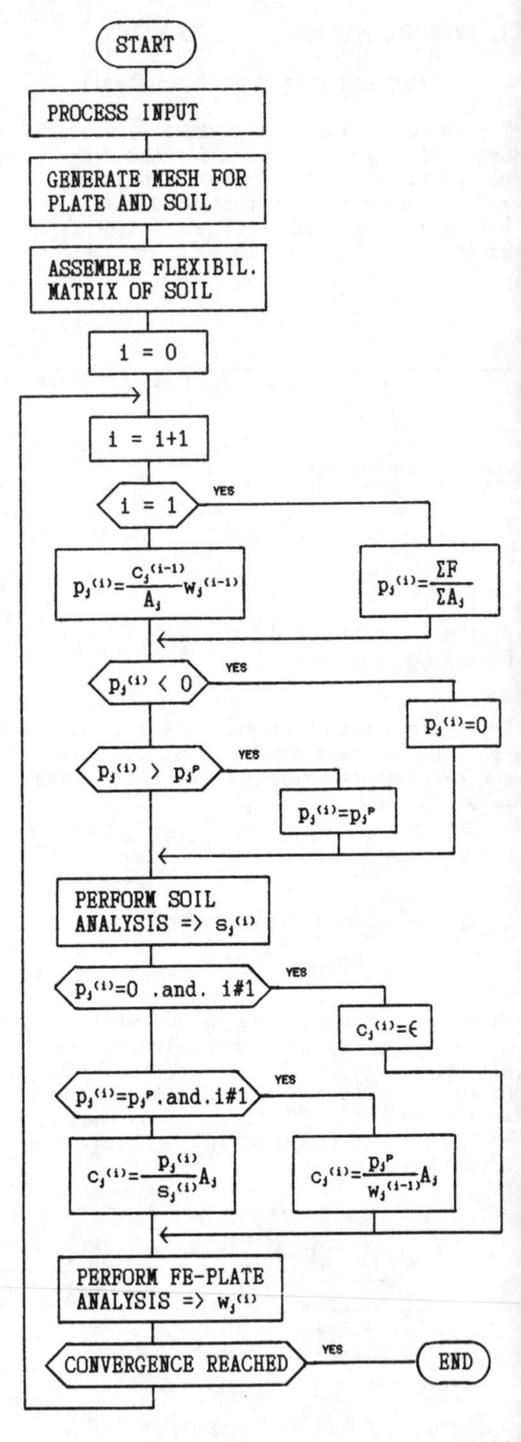

Figure 4. Flowchart of the iteration process.

4 NUMERICAL RESULTS

4.1 Assumptions for the numerical calculation

A rectangular plate with the dimensions 12 x 6 m is chosen and subdivided into 12 x 12 elements yielding to 13 x 13 nodal points for the plate and 13 x 13 soil elements too. Taking advantage of the double symmetry the computation cost is extensively reduced. For the elastic response of the subsoil, the model of the isotropic half-space is used. The elastic constants for the plate and the soil are assumed to be: E_s=10 MN/m², E_p=260000 MN/m², μ_s=0, μ_p=0.15.

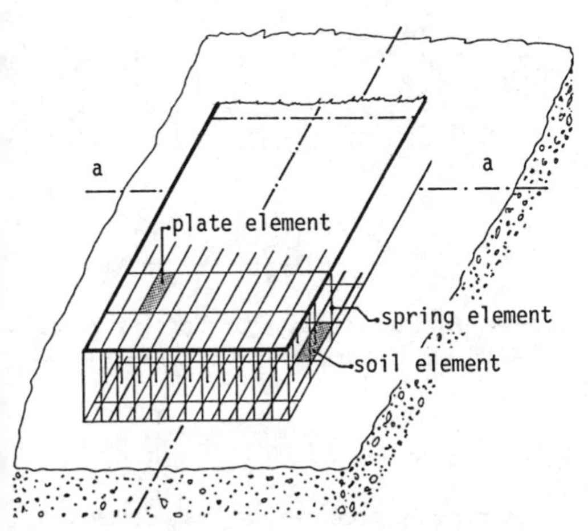

Figure 5. Discretization of the plate and the soil-foundation interface.

4.2 Elastic response of the subsoil

For a judgement of the proposed method, the numerical results of the middle section a-a of the above example are compared with an analytical solution. This specific analytical solution was given by Borowicka [1] for the contact pressure beneath a uniformly loaded smooth strip lying on the elastic isotropic half-space. Figures 6, 7 and 8 show the comparison of the present method with Borowicka's solution for several values of the relative stiffness K, which is defined as

$$K = \frac{E_p(1-\mu_s^2)}{6E_s(1-\mu_p^2)} \quad \frac{h^3}{a^3} \tag{10}$$

where 2a is termed the breadth of the strip. Using the above-mentioned assumptions, equation (10) was evaluated for K = π/30, π/10, π/3.

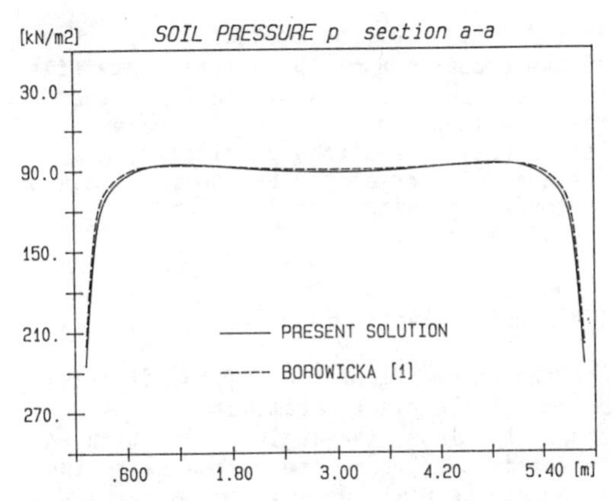

Figure 6. Contact pressure distribution K=π/30, h=18.5 cm.

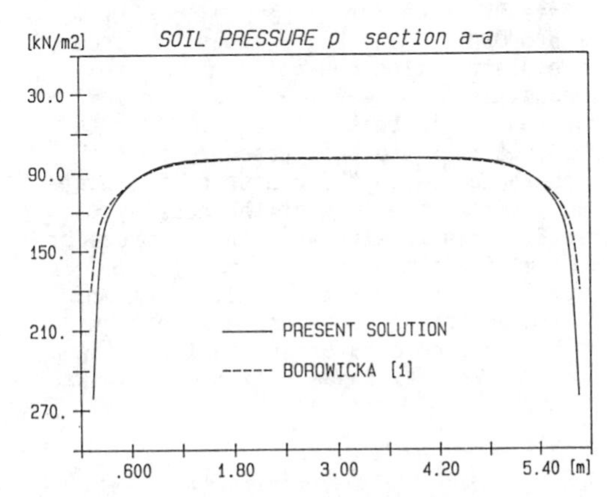

Figure 7. Contact pressure distribution K=π/10, h=26.7 cm.

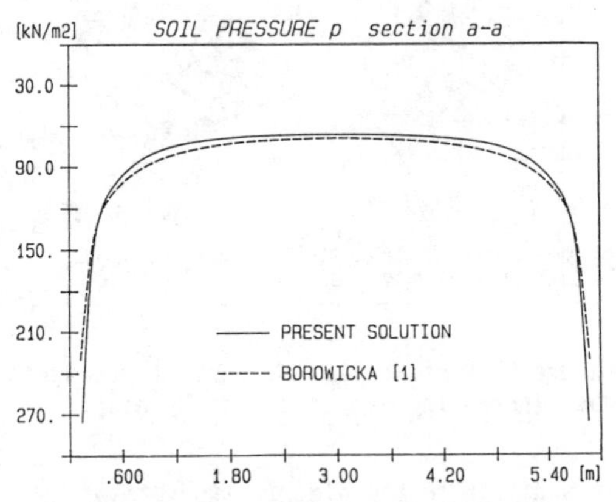

Figure 8. Contact pressure distribution K=π/3, h=40.0 cm.

An indication for the sufficient accuracy of the comparison of the actual plate with the strip is furnished by the fact, that the numerical results do differ very little only in sections parallel to a-a. The analysis for comparing the results was performed applying a uniform load $q = 100$ kN/m².

4.3 Elastic-plastic response of the soil

For the prior calculation of the distribution of the plastic ultimate stresses under the plate, the angle of friction was selected at $\Phi=27.5°$. The stress along the edges of the plate was picked to be 150 kN/m² corresponding to a certain depth under the ground surface, on which the foundation is assumed to be located. The gradient chosen for p^p is in accordance with ÖNORM B 4432. Fig.9 shows the pressure distribution beneath the uniformly loaded plate for several load steps. When the load is increased it can clearly be observed that the soil pressure shifts from the edges to the center of the plate, and leads to the loss of the bearing capacity. This is simulated in the procedure by limiting the soil pressure p_j to p_j^p, noticing that p_j^p is a function of x and y. The numerical procedure was performed using the same data as in the foregoing example. The thickness h of the plate was 40 cm.

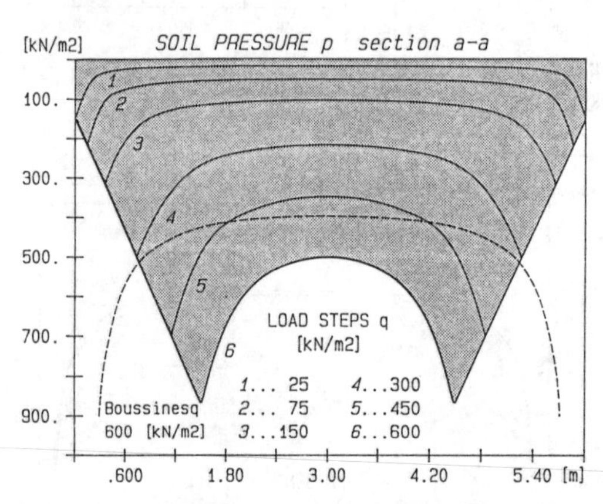

Figure 9. Distribution of contact pressure when increasing the load step by step.

In addition to the distribution of the pressure acting in the soil-foundation interface, this analysis yields the bending moments, required for the design of the plate. The chart given in Fig.10 shows the distribution of the bending moments $m_x/q*100$ versus x. q denotes the applied uniform load. It may be stated that increasing the load changes the kind of stress substantially and even the sign of the bending moment may change. Thus a frequently used technique in engineering practice, reinforcing foundation slabs on both sides equally, is well recommended for this case.

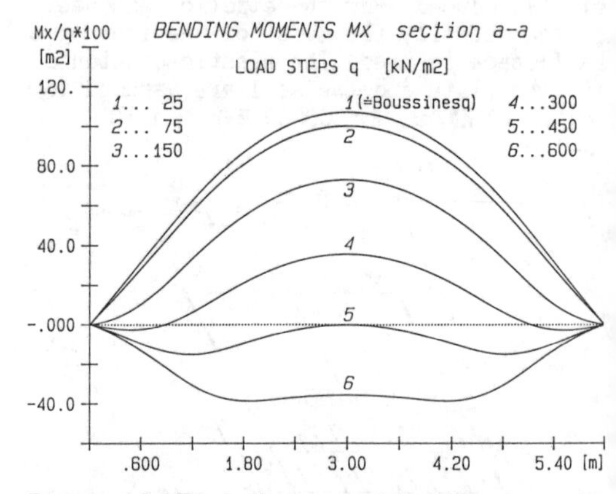

Figure 10. Distribution of bending moments $m_x/q*100$ when increasing the load step by step.

5 CONCLUSION

A numerical procedure has been described which enables consideration of arbitrarily shaped foundation mats. For the analysis of the plate a common finite element program may be used and for the response of the soil any linear model may be incorporated. The comparison of the results between this present method and an analytical solution computing the contact pressure distribution beneth a uniformly loaded smooth strip lying on the surface of the isotropic half-space, shows good conformity after only few iteration steps. Since the plate is supported by discrete springs connected to the nodal point of the plate, no updating of the stiffness matrix is required during the whole iteration process which makes the method more economical. Besides the evaluation of the plate bending moments this technique also may be used for a reliable settlement analysis. While choosing an appropriate model for the soil response, it is also feasible to calculate the mutual influence between neighbouring structures.

Extending beyond the assumption of purely elastic behaviour of the soil, this method offers the possibility to handle the interaction problem, even considering non linear soil response, where this non linearity of the soil is related to the loss of bearing capacity of the soil, according to the common assumptions used in soil mechanics.

BIBLIOGRAPHY

[1] H.Borowicka, Druckverteilung unter elastischen Platten, Ing.-Archiv, 10(2), 113-125 (1939).

[2] J.Boussinesq, Application des Potentieles à l'ètude de l'èquilibre et du mouvement des solides èlastiques, Gauthier-Villard, Paris, 1885.

[3] P.T.Brown and R.E.Gibson, Rectangular loads on inhomogeneous elastic soil, J.Soil Mech.Found.Div., Proc.ASCE, 99(SM10), 917-920 (1973).

[4] Y.k.Chow, Vertical deformation of rigid Foundations of arbitrary shape on layered soil media, Int.J.numer. anal.methods geomech., 11, 1-15 (1987).

[5] R.A.Fraser and L.J.Wardle, Numerical analysis of rectangular rafts on layered foundations, Geotech., 26(4), 613-630 (1976).

[6] H.Graßhoff, Setzungsberechnungen starrer Fundamente mit Hilfe des 'kennzeichnenden Punktes', Bauingenieur, 30(2), 53-54 (1955).

[7] J.Majer, Ein allgemeiner Ansatz zur Berechnung des elastisch gebetteten Trägers, Österr.Ing.Zeitschrift, 13(2), 119-126 (1970).

[8] P.L.Pasternak, On a new method of analysis of an elastic foundation by means of two foundation constants, Gosudarstvennoe Izdatelstro Liberaturi po Stroitelstvui Arkhitekture, Moscow, 1954.

[9] H.G.Poulos and E.H.Davis, Elastic solutions for soil and rock mechanics, Wiley, New York, 1974.

[10] A.P.S.Selvadurai, Foundamental results concerning the settlement of a rigid foundation on an elastic medium due to an adjacent surface load, Int.J.numer.anal.methods geomech., 7, 209-223 (1983).

[11] R.Stark, Berechnung rechteckiger punktförmig belasteter Fundamentplatten, Diplomarbeit Univ. Innsbruck, Innsbruck, 1979.

[12] W.Steinbrenner, Tafeln zur Setzungsberechnung, Straße, 1, 121-124 (1934).

[13] K.Wieghart, Über den Balken auf nachgiebiger Unterlage, Z.angew.Math. Mech., 2(3), 156-184 (1922).

[14] E.Winkler, Die Lehre von der Elastizität und Festigkeit, Dominicus, Prag, 1867.

Numerical Methods in Geomechanics (Innsbruck 1988), Swoboda (ed.)
© 1988 Balkema, Rotterdam. ISBN 90 6191 809 X

Numerical analysis of slab foundations

F.Rodríguez-Roa & C.Gavilán
School of Civil Engineering, Catholic Universty of Chile

ABSTRACT: A Finite Element Model was used in order to study the settlements and internal forces induced in slab foundations. The analysis considered slab foundations of different flexural stiffness under framed structures with finite and infinite basal dimensions. The stiffness of the superstructure was not taken into account.

It was concluded that in slab foundations of framed structures of infinite extent, the simplified Winkler's solution is a very close approximation to the continuum-medium solution. However in slabs of finite extent, large differences can result in the values of settlements and internal forces computed by both procedures.

When modelling the soil as an anisotropic linear elastic continuum medium, the maximum angular distortion and internal forces induced over the slab foundation are not significantly affected by variations of some of the supporting-soil properties, such as Anisotropic Ratio, Poisson's Coefficients, and the Shear Modulus associated to a vertical plane. In any case the Shear Modulus is the parameter that can originate the most significant differences.

1 INTRODUCTION

To design a raft foundation the engineer must know the magnitude and distribution of its internal forces. However, practice has shown that traditional methods to evaluate such forces often yield results far away from reality; Terzaghi and Peck (1967) pointed out: "Since the difference between the theoretical and the real distribution of the bending moments in the raft can be very large, it is commonly advisable to provide the raft with twice the theoretical amount of reinforcement". This uncertainty frecuently found in practice was the main reason to undertake this study, whose objective is to get a better understanding of some aspects related to this type of foundation design.

The subgrade-reaction-coefficient method, first proposed by Winkler (1867), assumes as it is known, that the intensity of the vertical contact pressure is proportional to the settlement at every point of the raft foundation. The subgrade reaction coefficient to be used depends largely on the properties of the soil and on the geometrical characteristics of the foundation (Terzaghi (1955), Bowles (1974), Selvadurai (1979)).

Another method often used for designing slab foundations has been to assume the soil as a linear elastic continuum-medium (Cheung and Zienkiewicz (1954), Cheung and Nag (1968), Fraser and Wardle (1976)). It is generally accepted that the continuum model provides a better physical representation of the supporting soil (Hain and Lee (1974), Horvath (1983)). However, mainly due to the simplicity of the Winkler's solution this method is still used in practice for the design of raft foundations.

This paper presents results obtained for foundation slabs of different flexural stiffnesses under framed structures with finite and infinite basal dimensions, in order to compare the slab foundation behavior when the soil is modelled as a continuous medium or as a Winkler's material. The behavior of such foundations was also studied considering changes in some soil properties such as the Anisotropic Ratio, Poisson's Coefficients, and the Shear Modulus associated to a vertical plane.

2 FINITE ELEMENT MODEL

The model subdivides the slab, within its own plane, into a mesh of twelve-degree-of-freedom rectangular elements (Clough and Tocher (1965), Zienkiewicz (1977), Desai and Abel (1972), Przemieniecki (1968)). An isotropic linear elastic constitutive law was considered for the slab's material.

The ground was simulated as a linear elastic and horizontally stratified medium and discretized in a number of rectangular prism elements; these elements are further divided by the computer program into five constant-strain tetrahedra (Zienkiewicz (1977)). Through the combination of the five tetrahedral stiffness matrices the program constructs the hexahedral stiffness matrix. The mean value of the stresses computed for each one of the five tetrahedra, is assigned to the centroid of each hexahedral element.

Only the terms of the stiffness matrix associated to the vertical displacement of the nodes of the tetrahedral elements were added to the corresponding terms of the stiffness matrix of the slab elements in their common nodes. Since the nodes of the slab elements were assumed to have no horizontal displacements it was necessary to impose the same restriction on the nodes of the soil elements which were in contact with the slab. This restriction is equivalent to have an infinitely rough soil-slab interface.

Alternatively, the computer program may also use the Winkler's solution to model the subsoil.

The superstructure stiffness was not included in this study.

The computer program, written in FORTRAN, was developed for a DEC-10 Digital computer. The set of linear equations was solved using a direct Gauss method and there were no storage capacity problems in the cases herein analyzed. However, despite the effort made to minimize the storage requirements, the computer memory might not be large enough to analyze a more general situation. In such case it would be necessary to modify the program and use secondary storage units.

3 MODEL VERIFICATION

To check the results of the computer program, a number of particular cases with previously known solutions were analyzed (Rodríguez-Roa and Gavilán (1985 and 1986)), one of which is reported here. This is the study of the displacements and stresses induced in the ground by a rigid square plate under concentrated vertical loads located at each of the four corners. The plate was modelled using a single finite slab element loaded by a constant vertical force applied at each node (Fig. 1). The ground was modelled using a single cubic element made up of an isotropic linear elastic material.

The parameters used were as follows:

Vertical load (P)	=	6.13 kN
Cube side (a)	=	0.50 m
Soil properties:		
Young's Modulus	=	9806.65 kPa
Poisson's Coefficient	=	0.25
Unit Weight	=	0.0 kN/m^3
Plate properties:		
Young's Modulus	=	205.94x10^6 kPa
Poisson's Coefficient	=	0.2
Thickness	=	0.05 m
Unit Weight	=	0.0 kN/m^3

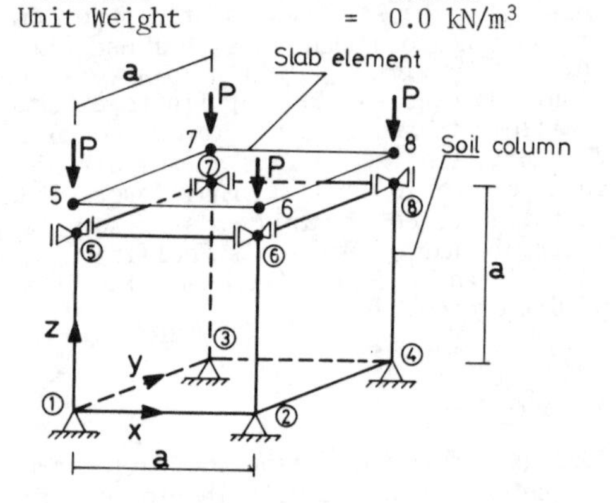

Fig. 1 Infinitely rigid slab over a soil column of height "a"

The finite element solution yielded a mean settlement value in the soil surface (nodes 5,6, 7 and 8 in Fig. 1) of 4.178×10^{-3} m and a vertical compressive stress of 98.16 kPa at the centroid of the cubic soil element.

Based on the Theory of Mechanics of Solids the analytical solution concerning the shortening of the soil column was determined using the soil's Oedometric Modulus and not its Young's Modulus because of the restrictions or boundary conditions imposed to the soil horizontal displacements (Fig. 1). This theoretical procedure yielded a settlement value of 4.166×10^{-3} m, which is very close to the numerical model solution. Furthermore, considering a mean vertical stress equal to $4P/a^2$, a value of 98.07 kPa is determined, which is almost exactly equal to the numerical model result.

4 SLAB FOUNDATION OF INFINITE EXTENT

In order to compare the results obtained using the continuum-medium solution with the results obtained using the Winkler's solution, a slab foundation for a framed structure of infinite extent was analyzed. It was assumed that the framed structure had a constant-distance distribution of columns in each one of the two orthogonal directions as it is seen in Fig. 2. Each column only transmits a constant concentrated vertical load to the foundation. Since the problem is symmetric the extent of the finite element mesh was reduced to the area hatched in Fig. 2. The results that follow were derived using a mesh made up with 16 rectangular slab elements and 112 rectangular-prism soil elements (Fig. 3).

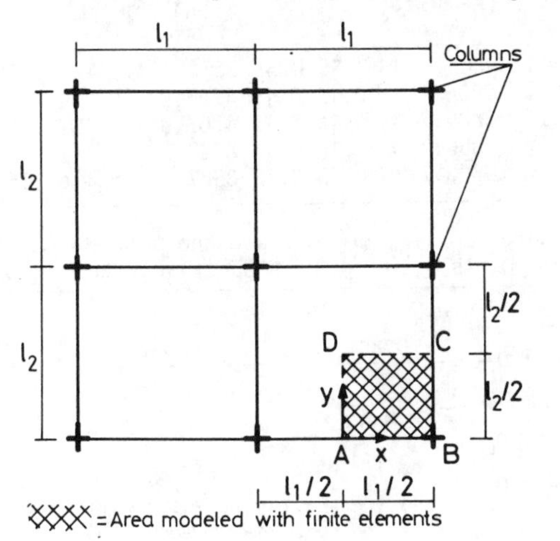

Fig. 2 Slab foundation of infinite extent studied.

The data used was:

Column separation: $\ell_1 = \ell_2 = 5$ m
Vertical load for each column: P=980.6 kN
Young's Modulus of the slab foundation = 196.13x10^5 kPa
Poisson's Coefficient of the slab foundation = 0.2
Unit weight of the slab = 0.0 kN/m^3
Thickness of the compressible homogeneous soil stratum = 5.0 m
Soil unit weight = 0.0 kN/m^3
$E_1 = E_2$ = 9806.7 kPa
$\nu_1 = \nu_2$ = 0.35
$G_1 = G_2$ = 3632.4 kPa

where,

E_1, G_1, ν_1: Young's Modulus, Shear Modulus and Poisson's Coefficient of the soil associated to the horizontally stratified plane.
E_2, G_2, ν_2: Young's Modulus, Shear Modulus and Poisson's Coefficient of the soil associated to the direction normal to the bedding plane.

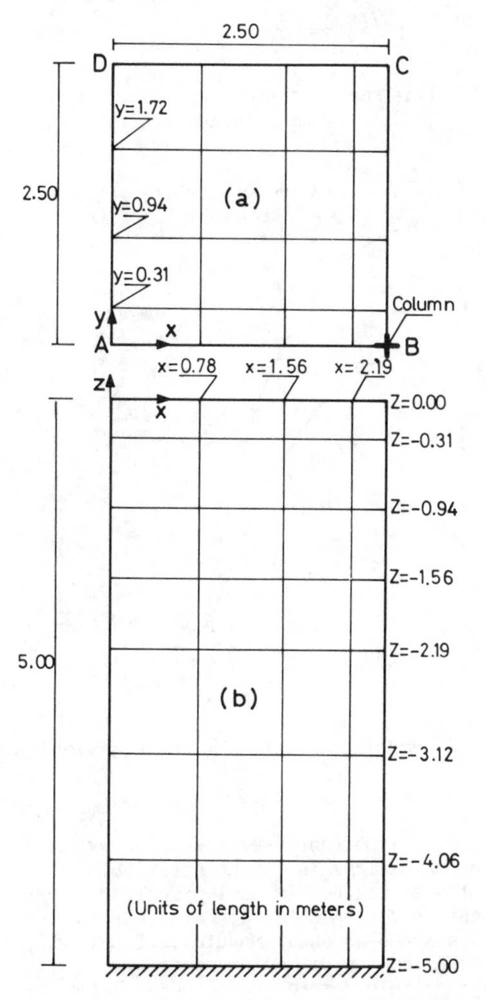

Fig. 3 Finite element mesh for the slab foundation of infinite extent

a) Plan view
b) Vertical cross section

Figure 4 illustrates the continuum-medium results obtained for a 0.80 meter-thick rigid slab. This figure shows the bending moments M_x and M_y along the line y = 0.155 m (see Fig.3). It also includes the results obtained using an equivalent Winkler's model, defined for a subgrade reaction coefficient k_0, which was determined so that the settlement of the rigid slab was the same as in the continuum-medium solution:

k_0 = (Mean vertical unit load on the slab)/(Rigid-slab settlement obtained for the continuum-medium solution)

= (980.6 kN/(5.0 m x 5.0 m))/(0.0128 m)
= 3064.4 kN/m^3

As it is seen in Figure 4, the results obtained using both methods are very close.

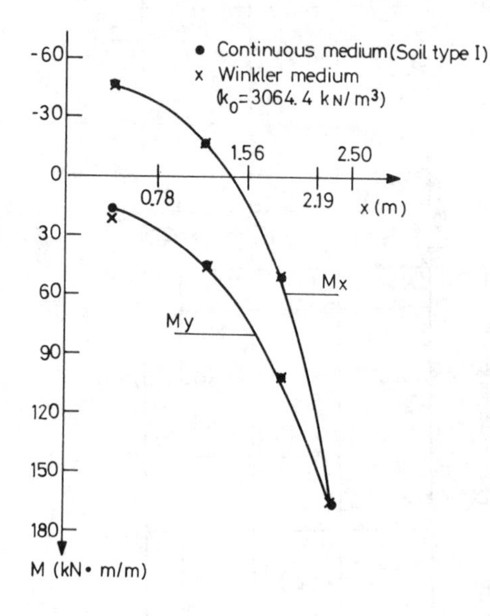

Fig. 4 Bending moments, M_x and M_y, along y = 0.155 m, in the 0.80 meter-thick rigid infinite slab

Another important aspect when analyzing slab foundations is the sensitivity of the results to changes in soil properties such as the Anisotropic Ratio, Poisson's Coefficients and/or Shear Modulus. Four additional continuum-medium cases were examined in order to study the influence of the parameters mentioned above. The soil types used and the settlements obtained are shown in Table 1.

Table 1 shows an important variation of the settlement values of the slab for the different types of soil considered. The maximum difference of 46% occurs when comparing the influence of soils I and II. However, the results also showed that in all cases bending moments M_x and M_y do not present significative changes from those illustrated in Figure 4.

In order to get a better understanding of the structural behaviour of the slab foundation, the case of a 0.15 m-thick flexible infinite slab was also modelled. The settlements obtained for this slab using both Winkler's solution (k_0=3064.4 kN/m^3) and the continuum-medium solution (Soil type I) are presented in Figure 5. It is possible to conclude that in this type of flexible foundation the Winkler's model is also a good approximation to the equivalent continuum-medium solution. This conclusion is also confirmed by the results shown in Figure 6 for M_x and M_y values.

Table 1. Settlements (mx10^{-2}) of selected points of the 0.80 meter-thick rigid infinite slab.

Soil type	E_1 (kPa)	$\frac{E_1}{E_2}$	ν_1	ν_2	G_2 (kPa)	ω_B	$\omega_A=\omega_C$	ω_D
I	9806.7	1	0.35	0.35	3632.4	1.3	1.3	1.3
II	9806.7	1	0.15	0.15	4263.9	1.9	1.9	1.9
III	9806.7	3	0.15	0.15	4263.9	1.7	1.7	1.7
IV	9806.7	3	0.15	0.15	8527.9	1.7	1.7	1.7
V	9806.7	3	0.35	0.15	8527.9	1.6	1.6	1.6

where, ω_A, ω_B, ω_C, ω_D, are the settlements of points A, B, C, D, respectively (Fig. 2).

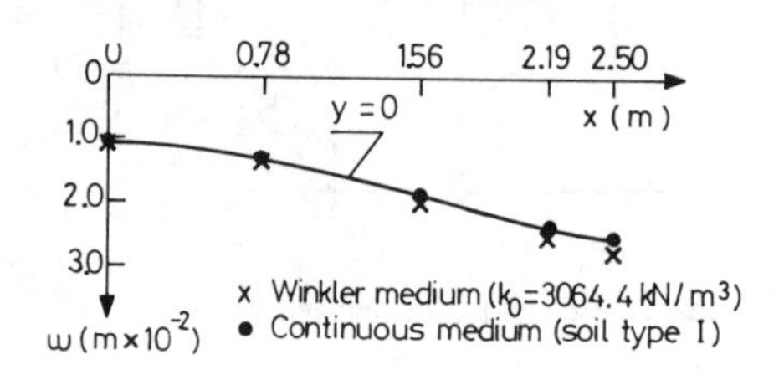

Fig. 5 Settlements of the 0.15 meter-thick flexible infinite slab

Table 2. Settlements (mx10^{-2}) of selected points, and maximum angular distortion, δ, of the 0.15 meter-thick flexible infinite slab.

Soil type	ω_B	$\omega_A=\omega_C$	ω_D	δ
I	2.51	1.09	0.83	1/176
II	3.16	1.74	1.47	1/176
III	2.93	1.53	1.27	1/179
IV	2.81	1.53	1.31	1/195
V	2.71	1.44	1.21	1/197

Table 2 summarizes the settlements of points A,B,C,D, and the maximum angular distortion, δ, between them, obtained for the flexible infinite slab foundation.

This Table shows that the largest settlement is 3.16×10^{-2} m for soil type II and it occurs under point B. This value is 66% larger than the one obtained for the infinite rigid slab at the same point.

The angular distortion δ has a mean value of 1/177 for soils I, II and III, and 1/196 for soils IV and V. The latter soils have a Shear Modulus G_2 approximately twice the Shear Modulus of the former. This would indicate that the parameter G_2 is the only one that influences the magnitude of δ to some significative degree. This conclusion can also be extended to M_x and M_y, because it was observed that they are practically the same for soils I, II and III, and also for soils IV and V. Anyway, the largest variation in the slab bending moments produced by the different types of soils was only about 18%.

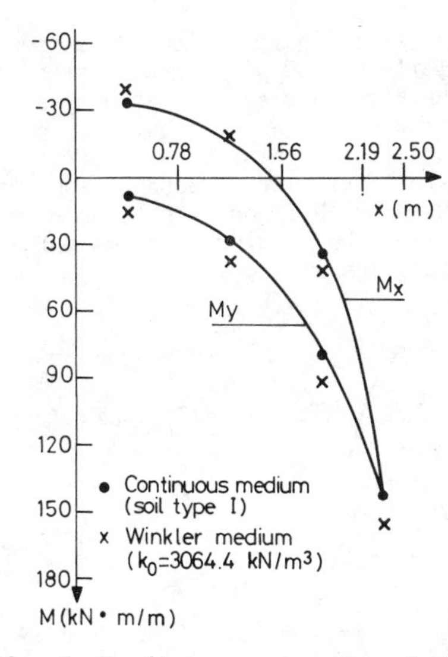

Fig. 6 Bending moments, M_x and M_y, along y = 0.155 m, in the 0.15 meter-thick flexible infinite slab

5 SLAB FOUNDATION OF FINITE EXTENT

The slab foundation of finite extent shown in Fig. 7 was studied. In this case it has also been assumed that the foundation soil is an homogeneous linear-elastic, weightless, 5 meters-thick stratum.

The soil mesh was extended 15.00 m and 12.50 m beyond the slab along each direction, x and y (Fig. 7). These distances can not be too large because of the computer memory requirements. Also, in order to minimize the effects of the boundary conditions of the soil mesh in the computed slab results, such distances can not be too small.

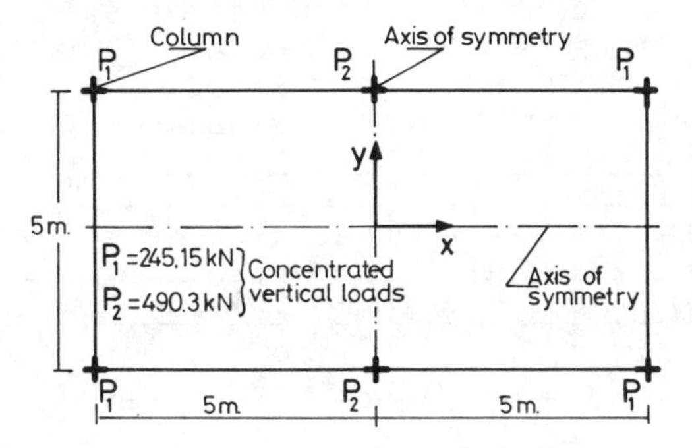

Fig. 7 Slab foundation of finite extent studied

Using the two-axis of symmetry, only a quarter of the slab and foundation soil were modelled with finite elements.

In order to better simulate the problem under the continuum-medium hypothesis, different finite element meshes were analyzed separately; this analysis showed that a mesh made up with 8 rectangular slab elements and 252 rectangular-prism soil elements, was accurate enough. Such mesh was used to derive the results that will be presented herein (Figs. 8 and 9).

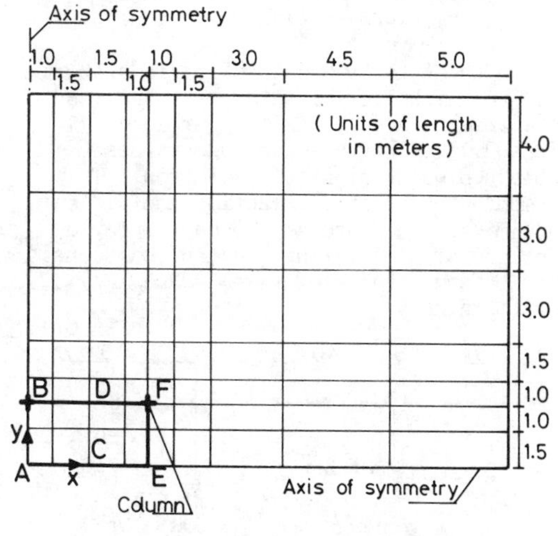

Fig. 8 Finite element mesh for the slab of finite extent: Plan view

The material of the slab was supposed to have the following parameters:

Young's Modulus = 196.13×10^5 kPa
Poisson's Coefficient= 0.2

Unit Weight = 0.0 kN/m^3

Initially the soil was modelled as a continuum medium using soil-type I properties (see Table 1).

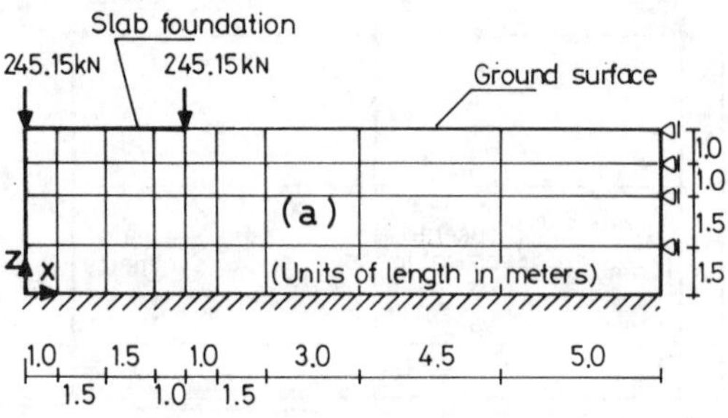

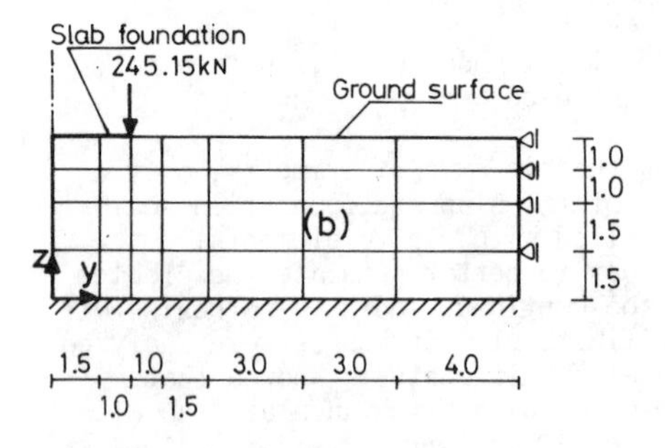

Fig. 9 Finite element mesh for the slab of finite extent:

a) Cross section (xz plane)
b) Cross section (yz plane)

Two cases, a 0.80 m-thick rigid slab and a 0.15 m-thick flexible slab were analyzed. The results are shown in Figures 10, 11, 12 and 13. They also include values obtained using Winkler's solution. To compute the subgrade reaction coefficient a similar procedure as the one employed for the slab of infinite extent was considered, and a value of k_0 = 5099.5 kN/m^3 was obtained.

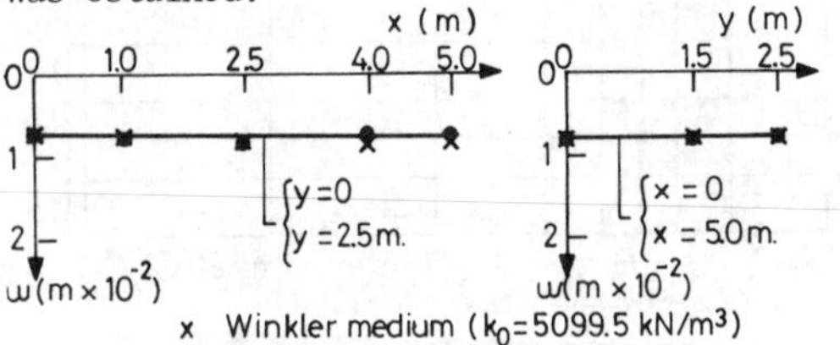

Fig. 10 Settlements of the rigid slab of finite extent studied

From Figures 11, 12 and 13, it is clear that Winkler's solution for slabs of finite extent presents a poor approximation both in settlements and internal forces.

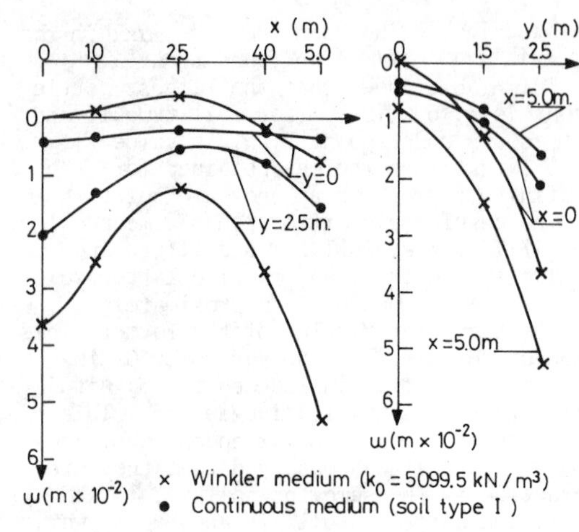

Fig. 11 Settlements of the flexible slab of finite extent studied

Moreover, from Figure 11 it can be seen that Winkler's method gives upward slab displacements in some zones, reaching a maximum value of - 0.004 m. Regarding the slab internal forces, Winkler's theory overestimates the negative bending moments induced on the slab foundation by as much as three times the continuum-medium solution values (Figs. 12 and 13).

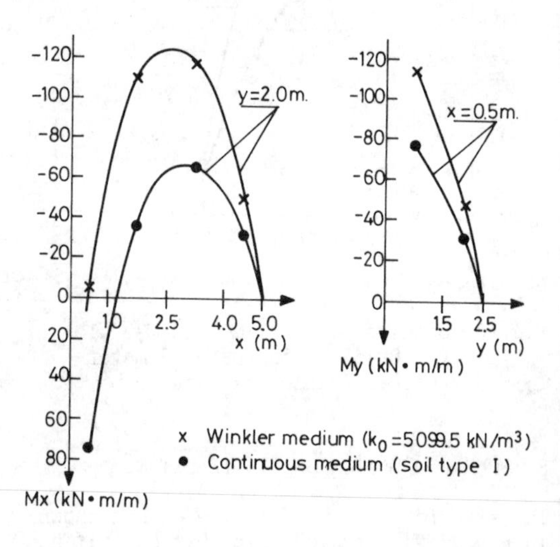

Fig. 12 Bending moments in the rigid slab of finite extent studied

Another interesting aspect is to consider the anisotropy of the medium and to analyze its influence on the behaviour of the slab foundation. Tables 3 and 4 show the settlements on the rigid and the flexible slab respectively, assuming for the soil five different continuum media with geotechnical properties as those defined in Table 1.

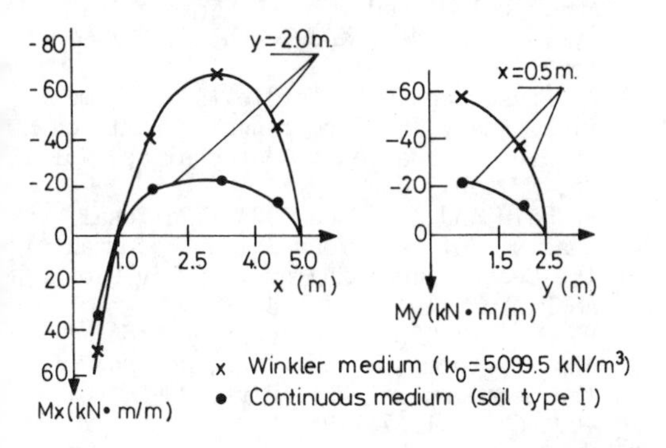

Fig. 13 Bending moments in the flexible slab of finite extent studied

As expected, maximum values of settlement and angular distortion are larger in the flexible slab foundation. It is also observed from Tables 3 and 4 that the resulting angular distortions are grouped in both cases around two different mean values, one corresponding to soil types I, II and III, and the other to soil types IV and V. This is the reason why the internal forces computed for soil types I, II and III, are practically equal. The same result is true for soil types IV and V.

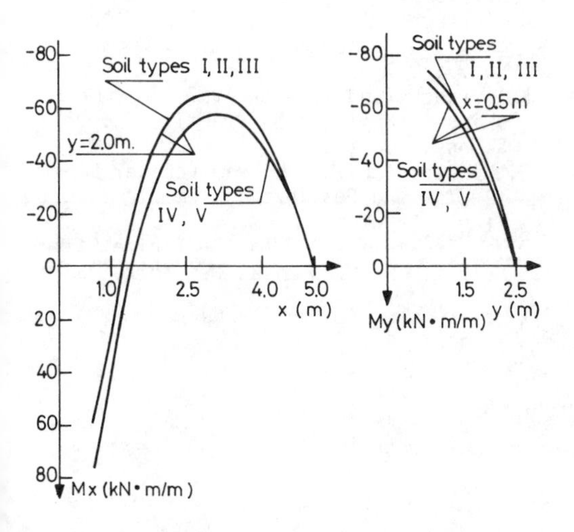

Fig. 14 Bending moments, rigid slab of finite extent over different continuous media

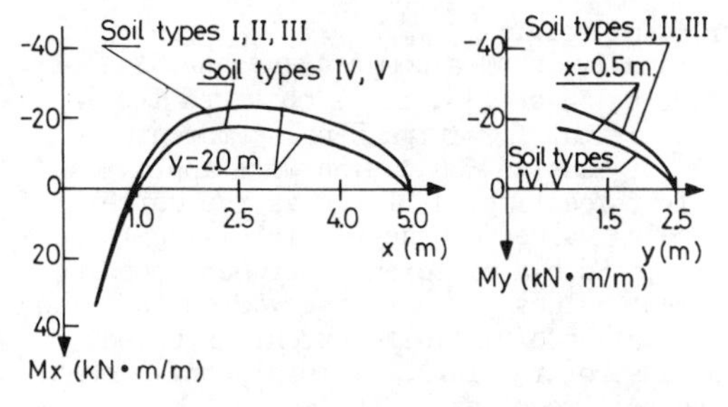

Fig. 15 Bending moments, flexible slab of finite extent over different continuous media

Table 3. Settlements (m x 10^{-2}) of selected points, and maximum angular distortion, δ, of the 0.80 meter-thick rigid finite slab

Soil type	ω_A	ω_B	ω_C	ω_D	ω_E	ω_F	δ
I	0.76	0.79	0.75	0.77	0.75	0.78	1/7669
II	0.92	0.95	0.91	0.93	0.91	0.94	1/7622
III	0.85	0.88	0.84	0.86	0.83	0.87	1/7669
IV	0.72	0.75	0.70	0.72	0.69	0.72	1/7886
V	0.70	0.73	0.69	0.70	0.67	0.71	1/7911

where ω_A, ω_B, ω_C, ω_D, ω_E, ω_F, are the settlements of points: A,B,C,D,E,F, respectively (Fig. 8).

Table 4. Settlements (m x 10^{-2}) of selected points, and maximum angular distortion, δ, of the 0.15 meter-thick flexible finite slab

Soil type	ω_A	ω_B	ω_C	ω_D	ω_E	ω_F	δ
I	0.44	2.10	0.20	0.52	0.31	1.55	1/151
II	0.60	2.31	0.34	0.67	0.43	1.74	1/146
III	0.55	2.19	0.30	0.60	0.38	1.63	1/153
IV	0.51	1.80	0.33	0.50	0.34	1.26	1/193
V	0.49	1.77	0.31	0.48	0.32	1.23	1/194

Figures 14 and 15 show the changes in bending moment values of the rigid and the flexible foundation respectively. It can be concluded that modifications in the Anisotropic Ratio, or Poisson's Coefficients of the soil, practically do not affect internal force values. In connection with the influence of the Shear Modulus, G_2, associated to a vertical plane, it is interesting to observe that when its value is doubled, the corresponding bending moments show variations that are under 15% and 30%, for the rigid and the flexible slab respectively.

Finally, from a comparative analysis of figures 14 and 15, it is observed that the maximum negative bending moments on the rigid slab foundation were approximately three times as large as the corresponding values on the flexible slab foundation. Such a difference between bending moment values was not observed on the slab foundation of infinite extent studied because of the larger lateral confinement the soil receives in that case.

6 CONCLUSIONS

On slab foundations for framed structures of infinite extent the simplified Winkler's solution is very close to the continuum medium solution. However, in slabs of finite extent, the differences produced on evaluating settlements and internal forces when using both procedures can be significantly large.

When modelling the soil as an anisotropic linear elastic continuum medium, it is concluded that the maximum angular distortion and internal forces induced over the slab foundation do not present very significant changes when different values of Anisotropic Ratio, Poisson's Coefficients, and Shear Modulus associated to a vertical plane, are used. In any case the Shear Modulus is the parameter that can originate the most significant differences.

ACKNOWLEDGMENTS

The authors would like to thank the "Dirección de Investigación de la Pontificia Universidad Católica de Chile" for their financial support to the research project DIUC 7/84, that originated the results presented in this paper.

REFERENCES

J.E. Bowles 1974. Analytical and Computer Methods in Foundation Eng. McGraw Hill, N.Y.

R.W. Clough and J.L. Tocher 1965. Finite element stiffness matrices for analysis of plates in bending. International Conference on Matrix Methods in Structural Mechanics, Air Force Institute of Technology, Ohio.

Y.K. Cheung and D.K. Nag 1968. Plates and beams on elastic foundations-Linear and nonlinear behaviour. Geotechnique, Vol. 18:250-260.

Y.K. Cheung and O.C. Zienkiewicz 1954. Plates and tanks on elastic foundations - An application of the finite element method. International Journal of Solids Structures, Vol. 1:451-461.

C.S. Desai and J.F. Abel 1972. Introduction to the Finite Element Method. A Numerical Method for Engineering Analysis. Van Nostrand Reinhold, New York.

R.A. Fraser and L.J. Wardle 1976. Numerical analysis of rectangular rafts on layered foundations. Geotechnique, Vol. 26, N° 4:613-630.

S.J. Hain and I.K. Lee 1974. Rational analysis of raft foundation. Journal of the Geotechnical Engineering Division of ASCE, Vol. 100, GT7: 843-860.

J.S. Horvath 1983. New Subgrade Model Applied to Mat Foundations, Journal of the Geotechnical Eng. Div., ASCE, Vol. 109, GT12: 1567-1596.

J.S. Przemieniecki 1968. Theory of Matrix Structural Analysis. Mc Graw-Hill, New York.

F. Rodríguez-Roa and Gavilán 1985. Losas de Fundación. DIE-85-2, Departamento de Ingeniería Estructural, Pontificia Universidad Católica de Chile, Chile.

F. Rodríguez-Roa and C. Gavilán 1986. Influencia de la rigidez de la superestructura en los esfuerzos y deformaciones de una losa de fundación. II Simposium sobre Aplicaciones del Método de los Elementos Finitos en Ingeniería, U. Politécnica, Barcelona, Spain.

A.P.S. Selvadurai 1979. Elastic Analysis of Soil-Foundation Interaction. Elsevier, New York.

K. Terzaghi, 1955. Evaluation of coefficients of subgrade reaction. Geotechnique, Vol. 5:297-326.

K. Terzaghi and R. Peck 1967. Soil Mechanics in Engineering Practice. J. Wiley & Sons, New York.

E. Winkler 1867. Die Lehre von der Elastizitat und Festigkeit, Dominicus Verlag, Prag.

O.C. Zienkiewicz 1977. The Finite Element Method. Mc Graw Hill, New York.

Numerical Methods in Geomechanics (Innsbruck 1988), Swoboda (ed.)
© 1988 Balkema, Rotterdam. ISBN 90 6191 809 X

The engineering design of pavement units developed from finite element research methods

John W.Bull
University of Newcastle upon Tyne, UK

ABSTRACT. The use of precast concrete pavement units ranging in size from two metres square to 300 mm square is a well established, empirically based, road paving technique in Europe, North America and Japan. A finite element based method using two metre square paving units has been successfully developed for the design of container terminal pavements, where 900 kN axle-loads and 2% CBR sub-grades, subjected to large settlements are not uncommon. A design example is given. For the smaller units less than 600 mm square, further research at Newcastle upon Tyne University has developed finite element methods, linked to laboratory and on-site testing, that can relate the field measurements due to standard axle-loading and pavement unit movement to the stresses and fatigue life of the pavement units. The basis of the design method is explained and an example given. For both of the design methods, the pavement life and the sub-grade stress can be calculated and related to the stresses on underground services such as gas and water supply pipes.

1. INTRODUCTION

The use of precast concrete pavement units ranging in size from two metres square down to 300 mm square is a well established road paving technique in Europe as well as in North America and Japan (Rollings 1981).

The range of manufactured sizes and manufacturing techniques is considerable. The smaller precast concrete paving units, up to 600 mm by 900 mm are manufactured using either the wet press process or the dry press process. Larger sizes, which may be up to 3.2 metres by 5.3 metres and even 2.29 metres by 10.0 metres are cast in moulds using high strength concrete. The larger sizes are normally reinforced with steel mesh although prestressing wires have been used, (Rollings 1981). The thicknesses range from 50 mm for the smaller sizes up to 220 mm.

The advantages of using precast concrete are its economy of scale in mass production, guaranteed quality control during manufacture and the rapid construction of the pavement.

In the early 1980s a major research effort was initiated at the University of Newcastle upon Tyne to develop a design method for the use of precast concrete pavement units subjected to vehicular overrun. The initial research was into the use of the two metre square pavement units. With the successful completion of the finite element numerical analysis, a similar finite element programme of work was developed for the smaller paving flags, having a plan area less than 0.36 m^2.

2. REPAIR CRITERIA FOR PAVEMENT UNITS

Existing pavement design in the United Kingdom (Croney 1977) converts the wheel loading effects into the cumulative number of 80 kN standard axles, to obtain the pavement thickness and strength. Unlike normal highway pavements, where large amounts of empirical data is available, the design method for pavement units relies heavily upon the use of the finite element method, backed up by the results of full-scale laboratory testing.

For precast concrete pavement units neither the deflection criteria of flexible pavements nor the cracking criteria for rigid pavements are adequate repair criteria. For example, in flexible pavement design, a pavement would be considered to have failed if the surface deformation exceeds 25 mm (Croney 1977). For two metre square pavement units, a 25 mm differential settlement between units is overcome by lifting and relevelling of the units. For units less than 600 mm square, the repair criteria is either the cracking of the unit, of the differential movement between the units. The author decided that the critical repair parameters in the pavement were the maximum tensile stress in the pavement unit (Vesic 1969) and the maximum compressive stress on the sub-grade (Bull 1986). The pavement unit life can then be expressed as the number of standard axle-load applications, related to the pavement unit tensile stress and the point at which pavement relevelling is required related to the sub-grade verticle stress.

3. FINITE ELEMENT MODELLING

To investigate the pavement and sub-grade stresses the finite element package PAFEC (PAFEC 1978) was used. The use of the finite element method enabled the complexities of the foundation together with the shape and size of the loaded area to be overcome. For example, the initial analytical work by Westergaard (Westergaard 1926) identified three critical load cases for concrete pavements, namely the central load position, loading at the middle of an edge and the loading at the corner of the unit. These three load positions were analysed and resulted in many pavement designers having to develop the equivalent single wheel load concept, where more than one wheel load of a vehicle, was represented as a single wheel load. This led to inaccurate stresses in the pavement units (Bull 1987).

The PAFEC finite element system has available, a number of finite element types. These types include thin shell and brick elements, which allow the analysis of, the smaller pavement units, the layered foundations and loads distributed over a tyre contact area at various positions on the unit.

Comparative trials to determine the most suitable combination of elements for the two metre square units, based on a simply supported plate and a plate with free edges were run. It was found that provided the elastic foundation layer was thin, the three dimensional brick finite elements overlaid by four noded thin shell elements gave the best agreement with the analytical solutions, (Ackroyd 1985). The accuracy of the computer model was assessed using either the continuity of stresses across element boundaries, or its convergence to a unique solution as the mesh density was increased.

The four node thin shell element was used to model the pavement unit and the eight node brick elements were used for the bedding sand, the sub-base and the sub-grade. It was also found that the central loading position was not as critical as the corner loading position or the centre of the edge position and was deleted from the analysis.

The variables investigated in the analysis and their values are given in Table 1. The values in brackets apply to the units 600 mm square and smaller.

TABLE 1 : Pavement Design Parameters

Pavement layer	Value
Loaded area	200 mm x 200 mm
Pavement unit	2 m x 2 m x 150 mm (see Table 2) 50 MPa concrete (32 MPa)
Bedding layer	50 mm thick (25 mm) 7.5% CBR
Sub-base	300 mm thick (75 mm) 20% CBR
Sub-grade	1200 mm thick (200 mm) 0.3% CBR

Varying each parameter in turn, with all other parameters remaining at their standard values, the following conclusions were drawn.

1. The thickness and the CBR values of the sand bedding layer altered the concrete stresses and sub-grade stresses by less than one per cent and were deleted from the design charts.
2. The effect of varying the sub-grade thickness, reduces as the thickness reaches 1200 mm. Few sub-grades are thinner than this value and the charts assume the thickness to be 1200 mm.
3. The sub-grade stresses depend principally upon the sub-grade CBR, the sub-base thickness and sub-base CBR. The combined effects of the other parameters cause variations of about 10%.
4. The maximum sub-grade stress occurred when the unit was loaded at a corner and the maximum pavement unit stress occurred when the unit was loaded at the centre of one edge.

4. DESIGN CHARTS FOR TWO METRE SQUARE UNITS

The design charts were produced from computer runs for corner and edge loading.

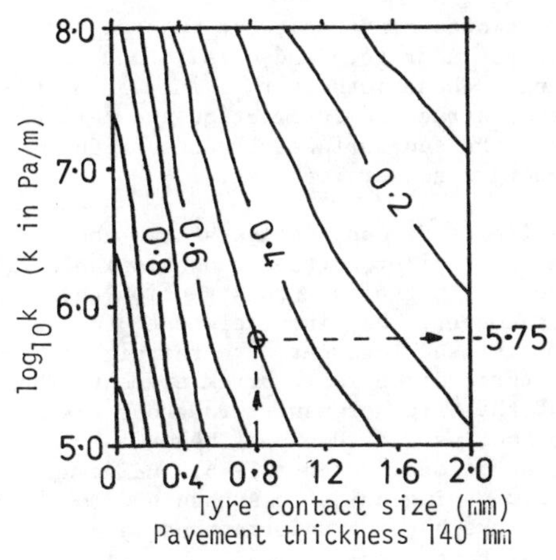

Figure 1: Pavement Unit Stress

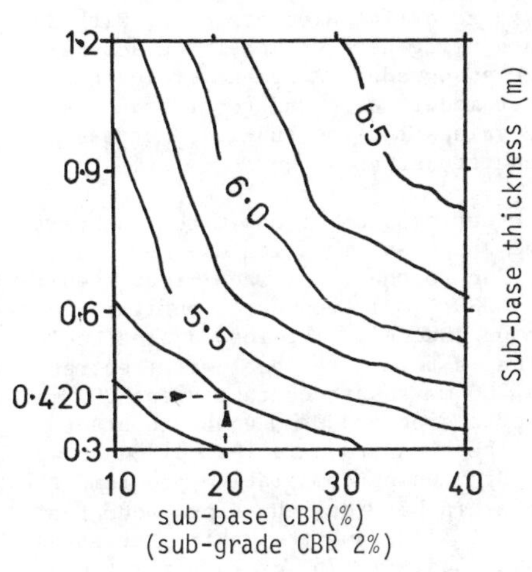

Figure 2: Sub-base thickness

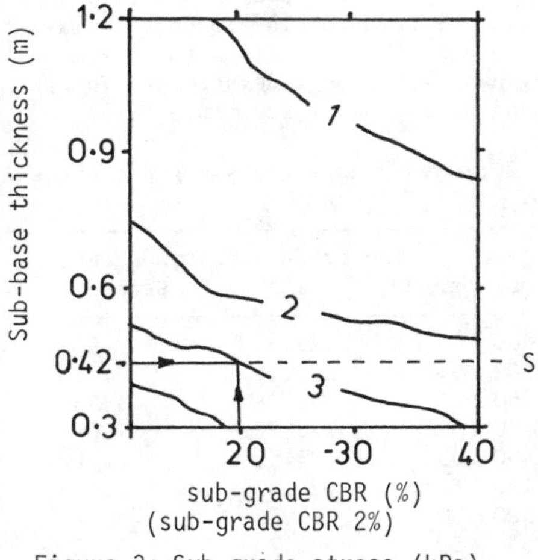

Figure 3: Sub-grade stress (kPa)

Sub-grade stress 3 kPa x $\frac{100\ kN}{10kN}$

= 30 kPa

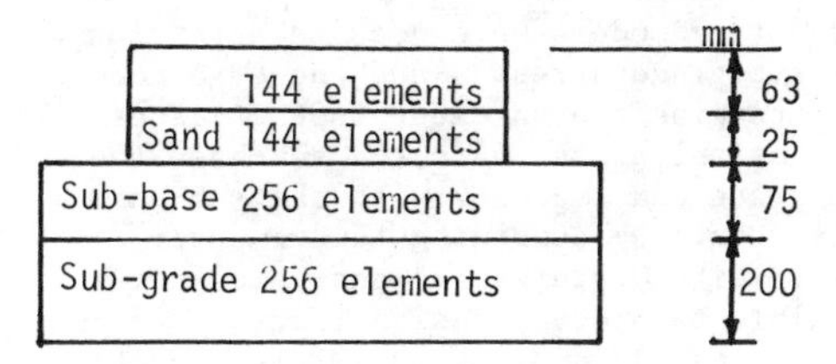

Figure 4: Typical pavement cross-section

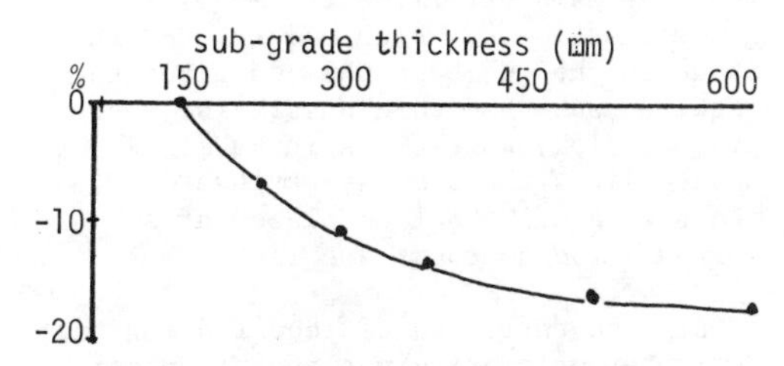

Figure 5: Concrete stress related to sub-base thickness

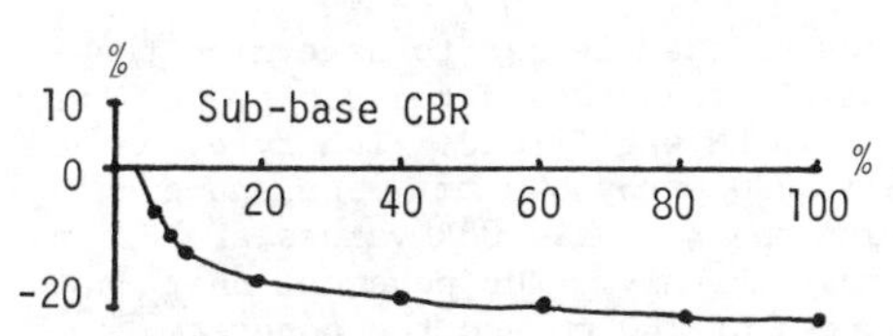

Figure 6: Concrete stress related to sub-grade CBR

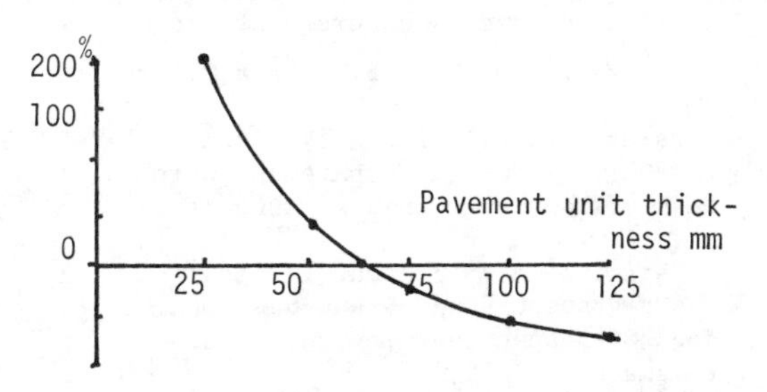

Figure 7: Concrete stress related to pavement unit thickness

The corner models were used to determine the sub-grade stresses and the edge model to determine the pavement unit stresses. The variables used were the sub-base and sub-grade CBR together with the sub-base thickness. By combining the computer result with Westergaards analysis it is possible to produce design charts as shown in Figures 1,2 and 3. Figure 1 is used to determine the maximum pavement unit stress for a standard 10 kN load at the centre of one edge. The contours are in MPa and the verticle axis determine the modulus of sub-grade reaction k. Figure 2 relates the k value to the sub-base CBR and gives the required sub-base thickness. The contours give Log 10k, where k is in Pa/m. Figure 3 determines the sub-base stress due to the standard 10 kN load placed at one corner. The contours are in Pa.

Only the three charts required for the design example are given in this paper.

4.1 Design Example

The design example has to determine the sub-base thickness and the sub-grade stress for a 100 kN wheel-load. The tyre contact size is 800 mm by 800 mm. The sub-grade CBR and the sub-base CBR values are 2% and 20% respectively. The pavement unit has a thickness of 140 mm and the concrete stress is not to exceed 5 MPa.

Relating the applied load of 100 kN and the concrete stress of 5 MPa to the basic 10 kN load gives a concrete stress of

$$(10\ \text{kN} \div 100\ \text{kN}) \times 5.0\ \text{MPa} = 0.5\ \text{MPa} \quad (1)$$

Using Figure 1 with a tyre contact area of 800 mm x 800 mm, a pavement unit thickness of 140 mm gives a k value of 5.75.

Using the 5.75 contour in Figure 2 with a sub-grade CBR and a sub-base CBR of 2% and 20% respectively, gives a sub-base thickness of 420 mm.

In Figure 3 the CBR values and sub-base thickness value found from Figure 2 are used to determine the sub-grade stress of 3 kPa for a 10 kN basic load. For the 100 kN load the sub-grade stress is

$$((3\ \text{kPa} \times 100\ \text{kN}) + 10\ \text{kN})) = 30\ \text{kPa} \quad (2)$$

5. FINITE ELEMENT MODELLING OF THE SMALL PAVEMENT UNITS

For the finite element analysis of pavement units 600 mm square and less, research at the University of Newcastle upon Tyne has developed finite element methods, linked to laboratory and on-site pavement testing. These methods take into account the soil/structure interaction and relate this to the sub-surface stresses produced on underground services.

The finite element analysis used the basic material properties shown in Table 1. Figure 4 is a typical cross section through the pavement. The finite element package PAFEC was used together with the eight node, three dimensional brick element. The use of the brick element throughout was preferred, as the theory of beams on elastic foundations (Hetenyi 1946) indicated that for paving units less than 600 mm square, the bending deformations were so small as to be negligible.

The paving unit and the bedding sand were represented by 144 elements each, with 256 elements being used to model the sub-base and the sub-grade. A number of further computer models were run to provide assymptotic values for the sub-grade stress and the concrete stress.

A further consideration in the analysis was the shape and location of the loaded area. Due to the legal wheel-load requirements (HMSO 1982) two wheel configurations had to be checked. The loaded area could be either a twin-tyre wheel with centres of the 200 mm square contact areas 300 mm apart or a single 300 mm square contact area. The standard axle-load of 80 kN was used. By running comparative programs the relationship between the single and the twin wheel loads and the maximum stresses were obtained. In 75% of loadings the single wheel load gave the maximum sub-grade stress and the maximum paving unit stress. Only for the 300 mm square paving unit did the twin wheels produce the maximum stresses. Table 2 gives the maximum paving unit and sub-grade stresses for the standard paving unit sizes (BS368:1971).

TABLE 2 : Paving unit and sub-grade stresses

Paving unit size mmxmmxmm	Sub-grade stress kPa	Concrete stress MPa
600 x 600 x 63	122.48	7.94
600 x 600 x 50	125.73	10.71
450 x 450 x 70	123.38	4.53
450 x 450 x 50	125.32	7.69
400 x 400 x 65	124.58	3.81
400 x 400 x 50	125.67	5.88
300 x 300 x 60	163.42	3.02
300 x 300 x 50	163.78	4.17

For each of the variables in Table 1 for the 600 mm square units, a series of finite element programmes were run. The results were drawn in % form, relating the alteration in the variable to the change in the paving unit stress and sub-grade stress. In the interests of clarity only the % alteration in paving unit stress is shown on Figures 4,5 and 6. Figure 4 shows how varying the sub-base thickness reduces the paving unit concrete stress, while Figure 5 relates the paving unit concrete stress to the sub-grade CBR. In Figure 6 the relationship between paving unit concrete stress and paving unit thickness is explored.

Two further pieces of research were drawn together to produce the small element paving unit design method. Firstly, the relationship between the load induced paving unit concrete stress, divided by the concrete modulus of rupture (The stress ratio), and the number of load repetitions has been investigated by many researchers, but the initial design method used here uses Portland Cement Association (PCA 1966) relationships. Work is progressing to assess other fatigue relationships and their viability in our design method. Secondly, the British Standards Institute (BS368:1971) failure stress values for the paving units are used to determine the paving unit concrete modulus of rupture.

6. DESIGN CHARTS FOR SMALL ELEMENT PAVING UNITS

The design of a pavement using 300 mm x 300 mm x 60 mm thick paving units is requested. The required number of standard axle over-runs is 130,000 and no alterations in the standard values of Table 1 are allowed except for variations in sub-base thickness, sub-grade CBR and if necessary, the paving unit thickness.

From Table 2 the paving unit concrete stress for a standard axle-load is 3.02 MPa. The failure flexural stress is 4.79 MPa (BS368:1971), giving a stress ratio of:

$$3.02 \text{ MPa} \div 4.79 \text{ MPa} = 0.63 \quad (3)$$

To sustain 130,000 standard axle overruns a stress ratio of 0.55 is required (PCA 1966) and the applied stress must be reduced to:

$$4.79 \text{ MPa} \times 0.55 = 2.63 \text{ MPa} \quad (4)$$

a reduction of 12.9% on the standard axle applied stress of 3.02 MPa. The 12.9% reduction can be achieved in the following way:

1) A 10% reduction in stress can be achieved by increasing the sub-base thickness from its standard value of 75 mm to 225 mm.
2) Increasing the sub-grade CBR from 2% to 4% would reduce the stress by a further 3%.

Adding the 10% to the 3% gives a 13% stress reduction, which is more than required to give 130,000 standard axle over-runs.

The alteration in the sub-grade stress can be calculated in a similar way. Increasing the sub-base thickness, reduces the sub-grade stress by 43.55%, but increasing the sub-grade CBR, increases the sub-grade stress by 22.5%. The final alteration in sub-grade stress is as follows:

$$\frac{163.78 \text{ kPa} \times ((100\% + (-43.5\% + 22.5\%))}{100\%}$$

$$= 129.39 \text{ kPa} \quad (5)$$

The value of this sub-grade stress can be used to determine the verticle pressures on underground services.

An alternative way of reducing the paving unit stress by the required 12.9% is to increase the unit thickness from 60 mm to 67 mm (Figure 7). However, the sub-grade stress is only reduced by 2% to 160.15 kPa, but it does give the pavement designer the choice to optimise his design.

7. CONCLUSIONS

1. A design method for two metre square precast concrete pavement units has been described and illustrated.
2. A design method for square precast concrete pavement units ranging in size from 600 mm square down to 300 mm square is being developed, and has been described. A design example has been given. The method allows an optimum pavement design for standard axle-loading and for the stresses on underground services to be determined.
3. Research is continuing to develop and validate the design method for small element paving units. Work is progressing both in the laboratory and at four sites in the United Kingdom.

8. ACKNOWLEDGEMENTS

This work is supported by the United Kingdom Science and Engineering Research Council (SERC), Redland Aggregates Ltd, Redland Precast Division and the National Paving and Kerb Association (NPKA).

9. REFERENCES

Ackroyd, R.F. and Bull, J.W. (1985) The design of precast concrete pavements on low bearing capacity sub-grades. Computers and Geotechnics, 1, 279-291.

BS368:1971 Precast concrete flags, BSI, London.

Bull, J.W. (1986) An analytical solution to the design of precast concrete pavements. Int J Num Methods Geomechanics, 10, 115-123.

Bull, J.W. and Salmo, S.H. (1987) The use of the equivalent single wheel load concept for discrete raft type pavements. Computers and Geotechnics, 3, 29-35.

Croney, D. (1977) The design and performance of road pavements. Department of the Environment, TRRL, HMSO, London.

Hetenyi, M. (1946) Beams on elastic foundation. Ann Arbor: University of Michigan Press, USA.

HMSO, (1982) Motor vehicles (Contruction and use) (Amendment) (No.7) Regulation, HMSO, London.

PAFEC, (1978) Theory, Results, PAFEC Ltd, Nottingham.

Vesic, A.S. and Saxena, S.K.(1969) Analysis of the structural behaviour of road test rigid pavements, Highway Research Record No.291.

Westergaard, H.M. (1926) Stresses in concrete pavements computed by theoretical analysis. Public Roads 7 (2), 25-35.

Numerical Methods in Geomechanics (Innsbruck 1988), Swoboda (ed.)
© 1988 Balkema, Rotterdam. ISBN 90 6191 809 X

Designing of spread footings on karst-prone sites

V.I.Solomin
Chelyabinsk Polytechnical Institute, USSR
T.A.Malikova
Research Institute of Bases and Underground Structures, Moscow, USSR
P.A.Karimov
Magnitogorsk Mining and Metallurgical Institute, USSR

Abstract:In the designing of foundations on karst areas reinforced concrete slabs were used as one of efficient types of foundations.Winkler model with variable stiffness factor has been adopted.This made possible accounting heterogeneity of soil base in plan as well as funnels and their boundary weakened zones.

The problem of designing buildings on karst hazardous sites in gaining extreme urgency nowadays due to extensive construction within areas affected by karst and suffosion phenomena as well as to frequent karst occurence on sites that have not been classified as karst hazardous as yet.Location and dimensions of karst subsidence are inherently statistical that makes probabilistic approach indispensable for solving the problem.

The probabilistic approach makes it possible:to assess the feasibility of construction on the karst site in question,to select the type of the building andits footings that would ensure the minimal cost for the stipulated reliability level; to assess probability of the subsidence under existing buildings and its implications respective(damage rate of the buildings and rebuilding costs).

We define "refusal" as the building's failure to carry on its normal functions.Three main types of refusals caused by karst subsidence may be singled out:excessive tilts and settlements;loss of strength and stiffness of the footing slab leading to the damage of some components of the building;destruction either of the whole building or its portion.

We assume that a refusal is caused by a karst subsidence whose diameter is greater than a certain theoretical limit value related to the kind of the refusal.The probability of the refusal is determined from the following formula:

$$p_o = p_1(1-p_d), \qquad (I)$$

with p_I as probability of karst subsidence occurence;p_d as probability that the subsidence would have smaller diameter than the theoretical limit value.

To determine p_I and p_d we apply the method proposed by V.V.Tolmachev (Tolmachev V.V. 1980).

The assumptions given above enable in principle designing buildings in karst conditions with any predetermined reliability.But the respective requirements are to comply with the feasibility conditions.

Fig.I shows the diagram of C_I,i.e. capital investment for the footing slab karst protective measures,versus refusal probability.This diagram has been plotted from the data obtained by numerical analysis of a number of buildings with framed superstructure.

Consider the characteristic points of the diagram.Point A corresponds to a slab designed without account of karst subsidence and having,therefore, minimal cost.The probability of such slab refusal is equal to p_I since a subsidence of any size under it would cause at least one type of refusal.

To reduce the refusal probability it is necessary to take into account the karst subsidence occurence under the building.To this end, minimal dimensions of the footing

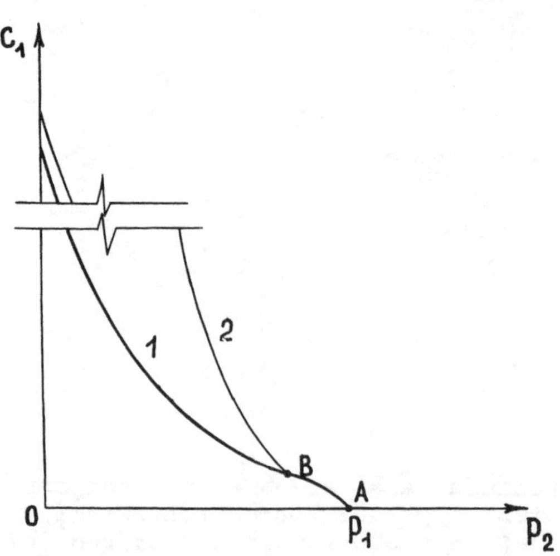

Figure I. Anti-karst measures cost versus refusal probability curve (I-curve of refusals caused by failure of the slab,2-curve of refusals caused by excessive tilts and settlements with no failure of the slab).

slab are initially assumed .
Point B corresponds to a slab whose dimensions and reinforcement abide by the requirements for its strength and stiffness.The maximum diameter of the subsidence is such that there is no need to enlarge the footing for ensuring allowable values of settlements and tilts.In this particular case the occurence of a subsidence whose diameter is greater than d_B would lead to refusal.Further reduction of probability of refusals of any kind makes it necessary to enlarge the dimensions of the footing slab so that mean and maximum contact pressures as well as its settlements,tilts and strehgth would not surpass allowable limits.Extension of the edge consoles of the slab gives rise to considerable increase of internal stresses in it and, hence to its increased reinforcement. The cost of the slab grows dramatically if the probability of the refusal of any kind is being decrease (line 2).However,if we allow excessive tilts and settlements and stipulate integrity of the structure as the main objective than the slab's strength is ensured,for the case when the diameter of the subsidence is greater than d_B ,at lower cost than that of the slab designed to withstand the same subsidences and, simultaneously,to have tilts and settlements within allowable limits (lineI).Therewith,we should bear in mind that for a slab of particular size and a soil of particular properties the tilts and settlements might be below allowable limits any way. In this case lines I and 2 coincide while point B disappears from the diagram altogether.
Assessment of a footing slab cost is linked up with its static analysis.Evidently,the cost of the slab depends on its dimensions,thickness and reinforcement rate.
When anti-karst measures scope is specified for a project design the question arises whether they are economically feasible.The less are the expenditures for anti-karst measures the greater are the investments to restore the damaged buildin Karst control feasibility analysis could be based on the same principles as are applied to seismic protection design analysis.
The optimization criterion could be the mean cost for the karst control measures(Aizenberg ,Neiman I973):

$$C = C_1 + C_g = min, \qquad (2)$$

where C is the total cost linked up with karst hazard;C_I is initial karst control cost; C_B is expected los due to temporal delay in service.
Social factors should also be taken into account,beside economic ones,when designing residential, public and the majority of industrial buildings.Most authors hold the view that the scope of karst control measures should stem from the allowable destruction probability for the whole building or its components i.e. this scope is determined from curve I.But the building destruction probability value has not been standardized as yet.It is only evident that its value (P) is to make allowance for the class of the buildings.Thus,the karst control cost optimization problem solution boils down to solving equation (2) with constraint $P_{destr.} \leq (P)$ with P_{destr} standing for probability of destruction.By modifying formula (2) as is done in (2) we obtain

$$C = C_1 + p_o A/(ET) = min, \qquad (3)$$

with A as damage cost due to the building refusal;E as a norm for reduced expenditures which is now taken to be equal to 0.08; T as the service life term of the building.Clearly,damage cost A due to the building refusal ia a stochastic value and depends on many factors such as the cost of the building,anti-karst measures scope,the diameter of the subsidence,its location with respect to the building,stiffness of the superstructure,the number of residents and other factors.Mathematical expectancy of value A is to be incorporated in formula (3).

For this purpose, statistical data on buildings karst damage costs is necessary.But such data has not been published yet.The following considerations may help determine an approximate value of A.Assume that a building designed to withstand a karst subsidence with diameter d_d (d_d stands for the karst diameter assumed in the design) suffers some irreparable damage if the occured subsidence diameter is less than d_{lim} where d_{lim} is the subsidence diameter for which the subsoil may loose its stability.

Assume the building repair cost dependence on the occured subsidence diameter depicted on Fig.2. to be valid.The diagram may be broken down into three segments.Segment I corresponds to the cost for restoring the building to its original state if the occured subsidence diameter is less than that taken in the design analysis.In this particular case the building does not suffer severe damage and the mitigation cost only covers activities for filling the gap and for taking measures to prevent possible future subsidence.Segment 2 corresponds to the building restoration cost in the case when the occured subsidence exceeds that assumed in the design but there does not happen total destruction.The cost of restoration and repair operations might be quite high and depends on the occured subsidence diameter and its location with respect to the footing slab.

Consider tentatively that for this segment the damage loss cost assumes values from 0,2 to 0,9 fraction of the estimate building cost (S). Segment 3 corresponds to the irrepairable damage of the building.

We tentatively assume the resulting damage cost to be equal to the doubled estimate cost of the building because it includes the expenditures for rehousing the people,compensation for the property damage,and the building demolishing cost.

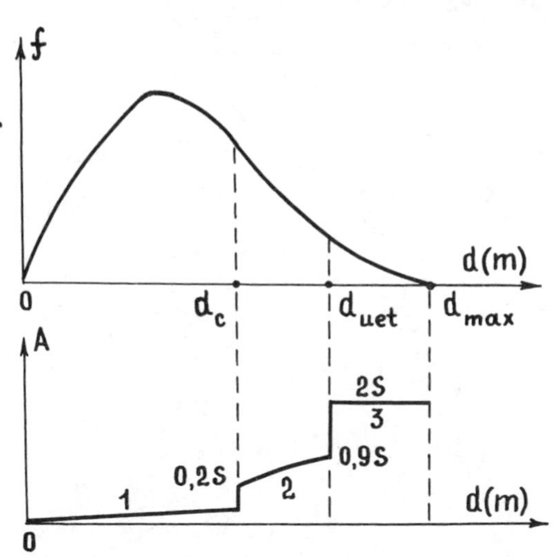

Fig.2. Expected damage versus karst subsidence curve.

Therefrom if the subsidence diameter density distribution function f(d) is available for the construction site area then the expected damage mean value can be determined from the following formula:

$$A=\int_0^{d_p} f(d)A_1(d)d(d)+\int_{d_p}^{d_{np}} f(d)A_2(d)d(d)+\int_{d_{np}}^{d_{max}} f(d)A_3(d)d(d). \quad (4)$$

Computer analysis carried out with the help of the programs based on the method described above has yielded the following conclusions: anti-karst measures scope may be specified on the basis of economic considerations for the sites where only small size subsidences may occur that do not involve human risk;reliability requirements must be the priority when specifying anti-karst measures scope for the sites where large size subsidences may occur.

REFERENCES

Tolmatchev V.V.I980.A probablistic approach to the assessment of stability of karst prone sites and to the design of anti-karst measures.Inzhenernaya Geologia 3: 98-I07.

Aizenberg Ja.M.,Neiman A.J.Economic assessment of seismo-resistant structures optimality and the balanced risk principle. Stroitelnaya Mekhanika i Raschet Sooruzhenyi 4:6-IO.

Numerical Methods in Geomechanics (Innsbruck 1988), Swoboda (ed.)
© 1988 Balkema, Rotterdam. ISBN 90 6191 809 X

Simulation of behaviour of a gravity structure subject to lateral loading

V.G.Fedorovsky
Research Institute of Bases and Underground Structures, Moscow, USSR
S.V.Kurillo, S.N.Levachev & A.F.Lunin
Moscow Civil Engineering Institute, USSR

ABSTRACT: Experiments have yielded that models of offshore gravity platforms subject to lateral loads experience additional settlements, tilts and redistribution of contact stresses. A contact model of the soilbase is proposed that predicts these nonlinear phenomena.

1 INTRODUCTION

Offshore developments in northern seas require installation of gravity platforms that can resist extreme environmental impacts, ice inclusive. The main structural features of such platforms, footings and the conditions which they bear may be conceived from the following factors. The spread footing slab may have up to 15000 m² area and can bear vertical load up to 2 GN, lateral load up to 0,8 GN, and overturning moment up to 15 GNm and more. In such conditions the offshore platforms are to be designed to bear considerable lateral and vertical movements that trigger nonlinear soil behaviour.

In order to investigate the specifics of the footing-soil interaction experiments have been staged on models of various sizes and design. On small size (F=0.3-0.6 m²) models the metodology was studied and then applied to large-scale experiments with a rigid 4 m² test plate.

2. EXPERIMENTS

The experiments were carried out in 8x5x3.2 m sandbank equipped with mechanical and hydraulic facilities for replacing the soil and with a loading hydraulic system that ensured central and eccentric independent application of vertical and lateral static forces up to 1MN following various patterns and modes of loading. Medium grain sand was put down to different densities with natural water content (4-6%) and then in some cases was saturated with water. Up to 22 soil contact dynamometers were installed in the plate for measuring normal and shear stresses at the bottom of footing (see [4]). Settlement, lateral displacement and tilt were measured with the help of benchmark system.

The general purpose facility enables testing a footing with additional deep units, namely piles or skirt. The results of this work are, however, not presented here.

Below brief description of methodology of the experiments is given to assess its effectiveness.

1. There has been tested a small-size model footing of a gravity platform on saturated (or of natural water content) sand of medium density that frequently occurs on many offshore sites. The plate is classified as rigid one as the actual footing in compliance with flexibility index r (see [3]).

2. The experiments were targeted to clarify plate-soil interaction and to verify the new method of analysis laid out below.

3. Combination of loads (vertical, lateral forces, and overturnind moment), their application (central, eccentric) and their lev-

el (value of contact stresses along the bottom) corresponded to static loading of ice-resistant platforms. The loading mode also imitated real conditions: firstly, vertical load was applied stepwise, then also stepwise (not more than 10% of the maximum load per one step) lateral load was applied to achieve the ultimate state. Each subsequent step was maintained until deformations stabilized.

4. The main requirement posed to the tests was to get maximum possible range of data, primarily, on soil resistance at the contact for the cross check-up of measurements and for obtaining reliable experimental data.

5. Special attention was put to check-up soil compaction and to determination soil strength and deformation parameters in water-saturated state.

The experiments (their number was over 40) have yielded the following main results (Fig. 1,2):

- test plate settlements w, after the pressure normal to the plate bottom achieved a certain value, are nonlinear versus vertical load P;
- lateral displacement u versus lateral load H dependence is highly nonlinear over the whole range of H-u curve up to critical value of H;
- test plate settlements grow if lateral load is applied:,
- normal contact stresses profile induced by vertical load becomes parabolic when the lateral force grows with a maximum in the center of the plate and lowering stresses at the edges;
- shear stresses, that are directed towards the center of the plate if H=0, grow at the back part of the plate and lower at the front part when the lateral load increases;
- application of the lateral force H tilts the plate i.e. the front edge settlement becomes greater than that of the back edge. This phenomenon occurs even if h=0, h as force H application point height above the plate bottom (Fig.1);
- water saturation of soil leads to its considerable softening.

It turned out that the apparent Young modulus E_0 of the soilbase determined from the slope of the P-w curve linear segment is roughly proportional to $\gamma' B$ with γ' as the sand specific gravity with water displacement taken into account, B as the plate width. This is evident from the fact that settlement of the plate resting on "dry" sand does not depend on its size while said settlement is doubled for the case of water saturated soil base (Fig. 2).

3. THEORY

The proposed method of analysis is based on the concept of soilbase contact model (CM) [1,2] in which deformativity of the soilbase is expressed in an integral form as stresses versus displacements relationship at the footing-soil contact plane. The assumed CM, same as the one proposed by Barvashov and Fedorovsky [1,2], is a serial arrangement of two layers. The upper layer (all its parameters are indexed with 1 below) is a generalized Winkler CM whose elasto-plastic springs are independently deformed vertically and laterally. The lower layer (all its parameters are indexed with 2 below) is a generalized Filonenko-Boroditch - Pasternak - Vlasov CM. The upper Winkler layer carries the footing and transmits the contact stresses to the lower layer unchanged. Total displacements of the footing are defined by the following equations

- vertical $w = w_1 + w_2$
- lateral $u = u_1 + u_2$ (1)

Constitutive equations of the lower CM are obtained with the help of Vlasov's variational technique (see [5]). Displacements of the elastic layer with the thickness H_2 and elastic parameters E_2, μ_2 on a stiff base in plane strain are expressed in the following form

$$u(x,z) = \sum_{i=1}^{m} \bar{u}_i(x)\varphi_i(z)$$

$$w(x,z) = \sum_{j=1}^{n} \bar{w}_j(x)\psi_j(z) \qquad (2)$$

$$-\infty < x < \infty; \; 0 \le z \le H_2$$

where φ_i and ψ_j are fixed functions of z while $\bar{u}_i(x)$ and $\bar{w}_j(x)$ are unknown functions. If we assume $m = n = 1, \; \varphi_1(0) = \psi_1(0) = 1$

then $\bar{u}_1$ and $\bar{w}_1$ are the layer surface displacements. Thereupon, functions $\bar{u}$ and $\bar{w}$ (index 1 may be omitted) can be found from the equations

$$A\bar{u}_{xx} - B\bar{u} + C\bar{w}_x + \tau = 0$$
$$-C\bar{u}_x + D\bar{w}_{xx} - F\bar{w} + \sigma = 0 \qquad (3)$$

where τ and σ stand for shear and normal stresses on the surface z=o, respectively, and

$$A = \frac{E_2}{1-\mu_2^2}\int_0^{H_2} \varphi_1^2 dz;$$

$$B = \frac{E_2}{2(1+\mu_2)}\int_0^{H_2} \varphi_1'^2 dz;$$

$$C = \frac{E_2}{1-\mu_2^2}\int_0^{H_2}\left(\mu_2\varphi_1\psi_1' - \frac{1-\mu_2}{2}\varphi_1'\psi_1\right)dz; \qquad (4)$$

$$D = \frac{E_2}{2(1+\mu_2)}\int_0^{H_2} \psi_1^2 dz;$$

$$F = \frac{E_2}{1-\mu_2^2}\int_0^{H_2} \psi_1'^2 dz.$$

The simplest option corresponds to linear functions φ_1 and ψ_1

$$\varphi_1(z) = \psi_1(z) = \frac{H_2 - z}{H_2} \qquad (5)$$

Functions $\bar{u}(x)$ and $\bar{w}(x)$ found by solving system (3) are employed for determining displacements in accordance with equation (1) for u_2 and w_2 respectively. Displacements of the upper CM in elastic state are expressed as follows

$$u_1 = \tau/K_u; \quad w_1 = \sigma/K_w \qquad (6)$$

with K_u and K_w as constant parameters (moduli of subgrade reaction). After achieving a certain level of loading a limit state is generated a plastic flow begins at a given point of the upper CM. Possible forms of limit soilbase behaviour when it interacts with a flat plate are as follows: a) superficial plane shear; b) deep upthrust; c) combinations of both.

Plane shear is described by Coulomb law

$$\tau = \sigma\, tg\varphi + c \qquad (7)$$

where φ and c are the angle of internal friction and cohesion of the soil respectively.

The condition of unthurst for any point on the contact area at a distance x from front or back edge of the plate is described by means of integrating Kötter equation along the combined slip-line shaped as straight line-logarithmic spiral-straight line sequence for different inclinations α of the latter straight line

$$\sigma = \frac{\sigma_n + H}{\cos^2\varphi}[1 + \sin\varphi \sin(2\alpha - \varphi)] - H;$$

$$\tau = \frac{\sigma_n + H}{\cos^2\varphi}\sin\varphi\cos(2\alpha - \varphi) \qquad (8)$$

where

$$H = c\,ctg\varphi; \quad \sigma_n = \bar{A} + \bar{B}\gamma' x;$$

$$\bar{A} = [\lambda^2(1 + \sin\varphi) - 1]H;$$

$$\bar{B} = -\frac{\sin 2(\alpha - \varphi)}{2} + \frac{\sin\alpha}{\cos\varphi}\{\cos\psi\cos(\alpha - \varphi - \psi) +$$
$$+ \lambda^3[\sin\Phi\sin\varphi + \sin\psi\cos(\Phi - \psi)]\};$$

$$\lambda = exp[(\alpha + \Phi)tg\varphi];$$

$$\psi = arctg(3tg\varphi); \quad \Phi = \frac{\pi}{4} - \frac{\varphi}{2}.$$

Thereby, equations (7) and (8) cover all virtual cases of the limit state in any point on the contact surface of the soilbase under gravity structure and make it possible to establish the condition for transition of the upper CM from elastic to plastic state. The yield condition may be given in the form

$$\Phi(\sigma, \tau) = 0 \qquad (9)$$

and its graphic interpretation is shown on Fig.3. In elastic domain $\Phi < 0$ and the upper CM is described by equations (6). If $\Phi(\sigma, \tau) = 0$ ideal plastic flow occurs for which the stress point may only move along the yield surface.

For the obtained yield surface the associated flow rule has been adopted for segments 4-5 and 5-6 (see Fig. 3)

$$du_1^p = d\lambda \cdot \Phi_\tau; \quad dw_1^p = d\lambda \cdot \Phi_\sigma \qquad (10)$$

where du_1^p and dw_1^p are plastic increments of lateral and vertical displacements of the upper CM.
For a singular point 5 a Koiter's rule is used, and for segments 3-4, 6-7 as well as 2-3 and 7-8 the non-associated flow rule is postulated

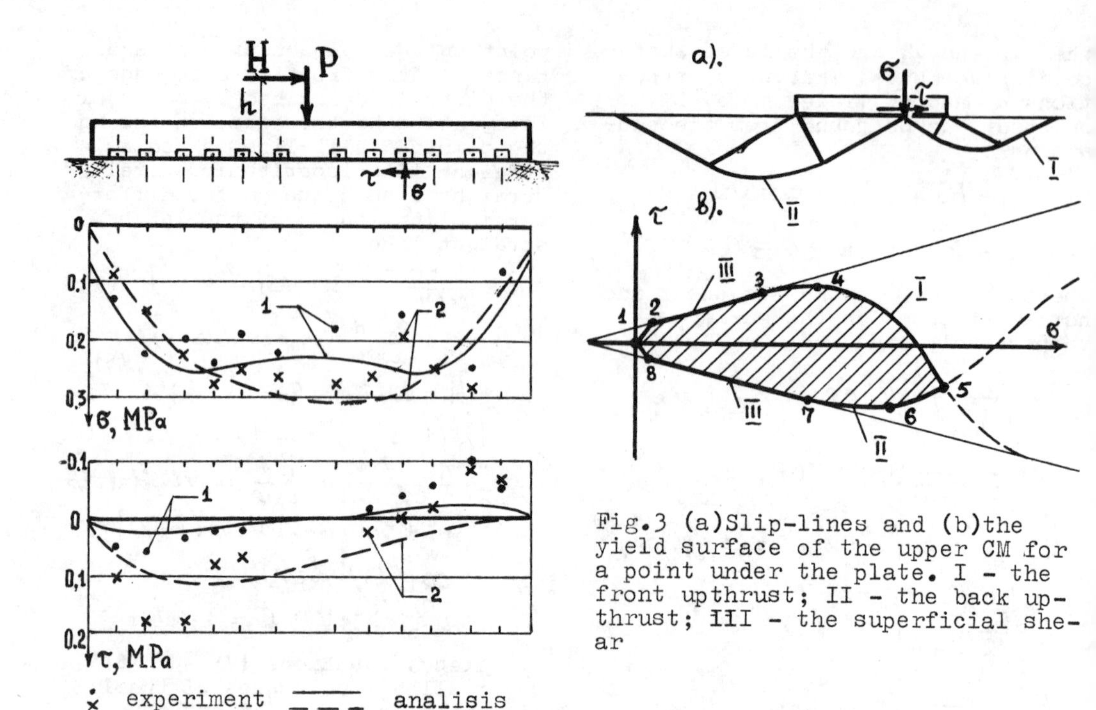

Fig.3 (a)Slip-lines and (b)the yield surface of the upper CM for a point under the plate. I - the front upthrust; II - the back upthrust; III - the superficial shear

Fig.1 Contact stresses under $4m^2$ square plate for vertical load P=800kN and various lateral loads (1 - H=0; 2 - H=200kN); the load application height h=0. Analytical results are given for a plane problem (the plate width B=2m) with the same mean cohtact stresses

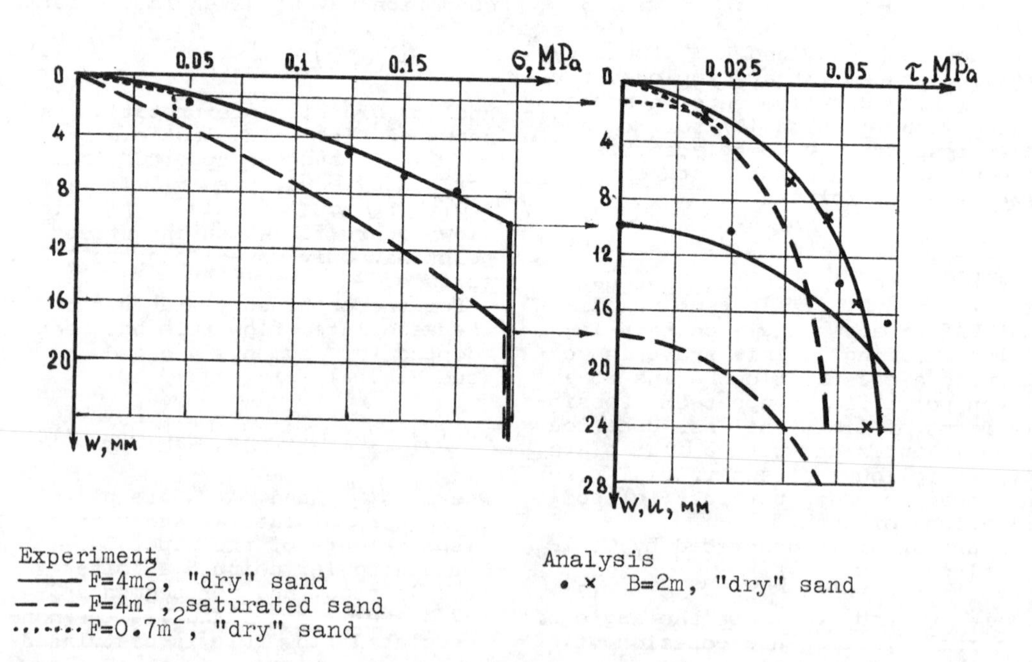

Experiment
—— $F=4m^2$, "dry" sand
– – – $F=4m^2$, saturated sand
.... $F=0.7m^2$, "dry" sand

Analysis
• × B=2m, "dry" sand

Fig.2 experimental and theoretical load-displacement curves; σ - mean normal stress, τ - mean shear stress under the plate

Table 1. Expressions of stress component increments for equations (13)

stress component	physical state of i-th node			
	elastic	plastic at the yield surface segment		
		2 - 3	3 - 4	4 - 5
$\Delta\sigma_i$	$K_w \Delta\bar{w}_i$ *	$K_w \Delta\bar{w}_i$	$K_w \Delta\bar{w}_i$	$\dfrac{\Phi_\tau^2 \Delta\bar{w}_i - \Phi_\sigma \Phi_\tau \Delta\bar{u}_i}{\Phi_\tau^2/K_w + \Phi_\sigma^2/K_u}$
$\Delta\tau_i$	$K_u \Delta\bar{u}_i$ **	$K_w tg\varphi \Delta\bar{w}_i$	$-\dfrac{\Phi_\sigma K_w \Delta\bar{w}_i}{\Phi_\tau}$	$\dfrac{\Phi_\sigma^2 \Delta\bar{u}_i - \Phi_\sigma \Phi_\tau \Delta\bar{w}_i}{\Phi_\tau^2/K_w + \Phi_\sigma^2/K_u}$

* $\Delta\bar{w}_i = \Delta w_c + l_i \Delta\theta - \Delta w_i$; ** $\Delta\bar{u}_i = \Delta u_c - \Delta u_i$

$$du_1^p = d\lambda \cdot \Phi_\tau ; \quad dw_1^p = 0. \tag{11}$$

Total displacement increments are as follows

$$du_1 = du_1^e + du_1^p$$
$$dw_1 = dw_1^e + dw_1^p \tag{12}$$

At the points 4 and 6 of associated and non-associated flow rules interface $\partial\Phi/\partial\tau = 0$ and, hence, there are no discontinuities. At the points 3 and 7 connection of different segments of the yield surface (deep and superficial shear) is also smooth.

The described combined contact model was used to develop a method of analysis and a computer program for evaluation of soilbase behaviour when the gravity structure is subjected to combined action of the vertical and lateral loads as well as of the moment produced by the lateral force. The analysis takes care of the loading history, therefore incremental iterative method has been employed to obtain the numerical solution of the plane problem.

Deformation zone T that included plate width B and free surface of the soilbase on both sides of the plate was divided into N equal segments $\Delta x = T/N$ long by nodes $x_i = i\Delta x$, $i = 0, 1, 2, \ldots N-1, N$. For each node of the grid incremental form of system (3) has been expressed in finite differences

$$\frac{A}{(\Delta x)^2}(\Delta u_{i-1} - 2\Delta u_i + \Delta u_{i+1}) - B\Delta u_i + \frac{C}{2\Delta x}(\Delta w_{i+1} - \Delta w_{i-1}) + \Delta\tau_i = 0;$$
$$-\frac{C}{2\Delta x}(\Delta u_{i+1} - \Delta u_{i-1}) - \frac{D}{(\Delta x)^2}(\Delta w_{i-1} - 2\Delta w_i + \Delta w_{i+1}) - F\Delta w_i + \Delta\sigma_i = 0; \tag{13}$$
$$i = 1, \ldots N-1$$

where $\Delta f_i = f_i^j - f_i^{j-1}$; $j, j-1$ are respective loading steps; $\Delta\tau_i$ and $\Delta\sigma_i$ at the nodes under the plate are expressed in terms of respective displacement increments Δu_i and Δw_i depending on physical behaviour of the i-th node in accordance to Table 1. In this table u_c, w_c and θ denote, respectively, lateral and vertical displacements of the center of the plate and its tilt, $l_i = x_i - x_c$ - the distance from the plate center to the i-th node. In the nodes outside the plate (on free surface) $\sigma_i = \tau_i = 0$. At the edge nodes (at the edges of the plate) values of $\Delta\sigma_i$ and $\Delta\tau_i$ are incorporated equal to half of those computed with the help of Table 1. At both ends of deformation zone ($i = 0, N$) $u_i = w_i = 0$.

Moreover, system (13) is supplemented with three equations of the plate equilibrium that makes the system confined. The solution is obtained by subsequent elimination of unknowns.

The loading step is selected equal tc the minimal one that ensures transition of one node (or two symmetrical nodes) into plastic state.

Due to curvilinear shape of the yield surface stress points (σ_i , τ_i) may move out of this surface. The correction is done with the help of the initial stress method using the current elasto-plastic (rather than elastic) stiffness matrix of the system. Comparison of theoretical and experimental data presented on Fig. 1,2 becomes somewhat difficult since the analysis was done for the plane problem conditions while a square plate was used for the experiment. For the comparison contact stresses along the central longitudinal line were taken. This is probably cause of the fact that initial shear stress (for H = 0) in the experiment are considerably greater than the theoretical ones. This difference is maintained later on.

4. CONCLUSIONS

Computations have yielded data showing that within the framework of the proposed CM all phenomena generated by lateral loading of gravity structures are described such as: additional settlement, redistribution of contact stresses with their reduction at the edges and tilt towards the applied lateral force. The results given above show that the proposed contact model of soil base can be applied to the analysis both of offshore gravity platforms and of onshore structures subject to lateral loads.

REFERENCES

[1] V.A.Barvashov, V.G.Fedorovsky. Three-parametered model of soil base and pile field with consideration of irreversible structural deformations of soils. Soil Mech. and Found. Eng., 4, 17-20 (1978) (in Russian)

[2] V.A.Barvashov, V.G.Fedorovsky. Contact models of soilbases. Archiwum Hydrotechniki, XXY11, 4, 629-645 (1980)

[3] M.I.Gorbunov-Possadov, T.A.Malikova, V.I.Solomin. Design of structures on elastic foundation. Strojizdat, Moscow, 1983 (in Russian)

[4] S.N.Levachev, V.G.Fedorovsky, Yu.M.Kolesnikov, S.V.Kurillo. Design of pile foundations of hydrotechnical structures. Energoatomizdat, Moscow, 1986 (in Russian)

[5] V.Z.Vlasov, N.N.Leontjev. Beams, plates and shells on elastic foundation. Physmathgiz, Moscow, 1960 (in Russian)

Numerical Methods in Geomechanics (Innsbruck 1988), Swoboda (ed.)
© 1988 Balkema, Rotterdam. ISBN 90 6191 809 X

Research of the interaction between circular foundation and layered ground with the method of finite elements

J.Nickolov & L.Ignatova
HTA 'A.Kunchev', Rousse, Bulgaria
P.Karachorov & K.Hamamdgiev
Geotechnics Laboratory to BAS, Sofia, Bulgaria

ABSTRACT: In the present work, the conditions are described and in addition, the results of the linear experimentation are discussed for clarification the shifting and stresses in a two-layer model of ground under ideally hard circular foundation at different correlations of the thickness of the upper harder layer to the foundation's diameter.

The calculations are carried out according to the finite elements at two models of the material. The first one is based on a precised model of Mohr-Coulomb and possibilities are created for isotropic strengthening or softening of the material and for simplified modelling of crack's appearing. The second one is experimented with hypoelastic non-linear behaviour of the material. Valuations of the used models are made on the base of comparison between the received in laboratory conditions and in field researches dealing with experimental results.

1. INTRODUCTION

The studying and the adequate modelling of the non-linear behaviour of the separate soil layers at vertical loading of the ground is of decisive importance for the full usage of the resources for the foundation projection.

The method of the finite elements in combination with powerful modern computer technics creates optimum premises for wide digital experimentation with different theoretical models for the material behaviour and consequent comparison with experimental results.

When the Geotechnics laboratory to BAS-Sofia carries out a research of the grounds consisting of cement-soil layer and loess layer, it is important that a sufficiently reliable digital approach is found, which has to be based on the possibly simplest model for non-linear behaviour of the object, and to ensure good correspondence between the calculation results and the field measurements.

A two-layer axi-symmetrical model was accepted for the object of research. The upper thinner layer studies the strengthening with cement loess soil and the lower of loess soil. The loading is from the foundation, which can be examined as an ideally hard plate.

As it is known, the loess soil is cohesion material with inner friction, which non-linear behaviour at unconvertible deformation is moulded sufficiently well with Mohr-Coulomb' model. Also it is known, that it is natural that there is the deformational strengthening or softening of the loess soil which this model does not envisage. The deformational softening most oftenly is explained with the appearance of cracking, provided by the shearing efforts. An important peculiarity of the soils is that they can not bear essential tension stresses and its appearance is accompanied by the cracks formation. The modelling of this phenomenon with the finite elements method meets serious difficulties.

In many cases of researches with the finite element method, a strongly simplified solution is used, which is based on the non-linear elastic material's behaviour during the time of the whole deformation /Duncan-Chang's model/.

In the present work, the experimental results are compared with the calculated results of a precised Mohr-Coulomb's model and Duncan-Chang's model and conclusions are made about the model qualities.

2. METHOD OF SOLUTION

The calculations are carried out with the developed HTS "A. Kunchev"-Rousse finite element program system VIMKE for strengthen, deformational and dynamical analyses. This subsystem for solving physically non-linear problems is used accepting a geometrical linearity of the problem. The subsystem works after the initial stress approach on the base of the incremental associated and non-associated flow rules of the plastic flow. The convergence of the iterational process is normally controlled by displacement changes and by of residual forces /1/. The expedience of the similar acceleration in every iteration, is controlled by prophilactically acting algorithm.

The acceleration is made according to a specially developed procedure /2/.

Apart from the ordinary the metals criteria of plasticity some criteria are included suitable for examining of soils Drukker-Prager, Mohr-Coulomb, Duncan-Chang.

When examining the ground, also the field of the initial stresses is marked and also is brought by the base's weight. It is accepted that the process of deformation as a result of the initial stresses, has faded and the displacements are established. The initial stresses are supergathered with the additionally appearing stresses, which are caused by the changing loading, and the general stresses are responsible for irreversible deformation in the different points of the object.

3. PRECISED MODEL OF MOHR-COULOMB

According to the Mohr-Coulomb's model, the plastic criteria the plastic potential, is expressed with the function:

$$\frac{\sigma_1 - \sigma_3}{2} + \frac{\sigma_1 + \sigma_3}{2} sin\varphi = c \cdot cos\varphi \qquad (1)$$

where σ_1 and σ_3 are the biggest and respectively the smallest algebraic value for the principal stresses, c is the cohesive strength of the material and φ is the angle of internal friction in it.

In the associated theory the flow potencial is expressed with the same function /1/. Imperative with loess soils is the usage of the non-associated theory of plasticity. In addition, for the flow potencial it is appropriate that the analogical function to be accepted:

$$\frac{\sigma_1 - \sigma_3}{2} + \frac{\sigma_1 + \sigma_3}{2} sin\psi = c \cdot cos\psi \qquad (2)$$

in which the dilatancy angle ψ is accepted

$$\psi = \gamma\varphi \qquad 0 \leq \gamma < 1 \qquad (3)$$

As it is known in the principal stress space, that the geometrical interpretration of /1/ is sixgram pyramid. The interpretation of /2/ is analogical. The difficulties in finding the gradicut vextor along the corners of the pyramid are overcoming with the usual rounding off. The experience shows that $\gamma \approx 0$ is most suitable for the loess soils.

With ideally plastic material, the cohesiveness c and the angle of friction do not depend on the degree of deformation and so remain permanent during the time of deformation. For creating the simplest model of isotropic strain hardening or softening, we will accept that c and φ change mutually binded and it is in such a way, that the top of the pyramid has to remain on the same place:

$$c \cdot cot\varphi = const. \qquad (4)$$

If the concept Yield stress is introduced conditionally

$$Y = \frac{\sigma_1 - \sigma 3}{2} + \frac{\sigma_1 + \sigma_3}{2} sin\varphi \qquad Y = c \cdot cos\varphi \quad (5)$$

and the limits of the change of the angle of friction are known, φ_0 /initial at first, plasticity/ and φ_1 /finite at deep plastic deformation, Y will change in the limits

$$Y_0 = c_0 \cdot cos\varphi_0 \leq Y \leq Y_1 = c_1 \cdot cos\varphi_1 \qquad (6)$$

according to the degree of plastic deformation.

As a criteria for the degree of plastic deformation, clearly deviatoric equivalent plastic strain $\bar{\varepsilon}_p$ is accepted and then for the running Y the hiperbolic relationship can be accepted:

$$Y = Y_0 + \frac{Y_1 - Y_0}{a + \bar{\varepsilon}_p} \bar{\varepsilon}_p \qquad (7)$$

The constant a can be expressed for example in the form:

$$a = \frac{10 c_0}{E} \qquad (8)$$

where E is the module of the material elasticity. For the practical usage of that model for isotropic strengthening, the initial angle of friction φ_0, the initial cohesiveness c_0 and the most possible change of the angle of inner friction is sufficient to be known:

$$\Delta\varphi = \varphi_1 - \varphi_0 \qquad (9)$$

The moulding of crack forming, due to exceeding of the limit tension stresses in the material, is quite difficult because of the preliminary chosen discretisation of the object. That is the reason for the usage of strongly simplified approaches, modelling only the elastic unloading, appearing at the forming of crack. With the components of the main tensor in every point the principle stresses are determined and these which exceed the limit tension of the material σ^* are put to zero. After back transformation to the corresponding main tensor, the discrepancies between the real and corrected stresses are expressed as unbalanced nodal forces, independently of the fact is that the point before this moment is in an elastic or plastic condition. With the successive, itterations these forces bring suitable new distribution of the stresses, in such a way, that normally tension normal interaction is to be absent to the crack. This limiting is kept futher on in all following degrees of loading.

4. NON-LINEAR ELASTIC SOLUTION

Some calculations resembling the approach, described in /3/ were carried on for checking the fitness of the simplified Duncan-Chang's model, for studying the two-layer model of ground.

Elastic solution, dependent on the degree of deformation tangential module E_T and constant Poisson's ration μ, are used in this approach. The results of the triaxial test in laboratory conditions, were worked up for finding an analogical dependence:

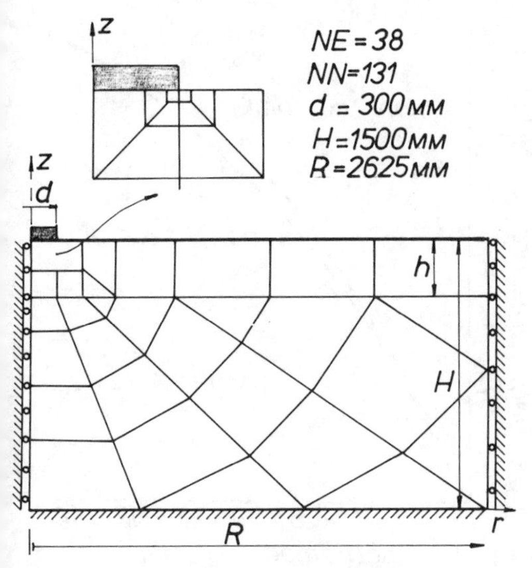

Fig.1 Numerical model

$$\bar{\sigma} = \sigma_1 - \sigma_3 = f(\bar{\varepsilon}),$$

$$\sigma_1 = \sigma_2 = \sigma_r \qquad \sigma_3 = \sigma_z \qquad \bar{\varepsilon} = \varepsilon_3 = \varepsilon_z \quad (10)$$

through which

$$E_T = \frac{d\bar{\sigma}}{d\bar{\varepsilon}} \quad (11)$$

It is essential for this dependence to be described with the curve:

$$\bar{\sigma} = \frac{E_0 \cdot \bar{\varepsilon}}{[1 + (\frac{E_0 \cdot \bar{\varepsilon}}{\bar{\sigma}_0})^n]^{\frac{1}{n}}} \quad (12)$$

which has for tangent the straight line of the ideal elasticity

$$\bar{\sigma} = E_0 \cdot \bar{\varepsilon} \quad (13)$$

and for the asymptote - the straight line of the ideal plasticity

$$\bar{\sigma} = \bar{\sigma}_0 \quad (14)$$

where E_0 is the initial elastic module of the elasticity.

The degree index n and the limit stress $\bar{\sigma}_0$ are approximated with simple functions of the confining pressure $\bar{\sigma}_r$. At the solution for σ_r, the average arithmetical of the radial and circle stress, which in the general case are of no importance, was accepted.

It was received for the studying loess soil after the working up of the results:

$$n = 0,00368\bar{\sigma}_r^2 - 0,355\bar{\sigma}_r + 0,64$$

$$\bar{\sigma}_0 = -0,0324\bar{\sigma}_r^2 + 0,606\bar{\sigma}_r + 9,66 \quad (16)$$

and for the cement soil

$$n = 0,00368\bar{\sigma}_r^2 - 0,355\bar{\sigma}_r + 0,64$$

$$\bar{\sigma}_0 = -0,220\bar{\sigma}_r^2 + 4,11\bar{\sigma}_r + 65,5 \quad (17)$$

as the stresses are in MPa.

5. RESEARCH RESULTS

The researches are carried out on three vertions of two-layers ground model, formed at different correlations of the thickness n of the cement-soil layer to the foundation's diameter d.

model	ZOS1 :	$h/d = 0,5$
model	ZOS2 :	$h/d = 0,75$
model	ZOS3 :	$h/d = 1$

as also on one layer model only of loess soil /ZOS4/.

At the solutions, a body of revolution was studied with the sizes of $H = 5d = 1500mm$, $R = 8,75d = 2625mm$, $d = 300mm$. At the discretisation of the object, isoparametric finite elements with 8 or 6 nodals at 3 x 3 Gauss points are used. The accepted discretisation /Fig.1/ at different versions discerns only with some sizes of the finite elements.

Brief description of the measurements

A great experimental program is fulfilled for proving the reliability of the calculating model and method of work. The experiments foresee research of the deformational and strengthening behaviour of two layers ground, consisting of lower layer - natural loess and upper - strain reinforsed loess with cement, cement-ground cushion. The experiments can be conducted at different correlations of h/d in the limit of $0,33 < h/d < 2,00$ and at two different correlations of the modules of the two layers.

Data are received for the dependences between the stresses and deformations, as for the mechanism of destruction of the ground.

The following material characteristics were accepted on the base of the carried out laboratory measurements, dealing with the calculations of the precised Mohr-Coulomb's model:

loess soil:

$$E = 14MPa \quad c_0 = 0,015MPa \quad \varphi_0 = 25^o$$

$$\Delta\varphi = -3,5^o \mu = 0,30 \quad \sigma^* = 0,003MPa$$

$$\gamma = 0 \quad \rho = 1900gr/cm^3$$

cement soil:

$$E = 95MPa \quad c_0 = 0,28MPa \quad \varphi_0 = 32^o$$

$$\Delta\varphi = 0^o \mu = 0,13 \quad \sigma^* = 0,45MPa$$

$$\gamma_0 = 0 \quad \rho = 1900gr/cm^3$$

Poisson's ratio $\mu = 0,485$ was accepted for the calculations after the model of "non-linear elasticity" for the two layers. The initial module of elasticity had

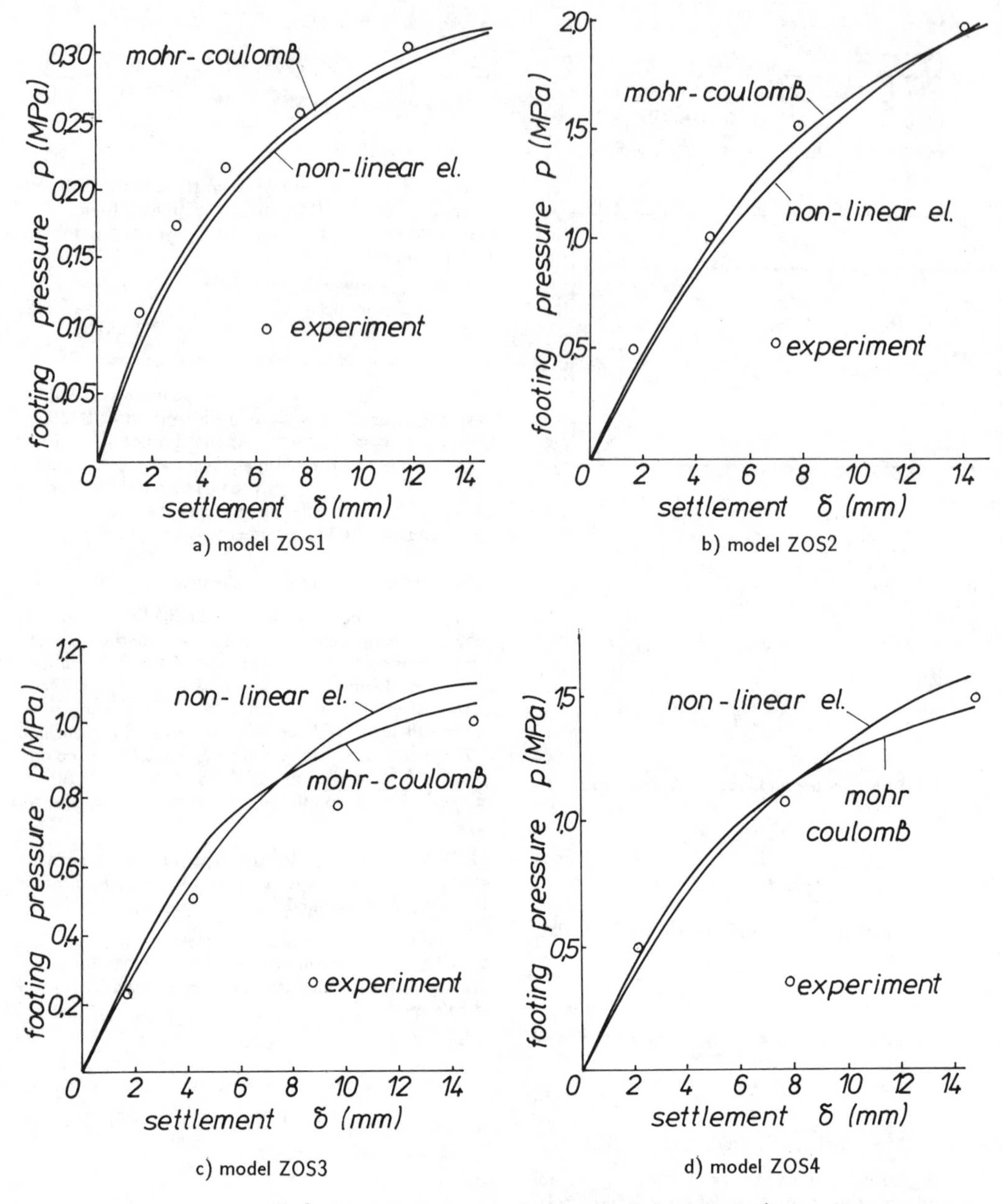

Fig.2 Load-displacement curves and experimental results

to be increased: with loess soil $E_0 = 18MPa$, and with cement soil $E_0 = 125MPa$.

The foundation was hard to accept and instead of loading displacements, the contact surface between it and the ground were given along. The respective footing pressure is calculated by integration of the stresses in the finite elements.

On Fig. 2, the experimental and calculated with the precised Mohr-Coulomb's model results, are correlated for the dependence between the maximal displacement and the exercised pressure. It is noticed that at general good correspondence, the graphics at ZOS3 and ZOS4 remain a little bit under the experimental results, while at ZOS1 the inverse effect exists. At small pressure, the

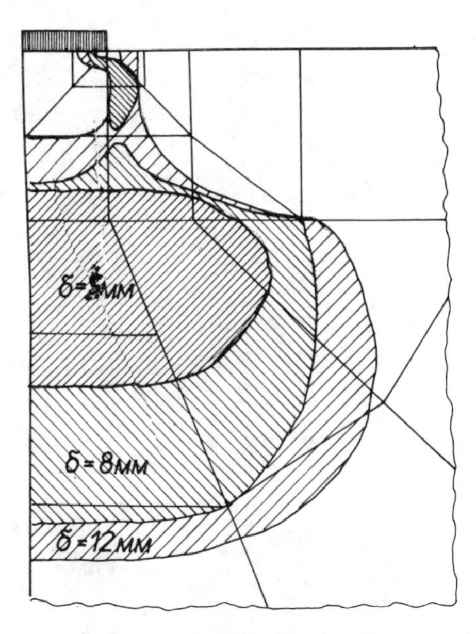

Fig.3 Model ZOS3: Plastic region

behaviour of the object is close to the elastic, plastic deformation appears, (Fig. 3) in small places around the foundation's periphery and to a certain level of depth around the axis of rotation. With the increase of loading these zones quickly increase and when they join a well expressed change of the object's behaviour appears - the displacements increase quickly with the increase of the loading. On Fig. 4 and 5, the development of the zone is shown with plastic deformational material, and the crack development (zone with elastically deformational material but with exclusive tension stresses) with the model ZOS4.

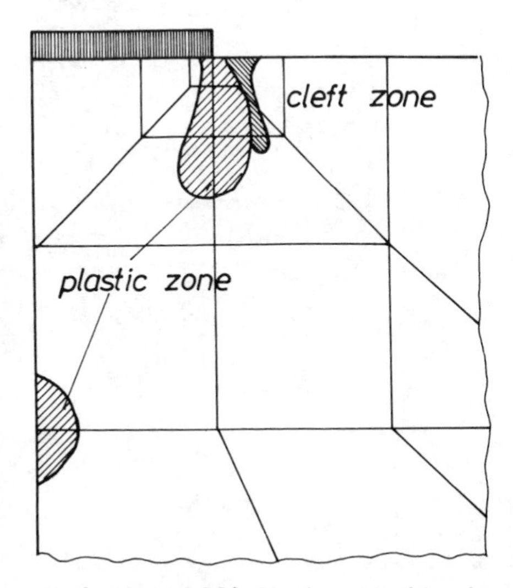

Fig.4 Model ZOS4: Plastic and 'split' region ($\delta = 1,5mm$)

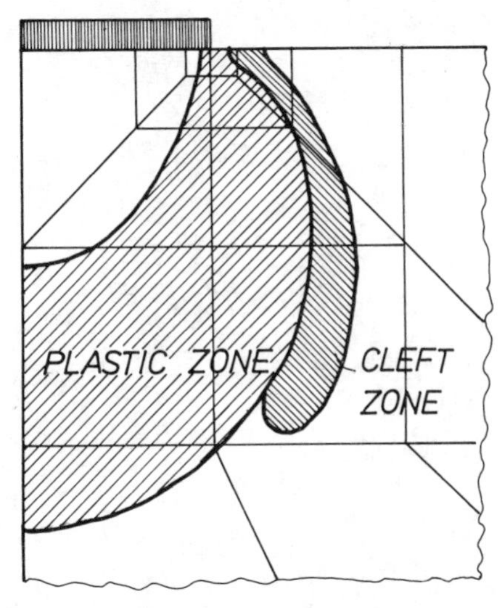

Fig.5 Model ZOS4: Plastic and 'split' region ($\delta = 3mm$)

The results about the dependence pressure-displacement, received after the model "non-linear elasticity" are similar to the described (Fig. 2). However, this is received with the increase of the initial module of elasticity with a factor of 30 appears considerably softer than the real object. Right results are not received with that model about the volume widening, also correct results are not received with the ground surface in proximity to the foundation, and the supporting ability of the ground is underestimated.

6. CONCLUSION

The carried numerical experiments show that after its precisesness, Mohr-Coulomb's model ensure sufficiently good and reliable correspondence between the calculations and the reality. The calculations after the model "non-linear elasticity", give the possibility for orientated values, but only after a respective increase of the initial module of elasticity (with 28 - 32 %). The project with supported by the scientific technical contract between Austria and Bulgaria, between the Technical University "Angel Kunchev" and the University of Innsbruck.

LITERATURE

[1] Nayak G.C., Zienkiewicz O.C., Elasto - plastic Stress Analysis - Int. J. Num. Meth. Engng. Vol. 5, 1972, 113 - 135

[2] Nickolov, J. Ob odnoj procedure dlja uscorenija schodimosti iteractionogo processa pri reshenii fizicheski nelinejnih zadach Proc. of the Intern. Conf. on Num. Meth. and Appl., Sofia, 1985, 459 - 464

[3] Desai C.S. Reese L.C. Analysis of circular footings on layered soils - J. Soil Mech. and Found. Division,1970, 1289 - 1309

Numerical Methods in Geomechanics (Innsbruck 1988), Swoboda (ed.)
© 1988 Balkema, Rotterdam. ISBN 90 6191 809 X

Loadings on box-type structures buried in elasto-plastic soil

Xiong Jian-guo & Gao Wei-jian
Institute of Engineering Mechanics, State Seismological Bureau, Harbin, People's Republic of China

ABSTRACT, The analysis of loads acting on a buried box-type structure subjected to ground surface air-blast loading is described. In the analysis the equivalent-layer model proposed by the first author in a previous paper is used and the soil is assumed to be a linear hardening medium. Through parametric studies the effect of the wave velocity of the soil, the buried depth and the dimension of the structure etc. are evaluated, and the expressions for estimating the pressures on the structure are given. Besides, procedure of dynamic response analysis of buried box-type structures is recommended.

1 INTRODUCTION

Box-type structure is one of the widely used shallow-buried structures in soils. For the analysis of its response to air-blast loading a equivalend-layer model and a two-step procedure of analysis were proposed by the first author in a previous paper[1]. The proposed model has many advantages, for instance, it reduces the dimensions of the problem by one and it eliminates the estimation of the natural frequency of the system. The applicability of the model has been examined in a subsequente paper[2]. However, the previous works[1,2] are restricted to the elastic soil behavior. As well known in many cases, especially for the air-blast loading problem, it is very important to take the nonlinear behavior of the soil into account. In seismic response analysis of soil deposit the "equivalent" linearization technique is widely used. In using this technique for soil-structure interaction analysis the nonlinear effect is generally divided into two categories, the primary effect and the secondary effect. The former is referred to as the nonlinear effect on the response of the free field to seismic excitation, while the latter to the additional effect due to the vibration of the structure in the soil and is estimated approximately in linear state using the values of the soil characteristics obtained from the analysis of the primary effect. It's suggested that the elasto-plastic response analysis by using the equivalent-layer model is more reasonable than the "equivalent" linearization technique used in the seismic response analysis due to the fact that in the equivalent-layer model the nonlinear effect has been involved not only in the free field but also in the structure vibration in soil.

Basing on the equivalent-layer model results of extensive parametric study on the loadings acting on a box-type structure buried in elasto-plastic soils and subjected to ground surface airblast loading are given in the paper, and procedure of dynamic response analysis of buried box-type structure is recommended. A detailed description can be found in Ref. [3].

2 FUNDAMENTAL RELATIONS AND EQUATIONS, BOUNDARY TREATMENT

For simplicity the linear hardening model and the Drucker-Prager criterior of yielding are used for the soil medium.

Applying the virtual work principle, one may derive the dynamic equilibrium equation for nonlinear material behavior condition

$$[M]\ \{\ddot{u}_{t+\tau}\} + [K]_{NL}\{\Delta u\} + \{F_t\} = \{R_{t+\tau}\} \tag{1}$$

in which $[M]$, and $[K]_{NL}$ denote the mass and stiffness matrix respectively, $\{F_t\}$ and $\{R_{t+\tau}\}$ — the node internal resistant force vector at time t and the equivalent node force vector due to external loading at time $t+\tau$ respectively, $\ddot{u}$ and Δu — acceleration and displacement increment respectively. The integration of eq.(1) was performed by using Newmark method. It has been shown the Newmark method is also non-conditional stable in case of piece-wise linear loading-displacement relation.

The treatment of the boundary is one of the problems encountered in using finite element method. The viscous-absorbing boundary proposed by Lysmer[4] has been successfully adopted as the bottom boundary in the dynamic response analysis for elastic soil condition. With elasto-plastic wave propagation from the stress wave theory the condition of dynamic compatibility at the front of weak discontinuity yields

$$\Delta\sigma = \rho\, C(\sigma)\, \Delta v \qquad (2)$$

where $\Delta\sigma$ and Δv are increment of stress and particle velocity respectively, ρ is mass density, C — wave velocity depending on stress level.

From eq.(2) we see that the viscous-absorbing boundary is also available for one-dimensional elasto-plastic wave propagation in the incremental sense. Thus in the present work the viscous-absorbing boundary is located at the bottom of the model to simulate the infinite extending of the medium.

3 NUMERICAL RESULTS

In order to verify the accuracy of the numerical calculation and the validity of the viscous-absorbing boundary stress time histories at various depths in homogeneous soil under impulsive loading on the ground surface with model length L equaling to 2.5m (solid line) and 30m (dot) are plotted in Fig.1, where theoretical results are also given. It shows that the numerical calculation agrees well with analytical solution and if only the response in the near surface region is of interest, then only a very small model length is necessary.

In the parametric study except the two-dimensional effect being not considered the influence of wave velocity in soil, buried depth and dimension(the height and the thickness of the roof and base slab) of the box structure, the ratio of the wave velocity in the overburden soil (i.e. the back-fill soil above the structure)to that in the soil beneath the structure (i.e. the subsoil) and the location of a stiff layer in region beneath the structure etc. has been taken into account. For comparison there are models associated with elastic soil condition and free field with elasto-plastic soil.

From the calculated results the variation of the pressure on the roof (σ_R) and base slab (σ_B) of the structure with wave velocity in the soil (as shown in Fig.2) can be expressed as

$$\sigma_R = 1.21\, P_m \exp[-0.055(\xi - 1)] \qquad (3)$$

and

$$\sigma_B = \sigma_R - 0.55\, P_m \exp[-0.382(\xi - 1)] \qquad (4)$$

where P_m— the peak value of the impulsive loading on the ground surface, $\xi = C_e/120$, here C_e is the elastic wave velocity in soil in m/sec. In Fig.2 corresponding results for elastic soil condition are also given, we see that compared with elastic soil condition consideration of the elasto-plasticity of the soil reduces the pressure on the base slab of the structure by 13-25%. The pressure on the roof of the structure is increased to some extent for medium stiff soil condition and does not change obviously in cases of very soft or stiff soil due to consideration of the soil elasto-plasticity.

The variation of pressures on the roof and base slab of the structure with its buried depth is plotted in Fig.3 and may be expressed as follows

$$\left.\begin{aligned}\sigma_R &= \sigma_R\big|_{H=3} - 0.021\, P_m (H-3)\\ \sigma_B &= \sigma_B\big|_{H=3} - 0.018\, P_m (H-3)\end{aligned}\right\} \quad H \geqslant 3\text{m} \qquad (5)$$

and

$$\left.\begin{aligned}\sigma_R &= P_m + (\sigma_R\big|_{H=3} - P_m)\, H/3\\ \sigma_B &= \sigma_B\big|_{H=3}\end{aligned}\right\} \quad H < 3\text{m} \qquad (6)$$

where $\sigma\big|_{H=3}$ is the pressure on the structure with buried depth H=3m.

Considering the fact that the stresses in the free field are more easy to be determined from calculation or measurement, it is more reasonable for practical purposes to link the pressure on the structure with the stress at the corresponding depth in the free field. From numerical results obtained for free field medels with combination of characteristics theory of stess wave propagation we may deduce the attenuation of the peak stress (σ_F)

in free field with depth (H)

$$\sigma_F = (1-7.14\ H/C_e)\ P_m \qquad (7)$$

By using eqs. (3) — (7) we arrive at the coefficient of pressure on the roof (α_R) and on the base slab (α_B) versus wave velocity of soil for different buried depth of the structure as shown in Fig.4. Coefficient α_R is defined as the ratio of the peak pressure on the roof of the structure to the peak value of the stress at the corresponding depth in the free field, while coefficient σ_B is defined as the ratio of the pressure on the base slab to that on the roof of the structure.

The influence of the back-fill soil above the structure on the pressure under different subsoil condition is shown in Fig.5. It is seen that generally the more stiff of the backfill soil is, the small is the response of the buried structure, however, for soft subsoil condition abnormal variation happens to the response of the buried structure.

The existing of a stiff layer with thickness of 10m. in the subsoil tends to increase the pressure on the structure due to mismatch of the impedance of the adjacent layers (see Fig.6). Nevertheless, the effect of the stiff layer can be neglected when it is more than 5m. apart from the base slab of the structure.

Results shows that the dimension of the structure leads to minor effect on the response of the structure.

The substructure analysis[1] indicates that the acceleration due to the monolithic motion of a structure is an important parameter in determining the response of a buried structure besides the pressure acting on it. Parametric study indicates that the maximum acceleration varies in the range of 3 to 10 m/sec^2 in cases considered. Supposing a maximum acceleration equaling to 10 m/sec^2 we have additional equivalent loading due to inertial force being equal to about $0.1\sigma_R$ for the roof and $0.2\sigma_B$ for the base slab. Due to the fact that peak pressure and the peak acceleration occur not at the same time and there are a lot of uncertainties in the parameters affecting the response of a buried structure, therefore we assume

$$P_R = 1.1\,\sigma_R \qquad P_B = 1.2\,\sigma_B \qquad (8)$$

where P_R and P_B are the loading on the roof and the base slab of a buried box structure taking into account of the monolithic motion.

4 RECOMMENDED PROCEDURE

Basing on the proposed equivalent-layer model and the two-step procedure and the results obtained in the present work the following procedure is recommended for the response analysis of the buried box-type structure to air-blast loading,

a. Estimation of the stresses in free field can be conducted by using Eq.(7) if no data are available.

b. The loadings on the roof and base slab of the structure are determined from Fig.4 multiplying coefficient 1.1 and 1.2 respectively in order to includ the inertial effect. The loading on the sidewall of the structure (P_s) may be evaluated from the stress in the free field multiplying coefficient of lateral pressure which can be found anywhere.

c. The dynamic response analysis of the structure is carried out according to the scheme shown in Fig.7, where hinge supports are introduced in order to consider shear force on the side wall resulting from the surrounding soil.

5 SUMMARY

a. Predominant parameters affecting the loading on a buried structure are the wave velocity in the soil and the buried depth of the structure. The influence of the dimension of the structure for the practical purposes can be neglected.

b. The elasto-plasticity of the soil reduces the pressure on the base slab of the structure, it may increase or may not affect the pressure on the roof of the structure depending on the stiffness of the soil.

c. The recommended procedure for the dynamic response analysis of box-type structure buried in elasto-plastic soil is very convenient for practical use.

d. The viscous-absorbing boundary used in the elastic wave propagation is also valid for elasto-plastic wave case.

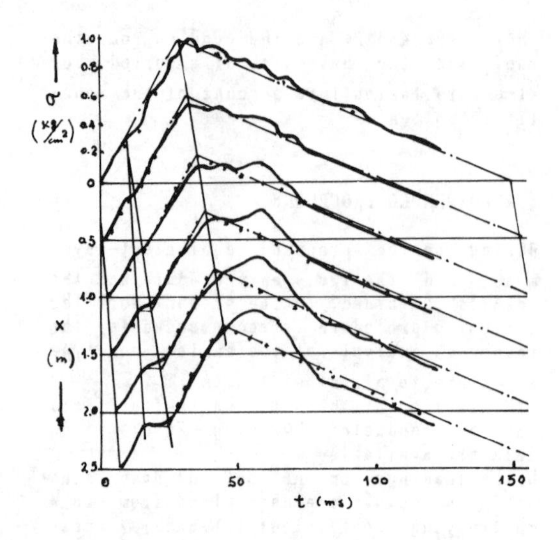

Fig. 1 Stress time histories at various depth(x) under triangular impulsive loading

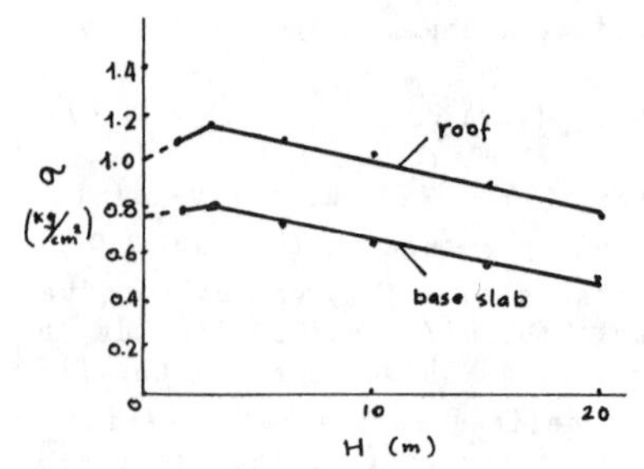

Fig. 3 Peak pressure on bur. str. vs. its buried depth

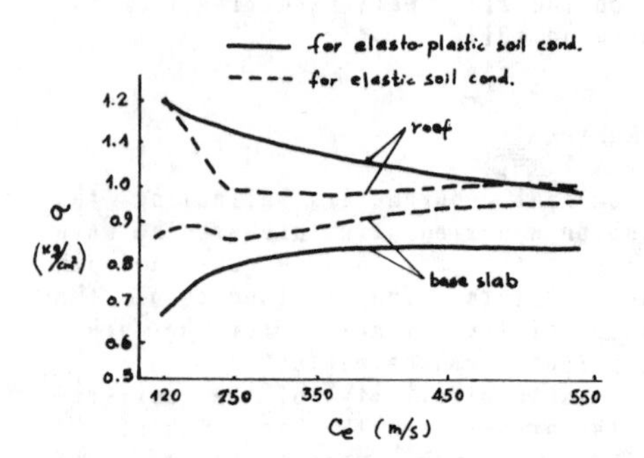

Fig. 2 Peak pressure on buried. str. vs. wave velocity

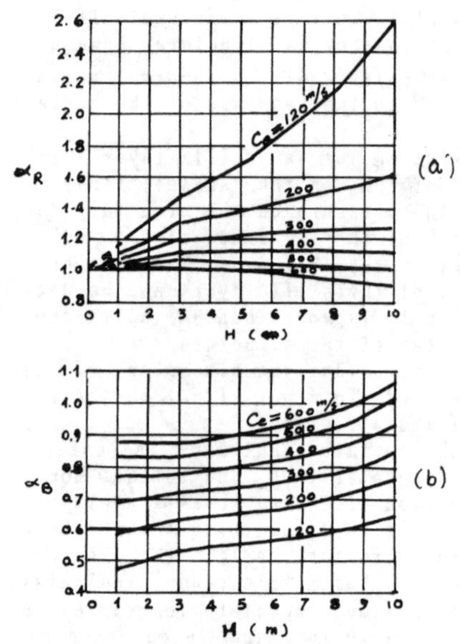

Fig. 4 Coeff. of pressure on roof (a) & base slab (b) vs. buried depth of structure

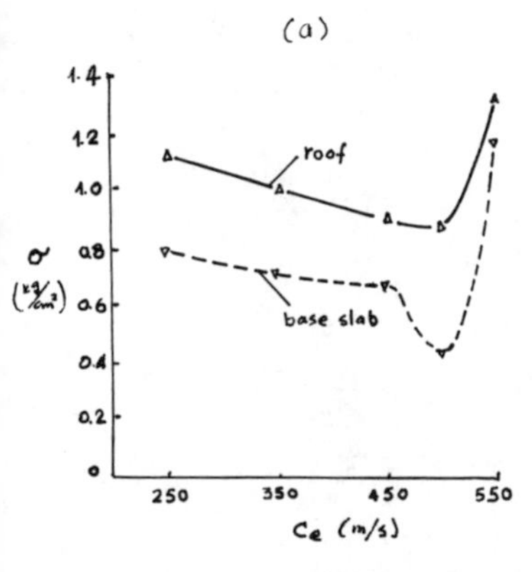

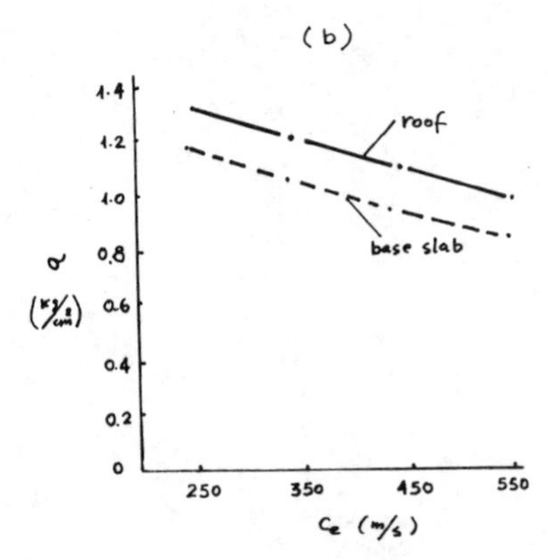

Fig.5 Effect of backfill soil on roof pressure of the bur.str. for subsoil with C_e=250m/s (a) and 550m/s (b)

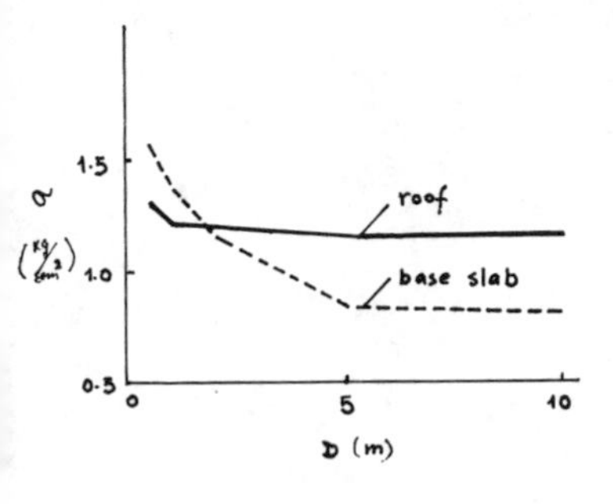

Fig.6 Pressure on the structure vs. distance from the stiff layer to the base slab of the buried structure

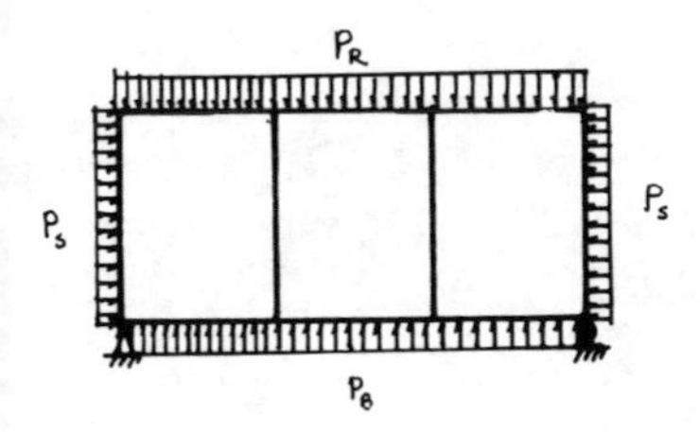

Fig.7 Scheme for response analysis of buried structure

REFERENCES

[1] Xiong Jianguo & Gao Weijian. Dynamic Respose Analysis of Buried Box Structures.Proc. 1st Int. Conf.on Computing in Civil Eng.,New York,N.Y.pp.251-261, 1981

[2] Xiong Jianguo,Gao Weijian & Xu Yiyan. Loading on Buried Box Structure.Chinese J. of Geotechnical Engineering,V.6, N.4 pp.13-23,1984

[3] Xiong Jianguo & Gao Weijian.Loadings on Box-Type Structure Buried in Elasto-Plastic Soil. Research Report, Inst. of Eng. Mech., State Seis. Bureau, Harbin, China,No 86-063,39 p., 1986

[4] Lysmer, J. & Kuhlemayer, R. L. Finite Dynamic Model for Infinite Media. ASCE J. Engng. Mech. Div., ASCE, 95, EM 5, pp. 859-877,1969

Numerical Methods in Geomechanics (Innsbruck 1988), Swoboda (ed.)
© 1988 Balkema, Rotterdam. ISBN 90 6191 809 X

Influence of neighbouring structures on deep foundations

O.Puła & W.Pytel
Geotechnical Institute, Wrocław, Poland

ABSTRACT: For external loads acting either at or below ground surface, the lateral distribution of pressure /on the surface of diaphragm wall protecting excavation/ is thus based on both elastic and limit theory. The effect of prestressed anchors causes the wall to displaced towards the soil which in turn may lead to passive limited state in soil behind the wall outside the excavation and under building's foundation. The method of calculation editional passive earth pressure as well as bearing capacity of strip footing in such conditions is presented in article.

1 INTRODUCTION

The constuction of engineering objects such as large-diameter pipelines, underground tunnels or garages in the vicinity of the already existing buildings is possible only when there is a proper retaining wall design emploed. The type and way in which such a retaining wall is constructed should dependend on the distance between foundation and the excavation wall and on the depth of foundation settling as well as value of the load applayed to the foundation /2,3/. If the distance is small, and external loads are considerable almost always slurry walles anchored in soil are used.

The effect of prestressed anchors causes the wall to displaced towards the soil which in turn may lead to passive limited state in soil behind the wall outside the excavation and under building foundation. In such a situation, the distribution of pressure on the wall as well as bearing capacity of foundation, changed.

In the theoretical solution presented analized was also the effect of soil type on the variability of these quantities.

2 THE SCHEME OF THE PROBLEM AND ITS SOLUTION

The scheme for calculating the bearing capacity of strip footing beside the slurry trench wall is schown in fig.1 .

It is assumed that all conditions in the halfplane are linear, elastic, isotropic, homogeneous and this plane is in failure state, Culombs criterion must be fulfilled at each point of the failure zone.

The plane of footing as well as the surface of the slurry wall are rough and the friction angle at the contact with soil is δ_F or δ_W. Other parametrs which are used ϕ, γ, c=0, D- depth of the base of the footing below ground surface

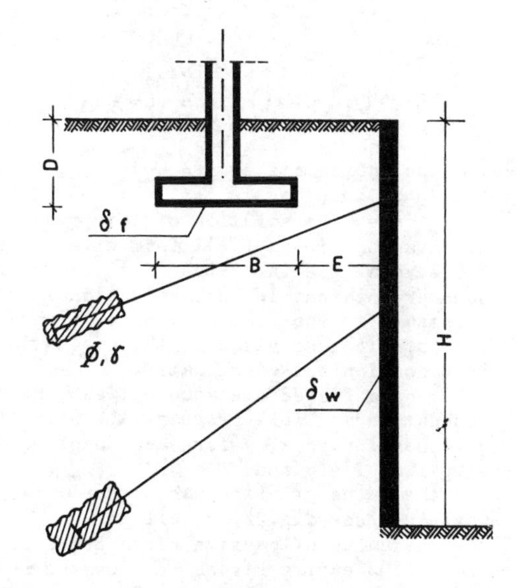

Figure 1. The scheme of the problem

and two values of δ_F and δ_W =1/3 or 2/3

In the solution the following is assumed:

- the existence of curvilinear stiff wedge limited by slip line field under the base,
- passive Rankine state on vertical wall

The calculation of state of stress in two dimensions, in each point of the area considered, leads for Coulomb's medium, to the solution of the system of three equations:

$$\frac{\partial \sigma_x}{\partial x} + \frac{\partial \tau_{xy}}{\partial y} = 0 \qquad \frac{\partial \sigma_y}{\partial y} + \frac{\partial \tau_{xy}}{\partial x} = \gamma$$

$$(\sigma_x - \sigma_y)^2 + 4\tau_{xy}^2 = (\sigma_x + \sigma_y + 2H)^2 \sin\Phi \qquad /1/$$

Sokołowski suggested introducing two auxiliary unknows σ and φ insted of tangential stress τ_{xy} as well as normal stress σ_x, σ_y for makeing the solution of system of equations easier. The basic system of equations could be expressed by unknows σ and φ as well as through two new variables ξ and η. We obtain:

$$\frac{\partial \xi}{\partial x} + tg(\varphi+\mu)\frac{\partial \xi}{\partial y} = a = -\frac{\gamma \sin(\varphi-\mu)}{2\sigma \cos 2\mu \cos(\varphi+\mu)}$$

$$\frac{\partial \eta}{\partial x} + tg(\varphi-\mu)\frac{\partial \eta}{\partial y} = b = \frac{\gamma \sin(\varphi+\mu)}{2\sigma \cos 2\mu \cos(\varphi-\mu)} \qquad /2/$$

As a result of farther transformations of the system of equations we obtain differential equation in the form:

$$\frac{dy}{dx} = tg(\varphi+\mu) \qquad \frac{d\xi}{dx} = a$$

$$\frac{dy}{dx} = tg(\varphi-\mu) \qquad \frac{d\eta}{dx} = b \qquad /3/$$

Since the equations are hyperbolic type, their solution is two families of slip-lines. The solution of this system of equations in explicit form exists only for several special cases of the task of boundary states. In such situation it was necessary to work out a computer program for numerical solution of those equations in dimensionless coordinates.

For established distance between the foundation and wall program WALL prints:

- coordinates x, y for each nodex of slip-line field and
- the value of principal stress σ_1 and angle φ /see fig.2/ as well as
- the value of passive earth pressure σ_p and shearing stress τ_p on the surface of a slurry wall,
- bearing capacity of foundation under conditions as in fig. 1.

3 THE RESULTS OF NUMERICAL CALCULATIONS

The calculation was performed for the following:

- three values of friction angle in soil Φ = 26°, 35° and 44° and two values of δ_F as well as δ_W = 1/3 and 2/3 Φ,
- proportions of the depth of the base of the footing below ground surface and width of strip footing D/B = 1/100, 1/2 as well as 1,
- four distances between strip footing and rigid, rough retaining construction.

The complete calculation procedure was given in /1/. In figure 2 some slip-line field are presented obtained from calculations by the method of Sokołowski.

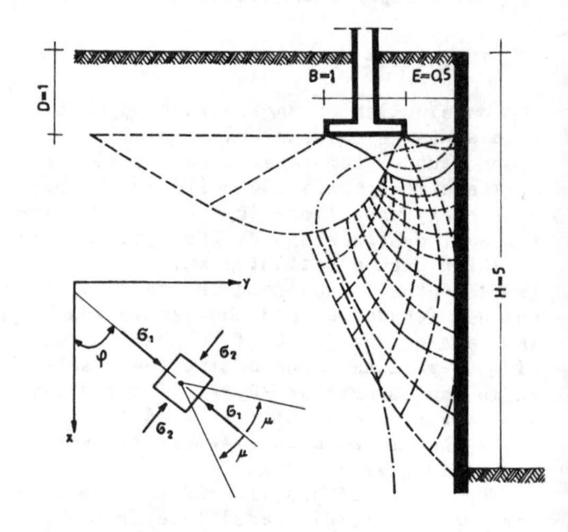

Figure 2. Slip-line field for Φ = 26°, δ_W = 2/3 Φ, D/B=1 and E=B/2 broken line, normal stresses σ_p on the wall - solid line.

On the basis of the calculated examples, the nomograms for reading the passive earth pressure σ_p on the rigid slurry wall dependent on the parameters mentioned above were developed.

In fig. 3 value and distribution passive earth pressure σ_p dependent on the depth of foundation settling were compared. At a depth of ~1.5 B below the level of loading, the maximum of passive earth pressure for all Φ and D/B parameters occurs. Increasing of the depth

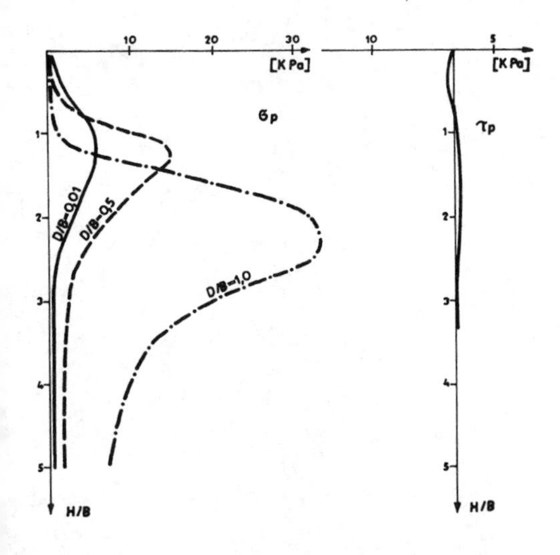

Figure 3. The distribution of passive earth pressure σ_p and shearing stress τ_p on the wall retaining construction for $\Phi = 35^{\circ}$ and $\delta_F = \delta_W = 2/3\ \Phi$.

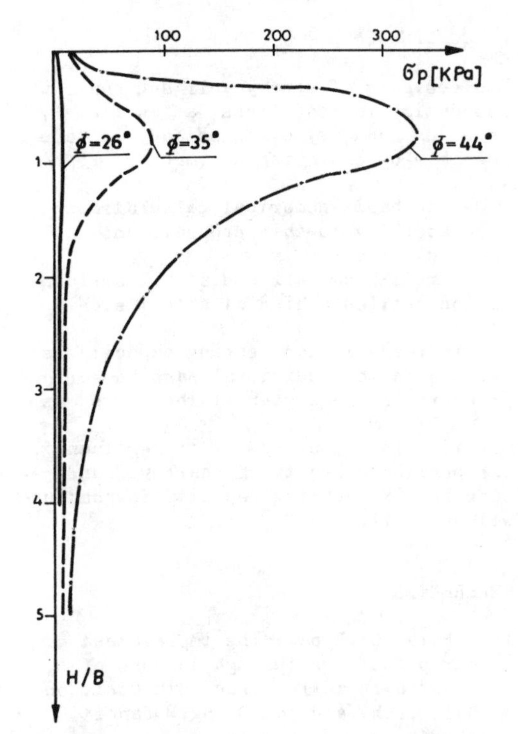

Figure 4. Increasing value of passive earth pressure on the surface of slurry wall depending on Φ for D/B = 1, E=0.5B

of foundation settling causes significant increasing of passive earth pressure on the wall.
The magnitude of passive earth pressure depends also on angle Φ characteristic for subsoil. Increasing friction angle of soil Φ and, related to it, increasing angle $\delta_F = \delta_W = 2/3$ or $1/3\ \Phi$ caused more than proportional increasing of passive earth pressure. Ilustrative values are shown in fig. 4. On the depth H more than 3 B increasing of editional passive earth pressure could be negligible.

The bearing capacity of shallow foundation increases three times for $\Phi = 44^{\circ}$ if the distance between the wall and the foundation decreased two times.

The increasing bearing capacity is greater, the greater is:

- friction angle in soil, depth of the base of the footing below ground D, roughness of the base of strip footing and slurry walls, or
- distance E is smaller.

Illustrative results of calculation are shown in fig. 5.

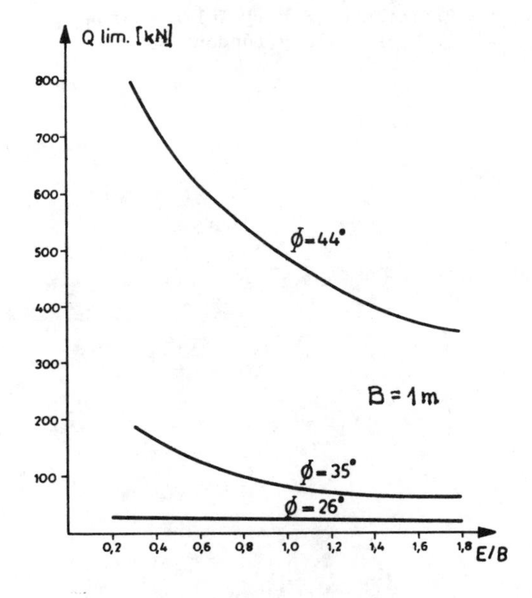

Figure 5. Increasing of the bearing capacity of shallow foundation caused by decreasing distance E for $\Phi = 35^{\circ}$ and D/B = 1.

4, CONCLUSION

The designig of slurry wall and shallow foundation in conditions, as in fig. 1, to using boundary state method, is made possible by employing a computer program WALL.

On the basis nemerical calculations made acording to this program, one can conclude, that:

- a retaining wall and a shallow foundation settled behind it affect each other,
- increase in the bearing capacity as well as in the additional passive earth pressure is the bigger if the bigger is ϕ, D/B, δ_F, δ_W,
- if velocity of E/B is bigger than 1.8 the bearing capacity of shallow foundation is like bearing capacity foundation without wall.

REFERENCES

1. O.Puła, Cz.Rybak, The improvement of strip footings through the use of se sheet pile walls. Proc. 6th Conf. on Soil Mech. and Found.Eng.Budapest, /1984/
2. Von W.Rossner and K.Winter, Normierte Verbau-und Abdeck-konstructionen, Verlag von Wilhelm Ernst and Sohn, Berlin Munchen, /1982/
3. P.P.Xanthakos,Slurry Walls,McGrow - Hill Book Company,London,/1979/

Numerical Methods in Geomechanics (Innsbruck 1988), Swoboda (ed.)
© 1988 Balkema, Rotterdam. ISBN 90 6191 809 X

The variant of Rayleigh's method applied to Timoshenko beams on Winkler-Pasternak foundations

C.P.Filipich & M.B.Rosales
Universidad Nacional del Sur, Argentina

ABSTRACT: The determination of the fundamental frequencies of Timoshenko beams in a Winkler-Pasternak medium is presented. It is proposed the use of the variant of Rayleigh's method which allows an optimization of the approximate modal functions through a non-integer exponential parameter. The case under study gives rise to a two-variable problem. Numerical results are given for hinged-hinged and clamped-clamped beams of uniform and variable cross section. Comparisons with available exact values yield very good agreement.

1 INTRODUCTION

Beams resting on elastic media have been studied quite extensively, both the static /1/ and lately the dynamic cases. These have been mainly approached for Bernoulli-Euler beams and Winkler foundations /2/ or isotropic, homogeneous semi-infinite elastic continuum /3/. But Kerr /4/ and Soldini /5/ have shown other models of soil behaviour which represent a large class of foundation materials. For example this is the case of the Winkler-Pasternak model (W-P) /6/ which admits that there is a reaction to the boundary layer rotation in addition to the Winkler effect.

Rades /7/ used this model in the study of the steady-state response of a beam, emphasizing the insufficiency of the Winkler model. Wang and Stephens /8/ showed the effect of rotatory inertia and shear deformation on the natural frequencies of the beam for various end restraints. This work was extended by Maurizi and Rosales /9/ to rotating elastically restrained ends. Wang and Gagnon /10/ studied the vibration of continuous Timoshenko beams.

In the present paper an approximate yet accurate and simple analytical solution (i.e. the variant of Rayleigh's method) is used to determine the fundamental frequencies of Timoshenko beams in a W-P medium. The use of a non-integer exponential parameter was originally suggested by lord Rayleigh /11/. This variant has been lately used in various eigenvalues and field problems by many authors /12,13,14,15/.

The case under study -a Timoshenko beam- gives place to a two-variable problem which to the authors' knowledge, has not been approached yet with this methodology. It supposes then a further application of this technique which, as it will be shown herein, yields very good results. The hinged-hinged and clamped-clamped beams are presented. The numerical results of a uniform cross section beam are compared with existing values yielding almost negligible errors.

In fact the actual authors' motivation has been the practical application of the method to linearly tapered beams of which the exact solutions are not available in the open literature. Both Timoshenko and Bernoulli-Euler beams with various section height ratios were computed for different foundation conditions.

2 MATHEMATICAL STATEMENT OF THE PROBLEM

The numerical values presented in this paper will be derived from an approximated method. Nevertheless the authors have believed convenient to include herein the exact solution of a Timoshenko beam in a W-P medium. They understand that there is an unsuitable interpretation of the physical sense given to the W-P model in the work known at present. In effect, in previous works such model is simulated as the medium shear reaction. In the present study it is shown that such effect is not more than the massive reaction to the boundary layer rotation as well as the Winkler effect is the reaction to the deflection.

Thus, it is concluded that C^1 continuity of the foundation is accepted and that is possible to admit other reactions based on subsequent derivatives of the boundary layer.

It should be pointed out that there is -at the very beginning- an approximation involved in assuming the response of a medium -massive in essence- reduced to the elastic effect of a massless boundary layer; i.e., taking into account only the elastic response (and neglecting the inertial behaviour). However such an approach allows a significant analytical simplification.

Figure 1 shows a beam embedded in an elastic surrounding medium of infinite extension to both sides of it.

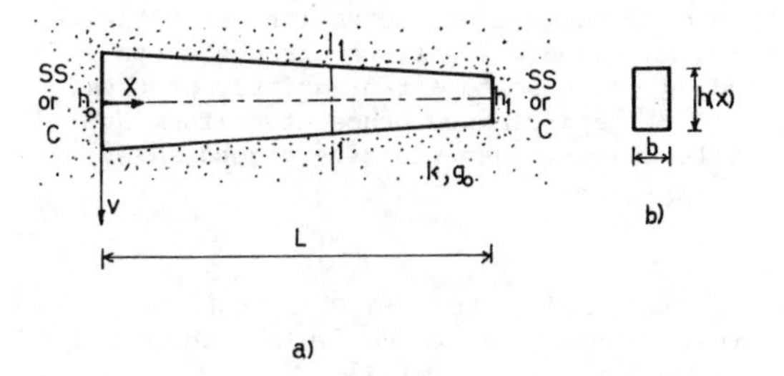

Figure 1. a) Timoshenko beam embedded in elastic surrounding medium; b) 1-1 cross section.

Being X the distance from left end of beam, t the time, L the length of beam, x = X/L and accepting normal modes of vibration with ω circular frequency, one has:

$$\bar{V}_L(X,t)=V_L(x,t)=v_L(x)\ e^{i\omega t}$$

$$\bar{V}(X,t)=V(x,t)=v(x)\ e^{i\omega t} \qquad (1)$$

$$\bar{\phi}(X,t)=\phi(x,t)=L\ \varphi(x)\ e^{i\omega t}$$

$$\bar{V}_R(X,t)=V_R(x,t)=v_R(x)\ e^{i\omega t}$$

denoting by $\bar{V}_L$, the foundation layer deflection to the left of the beam; $\bar{V}$, beam and foundation layer total deflection in the central zone; $\bar{\phi}$, bending slope in the central zone and V_R, deflection to the right of the beam.

The application of Hamilton's principle gives rise to the following differential equation system:

$$v_L'' - (w/p)^2\ v_L = 0, \qquad -\infty < x \leqslant 0 \qquad (2)$$

$$v_R'' - (w/p)^2\ v_R = 0, \qquad 1 \leqslant x < +\infty \qquad (3)$$

$$s^2\varphi'' - (1-\Omega^2 r^2 s^2)\varphi + v' = 0 \qquad (4a)$$

$$0 \leqslant x \leqslant 1$$

$$(1+s^2p^2)v'' + s^2(\Omega^2 - w^2)v - \varphi' = 0 \qquad (4b)$$

where the prime denotes differentation with respect to the spatial variable and

$$\Omega^2 \equiv \omega^2 L^4 \rho A/EI; \quad r^2 \equiv I/(AL^2); \quad s^2 \equiv EI/(KGAL^2)$$

$$w^2 \equiv kL^4/EI; \quad p^2 \equiv g_o L^2/EI$$

with A the cross sectional area, E modulus of elasticity, g_o Pasternak foundation constant, G modulus of rigidity, I moment of inertia of beam section, k Winkler foundation constant, K cross sectional shear factor and ρ mass density of beam material.

Actually in the present work an approximation through Rayleigh's quotient is used. For this reason it will not be neccessary to solve the differential equations. We only explicit the v_L and v_R solutions.

$$v_L(x) = A_L \sinh wx/p + B_L \cosh wx/p$$

$$v_R(x) = A_R \sinh wx/p + B_R \cosh wx/p \qquad (5)$$

As $x \to -\infty$, $v_L \to 0$ and $\delta v_L \to 0$ and as $x \to +\infty$ $v_R \to 0$ and $\delta v_R \to 0$, thus $A_L = B_L$ and $A_R = -B_R$ from which one can verify that

$$v_L'(x) = (w/p)v_L(x) \quad \text{and} \quad v_R'(x) = -(w/p)v_R(x) \qquad (6)$$

and the boundary conditions at $\pm\infty$ are satisfied.

In this paper both simply supported (SS) beams ($\varphi'(0)=\varphi'(1)=0$) and clamped-clamped (CC) beams ($\varphi(0)=\varphi(1)=\delta\varphi(0)=\delta\varphi(1)=0$) were considered and in both cases $v(0)=v(1)=\delta v(0)=\delta v(1)=0$. Consequently it is deduced that the boundary condition should be either $v_L'(0)=0$ or $\delta v_L(0)=0$. For any case it is $A_L=B_L=0$ and

$$v_L(x) = v_L'(x) = 0 \qquad \forall x \qquad (7)$$

and similarly

$$v_R(x)=v'_R(x)=0 \qquad \forall x \tag{8}$$

In general this will not be the case of beams with free ends.

Finally, Rayleigh's quotient may be written as follows:

$$\Omega^2 = \left[\int_0^1 \varphi'^2 dx + s^{-2}\int_0^1 (v'-\varphi)^2 dx + w^2 \int_0^1 v^2 dx + p^2 \int_0^1 v'^2 dx\right] / \left[\int_0^1 v^2 dx + r^2 \int_0^1 \varphi^2 dx\right] \tag{9}$$

which stands for both SS and CC beams.

In the case of variable cross section beams with A=A(x) and I=I(x), Rayleigh's quotient (9) is written as:

$$\Omega^2 = \left[\int_0^1 i(x)\varphi'^2 dx + s^{-2}\int_0^1 a(x)(v'-\varphi)^2 dx + w^2 \int_0^1 v^2 dx + p^2 \int_0^1 v'^2 dx\right] / \left[\int_0^1 a(x)v^2 dx + r^2 \int_0^1 i(x)\varphi^2 dx\right] \tag{10}$$

where $A(x)=a(x)A_o$; $I(x)=i(x)I_o$; A_o and I_o are the left cross sectional area and left moment of inertia, being a(x) and i(x) arbitrary functions.

3 EXPONENTIAL VARIANT OF RAYLEIGH'S METHOD

Approximate polynomial functions for v(x) and $\varphi(x)$ are introduced. These functions depend upon an exponential parameter γ which makes possible an optimization of the value Ω^2 obtained through Rayleigh's quotient.

For the problem under study, as it can be observed from the differential system (4), the two basic functions v and φ are not absolutely independent from each other. However and provided Rayleigh (or Rayleigh-Ritz) problems convergence condition requires to satisfy only the essential or principal boundary conditions and not the natural ones /16/, it would be enough to choose v and φ in this way and varying independently. But despite this, the static relationship that connects v and φ for a Timoshenko beam is imposed. This makes easier the determination of the functions constants. Observing the differential equation (4a) and neglecting the term in Ω^2 one assumes:

$$v'=-s^2\varphi''+\varphi \tag{11}$$

It is defined $\gamma_j=(\gamma+j)$, $j \geq 1$.

For both uniform and non-uniform beams the polynomial functions are chosen in such a manner that they satisfy the boundary conditions that one imposes:

*SS beam:

$$\varphi(x) \sim A(x^{\gamma_1}/\gamma_1)+B(x^3/3)+1$$
$$v(x) \sim A\left(-s^2 x^{\gamma}+\frac{x^{\gamma_2}}{\gamma_1\gamma_2}\right)+B\left(-s^2 x+\frac{x^4}{12}\right)+x \tag{12}$$

subject to the B.C.:

$$\varphi'(0)=\varphi'(1)=v(0)=v(1)=0 \tag{13}$$

which enables to fix A and B.

*CC beam:

$$\varphi(x) \sim A(x^{\gamma_1}/\gamma_1)+B(x^2/2)+x$$
$$v(x) \sim A\left(-s^2 x^{\gamma}+\frac{x^{\gamma_2}}{\gamma_1\gamma_2}\right)+B\left(-s^2 x+\frac{x^3}{6}\right)+\frac{x^2}{2} \tag{14}$$

and the B.C.:

$$\varphi(0)=\varphi(1)=v(0)=v(1)=0 \tag{15}$$

which enables to fix A and B.

Note that in (13) a non-essential B.C. ($\varphi'(0)=\varphi'(1)=0$) is imposed. This is unneccessary from the theoretical view-point but on the other hand it improves the rate of convergence.

Also, it should be pointed out that condition (11) could have been changed directly for (4a) without dropping the Ω^2 term. But this would increase the algebraic complexity and Rayleigh's quotient would yield an algebraic second order equation in Ω^2.

4 NUMERICAL RESULTS

After substituting equations (12) or (14), for each case, in Rayleigh's quotient (9) or (10), one obtains the fundamental frequency coefficient Ω^2 as a function of γ. Through a numerical minimization with respect to γ the optimum value of Ω^2 is found.

Numerical results were obtained for both SS and CC beams under various conditions.

Fundamental frequency coefficients for SS and CC Timoshenko beams of uniform cross section were included in reference /19/. They were studied with varying slenderness r and three media conditions (i.e., 1. $w^2=0$; $p^2=0$, 2. $w^2=25$; $p^2=0$, and 3. $w^2=25$; $p^2=25$). They were compared with values obtained by means of the exact expressions taken from /8/ yielding almost negligible errors. For instance, the largest error (1,2%) was found for a SS beam with $w^2=25$; $p^2=25$ and $r^2=0.02$. The average error for both SS and CC beams with slenderness r^2 ranging from 0.02 to 0 was of 0.02% approximately.

As a particular case, fundamental frequency coefficients were computed for a Bernoulli-Euler beam of variable cross section. The numerical values are shown in Table 1. Some exact solutions results are in brackets. A fair agreement can be observed.

Table 1. Fundamental frequency coefficients Ω for a Bernoulli-Euler linearly tapered beam (SS and CC) ($r^2=0$; $s^2=0$): Effect of height ratio variation.

	SS beam								
	$w^2=0$; $p^2=0$				$w^2=25$; $p^2=0$		$w^2=25$; $p^2=25$	
h_1/h_o	γ	Ω	Exact solution	‰ error	γ	Ω	γ	Ω
0.40	4.10	6.49629			3.58	8.90172	1.22	20.9577
0.50	2.85	7.14269	[7.12154]*	2.97	2.68	9.22262	1.28	20.4717
0.60	2.05	7.30196			1.99	9.43889	1.28	20.0762
0.66	1.92	8.11900	[8.11076]*	1.01	1.81	9.80626	1.25	19.8582
1.00	1.05	9.87568	[9.86960]**	0.61	1.05	11.0693	1.04	19.2173

	CC beam							
	$w^2=0$; $p^2=0$				$w^2=25$; $p^2=0$		$w^2=25$; $p^2=25$	
h_1/h_o	γ	Ω	Exact solution	‰ error	γ	Ω	γ	Ω
0.40	8.48	15.0608			8.34	16.2715	5.24	27.0765
0.50	6.09	16.3896			6.02	17.4130	4.46	26.9788
0.60	4.64	17.6729			4.60	18.5538	3.82	27.0886
0.66	3.96	18.5047			3.94	19.3087	3.42	27.2525
1.00	2.07	22.4472	[22.3733]**	3.30	2.07	22.9973	2.02	28.7922

*Values taken from reference /18/
**Values taken from reference /17/

Table 2. Fundamental frequency coefficients Ω for a Timoshenko linearly tapered beam (SS and CC) ($r^2=0.005$; $s^2=28\ r^2/9$): Effect of height ratio variation.

	SS beam					
	$w^2=0$; $p^2=0$		$w^2=25$; $p^2=0$		$w^2=25$; $p^2=25$	
h_1/h_o	γ	Ω	γ	Ω	γ	Ω
0.40	2.07	6.67327	1.95	8.96240	1.03	20.6985
0.50	1.75	7.08059	1.70	9.12133	1.07	20.1741
0.60	1.57	7.47958	1.52	9.31069	1.08	19.7255
1.00	1.06	9.03534	1.06	10.2832	1.05	18.5417
1.50	0.78	10.9066	0.78	11.7418	0.88	18.0248
2.00	0.67	12.7102	0.67	13.3002	0.74	18.1478

	CC beam					
	$w^2=0$; $p^2=0$		$w^2=25$; $p^2=0$		$w^2=25$; $p^2=25$	
h_1/h_o	γ	Ω	γ	Ω	γ	Ω
0.40	3.15	14.5205	3.10	15.7267	2.35	24.9717
0.50	2.92	14.8490	2.87	15.9414	2.32	24.5255
0.60	2.71	15.1872	2.68	16.1833	2.27	24.1791
1.00	2.09	16.7470	2.09	17.4651	2.00	23.6060
1.50	1.58	19.1354	1.58	19.4655	1.64	24.2105
2.00	1.24	21.8101	1.24	22.1830	1.35	25.7137

Table 2 illustrates an application of the methodology to Timoshenko beams of variable cross section with b=constant; $h(x)=h_o[1-(1-h_1/h_o)x]$; $a(x)=bh(x)$; $i(x)=b[h(x)]^3/12$ (see Figure 1) and $r^2=0.005$; $s^2=28r^2/9$ (i.e. taking into account rotatory inertia and shear deformation). The values show the notable influence of the W-P effect. It should be pointed out that for each height ratio the dimensionless parameters r^2, s^2, w^2, p^2, and Ω^2 were all referred to the left section (i.e. depending on h_o, A_o, I_o).

5 CONCLUSIONS

It has been shown that the use of the exponential variant of Rayleigh's method extended to a two-variable problem yields results of remarkable accuracy.

For variable cross section beams the results are presumingly acceptable. Observing Table 2 it is noticeable that the values corresponding to Timoshenko beams without foundation and in a Winkler medium vary monotonously. This is also the case of tapered Bernoulli-Euler beams as shown in /17/. On the other hand the values representing the fundamental frequency coefficients of Timoshenko beams in a W-P medium have not the same characteristic: there is a given height ratio for which the frequency coefficient is minimum.

It is important to note that this result arose from the use of a simple methodology and it reveals new information about the notable influence of the Pasternak model on the dynamic behaviour of Timoshenko beams.

Finally it is felt that further investigation is needed regarding the dynamic response of Timoshenko beams of variable cross section.

ACKNOWLEDGEMENT

The second author has been supported by a grant from the Consejo Nacional de Investigaciones Científicas y Técnicas (Argentina).

REFERENCES

/1/ M.Hetényi, Beams on elastic foundation, The University of Michigan Press, Michigan, 1946

/2/ M.Hetényi, Beams and plates on elastic foundations and related problems, Applied mechanics reviews 19, 95-102 (1966)

/3/ R.E.Richart, J.R.Hall and R.D.Woods, Vibrations of soils and foundations, Prentice-Hall, New Jersey, 1970

/4/ A.D.Kerr, Elastic and viscoelastic foundation models, Journal of applied mechanics 31, 491-498 (1964)

/5/ M.Soldini, Contribution a l'étude théorique et expérimentale des déformations d'un sol horizontal élastique a l'aide d'une loi de seconde approximation, Publication laboratoire photoélasticité, Ecole Polytechnique Fédérale, Zurich, 9 (1965)

/6/ G.Menditto, Su alcuni problemi di elementi di fondazione riposanti su un suolo alla Pasternak, Tecnica italiana, 32, 865-874 (1967)

/7/ M.Rades, Steady-state response of a finite beam on a Pasternak-type foundation, International Journal of Solids and Structures 6, 739-756 (1970)

/8/ T.M.Wang and J.E.Stephens, Natural frequencies of Timoshenko beams on Pasternak foundations, Journal of Sound and Vibration 51,149-155 (1977)

/9/ M.J.Maurizi and M.B.Rosales, Free vibrations of Timoshenko beams on a W-P foundation, Department of Engineering, Univ. Nacional del Sur, (1986)

/10/ T.W.Wang and L.W.Gagnon, Vibrations of continuous Timoshenko beams on W-P foundations, Journal of Sound and Vibration 59, 211-220 (1978)

/11/ Lord Rayleigh, Theory of Sound (two volumes), Dover publications, New York, 1877 (1945 re-issue)

/12/ A.Stodola, Steam and Gas turbines with a supplement on the prospects of the thermal prime mover (Vol.II), McGraw-Hill Book Co, New York, 1927

/13/ C.W.Bert, Application of a version of Rayleigh technique to problems of beams columns, membranes and plates, 2nd. Conf.on Applied Mathematics, Edmonton, Oklahoma, April 18-19, (1986)

/14/ P.A.Laura and R.H.Gutierrez, Vibrating non-uniform plates on elastic foundation, Journal of Engineering Mechanics (ASCE) 111 (9), 1185-1196 (1985)

/15/ R.Schmidt, Towards resurrecting the original Rayleigh method, Journal of the Industrial Mathem. Society 35 (1), 69-73 (1985)

/16/ S.G.Mikhlin, Variational methods in mathematical physics, Pergamon Press, New York, 1964

/17/ R.D.Blevins, Formulas for natural frequency and mode shape, Van Nostrand Reinhold Co, New York, 1979

/18/ R.P.Goel, Transverse vibration of tapered beams, Journal of Sound and Vibration 47(1), 1-7 (1976)

/19/ C.P.Filipich and M.B.Rosales, A variant of Rayleigh's method applied to Timoshenko beams embedded in a Winkler-Pasternak medium, Journal of Sound and Vibration (to be published)

Numerical Methods in Geomechanics (Innsbruck 1988), Swoboda (ed.)
© 1988 Balkema, Rotterdam. ISBN 90 6191 809 X

Elastic-plastic pile-frames interaction

Ragnar Larsson & Nils-Erik Wiberg
Department of Structural Mechanics, Chalmers University of Technology, Gothenburg, Sweden

ABSTRACT: A FE-algorithm for small deformation incremental plasticity is presented for the analysis of a plane frame interacting with a piled foundation. Both the frame and the piles, consisting of an elasto-plastic material, are analysed by using beam variables. For the interaction with the soil medium, non-linear load transfer functions are used. The solution of the non-linear equations arizing from the discretization is obtained by a Newton iteration procedure optimized by 'slack' line search. A few examples show the structural response for proportional, non-proportional and cyclic loading.

1 INTRODUCTION

Every building interacts with a foundation of one type or another. The structural response from different loading conditions depends strongly on the relative non-linear stress-strain relationships of the frame, the foundation and the soil. Unfortunately, it is a hard task to take into account all the non-linear characteristics as these would involve extensive requirements to be made on the solution procedure, and hence the problem has been approximated in different ways. In Wiberg [1] a plane frame was analysed according to second order - yield hinge theory with a foundation modelled on non-linear generalized springs, and in Larsson and Wiberg [2] a second order linear elastic analysis is described with the aid of exact solutions for both the elements of the frame and for the foundation.

Concerning the interaction with the soil medium the situation seems somewhat clearer for methods based on load transfer functions, [3], [4], both on empirical and experimental grounds.

In the present approach beam variables are used to analyse the frame and the piles. The constitutive relations are then enforced by a layer technique similar to the one adopted by Corradi and Poggi [5]. As shear deformations are neglected only the axial normal stress needs to be considered in the yield condition. This approach competes favourably with the plastic hinge idealization, which in many cases provides solutions acceptable for many engineering purposes, when plastic hardening and non-proportional loading are considered. For the interaction with the soil non-linear, hardening or softening, load transfer functions in the axial direction of the piles are adopted. Another interesting idea is to use a thin-layer element as proposed by Desai and Zaman [6]. The real constitutive parameters of the soil can then be incorporated more accurately.

The incremental constitutive relations are integrated by using an implicit scheme which has been extensively investigated by Runesson [7]. For hardening plasticity the displacement rate in each time increment is defined as an extremal point of a certain associated convex functional. For softening plasticity, however, the functional becomes non-convex and several stable and unstable solutions are available. By guidance of that stable solutions are localized, a computational procedure is outlined by Runesson and Larsson [8]. A discussion of the efficiency of the algorithm is beyond the scope of this paper as it already has been assessed in a previous work [9]. This paper is concluded with a few examples with proportional, non-proportional and cyclic loading. It is worth mentioning that the examples are intended to indicate the potentiality of the formulation.

2 THEORETICAL BACKGROUND

2.1 Preliminaries

A plane frame, such as in Fig. 1, consisting of straight members with a monosymmetric cross-section, is considered in the elastic-plastic analysis. The loading of the frame, which may be line loads, point loads and prescribed displacements, vary slowly enough to make dynamic effects negligible.

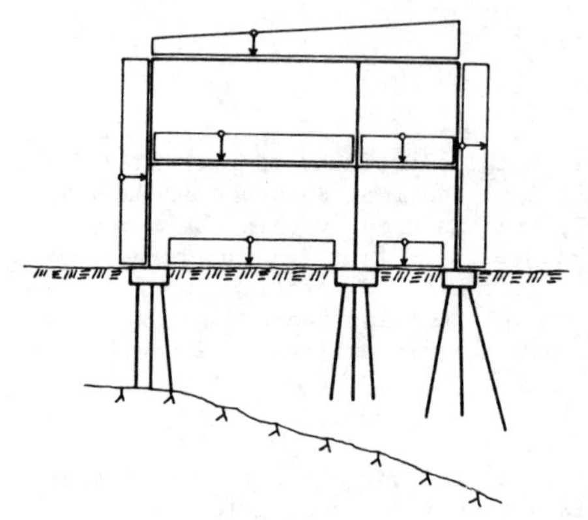

Fig. 1 Elasto-plastic frame considered

The governing equations of the problem concern a member, defined in a one-dimensional interval I of length L, cut out of the frame. The structure problem is then established by incorporating non-homogeneous boundary conditions at the ends of each member into a set of coupled boundary conditions. The boundaries, i.e. the connective points of the members, are prescribed by points forces and/or prescribed displacements.

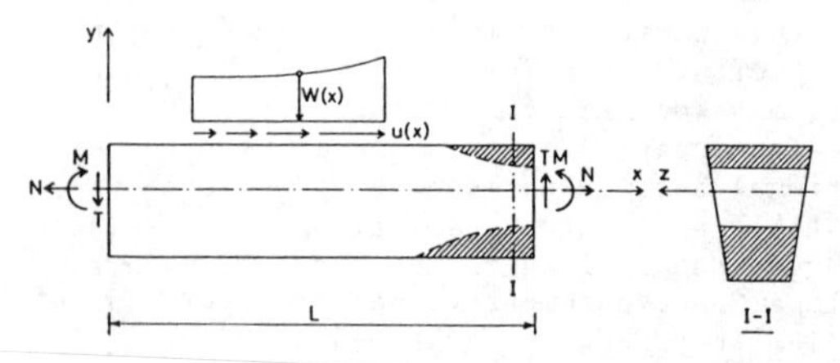

Fig. 2 Elastic-plastic state of a member cut out of the structure

For the interaction with the soil-medium the elements of the foundation, i.e. the piles, are analysed by using non-linear load transfer functions which may describe hardening or softening behaviour of the soil, see Fig. 3. At the end of the piles a spring element can be added.

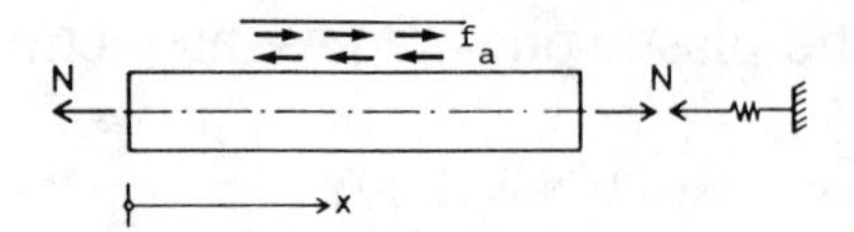

Fig. 3 Pile interacting with soil

2.2 Governing equations for a member

2.2.1 Plasticity problem

The plastically permissible axial stresses σ of the member are defined by the set B

$$B = \{(\sigma,\kappa,\alpha):F(\sigma,\kappa,\alpha) \leq 0\} \qquad (1)$$

where the yield function $F(\sigma,\kappa,\alpha)$ is a function of the uniaxial stress σ, the hardening parameter κ and the parameter α accounting for the kinematic hardening. As shear deformations are neglected the yield condition can be defined as

$$F(\sigma,\kappa,\alpha) = |\sigma| + \alpha_y(\kappa) - \sigma_y(\kappa) \qquad (2)$$

where $\sigma_y(\kappa)$ and $\alpha_y(\kappa)$ are functions defining the isotropic and the kinematic hardening, respectively. The functions σ_y and α_y are obtained from an experimental test curve as in Fig. 4.

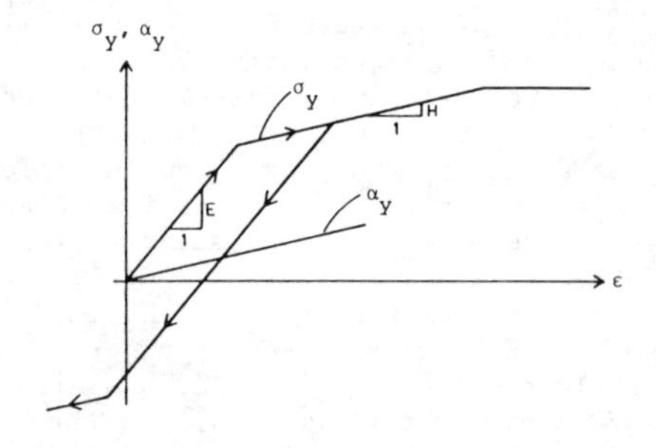

Fig. 4 Experimental test curve defining isotropic and kinematic hardening

The hardening parameter κ is related to the rate of the plastic part of the uniaxial strain ε as

$$\kappa = \int_0^t |\dot{\varepsilon}^P| dt \qquad (3)$$

According to classical beam theory the uniaxial strain ε is connected to the beam variables, i.e. the curvature γ and the generalized uniaxial strain ν, as

$$\varepsilon = \underset{\sim}{y}^T \underset{\sim}{\varepsilon};\ \underset{\sim}{y}^T = [-y,1];\ \underset{\sim}{\varepsilon}^T = [\gamma,\nu] \quad (4)$$

In small deformation theory the vector $\underset{\sim}{\varepsilon}$ is related to the displacements $\underset{\sim}{u}$ with the aid of the strain operator $\bar{\nabla}$ as

$$\underset{\sim}{\varepsilon} = \bar{\nabla}\underset{\sim}{u};\ \begin{bmatrix}\gamma\\ \nu\end{bmatrix} = \begin{bmatrix} d^2/dx^2 & 0 \\ 0 & d/dx \end{bmatrix}\begin{bmatrix} v \\ u\end{bmatrix} \quad (5)$$

Further, the beam stress $\underset{\sim}{\sigma}$, i.e. the moment M and the axial force N, is defined by

$$\underset{\sim}{\sigma} = \int_A \underset{\sim}{y}\sigma dA,\ \underset{\sim}{\sigma}^T = [M,N] \quad (6)$$

where A is the cross section area and σ is the uniaxial stress of a lamina.

Introducing the elastic stress-strain relation $\sigma = E\varepsilon$, where E is the modulus of elasticity, the elastic stress-strain relation in beam variables is obtained

$$\underset{\sim}{\sigma} = D\underset{\sim}{\varepsilon},\ D = E\int_A \underset{\sim}{y}\underset{\sim}{y}^T dA \quad (7)$$

Finally, the problem in incremental plasticity is to find the stresses and displacements which satisfy

$$\begin{cases} -\bar{\nabla}\underset{\sim}{\sigma} = \underset{\sim}{U} \text{ in I},\ \underset{\sim}{U} = [W, U - f_a]^T \\ \dot{\varepsilon} = \frac{1}{E}\dot{\sigma} + \dot{\varepsilon}^P \\ u(0) = u(L) = 0 \\ v(0) = v'(0) = v(L) = v'(L) = 0 \end{cases} \quad (8)$$

The rate of plastic strain is defined by the constitutive inequality

$$(\sigma - \tau)\dot{\varepsilon}^P \geq 0 \quad \forall\ \tau \in B^*(\kappa,\alpha) \quad (9)$$

where the state $B^*(\kappa,\alpha) \in B$ is defined by

$$B^*(\kappa,\alpha) = \{\sigma : F(\sigma) \leq 0\} \quad (10)$$

2.2.2 Time integration procedure

The corresponding discrete problem in time of (8) can be obtained by applying the backward Euler method for the time integration. Thus, by assuming the solution $(\underset{\sim}{\sigma}_n, \underset{\sim}{u}_n)$ to be known the state $\underset{\sim}{u}_{n+1}$ at $t_{n+1} = t_n + \Delta t$ can be obtained as

$$\underset{\sim}{u}_{n+1} = \underset{\sim}{u}_n + \Delta t\dot{\underset{\sim}{u}}_{n+1} \quad (11)$$

where $\underset{\sim}{\sigma}_{n+1}$ satisfy

$$\begin{cases} -\bar{\nabla}\underset{\sim}{\sigma}_{n+1} = \underset{\sim}{U}(t_{n+1}) \text{ in I} \\ (\sigma_{n+1} - \tau)\frac{1}{E}(\sigma^E_{n+1} - \sigma_{n+1}) \geq 0, \\ \forall\ \tau \in B^*(\kappa_{n+1}, \alpha_{n+1}) \\ u(0) = u(L) = 0 \\ u(0) = v'(0) = v(L) = v'(L) = 0 \end{cases} \quad (12)$$

where σ^E_{n+1} is a fictitious elastic stress defined by

$$\sigma^E_{n+1} = \sigma_n + E\Delta\varepsilon \quad (13)$$

When the incremental constitutive inequality of (8) is considered the elastic state (E) and the plastic state (P) must be distinguished; the states are defined by

$$\begin{cases} \Delta\varepsilon^P = \frac{1}{E}(\sigma^E_{n+1} - \sigma_{n+1}) = 0 & \text{(E)} \\ \Delta\varepsilon^P = \frac{1}{E}(\sigma^E_{n+1} - \sigma_{y,n+1}\cdot s) & \text{(P)} \end{cases} \quad (14)$$

The incremental constitutive relations can now be integrated as

$$\sigma_{n+1} = \begin{cases} \sigma^E_{n+1} & \text{(E)} \\ \sigma_{y,n+1}\cdot s & \text{(P)} \end{cases} \quad (15)$$

$$\alpha_{n+1} = \begin{cases} \alpha_n & \text{(E)} \\ \alpha_{y,n+1}\cdot s & \text{(P)} \end{cases}$$

$$\varepsilon^P_{n+1} = \begin{cases} \varepsilon^P_n & \text{(E)} \\ \varepsilon^P_n + \frac{1}{E}(\sigma^E_{n+1} - \sigma_{y,n+1}\cdot s) & \text{(P)} \end{cases}$$

$$\kappa_{n+1} = \begin{cases} \kappa_n & \text{(E)} \\ \kappa_n + \frac{1}{E}(|\sigma^E| - \sigma_{y,n+1}) & \text{(P)} \end{cases}$$

where $s = \text{sign}(\sigma^E_{n+1})$, $\sigma_{y,n+1}$ and $\alpha_{y,n+1}$ are updated stresses after the projection of the fictitious elastic stress onto the yield line. Thus, for a given σ^E_{n+1} one may solve (15) for κ_{n+1} if the experimental relations $\sigma_y(\kappa)$ and $\alpha_y(\kappa)$ are known. Due to the implicit character of $\sigma_y(\kappa)$ and $\alpha_y(\kappa)$, however, the solution has in general to be found iteratively. A simple exception is offered by perfect plasticity defined by

$$\sigma_{n+1} = \begin{cases} \sigma^E_{n+1} & \text{(E)} \\ \sigma_{y,n} & \text{(P)} \end{cases} \quad (16)$$

The similar arguments as for the beam variables can be used to integrate the constitutive relations of the load transfer function. It is then assumed that a test curve for the load transfer function is given as in Fig. 5.

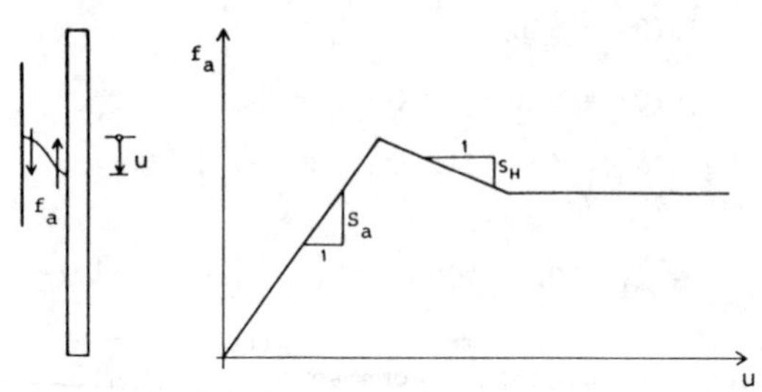

Fig. 5 Experimental test curve for the load transfer function

3 FINITE ELEMENT FORMULATION

The differential equation in (12) is recast into its variational form by multiplying a test function $w = [v',u']$ and then integrating in parts over the interval I. By introducing some basic concepts from functional analysis we obtain

$$(\bar{\nabla}\underset{\sim}{w},\underset{\sim}{\sigma}) + (u',f_a) = (\underset{\sim}{w},\underset{\sim}{U}) \quad \forall \underset{\sim}{w} \in V \tag{17}$$

where

$$(\bar{\nabla}\underset{\sim}{w},\underset{\sim}{\sigma}) = \int_L (\bar{\nabla}\underset{\sim}{w})^T \underset{\sim}{\sigma} dx; \quad (u',f_a) = \int_L u' f_a dx;$$

$$(\underset{\sim}{w},\underset{\sim}{U}) = \int_L \underset{\sim}{w}^T \underset{\sim}{U} dx$$

and V is a Hilbert space introduced by

$$V = \{\underset{\sim}{u} : v \in H_0^2(I),\ u \in H_0^1(I)\}$$

The variational problem (17) is solved by construction of a finite dimensional subspace V_h to V. A subdivision of I into subintervals $I_k = (x_{k-1},x_k)$, $k = 1,\ldots,n$ with the length $l_k = x_k - x_{k-1}$, is made. The basis functions are then defined as a set of piecewise continuous functions on each subinterval I_k so that the solution $\underset{\sim}{u}_h$: u_h, v_h can be represented as a linear combination of the basis functions as

$$\underset{\sim}{u}_h : \begin{cases} v_h = \sum_{k=0}^{n} (v_k\psi_k + v_k'\chi_k), & v_k = v_h(x_k), \\ & v_k' = v_h'(x_k) \\ u_h = \sum_{k=0}^{n} u_k\varphi_k, & u_k = u_h(x_k) \end{cases} \tag{18}$$

In Fig. 6 the basis functions ψ, χ, and φ are depicted.

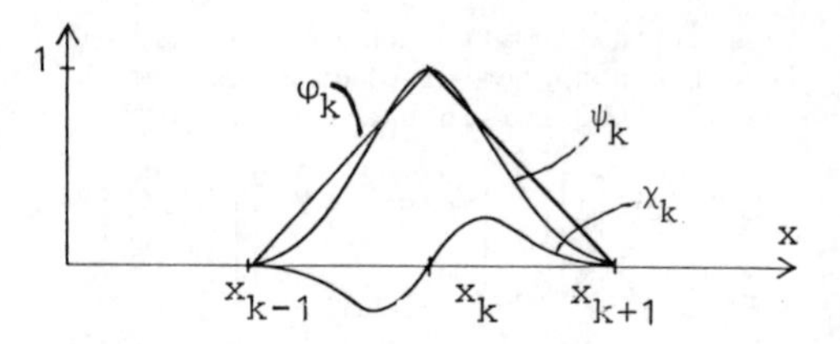

Fig. 6 Basis functions for the member

A set of nonlinear equations is obtained by combining (17) with (18), which gives

$$b - F = 0 \tag{19}$$

where b is a vector of elements b_k containing the internal forces and F is a vector of elements F_k containing the external forces; the elements b_k and F_k become

$$b_k = [\int_L M\psi_k'' dx,\ \int_L M\chi_k'' dx,\ \int_L (N\varphi_k' + f_a\varphi_k) dx]^T \tag{20}$$

$$F_k = [\int_L W\psi_k dx,\ \int_L W\chi_k dx,\ \int_L U\varphi_k dx]^T$$

4 NUMERICAL PROCEDURES

4.1 Iterative procedure

The out-of-balance forces g defined as

$$g = b - F \tag{21}$$

may be shown to be the gradient of a certain associated functional. An extremal point of the functional is reached when the gradient vanishes. When hardening plasticity is considered the functional becomes convex, and hence, there exists a unique solution. In softening, however, the functional becomes non-convex and several stable and unstable solutions are available. Some guidance to trace the most stable one can be obtained from the principal belief that softening materials exhibit localization of plastic strains. Discontinuous displacements have also been suggested for finite elements in perfect plasticity [10].

In order to solve $g = 0$ a variety of methods can be employed, e.g. conjugate gradient or 'quasi Newton' methods. In this paper, however, a true Newton iteration procedure is defined; the procedure can be outlined as:

The solution $\underset{\sim}{u}_h^n$ at $t = t_n$ is known so that

$$\underset{\sim}{u}_h^{n+1} = \underset{\sim}{u}_h^n + \Delta t \dot{\underset{\sim}{u}}_h^{n+1}$$

$$\underset{\sim}{v} = [v(x_0), v'(x_0), u(x_0), \ldots, v(x_n), v'(x_n), u(x_n)] \tag{22}$$

where $\underset{\sim}{v}$ is a vector containing the nodal parameters of $\underset{\sim}{u}_h$, then $\dot{\underset{\sim}{v}}^{n+1}$ is solved from

$$K_{i-1} d_i = -g(\dot{\underset{\sim}{v}}_{i-1})$$
$$d_i = \dot{\underset{\sim}{v}}_i - \dot{\underset{\sim}{v}}_{i-1} \tag{23}$$

where d_i is the search direction and K_{i-1} is the iteration matrix. The solution $\underset{\sim}{v}^{n+1}$ is then obtained as

$$\underset{\sim}{v}^{n+1} = \underset{\sim}{v}^n + \Delta t \sum_{i=1}^{N} \rho_i d_i$$

where N is the number of iterations required to satisfy the convergence criterion and ρ_i is an acceleration factor obtained from line search. The iteration matrix K_{i-1} contains the elements

$$K_{i-1} = \begin{bmatrix} H(\psi_j,\psi_k) & H(\varphi_j,\chi_k) & H(\psi_j,\varphi_k) \\ H(\chi_j,\psi_k) & H(\chi_j,\chi_k) & H(\chi_j,\varphi_k) \\ H(\varphi_j,\psi_k) & H(\varphi_j,\chi_k) & H(\varphi_j,\varphi_k) \end{bmatrix}$$

which are defined by

$$H(\underset{\sim}{w},\dot{\underset{\sim}{u}}) = \int_L [(\bar{\nabla}\underset{\sim}{w})^T \frac{\partial \underset{\sim}{\sigma}}{\partial \dot{\underset{\sim}{\varepsilon}}} \frac{\partial \dot{\underset{\sim}{\varepsilon}}}{\partial \dot{\underset{\sim}{u}}} + u' \frac{\partial f_a}{\partial \dot{u}} \frac{\partial \dot{u}}{\partial \dot{\underset{\sim}{u}}}] dx$$

where

$$D_H = \frac{\partial \underset{\sim}{\sigma}}{\partial \dot{\underset{\sim}{\varepsilon}}} = \int_A H^*(y) \underset{\sim}{y}\underset{\sim}{y}^T dA; \quad S = \frac{\partial f_a}{\partial \dot{u}}$$

$$H^*(y) = \begin{cases} E & (E) \\ H & (P) \end{cases}; \quad S = \begin{cases} S_a & (E) \\ S_H & (P) \end{cases}$$

Note that if the entire cross section is elastic then clearly $D_H = D$.

Line search is used to optimize the algorithm, i.e. we seek the root of the gradient in the search direction. Thus, we seek the root of $\psi(\rho_i) = g^T(\dot{\underset{\sim}{v}}_{i-1} + \rho_i d_i) d_i$; this is approximated by

$$\rho_i = \frac{-\psi(0)}{\psi(1) - \psi(0)}$$

In order that d_i is accepted it must be required that $\psi(0) < 0$, i.e. $K_{i-1} d^T K_{i-1} d \geq 0$. This condition is trivially satisfied for convex problems but not for non-convex problems. The next requirement is that $0 < \rho_i < 2$ in order to ensure convergence for a strictly convex problem.

4.2 Integration of the cross section

For the integration of the stiffness moduli D_H a mono-symmetric cross section as in Fig. 7 is considered. The cross section is then subdivided into a finite number of layers as $y = 0 = y_0 < \ldots < y_n = h$, with the length $h_j = y_j - y_{j-1}$. Furthermore, a shift in origin β is introduced in order that D_H be uncoupled when the cross section is elastic.

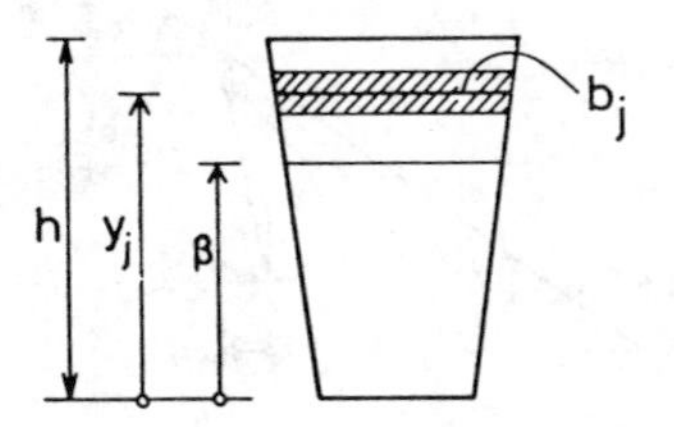

Fig. 7 Mono-symmetric cross section

Then, the stiffness moduli D_H is integrated as

$$D_H = \sum_{j=1}^{n} H^*(y_j) \underset{\sim}{y}_j \underset{\sim}{y}_j^T b_j h_j \tag{25}$$

where

$$\underset{\sim}{y}_j = [(y_{j-1} - \beta + h_j/2), 1]^T$$

5 STRUCTURAL COUPLING

In order to analyse the entire frame the members must be coupled at the nodes. A variational formulation of the entire frame with non-homogeneous boundary conditions included yields

$$\sum_e [(\nabla \underset{\sim}{w}, \underset{\sim}{\sigma}) + (u', f_a)] = \sum_e (\underset{\sim}{w}, \underset{\sim}{U}) + \sum_e n_1^{*T} N_1 \tag{26}$$

$$n_1^* = [n_1, t_1, m_1, n_2, t_2, m_2]^T$$

$$N_1 = [N_1, T_1, M_1, N_2, T_2, M_2]^T$$

where the sum is over all members of the frame.

A set of coupled boundary conditions have now been obtained. This is represented by the term

$$\sum_e n_1^{*T} N_1$$

containing the end forces N_1 and the end displacements n_1^*, which are defined in the local coordinate system of each member.

A very general coupling of the members, allowing for stiff areas, hinges and rollers, can be obtained if transformations of the displacement variables are made from the local coordinate system $x_1 - y_1$ to a specified system $x_s - y_s$ and/or a global system $x - y$, see Fig. 8.

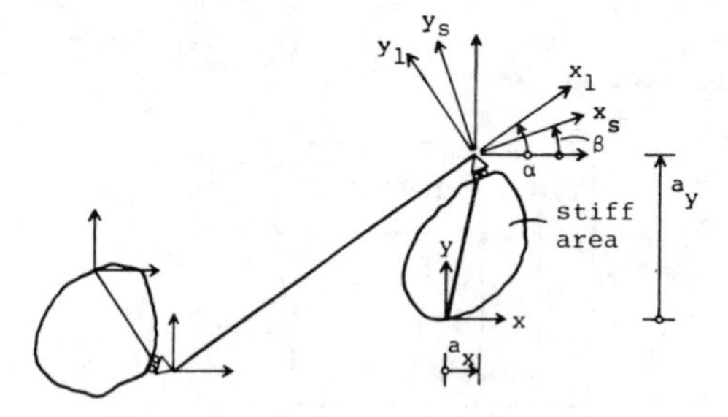

Fig. 8 Member with coordinate systems introduced

Displacement variables defined in the various systems are related to each other by the following transformations, see Ref. [2]:

$$n_e = Bn_s, \quad n_s = ACn_s \tag{29}$$

The coupled boundary conditions can then be defined with the aid of the diagonal matrices Λ_s and Λ with the relation $\Lambda = I - \Lambda_s$. The matrices Λ_s and Λ refer to element forces defined in the specified system $x_s - y_s$ and in the global system $x - y$, respectively.

The subsidiary conditions, i.e. definition of hinges and rollers, are defined in Λ_s as

$$\sum_e \Lambda_s B^T N_e = \underset{\sim}{P}_s \tag{30a}$$

and the remaining part are defined via Λ as

$$\sum_e C^T A^T \Lambda B^T N_e = \underset{\sim}{P} \tag{30b}$$

where $\underset{\sim}{P}_s$ and $\underset{\sim}{P}$ are applied point forces.

6 NUMERICAL EXAMPLES

6.1 General

For the evaluation of the gradient g and the iteration matrix K basis functions obtained from the homogeneous solution of the corresponding elastic problem were used, see Ref. [11]. The system was then integrated by using Gauss point integration over the length and the layer technique was used for the integration of the cross section (3 Gauss points and 6 layers were used for all elements for convenience) Convergence in the iteration procedure was checked by comparing the error e with the tolerance $\delta = 0.001$ as

$$e = ||d|| / ||v_E|| \leq \delta \tag{31}$$

where $||\cdot||$ is the Euclidean norm and v_E is the elastic response.

6.2 Plane frame resting on a piled foundation

The plane frame in Fig. 9 is resting on a piled foundation. The foundation, which is embedded in a cohesive subsoil, consists of 2 pile groups with 2 piles attached to the pile caps; the pile caps are rigid and are not in contact with the soil. The non-linear material characteristics of both the frame and the soil are shown in Fig. 10.

The loading consists of a uniformly distributed load acting on top of the frame, see Fig. 9, and is subjected according to three different loading programs, of proportional, non-proportional and cyclic loading, as shown in Figs. 11-12.

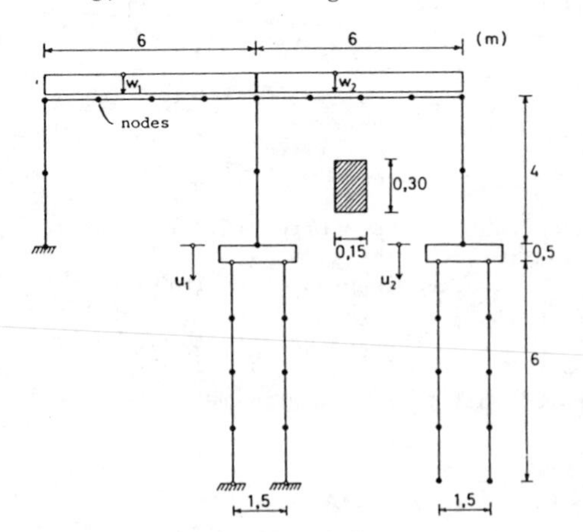

Fig. 9 Analysed frame resting on a piled foundation

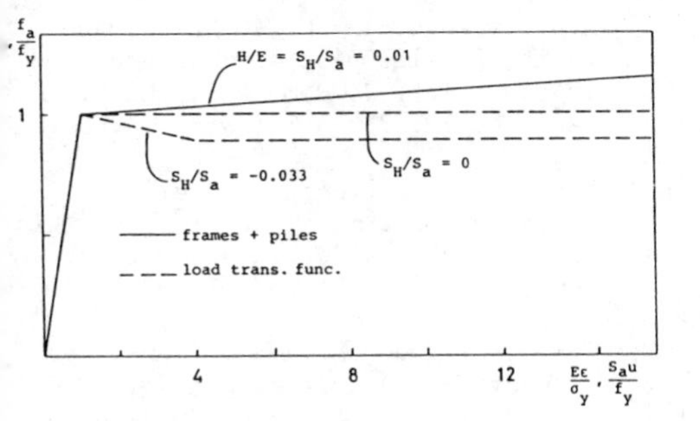

Fig. 10 Material characteristics of the frame, the piles and the soil

The properties in the elastic range for the frame, the piles and the soil are as follows: E = 210 GPa for $\sigma_{y0} < 220$ MPa, E = 21 GPa for $\sigma_{y0} < 22$ MPa and S_a = 3 MPa for $f_{y0} < 10$ kN/m, respectively.

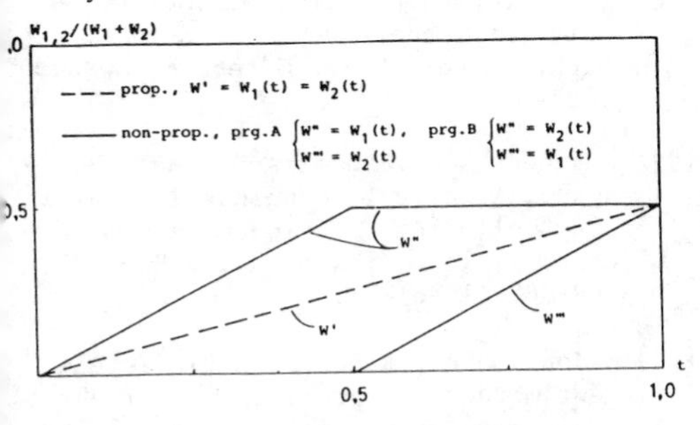

Fig. 11 Loading programs displaying proportional and non-proportional load

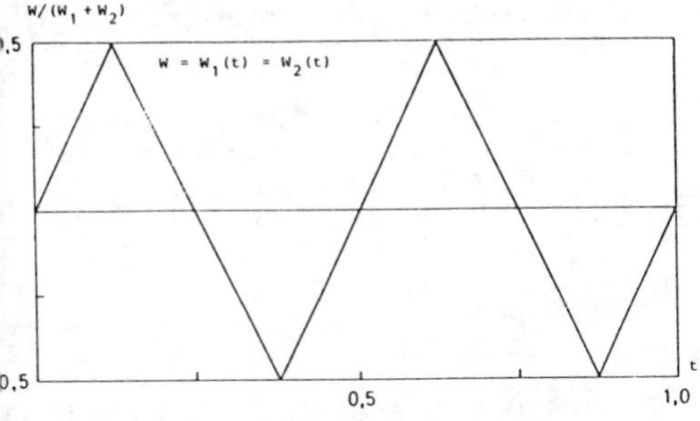

Fig. 12 Loading program displaying cyclic load

In the first example proportional loading (W(1) = 100 kN/m) is considered with hardening, perfect elastic-plastic and softening behaviour of the load transfer function, see Fig. 10. The solutions were obtained by using 10 timesteps and linear extrapolation of the displacement rate from the previous timestep was used. Fig. 13 significantly displays a more flexible behaviour of the structure when the ratio S_H/S_a decreases. The figures ①, ② and ③ denote where and in which order "plastic elements" occur.

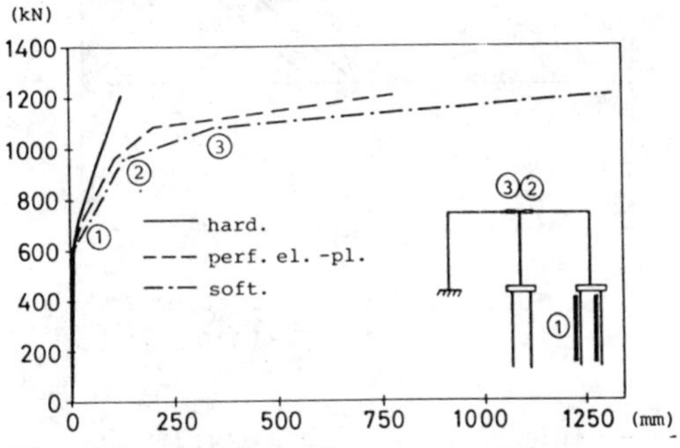

Fig. 13. Total load versus settlement u_2 of the right pile cap

In the second example non-proportional loading according to Fig. 11 is considered with perfect elastic-plastic behaviour of the load transfer function. For the cases of non-proportional loading the use of linear extrapolation from the previous timestep must be used with some care. In this example, however, linear extrapolation was used as the loading is "almost" proportional. Fig. 14 shows the settlements of the left and the right pile cap in 20 timesteps for different loading histories. It can be observed that the stress-strain dependency makes a difference when loading programs A and B are compared. No difference is obtained at W(1) = 48 kN/m when program A is compared with proportional loading as the yielding takes place only in the shear layer of the right pile group.

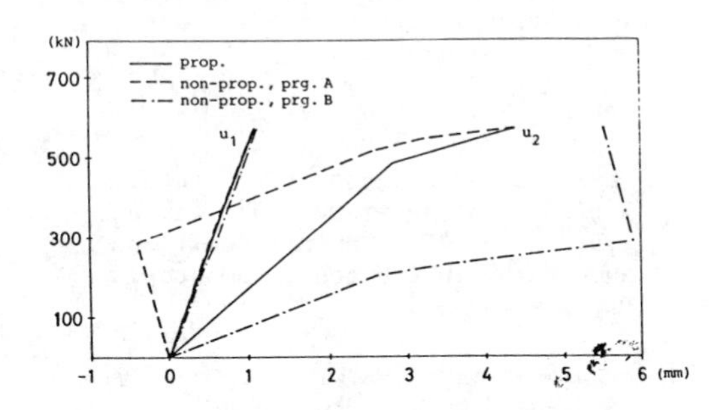

Fig. 14 Total load versus settlement u_1 of the left and u_2 of the right pile cap

In the third example cyclic loading is considered. Perfect elastic-plastic behaviour is used for the load transfer function. A "zero predictor" was used in this case as no convergence with the linear extrapolation was obtained. Fig. 15 shows the settlement of the left and the right pile cap in 40 timesteps for two load

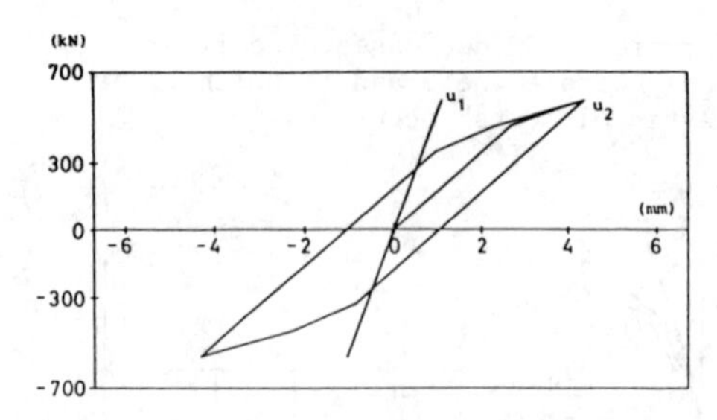

Fig. 15 Total load versus settlement u_1 of the left and u_2 of the right pile cap

cycles (see Fig. 12) with an amplitude W = 48 kN/m.

7 CONCLUSIONS

A FE-algorithm in small deformation incremental plasticity was presented for elastic-plastic pile-frame interaction. The theory is based on consistent convex analysis, and hence the problem was considered on a sound theoretical ground. The use of the backward Euler method for the time integration seems appealing as the projected stress σ_{n+1} is always assured plastically admissable.

In the numerical examples hardening, perfect and softening plasticity was considered for proportional loading. In softening plasticity much more research needs to be made due to the non-uniqueness of the solution. For the cases of non-proportional loading the use of a good predictor is strongly recommended.

8 REFERENCES

[1] N.-E.Wiberg, Soil-structure interaction in plane frame stability analysis, Proc. Fourth Int. Conf. on Num. Meth. in Geomech., Edmonton, May 31 - June 4, 781-786, 1982.

[2] R.Larsson and N.-E.Wiberg, Stability of plane frames including soil structure interaction, Chalmers Univ. of Techn., Dept. of Struct. Mech., Report 87:1, Göteborg, Sweden, 1987

[3] Y.K.Chow, Analysis of vertically loaded pile groups, Int. J. Num. Anal. Meth. Geomech. 10, 59-72 (1986)

[4] A.G.Heydinger and M.W.O'Neill, Analysis of axial pile-soil interaction in clay, Int. J. Num. Anal. Meth. Geomech. 10, 367-381 (1986)

[5] L.Corradi and C.Poggi, A refined finite element model for the analysis of elastic-plastic frames, Int. J. for Num. Meth. in Eng. 20, 2155-2174 (1984

[6] C.S.Desai and M.M.Zaman, Thin-layer element for interfaces and joints, Int. J. Num. Anal. Meth. Geomech. 8, 19-43 (1984)

[7] K.Runesson, Implicit integration of elasto-plastic relations with reference to soils, Int. J. Anal. Num. Meth. Geomech. 11, 315-321 (1987)

[8] K.Runesson and R.Larsson, Characteristics and computational procedure in softening plasticity, Chalmers Univ. of Techn., Dept. of Struct. Mech., Report 87:8, Göteborg, Sweden, 1987

[9] K.Runesson, A.Samuelsson, and L.Bernspång, Numerical technique in plasticity inlcuding solution advancement control, Int. J. Num. Meth. Eng. 22, 769-788 (1986)

[10] C.Johnson and R.Scott, A finite element method for problems in perfect plasticity using discontinuous trial functions, Nonlinear Finite Element Analysis in Structural Mechanics (Eds. Wunderlich et al.), Springer, Berlin, 1981

[11] R.Larsson and N.-E.Wiberg, On exact and hierarchical finite elements for frame structures, NUMETA 87, 1, Martinus Nijhoff Publ., 1987

9 Earth structures, slopes, dams, embankments

Numerical Methods in Geomechanics (Innsbruck 1988), Swoboda (ed.)
© 1988 Balkema, Rotterdam. ISBN 90 6191 809 X

Layered analysis of embankment dams

D.J.Naylor
University College, Swansea, UK

D.Mattar, Jr.
PROMON, Sao Paulo, Brazil
(Previously research assistant at Swansea)

ABSTRACT: The accuracy of the idealisation of the construction of embankment dams using the finite element method is examined. It is found that good accuracy can be obtained using many less than the ten layers commonly used. To achieve this it is necessary to correctly interpret the computed displacement and also apply a stiffness reduction factor of about 4 to new layers. The results of a parameter study on a hypothetical dam section and also an application to a real dam are presented to justify the findings.

1 INTRODUCTION

This paper concerns the modelling of the construction of large embankment dams by numerical methods, especially the finite element method. These fills are normally constructed in substantially horizontal layers ranging in thickness from about 0.2 to 1.0m. Consequently there will be a large number of layers, perhaps 100 or more in a large dam. The limitations of computer modelling require that the real situation be idealised as the placement of relatively few thick layers. Each is conceived as being constructed under conditions of zero gravity subsequently "switched on" to each layer in turn. For major dams when the interest lies in predicting displacements within the dam typically about ten layers have been used. Not so many are considered necessary for stress prediction. Table 1 gives some examples. These have established a precedent for the number of layers required.

It is time consuming and therefore expensive to carry out a multi-layer analysis using the finite element method as each layer requires a separate analysis. It is, therefore, desirable to minimise the number of layers consistent with obtaining appropriate accuracy. A high accuracy will not be justified as the source of error in modelling the material properties of the soil is likely to be much greater than that due to the number of layers.

The broad objective then of this paper is to look into the question of how many layers are appropriate. This is achieved by looking at first one dimensional case of a laterally extensive fill and then at fills in general

Table 1. Number of layers (n) used in major dam analyses

Dam	n	Source
"Standard" (hypothetical)	7,10,14	
Otter Brook, U.S.A.	14	Clough and Woodward (1967)
Oroville, U.S.A.	9, 12	Kulhawy and Duncan (1972)
Scammonden, U.K.	9	Penman et al (1971)
Llyn Brianne, U.K.	10	Penman and Charles (1973)
Llyn Brianne, U.K.	10	Cathie and Dungar (1978)
Alvito, Portugal	10	Seco Pinto (1983)
Beliche, Portugal	11	Veiga Pinto (1983)

Two techniques are identified for obtaining good accuracy with few layers. The first concerns the interpretation of the displacement output from finite element analyses. The second provides criteria for reducing the stiffness of the relatively thick analytical layers as they are placed to more closely approximate the real situation.

Before proceeding further it must be made

quite clear what is meant by "displacement" in the context of a rising fill. It is: The movement of a marker placed on the surface of the fill. The important point is that the zero datum for displacement is established the moment the marker is installed when it is assumed to be flush with the current fill surface. Its subsequent movement will be due to the compression of the underlying material by the material added above the marker. This definition holds good whether or not there are time effects, i.e. consolidation or creep. In the analyses carried out here, however, time effects are not considered. All movements are assumed to occur immediately upon application of load.

2 THE ONE DIMENSIONAL CASE

2.1 The irrelevence of the number of layers

The modelling of a continuously placed laterally extensive fill by discrete layers is unaffected by the number of layers provided the layer boundaries coincide with the elevations at which settlement predictions are required. The reason is that the settlement of a point in the fill depends solely on the weight of material placed above it. This will not depend on how many layers are used to place that material.

This has an important implication for nonlinear analyses. In these it is usual for loads to be applied incrementally. Suppose n increments are required to model a fill construction. For the one dimensional case there would be a choice between building the fill up in n layers, applying gravity to each in turn, or alternatively building the fill in one layer and applying its self weight in n increments. Both would have precisely the same effect on the underlying material. The former, however, would be less convenient computationally due to the changing geometry. A compromise involving ℓ layers with m increments in each so that very roughly $\ell \times m = n$ may be used for two and three dimensional analyses. If there is little differential settlement (as in a wide dam of homogeneous stiffness) then it will be expedient to make ℓ relatively small).

2.2 The interpretation of settlement

A significant source of error in the finite element analysis of fills can arise due to the incorrect interpretation of displacements at nodes located between fill boundaries. This is due to failing to appreciate that the datum for displacement is undefined between layer boundaries.

This is illustrated by the one dimensional example shown in Figure 1. The 6m

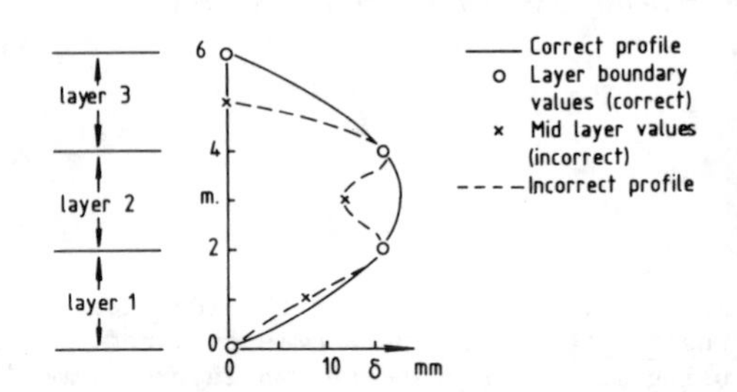

Fig. 1 Correct and incorrect interpretations of settlement in one-dimensional fill

high fill is taken as linear elastic. The confined modulus, D, is 10 MPa and the unit weight, γ, is 20 kN/m^3. The incorrect mid layer values shown by the crosses were obtained by assuming the datum for the mid layer point to be zero when the layer above it was placed. The settlement at the 1m height due to the placing of layer 2 is calculated on this basis as 4mm. This is subsequently increased to 8mm due to placing the third layer. The correct values (which would have been obtained had a layer boundary occurred at the 1m level) are 6 and 10mm respectively. Thus an error of 2mm is 'locked in' due to the incorrect datum assumption.

It is suspected that earlier workers used as many layers as they did to smooth the settlement profile which, due to this factor, tended to be irregular. The correct procedure is to ignore the mid layer values as there is no settlement datum for them. If this is done a smooth (and in the one dimensional case exact) settlement profile results.

3 THE GENERAL CASE

3.1 Stiffness reduction factor

The layer thickness is irrelevant in the one dimensional case because there is no differential settlement and therefore no bending in the layers as they are placed. The existence of bending in two and three dimensional analysis is therefore the cause of error due to the analytical layers being thicker than the actual. Kulhawy et al (1969) recognised this. They found in the analysis of the 'standard dam' described by Clough and Woodward (1967) that similar accuracy was obtained using 7 layers with the stiffness of new layers reduced to a small value as compared with 14 layers without reduction.

The question is: is there an optimum stiffness reduction factor, and if so what is it? Intuitively an optimum value between 1 and the large value used by Kulhawy would be expected to exist. Naylor et al (1984) derived theoretically an optimum reduction factor of 4 for a linear elastic fill material. This was derived by considering a new layer as a beam. The flexural rigidity of the analytical layer had to be such that when loaded under its self weight the same displacement profile along its base resulted as in the actual situation when the fill was built up in relatively thin layers. The theory is given in full in the Appendix to the 1984 paper.

For nonlinear materials in which the stiffness increases as the stress level increases some of the stiffness reduction needed is anticipated as the average stiffness during the building up of the gravity load will be less than the final. Consequently it would appear that a lower reduction factor should be used for such materials. On this basis f = 3 was used for the nonlinear analyses described below with f = 4 for the elastic. This assumption may not, however, be valid, in which case the optimum f for nonlinear materials may exceed 4. See below.

3.2 Parameter study

In order to assess the validity of the foregoing a parameter study was carried out of a situation resembling that of a central clay core dam but exaggerating the differential settlement across a horizontal section. This was achieved by assuming the clay core to be bounded by rigid shoulders. The section is shown on Figure 2. It was chosen with the Beliche dam in southern Portugal in mind as the authors' were involved in modelling the construction of this dam by the finite element method (Naylor et al, 1986).

The mesh of Figure 2(b) was used to model the fill construction. Reduced integration (i.e. 2x2 Gauss quadrature) was used for the eight noded elements.

The parameters varied comprised the number of layers, the material model and the new layer stiffness reduction factor (f). 3, 6 and 12 layers were compared. The material models were linear elastic and the variable elastic 'K-G' model described in Chapter 4 of Naylor et al, 1981. f was

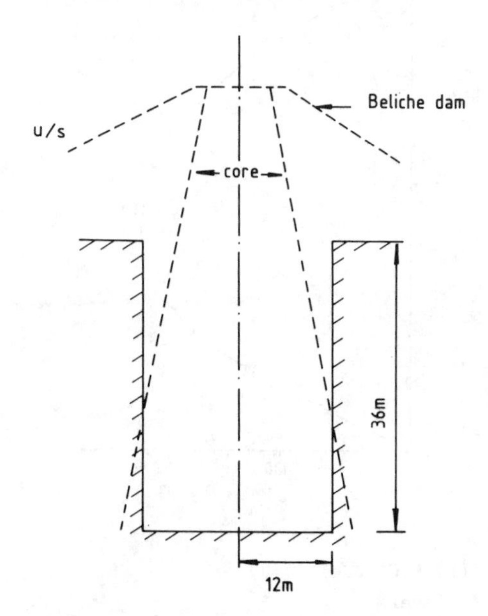

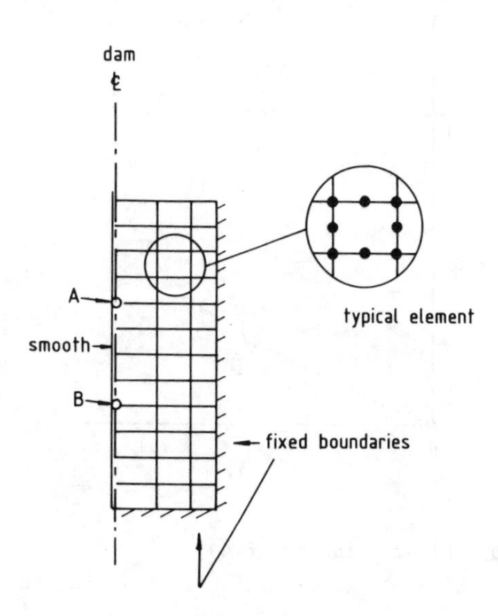

Fig. 2 Geometry for rigid shoulders idealisation

varied from 1 to 1000 with 4 being used as a 'central' value for the elastic analyses and 3 for the K-G.

For the elastic analyses Young's modulus was taken as 10 MPa and Poisson's ratio 0.4. Pore pressures were not included in the analyses, it being assumed that drained conditions applied.

A plane strain version of the K-G model was used in the nonlinear analyses. In this the tangential bulk modulus K was replaced by $\bar{K} = K + G/3$, G being the tangential shear modulus. The model then defines stress dependent tangential (or incremental) moduli as follows:

$$\bar{K} = \bar{K}_1 + \bar{\alpha}_K \sigma_s \quad (1)$$

$$G = G_1 + \alpha_G \sigma_s + \beta_G \sigma_d \quad (2)$$

in which $\bar{K}_1$, G_1, $\bar{\alpha}_K$, α_g, and β_G are empirical constants (β_G negative), and σ_s and σ_d are mean and deviatoric stress measures: $\sigma_s = \frac{1}{2}(\sigma_1 + \sigma_3)$, $\sigma_d = (\sigma_1 - \sigma_3)$ where σ_1 and σ_3 are the major and minor principal stresses. The values of the constants are given in Table 2. They are the same as those used for the Beliche dam clay core.

Table 2. Clay core K-G model parameters

$\bar{K}_1$	G_1	$\bar{\alpha}_K$	α_G	β_G
10.7	6.4	47	28	-43

Units: $\bar{K}_1$, G_1 MPa, $\bar{\alpha}_K$, α_G, $-\beta$ dimensionless

Displacements on the core centre-line from the parameter study are presented in Figures 3, 4 and 5.

Figure 3 compares the settlement profile for different numbers of layers using the preferred f values of 4 for linear elastic and 3 for the K-G model. The differences are so small as to be barely distinguishable.

Figure 4 shows the effect of varying f the number of layers being kept constant at 6. As would be expected the unreduced new layer stiffness (f=1) showed the least settlement. Interestingly the preferred f value profiles were close to the large f profile for the elastic analysis whereas they were close to the f=1 profile for the K-G. No explanation for this is offered. The effect of varying f was greater for

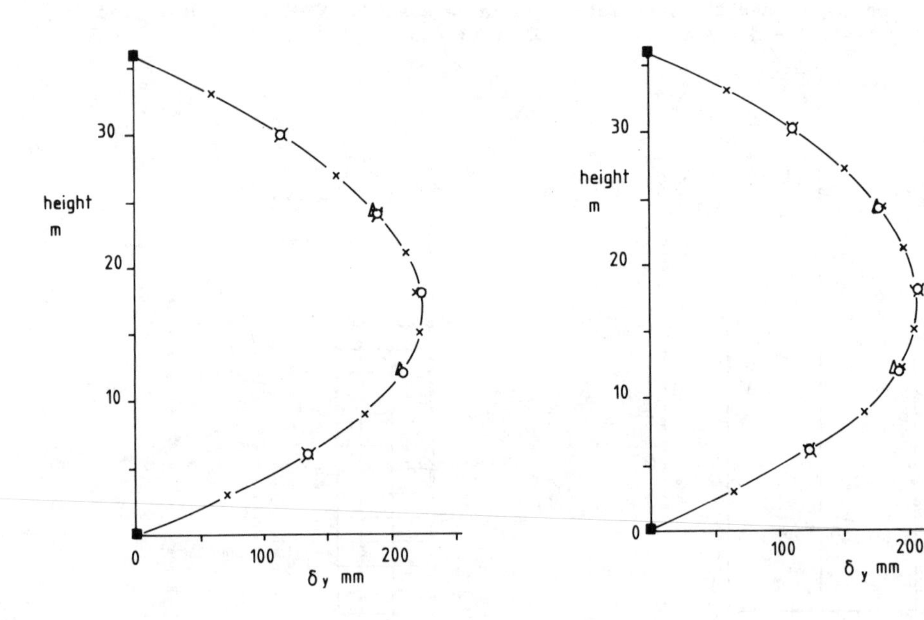

(a) Linear elastic (f=4) (b) K-G model (f=3)

Legend Δ 3 layers
O 6 layers
—X—12 layers

Fig.3 Rigid shoulders study - effect of number of layers on settlement

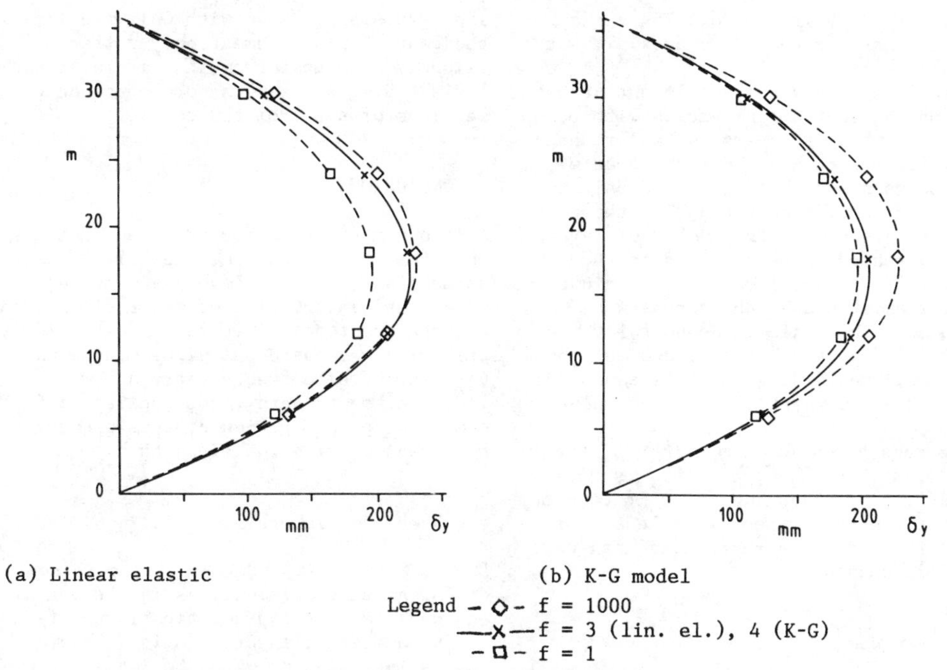

Fig.4 Rigid shoulders study - effect of f, 6 layer analyses

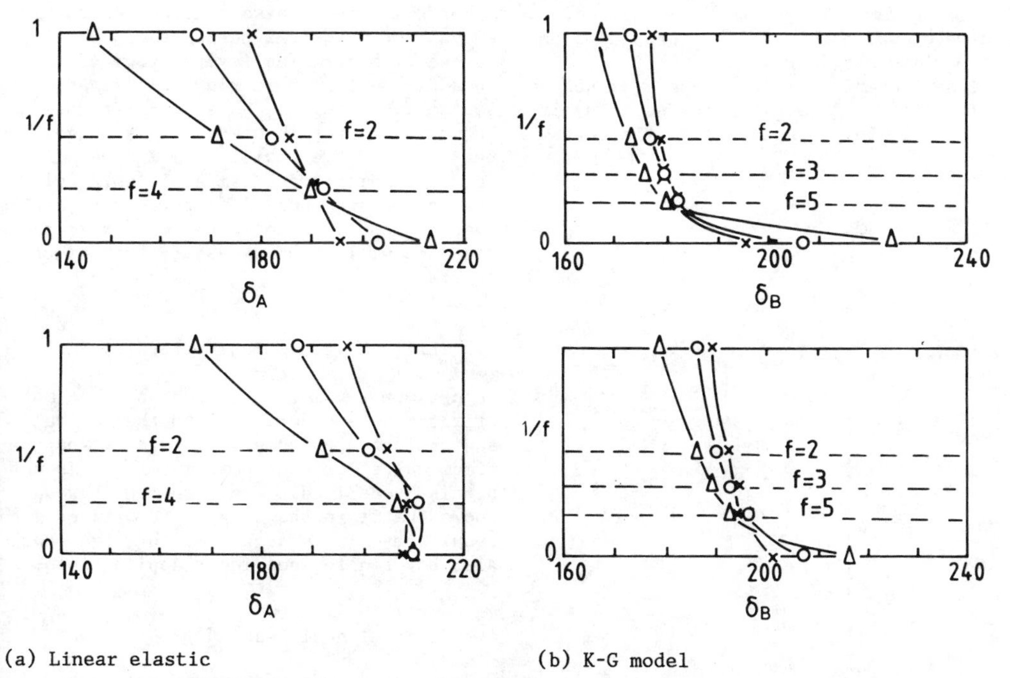

(a) Linear elastic

(b) K-G model

Location of points A and B shown on Fig.2. Legend as in Fig.3

Fig.5 Rigid shoulders study - f:settlement:number of layers relation for two points on centre-line

the 3 layer analysis (approx. 25% variation) and less for the 12 (approx. 7% variation).

Figure 5 shows how the settlement at two points on the centre-line varies with both f and the number of layers. The left hand pair of figures relates to the linear elastic analyses and the right hand pair to the K-G. It should be noted that the curves become steeper as the number of layers (n) is increased and would tend to the vertical as n became large. If a unique optimum f existed then the curves would all cross at this value. Support for the theoretical f=4 value is provided for the elastic analyses. However, this cannot be said for the K-G where the optimum value appears to be nearer to 5 than 3.

Space does not allow comparison of the stress distributions here. This is given in Naylor et al, 1984. Suffice it to comment that generally these were less sensitive to the parameter variations than were the displacements.

3.3 Beliche dam

An investigation into the effect of using different numbers of layers was made as part of the modelling of the construction of Beliche dam. The K-G model was used. f, however was taken as 5 rather than 3 as in the rigid shoulders study.

The settlement profile on the core centre-line (see Figure 2 for the section) is shown on Figure 6. Again the differences are very small. Even with only two layers the central displacement only differs slightly from that from the six layer analysis. Similar results were obtained at sections offset from the centre.

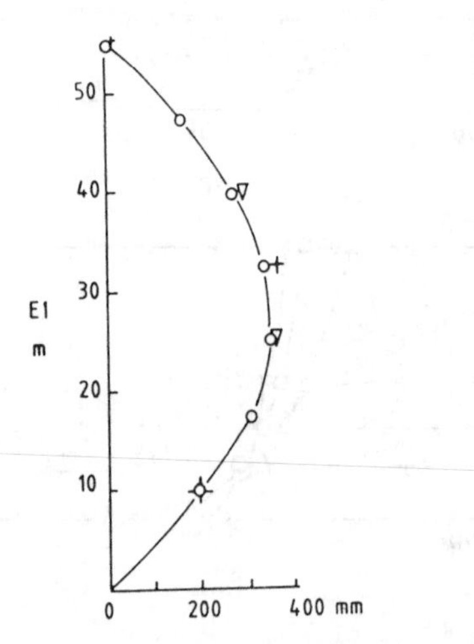

Fig. 6 Settlement on Beliche dam centre-line

4 CONCLUSIONS

A strong motivation for this research has been the need to examine the belief that as many as about ten layers are needed to model the construction of major fills such as embankment dams. it seems that this criterion was based on pioneering work carried out some twenty years ago when finite elements were first applied to fill dams. No proper review of this appears to have been carried out since then.

4.1 Number of layers

It has been shown that when there is no differential settlement, as in the one dimensional case, the prediction is unaffected by the number of layers provided layer boundaries coincide with the levels at which settlement predictions are required. This implies that the computed values between layer boundaries are ignored.

For the general case accuracy quite high enough for practical purposes can be obtained with very few layers provided that in addition to considering only layer boundary settlements an appropriate stress reduction factor is used. It is suggested that 4 to 6 layers will normally be sufficient in embankment dam analyses. Only when there are special features such as the need to model a complicated construction sequence will it be necessary to use more layers.

4.2 Stiffness reduction factor (f)

An optimum f value for linear elastic materials has been derived both theoretically and validated by numerical trials. The optimum value for nonlinear materials is not precisely defined. The research, however, shows that f in the range 3 to 5 gives good results, so it is suggested that f=4 should also be used for nonlinear applications.

4.3 Material nonlinearity

The extent to which the material is nonlinear is believed to have no bearing on the question of how many layers are required for a given accuracy. The relevant factor here - at least in the one dimensional case - is the total number of increments used in modelling the construction. Thus if few layers are required more increments per layer will be required. It will generally be more convenient to use fewer layers although the cost savings will be less than in a linear elastic analysis where only one solution per layer is necessary.

ACKNOWLEDGEMENTS

Thanks are due to the then director, J. Ferry Borges and senior engineer, E. Maranha das Neves, of the Laboratório Nacional de Engenharia Civil (L.N.E.C.), Lisbon, Portugal for their co-operation and assistance with the work on the Beliche dam. The support of S.E.R.C., who financed the second author under grant GR/C/66510, is also acknowledged. His contribution to this paper formed part of the work covered by this grant.

REFERENCES

Cathie, D.N. and Dungar, R. 1978. Evaluation of finite element predictions for constructional behaviour of a rockfill dam. Proc. Instn. Civ. Engrs., Pt. 2, 65(Sept.): 551-568.

Clough, R.W. and Woodward, J.R. 1967. Analysis of embankment stresses and deformations. Proc. Am. Soc. Civ. Engrs., 93, SM4: 529-549,

Kulhawy, F.H., Duncan, J.M. and Seed, H.B. 1969. Finite element analyses of stresses and movements in embankments during construction. Geotechnical Engineering Report No. TE-69-4, University of California, Berkeley, California.

Kulhawy, F.H. and Duncan, J.M. 1972. Stresses and movements in Oroville Dam. Proc. Am. Soc. Civ. Engrs., 98, No. SM7: 653-665.

Naylor, D.J., Maranha das Neves, E., Mattar, Jr., D. and Veiga Pinto, A.A. 1986. Prediction of construction performance of Beliche Dam. Geotechnique 36, No. 3: 359-376.

Naylor, D.J., Mattar, Jr., D. and Engmann, F.O. 1984. Layered analysis of embankment construction. Internal report C/R/480/84, Dept. of Civil Engng., Univ. College, Swansea.

Naylor, D.J., Pande, G.N., Simpson, B. and Tabb, R. 1981. Finite elements in geotechnical engineering, Pineridge Press, Swansea.

Penman, A.D.M., Burland, J.B. and Charles, J.A. 1971. Observed and predicted deformations in a large embankment dam during construction. Proc. Instn. Civ. Engrs., 49(May): 1-21.

Penman, A.D.M. and Charles, J.A. 1973. Constructional deformations in rockfill dam. Proc. Am. Soc. Civ. Engrs., 99, SM2: 139-163.

Sêco e Pinto, P.S. Fracturacão Hidráulica em Barragens de Aterro Zonadas. L.N.E.C. internal report, March 1983 (equivalent to doctoral thesis).

Veiga Pinto, A.A. Previsão do Comportamento Estrutural de Barragens de Enrocamento, L.N.E.C. internal report, July, 1983 (equivalent to doctoral thesis).

Numerical Methods in Geomechanics (Innsbruck 1988), Swoboda (ed.)
© 1988 Balkema, Rotterdam. ISBN 90 6191 809 X

Slope analysis and theory of plasticity

P.De Simone
Università di Napoli, Italy
L.Mongiovì
Università di Roma, 'Tor Vergata', Italy

ABSTRACT : The slope collapse problem is studied by means of the method of characteristics by introducing two lines of stress discontinuity which originate from the vertex of the slope. It is shown that the problem is composed of two Cauchy domains truncated by the lines of discontinuity and connected by a free boundary generalized Goursat domain, and that the Cauchy domain under the top surface has to be in a passive state in order to obtain an admissible solution.

1 INTRODUCTION

The slope collapse problem is generally studied by the limit equilibrium method owing to its applicability to heterogeneous materials and complex geometrical conditions. However, due to the lack of rigorous theoretical formulation, it is generally impossible to evaluate the reliability of the results (Mongiovì, 1981).

On the other hand, to date there are very few solutions obtained using rigorous methods of the theory of plasticity. These are restricted to particular situations such as the infinite slope and the slope of undefined shape in a homogeneous Mohr-Coulomb medium (Sokolovskii, 1965), or the slope in a Tresca medium whose cohesion increases linearly with depth (Booker and Davis, 1972).

From the view-point of the method of characteristics the slope stability problem differs remarkably from other collapse problems, such as the bearing capacity of a foundation or the limiting earth pressure against a retaining structure, where the solution can be obtained considering a sequence of a Cauchy, a Riemann and a Goursat domain. Such a sequence is not well-posed from the phisical-mathematical viewpoint, being tractions fixed both on slope and top surfaces.

In this paper, the slope stability analysis is performed by means of the method of characteristics using a different sequence of initial value problems, which is based on introduction of lines of stress discontinuity originating from the vertex of the slope.

2 BEARING CAPACITY AND SLOPE STABILITY

The bearing capacity of a strip footing, from the viewpoint of the method of characteristics, is a sequence of a Cauchy (initial line), a degenerated Riemann (initial characteristics) and a Goursat (mixed) problem, as shown in fig. 1a with reference to a footing on a slope of a Tresca medium (e.g. Sokolovskii, 1965).

This sequence is not suitable for the analysis of the collapse of a slope, owing to the lack of one degree of freedom in the problem caused by the absence of the footing (De Simone, 1986).

In order to handle the problem by means of the a aforementioned sequence, Sokolovkii considered a slope with a free shape. In this way one degree of freedom is transferred from stress to geometry, and a free surface Goursat problem is defined on the slope, the bearing capacity sequence (C-R_D-G) being retained (fig. 1b).

The scheme analyzed by Sokolovskii is completed by a surcharge acting on the top surface of the slope, whose intensity depends on the inclination of the slope in the vertex O (fig. 1b). In the following the essential role played by this hypothesis on the solution of such a scheme will appear evident.

On the other hand, the problem of the collapse of a slope with a fixed shape (in particular rectilinear) and without any limitation on the value of the surcharge (in particular zero traction condition) is completely different from the one analyzed by Sokolovskii. As a matter of fact both, the top surface OA and the slope surface OB (fig. 1c) define a Cauchy domain. The problem can be now posed as follows : "Is it possible (and how) to connect such two domains by means of some a third one?"

3 SLOPE ANALYSIS USING DISCONTINUITIES

In a previous paper (De Simone, 1986), owing to the

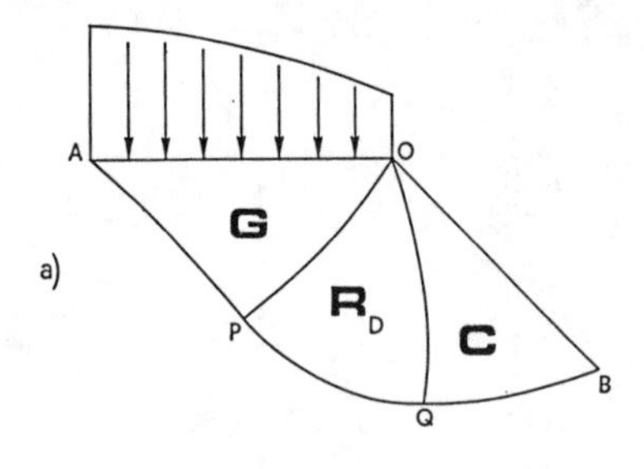

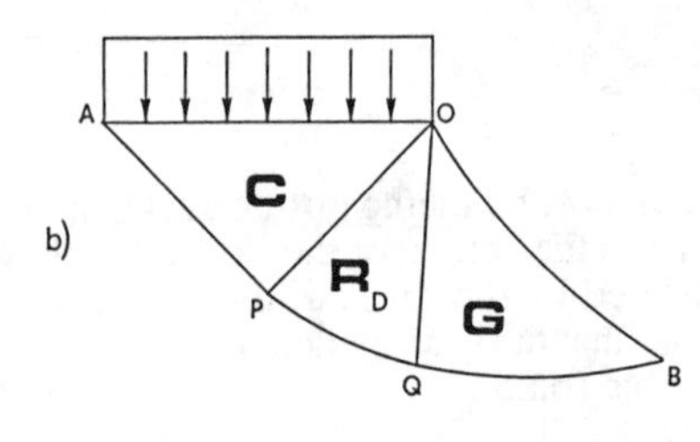

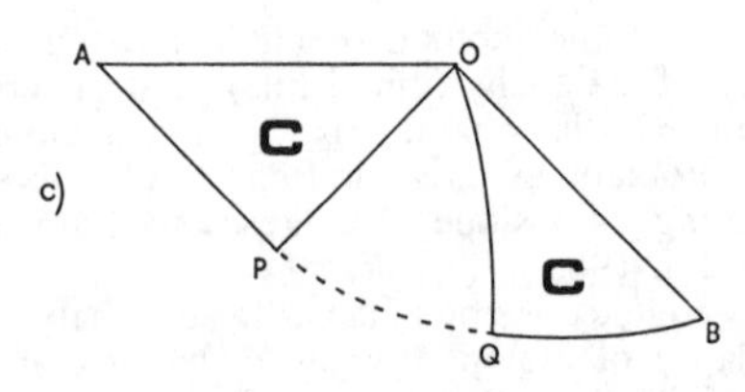

Fig. 1 Cauchy (C), degenerated Riemann (R_D) and Goursat (G) domains in the bearing capacity problem of a strip footing a), and in the Sokolovskii solution of the slope collapse b); Cauchy domains in the slope collapse c)

impossibility of connecting (in general) two Cauchy domains by means of a Riemann one, it was concluded that the solution of the slope collapse problem should be sought in a more general context (elliptic and hyperbolic domains) than the one of the method of characteristics (hyperbolic domains only).

A more careful examination of the problem shows that this is not necessarily the case, as it is possible to connect two Cauchy domains by means of suitable generalized Goursat domains, if lines of discontinuity of stresses are introduced into the problem, as it will be shown afterwards.

A line of discontinuity for stresses is a curve across which stress exhibits some discountinuity. Equilibrium requirements (i.e. both the normal stress σ_n and tangential stress τ_{nm} components along the line of discontinuity η must be continuous across the line) give the jump conditions to be satisfied by the stress on both sides of this line (fig. 2a).

In the case of a Tresca medium, which is considered in this paper, the jump conditions where

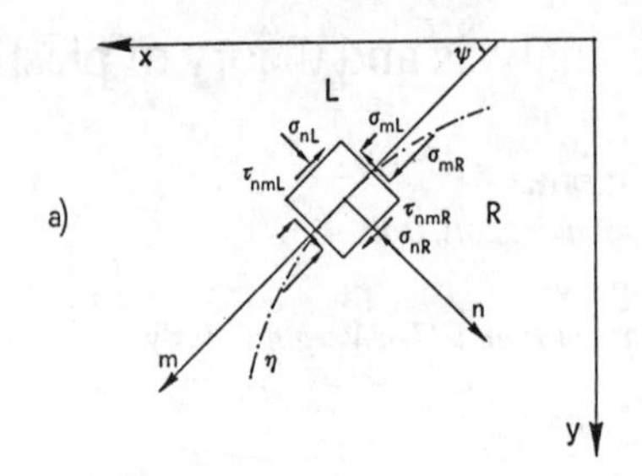

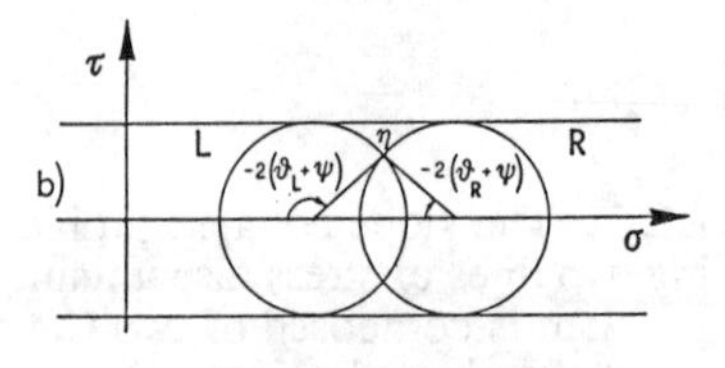

Fig. 2 Line of discontinuity for stresses in physical a), and in Mohr b) planes

first obtained by Prager (1948). Characterizing the state of stress at a point by means of the angle θ from the maximum principal stress σ_1 to the x-axis and the "mean" stress $p = (\sigma_x+\sigma_y)/2$, it is easy to obtain the two jump conditions (fig. 2b) :

$$\theta_L+\theta_R=\pi/2-2\psi \pm n\pi \qquad (n = 0, 1, 2, ...) \qquad (1)$$

$$p_L-p_R=2c\cos(2\theta_L+2\psi) \qquad (2)$$

where L and R mean respectively left and right side, and ψ is shown in fig. 2a. Introducing the parameter χ defined by Sokolovskii (1965) as $\chi = (p-\gamma y)/2c$, eq. (2) becomes:

$$\chi_L-\chi_R=\cos(2\theta_L+2\psi) \qquad (3)$$

In order to introduce discontinuities into the problem, we have first to decide the kind of each Cauchy domain (i.e. active or passive state). As a matter of fact, while the Cauchy domain below the slope (OBQ in fig. 3) is in a passive state, it is not a priori evident the state of the Cauchy domain below the top surface (OAP in fig. 3).

The stress state in the Cauchy domain below the slope OBQ is the same as in the infinite slope problem; along the initial line OB, from the condition of zero traction, it follows that $\theta = \omega$ and $\chi = 1/2 - \gamma y/2c$, while at a generical point x', y' :

$$\theta = \omega + 1/2\sin^{-1}(\gamma y'/c\sin\omega) \qquad (4)$$

$$\chi = 1/2\cos(2\theta-2\omega) + \gamma x'/2c\sin\omega \qquad (5)$$

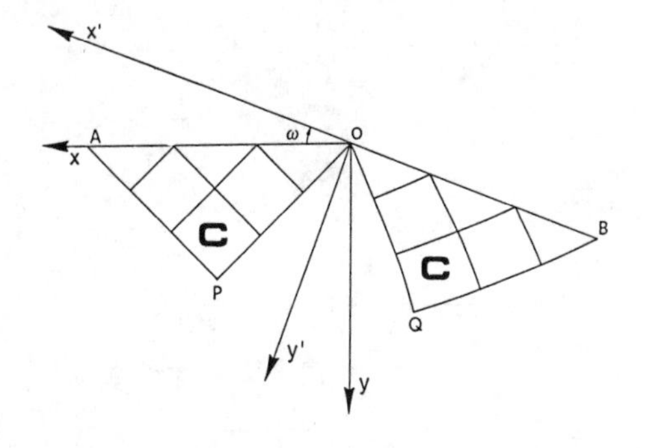

Fig. 3 Cauchy domain in the slope collapse problem

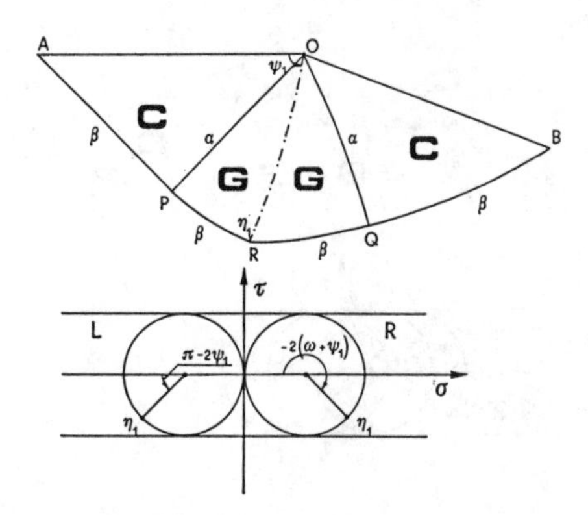

Fig. 4 Slope collapse. Active case, one discontinuity line

the characteristics being arcs of cycloids (Sokolovskii, 1965).

In the Cauchy domain below the top surface the stress state is very simple both for active and passive state; the characteristics are rectilinear and inclined at $\pm\pi/4$ to the x-axis, and $\theta = \pm\pi/2$, $\chi = -1/2$ in the active case, and $\theta = 0$, $\chi = 1/2$ in the passive case.

4 THE CASE OF ACTIVE STATE

The first case analyzed is the one of active state below the top surface; this case corresponds to the one studied by Sokolovskii, and in fact it seems the more logical extension of the bearing capacity pattern to the slope collapse problem.

4.1 One discontinuity

Introducing one line of discontinuity through the vertex of the slope (OR in fig. 4), the two Goursat domains OPR and OQR can be considered, having respectively OP and OQ as initial characteristics, and the line of discontinuity itself as a common non characteristic initial line.

However, it is easy to verify that such a pattern is impossible. In fact, writing the jump conditions (1) and (3) at the point O, it can be seen that the initial value of ψ_1 at that point must verify simultaneously the two equations:

$$\psi_1 = -\omega/2 \tag{6}$$

$$\psi_1 = -2\omega \tag{7}$$

and that it is possible only for $\omega = 0$.

4.2 Two discontinuities

Introducing instead two lines of discontinuity, it is to expect that the number of equations will equate the number of unknowns. In this case each line of discontinuity must cut one Cauchy domain, which actually results in truncated Cauchy domains (fig. 5a), and the lines of discontinuity comprise two Goursat domains with a common stress state at point O. Thus, at the point O there are four unknowns (θ and χ in the Goursat domains, and the inclinations of the two lines of discontinuity ψ_1 and ψ_2) and four equations (two jump conditions for each line of discontinuity); hence the problem is determined, at least in principle.

Eliminating θ and χ from the four jump conditions the two equations for ψ_1 and ψ_2 are obtained:

$$\psi_1 - \psi_2 = \pi/4 + \omega/2 \tag{8}$$

$$\cos(2\psi_1) + \cos(2\psi_2 + 2\omega) = 1 \tag{9}$$

Unfortunately, this system gives results which are unaccetable, since one of the discontinuities lies

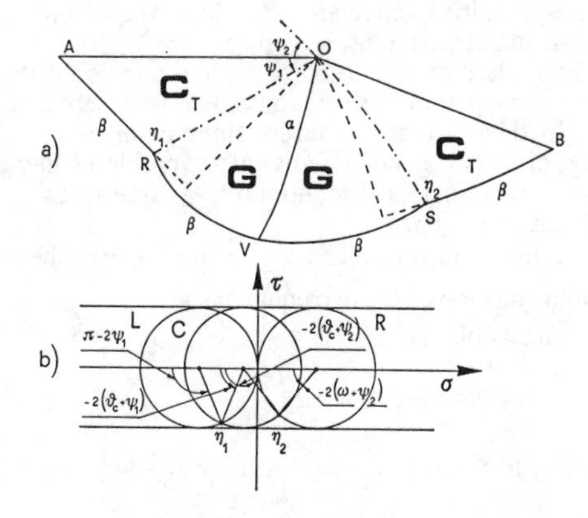

Fig. 5 Slope collapse. Active case; two discontinuity lines

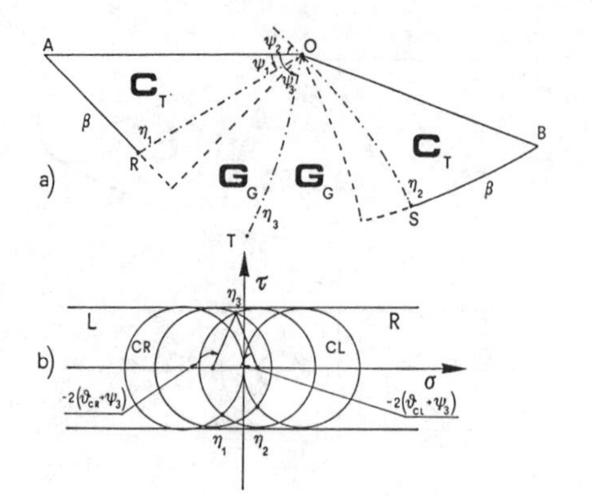

Fig. 6 Slope collapse. Active case, three discontinuity lines

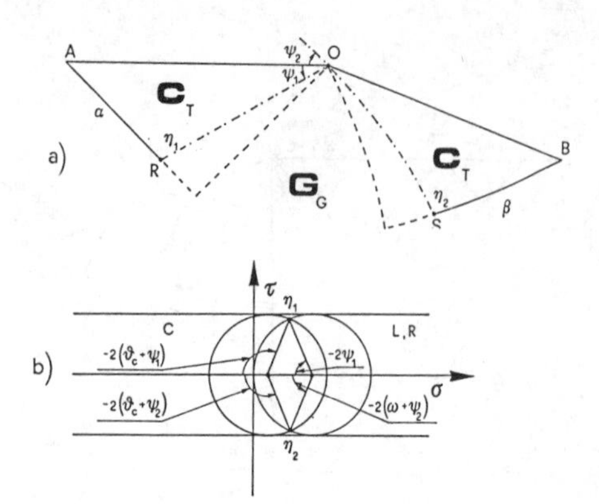

Fig. 7 Slope collapse. Passive case, two discontinuity lines

beyond the boundary of the medium.

On the other hand, the problem would give acceptable solutions if the inclination ω were negative (upward ground inclination). In this case, owing to the appearance in the vertex of a (non degenerated) Riemann zone, eq. (8) becomes :

$$\psi_2 - \psi_1 = \pi/4 - \omega/2 \quad (10)$$

and the system gives accettable results.

It is to observe that a similar problem has been analyzed by Chen (1975), from the view point of limit analysis, only in the case of negative ω values; obviously, its results are actually coincident with the ones obtained solving eqs. (9) and (10).

4. 3 Three discontinuities

Increasing the number of discontinuities, the number of unknowns overcomes the number of equations.

As a matter of fact, if d is the number of discontinuities, there are 2d (jump) equations and 3d-2 unknowns; thus, selecting d = 3 (see fig. 6) a whole class of solutions exists, and one could think that some part of this solutions could be acceptable.

In this case the three lines of discontinuity comprise two generalized Goursat problems having each two lines of discontinuity as free initial non characteristic lines.

Eliminating the θ and χ unknowns from the six jump equations, the two equations for ψ_1, ψ_2 and ψ_3 are finally obtained:

$$\psi_1 + \psi_2 - \psi_3 = -\pi/2 - \omega/2 \quad (11)$$

$$\cos(2\psi_1) + \cos(2\psi_2 + 2\omega) - \cos(2\psi_3 - 4\psi_1) = 1 \quad (12)$$

After some calculations, it can be seen that also this case yields unacceptable results.

5 THE CASE OF PASSIVE STATE

In the (perhaps) less evident case of the Cauchy zone below the top surface in a passive state, and with reference directly to two discontinuities, the geometry of the problem is as shown in fig. 7a, with one only generalized Goursat problem comprised between the two lines of discontinuity.

From the jump conditions at the vertex point the following equations for ψ_1 and ψ_2, are obtained :

$$\psi_2 - \psi_1 = -\pi/2 - \omega/2 \quad (13)$$

$$\psi_2 + \psi_1 = -\omega \quad (14)$$

whose solutions is :

$$\psi_1 = (\pi - \omega)/4 \quad (15)$$

$$\psi_2 = -3(\pi - \omega)/4 \quad (16)$$

In order to determine completely the two lines of discontinuity we have to solve the generalized Goursat problem defined by such lines together with their jump conditions. This problem, which is more complex than the classical Goursat one and scarcely referred to literature, is named by Sokolovskii the fourth boundary value problem. It is obtained by the classical Goursat problem, substituting the initial characteristic by an initial non characteristic line, along which only one unknown (or an equation) is given.

In this study the problem is even more complex, the initial lines being not known (free boundaries). Obviously, the degree of freedom connected with this indefiniteness is eliminated introducing one more equation along the discontinuity itself; as a matter of fact, along each discontinuity there are two equations to be satisfied (i.e. the jump conditions).

This problem, which is a generalized free

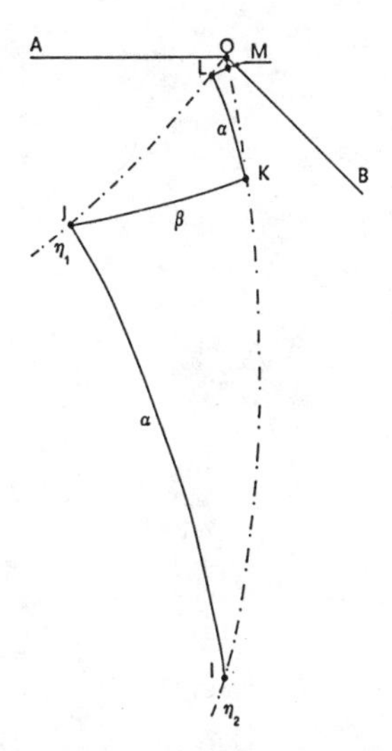

Fig. 8 Free boundary generalized Goursat problem. Cassiopeia-like characteristics

boundaries Goursat problem, can be looked at as a domain in which a characteristic of one family starting from one initial line, reflects on the opposite initial line into a characteristic of the other family in a way very similar to Cassiopeia, the well know constellation in the Milky Way, near the Polar Star.

Preliminary calculations show that the problem can be easily solved using the finite difference approximation.

6 CONCLUSIONS

The method of characteristics, previously applied by Sokolovskii to slope collapse problems under strong geometry and load restrictions, is shown to be applicable to such kind of problems without any limitation, introducing two lines of discontinuity for stresses. It is shown that the Cauchy domain under the top surface must be in a passive state, and that it is connected to Cauchy domain under the slope surface by a free boundaryies generalized Goursat domain.

Some preliminary finite difference calculations show that the solution can be extended from the vertex to some depth into the medium. A subsequent work will investigate in detail such a depth and its connection with the critical height of the slope.

REFERENCES

Booker, J.R. and Davis, E.N. 1972. A note on a plasticity solution to the stability of slopes in inhomogeneous clays. Geotechnique, 22 : 509 - 513.

Chen, W.F. 1986. Limit analysis and soil plasticity. Elsevier.

De Simone, P. 1986. Considerazioni sulla verifica di stabilità delle scarpate secondo le teorie della plasticità e della frattura. Atti XVI Convegno Nazionale Geotecnica, I : 275 - 281.

Mongiovì, L. 1981. Considerazioni sui metodi dell'equilibrio limite. Atti Riunione Gruppo Ingegneria Geotecnica C.N.R., 245 - 253.

Prager, W. 1948. Discontinuous solutions in the theory of plasticity. Courant Anniversary Volume, 289 - 299.

Sokolovskii, V.V. 1965. Statics of granular media. Pergamon Press.

Numerical Methods in Geomechanics (Innsbruck 1988), Swoboda (ed.)
© 1988 Balkema, Rotterdam. ISBN 90 6191 809 X

Three-dimensional finite element collapse analysis for foundations and slopes using dynamic relaxation

Tadatsugu Tanaka
Meiji University, Kawasaki, Japan

Osamu Kawamoto
National Research Institute of Agricultural Engineering, Ibaraki, Japan

ABSTRACT: A numerical procedure for prediction of the collapse loads of soil structure is presented. Use was made of dynamic relaxation with a one-point integration of linear isoparametric elements. The constitutive model used was the elastic, perfectly plastic Mohr-Coulomb model. Three-dimensional elasto-plastic problems such as surface footings and slopes were analysed and collapse loads were obtained for soil in which the difference between the angle of internal friction and the angle of dilatancy exceeds 30 degrees.

1 INTRODUCTION

One important problem related to soil mechanics is the accurate prediction of collapse loads of soil structures. Heretofore, attention has been mostly confined to two-dimensional soil structures without friction or with a low friction angle. A study of collapse loads of granular materials with a high friction angle was made by de Borst & Vermeer(1984), however, a numerical instability appears when the difference between the angle of internal friction and angle of dilatancy exceeds 30 degrees. As discussed later, the numerical instability disappears when the round-off error in the evaluation of variables near the singular points of the Mohr-Coulomb constitutive model is minimized. Even in case of the two-dimensional collapse analysis, a large amount of time at the computer is required and the efficient numerical procedure for the use of three-dimensional analysis has to be designed. Dynamic relaxation was proposed by Otter (1965) and Day (1965) in connection with the finite difference method. In the early stage, this procedure had disadvantages such as slow convergence etc. Underwood (1983) obtained reasonable solutions of nonlinear problems, using the finite difference method and the explicit dynamic relaxation scheme. The capability of this scheme for large-scale nonlinear problems using a finite element method has not been fully clarified.

For the collapse analysis of frictional, dilatant solids , methods such as 8-noded isoparametric elements with a reduced integration rule and 15-noded triangular elements seem to be reasonably accurate (see de Borst & Vermeer (1984)). In addition, linear isoparametric elements with a special integration rule show promise for efficient use in the collapse analysis. Kelly (1980) has shown that special reduced integration of the 4-noded Lagrange elements can be efficiently used to evaluate limit loads of materials without a friction, in the two-dimensional analysis. Tanaka & Yasunaka (1982) showed that in dynamic analysis an explicit scheme with the reduced integration technique for linear isoparametric elements yielded good results. Malkus & Hughes (1978) developed selective integration schemes for linear isoparametric elements, using a one-point integration for volumetric stress and 2×2 Gauss quadrature for deviatoric stresses. For an efficient evaluation of the variables, a one-point integration for both volumetric and deviatoric stresses is preferable. As for linear isoparametric elements, the use of such one-point integration results in deformation modes called hourglass or zero energy modes, and singularity appears in the related equations. Hourglass control schemes such as developed by Flanagan & Belytschko (1981) proved to be efficient, but use of such schemes is restricted to linear problems.

In this paper, a numerical procedure using dynamic relaxation with a one-point integration of linear isoparametric elements is presented. Using this procedure, three-dimensional elasto-plastic problems such as surface footings and slopes of soils in which the difference between the angle of internal friction and the angle of dilatancy exceeds 30 degrees were analysed. The effect of hourglassing was also given attention.

2 FORMULATION BASED ON DYNAMIC RELAXATION

Governing equations of the elasto-plastic collapse analysis in the static case are given by

$$\{P\}-\{P\}^{init}=\{F\} \tag{1}$$

and

$$\{P\}=\sum^{N}\int_{Ve}[B]^{T}\{\sigma\}dV \tag{2}$$

where $\{P\}$ is the vector of internal forces, $\{P\}^{init}$ is the vector of nodal forces due to initial stress, $\{F\}$ is the vector of external forces, $[B]$ is the displacement-strain transformation matrix, N is the number of elements, $\{\sigma\}$ is the vector of stresses in each element, and V_e is the volume of each element. To redistribute the unbalanced residual forces expressed as the difference between the left and right side of eq.(1), the stiffness matrix is often used. It offers computational disadvantages even in the two-dimensional analysis because a large amount of time at the computer is required. To overcome this difficulty, the solutions are obtained as final (steady state) ones of the transient dynamic problem

$$[M]\{\ddot{u}\}+[C]\{\dot{u}\}+\{P\}-\{P\}^{init}=\{F\} \tag{3}$$

where $[M]$ is the mass matrix, $[C]$ is the damping matrix, $\{\ddot{u}\}$ is the acceleration vector, and $\{\dot{u}\}$ is the velocity vector. The following central difference expressions are introduced for the derivatives in eq.(3)

$$\{\dot{u}\}^{t}=\frac{\{u\}^{t+\Delta t}-\{u\}^{t-\Delta t}}{2\Delta t} \tag{4}$$

and

$$\{\ddot{u}\}^{t}=\frac{\{u\}^{t+\Delta t}-2\{u\}^{t}+\{u\}^{t-\Delta t}}{(\Delta t)^{2}} \tag{5}$$

where Δt is the time increment and $t+\Delta t$, $t, t-\Delta t$ denotes time. Using mass matrix $[M]$, damping matrix $[C]$ is determined by

$$[C]=\alpha[M] \tag{6}$$

where α is the damping ratio. Substituting eq.(4), (5), and (6) into (3) gives the nodal displacements as

$$\{u\}^{t+\Delta t}=\frac{1}{1+\frac{1}{2}\alpha\Delta t}[M]^{-1}\{(\Delta t)^{2}(\{F\}-\{P\}+\{P\}^{init})+2[M]\{u\}^{t}-(1-\tfrac{1}{2}\alpha\Delta t)[M]\{u\}^{t-\Delta t}\} \tag{7}$$

Solutions are easily obtained by introducing the lumped mass into the mass matrix $[M]$ in eq.(7). Optimum convergence is achieved for the maximum value of the time increment within the stability limit of the central difference time integrator, and the approximate value of the critical dampimg ratio. Stability for the time increment is expressed by

$$\Delta t\leqq\beta\ell\sqrt{\frac{\rho(1+\nu)(1-2\nu)}{E(1-\nu)}} \tag{8}$$

where $\beta\leqq 1.0$ is a constant, ℓ is the minimum distance among nodal points, E is the Young's modulus, ν is the Poisson's ratio, and ρ is the density. From eq.(8), taking $\Delta t=1$, the fictitious density for the pseudo mass matrix is obtained

$$\rho=\frac{E(1-\nu)}{\beta^{2}\ell^{2}(1+\nu)(1-2\nu)} \tag{9}$$

For the linear problem, the critical damping ratio is expressed by

$$\alpha=2\sqrt{\frac{\{u\}^{T}[K]\{u\}}{\{u\}^{T}[M]\{u\}}} \tag{10}$$

where $\{u\}$ is the eigenvector associated with the lowest eigenvalue. For the non-linear problem, the approximate value of the critical damping ratio can be given by the same formula as eq.(10) selecting the nodal displacement vector $\{u\}$ and the tangent stiffness matrix $[K]$ instead of the eigenvector and the stiffness matrix in the linear problem respectively. The tangent stiffness matrix $[K]$ is approximated by the diagonal form (Underwood (1983))

$$K_{ii}=\frac{{}^{t}p_{i}-{}^{t-\Delta t}p_{i}}{\Delta t\ {}^{t}\dot{u}_{i}} \tag{11}$$

where K_{ii} is a diagonal term of approximated tangent stiffness matrix, and ${}^{t}p_{i}$, ${}^{t-\Delta t}p_{i}$ are components of the internal force vector at time t and $t-\Delta t$, respectively, and ${}^{t}\dot{u}_{i}$ is a component of the velocity vector at time t. As a measure of the satisfaction of governing equations, the norm of residual forces and energy are used. Convergence is declared when

$$\frac{\|\{F\}-\{P\}+\{P\}^{init}\|^{2}}{\|\{F\}\|^{2}}<\varepsilon_{1} \tag{12}$$

and

$$\frac{\Delta t\{\dot{u}\}^T[\{F\}-\{P\}+\{P\}^{init}]}{\Delta t\{\dot{u}_0\}^T\{F\}}<\varepsilon_2 \quad (13)$$

where $\{\dot{u}\}$ is the velocity vector at each time, $\{\dot{u}_0\}$ is the velocity vector at time=0, ε_1 is the tolerance for force norm, and ε_2 is the tolerance for energy norm. If eq.(12) or (13) is not satisfied in predetermined iteration cycles (1000 in our case), a collapse condition is considered to have been attained.

3 CONSTITUTIVE MODEL

The constitutive model employed is the elastic, perfectly plastic Mohr-Coulomb model. Plastic flow occurs when the following yield condition is satisfied

$$f=\sigma_m \sin\phi+\sqrt{J_2}\cos\alpha-\frac{\sqrt{J_2}}{\sqrt{3}}\sin\alpha\sin\phi-C\cos\phi=0 \quad (14)$$

where ϕ is a friction angle, C is cohesion, σ_m is the mean stress, J_2 is the second invariant of deviatoric stresses, and α is the Lode angle. The plastic potential function which defines the plastic strain increment is given by

$$\psi=\sigma_m \sin\phi'+\sqrt{J_2}\cos\alpha-\frac{\sqrt{J_2}}{\sqrt{3}}\sin\alpha\sin\phi'-C\cos\phi' \quad (15)$$

where ϕ' is a dilatancy angle. The elastic and plastic matrix including derivatives of these functions are singular when the Lode angle equals to ±30°. In the case of a non-associated flow with high friction angles under a plane strain condition, the numerical instability which appears is attributed to the fact that under this condition the Lode angle tends to coincide with ±30° (Tanaka (1985)a,b). Therefore, the round-off error in the evaluation of variables in the neighbor of these singular points should be minimized. For this reason, variables in the neighbor of singular points ($30° \geq \alpha \geq 29.99°$ or $-29.99° \geq \alpha \geq -30°$) are evaluated, using Drucker-Prager yield criterion.

4 ONE POINT INTEGRATION

In this paper collapse analysis is executed using linear isoparametric hexahedron elements with a one-point integration. Use of these elements usually results in deformation modes called hourglass or zero energy modes and singularity appears in the related equations. Therefore, the technique developed for hourglass control (Artificial Stiffness developed by Flanagan & Belytschko (1981)) was examined. Effects on the numerical solution are given attention in the latter part of the paper.

5 RESULTS

5.1 SURFACE FOOTING

This example compares the numerical and the analytical solutions of a smooth surface footing. For a smooth surface footing on a cohesive-frictional and weightless subsoil under plane strain condition, the solution for the associated flow rule is given by Prandtl. Numerical solutions are obtained using 3D finite element model, under a plane strain condition. The finite element mesh (54 elements, 140 nodal points) is shown in Fig.1. Fig.2 shows the load-displacement curves obtained. The difference between the numerical solution given by the associated flow rule and the analytical solution is within 3%. This result shows that a reasonable solution has been obtained even for a coarse finite element mesh. Fig.2 also shows

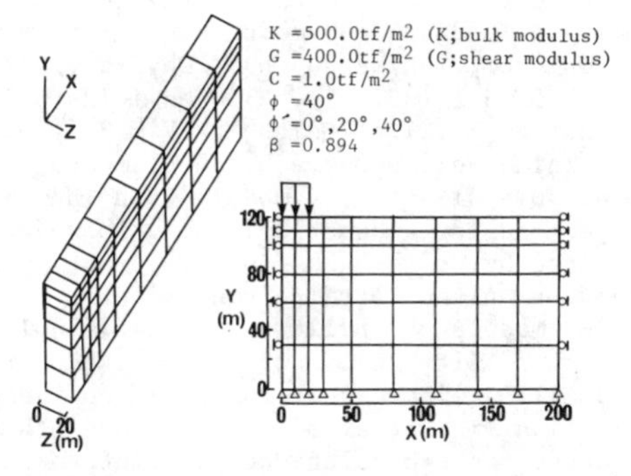

Fig.1 Finite element mesh for a smooth surface footing

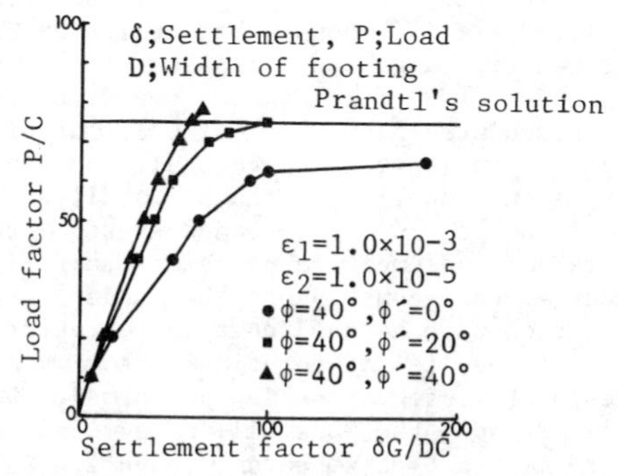

Fig.2 Load-displacement curves for a smooth surface footing

Table 1. CPU time for a smooth surface footing (including compilation time)

Dilatancy angle (ϕ',°)	Total number of iteration	CPU time (sec)
0.0	2289	156
20.0	2742	198
40.0	3786	269

that a stable solution can be obtained for the non-associated flow rule with a high degree of non-normality (ϕ=40°, ϕ'=0°). Table 1 shows the computer time required for this calculation. The computational work has been carried out on the ACOS 850-MODEL 20 computer (16 MIPS), and double-precision arithmetic operation has been implemented. It is clear from this table that collapse loads can be efficiently obtained by presented method.

5.2 SLOPE STABILITY

This example compares with the numerical solution and the experimental result of the slope stability problem. The experimental results of tilting tests on small-scale 2D and 3D model were reported by Ohmachi, Tokimatsu and Ohyama (1981) for 1.5 on 1 (33.69°), 1.8 on 1 (29.05°) and 2.5 on 1 (21.8°) slopes. Fig.3 shows the finite element mesh (30 elements, 92 nodal points) of 2D model with 33.69° slope angle. Numerical solutions are obtained by using a 3D finite element mesh under a plane strain condition. Small scale models are built on a flat table, and the tilt angle of this table is gradually increased until collapse condition is attained. To trace this process, the equivalent nodal point forces of body forces acting in the horizontal and vertical direction are evaluated according to the tilt angle of the flat table. Fig.4 shows the relationship between tilt angle and displacements at a top node of 2D model tests given by the non-associated flow rule (ϕ'=0°). The tilt angle at collapse is successfully obtained for each slope. For the 3D small-scale model built on a V-shaped base, the tetrahedron element is necessary to follow exactly the geometry of the model. But reduction of a hexahedron to a tetrahedron element results in numerical error when a one-point integration is performed. For this reason, the 3D small-scale model is approximated by hexahedron elements. Half space is analysed because of the symmetric geometry. Fig.5 shows the finite element mesh (70 elements, 166 nodal points) of the 3D model tests. The relationship between tilt angle and displacements of 3D model tests is plotted in Fig.6. Collapse loads are successfully obtained also in 3D model tests. The relationship between tilt angle at collapse and slope angle is plotted together with experimental results in Fig.7. Numerical results are slightly lower than experimental results in the 2D model case and higher in the 3D model with a high slope angle. In general, the agreement is satisfactory. Investigations on hourglassing are indispensable when linear isoparametric elements with a one-point integration are used. For this reason, the effect of hourglass control was examined using an artificial stiffness. In this procedure, an appropriate value of the stiffness parameter κ has to be chosen. Fig.8 shows the relationship between the tilt angle and the displacement given by typical values of the stiffness parameter κ in the case of 1.5 on 1 slope of 3D model. The tilt angle at collapse gradually increases with the stiffness parameter. For the case of κ=0.1, a collapse load is not obtained when the tilt angle has reached 70°, which largely exceeds the experimental result. In Fig.9 is plotted the displacement pattern along the nodal points shown in Fig.5. This is the same case as for Fig.8. An irregular distribution of nodal displacements appears with no hourglass control. The displacements should be evaluated at the mid points of the nodes, because linear isoparametric elements with a one-point integration have the same effect as pseudo-equilibrium elements. No irregularity appears in the dotted lines in Fig.9 drawn through the mid points of the nodes. It is clear from these results that normal solutions can be obtained without an hourglass control. With use of an approximate value of critical damping ratio in dynamic relaxation, hourglass modes of linear isoparametric elements with a one-point integration can be suppressed. In previous studies on slope stability, the difference of numerical results between associated and non-associated formulation was slight and is attributed to the low degree of constraint on the plastic flow (Zienkiewicz, Humpheson and Lewis(1975)). These results were based on a two-dimensional condition and a low friction angle. Fig.10 compares solutions based on associated and the non-associated flow rule in the 2D and 3D cases. This figure shows that in case of a high friction angle, the difference between the collapse loads for associated and non-associated plasticity is not slight even in case of 2D. It is also clear that the differences between the results given by the associated and non-associated flow in the 3D model are similiar in some cases, and in others, greater than those of the 2D model. Table 2 shows the computer time in this calculation. It can be seen in this table that solutions were efficiently obtained.

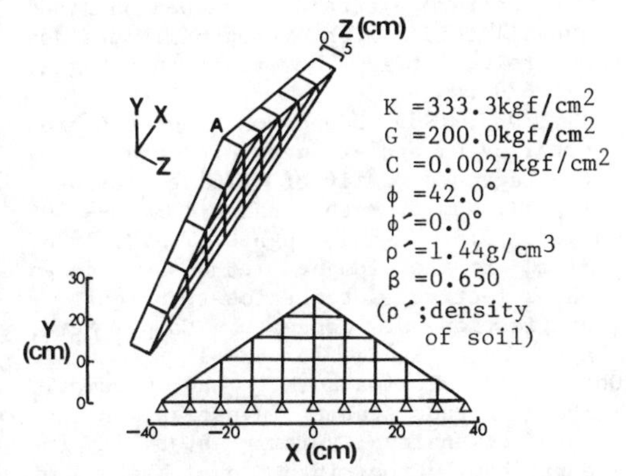

Fig.3 Finite element mesh for a 2-D slope model

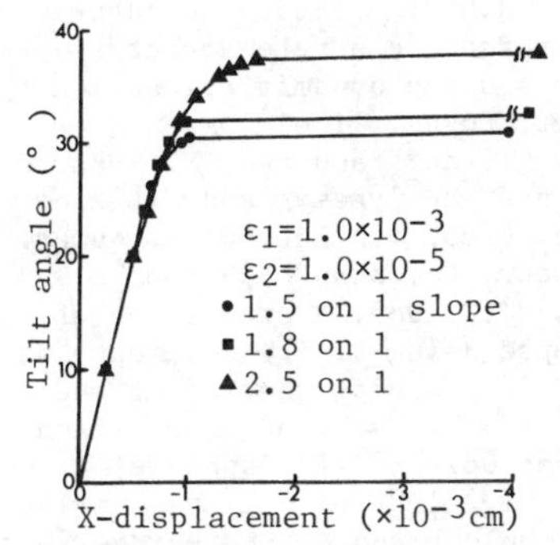

Fig.4 Tilt angle-displacement curves for a 2-D slope model (point A in Fig.3)

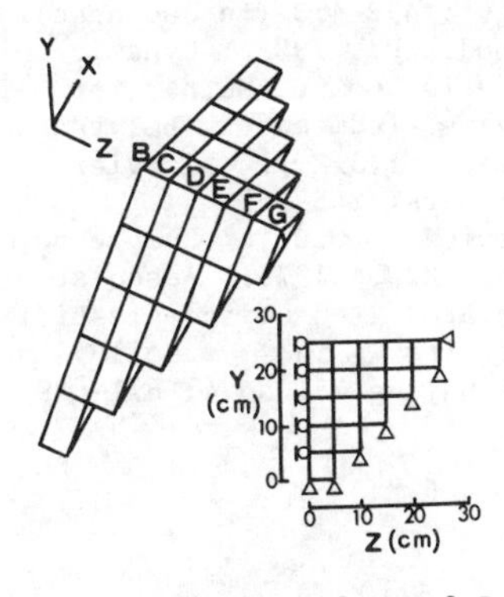

Fig.5 Finite element mesh for a 3-D slope model (material properties and the cross section in X-Y plane are shown in Fig.3)

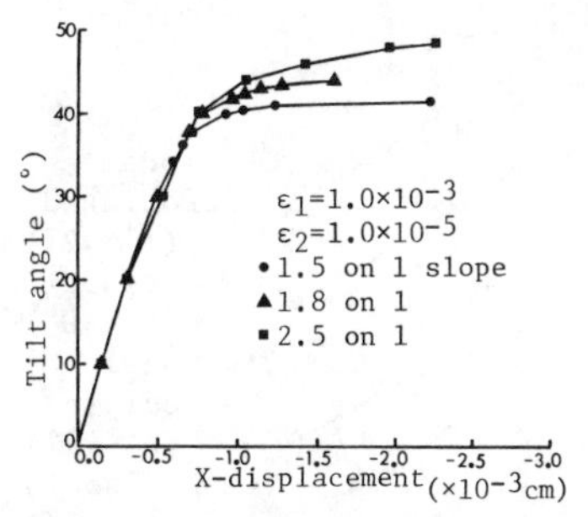

Fig.6 Tilt angle-displacement curves for a 3-D slope model (point B in Fig.5)

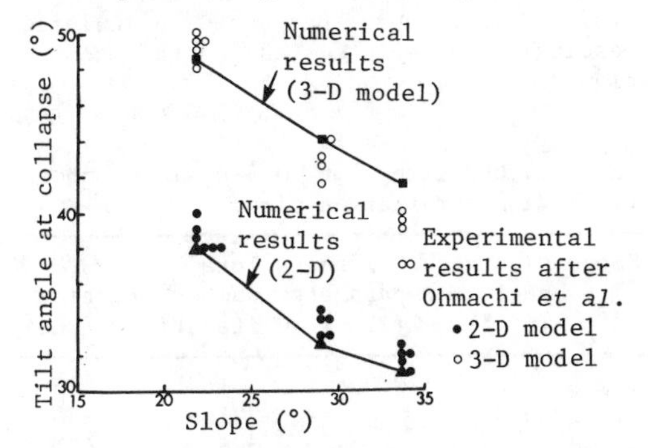

Fig.7 Relationship between tilt angle at collapse and slope angle

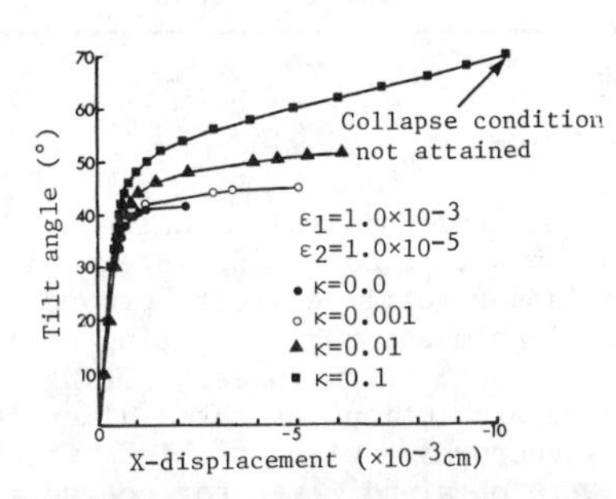

Fig.8 Tilt angle-displacement curves for various values of the stiffness parameter

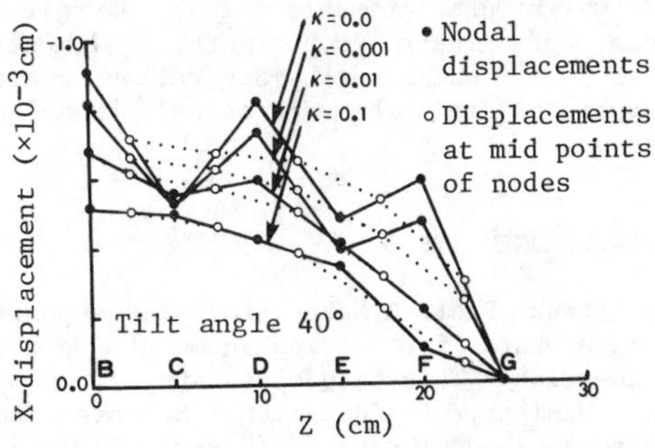

Fig.9 Displacements across the mesh for various values of the stiffness parameter

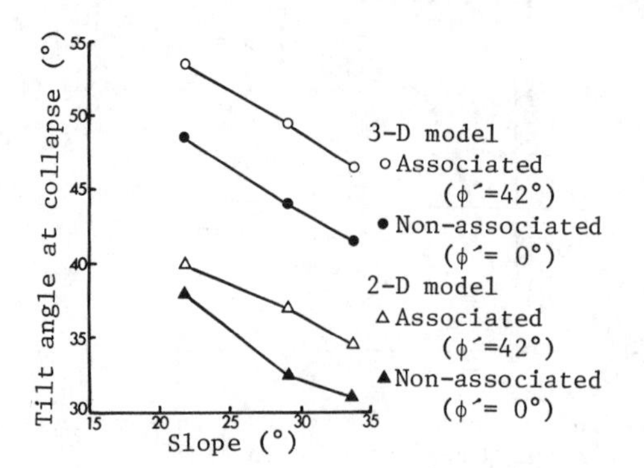

Fig.10 Comparison between calculated results - associated and non-associated flow rule

Table 2. CPU time for a 3-D slope model (including compilation time)

Slope	Dilatancy angle (φ´,°)	Sfiffness parameter (κ)	Total number of iteration	CPU time (sec)
1:1.5	42.0	0.0	3314	275
1:1.5	0.0	0.0	3272	286
1:1.8	42.0	0.0	5065	421
1:1.8	0.0	0.0	3501	303
1:2.5	42.0	0.0	5411	453
1:2.5	0.0	0.0	3606	312

6 CONCLUSION

Finite element analysis using dynamic relaxation with a one-point integration of linear isoparametric elements proved to be an efficient means for predicting collapse loads of soil structures. Using this procedure, solutions of three-dimensional soil structures with a high friction angle were obtained even for coarse mesh. An approximate value of critical damping ratio is used for the dynamic relaxation and with use of this value, hourglass modes of linear isoparametric elements with a one-point integration are suppressed in the numerical examples presented.

ACKNOWLEDGEMENTS

We thank Prof. R.Nakano for kind support and M.Ohara for critical comments on the manuscript. This study was supported by The Ministry of Education, Science and Culture, Japan (Grant No.(B)-60460213).

REFERENCES

de Borst, R. and Vermeer, P.A. 1984. Possibilities and limitations of finite elements for limit analysis. Geotechnique 34, No.2: 199-210

Day, A.S. 1965. An introduction to dynamic relaxation. The Engineer 219: 218-221

Flanagan, D.P. and Belytschko, T. 1981. A uniform strain hexahedron and quadrilateral with orthogonal hourglass control. Int. J. Numer. Meth. Engng. 17: 679-706

Kelly, D.W. 1980. Bounds on discretization error by special reduced integration of the Lagrange family of finite elements. Int. J. Numer. Meth. Engng. 15:1489-1506

Malkus, D.S. and Hughes, T.J.R. 1978. Mixed finite element methods - reduced and selective integration techniques: a unification of concepts. Comp. Meth. Appl. Mech. Engng. 15: 63-81

Ohmachi,T., Tokimatsu,K. and Ohyama,T. 1981. Pseudo-static investigation of three-dimensional seismic stability for earth dam. Proc. 16th Annual Meeting of the Japanese Soc. of Soil Mech. and Foundation Engng.,Kanazawa: 1101-1104

Otter, J.R.H. 1965. Computations for prestressed concrete reactor pressure vessels using dynamic relaxation. Nucl. Struct. Engng. 1: 61-75

Tanaka, T. and Yasunaka, M. 1982. Dynamic response analyses and shaking table tests. Proc. 4th Int. Conf. Numer. Meth. Geomech., Edmonton: 459-468

Tanaka, T. 1985. A prediction of plastic collapse using finite element methods. Proc. 20th Annual meeting of the Japanese Soc. of Soil Mech. and Foundation Engng., Nagoya: 967-968 (In Japanese)

Tanaka, T. 1985.b Collapse analysis of Mohr-Coulomb and Drucker-Prager material. Proc. Annual Meeting of the Japanese Soc. of Irrigation, Drainage and Reclamation Engng. : 412-413 (In Japanese)

Underwood, P. 1983. Dynamic relaxation. Ch.5 in Comp. Meth. for Transient Analysis (edited by Belytschko,T. and Hughes, T.J.R.). Elsevier Science Publishers: 245-265

Zienkiewicz, O.C., Humpheson, C. and Lewis, R.W. 1975. Associated and non-associated visco-plasticity and plasticity in soil mechanics. Geotechnique 25, No.4: 671-689

Numerical Methods in Geomechanics (Innsbruck 1988), Swoboda (ed.)
© 1988 Balkema, Rotterdam. ISBN 90 6191 809 X

Numerical methods for locating the critical slip surface in slope stability analysis

Venanzio R.Greco
University of Calabria, Arcavacata di Rende, Italy

ABSTRACT: This paper compares in terms of their speed and efficiency some of the more commonly used Nonlinear Programming Methods for locating the critical slip surface in slope stability analyses. In all these methods the slip surface is approximated bv a broken line whose coordinates at the vertices are yielded by the different minimization procedures used. Although all the methods tested require a large number of potential slip surfaces in searching for the minimum, they are all efficient enough to be used in practical applications.

1 INTRODUCTION

Until comparatively recently the methods proposed for analyzing slope stability were developed in connection with prefixed slip surfaces only. Consequently no procedures existed to search for a critical slip surface whenever its shape was not already assigned. In the particular case of circular slip surfaces, the search technique most commonly used was the grid method which involved assuming that the vertices of a network represented the centres of possible slip circles. The main disadvantage of this particular method is given by the large number of potential slip surfaces to be examined. Moreover, it is only usable in cases such as the circle, when the shape of the slip surface can be described by a limited number of parameters.

In the case of surfaces which are general in shape, the most widely used technique involved examining a large number of potential slip surfaces and assuming the lowest safety factor to be that of the slope. Since all the possible surfaces are not in fact examined, this approach can obviously lead to inaccuracies or errors.

In order to overcome this problem, in the late 1950s and early 1960s (Kopacsy 1957, 1961) and once again from the second half of the 1970s onwards, methods based on Variation Calculus were used to find the critical slip surface. However, it has been pointed out by De Josselin De Jong (1981) and Castillo & Luceño (1982) that many of these methods are not very rigorous from a mathematical point of view.

Towards the end of the 1970s numerical methods were introduced in slope stability analysis: Celestino & Duncan (1979) proposed a simple nonlinear programming procedure which is fairly similar to the Univariate Method, whereas Baker (1980) developed a method based on Dynamic Programming.

Although the majority of Nonlinear Programming Methods have been available for the past twenty years, it is only recently that they have been used to any great degree in Soil Mechanics. As a result of the decreased interest in analytical methods based on Variation Calculus, other numerical methods based on Nonlinear Programming have also been proposed: Nguyen (1985) has used the Simplex Method, Arai & Tagyo (1985) the Conjugate Gradient Method and Greco & Gullà (1985) the Pattern Search Method.

At the same time probabilistic procedures have been presented by Boutrup & Lovell (1980) and Siegel, Kovacs & Lovell

(1981), who have proposed methods of the random jumping type, and Cherubini & Greco (1987), who have used a random walk type technique.

No attempts have been so far made to compare the speed and efficiency of the various numerical search methods. This paper presents a comparison between the following:

1. the Pattern Search Method;
2. the Simplex Method;
3. the Univariate Method;
4. the Steepest Descent Method;
5. the Conjugate Gradient Method.

2 NONLINEAR PROGRAMMING METHODS IN SLOPE STABILITY ANALYSIS

In order to use the aforementioned numerical Nonlinear Programming Methods in slope stability analysis, the slip surface must be approximated by a broken line, whose coordinates at the vertices are the variables of the problem.

To each slip surface, represented in a m-dimensional vectorial space by the point:

$$\boldsymbol{x} = \{x_1, x_2, \ldots, x_n, y_1, \ldots, y_n\}^t , \quad (1)$$

a safety factor $F(\boldsymbol{x})$ is associated.

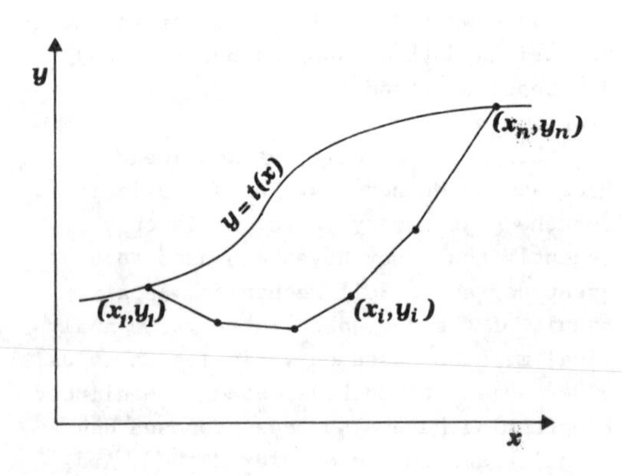

Figure 1. Cross section of slope and broken line approximating the slip surface.

The critical slip surface is then obtained by minimizing the objective function, i.e. the safety factor, with respect to $\boldsymbol{x}$:

$$F(\bar{\boldsymbol{x}}) = \min F(\boldsymbol{x}). \quad (2)$$

The minimization procedure is generally carried out as follows:

1. Calculate the safety factor F_o of a trial slip surface, represented by $\boldsymbol{x}_o$;

2. Assume a suitable direction, $\boldsymbol{d}_k$ (k=1 to start with), which points in the general direction of the minimum;

3. Find an appropriate step length, α, so that the new point, $\boldsymbol{x}_{k+1}$, given by:

$$\boldsymbol{x}_{k+1} = \boldsymbol{x}_k + \alpha \boldsymbol{d}_k , \quad (3)$$

is characterized by a safety factor, F, which is smaller than that of the previous slip surface.

4. Stop the procedure when $\boldsymbol{x}_{k+1}$ is a minimum. Otherwise, set k=k+1 and go to step 2.

Although the search for a critical slip surface involves obtaining a constrained minimum, the problem is currently solved by using unconstrained minimization methods. These methods can be divided into two different classes.

1. Descent Methods where the derivatives of $F(\boldsymbol{x})$ have to be calculated;

2. Direct Search Methods where no derivatives are involved.

The Random Search, Pattern Search, Rotating Directions, Univariate and Simplex methods represent some examples of the second class, whereas the first includes the Steepest Descent, Conjugate Gradient and Variable Metric methods.

All Nonlinear Programming methods are iterative and solutions are obtained by a step to step procedure. In some cases however, the minimization procedure can slow down and stop without converging at a minimum. When this happens the procedure must obviously be considered inefficient.

The path from trial to critical slip surface differs from one method to another.

The Steepest Descent Method puts $\boldsymbol{d}_k = -\boldsymbol{g}_k$ where $\boldsymbol{g}_k$ is the gradient of the safety factor calculated for the slip surface $\boldsymbol{x}_k$, and α is such that $F(\boldsymbol{x}_{k+1})$ is minimized with respect to α.

The Univariate Method assumes that the direction $\boldsymbol{d}_k$ coincides with each of the coordinate axes in turn.

The Conjugate Gradient Method puts $\boldsymbol{d}_k = -\boldsymbol{g}_k$ in the first step and

$$\boldsymbol{d}_k = -\boldsymbol{g}_k + \frac{\boldsymbol{g}_k^t \boldsymbol{g}_k}{\boldsymbol{g}_{k-1}^t \boldsymbol{g}_{k-1}} \boldsymbol{d}_{k-1}. \qquad (4)$$

at the k-th step.

The Pattern Search Method alternates exploration and extrapolation at each step. In the exploration phase the displacement is assumed to be fixed and parallel to each of the coordinate axes in turn. In the extrapolation phase it is equal to the entire displacement found by exploration.

The Simplex Method initially evaluates the safety factor in the m+1 points constituting the simplex, where m=2n-2 are the free variables. At each step the search is carried out in the direction from the point having the largest safety factor to the baricenter of the simplex.

Descent Methods are generally faster and more efficient than Direct Searches since they utilize a larger body of data concerning the objective function. There are however some cases where the Steepest Descent Method proves inefficient in its search for a minimum. Where the Direct Search Methods are concerned, it must be pointed out that the Simplex Method is only efficient when the number of variables is small, while the efficiency of the Univariate Method is directly dependent on the shape of the objective function. However, it must be underlined that these conclusions were reached by researchers working in much different fields from slope analysis.

In slope stability analysis the iterative procedure used in calculating the objective function, $F(\boldsymbol{x})$, inevitably introduces a rounding-off error which can have considerably influence on the value obtained for the partial derivatives of $F(\boldsymbol{x})$, even when incremental ratios are used. In fact, when the finite increment in a variable is very small, the rounding-off error can create problems in correctly evaluating the difference in $F(\boldsymbol{x})$ from a point to another. If, on the other hand, the increment is very large, even the sign of the derivative can be wrong. It is essential therefore in using Descent Methods to keep these particular problems in mind.

The reliability of the solution obtained will also depend on the following:

1. The number of vertices of the broken line approximating the slip surface;
2. The position of the vertices of the trial slip surface;
3. The particular way in which a minimization procedure is implemented;
4. The existence of more than one minimum.

3 APPLICATION

Although the aforementioned procedures have been used in different slope stability analyses, the present investigation only concerns an application to the slope shown in fig.2. However, since the results obtained in the different analyses are all fairly similar, the example illustrated does appear to be fairly representative. The slope, whose geotechnical parameters are given by $c'=4.25$ t/m^2, $\varphi'=15^0$, $\gamma=1.92$ t/m^3, has been taken from Arai & Tagyo (1985). The dotted and solid lines in fig.2 show the trial and critical slip surfaces whose safety factors are respectively given by 1.433 and 1.260. In both cases the surfaces are characterized by 7 vertices and the simplified Janbu method has been used in calculating the safety factors.

The derivatives used in the Descent Methods have been approximated by the incremental ratios with increments of one metre in the variables. The initial simplex has been made by increasing or decreasing each of the coordinates in turn by one metre. The procedure used in the Univariate Method is that proposed by Celestino & Duncan (1979).

Fig.2 also plots the safety factor of the slope versus the number of potential slip surfaces examined and can provide useful data concerning the speed and efficiency of the various minimization methods described. Since the Simplex Method shows a certain tendency to stop prematurely, the results shown in the figure can only be obtained by starting the procedure over and over again. Although the Univariate Method seems to be very fast at the beginning, in the later stages it very often slows down and comes to premature halt.

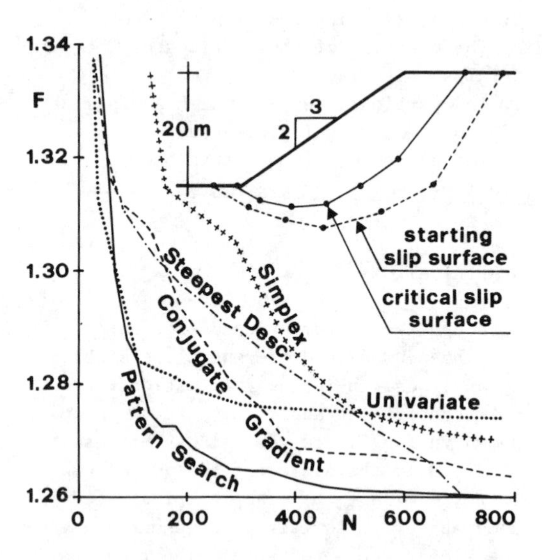

Figure 2. Safety factor (F) versus the number of examined potential slip surfaces (N) for the various minimization methods.

The results obtained by using the Conjugate Gradient and Steepest Descent Methods are not very different. Although in the case reported the former is initially the speeder of the two, it has been found that the value of the incremental step, Δ, exerts considerable influence on the results as can be easily seen from fig.3 where F is plotted against N with Δ= 0.5, 1, 2 and 3 m.

The Pattern Search Method almost always proves to be the fastest and the most efficient of the procedures examined. Moreover, its speed and efficiency is not very strongly in...enced by the length of the incremental step.

However, from a practical point of view it can be said that no great differences exist between the efficiency levels of the various methods since the values of the minima are found to differ by less than 2

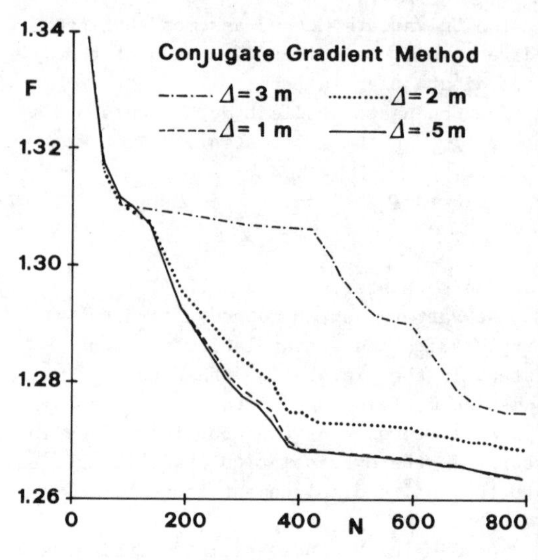

Figure 3. Influnce of the incremental step (Δ) on the convergency speed of the Conjugate Gradient Method.

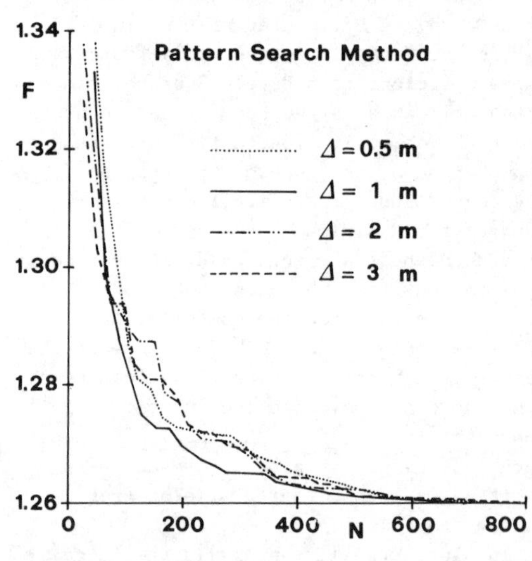

Figure 4. Influence of the incremental step (Δ) on the convergency speed of the Pattern Search Method.

percent.

Where the number of vertices of the slip surface are concerned, it must be remembered that greater the number of vertices, the better the approximation obtained. However, this advantages is counterbalanced by the increase in the computation time required. A broken line with 7 vertices does seem to represent a reasonable compromise.

The position of the vertices in the trial slip surface can also exert considerable influence on the solution not only because the computation time can be either doubled or trebled but also because the solution can become less precise. It is therefore essential to check that the vertices of the critical slip surface are in fact distinct and that they are primarily situated in the curved portion of the slip surface.

The way in which the minimization procedure is implemented can also have important effects on its effiency and particularly on its speed. For example, The Conjugate Gradient Method used by Arai & Tagyo (1985), where the abscissae of the vertices are assumed fixed, can require very different computation times than those needed by the present implementation.

Finally, it must be remembered that Nonlinear Programming Method yield a local rather than a global minimum. However, in the particular case of slope stability analysis, it is generally sufficient to check that the results from different trial slip surfaces are essentially the same.

4 CONCLUSIONS

The Nonlinear Programming Methods examined have proved reasonably efficient in locating the critical slip surface in slope stability analyses. Although the Simplex Method shows a tendency to halt prematurely, this particular difficulty can be overcome by starting the procedure over and over again. The speedy of the Descent Methods are strongly influenced by the incremental ratios employed in obtaining the derivatives, while the influence of the incremental step is practically negligible in the Pattern Search Method which seems to be slightly faster and more efficient than the other methods examined. However, whatever the disadvantages, the results yielded by all these methods are perfectly acceptable in practical applications.

REFERENCES

Arai, K. & Tagyo, K. 1985. Determination of noncircular slip surface giving the minimum factor of safety in slope stability analysis. Soils & Foundations. 35: 43-51

Baker, R. 1980. Determination of the critical slip surface in slope stability computations. Int. Jou. Num. Anal. Meth. Geom. 4:333-259.

Boutrup, E. & Lovell, C.W. 1980. Searching techniques in slope stability analysis. Engineering Geology. 16: 51-61.

Castillo, E. & Luceño, A. 1982. A critical analysis of some variational methods in slope stability analysis. Int.Jou.Numer. Meth. in Geom. 6: 195-209.

Celestino, T.B. & Duncan, J.M. 1979. Simplified search for noncircular slip surfaces. Proc. X ICSMFE. 3:391-394

Cherubini, C. & Greco, V.R. 1987. A probabilistic method for locating the critical slip surface in slope stability analysis. Proc. ICASP5. 2:1182-1187.

De Josselin De Jong, G. 1981. A variational fallacy. Geotechnique. 31:289-290.

Greco, V.R. & Gulla`, G. 1985. Critical slip surface search in slope stability analysis. Rivista Italiana di Geotecnica. 189-198.

Kopacsy, J. 1957. Three-dimensional stress distribution and slip surface in earth works at rupture. Proc. 4th ICSMFE.3a: 17.

Kopacsy, J. 1961. Distribution des contraintes à la rupture, forme de la surface de glissement et hauteur théorique des talus. Proc. 5th ICSMFE. 6:23

Nguyen V.U. 1985. Determination of critical slope failure surfaces. Jou. Geot. Eng. Div. ASCE. 11:238-250.

Siegel, R.A., Kovacs, W.D. & Lovell, C.W. 1981. Random surface generation in stability analysis. Jou.Geot.Eng.Div. ASCE. 107:GT7, 996-1002

Numerical Methods in Geomechanics (Innsbruck 1988), Swoboda (ed.)
© 1988 Balkema, Rotterdam. ISBN 90 6191 809 X

Reliability analysis of propped embedded cantilever walls

Kanagasabai Ramachandran
Department of Civil Engineering, Imperial College, London, UK

ABSTRACT: Probabilistic methods are now increasingly used in geomechanics to assess the safety of earth slopes, retaining walls and dams under seismic excitations. In this paper, the results of a preliminary study carried out at Imperial college on the safety assessment of propped embedded cantilever retaining walls are presented. The problems associated with such probabilistic analyses are identified and solutions are suggested for a few selected problems.

1 INTRODUCTION

Probabilistic methods are slowly gaining acceptance with the majority of geotechnologists in north America and Europe. However, the level of understanding varies tremendously across these continents. While very refined approaches are suggested by researchers in USA, Germany and Scandinavia, the research papers published in Britain lack basic understanding and apprecition of probabilistic concepts. It is not surprising that there is a good deal of controversy over the use of probabilistic methods amongst geomechanists in Britain. The researchers in Britain, instead of giving attention to problems pertaining to geomechanical designs and developing suitble methodologies to deal with specific geotechnical problems, tend to discuss the basic ideas in reliability theory developed a decade earlier.

This paper attempts to remedy some of these shortcomings and to highlight the areas where research effort should be directed to, in the near future. The research resources available for such studies are very scarce in Britain and therefore these should be properly channelled to make real progress in the application of probabilistic concepts to geotechnical analysis and design.

Reliability theory is now well developed and can be directly applied to geotechnical analyses and designs. There is no need for geotechnologists to understand the complex mathematics inherent in the advanced reliability theory; but it is essential for them to understand and to appreciate the basic concepts in the reliability theory thoroughly before applying it to geotechnical problems. Partial or improper understanding can lead to unsafe and uneconomical designs as pointed out in an earlier discussion (Ramachandran 1986). A clear distinction should be made here between concepts and complex mathematics. Complex mathematics is not easy to comprehend by every one but an average geotechnologist should not have any difficulty with concepts. For example, consider the problem of constrained minimisation. It is quite easy to formulate an objective (cost) function and a set of constraint equations, although the mathematics involved in the minimisation process can be, and is, complex. Of course, it is better if the user of the minimisation algorithm understands the logic; however it is not necessary.

All engineering designs, whether structural or geotechnical, are made under a great deal of uncertainty in material properties, loadings and mathematical models used in the analysis. Although the mathematical models generally play a major role in the resulting design parameters, the uncertainty in material properties and loadings cannot be given less prominence as suggested by a few geotechnologists in Britain. With refined analyses like the finite element method available on micros, more attention should be given to the uncertainties on these variables and their treatment in the design process. It is illogical to use the worst possible values in the design.

At present, most geotechnical design decisions are made on the basis of deterministic analyses with one or more factors of safety to cover the uncertainty and inaccuracy, and to ensure an 'acceptable' level of confidence in the design. The term 'acceptable' is very subjective and is usually decided and recommended by code writers. There are serious difficulties as described by Hoeg and Murarka (1974), Burland,Potts and Walsh (1981) and others with the use of these safety factors in the design of retaining walls. The recommended factors of safety for different design approaches do not give the same 'notional' level of safety.

On the other hand, a probabilistic analysis will give the same value for the 'notional' level of safety. This is the basis of the reliability approach developed in the structural engineering field, which is now widely used in other branches of engineering as well. Hoeg and Murarka (1974) appear to be the first to use the reliability approach to the analysis and design of retaining walls, although Meyerhof (197o) gave some relationship between the factors of safety and probability of failures in a publication appeared a few years earlier (meyerhof 197o). Hoeg and Murarka used the so-called mean value first order second-moment method of reliability analysis. In the period 1974 - 1986 a number of publications have been published on the application of reliability theory to various geotechnical problems. Very recently there was a paper in Geotechnique (Smith 1985) on the probabilistic analysis of retaining walls somewhat similar to the work of Hoek and Murarka using the so-called advanced first order second-moment method of reliability analysis. This paper failed to deal with the problem in a logical and an acceptable manner and evaded a number of critical issues as discussed by the author in Geotechnique (Ramachandran 1986).

In this paper, a brief discussion on the reliability theory is first given for completeness and clarity. The important issues such as the selection of basic variables, correlation between the basic variables, possible probability distribution functions for these basic variables and their effect on the outcome of the reliability analysis, and uncertainty associated with mathematical models used in the analysis are then given attention. The retaining wall used by Smith (1985) for his analysis has been used for this study in order to highlight the importance of proper understanding of probabilistic concepts. It should be noted that the reliability analysis, unless performed by or with the help of an experienced probabilistic mechanist can lead to unsafe and uneconomical designs. The discussion on this paper avoids the controversy of the suitability or otherwise of existing design formats as the probabilistic methods are not affected by them.

2 RELIABILITY ANALYSIS THEORY

First step in the reliability analysis is to choose carefully the basic variables to be used in the analysis. Basic variables are those variables which have a profound influence on the design. If in doubt, sensitivity study should be done with suspected variables to decide on their influence on the failure of the design (structure). In certain cases, the basic variables may not be apparent at the outset and further study may be necessary. For this presentation, it is assumed that the basic variables can be easily identified. The next step is to assign suitable probability distributions for these basic variables with the help of data if available, or by judgement. As can be seen from the results reported later, the assumption of normal (Gaussian) distribution is acceptable in most cases. Investigation of possible correlation between these basic variables should then be undertaken. If it is not possible to calculate the correlation coefficient between two variables, the author suggests that the reliability analysis should be repeated for a range of possible correlation coefficients and a conservative decision should be taken as to the reliability of the structure under consideration. However, as shown later in this paper it is advisable to avoid such situations and to select uncorrelated (independent) basic variables.

The author's algorithm for the evaluation of the so-called reliability (safety) index β is very simple and is given below. The probability of failure of the structure in the given limit state is then given by $\phi(-\beta)$, where ϕ is the standard normal distribution function, the values of which are tabulated in text books.

$$\sigma = \left(\sum_{j=1}^{n} \sum_{i=1}^{n} g_i g_j \operatorname{cov}(X_i, X_j) \right)^{1/2}$$

$$\beta = \left[\sum_{i=1}^{n} g_i(\mu_i - x_i) + Z \right] \Big/ \sigma$$

$$x_i^n = \mu_i - \frac{\beta}{\sigma} \sum_{j=1}^{n} g_j \operatorname{cov}(X_i, X_j)$$

where $g_i = \partial Z / \partial x_i$, μ_i is the mean value of the ith basic variable, $\operatorname{cov}(X_i, X_j)$ is the covariance between the variables X_i and X_j and x_i^n is a new value of x_i (in the iterative process).

3 APPLICATION TO RETAINING WALL ANALYSIS

Figure 1 shows the details of a sheet-piled propped cantilever wall in a cohesionless (c=o) soil. The ground water level may be assumed to be at dredge level

3.1 Safety margin Z

Method recommended by the civil engineering code of practice no 2: earth retaining structures (Institution of structural engineers 1951) may be used to to obtain an expression for factor of safety F against failure by overturning. The safety margin Z is then simply given by (Ramachandran and Hosking 1985)

$$Z = F - 1$$

where F contains all the basic variables and other parameters (deterministic variables and constants). For the given dimensions in Figure 1, Z takes the following form.

$$Z = 38(\gamma - 1o)K_p / 3o3\gamma K_a + 38(\gamma - 1o)K_a - 1$$

where K_a = active pressure coefficient
K_p = passive pressure coefficient
γ = unit weight of soil

Note that the depth of the cantilever wall and embedment length are assumed to be deterministic for simplicity, but may be included in the analysis if required.

3.2 Selection of basic variables

The variables that appear in the simplified form of the safety margin Z are K_a, K_p and γ. It is well known that both K_a and K_p are functions of the angle of internal friction Φ and the angle of wall friction δ, and there is not enough evidence to suggest that the unit weight has correlation with Φ or δ. It is now quite easy to see that δ, Φ and γ are the basic variables for this particular reliability analysis. In the earlier publication (Smith 1985) δ is assumed to be equal to $\Phi/2$. Although it is not reasonable to make this assumption, δ has been taken as $\Phi/2$ in this study just for comparison purposes. δ will be treated as a correlated variable in a future publication (Ramachandran and Frangos 198 An experienced analyst should know that the effect of δ on the factor of safety F is very minimal compared to that of Φ and thereby reduce the number of basic variables to one, namely Φ. However for illustrative purposes, both δ and Φ are retained as basic variables in this study.

3.3 Possible probability distributions

The unit weight can be safely assumed to have a normal distribution by central limit theorem. If necessary, calculations may be repeated with lognormal distributions to check for any variation in the final result. Is the angle of internal friction Φ or the shear strength parameter tan Φ a basic variable?. It appears that geotechnologists have taken tan Φ as the basic variable and studied its probability distribution (Lumb 197o). As the author is not aware of any other probability study on Φ or tan Φ normal distribution was assumed for tanΦ. The study was however repeated with a normal distribution for Φ instead of tanΦ as well. There may be objections for the use of normal distributions on the grounds that Φ or tan Φ cannot take negative values. To satisfy those doubters, the normal distributions may be truncated at zero values and suitable modifications can be made to the distribution function (Ramachandran and Hosking 1985).

3.4 Correlation between the basic variables

By using Φ and γ as basic variables, the problem of calculating the correlation coefficients is avoided. If K_a and K_p are used as basic variables as wrongly suggested by Smith (Smith 1985), we will face with the following two unsolvable problems:

1. What are the probability distribution functions for K_a and K_p ?.
2. What is the correlation coefficient between K_a and K_p ?.

If someone wishes to use K_a and K_p as basic variables, reliability analysis should be repeated for correlation coefficients varying from zero to -o.99 (K_a and K_p are negatively correlated - an increase in K_a corresponds to a decrease in K_p and vice versa.) and the most unfavourable result should be taken as the reliability index corresponding to the safety margin under consideration. The assumption of normal distributions for K_a and K_p, though not exact, seems to be reasonable for the cases studied. However, the author disapproves the use of K_a and K_p as basic variables.

There is no evidence so far to suggest that there may be any correlation between the unit weight γ and and the angle of internal friction Φ (or shear strength tan Φ). If the wall friction angle δ (or tan δ) is used as a basic variable, then the correlation between δ and Φ (or tan δ and tan Φ) should be given consideration. In the authors opinion, δ and Φ appear to be uncorrelated and geotechnologists may be able to provide some guidance on this aspect in the near future.

3.5 Mathematical model uncertainty

Evaluation of the uncertainty associated with the mathematical models used in the analysis is the most difficult task in the whole process. In some cases, it may be possible to collect data to fit a probability distribution for this model uncertainty by performing a series of tests (Ramachandran 1986, Ramachandran and Baker 1981).

In geotechnics it may not be possible to perform such experiments to compare the actual value of K_a (or K_p) and the value predicted by our mathematical model. In the absence of relevant data, it is not very unreasonable on a practical point of view to assume a normal distribution for the model uncertainty in K_a and K_p.

4 MATHEMATICAL MODEL AND DATA

To get reliable answers from our analysis, physical process should be properly modelled. In the case of an embedded retaining wall, the active and passive earth pressures are difficult to model. This paper in the main follows the procedures adopted by Burland et al (Burland, Potts and Walsh 1981) for modelling active and passive earth pressures. The active earth pressure coefficient K_a is

$$K_a = \frac{\cos^2(\phi - \alpha)}{\cos^2\alpha \cos(\delta - \alpha)} \left\{1 + \left[\frac{\sin(\phi + \delta)\sin\phi}{\cos(\delta - \alpha)\cos\alpha}\right]^{1/2}\right\}^{-2}$$

For passive pressure coefficient K_p, values given by Caquot and Kerisel (1948) have been used. Author has fitted a sixth order polynomial for the value of K_p in terms of Φ and the results are shown in Table 1. The polynomial is given by

$$K_p = .02469-34.3753\Phi+7.45417\Phi^2-.604717\Phi^3+.024377\Phi^4-.0004823\Phi^5+.0000038667\Phi^6$$

Similarly another fifth order polynomial has been fitted for the correction factor C_f for $\delta = \Phi/2$ and the polynomial is

$$C_f = .000233+1.007949\Phi-0.005576\Phi^2-0.0000382\Phi^3-0.0000028\Phi^4+1.82\times10^{-8}\Phi^5$$

Table 1. Polynomial fitted to K_p

Φ	Actual value	Polynomial value
15	2.19	2.17
20	3.01	3.03
25	4.29	4.27
30	6.42	6.44
35	10.20	10.18
40	17.50	17.52
45	33.50	33.48

Table 2. Polynomial for correction factor ($\delta = \Phi/2$)

Φ	Actual value	Polynomial value
10	0.946	0.946
15	0.907	0.907
20	0.862	0.862
25	0.808	0.808
30	0.746	0.746
35	0.674	0.674
40	0.592	0.592
45	0.500	0.500

4.1 Approximate model for K_p

For simplicity, Coulomb's equation may be used for K_p with model uncertainty factc M_p. The relevant equation for K_p is then

$$K_p = M_p \frac{\cos^2(\phi - \alpha)}{\cos^2\alpha \cos(\delta - \alpha)} \left\{1 - \left[\frac{\sin(\phi + \delta)\sin\phi}{\cos(\delta + \alpha)\cos\alpha}\right]^{1/2}\right\}^{-2}$$

4.2 Data used in the analysis

Normal distributions are assumed for both $\bar{\gamma}$ and $\tan\Phi$. Exercise was repeated with normal distribution for Φ and truncated normal distribution for $\tan\Phi$. Mean valu of Φ and γ are kept at 35° and 20 kN/m^3 to be consistent with the design. The corresponding value of $\tan\Phi$ is 0.7. 5% and 10% coefficient of variations were used in the analysis. The author recommends a value o 10% for the coefficient of variation for Φ and 5% for γ. Uncertainty in the mathematic models for K_a and K_p are not used in the results presented here. These refinements are under investigation and the results will be presented later (Ramachandran and Frangos 1987).

RESULTS AND DISCUSSION

'he results obtained with the author's general purpose reliability program (Ramachandran 1982, Ramachandran 1986) are tabulated in Tables 3-7. TanΦ and γ were used as the basic variables for the results in Table 3.

Table 3. With tan Φ and γ as basic variable

Variable name	Mean val	Std dev	Design value	Sensitivity coefficient
Tan Φ	0.70	0.039	0.600	0.9589
γ	20.0	1.00	19.24	0.2838

Reliability index = 2.6723
Probability of Failure = 0.003767

Results in Table 4 were obtained when the above data was slightly modified (standard deviation of tan Φ = 0.09)

Table 4. Modified std dev for tan Φ

Variable name	Mean val	Std dev	Design value	Sensitivity coefficient
Tan Φ	0.70	0.09	0.593	0.9934
γ	20.0	1.0	19.86	0.1150

Reliability index = 1.1940
Probability of Failure = 0.1162

In order to check the validity of using Φ as a basic variable instead of tan Φ, program was rerun with Φ and γ as basic variables. Results are shown in Table 5 and the effect of truncating tan Φ (or Φ) at zero value is given in Table 6.

Table 5. With Φ and γ as basic variables

Variable name	Mean val	Std dev	Design value	Sensitivity coefficient
Φ	35.0	1.75	30.91	0.9648
γ	20.0	1.0	19.36	0.2631

Reliability index = 2.4234
Probability of failure = 0.007687

Table 6. With truncated tan Φ

Variable name	Mean val	Std dev	Design value	Sensitivity coefficient
tan Φ	0.70	0.039	0.600	0.9588
γ	20.0	1.0	19.42	0.2841

Reliability index = 2.6727
Probability of failure = 0.003763

The inaccurate results obtained with the wrong assumption of zero correlation between K_a and K_p (Table 7) clearly shows the need for proper understanding of the concept of correlation between the basic variables.

Table 7. With independent K_a and K_p

Variable name	Mean val	Std dev	Design value	Sensitivity coefficient
K_a	0.235	0.02	0.2573	0.3554
K_p	7.100	0.90	4.5570	0.9027
γ	20.00	1.00	19.24	0.2428

Reliability index = 3.1307
Probability of Failure = 0.00087

Summary of the results for various assumptions is given below in Table 8.

Table 8. Summary of results

Basic variables	Reliability index	Failure prob.
γ and tan Φ (Normal distribution for γ and tan Φ)	2.6723	0.0038
γ and tan Φ (truncated normal distribution for Φ)	2.6727 2.6727	0.0038
K_a and K_p with γ (normal distribution for all)	3.131	0.00087

6 CONCLUSIONS

The use of reliability theory in the safety assessment of propped cantilever retaining walls has been investigated and some preliminary results are given. These results show that serious errors can result if the basic variables are not properly selected and their correlations not considered in the reliability analysis. Author recommends that K_a and K_p should not be used as basic variables. Results obtained with truncated normal distribution for tan Φ (or Φ) do not differ from that obtained with normal distribution for tan Φ (or Φ). This suggests that normal distribution can be used for tan Φ in future investigations.

It has been assumed implicitly that there is only one dominant failure mode for the cantilever investigated. In fact, there may be a number of equally likely failure modes. It is therefore necessary to extend the reliability analysis for other possible failure mechanisms. Once the probability of failure of all possible modes are calculated, bounds on the total failure probabability of the cantilever can easily be evaluated by the author's bounds (Ramachandran 1986)

7 ACKNOWLEDGEMENTS

The author thanks Dr S K Sarma for the useful discussions and Helen Frangos for her computational assistance.

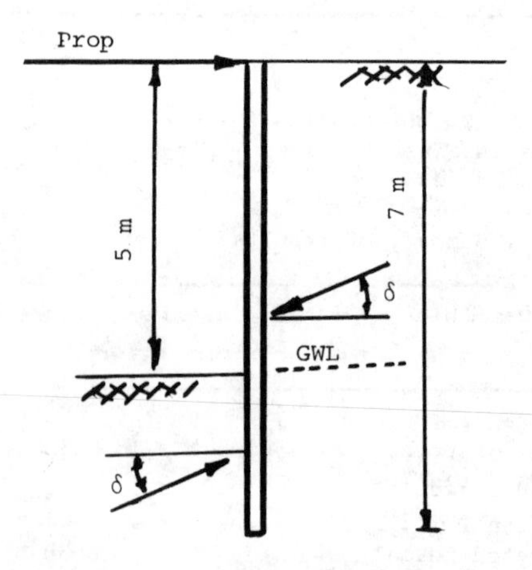

Figure 1. Propped cantilever wall

8 REFERENCES

Burland,J.B.,Potts,D.M. & Walsh,N.M. 1981. The overall stability of free and propped embedded cantilever retaining walls. Ground Engineering, 5,28-38

Hoeg,K & Murarka,R.P. 1974. Probabilistic analysis and design of a retaining wall. Am. Soc of civil engrs., geotech. div.,1oo.

Lump, P. 197o. Safety factors and the probability distribution of soil strength. Canadian geotechnical journal, 7,225-242

Meyerhoff,G.G. 197o. Safety factors in soil mechanics. Canadian geotechnical journal,7.

Ramachandran,K. 1982. Relyiç - reliability analysis program, Civil engineeing department, Imperial college,London.

Ramachandran,K & Hosking,I. 1985. Reliability approach to stability analysis of soil/ rock slopes. Fifth int. conf. on num. meth. in geomechanics, Nagoya, Japan. 1o19-1o28.

Ramachandran,K. 1986. Bound for trivariate integrals in system bound. Journal of structural engineering, ASCE,112,9o7-923.

Ramachandran,K. 1986. Discussion on the paper by G.N.Smith. Geotechnique, 36,617-61

Ramachandran,K & Frangos,H. 1977. Safety assesment of retaining walls by second-moment reliability methods. under preparation.

Smith,G.N. 1985. The use of probability theory to assess the safety of propped embedded cantilever retaining walls. Geotechnique,35,451-461.

Numerical Methods in Geomechanics (Innsbruck 1988), Swoboda (ed.)
 ISBN 90 6191 809 X

Analysis of concrete dam-rock base interaction by numerical methods

S.B.Ukhov, V.G.Orekhov, M.G.Zertsalov, V.V.Semenov & G.I.Shimelmits
MICI Kuibyshev V.V. Moscow, USSR

ABSTRACT: The paper presents materials of investigations made for design solutions substantiation of a series of dams. It is shown that the methods of finite elements allows to take into account the features of rock masses: heterogeneity, jointing, anisotropy, complex interaction of various physical-mechanical processes, when analysing concrete dams - rock bases interaction. Close links between geologic, geomechanic and analytic aspects of the problem are emphasized. The effect of the methods of schematization and solving problems on the final result is also under study.

Concrete dam and rock base represent an indivisible system. Such a system always has complex geometrical inner and outer contours, it is heterogeneous in its mechanical and filtration properties and may be anisotropic at separate areas. This conditions the advisability of making the system analysis by numerical methods. Our experience shows that the method of finite elements allows to take into account structural features, physical and mechanical properties of rock masses and to obtain exhaustive information about stress-strain state, strength and stability of both the system elements and system as a whole.

Lower the examples of calculation analysis of real hydrotechnical objects are given. Here we tried to stress close links between different aspects of the problem, the effect of methods of solving problems and schematization of bases on final result.

1. A base of a dam, 50 high (Fig.1) is represented by alternating layers of sandstones and argillites with seam thickness up to several meters. Clearly seen anisotropy of deformation properties is characteristic of the base as a whole. Since the extent of geomechanic study was not enough to specify calculation indexes of base rock properties exactly, three variants of the problem were examined, with the index of rock deformation anisotropy D_1/D_2 varying from 1 up to 3,5,4,5, where D_1 and D_2 are strain modulus, directed along main axes of anisotropy.

The table gives safe sliding factor values along the most probable slide surfaces (Fig. 1), determined on the basis of engineering-geologic analysis of the situation.

It is characteristic that when the index of deformation anisotropy of base rocks changed by a factor of 4.5,

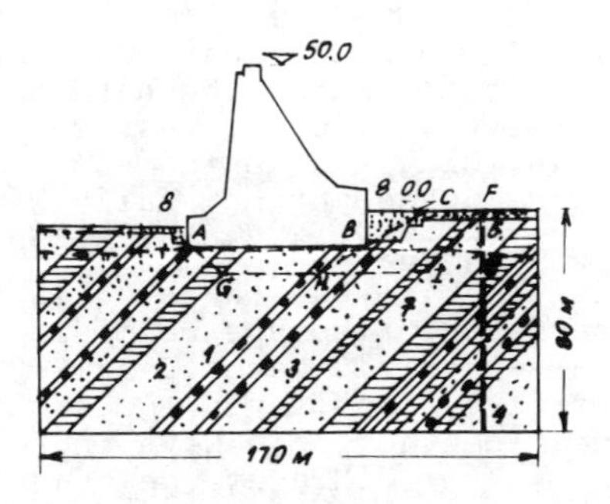

Figure 1. Calculation scheme of gravity dam and laminated rock foundation: 1,2,3 - different types of the rock, 4 - subvertical fracture, 5,6, 7 - zones of the effect of rock weathering, 8 - backfill of the foundation pit, A-I - letter symbols of slide surfaces.

Table 1

Intervals of / values for different rock seams	Safe sliding factor values for slide surfaces (Fig. 1)				
	AB	ABC	ABEF	AGHBC	AGIF
*1	1.18	1.41	1.56	3.66	4.01
1.5-2.0	1.08	1.28	1.42	3.42	3.83
3.5-4.5	1.01	1.12	1.27	3.21	3.51

the safe sliding factor changed much less (about 10-30%). This important conclusion demonstrates, that in some engineering applications the accuracy of determining rock masses deformation properties indexes is not of decisive importance. The obtained results of calculations made it possible not only to evaluate dam stability, but also to determine optimum direction of further base analysis and design of the structure. The detailed description of these results is given in (Ukhov 1979; Ukhov et al. 1981; Ukhov & Semenov 1984).

2. The effect of various structural types of the rock mass consisting of tuff-sandstones, tuff-aleurolites, spilites, forming seams of different deformability, on stress-strain state of the structure and the base was studied for a concrete gravity dam 170 m high. Calculations were made for different schemes of heteromodulus rocks location under the dam footing section.

The variant with the most rigid base (D = 25000 MPa) was examined on the first model. In this case in the near-contact zone at the head wall compressive vertical stresses σ_y were 0.87 MPa, and in the zone of downstream side maximum basic stresses were σ_{max} = 4.7 MPa. A zone of biaxial tension was registed in the dam base under the upstream side. With a less rigid base (D = 12000 MPa) compressive stresses in the dam in the near-contact zone have increased up to σ_y = 1.0 MPa at the head wall, and up to σ_{max} = 4.91 MPa in the zone of downstream bucket. The zone of biaxial tension was preserved in the dam base under the upstream side. The presence of cracks in the rock mass decreases tension zone in the base under the upstream side considerably. So, when there is a vertical crack in the base in the head wall area, which does not take up tensile stre ses, a zone of biaxial tension char ges into uniaxial one. Rocks of dif ferent rigidity being located under the dam footing area, transformatic of stress state takes place in the near-contact zone of the structure, depending on specific position of rigid and soft rock seams. For the engineering-geological model refine for the present period, compressive stresses in the contact area near the head wall were σ_y = 0.97 MPa, and in the zone of downstream bucke basic stresses were σ_{max} = 4.8 MP A zone of biaxial tension was regis tered in the dam base under the upstream side.

The investigations showed that di ferent deformability of rocks may change stress state considerably ir the contact zone of the dam and the base. The location of less rigid rocks in the upstream side area of the dam and rigid rocks - in the downstream side is most favourable. Cracks opening in the base near the upstream side being taken into account, tensile stresses values are decreased considerably and the dimensions of seal failure zone are reduced.

Calculation analysis made it possible to determine the most favourable position of dam location and to recommend grouting of several base areas to achieve more uniform distribution of stresses in the near-contact zone.

3. Rock base of a concrete dam, about 100 m high (Fig. 2), formed by a complex of paleozoic metamorph shales, was schematized as heteroge neous fissured-block medium. The mo dulus of deformation of the base within the blocks changed from D = = 4000 MPa on the surface D =

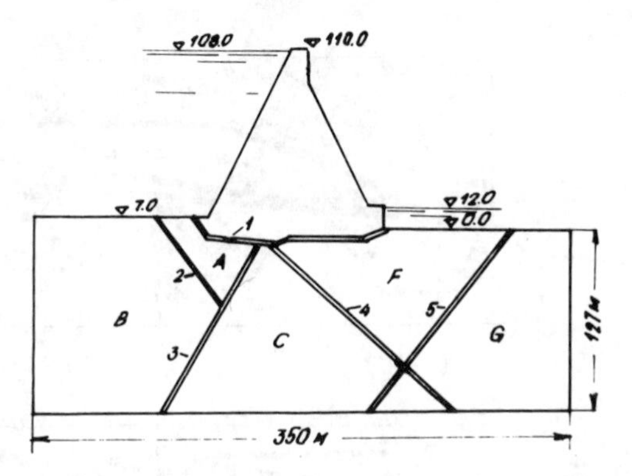

Figure 2. Calculation scheme of gravity dam and fissured-block base: 1 - contact zone of the dam and the base; 2-5 - isolated large cracks; A-C - blocks of solid heterogeneous rock

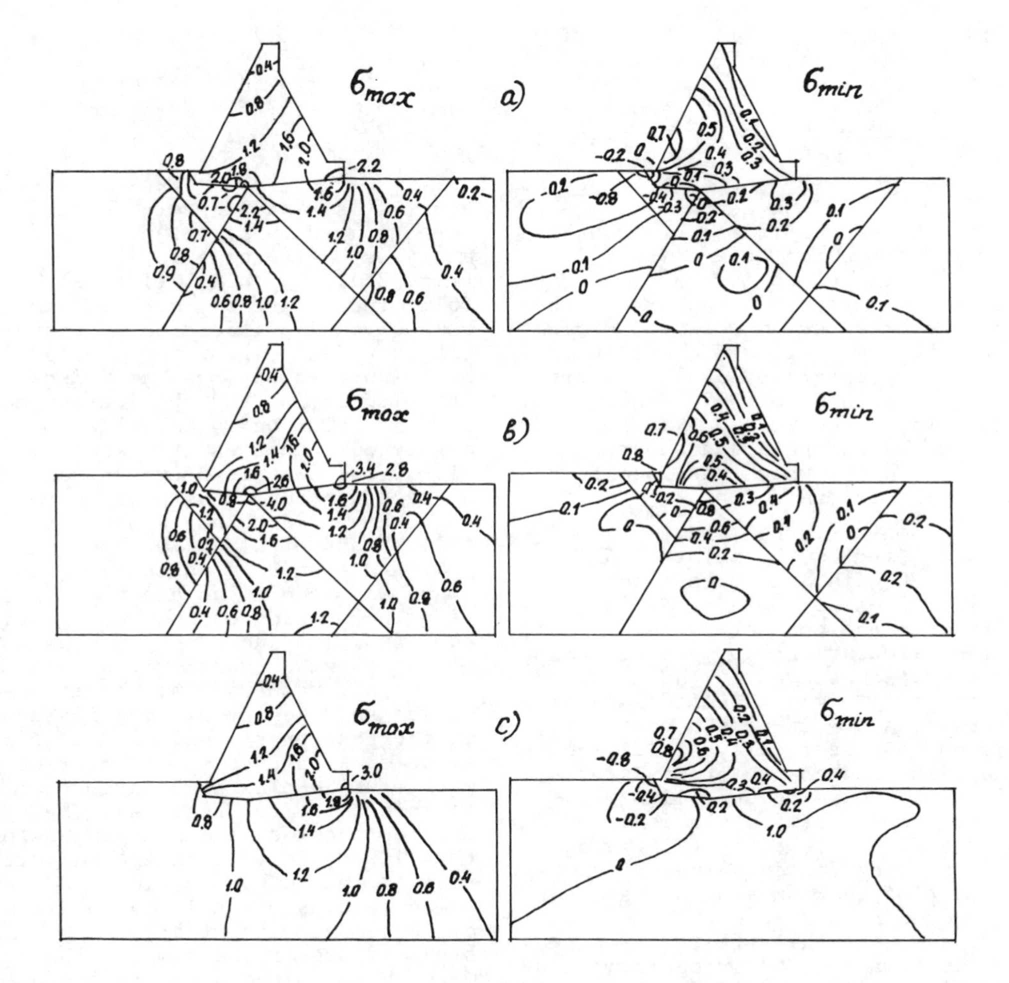

Figure 3. Isolines of maximum (σ_{max}) and minimum (σ_{min}) basic stresses, MPa: a) fissured-block base (elastic problem); b) the same (7-th iteration); c) solid base.

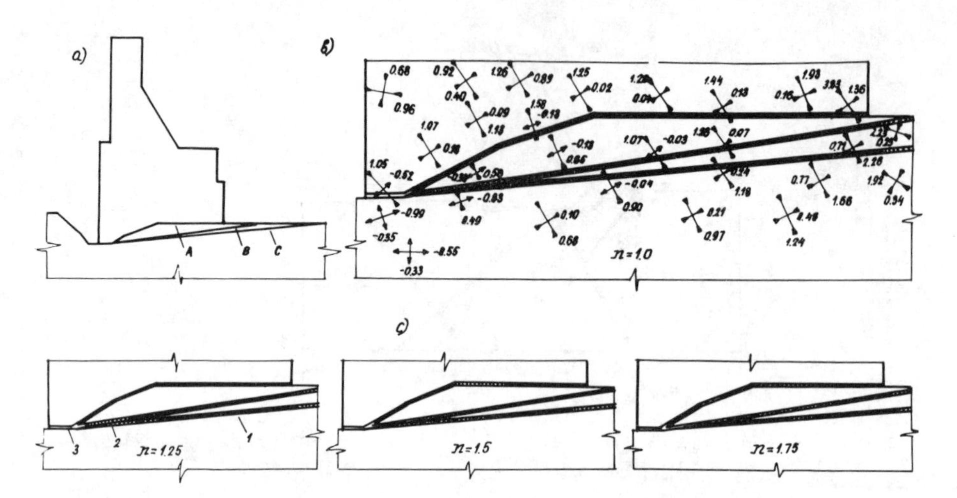

Figure 4. The analysis of a gravity dam 90 m high and fissured rock base: a) calculation scheme; b) stress state (MPa) of contact zone with overload factor n = 1.0; c) state of "A" contact and fissures "B" and "C" with different overload factors n; 1 - undisturbed contact; 2 - shear zone; 3 - open crack.

= 12000 MPa at the depth of 25 m and lower, with coefficient of transverse deformation being ν = 0.1. It was singled out in the base: a zone of contact between the base and the dam, represented by rocks of medium and high jointing, different in strength properties, and large cracks with the opening of 0.7 m and 1.0 m. Strength of contacts was determined by tensile strength and shear strength τ_{lim}, which was specified by non-linear relation $\tau_{lim} = f(\sigma)$ with a non-disturbed state of the contact, and by the relation $\tau_{lim} = \sigma \cdot tg\,\varphi$ with a disturbed due to chear contact, where σ is stress normal to the contact plane.

The calculation was made according to iteration scheme by the method of alternating rigidity. After seven iterations stress-strain state of the system was practically stabilized, the value and character of stresses distribution being appreciably different from elastic solution, obtained at the first iteration (Fig. 3 a,b). Safe sliding factor value along the contact of the dam and the base has decreased from 2.9 to 2.7.

To evaluate the effect of large cracks, forming the irregular fissured-block structure of the base, the analysis of the same problem was made for the model of solid heterogeneous base. The difference in the results is rather great, which is corroborated by stress distribution analysis in the calculation area (Fig. 3c). The calculation data are fully indicative of the fact, that excessive simplification of geomechanic model (e.g. representation of fissured-block base as a model of solid medium and application of elastic solution in calculations) may lead to the results far from re ality. The details of geomechanic and calculation results are given in (Ukhov 1979; Ukhov & Semenov 1984; Ukhov et al. 1978).

4. A dam under construction, 90 m high, was used as an example for so ving the problem of concrete gravit dam stability and strength. The bas is represented by dolerites mass of different preservation. The effect of weakened gently sloping fissures in the base on strength and stabili ty of dam-base system was considere with elastic-plastic behaviour of fissure's filler (Fig. 4a). The bas was represented as a homogeneous on (D = 10000 MPa). Strength characteristics in the base varied.

The results of basic calculation cycle demonstrate, that stress-strain state of the dam and the bas with the examined strength and defo mation characteristics of the conta and the fissures changes only sligh Tension zone in the upstream side o the dam (Fig. 4b) causes opening of

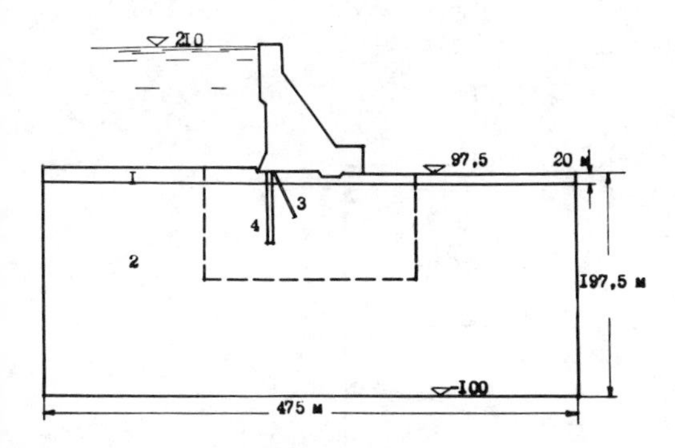

Figure 5. Calculation scheme of gravity dam and base: 1 - a zone of high jointing; 2 - a zone of preserved rock; 3 - drainage holes; 4 - cement-grout curtain.

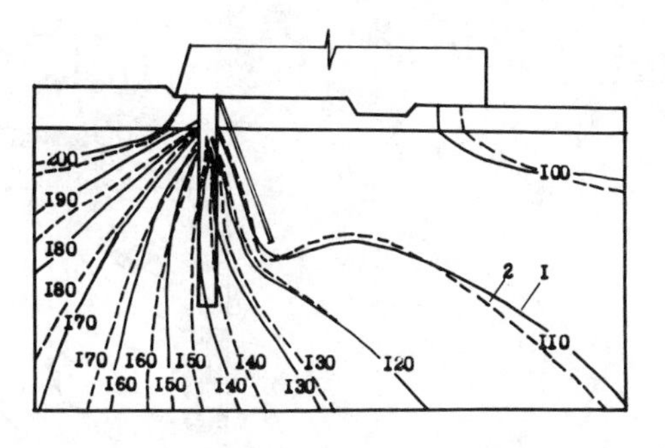

Figure 6. Lines of equal potentials, m: 1 - elastic model of the base; 2 - elastic-plastic model of the base, mutual effects of stress-strain state and filtration regime being taken into account.

contact in the area 8-10 m away from the dam. It also causes opening of fissures wedged in near the dam's toe. The basic part of the contact between the dam and the base is not disturbed. It may be concluded, that, when servife loads are acting, dam-base system has a stability margin.

The second series of calculations was made when the structure was overloaded by a horizontal force T exceeding the design value T_c. The factor of overloading was determined as $n = T/T_c$. The increase of horizontal load leads to further opening of fissures and to broadening of the shear zones along the fissures and the contact section (Fig. 4c). With n = 1.25 stability of the system is preserved, although some contact areas are in limiting state. With n = 1.5 stability of the system is still preserved, but a far greater part of fissures and contact areas is in limiting state. And finally, with n = 1.75 it may be assumed, that the system has practically exhausted its load-carrying capacity, and stability has been disturbed along the fissures' "B" and "C" planes.

The calculation analysis showed, that the system preserves its stability with overloads of about n = 1.25, parameters of the fissures being rather low (C=0, tg φ = 0.65). In addition it was determined, that the toe of the dam is very important for providing structure stability. The investigations results made it possible to work out a complex of constructive procedures to increase a load-carrying capacity of dam-base system, including (for some sections) installation of the toe, grouting of the base under the downstream wedge of the dam and setting the structure against the rock mass at the downstream side.

5. The problem of interaction of concrete dam 110 m high and rock base was solved subject to such factor as heterogeneity and anisotropy of the base, elastic-plastic character of rock deformation, interrelation of stress-strain state and water permeability characteristics, the effect of counter-filtration devices on dam-base interaction.

Calculations were made by the iteration scheme. At every step the problems of filtration and statics were solved consecutively and correction of deformability and water permeability of base rock masses was accomplished. Elastic-plastic problem of statics was solved in plane variant. In solving the filtration problem three-dimensional base area was considered, which is connected with the necessity of modelling space character of stream motion through a number of drainage holes. The details of formulation and methods of solution of the problem are given in papers (Ukhov & Semenov 1984; Semenov 1986).

Dam location area consists of flych deposits of stone coal system represented by alternating layers of sandstones and argillites. Rock lamination is characterized by a complex relationship, forming a relatively

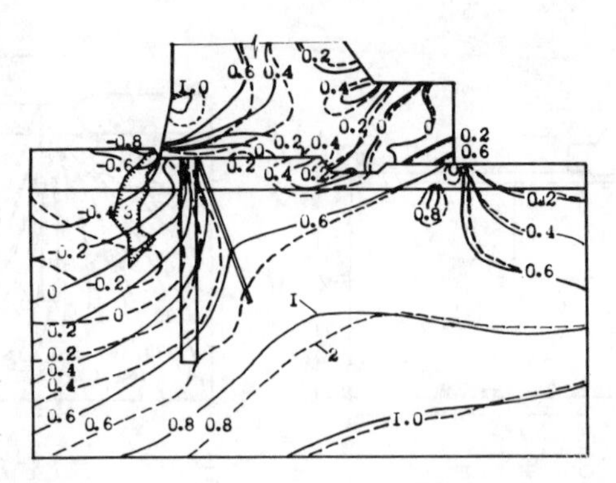

Figure 7. Isolines of minimum basic stresses σ_{min}, MPa: 1 - elastic model of the base; 2 - elastic-plastic model of the base, mutual effect of stres strain state and filtration regime being taken into account; 3 - cracking zone

homogeneous thickness as a whole. The area assigned for further analysis is shown by a dotted line in Fig. 5.

Fig. 6 illustrates the effect of stress-strain state on potentials distribution. Rearrangement of potentials field is most considerable in the base area in front of cement-grout curtain and a number of drainage holes. These changes are less considerable under the footing of the structue due to regulating effect of counter-filtration devices. The prevalence of compressive strains in the base under the downstream side of the dam caused reduction of filtration coefficients to 1.5-2 of their initial values. At the upstream side where tensile deformations predominate, filtration coefficients have increased 4 fold and more. The base, being isotropic in its natural state, acquired filtration anisotropy.

The analysis of stress state shows, that the field of minimum basic stresses σ_{min} differs greatly from elastic solution results (Fig. 7). In elastic solution absolute values of tensile stresses in the base under the upstream side of the dam exceed several time the tensile strength of the rock. The tensile area covers the near-contact area in the body of the dam. Development of cracking in the base at the upstream pool (Fig. 7) has unloaded the adjacent mass area and completely eliminates stresses in the body of the dam.

The opportunities of the proposed calculation approach make high demands of the volume of information characterizing the model of rock bases filtration properties. It is generally accepted when working out such models, to single out zones of different filtration in the base an set filtration coefficients. These propositions are supplemented by th necessity to give water permeabilit change forecast, depending on strai state of the mass.

REFERENCES

Ukhov, S.B. 1979. Principles used in developing geomechanical model of rock masses for solving engine ering problems. Proc. 4th Congres of the ISRM, Montreaux, vol. 2.

Ukhov, S.B. Gasiev, E.G. & A.G. Lycoshin 1981. Construction of rock masses engineering-geological and geomechanical models for solving engineering problems. Hydrotechni cheskoye stroitelstvo, No 3.

Ukhov, S.B. & V.V. Semenov 1984. Us of numerical methods for studies on the interaction between concrete dams and heterogeneous fissured block foundations. Proc. Indo-Soviet workshop on rock mechanics. New Delhi.

Ukhov, S.B., Semenov, V.V., Kotov, P.B. & N.N. Shevarina 1978. Analysis of interaction between a concrete dam and heterogeneous fissured-block rock base. Hydrotechnicheskoye stroitelstvo, No 9

Semenov, V.V. 1986. Joint static an filtration calculations of concre dams rock bases. In the book "Aplication of Numerical Methods to Geomechanic Problems", Moscow, MI

Numerical Methods in Geomechanics (Innsbruck 1988), Swoboda (ed.)
 ISBN 90 6191 809 X

Failure criteria for slope analysis using initial stress method

P.K.K.Lee & K.L.Ng
University of Hong Kong

ABSTRACT: In the past, finite element method incorporating initial stress method has been applied to various problems in geomechanics. The prediction of collapse load of an earth structure is a useful application. It was pointed out by Zienkiewicz that a slowness or lack of convergence of iteration in the computation can be regarded as a symptom of structural failure. Numerical stability of the failure criterion has been studied and an alternate failure criterion considering the number of yielded elements has been developed which gives agreeable results to those derived from classical soil mechanics. Examples on problems related to soil slopes have been analysed with this proposed criterion. Comparison of results with those analysed by other criteria helps to demonstrate the differences in predicting the collapse of earth structures.

1 INTRODUCTION

Collapse loads of various earth structures have been successfully predicted using finite element method incorporating initial stress method. With the elasto-plastic model, Smith (1973,1974) analysed some typical bearing capacity problems and idealized slopes with angles ranging from 30° to 90°. Following the failure criterion suggested by Zienkiewicz (1971) that a slowness or lack of convergence of iteration in the computation can be regarded as a symptom of structural failure, excellent agreement were achieved between the predicted collapse loads and those given by conventional limit analysis methods. Snitbhan (1976) also studied the stability of a vertical cut using large deformation analysis and the failure criterion was based on 'loss of ground'.

In this paper, typical slope stability problems are analysed and the collapse loads are predicted by using various failure criteria. A simple criterion considering the extent of yield region within the soil mass is introduced. It can be shown that the extent of yield region has some significance on the behaviour of the earth structure towards failure. The collapse loads so predicted are also compared with those given by others and the suitability of the proposed failure criterion for slope stability problems are evaluated.

2 INITIAL STRESS METHOD

The constitutive relationship for an elasto-plastic material is described by

$$\{\delta\sigma\} = (D-p)\{\sigma\varepsilon\} \qquad (1)$$

where D is the elastic component of the constitutive matrix which only depends on the two parameters E & ν if the material is isotropic and linear elastic. P is the plastic component of the constitutive matrix and is dependent on the stress state of the element, yield criterion, flow rule and any hardening law adopted.

Since P is stress dependent, the stress-strain relationship is therefore non-linear. The non-linear characteristics can be dealt with by techniques such as the initial stress method, variable stiffness method etc.

In the initial stress method, the global stiffness of the structure in the finite element analysis is kept constant. By assuming that the properties of soil are purely linear elastic, the stress and strain for each element are first predicted. Stresses exceeding the yield stress of each element are converted to a set of nodal forces acting at the corresponding nodes of the element. With the updated load vector, the stress and strain for each element are computed the second time. Excess stresses resulted from the second computation are again converted to nodal forces and this iteration procedure continues. During the iteration process, the excess stresses of the elements are distributed to neighbouring elements and the yield region may extend to cover more elements. The iteration will continue until the nodal forces are diminished to a tolerable magnitude.

3 STABILITY OF SLOPES PREDICTED BY VARIOUS METHODS

3.1 Numerical model

A series of idealized homogeneous saturated clay slopes with slope angles ranging from 15 to 75 , are analysed. The elasto-plastic model is used together with Tresca yield criterion and non-associated flow rule for no plastic volume change during yielding. The elastic properties of the soil have been taken as constant before yielding occurs. Plane strain condition is assumed.

A mesh of 273 elements consisting of 4-node quadrilateral and 3-node triangular elements have been used. The arrangement of the elements, the boundary conditions and the dimensions of the slope together with the elastic properties of soil are given in Figure 1.

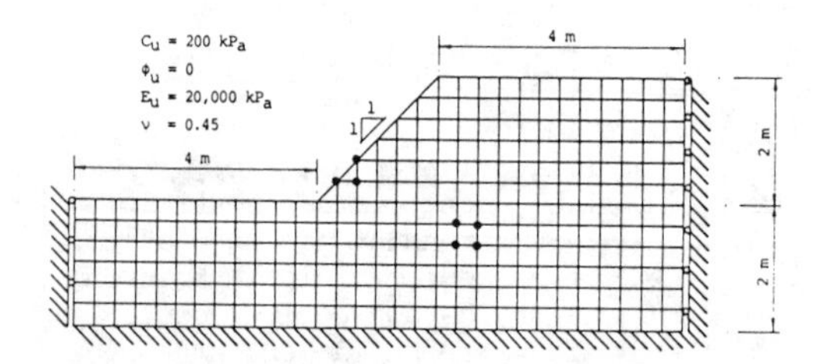

Figure 1. Finite element mesh and boundary conditions of the idealized slope.

3.2 Loading condition

Taylor (1948) suggested that the stability of such slopes depends on the stability number, S, where

$$S = \frac{C_u}{N\gamma H} \quad (2)$$

S is a dimensionless parameter. C_u is the cohesion and γ is the specific weight of soil. H is the height of the slope and N is a dimensionless factor. To obtain S, a slope can be loaded by decreasing C_u by N times or increasing γ by N times until the slope fails.

In this example the slopes are in a stress-free condition before loading and the method of increasing γ is used. The loading increment is kept at 1.6% of the ultimate value of Nγ but for comparison purpose, loading increment of 8% which is 5 times larger have also been used in the computation.

3.3 Load increment and number of iterations

Failure of a slope can be defined as a failure in convergence of nodal loads within a certain criterion. The convergence criterion adopted by Smith (1974) was that the change in any term in the incremental load vector between two successive iterations should not exceed one thousandth of the maximum term in that load vector and the maximum allowable number of iterations was 30. However, the number of iterations specified is arbitrary and highly dependent upon the convergence criterion.

In the present example, the computation becomes numerically unstable after 30 iterations and oscillates within a wide range beyond 30 iterations (Figure 2).

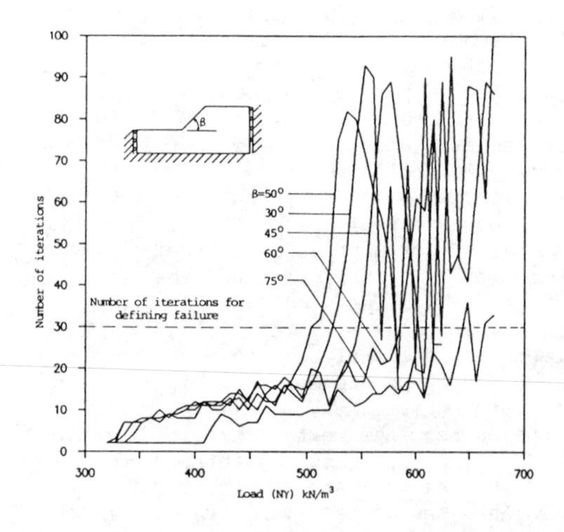

Figure 2. Number of iterations against load.

If the spread of the yield region developed within the soil mass corresponds to higher number of iterations needed to converge the nodal forces, oscillation may mean that the spread is not extending smoothly enough after the specified number of iteration is reached.

Figure 3a shows the yield pattern given by Smith (1974) for a slope with an angle equals to 45^{0}. A similar yield pattern can be obtained for the same slope if the magnitude of load increment is 5 times the proposed 1.6% of maximum Nϒ. (Figure 3b). However, a rather different pattern is obtained when the proposed loading increment is used (Figure 3c) and a complete failure mechanism can be observed. The formation of the mechanism suggests that the maximum number of iterations of 30 is appropriate. The collapse loads predicted using the finer loading increment agree closely with the Talyor's solution as shown in Figure 4.

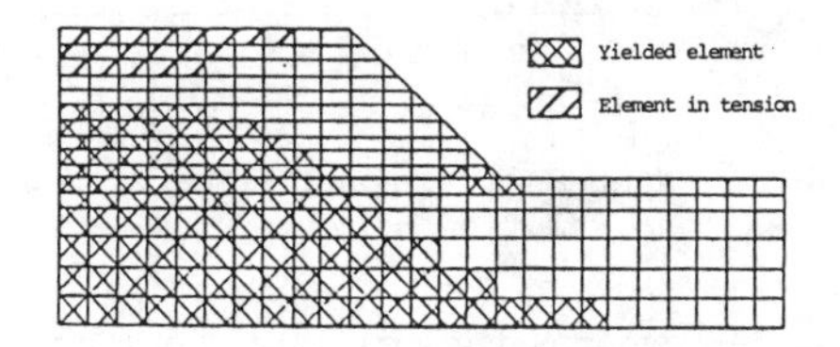

Figure 3a. Smith's yield pattern.

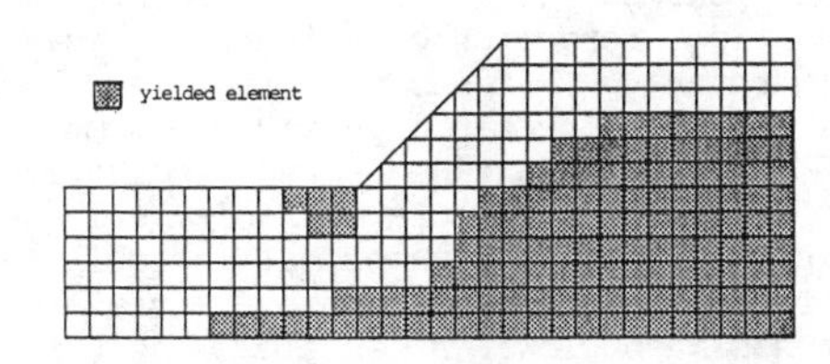

Figure 3b. Yield pattern when δNϒ = 40 kPa.

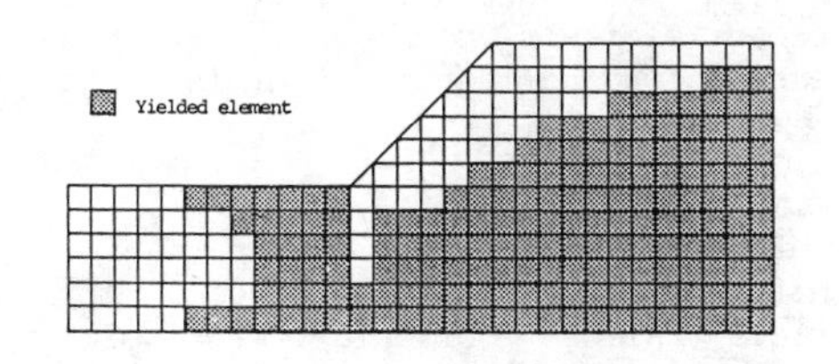

Figure 3c. Yield pattern when δNϒ = 8 kPa.

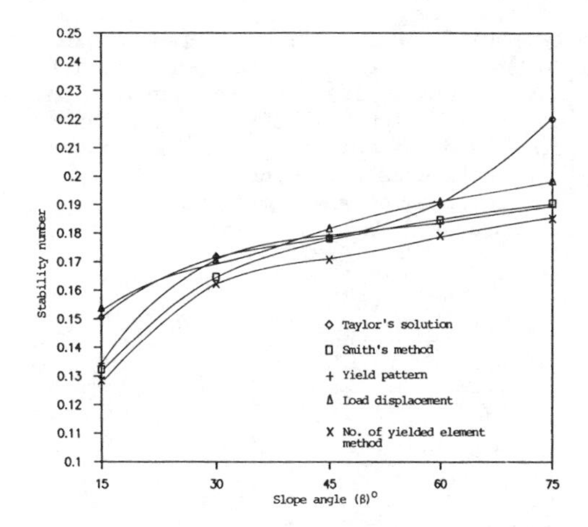

Figure 4. Comparison of stability number predicted by various methods.

3.4 Load-displacement relationship

Failure can also be defined by excessive movement at a particular point within the slope. In some earth structures, it may be difficult to select a representative point in particular with failures involving sliding of soil masses. Zienkiewicz (1975) has considered the excessive movement at the crest of an idealized homogeneous embankment to define the collapse loads.

For the slopes in this example, collapse loads are predicted by using this approach and plotted in Figure 5.

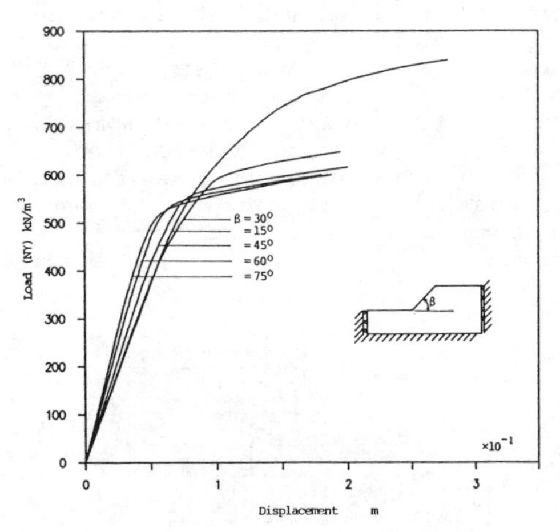

Figure 5. Load-displacement curves for the idealized slopes.

From this figure, slopes with angles above 30^{0} show a more progressive failure. Excessive movement can also be observed in the velocity field just before and at failure of the slope as shown in Figures 6a and 6b. The collapse loads so obtained also agree very closely with standard Taylor's solution (Figure 4).

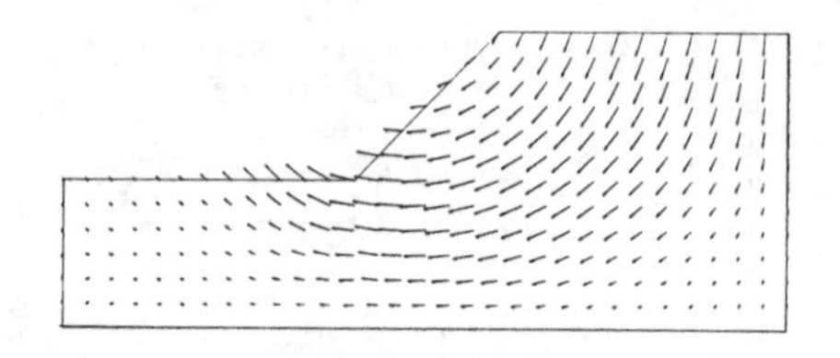

Figure 6a. Displacement field of the load increment just before failure. ($\beta = 45^{0}$).

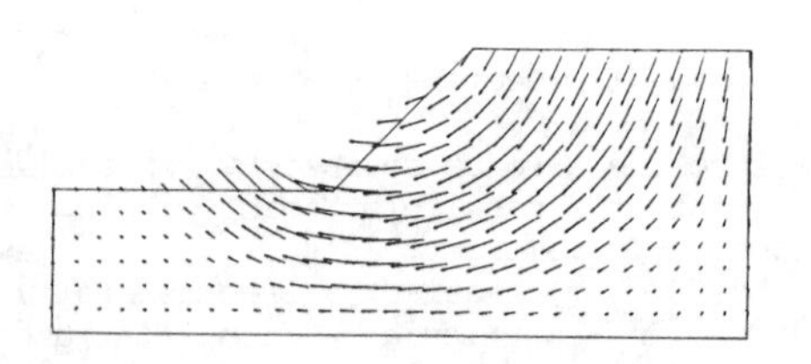

Figure 6b. Displacement field of the load increment at failure. ($\beta = 45^{0}$).

3.5 Yield pattern

When the slopes are loaded to a certain extent, yielding occurs and spreads through the soil. At certain point of loading, the yield region forms a complete failure mechanism as shown in Figure 7. The collapse loads obtained by observing the completion of a failure mechanism agree closely with the standard solution, although it is the upper bound solution.

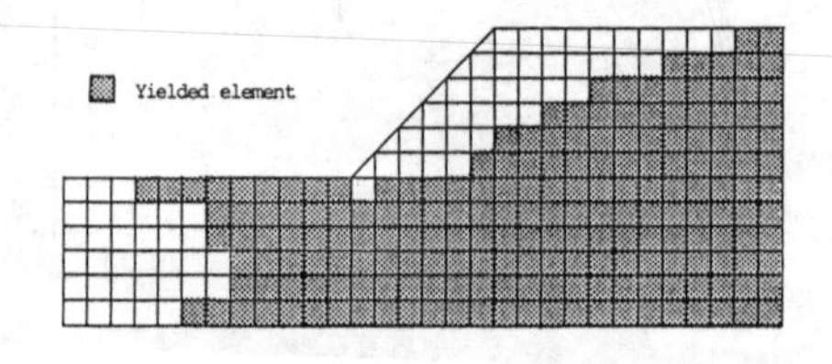

Figure 7. Yield pattern at failure (β=45^{0}).

3.6 Number of yielded elements - the proposed criterion

The extent of the yield region can be represented by the increase in yielded elements if the mesh is uniform. The number of yielded elements for the slopes increases until at certain point, the increase diminished as shown in Figure 8.

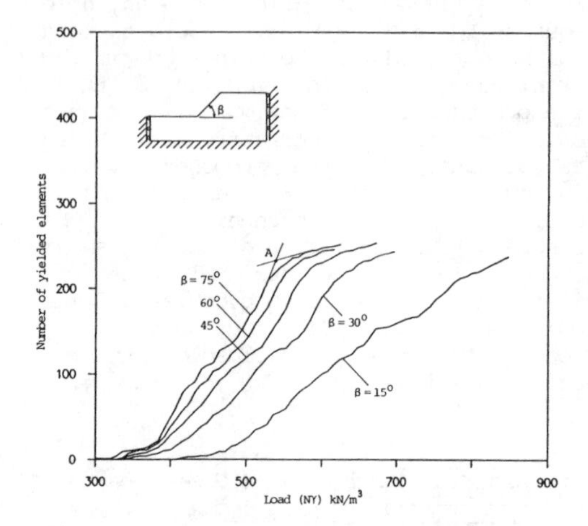

Figure 8. Number of yielded elements and applied load relationship.

Point A denotes the point at which the increase in the number of yielded element drops. The loads at these points for various slopes agree closely with the corresponding collapse loads predicted by observing formation of failure mechanism (Figure 4). A sudden drop in the increase of number of yielded elements may therefore be taken as an indication of formation of failure mechanism. These help to explain the numerical instability in the Smith's approach when the yielding extends much slower after the slope fails and does not spread smoothly through the medium. However, the correlation between the formation of failure mechanism and a drop in the increase of the number yielded elements becomes less obvious when the loading increment $\partial N\gamma$ is large.

4 CONCLUSION

Finite element analysis incorporating various failure criteria described in this paper all give consistant collapse loads for slope stability problems. The results differ by less than 10% provided the load increment is small enough.

A fine mesh is definitely required for

the number of yielded element approach because the accuracy of analysis depends on the smoothless of the spread of the yielded region. However, the fineness of the mesh does not necessarily lead to convergent results irrespective the magnitude of the loading increment.

When using Smith's method, an arbitrary collapse load has to be predicted and previous experience on the technique is necessary.

The yield pattern and number of yielded element methods tend to overestimate the collapse loads as the definition of the load is based on the formation of a failure mechanism. On the other hand, the load-displacement relationship is a straight foreward method for prediction of collapse load although some difficulty may arise when the failure mechanism cannot be clearly identified.

5 REFERENCES

Almeida, M.S.S., Britto, A.M. and Parry, R.H.G. 1986. Numerical modelling of a centrifuged embankment on soft clay. Canadian Geotechnical Journal, Vol.23, 103-114.

Griffiths, D.V. 1980. Finite element analysis of walls, footings and slopes. Proceedings, Symposium on Computer Application in Geotechnical Problems in Highway Engineering, Cambridge.

Hill, R. 1950. The mathematical theory of plasticity. Clarendon Press, Oxford.

Naylor, D.J., Pande, G.N., Simpson, B. and Tabb, R. 1981. Finite elements in geotechnical engineering. Pineridge Press, Swansea, U.K.

Simpson, B. 1973. Finite element computations in soil mechanics, Ph.D. thesis, Engineering Department Cambridge University, Cambridge, England.

Smith, I.M. 1973. Numerical analysis of plasticity in soils. Proceedings, Symposium on the role of plasticity in soil mechanics, Cambridge, 279-289.

Smith, I.M. and Hobbs, R. 1974. Finite element analysis of centrifuged and built-up slopes, Geotechnique, 24, No.4, 531-559.

Smith, I.M. 1982. Programming the finite element method with application to geomechanics, John Wiley & Sons.

Snitbhan, N. and Chen, W.F. 1976. Finite element analysis of large deformation in slopes. Numerical Methods in Geomechanics, ed. C. Desai, 744-758, A.S.C.E.

Taylor, D.W. 1948. Fundamentals of soil mechanics, John Wiley and Sons, New York.

Terzaghi, K. 1943. Theoretical soil mechanics, John Wiley and Sons, New York.

Terzaghi, K. and Peck, R.B. 1967. Soil mechanics in engineering practice, 2nd ed. John Wiley and Sons, New York.

Zienkiewicz, O.C., Valliappan, S. and King, I.P. 1969. Elasto-plastic solution of engineering problems: initial stress, finite element approach. International Journal for Numerical Methods in Engineering, Vol.1, 75-100.

Zienkiewicz, O.C. 1971. The finite element method in engineering science, McGraw Hill, New York.

Zienkiewicz, O.C., Humpheson, C. and Lewis, R.W. 1975. Associated and non-associated viscoplasticity and plasticity in soil mechanics, Geotechnique, 25, No.4, 671-689.

Zienkiewicz, O.C. 1977. The finite element method, 3rd ed. McGraw Hill, New York.

Numerical Methods in Geomechanics (Innsbruck 1988), Swoboda (ed.)
© 1988 Balkema, Rotterdam. ISBN 90 6191 809 X

The nonlinear behaviour of ground anchors and its consideration in finite element analysis of tied-back walls

M.Matos Fernandes
Faculdade de Engenharia, Universidade do Porto, Portugal
J.C.-B.Falcão
Laboratório Nacional de Engenharia Civil, Lisboa, Portugal

ABSTRACT: Over 60 in situ anchor tests are analysed by fitting the load-displacement curves by a hyperbolic model similar to the one used for soils. This allows for an easy quantification of the discrepancies between experimental diagrams and the elastic extension of the free anchor lengths. Finite element analyses of an excavation supported by a tied-back wall are performed, employing the above mentioned model for the anchors. It is concluded that reasonable estimates of anchor loads can only be achieved if anchor behaviour is accurately modelled.

1 INTRODUCTION

Ground anchors have been widely used over the last 25 years, specially in retaining structures for deep excavations in urban areas. In these problems, the use of anchors allows for unobstructed excavations and reduces movements.

On the other hand, the use of finite element (f.e.) models in recent years has allowed for a safer design of those excavations, namely by means of a better understanding of the factors which influence movements and by increasing the designer's ability to predict and control them within acceptable limits.

The use of those numerical models requires realistic values of the parameters which command the mechanical behaviour of the soil and the structure.

The paper analyses the mechanical behaviour of ground anchors with reference to experimental results obtained by the Portuguese National Laboratory for Civil Engineering (LNEC). A method is proposed to incorporate load anchor test results into f.e. models.

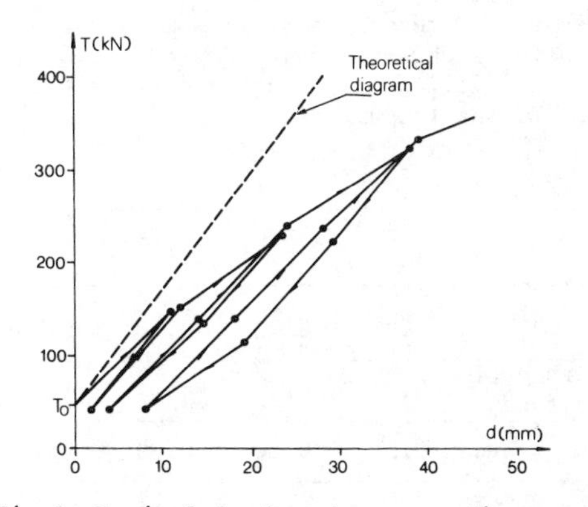

Fig.1 Typical load anchor test diagram

2 LOAD-DISPLACEMENT ANCHORS BEHAVIOUR FROM IN SITU TESTS. HYPERBOLIC APPROXIMATION

Up to 1983, LNEC had carried out about 70 field load tests of ground anchors included in the design (preliminary tests) and construction (acceptance tests) of some important geotechnical works.

Figure 1 shows a typical test diagram relating the displacements, d, measured at the anchor head, to the tensile loads, T. The dashed line corresponds to the elastic deformation of the free portion of the anchor tendon. Its slope represents the theoretical stiffness of the anchor:

$$K_t = \frac{EA}{L} \tag{1}$$

in which E is the Young modulus of the steel, A is the cross sectional area of the tendon and L the so called free anchor length.

An attemp has been made to fit the load-displacement test diagrams by an hyperbolic law similar to that proposed for soils by

Kondner (1963). The approach had already been applied to ground anchors by Evangelista and Sapio (1977) and Davis and Plumelle (1982).

The procedure is as follows:

i) the results from each test are plotted using as axes d and $T'=T-T_o$ (T_o is the initial load usually applied to bed the tendon, Figure 2a);

ii) the experimental points are fitted by a straight line, whose equation is:

$$\frac{d}{T'} = a + bd \qquad (2)$$

iii) the hyperbola which approximates the test diagram (Figure 2b) is then obtained by expressing T' as a function of d:

$$T' = \frac{d}{a+bd} \qquad (3)$$

The parameter a represents the inverse of the initial tangent stiffness:

$$K_i = \frac{1}{a} \qquad (4)$$

As for very low values of T' only the elastic deformation of the free anchor length significantly contributes to the displacements, it is reasonable to expect similar values for K_i and K_t.

On the other hand, b is the inverse of the tensile load for an infinite displacement, that is, the asymptote of the hyperbola:

$$T'_{ult} = \frac{1}{b} \qquad (5)$$

Increasing values of parameter b, therefore mean worse conditions of fixity offered by the grouted length, larger curvature of the test diagram and larger discrepancy between this one and the theoretical line.

By differentiating equation (3), the current tangent stiffness, K_{tg}, is obtained as:

$$K_{tg} = \frac{\partial T'}{\partial d} = (1 - \frac{T'}{T'_{ult}})^2 K_i \qquad (6)$$

This methodology has been applied to the results of 62 load anchor tests performed at 8 different sites. The tests carried out at each site, involving anchors with the same free and grouted lengths, still tendon and initial load, have been analysed togethe[r]. Figure 3 shows the results from some of thes[e] tests as well as the hyperbolic curves obtained as described above. As can be seen the approximation achieved if fairly good.

The tangent stiffness, K_w, corresponding to the design anchor working load has been computed for each case by using equation (6). It should be noted that the anchors tested, with the sole exception of those of Fig. 3b, were permanent anchors. For temporary anchor[s] it is common practice to establish working loads 30 to 50% greater than those for permanent anchors. Therefore, a computation was also made of K_w for a load 1.3 times the one referred to above. So, K_{wp} and K_{wt} will be used to denote the tangent stiffness for the working load, respectively, for permanent and temporary anchors.

The study has also included the analysis of the mechanical behaviour of the anchors for unloading-reloading situations. For this type of load path, the anchors, similarly to the soils, approximately exhibit a linear elastic behaviour, as can be seen in Fig. 1. As the unloading-reloading stiffness K_{ur}, does not seem to depend on the load level at which unloading-reloading cycle is performed, an average value of K_{ur} was adopted for each test or group of tests analysed together.

Table I summarizes the results of the study carried out by comparing the differen[t] anchor stiffness defined in the paper. This table shows some interesting trends.

Looking at the first column, one can notice that the initial tangent stiffness, K_i, may substantially diverge from the theoretical one K_t. When K_i is higher than K_t the reason is usually one of the following: i) difficulty in controlling the length along which the tendon is sealed under pressure to the ground; ii) some shear transfer along the "free length", induced by backfilling the hole without preventing bond to the tendon.

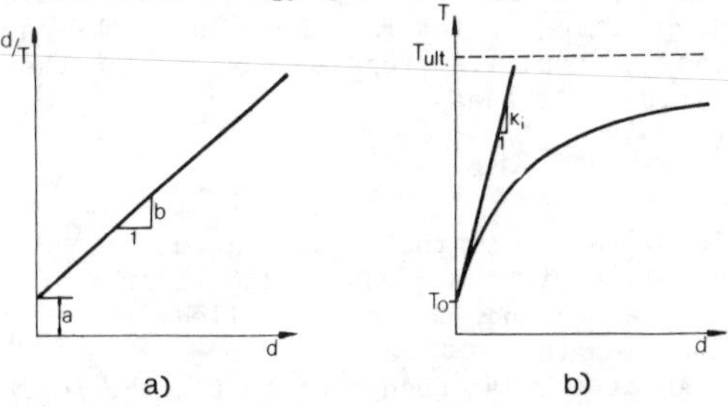

Fig. 2 Hyperbolic approximation to T-d diagram

TABLE I

	K_i/K_t	K_{wp}/K_i	K_{wt}/K_i	K_{ur}/K_i
Average of 62 tests	0.93	0.72	0.64	1.00
Average of 8 sites	0.94	0.70	0.62	1.00
Upper bound	1.27	0.84	0.80	1.14
Lower bound	0.64	0.34	0.21	0.84

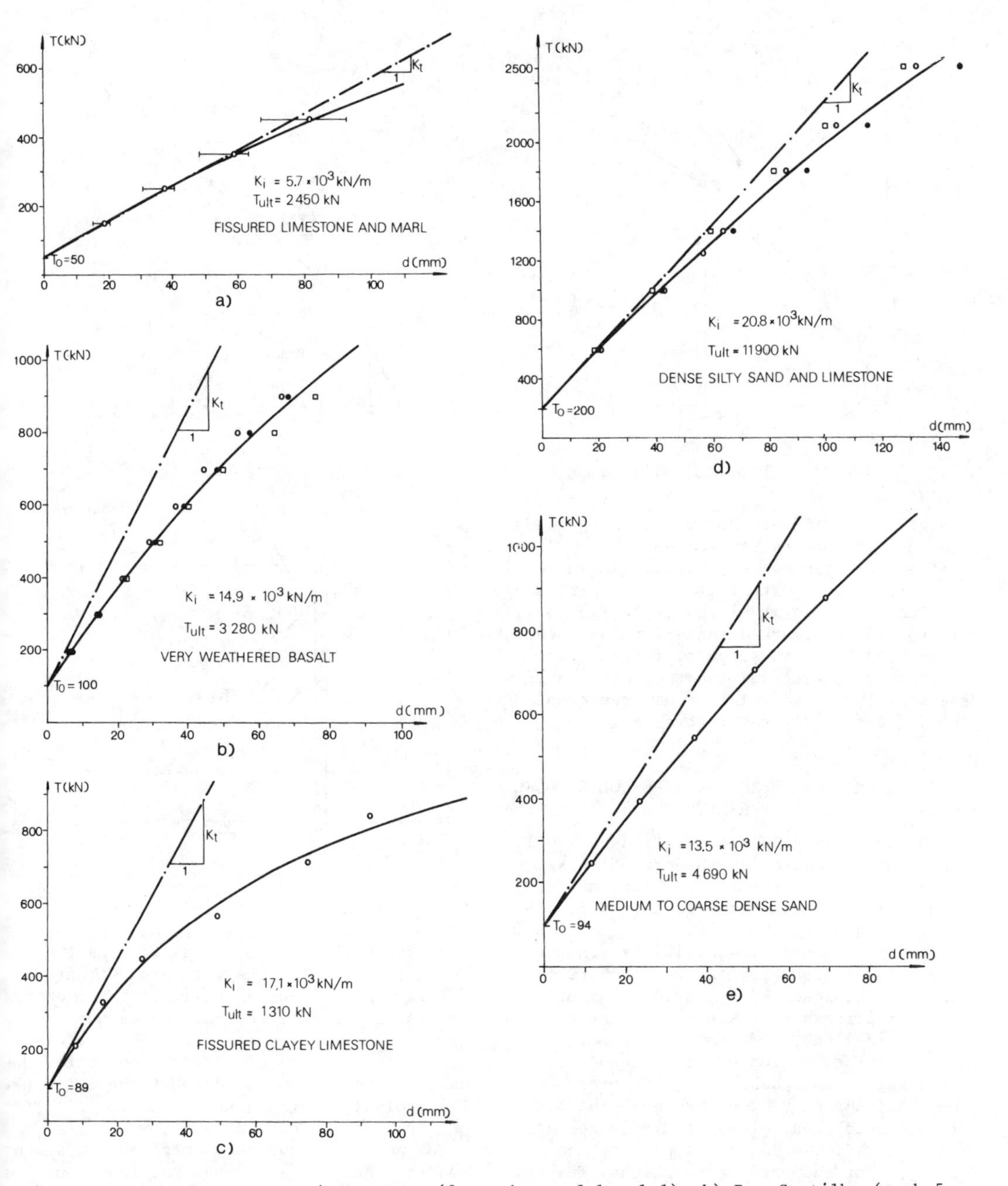

Fig. 3 Load anchor tests: a) Monsanto (8 anchors of level 1); b) Rua Castilho (anch.5, 5A, 18A); c) Casal Ventoso (anch. E1); d) Almada (anch. 27, 37, 49); e) Barreiro (anch. 1)

On average, K_i is seen to be lower than K_t, as should be expected. According to Pinelo and Matos Fernandes (1981), this is due to the (usually small) contribution from the elastic displacements of the grout and of the soil. This conclusion is further supported by the average of K_{ur}/K_i being equal to 1.00 (see column 4). As a matter of fact, if only the elastic displacements are recoverable, and if the unloading-reloading and the initial tangent stiffness coincide, this proves that the latter represents the elastic component of the recorded displacements when the tensile load increases.

Turning now to columns 2 and 3, we detect major differences between (average)

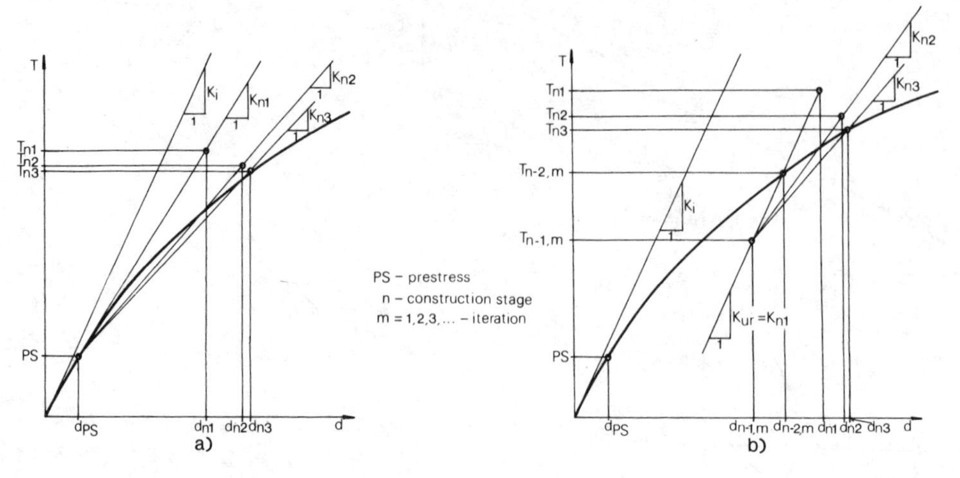

Fig. 4 Iterative process: a) primary loading; b) reloading-primary loading

tangent stiffness for the working loads and K_i. In some extreme cases, those differences can be very high, like in the anchors built to stabilize the natural slope of Casal Ventoso at Lisbon (see Fig. 3c). It should be remarked that if these anchors have been built to support, for example, a temporary excavation, the tangent stiffness at the working load would represent only about 20% of the theoretical one.

3 CONSIDERATION OF ANCHOR HYPERBOLIC MODEL IN FINITE ELEMENT ANALYSES

In excavations supported by tied-back walls, the anchor grouted lengths are often set within the stiff formation which underlies the soft excavated deposit. As the displacements induced by the excavation in the stiff ground are practically negligible, in f.e. analysis of such problems the anchors are usually represented by spring elements whose stiffness is taken equal to the theoretical one.

In order to evaluate the influence of nonlinear anchor behaviour, the proposed hyperbolic anchor model has been introduced in a conventional plane strain incremental nonlinear f.e. model developed for the analysis of supported excavations. The consideration of such a behaviour for the anchors requires, at each loading increment, an iterative process running concurrently to the one necessary to account for the soil and joints nonlinearity. Fig. 4 illustrates the procedure for two types of load paths.

In order to test the procedure, we have considered a symmetric cut, 9.3m deep and 12.0m wide, excavated in a 15.0m thick soft clay deposit. Soil parameters are shown in Figure 5 (the hyperbolic model has also been

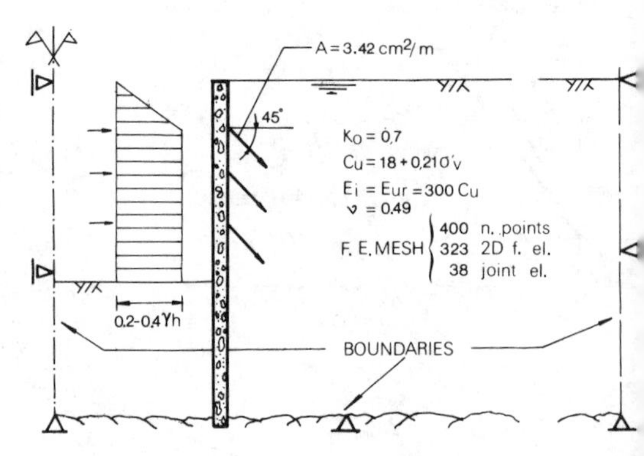

Fig. 5 Numerical case study

used for soil behaviour), as well as the characteristics of the retaining structure, a 0.6m thick diaphragm wall with three anchor levels. It is assumed that the anchor grouted lengths are set into bedrock, which allows for the use of springs connecting the wall to the bottom boundary of the f.e. mesh Trapezoidal diagrams were used to calculate prestressing anchor loads.

The analyses have been performed in seven loading increments or construction steps: 1, 3, 5 and 7, corresponding to the excavation stages, and 2, 4 and 6, concerning to the three anchor levels prestressing.

Two different anchor models have been employed: i) linear elastic ($K = K_t$), denoted by LE; ii) nonlinear hyperbolic ($K_i = K_{ur} = K_t$), for which two choices have been considered: HM (hyperbolic maximum) and HI (hyperbolic intermediate) according the range of experimental results (see Table I) used for the definition of the hyperbola modelling the anchor response for primary loading.

Fig. 6 represents final wall deflections and ground surface settlements computed by adopting a design prestress anchor diagram whose maximum ordinate pressure is 0.2γh. The curves are similar for the three analyses but one can see that the consideration of nonlinear anchor behaviour moderately increases displacements.

Fig. 7 shows the load-displacement diagrams assumed for the first anchor level in analyses LE and HM. The figure also includes the points corresponding to the calculated tensile loads and displacements of the anchor head in the stages subsequent to the prestressing (which is applied at stage 2). It is interesting to see how these points follow the model which had been imposed for T-d diagram in analysis HM: i) at the excavation stages (3, 5 and 7) new maximum values are achieved for the tensile anchor loads, the corresponding points being located on the hyperbola; ii) in stages 4 and 6 prestressing is applied to anchor levels 2 and 3, respectively, an elastic unloading being induced to anchor level 1 in both cases.

As can be easily evaluated from Fig. 7, the differences between maximum final anchor loads are very substantial. This suggests a strong dependence of the loads carried by the anchors on their load-extension relationships.

Three more analyses have been performed by duplicating anchor prestress loads. Fig. 8 illustrates final wall displacements and surface settlements from the analyses. In Fig. 9 one can observe the evolution during the construction of the anchor loads, expressed as a percent of their initial value.

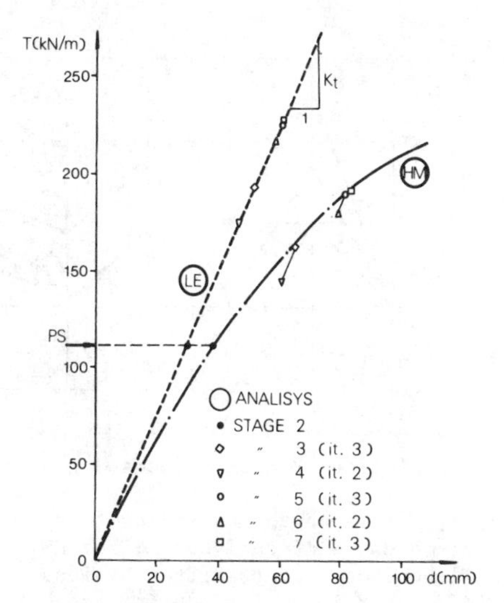

Fig. 7 T-d values and diagrams for the first anchor level (analyses LE and HM, 0.2γh trapezoidal prestress diagram)

The examination of Figures 8 and 9 shows that increasing anchor prestress has allowed for a much better overall performance, namely by reducing soil and wall displacements, as well as the differences between prestress load and maximum working anchor load, which now occurs at the excavation stages imediately after anchor prestressing. This means that, for each anchor level, nonlinear behaviour is now limited to a single construction stage. Nevertheless, the influence of anchor nonlinearity remains relevant because increasing prestress loads reduces the anchor stiffness for the subsequent construction stage (see equation 6).

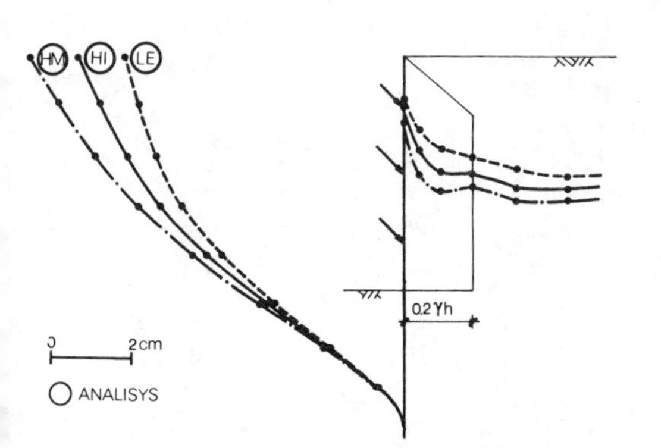

Fig. 6 Final wall deflections and surface settlements (0.2γh trapezoidal prestress diagram)

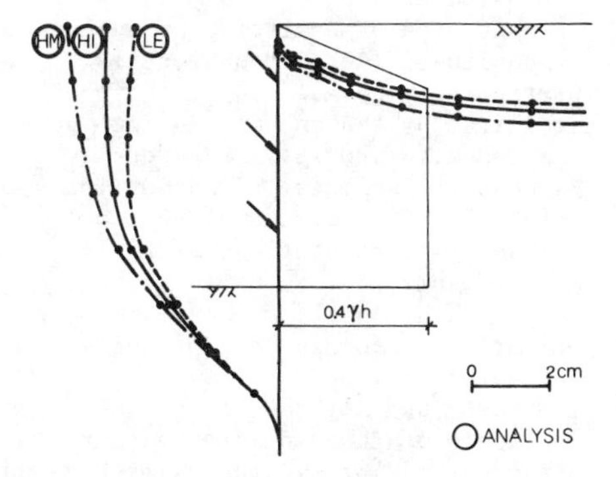

Fig. 8 Final wall deflections and surface settlements (0.4γh trapezoidal prestress diagram)

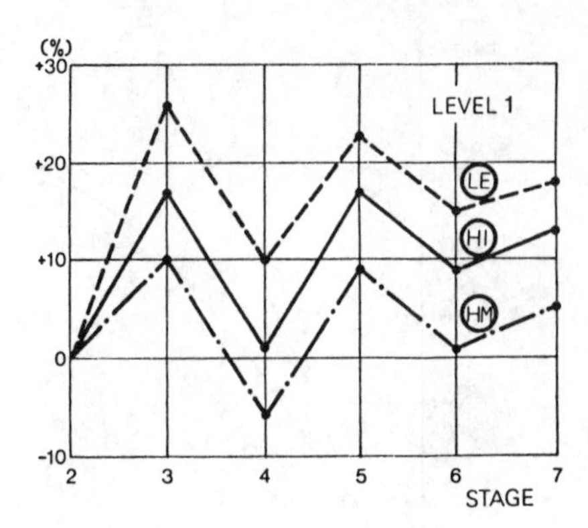

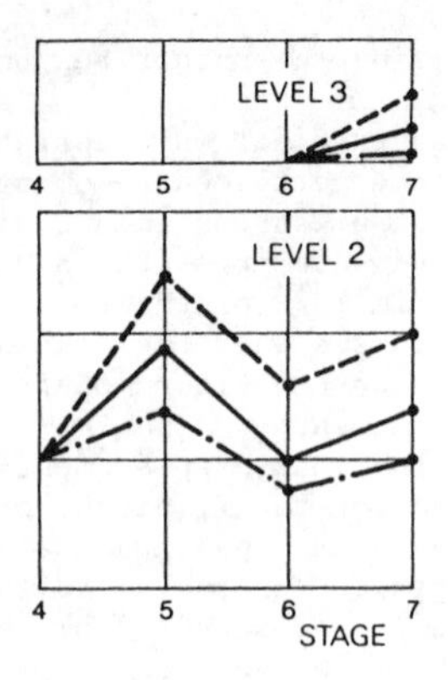

Fig. 9 Evolution of anchor loads during the construction (0.4γh trapezoidal diagram)

Finally, it is worth mentioning that for the extreme cases of analyses LE and HM the wall bending moments differ by less than 10%, regardless of the prestress level selected.

4 CONCLUSIONS

i) Field load tests show that the mechanical behaviour of ground anchors is highly nonlinear, even when they are built into very stiff or dense soils or into rocks of low to medium strength.

ii) The load-displacement behaviour of the anchors may be approximately represented by a hyperbolic model of the type used for nonlinear soil analysis; the initial tangent stiffness is generally close to the theoretical value, corresponding to the elastic extension of the free anchor length.

iii) The above mentioned model allows for a simple quantification of the discrepancies between the actual and the theoretical anchor stiffness, for a given load level; furthermore, it may be easily introduced in f.e. programs for supported excavation analysis.

iv) In this type of analysis, the influence of nonlinear anchor behaviour is particularly relevant upon the loads mobilized by the anchors during the construction process. Therefore, reasonable estimates of anchor loads in preconstruction f.e. analyses, or reliable interpretations of field measurements in f.e. backanalyses, may only be achieved if anchor load-displacement relationships are properly accounted for.

v) The last, but not the least, the influence of the nonlinear anchor behaviour can be substantially reduced by tensioning the anchors to the predicted maximum anchor load before locking the design prestress load.

REFERENCES

Davis, A.G. and Plumelle, C. 1982. Full-scale tests on ground anchors in fine sand. J. Geotech. Eng. Div., ASCE, GT3, pp. 335-353.

Evangelista, A. and Sapio, G. 1977. Behaviour of ground anchors in stiff clays. Revue Française de Géotechnique, No.3, pp. 39-47.

Kondner, R. 1963. Hyperbolic stress-strain response: cohesive soils. J. Soil Mech. Found. Div., ASCE, SM1, pp. 115-143.

Pinelo, A. and Matos Fernandes, M. 1981. Anchor behaviour and tied-back walls analysis. Proc. 10th ICSMFE, Stockholm, Vol. 2, pp. 219-225.

ACKNOWLEDGEMENTS

The authors are indebted to LNEC for providing the necessary support and to Mr. J.M.M.C. Marques for his suggestions about the presentation of the paper.

Numerical Methods in Geomechanics (Innsbruck 1988), Swoboda (ed.)
© 1988 Balkema, Rotterdam. ISBN 90 6191 809 X

Stability of vertical corner cuts

D.Leshchinsky & T.Mullett
University of Delaware, Newark, USA

ABSTRACT: Outlined are the analytical results of a 3-D slope stability method which is based on the variational limit-equilibrium approach. Also outlined is a numerical procedure utilizing these results. Limitations of the variational analysis and of the numerical procedure performance are discussed. Numerical study of the 3-D stability of homogeneous vertical corner cuts with variable angle, considering pore-pressure, is conducted. The results of this study are condensed into design charts enabling one to determine the factor of safety as well as the volume of a potentially sliding mass. Comparison of the variational results with published results obtained from a limit-analysis indicates the former ones to be more critical.

1 INTRODUCTION

Vertical cuts are often required as part of a construction process. These cuts are comprised of straight segments. When intersecting, these segments define corner cuts. A rational and economical design of such corners requires knowledge of their stability against catastrophic collapse.

The most simplified, and perhaps practical, stability analysis is the one based on limit-equilibrium approach. Since the corner is of a 3-D nature, however, a 3-D analysis is required. Furthermore, in some cases cuts are left open over prolonged periods (e.g., open pit mines) justifying an "effective-stress" stability analysis. Leshchinsky and Baker (1986) reviewed the few available 3-D limit-equilibrium methods of slope stability and concluded that each, to a certain extent, appears to be unsatisfactory. The analysis outlined in this work is based on a variational formulation of the limit-equilibrium slope stability problem. In this approach the failure mechanism is unknown a-priori and it is being determined as part of the solution procedure. It should be stated that the original statement of this variational approach was presented by Baker and Garber (1978) and Garber and Baker (1979). This approach was extended to 3-D problems by Leshchinsky et al. (1985).

The outlined analysis is limited to homogeneous soil and symmetrical corners. The significance of the analysis as well as the existence limit of the numerically derived results are clearly stated. The results are presented in a condensed form of stability charts which enable one to estimate the safety factor and the volume of the potentially sliding mass as a function of the corner's angle, strength parameters of the soil and an assumed homogeneous pore pressure ratio r_u.

2 VARIATIONAL LIMITING-EQUILIBRIUM

2.1 Results of Analysis

The following is just an outline of the results obtained from a 3-D variational analysis. The mathematical details of the analysis are presented elsewhere [Leshchinsky et al., (1985) combined with Leshchinsky and Baker (1986)].

It is assumed that the linear Mohr-Coulomb failure criterion is valid. Thus, the common approximation of mobilized shear strength can be stated as

$$\tau = [c' + (\sigma - u)\psi]/F_s \quad (1)$$

where c' and ϕ' ($\psi = \tan\phi'$) are the effective cohesion and internal angle of friction, respectively; σ and u are the total normal stress and pore-pressure, respectively; τ is the mobilized shear strength at a point; and F_s is the strength reduction coefficient termed the factor of safety.

To invoke the limit-equilibrium approach one assumes that there will develop a 3-D

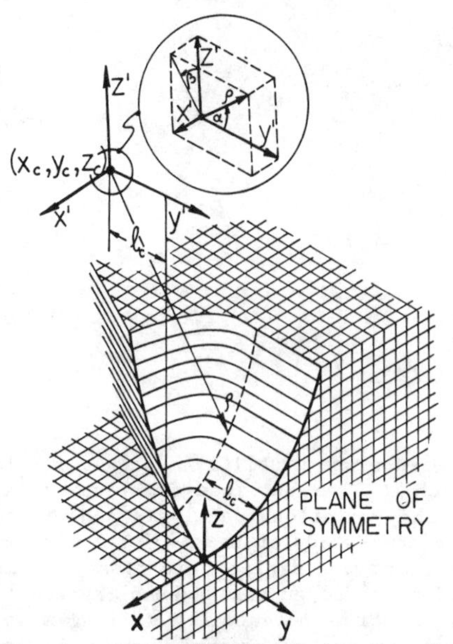

Fig. 1. Notation Used in the Vertical Corner Cut Problem.

surface over which Eq. (1) will realize; i.e., over this surface F_s is constant. Combining this assumption with the limiting equilibrium equations produces a problem which includes three unknown functions: the slip surface function z(x,y), the direction of shear stress τ acting over this surface $\theta(x,y)$, and normal stress function $\sigma(x,y)$. Utilizing a variational technique, in which the functions z(x,y), $\theta(x,y)$ and $\sigma(x,y)$ that produce $\min(F_s)$ are sought, Leshchinsky and Baker (1986) obtained the following 3-D failure mechanism:

$$\rho = \begin{cases} A \exp(-\psi_m\beta)/\sin\alpha \text{ for } |y|\le\ell_c & \text{(2a)} \\ A \exp(-\psi_m\beta)\cdot\sin\alpha \text{ for } |y|\ge\ell_c & \text{(2b)} \end{cases}$$

where $\rho=\rho(\alpha,\beta)$ is the slip surface in spherical coordinates (ρ,α,β), analogous to z(x,y), see Fig. 1; A is an unknown constant; $\psi_m = (\tan\phi')/F_s$; and y is the ordinate illustrated in Fig. 1, where y=0 defines the symmetry plane for the problem. Eq. (2a) defines a cylinder, $2\ell_c$ long, while Eq. (2b) symbolizes end caps attached smoothly to the central cylinder. The function $\rho(\alpha,\beta)$ is a 3-D expansion of the 2-D log-spiral family of slip surfaces.

The direction of elemental shear stress acting over ρ is

$$\tan\theta = \begin{cases} \dfrac{-\psi_m \sin2\alpha}{\psi_m\cos2\alpha \sin\beta+\cos\beta} & \text{over end caps (3a)} \\ 0 & \text{over cylinder (3b)} \end{cases}$$

where $\theta=\theta(\alpha,\beta)$ is measured on a horizontal plane z=const. as shown in Fig. 2b. It should be pointed out that since each elemental shear force opposes the sliding body movement, $\theta(\alpha,\beta)$ can be viewed also from a kinematic standpoint thus constituting part of the failure mechanism.

Attempts to determine $\sigma(x,y)$ failed because of overcomplicated expressions. This, however, is not an obstacle in estimating F_s via a numerical procedure as is shown later on.

2.2 Significance of Results

The failure mechanism [Eqns. (2) & (3)] is a result of variational extremization of a safety functional. The minimal value of this functional is the factor of safety and it corresponds to a single critical surface. Ideally, this critical surface should produce a genuine minimum considering infinite possible mechanisms. However, some researchers [e.g., DeJong (1980,1981), Luceno and Castillo (1981)] have questioned the existence of minima in the 2-D variational limit-equilibrium analysis as formulated by Baker and Garber (1977,1978). Furthermore, it has been shown [e.g., Castillo and Luceno (1983), Leshchinsky et al., (1985)] that variational extremization of limit-equilibrium problems is equivalent to a rigorous determination of an upper-bound in the strict framework of limit-analysis. Consequently, although the variationally obtained failure mechanism signifies a stationary value, among other possible 3-D failure mechanisms it may actually yield a "saddle" value for F_s.

Under the above mathematical circumstances there is clearly a question as to the practical justification of using the variational analysis. The following provides such justification. Baker (1981) showed that the 2-D variational results are identical to Taylor's (1948) which, in turn, are considered acceptable by practitioners. Moreover, it has been demonstrated [e.g., Chen et al. (1969,1971)] that good agreement exists between results obtained from limit-analysis and from well-established limit-equilibrium techniques for 2-D problems. It is expected, therefore, that a rigorous extension of a limit-analysis approach to include the third dimension, will yield also reasonable results. The variational approach is particularly valuable in 3-D problems where an intuitive choice for a failure mechanism, typically done in the conventional limit-equilibrium analysis, is much more difficult then in the 2-D case. Most importantly, Leshchinsky and Baker (1986) compared predictions based on the variational approach with those obtained from other 3-D limit-equilibrium methods and showed that essentially, for $c-\phi$ soil, the variational F_s was the smallest, exhibiting consistent behavior when boundary problems were analyzed.

3 OUTLINE OF NUMERICAL PROCEDURE

Based on Eqns. (1) thru (3), the moment limit-equilibrium equation for the sliding mass, written about the axis of rotation (see Fig. 1), can be assembled rigorously without specifying the normal stress function $\sigma(x,y)$. This phenomenon is typical to log-spiral slip surfaces. Using this moment equation [e.g., Leshchinsky and Baker (1986)], the mobilized cohesion can be written as follows

$$c'_m = \left\{\gamma\left[\iint_{D_1} M_{12}dxdy + \iint_{D_2} M_{22}dxdy\right] + \psi_m\left[\iint_{D_1} uM_{11}dxdy + \iint_{D_2} uM_{21}dxdy\right]\right\} \Big/ \left\{\iint_{D_1} M_{11}dxdy + \iint_{D_2} M_{21}dxdy\right\} \quad (4)$$

where c'_m is the mobilized cohesion (i.e., $c'_m = c'/F_s$); γ is the unit weight of the soil; u is the pore pressure at (x,y); D_1 and D_2 are the projections of the cap and cylindrical surface portions, respectively, onto the x-y plane; and

$$M_{11} = \frac{\exp(-\psi_m\beta)\ \sin^2\alpha}{(\cos^2\alpha - \sin^2\alpha)\ \cos\beta + \psi_m\ \sin\beta} \quad (5a)$$

$$M_{12} = [A\ \exp(-\psi_m\beta)\ \sin^2\alpha\ \cos\beta - (Z-z_c)]\ \exp(-\psi_m\beta)\ \sin^2\alpha\ \sin\beta \quad (5b)$$

$$M_{21} = \frac{\exp(-\psi_m\beta)}{\psi_m\ \sin\beta - \cos\beta} \quad (5c)$$

$$M_{22} = [A\ \exp(-\psi_m\beta)\ \cos\beta - (Z-z_c)]\ \exp(-\psi_m\beta)\ \sin\beta \quad (5d)$$

and $Z = Z(x,y)$ is the corner surface.

For a given problem [i.e., Z(x,y), u(x,y), γ, ϕ'_m ($\psi_m = \tan\phi'_m$)] and an arbitrarily selected slip surface (i.e., A, x_c, y_c, z_c), Eq. (4) enables one to determine the required mobilized cohesion c'_m. The limiting equilibrium analysis, however, seeks the minimum factor of safety, F_s, produced by all possible potential slip surfaces for a given problem. Since c'_m equals c'/F_s, minimization of F_s is equivalent to maximization of c'_m. Hence, Eq. (4) can be rewritten in the following symbolic format

$$c'_m = \max_{(A,x_c,y_c,z_c)} f(A,x_c,y_c,z_c | Z,u,\phi'_m,\gamma) \quad (6)$$

Eq. (6) states that the maximum value of c'_m is sought for all possible slip surfaces and for a given corner, pore pressure, γ and ϕ'_m. Such a statement is particularly convenient when production of stability charts is desired. The safety factor for an actual problem (i.e., for given Z(x,y), u(x,y), γ, ϕ', c') can be estimated using these charts by trial and error.

The numerical steps required in executing the statement in Eq. (6) are, in principle, as follows:

a) Assume the value of A, x_c, y_c and z_c.

b) Carry out a numerical integration of the terms shown in Eqns. (4) and (5), and determine the corresponding c'_m. Note that a mixed coordinate system is involved.

c) Change the values of A, x_c, y_c and z_c and repeat steps (a) and (b) until maximum c'_m is located.

The results presented in the next section were obtained using the above outlined numerical procedure.

4 RESULTS

4.1 Normalization

To enable an efficient results presentation, the following nondimensional parameters are used

$$N_m = \frac{1}{\gamma H} \frac{c'}{F_s} \quad (7a)$$

$$r_u = \frac{u}{\gamma h} \quad (7b)$$

where N_m and r_u are the well known stability number and pore pressure ratio, respectively; H is the height of the corner cut; and h is the depth of the point (x,y) on the slip surface as measured from the corner surface [i.e., from Z(x,y)]. It can be verified that through substitution of Eq. (7) into (4) and considering Eq. (6), one can express, for the critical case, N_m as function of ϕ'_m, r_u and ζ where 2ζ is the central angle of the vertical corner cut (see Fig. 1).

The design value of the ratio r_u, assumed to be constant, has significant influence on stability; however, based on the current state of knowledge, its determined value

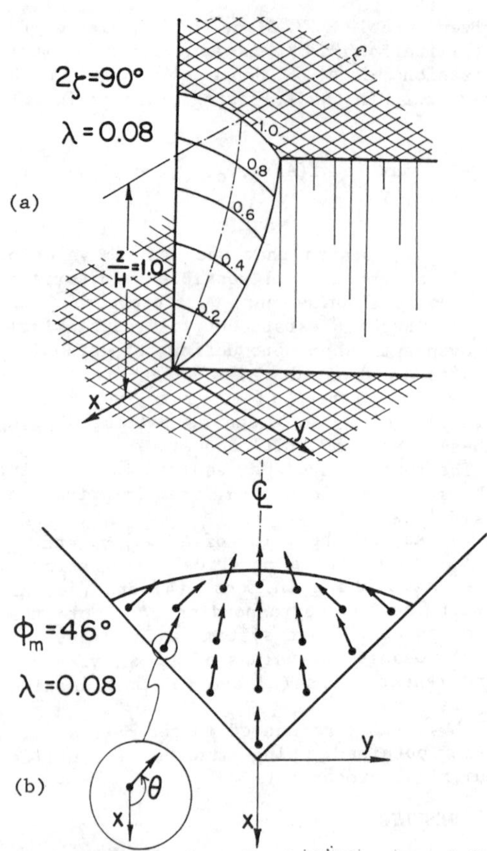

(a)

(b)

Fig. 2. Slip Surface for λ=0.08 and r_u=0.0: (a) 3-D View; (b) Projection of Shear Resisting Force Vectors.

for the present problem can only be speculated. In any event, inclusion of r_u in the results gives a rather comprehensive insight into the solution behavior.

4.2 Design Charts

Fig. 2a illustrates a 3-D view of a typical slip surface for r_u=0 and 2ζ=90°. Fig. 2b shows the plan view of this surface and the direction θ. Fig. 3 presents the plan view of various slip surfaces as a function of λ (again r_u=0 and 2ζ=90°). It is interesting to note that in all analyzed cases for which 2ζ<180°, the critical slip surface was comprised of a cylinder with end caps attached to it; i.e., not a cylinder or a cap alone.

Figs. 4 thru 7 are design charts relating N_m, the volume of the potentially sliding mass, ϕ'_m, ζ and r_u. Values in these charts constitute solution of Eq. (6). Utilization of the charts requires a trial and error approach. As an example consider the

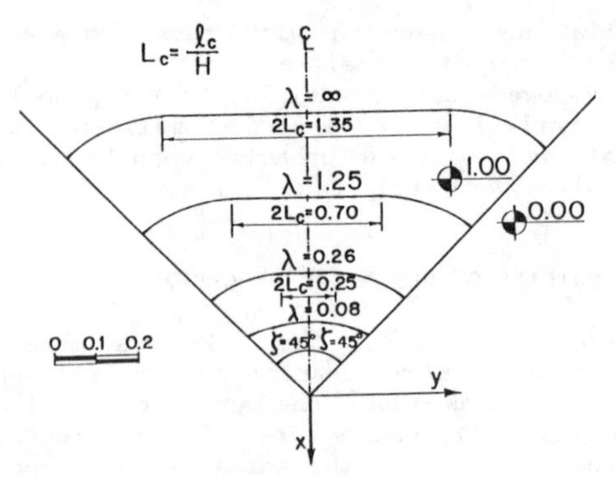

Fig. 3. Plan View of Slip Surface as a Function of λ.

following problem: 2ζ=90°, γ=18kN/m^3, H=5m, ϕ'=25°, c'=18 kPa and r_u is estimated to equal 0.1; determine F_s. Assuming F_s=1.0, one calculates ϕ'_m=25° and N_m=18/(18•5)=0.2. Using Fig. 5 it follows that the point defined by (25°,0.2) is way above the curve for 2ζ=90°. By trial and error it can be verified that only when F_s=1.3, the defined point (i.e., ϕ'_m=19.7° and

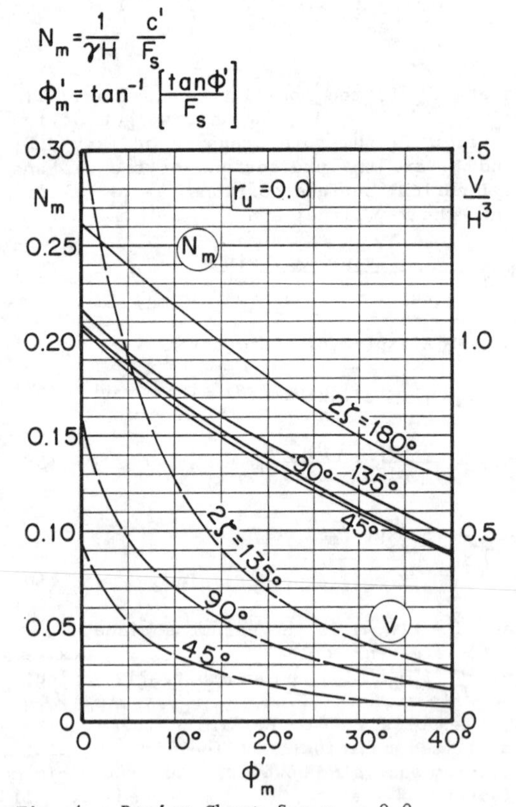

Fig. 4. Design Chart for r_u = 0.0.

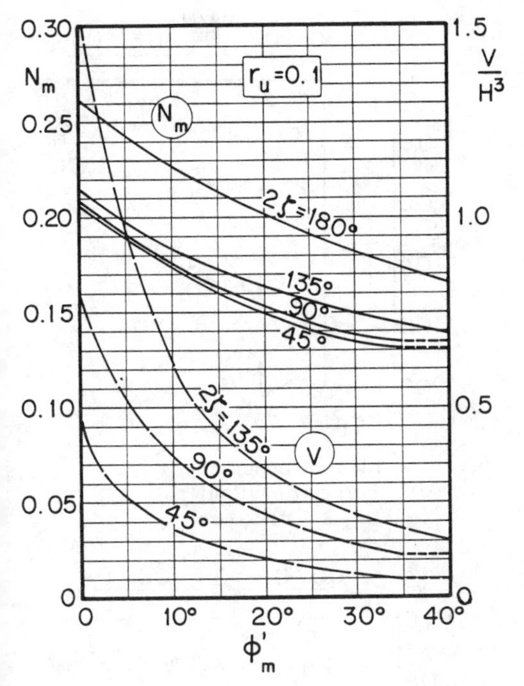

Fig. 5. Design Chart for r_u = 0.1.

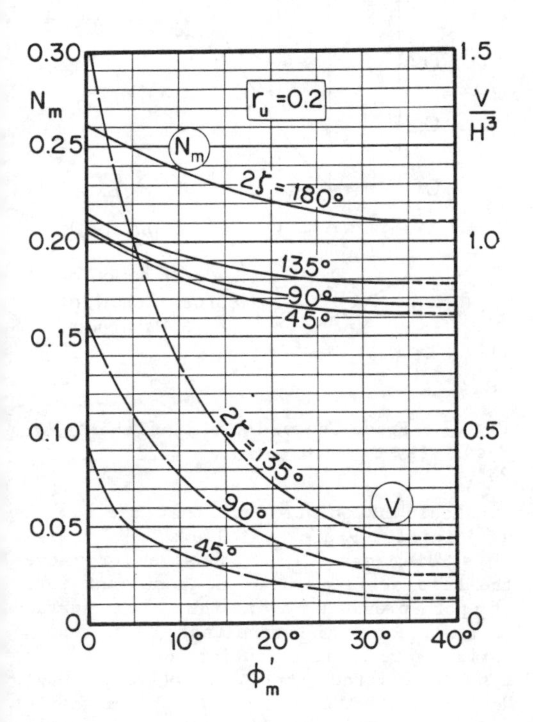

Fig. 6. Design Chart for r_u = 0.2.

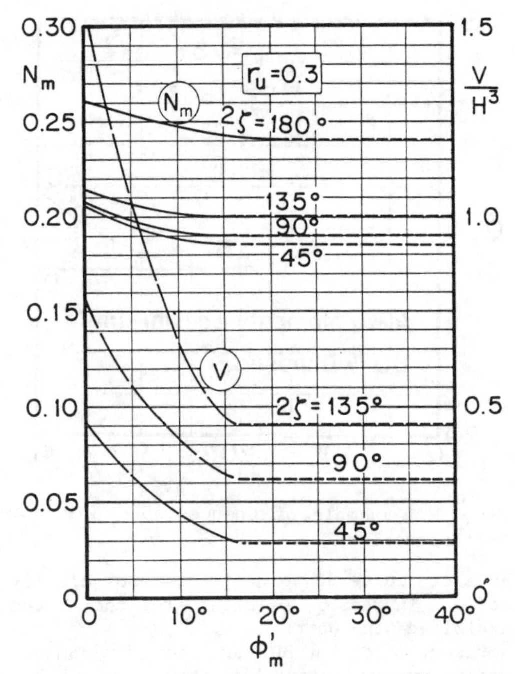

Fig. 7. Design Chart for r_u = 0.3.

N_m=0.1538) lays on the curve. From Fig. 5 it also follows that (v/H^3) = 0.22; i.e., the volume of the potentially sliding mass equals 27.5 m^3. It should be pointed out that estimation of the volume of potential failure is essential in any serious risk cost-analysis of earth structures.

Notice that the design charts include also values for 2ζ=180°. This case signifies actually a 2-D boundary and therefore no volume is associated with it. In fact, a slip surface comprised on a very long cylinder relative to the end caps was analyzed and it can be verified that the computed values of N_m as function of ϕ_m for r_u=0.0 are identical to Taylor's.

4.3 Existence Limit

Observing Figs. 5 thru 7 one sees that some of the curves become horizontal. These horizontal portions are marked by short dashed lines and are more frequent as r_u and ϕ'_m increase. These segments do not signify a solution of Eq. (6) but are arbitrarily imposed. For the sake of discussion consider Fig. 8 which represents essentially a 2-D problem (i.e., 2ζ=180°) and r_u=0.3. Curve ABC on this figure constitutes a solution of Eq. (6). Clearly, portion BC signifies a paradoxical situation; i.e., as ϕ'_m increases the required c'_m (or N_m) increases too. One would expect,

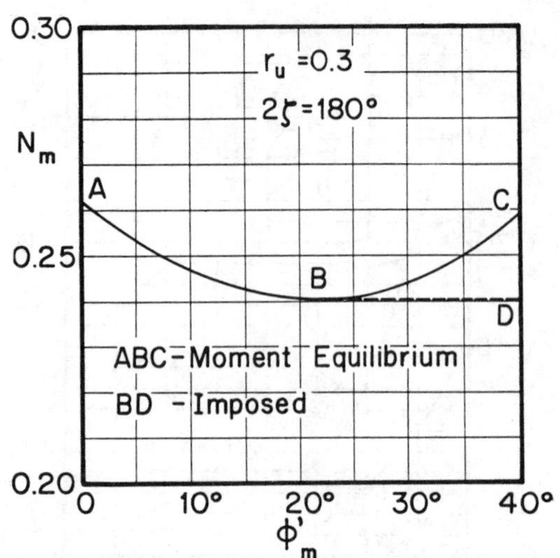

Fig. 8. Example of Unacceptable Solution Performance.

however, curve BC to decrease monotonically as ϕ'_m increases, similar to the trend exhibited in portion AB. Imposing the horizontal segment BD therefore is a conservative approximation to the expected behavior. Of course, an explanation for the solution's awkward behavior and the subsequent elimination of curve BC is required.

Recall that points on curve ABC were essentially obtained through the satisfaction of the moment limit-equilibrium equation written about the axis of rotation. Once the critical surface has been determined, however, one can define the corresponding global force polygon using the principles of statics; e.g., see Fig. 9a.

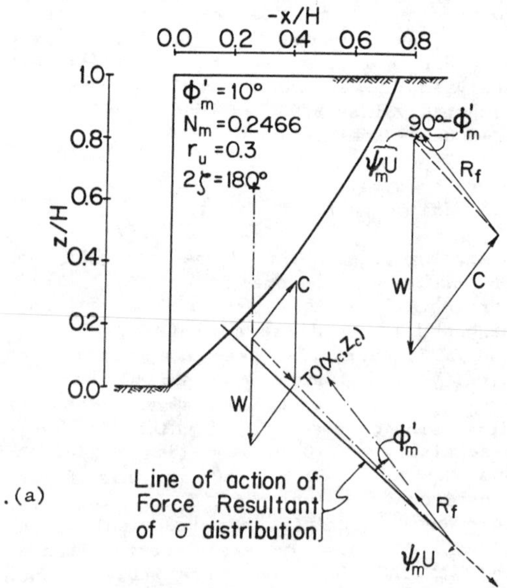

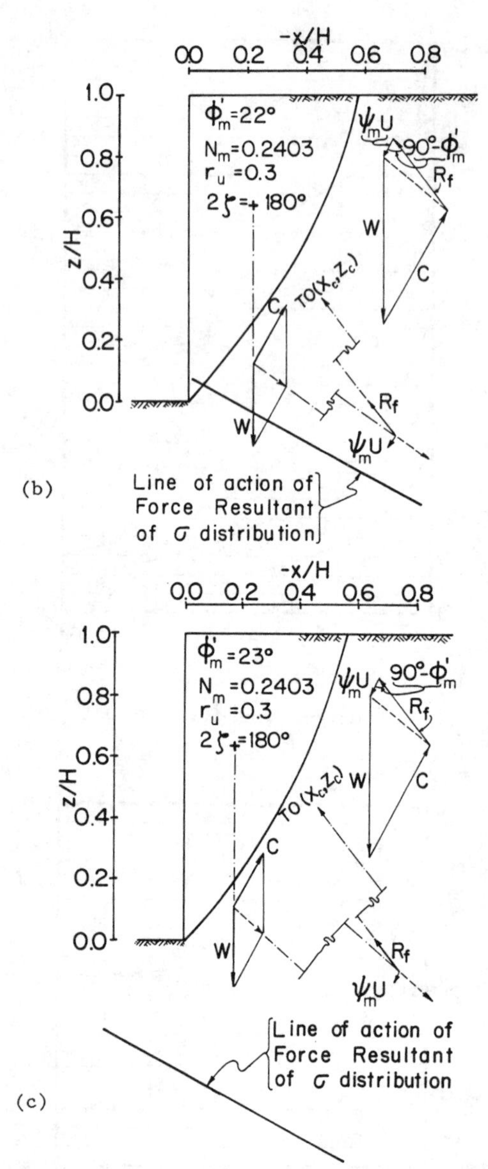

Fig. 9. Force Vectors Acting on the Sliding Mass: (a) ϕ'_m = 10°; (b) ϕ'_m = 22°; (c) ϕ'_m = 23°.

It should be pointed out that the symbols used in Fig. 9 are as follows: W = weight of sliding mass, C = cohesive force over the slip surface, U = the resultant force of pore-pressure over the slip surface ($U=r_uW$), R_f = the resultant force due to normal stress distribution (i.e., σ) and its associated friction (i.e., $\psi_m\sigma$). Notice that the vector ($-\psi_m U$) signifies reduction in shear strength as basically stated in Eq. (1) [i.e., $(\sigma-u)\psi_m$] and is inclined at $(90°-\phi'_m)$ to R_f. As can be seen from Fig. 9a, for the case of ϕ'_m=10° the line-of-action of the resultant force resulting from σ distribution (inclined at ϕ'_m relative to R_f), represents a feasible normal stress distribution. Fig. 9b indicates that as ϕ'_m increases the line-of-action approaches the toe. Fig. 9c shows that increasing ϕ'_m from 22° to 23° results with a line-of-action which exits the sliding mass and thus represents a resultant of impossible distribution of σ. Consequently, although points on curve BC in Fig. 8 satisfy moment equilibrium, force equilibrium then is unfeasible (i.e., requiring impossible distribution of σ) and therefore, this portion of the curve should be eliminated.

For simplicity, Figs. 8 and 9 represent a 2-D boundary problem. It can be verified, however, that the conclusion regarding possible statical inadmissibility of the results holds also for symmetrical 3-D problems.

4.4 Comparison

Fig. 10 shows predictions by the variational analysis and by limit-analysis which employs an intuitively selected 3-D admissible failure mechanism as introduced by Giger and Krizek (1976). The variational results are more critical (i.e., for a given ϕ_m the variationally required N_m is greater), especially as 2ζ increases. The conclusion of this comparison supports the arguments raised in the section "Significance of Results."

5 CONCLUSION

Outlined are the analytical results of a 3-D slope stability method which is based on the variational limit-equilibrium approach. Also outlined is a numerical procedure utilizing these results. The variational method is equivalent to an upper-bound approach in the framework of limit-analysis where a rigid body movement is considered. Consequently, a discussion regarding the practical significance of the results is presented.

The variational approach was applied to study the stability of homogeneous vertical corner cuts with variable corner's angle. Subsequently stability charts which enable the prediction of the factor of safety and the volume of the potentially sliding mass, as function of ϕ', c', γ, H, r_u and ζ, are presented. It is shown that the utilized numerical procedure may yield inadmissible results. A conservative remedy to this problem is suggested.

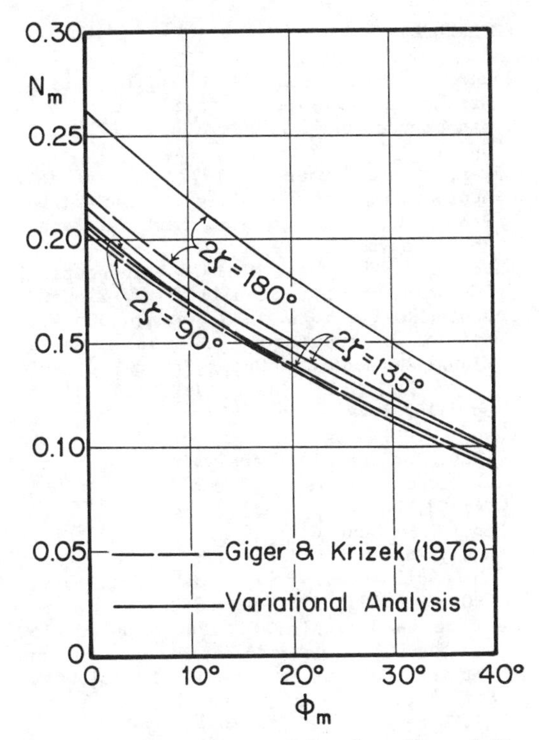

Fig. 10. Comparison of Results (r_u = 0.0).

Results indicate that as ζ decreases, stability increases. However, a typical consequence of construction is disturbance of soil, especially in sharp corners. Such disturbance may lead to a reduced strength resulting in decreased stability.

The presented analysis can be modified to deal with situations such as external loads, nonhomogeneous pore-water pressure distribution, cut of limited extent [e.g., Leshchinsky and Mullett, (1987)], etc. Modifications are straightforward since the failure mechanism remains unchanged; only the moment equilibrium requires adequate alteration.

ACKNOWLEDGMENT

This material is partially based upon work supported by the National Science Foundation under Grant No. ECE-8503572. Ms. T. L. Mullett's help with some of the computer runs is appreciated.

REFERENCES

Baker, R. 1981, "Tensile Strength, Tension Cracks, and Stability of Slopes," Soils and Foundations, JSSMFE, Vol. 21, No. 2, pp. 1-17.

Baker, R. and Garber, M. 1977, "Variational Approach to Slope Stability," Proc. of the 9th Intl. Conf. on Soil Mech. & Found. Eng., Tokyo, Vol. 2, pp. 9-12.

Baker, R. and Garber, M. 1978, "Theoretical Analysis of the Stability of Slopes," Geotechnique, Vol. 28, No. 4, pp. 395-411.

Castillo, E. and Luceno, A. 1983, "Variational Methods and Upper Bound Theorem," J. of Eng. Mech., ASCE, Vol. 109, No. 5, pp. 1157-1174.

Chen, W.F., Giger, M.W. and Fang, H.Y. 1969, "On the Limit Analysis of Stability of Slopes," Soils and Foundations, JSSMFE, Vol. 9, No. 4, pp. 23-32.

Chen, W.F. and Giger, M.W. 1971, "Limit Analysis of Stability of Slopes," J. of the Soil Mech. and Found. Eng. Div., ASCE, Vol. 97, SM1, pp. 19-26.

De Jong De Josselin, G. 1980 "Application of the Calculus of Variations to the Vertical Cut-off in Cohesive Frictionless Soil," Geotechnique, Vol. 30, No. 1, pp. 1-16.

De Jong De Josselin, G. 1981, "Variational Fallacy," Geotechnique, Vol. 31, No. 4, pp. 289-290.

Garber, M. and Baker, R. 1979, "Extreme-Value Problems of Limiting-Equilibrium," J. of the Geotechnical Eng. Div., ASCE, GT-10, pp. 1155-1172.

Giger, M.W. and Krizek, R.J. 1976, "Stability of Vertical Corner Cut with Concentrated Surcharge Load," J. of the Geotechnical Eng. Div., ASCE, Vol. 102, GT-1, pp. 31-40.

Leshchinsky, D., Baker, R. and Silver, M.L. 1985, "Three-Dimensional Analysis of Slope Stability," Intl. J. for Num. and Ana. Methods in Geomechanics, Vol. 9, pp. 199-223.

Leshchinsky, D. and Baker, R. 1986, "Three-Dimensional Slope Stability: End Effects," Soils and Foundations, JSSMFE, Vol. 26, No. 4, pp. 155-167.

Leshchinsky, D. and Mullett, T.L. 1987, "Design Charts for Vertical Cuts," J. of the Geotechnical Eng. Div., ASCE, in press.

Luceno, A. and Castillo, E. 1981, Discussion of Garber and Baker "Extreme-Value Problems of Limiting-Equilibrium," J. of the Geotechnical Eng. Div., ASCE, Vol. 107, pp. 118-121.

Taylor, D.W. 1948, Fundamentals of Soil Mechanics, John Wiley and Sons.

Numerical Methods in Geomechanics (Innsbruck 1988), Swoboda (ed.)
© 1988 Balkema, Rotterdam. ISBN 90 6191 809 X

Consolidation analysis of an experimental fill in a reclamation area

R.F.Azevedo & W.T.Pinto
Pontifícia Universidade Católica do Rio de Janeiro (PUC/RJ), Brazil

ABSTRACT: As part of a large urbanization scheme, circular artificial islands were hydraulically sandfilled at the border of Itaipu lagoon, near Rio de Janeiro city in Brazil. To investigate the mode of settlement of the soft peaty deposit where the islands were to be created, an instrumented test fill was constructed. This paper compares the observed fill behaviour which results obtained using the classical one-dimensional Terzaghi's theory, the finite strain one-dimensional consolidation theory proposed by Gibson et al. (1981) and the two-dimensional coupled consolidation theory developed by Biot's (1941).

1 INTRODUCTION

In 1977 a large brazilian construction company started developing a reclamation area near Niteroi city, Rio de Janeiro, Brazil. The construction consisted of making circular islands in the border of Itaipu lagoon by using a technique in which, initially, the whole reclamation area was covered by a fill hidraulically made by dredging sand from the bottom of the lagoon and, susequently, forming the islands by excavating channels between them while retaining the excavated slopes with crib-walls (Figure 1).

The construction started but two problems arose. First, the planned fill elevation was not being achieved at the desired time due to settlement of the underlying soft layer. Second, during the excavations, the island slopes failed before reaching the planned channel depth, preventing the crib-wall instalations.

The Geotechnical Group of the Civil Engineering Department of PUC/RJ was then invited to study these problems. An extensive research program was established involving the construction of an experimental fill for settlement and progress of consolidation studies, an experimental unsupported excava-

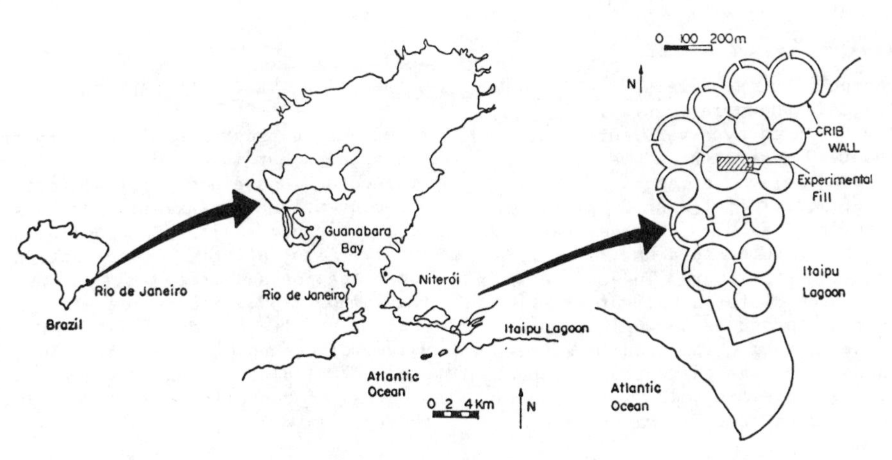

Figure 1 - General View.

tion taken to failure for strength studies, and laboratory and field tests (Russo Neto, 1980; Silva, 1979; Pinheiro, 1980; Carvalho 1980).

The observed experimental fill behavior was published by Sandroni et al. (1981), whereas the whole reclaimed area construction and research as well as the behavior of the experimental excavation were published by Sandroni et al. (1984).

Einsentein and Sandroni (1979) published an analysis of settlement and progress of consolidation of reclamation fills. They proposed using conventional Terzaghi's theory with a modified value of c_v (coefficient of consolidation) to account for the loading variation with time due to sinking below water of the fill. They also considered the drainage path decreasing with time due to large deformations. At the end, an example of application was made using data corresponding to the Itaipu experimental fill. No comparisons were made between field and analytical results.

This paper presents another analysis of the experimental fill, using conventional Terzaghi's theory without any modification, a simplified version of the finite strain theory (Gibson et al., 1967) proposed by Gibson et al. (1981) and a two-dimensional finite element implementation of the coupled consolidation theory (Biot, 1941). Laboratory results available were used to calibrate all these theories. Comparisons between field and analytical results with the different theories are made and discussed.

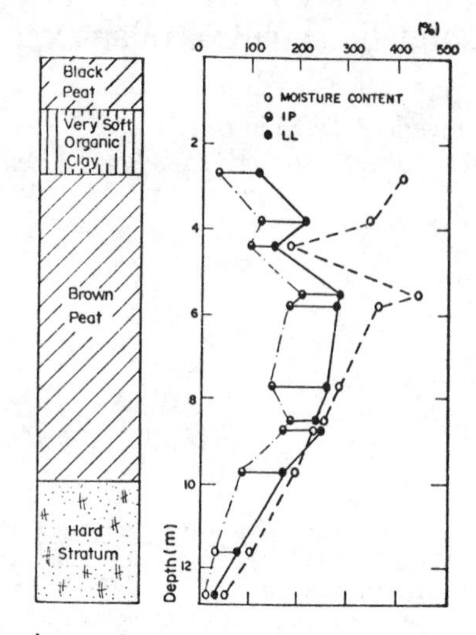

Figure 2 - Geotechnical Profile.

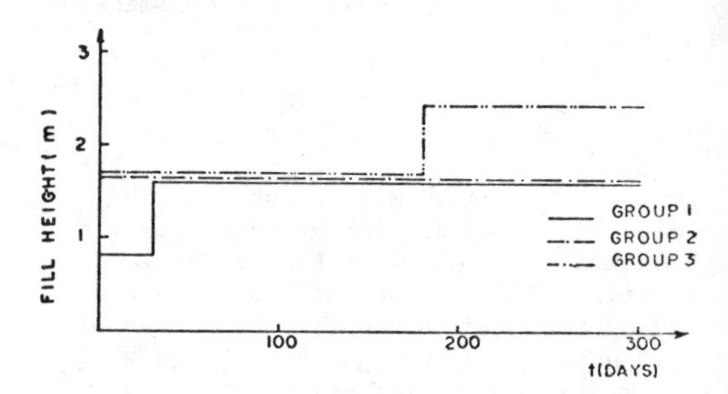

Figure 3 - Fill height vs. time curves.

2 THE EXPERIMENTAL FILL

Most of the information given in this section were obtained in the publications mentioned above and will be briefly repeated here only to insure completeness of the paper.

The geotechnical profile at the experimental site is shown in Figure 2. It is basically composed by a very soft peaty layer about 10 meters thick overlying a harder sandy-clay stratum.

The experimental fill was 50 meters wide and 150 meters long. Its thickness was 2.7 meters in the middle and about 1.7 meters in the rest of its length. To achieve this thickness three different height vs. time curves were observed as shown in Figure 3.

The instrumentation included 14 settlement plates placed at the top of the peat layer, 5 open-tube piezometers, 7 magnetic settlement indicators, one water-level observation well and a deep-seated benchmark (Figure 4). Only movements of the settlement plates are used for comparisons in this paper.

3 LABORATORY TESTS AND CALIBRATION

An extensive program of laboratory tests was carried out involving conventional drained and undrained triaxial tests and non-conventional triaxial stress-path tests (Pinheiro, 1980; Sandroni et al., 1984). Also, a series of conventional consolidation and creep tests was performed (Carvalho, 1980; Sandroni et al., 1981).

Using results of consolidation tests, values of m_v and c_v were evaluated and used to calculate the final settlement and progress of consolidation by Terzaghi's theory (Table 1).

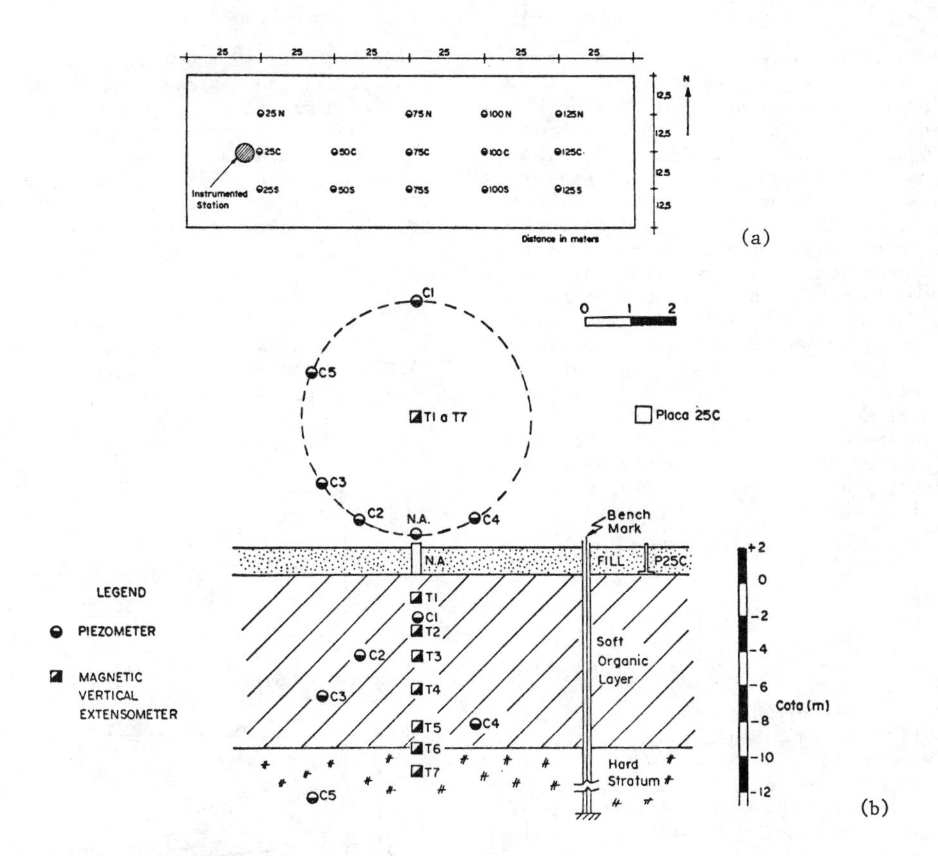

Figure 4 - Fill Instrumentation (a) Top View. (b) Cross-Section.

For Biot's analysis, the parameter E was obtained with the expression:

$$E = \frac{(1 + \nu)\,(1 - 2\nu)}{m_v\,(1 - \nu)} \qquad (1)$$

in which ν was assumed equal to 0.35, and k was obtained according to:

$$k = m_v \cdot c_v \cdot \gamma_w \qquad (2)$$

The values of these three parameters are also presented in Table 1.

TABLE 1

water content	w = 240%
total unit weight	γ_t = 1.15 t/m³
water unit weight	γ_w = 1.0 t/m³
solids unit weight	γ_s = 1.8 t/m³
initial void ratio	e = 4.31
coeff. vol. compressiblity	m_v = 0.03 m²/t
coefficient consolidation	c_v = 6.5 x 10⁻³cm²/seg
compression index	C_c = 0.78
elasticity modulus	E = 20.8 t/m²
Poisson's coefficient	ν = 0.35
permeability	k = 1.9 x 10⁻⁶cm/seg

In the Gibson et al. (1967) finite strain consolidation theory the governing equation is

$$\frac{\partial}{\partial z}\left(g(e)\,\frac{\partial e}{\partial z}\right) \pm f(e)\,\frac{\partial e}{\partial z} = \frac{\partial e}{\partial t} \qquad (3)$$

where

$$g(e) = -\,\frac{k\,(e)}{\gamma_w\,(1 + e)}\;\frac{d\sigma'}{de} \qquad (3a)$$

$$f(e) = -\left(\frac{\gamma_s}{\gamma_w} - 1\right)\frac{d}{de}\;\frac{k\,(e)}{1 + e} \qquad (3b)$$

where z is the reduced coordinate that is equal to the volume of solids contained in a volume lying between the datum plane and the point in question.

In this theory the input parameters are the relationships between the void ratio (e) with the effective stress (σ') and the permeability. Znidarcic et al. (1986) presented a new analysis, based on this theory, of the constant rate of deformation consolidation test from which these relationships can be obtained without any a priori assumptions regarding their form. Pane (1981) developed a finite difference computer program to integrate (3) with various boundary conditions and any form of these relationships. Croce et al. (1984) validated the new theory, laboratory technique and program by comparing analytical and centrifugal results. This approach seems to be the most appropriate for consolidation of very soft material. However, it requires non-conventional laboratory data that are not usually available and indeed were not available for the Itaipu experimental fill.

The non-linear equation (3) was rendered linear by Gibson et al. (1981) by assuming, first, g(e) as a constant and, second, by making

$$\lambda(e) = -\frac{d}{de}\frac{de}{d\sigma'} = \text{constant} \quad (4)$$

therefore

$$e = (e_{00} - e_{\infty}) \exp(-\lambda\sigma') + e_{\infty} \quad (4a)$$

where e_{∞} and e_{00} are assumed to be the void ratio at end and beginning of consolidation, respectively.

Values of e_{∞}, e_{00}, λ and g evaluated voith the results of the oedometer tests, as recommended by Gibson et al. (1981), are shown in Table 2.

4 COMPARISONS BETWEEN ANALYTICAL AND FIELD RESULTS

In all analyses the soft layer was assumed to be double-drained and the total unit weight of the fill was considered constant and equal to 1.75 t/m³. According to the height vs. time curves (Figure 3) the analysis was divided into three groups:

- Group 1 corresponds to plates 25C, 25N 125C and 125N. In this group the fill was 80cm thick during the first 30 days and from there on the average thickness was 160 cm. Figure 5 shows comparisons between analytical and observed settlement curves for this group.
- Group 2 corresponds to plates 25S, 50S and 50C. Since the beginning, the average thickness of the fill in these plates was 165cm remaining constant all the time. Figure 6 shows comparisons between analytical and observed settlement curves for this group.
- Group 3 corresponds to plate 75C where a thickness of 170cm was observed during the first 180 days and then it was increased to 270cm. Figure 7 shows similar comparisons for this plate.

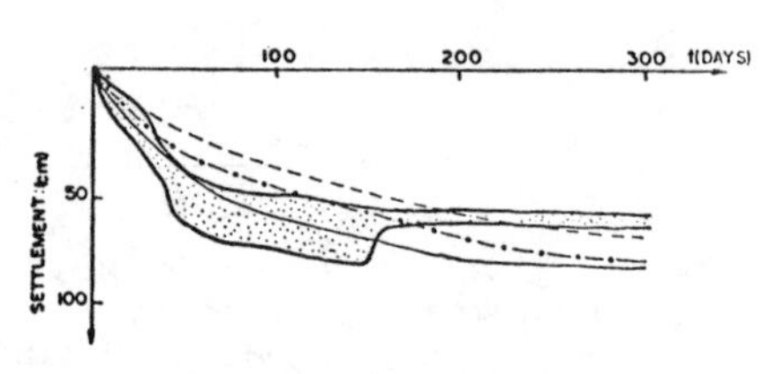

Figure 5 - Settlement Curves for Group I.

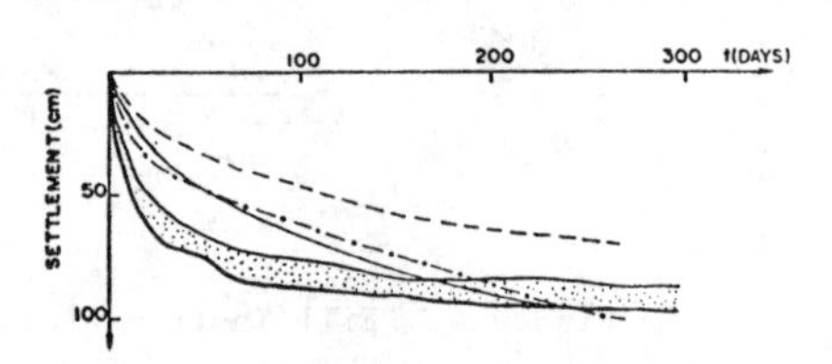

Figure 6 - Settlement Curves for Group II.

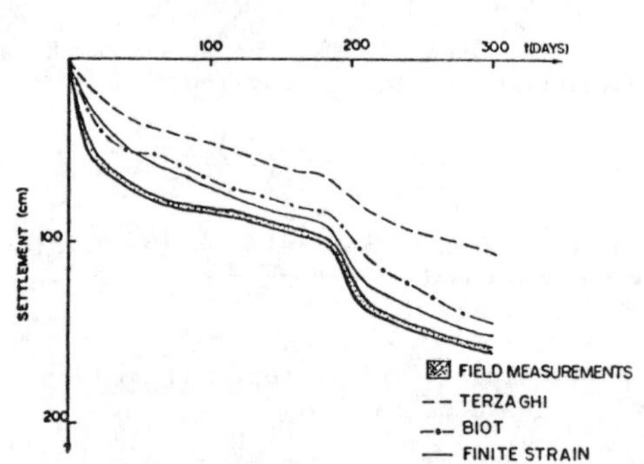

Figure 7 - Settlement Curves for Group III.

TABLE 2

GROUP	LOAD STEP	$\Delta_q (t/m^2)$	$\delta'_a (t/m^2)$	e_a	e_{00}	e_{∞}	$\lambda (m^2/t)$	$g (m^2/day)$
I	1	1.4	1.45	4.08	4.75	3.85	0.885	21.8×10^{-2}
	2	2.8	2.88	3.85	4.75	3.72	0.694	23.9×10^{-2}
II	1	2.9	2.19	3.94	4.75	3.71	0.657	23.0×10^{-2}
III	1	2.98	2.24	3.94	4.75	3.70	0.628	23.0×10^{-2}
	2	4.73	4.6	3.69	4.75	3.59	0.519	25.6×10^{-2}

5 CONCLUSIONS

Based on the comparisons shown the following observations can be made:

- for Group 1 (Figure 5), the scatter in the field results is quite large because of differences in the load carried by each plate. The loading process of the plates in this Group was sufficiently similar to group them together, however some differences ocurred that justify the scatter in the field results. Besides that, the settlement of the plate that was carrying more load was not measured until the end. For this Group, at the beginning, results given by the finite strain consolidation (FSC) and Biot's theories were in better agreement than the one given by Terzaghi's. However, the final settlement given by this last theory was in better agreement than the one given by the others, although the settlement of the plate that was setling more could not be measured until the end and until the moment it was being measured it agreed very well with the FSC theory.
- for Groups 2 (Figure 6), the theories were not able to match the progress of settlement, although the final settlement was accurately matched by the FSC theory.
- for Group 3 (Figure 7), the FSC theory, although in the beginning predicting smaller settlement, was in close agreement with the observed final settlement and with the shape of the observed settlement curve.

From these observations and having in mind that one of the objective of the research was to investigate which assumptions on Terzaghi's theory is less in agreement with reality, it can be concluded that:

- the simplified version of the finite strain consolidation (FSC) theory could represent the observed field behavior better than the other theories.
- the settlement curves obtained with the FSC and Biot's theories were very close until the point analysed (300 days). However, at this time, the degree of settlement indicated by the FSC analysis (more than 90%) was much more in agreement with the field one than Biot's.
- for the case analysed, the consideration of finite strain and non-linear properties on the constitutive relations between the void ratio and effective stress and the void ratio and permeability was more important than the consideration of the two-dimensionality of the problem.

REFERENCES

Biot, M.A. 1941. General theory of three-dimensional consolidation Journal of Applied Physics, 12; 155-164.

Carvalho, J. 1980. Study of secondary compression of Itaipu soft deposit (in Portuguese). M.Sc. Thesis, Pontifícia Universidade Católica do Rio de Janeiro, Rio de Janeiro, Brazil, 213p.

Croce, P.; Pane, V.; Znidarcic, D.; Ko, H.-Y.; Olsen, H.W. & Schiffman, R.L. 1984. Evaluation of consolidation theories by centrifugal modelling. Proc. Conf. Applications of Centrifuge Modelling to Geotechnical Design, Manchester University, p. 380-401.

Einsenstein, Z. & Sandroni, S. 1979. Settlement analyses for a reclamation fill on very soft deposits. 32nd Canadian Geotechnical Conference, Quebec, 2, 3/104-3/115.

Gibson, R.E.; England, G.L. & Hussey, M.J. L. 1967. The theory of one-dimensional consolidation of saturated clay, I. Finite non-linear consolidation of thin homogeneous layers. Geotechnique, 17:261-273.

Gibson, R.E.; Schiffman, R.L. & Cargill, K. W. 1981. The theory of one-dimensional consolidation of saturated clays, II. Finite non-linear consolidation of thick homogeneous layers, Canadian Geotechnical Journal, 18:280-293.

Pane, V. 1981. One-dimensional finite strain consolidation, M.Sc. Thesis, University of Colorado at Boulder, Colorado, USA, 148p.

Pinheiro, J.C.N. 1980. Triaxial tests in soft peaty soil of the Itaipu Lagoon (in Portuguese). M.Sc. Thesis, Pontifícia Universidade Católica do Rio de Janeiro, Rio de Janeiro, Brazil, 253p.

Russo Neto, L. 1980. Instrumented settlement fill on the soft peaty deposit of the Itaipu Lagoon (in Portuguese). M.Sc. Thesis, Pontifícia Universidade Católica do Rio de Janeiro, Rio de Janeiro, Brazil, 169p.

Sandroni, S.; Russo Neto, L. & Carvalho, L. 1981. Instrumented settlement fill in a peaty deposit. Proc. 10th. Int. Conf. on Soil Mechanics and Foundation Engineering, Stockholm, 1/40, 229-232.

Sandroni, S.; Silva, J.M.J. & Pinheiro, J. C.N. 1984. Site investigations for unretained excavations in a soft peaty deposit. Canadian Geotechnical Journal, 21:36-59.

Silva, J.M.J. 1979. Instrumentation, observation and preliminary analysis of an experimental excavation on the soft organic soil of the Itaipu Lagoon (in Portuguese). M.Sc. Thesis, Pontifícia Universidade Católica do Rio de Janeiro, Rio de Janeiro, Brazil, 145p.

Znidarcic, D.; Schiffman, R.L.; Pane, V.; Croce, P.; Ko, H.-Y. & Olsen, H.W. 1986. The theory of one-dimensional consolidation of saturated clays: part V, constant rate of deformation testing and analysis. Geotechnique, 36:227-237.

Numerical Methods in Geomechanics (Innsbruck 1988), Swoboda (ed.)
© 1988 Balkema, Rotterdam. ISBN 90 6191 809 X

The interaction mechanism of granular soils with geogrids

Gunther E.Bauer
Department of Civil Engineering, Carleton University, Ottawa, Canada
Yousry M.Mowafy
Department of Civil Engineering, Al-Azhar University, Nasr City, Cairo, Egypt

ABSTRACT: The transfer of stresses between soil and geogrids is investigated experimentally for different types of meshes and various soil materials. The obtained results showed that this interaction mechanism depends on the soil type, grain size and mesh type. The results are presented in graphical form and are compared to those obtained from a nonlinear F.E. analysis.

1 INTRODUCTION

The use of natural fibres to reinforce soils is many thousands of years old. In recent years man-made materials, such as wood, steel, concrete and plastics, have been used to build soil support and earth retaining structures. Since the late 1960's, galvanized smooth and ribbed steel strips have been commonly used to build more than 5,000 reinforced earth structures in over 30 countries. Credit for developing this concept on a rational basis belongs to the French engineer, Henri Vidal. The vast majority of these structures were built with galvanized mild steel strips and granular backfill materials and the service life is governed by the rate of corrosion of the reinforcing elements. This problem of corrosion can be severe, especially in countries where salting of roads during the winter months has become a standard practice in keeping roads trafficable. It was therefore, a natural step to turn to the use of plastic grids or meshes made from non-corrodible materials.

The use of smooth strips or sheets as reinforcing elements have also a major drawback in that they can only transfer stresses by pure surface friction mobilized between the soil and the inclusion. In contrast meshes or geogrids fulfill the additional function in that they confine the granular material within the grid apertures and also interlock with the soil. For a given granular aggregate and geogrid, the degree of confinement and the amount of interlock depends on the apertures, i.e. windows, and the size, shape and hardness of the soil particles. It stands to reason that for a certain geogrid having given grid apertures, there should be a granular soil type to produce the best interlock and confinement.

The objectives of this investigation were to determine experimentally the maximum pull-out resistance of several mesh-types with various granular materials. At the same time it was attempted to separate quantitatively the components of interlock and surface friction which, in general, determine the total pull-out resistance of a geogrid or mesh.

A non-linear F.E. program called REA (Reinforced Earth Analysis) was used to simulate the pull-out/displacement response of the grids investigated. The agreement between experimental and analytical results is quite good.

2 SOIL/MESH INTERACTION

The basic interaction mechanism in reinforced soil is governed by the mobilization of frictional resistance between the reinforcement element and the soil. This basic concept is quite simple when flat sheet elements are used as reinforcement. The surface or sliding friction can easily be determined experimentally and the coefficients of friction can be considered a constant for a given soil and reinforcing material. The problem becomes increasingly more complex when the reinforcing elements are ribbed strips or meshes. In this paper the terms grid and mesh are used interchangeably and apply to any open net-like structure, whereas the word geogrid is reserved for plastic grids only. For the case of mesh reinforced soil the main mechanism of soil/mesh interaction depends on the following factors:

(1) the geotechnical properties of the soil (i.e. density, particle sizes and distribution, etc.), and,

(2) the material properties as well as the geometry and configuration of the mesh.

Then for a given soil and a specified mesh or grid type, the interaction between the reinforcement and the aggregate is a function of three mechanisms:

(1) the frictional resistance mobilized between the plane surfaces of the mesh and the soil particles in contact with them,

(2) the shearing resistance of the soil particles which have interlocked through the apertures of the mesh, and

(3) the passive resistance of the soil that bears against the reinforcement elements which are normal to the direction of the shear displacement. These three possible interaction mechanisms are illustrated in Figure 1. The stresses mobilized in the three mechanisms are difficult to quantify analytically, but their influence on the total pull-out resistance of a mesh can be determined experimentally. It is obvious that in a very fine sand or loose silt the interlock mechanism is negligible, whereas in a well graded gravelly soil, this factor will be the major component contributing to the pull-out resistance. Flat sheets or strips used as reinforcements will generate their pull-out resistance totally from the interface friction. If complete interlock occurs between a grid and soil particles, that is, no relative displacement between the soil particles confined within the mesh apertures and the mesh itself will take place, the passive shear resistance and surface friction are not brought into action. These latter two concepts are easy to understand but the concepts of interlock and particle confinement need further discussion since they are key elements in mesh or grid reinforcement techniques.

2.1 Interlock and Confinement

Interlock and soil confinement can only be present where soil particles fully or partly penetrate through the mesh apertures and at the same time are confined within such an opening. If full interlock occurs, which means that the particles within the mesh opening are completely locked in place and are confined by the surrounding mesh, then shear will occur within the soil on surfaces above and below the grid (Figure 1(b)). Therefore, the pull-out resistance is thus the sum of the shear strength components acting on shear planes above and below the mesh. Then the pull-out resistance can be given by the following equation:

$$P_R = 2\sigma'_n \cdot A_n \cdot \tan\phi_m \qquad (1)$$

where P_R = pull-out resistance (force units)

σ'_v = effective normal stress on shear plane (for a horizontally placed mesh, σ'_n can be taken as the effective overburden stress including surcharge effects)

A_n = nominal or overall area of the mesh, and

ϕ_m = mobilized internal friction angle of the soil.

The pull-out force will be greatest for a dense, well-graded granular soil with angular particles, giving a high degree of soil confinement within the mesh apertures. Theoretically then the maximum value the mobilized friction angle, ϕ_m, can assume is equal to the internal friction angle, ϕ, of the soil.

The results published by Ingold [1] and Morgado [2] have given values of mobilized friction angles greater than the values corresponding to the internal friction.

This apparent paradox phenomenon can be explained by the fact that the mobilized friction angles were back-calculated using equation (1). Two factors which are not considered by this equation are that the actual shear planes during a pull-out test might be greater than the nominal planes assumed. Secondly, the measured pull-out resistance, P_R, could also contain a component of passive resistance which the mesh elements normal to the pull-out direction may generate.

The degree of particle confinement within the mesh apertures depends on the dimensions of the opening and on the shape, size and hardness of the soil particles. It stands to reason that for a certain mesh material, with given apertures, there should be a granular soil type to produce the best interlock and confinement. This topic was the main objective of the experimental investigation which consisted of several series of pull-out tests which are discussed in the following sections.

3 EXPERIMENTAL PROGRAMME

The experimental programme consisted of several series of pull-out tests which were performed with different meshes and soil types in a steel tank. The tank was 1020 mm long, 300 mm wide and 400 mm deep. A 10 mm wide slot was cut into the end wall at midheight of the tank to allow the reinforcing mesh to be pulled through. The soil was placed and compacted inside the tank to midheight, i.e. to the elevation of the slot, then the reinforcing mesh or geogrid was placed and the tank was continued to be filled with the granular aggregate. The mesh protruded through the slot and was gripped by a pulling mechanism. This pulling mechanism consisted of a mechanical gear box screw jack assembly which was driven by an electrical motor. The pulling force and displacement were monitored with a load cell and a linear-voltage-displacement-transducer (LVDT), respectively. Both recording transducers were connnected to a x-y plotter. A detailed description of the testing facility and test procedure were given by Mowafy [3].

In a separate series of tests, the tank was adapted to function as a large direct shear box in order to determine the mechanical properties of the various granular soils used in this investigation. In either test series, i.e. pull-out and soil strength tests, the surface of the soil could be surcharged in order to increase the normal stress on the horizontal shear plane(s).

3.1 Soil Aggregates

The soils used as granular backfill material consisted of (1) a well graded crushed limestone aggregate, (b) a uniformly graded well rounded silica sand (medium Ottawa sand) and (c) three sizes of crushed limestone. The wellgraded limestone aggregate was placed, for different test series, at unit dry weights of 16, 19 and 22 kn/m^3 which corresponded to internal friction angles of 38, 45 and 53°. The three uniformly crushed limestone aggregates had average grain size diameters of 10 mm (1/4 in), 16 mm (5/8 in) and 19 mm (3/4 in). At a unit dry weight of 16 kn/m^3 the corresponding friction angles were 37, 38 and 41°. As mentioned earlier the reason for using different size aggregates was to investigate the effect of particle size on confinement and interlock for a given aperture of a mesh. The uniform Ottawa sand, i.e. well rounded uniform quartz particles, was rained into the test tank at a unit dry weight of 16 kn/m^3 which corresponded to a friction angle of 26°.

3.2 Reinforcement Meshes

Seven steel meshes having different apertures, or openings were investigated. A smooth galvanized steel sheet was also tested in pull-out in order to determine the pure surface friction component. Three types of TENSAR geogrids were choosen in this study, SS1, SS2 and AR1. These geogrids were selected because of their excellent strength and interlock properties as determined in an earlier study [3]. TENSAR geogrids are made from high strength polypropylene material and are well suited as soil reinforcements. In order to determine the pure frictional resistance of polypropylene, a flat sheet of this material was tested in pull-out with the various soil aggregates and soil densities.

4 TEST RESULTS

A summary of the number and type of tests for the steel meshes is presented in Table 1. A total of 144 tests were performed. In addition several repeat tests in each category were carried out to check the repeatability of results. The most tests were performed with the well graded limestone aggregate. A typical pull-out/displacement response is given in Figure 2. The steel mesh opening was 51 by 51 mm which corresponded approximately to aperture of the TENSAR geogrid SS2. At small displacements the behaviour is quite linear and the various curves coincide regardless of surcharge intensity. The total pull-out force is of course dependent on the normal stress, i.e. surcharge intensity, acting on the shear surface(s). Figure 2 gives a summary of the pull-out behaviour for several steel meshes tested in the 19 mm diameter limestone aggregate. This figure shows the results for a surcharge of 20.7 kn/m^2 and a unit dry weight of 16 kn/m^3. Also shown on this figure is the behaviour of the sheet metal. It is quite obvious that a large portion of the pull-out resistance is attributed to interlock. The phenonemon of interlock is further illustrated in Figure 4, where the interlock and frictional components of the pull-out resistance are given with regard to surcharge intensities and dry unit weight of the aggregate. The other tests showed similar results. In general it can be said that the interlock contributed between 85 to 95 percent to the total pull-out resistance. This was true regardless of mesh aperture and soil density. The steel meshes with apertures of 25 by 25 mm and 51 by 51 mm exhibited the greatest resistance to pull-out.

The test series for the TENSAR geogrids are summarized in Table 2. A total of 42 tests were performed. All geogrids were tested in both directions, that is, along and across the direction of manufacture. The geogrids are continuous along rolling direction. In all cases the grids tested across roll direction showed a greater pull-out resistance and TENSAR geogrid SS2 exhibited the highest resistance regardless of soil type or surcharge intensity. Typical load-displacement responses are given in Figure 5 for the SS2 grid and the well-graded limestone aggregate. The results for interlock and friction with regard to the size of mesh opening is presented in Figure 6. Again the fact that interlock contributes the largest component to total pull-out resistance is quite clear from this figure regardless of aggregate size and mesh opening.

In summary it can be concluded from the test results obtained on the steel meshes and geogrids that the maximum grain size should not exceed one-third of the mesh opening to give best interlock. Also a well graded aggregate where the maximum size aggregate size does not exceed the criterion stated in the previous sentence will provide the best particle confinement within the grid apertures. A certain amount of strain is necessary within the reinforced soil in order to mobilize the soil/grid interaction mechanism as illustrated in Figures 2, 3 and 5. A larger strain, i.e. displacement, is needed for the TENSAR geogrids as compared to the steel meshes in order to mobilize the full pull-out resistance.

5 FINITE ELEMENT ANALYSIS OF TEST RESULTS

The results from the experimental investigation have shown that the pull-out resistance of a geogrid or mesh depends on the mechanical properties of the soil and the grid. Since very few research establishments have test facilities to carry out pull-out tests, an analytical model was needed in order to predict the behaviour of open netlike inclusions in a granular soil media. The authors have previously investigated the stress and deformation behaviour of reinforced earth structures subjected to selfweight and external loading using a F.E. analysis [3,4,5]. The non-linear finite element program REA which was developed by Hermann and his colleagues [6,7,8] was adapted in the present study to simulate the behaviour of mesh type reinforcements. The program takes into account the nonlinear inelastic soil behaviour and slippage and yielding of the reinforcement. The program uses isoparametric elements with four nodes and an incremental iterative procedure. Detailed description of the program and its application have been given elsewhere [3,6,7,8] and will not be repeated here. Figure 6 shows the finite element mesh and the embankment geometry which was used in predicting the pull-out behaviour of the reinforcing grid. It should be noted that the slope geometry shown in this figure is the same as the embankment in the test tank.

The mesh given in Figure 7 contains 139 elements and 145 nodal points. One-dimensional elements, i.e. elements 125 to 139 were employed to model the reinforcing sheet or grid. It was assumed that the bending stiffness of the reinforcing was negligible. Elements 1-124 are continuum elements representing the soil, i.e. the well graded limestone aggregates. The elastic and the plastic moduli were respectively 200,000 and 1400 MPa for steel mesh and sheet metal; and were 4500 and 1400 MPa, respectively for the SS2-Tensar grid and the polypropylene sheet. The yield stresses were 320 and 230 MPa for steel and Tensar, respectively. To simulate the relative movement between the reinforcement and the surrounding soil, fictitious bond springs were used. The appropriate selection of the spring coefficient is paramount in order to model the pull-out behaviour. The movement of the reinforcement relative to the soil is denoted by δ' and the resulting resisting force (provided from the load-displacement curves of the pull-out tests) is given as F. Then the spring coefficient k is defined as $k=F/\delta'$. The coefficient k' (which was used as input parameter in the program) is defined as

$k'=k/A$ (where A is the cross-sectional area of the reinforcement element). Using the above expression for the spring coefficient, average values of k' equal to 2 and 10 $N/mm^2/mm$ were obtained for the sheet metal and the steel mesh respectively. Values of k' equal to 0.8 and 15 $N/mm^2/mm$ were selected for polypropylene and SS2-Tensar grid, respectively.

5.1 Sheet Metal Reinforcement

The results of the analysis using sheet metal reinforcing under three different surcharge intensities, i.e. 7.2, 13.9 and 19.7 kN/m^2, were analyzed using the finite element technique. The coefficient of friction between the sheet metal and the well graded crushed limestone was found from the results of the pull-out tests and were 1.1, 0.8 and 0.7 for surcharge intensities of 7.2, 13.9 and 19.7 kN/m^2, respectively. A total of nine increments were employed for each pull-out test. Two increments were used to construct and surcharge the embankment, and seven displacement increments were applied at node A (Figure 7).

The comparison between the experimental and the finite element results for the different surcharge values are shown in Figure 8a. The results of this figure show a good agreement between the experimental and the analytical values from the finite element analysis. The displacement at maximum pull-out force were 2.3 mm and 1.7 mm for the experimental and the finite element analyses, respectively.

5.2 Steel Mesh Reinforcement

The experimental pull-out test results from the steel mesh series (51 x 51 mm openings) were also analyzed with the nonlinear finite element program REA. The coefficient of friction between the mesh and the limestone aggregate was found to be 2.3. Ten incremental steps were used to determine the load-displacement response. A total displacement of 6 mm at nodal point A (Figure 6) was needed to completely define the load-displacement behaviour. Two different surcharge increments, i.e. 5.9 and 11.2 kN/m^2, were analyzed. Since the steel mesh was treated as a sheet, the results of such an analysis could be sensitive to the value of the spring coefficient. The calculated spring coefficient for the steel mesh was 10 $N/mm^2/mm$.

Figure 9 shows the results of this study under a surcharge intensity of 5.9 kN/m^2. This figure indicates an increase of the pull-out force with an increase of the spring coefficient. The results of Figure 9 show that a value of 25 $N/mm^2/mm$ for the spring coefficient gives the best agreement between the measured and analytical pull-out force. For instance, the difference between the experimental and the finite element values for the maximum pull-out loads were found to be 15 and 11% for spring coefficients of 10 and 20 $N/mm^2/mm$ respectively. This difference was found to be 4% for a spring coefficient of 25 $N/mm^2/mm$. Figure 9 also indicates that the finite element analysis yielded an average displacement of 5.4 mm corresponding to the maximum pull-out load while in the test displacement of 6.5 mm was recorded.

Figure 9 shows a comparison between the finite element and the experimental results for the steel mesh for a surcharge intensity of 11.2 kN/m^2. This figure shows a good agreement. The maximum pull-out load in the experimental test was 5750 N compared to 5100 N for the finite element analysis. The corresponding displacements were 5.4 and 5.2 mm, respectively.

5.3 Polypropylene Sheet

A comparison of results for the polypropylene sheet reinforcing is given in Figure 8b. This figure shows the load-displacement curves of the experimental results and the finite element analysis for three surcharge intensities, i.e. 7,15.5 and 24 kN/m^2. The coefficient of friction between the sheet and the granular material was found to be 0.17.

5.4 Tensar Geogrid (SS2)

A comparison of results from the experimental tests and the F.E. analysis are given in Figure 10. A total displacement of 11 mm had to be applied at the nodal point A (Figure 6) in order to define completely the theoretical load-displacement response of the geogrid subjected to a surcharge intensity of 7 kN/m^2. Other surcharge intensities yielded similar good agreements between the experimental and analytical results. It should be noted that the input parameters for the F.E. analysis were predetermined quantities based on the material properties of the soil and reinforcements. These quantities were determined experimentally from independent tests. The maximum difference in the pull-out force between tests and analysis was 20 percent. The mean difference was in the order of 9 percent. A better agreement could, of course, be obtained by adjusting the spring coefficients. A slightly lower spring coefficient would have given a better congruence between the experimental and analytical force-displacement response.

In summary it can be said that the nonlinear finite element program was found to be capable to simulate the load-displacement behaviour of the experimental curves. The values of the spring coefficient and the coefficient of friction between the reinforcing elements and soil aggregates can be predetermined from independent laboratory tests. The load-displacement curves for various reinforcing elements, such as sheet metal, polypropylene sheet, steel mesh and plastic geogrid, could be modelled realistically with the nonlinear finite element analysis.

6 CONCLUSIONS

The following conclusions can be drawn based on the experimental and analytical results presented in this treatise.

Meshes and geogrids derive most of their pull-out resistance from interlocking with the granular soil aggregates. For the granular backfill and the different meshes investigated up to 95 percent of the total resistance to pull-out can be attributed to interlock.

Both the size of the grid opening and the average diameter of the soil particle affect the interlock and confinement of the soil with the mesh. A well-graded crushed rock aggregates with maximum particle size not exceeding one-third of the mesh opening provided the best interlock and confinement and hence the greatest pull-out resistance.

Tests on the three TENSAR geogrids (SS1, SS2 and AR1) have yielded higher pull-out resistances when tested across rather than along the direction of roll. Moreover, TENSAR geogrid SS2 exhibited the highest resistance of the three geogrids investigated. The plastic geogrids need to be strained more than the steel meshes in order to develop their full pull-out resistance.

The nonlinear finite element program REA was found to be able to simulate the load-displacement response to the experimental pull-out test results.

7 ACKNOWLEDGEMENTS

The financial assistance which was provided for this study by the Natural Sciences and Engineering Research Council of Canada (NSERC) was greatly appreciated.

8 REFERENCES

1. Ingold, T.S. 1983. A laboratory pull-out testing on grid reinforcement in sand. Geotechnical Testing Journal, Vol.6, No.3, pp.101-111.
2. Morgado, M.M. 1983. Investigation of steel mesh reinforcement for earth structures. M.A.Sc. Thesis, Department of Civil Engineering, University of Ottawa, Ottawa, Canada.
3. Mowafy, Y.M. 1986. Analysis of Grid Reinforced Earth Structures, Ph.D. Thesis, Department of Civil Engineering, Carleton University, Ottawa, Canada, p.384.
4. Bauer, G.E. and Mowafy, Y.M. 1985. A nonlinear F.E. analysis of reinforced embankments under external loading. 5th Int. Conf. on Numerical Methods in Geomechanics, Nagoya, Japan, p.12.
5. Herrmann, L.R. 1978. Finite element analysis of contact problems. ASCE Journal of Engineering Mech. Div. 104, EM5, pp.1043-1054.
6. Herrmann, L.R. 1978. User's manual for REA general two-dimensional soils and reinforced earth analysis program. Department of Civil Engineering Report, University of California, Davis.
7. Herrmann, L.R. and Al-Yassin, Z. 1978. Numerical analysis of reinforced soil system, ASCE Symposium of Earth Reinforcement, Pittsburgh, pp.429-457.
8. Duncan, J.M. and Chang, E.Y. 1970. Nonlinear analysis of stress and strain in soils. ASCE Journal of Soil Mech. and Fdn. Div., 96 (SMS) pp.1629-1653.

Table 1. Summary of Pull-Out Tests On Steel Mesh

Soil Type	Soil Unit Weight (RN/m^3)	Apertures of Steel Mesh (mm x mm)	Number of Surcharge Increments	Total Number of Tests
Well graded	16.0	13x13; 25x25; 25x25; 25x51; 51x25; 51x51; 51x102; 102x51 sheet metal	5 each	80
Well graded	12.0	25x25; 51x51 sheet metal	4 each	12
Well graded	22.0	25x25; 51x51 sheet metal	4 each	12
Uniform 10mm (3/8")	16.0	13x13; 25x25 51x51; 51x102 sheet metal	2 each	10
Uniform 16mm (5/8")	16.0	13x13; 25x25 51x51; 51x102 sheet metal	2 each	10
Uniform 19mm (3/4")	16.0	13x13; 25x25 51x51; 51x102 sheet metal	2 each	10
Quartz sand	16.0	13x13; 25x25 51x51; 51x102 sheet metal	2 each	10

Grand Total: 144 Tests

Table 2. Summary of Pull-Out Tests On Tensar Geogrids

Soil Type	Soil Unit Weight (kN/m^3)	Tensar Geogrid	Number of Surcharge Increments	Total No. of Tests
Well	16.0	SS1 (1)	5	5
Graded	16.0	SS1 (2)	5	5
	22.0	SS1 (1)	4	4
Well	16.0	SS2 (1)	5	5
Graded	16.0	SS2 (2)	5	5
	22.0	SS2 (1)	4	4
Well	16.0	AR1 (1)	5	5
Graded	16.0	AR1 (2)	5	5
	22.0	AR1 (1)	4	4

Grand Total: 42 Tests

(1) Grid tested across roll direction
(2) Grid tested along roll direction

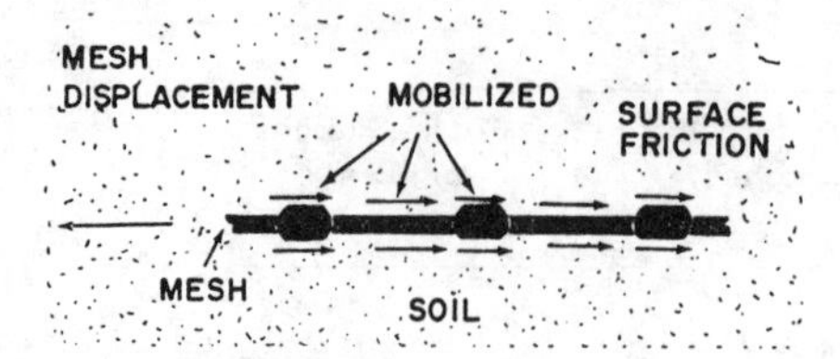

a) SKIN OR SURFACE FRICTION

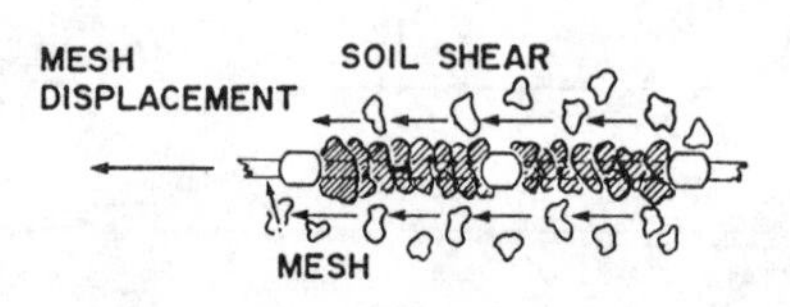

b) FULL SOIL SHEARING DUE TO INTERLOCK

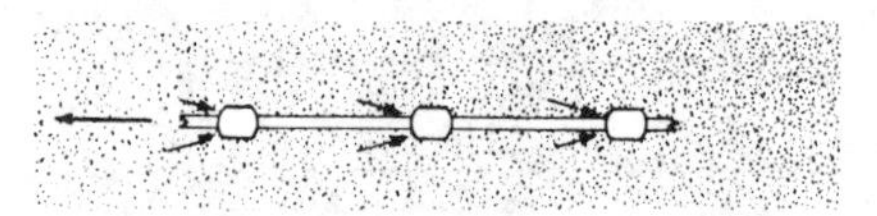

c) PASSIVE SHEAR RESISTANCE

Fig. 1. Interlock and Friction

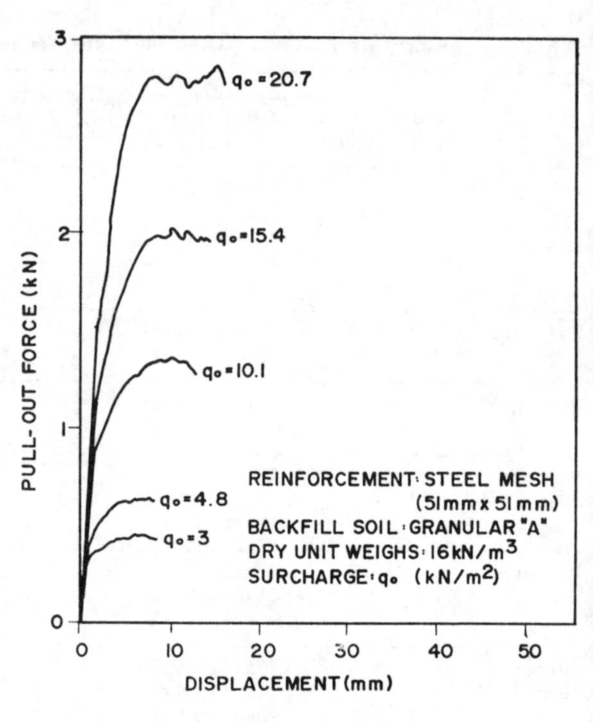

Fig. 2. Effect of Surcharge on Pullout

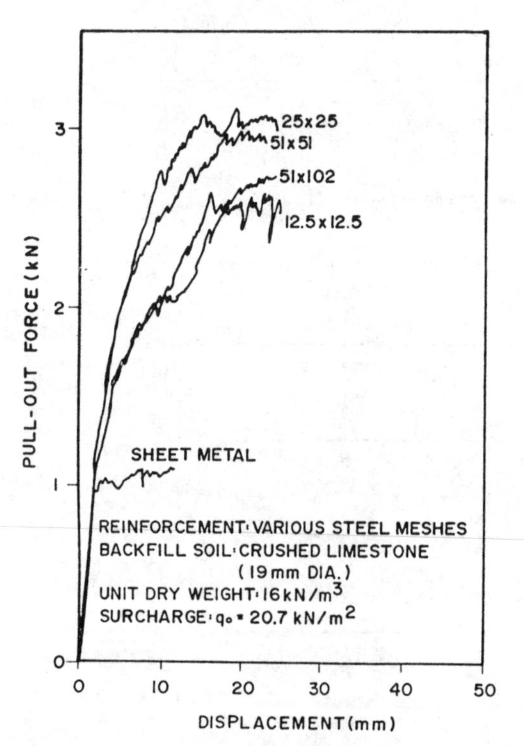

Fig. 3. Pull-out Forces vs. Displacement

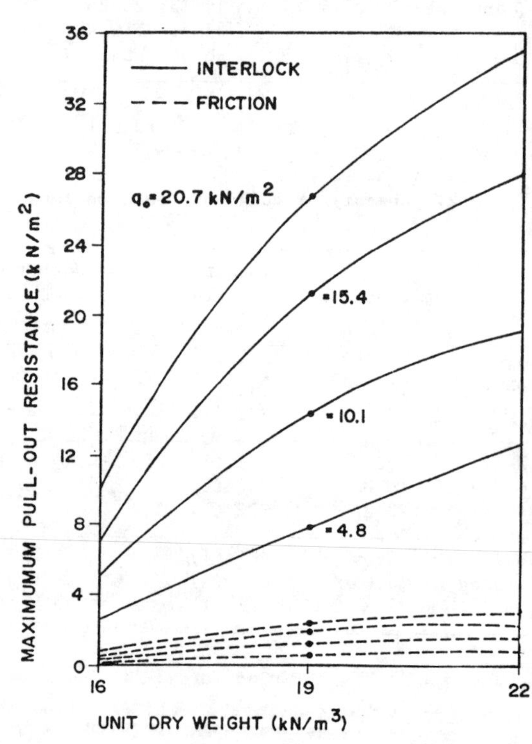

Fig. 4. Pull-out Resistance vs. Soil Density

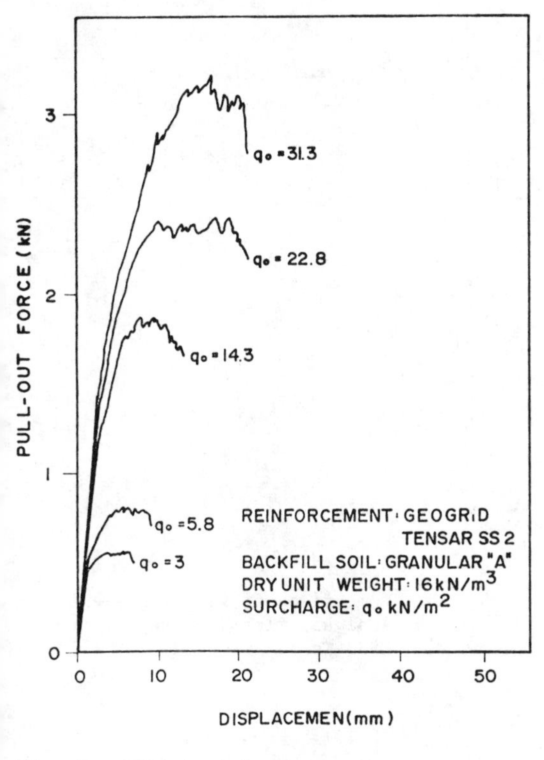

Fig. 5. Effect of Surcharge on Pull-out Force-TENSAR SS2

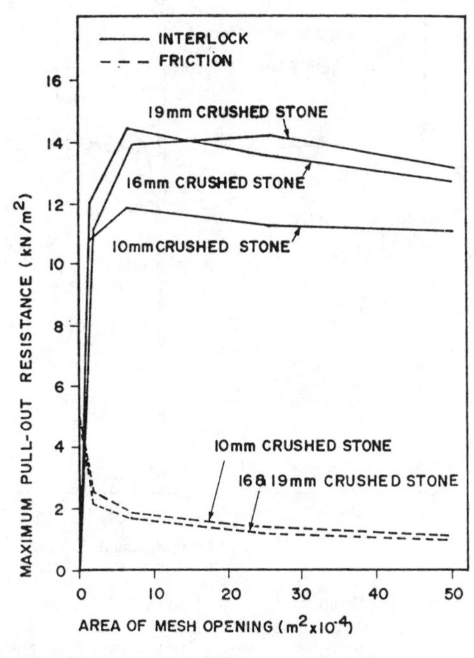

Fig. 6. Pull-out Resistance for Different Aggregates

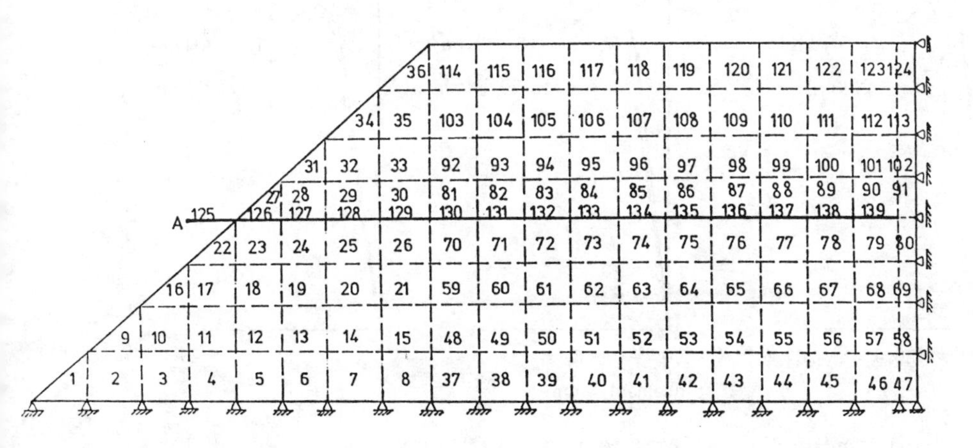

Fig. 7. F.E. Mesh Used to Simulate Pull-out Resistance of Geogrid

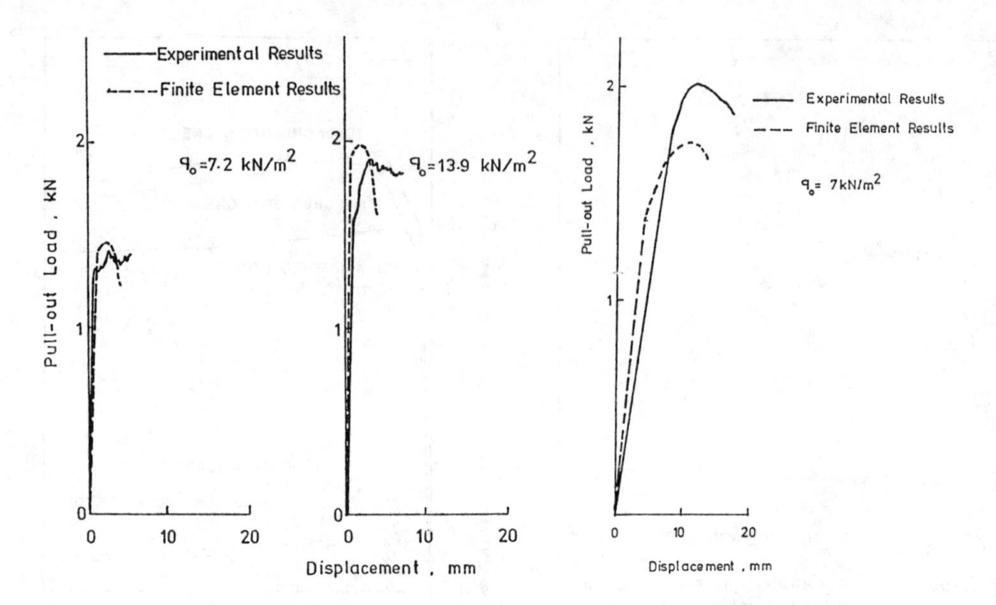

Fig. 8. Comparison of Finite Element Analysis with Exp. Results
a) Sheet Metal b) Polypropylene

Fig. 10. Comparison of Finite Element Analysis with Exp. Results (SS2 Tensar Geogrid)

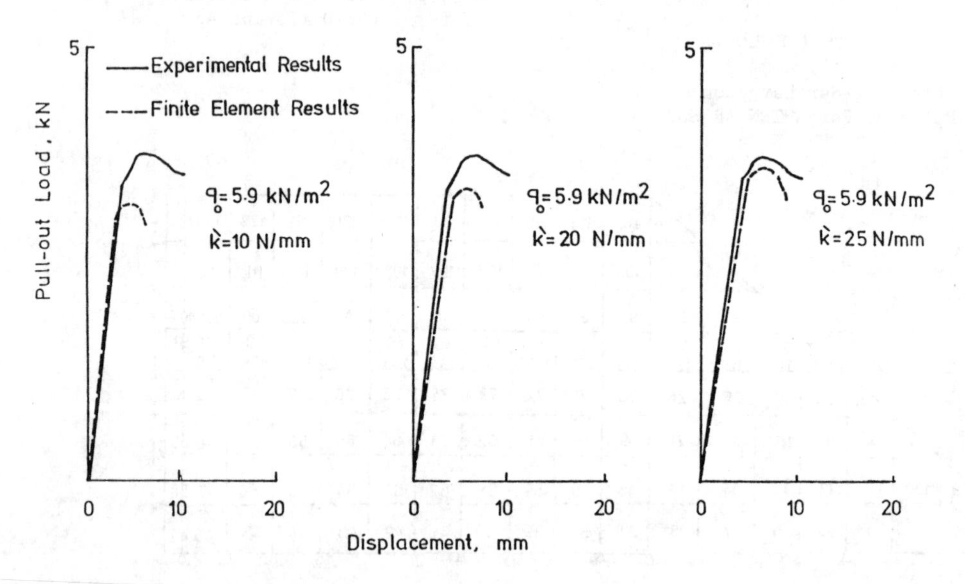

Fig. 9. Comparison of Finite Element Analysis with Exp. Results (Steel Mesh - 51 x 51 mm Apertures)

Numerical Methods in Geomechanics (Innsbruck 1988), Swoboda (ed.)
© 1988 Balkema, Rotterdam. ISBN 90 6191 809 X

The analysis of steel-reinforced embankments on soft clay foundations

R.K.Rowe & B.L.J.Mylleville
Geotechnical Research Centre, Faculty of Engineering Science, University of Western Ontario, London, Canada

ABSTRACT: The factors to be considered when using numerical models for studying the behaviour of steel reinforced embankments are discussed. Some results of finite element analyses are presented which illustrate that the mode of failure may vary substantially depending on the amount of reinforcement used and the properties of the foundation soil. It is suggested that results obtained from finite element analyses may form a starting point from which to develop design guidelines for steel reinforced embankments.

1 INTRODUCTION

The past several years have seen a significant increase in the use of reinforcement as a means of ensuring the stability of embankments on soft soil foundations. The behaviour of reinforced embankments on soft foundations is dependent on non-linear soil-structure interaction and the problem lends itself to being analyzed using the finite element technique. The objective of the present paper is to discuss the factors which the authors consider to be important in modelling the behaviour of reinforced embankments. Emphasis will be placed on the practical application to steel reinforced embankments on soft clay foundations.

Results are presented to illustrate the importance of numerical modelling as a tool for improving understanding of steel reinforced embankment behaviour. The importance of some of these results with regards to design considerations is also discussed.

2 NUMERICAL MODELLING AND PARAMETER SELECTION

To understand the effects of using a particular form of reinforcement, it is necessary to model the response of the entire reinforced system up to failure. This involves modelling of plastic failure within the soil using an elasto-plastic formulation (eg. see Zienkiewicz, 1977). The specific formulation adopted by the authors assumes a Mohr-Coulomb failure criterion together with a flow rule of the form proposed by Davis (1968).

In the construction of embankments on soft clay foundations, the most critical situation generally corresponds to that at the end of construction. In predicting the short term behaviour of embankments on relatively soft cohesive soils, the undrained shear strength c_u (ϕ=0) and undrained modulus E_u (taking Poisson's ratio to be 0.48) may be used. The results of some analyses reported herein are for the case of strength and modulus increasing linearly with depth, which is typical of soft, normally or slightly overconsolidated clays. The ratio of undrained modulus to undrained strength used in the analyses was 125 (E_u/c_u=125). The depth of clay deposit considered was 15 m, underlain by a rigid base. Figure 1 shows the general arrangement of the physical problem being analyzed.

In order to obtain reasonable stresses and strains in the embankment fill, it is necessary to consider both the stress dependent stiffness characteristics of granular materials as well as plastic failure. A quite simple non-linear elastic model which accounts for stress-dependent stiffness based on Janbu's equation viz.

$$(E/p_a) = K(\sigma_3/p_a)^m \qquad (1)$$

has proven to be very successful for modelling the behaviour of granular embankments and dams at load levels well below failure (eg. Duncan, 1980). Thus, for the purposes of the analyses reported herein, the non-linear stiffness characteristic of the soil was modelled using Janbu's equation using parameters determined to be relevant for typical fill materials while plastic failure was modelled using a Mohr-Coulomb failure criterion and a non-associated flow rule. The parameters adopted for the fill were ϕ=36°, γ=20 kN/m^3, ν=0.35, K=100, m=0.5.

The reinforcing elements in the embankment were modelled as bar elements where the axial stiffness is a representative value per unit width of embankment.

One practical consideration which had to be addressed when using reinforcement in the form of Reinforced Earth (R) (eg. Vidal, 1968) type steel strips is the location of the reinforcement within the system. It is generally accepted that the strips be placed within the fill to develop frictional resistance on both faces of the reinforcement. The analyses performed considered a single layer of steel strips 375 mm above the clay/fill interface (see Figure 1).

The ribbed reinforcing steel strips modelled in these analyses were 50 mm wide and 5 mm thick with a yield strength of 350 MPa and an ultimate tensile strength of 490 MPa. Centre to centre strip spacings (S) of 125 mm and 375 mm were considered.

In modelling the behaviour of reinforced soil masses using strip reinforcement, it is necessary to consider two possible mechanisms of failure at the soil-reinforcement interface as indicated below.

(a) If there is insufficient anchorage capacity, failure will occur at the soil reinforcement interface above and below the reinforcement as the reinforcement is pulled out of the soil. This "pullout" mode involves displacement of the reinforcement relative to the soil on both sides of the reinforcement. This rarely occurs for sheets of reinforcing material however it is important when examining strip reinforcement as considered here.

(b) If the shear strength of the soil reinforcement interface is less than the shear strength of the soil alone, then failure may occur by sliding of the soil along the upper surface of the reinforcement and the upper soil mass moves relative to both the reinforcement and the underlying soil. This rarely occurs.

In general, it is logical to locate strips entirely within the fill, however if the reinforcement is placed on top of the soft foundation then a third failure mechanism can involve the soft foundation material being squeezed out from beneath the reinforcement layer and the entire reinforced embankment.

When reinforcement consists of separate reinforcing strips (eg. steel strips), then special care is required to correctly model the failure mechanism. For these materials, the interface shear resistance in direct shear (eg. if there is sliding of the soil along the upper surface of the reinforcement) will be substantially higher than the interface resistance in pullout. In modelling these materials, it is necessary for the formulation of the interface element to be such that it can detect whether it is in a direct shear or pullout mode and to then select the appropriate interface parameters to model this mode of shearing. Thus the behaviour of the interface element on one side of the reinforcement is related to the behaviour of the interface element on the other side (since the mode of shearing can only be assessed by consideration of the direction of shear on either side of the reinforcement).

For strip reinforcement, independent movement of the soil above and below the plane of reinforcement can only occur during a direct shear mode of failure. Pullout of strips is really a three-dimensional phenomenon in which the strips move relative to the soil around them but the soil between strips remains continuous. As noted by Naylor and Richards (1978), the approach of using a conventional joint element (or nodal compatibility element) to model this slip implicitly treats the strips as an equivalent two-dimensional sheet and will cause serious error since it interrupts the transfer of shear stress through the soil.

Since pullout of strips does represent a truly three-dimensional situation, it can only be approximately modelled in a two-dimensional analysis. A number of different approaches can be adopted. For example, Naylor and Richards (1978) proposed a composite formulation which ensured

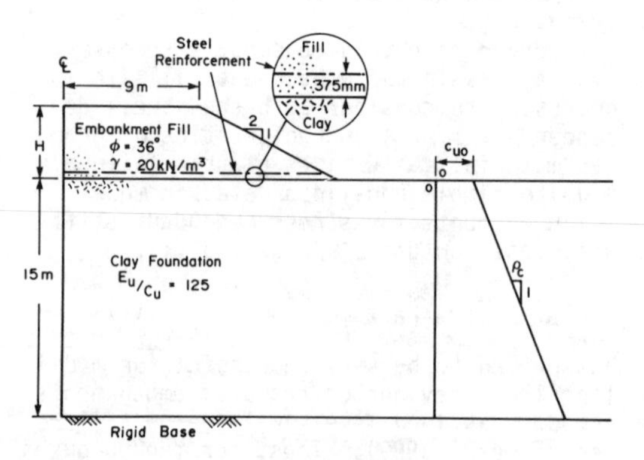

Figure 1. General arrangement of the physical problem being analyzed

continuity of shear stress in the soil after pullout by introducing a "conceptual shear zone". An alternative approach implemented by the authors in their formulation involves an interface element which has a node above the reinforcement, a node on the reinforcement and a node below the reinforcement. Prior to slip, normal and tangential compatibility between the soil and reinforcement is enforced by means of very stiff springs. The normal and shear stresses "above" and "below" the reinforcement are automatically monitored. If a pullout mode of failure occurs (as inferred by the direction of shear above and below the reinforcement together with a Mohr-Coulomb failure criterion), then the computer program automatically enforces compatibility between the soil nodes "above" and "below" the reinforcement (thereby maintaining continuous transfer of shear stress in the soil) while allowing slip between the reinforcement node and the two soil nodes. The normal force between these nodes is used to assess the normal forces acting on the strip; the corresponding shear resistance (based on a Mohr-Coulomb failure criterion) between the strip and soil is applied to both the upper and lower soil node, and as the equilibrating force to the node on the soil strip.

Since the strip only covers a small area of the soil, the Mohr-Coulomb parameters must be adjusted to take account of the actual surface area, per unit width of the embankment, which is in contact with the soil.

Research in the area of reinforced embankment behaviour (eg. Rowe and Soderman, 1987) has shown that the amount of force which can be developed in the reinforcement is in some instances limited by slip at the clay/fill interface immediately beneath the embankment. In order to allow for this condition, slip nodes were introduced at the clay/fill interface, with capacity limited to the undrained shear strength of the clay at the surface.

Small strain analyses were performed utilizing a finite element mesh with 4247 degrees of freedom. Refinement of the mesh was checked using the case of a rough rigid footing. Comparison between collapse loads for highly reinforced embankments agreed with collapse loads for rigid footings, obtained from plasticity theory, to better than 7%. (The technique used to perform this comparison has been described in detail by Rowe and Soderman, 1987b).

The embankments were "constructed" by turning on gravity within rows of elements and involved up to 14 lifts and a total of up to 250 loadsteps.

3 RESULTS OF SOME STEEL REINFORCED EMBANKMENT ANALYSES

The input parameters used in the finite element analyses reported herein were factored according to a limit states design type philosophy. The actual partial factors used were taken from the Ontario Highway Bridge Design Code, 1983. For example, a factor of 0.65 was applied to the clay foundation strength properties and factors of 0.8 and 1.25 were applied to the tangent of the angle of internal friction and unit weight of the fill, respectively. The values of clay foundation strengths quoted in the cases presented are nominal (unfactored) values.

For purposes of further discussion, failure of the reinforced embankment has been defined as the height at which the increment in vertical displacement is equal to or exceeds the increment in fill thickness. This corresponds to the height beyond which the addition of more fill will not result in a net increase in embankment height.

For some geotextile reinforced embankments this practical definition of failure may result in failure heights considerably less than the plasticity "collapse height" (Rowe and Soderman, 1987a). However, for heavily reinforced embankments such as those examined here, the failure height is in fact very close to the plasticity collapse height at which uncontained plastic flow occurs.

Figure 2 shows the plastic region and velocity field at failure for an embankment constructed on a clay foundation with a nominal undrained shear strength at the surface of c_{uo} = 15 kPa, and a rate of increase with depth of ρ_c = 2.5 kPa/m. (Note that factored parameters were actually used in the analysis.) The failure height for this reinforced embankment with a steel spacing of 375 mm was found to be 4.4 m, as compared to 2.8 m for the unreinforced case. With the limit state philosophy adopted in these analyses, the failure height obtained using factored parameters corresponds to the allowable working height for the nominal conditions and so in this case the reinforcing steel at a 375 mm spacing gives rise to a 57% increase in the allowable height to which an 18 m wide (crest width) embankment could be constructed on this deposit.

Careful examination of the velocity field shown in Figure 2 reveals an extensive lateral component to the deformation of both the fill and foundation material. In this case, the failure mechanism is controlled by yield in the reinforcement,

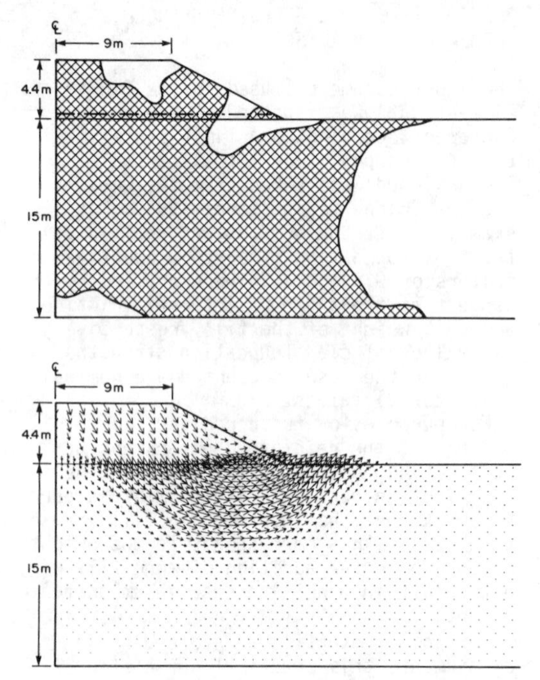

Figure 2. Plastic region and velocity field at failure for nominal parameters c_{uo}=15 MPa, ρ_c=2.5 kPa/m and S=375 mm

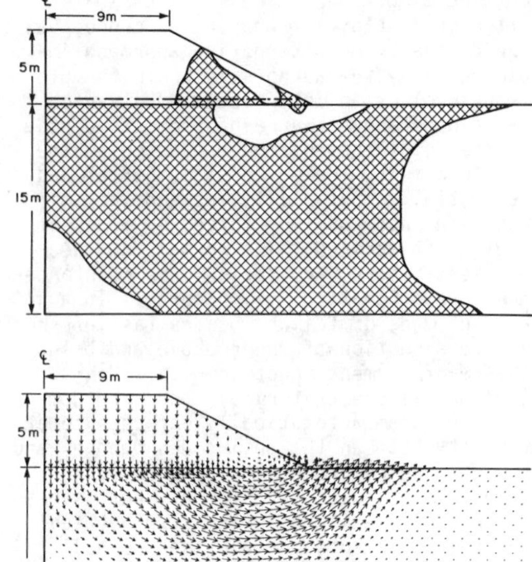

Figure 3. Plastic region and velocity field at failure for nominal parameters c_{uo}=15 kPa, ρ_c=2.5 kPa/m and S=125 mm

general foundation failure and limited reinforcement pullout.

The effect of a three-fold increase in the amount of reinforcement, S = 125 m, can be seen in Figure 3. Assuming the same clay foundation properties as in the previous case, the increase in steel prevented yield of the reinforcing strips in this case. The reinforced failure height of 5.0 m corresponds to a 79% improvement over the unreinforced case. In this case, the collapse mechanism is a function of general foundation soil failure and some reinforcement pullout. As might be expected, increasing the amount of steel by a factor of three has improved the embankment performance and increased the failure height relative to that for S = 375 mm, however it is noted that this case with S = 125 mm is approaching the maximum height which can be achieved for a perfectly reinforced embankment and hence the provision of additional steel (i.e. adopting a spacing closer than 125 mm) is unlikely to achieve significant additional improvement in failure height for these conditions.

Apart from the obvious increase in failure height, there are some important differences in the failure mechanism when compared to the S = 375 mm case. Examining the velocity field in Figure 3, it can be seen that the displacement vectors in the embankment fill between the centreline and shoulder show essentially no lateral component of displacement. This coupled with the fact that the plastic region shows no plasticity in the same area, would suggest that the more heavily reinforced (S = 125 mm) embankment moves downward as a rigid block over a width of approximately 30 m. It is interesting to note that the point about which rotation appears to be occurring (i.e. the edge of the approximate rigid footing) corresponds to the point at which the applied pressure γh is equal to the surface bearing capacity of 5.14 c_{uo} when the factored values of γ and c_{uo} are considered. This provides evidence for the approximate approach for estimating collapse heights proposed by Rowe and Soderman (1987b).

Figure 4 shows the shear stress distribution above and below the reinforcing strips per metre width of embankment, as obtained from finite element analysis for the case of c_{uo} = 15 kPa, ρ_c = 2.5 kPa/m, and S = 125 mm. Between the centreline and halfway to the shoulder, the reinforcing strips

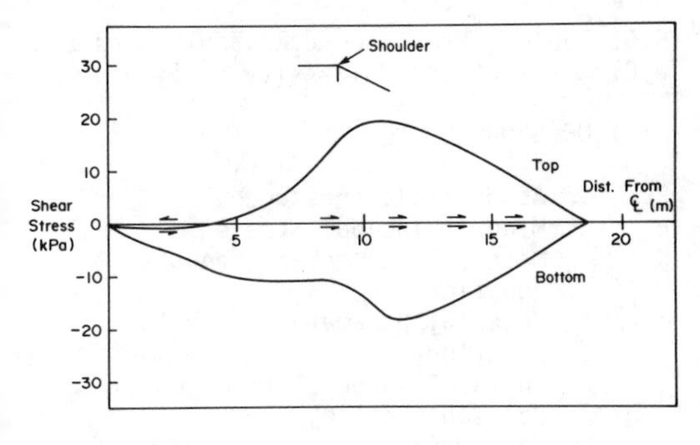

Figure 4. Shear stress distribution along reinforcement at failure for nominal parameters c_{uo}=15 kPa, ρ_c=2.5 kPa/m, S=125 mm

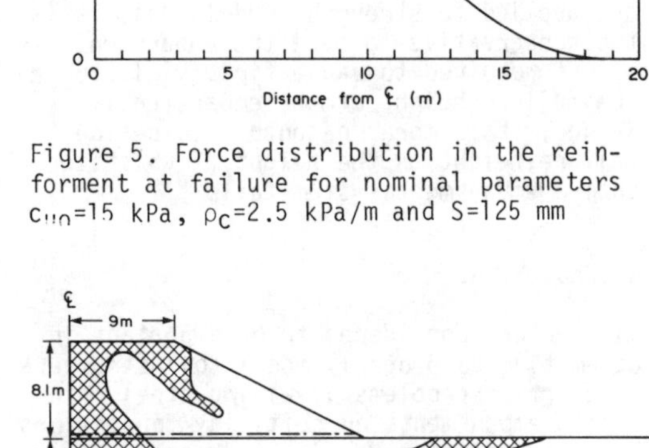

Figure 5. Force distribution in the reinforment at failure for nominal parameters c_{uo}=15 kPa, ρ_c=2.5 kPa/m and S=125 mm

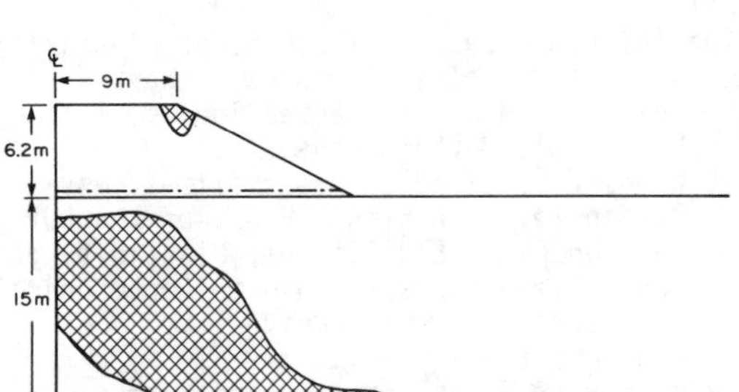

Figure 6. Plastic region at first yield of reinforcement for nominal parameters c_{uo}= 30 kPa, ρ_c=2.5 kPa/m and S=125 mm

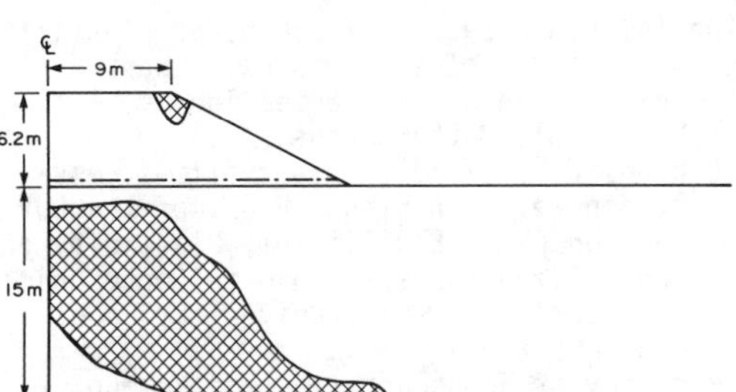

Figure 7. Plastic region at failure for nominal parameters c_{uo}=30 kPa, ρ_c=2.5 kPa/m and S=125 mm

are subjected to direct shear whereas the remainder of the reinforcement experiences "pullout" type shear although even at failure pullout is only occurring along the outer 10 m of the reinforcement (i.e. between the shoulder and toe of the embankment).

The reinforcement force distribution corresponding to the shear stress distribution of Figure 4 is shown in Figure 5. For this case, the maximum reinforcement force occurs at the centreline of the embankment.

Since the reinforcing strips commercially available are of limited length, the distribution of forces developed in the reinforcement would have definite implications when selecting splice locations in a design application. Ideally, one would like to locate splices in regions of minimum reinforcement force. Plots such as Figure 5 give an indication of what magnitude of reinforcement forces might be developed, and may aid in determining whether a given splice location is feasible.

In many instances, collapse of a reinforced embankment on a soft clay foundation is preceded by yield in the reinforcement. Once this yield has occurred, additional fill can be added prior to collapse of the entire reinforced embankment system. In some cases, the quantity of this additional fill is significant and the finite element analysis is particularly useful for establishing the circumstances under which this situation might occur.

Figures 6 and 7 show results for a reinforced embankment with a steel spacing of 125 mm, constructed on a clay foundation with nominal parameters c_{uo} = 30 kPa and ρ_c = 2.5 kPa/m.

Figure 6 shows the extent of plasticity in the soil at the occurrence of first yield in the reinforcing strips. It can be seen that up to this point, yield in the fill and foundation has been limited by the stiff reinforcement and that what failure there is in the foundation is contained by a large region of elastic soil. Clearly, the embankment is still stable. Figure 7 shows the plastic region for the embankment at failure. Following first yield, 1.9 m of fill was added before failure occurred. This represents an additional 30% increase in height beyond the point of first yield in the reinforcement.

This raises an important point in regards to what one defines as failure of the reinforced soil system. When designing steel reinforced embankments using limit state design (i.e. where partial factors are applied to strengths and loads), it is too conservative to take the embankment height required to cause first yield to be the failure height of the embankment. Rather, it is more reasonable to define the failure height as the height at which the entire embankment system fails.

4 CONCLUSIONS

The factors considered to be important in attempting to properly model soil-structure interaction problems involving steel reinforced embankments on soft clay foundations have been discussed. It has been shown that the mode of failure may vary substantially depending on the amount of reinforcement used and the properties of the foundation. It has been demonstrated that a substantial improvement in collapse height can be achieved using reinforcement in the form of steel strips.

It has been noted that in some cases yield of the reinforcement occurs well before collapse of the entire embankment system and it is suggested that when using limit state design for reinforced soil systems, failure should be associated with failure of the entire system rather than a component of the system, such as the steel. The finite element results also show that large forces can develop along a significant length of the reinforcement and hence care is required in the selection of location for splices in the reinforcement.

In general, it has been shown that numerical modelling using the finite element technique can be a powerful tool for use in developing design guidelines for steel reinforced embankments on soft clay foundations.

5 ACKNOWLEDGEMENT

The work presented in this paper is funded by the Natural Science and Engineering Research Council of Canada under grant A1007. Additional funding was provided by the Reinforced Earth Company (Australia). The authors gratefully acknowledge the value of discussions with Mr. M. Boyd of the Reinforced Earth Co. and with Professor H.G. Poulos, Professor J.R. Booker and Dr. J.C. Small of the University of Sydney.

LIST OF SYMBOLS

P_a	Atmospheric pressure
σ_3	Minor principal stress
K	Material parameter, Janbu's equation
m	Material parameter, Janbu's equation
ϕ	Angle of internal friction
ν	Poisson's ratio
γ	Unit weight
S	Centre to centre reinforcing strip spacing

REFERENCES

Davis, E.H. (1968). Theories of plasticity and failure of soil masses. Chapter 6 in Soil Mechanics - Selected Topics, I.K. Lee (Ed.), Butterworths.

Duncan, J.M. (1980). Hyperbolic stress-strain relationships. Proc. of the Workshop on Limit Equilibrium, Plasticity and Generalized Stress-Strain in Geotechnical Engineering. McGill University, Montreal, Canada.

Ministry of Transportation and Communications. Ontario Highway Bridge Design Code (and Commentary). Toronto: Highway Engineering Division, 1983.

Naylor, D.J. and Richards, H. (1978). Slipping strip analysis of reinforced earth. Int. J. for Num. and Analyt. Meth. in Geomech., 2, pp. 343-366.

Rowe, R.K. (1986). Numerical modelling of reinforced embankments constructed on weak foundations. Proc., 2nd International Symposium on Numerical Models in Geomechanics, Ghent, Belgium.

Rowe, R.K. and Soderman, K.L. (1987a). Reinforcement of embankments on soils whose strength increases with depth. Proc., Geosynthetics '87 Conference, New Orleans, U.S.A.

Rowe, R.K. and Soderman, K.L. (1987b). Stabilization of very soft soils using high strength geosynthetics: the role of finite element analysis. 1st Geosynthetics Research Institute Seminar, Philadelphia, U.S.A., October 1987.

Vidal, H. (1969). The principle of reinforced earth. Highway Research Record No. 282, pp. 1-16.

Zienkiewicz, O.C. (1977). The Finite Element Method in Engineering Science, 3rd Edition. London: McGraw-Hill.

Numerical Methods in Geomechanics (Innsbruck 1988), Swoboda (ed.)
© 1988 Balkema, Rotterdam. ISBN 90 6191 809 X

Experience with numerical modelling of dams

M.Doležalová, A.Hoření & V.Zemanová
Hydroprojekt, Prague, Czechoslovakia

ABSTRACT: Experiences with FEM analyses of earth - and rockfill dams aiming at the safe design of their sealing elements against cracking and hydraulic fracturing are summarized in the paper. Particular attention is paid to field measurement results which were used to select design criteria for impervious core and basic assumptions for the computational and constitutive models. For this purpose strain paths in rockfill dams were derived from field measurements and interpreted by laboratory tests. The frequent occurrence of the stress path causing unloading in shear and the significance of displacements due to wetting and creep are emphasized and the corresponding algorithms described. Further the experiences are discussed and some selected applications presented.

1 INTRODUCTION

Up to early seventies the slope stability was considered as the basic design criterion for earth- and rockfill dams. Applying it also to the impervious cores, stiff core materials placed at water content below optimum were preferred aiming at the reduction of the porewater pressure. As a safety measure the water content of the clay cores was often reduced during construction /1/ or even the core itself was drastically narrowed /2/.

At this time, however, field performance records have been already available showing cracking and piping /3,4,5,6,7/ and later hydraulic fracturing as a very frequent, if not predominant, failure mode of earth- and rockfill dams /2,5,6,7,8,9,10,11/. As a rule, dry, stiff cores were demaged, while flexible, wet cores exerting high pore pressures performed well /1,12,13,14/. These records clearly showed that depending on the supposed failure mode, the requirements to the core could be quite different, even opposite ones.

In the light of these findings it becomes evident that the prevention against cracking and hydraulic fracturing is to be considered as a decesive design criterion for impervious cores and methods for predicting these phenomena are necessary.

For this purpose a design concept based on the analysis of field measurement results and on the use of numerical methods for predicting cracking and arching effect in the clay cores was suggested by the first author /5,6,16/. Along with that a soil mechanics oriented softwere including different 2D and 3D FEM codes with path dependent constitutive model and pre- and postprocessing programs has been developed at Hydroprojekt Prague.

Using this softwere, numerous 2D and 3D FEM analyses have been performed for high rockfill dams built in Czechoslovakia, Bulgaria and Germany: Římov (45m), Stanovice (60m) /16/, Jirkov (50m) /17/, Dalešice (90m) /14/, Nová Bystrica (60m), Slezská Harta (70m), Pades (90m), Asenovec (70m) /18/ and Goldisthal (90m) /19,20/.

These analyses resulted in practical recommendations which were directly implemented in the design practice. The recommendations concerned not only the prediction of thegeneral dam behaviour and the cracking and hydraulic fracturing potential /16,14, 18/ but also the interpretation of the field measurement results /14,17/ and suggestions for dam reconstruction /17/.

None of these solutions, however, were used for slope stability calculations as the stress state computed for a safe slope is neither identical nor proportional with the stress state at the slip failure. More sophisticated constitutive models /21,22/ or new numerical methods are necessary for this purpose /23,24/. So far the use of them, however, has not brought considerable improvements in comparison with the currently applied limit state methods /22,24/.

The findings, experiences and algorithms

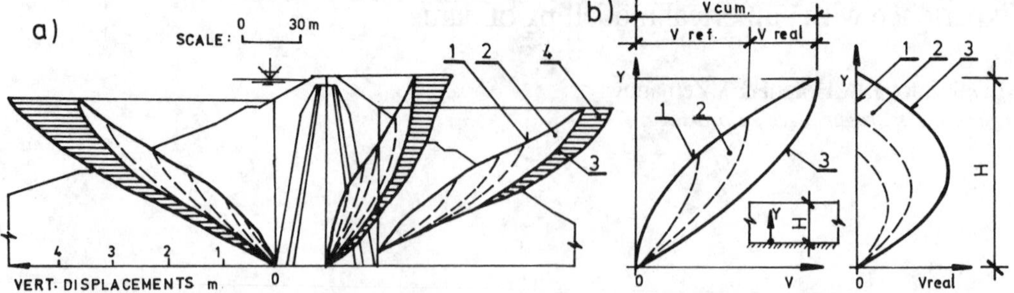

Fig.1. a) Vertical displacements registered by inclinometers during the dam construction and reservoir filling (after [26])

b) Vertical displacements during the multilift construction of a homogeneous infinite layer

described in the following refer mostly to the earth- and rockfill dams built in narrow valleys on imcompressible bedrock and loaded by static loads.

2 FIELD PERFORMANCE OF DAMS

According to field evidence [1-15,25,26] the characteristic features of the dam behaviour which should by simulated by a relevant computational model are as follows.

1. The specific distribution of displacements during the multilift construction procedure shown in Figs. 1 and 2. In these Figures the cumulative and references curves are notable which indicate the displacements compensated during construction and the position of the measuring device.

The magnitude of the real displacements during construction is determined by the difference between these curves. Multilift construction sequence produces a parabolic distribution with a maximum displacement in the middle of the layer. A hypothetical single lift loading of the dam produces a curve with the maximum at the crest which corresponds to the cumulative curve.

For a rough estimation it is useful to know the analytical expressions for these curves which are valid for a homogeneous, linear, infinite layer on imcompressible foundation (Fig.1b):

$$V_{cum} = \frac{\gamma}{E}\left(Hy - \frac{y^2}{2}\right); \quad V_{ref} = \frac{\gamma}{2E}\, y^2 \qquad (1)$$

$$V_{real} = \frac{\gamma}{E}\left(Hy - y^2\right),$$

where V_{cum}, V_{ref}, V_{real} are the cumulative, compensated and real displacements, and γ, E, H are the bulk density, deformation modulus and the height of the layer.

2. The post-construction displacements

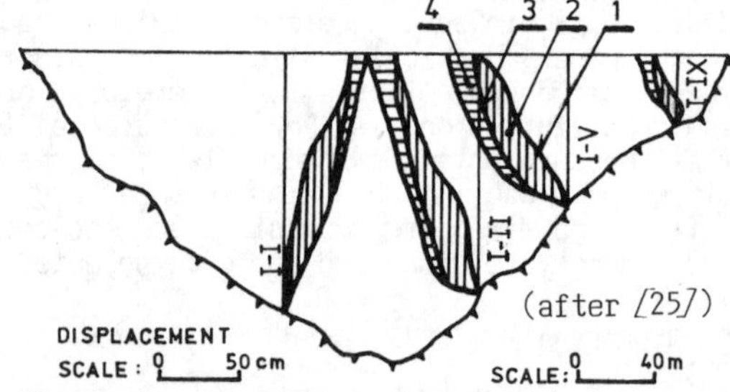

Fig.2. Longitudinal horizontal displacements registered by inclinometers during dam construction and reservoir filling

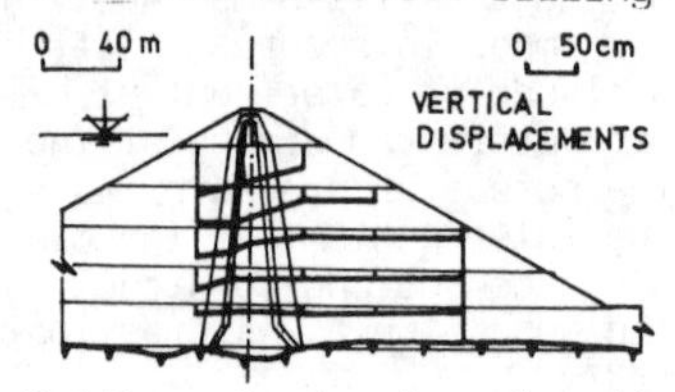

Fig.3. Settlements due to saturation at the first filling of the reservoir [27]

which are considerably influenced by the settlements due to saturation and creep. Sudden settlement of the upstream rockfill due to wetting at the first filling of the reservoir is a typical phenomenon which unfavourably influences the displacement pattern (Figs.1 and 3) and often causes cracking.

According to the field records the pore pressure dissipation in the clay cores of zoned dams is very slow and it continues several years after construction [1,13,14, 15,27]. Hence, the longtime settlement of the crest of rockfill dams occurs mostly due to creep (Fig.4). Note that all post-construction movements are concentrated on the crest (Fig.1,2,3).

3. Differential settlements which develop during the dam construction and first reservoir filling due to following factors:

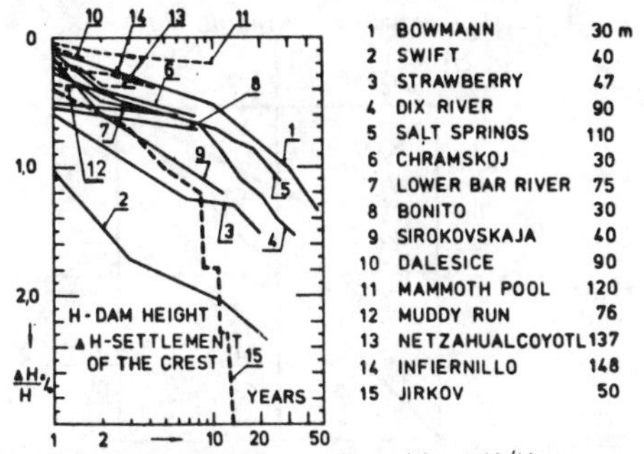

Fig.4. Crest settlement ratio ΔH/H versus time for rockfill dams (—— rockfill; --- earth-rockfill)

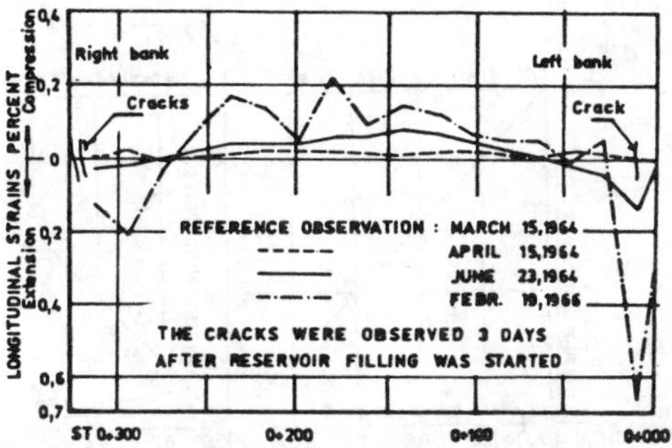

Fig.5. Longitudinal strains and cracking of the crest uppon filling of reservoir /27/

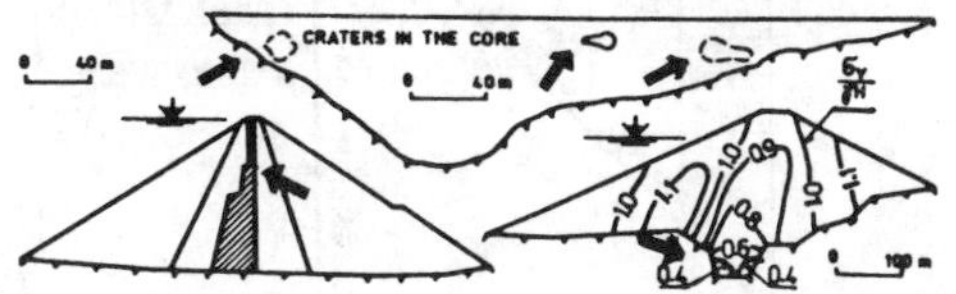

Fig.6. Zones of distress in earth-rockfill dams according to field measurements

steep abutments causing large longitudinal horizontal displacements (Fig.2), different stiffnesses of the core and shoulders (Fig.1), settlements due to wetting (Figs. 1,3) and compressible foundation.

4. Tensile zones caused by differential settlements in longitudinal directions. A typical pattern which brought about the cracking of the crest is shown in Fig.5.

5. Zones of low stresses produced by load transfer resulting in arching effect. As a rule, these zones are registered in narrow cores /2,7/, narrow key trenches /10,13/, along the crests /2,9/ and steep abutments /5/, at bedrock irregularities and around constructions embedded into the dam /8,12/ (Fig.6).

The zones of low stresses allow for hydraulic fracturing when an unfavourable cumulation of the following factors occur:
- combined arching effect in transversal and longitudinal directions;
- rapid first filling of the reservoir;
- use of dry, stiff, broadly coarsed, erodable core materials;
- imperfect contact of the core with the adjacent fills or bedrock (absence of effective filters, open fissures, etc.)

6. Some field evidences indicating a considerable overestimation of displacements computed by FEM when using standard triaxial tests /28,29/, and good agreement with using oedometer tests. An other finding concerns rockfill dams with membranes where displacements due to water pressure are much more smaller than would be indicated by the deformation modulus of the rockfill determined by field tests /30/.

From these findings the following inferences were drawn for the computational model:

1. The decisive loading stages are the end of construction and the first rapid filling of the reservoir. An accurate simulation of the construction sequence is necessary for computing real stresses especially for the crest.

2. The analysis of the dam in longitudinal direction is necessary either by 3D solution or an approximate 2D solution. For predicting cracking, the right simulation of post-construction settlements is of great importance.

3. The behaviour of the core material in tension must be tested and considered by the constitutive model. The bending test was found out as the most appropriate one /3,5/.

4. Regarding the slow pore pressure dissipation and the rapid reservoir filling, it is not inevitable to apply a coupled approach for predicting cracking and hydraulic fracturing. Moreover, even a total stress approach is acceptable for a wet core, as the cracking and hydraulic fracturing criteria are formulated in total stresses.

5. The settlements due to wetting and creep and the stress paths followed by fills must be taken into account.

3 STRAIN PATHS PROCEDURES

3.1 Strain path concept, graphical presentation

To find out the typical stress paths followed by the fills in a rockfill dam, first the typical strain paths were derived using field measurement results.
More detailed information on these works are contained in /31,32,19/, there a brief summary of the main points is given.

The strain paths are defined in the 2D space of principal strain ratios $e_2 = \varepsilon_2/\varepsilon_1$

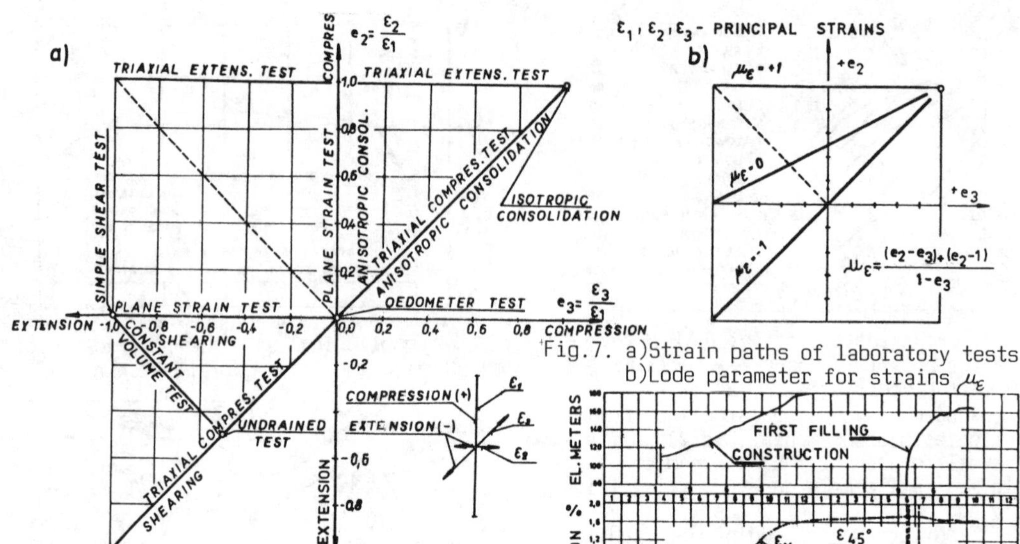

Fig.7. a)Strain paths of laboratory tests
b)Lode parameter for strains μ_ε

and $e_3 = \varepsilon_3/\varepsilon_1$ as the time sequence of strains in a given point ($\varepsilon_1, \varepsilon_2, \varepsilon_3$ - principal strains, compression (+)). In this space the strain paths of various laboratory tests and the Lode parameter for strains μ_ε can be plotted (Fig.7). The latter expresses the influence of the third strain invariant and gives information about the type of loading:
$\mu_\varepsilon = -1$(compression), $\mu_\varepsilon = 0$(shearing), $\mu_\varepsilon = +1$ (extension).

3.2 Strain paths in rockfill dams

The strain paths in various rockfill dams were evaluated using inclinometer and extensometer records and identifying one of the strain components with the corresponding principal strain. Typical examples of the measurements results and the computed strain paths are shown in Figs.8 and 9.

Based on similar plots for 5 rockfill dams the following was concluded:

1. The starting points of paths are mostly in the zone of anisotropic consolidation.
2. The directions and lengths of the paths are different: some paths are long with decreasing e_2, e_3 (cores, transits) while the others are short with increasing e_2, e_3 (central parts of shells).
3. Regardless to the permanently increasing external loads (embankment construction, reservoir filling), the in situ strain paths frequently change their directions. This indicates the change of anisotropy of the principal strains in a given point.
4. All paths are close to the oedometer and plane strain paths.

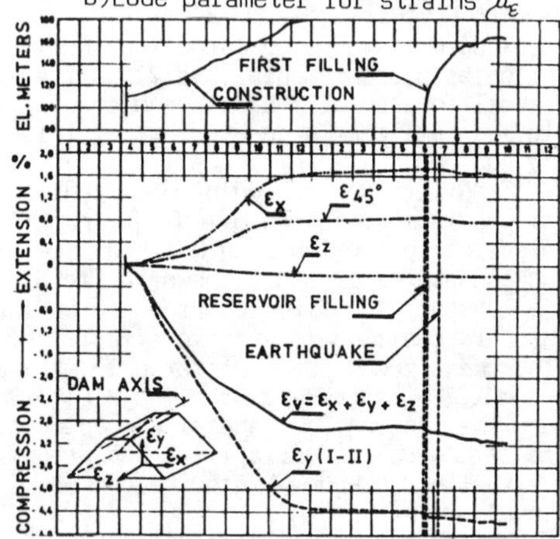

Fig.8. Strains versus time recorded by inclinometers [25,27]

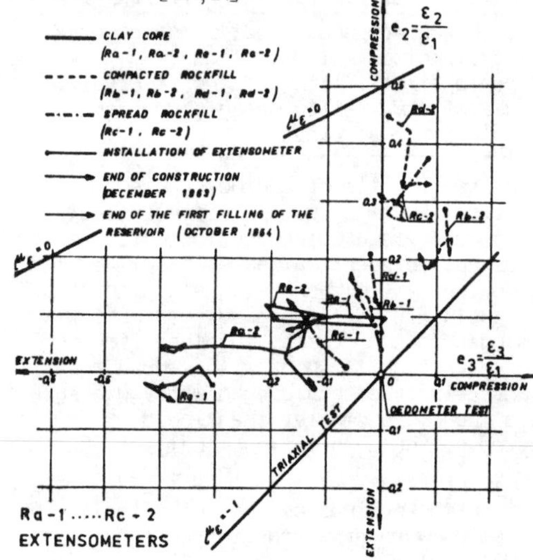

Fig.9. Strain paths in Infiernillo Dam [31]

5. Shearing at constant volume but no dilatancy was found out (see Fig.10).
6. The approximate range of the Lode parameter change is $-1,0 < \mu_\varepsilon < -0,4$.

3.3 Strain paths of laboratory tests

The analysis of the CR tests of isotropically consolidated and anisotropically consolidated samples with $K=\Delta\sigma_3/\Delta\sigma_1 =$ = const ($\sigma_1, \sigma_2, \sigma_3$ - principal stresses) resulted in following findings (Fig.10):

1. The long strain paths with decreasing $e_3=e_2$ ($\Delta e_3=\Delta e_2 < 0$) belong to the CR tests where $\Delta K < 0$ ($\Delta K=K-K_c$; K_c - stress ratio at the consolidation).
2. The short strain paths with increasing e_3 ($\Delta e_3 > 0$) belong to the CR tests where $\Delta K > 0$.
3. By relating $e_3= \varepsilon_3/\varepsilon_1$ to the shear strength mibilization $i=\tau_{oct}/\tau_{oct}^{lim}$ (τ_{oct} -octahedral shear stress, τ_{oct}^{lim}-the limit τ_{oct} according to failure hypothesis) a nearly unique relation can be obtained for CR tests with different K (Fig.11).
4. According to this relation the paths cluster into the three groups:

1st group: paths with increasing shear strength mobilization ($\Delta i > 0$; $\Delta K < 0$; $\Delta e_3 < 0$);

2nd group: paths with constant shear strength mobilization ($\Delta i \doteq 0$; $\Delta K \doteq 0$; $\Delta e_3=0$, i.e. oedometer paths);

3rd group: path with decreasing shear strength mobilization ($\Delta i<0$; $\Delta K > 0$; $\Delta e_3 > 0$).

3.4 Path dependence of deformation parameters

Evaluating the tangent values of deformation parameters E_t, μ_t (E_t - tangent deformation modulus, μ_t - tangent Poisson ratio) for CR tests with different K and relating them to i, another, almost unique, relations were found out for the paths of the 1st group (Fig.12a). This relations indicates a decrease of E_t (and increase of μ_t), i.e. the loosening of the soil.

Similar relations for anisotropically consolidated samples indicate the same for the paths of the 1st group with $\Delta i > 0$ (right side of Fig.12b). For the paths of the 2nd and 3rd groups with $\Delta i \leqq 0$, however, a sharp, increase of E_t (and decrease of μ_t) was found out (left side of Fig.12b). The latterindicates the hardening of the soil caused by unloading in shear.

Thus the deformation parameters E_t and μ_t depend not only on the stress level determined by σ_{oct} (octahedral normal

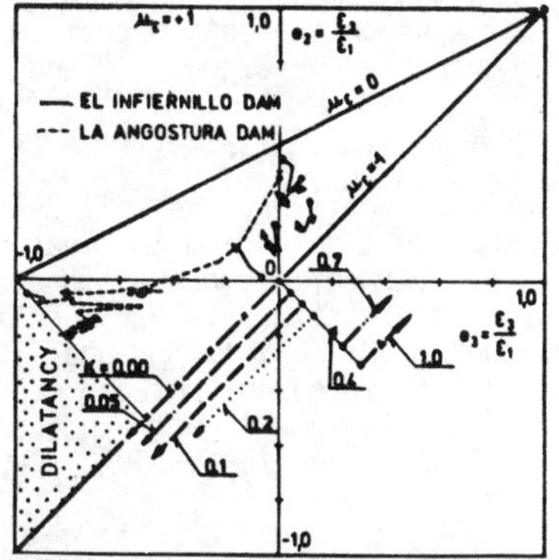

Fig.10. Comparison of the in situ strain paths with the paths of triaxial CR tests for sand

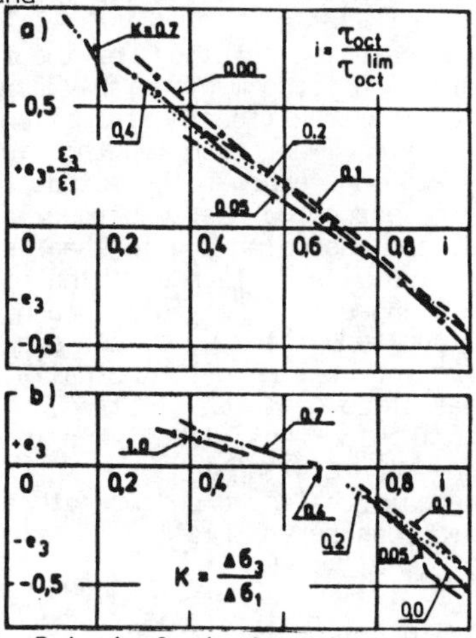

Fig.11. Principal strain ratio e_3 versus shear strength mobilization $\underline{i}$ for CR tests with different K

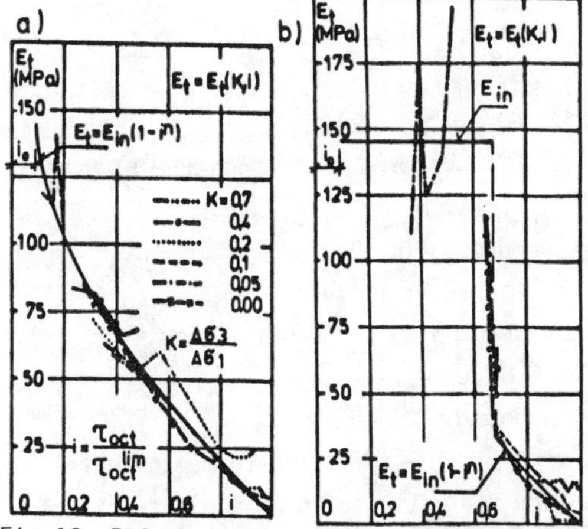

Fig.12. Deformation modulus E_t versus shear strength mobilization $\underline{i}$ for isotropically (a) and anisotropically (b) consolidated samples

stress) and the shear strength mobilization i but also on the path direction changes controlled by sign $\Delta\sigma_{oct}$ and sign Δi.

The analysis performed, as well as the Fig.13 show that unloading in shear is a frequent phenomenon in embankments and the majority of the strain paths in rockfill dams belong to the last two groups. This explains the disagreement between measurements and computations when the unloading in shear is neglected [28,29,30].

4 ALGORITHMS

4.1 Path dependent constitutive relations

Simple hypoelastic relations of the first order are used for computing the tangent values of the traditional deformation parameters E_t and μ_t for each loading step. These relations express the mutual influence of hardening due to stress level increase and loosening due to shear strength mobilization. The influence of the path direction changes is taken into account by switch functions: sign $\Delta\sigma_{oct}$ and sign Δi. According to the sign of principal stresses, the zones of compression (isotropy), compression/tension (anisotropy) and tension (isotropy) are distinguished.

In the compression zone four path directions are considered (Fig.14):

1. $\Delta\sigma_{oct} \geq 0;\ \Delta i > 0$

$$E_t = E_{in}(1-(1-\delta)i_*^{\,n}) \qquad (2)$$

$$\mu_t = \mu_p + (\mu_{max} - \mu_p)i_*^{\,m} \qquad (3)$$

2. $\Delta\sigma_{oct} \geq 0;\ \Delta i \leq 0$

$$E_t = E_{in} = E_p\left(\frac{\sigma_{oct}}{\sigma_o}\right)^p \qquad (4)$$

$$\mu_t = \mu_p \qquad (5)$$

3. $\Delta\sigma_{oct} < 0;\ \Delta i > 0$

E_t according to (2) substituting E_{unload} for E_p

μ_t according to (3)

4. $\Delta\sigma_{oct} < 0;\ \Delta i \leq 0$

E_t according to (4) substituting E_{unload} for E_p

μ_t according to (5)

where E_p, μ_p - initial deformation parameters for unit stress σ_o and for initial shear strength mobilization i_o

p - exponent controlling hardening by σ_{oct}

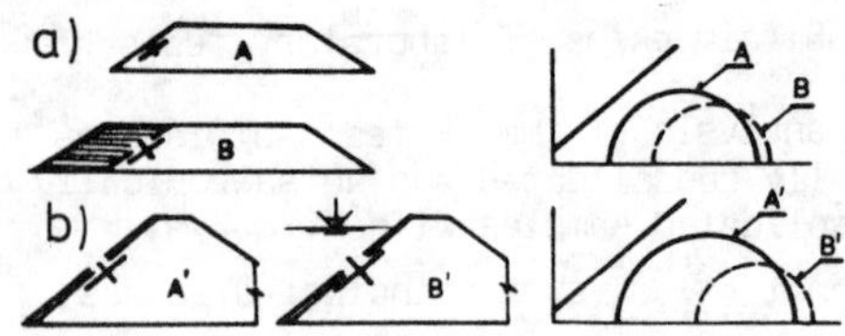

Fig.13. The occurrence of unloading in shear in embankments during the construction (a) and first filling of the reservoir (b)

n,m - exponents controlling loosening by τ_{oct}

δ - ratio determining the minimum value of E_t at failure

$$i_* = \frac{i - i_o}{1 - i_o}\ .$$

All parameters are determined by curve fitting procedure using standard triaxial, eventually simple shear and oedometer tests [19].

In the zone of compression/tension four similar paths and the anisotropy due to different response of compacted fills to compression and tension are taken into account. In the direction of the major principal stress the above relations are used, while in the direction of minor principal stress a bilinear behaviour governed by the deformation modulus in tension andthe tensile strength is assumed. For the tension zone the same assumption holds and the required parameters are determined by bending test (Fig.15).

Each loading step is repeated for adjusting the parameters to the stress state.

As a failure criterion the generalized Von Mises criterion is used which is more realistic in the estimated range of Lode parameter change ($-1 < \mu_\varepsilon < -0{,}4$) than the Mohr-Coulomb one, being too much on the safe side [33]. A similar conclusion is made in [22] but the recommended range is only $-1 < \mu_\sigma < -0{,}8$.

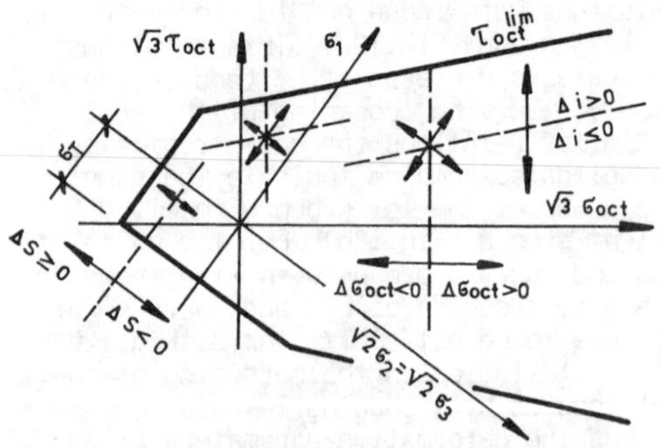

Fig.14. Stress paths considered by the constitutive model

4.2 Settlements due to wetting

A more detailled description of the algorithm using initial strain method [35] for calculating settlements of rockfill due to wetting is given in [14]. Based on experimental results of Hayashi [38] and Marsal [36] regarding the influence of the void ratio e and moister content w, and Žurek [37] and Kolosov concerning the effect of the stress level, the following relation was suggested for the vertical submerging strain component ε_s:

$$\varepsilon_s = /a(w)+b(w)e+c(w)e^2/ \ . \ /\frac{\sigma_y}{\sigma_{exp}}/^{0,5}, \quad (6)$$

where σ_{exp}= 0,9 MPa - experimental load for which the subsequent relations were verified [38]:

a = -0,082 + 0,009 w
b = 0,550 + 0,080 w
c = -0,500 + 0,052 w

The relation (6) is valid in the range $0,2 < e < e_{max}$ and $0,5 < w < 3,5\%$. The value of e_{max} depends on w:

w%	0	1	2	3	3,5
e_{max}	0,55	0,525	0,491	0,472	0,35

Vector $\underline{\varepsilon}_s = (0,\varepsilon_s,0,0,0,0)^T$ is used as an initial strain vector for calculating equivalent nodal forces which form the fictitious load vector when computing submerging settlements by FEM.

4.3 Creep

Displacements due to creep are also computed by initial strain method and the creep laws suggested by Feda [39] are applied. Evaluating the results of oedometer and ring shear tests carried out for different materials, some important relations between the uniaxial creep rate $\dot{\varepsilon}_a$, volumetric creep rate $3\,\dot{\varepsilon}_{oct}$ and distortional creep rate $\dot{\gamma}_{oct}$ were found out:

1. Relation $\dot{\varepsilon}_a$ versus log t (t-time) is linear (Fig.16)
2. $3\,\dot{\varepsilon}_{oct}/\dot{\varepsilon}_a = \varkappa_1 = 0,12 \div 0,23$ (7)
3. $\dot{\gamma}_{oct}/\dot{\varepsilon}_a = \varkappa_2 = 0,82 \div 2,88$ (8)
4. $\dot{\varepsilon}_{oct}$ and $\dot{\gamma}_{oct}$ depends on $i = \tau_{oct}/\tau_{oct}^{lim}$

Using these findings the following creep laws were suggested [39]:

$$\varepsilon_a = \varepsilon_a^o + 10^{a\sigma_{oct}-b} \ . \ln t \quad (9)$$

$$\varepsilon_{oct} = \varepsilon_{oct}^o + 10^{a\sigma_{oct}-c}(1-i)^{-d} . \ln t \quad (10)$$

$$\gamma_{oct} = \gamma_{oct}^o + 10^{a\sigma_{oct}-g}(1-i)^{-f} . \ln t \quad (11)$$

where ε_a^o, ε_{oct}^o, γ_{oct}^o - uniaxial and octahedral strains for t=0 and a,b,c,d,f,g - empirical parameters which could be calibrated using creep test results or estimated basing on the generalized results of the above mentioned experiments.

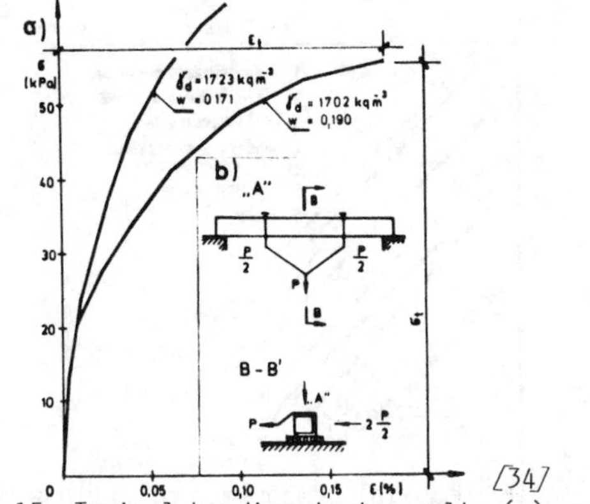

[34]

Fig.15. Typical bending test results (a) and the draft of the bending test apparatus (b)

Displacements due to creep are computed for a given loading step assuming constant stress level during a time increment Δt. The calculation consist of the following steps:

1. Calculation of the creep vector related to the time increment Δt using relations (9),(10),(11).
2. Calculation of the equivalent nodal forces forming the fictitious load vector.
3. FEM solution using this fictitious load vector.

As input data the stress tensor $\underline{\sigma}^o$ corresponding to the load increment, the parameters a,b,c,d,f,g and the time increment Δt are given.

The creep vector $\Delta\underline{\varepsilon}^r = (\Delta\varepsilon_x^r, \Delta\varepsilon_y^r, \Delta\varepsilon_z^r, \Delta\gamma_{xy}^r, \Delta\gamma_{xz}^r, \Delta\gamma_{yz}^r)^T$ is computed by following steps:

1. Treatment of special cases: if $\sigma_{oct}^o > 0$ (tension) then $\Delta\underline{\varepsilon}^r = (0,0,0,0,0,0)^T$; if $\varepsilon_2^o = \varepsilon_3^o \leq 0$ then $\Delta\varepsilon_1^r = 10^{a\sigma_{oct}-b} \ . \ln \Delta t$ and $\Delta\varepsilon_2^r = \Delta\varepsilon_3^r = 0$

2. Computation of the volumetric and distortional creep according to (10) and (11):

$$\Delta\varepsilon_{oct}^r = 10^{a\sigma_{oct}-c}(1-i)^{-d} \ln \Delta t \quad (12)$$

$$\Delta\gamma_{oct}^r = 10^{a\sigma_{oct}-g}(1-i)^{-f} \ln \Delta t \quad (13)$$

3. Computation of θ in the octahedral plan:

$$\theta = \frac{1}{3}\sin^{-1}\left(-\frac{3\sqrt{3}}{2}\frac{J_3^o}{(J_2^o)^{3/2}}\right); \ -\frac{\pi}{6} < \theta < \frac{\pi}{6} \quad (14)$$

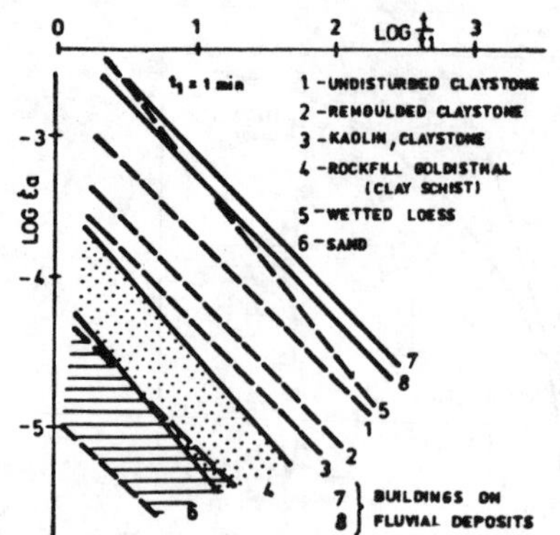

Fig.16. Uniaxial creep rate of various materials (by Feda /39/)

J_2^o - 2nd invariant of the stress deviator

J_3^o - 3rd invariant of the stress deviator

4. Computation of the principal creep components:

$$\begin{bmatrix}\Delta\varepsilon_1^r\\ \Delta\varepsilon_2^r\\ \Delta\varepsilon_3^r\end{bmatrix} = \frac{1}{\sqrt{2}}\Delta\gamma_{oct}^r \cdot \begin{bmatrix}\sin(\theta + \frac{2\pi}{3})\\ \sin\theta\\ \sin(\theta + \frac{4\pi}{3})\end{bmatrix} + \Delta\varepsilon_{oct}^r \quad (15)$$

5. Computation of the principal angles using the components of the stress tensor $\underline{\sigma}^o$.

6. Computation of the components of the creep vector $\underline{\Delta\varepsilon}^r$ using the principal angles and principal creep components.

In this way all the six creep components can be obtained which are considered further as initial strains. Using these strains the fictitious load vector $\underline{F}_n$ is calculated according to initial strain procedure /35/.

In case of incremental loading a total fictitious load vector $\underline{F}_n$ corresponding with the cumulated stresses and time is computed for each loading step and the difference $\underline{\Delta F}_n = \underline{F}_{n-1} - \underline{F}_n$ is applied as a loading vector for the step under consideration.

4.4 Experiences

1. The path dependent constitutive model using incremental tangent stiffness approach requires small loading steps, especially when failure zones arise. In the opposite case the equilibrium fails and uncorrect stresses can be obtained. To avoid that the equilibrium is to be checked for each loading step using initial stress method.

2. No simple constant strain triangles can be used regarding the sharp changes of the stress components. Particularly, the contact of the triangle with joint elements can cause numerical difficulties. As the simplest element the quadrilateral element consisting from four triangles can be recommended.

3. Modelling the initial geostatic stress state of the subsoil a right selection of the Poisson ratio is necessary. A too low Poisson ratio brings about low lateral pressure and large stress deviator which could result in an unrealistic stress state being close to the failure. A check by the formula $\mu_t > \frac{1-\sin\varphi}{2-\sin\varphi}$ derived from the Jáky's relation could be recommended (φ-internal friction angle).

4. Similarly, at a particular loading step an error in Poisson ratio influences considerably the minor and middle principal stresses and hence the stress deviator and shear strength mobilization. In this way uncorrect deformation moduli and uncorrect displacements due to creep could be obtained. Analogical errors could be brought about by a failure hypothesis which is too much on the safe side.

5. This simple, incrementally linear model cannot be recommended for analysis of structures which are in the limit state. Neither the direction of strain increments nor the volume strains can be correctly computed for this state and numerical difficulties likely arise. For these problems more sophisticated models are to be applied. When selecting a new model the graphical comparison shown in /40/ is of great value. The rate type relation suggested by Kolymbas /21/ was tested by us with good results and it was selected for the further use. A problem arise, however,with the nonlinear system of eguations which results from the incremental nonlinearity of this model.

5 APPLICATIONS

In this part some selected results of FEM analyses where the foregoing algorithms were applied are given. The prediction of deformations, distress zones, cracking and hydraulic fracturing was aimed at and IBM 370 and HP 1000 computers were used. Computational models with nodal point number about 1500 and quadrilateral, isoparametric and joint elements were applied.

5.1 Dalešice Dam

For this 90 m high dam - the highest earth-rockfill dam in Czechoslovakia - extensive 2D and 3D analyses were carried out /14/ in the early stage of construction (embankment height 30 m) when an unusual behaviour of the dam was indicated by the measuring devices. In the sloped clay core (w_L=0,35 , I_p=0,15 , E_p=15 MPa , μ_p=0,45 , c_u=0,12 MPa, φ_u=7°) compacted at $w > w_{opt}$ and supported by rockfill shoulders (amphibolite, porosity n = 17%, E_p= 50 MPa, μ_p=0,26; φ= 42°) considerable arching effect (50-60%) and high pore pressures arised. The 2D and 3D computational models were adjusted to the performance of the dam in this initial stage and the behaviour at the end of construction and first filling of the reservoir was predicted. A fair agreement for the downstream horizontal movements (computed: 13 cm, measured 10 cm for water level increase 14 m) and for the upstream submerging settlements (calculated 12 cm, measured 11 cm, Fig.6 in /14/) was obtained.

The safety of the core against hydraulic fracturing and cracking was checked using criteria (16) and (17):

$p < \sigma_i + \sigma_t$ (16); $\varepsilon_y^{crest} < \varepsilon_t$ or $\sigma_y^{crest} < \sigma_t$ (17)

where p-hydrostatic pressure on the upstream face of the core, σ_i - total principal stresses in the stream direction (i= 1,2); ε_y^{crest}- longitudinal principal strain at the crest; σ_t, ε_t - tensile strength and maximum elongation of compacted clay according to bending test. A low safety factor against hydraulic fracturing (σ_2/p = 1,1 taking σ_t= 0, see Fig.3 in /14/) but a high one against cracking (1,8--2,7) was found out (Fig.17). Due to construction pore pressures which were higher than the reservoir water pressure no hydraulic fracturing occured and hitherto the performance of the dam is fairly good.

5.2 Goldisthal Dam

Some results of 3D solution for this 90 m high rockfill dam being to be built in GDR have been already published in /19/. The analysis aimed at prediction of asphaltic concrete facing deformations due to reservoir filling and creep. During the reservoir filling the unloading in shear (see Fig.13) manifested itself and the path dependent model computed almost 3 times higher deformation moduli than at the end of the construction (Fig.18 in /19/). The displacements due to reservoir filling and creep are shown in Fig.18. The creep parameters derived from the oedometer creep tests were as follows: a=2,84 , b=5,55 , c=6,25 , d=0,3 , f=0,9 , g=5,55 /39/. The maximum deflection of the facing is 60 cm from which 35 cm occurs due to creep. Near to the bottom a smaller but still significant influence of the creep was computed: 17 cm from a total deflection 44 cm.

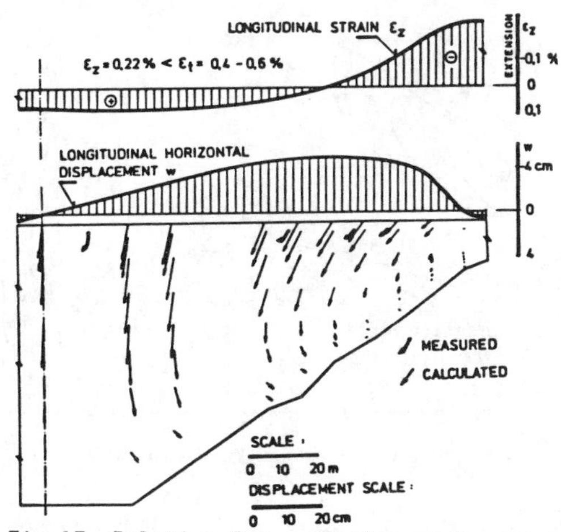

Fig.17. Dalešice Dam - displacements and tensile strains due to wetting and creep by 3D FEM /41/

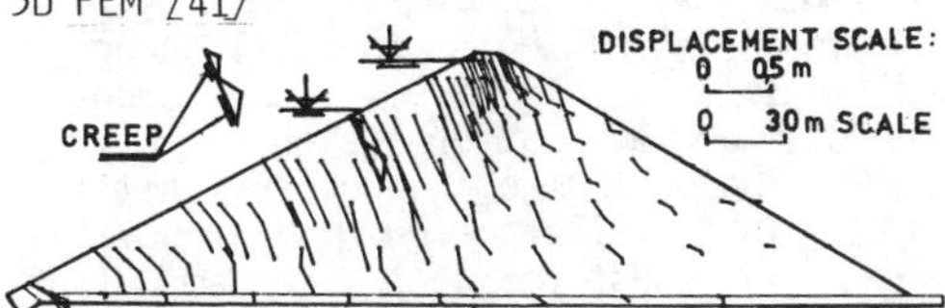

Fig.18. Goldisthal Dam - displacements due to reservoir filling and creep by 3D FEM

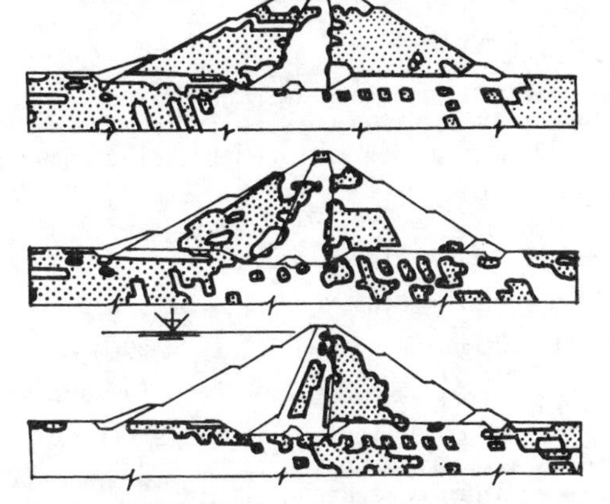

Fig.19. Nová Bystrica Dam - zones where unloading in shear and hardening of the fill occur during the construction and reservoir filling

5.3 Nová Bystrica Dam

This 60 m high dam with central core (clay with gravel, w_L=0,36, I_p=0,17 , E_p=9,7 MPa, μ_p=0,45 , c_u=0,05 MPa , φ_u = 5°) and rockfill shoulders (sandstone with claystone, E_p= 48 MPa, μ_p=0,27 , φ= 40°) is under construction in the region of Carpathian

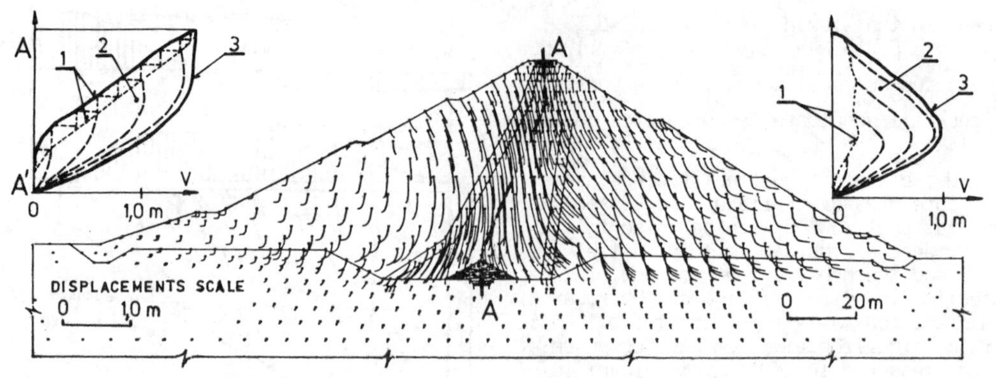

Fig.20. Nová Bystrica Dam - displacements at the end of the construction; 1,2,3-see Fig.1

Flysch. By modelling the multilift construction of the dam and the reservoir filling, the zones where unloading in shear resulting in the hardening of the fill occurs were registered (Fig.19). The corresponding displacement pattern is shown in Fig.20. The maximum constructional settlement is 76 cm while the cumulativ settlement at the crest indicating the amount of the fill compensated during the construction is 148 cm.

5.4 Slezská Harta Dam

Only two 2D solutions instead of the planned 3D solution were performed for this 71 m high dam with central clay core (w_L=0,44 , I_p=0,215 , E_p= 8 MPa , μ_p= 0,45 , c_u= 0,06 MPa, φ_u=15,3^o , CIU tests), wide transit zones (clayey gravel, E_{p_o}= 30 MPa , μ_p= 0,30 , c´= 0,05 MPa , φ´= 31°) and rockfill shoulders (basalt, E_p = 50 MPa , μ_p= 0,26 , φ=34°), unless the unfavourable geological conditions of the site (fractured basalt stream on the right side of the valley, bedrock weakened by wide, compressible faults) should require the 3D analysis. The displacements and principal stresses along the face of the core computed for the maximum cross section at the end of construction are shown in Fig.21a. The predicted settlements due to saturation and creep are presented in Fig.21b. Displacement pattern computed for the longitudinal section (Fig.21c) proves plane strain conditions in the middle part and considerable settlements of the bedorck (up to 15 cm) at the right side of the valley. The corresponding stress state (Fig.21d) shows remarkable stress transfer in the bedrock due to faults and only small tensile and distress zones in the core. The solution approved the effectivness of the suggested measures against cracking and hydraulic fracturing: the concrete wall separating the parts of the dam and the bedrock with opposite deformational trends, the flexible wet core with higher lateral pressures, the larger difference between the maximum water level and crest (3,80 m), the effective downstream filters along all zones of stress decrease, the higher requests to the compaction of the sluiced upstream rockfill to lower the submerging settlements and the slow reservoir filling to allow the swelling of the clay.

6 CONCLUSION

According to our experiences gained from field measurements and numerical solutions, the safety of sealing elements of rockfill dams can and should be assessed by numerical methods. Along with the current slope stability analysis this assessment should be included into the standard design procedures using PC. For this purpose the findings and algorithms presented in the paper could be useful.

ACKNOWLEDGEMENT

The FEM codes were developed and the tasks were solved in collaboration with Dr.V.Ulrich at Hydroprojekt Prague.

REFERENCES

/1/ A.D.Hosking, Monitoring of Earth-Rock Dams, Australian Geomechanics Journal, pp. 1-12, 1974

/2/ D.M.Wood, B.Kjaernsli and K.Höeg, Thoughts concerning the unusual behaviour of Hyttejuvet Dam, Proc.of 12ve ICOLD,

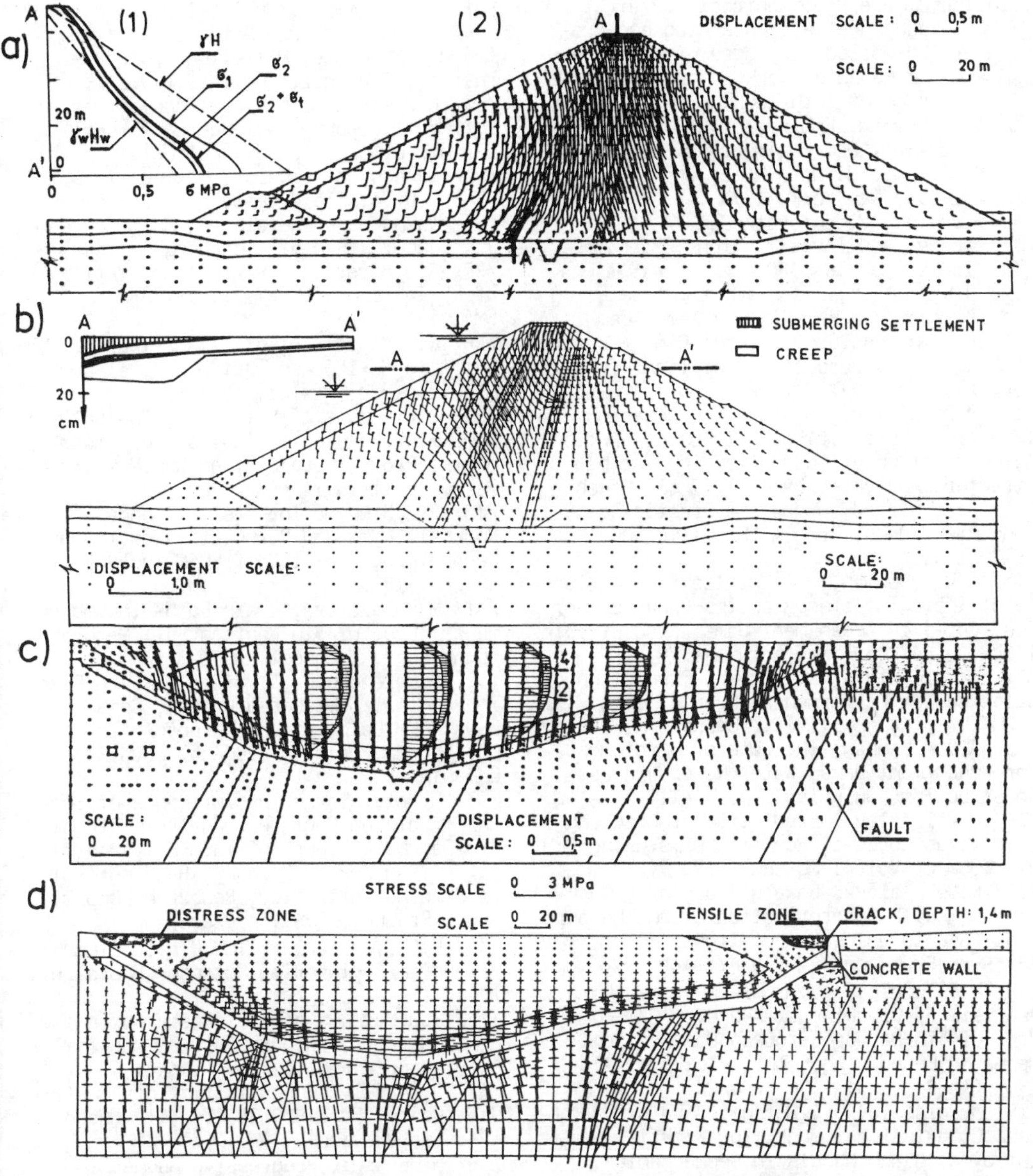

Fig.21. Slezská Harta Dam - performance of the dam with wet soft clay core
a) prediction of hydraulic fracturing (1) and displacement pattern (2) at the end of the construction
b) displacements due to wetting at the first filling of the reservoir and due to creep
c) construction and postconstruction displacements in longitudinal direction
d) stress distribution in longitudinal direction

pp. 391-414, Mexico, 1976
/3/ G.A.Leonards, J.Narain, Flexibility of clay and cracking of earth dams, Proc.ASCE, SM2, 1963
/4/ J.L.Sherard et al., Earth and Earth-Rock Dams, New York, 1963
/5/ M.Doležalová, The Influence of Abutment Steepness on Cracking of Clay Cores of Rockfill Dams, Ph.D.Thesis, Moscow, 1968
/6/ M.Doležalová, The Effect of Steepness of Rock Canyon Slopes on Cracking of Clay Cores of Earth-and Rockfill Dams, Proc. of

10th ICOLD, Q36/R13, Montreal, 1970
/7/ P.R.Vaughan et al., Cracking and Erosion of the Rolled Clay Core of Balderhead Dam and the Remedial Works Adopted for its Repair, Proc. of 10th ICOLD,Q.36/R.5, 1970
/8/ J.L.Sherard, Embankment Dam Cracking, in: Embankment Dam Engineering, S.Poulos and R.Hirschfeld, John Wiley and Sons, pp. 272-353, New York, 1973
/9/ H.Vestad, Viddalsvatn Dam, A History of Leakages and Investigations, Proc.of 12ve ICOLD, pp. 369-389, Mexico, 1976
/10/ H.B.Seed et al., Hydraulic Fracturing and its Possible Role in the Teton Dam Failure, in: Failure of Teton Dam. A report of Findings by Buro of Reclamation, Interior Review Group, 1977
/11/ M.Doležalová, Hydraulic Fracturing of Impervious Cores of Rockfill Dams, Conf.on Automatization of Foundation and Earth Structure Analysis, Brno, 1979 (in Czech)
/12/ J.L.Sherard, Hydraulic Fracturing in Embankment Dams, Proc.ASCE, GT 10, pp.905--927, 1986
/13/ A.D.M.Penman, J.A.Charles, The Influence of their Interfaces on the Behaviour of Clay Cores in Embankment Dams, Proc.of 13th ICOLD, Q.48/R.39, New Delhi, 1979
/14/ M.Doležalová, F.Leitner, Prediction of Dalešice Dam Performance, Proc.of 10th ICSMFE, 1/16, pp.111-114, Stockholm, 1981
/15/ A.D.M.Penman, The waterproof element for embankment dams, Water Power and Dam Construction, July 1985, pp.33-112
/16/ M.Doležalová, Prediction of Cracking in Impervious Cores of Rockfill Dams using FEM, 8th ISSMFE, Vol.4, Moscow, 1973
/17/ M.Doležalová, Case History of a Rockfill Dam with Interpretation of the Measurement Results using Computational Model, Int.Symp.FMGM, pp.827-838, Zurich, 1983
/18/ M.Doležalová, V.Mikulášková, V.Škoch, Analysis of Cracking at Asenovec Dam, Proc.of 5th Danube European Conf.on Soil Mech.and Found. Eng., pp.31-45, Bratislava, 1977 (in Russian)
/19/ M.Doležalová, A.Hoření, A Path Dependent Computational Model for Rockfill Dams, Int.Symp.on Num.Models in Geomechanics, pp.577-587, Zurich, 1982
/20/ M.Doležalová, Computation of Dams by Finite Element Method. State of Experience, Lecture, University Stuttgart, 1983
/21/ D.Kolymbas, A Constitutive Law of the Rate Type for Soils and Other Granular Materials, Proc.of 1st Czechoslovak Conference on Num.Meth. in Geomech., High Tatras, 1987
/22/ J.K.Zareckij, V.N.Lombardo, Statics and Dynamics of Embankment Dams, Energoatomizdat, Moscow, 1983 (in Russian)
/23/ T.Tamura, S.Kobayashi, T.Sumi, Rigid Plastic Finite Element Method for Soil Structure, Proc.of 5th Int.Conf.on Num. Meth. in Geomech., pp.185-192, Nagoya,1985
/24/ P.Gussmann, Die Methode der Kinematischen Elemente, Mitteilung 25, Baugrundinstitut Stuttgart, pp.1-97, 1986
/25/ R.J.Marsal, L.R.Arellano, Presa el Infiernillo, observaciones en la cortina durante el periodo de construction y primer Llenado del embalse, Mexico, 1965
/26/ B.Gilg, Field tests and control measurements on Gäscheneralp and Mattmark dams, 8th ICOLD, R.8/Q.31, Edinburgh, 1964
/27/ R.J.Marsal, L.R.Arellano, Performance of El Infiernillo Dam, 1963-1966, Proc.ASCE SM4, 1967
/28/ R.J.Marsal, et al., Behaviour of Dams Built in Mexico, Comission Federal de Electricidad, Mexico, 1977
/29/ Z.Eisenstein, S.T.C.Law, The Role of Constitutive Laws in Analysis of Embankments, Proc.of 3rd Conf.on Num.Meth.in Geomech., pp.1413-1430, Aachen, 1979
/30/ V.F.B.de Mello, Design Trends on Large Rockfill Dams and Purposeful Monitoring Needs, Proc.of Int.Conf.FMGM, pp.805-826, Zurich, 1983
/31/ M.Doležalová,Strain Paths in Embankments, Proc.of 5th Conf.on Soil Mech.and Found.Eng., Budapest, 1976
/32/ M.Doležalová, A.Hoření, Strain Paths in Rockfill Dams-Measurements, Constitutive Laws, FEM Calculations, Proc.of 4th Int. Conf.on Num.Meth.in Geomech.,pp.679-689, Edmonton
/33/ M.Doležalová, The Influence of Some Input Data on Nonlinear Analyses of Dams, 6th Danube-Europe Conf.ISSMFE, Varna, 1980
/34/ I.Vaníček, J.Záleský, Bending tests of Slezska Harta clay, Research Report, ČVUT, Prague, 1986 (in Czech)
/35/ O.C.Zienkiewicz, The Finite Element Method in Engineering Science, McGraw-Hill, 1971
/36/ R.J.Marsal et al., Research on the Behaviour of Gran.material and Rockfill Samples, CFE Publication, 1965
/37/ J.Žurek, Compressibility of Rockfill at High Loads, Proc.VODGEO, Vol.11, Moscow, 1965 (in Russian)
/38/ M.Hayashi, Progressive Submerging Settlement During Water Loading to Rockfill Dam, Proc.of Int.Symp."Criteria and Assumptions for Num.Analyses of Dams", Swansea, 1975
/39/ J.Feda, Creep of Soils, Academia, Prague, 152 pp., 1983 (in Czech)
/40/ G.Gudehus, A comparison of some constitutive laws for soils under radially symmetric loading and unloading, Proc.of 3rd Int.Conf.on Num.Meth.in Geomech., Aachen, 1979
/41/ F.Leitner, M.Doležalová, Statics of embankment dams, Research Report, Hydroprojekt, Prague, 1980 (in Czech)

Numerical Methods in Geomechanics (Innsbruck 1988), Swoboda (ed.)
© 1988 Balkema, Rotterdam. ISBN 90 6191 809 X

Finite element modelling of rapid drawdown

C.O.Li & D.V.Griffiths
Simon Engineering Laboratories, University of Manchester, UK

ABSTRACT: Aspects of finite element modelling of rapid-drawdown phenomena in soil slopes are discussed. The analyses involve finding the position of the steady-state phreatic surface through an earth dam, followed by sudden unloading due to reduction in water level on the upstream face. The analyses are performed in plane-strain using 8-noded quadrilateral elements, and it is shown that in the undrained analyses, oscillations in the computed pore pressures are considerably reduced by a form of selective reduced integration. It is found that the residual pore pressure distribution after the drawdown is not uniform with higher pore pressure concentration near the toe.

1. Introduction

A list of failures of earth dams between 1936-61 attributed to conditions set up during drawdown has been documented by Morgenstern (1963). Sudden drawdown leads not only to possible stability problems on the upstream slopes of earth dams, but may also induce slides in the natural slopes of the reservoir area. In addition, Koppejan et al(1948) suspected that the drawdown mechanism during tidal recession was one of the causes of coastal flow slides. Jones et al (1961) also recorded that landslides on the banks of the Franklin D. Roosevelt lake were more numerous after a lowering of the level of water impounded behind the Grand Coulee Dam.

The mechanism of drawdown may be considered in stages. First the pore pressure distribution in an earth dam prior to lowering of the reservoir level is governed by the steady seepage condition where the soil has consolidated under its own self weight. Following this, the lowering of the water level causes changes in the pore pressure as a result of the unloading during drawdown establishing new boundary conditions for seepage through the dam. During this transient phase, the free-surface gradually adjusts to a new equilibrium position by dissipating the excess pore pressures. As a result, the effective stresses increase with time as does the factor of safety of the upstream slope.

In fill materials of 'low' permeability, a significant amount of time is required for equilibriumn to be reached after drawdown. The most critical time however, occurs immediately after drawdown hence an 'undrained' assumption is conservative. The 'rapid drawdown' mechanism can therefore be reviewed as a consolidated (steady seepage) - undrained (rapid drawdown).

2. Analyses of Drawdown

Early attempts to analyse embankment stability during drawdown were carried out by Mayer (1936). An assumed failure plane in conjunction with the Method of slices led to a resultant shear force which implied failure if it was greater than the shear resistance of the soil. Lowe (1960) proposed a slip circle again in conjunction with the Method of slices. This more refined approach obtained an undrained shear strength (C_u) accounting for the anisotropically consolidated soil such as that occuring in elements of soil under high reservoir level prior to drawdown. Morgenstern (1963) recognised that the change in loading due to drawdown will induce immediate changes in pore pressure prior to consolidation. This concept is consistent with Skempton (1954) who first established the well-known equations in which changes in pore pressure under undrained conditions are related to changes in total stresses via the pore pressure parameters A and B. Using this approach, Morgenstern used slip circle methods to produce a set of stability charts for upstream slopes with various inclination and shear strength parameters (c', ϕ') (Morgenstern 1963).

Very few finite element analyses of the rapid drawdown problem have been reported. Desai (1977) and Li (1983) proposed a finite element procedure for seepage and stability of earth dams during transient flow where pore pressures begin to dissipate following drawdown. In these analyses, the effects of the fluid and solid phases were not coupled, but superimposed during the analyses. In the present work, the authors present a more rigorous approach in which coupling between the phases is considered. The analysis can be considered in two parts; a steady state solution followed by undrained unloading.

3. Method of Analysis

3.1 Steady Seepage

Initially, a finite element procedure is derived to locate the position of the free surface prior to drawdown. There are two basic methods for doing this; the mesh adaptive methods e.g. Taylor and Brown (1967), Finn (1967), Witherspoon and Neuman (1970), Smith and Griffiths (1987) and the fixed mesh methods e.g. Desai (1976), Bathe (1979). The fixed mesh approach has been used in the present study for the following reasons:

(i) The same mesh can be used for both the seepage and stress analyses.

(ii) The mesh adaptive methods can run into difficulties in the case of layered soils with horizontal interfaces and for soils involving non-homogeneous permeabilities. It is also difficult to assign appropriate material properties when the free surface crosses on interface. (Li 1983).

(iii) The mesh may become so distorted that the error due to the finite element approximation becomes significant.

3.2 Slope Stability

Gravity loads are applied to the slope and constant stiffness iterations (modified Newton-Raphson) are used to obtain a converged solution. The viscoplastic iterative technique is described elsewhere (e.g Zienkiewicz 1977, Grifiths 1980).

In the present analyses, after location of the steady state seepage line, the steady residual pore pressure is applied as nodal water force vector together with the gravity loading due to soil self weight and the upstream force corresponding to the water level prior to drawdown. This generates the consolidated stresses corresponding to the steady state condition. These stresses then act as the intial stress state prior to rapid (undrained) drawdown unloading.

4. The Fixed Mesh Method

Consider 2-D steady seepage through a dam as shown in Figure 1.

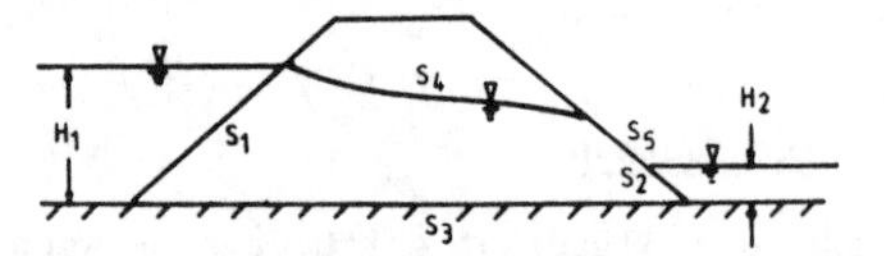

Figure 1: Boundary Conditions

For no sources or sinks within the system, the governing equation is due to Laplace of the form:

$$\frac{\partial}{\partial x}\left(k_x \frac{\partial \phi}{\partial x}\right) + \frac{\partial}{\partial y}\left(k_y \frac{\partial \phi}{\partial y}\right) = 0 \qquad (1)$$

where ϕ = potential, k_x, k_y = permeabilities in x and y directions. The boundary conditions from Figure 1 can be summarised as follows:

$$\begin{aligned} &S1 \quad \phi = H_1 \\ &S2 \quad \phi = H_2 \\ &S3 \quad \frac{\partial \phi}{\partial n} = 0 \\ &S4 \quad \phi = y, \ \frac{\partial \phi}{\partial n} = 0 \\ &S5 \quad \phi = y \end{aligned} \qquad (2)$$

Boundary conditions S1, S2, S3 can be easily satisfied in the finite element formulation (e.g. Smith and Griffiths 1987). On the free surface, the location of S4 is unknown 'a priori' and condition S5 depends on S4. The method used is to set the permeability to zero, or in practice a very small number, when the computed pressure becomes negative (i.e. above the free surface) during each iteration (Bathe 1979, Werner 1986). A smoothing technique proposed by Werner (1986) is also used to avoid undesirable oscillation which occurs when a sudden change in permeability is encountered (Figure 2).

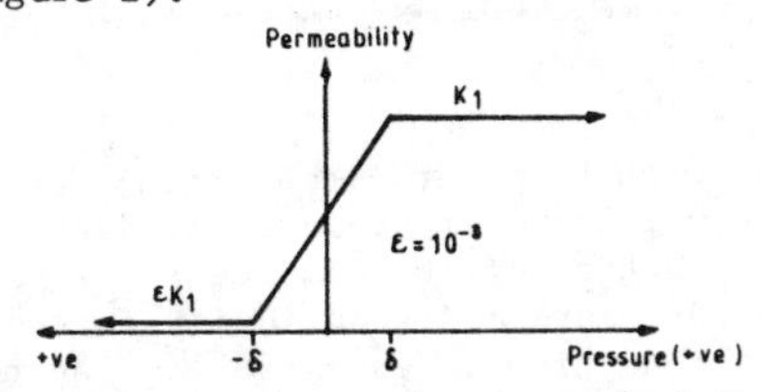

Figure 2: Permeability–Pressure Relationship

The length of the seepage surface S5 is estimated on each iteration by minimising the net outflow given by the expression

$$\frac{\partial}{\partial x}\left(k_x \frac{\partial \phi}{\partial x}\right) + \frac{\partial}{\partial x}\left(k_y \frac{\partial \phi}{\partial y}\right) \quad (3)$$

5. Simplified Biot formulation for 2-D undrained Analysis

Assuming for the moment an elastic soil skeleton and an incompressible pore fluid we get the following equilibrium equation:

$$\frac{\partial \sigma'x}{\partial x} + \frac{\partial \tau xy}{\partial y} + \frac{\partial u}{\partial x} = F_x \quad (4)$$

$$\frac{\partial \sigma'y}{\partial y} + \frac{\partial \tau xy}{\partial x} + \frac{\partial u}{\partial y} = F_y \quad (5)$$

where u = excess pore pressure, F_x, F_y = body forces.

With reference to the pore fluid, Darcy's law and continuity conditions can be summarised thus:

$$\frac{\partial}{\partial x}\left(k_x \frac{\partial \phi}{\partial x}\right) + \frac{\partial}{\partial y}\left(k_y \frac{\partial \phi}{\partial y}\right) = \frac{\partial}{\partial t}(\varepsilon_x + \varepsilon_y)$$

(Where plane strain conditions are implied). For saturated undrained conditions, no volume change takes place hence only the stress equilbrium equation is required which can be written as:

$$\underline{B}^T \underline{\sigma}' + \underline{B}^T \underline{U} = \underline{F} \quad (6)$$

where $\underline{B}^T = \begin{vmatrix} \frac{\partial}{\partial x} & o & \frac{\partial}{\partial y} \\ o & \frac{\partial}{\partial y} & \frac{\partial}{\partial x} \end{vmatrix}$ $\quad \underline{u} = \begin{vmatrix} u \\ u \\ o \end{vmatrix}$

$\underline{\sigma}' = \begin{vmatrix} \sigma'_x \\ \sigma'_y \\ \tau_{xy} \end{vmatrix}$

Relating stress to strains we get:

$$\underline{\sigma}' = \underline{D}'\underline{\varepsilon}, \quad \underline{u} = \underline{D}^u \underline{\varepsilon}^u \quad (7)$$

where $\underline{D}^u$ = stress strain matrix of the pore fluid

$\underline{D}'$ = effective stress strain matrix of the soil skeleton

For undrained conditions, there is no relative movement between the pore water and soil skeleton, hence:

$$\underline{\varepsilon} = \underline{\varepsilon}^u \quad (8)$$

In addition, we assume that the pore pressures contribute only to direct stresses (and not at all to shear stresses) hence:

$$\underline{D}^u = K_A \begin{vmatrix} 1 & 1 & 0 \\ 1 & 1 & 0 \\ 0 & 0 & 0 \end{vmatrix} \quad (9)$$

where K_A is the apparent Bulk Modulus of the pore fluid, usually assigned a large, but finite value (Griffiths 1985). The analysis now amounts to a 'penalty' formulation in which the total Poisson's ratio of the soil/fluid system approaches 0.5.

We can now write a simplified version of the Biot formulation for an undrained analysis. From equations (6) and (7).

$$\underline{B}^T (\underline{D}' + \underline{D}^u)\, \underline{\varepsilon} = \underline{F} \quad (10)$$

or

$$\underline{B}^T\, \underline{D}\, \underline{\varepsilon} = \underline{F} \quad (11)$$

where $\underline{D} = \underline{D}' + \underline{D}^u$

represents the total stress/strain matrix.

A similar formulation was first obtained by Naylor (1974). Although this method can be used to estimate pore pressures from equation (7), it has been found that the values obtained at the Gauss points tend to oscillate. A technique for smoothing these oscillations is now described.

6. An Integration Technique for the Pore Pressure Smoothing

For nearly incompressible material such as saturated soils during undrained loading, 'reduced' integration is commonly used in conjunction with 8-node elements to avoid the troublesome volumetric portion of the strain energy and to overcome 'mesh locking' (e.g. Fried 1974, Zienkiewicz 1977). Reduced integration relaxes the number of constraints imposed by satisfying the incompressibility condition only at the Gauss points. As a result, the overall performance of the element may be improved, but errors are introduced into the volumetric strain energy by relaxing the 'zero' volume change condition throughout the element. Although these errors may be small in magnitude, they may become significant as pore pressures after multiplication by a 'large' fluid bulk modulus.

In the present work, it is proposed that exact integration is used to obtain the total strain energy, but stresses will still be sampled at the 'reduced' integration point locations. The reduced integration points for stresses will continue to be used because:

(i) Computation time is reduced when using constant stiffness iterations in which the global stiffness matrix is formed once only.

(ii) The reduced integration points are the best positions for assembling stresses for either reduced or exact integrations schemes as observed by Zienkiewicz (1977).

The integration technique described above will be confirmed by numerical examples in the next section. It can be shown that the particular application described in this paper does not suffer from 'mesh locking' when using exact integration.

Discussions

Examples of the proposed fixed mesh method using 8-node quadrilateral elements with 'reduced' (2 x 2) integration are presented in Fig. 3 and 4. The free surface solutions are observed to converge rapidly in all the cases and agree very well with other published solutions. With the given unit permeabilities in x- and y- directions, the calculated flow rate for the vertical homogeneous dam in fig 3 is 4.59 compares favourably with the result 4.55 from Ligget (1977). For the

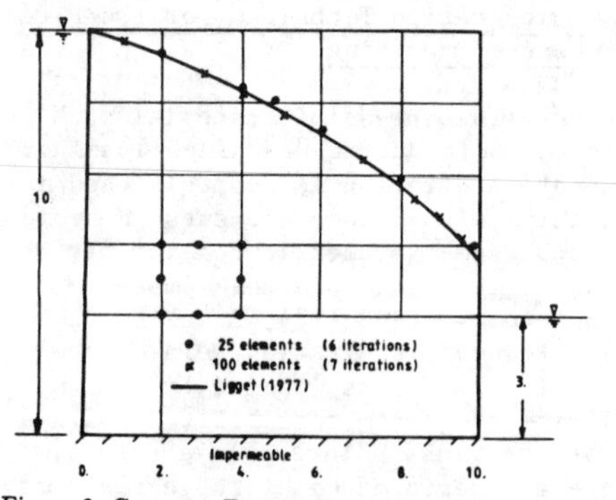

Figure 3: Compare Free Surface with published solution

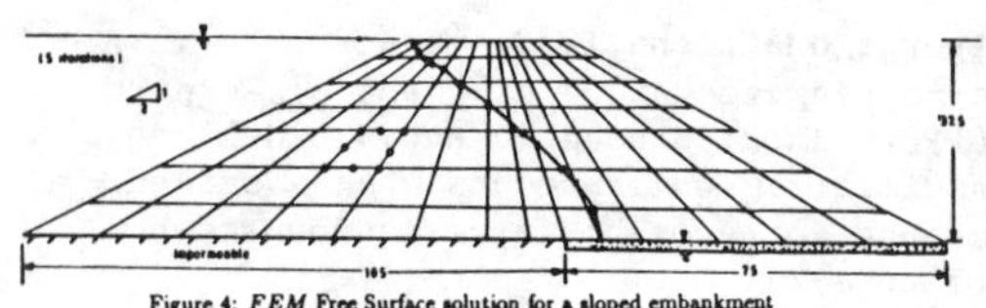

Figure 4: *FEM* Free Surface solution for a sloped embankment

example with horizontal toe drain, the point at which the impervious boundary along the bottom of the dam meets the drain will create a discontinuity in the boundary conditions. This problem can be avoided by representing the boundary meeting point with two nodes, one connected to the element above the impervious boundary with the other, with zero prescribed pressure, connected to the element above the drain.

In stress analysis for slope stability during rapid drawdown, the element is assumed to be elastic-perfectly plastic with the Mohr-Coulomb failure criterion. Non-linearity introduced by the plasticity is accounted for using the viscoplastic method. Non-dilatant soil behaviour is assumed and non-associated flow rule is used. The factor of safety (FOS) is defined as $C_f' = C'/FOS$, $\phi_f' = \text{Arc tan}(\tan \phi'/FOS)$. An example of a fully submerged slope under complete drawdown is used to study the effect of the magnitude of KA (Fig. 5). It shows that the FOS remains unchanged throughout the range of KA = 25 E' to 1000E'. However, the pore pressure response due to the drawdown unloading suffers severe oscillations. The magnitude of these oscillations increases with increasing KA (Fig. 6). This is thought to be due to the errors involved in the volumetric strain ($\Delta\varepsilon_v$).

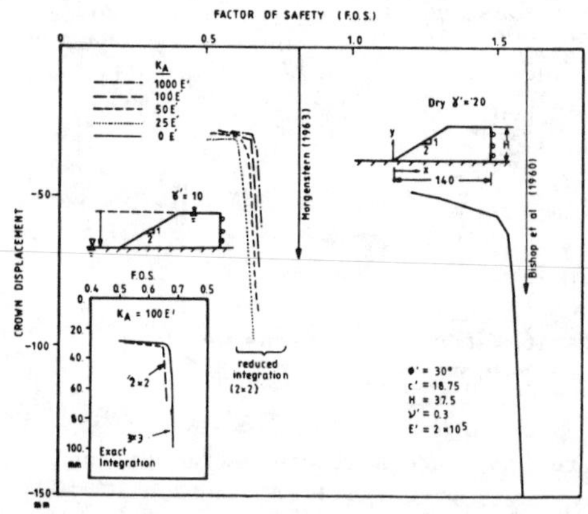

Figure 5: Effect of K_A on *F.O.S.* : Compare with published results

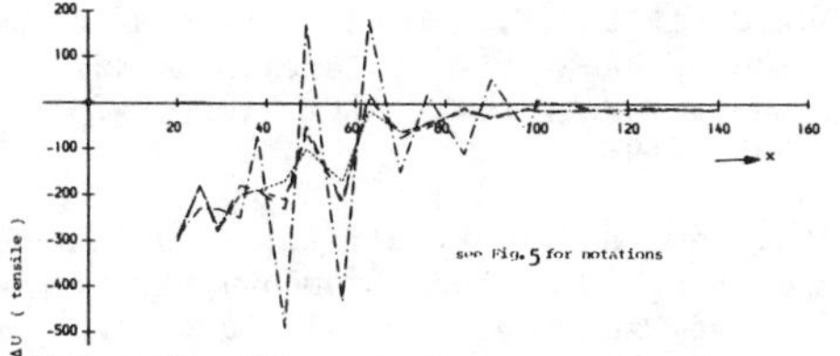

Figure 6: Effect of K_A on pore pressure response at failure at $y = 8.9$.

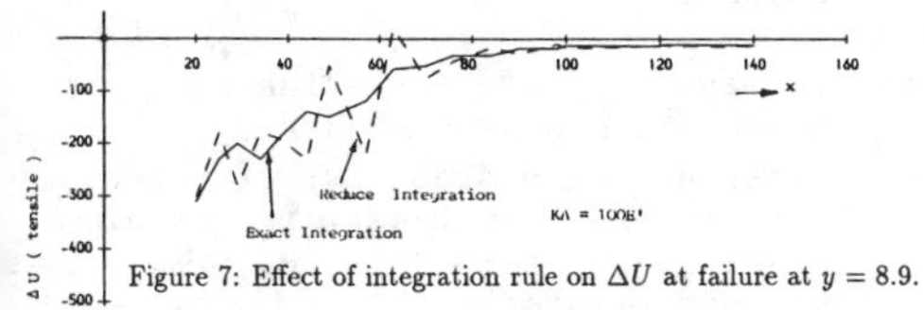

Figure 7: Effect of integration rule on ΔU at failure at $y = 8.9$.

Since the pore-pressure is $\Delta U = KA . \Delta\varepsilon_v$, increasing KA will magnify the error in the pore pressure and hence the oscillations. The error in the volumetric strains is a consequence of 'relaxing' the element using 'reduced' integration, and can be reduced if exact integration is used. This is confirmed in Fig. 7. Comparing the FOS of the fully submerged slope under complete rapid drawdown with published results, the coupled FEM solution (FOS = .67) underestimates that from Morgenstern (FOS = .82) by about 18% (Fig. 5). The 'dry' solution however, agrees exactly with that by Bishop and Morgenstern (1960). A close examination of the residual pore pressure distribution after complete drawdown of the coupled solution shows a non-uniform distribution with higher value concentrated near the toe areas (Fig. 8). The pore pressures obtained in the present work correspond to an elastic, perfectly plastic (non-dilative) soil model. Morgenstern (1963) assumed a uniform residual pore pressure distribution given by $\gamma_w . h$ (Fig. 8) and this simplified approach leads to an overestimation of the factor of safety. In very loose materials during undrained shear, the pore pressures would be greater still, leading to even lower factors of safety. Fig. 9 shows the results of the coupled analysis for upstream failure due to different amounts of drawdown. Both a single slope and a double sided embankment have been considered and the results agree well with Morgenstern (1963).

Conclusions

The paper has discussed coupled FEM and fixed mesh methods for rapid drawdown analysis. It has been shown that the residual pore pressure distribution after the drawdown is not uniform with pore pressure tending to the highest near the toe. This led to the Factors of Safety lower than those obtained using traditional methods. The oscillations in the pore pressure response due to using 'reduced integration' can be greatly improved by using exact integration to form the global stiffness matrix.

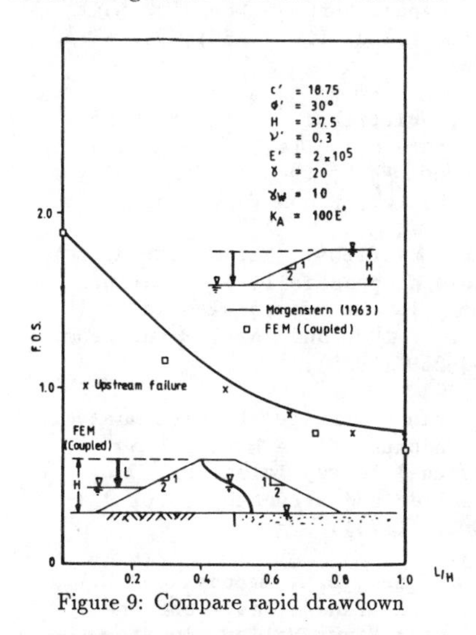

Figure 9: Compare rapid drawdown solutions with published results.

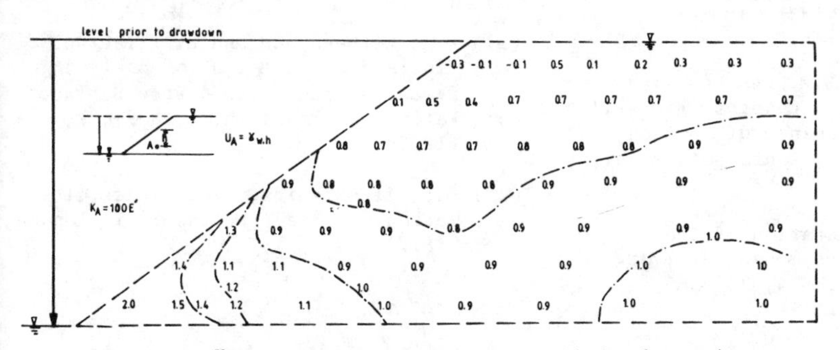

Figure 8: $\frac{U_{FEM}}{\gamma_w . h}$ distribution after complete rapid drawdown using exact integration rule.

References

(1) A.W. Bishop, and N. Morgnestern, :Stability coefficients for earth slopes, Geotech. 10, 129-150 (1960).

(2) K.J. Bathe and M.R. Khoshgoftaar, :Finite Element Free Surface Seepage Analysis without mesh iteration, Int. J. Num. Anal. Meth. Geom., 3, 13-22 (1979).

(3) C.S. Desai: Finite Element Residual Schemes for Unconfined flow, Int. J. Num. Meth. Engng, 10, 1415-1418 (1976).

(4) C.S. Desai: Drawdown Analysis of slopes by Numerical Method", J. Geot. Engng, Div., ASCE, 103, 667-676 (1977).

(5) W.D.L. Finn: Finite-Element analysis of seepage through Dams, J. Soil Mech. Found. Div., ASCE, 93, 41-48 1967.

(6) D.V. Griffiths: Finite Element analysis of walls, footings and slopes, Ph.D, thesis, University of Manchester, Simon Eng. labs. (1980).

(7) D.V. Griffiths: The effect of pore-fluid compressibility on failure loads in the Elasto-plastic soil" Int. J. Num. and Anal. Meth. Geom. 9, 253-259 (1985).

(8) I. Fried: Finite Elements Analysis of Incompressible Material by Residual Energy Balancing, Int. J.Solids and Structures, 10, 993-1002, (1974).

(9) F.O. Jones, D.R. Embody, and W.L. Peterson: Landslides along the Columbia River Valley, Northeaston, Washington, Professional Paper No 367, U.S. Geological Survey, Washington (1961).

(10) A.W. Koppejan, B.M. van Wamellen, and L.J.H. Weinberg: Coastal flow slides in the Dutch province of Zeeland, Proc. Second Int. Conf. Soil Mech., 5, 89-92, (1948).

(11) J.A. Ligget: Location of free surface in porous media, J. Hydr. Div. proc. ASCE, HY4, 353-365, (1977).

(12) G.C. Li and C.S. Desai: Stress and Seepage Analysis of Earth Dams, J. Geot. Engng. Div., ASCE, 109, 947-960, (1983).

(13) J. Lowe and L. Karafiath: Stability of Earth Dams upon Drawdown, Proc. First Pan. Conf. on Soil Mech. and Finite Element, Mexico City, 537-552, (1960).

(14) A. Mayer: Characteristics of Materials used in Earth Dam Construction - stability of Earth Dams in cases of Reservoir Discharge, Proc. Second Congr. Large Dams, 4, 295-327, (1936).

(15) N. Morgenstern: Stability Charts for Earth Slopes during Rapid Drawdown. Geotech. 13, 121-131, (1963).

(16) D.J. Naylor: Stresses in nearly incompressible materials by Finite Elements with application to the calculation of excess pore pressure. Int. J. Num. Meth. Engng. 8, 443-460 (1974).

(17) I.M. Smith and D.V. Griffiths: Programming the Finite Element Method, 2nd Edition, Wiley and Sons, to be published in 1987.

(18) A.W. Skempton: The pore-pressure coefficient A and B, Geotech. 4, 143-147, (1954).

(19) R.L. Taylor and C.B. Brown: Darcy flow solutions with a Free Surface. J. Hydr. Div. Proc. ASCE, 93, 25-33, (1967).

(20) H. Werner and E. Rank: An Adaptive Finite Element approach for the Free Surface Seepage problem. Int. J. Num. Meth. Engng, 23, 1217-1228 (1986).

(21) P.A. Witherspoon and S.P. Neuman: Finite Element Method of Analysing Steady Seepage with a Free Surface. Water Res. Research, 6, 889-897, (1970).

(22) O.C. Zienkiewicz, The Finite element method. 3rd ed. McGraw-Hill, London, (1977).

Numerical Methods in Geomechanics (Innsbruck 1988), Swoboda (ed.)
© 1988 Balkema, Rotterdam. ISBN 90 6191 809 X

Analysis of gravity dams on complex rock formations with effects of seepage

C.Chinnaswamy
Central Water Commission, New Delhi, India
K.G.Sharma & A.Varadarajan
Department of Civil Engineering, I.I.T. Delhi, India

ABSTRACT : In many of the geotechnical engineering problems, particularly analysis of dam-foundation structures, the water flow induced forces have to be taken into consideration in order to arrive at a more realistic solution. The water flow induced forces viz. seepage forces and hydrostatic forces, which can be calculated from the seepage analysis, mainly depend upon the flow properties of the media. For analysis of flow through dam foundation with seams or faults, a new joint element formulation has been suggested to simulate flow through joints. By using the joint element, results of the seepage analysis of a dam-foundation with sub-horizontal seams have been presented. With the calculated water flow induced forces from the seepage analysis, elastic and elasto-plastic behaviour of a dam-foudation system with seam have been compared.

1. INTRODUCTION

In recent years, several catastrophic and many more less severe dam failures have occured and in most of the cases (34% of the total dam failures) geological factors like presence of seams, bedding planes, faults etc. , have been responsible for damage or the ultimate failure of the structures. The failure of Malpasset dam in 1969 was due to the development of the uplift pressure at depth caused by the geological factors like presence of rock faults and high sensitivity of their permeability to stress. The presence of seams or faults are liable to be the potential sliding paths leading to the most insiduous weakness in the foundation and hence their presence in the foundation necessitates special measures in the design. In cases, where the discovery of the weaknesses has not been made until the construction is underway, methods of dealing with the problem have usually been very expensive . Therefore, it is important to analyse the dam-foundation by considering the seams and faults in the foundation and also with water flow induced forces.

2. FINITE ELEMENT METHOD (FEM)

Among the three possible approaches, viz. (i) limit equilibrium method (LEM), (ii) Model studies (MS), and (iii) Numerical method (NM) particularly in complex rock formations, numerical method is a very convenient tool because of less limitations and more flexibilities. In problems of stress analysis with effects of seepage the finite element method among the numerical methods, is the best suited one. The solution procedure for both the stress and seepage analyses by FEM is almost same except in the element stiffness metrix formulations. There will be some problems in getting the load-vector due to water flow induced forces from seepage analysis corresponding to the finite element mesh of the stress analysis. However, these difficulties can be alliviated by using the same finite element mesh for stress and seepage analyses.

3. FINITE ELEMENT REPRESENTATION

3.1 Stress Analysis

In this paper, eight noded parabolic isoparametric elements are used to represent the continuum. The characteristics and applications of this element have been reported in several texts [1] and so is not described here. Interface elements are used to model the effect of differential displacements along the seams or faults. A six noded iso-parametric interface element with quadratic variation of both geometry and relative displacements is used to represent the seams. The details of this element formulation has been already reported by Sharma et. al. [2].

3.2 Seepage Analysis.

In this work, flow analysis is to be carried out first during the stages of reservoir filling to calculate the seepage forces in the foundation and uplift pressure at dam-base. With these water flow induced forces and other external loads, stress-strain behaviour is to be investigated. Unless the finite element mesh is same for both the analyses, the procedure of determining the seepage forces at nodes of mesh used in the stress analysis will become complicated. With this point in view, the same eight noded iso-parametric elements are used to represent the continuum and its formulation can be found elsewhere [3].

For a long time, research works have been done on the study of flow through porous media which may be hetrogeneous and anisotrophic. On the other hand, flow through jointed media, particularly discontinuous rock is not well understood. In a seepage analysis, the rock-mass can be considered as a continuum or discontinuum. In the discontinuum approach, joints are commonly modelled as equivalent parallel plates with flow governed by Darcy's law. Graphical and analytical methods applicable to simple two dimensional laminar or turbulant flow analysis has been presented by Louis [4]. The finite element technique as applied to two dimensional laminar flow in rigid joint systems has been studied by Wilson and Witherspoon [5]. The method has been extended to three dimensions and turbulant flow by Wittke et. al. [6]. Ohnishi et. al [7] presented a two noded line element to simulate the flow through joints. However, these existing models do not suit in case of stress and seepage coupling problems.

3.2.1 New joint element formulation.

A new iso-parametric joint element with quadratic variation of geometry, relative potential head $\Delta\bar{U}$ and mean potential head $\bar{U}m$ is described by a six noded element of constant thickness 't' . The behaviour of this element is characterised by the relationship between the mean and relative potential heads at the mid-point of each sets of nodes (top and bottom) and corresponding flow velocities. The shape functions $N_i(\xi)$ associated with nodes 1 to 6 are same as that for the joint elements in stress analysis.

At any point in the element,

$$x = \sum_{i=1}^{3} N_i x_i \quad ; \quad y = \sum_{i=1}^{3} N_i y_i \tag{1}$$

$$\Delta\bar{U} = \sum_{i=1}^{3} N_i \Delta\bar{U}_i \quad ; \quad \bar{U}_m = \sum_{i=1}^{3} N_i \bar{U}_m$$

The hydraulic gradient at any point along the element line can be written as,

$$[g] = \sum_{i=1}^{3} \begin{bmatrix} N_i & 0 \\ 0 & \frac{dN_i}{dS} \end{bmatrix} \begin{Bmatrix} \Delta\bar{U}_i / t \\ \bar{U}_{mi} \end{Bmatrix} \tag{2}$$

$$= \sum_{i=1}^{3} \begin{bmatrix} N_i & 0 \\ 0 & \frac{dN_i}{dS} \end{bmatrix} \begin{Bmatrix} \frac{\bar{U}^T - \bar{U}^B}{t} \\ \frac{\bar{U}^T + \bar{U}^B}{2} \end{Bmatrix} \tag{3}$$

where t=thickness of the element and suffixes T and B stand for the top and bottom nodes.

Equation (3) can be re-written as,

$$[g] = \sum_{i=1}^{3} \begin{bmatrix} \frac{-N_i}{t} & \frac{N_i}{t} \\ \frac{1}{2}\frac{dN_i}{dS} & \frac{1}{2}\frac{dN_i}{dS} \end{bmatrix} \begin{Bmatrix} \bar{U}^B \\ \bar{U}^T \end{Bmatrix} \tag{4}$$

By expanding the summation terms the gradient matrix becomes,

$$[g] = \begin{bmatrix} \frac{-N_1}{t} & \frac{-N_2}{t} & \frac{-N_3}{t} & \frac{+N_3}{t} & \frac{N_2}{t} & \frac{N_1}{t} \\ \frac{1}{2}\frac{dN_1}{dS} & \frac{1}{2}\frac{dN_2}{dS} & \frac{1}{2}\frac{dN_3}{dS} & \frac{1}{2}\frac{dN_3}{dS} & \frac{1}{2}\frac{dN_2}{dS} & \frac{1}{2}\frac{dN_1}{dS} \end{bmatrix} \begin{Bmatrix} \bar{U} \end{Bmatrix} \tag{5}$$

where $[\bar{U}^T] = [\bar{U}_1 \; \bar{U}_2 \; \bar{U}_3 \; \bar{U}_4 \; \bar{U}_5 \; \bar{U}_6]$

Symbolically, equation (5) can be written as,

$$[g] = [B] \; [U] \tag{6}$$

Matrix (B) can be written as a product of matrices [A] and [B*] as,

$$[B] = [A] \; [B^*]$$

$$\text{where } (A) = \begin{bmatrix} 1/t & 0 \\ 0 & 1 \end{bmatrix} \tag{7}$$

$$[B^*] = \begin{bmatrix} -N_1 & -N_2 & -N_3 & N_3 & N_2 & N_1 \\ \frac{1}{2}\frac{dN_1}{dS} & \frac{1}{2}\frac{dN_2}{dS} & \frac{1}{2}\frac{dN_3}{dS} & \frac{1}{2}\frac{dN_3}{dS} & \frac{1}{2}\frac{dN_2}{dS} & \frac{1}{2}\frac{dN_1}{dS} \end{bmatrix} \tag{8}$$

The element stiffness matrix $[K^e]$ is given by,

$$[K^e] = \int_V [B]^T [D] [B] dv$$
$$= \int_S [B]^T [D] [B] t.1 dS \qquad (9)$$

where [D] is the permeability matrix.Since dS= 1.dξ and [B]=[A] [B*],

$$[K^e] = \int_{-1}^{1} [B^*]^T [A]^T [D] [A] [B^*] t.1 d\xi$$
$$= \int_{-1}^{1} [B^*]^T [D^*] [B^*] t.1 d\xi \qquad (10)$$

where $[D^*] = t [A]^T [D] [A]$.

The above integral can be evaluated by using Gauss quadrature numerical integration and in the process, three point integration is found essential.

In all the seepage problems, it is convenient to determine the element stiffness matrix with respect to the principal axes of permeability. The directions along and normal to the element line can be considered as the principal directions of permeability and hence,

$$[D] = \begin{bmatrix} k_n & 0 \\ 0 & k_s \end{bmatrix} \qquad (11)$$

where k_n -permeability in the direction normal to the element line.

k_s -permeability along the element line.

Substituting for [D] in equation (10).

$$[D^*] = \begin{bmatrix} k_n/t & 0 \\ 0 & t.k_s \end{bmatrix} \qquad (12)$$

4.FLOW THROUGH DAM-FOUNDATION WITH SEAMS

With reference to flow through dam-foundation, there are two distinct aspects to be taken care of in the design of gravity dams. The first one is the flow pattern which brings out the distribution of hydraulic gradients and the direction of seepage forces. The other aspect is the uplift pressure at the dam-base which plays a major role as a destabilising force (activating force) in the stability of the dam. Hence, accurate estimation of seepage forces and uplift pressures is very important in the design of dams. The geological complexities may change the flow pattern and the flow induced forces which may possibly lead to even severe catastrophic failures [9].

By using the finite element formulations presented in the previous section, seepage analyses have been carried out for foundation with seams. The flow parameter of the seam is considered in the form of permeability ratio,

$$P_r = \frac{K_{js} \cdot t}{K_r \cdot 1} \qquad (13)$$

where k_{js} is the permeability of the seam; k_r is the permeability of the rock and t is the thickness of the seam. The results of seepage analysis in the form of flownets are shown in figures 1, 2 and 3. By comparing figures 1 and 2,it is seen that the flow length through seam increases with the increase in the permeability ratio.

From the results of seepage analysis, hydraulic gradients, i are calculated at each Gaussion point and the body forces are calculated by taking $-i\gamma_\omega$ as the density to have the seepage forces[10]. The seepage forces in the seam portion are neglected.

5.ELASTO-VISCOPLASTICITY

In the present study, elasto-viscoplastic theory [11] has been adopted and used as an artifice to obtain elasto-plastic solution. Both for continuum and interface elements, Mohr-Coulomb yield criterion has been used in conjunction with 'No Tension' yield criterion.

6.ANALYSIS

In the present study, a dam section with a base width of 100m and a height of 125m with a downstream face slope of 0.8 horizontal to one vertical has been considered. In order to take the foundation interaction into account with effects of seepage, 7 times the base width on either side and 4 times the base width in the depth direction have been considered. The details of the material properties are as follows:

a) Intact rock:

Young's modulus 1.02×10^7 kN/m^2 (1.04×10^6 t/m^2)

Poisson's ratio 0.15
c 2450 kN/m^2 (250 t/m^2)
ϕ 42°

Tensile strength 452 kN/m^2(46 t/m^2)

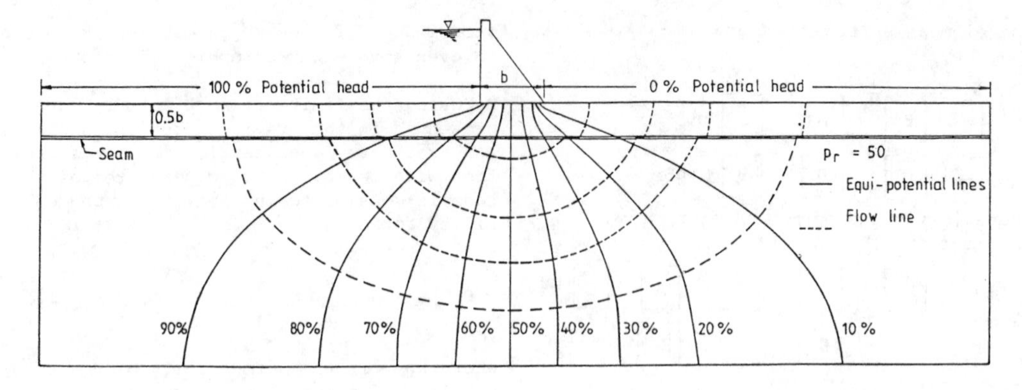

Fig.1. Flow through dam-foundation with a Horizontal seam, P_r=50.

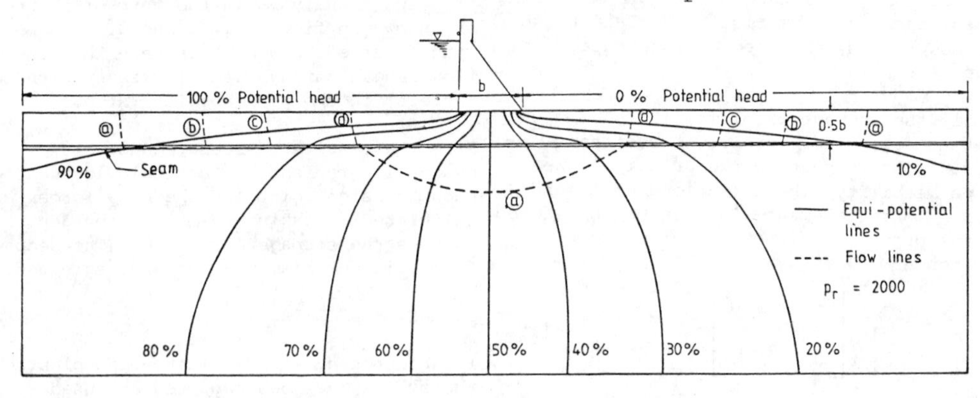

Fig 2 Flow through dam-foundation with a Horizontal seam, P_r=2000.

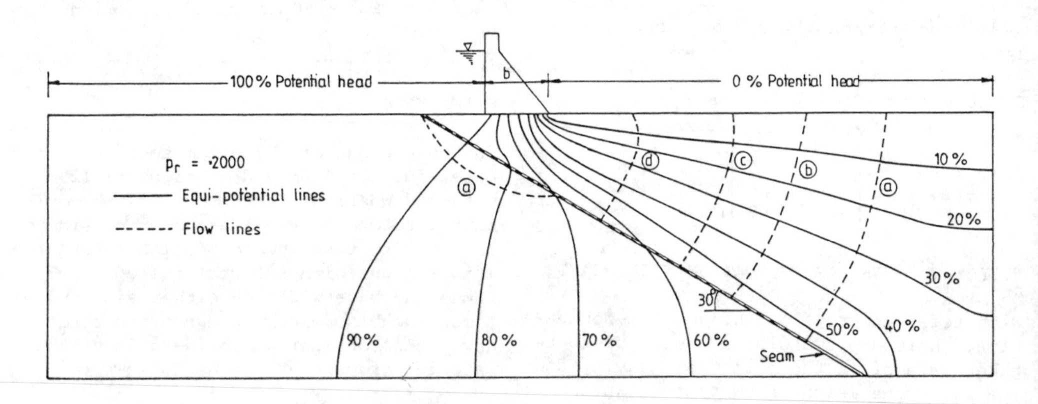

Fig 3. Flow through dam-foundation with a seam of 30° orientation dipping D/S at location ·1b from the heel of the dam.

Permeability 0.1×10^{-6} m/sec

Density 27 kN/m^3 (2.75 t/m^3)

In-situ stress ratio 1

b) Concrete:

Young's modulus 1.02×10^7 kN/m^2 (1.04×10^6 t/m)

Poisson's ratio 0.30

c 2400 kN/m^2 (244 t/m^2)

ϕ 40°

Tensile strength 750 kN/m^2 (76.45 t/m^2)

Density 24 kN/m^3 (2.44 t/m^3)

c) Seam:

Shear stiffness, k_s 0.1×10^6 $kN/m^2/m$ (10193 t/m^3)

Normal stiffness, k_n 0.5×10^6 $kN/m^2/m$ (50968 t/m^3)

c 0

ϕ 25°

Tensile strength 1 kN/m^2 (0.102 t/m^2)

Insitu stress ratio 1.0

Permeability ratio 50

With the external loads and the waterflow induced forces during the stages of reservoir filling, elastic and elasto-plastic analyses have been carried out. Two cases of seams in the foundation viz. a horizontal seam at a depth of 0.1b from the dam-base and an inclined seam with orientation of 30° dipping downstream and day lighting at the heel of the dam have been studied.

7. RESULTS AND DISCUSSION

From the analyses carried out, the relative displacements, ΔR and the local factor of safety, F_{sl} $\left(= \frac{c + \sigma_n \tan\phi}{\tau}\right)$ at various locations along the seam are calculated. The variations of ΔR and F_{sl} at the full reservoir level condition (FRL) are shown in fig.4. It is seen that at FRL condition, the maximum ΔR occurs below the toe of the dam for both elastic and elasto-plastic analyses and the maximum value of ΔR in elasto-plastic analysis is twice that in elastic analysis. Relative displacement is positive (movementof top node is towards right with respect of bottom node) all along the seam in case of elastic analysis where as in elasto-plastic analysis, ΔR is negative along the seam on the upstream side of the heel of the dam (fig.4a).

The variation of the local factor of safety, F_{sl} is plotted in fig.4b. At location along the seam below the heel, $F_{sl} = 0$ because of tension zone below the heel of the dam. In the elastic analysis, F_{sl} is less than unity almost at all points below the dam-base, indicating shear stress induced is more than the strength mobilized. In the elasto-plastic analysis, the excessive

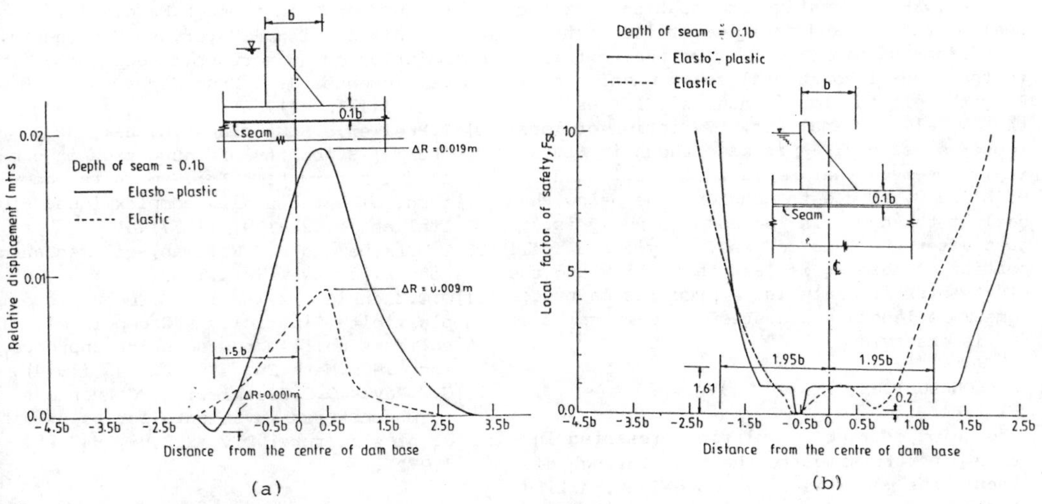

Fig.4, Variation of ΔR and F_{sl} at FRL condition for depth factor of 0.1.

shear stress are redistributed to have $F_{sl}=1$. The length of the seam with F_{sl} equal to unity is three times the base width of the dam.

In a similar way, variation of ΔR and F_{sl} are plotted in fig. 5 for a case of dam foundation with a seam inclined at 30° dipping

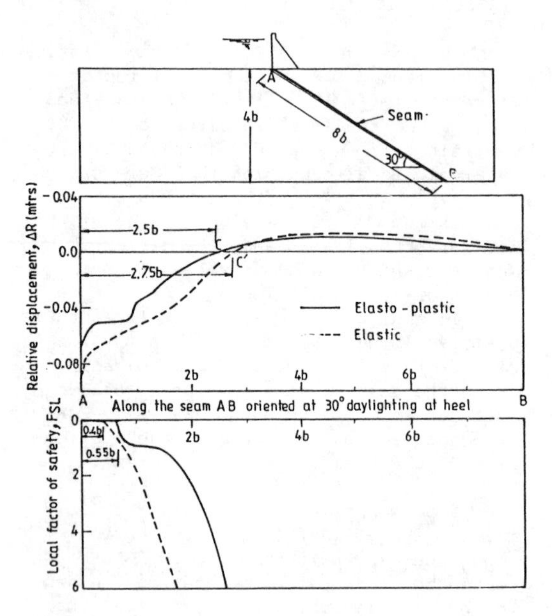

Fig, 5. Variation of ΔR and F_{sl} along the seam oriented at 30° day lighting at heel.

downstream and day-lighting at the heel of the dam. In both elastic and elasto-plastic analyses, ΔR is negative at location near the heel of the dam and positive at locations away from the dam (Fig.5). On the contrary to the case of horizontal seam, magnitude of ΔR by the elasto-plastic analysis is less than that by elastic analysis. Variation of local factor of safety, F_{sl} is also shown in Fig.5 It is observed that F_{sl} is zero for length of 0.4b and 0.55b due to tension zone below the heel of the dam. In the elastic analysis, just after the tension zone, only for a small portion of seam F_{sl} is less than unity and the elasto-plastic analysis, F_{sl} remains unity for some more length and a steep rise in value of F_{sl} is observed.

8. CONCLUSIONS

The joint element formulation presented in this paper to simulate the flow through discontinuous media is very convenient, particularly in the stress and seepage coupled problems. The effect of seam in the foundation on the flow pattern for three cases have been presented. From the stress analysis with the effects of seepage, variation of relative displacements and points of local failures have been presented. This helps to identify the places where to provide the corrective measures like shear keys, vertical plugs, etc. However, more detailed elasto-plastic analysis is necessary to determine the overall factor of safety.

REFERENCES

[1] K.J. Bathe and E.L. Wilson, Numerical methods in finite element analysis. Prentice hall inc, New Jersey, 1976.

[2] H.D.Sharma, G.C.Nayak and J.B.Maheswari Generalisation of sequential non-linear analysis - A study on rockfill dam with joint elements, C.S.Desai(Ed) Numerical methods in geomechanics, Vol.II, 662-685.

[3] C.S.Desai and J.F.Abel, Introduction to finite element method, Von Nostrand Reinhold, New York, 1972.

[4] C.Louis, A study of ground water flow in jointed rocks and its influence on the stability of rock masses. Rock mechanics Research report No.10, Imperial College, 1969.

[5] C.R.Wilson and P.A.Witherspoon, Steady state flow in rigid networks of fracture Water Resources Research, 10, No.2 (1970)

[6] W.Wittke, P.Rissler, S.Semprich, Three dimensional laminar and turbulant flow through fissured rock according to continuous models, Proc. of the Symposium on Percolation through fissured rock, Stuttgart (1972).

[7] Y.Ohinishi, H.Shibata and M.Nishigaki Finite element analysis of seepage flow in regularly jointed rockmass, 5th Int.Conf. on Num.meth. in geomech. Nagoya, II, (1985)

[8] O.C.Zienkiewicz, P.Mayer, and Y.K.Cheung, Solution of anisotrophic seepage by finite elements, J.of Engg.Mech.Div. ASCE, EM1, 111-120 (1966).

[9] L.Mueller, H.B.Muehlhaus, G.Reik, and B. Sharma, Stability of foundation in comlex rock formation, Int.Sym.on the geotech. of structurally complex foundation Italiana, 2, 127-139, (1977).

[10] T.W.Lambe and R.V.Whitman, Soil mechanics John Wiley, New York, 1969.

[11] O.C.Zienkiewicz and I.C.Cormeau, Viscoplasticity, Plasticity & Creep in elastic solids, A unified num.solution approach, Int.J.Num.Meth.Engg., 8, 821-845 (1974).

[12] G.C.Nayak, O.C.Zienkiewicz, Convenient form of stress invariants for plasticity J. of structural Div.ASCE, 98, 949-954 (1972).

Numerical Methods in Geomechanics (Innsbruck 1988), Swoboda (ed.)
© 1988 Balkema, Rotterdam. ISBN 90 6191 809 X

Consolidation analysis of partially saturated soils – Application to earthdam construction

E.E.Alonso, F.Batlle, A.Gens & A.Lloret
Technical University of Catalunya, Barcelona, Spain

ABSTRACT: An earth dam analysis is carried out using a formulation, described in the paper, that allows the solution of the combined flow and deformation problem for partially saturated soils. The formulation takes into account the basic features of behaviour of partially saturated soil. The analysis is performed incrementally, each increment having two stages. The first one involves the solution of the (transient) flow equation whereas the second one involves the solution of the equilibrium equations. Undrained events can also be analyzed using the same basic formulation which does not require any modification when soil reaches full saturation. The construction of a 90 m. high zoned earth dam has been simulated. The effects of varying the compaction water content are analyzed. Particulary relevant is the response of the clay core in terms of generated pressure during step by step construction.

1. INTRODUCTION

In earthdam design, pore pressure generation in clay cores during embankment construction is an important consideration since it is directly related to the deformation and stability of the structure. The magnitude of these water pressures will be governed by the intensity of the total stresses induced by the accumulation of soil layers and by the rate of pore pressure dissipation. Both aspects are closely related to the as-compacted water content conditions of the soil. In general terms, compacted soils wet of optimum experience a significant increase of pore pressure during construction whereas soils compacted dry to optimum do not show a comparable increase in pore pressure and, in addition, pore pressure dissipation is , in the latter case, considerably faster than in wet of optimum conditions [1], [2].
Compacted soils are partially saturated soils. Their behaviour, which has been reviewed recently [3] , is, in many respects, considerably different from the behaviour of saturated soils. Particularly important are the volumetric deformations (either collapse or swelling) and changes in stiffness induced by changes in pore water suction and the strong dependence of permeability (both to water and air) on degree of saturation. Unsaturated flow has deserved considerable attention in recent years. However, there is a lack of stress-strain models of general applicability for these soils.
Several techniques have been used to predict pore pressure built-up and deformations during embankment construction. The classical Biot analysis for saturated conditions [4] ,the variability of Skempton B parameter, elastic modulus and coefficient of consolidation with stress intensity [5] and the use of hyperbolic stress-strain models [6] were some of the early assumptions.
More recently formulations for partially saturated soil have been introduced [7],[8]. They consider the compressibility of the pore fluid and rely in the Terzaghi effective stress principle. Other authors [9] use a simplified version of Bishop effective stress equation for partially saturated conditions to analyze the behaviour of an earthdam clay core.
As far the flow in earthdams under partially saturated conditions is concerned solutions which use finite difference or finite element techniques , [10], [11], have been developed. The soil is however assumed to be rigid in these cases and the initial conditions (distributions of pore water pressure and degree of saturation) are an input of the model.
The purpose of this paper is to present a general formulation for the full coupled stress-strain-flow behaviour of partially saturated soils which can be applied to either drained and undrained conditions. The model takes into account the effect of suction in stress-strain behaviour, strength and degree of saturation of the soil. In addition,permeability to both air and water is made dependent on porosity and degree of saturation. The present paper extends and completes some previous work done by the authors [12],[13]. The formulation developed is finally applied to the step by step construction of an earthdam under plain strain conditions. The effect of varying the initial water content conditions either to the wet or dry side of the optimum is also described in the paper.

2 CONSTITUTIVE RELATIONS

The swelling and collapse behaviour of partially saturated soil due to a reduction in suction (wetting) cannot be explained by an effective stress relationship such as Bishop's equation [14],[15]. A separate consideration of total stress , $\boldsymbol{\sigma}$, (in excess of air pressure, p_a) and suction - $(p_a - p_w)\delta_{ij}$, where p_w is the water pressure - seems to be necessary to describe properly the behaviour of partially saturated soils. A number of null tests [16] have provided experimental evidence in this regard.

A practical stress-strain relationship which may take properly into account the relevant behaviour of partially saturated soils is:

$$d\boldsymbol{\varepsilon} = \boldsymbol{D}^{-1}\, d\boldsymbol{\sigma}^* + d\boldsymbol{\varepsilon}_0 \tag{1}$$

where the strain increments $d\boldsymbol{\varepsilon}$ are obtained as a sum of the total stress changes effects ($\boldsymbol{\sigma}^* = \boldsymbol{\sigma} - \boldsymbol{m}\, p_a$; $\boldsymbol{m}^t = [1,1,1,0,0,0]$) and the volumetric strains $d\boldsymbol{\varepsilon}_0$ induced by suction changes. $\boldsymbol{\varepsilon}_0$ may be determined by means of laboratory tests. An expresion such as :

$$d\boldsymbol{\varepsilon}_0 = \boldsymbol{D}_s^{-1}\, d(p_a - p_w) \tag{2}$$

may provide a suitable relationship for computational purposes. In general, the terms of matrix $\boldsymbol{D}_s$ will depend on stress lavel and existing suction. An approximate procedure to obtain matrix $\boldsymbol{D}_s$ is to use the concept of state surface for void ratio [17]. This surface provides the variation of void ratio ,e, with applied total stress and suction for a given experimental procedure (oedometric or triaxial conditions in most cases). For instance,for isotropic or oedometric stress conditions the following expressions have been suggested for these state surfaces [18] :

$$e = d + a\, log(\sigma - p_a) + b\, log(p_a - p_w) + c\, log(\sigma - p_a)\, log(p_a - p_w) \tag{3}$$

where a,b,c,d are constants and σ the relevant stress which defines the stress state (i.e. the vertical stress in the oedometric cases). The state surface given by (3) does not consider explicitly the influence of deviatoric stresses on volumetric change induced by suction changes. Fig.1 illustrates a state surface for a low plasticity clayly silt used in the construction of a dam core.

The coefficients of matrix $\boldsymbol{D}$ may be defined through a nonlinear elastic model. The compressibility modulus K is obtained from equation (3) (for isotropic test conditions). The shear modulus can be obtained from a hyperbolic shear stress- shear strain law.

$$G = [G_0 + M(p_a - p_w)]\, [1 - (\sigma_1 - \sigma_3)R/(\sigma_1 - \sigma_3)_f]^2 \tag{4}$$

Note that the stiffenning effect of suction increase has been introduced in equation (4), through the initial G modulus, by means of a linear relationship with parameter M . This is accordance with some experimental evidence [19] . The following expression , suggested in [20] , may be used as a failure criteria :

$$(\sigma_1 - \sigma_3)_f = c' + (\sigma_n - p_a)\, tan\,\phi' + (p_a - p_w)\, tan\,\phi^b \tag{5}$$

where c', ϕ' and ϕ^b are constants.

Besides stress-strain behaviour, the description of the state of a partially saturated soil requires the knowledge of the variation of current degree of saturation, S_r (or water content) with applied stress and suction. Again for cases in which the stress state could be defined by a single state parameter, σ , the following relationship was proposed on the basis of experimental data [18] :

$$S_r = 1 - [a' + d'(\sigma - p_a)]\; Th\, [b'(p_a - p_w)] \tag{6}$$

where a' , b' and d' are constants. Fig. 2 shows this relationship for the same clayey silt whose volumetric behaviour was presented in Fig. 1.

3. AIR AND WATER FLOW

The motion of air and water can be described by a generalization of Darcy's law:

$$\boldsymbol{v}_w = -\boldsymbol{K}_w\, \nabla\, (z + p_w/\gamma_w) \tag{7}$$

$$\boldsymbol{v}_a = -\boldsymbol{K}_a\, (1/\gamma_a\, \nabla\, p_a + \nabla\, z) \tag{8}$$

where $\boldsymbol{v}_w$ and $\boldsymbol{v}_a$ are the velocities of water and air in Darcy's sense. $\boldsymbol{K}_w$ and $\boldsymbol{K}_a$ are coefficients of permeability and γ_a and γ_w the specific weights of air and water.

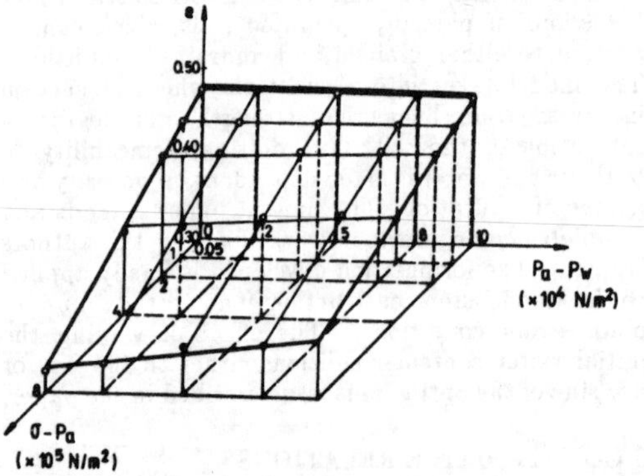

Fig.1 State surface for void ratio of a low plasticity clayey silt

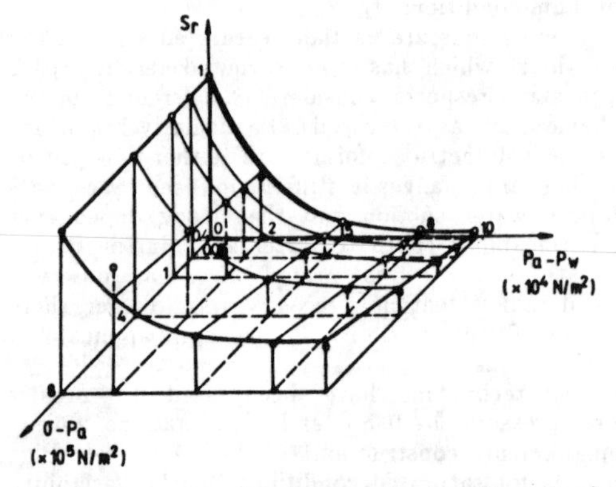

Fig.2 State surface for degree of saturation of a low plasticity clayey silt

The coefficients of permeability vary strongly with degree of saturation and, to a lesser extent, with the porosity of the soil. The following expresions proposed in [21] and [22] for the water flow and in [23] for the air flow have been used in this work:

$$K_w(e, S_r) = A\left(\frac{S_r - S_{ru}}{1 - S_{ru}}\right)^3 10^{\alpha e} \qquad (9)$$

$$K_a(e, S_r) = B\,\frac{\gamma_a}{\mu_a}[e(1 - S_r)]^\beta \qquad (10)$$

In these expresions, A, B , α, β and S_{ru} are constants and μ_a the viscosity of the air. The variation of permeability with stress and fluid pressures may now be obtained with the aid of the constitutive relationships previously described.

4 BASIC FORMULATION

The flow of water and air will induce changes in the fluid pressures which will result , in turn , in soil deformations and changes in degree of saturation. Both effects will alter the stress state. On the other hand, any stress and fluid pressure change will modify the soil permeabilities. In short, flow and deformation phenomena are highly coupled. Three basic set of equations must be solved : continuity of water and air flow and mechanical equilibrium. These equations are :

$$\frac{\partial(\rho_w n S_r)}{\partial t} + div(\rho_w \boldsymbol{v}_w) = 0 \qquad (11)$$

$$\frac{\partial}{\partial t}[\rho_a n(1 - S_r + HS_r)] + div[\rho_a(\boldsymbol{v}_a + H\boldsymbol{v}_w)] = 0 \qquad (12)$$

$$\frac{\partial(\sigma_{ij} - \delta_{ij}p_a)}{\partial x_j} + \frac{\partial p_a}{\partial x_i} + b_i = 0 \qquad (13)$$

where b_i are body forces, ρ_w and ρ_a the densities of water and air , n the porosity and H the Henry's constant.
Application of a Galerkin discretization procedure reduces the preceding set of second order differential equations to the following set of first order differential equations :

$$\begin{bmatrix} 0 & 0 & 0 \\ 0 & S_a & S_{aw} \\ 0 & 0 & S_w \end{bmatrix} \begin{bmatrix} \boldsymbol{u} \\ p_a \\ p_w \end{bmatrix} + \begin{bmatrix} K_T & L_a & L_w \\ H_a & T_a & T_{aw} \\ H_w & T_{wa} & T_w \end{bmatrix} \frac{\partial}{\partial t} \begin{bmatrix} \boldsymbol{u} \\ p_a \\ p_w \end{bmatrix} = \begin{bmatrix} f \\ r_a \\ r_w \end{bmatrix} \qquad (14)$$

where $\boldsymbol{u}$ are displacements and

$$S_a = \int (\nabla\bar{N})^T \frac{1}{g} K_a \nabla\bar{N}\, dv \qquad (15.a)$$

$$S_{aw} = \int (\nabla\bar{N})^T \frac{\rho_a H}{\gamma\omega} K_w \nabla\bar{N}\, dv \qquad (15.b)$$

$$S_w = \int (\nabla\bar{N})^T \frac{1}{\gamma\omega} K_w \nabla\bar{N}\, dv \qquad (15.c)$$

$$K_T = \int B^T D B\, dv \qquad (15.d)$$

$$L_a = \int B^T(-DD_s^{-1} + m)\bar{N}\, dv \qquad (15.e)$$

$$L_w = \int B^T(DD_s^{-1})\bar{N}\, dv \qquad (15.f)$$

$$H_a = \int \bar{N}^T \rho_a [(1 - S_r + HS_r)m^T + n(H-1)g_1 D]\, B\, dv \qquad (15.g)$$

$$T_a = \int \bar{N}^T \beta n[(1 - S_r + HS_r) + p_a(H-1)(g_1(-DD_s^{-1}) + g_2)]\,\bar{N}\, dv \qquad (15.h)$$

$$T_{aw} = \int \bar{N}^T \rho_a n(H-1)(g_1 DD_s^{-1} - g_2)\bar{N}\, dv \qquad (15.i)$$

$$H_w = \int \bar{N}^T [n g_1 D + S_r m^T]\, B\, dv \qquad (15.j)$$

$$T_{wa} = \int \bar{N}^T n [g_2 - g_1 DD_s^{-1}]\,\bar{N}\, dv \qquad (15.k)$$

$$T_w = \int \bar{N}^T n [g_1 DD_s^{-1} - g_2]\,\bar{N}\, dv \qquad (15.l)$$

$$f = \int N^T \frac{\partial b}{\partial t} dv + \int_{S_1} N^T \frac{\partial \tau}{\partial t} dS \qquad (15.m)$$

$$r_a = -\int_v \rho_a(\nabla\bar{N})^T(K_a + HK_w)\nabla z\, dv - \int_{S_2} \bar{N}^T \pi\, dS \qquad (15.n)$$

$$r_w = -\int (\nabla\bar{N})^T K_w \nabla z\, dv - \int_{S_2'} \bar{N}^T \lambda\, dS \qquad (15.o)$$

In these equations $\beta = \rho_a/p_a$, g is the gravity constant, $g_1 = \partial S_r/\partial(\sigma - mp_a)$, $g_2 = \partial S_r/\partial(p_a - p_w)$, N are the shape functions and B a differential operator strains and nodal displacements $\bar{N}$ the shape functions used in the discretization of water and air pressures and τ the imposed stress in boundary S_1 . The imposed flows in boundaries S_2 and S_2' are:

$$air : \rho_a(v_a + Hv_w)n - \pi = 0 \qquad (16.a)$$

$$water : v_m n - \lambda = 0 \qquad (16.b)$$

n being the unit normal to the boundary.
Once discretized in time the system of equations (14) may be solved simultaneously. However, some benefits may be obtained from a two-stage solution approach, given the strong nonlinearities of the problem:

a) A separate consideration of the flow and stress-strain problems allows the use of existing programs developed to handle the "simpler" problems. It will only required a linking software between them (i.e. a program to solve flow in partially saturated soils and a conventional stress-strain analysis program may be used with limited modifications). The computational scheme is indicated in Fig. 3.
b) In general, the computational efficiency increases due to the smaller bandwith of the resulting systems of equations.

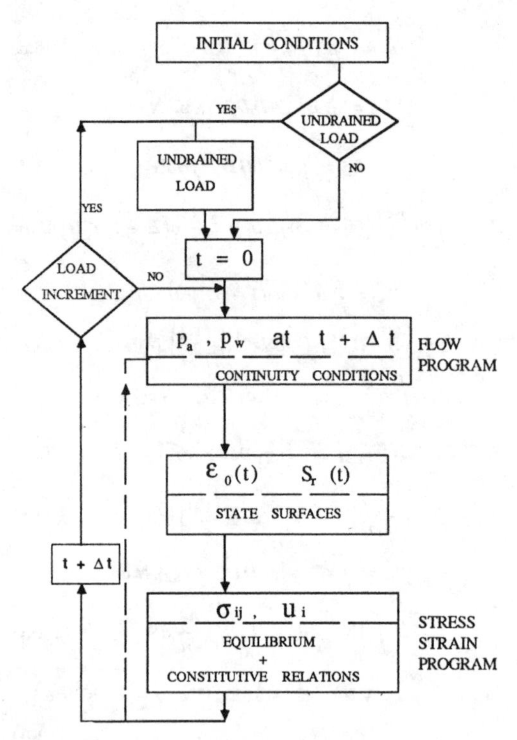

Fig.3 Layout of computational procedure

5 UNDRAINED LOADING

The formulation outlined above is also valid to solve undrained loading problems. Very small time increments will, however, be required and this may lead to numerical difficulties. To overcome this situation an specific formulation for undrained loading has been developed.

The continuity equations for air and water will be replaced by expressions which impose the conservation of air and water masses in the soil pores during the loading process. In this way the degree of saturation and volumetric deformation after loading may be expressed in terms of changes in air and water pressure:

$$\Delta\varepsilon_v = \frac{e_0}{1+e_0}(V_p - 1) \tag{17}$$

$$S_r = V_w / V_p \tag{18}$$

where

$$V_w = S_{r_0}\left[1 - C_w(p_w - p_{w_0})\right]$$

$$V_a = \frac{p_{a_0}}{p_a}(1 - S_{r_0} + H S_{r_0})$$

$$V_p = (1-H)V_w + V_a$$

C_w is the compressibility of the water and the suffix o denotes the value of a variable before loading.

Two compatibility conditions must be satisfied at every integration point: a) The volumetric strain given by equation (17) should be equal to the volumetric strain computed in the stress-strain part of the analysis and b) The degree of saturation (equation 18) must lie on the state surface for degree of saturation after loading (i.e. equation 6). Both conditions are required to obtain the air and water pressures after loading. The system of equations to be solved is therefore :

$$\frac{\Delta e}{1+e_o} = \boldsymbol{m}^t\left[\boldsymbol{D}^{-1}\Delta(\boldsymbol{\sigma} - \boldsymbol{m}p_a) + \boldsymbol{D}_s^{-1}\Delta(p_a - p_w)\right] \tag{19}$$

$$\frac{V_w}{V_p} = g(\boldsymbol{\sigma} - \boldsymbol{m}p_a \,,\, p_a - p_w) \tag{20}$$

$$\boldsymbol{K}_T\Delta\boldsymbol{u} + \boldsymbol{L}_a\Delta\boldsymbol{p}_a + \boldsymbol{L}_w\Delta\boldsymbol{p}_w = \boldsymbol{f} \tag{21}$$

The problem has to be solved in an iterative way since the stress change depends on the computed values of p_a and p_w . A value of $\boldsymbol{\sigma}$ is first assumed and equations (19) and (20) are solved for p_a and p_w ; the stress-strain problem (equation 21) is then solved for updated value of $\boldsymbol{\sigma}$.

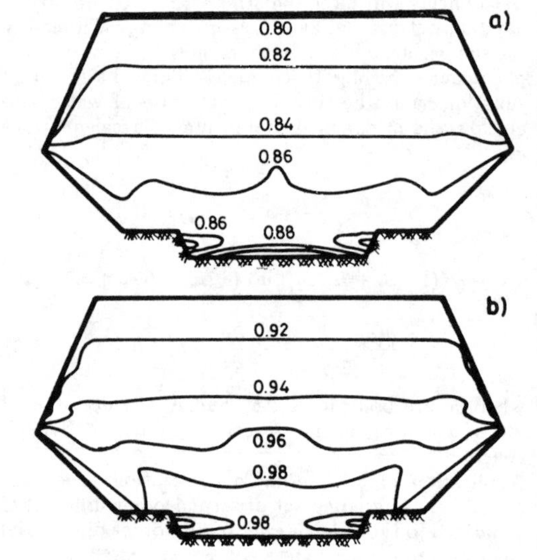

Fig.4 Contours for degree of saturation at an intermediate construction stage a) Drier soil, S_{r_0}=0.8; b) Wetter soil S_{r_0}=0.9

6 EARTHDAM CONSTRUCTION

The step by step construction of a 90 m. high earth dam has been analyzed by means of the preceding formulation. The dam is a zoned embankment with a relatively thick central core, upstream and downstream filters and rockfill shoulders. For the purposes of the simulation of the construction only two zones have been considered: The core and a zone which includes the filters and the shoulders. This second zone was assumed to be highly permeable and less deformable than the clay core.The stress-

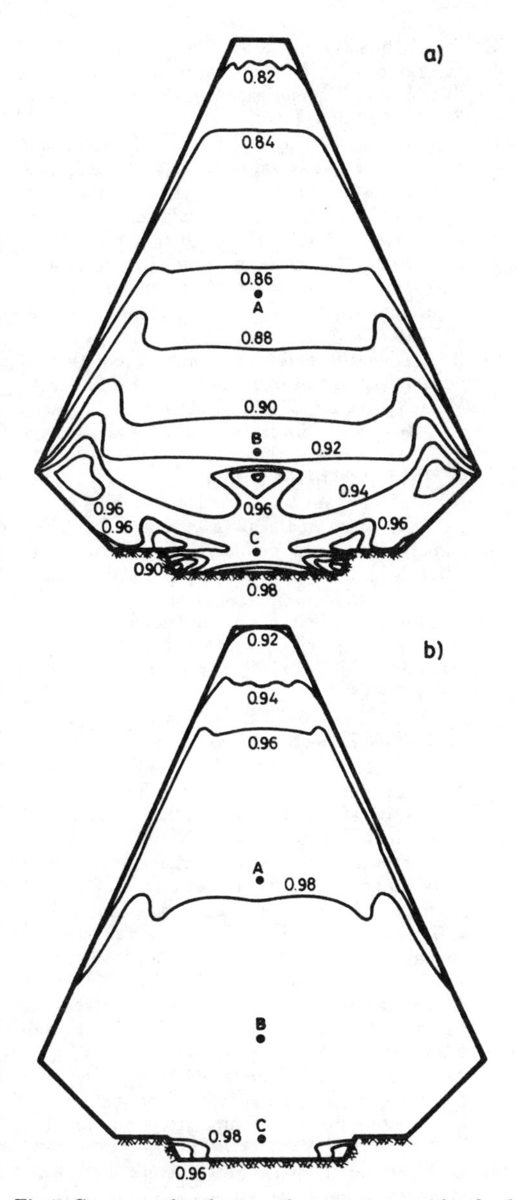

Fig.5 Contours for degree of saturation at the final construction stage a) Drier soil, $S_{r_0}=0.8$; b) Wetter soil $S_{r_0}=0.9$

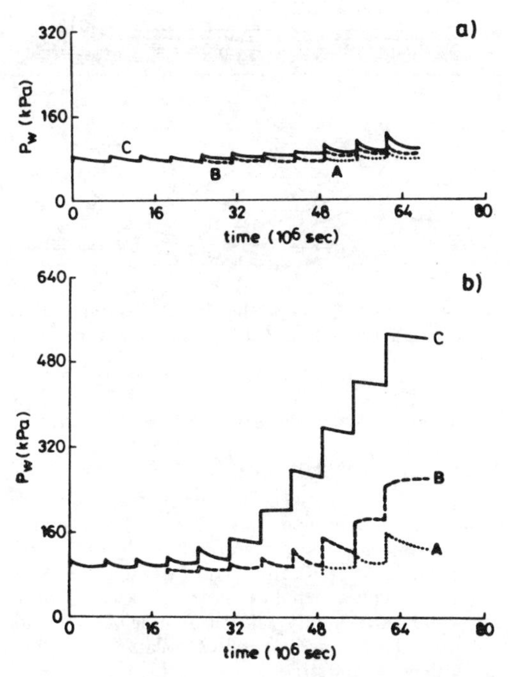

Fig.6 Evolution of pore water pressure in three locations. a) Drier soil, $S_{r_0}=0.8$; b) Wetter soil $S_{r_0}=0.9$

strain analysis involved the full cross-section of the dam. However the consolidation problem was solved in the core with the appropiate boundary conditions (full drainage allowed at the boundaries except for an impervious foundation zone, see Fig 4 and 5).The possibility of using different domains for the solution of the flow or the mechanical part of the coupled analysis is a particulary suitable feature of the model developed. In fact, it allows an optimization of computer resources in terms of memory size and computing time. Every new layer was applied under undrained conditions and a consolidation process was then initiated. The state surfaces for the core material are those shown in Figs. 1 and 2.

Two placement water content conditions were simulated. The "dry" soil has an initial degree of saturation $S_{r_0}=0.8$. For the "wet" soil $S_{r_0}=0.9$. Figs. 4a and 4b show the degree of saturation contours for both cases at an intermediate construction stage. The distribution of degree of saturation at the end of construction is shown in Fig. 5a and 5b. The increase in saturation towards the bottom of the core reflects the increasing stress due the increase in dam height. The reduction in degree of saturation at the core-shoulders contacts is a result of the drainage allowed at these boundaries.The evolution of water pressures at three selected points (indicated in Fig. 5) is shown in Figs. 6a and 6b. The dryer soil experiences a very slight increase in pore water pressure during the construction process. In the wet soil (Fig. 6b), a sudden increase in pore pressure, as a consequence of the placement of a new layer, is observed at some time during the construction process. This is a consequence of the higher degree of saturation of the soil and its low deformability under undrained conditions.Fig 6b. also shows the variation of pore pressure due to consolidation effects. The deformation of the dam can also be followed during the construction process. The deformed shapes of the whole mesh for an intermediate and the final construction stage is shown, for the wetter soil, in Fig. 7. These displacements include the effects of increasing dam

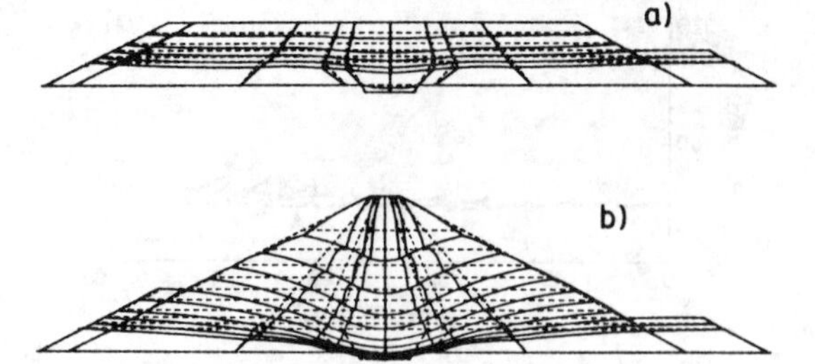

Fig.7 Deformed shape of the dam. a) Intermediate construction stage; b) Final construction stage.

height and the movements due to consolidation.

7 CONCLUDING REMARKS

In this paper, a formulation for flow and deformation analyses which takes into account the basic features of behaviour of partially saturated soil has been described. These include the strong variation of air and water permeability with the degree of saturation and the characteristic swelling/collapse behaviour which depends on stress level. In the analysis, both air and water continuity equations are satisfied. The linking between the flow and the stress-strain analyses allows the computation of strains and defotmations due to changes in suction. Undrained events can also be dealt with using the same basic formulation. The analysis of some aspects of the construction of an earthdam shows the capabilities of the formulation for the application to engineering design.

REFERENCES

[1] J.L. Sherard, R.J. Woodward, S.F. Gizienski and W.A. Clevenger. Earth and earth-rock dams.Engineering problems and construction. John Wiley. New York. (1963)

[2] A.D.M. Penman. Construction pore pressures in two earth dams. Clay Fills. Institution of Civil Eng. The Burlington Press. London. (1979)

[3] E.E. Alonso, A. Gens and D.W. Hight. Special Problem Soils: General Report. IX E.C.S.M.F.E. Dublin (1987)

[4] I.M. Smith and R. Hobbs. Biot analysis of consolidation beneath embankments . Geotechnique. 26 (1). 149-171 (1976)

[5] Z. Eisenstein and S.T.C. Law . Analysis of consolidation behaviour of Mica dam. J. Geotech. Eng. Div. A.S.C.E. 103 GT8 . 879-895 (1977)

[6] S. Cavounidis and K. Hoeg. Consolidation during construction of earth dams. J. Geotech. Eng. Div. A.S.C.E. 103 GT10 . 1055-1067 (1977)

[7] J. Ghaboussi and K.J. Kim . Analysis of saturated and partially saturated soils. Int. Symp. on Numerical Models in Geomech. Zurich. 377-390. (1982)

[8] C.S. Chang and J.M. Duncan .Consolidation analysis for partly saturated clay by using an elastic-plastic effective stress-strain model. Int. Jour. for Numerical and Analytical Meth. in Geomech. 7. 39-55 (1983)

[9] D. Aubry, O. Ozanam and J.P. Person . Ecoulements non satures en milieux poreux deformables. IX E.C.S.M.F.E. Dublin . Vol 2. 537-540 (1987)

[10] B.G. Richards and C.Y. Chan .Prediction of pore pressures in earth dams. VII I.C.S.M.F.E. Mexico . Vol 2. 355-362. (1969)

[11] S.P. Neuman . Saturated-unsaturated seepage by finite elements. J. Hydraul. Div. A.S.C.E. 99 Hy12 . 2223-2250 (1973)

[12] E.E. Alonso and A. Lloret . Behaviour of partially saturated soil in undrained loading and step by step embankment construction. I.U.T.A.M. Symp. on Def. and Fail. of Granul. Mat. Delft . 173-180. (1982)

[13] A. Lloret, A. Gens, F. Batlle and E.E. Alonso. Flow and deformation analysis of partially saturated soil. IX E.C.S.M.F.E. Dublin .Vol 2. 565-568 (1987)

[14] J.E.B. Jennings and J.B. Burland . Limitations to the use of effective stress in partly saturated soils. Geotechnique 12, (2) 125-144. (1962)

[15] A.W. Bishop and G.E. Blight . Some aspects of effective stress in saturated and unsaturated soils. Geotechnique 13, (3) 177-197. (1963)

[16] D.G. Fredlund and N.R. Morgenstern. Stress state variables for unsaturated soils. J. Geotech. Eng. Div. A.S.C.E. 103 GT5 . 447-466 (1977)

[17] E.L. Matyas and H.S. Radhakrishna. Volume change characteristics of partially saturated soils. Geotechnique 18 (4). 432-448 (1968)

[18] A. Lloret and E.E. Alonso . State surfaces for partially saturated soils. XI I.C.S.M.F.E. San Francisco. Vol 2 557-562 (1985)

[19] A. Brull . Caracteristiques mecaniques des sols de fondation de chaussees en fonction de leur etat d'humidite et de compacite. Coll. Int. sur le Compactage .Vol 1. 113-118 (1980)

[20] D.G. Fredlund, N.R. Morgenstern and R.A. Widger. The shear strength of unsaturated soils. Canad. Geot. Jour. 15. 313-321 (1978)

[21] S. Irmay .On the hydraulic conductivity of unsaturated soils. Trans. Amer. Geoph. Union. 35 . 463-468. (1954)

[22] T.W. Lambe and R.V. Whitman. Soil Mechanics. John Wiley. New York (1968)

[23] Y. Yoshimi and J.O. Osterberg. Compression of partially saturated cohesive soils. J. Soil Mech. Found. Div. A.S.C.E. 89 SM4. 1-24. (1964)

Numerical Methods in Geomechanics (Innsbruck 1988), Swoboda (ed.)
 ISBN 90 6191 809 X

The numerical analysis for the interaction of concrete core and rockfill shell of a dam and the prototype measurement and verification

Gao Lianshi
Department of Hydraulic Engineering, Tsinghua University, Beijing, People's Republic of China
Zhang Jinsheng
Ministry of Water Conservancy and Hydroelectric Power, Beijing, People's Republic of China

ABSTRACT: This paper presents a rockfilled dam with a concrete core of fifty eight meter height serving as upstream cofferdam of Longyang Xia Reservoir. The rockfilled dam began to prevent water in summer of 1980, since then it had worked normally for seven years. It stood a severe test of extreme flood in 1981. The systematical monitoring was performed in the flood and dry period. After that the FEM was used in back analysis for studying the interaction between concrete core and rockfilled shell and comparing with the measuring result. This paper concluded the experience of adopting the concrete core in rockfilled dam according to the FEM analysis and proto-type monitoring.

1. Introduction

Longyang Xia hydroelectric power station is a huge type project on yellow River just under construction. It is located in Qinghai Province, China, to be planned in installed capacity of 1280 MW. The dam site is in a narrow valley and the pedestal rock is diorite-granite. The dam type is a gravity arch dam with a height of 175 meters. Its diversion construction included the diversion tunnel in right bank, rockfilled coffer dam and temporary flood way. This type of diversion ensured that the construction in dam foundation trench could sustain in full year. Of the diversion construction, the upstream cofferdam is a main engineering, it is a rockfilled dam with a concrete core. Due to the extreme flood came across in 1981, it was raised in height from 54 meters of original design to 58 meters for the temporary water prevention.

In the summer of 1980, the diversion construction was completed. From that time it operated normally for seven years and encountered the extraordinary flood of 5570 m^3/s, of which the probability was once in two-hundred years. The storage level at that time rose up to 2494.8 m, it almost met the top of concrete core. The reservoir formed by the cofferdam reached to a capacity of 0.98 billion m^3. The cofferdam kept on high water level safely to operate for seven days. Fig. 1 showed the working situation of cofferdam in extraordinary flood period, 1981.

Fig.1 Upstream Cofferdam in Overflow

The cases of Longyang Xia diversion construction has been presented in details in references [1] [2] [3]. In this paper, the results of prototype observation and back analysis are emphasized and the working condition and the structure characteristic of the concrete core are discussed.

2. Design, Construction and Monitoring

2.1 Design and Construction

Fig. 2 shows the cross section of cofferdam. It is in height of 58 m, and the width of the concrete core is one meter at top and two meters at bottom, that makes the average ratio of height to width to be forty. The core is insurt in to the pedestal rock through a bottom seat. Between the bottom seat and the core there is a structure joint, and in the core there are three horizontal joints and six vertical joints else. All the joints are sealed with sealing strip, shown in Fig.2. The filter zones, which has a slope of ten to one, filled at both sides consist of sand and gravel.

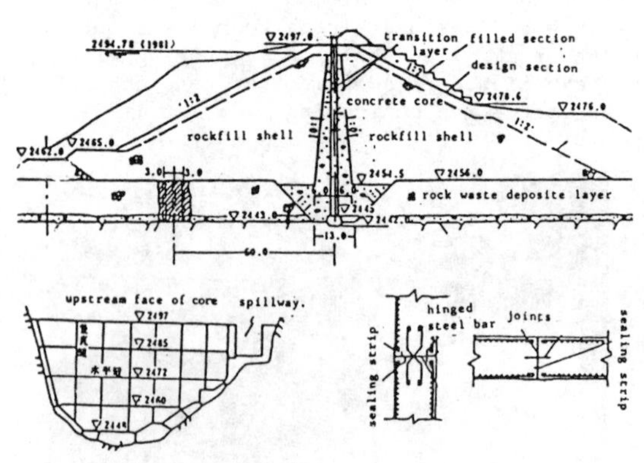

Fig.2 Sketch of Upstream cofferdam

During construction, each layer was covered with filled rock in thickness of one meter and compressed with plate roller of 8 tons except the closer part near the core, which was compacted with punner. The average dry density of the compacted filter layer reached to 20.4 KN/M^3, and that of the rock shell layer reached to 19.3 KN/M^3. For increasing the prestressed pressure in core, the construction order required the downstream dam shell to be filled higher than the upstream dam shell in two to three meters.

2.2 Proto-type Observation

During constraction, there were more than fifty pieces of measuring instruments such as steel bar meters, joint meters , earth pressure meters, pervious meters and displacement points to be embeded in cofferdam. They played an important role in proto-type monitoring and gave a lot of observation data cumulated for seven years. Fig. 3 shows their arrangement.

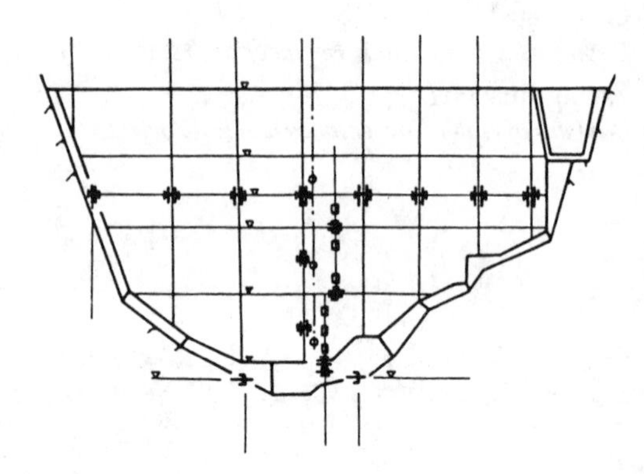

Fig.3 Arrangement of Observation Instruments

The moritoring cases are indicated as following:

(i) Joint meters and working condition of core joints

A total of 28 joint meters were embeded in horizontal and vertical joints at two sides of core. The monitoring result showed that the horizontal joints always subjected to pressure in spite of dry of flood period and the bending moment existed in arbirary height of core. So the effect of hinged joint did not take place actually. The vertical joints also subjected to pressure in river bed, but to tension at two banks, where the gap of vertical joints extended up to 0.75 cm in storage period, it gave play to expectant effects.

(ii) Pervious meters

Two pervious meters were laid on pedestal rock at center of river bed near downstream lateral of the core. The observation results showed that the water level at measuring point was lower either in flood period or in dry period. There was no case in leakage, that indicated the impervious core worked in normal and the drainage passage was unobstructed.

(iii) Earth pressure meters

Six earth pressure meters were arranged symmetrically on two sides of core at up and down stream and lined on three different level. Due to the failure of observation instruments, there was no completely result gained.

(iV) Steel bar meters

12 steel bar meters were fixed on steel reinforcement at two sides of core and arranged on six levels. It showed that the concrete core worked essentially in

compressive case, but there was a greater minus bending moment at bottom in flood period.

(V) Displacement monitors

The displacement measuring points on dam slope all were distroyed during construction. Only five line measuring points laid at top of the core gained the observation results. At the cross section on center line of river bed, the measured displacement on core top was 1.6 cm while the water level of reservior rose up to 2486.5 m . Due to raising the height of cofferdam in 1981, the measuring work had broken off later on. The settlement of core was very small, only being within the scope of observation error.

3. Back Analysis of Core and The Comparison With Proto-type Measuring

3.1 Back Analysis

It is of important significance to take a back analysis for rock filled dam which had such sysmetical proto-type measuring data. Its effects are: 1), according to the comparison between back analysis and proto-type measuring, we can further explain and check the monitoring results and analyze and judge their reality; 2), we can utilize the measuring results of displacement and stress back to calculate the proto-type deformation modulus of dam shell; and 3), through examining the measuring data we can correct the parameters of FEM.

The back analysis applied the variable stiffness method with piecewise linear form. In calculation, we took the maximum cross section in river bed and assumed the core as a fixed support continue structure due to the existance of bending moment on horizontal joints in core, that gained from measuring. The rock filled cofferdam was divided into 17 stages to bearing load and simulated the construction process which desired the downstream filled layer always being higher than the up stream layer in three meters. The finite element adopted was the arbitrary quadrangle isoparametric element and the contact element respectively for filled rock and contact face between concrete and rock material. The linear elastic model was used in concrete core and the hyperbolic model of Duncan-Chang was used in dam shell. The parameters of dam materials were chosen by refering the testing results of some dams and were then corrected by back analysis until reaching to agreement in deformation of FEM and proto-type monitor [4].

3.2 The Comparison Between FEM and proto-type observation

3.2.1 Core Displacement

In the light of the measuring results, the displacement of the concrete core was very small in construction period since the balance in lateral pressure in up and downstream. Fig. 4 showed the horizontal displacement on core top at various storage levels. It can be seen that the measured and calculated values of displacement on core top increased regularly along with the increase of storage level, the maximum values all closed to 2 cm. It can be thought that the model and parameters used in FEM through the correction of back analysis could reflect the actual condition of dam materials.

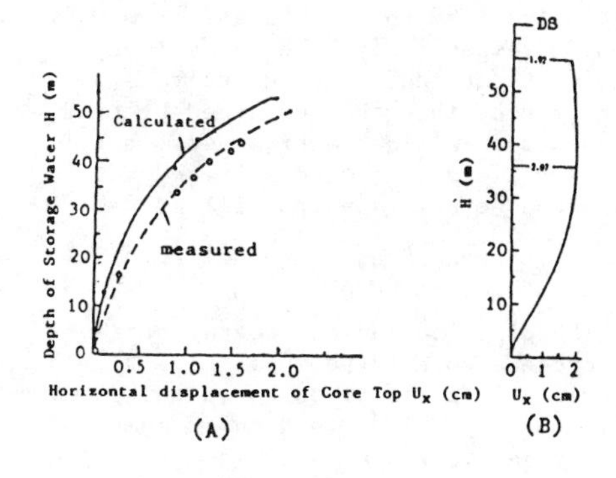

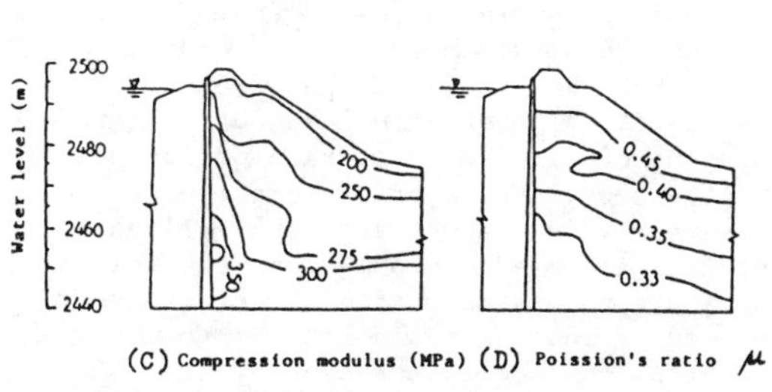

Fig.4 Horizontal Displacement of Core and Deformation Character of Back Analysis in Storage Stage

When the reservoir was full filled the distributive curve of horizontal displacement was shown in Fig. 4(B). The core moved toword downstream since it subjected to the loads of water and soil pressures from upstream, until it was balanced by the resistant force formed in downstream dam shell. Due to the resistance offered in dam shell being nonuniform, the horizontal diplacement curve of core tended to bend to upstream on its upper part, thus an anti-bending point appeared on its lower part 6 m from its bottom. The maximum value of core displacement at height of 36 m was 2.07 cm, but was 1.92 cm at core top. The deflection of concrete core was 0.037 percent, this value indicated the downstream dam shell offered a very strong resistance to it.

The distributions of elasticity modulus E and Poisson's ratio μ in downstream dam shell obtained by back analysis are shown in Fig. 4(C) and (D). It discribed that their scopes of distributions were associated with the rigidity of dam shell. In laboratory tests, it is difficut to simulate the actual stress path of dam loading, only through back analysis the elasticity constant which accords with the loading process can be calculated. Fig. 4 (C) and (D) showed that the elasticity constant E and μ in storage period ranged from 200 to 350 MPa and from 0.31 to 0.45 respectively. The elasticity modulus E obtained by back analysis was much greater than that obtained by traditional triaxial laboratory tests, but it was very close to some measuring results of some concrete face rock filled dam [5].

3.2.2. Lateral Pressure and Tangential Force on Core

The distribution of lateral pressure on concrete core is shown in Fig. 5. The lateral soil pressure on two sides of core only had a little difference in completion stage, it essentially was symmetric. The soil pressure increased from top to bottom, the maximum value reached to 363.0 KPa, but in foundation trench it redused to a lower value due to the transmission of stress. In storage period due to the hydro-static pressure acting on upstream face of core and the influence of floatation and deformation, the soil pressure on upstream face of core redused obviously and the resistance in downstream dam shell increased by some 40 percent except the place under anti--bonding point. The sum of water and soil pressure on upstream was in balance to the soil resistance offered by downstream shell. Therefore, the horizontal force subjected to by the core was small and the bending moment in core also offset likely by the downstream resistance, so that the horizontal displacement was limited to a small value.

The tangential force on surface of core was essentially symmetric, the mean value is some 70 KPa in completion stage, but it increased on downstream side and decreased on upstream side when the storage level rose up.

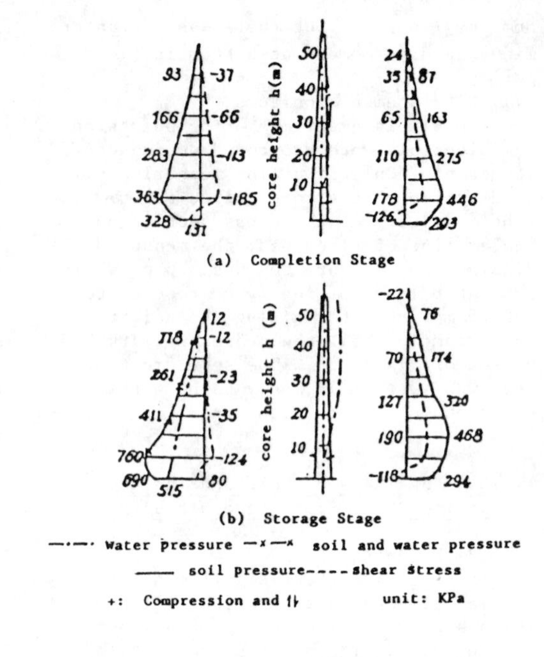

Fig.5 Stress Distributions on Core Sides

3.2.3. Stress in Core

Fig. 6 (A) shows the changing process of storage level from January, 1981 to October, 1983. The maximum storage level of 2494.8 m was appeared in September, 1981 and 2nd high storage level of 2483.0 m was appeared in July, 1983. The actual measuring value of normal stress at upstream and downstream σ_{US} and σ_{DS}, are shown in Fig. 6 (B) and 6 (C), which on different level of 2463 m and 2450 m.

From Fig. 6 (B) and (C), we knew that the stress in core was uniform in dry period of 1981, stage I, it almost existed no bending stress. In flood period of 1981, stage II, the bending stress in core reached to maximum value according with high storage level of that time. After that, in spite of dry period or flood period, stage III or IV, the bending stress could not recover to an appropriate value along with the changing of storage level, that indicated the restrain action of the plastic deformation of dam shell limited to recover the deformation of concrete core.

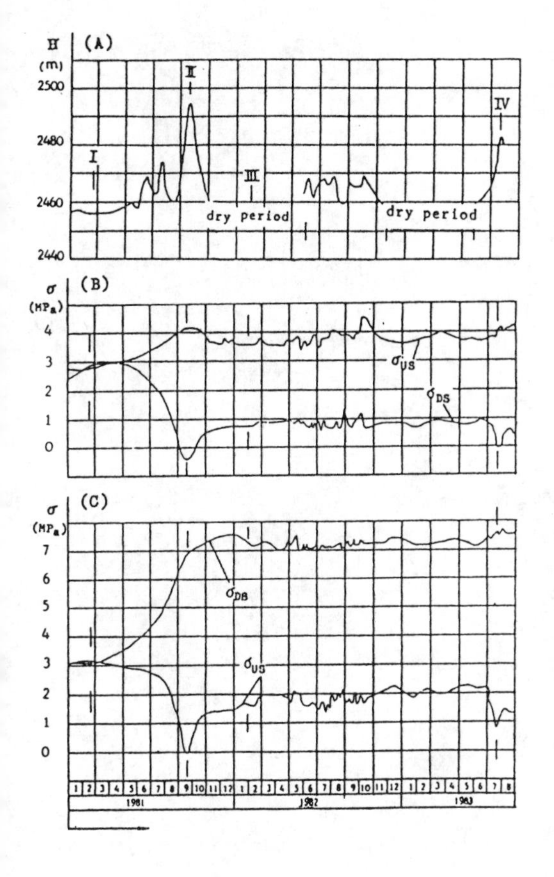

Fig.6 Measured Relation Curves Between Normal Stress and Storage Process

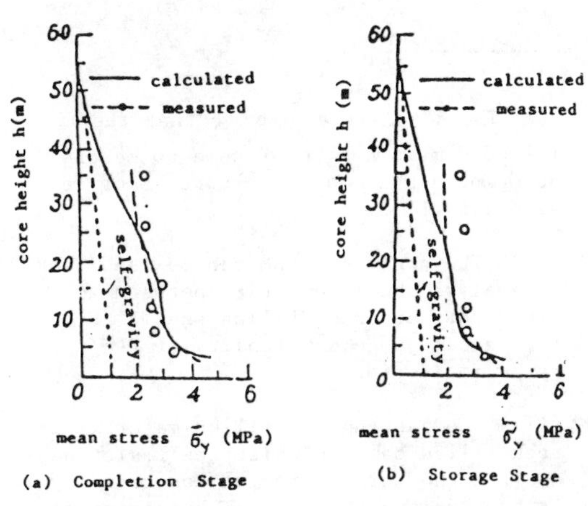

Fig.7 Average of Normal Stress in Core

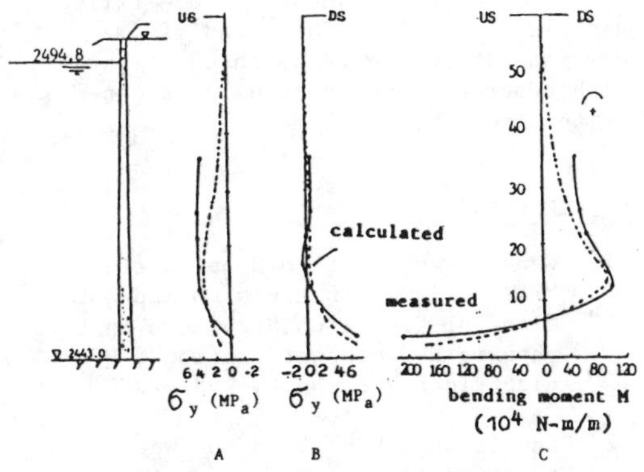

Fig.8 Calculated and Measured Core Edge Stress and Distribution of Bending Moment in Full Storage

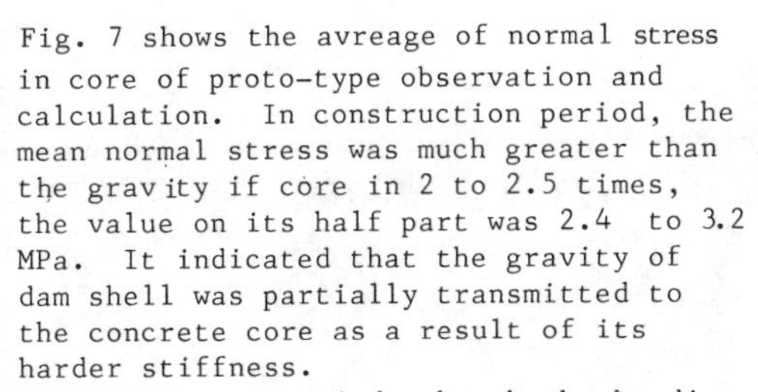

Fig. 7 shows the avreage of normal stress in core of proto-type observation and calculation. In construction period, the mean normal stress was much greater than the gravity if core in 2 to 2.5 times, the value on its half part was 2.4 to 3.2 MPa. It indicated that the gravity of dam shell was partially transmitted to the concrete core as a result of its harder stiffness.

In storage period, though the bending stress in core increased, the normal stress still kept in compression, the mean value on its half part trdused to 2.0 to 3.0 MPa, Fig. 7 (B) and 8.

Fig. 8 (A) and (B) show the edge stress on up and downstream faces of core in full storage level. The calculated value was in agreement with the measuring one. The edge stress likely was compressive, the maximum pressure at do-wnstream was some 6 MPa. The measuring tension was -0.1 MPa which occured at one third of the height, but in calculation there was no tension.

3.2.4 Bending moment in core

According to measuring and calculation the bending moment was given in Fig. 8(C). The minus bending moment at core bottom was 1700 KN-m in calculation and 2000 KN-m in measuring, it was very close each other. The values of them were rapidly down to zero at height of 6 to 7 meters from bottom. The maximum value of 900 KN-m appeared at height of 15 meters from bottom. Above the antibending point, the bending moment tended to a small value.

4. Conclusion

(i) The actualities approve that the rock filled dam in height of some 60 meters adopting a thin concrete core is of feasibility.

(ii) The nonlinear FEM can be used back to analyse the interaction between concrete core and rock filled shell. From that we can judge the reality of proto--type monitoring.

(iii) Compacting the filled material in rock filled dam can raise the resistance against the deformation of concrete core. It can improve the working condition of the concrete core.

(iv) Using concrete core to proof against the perviousness in rock filled dam is available. This kind of impervious structure is worth to compare with concrete face or bituminous concrete core.

REFERENCES

[1] General Bureau of Hydroelectric Power, "The Design, Construction and Anti-flood of Diversion Structure in Longyangxia Hydro-power Station", Hydroelectric Power, No.1, 1982, PRC.

[2] Pan Jiazung, "The Study on Working Condition of a Rock Waste Cofferdam With Concrete Core", Hydroelectric Power, No. 1, 1982, PRC.

[3] Yi Shinyian, Ma Raifen and Ma Kuifa, "The Report of Internal Observation in Concrete Core of Upstream Cofferdam of Longyangxia Hydro-power Station", Longyang Water Power, No.1, 1982, PRC.

[4] Gao Lianshi, "The Finite Element Analysis of Interaction between Concrete Core and Rock Filled Shell for Longyangxia Cofferdam", Chinese Journal of Water Conservancy, No.3, 1987, PRC.

[5] Marques, L., Maurer, E., Toniatti, B., "Deformation Characteristics of F_{02} Do Areia Concrete Face Rockfill Dam, as Revealed by a Simple Instrumentation System, 15th ICOLD, Vol.1, Q56, 1985.

Numerical Methods in Geomechanics (Innsbruck 1988), Swoboda (ed.)
© 1988 Balkema, Rotterdam. ISBN 90 6191 809 X

Finite element analysis of fill type dams – Stability during construction by using the effective stress concept

Y.Kohgo & T.Yamashita
National Research Institute of Agricultural Engineering, Ministry of Agriculture, Forestry and Fisheries, Tsukuba, Japan

ABSTRACT: A numerical method of general stability analysis against deformations and collapses of fill type dams during construction using the effective stress concept is given. It is based on a stress-seepage coupled analysis and a nonlinear elasto-plasticity finite element analysis after Nayak's solution technique. The numerical results for the stability analyses of an imaginary homogeneous dam show that the magnitude of collapse strength estimated by this method coincides well to ones estimated by the limite equilibrium methods (Morgenstern-Price and simplified Bishop methods) and the maximum rates of horizontal movements caused by the rise of fill influence the spread of plastic zones. It is concluded from these results that this method is valid for investigating the stability of fill type dams and the monitoring of horizontal movements during construction is a suitable index of the safety of dams.

1. INTRODUCTION

On consideration of the stability of soil structures, it is necessary to investigate their stability from two points of view. One is the stability for deformations, and the other is the stability for collapses. For the former, stress∿deformations analyses based on finite element methods have been adopted, where soil is usually regarded as a linear or nonlinear elasticity. For the latter, limite equilibrium methods have been used mainly. In the process of collapses of soil structures, firstly deformations become greater, nextly localized collapses are advanced and finally soil structures fail over all. Therefore, deformatins and collapses are closely related each other. Deformations and strength of soil should always be given in terms of the effective stress concept. According to the points of view described above, we should develop a method to investigate the stability continuously in the process from deformations to collapses in terms of the effective stress concept.

Finite element analysis simulating until the collapse of slopes has been achieved by Zienkiewicz et al. (1975), where the visco-plastic analysis method and the perfect elasto-plastic model were used. In their results, the collapse strength of slopes was exactly estimated, and slip surfaces were shown as those going through the concentrated area of shear strains. However, this method is not suitable and seepages of the pore water should be considered when the transient stability of soil structures (for example, stability of fill type dams during construction) is invetigated by using the effective stress concept.

In this paper, we will show the numerical method of general stability analysis against deformations and collapses of fill type dams during construction using the effective stress concept. The stability of an imaginary homogeneous dam will be examined by using this method, and we will discuss about its applicability and the behavior of the earth dam.

2. SOLUTION METHOD

2.1 Field equations

We need the equilibrium equations of the soil and continuity equations of the pore fluid as the field equations. In description below, it is assumed that inertia effects could be neglected.

If the soil is saturated with the water (which is assumed to be imcompressible), the field equations should be as follows (Biot, 1941).

$$\sigma'_{ij,j} + \delta_{ij}P_{,j} + \rho F_i = 0 \quad \cdots\cdots\cdots\cdots (1)$$

$$q_{i,i} + \dot{\varepsilon}_{ii} = 0 \quad \cdots\cdots\cdots\cdots (2)$$

where σ'_{ij} denotes the components of effective stress tensor, P denotes the pore water pressures, ρ denotes the mass density of the soil, F_i are the components of the body force vector, δ_{ij} is the Kroneker delta, q_i denotes the components of relative displacement velocity vector of the fluid with respect to the soil skeleton and ε_{ii} is the volumetric strain of the soil skeleton.

Now, we restrict our consideration to two typical unsaturated soil conditions. One is the unsaturated condition where the air bubbles are entirely surrounded by the water. The other one is the unsaturated condition where the pore air is connected with the atmosphere. Here, we call the former the occluded unsaturated condition and the latter the uniform unsaturated condition, respectively. In the occluded unsaturated case, the soil is regarded as the saturated one filled with the single homogeneous fluid combined with the air and the water. Then, Eq. (1) can be used without changing by treating P as the homogeneous pore fluid pressures. Eq. (2) should be rewritten as follows to take account of the compression of the fluid. (Ghaboussi et al., 1973)

$$q_{i,i} + \dot{\varepsilon}_{ii} - \dot{e}_{ii} = 0 \quad \cdots\cdots\cdots\cdots\cdots\cdots (3)$$

Where e_{ii} is the volumetric strain of the fluid.

Considering three cases, that is, the saturated, occluded unsaturated and uniform unsaturated cases, the mass conservation law of the pore water should be used as the field equation (Scott, 1965) and the equilibrium equations could be rewritten using the stress state variables (Fredlund et al., 1977). If soil particles and the water are assumed to be imcompressible, the field equations could be described as follows (Kohgo, 1986).

$$\sigma'_{ij,j} + \langle S-S_c \rangle \delta_{ij}P_{,j} + \rho F_i = 0 \quad \cdots\cdots (4)$$

$$q_{i,i} - \dot{e}_{ii} + S\dot{\varepsilon}_{ii} = 0 \quad \cdots\cdots (5)$$

Where S is the degree of saturation and S_c is the degree of saturation at the air entry value, which distinguishes between the occluded ($S \geqq S_c$) and uniform unsaturated conditions ($S<S_c$) (refer to Fig.1). The bracket < > denotes 1 and 0 when $S \geqq S_c$ and $S<S_c$, respectively. e_{ii} does not represent the volumetric strain of the fluid but the change of storage of the water due to the change of degree of saturation. Therefore, Eq. (5) consists with the Richards' formula which governs the unsaturated flow.

In the following, Eqs. (4) and (5) will be adopted as the field equations. The generalized Darcy's law, the assumption of small deformations, the stress∿strain relationships and the soil water retentivity relationship will be used. These relationships are expressed as follows.

$$q_i = k_{ij}(P_{,j} + \rho_f F_j)/\rho_f \quad \cdots\cdots\cdots\cdots (6)$$

$$\varepsilon_{ij} = (u_{i,j} + u_{j,i})/2 \quad \cdots\cdots\cdots\cdots (7)$$

$$d\sigma'_{ij} = D_{ijkl}d\varepsilon_{kl} \quad \cdots\cdots\cdots\cdots (8)$$

$$de_{ii} = n'dP/E_f = dP/E'_f \quad \cdots\cdots\cdots\cdots (9)$$

Where ρ_f, k_{ij}, u_i, ε_{ij}, D_{ijkl} and n' denote the mass density of the water, the permeability tensor, the displacement vector of the soil skeleton, the components of strain tensor of the soil skeleton, the components of elasticity tensor for the soil skeleton and porosity, respectively. $1/E_f$ is $-\partial S/\partial P$. Subscripts after a comma denote spatial differentiation in the standard indicial notation used. Repeated indices indicate summation and a superposed dot denotes differentiation with respect to time.

The boundary conditions are;

$$\left.\begin{array}{lll} u_i = \bar{u}_i & & \text{on } S_1 \\ P = \bar{P} & & \text{on } S_2 \\ T_i = (\sigma'_{ij} + \delta_{ij}P)n_j & & \text{on } S_3 \\ Q_i = q_i n_i & & \text{on } S_4 \end{array}\right\} \quad (10)$$

2.2 Formulation of finite element method

By using the weighted residual method, Eqs. (4) and (5) can be approximated as a set of integral equations.

$$\int_V (\sigma'_{ij,j} + \langle S-S_c \rangle \delta_{ij}P_{,j} + \rho F_i)u^*_i dV = 0 \quad \cdots\cdots (11)$$

$$\int_V (q_{i,i} - \dot{e}_{ii} + S\dot{\varepsilon}_{ii})P^* dV = 0 \quad \cdots\cdots (12)$$

Where u^*_i and P^* are weight functions.

If Green's theorem and the relationships of Eqs. (6) ∿ (10) are applied to Eqs. (11) and (12) and the finite elements discretization of the domain is carried out, Eqs. (11) and (12) can be represented as the following equations.

$$\Phi_1=\sum_{i=1}^{\ell}\int_v B^T\sigma'(t_n)dV+\sum_{i=1}^{\ell}\int_v \langle S-S_c\rangle C^T NP(t_n)dV$$

$$-\sum_{i=1}^{\ell}\int_{s_3} N_{ub}^T N_{ub}F_b(t_n)ds-\sum_{i=1}^{\ell}\int_v \rho N_u^T F dv = 0 \quad \cdots\cdots (13)$$

$$\Phi_2=\sum_{i=1}^{\ell}\int_v SN^T Cu(t_n)dv-\sum_{i=1}^{\ell}\int_v SN^T Cu(t_{n-1})dv$$

$$-\sum_{i=1}^{\ell}\int_v N^T e(t_n)dv+\sum_{i=1}^{\ell}\int_v N^T e(t_{n-1})dv$$

$$+\Delta t[\alpha\sum_{i=1}^{\ell}\int_{s_4} N_b^T N_b Q_b(t_n)ds$$

$$+(1-\alpha)\sum_{i=1}^{\ell}\int_{s_4} N_b^T N_b Q_b(t_{n-1})ds]$$

$$-\Delta t[\alpha\sum_{i=1}^{\ell}\int_v N'^T\{q_p(t_n)+q_g(t_n)\}dv$$

$$+(1-\alpha)\sum_{i=1}^{\ell}\int_v N'^T\{q_p(t_{n-1})$$

$$+q_g(t_{n-1})\}dv] \quad \cdots\cdots\cdots\cdots\cdots (14)$$

Where σ', u, P, N, N', N_u, B, C, e, F, Q_b, F_b, q and α denote the effective stress vector, the nodal displacement vector, the nodal pore water pressures, the interpolation function for pore water pressures, the gradient matrix for pore water pressures, the interpolation function for displacements, the nodal displacement ∿ strain matrix, the nodal displacement ∿ volumetric strain matrix, the change of storage of the water due to the change of degree of saturation, the body force vector, the prescribed flux on the boundary nodal points, the prescribed traction on the boundary nodal points, the flux vector and the coefficient characterizing single-step temporal discretization ($\alpha \geq 1/2$ for stability), respectively. Subscripts b, p and g denote the value on the boundary, the value due to pore water pressures and the value due to the gravitational potential. Φ and $\sum_{i=1}^{\ell}$ denote the residual vector and the sum of total elements, respectively.

If the material nonlinearities and some kinds of boundary conditions are taken into account, Eqs. (13) and (14) will become a set of the nonlinear simultaneous equations. Then, we need the iteration to obtain the solutions which are always satisfied with Eqs. (13) and (14). The procedure of the iteration will be shown in the next section.

2.3 Procedure of nonlinear solution

The solution procedure based on the modified Newton-Raphson method is adopted here, but we will previously show that based on the Newton-Raphson method. If the Newton-Raphson method is applied to Eqs. (13) and (14), the solution procedure can be expressed as follows.

$$\left.\begin{array}{l}
d\Phi=K^* d\delta \\
d\Phi=\begin{Bmatrix} d\Phi_1 \\ d\Phi_2 \end{Bmatrix},\quad d\delta=\begin{Bmatrix} du(t_n) \\ dP(t_n) \end{Bmatrix}, \\
K^*=\begin{bmatrix} K & A \\ G & -\alpha\cdot\Delta t\bar{K}-E \end{bmatrix} \\
K=\sum_{i=1}^{\ell}\int_v B^T DBdv,\quad A=\sum_{i=1}^{\ell}\int_v \langle S-S_c\rangle C^T Ndv \\
G=\sum_{i=1}^{\ell}\int_v SN^T Cdv,\quad \bar{K}=\sum_{i=1}^{\ell}\int_v N'^T RN'/\rho_f dv \\
E=\sum_{i=1}^{\ell}\int_v N^T N/E_f' dv
\end{array}\right\} \quad (15)$$

Where D denotes the elasticity matrix and R denotes the permeability matrix.

According to Eq. (15), the m+1th iteration δ_{m+1} can be evaluated by;

$$\delta_{m+1} = \delta_m + \Delta\delta_{m+1} \quad \cdots\cdots\cdots\cdots\cdots (16)$$

$$\Delta\delta_{m+1} = -K_m^{*-1}\Phi^m \quad \cdots\cdots\cdots\cdots\cdots (17)$$

If K^*_m is equal to K^*_0 and is constant, this method is equivalent to the modified Newton-Raphson method. Eq. (18) can be used as K^*_0 matrix despite the dissymmetry of K^* matrix (Kohgo, 1987). Therefore, the coefficients above (or below) the principal diagonal of K^*_0 matrix may be only stored because of the symmetry of K^*_0 matrix.

$$\left.\begin{array}{l}
K^*_0 = \begin{bmatrix} K & L \\ L^T & -\alpha\cdot\Delta t\bar{K}-E \end{bmatrix} \\
L = \sum_{i=1}^{\ell}\int_v C^T Ndv
\end{array}\right\} \quad \cdots\cdots\cdots\cdots (18)$$

Following the above description, in the modified Newton-Raphson method, Eq. (17)

should be rewritten into Eq. (19), and Eq. (16) can be used as it is.

$$\Delta\delta_{m+1} = -K_0^{*-1}\Phi^m \quad \cdots\cdots\cdots\cdots\cdots (19)$$

$$\left.\begin{aligned}
\Phi^m &= \begin{Bmatrix} \Phi_1^m \\ \Phi_2^m \end{Bmatrix} \\
\Phi_1^m &= \sum_{i=1}^{\ell}\int_v B^T\sigma'(t_n)_m dv \\
&+ \sum_{i=1}^{\ell}\int_v \langle S-S_c\rangle C^T NP(t_n)_m dv \\
&- \sum_{i=1}^{\ell}\int_{s_3} N_{ub}^T N_{ub} F_b(t_n) ds - \sum_{i=1}^{\ell}\int_v \rho N_u^T F dv \\
\Phi_2^m &= \sum_{i=1}^{\ell}\int_v SN^T Cu(t_n)_m dv \\
&- \sum_{i=1}^{\ell}\int_v SN^T Cu(t_{n-1}) dv \\
&- \sum_{i=1}^{\ell}\int_v N^T e(t_n)_m dv \\
&+ \sum_{i=1}^{\ell}\int_v N^T e(t_{n-1}) dv \\
&+ \Delta t[\alpha\sum_{i=1}^{\ell}\int_{s_4} N_b^T N_b Q_b(t_n) ds \\
&\quad + (1-\alpha)\sum_{i=1}^{\ell}\int_{s_4} N_b^T N_b Q_b(t_{n-1}) ds] \\
&- \Delta t[\alpha\sum_{i=1}^{\ell}\int_v N'^T\{q_p(t_n)_m + q_g(t_n)_m\} dv \\
&\quad + (1-\alpha)\sum_{i=1}^{\ell}\int_v N'^T\{q_p(t_{n-1}) \\
&\quad + q_g(t_{n-1})\} dv]
\end{aligned}\right\} (20)$$

The operations of Eqs. (16), (19) and (20) should be iterated until the solutions of δ have been satisfied with the convergence criteria. The criteria used here is;

$$\frac{\|\Delta\delta\|_2}{\|\delta\|_2} \leq \varepsilon , \quad \|\Delta\delta\|_2 = \sqrt{\Delta\delta^T\Delta\delta} \quad \cdot\cdot (21)$$

In the following analysis, the tolerance ε is 1×10^{-3}. If the soil is regarded as the elasto-plasticity, operations of plastic stress calculation should be inserted between operations of Eqs. (19) and (20). This operation technique follows Nayak's solution technique (1972) which pays attention to the exclusion of cumulative errors.

Embankment construction is simulated using the technique developed by Clough et al. (1967).

In only considering the saturated and occluded unsaturated conditions, Eqs. (1) and (3) might be adopted as the field equations. This is accomplished by replacing both $\langle S-S_c\rangle$ and S to unity in Eq. (20).

2.4 Material nonlinearity

Here, we will consider the nonlinearities for deformations of the soil skeleton, the permeability and the soil water retentivity.

In the nonlinearities of deformations of the soil skeleton, the soil is regarded as the nonlinear elasto-plasticity. In the elastic region, Eq. (22) represents the shear (G) and bulk (K) moduli using the stress invariants.

$$\left.\begin{aligned}
G_{ld} &= G_0 + \bar{\gamma}_{ld}\sqrt{J_2} - \gamma_{ld}P'_m \\
G_{un} &= G_{0u} + \bar{\gamma}_{un}\sqrt{J_2} - \gamma_{un}P'_m \\
K &= -2.3(1+e_0)P'_m/\kappa + K_0
\end{aligned}\right\} \cdots\cdots\cdots (22)$$

Subscripts ld and un denote the loading and unloading conditions, respectively. The stress invariants J_2 and P'_m can be expressed by;

$$\left.\begin{aligned}
P'_m &= \sigma'_{ii}/3 \\
J_2 &= S_{ij}S_{ij}/2 \\
S_{ij} &= \sigma'_{ij} - \delta_{ij}P'_m
\end{aligned}\right\} \cdots\cdots\cdots\cdots\cdots (23)$$

κ is the slope of overconsolidation line ($e \sim \log P'_m$). G_0, γ_{ld}, $\bar{\gamma}_{ld}$, G_{0u}, γ_{un}, $\bar{\gamma}_{un}$ and K_0 denote the material parameters.

In the plastic region, the yield and plastic potential functions are assumed to be Drucker-Prager and von Mieses models, respetively.

Although the permeability is affected by many factors, void ratio (e) and degree of saturation (S) will be only considered as its factors here. The coefficient of permeability is;

$$\left.\begin{aligned}
k &= k_s E_s H_s \\
E_s &= \{e^n/(1+e)\}/\{e_0^n/(1+e_0)\} \\
H_s &= \{(S-S_f)/(1-S_f)\}^m
\end{aligned}\right\} \cdots\cdot (24)$$

Where k_s is the saturated coefficient of permeability at $e=e_0$, S_f is the residual degree of saturation, where k is theoretically assumed to be zero, and both n and m are the material parameters.

In the uniform unsaturated condition, the nonlinearity for the soil water retentivity as shown in Fig.1 could be expressed using the modified Brooks and Corey's equation (1966). In the occluded unsaturated condition, its nonlinearity could be theoretically obtained by assuming that the pore water pressure is nearly equal to the pore air pressure. These are;

$$\left.\begin{aligned} & S = -(1-S_c+hS_c)\frac{(P_{ac}+P_a)}{(P+P_a)} + (1+hS_c) \\ & \frac{1}{E_f} = -\frac{\partial S}{\partial P} = -\frac{(1-S_c+hS_c)}{(P+P_a)^2}(P_{ac}+P_a) \\ & \qquad\qquad\qquad\qquad (\text{when } S>S_c) \\ & S = S_f + (S_c-S_f)\left(\frac{P_{ac}}{P}\right)^{\lambda_s} \\ & \frac{1}{E_f} = -\frac{\partial S}{\partial P} = \lambda_s\frac{(S-S_f)}{P} \qquad (\text{when } S<S_c) \end{aligned}\right\} \quad (25)$$

Where P_a, h and P_{ac} denote the atmospheric pressure, Henry's coefficient of solubility and the air entry pore water pressure, respectively, and λ_s is the material parameter related to the material parameter m as $m=3+2/\lambda_s$, which has been used to express the permeability nonlinearity. When $S=S_c$, the mean values of both ones are used.

When the saturated and occluded unsaturated conditions are only considered, S and E_f could be theoretically estimated as follows. In this case, E_f can be regarded as the bulk modulus of the air-water mixture.

$$\left.\begin{aligned} & S = \frac{S_0}{(1-h)S_0+(1-S_0+hS_0)\left(\frac{P_0+P_a}{P+P_a}\right)} \\ & E_f = -\frac{(P+P_a)^2}{(1-S_0+hS_0)(P_0+P_a)} \end{aligned}\right\} \cdots\cdots (26)$$

Where P_0 and S_0 denote the initial pore pressure and the initial degree of saturation, respectively.

3. APPLICATION TO HOMOGENEOUS EARTH DAM

Here, the stability of the homogeneous earth dam during construction is investigated using the solution method described above. The saturated and occluded unsaturated conditions are considered in this analysis, because the degree of saturation in homogeneous dams should be usually comparatively high. Consequently, both $\langle S-S_c \rangle$ and S in Eq. (20) are replaced to unity.

Two cases (Case A and B) were analyzed. Case A is the case of collapse during construction, Case B is the safely constructed case. The difference between two cases exists only in the magnitude of the saturated coefficient of permeability k_s. They are 1×10^{-4}m/day in Case A and 1×10^{-2}m/day in Case B. The dimension of the dam (H=50 m, B=305m), finite element mesh, the speed of construction and the boundary conditions are shown in Fig.2. As the boundary conditions, the pore water pressures P are zero at the up and down stream foundations and should be always less than or equal to the atmospheric pressure on the embankment surface. The base is rigid and impervious. The elements used here are quadratic isoparametric elements (8-noded serendipity elements) for the displacement field and superparametric elements for the pore water pressure field, which have the lower interpolation function by one order than elements used for the displacement field. The reduced integration (a 2×2 Gauss point integration) is also adopted here to estimate the collapse strength. The time interval Δt is 8 days. The construction of this dam is simulated in ten layers with 5m lift/layer.

The results of these analyses are shown in Figs.3 ∿ 8. Fig.3 shows the pore water pressure distributions at different stages in Case A. At the 6th and the 7th layer stages (The Nth layer stage represents the condition at the Nth point where the real curve of construction and its approximated stepwise incremental curve as illustrated in Fig.2 cross each other.), the zero pore water pressure line approximately corresponds to the embankment surface. At the 8th and 9th layer stages, this pressure line near the crest appears at the place which is slightly below the embankment surface. The pore water pressure caused by banking is about 80t/m^2 at the center base of the dam at the 9th layer stage, and this value is about 85% of the overburden pressure. Figs.4(a) and (b) show the development of plastic zones in Case A and B, respectively. The distributions of shear strains at the 9th layer stage are shown in Fig.5. The slip surface, which was estimated as the arc surface going through the concentrated area of shear strains using the procedure propossed by Zienkiewicz et al., is also represented in this figuer. The safty factors (F_s) in this surface calculated by means of the limite equilibrium methods were 0.84, 0.98 and 0.99 in Fellenius, simplified Bishop and Morgenstern-Price methods, respectively. In these calculations, the values of pore water pressures on the slip surface and between the slices (only for Morgenstern-Price method) were estimated from the pore water pressure distributions illustrated in Fig.3 when the pore water pressures $P \geq 0$. When $P<0$, P set to zero. By the way, F_s was 0.99 in simplified Bishop method considering both the negative and positive pore water pressures. According to the above results, Safety factors should be estimated as slightly safty side values. However, it might be concluded that this embankment is almost under the critical condition as shown in the analyses by means of simplified Bishop and Morgenstern-Price methods. The safty factor by Fellenius method gives fairly safty side value as it has been already pointed out. According to the fact that the calculation did not converge at the 10th layer banking in this analysis, this embankment is also estimated to be almost under the critical condition. Therefore, the magnitude of collapses strength and the location of the slip surface could be fairly exactly obtained by the method represented here.

The relationships between the height of dams and the maximum rates of horizontal movements (R_{hm}) caused by the rise of fill are plotted in Fig.6. The solid circles and triangles denote the values observed during construction of nine dams (Penman, 1986). The open triangles and circles denote the values obtained from our analyses in Case A and B, respectively. These maximum rates were obtained at C point (in Case A) and D point (in Case B) in Fig.2. The value of R_{hm} is 7 at the 6th layer stage, exceeds 20 at the 7th layer stage, becomes 103 at the 8th layer stage and really reaches 397 at the 9th layer stage, in Case A. Fig.4(a) indicats obviously that at the 7th layer stage, plastic zones fairly extend at the down stream side, however at 6th layer stage, the zones are limited at both up and down streams of dam base. On the other hand, in Case B, the value of R_{hm} is 8, 14, 20, 23 and 19 at the 6th, 7th, 8th, 9th and 10th layer stages, respectively. Though this value continuously increases up to the 9th layer stage, it decreases at the 10th layer stage. It is clear from Fig.4(b) that plastic zones progress toward the center of the dam with the embankment from the 6th to the 9th layer stage, while they shrink at the 10th layer stage. It might be concluded from above considerations that the development of plastic zones are fairly accelerated when the value of R_{hm} exceeds 20 in these analyses. This value corresponds to that on the warning line of failure illustrated in Fig.6.

Changes of vertical and horizontal displacements generated by the rise of fill are shown in Figs.7 and 8. Fig.7 represents the relationship between vertical displacements and height of embankment at AA' section in Fig.2. In Case A, few vertical displacements occur until the 8th layer stage, but they rapidly increase at the 9th layer stage. In Case B, vertical displacements gradually increase and are greater than those in Case A until the 8th layer stage. Fig.8 represents horizontal displacements at BB' section in Fig.2 where the maximum horizontal displacements occur. In both cases, a few horizontal displacements generates at 7th layer stage. As the embankment rises higher, they accelerative-ly increase in Case A, but do not so fairly increase in Case B.

Thus, the behavior of horizontal displacements directly expresses whether unstability of the embankment increases or not. This implies that R_{hm} pointed out by Penman et al. is valid as a warning index of safty of dams.

4. CONCLUSIONS

We obtained the following conclusions.

(1) Comparing with the results of the limite equilibrium methods, the magnitude of collapses strength and the location of the slip surface can be fairly exactly obtained by the method represented here.

(2) The magnitude of collapses estimated by this method is almost as same as the one estimated by Morgenstern-Price or simplitied Bishop method. The value estimated by Fellenius method gives fairy safty side value as it has been already pointed out.

(3) It is important to monitor the horizontal movements during construction in checking on the stability of embankments. The figure proposed by Penman, which is used to check on whether embankments are stable or not, is practical (see Fig.6).

We would like to our grateful thanks to Prof. T. Tanaka, Dr. S. Iwata, Dr. T. Miyazaki and Dr. S. Hasegawa for many helpful suggestions and encouragement during the course of this work.

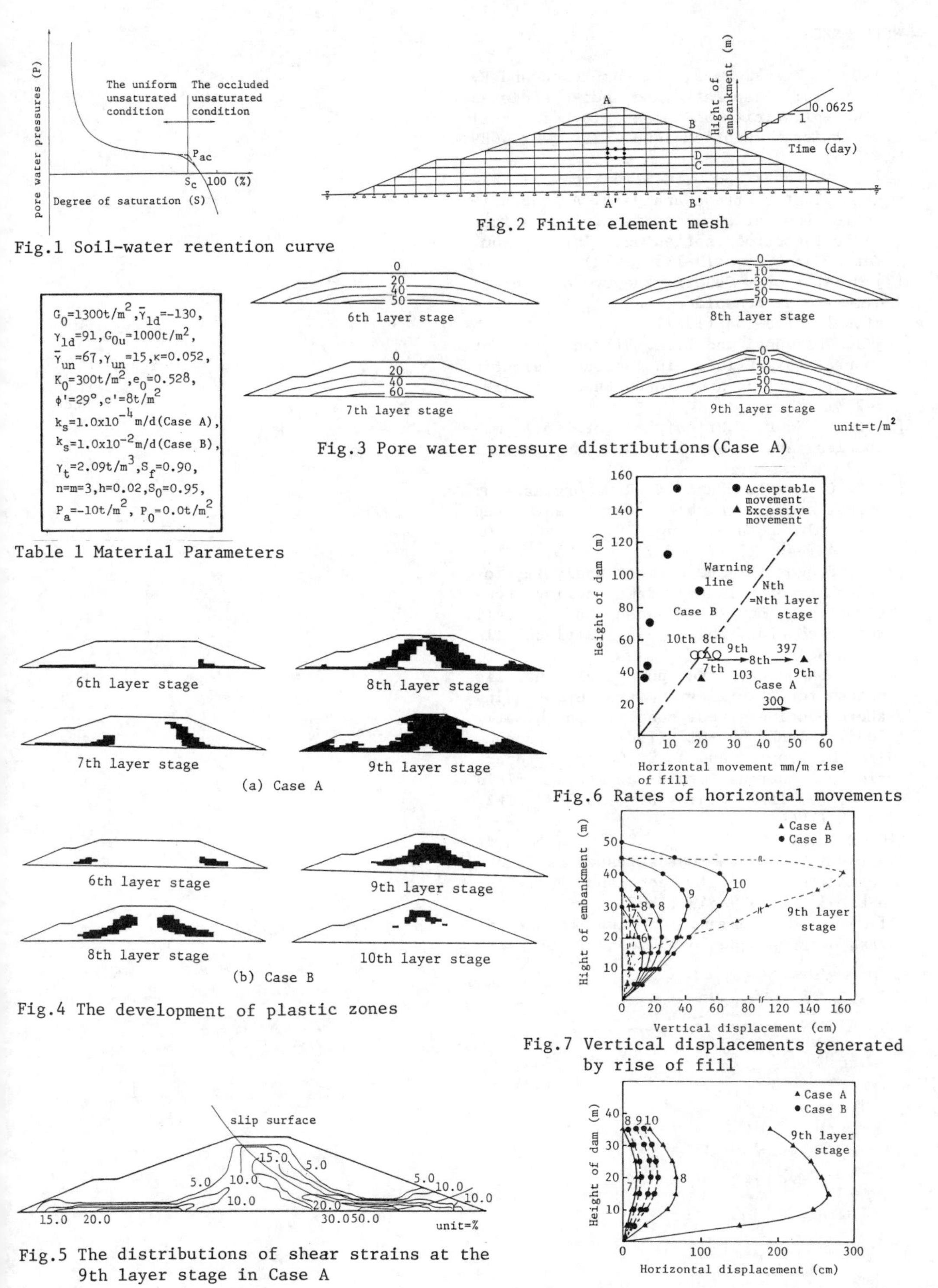

Fig.1 Soil-water retention curve

Fig.2 Finite element mesh

Fig.3 Pore water pressure distributions (Case A)

$G_0=1300t/m^2, \bar{\gamma}_{1d}=-130,$
$\gamma_{1d}=91, G_{0u}=1000t/m^2,$
$\bar{\gamma}_{un}=67, \gamma_{un}=15, \kappa=0.052,$
$K_0=300t/m^2, e_0=0.528,$
$\phi'=29^\circ, c'=8t/m^2$
$k_s=1.0\times10^{-4}m/d$ (Case A),
$k_s=1.0\times10^{-2}m/d$ (Case B),
$\gamma_t=2.09t/m^3, S_f=0.90,$
$n=m=3, h=0.02, S_0=0.95,$
$P_a=-10t/m^2, P_0=0.0t/m^2$

Table 1 Material Parameters

Fig.6 Rates of horizontal movements

Fig.4 The development of plastic zones

Fig.7 Vertical displacements generated by rise of fill

Fig.5 The distributions of shear strains at the 9th layer stage in Case A

Fig.8 Horizontal displacements generated by rise of fill

REFERENCES

[1] O. C. Zienkiewicz, C. Humpheson and R. W. Lewis; Associated and non-associated visco-plasticity and plasticity in soil mechanics, Geotechuique 25, 671-689 (1975)

[2] G. C. Nayak and O. C. Zienkiewicz: Elasto-plastic stress analysis. A generalization for various constitutive relations including strain softening, Int. J. Num. Meth. Engng. 5, 113-135 (1972)

[3] M. A. Biot: General theory of three-dimensional consolidation, J. Appl. Physics 12, 155-164 (1941)

[4] J. Ghaboussi and E. L. Wilson: Flow of compressible fluid in porous elastic media, Int. J. Num. Meth. Engng. 5, 419-442 (1973)

[5] R. F. Scott: Principles of soil mechanics, Addison-Wesley, Reading, Mass., 1963

[6] D. G. Fredlund and N. R. Morgenstern: Stress state variables for unsaturated soils, J. Geothch. Eng. Div. ASCE 103 GT5, 447-466 (1977)

[7] Y. Kohgo: Finite element analysis for stability of fill type dams during construction, The 21th Japan Natl. Conf. Soil Mech. Fdn. Eng., 591-592 (1986) (in Japanese)

[8] Y. Kohgo: On the numerical analysis method for stability of structures using the effective stress concept, Japan Natl. Conf. JSIDRE, 316-317 (1987)(in Japanese)

[9] R. H. Brooks and A. T. Corey: Properties of porous media affecting fluid flow, J. Irri. Drain. Div. ASCE 92 IR2, 61-88 (1966)

[10] R. W. Clough and R. J. Woodward: Analysis of embankment stresses and deformations, J. Soil Mech. and Fdn. Div. ASCE 93 SM4, 529-549 (1967)

[11] A. D. M. Penman: On the embankment dam, Geotechnique 36, 303-348 (1986)

Numerical Methods in Geomechanics (Innsbruck 1988), Swoboda (ed.)
© 1988 Balkema, Rotterdam. ISBN 90 6191 809 X

Three-dimensional movements of embankment ends on soft layer

Y.Yokoyama, O.Kusakabe & T.Hagiwara
University of Utsunomiya, Japan

ABSTRACT: A three dimensional stability analysis was carried out to examine the behavior of embankments on soft clay, with special reference to end parts of the embankments to which bridge abutments are attached. It was found that longitudinal movement beneath the abutments increases with increasing the width of the embankment.

1 INTRODUCTION

Networks of transportation such as highways form one of the most important infrastructures in any counties. In Japan, most of roads has been constructed on soft alluvial deposits, prevailing along coastal areas. Up to 1960s roads were constructed to have normally two lines. Main concern over the stability of road embankment had been stability of the embankment slopes, to which two dimensional stability analysis is applied.

Since 1960s when wider roads began to be constructed, an attention has been given to movements of bridge foundations situated at end parts of the embankment. Massive data were collected and a correlation was found between horizontal movement of the bridge foundation and the factor of safety against the failure, assuming two dimensional slip surface at an end of the embankment in longitudinal direction, as shown in Figure 1 (Kimura et al., 1986).

In design consideration for the bridge foundations of narrower roads, earth pressure acting on superstructures is considered, but normally no consideration is made for the lateral pressure acting on the substructures such as piers, piles and caissons. This practice has been satisfactory as long as roads are narrow. However, current trend of constructing wider road, say four or six lines, necessities detailed consideration of the effect of earth pressures onto the substructure.

There are currently a large number of plans to upgrade two lines roads to four lines roads. In this context, there is a need for studying three dimensional behavior near the end part of embankment.

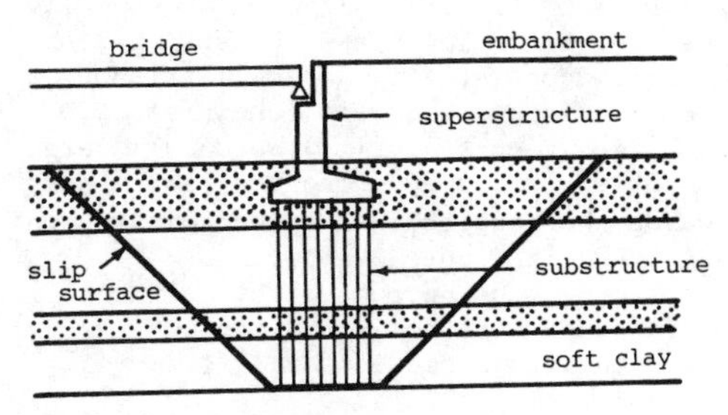

Fig. 1 Key sketch of the problem

This paper examines the effect of embankment width on the failure mechanism and design consideration for substructures beneath bridges by three dimensional upper bound analyses and centrifuge model tests.

2 THREE DIMENSIONAL UPPER BOUND CALCULATIONS

2.1 Soil profile and soil model

The soil is assumed to be saturated and normally consolidated, of which strength increases linearly with depth. The strength profile may be expressed of the form

$$Cu = Co + kz$$

where Cu : undrained shear strength at the depth of z
Co : undrained strength at the ground surface
k : the rate of strength increase with depth
z : depth measured from the ground surface.

Embankment material is treated as a flexible load and the shear strength of the material is not taken into account in the analysis, as usual practice. The soil is assumed to behave as a Tresca material and the upper bound formulation proposed by Drucker (1953) is used in this analysis.

2.2 Preliminary Considerations

Three dimensional nature at the end part of an embankment may be simplified as shown in Figure 2(a), in which there is a rectangular flexible load on a normally consolidated clay and the failure mechanism consists of two small triangular prisms sliding sideways (shown as BCDK-FGHL in Figure 2 (b)) and a large triangular prism pushing the soil away towards the longitudinal direction (shown as AEI-BFJ in Figure 2 (b)).

Adopting the proposal by Yamaguchi (1985) that the horizontal mean stress acting the cross section ABJI is

$$\bar{\sigma}_y = (\pi/2 + 1)*(Co + Z/2)$$

where $Z = kh/Co$,
the bearing capacity factor Nc is calculated by equating the energy dissipated to external work done according the upper bound theorem. The expression of Nc is of the form as

$$Nc=\min\{(1-m)\left(\frac{\pi}{2}+1+\frac{L}{h}+\frac{2h}{L}\right)\left(1+\frac{Z}{2}\right) + \frac{h}{B}\sqrt{1+\left(\frac{L}{h}\right)^2}\left(1+\frac{Z}{3}\right)+6m\left(1+\frac{mZ}{6}\cdot\frac{B}{h}\right) + \sqrt{2}\,m^2\frac{B}{L}\left(1+\frac{mZ}{6}\cdot\frac{B}{h}\right)\}$$

in which m is the ratio of the width of the small prism to the total width of the load, indicating how much sideway movement exists in the failure mechanism.

By minimizing the equation with respect to geometrical parameters, m and L/h for various values of kh/Co, the value of Nc can be obtained. The results of Nc value are plotted against B/h in Figure 3.

It is interesting to note that there exists three domains in Figure 3. For the values of B/h less than 1.0 where m = 1, there is a domain where the failure mechanism consists of the two small prisms only, indicating that the critical failure occurs totally in plane strain condition. On the other extreme (m < 0.1), the failure mechanism consists of dominantly the larger triangle with the very small side prisms, implying that the failure occurs mostly in longitudinal direction.

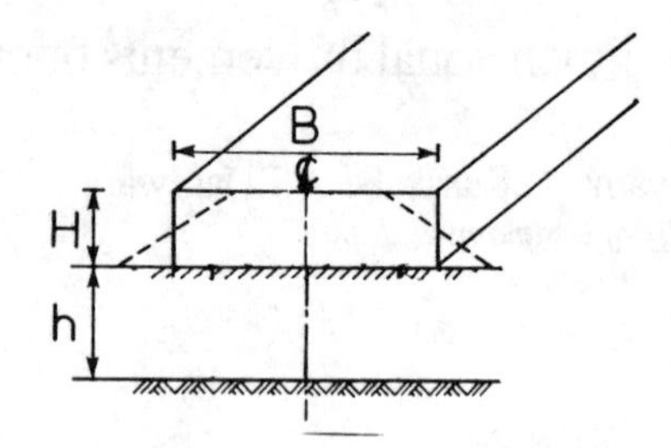

Fig.2 (a) Rectangular embankment

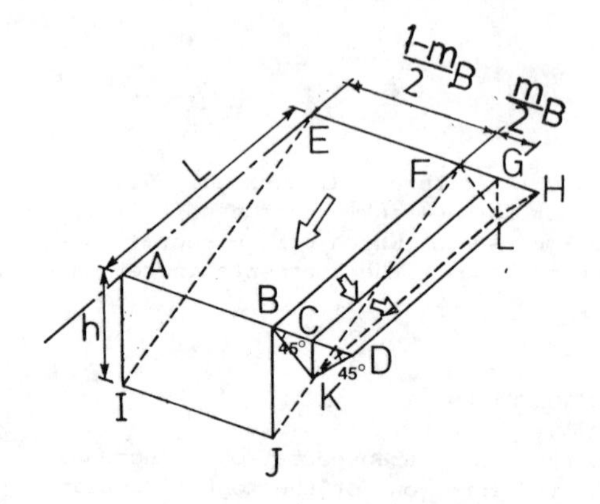

Fig.2 (b) Failure mechanism

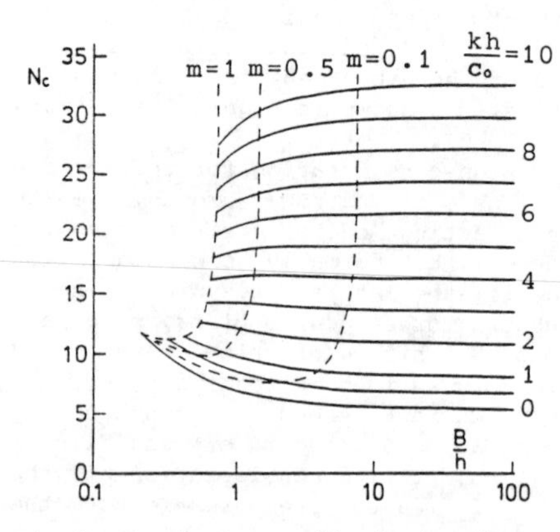

Fig.3 Bearing capacity factor Nc

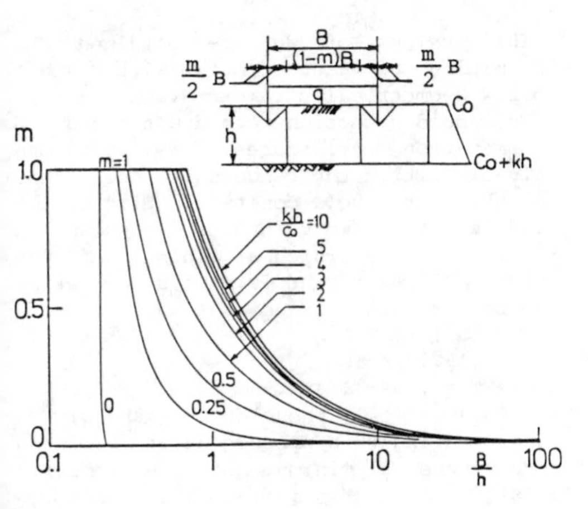

Fig.4 Change in m value for B/h

In between there is another domain where the failure mechanism is a mixture of both the two small side prisms and the larger triangle.

Figure 4 is the plot of the values of m giving the critical value of Nc against B/h. It becomes clearer in this figure how change in width of the load (embankment) influences on the movement of the soil beneath the area of load; namely the m value rapidly decreases with increasing B/h value. In other words, the soil moves more towards the longitudinal direction as the embankment width increases for a given depth of soft layer. This trend is more marked for the smaller value of kh/Co.

By the preliminary consideration, it is found that there is a general tendency that an abutment located at the end of the embankment is pushed by the movement of soil beneath the embankment when the width of the embankment becomes large. This tendency is generally agreed with field observation.

For example, if we had a road of of 4 meter wide on a soft clay of 4 meter thick(B/h = 1) and of kh/Co value of 1.0, and a bridge abutment was constructed on a caisson of 4 meter, the m value is then found to be 0.38 from Figure 4. That means that the caisson of 4 meter is being pushed by soil of 2.5 meter wide moving towards the longitudinal direction. Then we increased the width of the road and the caisson to be 24 meter(B/h = 6), the m value is going to be 0.06, which means that the caisson of 24meter is to be pushed by soil of 22.6 meter wide, more than 50% larger than before, in terms of unit width of the caisson.

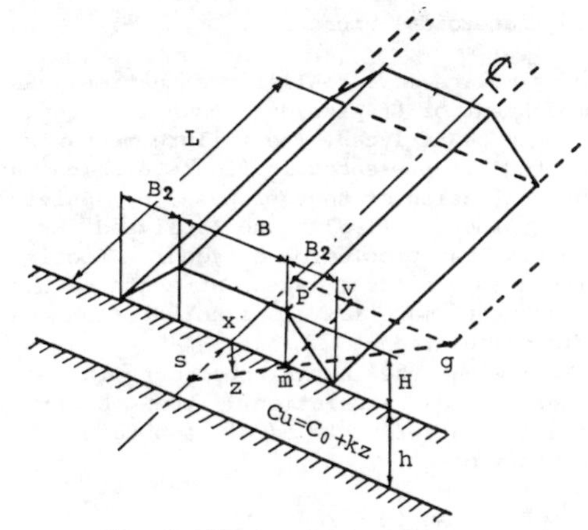

Fig.5 Trapezoidal embankment

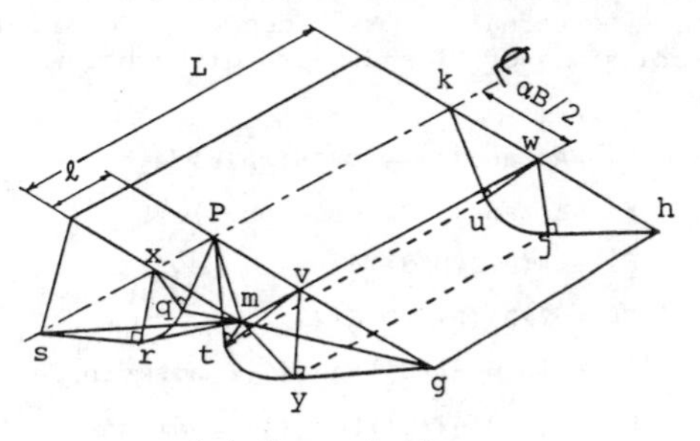

(a) General view

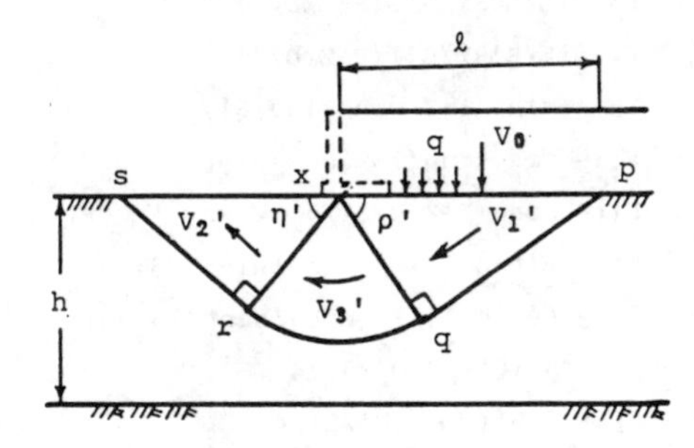

(b) Crosssection in longitudinal direction

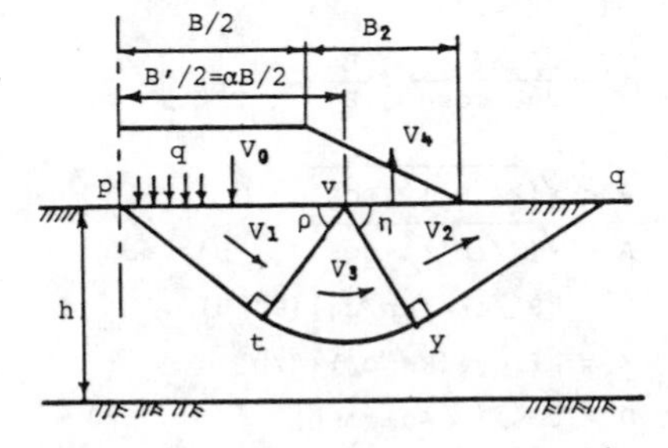

(c) Crosssection in lateral direction

Fig.6 Failure mechanism

2.3 Trapezoidal embankment

To simulate more realistic situation, an embankment of trapezoidal shown in Figure 5 is to be analyzed. The failure mechanism adopted is presented in Figure 6 which is the extension of that proposed by Shield and Drucker (1953) who analyzed the indentation problem of a square footing with a smooth base. Shear strength of the embankment material is ignored also in this calculation.

Following the normal procedure in the upper bound calculation, the bearing capacity factor Nc (= q/Co) is obtained in the form of

$$Nc = \min\ (\Sigma Fi/G)$$

where Fi are the energy dissipated and G the external work done. Detailed expressions of Fi and G are given below.

$$F_1 = A \tan\rho(1+Z_B\cos\rho\sin\rho/6)/4$$

$$F_2 = A \tan\eta(1+Z_B\cos\rho\sinh/6)/4$$

$$F_3 = A(P+Z_B D/3)/4$$

$$F_4 = \ell/B'(P+Z_B D/6)/2$$

$$F_5 = (L/B'-\ell/B')\tan\rho(1+Z\ \cos\rho\sin\rho/4)$$

$$F_6 = (L/B'-\ell/B')\tan\eta(1+Z_B\cos\rho\sin\eta/4)$$

$$F_7 = (L/B'-\ell/B')(P+Z_B D/2)$$

$$F_8 = (L/B'-\ell/B')(P+Z_B D/4)$$

$$F_9 = \sin\rho(1+Z_B\cos\rho\sin\rho/6)/4$$

$$F_{10} = \cos\rho\tan\eta(1+Z_B\cos\rho\sin\eta/6)/4$$

$$F_{11} = \cos\rho(P+Z_B D/3)/4$$

$$F_{12} = A'\tan\rho'(1+Z_\ell\cos\rho'\sin\rho'/3)/4$$

$$F_{13} = A'\tan\eta'(1+Z_\ell\cos\rho'\sin\eta'/3)/4$$

$$F_{14} = A'(P'+2Z_\ell D'/3)/4$$

$$F_{15} = \ell/B'(P'+Z_\ell D'/3)/4$$

$$G = \frac{L}{B'}\left[1-\frac{B}{B_2}\frac{(\alpha-1)^2}{4\alpha}-\frac{B}{\alpha B_2}\left(\frac{B_2}{B}-\frac{\alpha-1}{2}\right)^2\frac{\cos\eta}{\cos\rho}\left\{1-\frac{2\ell}{3\alpha L}\frac{\cos\eta}{\cos\rho}\left(\frac{B_2}{B}-\frac{\alpha-1}{2}\right)\right\}\right]$$

$$A = \sqrt{(2\ell/B')^2+\cos^2\rho}$$

$$A' = \sqrt{(2\ell/B')^2+\cos^2\rho'},\quad B' = \alpha B$$

$$Z = kB'/c_0 = (kh/c_0)(B'/h)$$

$$Z_\ell = k\ell/c_0 = (kh/c_0)(\ell/h)$$

$$D = \cos\rho(\cos\rho+\cos\eta)$$

$$D' = \cos\rho'(\cos\rho'+\sin\eta')$$

$$P = \pi-\rho-\eta\ ,\quad P' = \pi-\rho'-\eta'$$

The values of Nc are obtained by minimizing the value of $\Sigma Fi/G$ with respect to six geometrical parameters .

The golden section method was used in minimization procedures. We vary one valuable within the range while the other valuable are held constant. Firstly we find a set of two points(a, b shown in Figure 7) by which a minimum of the function f is bracketed. Then a new set of two points is specified as

$$q_1 = Ta + (1-T)b$$

$$q_2 = (1-T)a + Tb$$

where T = (5 - 1)/2.

If $f(q_1) > f(q_2)$, then let $a=q_1$ and $q_1=q_2$. By repeating this process n times, the zone where a minimum of the function exists is bracketed in $(a-b)T^n$. By looping over all the valuable one can find the minimum. Actual calculation was carried out by a supermini-computer VAX II and CPU time for obtaining the minimum value was normally 150sec.

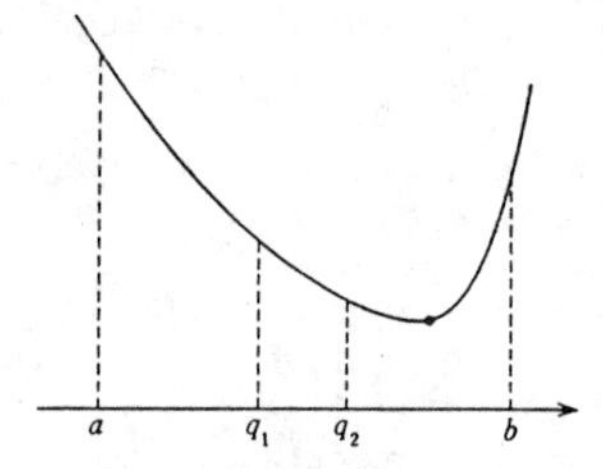

Fig.7 Minimization procedure

A typical example of Nc values is given in Figure 8 for the case of B2/h=0, L/B=10. For uniform clays, the Nc value is found to be 5.20 for B/h < 5, which indicates that if the aspect ratio L/B is more than 10, the bearing capacity is almost the same as that of two dimensional. Comparing the value of Nc given in Figure 3, it is said that this solution gives smaller value for the range of the B/h less than 5. Since it is known that smaller values are closer to the exact solution in the upper bound calculations, the envelope of the smaller values is to be used as shown in Figure 9.

Figure 10 is a plot of V/h^3 against B/h, where V denotes the soil volume moving towards the longitudinal direction. As seen in Figure 10, the soil volume increases rapidly with increasing B value for a given depth of soft layer, and decreases with increasing kh/Co value for a given B value.

The gradient of the relation between the soil volume and the width is more than unity, which implies that total earth

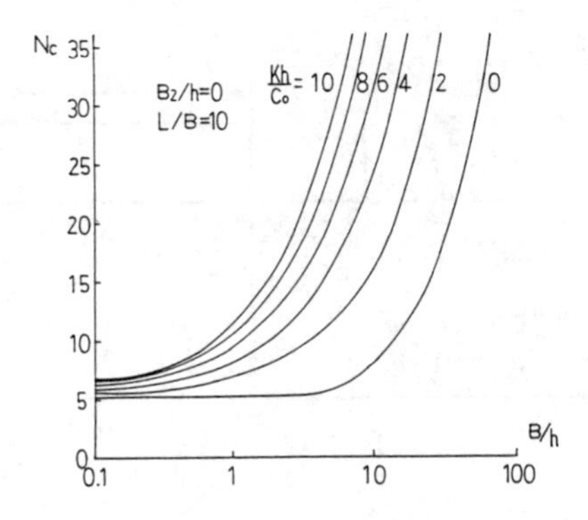

Fig.8 Bearing capacity factor Nc

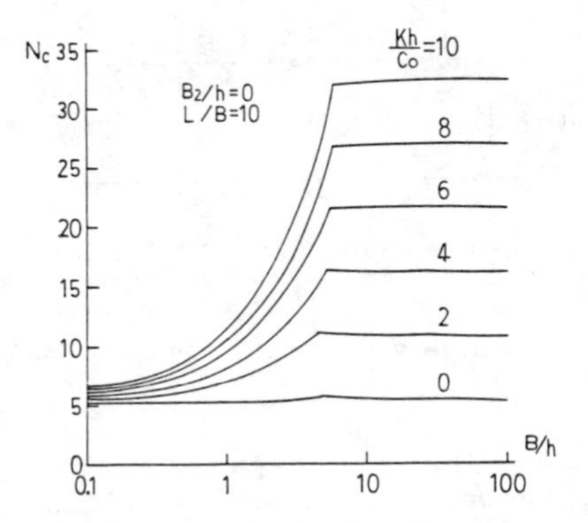

Fig.9 Enveplope of Nc value

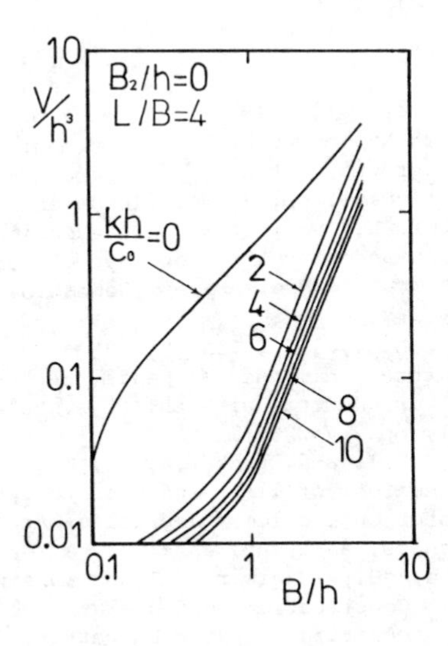

Fig.10 Change in soil volume moving longitudinal direction with B/h

pressure acting onto the substructure more rapidly increases than the width of the embankment, the same result as was obtained in the previous section.

From these two upper bound calculations, it has become clear that the soil beneath the end of embankment begins to move more to longitudinal direction as the width of embankment increases, implying that total earth pressure acting on substructures supporting abutments increases with increasing the width of embankment, creating the tendency of horizontal movement of abutments constructed at the end of wider embankments.

3 EXPERIMENTAL STUDY

3.1 Test procedures

A series of centrifuge model tests was conducted to verify the trend that the upper bound calculations suggested. The centrifuge used was a drum centrifuge at University of Utsunomiya, of which diameter is 0.8 meter and the maximum acceleration is 150 g. A normally consolidated clay layer was prepared by self weight consolidation technique (Kusakabe et al., 1988). Embankments were made of compacted clay and a series of noodles was inserted into the area beneath the end part of the embankment. The centrifuge was gradually spun up to 120 g. After stopping the centrifuge, deformations of the embankment and the ground were measured. The main purpose of the tests were to check the validity of the failure mechanism.

3.2 Comparison between the analytical prediction and the test results.

Figure 11 is the plane view comparison of the failure mechanism between the upper bound prediction and the experimental observation for two cases of B=50mm and 70mm. Continuous cracks are observed in the middle of the crest and at the end part larger movement takes place.

Virtually no movement is seen at the corners. The corner part seems to remain stational. The location of stationary zone is well predicted by the analytical predictions, in particular for the case of B=70mm.

In the case of B=50mm, another continuous cracks are seen in the middle of the slope surface. This is well compared with the thick dotted line indicating the location of point v of Figure 6(c).

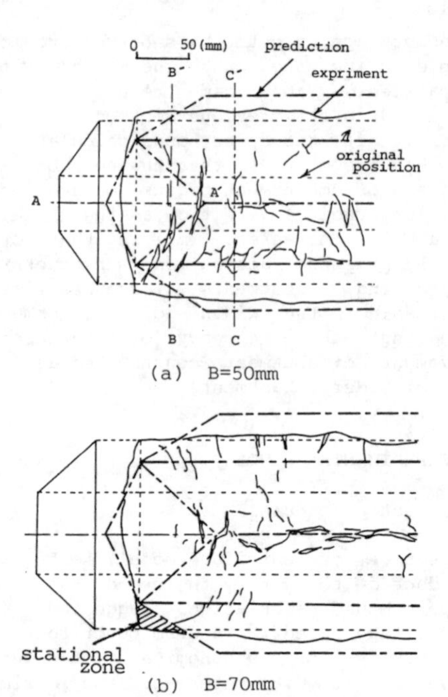

Fig.11 Plan view comparison of failure mechanism

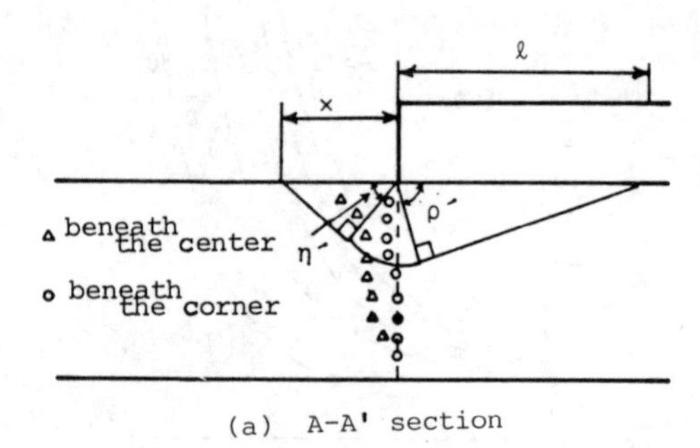

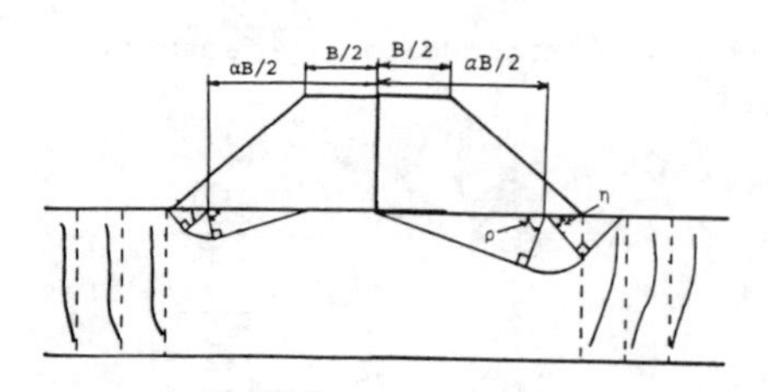

(b) B-B' section (c) C-C' section

Fig.12 Side view comparison of failure mechanism

Deformations at three different cross section shown in Figure 11 (a) of the ground measured by the movements of the noodles are drawn in Figure 12 for the case of B=50mm, together with the slip lines at the central section predicted by the upper bound calculations.

In Figure 12 (a), actual deformation reaches to greater depth, no lateral movement is observed below the corner. Figure 12 (b) and (c) shows the deformations at two different cross sections (B-B')(C-C') perpendicular to the A - A' plane shown in Figure 11 (a). These two figures clearly demonstrates that the failure becomes two dimensional as the distance from the end of the embankment becomes large, as is predicted by the analysis.

4 CONCLUSIONS

Three dimensional upper bound calculations were carried out to examine the behavior of the end part of embankments. It was found from the upper bound analyses that the soil movement towards longitudinal direction rapidly increases with increasing the width of the embankment. The failure mechanism adopted was supported by the centrifugal model tests.

Influence of the width of embankment on the soil movements at the end part of the embankment may be of practical importance in upgrading narrower roads to wider roads.

REFERENCE

Kimura et al. 1986, Prediction of horizontal movements of abutments on soft ground, Proc. of Symposium on Lateral Flow, pp. 77-84. (in Japanese).

Yamaguchi, H. 1985, Study on prediction of foundation deformation on layered soils, A report to Ministry of Education.(in Japanese)

Drucker, D.C., 1953, Limit analysis of two and three dimensional soil mechanics problems, Journal of Mech. Phys. Solids, Vol.1, pp 217-226.

Shield, R.T. and Drucker, D.C., 1953, The application of limit analysis to punch indentation problems, Journal of Applied Mechanics, ASME, pp. 453-460.

Kusakabe, O., Hagiwara, T and Kuroiwa,H 1988, Construction and operation of a drum centrifuge at University of Utsunomiya, Proc. of International Symposium on Centrifuge Modeling (to appear).

Numerical Methods in Geomechanics (Innsbruck 1988), Swoboda (ed.)
© 1988 Balkema, Rotterdam. ISBN 90 6191 809 X

Numerical analysis of a road embankment constructed on soft clay stabilized with stone columns

H.F.Schweiger
Institute for Soil Mechanics, Rock Mechanics and Foundation Engineering, Technical University of Graz, Austria
G.N.Pande
Department of Civil Engineering, University College of Swansea, UK

ABSTRACT: A new constitutive law for describing the behaviour of stone column reinforced clay is applied to the analysis of a road embankment constructed on soft clay stabilized by gravel piles. In this model the influence of the columns is assumed to be uniformly distributed over the reinforced region i.e. the constitutive law is formulated for an equivalent material consisting of gravel and clay. It combines the Mohr-Coulomb failure criterion and the Critical State model to represent the elasto-plastic behaviour of gravel and clay respectively. The compaction of the clay due to the installation process of the columns is taken into account by a simple approximation. The model has been implemented into a finite element code and it is emphasized that the finite element mesh is independent of column spacing and diameter thus parametric studies to determine optimum spacing are carried out very efficiently.

1 INTRODUCTION

Reinforced soft soils with stone columns is a widely used technique to improve settlement characteristics and bearing capacity of foundations. Despite the fact that many practical applications have been reported /1/,/2/ design is based on empirical rules in most cases because the complex soil - column interaction makes analytical and numerical treatment very difficult. Design charts and diagrams developed from experiments with a single pile allow an estimation of bearing capacity and settlements of foundations. In general this approach leads to conservative design /3/. Analytical solutions assuming a circular domain of influence and hence a "unit cell" are strictly applicable only for rigid rafts and face serious restrictions concerning boundary conditions /4/,/5/. If installed at toes of embankments an average increase of shear strength is usually taken into account when applying conventional slip circle analysis /6/ but settlements cannot be predicted using this approach. Finite element analysis modelling the columns by descrete elements turn out to be very costly and impractical for routine analyses.

The approach followed in this paper assumes an uniformly and homogeneously distributed influence of the columns over the reinforced region. It is further assumed that the total strains in soil and gravel are equal. The constitutive model is formulated for an equivalent material combining the Mohr-Coulomb criterion for the gravel and the Critical State model /7/ for the clay.

The radial support of the surrounding soil governs the load the column can carry and therefore an additional pseudo-yield criterion is introduced to establish equality of horizontal stresses in clay and columns.

A more detailed description of the model and the verification against analytical solutions have been presented in /8/, /9/ and only a short summary is given here. The simulation of the installation process is described briefly. In the last section the example of a road embankment constructed on stone column reinforced clay is analyzed.

2 SUMMARY OF BASIC EQUATIONS

The two basic assumptions of the constitutive model applied are:

1. The influence of the columns is uniformly and homogeneously distributed over the reinforced region (this assumption is reasonable because overall dimension are usually large compared to the spacing of

the columns).

2.Total strains are the same in columns and surrounding soil (this has been varified by in situ experiments at least for working loads /3/).

The model is incorporated into a finite element code and an elasto-viscoplastic algorithm is used to solve the nonlinear equations. Only time-independent behaviour is considered in this paper and "time" is purely an artifice to control convergence.

Applying the theory of viscoplasticity the basic equations describing the behaviour of the clay/column system are given in the following in incremental form.

Total strains are the sum of elastic and viscoplastic components.

$$d\underset{\sim}{\varepsilon} = d\,\underset{\sim}{\varepsilon}^{e} + d\,\underset{\sim}{\varepsilon}^{vp} \qquad (1)$$

In the following subscripts c, s and cs denote stresses, strains etc. related to clay, stone column and equivalent material respectively.

$$d\,\underset{\sim}{\varepsilon}_c = d\,\underset{\sim}{\varepsilon}_c^{e} + d\,\underset{\sim}{\varepsilon}_c^{vp} \qquad (2a)$$

$$d\,\underset{\sim}{\varepsilon}_s = d\,\underset{\sim}{\varepsilon}_s^{e} + d\,\underset{\sim}{\varepsilon}_s^{vp} \qquad (2b)$$

No slip condition implies

$$d\,\underset{\sim}{\varepsilon}_{cs} = d\,\underset{\sim}{\varepsilon}_c = d\,\underset{\sim}{\varepsilon}_s \qquad (3)$$

If ρ denotes the volume occupied by the columns per unit volume of the clay/column system which in turn represents the spacing of the columns the stress acting on the reinforced soil is

$$d\,\underset{\sim}{\sigma}_{cs} = (1-\rho)\,.\,d\,\underset{\sim}{\sigma}_c + \rho\,.\,d\,\underset{\sim}{\sigma}_s \qquad (4)$$

with

$$d\,\underset{\sim}{\sigma}_c = \underline{D}_c\,.\,d\,\underset{\sim}{\varepsilon}_c^{e} \qquad (5a)$$

$$d\,\underset{\sim}{\sigma}_s = \underline{D}_s\,.\,d\,\underset{\sim}{\varepsilon}_s^{e} \qquad (5b)$$

where $\underline{D}_c$ and $\underline{D}_s$ are matrices of elastic constants. The equivalent matrix $\underline{D}_{cs}$ is given by

$$\underline{D}_{cs} = (1-\rho)\,\underline{D}_c + \rho\,.\,\underline{D}_s \qquad (6)$$

The rates of viscoplastic strains are given by

$$\dot{\underset{\sim}{\varepsilon}}^{vp} = \gamma\,\langle F\rangle\,\frac{\delta Q}{\delta \underset{\sim}{\sigma}} \qquad (7)$$

where γ is taken as 1.0 and F and Q are yield functions and plastic potential functions respectively.

$$F = F(\underset{\sim}{\sigma}\,,\,\underset{\sim}{\varepsilon}^{vp}) = 0 \qquad (8)$$

$$Q = Q\,(\underset{\sim}{\sigma}) = 0 \qquad (9)$$

Equations (7), (8) and (9) are written for clay and column material seperately and any suitable functions can be used within the framework developed. Mohr-Coulomb criterion for the column material is employed in this study together with the Critical State model for the clay.

It has been mentioned that the lateral support of the surrounding soil is essential for the columns to carry any load. It follows that the radial stress in the clay and in the columns has to be the same at the clay-column interface. Although the interface between soil and column is fictitious in this model an additional pseudo-yield criterion has been introduced to ensure equality of radial stresses.

$$F = \sigma_{r,s} - \sigma_{r,c} = \Delta\sigma_r = 0 \qquad (10)$$

If $F > 0$ viscoplastic strains rates in radial directions are developed

$$\dot{\varepsilon}_{r,c} = \gamma\,.\,F \qquad (11a)$$

$$\dot{\varepsilon}_{r,s} = -\gamma\,.\,F \qquad (11b)$$

and added into the vector of viscoplastic strain rates obtained from yielding in columns or clay or both.

The vector of viscoplastic strain increments is obtained for clay and column material to

$$d\,\underset{\sim}{\varepsilon}_c^{vp} = \dot{\underset{\sim}{\varepsilon}}_c^{vp}\,.\,dt \qquad (12a)$$

$$d\,\underset{\sim}{\varepsilon}_s^{vp} = \dot{\underset{\sim}{\varepsilon}}_s^{vp}\,.\,dt \qquad (12b)$$

where dt is the time step length sufficiently small to ensure convergence.

Equivalent nodal forces are calculated and resolutions are carried out till convergence with respect to all three yield functions is achieved to a specified tolerance.

3 MODELLING OF INSTALLATION PROCESS

Construction of stone columns leads to compaction of the in situ soil the amount of which depends on various factors e.g. the installation technique applied, column diameter and spacing.

If a strain hardening plasticity model is used to describe the nonlinear behaviour of the clay the effect of installation of columns can be taken into account at least qualitatively by subjecting the soil to a radial strain before the external loads are applied.

Employing the Critical State model for normally consolidated clays the initial size of the yield ellipse in $\sigma_m - \bar{\sigma}$ - space depends on the preconsolidation pressure (P_{co}) i.e. the initial stress state. The expansion or contraction of the yield surface is governed by

$$P_{co} = P_{co} \exp \quad (-\chi . \varepsilon^{vp}_{vol}) \tag{13}$$

where

$$\chi = \frac{1 + e_o}{\lambda - \kappa} \tag{14}$$

e_o being the initial void ratio and λ and κ the compression and swelling indices respectively.

Due to space limitations only the basic steps which are carried out prior to the finite element analysis using a small computer program are outlined in the following:

1. Calculate initial stresses ($\sigma_y = \gamma . H$, $\sigma_x = \sigma_z = k_o . \sigma_y$) and corresponding preconsolidation pressures (P_{co}) for all Gauss points.

2. Apply radial strains related to volume occupied by columns to obtain stress state after installation.

3. Calculate average values for $\bar{\gamma}$ and $\bar{k}_o$ corresponding to the stress state after installation.

4. Input $\bar{\gamma}$ and $\bar{k}_o$ into the finite element program and calculate stresses and consolidation pressures before external loads are applied.

4 ANALYSIS OF A ROAD EMBANKMENT

To demonstrate the applicability of the proposed model an analysis of a road embankment is briefly presented.

The crude finite element mesh (Figure 1), sufficient for the purpose of this paper, consists of 63 8-noded isoparametric elements. The embankment, 7,5 m high and approximately 30 m wide at the top, is constructed on a 6 m soft clay layer on hard bedrock. The side slopes are inclined at a ratio of 1:1.6. The geometry and the soil parameters are similar to the problem described in /10/.
The material parameters are given in Table 1.

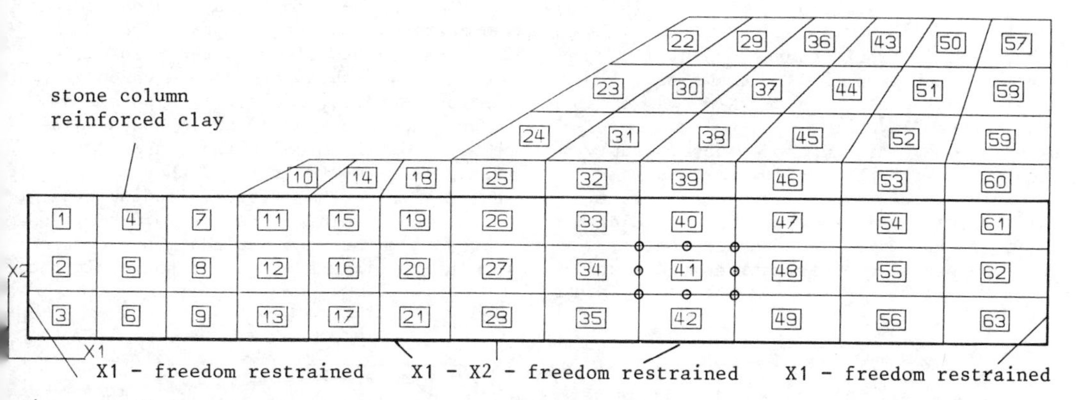

Figure 1. Finite element mesh

Table 1. Material parameters

	E (kN/m²)	ν (-)	γ (kN/m³)	k_o	ϕ (°)	c (kN/m²)	ψ (°)	χ (-)
Embankment	50 000	0.3	18.0	-	30	22	-	-
Clay	6 000	0.4	20.0	0.45	25	-	-	5.00
Column	72 000	0.3	20.0	0.45	40	0	0	-

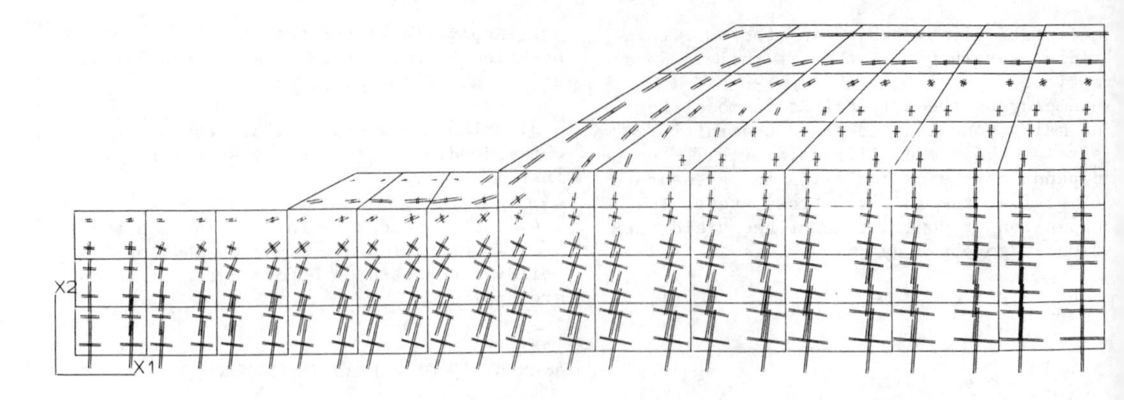

Figure 2. Principle stresses in clay at load level 40 % of the weight of embankment

No time-dependent behaviour is considered and drained conditions are assumed. The weight of the embankment is applied in 12 load increments.

In the first analysis the clay was assumed to be in its natural condition. The preconsolidation pressure (P_{co}) was derived from initial stresses ($\sigma_y = \gamma . H, \sigma_x = \sigma_z = k_o . \sigma_y$) i.e. it was assumed to increase with depth and was chosen as

$$P_{co} = P_{co,cal} \cdot 1.15 \qquad (15)$$

where $P_{co,cal}$ would cause the initial stress to lie exactly on the yield surface. A minimum value of P_{co} = 20 kN/m² was chosen for the top layers.

By solving this type of problems applying continuum mechanics safety factors comparable to those abtained from classical slip circle analysis cannot be calculated. However, monitoring the development of plastic zones and deformations the stability of the embankment can be judged. The analysis showed already at a load level of 40 % of the weight of the embankment widespread plastic zones and large displacements. It may be concluded that construction of the embankment without strengthening the clay is not possible. The result of this analysis is obvious if one considers the soil parameters and only the principle stresses are shown (Figure 2) for comparison with the reinforced case.

The second analysis assumes that 20 % of the clay has been replaced by stone columns. The change in the stress state due to the installation procedure is taken into account using the approach presented in the last chapter. The stresses obtained correspond to a $\bar{\gamma}$ of 30 kN/m³ and a value of $\bar{k}_o$ of 0.7 from which the consolidation pressures (i.e. the size of the yield surfaces) prior to external loading could be calculated. A minimum value of 50 kN/m² was assumed for the preconsolidation pressure for the top layers. Figure 3 shows plastic

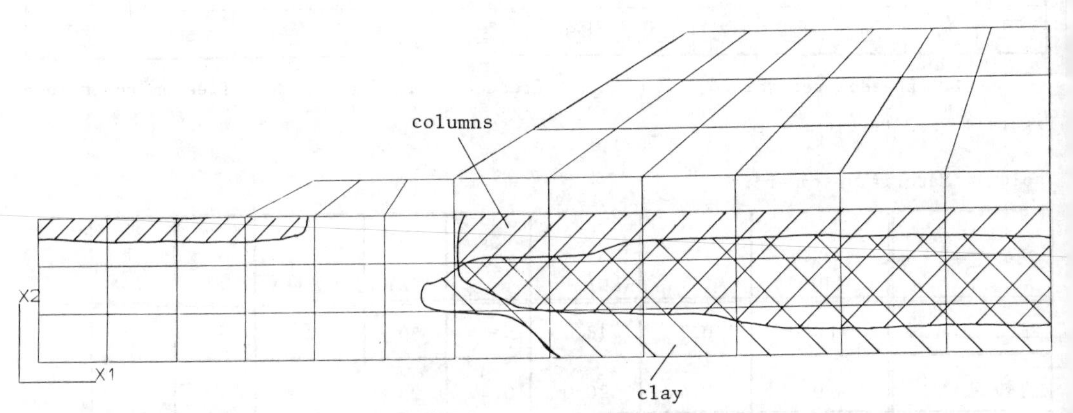

Figure 3. Plastic zones in clay and columns (reinforced case)

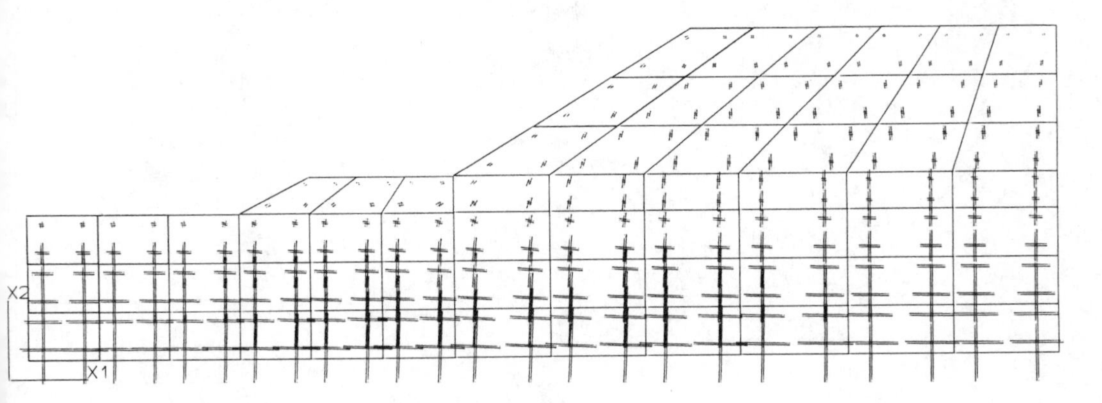

Figure 4. Principle stresses in clay (reinforced case)

zones for clay and columns when the full load is applied. If settlements for the reinforced case are compared to the unreinforced case the reductionfactor n (n = settlement reinforced / settlement unreinforced) varies from 0.35 at 10 % load to 0.04 at 40 % load. The significant difference in principle stress directions thus in the shear stresses is evident from Figure 4 which has to be compared with Figure 2. The load level is 40 % of the total weight of the embankment.

5 CONCLUSION

A new approach of modelling the behaviour of stone column reinforced clay has been applied to the analysis of a road embankment. A constitutive law is formulated for an equivalent material consisting of gravel and clay. It takes into account plastic deformations in both materials. It is incorporated into a finite element program which makes it readily applicable for the solution of boundary value problems thus being more versatile than methods which examine the behaviour of a single pile. A simple approach to take into account the installation process at least qualitatively has been presented. The example analyzed indicates that the developed framework is able to describe the behaviour of stone reinforced clays.

6 REFERENCES

/1/ G.A.Munfakh, Soil reinforcement by stone columns-varied case applications, Int.Conf.In Situ Soil and Rock Reinforcements, Paris, 1984

/2/ R.R.Goughnour and R.D.Barksdale, Performance of a stone column supported embankment, Int.Conf. on Case Histories in Geotechnical Engineering, St.Louis, USA, 735-742, 1984

/3/ R.D.Barksdale and R.C.Bachus, Design and Construction of Stone Columns, Fed. Highway Admin., U.S.Dept. of Transportation, Report No.FHWA/RD - 83/026, 1983

/4/ N.P.Balaam and S.R.Booker, Analysis of Rigid Rafts Supported by Granular Piles, Int.Jou.f.Num.and Analy.Methods in Geomech., Vol.5, 1981

/5/ N.P.Balaam and S.R.Booker, Effect of Stone Column Yield on Settlement of Rigid Foundations in Stabilzed Clay, Int.Jou.f. Num.and Analy.Methods in Geomech., Vol.9, No.4., 1985

/6/ E.Rathgeb and C.Kutzner, Some Applications Of The Vibro-Replacement Process, Geotechnique 25, 45-50, 1975

/7/ J.H.Atkinson and P.L.Bransby, The Mechanics Of Soils, An Introduction To Critical State Soil Mechanics, McGraw Hill, London, 1978

/8/ H.F.Schweiger and G.N.Pande, Modelling behaviour of stone column reinforced soft clays, Proc.2nd NUMOG, Ghent, 1986

/9/ H.F.Schweiger and G.N.Pande, Numerical Analysis of Stone Column Reinforced Foundations, Computers and Geotechnics 2, 347-372, 1986

/10/ P.S.Coupe, An extension of dynamic consolidation, Ground Engineering 18, No.6, 1986

Numerical Methods in Geomechanics (Innsbruck 1988), Swoboda (ed.)
© 1988 Balkema, Rotterdam. ISBN 90 6191 809 X

Search for noncircular slip surfaces by the Morgenstern-Price Method

Takuo Yamagami & Yasuhiro Ueta
University of Tokushima, Hanshin Consultants Co., Osaka, Japan

ABSTRACT: Four techniques for solving optimization problems have been investigated to search for a general noncircular critical slip surface by using the Morgenstern and Price method. These techniques are the Nelder and Mead simplex method, the Powell, the DFP and the BFGS methods. In the present study the problem of slope stability has been formulated as a simple, yet versatile procedure for solving an optimization process. A computer program has been developed and its basic characteristics have been revealed by investigating various aspects of example problems. It is shown that for slopes consisting of inhomogeneous soils, an initial estimate for the optimization process plays a particularly important role.

1 INTRODUCTION

In the analysis of slope stability it is quite difficult to search for a critical slip surface by using a general noncircular slip surface theory. One major difficulty arises from numerous possibilities of slip surface configurations. In recent years, many attempts have been conducted to overcome such difficulties: Baker[1] combined the Spencer method with dynamic programming; Boutrup and Lovell[2] and Siegel, Kovacs and Lovell[3] used random numbers to generate irregular shear surface; Celestino and Duncan[4] also combined the Spencer method with the alternating variable method; Nguyen[5] applied the simplex method to the Bishop, Wedge and Morgenstern-Price methods respectively; Arai and Tagyo[6] employed the simplified Janbu method and the Fletcher and Reeves conjugate gradient method; Greco and Gulla[7] showed a constrained minimization technique based on the Hooke and Jeeves method by coordinating it with the simplified Janbu method; Yamagami and Ueta[8] coupled the simplified Janbu method with dynamic programming involving Baker's ideas. As is obvious from the previous work, it is essential in some respects to determine noncircular critical slip surfaces with the aid of mathematical programming.

In general, slope stability research has been concerned with two aspects: one is to calculate a value of the factor of safety along a designated slip surface and the other is to efficiently locate the position of the critical slip surface. The Morgenstern and Price method[9] or the M-P method for short is one of the most popular methods employed in regard to the former above mentioned aspect, this being compared to other procedures [10,11,12]. In the latter aspect, however, the M-P method has hardly been employed, as compared with the Spencer or simplified Janbu methods, when noncircular slip surfaces are to be analysed. To the best of the author's knowledge, only Nguyen[5] has briefly showed a search result for a very simple example utilizing the M-P method. As Janbu[13] pointed out, however, when comparing different procedures, only the minimum factor of safety obtained from each method is meaningful. He also stated that "because the Fmin may occur for different shapes and locations, a comparison between two procedures for one surface only is in most cases misleading, at least." Considering these descriptions, it is of interest, in both the research and practical aspects, to construct a search method for Fmin-surfaces using the M-P method.

The purpose of this paper is to develop a search method by introducing four nonlinear programming algorithms into the M-P method. The first part of the paper deals mainly with a solution strategy and the four optimization techniques used.

This is followed by some example problems to examine the basic characteristics of the present method.

2 A BRIEF DESCRIPTION OF THE M-P METHOD

The M-P method is quite prominent and well documented elsewhere in the literature of slope stability analysis. Thus details of the method can be eliminated here. In this study the subroutine program to calculate the factor of safety for the M-P method has been developed with reference to that given in the textbook by Clayton and Milititsky[14]. The program requires us to input the following data:

(i) the shape of the f(x) distribution as well as (ii) the geometry of the problem, (iii) soil parameters, and (iv) the desired division of the slip mass into slices, where

$$f(x) = \frac{1}{\lambda} \cdot \frac{X}{E} \tag{1}$$

on any vertical plane within the slip mass. Eq.(1) relates the interslice shear force, X, and the interslice normal force, E, with a mathematical function f(x), where f(x) is called the interslice force function, and λ is a scaling constant representing the percentage of the f(x) function used.

The interslice force function may typically take the form of either (i) a constant, (ii) a half sine, (iii) a clipped sine, and (iv) a trapezoid [10, 15]. Some authors have indicated that the form of this function has little effect on the factor of safety, but Clayton and Milititsky[14] states that this depends upon the geometry of the problem and the parameters used. In this paper, however, only a constant distribution (f(x)=1.0) has been employed in the example problems below.

3 SEARCH STRATEGY FOR CRITICAL SLIP SURFACE

Consider n-vertical lines within a given slope to be analysed (see Fig.1). These lines represent the division of an assumed sliding mass into (n-1)-silces. These slice lines are now assumed to be fixed in space during the entire analysis. Under this condition any trial slip surface may be generated according to a combination of vertical coordinates, y_i (i=1~n), which are defined as the intersections of the slice lines and the trial slip surface. Each of the trial slip surfaces generated in this way ought to have a different

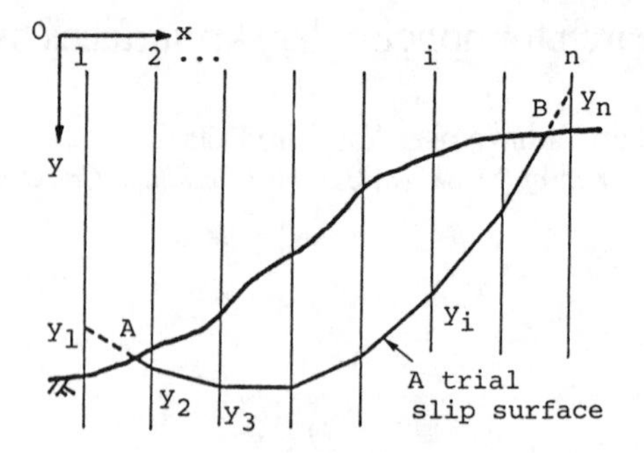

Fig.1 Generation of trial slip surfaces

value for the factor of safety. As a result we can consider the factor of safety F as a function of y_i:

$$F = F(y_1, y_2, \dots, y_n) \tag{2}$$

Thus, the current stability problem can be formulated as a nonlinear optimization problem which states the following: "find a set of independent variables, y_i, such that the objective function F is minimized":

$$\text{minimize} \quad F(y_i) \tag{3}$$

Solving this problem produces the location of the critical slip surface and the associated minimum value of the factor of safety. As mentioned in "INTRODUCTION", four techniques found in the nonlinear programming theory have been investigated to minimize $F(y_i)$.

One point to which special attention should be paid in implementing this procedure, is that the slice lines have to fully cover a region within which the correct solution is anticipated to exist. In particular, the end lines should be established at such a distance to incorporate the region under study. In this case a way to determine the real end points, A and B of the critical slip surface has already been given elsewhere [1, 8].

We must also notice the fact that, in general, a nonlinear optimization technique will offer only a local minimum of the objective function, not necessarily the global one. Therefore several different initial guesses should be made in order to finally obtain a desired result.

4 OPTIMIZATION TECHNIQUES USED

Consider a problem in which we find the

extreme point (the minimum) of a nonlinear function f(x) of the n-independent variables $x_1, x_2, \ldots, x_n$. If the independent, decision variables are restricted to values within a specified range or if other conditions are imposed, the problem is called a constrained optimization problem; if such restrictions or conditions are not imposed, it is called an unconstrained optimization problem. The function f(x) is the objective function as we have already used this term in conjunction with Eq.(2). Techniques for solving these nonlinear optimization problems can be classified into two major groups, although many have been presented so far. The first uses values of the objective function and those of its derivatives with respect to the independent variables in the optimization process. The second uses only values of the objective function. The latter, derivative-free techniques are in most cases called "direct search" methods. These direct search methods are favorable when the function f(x) does not have its derivatives or when computations of the derivatives are too complicated. On the contrary, it is generally said that these methods are inefficient as compared with gradient-based algorithms if the derivatives are easily obtained.

In this study, four nonlinear programming techniques have been employed to solve the unconstrained optimization problem, Eq.(3): the simplex method proposed by Nelder and Mead, Powell's (conjugate direction) method, the DFP (Davidon-Fletcher-Powell) method, and the BFGS (Broyden-Fletcher-Golfarb-Shanno) method. The simplex method is representative of the direct search methods. Powell's method requires only values for the objective function, but is not necessarily classified into the direct search method category. The DFP and the BFGS methods are both gradient-based methods and are included in the quasi-Newton methods; The BFGS formula is an extension of the DFP formula. For these two approaches, the first derivatives of f(x) are required and should be evaluated numerically as will be mentioned later.

For further details of the theoretical aspects the reader should refer to specialized text books (see e.g. Avriel[16], Reklaitis, et al.[17], Rao[18], Pike[19]). The reader may also refer to books regarding the computer program listings for the above techniques (e.g. Himmelblau[20], Siddall[21], Pike[19]).

5 NUMERICAL STUDIES

Basic characteristics of the present method will be elucidated in the following through numerical studies of two test problems: one problem is concerned with a simple slope involving homogeneous soil and the other a layered slope consisting of four different soils. Stability calculations herein were all made on the basis of the total stress approach. Soil properties used will appropriately be given in related figures.

As mentioned before, the DFP and BFGS methods require the first derivatives of the objective function. These derivatives, however, cannot be obtained analytically, thus resulting in simple finite-difference approximations:

$$\frac{\partial F}{\partial y_i} \doteqdot \frac{F(y_i+\Delta y_i) - F(y_i)}{\Delta y_i} \tag{4}$$

where Δy_i represents a small interval of each independent variable, y_i (i=1~n). Preliminary investigations have proved that a value of Δy_i on the order of $1.0\times10^{-4} y_{i,0}$ is preferable, where $y_{i,0}$ is an initial estimate for y_i.

Comparison of the four N. P. Algorithms. Fig.2 shows the final converged solutions to the homogeneous slope by the use of four N. P. algorithms, together with the initial estimate. The minimum factor of safety obtained from each algorithm and the computer time (CPU time) are also listed in the figure. For this a FACOM M-360 computer was used at the data processing center of the Univ. of Tokushima, Japan. In this problem, eight slices, i.e. nine independent variables have been employed as seen in Fig.2.

Effects of Initial Estimates. For the homogeneous slope, illustrations are given for the effects of initial estimates on the solutions in Fig.3 to Fig.6. Note that both the Powell and simplex methods involving the deepest initial estimate failed to give the converged solutions when only the nine slice lines were used. This was not due to a limited number of iterations for the N. P. techniques but because of the M-P method. Iterative computations are necessary for the M-P method to obtain the factor of safety; the maximum number of iterations were limited to forty in the present study. Adding an extra slice line, drawn as a dotted line on each side, however, we could reach the converged solutions as shown in Figs.5 and 6.

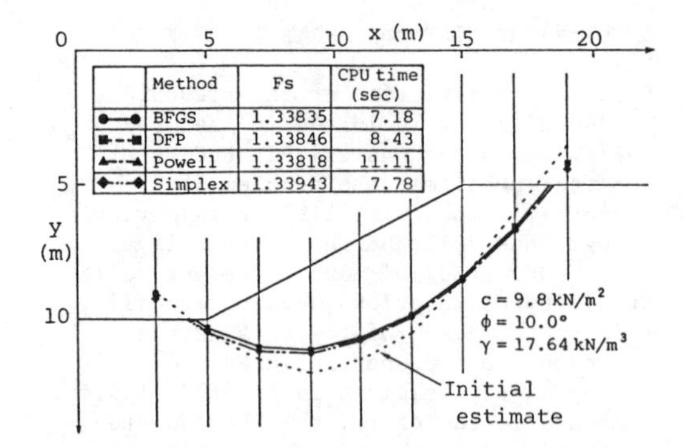

Fig.2 Comparison of the four NP algorithms

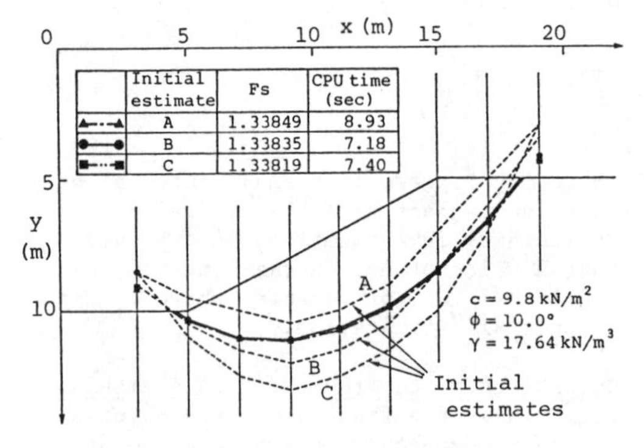

Fig.3 Effects of initial estimates (the BFGS method)

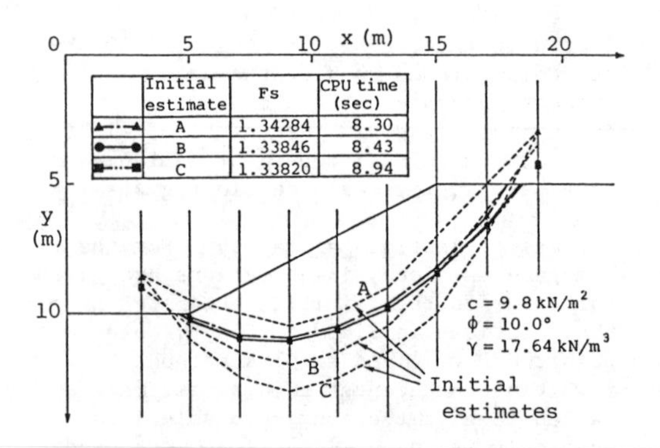

Fig.4 Effects of initial estimates (the DFP method)

In marked contrast to the homogeneous case, the effects of initial estimates are revealed to considerable degree in the case of the layered slope. These are shown in Figs.7 and 8. Each N. P. technique has reached a local optimum in Fig.7,

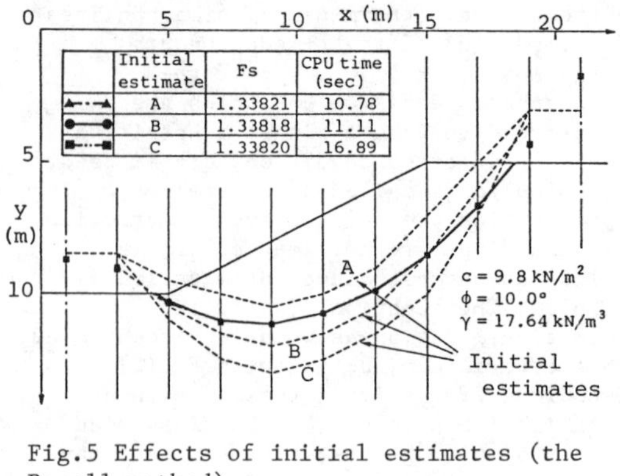

Fig.5 Effects of initial estimates (the Powell method)

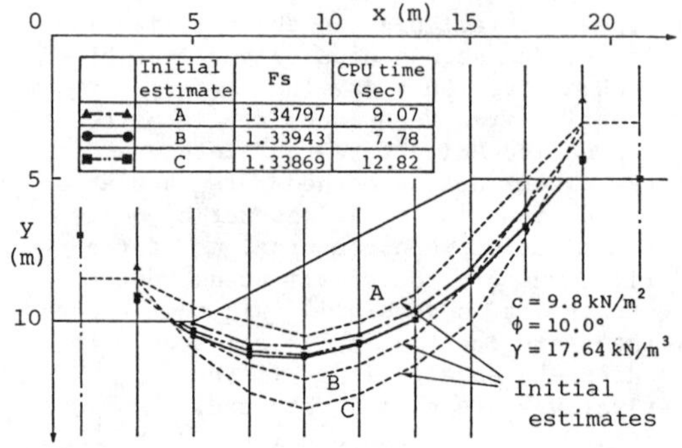

Fig.6 Effects of initial estimates (the simplex method)

thereby preventing us from approaching a preferable location. In Fig.8 the initial estimate was intentionally set up close to where the real critical slip surface should exist; this being deduced from the analyses based on the Bishop and the FEM + D. P. methods.

The above results show how important it is to investigate different initial estimates in order to obtain desired solutions for inhomogeneous slopes.

Effects of the Number of Independent Variables. The number of independent variables is a crucial factor in solving N. P. problems. In particular, the simplex method is said to be efficient only when solving problems involving a limited number of variables. Figs.2 and 8, and other data which has not been included, have supported this statement: the CPU time in the simplex method increases rapidly when the number of variables exceeds a certain limit.

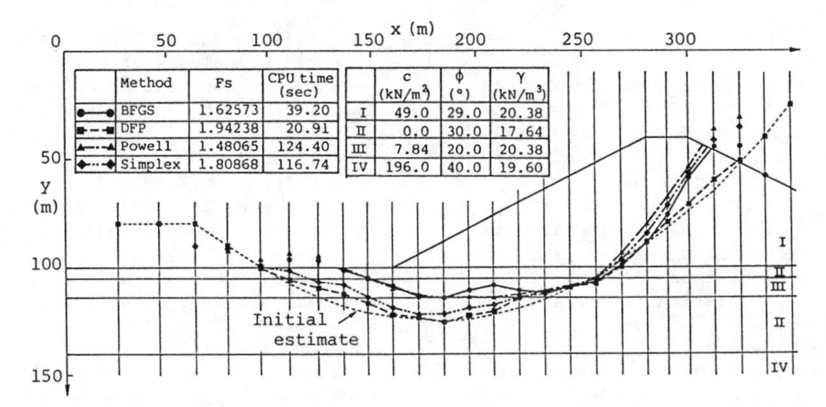

Fig.7 An example in which solutions reached their local optimum (Note the number of independent variables used)

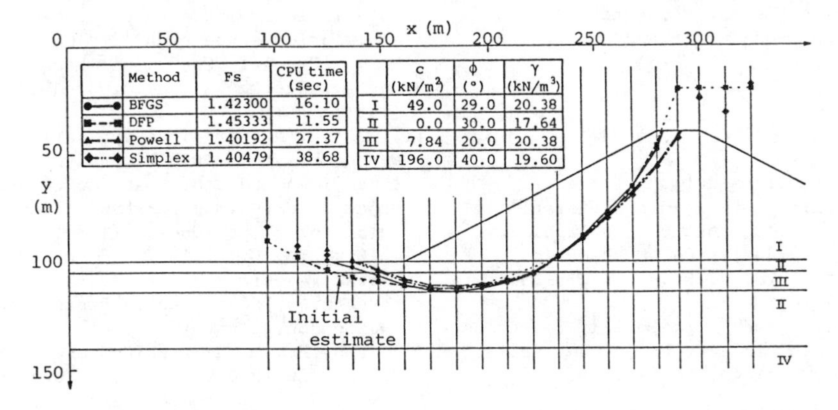

Fig.8 Final solutions obtained from the four NP techniques for the inhomogeneous slope

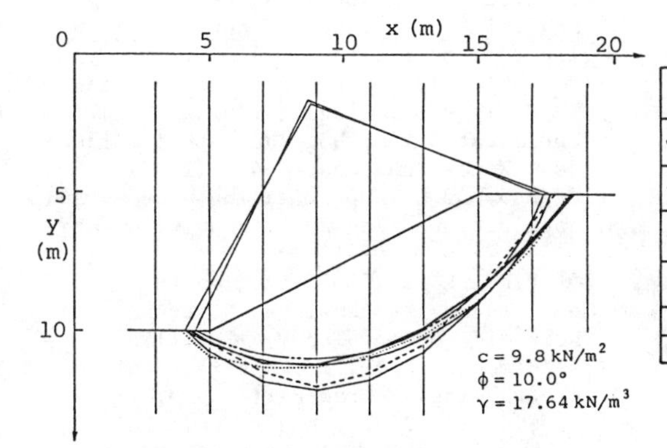

	Method	Fs	CPU time (sec)
——	Morgenstern-Price + Simplex method	1.339	7.78
······	Rigorous Janbu + Simplex method	1.346	8.22
——	Simplified Janbu + Simplex method	1.185	6.44
– – –	Simplified Janbu + D.P.	1.191	9.49
—·—	Fellenius	1.282	4.15
—··—	Simplified Bishop	1.348	10.09

Fig.9 Comparison of factor of safety equations

Comparison of Factor of Safety Equations. Fig.9 shows critical slip surfaces searched by using several slope stability analysis methods, together with the CPU times and associated values for the factor of safety. It is interesting to note that the M-P, Bishop and rigorous Janbu methods gave nearly the same factor of safety, although shapes and locations were slightly different.

6 CONCLUDING REMARKS

A search method has been investigated for general noncircular critical slip surfaces by coupling the M-P method with nonlinear optimization techniques. As in most other approaches which determine noncircular critical slip surfaces, the slope stability analysis was also formulated as a nonlinear optimization problem. A computer program was then developed to implement this formulation in terms of four NP techniques.

While the M-P method is quite popular when calculating the factor of safety for a given slip surface, it has hardly been employed in the search for the critical slip surface. Therefore, investigations viewed from various aspects were carried out to clarify the basic characteristics of the search procedure based on the M-P method.

REFERENCES

[1] Baker,R., Determination of the critical slip surface in slope stability computations, International Journal for Numerical and Analytical Methods in Geomechanics, Vol.4, No.4, pp.333-359 (1980)

[2] Boutrup, E. and C. W. Lovell, Searching techniques in slope stability analysis, Engineering Geology, Vol.16, pp.51-61 (1980)

[3] Siegel, R. A., W. D. Kovacs and C. W. Lovell, Random surface generation in stability analysis, ASCE, J. Geotech. Engng. Div., Vol.107, GT7, pp.996-1002 (1981)

[4] Celestino, T. B. and J. M. Duncan, Simplified search for noncircular slip surfaces, Proc. 10th Int. Conf. Soil Mech. Found. Engng., Stockholm, Vol.3, pp.391-394 (1981)

[5] Nguyen, V. U., Determination of critical slope failure surfaces, ASCE, J. Geotech. Engng. Div., Vol.111, No.2, pp.238-250 (1985)

[6] Arai, K. and K. Tagyo, Determination of noncircular slip surface giving the minimum factor of safety in slope stability analysis, Soils and Foundations, Vol.25, No.1, pp.43-51 (1985)

[7] Greco, V. R. and G. Gulla, Slip surface search in slope stability analysis, Rivista Italiana di Geotecnica, pp.189-198 (1985)

[8] Yamagami, T. and Y. Ueta, Noncircular slip surface analysis of the stability of slopes -an application of dynamic programming to the Janbu method-, J. of Japan Landslide Society, Vol.22, No.4, pp.8-16 (1986)

[9] Morgenstern, N. R. and V. E. Price, The analysis of the stability of general slip surfaces, Geotechnique, Vol.15, pp.79-93 (1965)

[10] Fredlund, D. G. and J. Krahn, Comparison of slope stability methods of analysis, Can. Geotech. J., Vol.14, pp.429-439 (1977)

[11] Duncan, J. M. and S. G. Wright, The accuracy of equilibrium methods of slope stability analysis, Engineering Geology, Vol.16, pp.5-17 (1980)

[12] Fredlund, D. G., J. Krahn and D. E. Pufahl, The relationship between limit equilibrium slope stability methods, Proc. 10th Int. Conf. Soil Mech. Found. Engng., Stockholm, Vol.3, pp.409-416 (1981)

[13] Janbu, N., Critical evaluation of the approaches to stability analysis of landslides and other mass movements, Proc. Inter. Sympo. on Landslides, New Delhi Vol.2, pp.109-128 (1980)

[14] Clayton, C. R. I. and J. Milititsky, Earth pressure and earth-retaining structures, Surrey Univ. Press, Glasgow and London, 300p. (1986)

[15] Fan, K., D. G. Fredlund and G. W. Wilson, An interslice force function for limit equilibrium slope stability analysis, Can. Geotech. J., Vol.23, pp.287-296 (1986)

[16] Avriel, M., Nonlinear programming, analysis and methods, Prentice-Hall Inc., 512p. (1976)

[17] Reklaitis, G. V., A. Ravindran and K. M. Ragsdell, Engineering optimization, methods and applications, John Wiley and Sons, 684p. (1983)

[18] Rao, S. S., Optimization, theory and applications (2nd Ed.), John Wiley and Sons, 747p. (1984)

[19] Pike, R. W., Optimization for engineering systems, Van Nostrand Reinhold Company Inc., 417p. (1986)

[20] Himmelblau, D. M., Applied nonlinear programming, McGraw-Hill, 497p. (1972)

[21] Siddall, J. N., Analytical decision-making in engineering design, Prentice-Hall Inc., 431p. (1972)

Numerical Methods in Geomechanics (Innsbruck 1988), Swoboda (ed.)
© 1988 Balkema, Rotterdam. ISBN 90 6191 809 X

Performance and analysis of anchored sheet pile wall in soft clay

H.Murakami, Y.Yuki & T.Tamano
Sewage Works Bureau, Osaka, Japan

ABSTRACT: This paper details the performance of anchored sheet pile wall supported by four rows of anchors, used for large-scale excavation (100m wide × 200m long × 10.4m deep) in soft clay in Osaka. The behaviour of the structure and the adjacent ground was observed carefully throughout the excavation work. Numerical analysis was carried out for different excavation stages, using the finite element method. Measured and computed behaviours are compared and discussed. As well, the mechanism of anchor load redistribution following anchor rupture was investigated for the same wall.

1. INTRODUCTION

The support system of the anchored wall, preloaded anchors, is rather complicated in interaction between structure and adjacent ground, especially in soft clay. To ensure the reliability of the anchored wall in soft clay, therefore, a case study, with careful observation, is required. Also important is to optimize wall design and to follow appropriate observational procedures during excavation, with consideration given both to the potential danger posed to field-installed anchors and to the difficulty in reinforcing them during excavation. In excavation work using the anchored wall, when an anchor already installed is ruptured for some reason, its load is redistributed to the adjacent anchors, possibly leading to the breakdown of the anchored wall. This is the most dangerous situation with the anchored wall method (Stille and Broms, 1976).

As an approach to these problems, this paper discusses the performance of anchored sheet pile wall in soft clay, together with the finite element analysis. The mechanism of load redistribution upon anchor rupture was also investigated for the same wall through an experiment involving anchor load release.

2. EXCAVATION WORK AND GROUND CONDITION

The anchored sheet pile wall was employed in a large-scale excavation (100m wide × 200m long × 10.4m deep in soft clay in Osaka), part of the construction of sewage treatment facilities. Fig. 1 shows the ground condition and the wall section at an instrument panel located in the center of the excavation width. Fig. 2 shows the subsoil profile and properties. The alluvial clay layer at G.L. -5.5 ∿ -12.0m comprises normally consolidated sensitive soft clay, with an undrained shear strength (Su) of 30 ∿ 70kN/m^2, a plastic limit (PL) of 30 ∿ 40%, a liquid limit (LL) of 80 ∿ 90%, and a natural water content (Wn) of 70 ∿ 80%. The ground-water level is G.L. -1m and the water head in sand layer at G.L. -16m is G.L. -8m. Anchors used are the pressure-grouted type, 14cm in diameter. The anchors have been designed to permit removal of steel wires after completion of the work, as necessitated by the future use of the

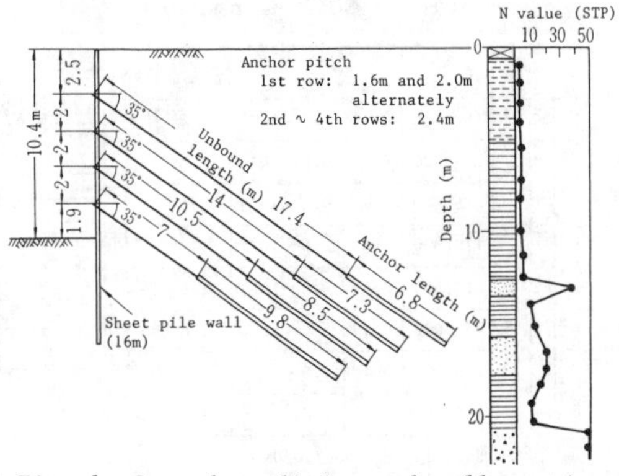

Fig. 1 Ground condition and wall section

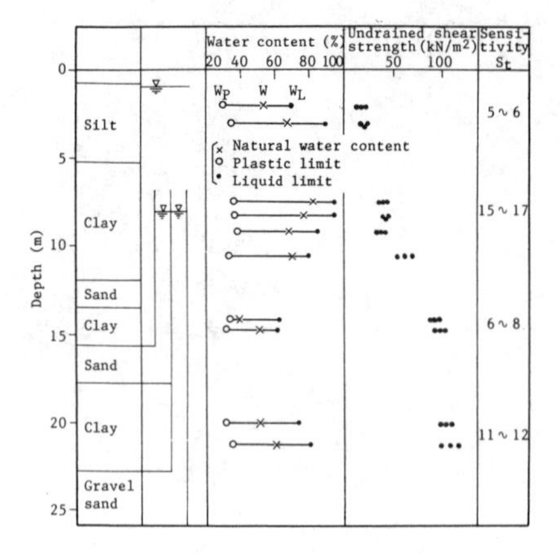

Fig. 2 Subsoil profile

ground below the adjacent road. Section modulus (Wz) of the sheet pile wall is 1,310cm^3 per meter. Two wale beams H-250 mm × 250mm × 9mm × 14mm (Wz: 867cm^3 × 2) were used against the anchor rows. The center distance of the anchors in the 1st row is alternately 1.6m and 2.0m, 2.4m in the 2nd, 3rd and 4th rows.

Although a dense sand layer with an N value of 50 or higher is generally used as the anchor binding layer, the anchors employed in this excavation were bound in medium dense sand layers with N values of 17 and 38 and a clay layer with an N value of 7. The construction process is shown in Fig. 4. The 1st to 9th stages were for excavation and anchor installation, the 9th to 10th stages (over a period of 480 days) for construction of the structure inside the wall, followed by the 10th to 19th stages for refilling and anchor removal (with steel wires removed after load release).

3. COMPUTATIONAL METHOD

Computation was carried out by means of coupled stress-flow finite element analysis of soil (program name: SIGNAS). In this analysis, problems involving seepage flow, such as interaction between saturated-unsaturated flow and stress-strain of the soil, can be solved (Akai et al., 1979). Equilibrium and continuous equations can be expressed as follows:

$$[k]\{u\} - [c]\{h - h_0\} = \{F\} \qquad (1)$$

$$[D]\frac{\partial\{u\}}{\partial t} + [s]\frac{\partial\{h\}}{\partial t} + A\{h\} = \{Q\} \qquad (2),$$

where [k]: stiffness matrix, [c]: matrix for conversion of pore pressure to body force, [D]: drainage matrix, [s]: porosity matrix, [A]: permeability matrix, {u}: displacement vector, {h}: total water head vector, {h_0}: initial water head vector, {F}: force vector, {Q}: hydraulic vector.

Finite element mesh and boundary conditions are shown in Fig. 3. Coefficient of permeability (k) of the sheet pile wall was set at 4.17 × 10^{-3}cm/s, a value obtained from the experimental study by Kubo and Murakami (1963). The right-hand boundary is allowed only vertical movement, with constant water head. The bottom boundary is fixed and undrained. The soil parameters using the computation are shown in Table 1. Moduli of elasticity (E) were calculated for the clay and sand layers respectively, with cohesion (C) and N value taken into account. Coefficients of permeability (k) were determined for the clay and sand layers respectively by laboratory consolidation and field permeability tests. Anchor preload to the ground was so applied as to transmit from the nodal points in the binding layer in consideration of the pressure-grouted anchor structures (Hata et al., 1985). Finite element analysis was conducted in the 1st to 10th stages.

Table 1 Soil parameters

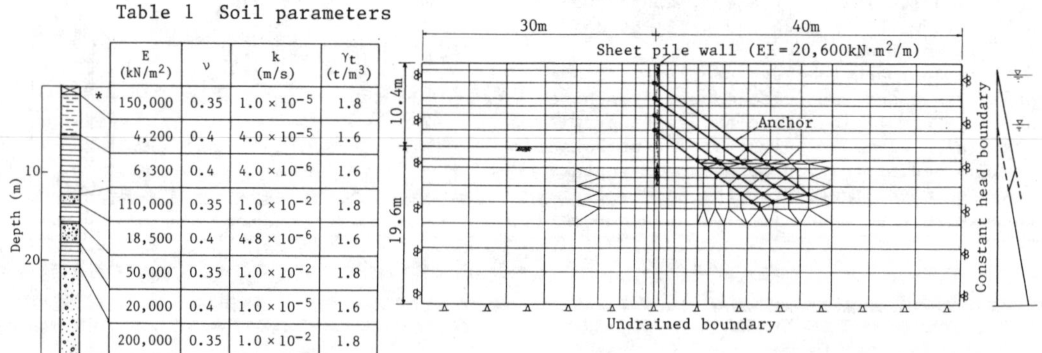

	E (kN/m^2)	ν	k (m/s)	γ_t (t/m^3)
*	150,000	0.35	1.0 × 10^{-5}	1.8
	4,200	0.4	4.0 × 10^{-5}	1.6
	6,300	0.4	4.0 × 10^{-6}	1.6
	110,000	0.35	1.0 × 10^{-2}	1.8
	18,500	0.4	4.8 × 10^{-6}	1.6
	50,000	0.35	1.0 × 10^{-2}	1.8
	20,000	0.4	1.0 × 10^{-5}	1.6
	200,000	0.35	1.0 × 10^{-2}	1.8

Sand layer E = 3,000N
Clay layer E = 210C
* Base course

Fig. 3 Finite element mesh and boundary conditions

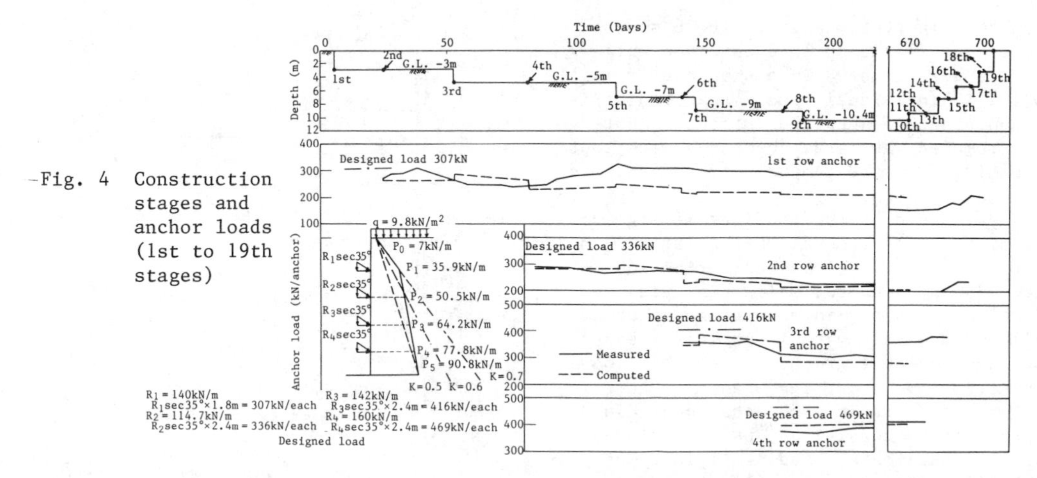

Fig. 4 Construction stages and anchor loads (1st to 19th stages)

4. MEASURED AND COMPUTED BEHAVIOUR

(1) Anchor load

Fig. 4 shows the measured anchor loads for the 2nd through 19 stages and those computed for the 2nd through 10th stages. Designed anchor loads were set as calculated, such that each row of anchors supported its lower row portion of the wall, in consideration of the lateral pressure distribution, shown in Fig. 4. This method, proven effective, permits simple calculation. In applying preload to anchors during their installation, anchor loads were set at 85% of the designed value, with a slight wall displacement allowance, and a reduction in earth pressure, from that at rest to the active earth pressure, taken into account. Each anchor had a pull-out resistance 1.5 times the designed load.

The measured and computed loads in the 1st through 9th stages coincide in that anchor load decreased in the 2nd and 3rd rows, with no significant change to the 4th row anchor load. In the 1st row, however, the measured load became higher than the initial setting, while the computed load tended to decrease. During the 10th stage, 480 days after the 9th stage, a reduction in anchor load measurements occurred in the 1st row, with an increase in the 3rd row. The increase in the 1st row anchor loads in the 1st through 9th stages were caused presumably by redistribution of earth pressure due to ground arching accompanying wall displacement; the subsequent decrease in the 10th stage was caused by outward displacement of the wall at the 1st row anchors, as shown in Fig. 7.

The following discusses an experiment conducted during anchor removal in the 10th through 19th stages to study load redistribution to adjacent anchors. Experimental procedures are described below, with removal of the 4th row anchors as an example. As indicated in Fig. 6, the measuring anchor (anchor equipped with load cell) ① was first removed and the resulting change in load of other measuring anchors (⑨ in the 4th row and ① and ⑨ in the 1st, 2nd and 3rd rows) was measured. Next, anchors ② through ⑨ were removed in that order, with measurement as above. All other 4th row anchors were then removed. Anchors in the other rows were also removed as above. Fig. 5 shows load redistribution after measuring anchor ① in each row was removed. When the 4th, 3rd and 2nd row measuring anchors ① were removed, load redistribution to adjacent anchors was as small as 0 ∿ 1.57%, the greatest portion being redistributed to the lower ground (96.87% and 91.44% being redistributed downward for 3rd and 2nd anchor removal, respectively). However, when the 1st row measuring anchor ① was removed, load redistributed to adjacent anchors was 9.13%, much larger

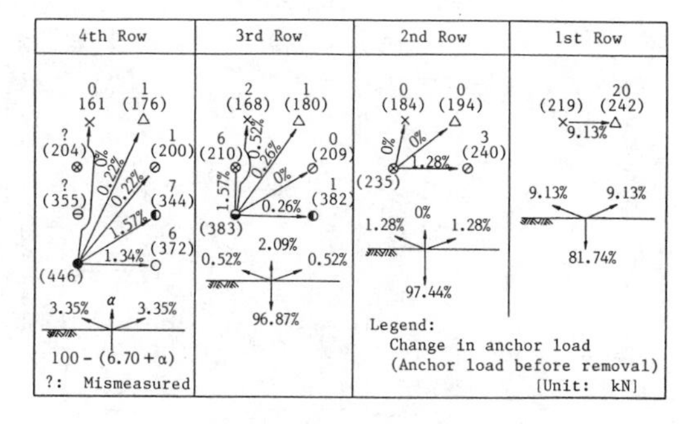

Fig. 5 Load redistribution upon removal of one anchor

than that in other cases. Fig. 6 shows the change in anchor load observed while anchors were removed in a sequence from ① to ⑨ in individual rows. Load redistribution observed when the 1st row anchors were removed was large, compared with that resulting from the removal of the 4th, 3rd and 2nd row anchors.

As evidenced by the findings in Figs. 5 and 6, that the removal of the 1st row anchors caused large load redistribution to the adjacent anchors, walls employing one row of anchors are in potential danger of collapse due mainly to insufficient support by the lower ground alone. The load redistribution occurring when an anchor is ruptured differs in each case, depending on the ground conditions, wall section and vertical/horizontal bending rigidity of the wall. The anchored wall studied in the present experiments proved to be highly safe, as the load redistribution occurring when an anchor in the wall ruptured was 20kN at most.

(2) Displacement and stress

Fig. 7 shows the measured and computed displacement of the wall in the 1st through 10th stages. Displacement occurred toward the inside during excavation, while return toward the outside was observed during preload application to the anchors. Measured and computed return toward the outside show similar patterns. The maximum measured displacement of the wall was approximately 4cm, smaller than that observed in other excavation works employing braced walls. Fig. 8 shows the measured displacement of the wall in the 9th through 19th stages. Steel wires were pulled out with anchor load released, after soil was replaced and roll-compacted up to 0.5m below each anchor installation point. As anchors in individual rows were removed successively, displacement of the wall gradually

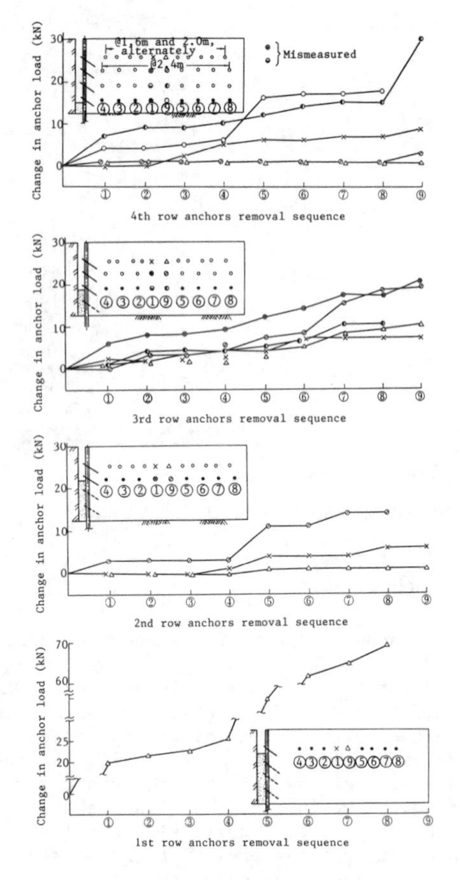

Fig. 6 Change in anchor load upon anchor removal

increased. When the 1st row anchors were removed, large displacement developed, reaching 20cm at the top of the wall. At the same time, a deep crack approximately 2cm in width occurred in the ground 9m behind the wall, proving that an earth wedge slide was caused by displacement due to removal of the anchors.

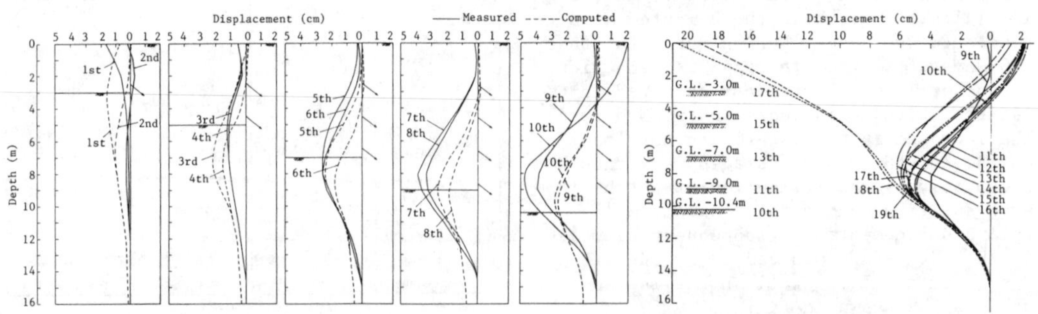

Fig. 7 Measured and computed wall displacement (1st to 10th stages)

Fig. 8 Measured wall displacement (9th to 19th stages)

Fig. 9 shows distribution of the stress over the outside of the wall as measured in the 1st through 9th stages and examples of the computed values in the 9th stage. As indicated by the distribution, the measured and computed stresses show high correspondence in the 9th stage. Unlike in ordinary braced walls, stress in anchored walls does not accumulate as excavation stages advance; during preload application to the anchors, stress shows a decrease as typically observed in the anchored wall method.

(3) Lateral pressure and water pressure

Fig. 10 shows measured lateral pressure (sum of earth and water pressure) and water pressure on the wall in the 1st through 9th stages. Water pressure shows decrease due to seepage of ground water from the wall during excavation and anchor installation. At the 4th row anchors, water pressure lies almost at $0 kN/m^2$. Prior to and after anchor preloading, excess pore pressure occurs due to displacement of the wall toward the outside. Water pressure above the 4th row anchors mostly comprises excess pore pressure. Lateral pressure shows an increase at the 1st row anchors and a decrease below them. A rise in lateral pressure at the 1st row anchors, corresponding to an increase in the 1st row anchor loads, is caused mainly by redistribution of earth pressure due to arching of the ground behind the wall.

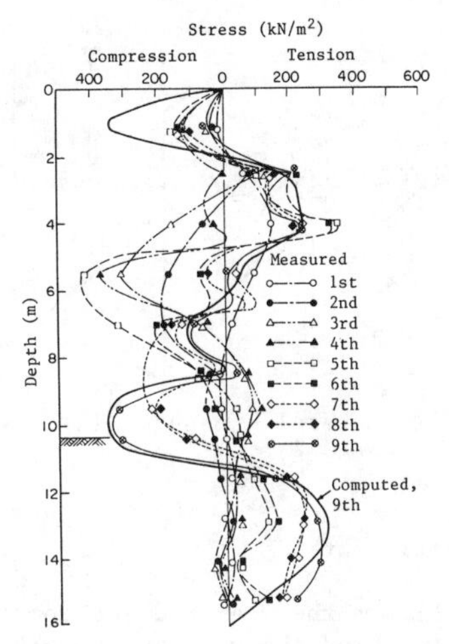

Fig 9 Wall stress (1st to 9th stages)

Fig. 11 shows the change in lateral pressure during removal of the anchors of individual rows. After the 4th and 3rd row anchors were removed, there was a decrease in lateral pressure due presumably to arching of the ground near the anchors removed or to an increase in the degree of mobilization of the shear strength of the same ground. After removal of the 2nd and 1st row anchors, the reduction in lateral pressure near the anchors removed and the increase in lateral pressure below them, caused presumably by the above-mentioned failure of the ground behind the wall, were remarkable.

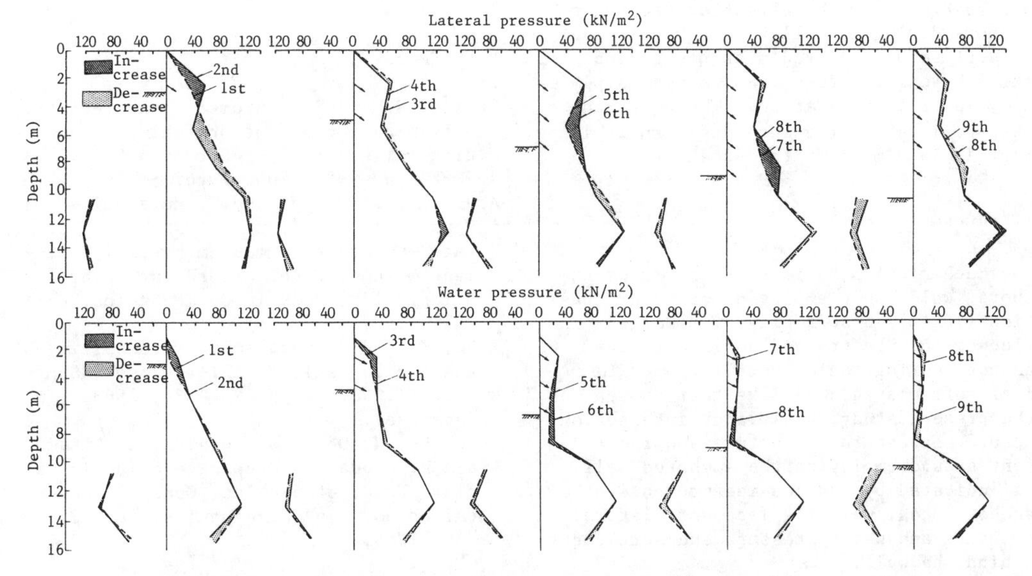

Fig. 10 Lateral pressure and water pressure (1st to 9th stages)

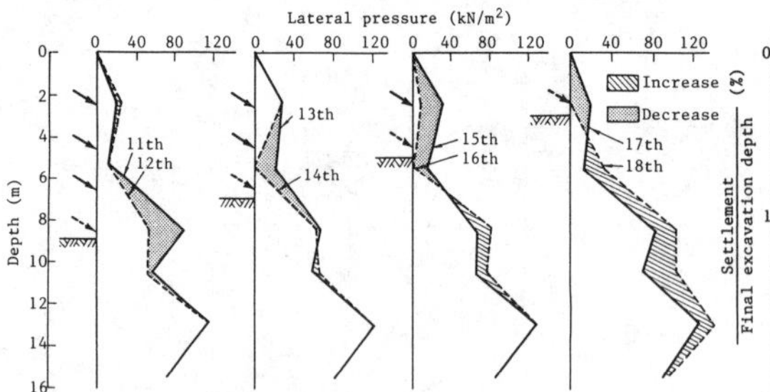

Fig. 11 Lateral pressure (11th to 18th stages)

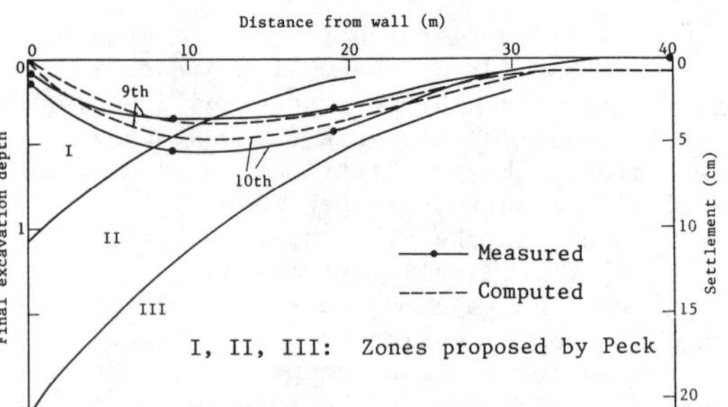

Fig. 12 Settlement behind wall (9th and 10th stages)

(4) Settlement

Fig. 12 shows the measured and computed settlement behind the wall in the 9th and 10th stages, with reference to the Peck's diagram. The maximum settlement, 3cm, or 0.35% of the excavation depth in the 9th stage, increased to 5.5cm or 0.55% in the 10th stage. Measured and computed values show relatively good correspondence.

Fig. 13 shows the measured ground-water level behind the wall and apparent level of seepage computed with pore pressure set at $0kN/m^2$ based on data obtained in the 3rd, 4th, 9th and 10th stages, for example. As indicated by the measured and computed values in the figure, the ground-water level behind the wall starts lowering, at around 40m behind the wall, toward the excavation bottom level. Between the 3rd and 4th stages, there was excess pore pressure caused by outward wall displacement occurring due to the effect of preload on anchors during installation. Development of settlement observed over the period from the 9th to 10th stages was caused mainly by clay layer consolidation. The computed seepage level in the 10th stage shown in Fig. 13 is almost in the steady state.

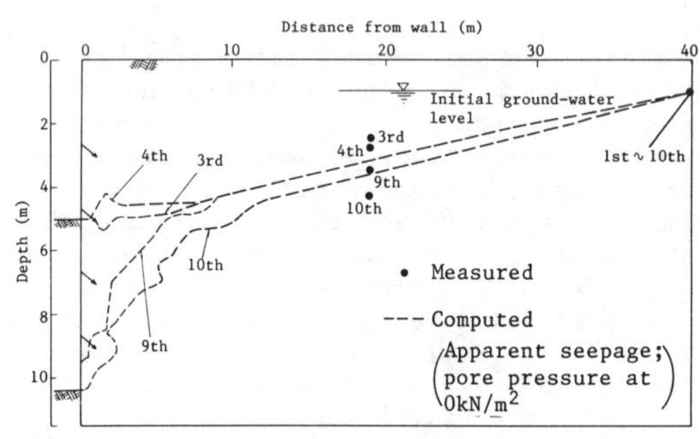

Fig. 13 Seepage behind wall (3rd, 4th, 9th and 10th stages)

4. CONCLUSIONS

The mechanical behaviour of four-rows anchored wall, when employed for an excavation (10.4m) in soft clay, was largely influenced by the preload applied to the anchors during their installation. In typical soft clay ground like that chosen for the present study, a distinct interaction occurs between the structure and the adjacent ground employing the anchored wall, as indicated by various measurements of anchor loads, wall displacement, lateral pressure and water pressure and settlement behind the wall.

ACKNOWLEDGEMENTS

We greatly appreciate the valuable advice extended to us for the present study by Professor K. Ueshita of Nagoya University.

REFERENCES

Stille, H. and B.B. Broms: Load redistribution caused by anchor failures in sheet pile walls, Proc. of European Conf. on SMFE, pp. 197 - 200, 1976.

Akai, K., U. Onishi and T. Murakami: Coupled stress flow analysis in saturated-unsaturated medium by finite element method, Proc. of 3rd Int. Conf. on Numerical Methods in Geomechanics, pp. 241 - 249, 1979.

Kubo, K. and M. Murakami: Permeability of sheet pile wall, Tsuchi-to-kiso, JSSMFE, Vol. 11, No. 2, pp. 25 - 31, 1963, (in Japanese).

Hata, S., H. Ohta, S. Yoshida, H. Kitamura and H. Honda: A deep excavation in soft clay, Proc. of 5th Int. Conf. on Numerical Methods in Geomechanics, pp. 725 - 730, 1985.

Numerical Methods in Geomechanics (Innsbruck 1988), Swoboda (ed.)
© 1988 Balkema, Rotterdam. ISBN 90 6191 809 X

Search for critical slip lines in finite element stress fields by dynamic programming

Takuo Yamagami & Yasuhiro Ueta
University of Tokushima, Hanshin Consultants Co., Osaka, Japan

ABSTRACT: Finite element stresses have been combined with dynamic programming to search for the critical slip line and the associated minimum factor of safety. Solutions obtained from the present method, although falling within the framework of limit equilibrium, can allow for constitutive laws for soils or the effect of initial stresses. No assumptions have been made on the shapes of slip lines except that they consist of a chain of line segments. The shape of the critical slip line is thus given as part of the solution.

The functions and features of a computer program developed for the present study are outlined. To show how this works, analyses perfomed with the computer program have been illustrated for two embankment problems.

1 INTRODUCTION

In recent years FE stress-deformation analyses have widely been performed for earthen structures. This trend will surely be increased with progress in the study on constitutive laws for soils. The FE stresses, however, do not directly show the stability condition of an earthen structure as a whole, although they yield valuable information on local factors of safety. On the other hand, limit equilibrium methods yield a value for the overall factor of safety of the structure but fail to account for the constitutive relationships of the soils or the effect of initial stresses. It is, therefore, very useful to combine the advantages of both approaches.

The purpose of this study is to make use of the FE stresses in stability analysis within the framework of limit equilibrium. For this purpose, we shall present a search method for the critical slip line in the FE stress field by employing a dynamic programming approach. Dynamic Programming (DP for short) is a technique used in solving multistage optimization problems[e.g. 1, 2]. The central concept of DP is "the principle of optimality" as it is called by Bellman. Recently Baker[3] and Talesnick and Baker[4] have developed a successful procedure for slope stability analysis which couples the DP technique with the Spencer method. In this study, the FE stresses and DP will be combined with the aid of Baker's pioneering work.

2 D.P. ALGORITHM FOR DETERMINING CRITICAL SLIP LINES

Let the curve $\widehat{AB}$ in Fig.1 be an assumed slip line connecting two arbitrary points, A and B, on the periphery of an earthen structure. We define the overall factor of safety for the assumed slip line as

$$F_s = \int_A^B \tau_f \, ds \Big/ \int_A^B \tau \, ds \tag{1}$$

where τ_f is the shear strength, and τ the shear stress. The integrations are carried out along the curve $\widehat{AB}$. The value of Fs in Eq.(1) depends on the shape and location of the curve $\widehat{AB}$; obviously our aim is to search for the critical slip line that has the smallest value for Fs.

A new auxiliary function G is introduced in order to rewrite Eq.(1):

$$G = \int_A^B (\tau_f - F_s \cdot \tau) \, ds \tag{2}$$

Now, it is known [3,5] that minimizing the function Fs in Eq.(1) is equivalent to minimizing the new function G in Eq.(2). In other words, the critical slip line has the minimum of not only the function Fs,

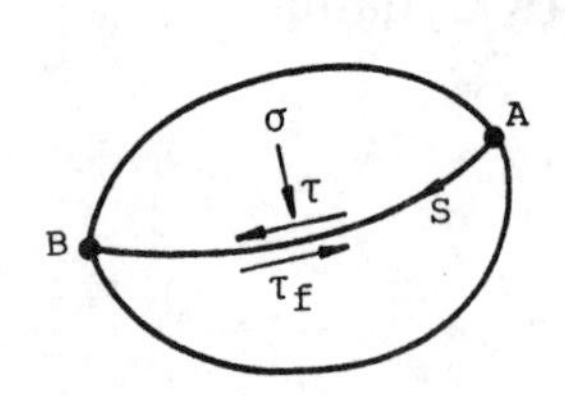

Fig.1 Definition of overall factor of safety

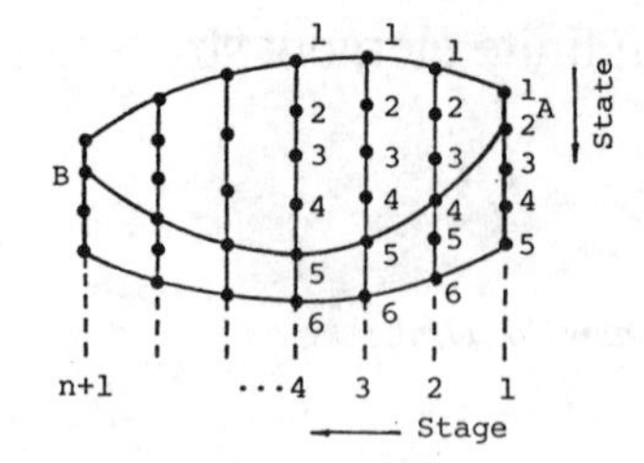

Fig.2 Search for critical slip line based on DP

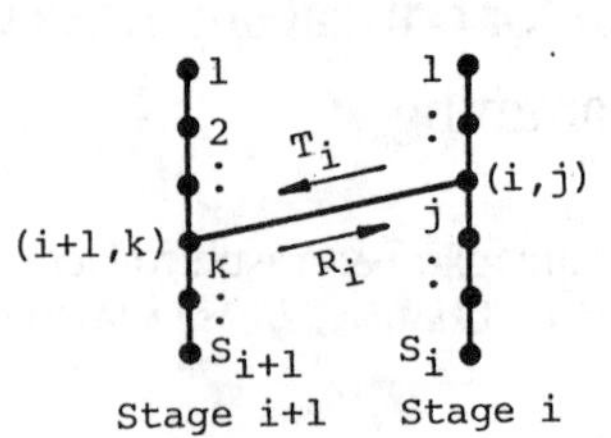

Fig.3 Application of the principle of optimality

but also the function G. For theoretical reasons, it is impossible to perform direct minimization of Eq.(1), therefore attempts must be made to minimize G in Eq.(2) instead:

$$G_m = \min G = \min \int_A^B (\tau_f - F_s \cdot \tau) \, ds \qquad (3)$$

where Gm is the minimum of the function G, which yields the critical slip line. Fs in Eq.(3) is not known in advance, so that starting with an initial assumed value of Fs, we must iterate the computation process which will be described later.

The DP algorithm is used to minimize G, provided that both the shear strength and the shear stress have been obtained from the FE analysis. To facilitate the DP implementation, Eq.(3) is rewritten in a discretized form as

$$G_m = \min \sum_{i=1}^{n} (R_i - F_s \cdot T_i) \qquad (4)$$

where Ri and Ti are relevant values of shear strength and shear force, respectively.

When applying the DP, the appropriate number of stages, the total number being (n+1), should be set up in the domain of analysis as shown schematically in Fig.2. Moreover, at each stage the appropriate number of states are created and are indicated by dots ● in Fig.2.

Let points(i,j) and (i+1,k) denote arbitrary states j and k at any two successive stages, respectively, as shown in Fig.3. Let us next consider the trajectory $\overline{jk}$ connecting the two points(i,j) and (i+1,k), and assume that the trajectory $\overline{jk}$ forms a part of a possible slip line. Then, the change in G, DGi(j,k), on passing from the point(i,j) to the point(i+1,k) is expressed as

$$DG_i(j,k) = R_i - F_s \cdot T_i \qquad (5)$$

where Ti and Ri, the respective shear force and shear strength along the trajectory, can be evaluated from the finite element stress field. The function DGi(j,k) is sometimes called the 'return function' in the literature on DP.

At this stage, we define a function Hi(j) which signifies the minimum value of G between the initial stage and the point(i,j); here the function Hi(j) is called the 'optimal value function'. Then, according to Bellman's principle of optimality, the recurrence relation for the present situation can be written as[3]

$$H_{i+1}(k) = \min_{j=1\sim S_i} [H_i(j) + DG_i(j,k)]_{k=1\sim S_{i+1}}^{i=1\sim n} \qquad (6)$$

with the boundary conditions:

$$H_1(j) = 0 \quad , \quad j = 1 \sim S_1 \qquad (6')$$

and

$$G_m = \min G = \min_{j=1\sim S_{n+1}} [H_{n+1}(j)] \qquad (6'')$$

where Si=the number of states at stage i.

The meanings of the above equations are briefly summarized as follows:

(a) The minimum value of G between the initial stage and state k at stage (i+1) is denoted by Hi+1(k); thus, Hi+1(k) must equal the minimum value of the sum of Hi(j) and DGi(j,k), where, as already defined, Hi(j) is the minimum value of G between the initial stage and any state j at the previous stage i, and DGi(j,k) is the return obtained on passing between the two states j and k. This fact is expressed by Eq.(6).

(b) There are no preceding stages for the states at the initial stage, therefore, all the optimal value functions are zero. This is given by Eq.(6').

(c) Application of the recurrence relation(6) to each state at each stage, starting with the initial stage, will eventually furnish the optimal value functions, that is, the values of G fc all states at the final stage. We must thus employ the minimum value of the optimal value functions. Eq.(6") expresses this requirement.

After the solution has reached the final stage and the minimum value Gm has been obtained with the above procedure, we can trace back the optimal path, e.g. the curve $\widehat{AB}$ in Fig.2; here the optimal path clearly represents the critical slip line. This formulation is identical with that of the shortest-path problems which are most common in the field of dynamic programming.

As stated before, since the value of Fs in Eq.(5) is initially unknown, we should begin the optimization procedure with an assumed value for Fs. The iteration process is thus needed to reach the correct critical slip line. This iteration process is as follows:

Suppose that the optimal path $\widehat{AB}$, which provides the minimum of G has been obtained on the basis of an assumed value $F_{s,i}$ for Fs. A new value $F_{s,it}$ for the factor of safety is then calculated along the curve $\widehat{AB}$:

$$F_{s,it} = \sum_{i=1}^{n} R_i \Big/ \sum_{i=1}^{n} T_i \qquad (7)$$

The convergence criterion is

$$|F_{s,i} - F_{s,it}| \leqq \delta \qquad (8)$$

where δ is the prescribed tolerance. If Eq.(8) is satisfied, then the path $\widehat{AB}$ represents the desired critical slip line, and $F_{s,i}$ or $F_{s,it}$ becomes the minimum factor of safety to be obtained; the iteration process will be completed at this stage. On the other hand, if Eq.(8) is not satisfied, a new assumed value $F_{s,i+1}$ for the factor of safety is evaluated by the following equation:

$$F_{s,i+1} = F_{s,i} + \alpha\,(F_{s,it} - F_{s,i}) \qquad (9)$$

where α is a factor to accelerate the convergence. The above process is repeated in terms of the newly assumed value for Fs until convergence is reached.

3 APPLICATION TO FINITE ELEMENT STRESS FIELDS

A computer program has been developed for the present study. The functions and features of the program will be briefly explained on the basis of the simple finite element mesh shown in Fig.4. For the sake of simplicity, only quadrilaterals are drawn in Fig.4, although the program can deal with both quadrilateral and triangular elements.

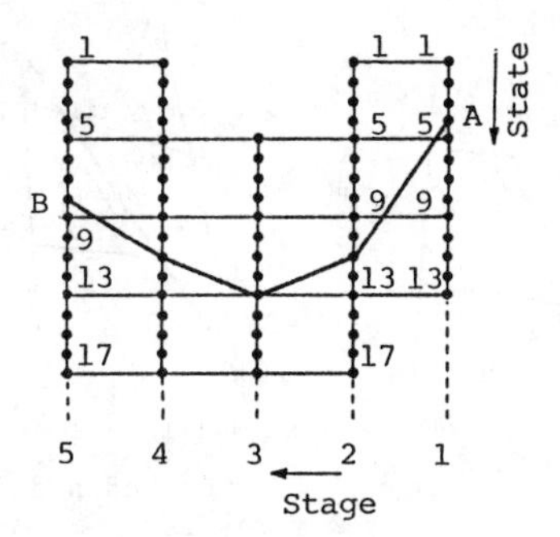

Fig.4 A schematic FE mesh used in the description of the computer program

Suppose that the stresses within each element in Fig.4 have been obtained from FE analysis carried out separately.

Setting up Stages and States. Vertical or near vertical lines in the finite element mesh are utilized as stages in this computer program. Furthermore, three additional points quadrisecting the side of each element, as well as nodal points are used as states at each stage. For example, there will be thirteen states altogether at stage one in Fig.4. Accomplishing the DP algorithm based on Eq.(6), after repeating the iteration process several times will thus lead to the critical slip line such as the broken line $\overline{AB}$ together with the minimum factor of safety.

Evaluation of Return Functions. The return function is defined by Eq.(5) with respect to a trajectory between stages i and i+1. When calculating the value for it, the computer program can handle either of the following two cases on stress distributions: (a) the stresses are assumed to be constant within each element, or (b) the nodal stresses which are smoothed or averaged by some measures are given.

a) Return Function for the Constant Element Stresses. - In this case, the return function may be simply calculated from the following expression (see Fig.5(a))

$$\left.\begin{aligned} R_i &= \Sigma\,\tau_f \cdot \ell = \Sigma\,(c+\sigma\tan\phi)\,\ell \\ T_i &= \Sigma\,\tau \cdot \ell \end{aligned}\right\} \qquad (10)$$

where the summation Σ is carried out over all the elements traversed by the trajectory, ℓ is the length of the intersection between the trajectory and an

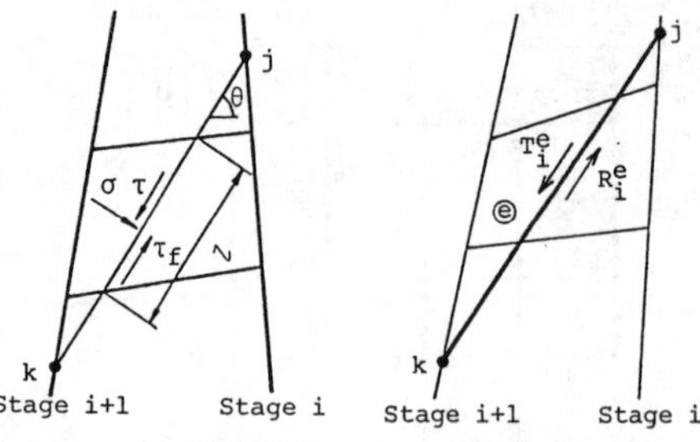

(a) Constant element stresses (b) Smoothed nodal stresses

Fig.5 Evaluation of return functions

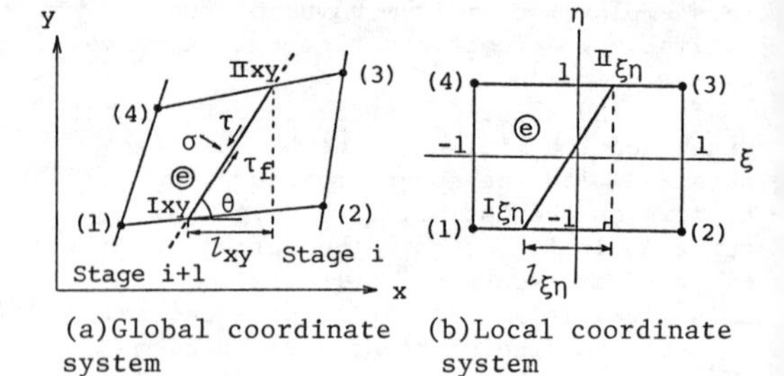

(a) Global coordinate system (b) Local coordinate system

Fig.6 Return function for smoothed stresses (Quadrilateral)

element, c and ϕ are the strength parameters and σ is the normal stress. The strength parameters and the normal stress should be properly distinguished according to the total and the effective stress approaches. Now, with the stress components σ_x, σ_y and τ_{xy} for any element, the assumption of constant stresses simply yields

$$\left.\begin{aligned} \sigma &= \sigma_x \sin^2\theta + \sigma_y \cos^2\theta - 2\tau_{xy}\sin\theta \cdot \cos\theta \\ \tau &= \tau_{xy}(\sin^2\theta - \cos^2\theta) - (\sigma_y - \sigma_x)\sin\theta \cdot \cos\theta \end{aligned}\right\} \quad (11)$$

where θ is the inclination of the trajectory to the x-axis.

(b) Return Function for the Smoothed Stresses. - Here we assume that the smoothed nodal stresses are given, based on an averaging technique [e.g. 6,7]. To allow a linearly varying stress within an element and to ensure inter-element continuity, stresses within a quadrilateral element are interpolated using the concept of iso-parametric elements (see Fig.6):

$$\tilde{\sigma} = \sum_{i=1}^{4} N_i \cdot \tilde{S}_i \quad (12)$$

where $\tilde{S}i$ = a smoothed nodal stress component(σ_x, σ_y or τ_{xy}), $Ni(i=1\sim4)$ = shape functions for the linear iso-parametric element, and $\tilde{\sigma}$ = the interpolated stress component corresponding to $\tilde{S}i$.

Now, as shown in Fig.5(b), let R_i^e and T_i^e represent the shear strength and the shear force, respectively, along the trajectory within an arbitrary element ⓔ. Then

$$R_i = \Sigma R_i^e \quad , \quad T_i = \Sigma T_i^e \quad (13)$$

where

$$R_i^e = \int_{I\,xy}^{II\,xy} \tau_f \, d\ell \quad , \quad T_i^e = \int_{I\,xy}^{II\,xy} \tau \, d\ell \quad (14)$$

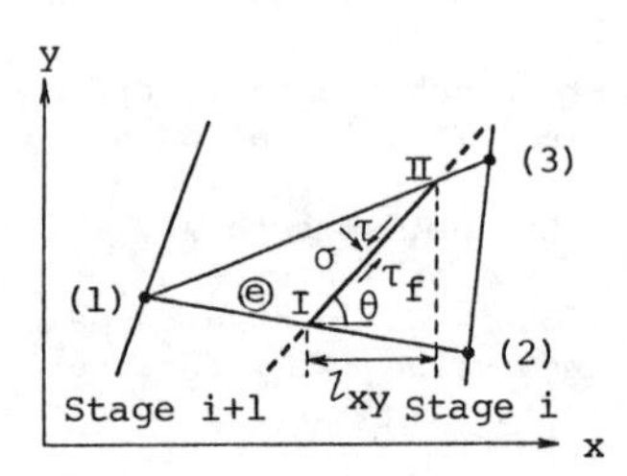

Fig.7 Return function for smoothed stresses (Triangle)

where points Ixy and IIxy are the intersections of the trajectory and element boundaries as seen from Fig.6(a), and the subscript xy denotes global coordinates. After the integrands are represented in terms of the interpolated stresses given by Eq.(12), the integrations above are performed in the local coordinate system shown in Fig.6(b). Consequently, each term related to integration must be transformed from the global coordinate system into the local system. However, details of the transformation process are not presented herein due to space limitations.

For a triangular element (Fig.7), shape functions $Ni(i=1\sim3)$ of the CST element are used:

$$\tilde{\sigma} = \sum_{i=1}^{3} N_i \cdot \tilde{S}_i \quad (15)$$

instead of Eq.(12), to consider a linear variation of stresses within the element. Eq.(15) automatically ensures stress continuity between elemets, despite the fact that Ni functions are defined in the global coordinate system.

Introduction of Fictitious Elemets. The elemets drawn with solid lines in Fig.8 represent a mesh used in FE analysis prior to the search for the critical slip line by DP. The DP analysis based only on this

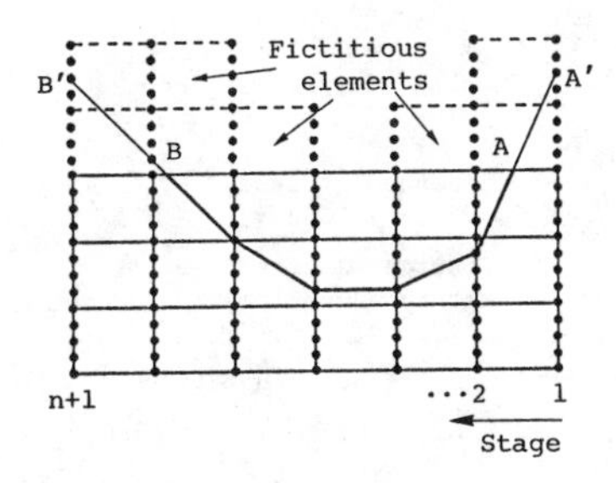

Fig.8 Schematic diagram showing fictitious elements

mesh cannot yield a slip line whose end points are located on the ground surface, such as $\overline{AB}$ in Fig.8. To solve this problem, fictitious elements may be added, as illustrated in the figure by dashed lines. Stresses within any fictitious element should be regarded as zero. The stresses so defined render return functions in the fictitious elements zero, this leading to a critical slip line, say $\overline{A'B'}$ in Fig.8. Thus, only the section $\overline{AB}$ of $\overline{A'B'}$ can be employed as the true critical slip line.

4 NUMERICAL ILLUSTRATION

Here, the computer program is applied to the nonlinear elastic stress fields in embankments. These stress fields are obtained by simulating the sequence of construction operations. The nonlinear incremental FE analysis of the embankments was performed by using the program ISBILD developed by OZAWA; this program incorporates the hyperbolic stress-strain relationship. The FE meshes have been formed from four-node linear isoparametric quadrilaterals and CST's.

The first example is concerned with a homogeneous embankment resting on a rough rigid base. The geometry of the embankment is shown in Fig.9 as Material No.I, and the soil properties used are listed in column I of Table 1. This hypothetical embankment was constructed in 10 sequential lifts.

The second example deals with the stability problem of an embankment on a compressible foundation ground. The whole structure in Fig.9 shows the geometry for this model; the foundation ground consists of two soil profiles, the material parameters of which are also given in Table 1. In this case, the construction was also performed in 10 lifts.

Fig.10 shows critical slip lines and associated factors of safety for the first

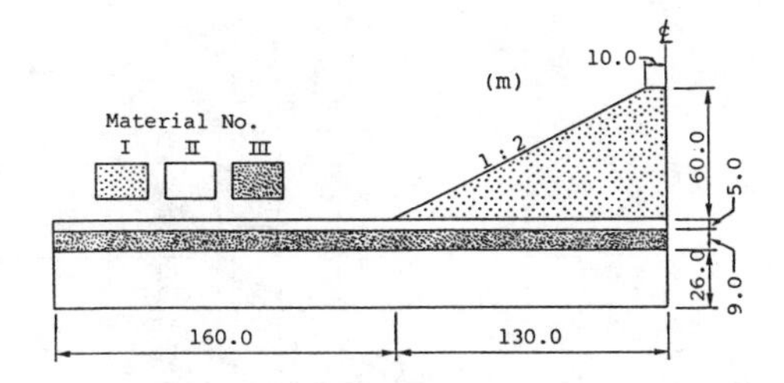

Fig.9 Embankment profiles

Table 1 Nonlinear hyperbolic model parameters

	I	II	III
ϕ (°)	29.0	30.0	20.0
c (kN/m^2)	49.0	0.0	7.84
K	580	300	2000
n	0.52	0.55	0.01
R_f	0.85	0.90	1.00
G	0.31	0.35	0.31
F	0.22	0.07	0.22
d	11.8	3.5	11.8
γ (kN/m^3)	20.38	17.64	20.38

example problem, determined by the present DP approach, together with those of the Bishop method. "Constant stresses within each element" in the figure signifies that each element was assumed to have constant stresses, the values of which were evaluated at their center. "Smoothed stresses" were obtained at each node in the FE mesh by simply averaging the stresses of all the elements surrounding the node under consideration. The elements drawn with dashed lines represent the fictitious element.

Fig.11 shows the search results for the second embankment case, in which a comparison was made among critical slip lines and factor of safety values, obtained from the present DP, Bishop and Morgenstern-Price methods.

Although little difference can be seen among the two stress conditions in Fig.10, a large discrepancy between the two is recognized in both the critical slip line and factor of safety in Fig.11. It should be noted, however, that these "Constant stresses" give a varying conservative value for the factor of safety in each case including the examples above. In Fig.11 differences between results from the proposed method and the Bishop or Morgenstern-Price methods are also greater. The authors can accept this fact, as the results are somewhat affected by the initial stresses in the foundation ground and because the proposed method can allow for such effects.

However, only simplified situations have

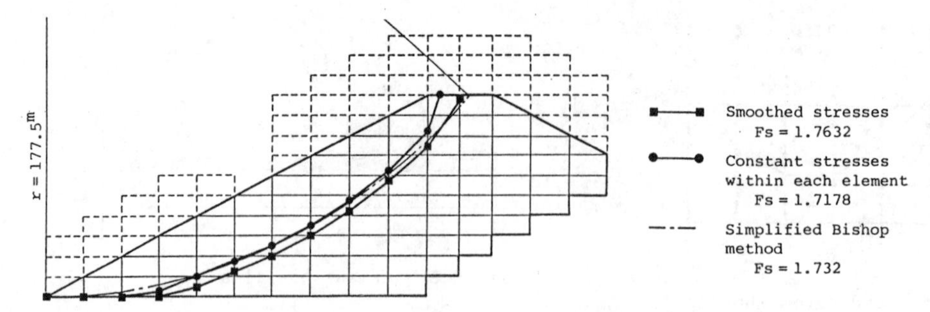

Fig.10 Search results for the embankment resting on a rigid base

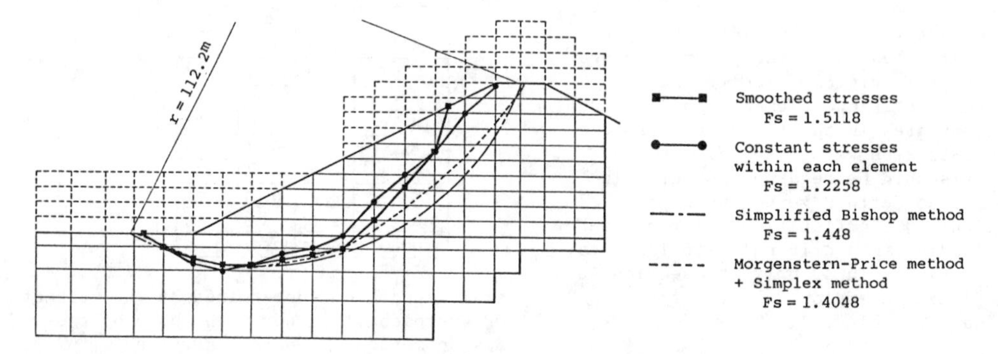

Fig.11 Search results for the embankment resting on the foundation ground

so far been employed, and the method used to obtain smoothed nodal stresses was too simple. Therefore, further proof is needed for final conclusions.

5 CONCLUDING REMARKS

A dynamic programming algorithm to search for the critical slip line has been presented; its purpose was to use the FE stresses in stability analysis within the framework of limit equilibrium. The greatest advantage of the present method is that it can incorporate the effects of the constitutive relations and/or initial stresses in the ground. Some theoretical aspects of the method were first described. This was followed by an explanation of a computer program developed to implement the present method.

Though further research may be needed for final conclusions, example problems show that the present method can provide a very promising tool for the stability analysis of earthen structures.

REFERENCES

[1] Dreyfus, S. E., Dynamic programming and the calculus of variations, Academic Press, 248p. (1965)

[2] Larson, R. E. and J. L. Casti, Principles of dynamic programming, part 1: Basic analytic and computational methods, Marcel Dekker Inc., 330p. (1978)

[3] Baker, R., Determination of the critical slip surface in slope stability computations, Inter. J. Numer. Ana. Meth. Geomech., Vol.4, No.4, pp.333-359 (1980)

[4] Talesnick, M. and R. Baker, Comparison of observed and calculated slip surface in slope stability calculations, Can. Geotech. J., Vol.21, No.4, pp.713-719 (1984)

[5] Petrov, I. P., Variational methods in optimum control theory, Academic Press, 216p. (1968)

[6] Oden, J. T. and H. J. Brauchli, On the calculation of consistent stress distributions in finite element approximations, Inter. J. Numer. Meth. Engng., Vol.3, pp.317-325 (1971)

[7] Hinton, E. and J. S. Campbell, Local and global smoothing of discontinuous finite element functions using a least squares method, Inter. J. Numer. Meth. Engng., Vol.8, pp.461-480 (1974)

Numerical Methods in Geomechanics (Innsbruck 1988), Swoboda (ed.)
© 1988 Balkema, Rotterdam. ISBN 90 6191 809 X

Reliability analysis of a test embankment by variance reduction and nearest-neighbor methods

D.T.Bergado* & N.Miura
Department of Civil Engineering, Saga University, Japan
*(*On sabbatical leave from the Asian Institute of Technology, Bangkok, Thailand)*
M.Danzuka
Department of Civil Engineering, Chiyoda Chemical Engineering and Construction, Yokohama, Japan

ABSTRACT: Probabilistic evaluation of settlement and stability of AIT test embankment has been carried out using three-dimensional variance reduction method and two-dimensional nearest-neighbor methods. Statistical data of the occurrence of fine sand and silt lenses in the soft Bangkok clay were obtained and probabilistically simulated to investigate their effects on the settlement prediction. The effect of radial drainage is taken into account under plain strain condition using the method of Lacasse (1975). Probabilistic settlement prediction using the method of Skempton and Bjerrum(1957) showed reasonable agreement with the observed values. Horizontal and vertical autocorrelation functions of the undrained shear strength were obtained using the field vane shear strength. Probabilistic slope stability analysis using the three-dimensional variance reduction method appeared to be more rational approach than the two-dimensional nearest-neighbor method. Including a 5% model error, the probabilities of slope failure of the test embankment at the final stage were 6.9×10^{-7} and 2.3×10^{-5} for the two- and three-dimensional analyses, respectively.

Introduction

This paper deals with the reliability analysis of AIT test embankment using the variance reduction and the nearest-neighbor methods. A test embankment with base dimensions of 18 by 30 m and top dimensions of 10.5 by 22 m was constructed and analyzed by Ohta and Ho(1982). Subsequent studies and deterministic predictions of its behavior were made by Poulos(1983), Jian(1983), Moy (1984) and Chen(1984). The initial and final stages of the test embankment are shown in Figs. 1 and 2, respectively. The profiles of the fine sand and silt lenses in the soft Bangkok clay were investigated and modeled. The autocorrelations of the vane shear strengths were obtained and applied. Most of the data presented herein were derived from the work of Danzuka(1985) under the guidance of the other author.

Soil profile and properties

The campus of the Asian Institute of Technology is located about 42 km north of Bangkok along the Phaholyotin Road. The soil profile at the site consists of the topmost, 2 m thick weathered clay layer underlain by a layer of soft, highly compressible, grayish clay from 2 to 9 m depth. Then, about 4 m thick, first stiff clay layer and about 10 m thick dense to very dense first sand layer with gravel can be found below the soft clay. Henceforth, alternating layers of stiff clay and very dense sand is encountered down to 400 m depth. The topmost 3 layers of the subsoil is shown in Fig.3 together with the basic properties. Jamshed (1975) and Ahmed(1977) studied the field and laboratory permeability of the soft Bangkok clay and showed that the permeability of a large size test specimen can be much higher than the small one as plotted in Fig.4.

Variance reduction method

Several pockets in the soil mass have either low or high strength. If the average strength is calculated over an interval of length, b, the variance will be less than the variance of the point values. Consequently, for a soil type j, the standard deviation of the interval values, $\tilde{S}bj$, is less than the standard deviation of the point values, $\tilde{S}pj$. Vanmarcke(1977) showed that the interval standard deviation of shear strength can be obtained from the corresponding point values by using variance reduction functions. Evaluations of the these functions in terms of cartesian coordinates has been made by Sharp et al(1981) as follows:

$$\tilde{S}bj = (\Gamma_{s,z}(b)_j)(\Gamma_{s,l}(L)_j)\tilde{S}pj \qquad ...(1)$$

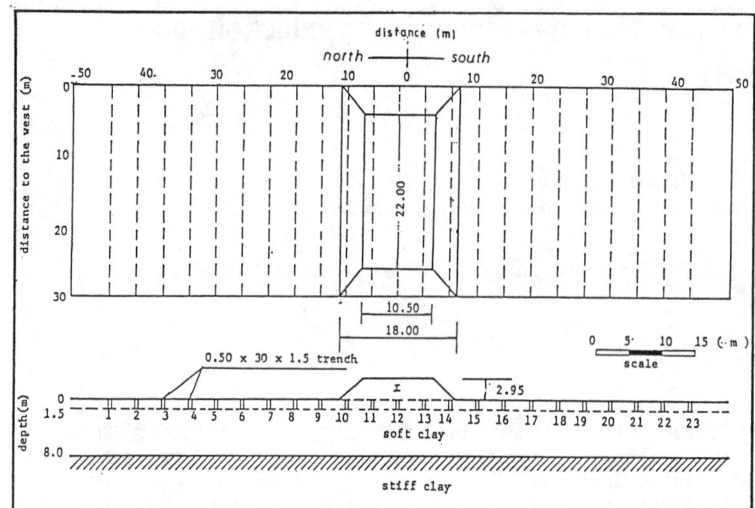

Figure 1. General view o
test embankment at initi
stages (Phase I).

(After Ohta and Ho, 1982

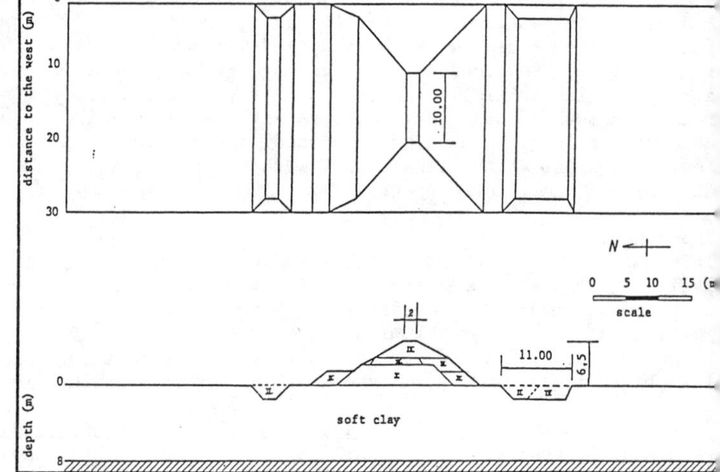

Figure 2. General view of
test embankment at final
stages (Phase IV).

(After Ohta and Ho, 1982)

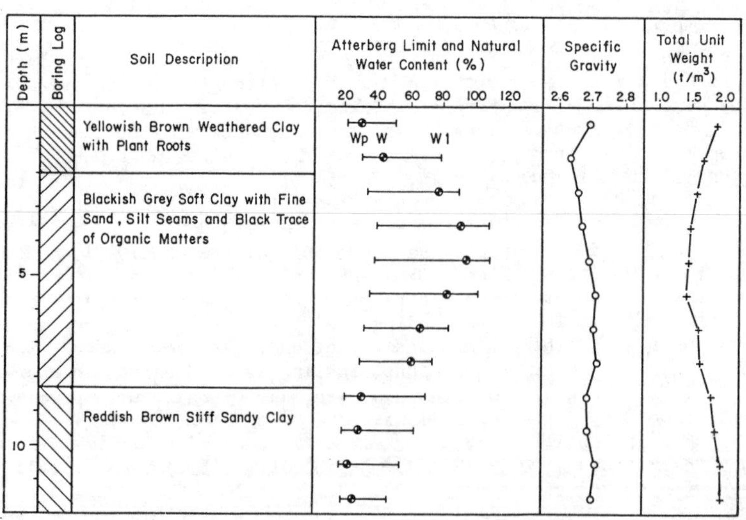

Figure 3. Soil profile ar
soil properties at the si

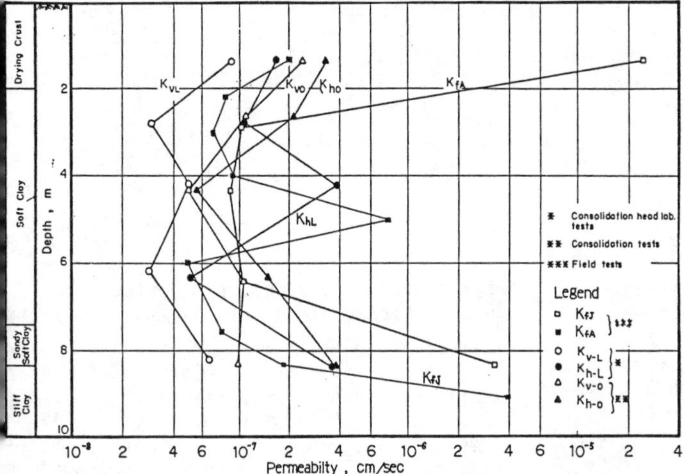

Figure 4. Laboratory and field soil permeability profile.

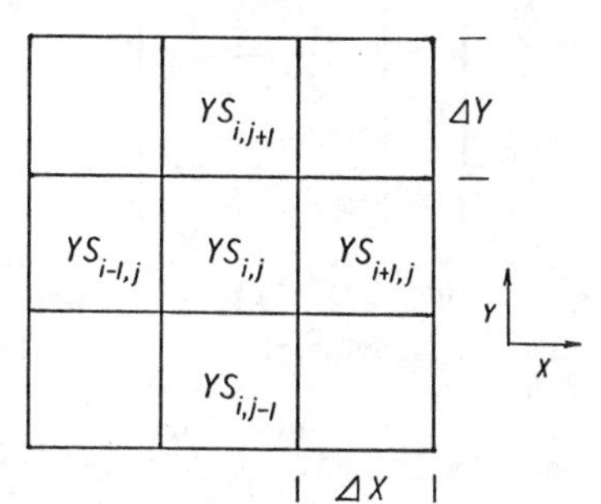

Figure 5. Schematic illustration of nearest-neighbor grid.

where $\Gamma_{s,z_j}(b)$ and $\Gamma_{s,1}(L)$ are the variance reduction along the embankment axis and along the failure surface, respectively, for the jth soil type. These variance reductions are functions of the autocorrelations as derived in Anderson et al(1981) and applied in Bergado and Anderson(1985). Autocorrelation function (ACF), $\rho(\Delta x)$ is a measure of the linear relationship between the soil parameter observed at any two points separated by a lag distance, Δx.
For exponentially decaying autocorrelation functions, the variance reduction factor can be obtained for interval length, b, as:

$$\Gamma^2(b) = \frac{1}{b^2}\int_{x}^{x+b}\int_{x'}^{x'+b} \rho(x)dxdx' \quad ...(2)$$

where $\Delta x = |x-x'|$. Thus, an autocorrelation function which decreases rapidly results in greater variance reduction than one which remains high over an extended lag distance.

The nearest-neighbor method
An autoregressive scheme extended into a spatial domain which is discrete block Monte Carlo technique was used by Smith and Freeze(1979a,b). Spatial dependence was built into a network of blocks by solving linear equations which relate the realization within one block to the realizations of the neighboring blocks with a known covariance structure. The simultaneous, nearest-neighbor autoregressive relation in two-dimensions can be written as:

$$YS_{ij} = \alpha_x(YS_{i-1,j} + YS_{i+1,j})$$
$$\alpha_y(YS_{i,j-1} + YS_{i,j+1}) + \xi_{ij} \quad ...(3)$$

where YS_{ij} is the random variable satisfying the nearest-neighbor, autoregressive relation, ξ_{ij} is an uncorrelated random variable associated with each realization, α_x and α_y are autoregressive parameters expressing the degree of spatial dependence of YS_{ij} in x and y directions, respectively. Figure 5 schematically illustrate this process equation for the block model. This method has been demonstrated by Bergado and Anderson(1985) and Bergado and Ju(1986).

Probability of failure
The value of the probability of failure can be computed from the mean ($\bar{F}b$) and standard deviation ($\tilde{F}b$) of the safety factor by means of the reliability index, β_b (Vanmarcke,1977). Assuming that the safety factor follows the normal distribution, the probability of failure, P_f, can be evaluated by the integration of the normal distribution(Dwight,1961). In this study, two infinite series as proposed by Anderson et al (1981) were used to evaluate the probability of failure. For the entire embankment of length B, the infinite combination of the overlapping failure masses of width b must be considered for the evaluation of P_f.

Probabilistic occurrence of sand and silt lenses
Five continuous boreholes were made at 1.2 m spacing. Shelby tubes with 7 cm diameter and 100 cm long were used for sampling of the soft Bangkok clay from 2 to 9 m depth. Each sample was cut up into two equal size along the axial direction to count and quantify the occurrence of the fine sand and silt lenses along the face of the sample. Full scale and detailed sketches were made for each sample. A typical occurrence of the lenses are shown in Fig.6. Two domains have been observed having different densities of the lenses, namely: (a) almost homogeneous clay with scattered fine sand and silt lenses amounting to an average of 0.2% by area from 2 to 7 m depth; (b) more heterogeneous clay containing larger fine sand and silt

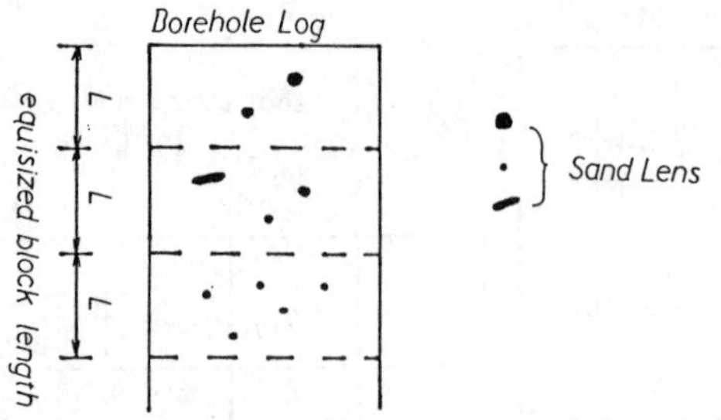

Figure 6. Occurrence of sand and silt lenses and illustration of equisized blocks.

lenses of averaging 20% by area from 7 to 9 m depth. For simplicity, only the vertical direction was considered. To carry out the one-dimensional modelling, each cut up sample face was divided into equisized blocks with length, L, as shown in Fig.6. The percent area of the lenses is calculated at each block. Subsequently, an equivalent sand seam at the middle of each block was assumed having the same percent area as the actual occurrence (Fig.7). Autocorrelation functions were calculated in domains 1 and 2. The resulting ACF averaged over 5 boreholes are shown in Figs. 8 and 9.

Effect of the fine sand and silt lenses on permeability

The presence of fine sand and silt lenses in the clay contribute much to the increase of its permeability. In this study, it is proposed that the increase in permeability is related to the percentage amount of the inclusions as shown in Fig.10. The permeability with no sand or silt should be that of the homogeneous clay (Kc) increasing to that of sand at 100% inclusion. The permeability of silty sand was assumed to be 1.5 x 10^{-4} cms. The data obtained by Jamshed (1975) and Ahmed(1977) shown in Fig.2 were utilized to correlate the field permeability to the density of fine sand and silt lenses in the clay. The value of x in Fig.10 was obtained to be 0.33 as illustrated in Fig.11 At 5.5 m depth, the field and laboratory permeability were found to be 1.05 x 10^{-7} cms and 5.3 x 10^{-8} cms, respectively, corresponding to a mean value of 0.25% fine sand and silt inclusions. At 8.0 m depth, more inclusions were found amounting to 19.7% and the field and laboratory permeability values were 4.3 x 10^{-6} cms and 4.5 x 10^{-8} cms, respectively.

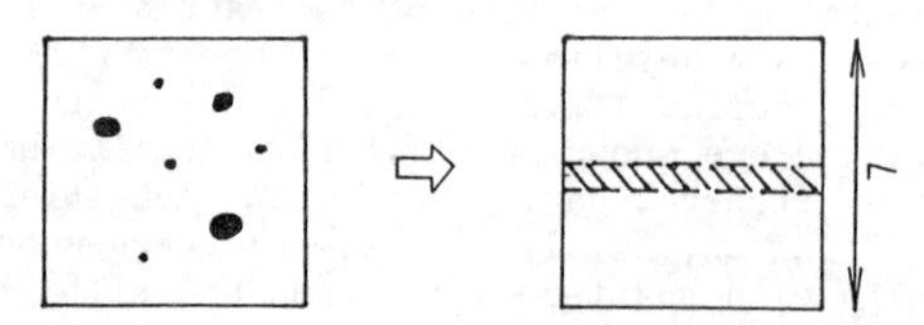

Figure 7. Concept of equivalent sand and silt lenses.

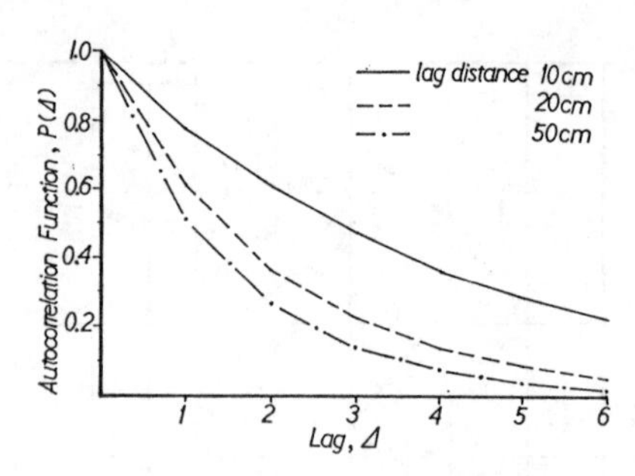

Figure 8. Autocorrelation function of sand and silt lenses in domain 1 (2 to 7 m depth

Probabilistic settlement analysis

Several investigators have applied probabilistic procedure in settlement analysis (Freeze,1977; Balasubramaniam et al, 1979; Somasundaran,1981). This study apply probabilistic settlement analysis of AIT test embankment during phase I and phase II. The seasonal groundwater table was assumed at 1.5 m depth similar to the assumption of Poulos(1983). The total and effective overburden pressures, the preconsolidation pressures as well as the static pore water pressures are plotted in Fig.12. Note that the piezometric drawdown due to groundwater pumping was neglected. The increase in vertical stresses due to the embankment load were calculated using the method of Janbu et al(1956). After generating the density of fine sand and silt lenses by the nearest-neighbor method, the permeability values were obtained using Figs.10 and 11 and assigned to each equisized block. Then, vertical and horizontal permeabilities were obtained as:

$$K_v = H/ \Sigma (Hi/Ki) \qquad ...(4)$$

$$K_h = \Sigma (KiHi)/H \qquad ...(5)$$

Next, the coefficients of horizontal and vertical consolidation were obtained assuming constant m_v value(Balasubramaniam et al,1979). Using the method of Murray(1971) the overall coefficient of consolidation fo the compressible layer was computed. Subsequently, the method of Lacasse(1975) was ap

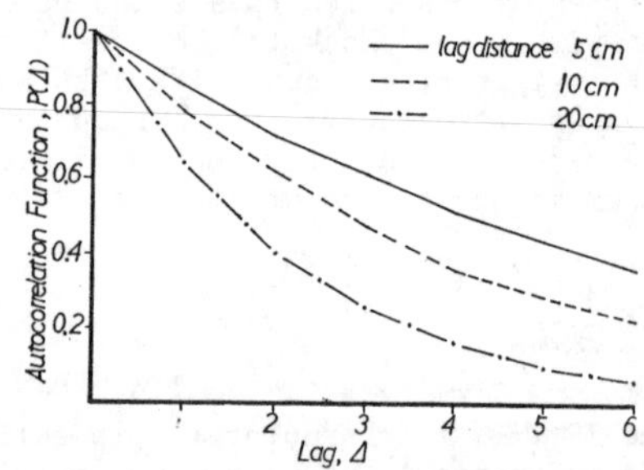

Figure 9. Autocorrelation function of sand and silt lenses in domain 2 (7 to 9 m depth)

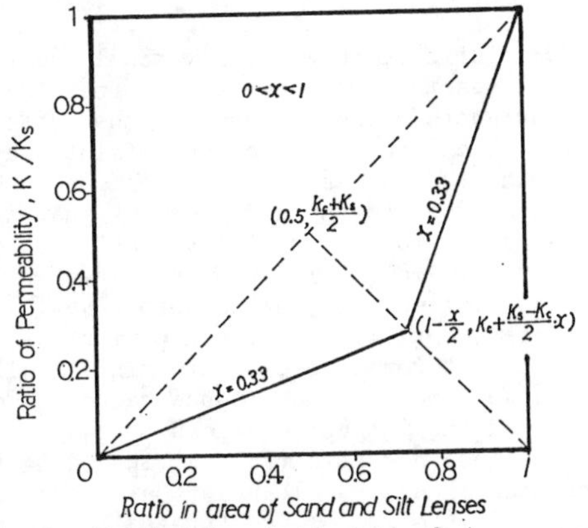

Figure 10. Model relationship between soil permeability and sand and silt lenses.

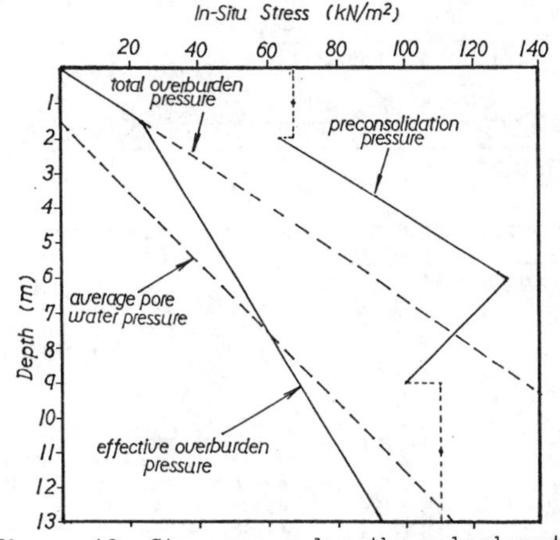

Figure 12. Stresses under the embankment and preconsolidation pressures.

plied to evaluate the time factor of anisotropic consolidation under the assumption of plane strain condition. Finally, the settlements were predicted using the method of Skempton and Bjerrum(1957). The procedure described above was followed for each realization and the settlements were averaged after 300 realizations.

Comparison of predicted and observed settlements

Jian(1983) reported the immediate settlement of 6.2 cm in Phase I of the test embankment and was applied in this study. It was found that the consolidation settlement was controlled predominantly by the overall averaged density of fine sand and silt lenses. The required parameters for the method of Skempton and Bjerrum(1957) were derived from Ohta and Ho(1982) and Moy (1984). In Fig.13, the settlements predicted by Skempton and Bjerrum method in conjunction with Lacasse method are plotted with the observed data. The predictions of Poulos (1983) and Jian(1983) are also plotted. The prediction of Poulos(1983) based on the one-dimensional theory with field C_v increased by a factor of 4 remarkably under predicted the settlements. Jian(1983) obtained and used field C_v values as much as 100 times the laboratory C_v. Skempton and Bjerrum method yielded fairly good agreement with the measured data.

Probabilistic characterization of the undrained shear strength, S_u

Field vane shear tests were conducted to obtain the statistical parameters of the undrained shear strength and the spatial variability in both horizontal and vertical directions. Ten points were spaced at 1.0 m. The tests were carried at 0.5 m interval from 2 to 7.5 m depth for a total of 120 tests. The resulting autocorrelation functions calculated after detrending the data are plotted in Fig.14 together with the theoretical ACF. The autoregressive parameters, α_x and α_z, were assigned to be 0.95 and 0.5, respectively. The vane test data were adjusted according to Bjerrum's correction factor.

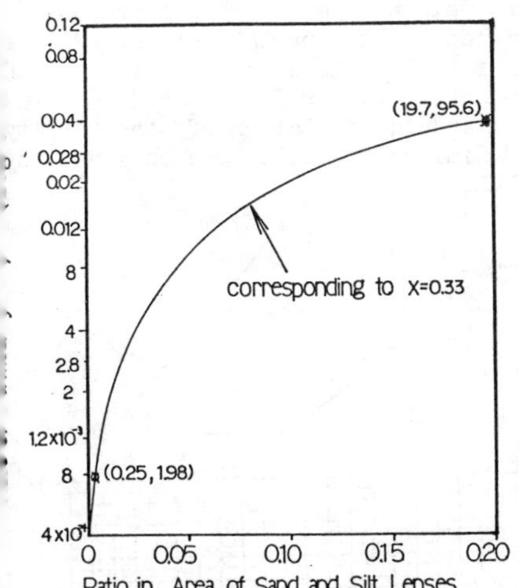

Figure 11. Plot of actual data to determine the value of parameter x in Figure 10.

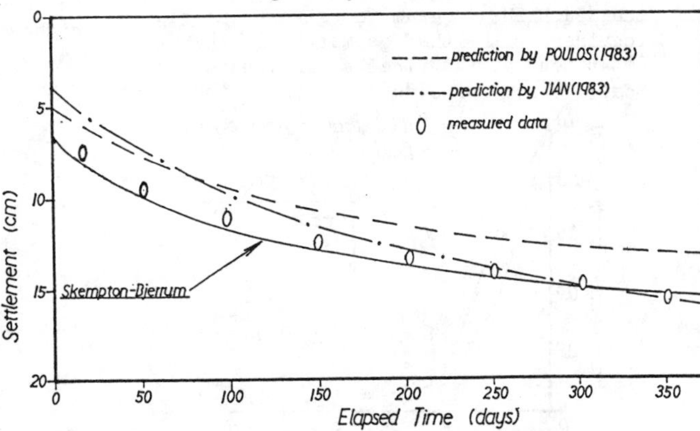

Figure 13. Actual and predicted settlements in initial stages (Phases I and II).

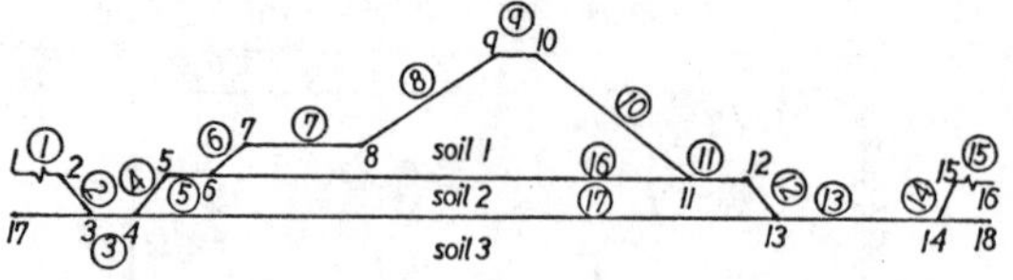

Point Data

No.	X-coord.	Z-coord.
1	5	8
2	35	8
3	37	6
4	39	6
5	41	8
6	44	8
7	46	9.6
8	52	9.6
9	59	14.5
10	61	14.5
11	69	8
12	72	8
13	74	6
14	82	6
15	83	8
16	100	8
17	5	6
18	100	6

Line Data

No.	P	P	S
1	1	2	2
2	2	3	2
3	3	4	3
4	4	5	2
5	5	6	2
6	6	7	1
7	7	8	1
8	8	9	1
9	9	10	1
10	10	11	1
11	11	12	2
12	11	12	2
13	13	14	3
14	14	15	2
15	15	16	2
16	17	18	3
17	19	20	4

Soil Data

No.	γ_T	S_u	ϕ_u	R_u	R_c
1	17.5	20.0	0	0	0
2	16.0	21.9	0	0	0
3	16.0	21.7	0	0	0

ACF Data

soil	C_x	C_y	C_z	T_x	T_y	T_z
1	0	0	0	0	0	0
2	0	0	0	0	0	0
3	0.5	0.5	1.5	0	0	0

Figure 14. Autocorrelation functions of shear strength in two coordinate directions.

Probabilistic slope stability analysis

Several researches have made contributions in the application of probability and statistics in the slope stability analysis (Alonso,1976;Catalan and Cornell,1976; Vanmarcke,1977;Sharp et al,1981;Bergado and Anderson,1985; Bergado and Ju,1986). To perform the probabilistic study, two methods were used, namely: the three-dimensional variance reduction method and the two-dimensional nearest neighbor method. The probabilistic slope stability analysis used the Bishop's simplified method(Bishop, 1955) contained in a computer program PROBISH developed at Utah State University (Anderson et al,1981). PROBISH is an interactive computer program for performing deterministic as well as probabilistic slope stability analysis. It is limited to circular failure surfaces. The critical failure surface used in the probabilistic analysis was first obtained in the deterministic analysis. The input data to the PROBISH program is given in Fig.15 together with the geometry of the test embankment at the final stage (Phase IV). Model error depends on the method of deterministic slope stability model and the assumptions that must be made. Cornel(1971) has suggested the variance of the safety factor which include model uncertainty may be computed as a combination of the variance due to the model error and the variance due to the spatial variability of the soil parameter. The probabilistic analysis was applied only in the soft clay layer. The embankment fill and the uppermost weathered crust were treated deterministically. The three-dimensional variance reduction method used the autocorrelations to evaluate the variance reduction(Equa.2) including the direction along the embankment axis and in that sense the model is three-dimensional. However, the mechanics of the model is two-dimensional except that the end resistance of the failure mass is considered. Including a model error of 5%, the following results were obtained:

Mean plane strain safety factor, $\bar{F}$ =0.96
Mean 3-D safety factor of width b, $\bar{F}b$=1.26
Probability of failure, $P_f=2.3\times10^{-5}$

The two-dimensional nearest-neighbor method used the critical failure surface obtained deterministically. The undrained shear strength is generated for each block. The location of the failure surface and the equisized blocks are shown in Fig.16. The factor of safety can be computed in each of the 300 realizations from which the mean and standard deviation can be computed. For a model error of 5%, the results are given below:

Deterministic mean safety factor, $\bar{F}$ =1.03
Mean 2-D safety factor, $\bar{F}r$=1.04
Probability of failure, $P_f=6.9\times10^{-7}$

The three-dimensional variance reduction model yielded higher probability of failure than the two-dimensional model even though the mean safety factor is larger due to the higher standard deviation of the safety factor. Similar results were obtained by Bergado and Ju(1986).

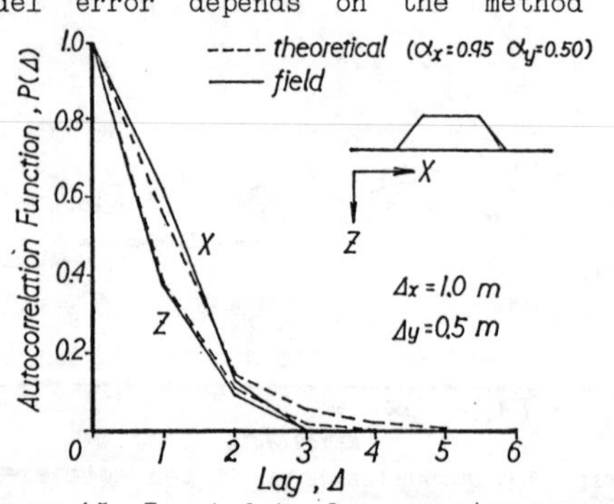

Figure 15. Input data for computer program PROBISH for slope stability analysis.

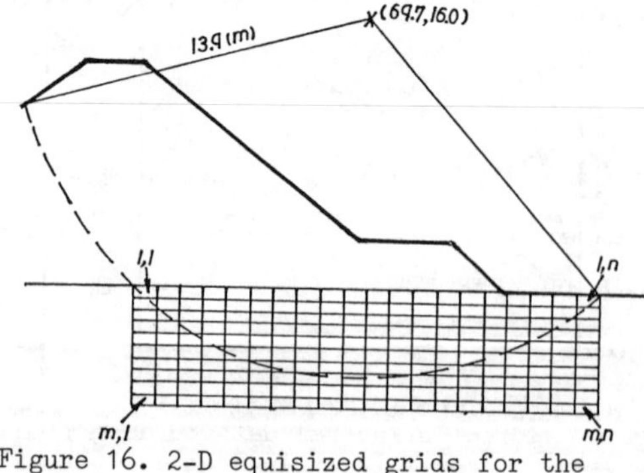

Figure 16. 2-D equisized grids for the nearest-neighbor method and failure surface.

Conclusions

1) The nearest-neighbor method can be used succesfully to predict the occurrence of fine sand and silt lenses in the soft Bangkok clay. Based on the simulation model the Skempton and Bjerrum method in conjunction with Lacasse's approximation predicted settlement magnitudes closely with the measured data in Phase I and Phase II.

2) The probabilistic study of the slope stability of the final embankment was done using the three-dimensional variance reduction and two-dimensional nearest-neighbor method resulting in probabilities of 2.3×10^{-5} and 6.9×10^{-7}, respectively, assuming a model error of 5%. The variance reductions were obtained from the autocorrelation functions. An autoregressive prediction equation was utilized in the nearest-neighbor method.

References

(1)M.M. Ahmed,Determination of permeability profile of soft Rangsit clay,M.Eng.Thesis No.1002, A.I.T., Bangkok, Thailand(1977).

(2)E.E. Alonso, Risk analysis of slopes and its application to slopes in Canadian sensitive clays, Geotechnique, Vol. 26, No. 3. pp.453-472 (1976).

(3)L.R. Anderson, D.S. Bowles, R.V.Canfield, and K.D. Sharp, Probabilistic modelling of tailings embankment designs, Research Report Submitted to U.S. Bureau of Mines, Contract No.J0295029, Utah State U.(1981)

(4)A.S. Balasubramaniam, C. Sivandran and Y. M. Ho, Stability and settlement analysis of soft Bangkok clay, Proc. 3rd ICONMIG, Aachen, West Germany (1979).

(5)D.T. Bergado and L.R. Anderson,Stochastic analysis of pore pressure uncertainty for the probabilistic assessment of earth slopes, Soils and Foundations, Vol. 25, No. 2, pp.53-76 (1985).

(6)D.T. Bergado and Y.C. Ju, Probabilistic computer modelling of a rockfill dam - A case of Khao Laem Dam, Soils and Foundations, Vol. 26, No. 4, pp.183-202(1986)

(7)A.W. Bishop, The use of slip circle in the stability analysis of slopes, Proc. Europ. Conf. on Stability of Slopes, Stockholm, Vol. 1, pp. 1-13 (1955).

(8)J.M. Catalan and C.A. Cornell, Earth slope reliability by level-crossing method, J. Geotech.Eng.Div.,ASCE, Vol. 103, No. GT7, pp.725-742 (1976).

(9)I.P. Chen, Settlement and stability analysis of embankment on soft Bangkok clay in AIT campus, M.Eng. Thesis,No.GT-83-22, A.I.T., Bangkok, Thailand (1984).

(10)C.A. Cornell, First-order uncertainty of soil deformation and stability, Proc. 1st Intl. Conf. Appl. Stat. Proba. Soil and Struct. Eng., HongKong, pp130-144 (1971).

(11)M. Danzuka, Reliability analysis of AIT test embankment, M.Eng. Thesis No. GT-84-10, A.I.T., Bangkok, Thailand (1985).

(12)H.B. Dwight, Probability integrals, tables of integrals and other mathematical data, The Macmillan Book Co., N.Y.(1961).

(13)R.A. Freeze, Probabilistic one-dimensional consolidation, J. Geotech. Eng. Div., ASCE, Vol. 103, No.GT7, pp.725-742 (1977)

(14)A.D. Jamshed, In-situ and laboratory permeability of Bangkok clay at Nong Ngoo Hao and Rangsit, M. Eng. Thesis No. 765, A.I.T., Bangkok, Thailand (1975).

(15)N. Janbu, L. Bjerrum and B. Kaernsli, Veiledning ved losning av fundamenteringsoppgaver, NGI Publication No.16, Oslo, pp. 30-32 (1956).

(16)Y.C. Jian, Analysis of settlement of an embankment on soft Bangkok clay in AIT campus, M. Eng. Thesis No.GT-82-21, A.I.T., Bangkok, Thailand (1983).

(17)S. Lacasse, Two-dimensional consolidation of a layer with double drainage under a strip load, Interim Report, Dept. of Civil Eng., Ecole Polytechnique, Montreal, Canada (1975).

(18)W.Y. Moy, Properties of subsoil related to stability and settlement of AIT test embankment, M. Eng. Thesis No.GT-83-25, A.I.T., Bangkok, Thailand (1984).

(19)R.T. Murray, Embankment constructed on soft foundations: settlement study at Avonmouth, TRRL Report No.LR-419 (1971).

(20)H. Ohta and Y. Ho, A trial embankment on soft Bangkok clay, phase I-III, AIT Research Report No.146, GTE Division, A.I.T., Bangkok, Thailand (1982).

(21)H.G. Poulos, Prediction of performance of AIT test embankment, Research Report No.R-453, Univ. of Sydney (1983).

(22)K.D. Sharp, L.R. Anderson, D.S. Bowles and R.V. Canfield, A model for assessing slope reliability, Proc. 16th Annual Meeting Trans. Res. Board, U.S.A. (1981).

(23)A.W. Skempton and L. Bjerrum, A contribution to the settlement analysis of foundation on clay, Geotechnique, Vol.7, No.4, pp.168-178 (1957).

(24)L. Smith and R.A. Freeze, Stochastic analysis of steady state groundwater flow in bounded domain: one-dimensional simulations, Water Resources Research,Vol.15, No.3, pp.521-528 (1979a).

(25)L. Smith and R.A. Freeze, Stochastic analysis of steady-state groundwater flow in bounded domain: two-dimensional simulations, Water Resources Research,Vol.15, No.6, pp. 1543-1559 (1979b).

(26)S. Somasundaran, A probabilistic approach to subsidence evaluation, M. Eng. Thesis No.GT-80-10, A.I.T., Bangkok(1981)

(27)E.H. Vanmarcke, Probabilistic modelling of soil profiles, J. Geotech. Eng. Div., Vol.103, No.GT11, pp.1127-1246 (1977).

Numerical Methods in Geomechanics (Innsbruck 1988), Swoboda (ed.)
© 1988 Balkema, Rotterdam. ISBN 90 6191 809 X

Static liquefaction of loose slopes

M.A.Hicks & S.W.Wong
University of Manchester, UK

ABSTRACT: Finite element computations for an underwater slope of loose sand, loaded undrained, are presented. These compare responses given by a linear elastic-perfectly plastic Mohr-Coulomb soil model, to those given by a nonlinear elastic-double hardening plastic constitutive law (Molenkamp (1981)). Large differences are found, and show the Mohr-Coulomb approach to be inadequate for modelling undrained soil behaviour. The Molenkamp results, however, provide a useful starting point for analysing a slope liquefaction test carried out at Delft Soil Mechanics Laboratory (Molenkamp and van Os (1987)).

1 INTRODUCTION

Groundwater conditions, and porepressures generated during deformation, dominate stability of loose granular fills. In particular, slopes formed underwater are especially vulnerable to disturbances and changes in porewater environment (Sladen et al. (1985)).

In Holland, erosion of dykes causes major concern. Here, changes in water level are often associated with slope failure, even in areas where gradients are small. Laboratory testing of a model underwater slope at Delft Soil Mechanics Laboratory, in conjunction with numerical analyses at Manchester University, have aimed to provide a better understanding of this problem.

Publications by Molenkamp (1982), Greeuw and Molenkamp (1986), and Molenkamp and van Os (1987), chart the progress of the experimental investigation. This ultimately involved inducing liquefaction in an underwater slope of loose sand, by allowing upward flow of water from fluidisation tubes attached to the base of the tank within which the slope and water were contained.

The numerical investigation has set out to independently compute model slope behaviour, using only laboratory data obtained from triaxial specimens, and without prior knowledge of actual test performance. Of prime consideration has been the numerical implementation of a suitable constitutive law, which can well represent the behaviour of soil during conditions of partial drainage.

The double hardening constitutive model proposed by Molenkamp (1981), and implemented in a general purpose finite element program by Hicks and Smith (1986) and Hicks (1988), is capable of simulating a wide range of stress paths (Griffiths et al. (1982), Griffiths and Smith (1983)), and forms the basis of the present work. Successful forward predictions of field performance for artificial islands (Hicks and Smith (1987)) and offshore foundations (Smith et al. (1988)) have already been obtained using this method of analysis.

This paper summarises a preliminary investigation into the suitability of the soil model and numerical algorithm for analysing the Delft liquefaction experiment. The Molenkamp model is used to investigate the undrained behaviour of a loose submerged slope subjected to different loading conditions, and the results compared with those obtained using an elastic-perfectly-plastic Mohr-Coulomb analysis.

These calculations also represent an extension of the previous work by Hicks and Smith (1986), where undrained and partially drained analyses of solid and hollow sand cylinders were used to demonstrate the differences between 'simple' (Mohr-Coulomb) and 'complicated' (Molenkamp) models when used in problems involving porepressures.

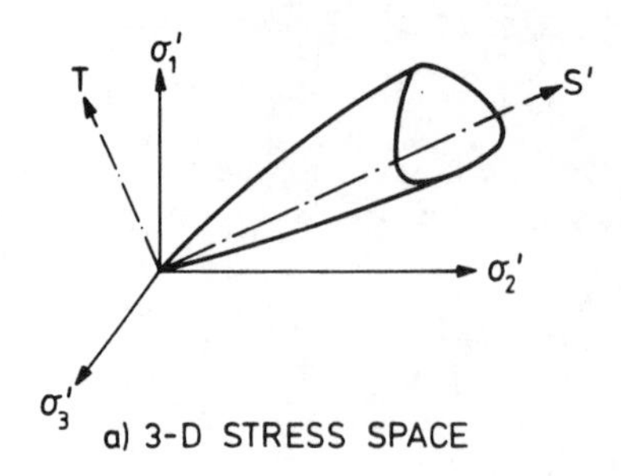

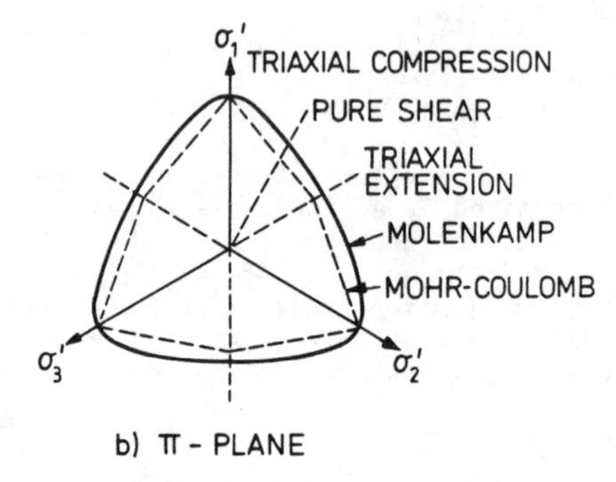

Figure 1. Molenkamp failure surface.

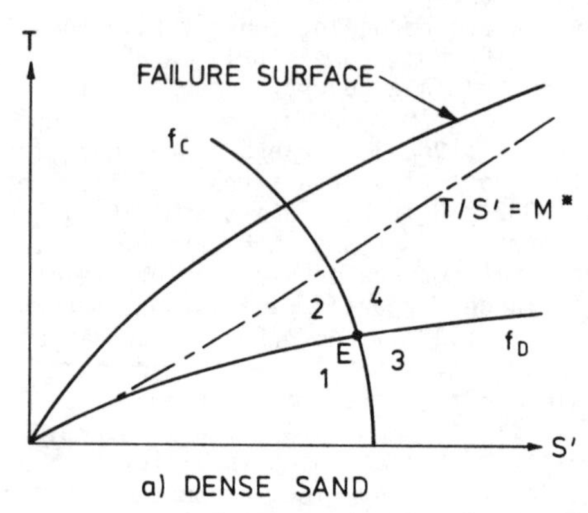

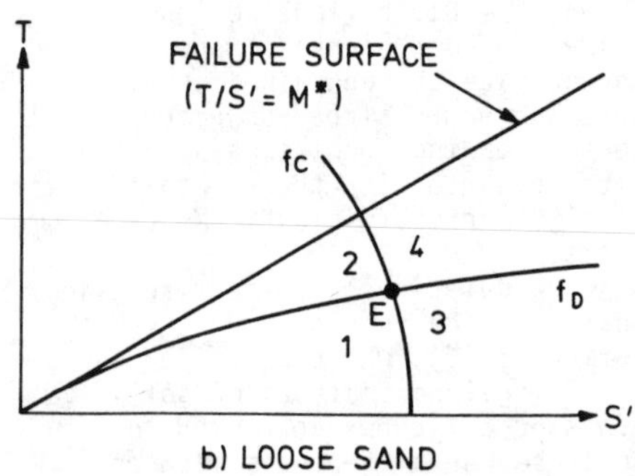

Figure 2. Molenkamp yield surfaces in octahedral plane for triaxial compression.

2 SOIL MODEL

Only a very brief review of the soil model can be included here. For a full description the reader should refer to Mokenkamp (1981). For the development and testing of the numerical algorithm, into which the model has been incorporated, see Hicks (1988).

Figure 1 shows 3-d stress space and π-plane representations of the Molenkamp failure surface. In the octahedral plane, it may be curved to account for decreased frictional resistance at increasing confining pressures. In the π-plane, it is shown to give higher strengths than Mohr-Coulomb, for the same strength in triaxial compression.

Figure 2 shows possible sections through the model in the octahedral plane corresponding to triaxial compression, for approximating (a) dense, and (b) loose, sands. Isotropic and deviatoric stresses are represented by invariants S' and T, respectively, where

$$S' = (\sigma_1' + \sigma_2' + \sigma_3')/\sqrt{3} \tag{1}$$

and

$$T = \sqrt{\{\tfrac{1}{3}[(\sigma_1 - \sigma_2)^2 + (\sigma_2 - \sigma_3)^2 + (\sigma_3 - \sigma_1)^2]\}}. \tag{2}$$

Stress space within the failure surface is divided into four distinct regions by the two isotropically hardening yield surfaces, f_C and f_D. Hence, starting from an initial stress state, E, four stress paths are possible, depending on how many, if any, of the yield surfaces are pushed out during loading. Three components of strain may occur, namely:

(i) elastic strains - due to elastic potential.

(ii) plastic strains caused by increasing mean effective stress - due to expansion of associated volumetric 'cap', f_C.

(iii) plastic strains caused by shear - due mainly to expansion of non-associated deviatoric yield surface, f_D.

Deviatoric yield surface expansion may result in contraction or dilation, depending on whether the stress path causing change in f_D is respectively inside or outside the no-volume change surface, shown chain dotted in Figure 2(a), and represented in the octahedral plane for triaxial compression by derived model parameter M*.

For a very loose material, such as that

used in the present analysis, Figure 2(b) shows that firstly, the failure surface can now be simplified to a straight line, and secondly, the criterion for no-volume change can now be taken as coincident with this limiting condition. Hence, all strains due to f_D become contractive. Furthermore, and of particular importance to the work presented here, because the deviatoric yield surface is always completely enclosed by the failure surface, expansion of f_D for very loose undrained soils can result in stress paths being driven down to very low effective stresses, and liquefaction in the sense of Casagrande (1976).

The soil model is fully defined by 21 input parameters, see Figure 3, although many of these are refinements which are taken to be constants. In practice, the number to be calibrated is somewhere between 5 and 12, depending on soil type.

3 ANALYSIS

Figure 3 shows mesh details for analysing a submerged slope of comparable dimensions to the laboratory model. Eight-node isoparammetric elements with 2 x 2 Gaussian integration have been used throughout. Discretisation of the problem area has been crude, although considered adequate for this preliminary investigation where only a qualitative comparison between 'simple' and 'complicated' models has been sought.

Parameter values for both soil models are also shown, and are those used by Hicks and Smith (1986). The Molenkamp parameters are typical for a very loose sand, while Mohr-Coulomb properties were previously derived to give similar shear stress-shear strain, and shear stress-volumetric strain, responses to the 'complicated' model in drained triaxial compression.

For each analysis, initial effective stresses are defined by a buoyant unit weight of 10 kN/m^3, and a coefficient of earth pressure at rest, K_0, of unity. Loading is assumed so rapid that no dissipation of excess porewater pressure can occur. This undrained condition is approximated using a large bulk modulus, K_A, to represent the porefluid, in the manner described by Naylor (1974). It has the advantage over a coupled Biot approach

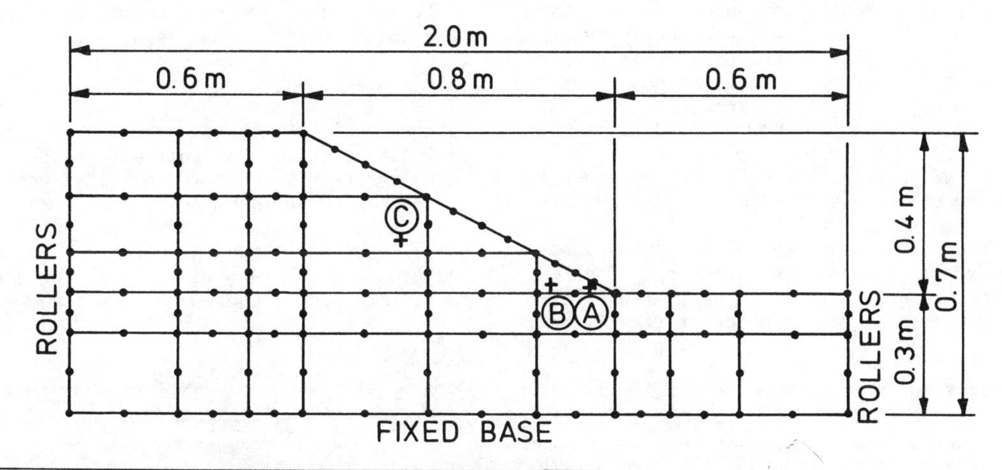

MOLENKAMP PARAMETERS										
ν	A	AP	B	BP	C	CP	DP	FIMU	FICV	SCV
.12	.0011	.42	.0023	.26	.5656	1.	.6	30.	30.	1.
VGC	VGP	NU	E	EP	LB	N	CG	CV	RT	
0.	1.	0.	.07	2.6	.3	7.	.8	.8	.3	

MOHR-COULOMB PARAMETERS				
c'	ϕ'	ψ'	E'	ν'
0. kN/m^2	30.°	0.°	10.^{4} kN/m^2	.3075

C (+) = GAUSS POINT C

$\rho' g = 10.$ kN/m^3

$K_0 = 1.$

$K_A = 2. \times 10^6$ kN/m^2

Figure 3. Slope mesh details and soil model parameter values.

using a small time step, in that problems associated with free draining boundaries are not encountered. However, oscillations in porepressures at the Gauss points can occur. To restrict this to an acceptable level, while maintaining essentially undrained conditions, the procedure described by Griffiths (1985) was adopted to derive the value for K_A. Contours of porepressure ratio, used in describing the following results, have been plotted using a smoothing process to remove remaining oscillations. This, however, has not affected the salient features of the contours.

Three loading cases have been considered, namely:

(i) gravity loading - the acceleration due to gravity is increased from its initial value, g, to higher levels given by g_{NEW}. Boundary conditions are a fixed mesh base, and rollers allowing only vertical movement at the two lateral boundaries.

(ii) surface loading - the horizontal surface at the top of the slope ($x \leq 0.6$ m) is surcharged by an increasing uniform pressure, q_B. Boundary conditions are the same as for gravity loading.

(iii) earthquake loading - this has been idealised quasi-statically by an increasing horizontal acceleration, g_H. Boundary conditions are as for (i) and (ii), except the left hand boundary is now allowed to move only in the horizontal plane.

For each load case, Molenkamp and Mohr-Coulomb models are compared using plots of crest settlement, ΔV_{CREST}, against increasing load, and using contours of excess porepressure ratio, r_u. Crest settlement is taken to be the vertical movement at the slope crest ($x = 0.6$ m).

4 RESULTS

Figures 4 to 7 summarise results for the gravity loaded slope. In Figure 4, crest settlement is plotted as a function of increasing gravity for both constitutive models. Although both cases show similar initial stiffness, the Molenkamp model (solid line) mobilises only 21% of the extra gravitational force carried by the simpler model (broken line).

Figures 5 and 6 illustrate why this is so, by considering effective stress behaviour at three Gauss points, A, B and C, see Figure 3, within the finite element mesh. For both figures, the deviatoric stress, T_{OCT}, is taken as the projection of invariant T on to the octahedral plane for triaxial compression. In Figure 5, this is plotted against mean effective stress. In Figure 6, T_{OCT} is shown as a function of γ, the deviatoric strain invariant, given by

$$\gamma = \sqrt{\{\frac{1}{3}[(\varepsilon_1 - \varepsilon_2)^2 + (\varepsilon_2 - \varepsilon_3)^2 + (\varepsilon_3 - \varepsilon_1)^2]\}}. \tag{3}$$

Due to the absence of shear-volume coupling in the 'simple' model, Mohr-Coulomb stress paths in the octahedral plane (Figure 5) climb vertically to the failure surface, leading to high strengths. Conversely, the contractive nature of the 'complicated' model causes stress paths to curve round and downwards to low effective stress states. Whereas the Mohr-Coulomb stress-strain response (Figure 6) shows elastic-perfectly-plastic behaviour, Molenkamp curves are characterised by considerable post-peak softening, and rapidly increasing strains, once the maximum shear strength has been attained.

Figure 7 shows contours of porepressure ratio for both constitutive models, at load levels approaching those for collapse. The porepressure ratio, r_u, at any point in the slope, is defined by

$$r_u = \frac{u_{xs}}{\rho' g_{NEW} h}, \tag{4}$$

where u_{xs} is the excess porewater pressure at that point, ρ' the soil effective density, g_{NEW} the updated gravitational acceleration defined in section 3, and h the depth below ground surface. Both analyses reveal high r_u contours originating at the toe, although for the Molenkamp model, these are localised in a

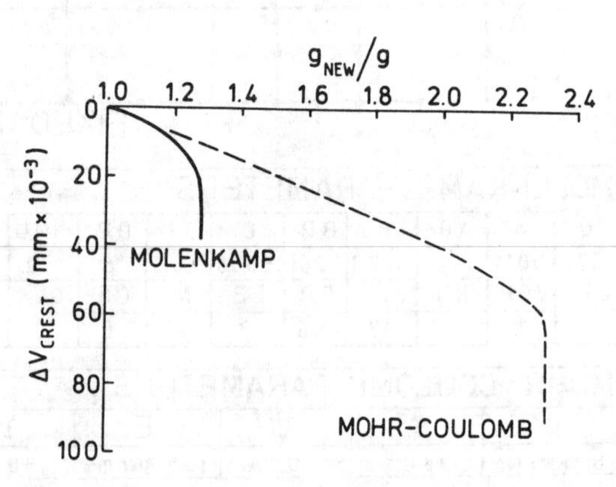

Figure 4. Gravity loading - load against settlement.

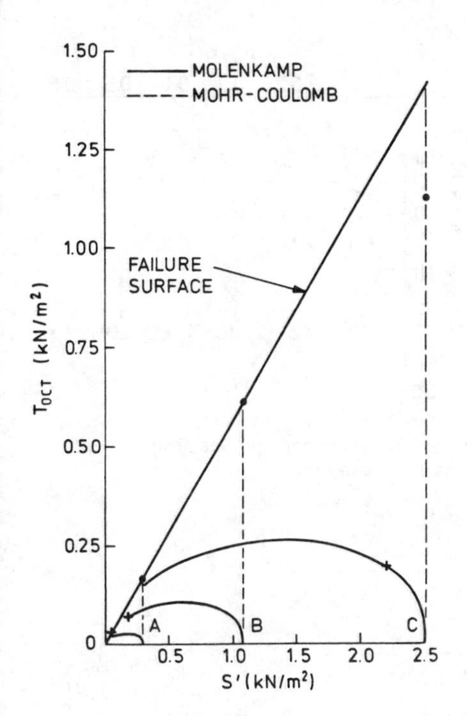

Figure 5. Gravity loading - octahedral effective stress paths.

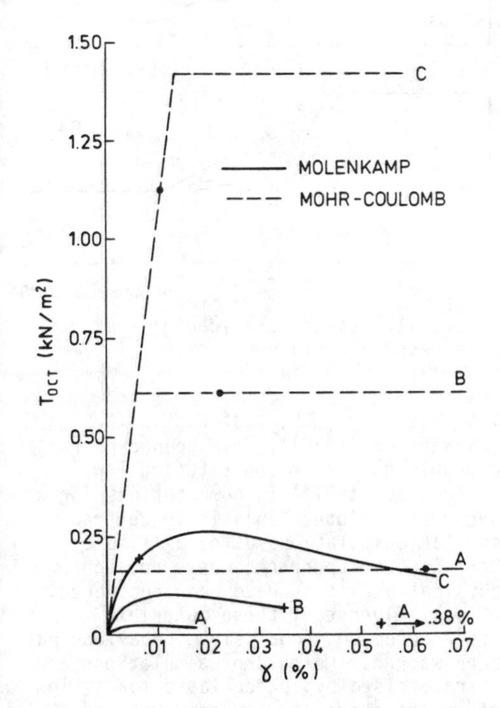

Figure 6. Gravity loading - stress-strain response.

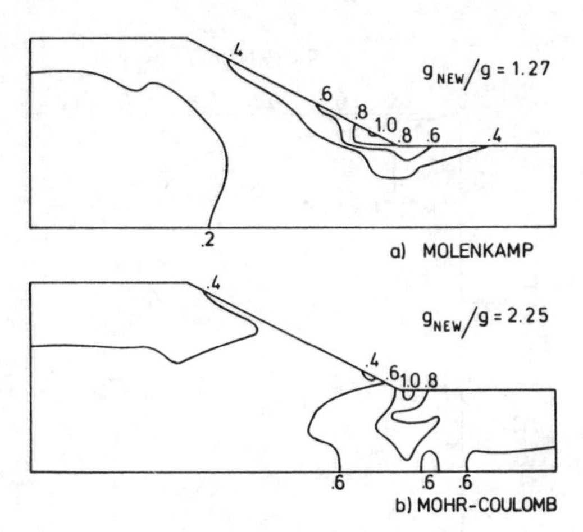

Figure 7. Gravity loading - excess porepressure ratio contours.

much shallower zone.

Returning to Figures 5 and 6, crosses and dots indicate stress and strain states at the Molenkamp and Mohr-Coulomb load levels, respectively, illustrated in Figure 7. These clearly show failure to be an unravelling process, with collapse spreading back up the slope from the toe once liquefaction has initiated. In Figure 6, localisation is pronounced for the Molenkamp model, where very large strains at Gauss point A have already taken place before the peak strength at Gauss point C has been fully mobilised.

Figures 8 and 9 show results for the surface loaded slope. This time, see Figure 8, surface load, q_B, is plotted against crest settlement. Again, Mohr-Coulomb mobilises significantly higher strengths than the Molenkamp equivalent.

In Figure 9, high r_u contours are localised in a narrow band beneath the foundation for the Molenkamp analysis, whereas high porepressure ratios for the Mohr-Coulomb analysis penetrate further beneath the foundation, as well as in the slope itself. In this figure, porepressure ratio is given by

$$r_u = \frac{u_{xs}}{\rho' gh + q_B} \qquad (5)$$

for all elements beneath the loaded area (x ⩽ 0.6 m), and by

$$r_u = \frac{u_{xs}}{\rho' gh} \qquad (6)$$

for all elements to the right of this zone (x ⩾ 0.6 m).

Finally, results for earthquake loading are given in Figures 10 and 11. In Figure

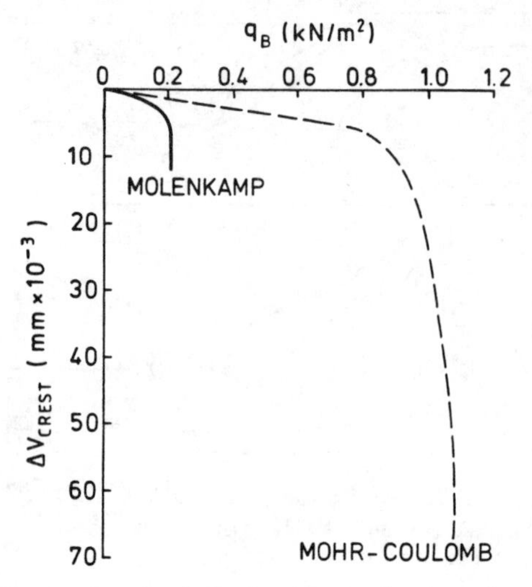

Figure 8. Surface loading - load against settlement.

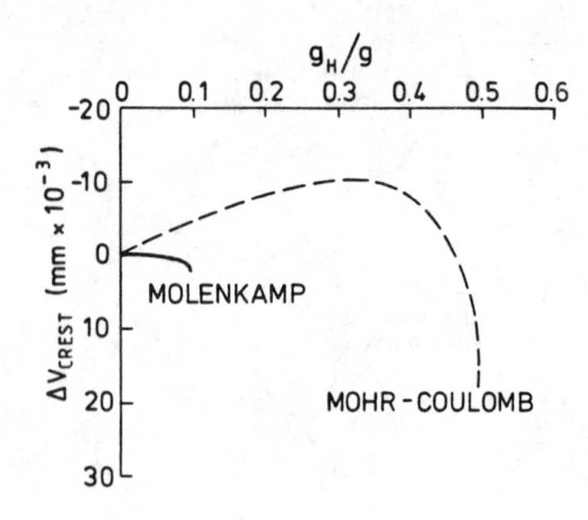

Figure 10. Earthquake loading - load against settlement.

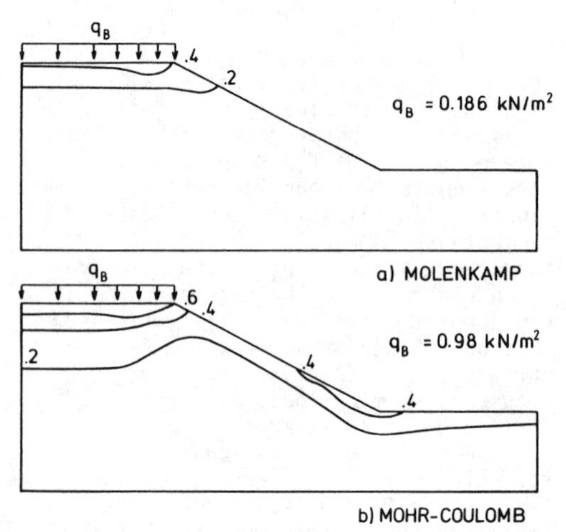

Figure 9. Surface loading - excess porepressure ratio contours.

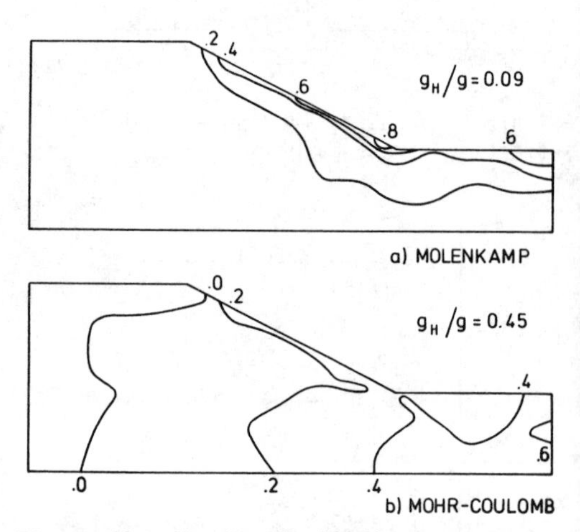

Figure 11. Earthquake loading - excess porepressure ratio contours.

10, the applied load is expressed as the ratio of horizontal acceleration to vertical gravitational acceleration. In Figure 11, the porepressure ratio is given by equation (6), shown above. For both figures, similar observations are made as for the two previous examples. Note, however, the initial rise in crest level for the Mohr-Coulomb analysis.

5 CONCLUSIONS

The essential features of very loose sand, loaded undrained, namely - contractancy, low shear strength, post-peak softening (Casagrande (1976)), have been captured by a double hardening constitutive law (Molenkamp (1981)) in computations for an underwater slope. This is in contrast to similar calculations using a 'simple' Mohr-Coulomb approach, where no such physical analogies have been recovered.

The influence of these material differences on global slope behaviour has been marked. Molenkamp calculations are characterised by low collapse loads, low effective stresses, and porepressure localisation, whereas Mohr-Coulomb calculations are characterised by high collapse loads, high effective stresses, and diffuse porepressure distributions.

It is clear that soil models which show essentially similar behaviour under drained loading conditions, may give very different responses for higher rates of loading. In the case of very loose soils, loaded undrained, collapse loads may be greatly overestimated using conventional methods of analysis. For undrained and partially drained problems, use of more sophisticated laws, which more closely reproduce 'real' soil behaviour, appears essential (Hicks and Smith (1986)).

This paper suggests the Molenkamp model (and encompassing finite element program) to be well suited for analysing the Delft liquefaction experiment (Molenkamp and van Os (1987)). Subsequent extension to modelling field slope problems seems likely (Smith and Hicks (1987)).

REFERENCES

Casagrande, A. 1976. Liquefaction and cyclic deformation of sands - a critical review. Harvard Soil Mechanics Series No. 88 (Presented at 5th Panamerican Conf. Soil Mech. Found. Eng., Buenos Aires (Nov. 1975)).

Greeuw, G. & Molenkamp, F. 1986. Calibrations for preparation of a liquefaction test in the BRUTUS-tank. Delft Soil Mechanics Laboratory SE-690500/2.

Griffiths, D.V., Molenkamp, F. & Smith, I.M. 1982. Computer implementation of a double-hardening model for sand. Proc. IUTAM Symp. Def. Failure Granular Mat., p. 213 - 221. Delft : Balkema.

Griffiths, D.V. & Smith, I.M. 1983. Experience with a double-hardening model for soil. Proc. Int. Conf. Constit. Laws Eng. Mat., p. 553-560. Arizona.

Griffiths, D.V. 1985. The effect of pore-fluid compressibility on failure loads in elasto-plastic soil. Int. J. Num. Anal. Meth. Geomech. 9 : 253 - 259.

Hicks, M.A. & Smith, I.M. 1986. Influence of rate of porepressure generation on the stress-strain behaviour of soils. Int. J. Num. Meth. Eng. 22 : 597-621.

Hicks, M.A. & Smith, I.M. 1987. 'Class A' prediction of Artic caisson performance. Submitted for publication.

Hicks, M.A. 1988. Numerically modelling the stress-strain behaviour of soils. PhD thesis to be submitted, University of Manchester, UK.

Molenkamp, F. 1981. Elasto-plastic double hardening model MONOT. Delft Soil Mechanics Laboratory CO-218595.

Molenkamp, F. 1982. Plan for liquefaction test in BRUTUS-tank and its numerical verifications. Delft Soil Mechanics Laboratory CO-218596/9.

Molenkamp, F. & van Os, R.C. 1987. Liquefaction test in the Brutus tank. Delft Soil Mechanics Laboratory SE-690504/2.

Naylor, D.J. 1974. Stresses in nearly incompressible materials by finite elements with applications to the calculation of excess pore pressures, Int. J. Num. Meth. Eng., 8 : 443-460.

Sladen, J.A., D'Hollander, R.D., Krahn, J. & Mitchell, D.E. 1985. Back analysis of the Nerlerk berm liquefaction slides. Can. Geotech. J. 22 : 578-588.

Smith, I.M. & Hicks, M.A. 1987. Constitutive models and field predictions in geomechanics. Proc. Int. Conf. Num. Meth. Eng. : Theory and Applications 2. Swansea : Nijhoff.

Smith, I.M., Hicks, M.A., Kay, S. & Cuckson, J. 1988. Undrained and partially drained behaviour of end bearing piles and bells founded in untreated calcarenite. Proc. Conf. Calcareous Soils. Perth, Western Australia : Balkema.

Numerical Methods in Geomechanics (Innsbruck 1988), Swoboda (ed.)
© 1988 Balkema, Rotterdam. ISBN 90 6191 809 X

Three-dimensional slope stability analysis by slice methods

Keizoh Ugai
Gunma University, Kiryu, Japan

ABSTRACT: In order to analyze three-dimensional (3-D) slope stability, new 3-D slice methods are proposed by extending Fellenius method, the simplified Bishop method, the simplified Janbu method and Spencer's method to three dimensions. Calculations are carried out to analyze 3-D effects on simple slopes that have a finite failure width. Applicability of these methods to practical cases and results of small-scale model tests are discussed.

1 INTRODUCTION

Recently, the author has presented 3-D stability analyses of cohesive slopes (see Ugai (1985) and Ugai et al. (1987)). However, for c-ϕ soil conditions new methods of analysis are needed to investigate 3-D slope stability. For problems with complicated geometry and soil conditions, slice methods are suited. Hovland (1977) presented a 3-D limiting equilibrium method for slopes of c-ϕ soils with arbitrary sliding surfaces by extending a 2-D (two-dimensional) slice method, assuming zero stresses on all the vertical intercolumn surfaces. He suggested that some situations exist where 3-D factors of safety (F_3) can be lower than 2-D factors of safety (F_2) in sandy slopes. This result has been questioned by Azzouz & Baligh (1978). Chen & Chameau (1983) extended Spencer's (1967) method for rotational sliding surfaces to three dimensions and showed results where the ratio F_3/F_2 is smaller than unity. These results are also questioned by Hutchinson & Sarma (1985). In 1987, two methods of 3-D stability analysis have been presented based on 3-D limiting equilibrium (Hungr and Cavounidis).

In this paper new 3-D slice methods are proposed by extending Fellenius method, the simplified Bishop method, the simplified Janbu method and Spencer's method to three dimensions.

2 EXTENSION OF 2-D SLICE METHODS TO THREE DIMENSIONS

Fig.1 shows an example of 3-D sliding

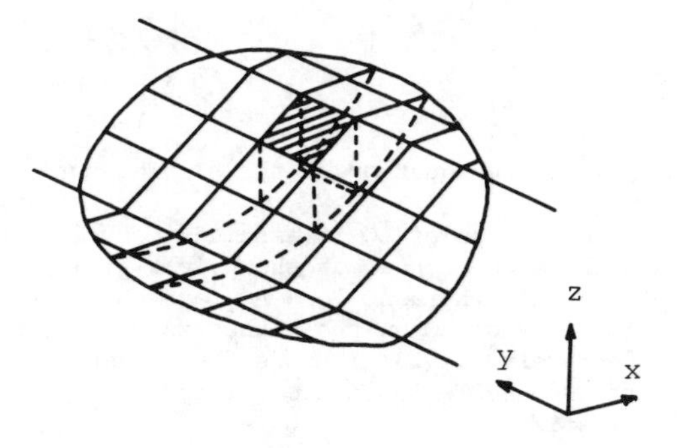

Fig.1 Sliding mass and divided columns

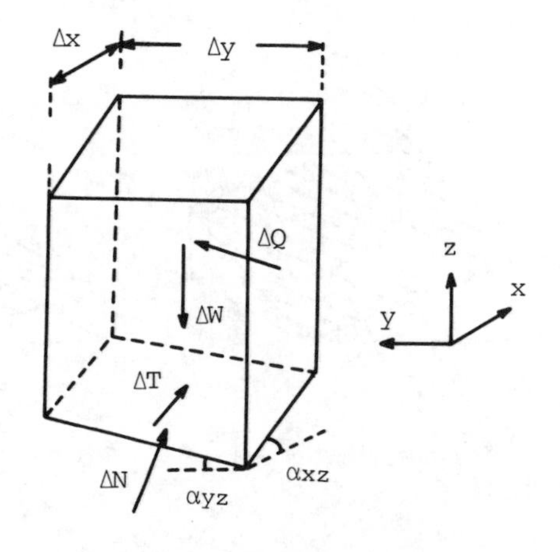

Fig.2 Forces acting on a column

condition. The sliding direction of the sliding mass is assumed to be in the xz-plane (perpendicular to the y-axis). The sliding mass is divided into several vertical columns. Fig.2 shows a single column and forces acting on it. ΔW is the weight of the column. ΔT and ΔN are the shear force and the total normal force acting on the base of the column. ΔQ is the resultant of intercolumn forces acting on the sides of the column. In Fig.2 α_{xz} and α_{yz} are the inclinations of the base of the column with respect to the x- and y-axes, respectively. Earthquake forces and pore-water pressures should be considered seperately. The directions of the forces ΔW, ΔT and ΔN are given by the following unit vectors:

$$\Delta W : \quad (0,\ 0,\ -1)$$

$$\Delta T : \quad (1/J',\ 0,\ \frac{\partial z}{\partial x}/J')$$

$$\Delta N : \quad (-\frac{\partial z}{\partial x}/J,\ -\frac{\partial z}{\partial y}/J,\ 1/J)$$

where

$$J'=\sqrt{1+(\frac{\partial z}{\partial x})^2} \tag{1}$$

$$J=\sqrt{1+(\frac{\partial z}{\partial x})^2+(\frac{\partial z}{\partial y})^2}=\sqrt{1+\tan^2\alpha_{xz}+\tan^2\alpha_{yz}} \tag{2}$$

$z(x,y)$ is the equation of the sliding surface.

The direction of ΔQ is assumed according to each slice method, as shown in Fig.3(a), (b) and (c), where ΔQ_1 and ΔQ_2 are the projections of ΔQ on the planes xz and yz, respectively. Fig.3(a) shows the assumption of the 3-D Fellenius method, where the inclinations of ΔQ_1 and ΔQ_2 are α_{xz} and α_{yz} with respect to the axes x and y, respectively. That is, in the 3-D Fellenius method ΔQ is assumed to be parallel to the base of the column. Fig.3(b) shows the assumption of the 3-D simplified Bishop and Janbu methods, where ΔQ_1 is parallel to the x-axis and ΔQ_2 is inclined by $\tan^{-1}(\eta\tan\alpha_{yz})$ with respect the y-axis. η is the unknown constant. Fig.3(c) is the assumption of the 3-D Spencer's method, where the inclinations of ΔQ_1 and ΔQ_2 are δ and $\tan^{-1}(\eta\tan\alpha_{yz})$ with respect to the x- and y- axes, respectively. δ is the unknown constant. Based on these assumptions and the failure condition, ΔN and ΔT (Fig.2) can be derived from the equilibrium of the forces in the direction perpendicular to the shadowed plane composed of ΔQ_1, ΔQ_2 and ΔQ (F-plane, B-plane or S-plane in Fig.3). The failure condition, in combination with the definition of the safety factor F, is written as

$$\Delta T=(cJ\Delta x\Delta y+\Delta N\tan\phi)/F \tag{3}$$

where c is the cohesion and ϕ is the angle of friction.

In the 3-D Fellenius method, F is defined by the following equation, based on the equation of moment equilibrium for the sliding mass.

$$F=\sum_{i=1}^{m}\sum_{j=1}^{n} Ri(cJ\Delta x\Delta y+\Delta Wij\tan\phi/J) \Big/ \sum_{i=1}^{m}\sum_{j=1}^{n} Ri\Delta Wij\sin\alpha xz \tag{4}$$

where Ri is the radius of the sliding surface and varies with y.

In the 3-D simplified Bishop method, F is defined by the simultaneous equations of moment and vertical force equilibrium for the sliding mass, as follows.

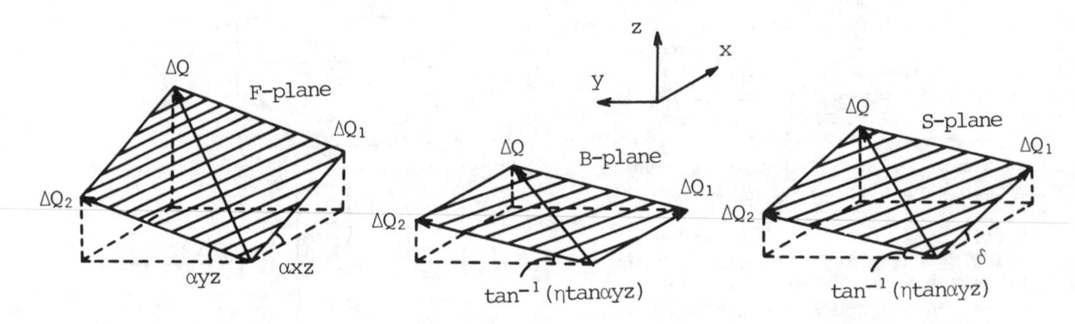

(a) 3-D Fellenius method (b) 3-D simplified Bishop and Janbu methods (c) 3-D Spencer's method

Fig.3 Assumptions of 3-D slice methods

$$F_m = \sum_{i=1}^{m}\sum_{j=1}^{n} [R_i\{c\Delta x\Delta y(1+\eta\tan^2\alpha_{yz})+\Delta W_{ij}\tan\phi\} / m_\alpha] / \sum_{i=1}^{m}\sum_{j=1}^{n} R_i\Delta W_{ij}\sin\alpha_{xz} \quad (5)$$

$$F_v = \sum_{i=1}^{m}\sum_{j=1}^{n} [\{c\eta\Delta x\Delta y\sin\alpha_{xz}\tan^2\alpha_{yz}+\Delta W_{ij}(F_v/J + \sin\alpha_{xz}\tan\phi)\}/m_\alpha] / \sum_{i=1}^{m}\sum_{j=1}^{n} \Delta W_{ij} \quad (6)$$

where

$$m_\alpha = (1+\eta\tan^2\alpha_{yz})/J+\sin\alpha_{xz}\tan\phi/F \quad (7)$$

F_m and F_v denote the safety factors derived from the equations of moment and vertical force equilibrium. From Eqs.(5) and (6) two unknowns F ($=F_m=F_v$) and η can be determined.

In the 3-D simplified Janbu method, F is defined by the simultaneous equations of vertical and horizontal force equilibrium for the sliding mass, that is, Eqs.(6) and (8), respectively. This method is applicable to arbitrary sliding surfaces.

$$F_h = \sum_{i=1}^{m}\sum_{j=1}^{n} [\{c\Delta x\Delta y(1+\eta\tan^2\alpha_{yz}\cos^2\alpha_{xz}) + (\tan\phi+\eta F_h\sin\alpha_{xz}\tan^2\alpha_{yz}/J)\Delta W_{ij}\} / (\cos\alpha_{xz}m_\alpha)] / \sum_{i=1}^{m}\sum_{j=1}^{n} \Delta W_{ij}\tan\alpha_{xz} \quad (8)$$

F_h denotes the safety factor derived from the equation of horizontal force equilibrium.

In the 3-D Spencer's method, three unknowns F, η and δ can be determined from the simultaneous equations of moment equilibrium and vertical and horizontal force equilibrium for the sliding mass (their equations are omitted).

These 3-D slice methods are reduced to the corresponding conventional 2-D ones for infinite slide width.

Finally, it should be noted that the unit vectors normal to the F-, B- and S-planes in Fig.3 have been taken into account in deriving Eqs.(4)-(8). These vectors are given by

$$\text{F-plane} : \left(-\frac{\partial z}{\partial x}/J,\ -\frac{\partial z}{\partial y}/J,\ 1/J \right)$$

$$\text{B-plane} : \left(0,\ -\eta\frac{\partial z}{\partial y}/J'',\ 1/J'' \right)$$

$$\text{S-plane} : \left(-\tan\delta/J_s,\ -\eta\frac{\partial z}{\partial y}/J_s,\ 1/J_s \right)$$

where

$$J'' = \sqrt{1+(\eta\frac{\partial z}{\partial y})^2} = \sqrt{1+(\eta\tan\alpha_{yz})^2}$$

$$J_s = \sqrt{1+\tan^2\delta+(\eta\tan\alpha_{yz})^2}$$

3 COMPUTATION AND RESULTS

3.1 EXAMPLE OF A SIMPLE SLOPE

As an example of 3-D failure, we consider a simple slope with the height H=25m and the inclination angle β=30°. The vertical section of the slope and the values of the other parameters are shown in Fig.4. The sliding surface is a rotational one having 100m width and is assumed to be composed of two parts: (1) a cylinder passing through the toe ($R_i=R_0$=61.42m) whose width is 20m; (2) semi-elliptical ends whose width is 40m. The sliding mass was divided into 76 columns.

The 3-D safety factors calculated by the 3-D slice methods proposed in this study and Hovland's method are shown in Table 1, where the corresponding 2-D safety factors and their ratios are also shown. From Table 1 it can be seen that: (1) In Fellenius and Hovland's methods, F_3 can be lower than F_2, as suggested by Hovland (1977) and Hosobori (1987). However, the other three methods do not give such a result. Therefore, the 3-D

Table 1 3-D and 2-D safety factors and their ratios of a simple slope

Methods	F_3	F_2	F_3/F_2
Fellenius	1.662	1.721	0.966
Bishop	1.851 (η=0.0076)	1.833	1.010
Janbu	1.736 (η=0.0082)	1.711	1.015
Spencer	1.849 (η=0.0062, δ=19.1°)	1.832 (δ=19.1°)	1.009
Hovland	1.658	1.721	0.963

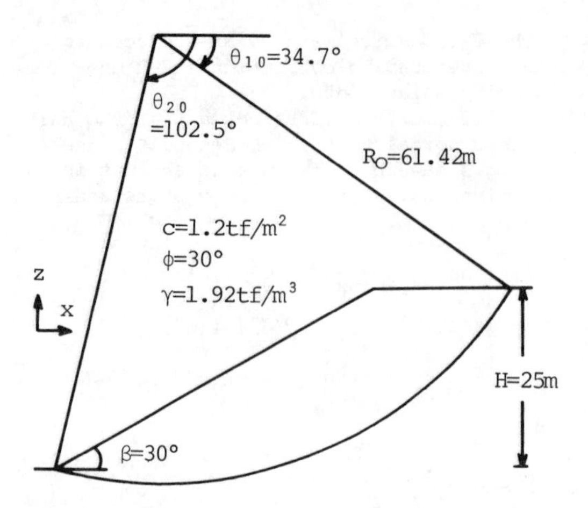

Fig.4 Cross-section of a simple slope

simplified Bishop, simplified Janbu and Spencer's methods are more reasonable and recommended; (2) The 3-D simplified Bishop and Spencer's methods have almost the same safety factors and give higher safety factors than the 3-D Janbu method. These results also hold in the 2-D case.

3.2 PARAMETRIC STUDIES ON SIMPLE SLOPES

Parametric studies on simple slopes were made, based on the 3-D Fellenius method. The sliding surface was assumed to consist of a cylinder of finite length with elliptical ends attached to it. The 3-D minimum stability factors were calculated as functions of the slope angle, the slope height and the failure length. From these calculations it was concluded that: (1) the 3-D effects become larger as the ratios of failure length to slope height become smaller; (2) the 3-D effects are large for cohesive slopes and small for sandy slopes; (3) F_3/F_2 can be smaller than unity for sandy slopes. To avoid this paradox in the 3-D Fellenius method, it is necessary to take into account the londitudinal stress in the slopes which acts perpendicular to the sliding direction. This stress gives the large effects on the 3-D stability of the slopes.

3.3 CASE STUDIES

A large landslide (the Ontake landslide) took place on the southern flank of Mount Ontake during the 1984 Nagano-seibu earthquake in Japan (Ishihara et al. (1987) and Sassa (1987)). Fig.5 shows the plan view

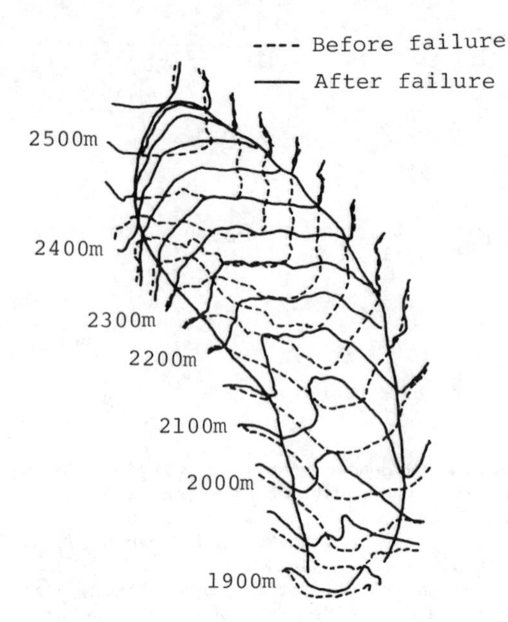

Fig.5 Plan view before and after the Ontake landslide

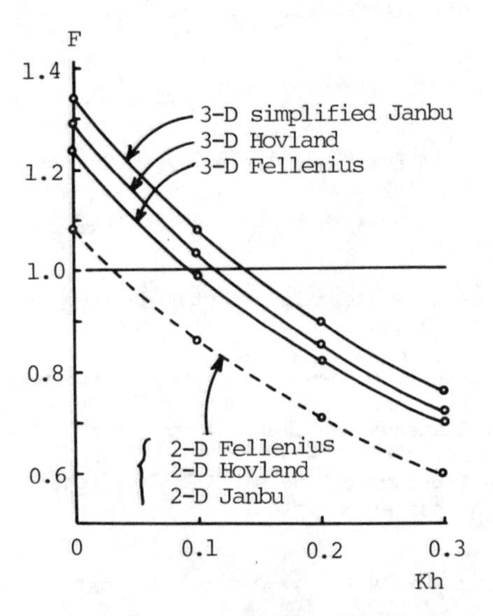

Fig.6 Relations between safety factor and seismic coefficient

before and after the landslide. Pseudo-static stability analyses were made for the landslide by means of the 3-D Fellenius, simplified Janbu and Hovland's methods. For the 3-D Fellenius method, the sliding surface was assumed to be a rotational one. The soil constants were determined as c=30 tf/m², φ=15°, γ=1.6tf/m³, referring to the

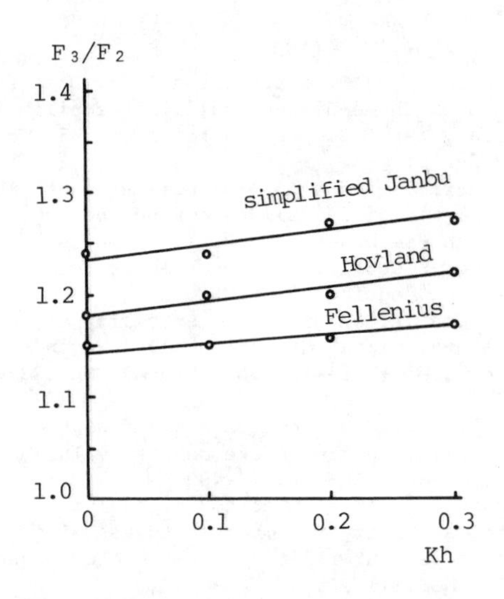

Fig.7 Relations between the ratio F_3/F_2 and seismic coefficient

experimental results by Ishihara et al. (1985). Fig.6 shows the relations between the 3-D safety factor and the seismic coefficient Kh. In this figure, the relations in the 2-D cases are also shown. Fig.7 shows the relations between the ratio F_3/F_2 and Kh. From Fig.6 and 7, it can be seen that the 3-D effects give 15%-25% higher values than the 2-D safety factor. Hence, if they are neglected, it can lead to an underestimation of back-figured Kh.

Three case studies including the Ontake landslide were carried out and it was found that the 3-D effects give 5%-30% higher values than the 2-D safety factor.

4 TEST RESULTS

Small-scale model tests were performed by inclining the 3-D simple slope models made of wet Toyoura sand, as shown in Fig.8. Fig.9 shows the shear test apparatus similar to that of the direct shear test, by which the strength parameters c and ϕ were determined. The shear test was done by inclining this apparatus until the sliding of its upper part occurred. Fig.10 shows one of the test results. The solid line is the observed sliding surface and the dotted line is that predicted by the 2-D Fellenius method. In this figure, the broken lines are those predicted by the 3-D Fellenius method, where κ is similar to the coefficient of earth pressure at rest and represents the magnitude of the londitudinal stress perpendicular to the sliding direction

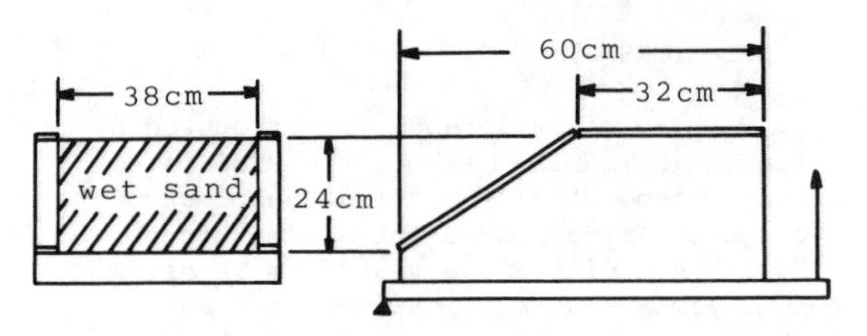

Fig.8 Slope model

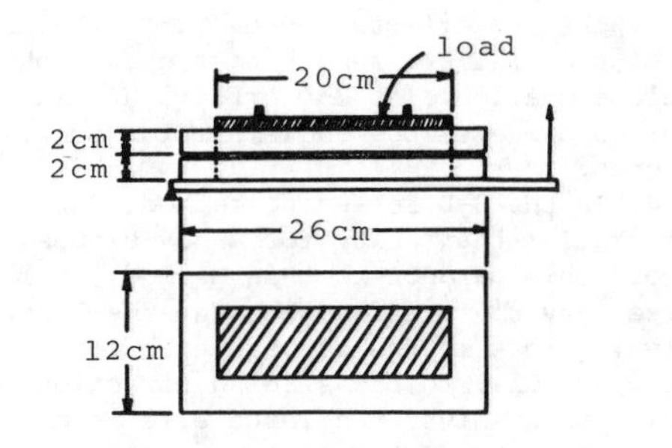

Fig.9 Shear test apparatus

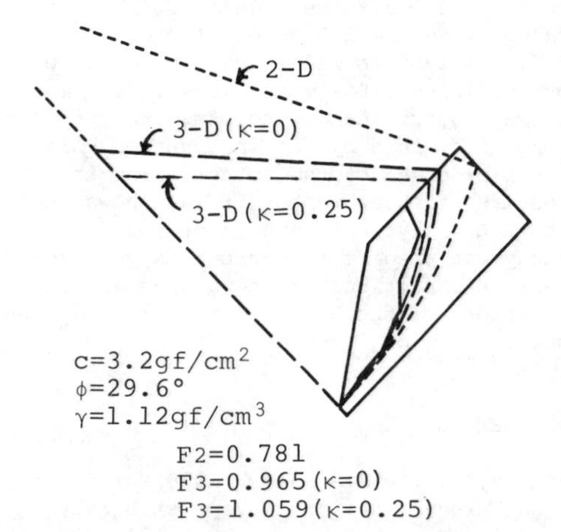

Fig.10 Comparison between observed and predicted sliding surfaces

stated in 3.2. Fom the results of these tests, it was shown that the 3-D analytical methods proposed in this study can predict more accurately the sliding conditions of the slopes and the positions of the 3-D sliding surfaces than the conventional 2-D methods.

5 CONCLUSIONS

1. New 3-D slice methods were proposed by extending Fellenius method, the simplified Bishop method, the simplified Janbu method and Spencer's method to three dimensions.

2. In Fellenius and Hovland's methods, F_3 can be lower than F_2. The remaining three methods do not give such a result. Therefore, the 3-D simplified Bishop, simplified Janbu and Spencer's methods are more reasonable and recommended.

3. The 3-D effects become larger as the ratios of failure length to slope height become smaller. The 3-D effects are large for cohesive slopes and small for sandy slopes.

4. In the 3-D Fellenius method, F_3/F_2 can be smaller than unity ror sandy slopes. To avoid this paradox in this method, it is necessary to take into account the londitudinal stress in the slopes which acts perpendicular to the sliding direction. This stress gives the large effects on the 3-D stability of the slopes.

5. Three case studies including the Ontake landslide were carried out and it was found that the 3-D effects give 5%-30% higher values than the 2-D safety factor.

6. Small-scale model tests were performed by inclining 3-D simple slope models made of wet Toyoura sand. The strength parameters were determined by the inclination shear tests proposed by the author. From the results of these tests, it was shown that the 3-D analytical methods proposed in this study can predict more accurately the failure conditions of the slopes and the positions of the 3-D failure surfaces than the conventional 2-D methods.

REFERENCES

Azzouz, A.S. & M.M.Baligh 1978. Discussion on Three-dimensional slope stability analysis method. Proc. ASCE 104, GT9:1206-1208.

Cavounidis, S. 1987. On the ratio of factors of safety in slope stability analysis. Geotechnique 37, No.2:207-210.

Chen, R.H. & J.L.Chameau 1983. Three-dimensional limit equilibrium analysis of slopes. Geotechnique 33, No.1:31-40.

Hosobori, K. 1987. Three-dimensional analysis of slopes. MSc thesis, Gunma University.

Hovland, H.J. 1977. Three-dimensional slope stability analysis method. Proc. ASCE 103, GT9:971-986.

Hungr, O. 1987. An extension of Bishop's simplified method of slope stability analysis to three dimensions. Geotechnique 37, No.1:113-117.

Hutchinson, J.N. & S.K.Sarma 1985. Discussion on Three-dimensional limit equilibrium analysis of slopes, Geotechnique 35, No.2: 215-216.

Ishihara, K. et al. 1985. Soil strength and analysis of landslides during the 1984 Nagano-seibu earthquake (In Japanese). Proc. Specialty Session held at the 20th Annual Meeting of JSSMFE.

Ishihara, K. et al. 1987. Stability of natural slopes during the 1984 Nagano-seibu earthquake. Proc. 8th ARCSMFE:237-240.

Sassa, K. 1987. The mechanism of high mobility in the Ontake debris avalanche. Proc. 8th ARCSMFE:487-490.

Spencer, E. 1967. A method of analysis of the stability of embankments assuming parallel interslice forces. Geotechnique 17, No.1:11-26.

Ugai, K. 1985. Three-dimensional stability analysis of vertical cohesive slopes. Soils and Foundations 25, No.3:41-48.

Ugai, K. et al. 1987. Three-dimensional stability analysis of cohesive slopes. Proc. 8th ARCSMFE:513-516.

Numerical Methods in Geomechanics (Innsbruck 1988), Swoboda (ed.)
© 1988 Balkema, Rotterdam. ISBN 90 6191 809 X

Finite element analysis of the stresses in slopes

Jue-guang Jiang & Qiang Xie
Southwest Jiaotong University, Sichuan, People's Republic of China

ABSTRACT: In this paper, the distribution and changes of the stresses in a slope was analysed with the method of the finite elements, when an artificial slope was excavated in a natural slope. The results obtained show that if natural slope angle is same, the vertical stress and the maximum shear stress increased obviously when the artificial slope angle increased in the slope foot, and a new tensile region formed and developed in the point of slope angle changed. The effects of results on the slope design were also discussed in this paper.

1 INTRODUCTION

In the China's railway construction, the method choosing a rock slope angle is mainly the method using experence data. In this method, theoretical basis is unsufficient. As one of research work in which the rock slope design in the China's railway construction will be standardization, it is necessary that researches the influences of the stresses distribution to a slope design and to a slope stability analysis when the slope is excavated.

Because of the complex geometric boundary in a slope, the analysis on the finite elements method is more simple and convenient. Some valuable results that involved such as the stresses distribution in slope, the influence of slope angle and the geometric shape of slope surface to the stresses, have be given by B. Hoyaux and B.Ladanyi(1), T. R. Stacey(2), and T. R. Stacey(3). But, when a natural slope will be excavated in part, the stresses distribution and changes was not still discussed. However, this condition is very common in the China's railway construction.

Basis on above consideration, the influences of the stresses distribution and changes when a natural slope will be excavated, will be analysed with the finite elements method by authors.

2 SLOPE MODEL USED IN CALCULATION

Fig.1 gives a slope model to calculate in this paper. Here, the top

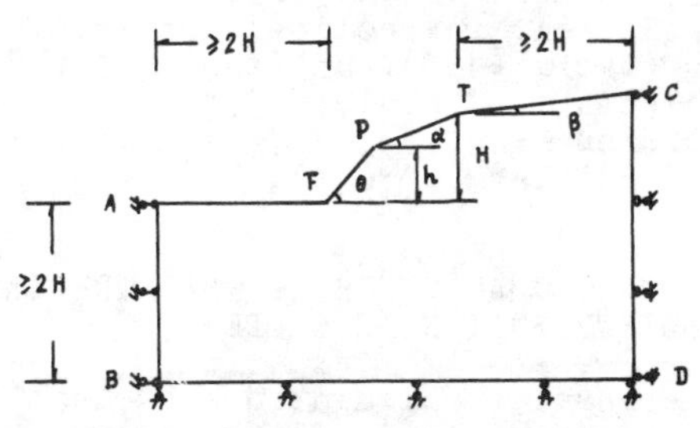

Fig.1 slope model usedin calculation

where: T--the slope top
P--the point of slope angle changed
F--the slope foot
H--the natural slope height
h--the artificial slope height
β--the top slope angle
α--the natural slope angle
θ--the artificial slope angle

face angle in the natural slope is β ($\beta=5^{0}$); the natural slope angle is α and the slope height is H(H=50^{m}); and the artificial (or excavated) slope angle is θ and slope height is h (h=30^{m}). According to Ref.1, the boundary size in the slope calculated is no less than 2H on both sides of the slope face and is no less than 3H in a total height.

On both vertical boundary, AB and CD sides, only vertical displacement allowed; and on horizontal boundary, BD side, only horizontal displacement allowed.

In the slope, the gravity is only applied.

Assumed that the material of the slope is a continuous, homogeneous and isotropic medium, and the mechanics model used for the finite elements calculation is linear-elastic, and linear-displacement. In calculating, the computer divided automatically the slope into about 300 trigonal elements in tatol. According to the test data in some sandstone, the parameters of the material of slope calculated is: the Young's modulus E=3 X 10^{4}MPa, the Poisson's ratio μ=0.25 and the unit weight γ=25kN/m^{3}.

Basis on above condition, the stresses distribution in slopes is calculated and the part results is discussed, when the natural slope angle α is respectively 35°, 55°, 75°, and 90°; as well as the artificial slope angle θ is respectively no excavation, 45°(tan θ =1:1), 53°(tan θ =1:0.75), 63°(tan θ =1:0.5), 68°(tan θ =1:0.4), 73°(tan θ =1:0.3), 78°(tan θ =1:0.2), and 84°(tan θ =1:0.1).

3 BASIC CHARACTERS OF THE STRESSES DISTRIBUTION IN SLOPE

The results calculated show that the basic characters of the stresses distribution in slope, natural or artificial, are as follows:

1. Vertical stresses In a natural slope which is no excavated, Fig.2a gives an example of the stresses distribution. It shows that there is a high stress region in the foot, and the incline which is parallel to the slope surface in the stress contours, has occured.

When the slope was excavated(Fig. 2b), the changes in the stress contours is that the stress in the foot has higher increased, and that the stress reduced in the point of slope angle changed.

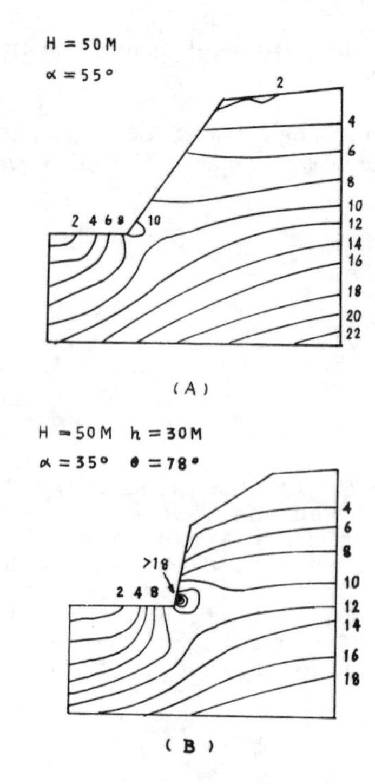

Fig.2 vertical stress(x10^{-1}MPa) contours in:
a. natural slope
b. artificial slope

2.Horizontal stress Fig.3a show the horizontal stresses distribution in a natural slope. In this figure, there is a high stress region in the foot and a tensile region in the top. The stress contours near to the slope surface is parallel to the surface.

When the natural slope was excavated(Fig.3b), the apparent changes on the stress distribution have appeared. A new tensile region occured in the point of slope angle changed, and on the artificial slope face. In addition, the tensile region also occured on the foundation for railway. And in the foot, the stress increased.

3. Maximum shear stress In Fig. 4a, the shear stresses distribution in the natural slope was given. It shows that the region in the foot is a region of the high shear stress. After excavated(Fig.4b), the changes of the shear stress is obviously increased in the foot and reduced

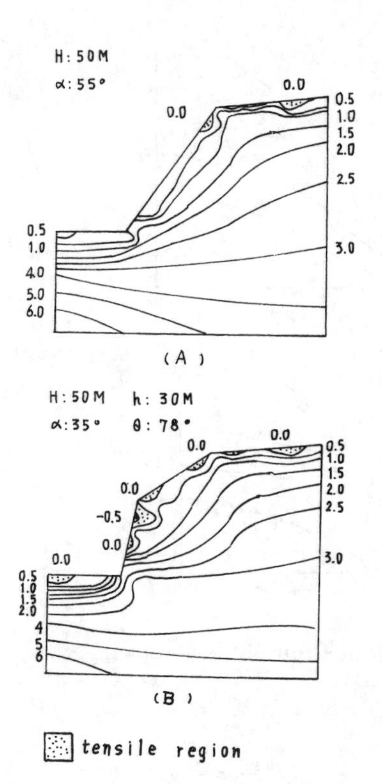

Fig.3 horizontal stress($\times 10^{-1}$MPa) contours in:
a. natural slope
b. artificial slope

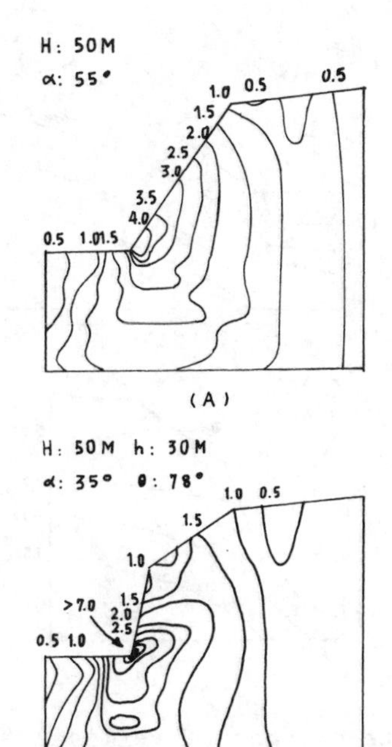

Fig.4 maximum shear stress ($\times 10^{-1}$ MPa) contours in
a. natural slope
b. artificial slope

in the point of slope angle changed.
Later, the meanings of above characters will be discussed.

4 THE CHANGES OF THE STRESSES DISTRIBUTION IN SLOPE WHEN SLOPE ANGLE CHANGED

In follows, the changes of the stresses distribution will be given when slope angle, natural or artificial, is changed.

4.1 Changes the artificial slope angle in a natural slope

When the natural slope angle is 35°, the stress conturs of four artificial slope angle, i.e., θ=35°, 43°, 63°, and 78°, is shown as Fig.5 to Fig.7. From these figures, we can obtain:

1. Vertical stress When the angle excavated increased, the stress obviously increased in the foot and on the face excavated. But in the point of slope angle changed, the stresses reduced.(Fig. 5)

2. Horizontal stress When the angle excavated increased, there are the apperent changes on the stresses distribution in slope. First, the tensile region in the top developed along the slope face. Second, in the point of slope angle changed, the compression stress turned into tensile stress, and the maximum tensile in the slope occured. Third, the tensile region also occured on the foundation for railway. And, the tensile increased in the foot.(Fig.6)

3. Maximum shear stress When the excavation deveioped, the shear stress rapidly increased in the foot only(Fig.7).

Above results can be summarized in Fig.8. The figure shows that the changes of stresses with the excavation slope angle increased in the top,

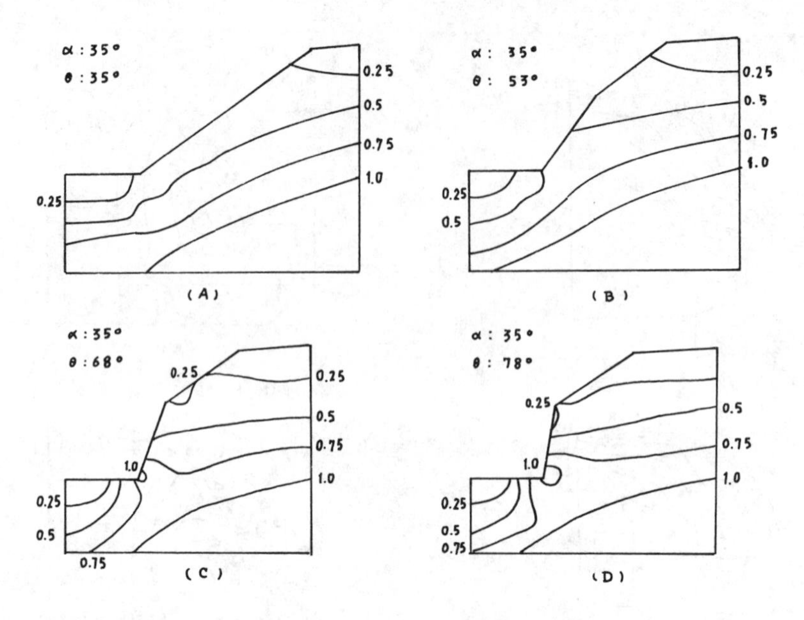

Fig.5 the changes of vertical stress($\sigma_v/\gamma H$) contours when the artificial slope angle(θ) increases

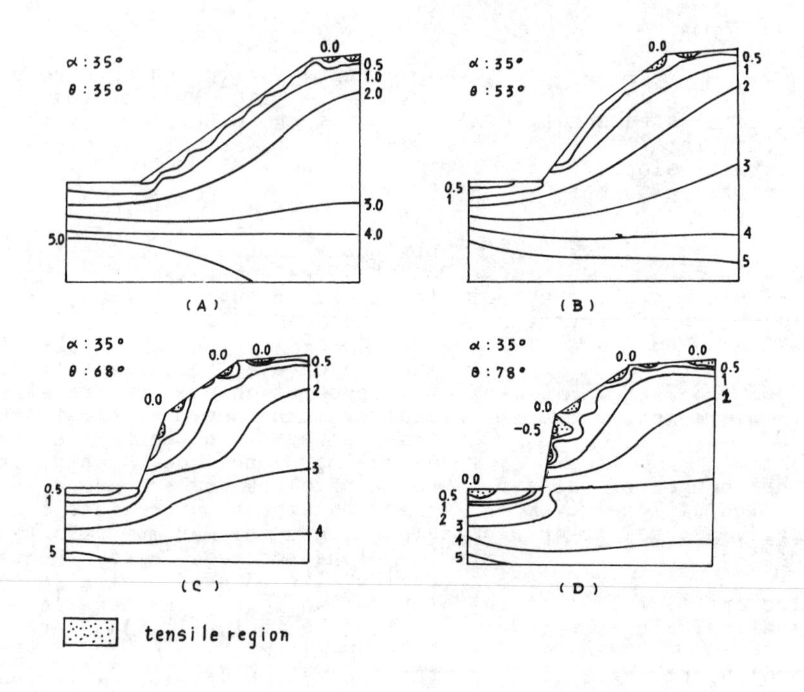

Fig.6 the changes of horizontal stress ($\times 10^{-1}$ MPa) contours when the artificial slope angle (θ) increases

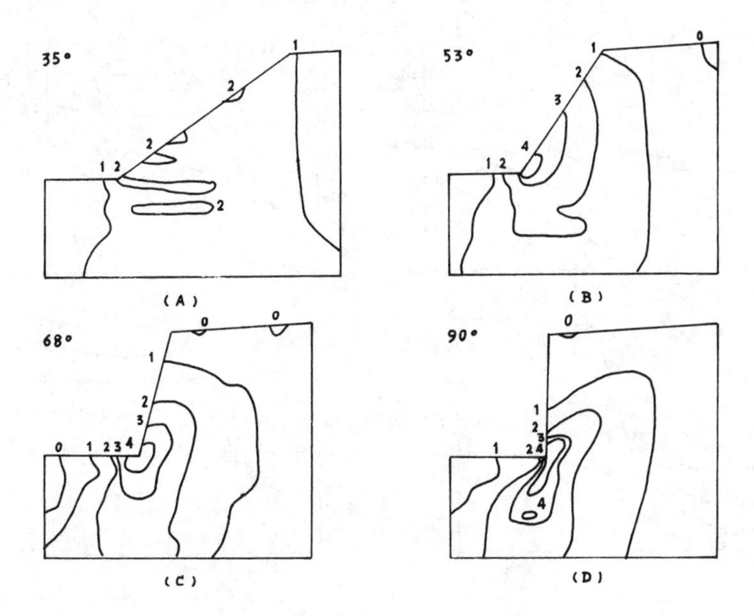

Fig.7 the changes of maximum shear stress($x10^{-1}$ MPa) contours when the artificial slope angle (θ) increases

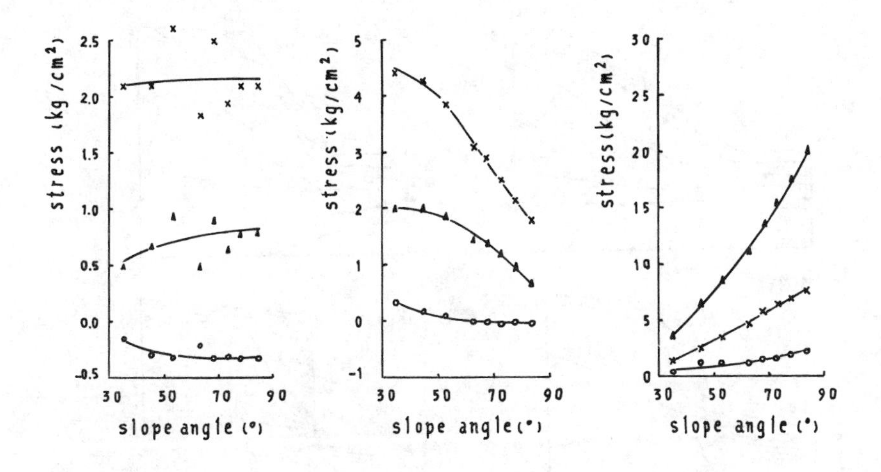

× vertical stress
o horizontal stress
▲ maximum shear stress

Fig.8 the relationships of the stress and the artificial slope angle in: a, top; b, point of slope angle changed and c, foot

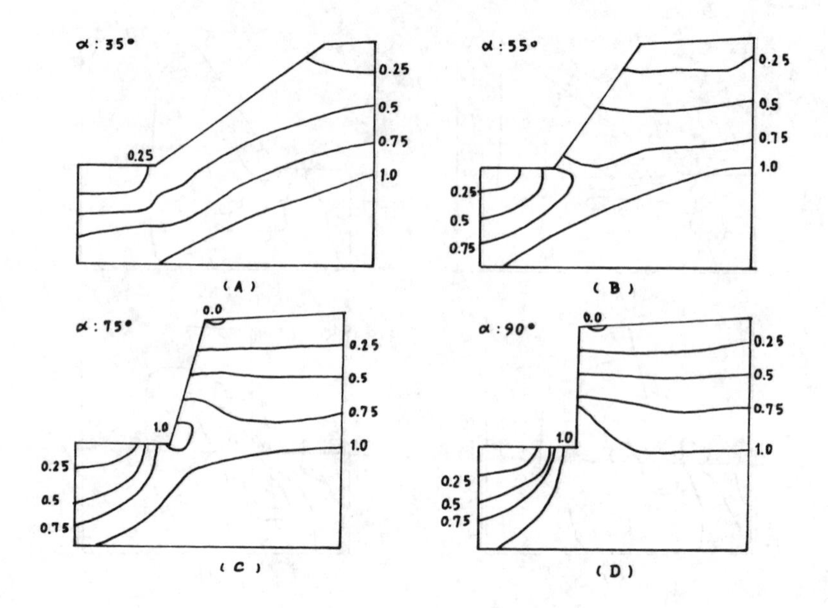

Fig.9 the changes of vertical stress ($\sigma_v/\gamma H$) contours when the natural slope angle (α) increases

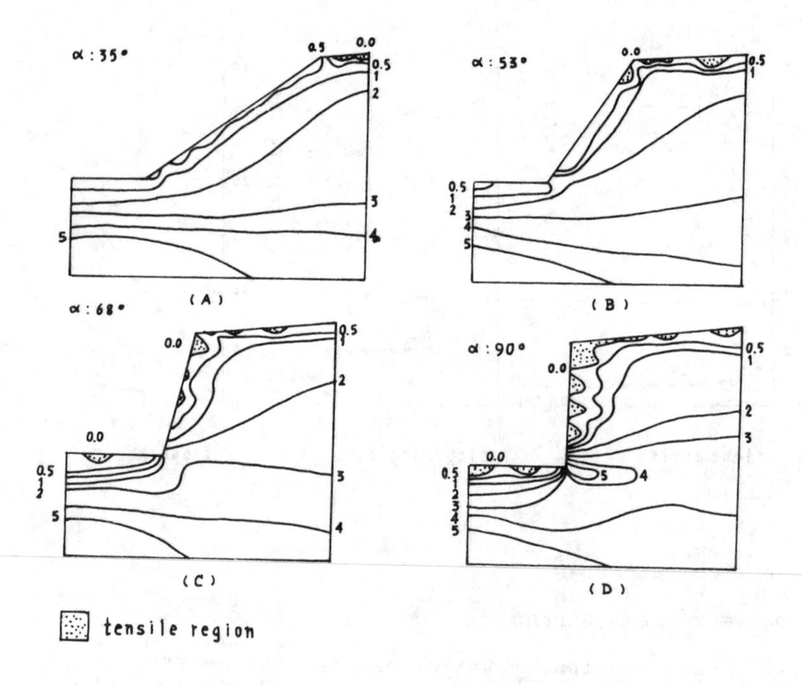

Fig.10 the changes of horizontal stress ($\times 10^{-1}$ MPa) contours when the natural slope angle (α) increases

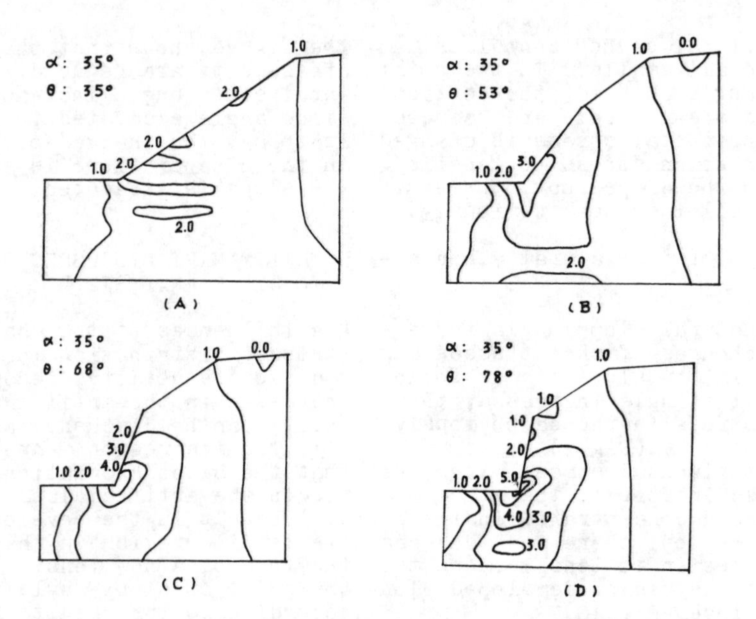

Fig.11 the changes of maximum shear stress ($x10^{-1}$ MPa) contours when the natural slope angle (α) increases

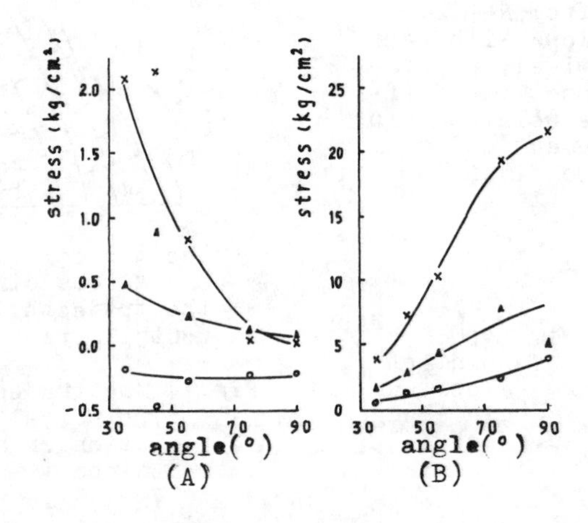

× vertical stress
o horizontal stress
▲ maximum shear stress

Fig.12 the relationships of the stress and the natural slope angle in: a. top, and b. foot either vertical or horizontal or shear, are all smaller. In the point of slope angle changed, the vertical stress and shear stress are reduced, and the horizontal stress is changed from compression stress to tensile stress. In the slope foot, three stresses are all increased obviously.

4.2 Changes of the natural slope angle

When the natural slope angle increased, the changes of the stresses distribution in the slope are as follows:

1. When the angle increased, the vertical stress increased obviously in the foot only.(Fig.9)
2. For horizontal stress, when the slope angle increased, the stress increased and concentrated in the foot. In the top, there is a changes from compression to tensile, and the tensile stress region developed along the slope face.(Fig.10)
3. The shear stress is obviously increased in the foot when the slope angle increased. But, the stress reduced appearly in the top.(Fig.11)

These results can be summarized in Fig.12. In addition, similar results can be also obtained from Ref.2.

If the artificial slope with same slope angle is respectively excavated in the natural slope with different slope angle, the relationship of the stress increasement and the natural slope angle are given in Fig.13.

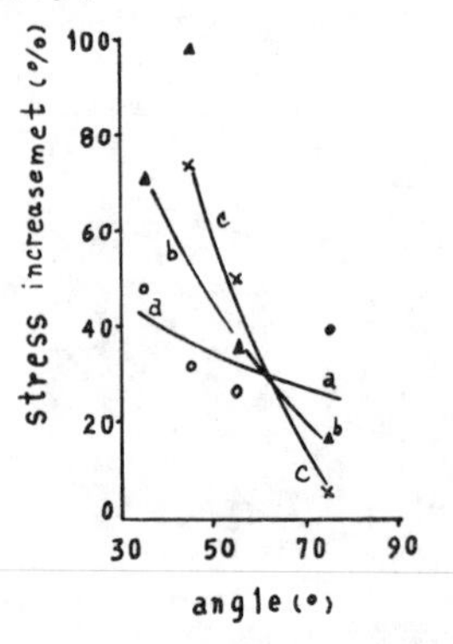

Fig.13 the relationship of the average stress increasement(i.e: $\Delta=((\Delta\sigma_V/\sigma_V+\Delta\sigma_H/\sigma_H+\Delta\tau_m/\tau_m)/3)$ and the natural slope angle

The figure shows that the stress increasements are reduced with the natural slope angle increased when the slope angle excavated is same. That is to say the changes of the stress in the steeper slope is smaller when the slope is excavated.

5 PRELIMINARY DISCUSSES TO APPLIED TO SLOPE ANALYSIS

The influences of the changes of the stresses distribution in a slope to the slope stability, because of the increased in the artificial slope angle, can be discussed as follows:

1. Tensile region Analysis shows that the slope excavation(i.e, increases in the artificial slope angle) will result in the development of the tensile region in the top, and the form of a new tensile region in the point of slope angle changed. According to the results from Fig.14

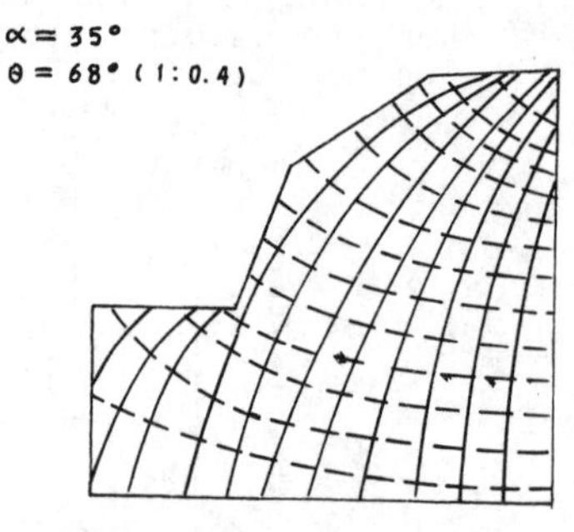

Fig.14 the trajectories of the principal stress direction in an artificial slope

and from Ref.4, the direction of tensile stress(i.e, the horizontal stress near slope face) is appoximately vertical to the slope face. When the tensile strength of the slope rock is reached, or when there were already the cracks in the slope, the cracks which are parallel to the slope surface would form or develop on the slope face excavated near the point of slope angle changed. Therefore, it is important that evaluates the possibility on slope failure along these cracks.

2. The stress concentration in the slope foot   Fig.8c shows that three stress all increased with the artificial slope angle increase in the foot. But the increase in the vertical stress and in the shear stress is more obvious. The slope foot would be a compression and shear failure region after excavated. Usually, the shear failure in a rockmass is a more important failure model.(5) Therefore, the shear failure in the foot could be carefully analysed in the slope design and in excavated.

3.The stress changes in the top, the point of slope angle changed and the foot   The relationships of the stress increasements and the angle of slope excavated show as Fig.15.

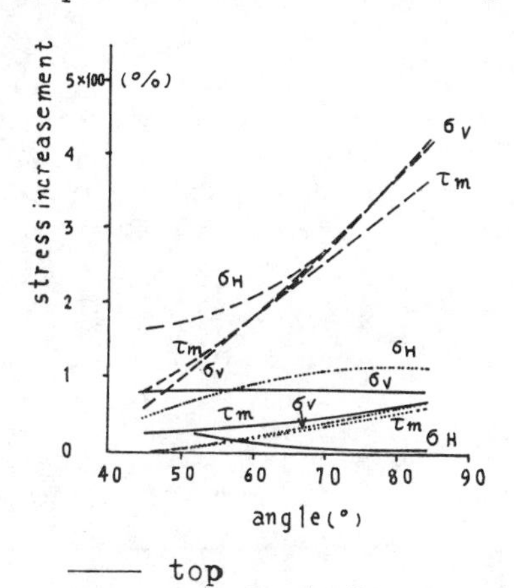

—— top
.... point of slope angle changed
--- foot

Fig.15 the relationships of the stress increasements and the artificial slope angle
where: σ_V--vertical stress
σ_H--horizontal stress
τ_m--maximum shear stress

It shows that the changes of three stress increasements is small in the top. In the point of slope angle changed, the increasement are appear increased. And in the foot, the increasements are rapidly arise. These facts confirm that in the foot and the point of slope angle changed, the influence of the stress change to a slope stability might be carefully considered in the slope design and the excavation.

6 CONCLUSIONS

1. In a natural slope, when the artificial slope angle increased, or when the slope angle change with excavated, the vertical stress and maximum shear stress will rapidly increase and will form a stress concentration region in the foot. And a new tensile region will occur and develop in the point of slope angle changed.

2. If the slope angle excavated is unchange, the influence of the excavation to the stress distribution will reduce with increase of the natural slope angle.

3. The results obtained show that the influence of the increase in the design slope angle, or the development of excavation, to three stresses distribution is small in the top, and is the greadest in the foot. Therefore, in slope design, for the point of slope angle changed, when the slope angle excavated is increased, the possibility of the tensile failure has to be carefully analysed. And for the slope foot, the shear failure will be research when the slope angle excavated is increased.

REFERENCES

1, B.Hoyaux & B.Ladanyi, Basic and Appied Rock Mech., Port City Press, Baltimore,U.S,621-631,1972
2, T.R.Stacey,Planning Open Pit Mines, J.G.INCE & SON(PTY.) LTD.,Johannesburg,South Africa,199-207,1970
3, T.R.Stacey,A three-dimensional consideration of stresses surrounding open-pit mine slope,Int. J.Rock Mech. & Min. Sci.,10,(1973)
4, Z.Y.Zhang et.al,Principles of analysis on engineering geology,Geology Press, Beijing,China,229-235, 1981
5, see Ref.4,68-69

Numerical Methods in Geomechanics (Innsbruck 1988), Swoboda (ed.)
© 1988 Balkema, Rotterdam. ISBN 90 6191 809 X

Comparison of the results of numerical and engineering stability analyses of the slopes of composite geological structures

S.B.Ukhov, V.V.Semenov, E.F.Gulko & M.G.Mnushkin
MECI Kuibyshev V.V., Moscow, USSR

ABSTRACT: The paper presents a geomechanic model and results of stability analysis of the slopes of composite geological structure. Stability analysis is accomplished by engineering method and by numeric calculations of stress-strain state in linear and non-linear formulation for slopes in natural state and for construction period: taking into account partial slope cutting for dam abutment and excavation of deep pit in the river bed.

To date engineers have to carry on construction of more and more main structures under complicated geological conditions. In doing so, they aften have to solve problems connected with rock masses stability in excavated slopes, in the edges of deep foundation pits, prisms of earth dams and other dams, in the banks of reservoirs. Reliability and efficiency of such solutions may be provided only with reliable engineering geological and geomechanical models of rock masses and tested calculation methods of geomechanics.

In many cases this task turned out to be rather complicated. Indeed, rock masses with complex rock formations of different composition and state must be schematized for calculations to relatively simple schemes. Even in this stage the engineering-geological models do not fit the real rock mass adequately. Experimental determination of mechanical properties characteristics of rock and semi-rock masses in particular and extention of the received data to the rock mass (geomechanical models construction) produce further rather uncertain errors. The calculation scheme, worked out this way, represents main characteristics of the real rock mass only to some extend.

Quite naturally, when working out a calculation scheme, engineer-geologist and geomechanic tries to take the best account of the factors, which are most dangerous for providing stability. However, the possibility to vary these factors and to evaluate the effect of their changes on the rock stability must be observed. Hence, it follows, that stability investigation of slopes and banks of composite geological structure must be of optimizing character. It becomes inevitable to do a large number of calculations with different parameters of calculation scheme.

Such a situation leads inevitably to an important contradiction: on the one hand, modern methods of numerical analysis in non-linear formulation allow to give a reliable forecast of the behaviour of the object under study, on the other hand, variant calculations may turn to be rather numerous and expensive. In this case simple and cheap in execution engineering calculation method may draw much attention. It would seem rather interesting to evaluate how the results, obtained by different methods of stability analysis match up, the object of investigation being the same. Lower the results of such comparison are given for the slope, the latter's scheme is shown in Fig. 1.

The basic part of the rock masses consists of marl blocks and clods, the interblock and interclod filler represented by rubble-loam material in frozen state. Underlying rocks

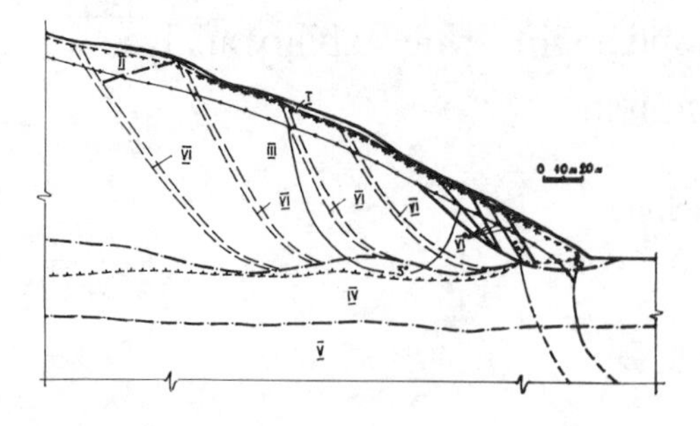

Figure 1. Engineering-geological model of the slope :—·—·-- lithologic and stratic boundaries; ┬┬┬- boundary of permafrozen rocks; —•—•— base of the bed of annual temperature fluctuations; ——3—— - isothermic lines; ⋛═ - slide-crush zones; ╲ - slide planes.

consist of crushed loam and separate blocks of preserved dolomites in melted state. Slope rock mass possesses ancient shift marks (slickensides and slide-friction zones). Using special engineering-geological and geomechanical investigations, an engineering-geological model was worked out, according to which the slope in question consists of six zones:

I - deluvial and solifluctional deposits in frozen and melted state: clod, rubble and landwaste grounds with sand loam filler;

II- gravitational (creep) formations in frozen and melted state, blocks, cut off from dolerite masses, dissected by cracks and sedimentation ditches, filled with clod rubble material;

III - marl blocks and clods, sizing from dozens meters to dozens centimeters with various orientation in space, bending of strata mainly from the river; the blocks and clods in question are of different preservation (about 60% are relatively preserved, about 30% - with wide cracks, filled with loam and partly with ice); interblock and interclod filler consists of loam-rubble material, rocks are salted (0.02% : 1.0%), frozen at the temperature of (-2° - -3°);

IV - heavily crumpled and crushed rocks, underlying zone III, represented by crushed loam, containing up to 30-40% of rubble from marls and aleurolites of low strength, and by separate blocks of preserved dolomites, rocks are heavily compacted: ρ = 1.8-1.9 g/cm^3; in the upper part - in frozen state, in the lower part - in melted state;

V - slightly deformed by the shift or preserved clay marls with occasional seams of dolomites, limestones and aleurolites in melted state;

VI - slide and friction zones (interblock ones) in zone III are filled with rubble-clod material with loam filler and ice; in the lower part of the slope they turn into slide planes VIa.

Investigation of above-mentioned rocks mechanical properties turned to be a very complicated task due to their heterogeneity and the effect of scale factor. Determination of mechanical properties effective characteristics was carried out by special laboratory and field experiments and by the method of mathematical modelling of the experiment (Ukhov et al. 1985). Mechanical behaviour of the rock masses and the slope on the whole was examined using conventional engineering models, i.e. deformability - using Hook's and Druccer-Prager's models, strength - by Coulomb-Mohr's equation. Design characteristics of mechanical properties, adopted as the basic variant are given in Table 1.

Slope stability analysis was carried out for natural profile and for construction period - taking into account partial slope cutting for dam abutment and excavation of deep pit in the river bed (Fig. 2). R.R.Chugaev's engineering method was used for stability analysis with fixed slide surfaces 1-9, corresponding to zone VI, VIa (Fig. 1). Calculations were also made by R.R.Chugaev's, K.Tertsagie's and A.V.Bishop's method. This method was applied for the search of round-cylindric slide surface, provided that the slope stability is the least.

Methods of K.Tertsagie and A.V.Bishop are well-known and do not need any commentary. R.R.Chugaev's method is recommended by normative literature for engineering practice and is applied for the analysis of heterogeneous rock masses with arbitrary slide surfaces contours (Designer's hand-book...1985). Calculations by Chugaev's method are based on the examination of the polygon of forces, acting on the element of

Table 1. Design characteristics of slope rocks

Zone NN	Rock state	E (MPa)	ν	φ(degreed)	C (kPa)	ρ (kN/m³)
I	frozen	75	0.3	14°	170	21.6
	melted	40	0.3	19°20'	5	22.0
II	frozen	80	0.3	19°20'	160	26.3
	melted	50	0.3	19°20'	5	26.3
III	frozen	150	0.35	16°	150	23.9
	melted	50	0.4	14°	40	23.9
IV	frozen	80	0.3	14°	30	22.6
	melted	60	0.3	14°	30	22.6
V	melting	180	0.25	24°	60	23.58
VI	frozen	-	-	16°	150	-
	melting	-	-	14°	40	-
VIa	frozen	-	-	13°30'	75	-

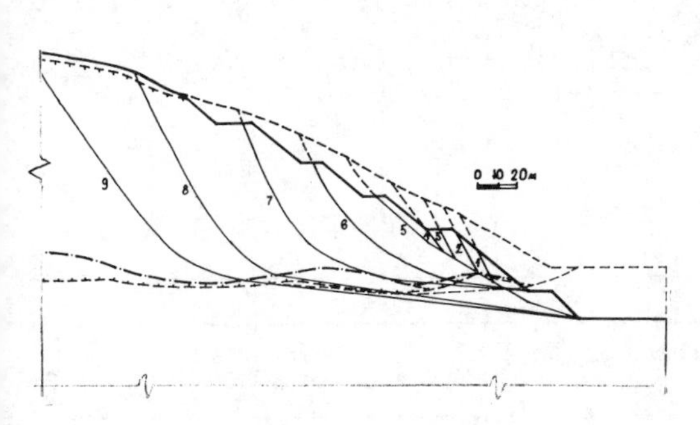

Figure 2. Scheme for slope stability calculations: 1-9 - analized slide surfaces, taken on the basis of geological research data (dashed line - location of the surfaces for the preconstruction period).

a potential slump body. It is assumed that in the limiting state the forces interacting with adjacent elements E_{li} and E_{ri} are directed at the angle of $\psi/2$ (where ψ is the angle of the shift) to the boundaries between the elements. In the absence of filtration and seismic effects application of equilibrium equation in projections on the coordinate axes and of additional condition $\Sigma \Delta E_{xi} = 0$, where ΔE_{xi} is the difference of forces $E_{ri} - E_{li}$, projected on the horizontal axis, leads to the factor of safety $F_1(K_{st}) - F_2(K_{st}) = 0$, where $K_{st} = \frac{tg\varphi}{tg\varphi'} \neq \frac{C}{C'}$. Here φ, C - are the true values of shear strength, φ', C' - values of the parameters, corresponding to the state of limiting equilibrium of the element.

Stress-strain analysis of the slope was carried out by the method of finite elements on the basis of elastic and elastic-plastic properties. In the latter case when deriving constraint equations between stress and strain, the flow surface was assumed to be associated with Coulomb-Mohr's strength law. Non-linear problem is solved in increments on the basis of initial stresses (Zenkevich 1975). The composed computer programme made it possible to describe slope heterogeneity: design area was approximated by 2386 elements.

As an example, the results of stress-strain analysis of the slope in construction period are shown in Fig. 3. Underworking of the slope led to a considerable change of stress field, particularly in the case of elastic-plastic solution. The calculation model (elastic and elastic-plastic solution) affects vertical stresses values σ_y to a lesser extend, to a greater extend it affects σ_x and τ_{xy} values. It should be noted, that elastic-plastic calculation being used, stabilization of stress fields was obtained for both variants, as the number of iterations increases,

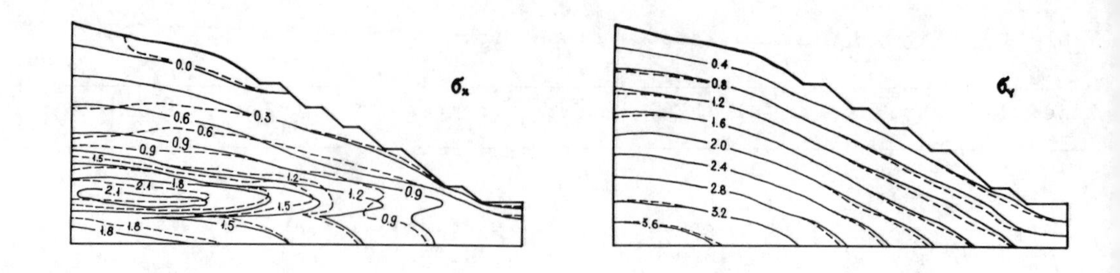

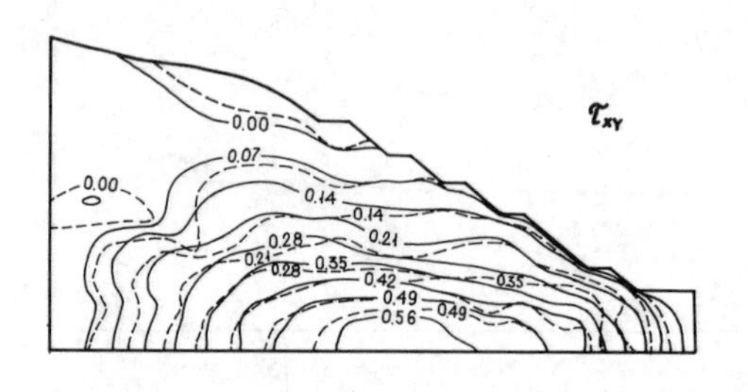

Figure 3. Results of numeric calculations of the slope stress-strain state in the construction period: a - isolines of horizontal stresses σ_x; b - isolines of vertical stresses σ_y; c - isolines of tangent stresses τ_{xy}; ——— - elastic-plastic solution; — — — - elastic solution

Table 2. K_{st} values for slide surfaces 1-9

Slide surfaces Nos	Calculations by R.R.Chugaev's method	Numeric calculations	
		Elastic solution	Elastic-plastic solution
1	1.616/2.259		
2	1.406/1.798		
3	1.291/1.392		
4	1.148/1.324		
5	1.089/1.394		
6	1.396/1.409	1.019/0.608	1.883/1.374
7	1.179/1.319	0.970/0.707	1.582/1.335
8	1.203/1.008	0.928/0.682	1.260/1.129
9	1.205/1.009	0.905/0.704	1.217/1.095

Note: in the numerator - slope variant in natural state: in the denominator-excavated slope (construction period).

which demonstrates general stability of the slope.

Stress field being known, slope stability was evaluated by the safety factor value $K_{st} = \frac{T_{str}}{T_{shear}}$, analogues to engineering methods, where T_{str}, T_{shear} represent correspondingly summed shear strength forces and shearing forces, acting along the surface. Table 2 gives these coefficients values for surfaces 1-9.

Stress-strain state being calculated, K_{st} values in elastic-plastic solution turned to be higher, than in plastic solution in all the cases. That fact corresponds to a natural process of slope deformation: in the areas approaching the limiting state strengths are redistributed to adjacent areas. The engineering method also demonstrated higher K_{st} values, than elastic solution did, but these values were lower than those, received by elastic-plastic solution.

The search for the most dangerous roud-cylindric slide surface (with $K_{st} \rightarrow min$) gave the following coordinates of the surface in question: the the method of R.R.Chugaev and K.Tertsagie: X = 130 m, Y = 180 m, R = 128.5 m (natural state), X = = 80 m, Y = 190 m, R = 158 m (construction period), by A.V. Bishop's method: X = 120 m, Y = 240 m, R = = 185.3 m (natural state), X = 70 m, Y = 260 m, R = 224.5 m (construction period). The location of these surfaces is similar to the surfaces 7 and 8, having the lowest K_{st} values (Table 2). K_{st} values for the most dangerous slide surfaces are given lower:

By R.R. Chugaev:	by K. Tertsagie:
$\frac{1.213}{1.053}$	$\frac{1.043}{0.917}$
By A.V. Bishop:	Elastic solution:
$\frac{1.119}{1.006}$	$\frac{0.917}{0.654}$
Elastic-plastic:	
$\frac{1.252}{1.110}$	

As before, in the numerator - for natural slope, in the denominator - for construction period.

Here elastics-plastic solution also demonstrates the highest K_{st} values, to which calculation results, obtained by R.R.Chugaev's and A.V.Bishop's methods are approaching.

It should be noted that according to precise test measurements, the slope is in the stable state, both in the natural state and after excavation. Consequently, within the limits of the accepted preconditions, all calculations, having demonstrated $K_{st} < 1$, do not represent actual slope behaviour.

Conclusions:

1. Elastic-plastic calculation of stress.strain state and safe sliding factor values, determined on its basis, give maximum K_{st} values. K_{st} values, determined by R.R. Chugaev's and A.V. Bishop's method, approach them to a considerable degree. K.Tertsagie's method and the elastic solution give excessive safety factor values, when slope stability is analysed.

2. R.R. Chugaev's and A.V. Bishop's engineering methods may be used for variant stability analysis of slopes. However, when K_{st} values, obtained by these methods, approach one, it is advisable to carry out control calculations on the basis of elastic-plastic properties.

REFERENCES

Ukhov, S.B., Chernyshov, S.N., Semenov, V.V. & E.F. Gulko 1985. The determination of effective characteristics of highly heterogeneous rock masses. Proceedings of the International Symposium "The role of mechanics in civil engineering and mining", Mexico, September, 1985.

Designer's hand-book. Foundations, bases and subsurface structures. 1985. Moscow, Stroiizdat, p. 340.

Zenkevich, O. 1975. Finite elements method in engineering. Moscow, Mir, 540 p.

Numerical Methods in Geomechanics (Innsbruck 1988), Swoboda (ed.)
© 1988 Balkema, Rotterdam. ISBN 90 6191 809 X

The stability of slopes in Champlain Sea Clays – A parametric study

Gunther E.Bauer
Ottawa-Carleton Institute for Civil Engineering, Carleton University, Ottawa, Canada

ABSTRACT: The Ottawa and St. Lawrence River Valleys have extensive deposits of post-glacial, sensitive clay, named Champlain Sea clay. These sensitive soils exhibit unusual geotechnical properties and are very susceptible to landslides. This paper reports on the results of an extensive parametric slope stability analysis considering several geometries, varying ground water conditions and non-linear strength envelopes. The results are summarized in graphical form.

1 INTRODUCTION

Slope failures in the Champlain Sea clay of Eastern Canada present a major hazard to human life and to property. This is particular true for the more densely populated areas of the Ottawa lowlands. There are many cases cited in the literature where loss of life and loss of property have occurred due to landslides in this type of soil (e.g. 1,2,3,4,5,6, etc.).

Past studies in this area have tried to deal with the slope stability problem in two ways: firstly, in a general way by publishing charts relating the height of a natural slope to the calculated factor of safety, and secondly, by back analyzing in detail failures using the best possible geometric and hydro-geological conditions assumed to have prevailed during the slope failures. The first approach was taken by Klugman and Chung [6] in an effort to allow a more efficient planning of land use and development of the area in the vicinity of Ottawa and it can be considered as an attempt to classify the failure potential of these slopes. The latter approach yielded detailed information for a particularly troubled site, but also showed that the analysis was particularly sensitive to the choice of shear strength parameter and the hydraulic conditions referred to during the failure of the slope [1,2,3,5,13 and 18].

This paper reports on the results of a parametric study where three slope geometries were chosen as representative for the areas under investigation. Different hydraulic groundwater conditions were investigated together with soil parameters from three actual sites where sliding had occurred during the past few years. This led to a total of 135 cases to be investigated. To accomplish this a special computer program had to be developed, based upon Bishop's modified analysis. Instead of the commonly used constant strength parameters ϕ' and c', variable effective strength and normal stress values τ' and σ_n' were employed, in order to describe more realistically the curved Mohr-Coulomb failure envelopes for this type of soil. The results of this study are summarized in the last three figures of the paper showing the effect of the groundwater conditions and slope geometry on the factor of safety.

2 GEOLOGY AND GEOTECHNICAL PROPERTIES

2.1 Geology

Near the end of the Wisconsin glaciation period, about 13,000 B.P. (B.P.=before present), the regression of the glaciers and the resulting meltwater, filled the depression of the Champlain Valley. Both the glaciation period and the Champlain Sea invasion are recorded in Eastern Canada by the characteristic topographies and sediments resulting from these events [1,7,8].

The inundation of the land was followed by estuarine and lacustrine periods due to the isostatic readjustment of the continential crust. During the uplight period, from approximately 13,000 to 8,000 B.P. [9,10], the land rebounded by more than 200 metres [11]. The clay materials which were deposited in the shallow sea environment are commonly referred to as "Leda clay" or "Champlain Sea clay" in the literature. This clay is characterized by its high sensitivity, it is termed a "quick clay", and its high natural water content and large void ratio. The engineering properties of the clay have been well investigated by several researchers [e.g. 1,12,13,14]. Most of the clay deposites are capped by a stiff highly weathered crust that possesses a higher shear strength than the underlying intact material. The exact transition from the oxidised brown clay to the softer grey clay varies, depending on the degree of erosion. The weathered crust is further characterized by a system of closely spaced fissures. From field observations it is known that in most cases the depth of initial slope failure occurs below the zone of active weathering [15,16,17].

The three most important factors which lead to slope failures in this soil, either singular or combined, are the porewater pressure conditions, surcharge loading and removal of toe support by water erosion.

The geotechnical properties of three sites were used in this study, namely Bearbrook, Castor River and South Nation River. All three sites are within 50 km of Ottawa and have a history of active sliding [1,2,3,4,5,6,18,19].

2.2 Soil Properties

As in any geotechnical engineering analysis, several soil parameters are needed in the investigation for the stability of slopes. The most important input in a stability analysis are the strength parameters and the porewater pressures which prevail in the field as well as the basic soil properties, such as density, natural water content and other index properties. The laboratory determination of shear strength parameters, from consolidated-drained and undrained triaxial compression tests, indicated that the effective strength for these clays cannot be represented by the classical Mohr-Coulomb strength envelope. For the three sites investigated this envelope is quite curved in the lower and medium normal stress ranges and it becomes linear only when the normal stress exceeds the preconsolidation pressure of the clay. Figure 1 shows a diagrammatic representation of the strength envelope of the three sites studied. It is obvious from this figure that the strength developed depends on the effective normal stress acting at the failure surface of the slope. Henceforth, the effective strength parameters c' and ϕ' are not constant for a particular soil but depend on the stress state of the soil.

In the analysis of natural slopes in Champlain Sea clays, the high normal stress range is of little practical importance since the mobilized strength is governed by the effective normal stresses which lie predominantly in the low stress ranges.

The actual effective strength envelopes for the Bearbrook, Castor and South Nation River sites were used in this study, as determined from triaxial soil specimens which were trimmed from undisturbed block samples. For all three sites the effective cohesion of the clays was in the order of 6 to 9 kPa. Therefore, in the computer analysis the actual effective shear stress and normal stress values, τ' and σ_n', were used. The strength envelopes, based on effective stresses were approximated by a nonlinear regression analysis. Where more than five data points, i.e. Mohr stress circles, were available, a fourth order polynomial of the following form was fitted:

$$S = A + B\sigma' + C\sigma'^2 + D\sigma'^3 + E\sigma'^4$$

where S = strength ordinate
σ' = effective normal stress ordinate, and A,B,C,D and E are regression coefficients

The coefficient of determination, R^2, is given as follows:

$$R^2 = \sum_{i=1}^{n} (z_i-\bar{S})^2 / \sum_{i=1}^{n} (S_i=\bar{S})^2$$

where z_i = approximate value for S_i
$\bar{S}$ = average S_i, and
S_i = the i-th dependent variable value

The coefficient of determination for all three strength envelopes was better than 0.9 indicating a good correlation between the experimental data and the derived polynomial.

2.3 Groundwater Conditions

In conventional slope stability analysis, where flow of water occurs under gravity flow and steady seepage conditions, porewater pressures within the soil mass can be evaluated from a flownet. These porewater pressures are then calculated either from the height of water column (piezometric surface) above a given point or the porewater pressure ratio. This ratio is defined as the porewater pressure divided by the total vertical stress at the same point of interest. Most computer programs for an effective stress analysis will handle steady seepage conditions due to gravity flow once the phreatic surface is given. But field investigations at the three sites have shown that several other flow conditions are possible [6,10,18,19]. Figure 2 shows a typical flow pattern which were observed in a natural embankment in the Ottawa area [13,20]. This figure illustrates an upward gradient at the toe of the slope and this upward flow is caused by a previous aquifer under an artesian pressure head underlying the clay deposit. In other regions a predominantly downward gradient was observed [18]. Therefore in this parametric study, on hand, three groundwater conditions were investigated. These flow regimes are graphically represented in Figures 3(a), (b) and (c). Figure 3(a) shows the flow pattern caused by gravity and the phreatic line becomes asymptotic to the upper and lower horizontal ground surfaces. In Figure 3(b) the flow is downward into a previous stratum underlying the clay slope representing infiltration from a slow snow melt or continuous rainfall. In Figure 3(c) an upward gradient causes the water to permeate through the clay. For this study only isotropic flow conditions were considered even though the computer analysis can handle layered soil deposits with anisotropic permeabilities.

2.4 Slope Geometry

Predicting the stability of a natural embankment is essentially an evaluation of the induced shear stresses and resisting shear strength. The mobilized shear stresses acting on a probable failure plane of a cohesive soil are governed by the height and steepness of the embankment. Observed failures in Champlain Sea clay occurred at slope inclinations between 15 and 30 degrees and heights varying between 8 and 24 metres [2,3,4,5,6,13]. In order to consider most practical cases and on the other hand to keep the computing effort in realistic proportions, the following slope heights and slope inclinations were analysed. Five heights, 5, 10, 15, 20 and 25 metres and three inclinations were considered. The three slope ratios were 3 horizontal to one vertical (i.e. $i=18.4^{\circ}$), 2.5 to 1 (i.e. $i=22^{\circ}$) and 1.75 to 1 (i.e. $i=30^{\circ}$).

In all cases the depth of soil below the level of the toe was taken as one third of the free slope height. This depth was sufficient in order that the lower boundary did not interfere with any base failures. Most critical circles cut the slope at the toe or just below the toe. In any case, the maximum depth below toe elevation did not exceed 1.2 times the slope height and none of these deep failure planes yielded a minimum factor of safety.

3 COMPUTER PROGRAMS

3.1 Slope Stability

The program which was used in this analysis was based on different existing programs, such as SB-slope, LEASE, ICES, SLOPE and E-Slope. The program uses the simplified Bishop method of slope stability analysis. It was written for IBM-PC's or compatibles and uses state-of-the-art computer graphics to facilitate geometric data checking, allows on-screen selection of failure surfaces and will print a hard copy graphics of the slope. The program is highly interactive and can handle the most situations facing a geotechnical engineer. The analysis options include : phreatic line or r_u-method for porewater pressures, free water surface, horizontal and vertical earthquake coefficients for a pseudo-static analysis, distributed surface loads and tension cracks. Up to 20 different sets of soil parameters can be handled. Data is entered following display prompts and may be corrected at any time. After all the data entry is complete, the data is displayed in an informative table for review and a menu of correction choices appears. The checked or corrected data is written to a data file and any analysis run can be easily recalled for further corrections, changes, or additional runs. There are four options for failure surface selection, as follows: (1) grid search for a specified circle centres and radii, (2) single circle analysis for specified centre and radius, (3) automatic search for an initial centre and radius and subsequent increment intervals, and (4) the interactive graphics may be used to specify failure surface by digitizing points. In this mode any area of the slope may be enlarged for better detail.

An extensive menu of output options allows the user a variety of output features, such as factor of safety contours from a grid search or automatic search, print graphic copy of critical circle and factor of safety, tabulation of factors of safety for all circles considered, printing only the overall minimum factor of safety, printing of the diagnostic slice data for the minimum factor of safety etc.

In order to make full use of the graphic possibilities, the personal computer must have, of course, an IBM graphics card or equivalent and a compatible dot matrix printer.

A total of 135 cases were analysed comprising of three sites (Castor, Bearbrook and South Nation River), three groundwater conditions, three slope inclinations and five slope heights. Generally 8 to 12 slip circles were needed before the overall minimum factor of safety could be computed. As mentioned earlier, the non-linear effective shear strength envelope for each site was used.

3.2 Seepage Program

The slope analysis program is able of handling several hydraulic conditions as mentioned in the previous section. For a strong downward or upward gradients as shown in Figures 3(a) and (b), respectively, the code had difficulty in determining the r_u values. In order to overcome this problem a finite element seepage program was used for these two flow conditions. The F.E. code calculated the hydraulic gradients at the required nodal points or elements. The program was originally developed by Desai in 1978 and a brief summary of this method was given by Desai and Christian [21]. The program used in this investigation was a modified version as given by Felio and Bauer [22] and is able to solve the following cases: steady seepage (confined and unconfined); anisotropic non-homogeneous soil systems and transient seepage (rapid drawdown and repeated hydrograph). The program uses 4-node elements and simple Euler integration.

In this study isotropic flow and homogeneous soil properties were assumed. The r_u-contours obtained from the seepage code were superimposed onto the geometry of slope stability analysis. Presently work is underway to fully integrate the two programs, i.e. slope stability and seepage analyses. The F.E. seepage program comes in two parts: (1) SEEPDATA.EXE: the interactive data pre-processor allows the user to input the parameters to run the main code. Corrections are made on the screen and the program saves the data in a format compatible with the main code. (2) SEEPAGE.EXE: this main program runs in a batch mode. The data can either be used from the pre-processors or by any text editor. The output from this investigation was transferred to a 1-2-3 Lotus spread sheet which was then used for plotting the results in the desired form. A typical flownet produced by the F.E. analysis is shown in Figure 4 for an upward gradient from a confined aquifier. Similar meshes were obtained for the other cases and boundary conditions.

4 RESULTS

As mentioned earlier there are five output options a user could choose from the stability program. In this study, the option to print the slice diagnostic for the lowest factor of safety of the critical circle was considered most appropriate. This information could be sent to both the screen for a visual check and to a disk file. The data was filed into the Lotus 1-2-3 program for storage and printing routine. Of course, the critical circle, or any circle, could be plotted directly on the plotter using the graphics printout option.

The final results are presented in graphical form relating the factor of safety to the slope height for a given slope inclination, porewater conditions and shear strength envelope. This representation was preferred over the conventional stability charts plotting a stability number against slope inclination since both shear strength parameters vary with the normal effective stress.

Figures 5, 6 and 7 summarize the results of this investigation as stability charts. An upward gradient with an artesian pressure head equal to the elevation of the crest of the slope (i.e. Figure 2) will yield the most critical stability conditions. Let us consider the case of a slope inclination ratio of 3 to 1 (i.e. i=30°), a desired factor of safety of 1.2 and an upward seepage pressure in the underlying previous aquifer corresponding to a piezometric level at the crest of the slope. From Figure 5 these conditions would yield slope heights of 13.5, 17.0 and 22.0 metres for the Castor River, South Nation and Bearbrook regions, respectively. Taken the same geometric conditions and soil properties but assuming a downward flow would yield a factor of safety of 1.8 for the three sites.

Figures 6 and 7 are similar graphs as presented in Figure 5 except for slope ratios 2.5 to 1 and 1.75 to 1, respectively.

The last three figures may therefore be useful to allow a more efficient planning and development of landuse for the three areas considered. The graphs also should be helpful in classifying the natural slopes in these areas for future failure potential.

Some of the results given in Figures 5, 6 and 7 were checked against those given by conventional stability charts proposed by Morgensterne and Price [23] and Cousins [24] and the agreement was quite good. Of course, one had to assume average soil strength parameters and porewater conditions in order to make a comparison with these charts.

The preceding analysis of slope stability did not consider the presence of tension cracks which usually occur as an early and important signal of impending failure. As mentioned before, the computer program can handle tension cracks, both dry and filled with water. Some preliminary runs were carried out with water filled cracks and the results indicated that the factor of safety may be reduced perhaps by as much as 20 percent.

5 CONCLUDING REMARKS

This analysis considered three unstable sites in the vicinity of Ottawa, Canada. The actual non-linear strength behaviour were used in this investigation. The slopes were subjected to three hydraulic flow conditions, all of which are possible and had been observed in the field, and the subsequent results of the analysis were given in graphical form relating the factor of safety to the slope height of a given steepness. These graphs or charts are considered useful in the evaluation of the stability for the three regions or any other regions with similar soil properties. The slopes were considered to be homogeneous but the soil strength properties vary depending on the effective normal stress acting on the probable failure planes. The author is quite aware of the fact that the soil in the field is desiccated to a certain depth and that this crust has strength properties similar to a granular, blocky material. But actual failures have shown that most of the sliding surfaces occur in the underlying softer intact clay. The influence of an overlying crust material on he stability of these slopes is the subject of a future investigation.

6 ACKNOWLEDGEMENTS

The author would like to acknowledge the contribution made by Mr. G.P. Deschamps who was involved in a previous study [3] and who has furnished the laboratory data used in this paper. Financial assistance under the author's NSERC Operating Grant is also acknowledged and was greatly appreciated.

7 REFERENCES

1. C.B. Crawford, Quick Clays of Eastern Canada, Engineering Geology, Vol.2, No.4, pp.239-265, (1968).

2. C.B. Crawford, and W.J. Eden, Stability of Natural Slopes in Sensitive Clay, JSMFD, ASCE, Vol.93, No.SM4, pp.419-436, (1967).

3. G.P. Deschamps, The Effect of Shear Strength and Pore Pressure Distribution on the Stability of Slopes, M.A.Sc. Thesis, University of Ottawa, Ottawa, (1979).

4. W.J. Eden, and P.M. Jarret, Landslide at Orleans, Ont. Tech. Paper No.321, NRC, DBR, Ottawa, (1971).

5. W.J. Eden, and Mitchell, Landslides in Sensitive Clays, Eastern Canada, Highway Research Rec. No.463, pp.18-27, (1978).

6. M.A. Klugman, and P. Chung, Slope Stability Study of the Regional Municipality of Ottawa-Carleton, Ontario, Canada, Ontario Geological Survey, Miscellaneous Report MP68, Ministry of Natural Resources, Ontario, (1976).

7. N.R. Gadd, Surfacial Geology of Ottawa Map Area, Ontario and Quebec, Geol. Surv. Can., Paper 62 (16), 4 pp. (1962).

8. J.E. Gillott, Mineralogy of Leda Clay, Can. Mineralogist , Vol.10, pp.797-811, (1971).

9. P.F. Karrow, The Champlain Sea and Its Sediments, in Soils in Canada, R.F. Legget, editor, Roy. Soc. Ca., Spec. Pub.3, pp.97-108, (1961).

10. S.H. Richard, Surfacial Geology Mapping: Ottawa-Hull Area (Parts of 31 F, F); in Report of Activities Part A, Geol. Surv. Can. Paper 75-1 A, pp.218-219, (1975).

11. T.C. Kenney, Sea Level Movements and the Geologic Histories of the Post-Glacial Marine Soils at Boston, Nicout, Ottawa and Oslo, Geotechnique, Vol.14, pp.203-230, (1964).

12. D.A. Sangrey, On the Causes of Natural Cementation in Sensitive Soils, Can. Geotech. J. Vol.9, pp.117-119, (1972).

13. G.E. Bauer, J.D. Scott and G.P. Deschamps, The Effect of Shear Strength and Pore Pressure Distribution on the Stability of Natural Slopes, Proc. Int. Symp. on Landslides, New Delhi, India, Vol.4, pp.297-302, (1980).

14. G.E. Bauer and S.K. Mital, Monitoring and Analysis of Pore Pressures During the Screw Plate Test, Symposium on Computer Aided Design and Monitoring in Geotechnical Engineering. AIT, Bangkok, Thailand, pp.150-166, (1986).

15. G.E. Bauer, J.D. Scott and D.H. Shields, The Deformation Properties of a Clay Crust, Proc. 8th Int. Conf. on SMFE., Moscow, U.S.S.R., Vol.1, pp.31-38, (1974).

16. G.E. Bauer and A. Tanaka, The Response and Analysis of Tower Silo Resting on a Fissured Clay Crust, 2nd Int. Conf. on Case Histories in Geotechnical Engineering, St. Louis, Mo., U.S.A., (1988) (Paper accepted).

17. G.E. Bauer, Analysis of a Failed Concrete Silo, 2nd Int. Conf. on Case Histories in Geotechnical Engineering, St. Louis, Mo., U.S.A., (1988) (Paper accepted).

18. J. Lafleur, Influence de l'eau sur la stabilite des pentes naturelles d'argile, These de doctorat, genie civil, Universite de Sherbrooke, (1978).

19. P. Larochelle, J.Y. Chagnon, and G. Lefebvre, Regional Geology and Landslides in the Marine Clay Deposits of Eastern Canada, Can. Geot. Jour. Vol.7, pp.145-157, (1970).

20. G.P. Deschamps, The Effect of Shear Strength and Pore Pressure Distribution on the Stability of Slopes, M.A.Sc. Thesis, University of Ottawa, Ottawa, (1979).

21. C.S. Desai, Flow Through Porous Media, Chapter 14 in C.S. Desai and J.T. Christian (eds.). Numerical Methods in Geotechnical Engineering, McGraw-Hill Book Co., New York, (1977).

22. G.Y. Felio and G.E. Bauer, Manuals: The Application of Microcomputers in Solving Geotechnical Engineering Problems, Short course, Carleton University, Ottawa, June 8-12, (1987).

23. N. Morgenstern, and V.E. Price, The Analysis of the Stability of General Slip Surfaces, Geotechnique, Vol.15, (1965).

24. B.F. Cousins, Stability Charts for Simple Earth Slopes, ASCE Journal, Geotechnical Engineering Division, Vol.14, February (1978).

SHEAR STRENGTH τ'
BEARBROOK
SOUTH NATION
CASTOR RIVER
φ'
LOW
INTERMEDIATE
HIGH
EFFECTIVE NORMAL STRESS σ'

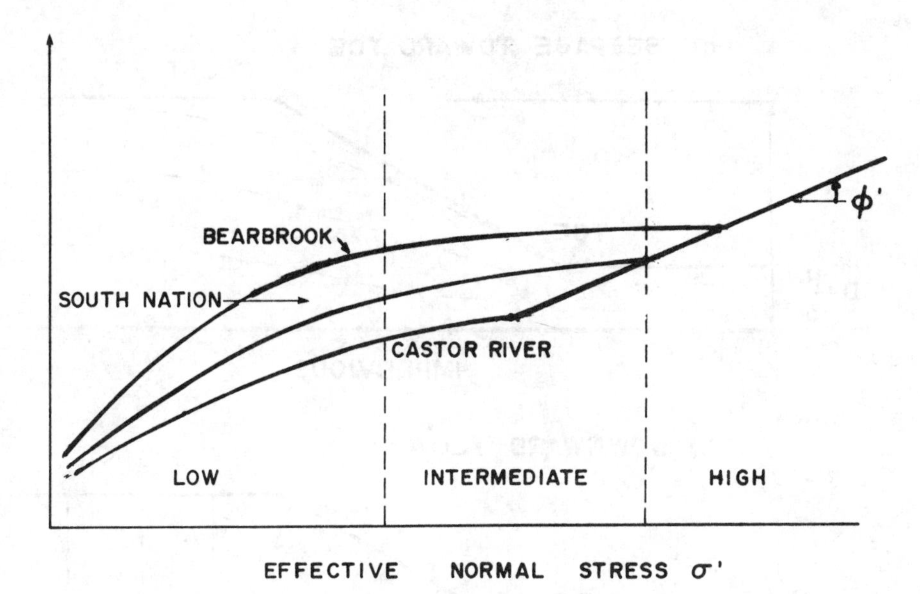

FIG. 1. MOHR-COULOMB FAILURE ENVELOPE
FOR CHAMPLAIN SEA CLAY

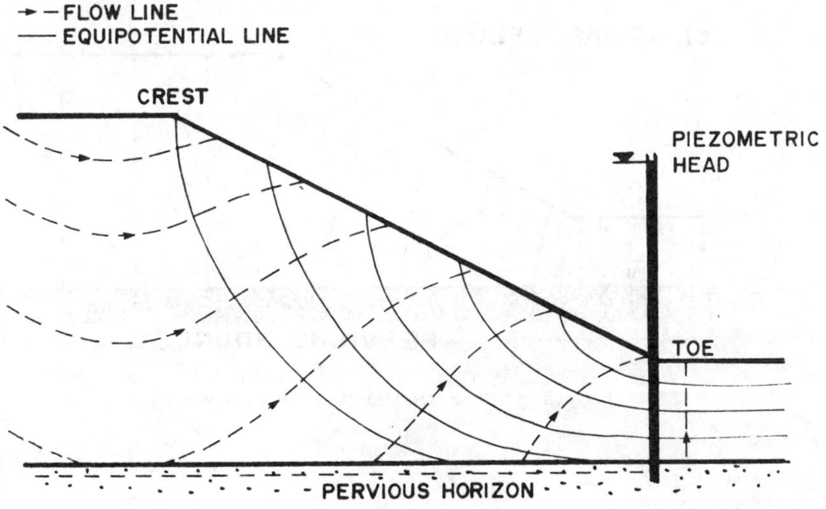

FLOW LINE
EQUIPOTENTIAL LINE
CREST
PIEZOMETRIC
HEAD
TOE
PERVIOUS HORIZON

FIG. 2. TYPICAL FLOW REGIME
IN THE OTTAWA AREA

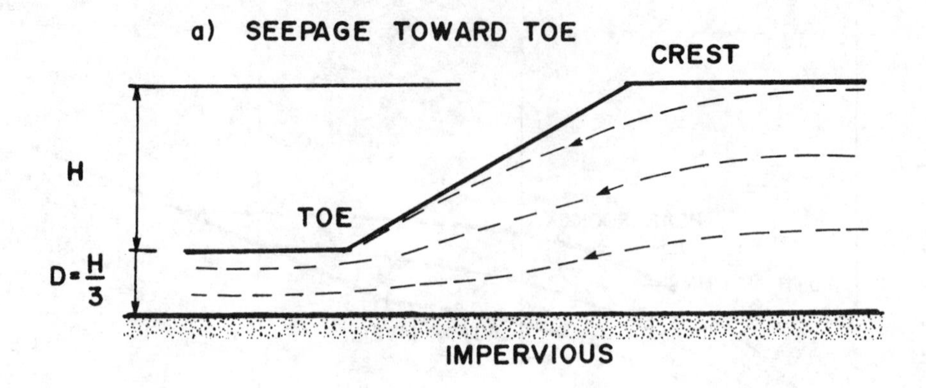

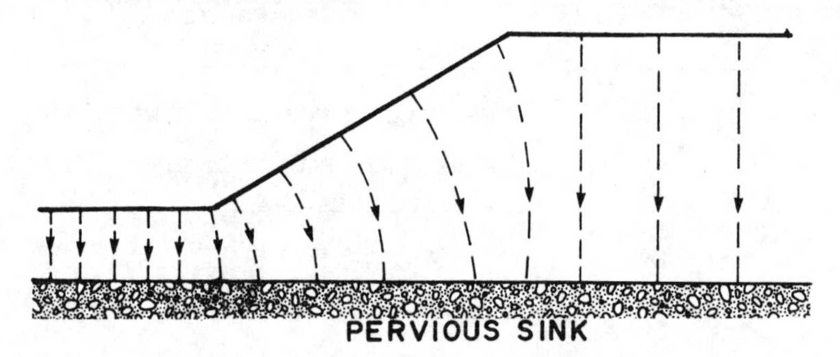

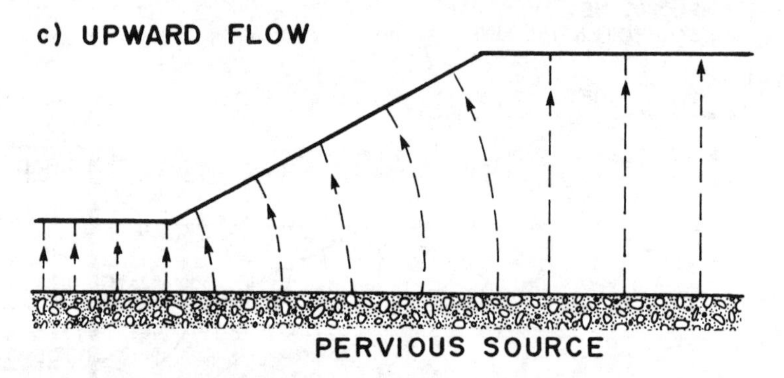

FIG. 3. THREE ISOTROPIC FLOW REGIMES

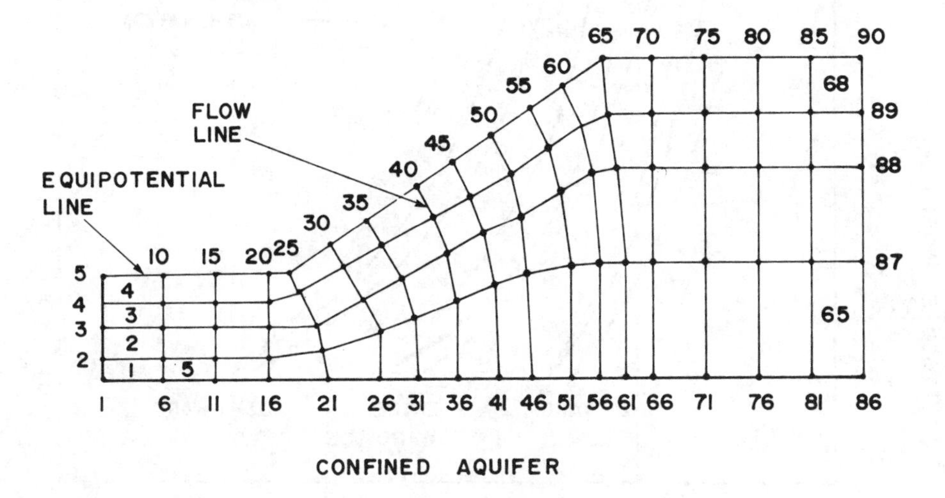

FIG. 4. F.E. FLOW NET FOR UPWARD FLOW

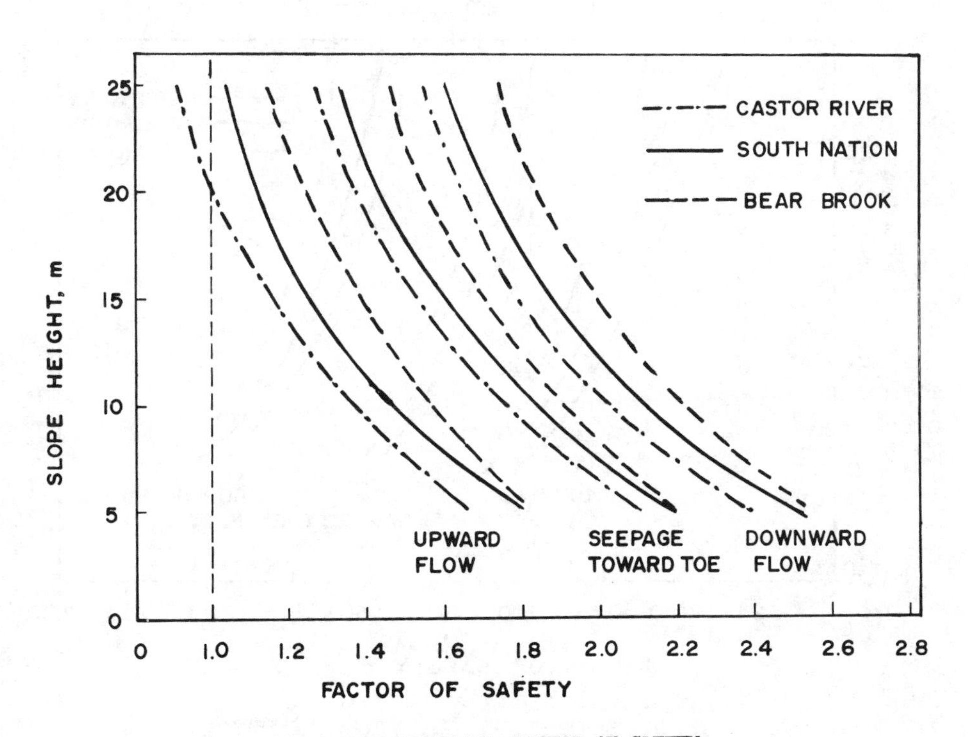

FIG. 5. SLOPE HEIGHT VS. FACTOR OF SAFETY
(SLOPE RATIO 3:1, $i = 18.4^{\circ}$)

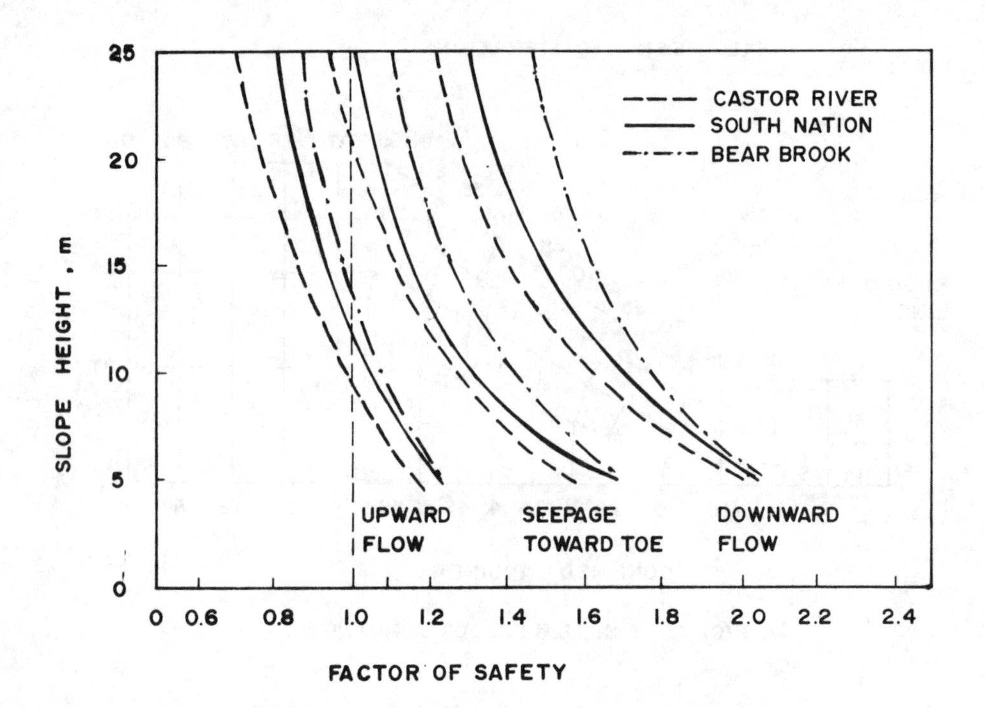

FIG. 6. SLOPE HEIGHT VS. FACTOR OF SAFETY
(SLOPE RATIO 2.5:1, i = 22°)

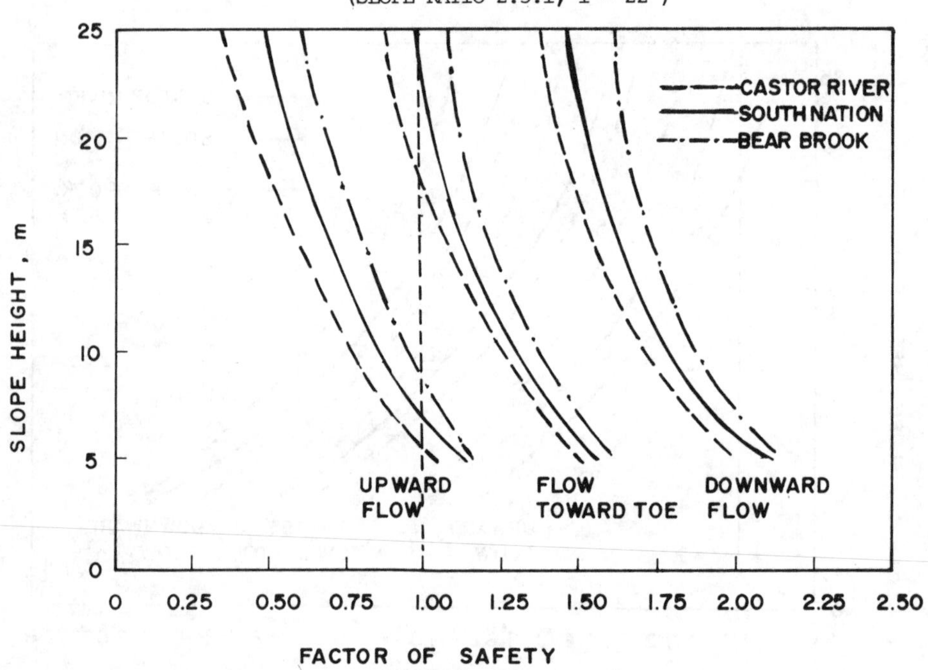

FIG. 7. SLOPE HEIGHT VS. FACTOR OF SAFETY
(SLOPE RATIO 1.75:1, i = 30°)

Numerical Methods in Geomechanics (Innsbruck 1988), Swoboda (ed.)
© 1988 Balkema, Rotterdam. ISBN 90 6191 809 X

Numerical modeling of the influence of seepage on foundation behaviour of gravity dams

P.Bonaldi
Ismes, Bergamo, Italy
D.Cricchi
Ismes Cons., Rome, Italy
R.Ribacchi
Department of Structural and Geotechnical Engineering, University of Rome, Italy

ABSTRACT: The deterministic models of dam behaviour currently utilized in dam monitoring require the effects of seepage body forces to be taken into account because they are not negligible with respect to the effects of water load on the dam face. For a full analysis, a finite element no-tension model allowing for coupling of seepage and state of stress should be adopted. On the basis of parametric numerical analyses, the effects of seepage on dam behaviour are discussed, taking into account the dishomogeneity of the rock mass resulting from the presence of a superficial loosened layer, the variation of permeability with depth, and the possible presence of a tensile region below the upstream heel.

1 INTRODUCTION

Monitoring of many Italian dams is performed by comparing the measured displacements in various points of the structure and of its foundation with the indications of a priori deterministic models [1,2,3].

Linear elastic models are commonly adopted because they allow superposition of the effects of the "external" causes (temperature, construction load and reservvoir level).

These models are characterized by elastic moduli to which reasonable values are initially assigned on the basis of in situ determinations or evaluations of the quality of the rock. More accurate values can be subsequently identified by a treatement of the monitoring results with statistical back-analysis techniques.

Although a non-linear behaviour can be observed in some cases, especially with accurate extensometer determinations, the linear models usually give a reasonably accurate representation of the global behaviour of the dam.

In various cases, mostly where gravity dams are concerned, an apparently abnormal behaviour could be ascribed to the influence of seepage forces within the foundation rock or also to the water load on the bottom of the reservoir [4,5].

It was therefore deemed useful to carry out an investigation on the seepage effects, taking into account also the possible presence of a tensile crack in the foundation below upstream face and a possible dishomogeneity of the rock mass. This latter factor is important because a detailed analysis of the behaviour of various dams [2,3] has shown a gradual increase of rock stiffness below the foundation surface.

2 NUMERICAL MODEL

The analysis was carried out by means of a finite element model which takes into account the effect of seepage as well as coupling between seepage and state of stress, resulting from the dependence of the permeability of rock masses on the state of stress [6]. The analysis can be carried out in terms of total or effective stresses. In the latter case, for instance, body forces acting in an opposite direction with respect to water pressure gradients and "initial" strains proportional to the intrinsic compressibility of the solid matrix, are introduced.

Second order triangular elements are adopted; the same mesh is used to solve both the stress problem and the seepage problem, and to find the nodal forces equivalent to the seepage gradients. This procedure, although incorrect from a purely theoretical standpoint, is convenient in practice and produces accurate results.

The coupled stress-seepage analysis is

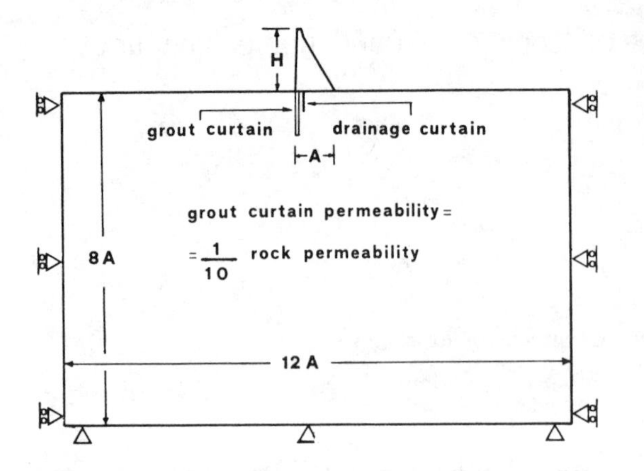

Fig. 1 Scheme of the finite element model

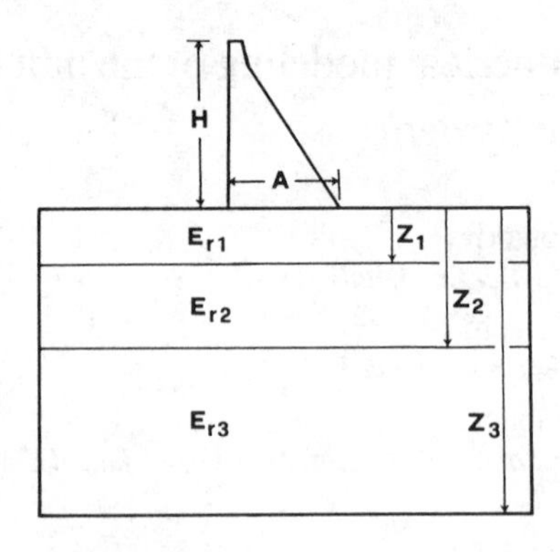

Fig. 2 Horizontal stratified model of the foundation rock

performed by means of an iterative technique, alternatively solving a stress and a flow problem. Usually less than 5 iterations are required to reach convergence.

The problem is treated as a plane strain; the errors introduced because of the finite length of the dam become smaller as the loosening and therefore the deformations in the superficial layers become more important.

To account for the presence of the drainage curtain in a plane strain simplified model, a method proposed by Andrade [7] was utilized. For an indefinite row of drains having well radius r and spacing a, crossing a confined aquifer of thickness D and permeability K, the water flow Q to each drain is given by:

$$Q = (\bar{h} - h_d) \cdot 2 \pi K \frac{D}{\ln \frac{a}{2 \pi r}} \quad 1)$$

where h_d is the piezometric height in the drain and $\bar{h}$ the average piezometric height along the alignment of the drains.

In the finite element model the alignment of drains should correspond to the sides of some elements of the mesh. In those elements the usual relationship between nodal flows $\underset{\sim}{F}^e$ and nodal potentials $\underset{\sim}{\Phi}^e$

$$\underset{\sim}{F}^e = \underset{\sim}{H} \underset{\sim}{\Phi}^e \quad 2)$$

is modified as

$$\underset{\sim}{F}^e = \underset{\sim}{\bar{H}} (\underset{\sim}{\Phi}^e - \underset{\sim}{\Phi}_d^e) \quad 3)$$

where $\underset{\sim}{\Phi}_d^e$ are the potential imposed on the drains and $\underset{\sim}{\bar{H}}$ depends on the element characteristics and on the geometry of the drains alignment, as expressed by rel. 1.

The numerical analyses were carried out for a gravity dam having the geometrical characteristics of the Passante dam [3], namely height H, base width A respectively equal to 66 and 47 m. The characteristics of the finite element model are shown in Fig.1.

3 REPRESENTATION OF THE RESULTS

The result of a deterministic analysis can be conveniently represented by means of the "influence function" [2,3].

The effect of the reservoir level w on the displacement δ (or rotations ϱ) at a point of the foundation can be written as:

$$\delta = \frac{1}{E_r} f_r \left(\frac{w}{H}\right) \quad 4)$$

Strictly speaking the influence functions depend on the ratio Ec/Er between the concrete and rock moduli, but we found that even strong variations of the ratio Ec/Er have only a very slight effect on fr . Our analyses were carried out for a ratio Ec/Er=2.

When the foundation rock is unhomogeneous, the representation of the influence functions is more complex, as they depend on the ratios between various rock moduli and on the thicknesses of the various layers.

In this investigation the analyses were carried out assuming the presence of a homogeneous deformable layer with thickness Z, over a perfectly rigid material. The variation of fr with the ratio Z/A allows a quick visualization of the influence of the more loosened superficial layers. Besides for a given stratified rock mass (Fig.2), the displacements can be approximately calculated by adding the

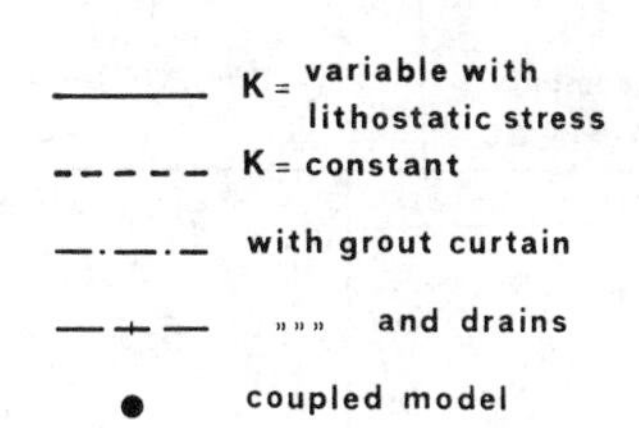

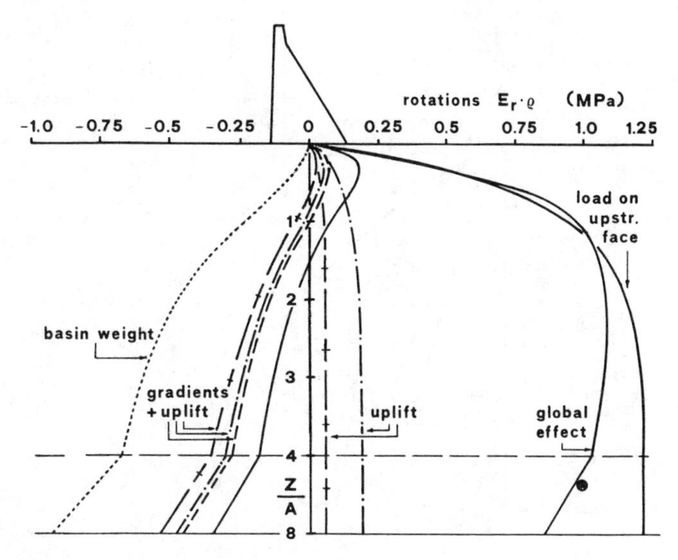

Fig. 3 Various components of the dam base rotations ϱ ,versus thickness of the deformable layer

contributes of the various layers with a method similar to that of Steinbrenner:

$$\delta(\frac{w}{H}) = (\frac{1}{E_{r1}} - \frac{1}{E_{r2}}) f_r (\frac{Z_1}{A}, \frac{w}{H}) + \qquad 5)$$

$$+ (\frac{1}{E_{r2}} - \frac{1}{E_{r3}}) f_r (\frac{Z_2}{A}, \frac{w}{H}) + \frac{1}{E_{r3}} f_r (\frac{Z_3}{A}, \frac{w}{H})$$

The function fr can be split into three terms fr1, fr2, fr3, which respectively account for the water load on the upstream face of the dam,the uplift pressure at the dam base and the body forces due to seepage within the rock mass.

In this study we have analyzed the rotations and the horizontal and vertical displacements in the central part of the dam base. These measurements, as in the case of the Passante dam, are usually obtained in a drift near the base of the dam.

3.1 Effects of water pressure on the dam face

This load, which can be represented with good approximation as an external load on the dam face, is usually considered the main factor contributing to dam displacements.

Fig.3 and 4 show the variation of the influence functions fr1 at full impounding (W/H=1) versus the thickness of the deformable layer. It is apparent that the rotations are influenced mainly by the characteristics of a superficial layer, whereas the horizontal displacements are also influenced to some extent by the deeper layers.

The results of the finite element model substantially corresponds to those which could be obtained by approximate analytical methods [8].

3.2 Effect of seepage

By adopting the simplified model shown in Fig.1 and by assuming uncoupling between state of stress and seepage, the influence functions are linear versus the water level. Their values depend not only on the thickness of the deformable layer, but obviously also on the seepage flow characteristic below the dam, which in turn are influenced by the presence of a grout and of a drainage curtain.

The seepage is also influenced by the variation of permeability in the rock mass. Two situations were analyzed: uniform permeability and a decreasing permeability with depth, according to a relationship of type [6]:

$$K = \frac{K_o}{(\sigma'_m + T)^n} \qquad 6)$$

in which σ'_m is the average effective lithostatic stress and T is a parameter of the rock.

In the grout curtain the permeability was assumed to be reduced to 1/10 of that of the rock; the drainage curtain was assumed to consist of holes 0.1 m in diameter and 3 m spacing.

Fig.3 shows the dependence of the rotations,at full impounding,on the thickness of the deformable layer, for various seepage conditions.

The uplift at the dam base causes a down

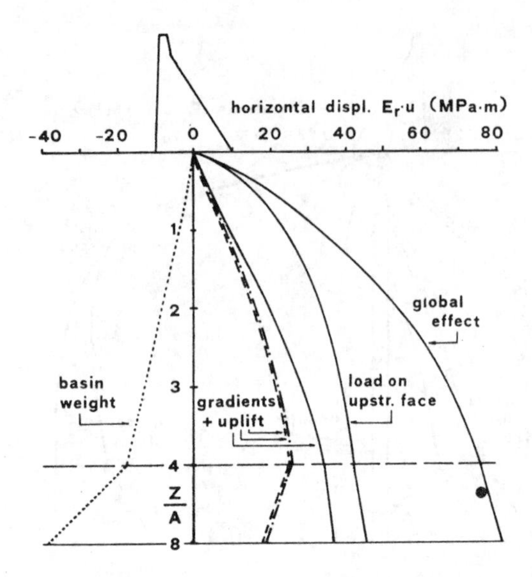

Fig. 4 Horizontal displacements, u , versus thickness of the deformable layer

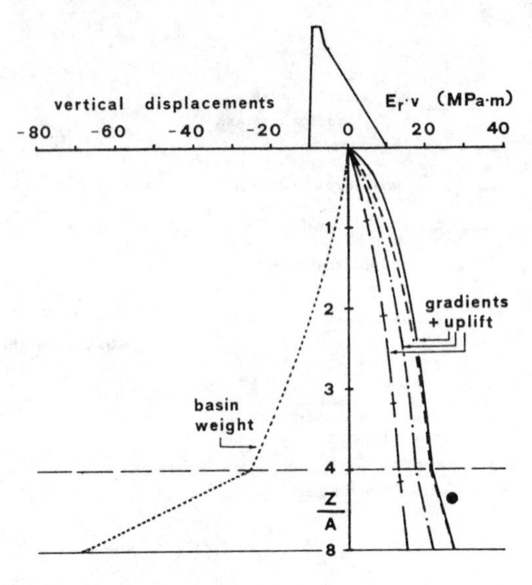

Fig. 5 Vertical displacements, v , versus thickness of the deformable layer

stream rotation, mainly influenced by the superficial layers; on the contrary, body forces due to seepage cause an upstream rotation, which shows a marked increase with the thickness of the deformable layer. Consequently the total effect of seepage can be a downstream or an upstream rotation, depending on the thickness of the deformable layer (that is in practice on the degree of loosening in the more superficial layer).

Horizontal displacements due to seepage are directed downstream and are not much lower than those deriving from the water load on the dam face; however they tend to decrease when the thickness of the deformable layer is very large.

When permeability varies with depth, vertical gradients both below the reservoir and in the downstream region decrease, whereas the horizontal gradients below the dam increse; consequently lower upstream rotations and higher horizontal downstream displacements are observed.

The vertical displacements of the dam base caused by the water load on the face are negligible, whereas those induced by seepage are directed upwards (Fig.5) ; by reducing uplift, the drains and grout curtain at the dam base also reduce the vertical displacements.

Figs. 6 and 7 show the rotations and the horizontal displacements, versus water level, for a given value of the thickness of the deformable layer. The effect of seepage varies linearly with the water

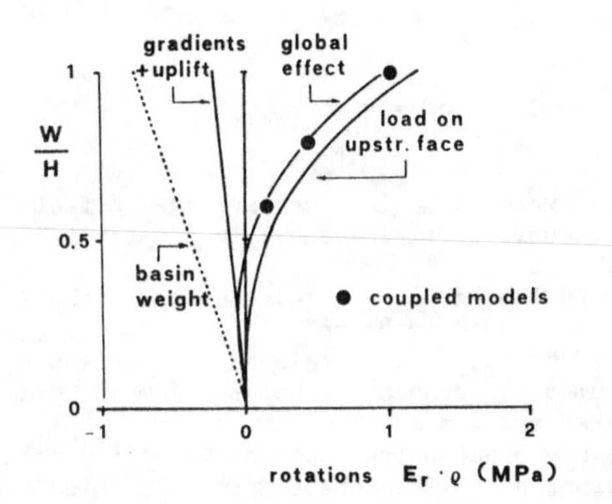

Fig. 6 Dam base rotations, ϱ ,versus water level (Z/A=5.5)

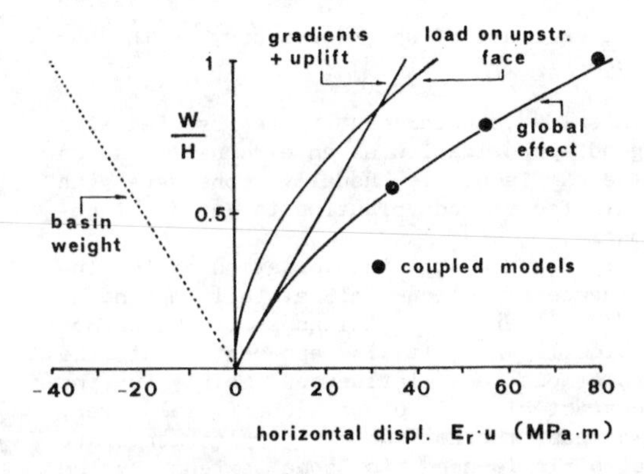

Fig. 7 Horizontal displacement, u ,versus water level (Z/A=5.5)

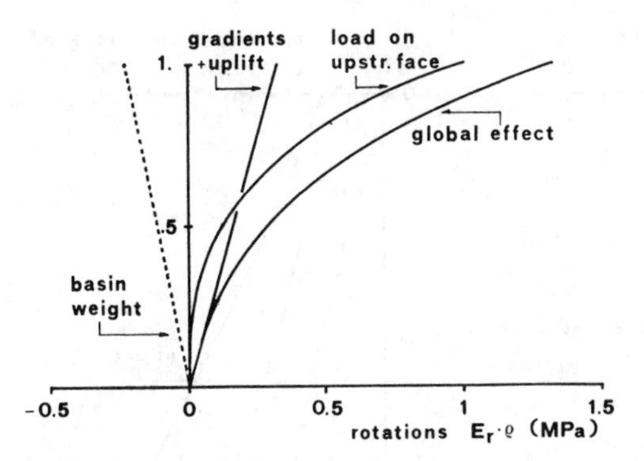

Fig. 8 Dam base rotations, ϱ ,versus water level (Z/A=1.0)

level, whereas that of load on the face becomes increasingly important at high water levels. Consequently in the case of Fig. 6 the global effect is initially a slight upstream rotation, followed by a larger downstream rotation at higher water levels. Such behaviour was noticed in many Italian gravity dams [4].

However an examination of the trend of the influence function in Fig.3 shows that the rotations will always be directed downstream if the superficial loosened layers are characterized by a very low modulus with respect to the deeper rock mass (Fig.8).

The results of some coupled analyses for a given thickness of the deformable layer are shown in Figs.3,4,5,6 and 7: we notice that displacements and rotations are not substantially different from those obtained with an uncoupled analysis, where a permeability varying only with natural lithostatic stresses is assumed.

The results concerning the uplift conditions, Fig.9, are however quite different: they show a substantial loss of efficiency of the grout curtain at high water levels, because of the tensile strains induced near the upstream heel of the dam.

In the foregoing the effects of a reservoir in the rock mass was represented by means of the body forces deriving from the gradients,assuming that steady state conditions had been reached.

In some previous studies [5,9,10] the effect of inpounding was represented as external loads applied on the bottom of the reservoir. This approach could represent for instance the transient situation occurring during a very rapid impounding, when the variations of water pressure do not propagate in the rock mass at depth.

The rotations and displacements which would be caused by this type of water load are also represented in Figs. 3,4,5,6 and

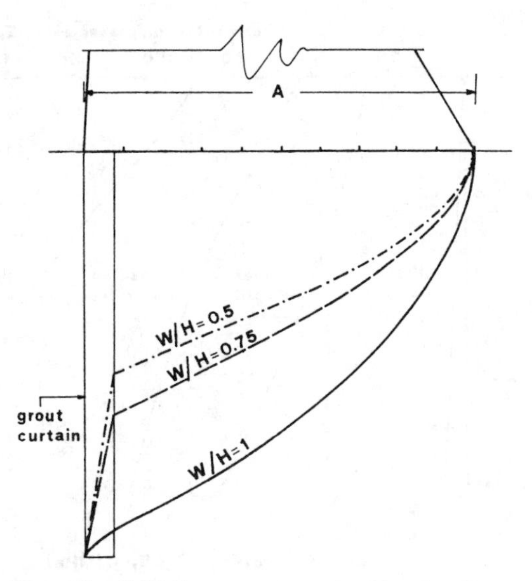

Fig. 9 Uplift pressure at the dam base versus water level (coupled model, no drains)

7. We notice that the rotations are directed upstream and are quite higher than those corresponding to the seepage effects; the horizontal and vertical displacements are directed respectively upstream and downwards,that is they are opposite to those corresponding to internal seepage effects.

3.3 Influence of cracking below the upstream heel

The load applied to the rock by the dam induces tensile stresses in the rock below the upstream heel and this may cause the formation of a vertical crack.Its presence was assumed for instance in some numerical models of real dams [9,10].

An investigation on its effects was performed assuming various depths of cracking

The results are presented in Fig. 10; the influence of the cracking is considerable and is equivalent to an apparent reduction of the stiffness of the rock mass;this should be taken into account in back-analyses of dam behaviour.

The indications in Fig. 10 refer to full impounding conditions; it is to be noted however that the depth of cracking could increse with increasing of water level, thereby introducing another non linearity condition.

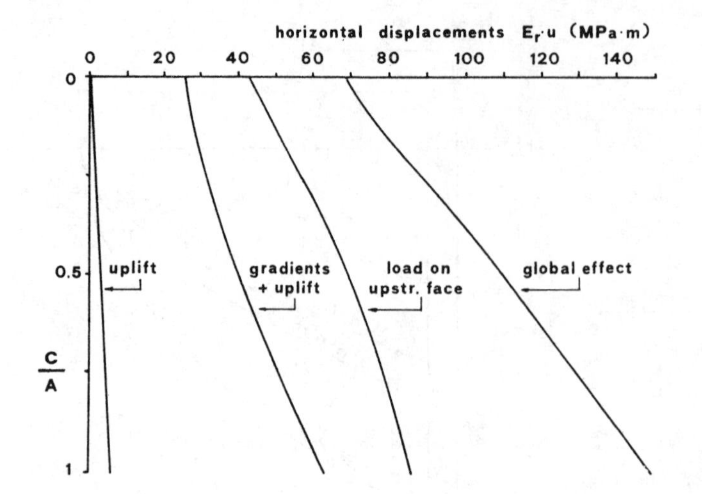

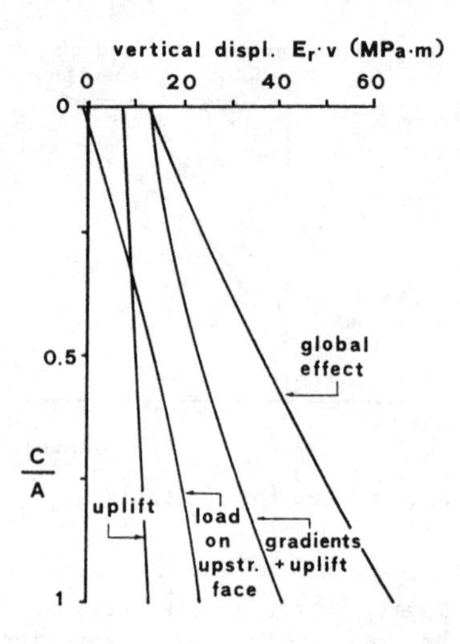

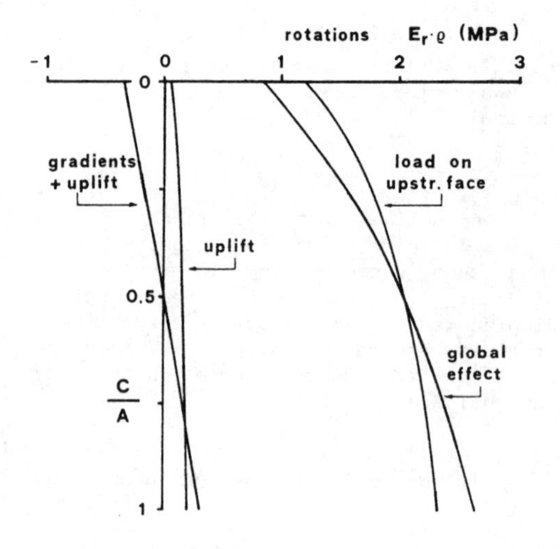

Fig. 10 Displacements and rotations of the dam base at full impounding as a function of the depth of the cracking below its upstream heel (Z/A=4)

REFERENCES

[1] M.Fanelli et al.,Experience gained du ring control of static behaviour of some large Italians dams, 13th ICOLD Congr., New Dehli,1982.

[2] P.Bonaldi et al., Evaluation of rock foundation behaviour for two dams in operation,14th ICOLD Congr.,q.32,927-942,Rio de Janeiro,1982.

[3] P.Bonaldi,et al., Foundation rock behaviour of the Passante dam, 5th Cong. ISRM,C,149-158, Melbourne,1983.

[4] A.Marazio,et al., Dèformabilitè de la roche de foundations dans le cas de quelques barrages italiens, Proc. 1st Congr. ISRM,2,603-616,Lisboa,1966.

[5] P.Bonaldi,et al., Pseudo 3-D analysis of the effects of basin deformations on dam displacements: comparison with experimental measurements, Proc. Int. Conf. Comp. Meth. and Exp. Measurements,329-341, Washington D.C., 1982.

[6] G.Manfredini et al., Mutual influence of water flow and state of stresses in the analysis of dam foundations, Proc.Int.Symp. Numerical Analysis of Dams,881-912,Swansea,1975.

[7] R.M. de Andrade,A drenagem nas fundacoes das estruturas hidraulicas,Engevix, Rio de Janeiro,1982.

[8] J.P.Giroud, Tables pur le calcul des foundations, Dunod,Paris, 1972.

[9] J.Bourbounnais, N.R. Morgenstern, An analysis of the deformation of three dam foundations,Proc. 3rd Congr.ISRM, 2B,685-690,Denver,1974.

[10] V.Souza Lima et al., Horizontal and vertical displacements of the Itaipu main dam. A study of fields measurements and theoretical predictions, 15th ICOLD Congr. 223-247,Lausanne, 1985.

(*) Research partly supported by CNR contribution.

Numerical Methods in Geomechanics (Innsbruck 1988), Swoboda (ed.)
© 1988 Balkema, Rotterdam. ISBN 90 6191 809 X

SUMSTAB – A computer software for generalized stability analysis of zoned dams

P.K.Basudhar, Yudhbir & N.S.Babu
Department of Civil Engineering, Indian Institute of Technology, Kanpur

ABSTRACT: The study pertains to the application of sequential unconstrained minimization technique in conjunction with Janbu´s generalized procedure of slices as a tool for autosearch of the critical slip surface and evaluation of the corresponding minimum factor of safety for zoned dams with geologic discontinuities in the foundation. Successful application of the developed computer software (SUMSTAB) has been demonstrated in the paper.

INTRODUCTION

Slope stability analysis is essentially a problem of optimization namely the determination of the slip surface that yields the minimum factor of safety.

Use of variational calculus and dynamic programming for stability analysis has been reported (Martin, 1982). Sequential Unconstrained Minimization Technique (SUMT) has been applied in slope stability computations (Bhowmick, 1984; Babu, 1986; Yudhbir and Basudhar, 1986).

In this paper an attempt has been made to highlight some aspects of the computer software (SUMSTAB) developed by the authors in analysing a zoned dam having geologic discontinuities in the foundation; here SUMSTAB stands for sequential unconstrained Minimization in Stability Analysis.

ANALYSIS

General

SUMSTAB has been developed by the authors to carryout pseudostatic stability analysis of zoned dams under steady state seepage and earthquake loading using the generalized procedure of slices (GPS) in conjunction with sequential unconstrained minimization technique (SUMT).

The developed method is capable of locating the critical shear surface corresponding to the minimum factor of safety without putting any prior restrictions on the nature of the slip surface. In this technique the stability problem is posed as an optimization problem wherein the factor of safety is minimized with respect to the coordinates of the slip surface and thus the critical surface is located.

To study the effect of earthquake it has been assumed that the quake imposes a horizontal acceleration having an amplitude that is equal to certain percentage of gravity and the resulting horizontal force acts at the centre of gravity of each slice.

The effect of pore water pressure on slope stability has been taken care of by considering an average value of pore pressure parameter, r_u, for the entire section.

The generalized procedure of slices developed by Janbu (1973) for homogeneous soils is assumed to be valid in analysing the dam with proper choice of the shear strength parameters and vertical stress at the base of each slice depending on its location. For reasons of space and brevity the method is not reported herein.

Statement of the Problem

Figure 1 shows the geometry of a zoned dam with a general potential slip surface with the sliding massd divided into N number of slices.

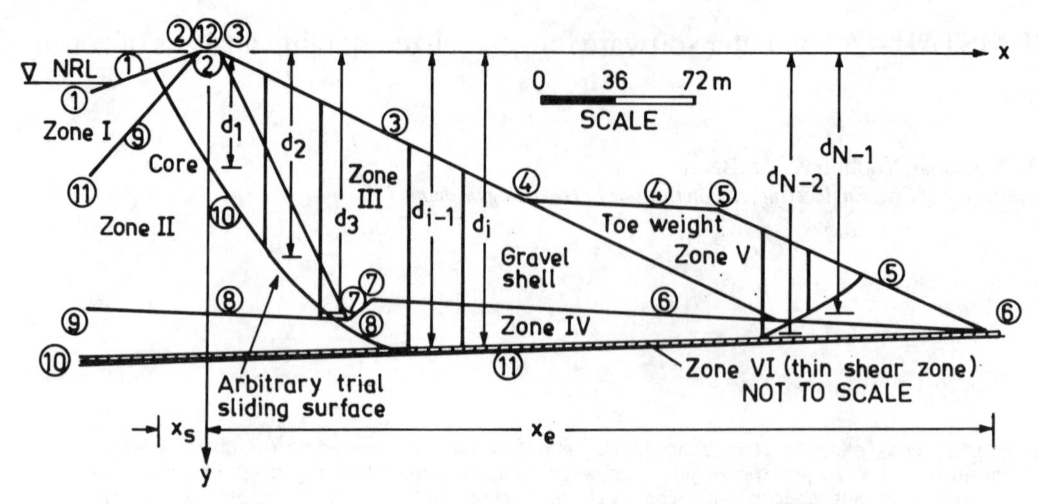

FIG. 1. IDEALIZED SECTION OF A ZONED DAM WITH THE POTENTIAL SLIDING MASS DIVIDED INTO SLICES.

For the given geometry of the dam section and soil properties the factor of safety is a function of the shape and location of the potential slip surface. The object is to determine the shape and location of the shear surface that gives the minimum factor of safety.

Design Variables and Objective Function

The coordinate system is shown in Fig.1. Let x_s and x_e be the x-coordinates of the starting and end points of the slip surface respectively and let $d_1, d_2, \ldots, d_{N-1}$ be the y-coordinates of the slip surface at the interfaces, which are equispaced on the slip surface. For a given number of slices the coordinates of the points along the slip surface are completely defined and the factor of safety can be expressed as a function of these coordinates. In the adopted procedure the factor of safety is minimized with respect to the y-coordinates of the interfaces of the slices and the x-coordinate of the starting and end points of the slip surface. Once the optimal design vector is found out, the design vector along with the x-coordinates of the slice interfaces will define the actual critical shear surface.

The elements of the design vector $\bar{D}$ are as follows

$$\bar{D}^T = (d_1, d_2, d_3, \ldots, d_{N-1}, x_s, x_e) \quad (1a)$$

writing $d_N = x_s$ and $d_{N+1} = x_e$ one obtains

$$\bar{D}^T = (d_1, d_2, d_3, \ldots, d_{N+1}) \quad (1b)$$

So the problem involves (N+1) variables where N is the number of slices into which the sliding mass is divided.

The objective function is the factor of safety and an expression for the same can be obtained from the original paper of Janbu. The factor of safety can be written in terms of the design vector as

$$F = f(\bar{D}) = f(d_1, d_2, d_3, \ldots, d_{N+1}) \quad (2)$$

Constraints

In order to ensure that the slip surface is physically reasonable and acceptable the following constraints are imposed on the shear surface:

1. The curvature of slip surface should be such that it is always concave upwards. This requires

$$d_{i+1} - 2d_i + d_{i-1} \leq 0 \quad (3)$$

2. The slip surface should be within the cross section of the dam. This requires

$$h_i - d_i \leq 0 \quad (4)$$

where h_i = y coordinate of the intersection point of top boundary line of the dam section with the vertical drawn through the point whose y-coordinate is d_i; i varies from 1 to N-1.

Adding all the above constraints the total number of constraints (M) is equal to (2N-2). So the problem is one of (N+1) design variables and (2N-2) side constraints.

Slices, Width of Slices and Zoning

For computational purpose in the idealized dam section (Fig.1) the different zones are numbered. All the intersection points defining the dam geometry are also to be numbered in order. The end points of any straight line whether or not intersected by any other straight line is also treated as intersection point (Fig.1, points 1, 9, 10 and 11). Similarly all the straight lines are also numbered in order. For clarity numbers of intersection points are encircled and numbers of straight lines are marked in semicircles drawn on respective straight lines.

Once the dam geometry is defined by the coordinates of the key intersection points, the coefficients of each straight line defining the dam section can be generated on the computer.

The optimal number of slices can be arrived at by a trade off study of computational efficiency and cost involved in slope stability computations. The total number of slices can also be controlled by proper choice of the slice width. In the present study for a large dam a width of slice of 24 m is found to be quite satisfactory. However, this is problem dependent.

One point to be highlighted is that for slice weight calculations or choosing the appropriate strength parameter along the base of a slice one need to locate the appropriate zone for different critical points. For finding the appropriate zone for any given point in the body of the dam the following procedure was adopted. For the considered point the coordinates are substituted in all the equations surrounding the zone, the inequality condition that are to be satisfied by the equations corresponding to the surrounding straight lines are noted. Hence, for a given point to lie in a particular zone it has to satisfy the inequality conditions of straight lines surrounding that zone. Due to round off errors the starting and end points of slip surface may not exactly satisfy the equations of straight lines on which they are lying. Hence instead of giving a strict equality condition a small margin of 0.0001 is provided. The end points of the slip surface cannot be beyond point 6 (Fig.1). This is put as an additional side constraints.

Presence of Thin Shear Plane in the Dam Foundation

If very thin shear zone is present in the foundation a reasonable thickness of the shear zone than the actual one is to be chosen so that during optimization process the intersection points of interfaces with the shear surface along the shear zone will not fall outside the shear zone. In the present study, while the actual thickness of shear zone is only 0.3 m, the value chosen on either side of shear zone is, however, 1 m. Though the thickness chosen is very high it has been observed that the critical shear surface does not move much from the exact shear plane, and, as such, the results are not affected in any way. But the scheme ensures that the shear surface does not lift off from the shear zone.

Optimization Formulation

The problem of finding the critical slip surface and the corresponding minimum factor of safety is stated as a mathematical programming problem as follows.

Find the design vector $\bar{D}_m$ such that $F = f(\bar{D}_m)$ is the minimum of $f(\bar{D})$ subject to

$$g_j(D_m) \leqslant 0; \quad j = 1, 2, \ldots, M$$

where M is the total number of constraints.

Minimization Procedure

The sequential unconstrained minimization technique using the interior penalty function formulation in combination with Powell's multidimensional search and quadratic fit for fiding the minimizing steps, has been used. The basic object of the penalty function method is to convert the original constrained problem into one of unconstrained minimization by blending the constraints into a composite function (Ψ). The detailed background of these methods are available in standard textbooks on optimization (Fox, 1971).

For problems with inequality constraints only, the ψ-function is defined as:

$$\psi(\bar{D}, r_k) = F(\bar{D}) - r_k \sum_{j=1}^{M} \frac{1}{g_j(\bar{D})}$$

where F is to be minimized over all D, satisfying

$$g_j(\bar{D}) \leq 0;\ j = 1, 2, \ldots, M$$

The penalty parameter r_k is made successively smaller in order to obtain the constrained minimum of F.

Under the senior authors´ guidance Babu (1986) developed the Sequential Unconstrained Minimization Stability Analysis Package (SUMSTAB) whose strength and weakness were further studied by Dhawan (1986).

Results and Discussions

The following design parameters have been used in the analysis:

Unit weight:	Core	20.40 KN/m
	Shell	24.00 KN/m
	Toe-weight	24.00 KN/m
Angle of shearing resistance (ϕ)	Core	26.5°
	Shell and toe-weight	35°
	Shear zone	18°

Numerical results have been obtained by using the DEC SYSTEM 1090 and the SUMSTAB package. For r_u= 0.35 and c´ = 0.0 KN/m typical results for g = 10% are presented in Fig.2.

It is observed from the figure that the critical surfaces crowd in a narrow zone in the core; the major portion of it ofcourse being controlled by the shear zone. These results indicate the location of any possible failure surface during an earthquake for the assumed psuedo-static analysis. The factor of safety for these slip surfaces ranges from 0.84 to 0.87.

It is interesting to observe that the nature of the critical failure surfaces is approximately similar to a wedge configuration.

It is well known that initial starting point plays a great role in getting the desired solution within reasonable number of iterations. If the obtained final solution is same irrespective of the starting points the obtained solution is the global minimum. This aspect has been studied for the given problem (Fig.2) and has been found that though the final solutions corresponding to different initial design vectors are different the value of the factor of safety does not differ much. It has also been observed that all the final optimal surfaces lie in a narrow band in the

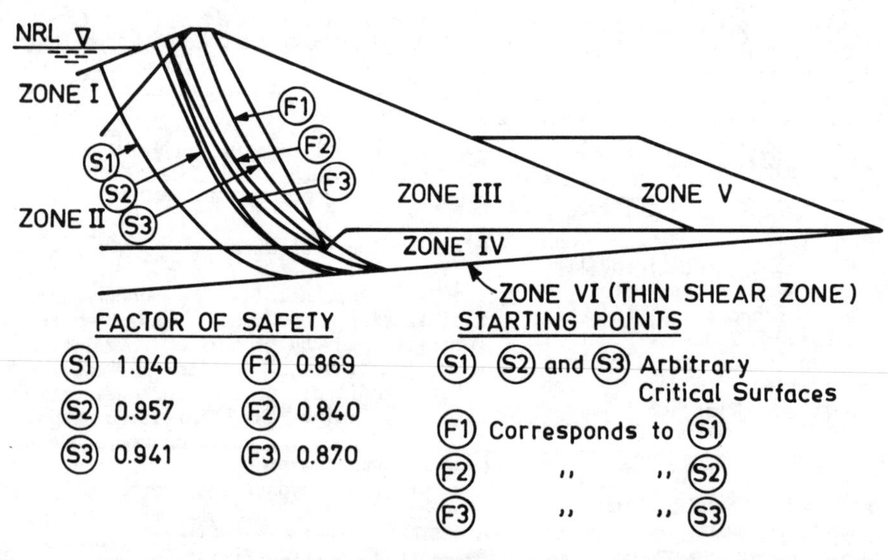

FIG. 2. COMPARISON OF EFFECT OF STARTING POINT ON THE ACCURACY OF COMPUTATION.

core and very near to the core in the foundation and over the rest of the portion they are the same and pass along the existing shear zone. As such, to arrive at the critical shear surface and the corresponding minimum factor of safety it is necessary to have solutions for different initial design vectors. It can be observed that the problem under consideration is unstable for the given parameters. All the final surfaces having factor of safety less than unity are hypothetical failure surface as failure will take place as soon as factor of safety approaches unity. To save computer time the search should be restricted to factor of safety equal to unity. The details of the special consideration needed to adopt SUMSTAB in the case of dams with geologic discontinuities in the foundation are reported elsewhere (Babu, 1986).

The influence of the number of function evaluations and the penalty parameter (r) on the solution have been studied and presented in Fig.3. Convergence of the objective function (F) and penalty function (ψ) with (a) the increase in number of function evaluations, and (b) decrease in r demostrates the efficiency of the proposed numerical scheme.

CONCLUSIONS

A software package (SUMSTAB) utilizing Sequential Unconstrained Optimization Technique has been described to enable

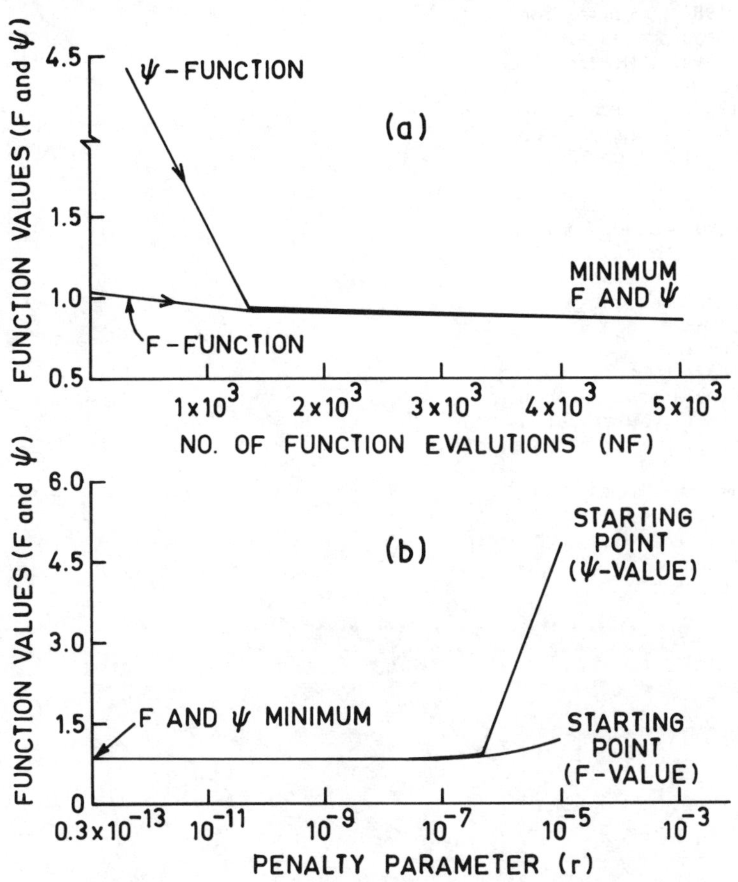

FIG. 3. (a) PATH OF F AND ψ FUNCTIONS FROM STARTING POINT WITH NF.
(b) DIAGRAM SHOWING VARIATION OF F AND ψ WITH r.

autosearch for critical slip surface and the corresponding minimum factor of safety using Janbu´s GPS method, without imposing a priori assumption regarding the slope of the potential failure surface. The method has been used for homogeneous and heterogeneous sections of embankments and has been extended to consider the influence of thin shear zones in the foundation materials.

REFERENCES

Babu, N.S. 1986. Optimization techniques in stability analysis of zoned dams, M.Tech.theses, Indian Institute of Technology, Kanpur, India.

Bhowmik, S.K. 1984. Optimization techniques in automated slope stability analysis, M.Tech. Thesis, I.I.T. Kanpur, India.

Dhawan, R.K. 1986. Parametric studies in slope stability analysis using SUMSTAB package, M.Tech.Thesis, I.I.T. Kanpur, India.

Fox, R.L. 1971. Optimization methods for engineering design. Addison Wesley, Reading, Mass.

Janbu, N. 1973. Slope Stability Computations. Embankment Dam Engineering, Casagrande Volume, John Wiley and Sons Inc.

Martin, J.B. 1982. Embankments and slopes by mathematical programming, J.B. Martin (Ed.), Numerical Methods in Geomechanics, D. Reidel Publishing Company, pp. 305-334.

Yudhbir and Basudhar, P.K. 1986. Revised stability analysis of down stream slope of Beas dam at Pong. A report submitted to the Beas Dam Design Directorate.

Numerical Methods in Geomechanics (Innsbruck 1988), Swoboda (ed.)
© 1988 Balkema, Rotterdam. ISBN 90 6191 809 X

Optimization of dam shape by boundary element method

Zhang Yang & Shen Jiayin
Hohai University, Nanjing, People's Republic of China

ABSTRACT: In this paper, the boundary element method is combined with the improved complex optimal method, and the complicated nonlinear implicit constraints are expanded point by point to linear constraints, thus, the check process of stress constraints at apexes is simplified and the volume of computation reduced. Optimal mathematical models are provided for the concrete gravity dam and the hollow gravity dam. Calculation is performed with the present method for the optimal shape of a gravity dam. Compared the original shape, the optimal shape can save the volume of concrete by 20%.

1 OPTIMIZATION OF STRUCTURES IN ENGINEERING PRACTICE

In engineering practice, the contradiction between safety and low cost is quite sharp. There are thousands of large dams, and the number of dams being built or to be built is quite large. Concrete dams are large in volume, the investment is also large, a part of which depends on the volume of concrete. Therefore, the optimal design of the shape of dams is very important for the reduction of the volume of concrete. The concrete gravity dam is a common dam type, in references [1] and [2], optimal design of the cross section of dams of this type was done by means of material strength method and finite element method. The calculation method used in [1] is in correspondence with that in the present code in our country, however, the accuracy is low. The method in [2] is accurate, but there is no corresponding code to refer to. Hollow dams are adopted to reduce the uplift pressure and to reduce the volume of concrete. In our country, more than 10 hollow dams have been constructed, most of which are hollow gravity dams. Because of the complexity of the structure of this type of dams, the material strength method is no longer applicable, and the finite element method, boundary element method etc. are needed in stress analysis. In the paper, optimal mathematical models are provided for the two types of dams, and the optimal shape calculation is done for a concrete gravity dam.

2 MATHEMATICAL MODEL OF DAM SHAPE OPTIMIZATION

It's necessary to set up the optimal mathematical model of the gravity dam and the hollow dam. And the general mathematical form of constrained optimization is

$$\underset{X \in \Omega}{\text{minimize}}\ F(X) \qquad (2\text{-}1)$$

Where $\Omega = \{X \mid g_i(X) \geqslant 0,\ i=1,\ 2,\ldots,\ m,\ x_j^l \leqslant x_j \leqslant x_j^u,\ j=1,\ 2,\ldots,\ k\}$ and $X=[x_1,\ x_2,\ \ldots,\ x_n]^T$.

2.1 Gravity dam optimal model

1. Design variables

The cross-section of a typical concrete gravity dam is shown in Fig.1, the height H and top width of which are given depending on the requirement of design. So there are only four variables to be considered, i.e.

$$X=[x_1,\ x_2,\ x_3,\ x_4]^T \qquad (2\text{-}2)$$

2. Objective function

For a two-dimensional problem, the area of the dam section can be considered as the objective function for the dam optimization, i.e.

$$F(X)=\tfrac{1}{2}[H(2B+x_4)+x_1x_2-x_3x_4] \qquad (2\text{-}3)$$

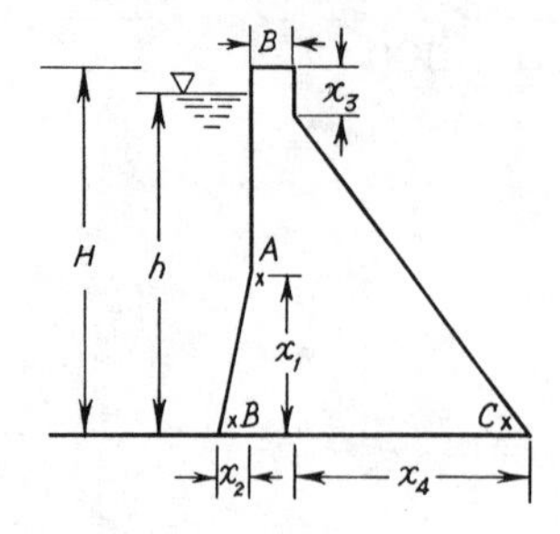

Fig.1

3. Constraints

(1) Geometric constraints

$$0 \leqslant x_1 \leqslant H$$
$$0 \leqslant x_2 \leqslant 0.2\,x_1 \qquad (2\text{-}4)$$
$$0 \leqslant x_3 \leqslant 2\,B$$
$$0.6(H-x_3) \leqslant x_4 \leqslant 0.8(H-x_3)$$

(2) Stability constraint

$$g_1(x) = fW/P - [Kc] \geqslant 0 \qquad (2\text{-}5)$$

where f is the friction coefficient between the dam bottom and the foundation rock, W and P are the total vertical force and horizontal force on the dam, respectively, [Kc] is the permissible safe sliding factor.

(3) Stress constraints

Through engineering experiences, we know that the biggest compression stress and tensile stress often appear where the structure section is sharply changed. So, for the gravity dam, the tensile stresses of points A and B (at the dam heel) and the compressive stress of point C (at the dam toe) should be considered as the governing stress. That is, the following constraints should be satisfied:

$$g_2(X) = [\sigma^-] - \sigma_A \geqslant 0$$
$$g_3(X) = [\sigma^-] - \sigma_B \geqslant 0 \qquad (2\text{-}6)$$
$$g_4(X) = [\sigma^+] + \sigma_C \geqslant 0$$

where $[\sigma^-]$ and $[\sigma^+]$ are tensile strength and compression strength respectively.

2.2 Hollow dam optimal model

1. Design variables

The height H of the hollow dam, the crest width B are given (Fig.2), and the semi-circle is adopted for the crest, which is tangential to two lines below, nine variables remain to be determined, i.e.

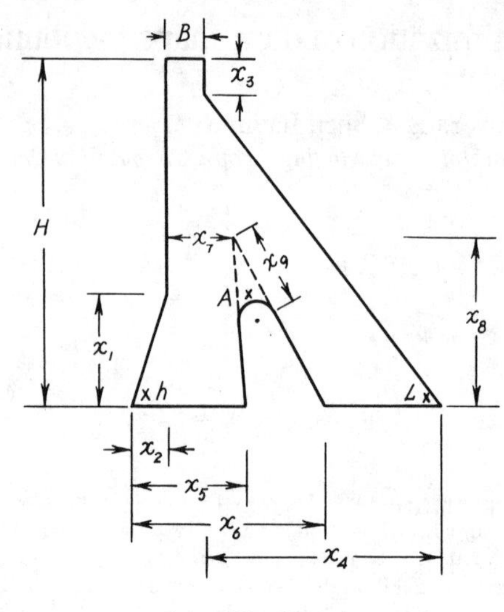

Fig.2

$$X = [x_1, x_2, \ldots, x_9]^T \qquad (2\text{-}7)$$

2. Objective function

Similar to the case of gravity dams, the area of the cross section of the hollow dam is taken as the objective function of optimization, i.e.

$$F(X) = Bx_3 + \frac{1}{2}(2B + x_2 + x_4)(H - x_3) + \frac{x_1 x_2}{2} - \frac{1}{2}(x_6 - x_5)x_8 + Rx_9 - \frac{\pi R^2}{2} \qquad (2\text{-}8)$$

$$R = x_9 \cdot tg\,\alpha$$
$$tg\,\alpha = \frac{(k_2 - k_1)(1 - k_1 k_2)}{(k_2 - k_1)^2 + (1 + k_1 k_2)^2} \qquad (2\text{-}9)$$
$$k_2 = \frac{x_8}{x_7 + x_2 - x_5}$$
$$k_1 = \frac{x_8}{x_7 + x_2 - x_6}$$

3. Constraints

(1) Geometric constrains

When considering the requirements of the upstream and downstream slopes and the limitation of the height, geometrical constraint conditions can be given as follows

$$\begin{aligned} 0 &\leqslant x_1 \leqslant H \\ -0.15x_1 &\leqslant x_2 \leqslant 0.2x_1 \\ 0 &\leqslant x_3 \leqslant 2B \\ 0.6(H-x_3) &\leqslant x_4 \leqslant 0.9(H-x_3) \\ 0 &\leqslant x_5 \leqslant x_4/2 \\ x_5 &\leqslant x_6 \leqslant (X_4+B)/4 \\ 0 &\leqslant x_7 \leqslant 5\,H \\ -H &\leqslant x_8 \leqslant H \end{aligned} \qquad (2\text{-}10)$$

$$\sqrt{(x_7-\frac{x_5+x_6}{2})^2+x_8^2}-\frac{H}{3}\leqslant x_9\leqslant\sqrt{(x_7-\frac{x_5+x_4}{2})^2+x_8^2}$$

(2) Stability constraint

Stability constraint condition is similar to the gravity dams, i.e.

$$g_1(X) = \frac{fW}{P} - [Kc] \geqslant 0 \qquad (2\text{-}11)$$

(3) Stress constraints

$$g_2(X) = -\sigma_h \geqslant 0$$
$$g_3(X) = 8\ kg/cm^2 - \sigma_A = 0 \qquad (2\text{-}12)$$
$$g_4(X) = 50\ kg/cm^2 + \sigma_L \geqslant 0$$

where σ_h is the maximum tensile stress at the dam heel, σ_A is the maximum tensile stress at the arch crown in hollow and σ_h is the maximum compressive stress at the dam toe.

3. IMPROVED COMPLEX METHOD

From above, we know that the mathematical model to be solved is a complex optimal problem involving nonlinear objective and nonlinear implicit constraints, so it is important to choose a feasible optimal method. There two kinds of optimal method, direct method and indirect method. The advantage of the former is that there is no need to differentiate the objective and constraints, therefore the calculation time can be saved, but, generally, no evident optimal effects can be obtained for 'one search'. For example, complex method is a direct search method, the principle of which is: in the n-dimensional space spanned by design variables, choose $K \geqslant n+1$ feasible points as 'top points' to form the polyhedron, or complex, then compare the values of the objective one by one so as to quit the worst point, which has maximum objective value, and replace it by a new feasible point. Then, the next step is performed: compare the objective value again, quit the worst point,..., as a result, the objective is being improved continuously. This method is generally used because of its simplicity and good results. But the constraints of the optimal problem in this paper are implicit and nonlinear, so it is not easy to check new top points to see if the stress constraints are satisfied. In order to improve the calculation efficiency and save the time on feasibility check of top points, the nonlinear stress constraints are expanded as linear constraints, point by point. This method, combining complex method with constraint linearization, is called improved complex method (ICM) in this paper, which is discussed below in detail.

3.1 Forming first complex

In the n-dimensional design space, choose $K \geqslant n+1$ vectors or top points (K=2n, in this paper). If these tops don't satisfy all constraints, then reform new tops until all top points become feasible. Now, the so called starting complex has been set up and their objective values can be calculated. Then the worst feasible point $X^{(h)}$ and the best feasible point $X^{(l)}$ can be easily found, which satisfy the following two Eqs.

$$F(X^{(h)}) = \max_{1\leqslant j\leqslant 2n}\{F(X^{(j)})\} \qquad (3\text{-}1)$$

$$F(X^{(l)}) = \min_{1\leqslant j\leqslant 2n}\{F(X^{(j)})\} \qquad (3\text{-}2)$$

respectively.

3.2 Finding reflecting point

Find the center of all top points except the worst point $X^{(h)}$, i.e.

$$X^{(0)} = \frac{1}{2n-1}\sum_{\substack{j=1\\ j\neq h}}^{2n} X^{(j)} \qquad (3\text{-}3)$$

(1) If $X^{(0)}$ is a feasible point, let

$$X^{(\beta)} = X^{(0)} + \beta\,(X^{(0)} - X^{(h)}) \qquad (3\text{-}4)$$

with β =1.3, 1.3/2, 1.3/4,..., till $X^{(\beta)}$ becomes a feasible point.

(2) If $X^{(0)}$ is not feasible, change the geometric constraints and then reform 2n top points or new complex. That is, return to the first step.

3.3 Comparing reflecting point with worst point

If $F(X^{(\beta)}) < F(X^{(h)})$, replace $X^{(h)}$ by $X^{(\beta)}$ to form new complex and return to the second step. If not, make β half smaller so as to satisfy the constraints. If the constraint can not be stisfied when the value β is very small, then replace $X^{(h)}$ by the next worst point and return to the second step to find new reflecting point.

3.4 Convergence standard

$X^{(l)}$ can be considered as the optimal solution if

$$\left\{\frac{1}{2n}\sum_{j=1}^{2n}\left[F(X^{(l)})-F(X^{(j)})\right]^2\right\}^{1/2}<\varepsilon$$

where $X^{(j)}$ are the 2n top points of the complex, ε is a small positive number.

4 BEM WITH BODY FORCE

There are stress constraints in the optimal model, so it is necessary to perform the stress analysis for the dam by numerical method and here BEM will be adopted.

4.1 Homogenization of governing Eq.

A elastostatic problem can be expressed mathematiclly as follows

$$\begin{aligned} N(U_j) &= b(y) \quad \forall y \in V \\ U_i &= \overline{U}_i(y) \quad \forall y \in S_1 \qquad (4\text{-}1) \\ t_i &= \overline{t}_i(y) \quad \forall y \in S_1 \end{aligned}$$

where $S_1 \cup S_2$ is the boundary of the region V, will $S_1 \cap S_2 = \phi$, N is Navier operator, which is nonlinear, $b_i(y)$ is a body force, U_i and t_i are displacement and traction with their given values U_i and t_i, on the boundaries, respectively.

Supose U_i^* is the special solution to the governing Eq., i.e.

$$N(U_i^*) = b_i(y) \quad \forall y \in V \qquad (4\text{-}2)$$

Let

$$\begin{aligned} U_i &= U_i^* + U_i^o \\ t_i &= t_i^* + t_i^o \end{aligned} \qquad (4\text{-}3)$$

where t_i^* is the special traction solution corresponding to U_i^*. Substituting Eq. (4-3) into Eq. (4-1), we can obtain the homogeneous Eq. and corresponding boundary conditions.

$$\begin{aligned} N(U_i^o) &= 0 \quad \forall y \in V \\ U_i^o(y) &= \overline{U}_i - U_i^* \quad \forall y \in S_1 \qquad (4\text{-}4) \\ t_i^o(y) &= \overline{t}_i - t_i^* \quad \forall y \in S_2 \end{aligned}$$

4.2 BEM formulation

Because of the homogenization of the governing Eq. the region integral is unnecessary. The BEM formulas for problem (4-4) are

$$CK_i\, U_i^o + \int_S T_{Ki}\, U_i^o\, ds = \int_S U_{Ki}\, t_i^o\, ds \qquad (4\text{-}5)$$

for the displacement, and

$$\sigma_{ij} = \int_S (D_{Kij}\, t_K^o - S_{Kij}\, U_K^o)\, ds - \sigma_{ij}^* \qquad (4\text{-}6)$$

for stress, where σ_{ij}^* is the special stress solution corresponding to U_i^*, U_{Ki} and T_{Ki} are fundamental solutions of displacement and traction, D_{Kij} and S_{Kij} are stress tensors corresponding to U_{Ki} and T_{Ki}, respectively, C_{Ki} are constants with different values for boundary points and inner points.

The inner point displacement and stress can be obtained by the discretization of Eqs. (4-5) and (4-6).

5 LINEAR STRESS EXPANSION

As we know, the stress constraints are nonlinear and implicit, it takes a lot of calculation time to check the top points to see if the constraints are satisfied. This difficulty can be overcome by the consideration of the linear expansion of the stress expression, which is based on the hypothesis that when the design variables $X=[x_1, x_2, .., x_n]^T$ are insignificantly changed, the boundary displacement and traction increments ΔU_i and Δt_i are small enough to be neglected.

Now discrete formula (4-6) and express it as follows

$$\sigma_{ij}(y,X) = E_m^{ij}\, l_m - \sigma_{ij}^*(y) \quad \forall y \in V \qquad (5\text{-}1)$$

$$\text{with } E_m^{ij} = [D_{Kij}\, t_K^o - S_{Kij}\, U_K^o]\, \phi W \qquad (5\text{-}2)$$

where index m means the m-th element, ϕ and W are the interpolation function and the numerical integral weight respectively. From Eq. (5-1) and the hypothesis above, the stress increment can be written as

$$\Delta\sigma_{ij} \doteq d\sigma_{ij} \doteq E_m^{ij}\frac{\partial l_m(X)}{\partial x_K}dx_K \qquad (5\text{-}3)$$

For general structure shape optimization, when element division is given, $l_m(X)$ usually has a simple relation with the variable set X, so it is easy to find the derivative $\partial l_m(X)/\partial X_K$.

The linear stress expression can be derived by means of formula (5-3). That is, if top point $X^{(b)}$ is in a small neighbourhood of $X^{(a)}$, then $\sigma_{ij}(y,X^{(b)})$ can be expressed as

$$\sigma_{ij}(y, X^{(b)}) = \sigma_{ij}(y, X^{(a)}) + E_m^{ij}\frac{\partial l_m}{\partial x_K}dx_K \qquad (5\text{-}4)$$

But if $X^{(b)}$ is far away from $X^{(a)}$, the stress analysis must be performed again in order to see whether the stress constraints are satisfied and then the stress can be expanded as a linear function of X at point $X^{(b)}$. This linearization of stress, point by point, brings about the linearization of constraints so that the optimization problem is simplified.

6 CALCULATION EXAMPLE

The cross-section of the concrete gravity dam of Yan Tan is shown in Fig.1, where the dam height H=80.0 m, the top width B=10.0 m, the upstream head h=76.0 m and downstream, h=0.0 m, the unit weight of concrete γ =2.4 kg/cm^3 , the concrete module and Poisson ratio are 2 x 10^5 Kg/cm^2 and 0.167 respectively, the rock module and Poisson ratio are 2.7 x 10^5 Kg/cm^2 and 0.2 respectively, the friction coefficient between the dam bottom and the foundation is 0.65, the permissible safe sliding factor [K]=1.05, tensile strength [σ^-]=1.0 Kg/cm^2, compressive strength [σ^+]=35 Kg/cm^2.

The dam shape optimization has been performed by means of our computer program (SOBEM) and the optimal results can be seen in table 1.

1. The best of the eight objectives in the first complex is F =2581.541 m^2 and the find optimal result is F_{113} =2053.104 m^2, which cuts down 20.5 percent of the amount of concrete.

2. From table 1, it can be seen that the stress constraint plays a governing role in the optimization. And in the end, the tensile stress σ_B is very close to its permissible value or tensile strength.

3. The smallest value of safe sliding factor is 1.109, which is as close as 5.6 percent to the permissible value.

As a conclusion of this paper, we would like to indicate the following valuable points:

1. It is adaptable to use BEM for the structure shape optimization, because boundary element mesh is much more suitable for the change of the structure shape than finite element mesh. Besides, it can simulate stress concentration.

2. In order to save the computing time and raise the optimal efficiency, the complex nonlinear stress constraints can be expanded as linear ones at some points.

3. In the optimization process, some variables e.g., x_3 in our example, can be adjusted experientially, which is more efficient and time-saving.

Tab. 1

	N						
	0	1	5	15	30	100	113
F	2581.54	2416.91	2306.13	2211.90	2160.35	2058.15	2053.10
x_1	35.401	18.832	14.818	11.274	46.750	21.236	19.397
x_2	2.573	0.794	2.025	1.899	4.007	1.948	3.867
x_3	9.218	16.075	17.509	18.030	19.832	19.907	19.853
x_4	49.052	50.354	47.723	45.222	42.105	41.185	40.421
σ_A	-2.81	-4.68	-5.29	-5.79	-1.60	-4.88	-4.95
σ_B	-1.46	-1.19	-0.63	-0.17	0.07	0.08	0.99
σ_C	-18.54	-18.43	-19.00	-19.67	-20.10	-20.81	-20.85
K	1.395	1.306	1.246	1.195	1.167	1.112	1.109

Notice : Length dimension is meter and stress dimension, Kg/cm^2, N is the times for iteration .

REFERENCE

Yang Zhonghou, Lu Tairen, Wang Dexin, Method of Geometric Programming and Its Application to optimal Design of Gravity Dam Section, Journal of East China Technical University of Water Resources, No. 4 (1982).

Hu Weijun, Li Duanmin, Optimal Aseismic Design of Gravity Dams, Journal of East China Technical University of Water Resources, No. 1 (1983).

The ministry of water conservancy and electric engineering, Design Standard of Concrete Gravity Dam (SDJ21-78), Publishing House of Hydraulic Engineering, 1978.

Numerical Methods in Geomechanics (Innsbruck 1988), Swoboda (ed.)
 ISBN 90 6191 809 X

Analysis of effective stress and movement of Lubuge Dam

Z.J.Shen & W.M.Zhang
Nanjing Hydraulic Research Institute, People's Republic of China

ABSTRACT: Lubuge dam is one of highest core type rockfill dams being constructed now in China. In this paper Biot's consolidation theory is used to predict the movement of various zones of dam body and pore pressure response in the core during its construction, reservoir filling and drawdown of water level. In the analyses three stress-strain models of soils, namely the revised Cambridge model, Duncan's nonlinear hyperbolic model and a new developed elasto-plastic hyperbolic model are compared with each others.

INTRODUCTION

Lubuge hydraulic electric plant with capacity of 600 MB is located in Xyangnixe river, a tributary of Nanpanjiang in Province Yunnan. The river with V-type valley is cut by a rockfill dam with a central core which has a maximum height of 100 m and a length of 200 m along the crest. The shell material is made up of dolomite and limestone excavated from foundation of the plant, while the core material is the weathered rocks with a quite large amount of clay particles. Owing to the absence of sand deposit near the construction site, the transition zones are filled with man-made sands crushed from rocks. The typical dam section is given in Fig.1.

The plant is now being constructed under the investment of The World Bank. The aim of this study is laid on the prediction of deformation and pore pressure response of the dam during its construction and reservoir impounding with emphasis of comparison of stress-strain models. The computation is executed with program BISS (Shen, 1983), which was originally written in ALGOL-60 and has been recompiled in FORTRAN-77 and transplanted in micro-computer of type IBM-PC.

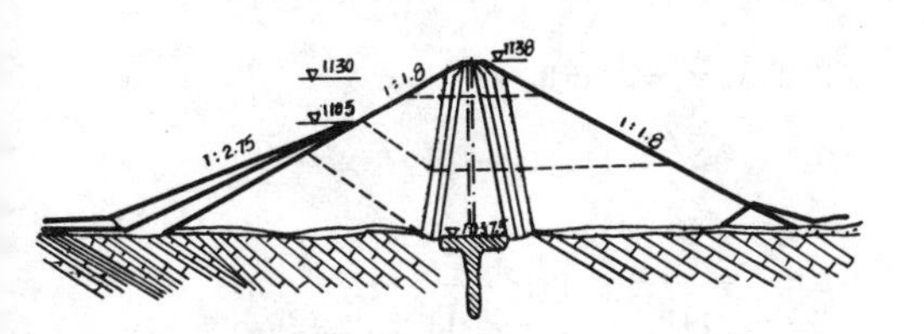

Fig.1 Cross-section of the dam

CONSTITUTIVE MODEL

Three stress-strain models, i.e. the variable elastic model proposed by J.M.Duncan et al.(1980), the Cambridge model revised also by J.M.Duncan et al.(1981) and a new elasto-plastic model suggested by the first author (Shen, 1987) have been used in the analyses. The brief discription of them is given blow.

1.Variable elastic model

For plane strain problem the incremental stress-strain relationship for variable elastic model can be written as follows :

$$\begin{vmatrix} \Delta\sigma_x \\ \Delta\sigma_y \\ \Delta\tau_{xy} \end{vmatrix} = \begin{vmatrix} d_1 & d_2 & 0 \\ d_2 & d_1 & 0 \\ 0 & 0 & d_3 \end{vmatrix} \begin{vmatrix} \Delta\varepsilon_x \\ \Delta\varepsilon_y \\ \Delta\gamma_{xy} \end{vmatrix} \quad (1)$$

If the tangential young's E_t and bulk modulus B_t are used as basic parameters, then

$$\left.\begin{aligned} d_1 &= 3B_t(3B_t+E_t)\div(9B_t-E_t) \\ d_2 &= 3B_t(3B_t-E_t)\div(9B_t-E^t) \\ d_3 &= 3B_tE_t\div(9B_t-E_t) \end{aligned}\right| \quad (2)$$

J.M.Duncan et al.(1980) proposed following formula for E_t and B_t :

$$E_t=E_i(1-R_fS_l)^2 \quad (3)$$

$$B_t=K_bp_a(\sigma_3/p_a)^m \quad (4)$$

where p_a —— atmospheric pressure, and

$$E_i=Kp_a\ (\sigma_3/p_a)^n \quad (5)$$

$$S_l=(\sigma_1-\sigma_3)\ /\ (\sigma_1-\sigma_3)_f \quad (6)$$

$$(\sigma_1-\sigma_3)_f=\frac{2\sin\phi}{1-\sin\phi}(\sigma_3+c*\cot\phi) \quad (7)$$

For materials of coarse particles, cohesion c=0,

$$\phi=\phi_1-\Delta\phi\ \log(\sigma_3/p_a) \quad (8)$$

K, K_b, n, m, ϕ_1, $\Delta\phi$, R_f are 7 constants obtained from conventional triaxial test. In addition, following loading function is used for loading-unloading criterion

$$f_l=S_l(\sigma_3/p_a)^{1/4} \quad (9)$$

The Young's modulus for unloading and reloading is calculated by

$$E_{ur}=K_{ur}p_a(\sigma_3/p_a)^n \quad (10)$$

The parameter E_t entering in Eq.2 is determined according to the ratio R_l of the updated value of f_l to its maximum value $(f_l)_{max}$ in the loading history

$R_l > 0.95 \quad (E_t)=E_t$

$R_l < 0.70 \quad (E_t)=E_{ur}$

$0.7<R_l<0.95 \quad (E_t)=E_t+4(E_{ur}-E_t)(0.95-R_l)$

2. Revised Cambridge model

The yield function of Cambridge model can be written as :

$$p_0=(1+\eta^2/M^2)(p+p_r)-p_r \quad (11)$$

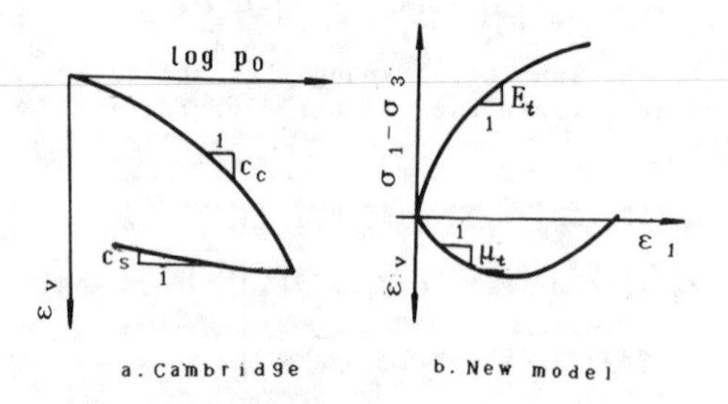

Fig.3 Parameter determination

where $\eta=q/p$, $p=(\sigma_1+\sigma_2+\sigma_3)/3$,

$q=1/\sqrt{2}\,[(\sigma_1-\sigma_2)^2+(\sigma_2-\sigma_3)^2+(\sigma_3-\sigma_1)^2]^{1/2}$.

The intersection p_r in Eq.11 was introduced by J.M.Duncan et al.(1981) to use the Cambridge model for soils other than normally consolidted clay (Fig.2a). The corresponding hardening function can be written as :

$$p_0=p_{0i}\exp[\varepsilon_v^p\ /\ (c_c-c_s)] \quad (12)$$

where ε_v^p —— plastic volumetric strain, c_c, c_s —— slope of compression and rebound branchs in $\varepsilon_v \backsim \ln p_0$ plot (Fig.3a).

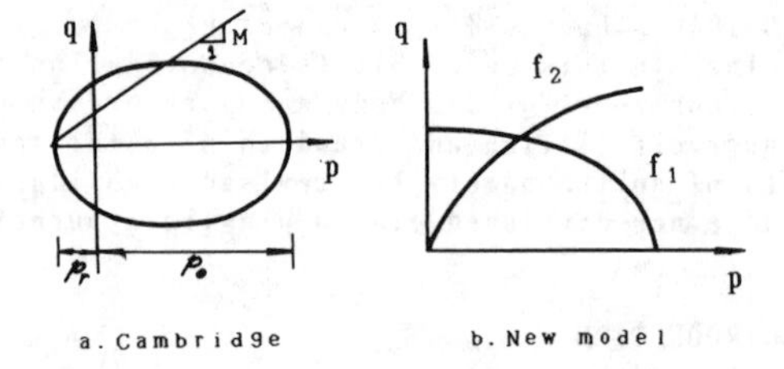

Fig.2 Yield locus in p-q plane

If in π plane Prandtl-Reuss flow law is used, the stress-strain relationship can be written as :

$$\begin{vmatrix}\Delta\sigma_x\\ \Delta\sigma_y\\ \Delta\sigma_z\\ \Delta\tau_{xy}\end{vmatrix} = \begin{vmatrix} M_1-P(s_x+s_x)/q-Qs_x{}^2/q^2 & M_2-P(s_x+s_y)/q-Qs_xs_y/q^2 & -Ps_{xy}/q-Qs_xs_{xy}/q^2\\ M_2-P(s_y+s_x)/q-Qs_ys_x/q^2 & M_1-P(s_y+s_y)/q-Qs_y{}^2/q^2 & -Ps_{xy}/q-Qs_ys_{xy}/q^2\\ M_2-P(s_z+s_x)/q-Qs_zs_x/q^2 & M_2-P(s_z+s_y)/q-Qs_zs_y/q^2 & -Ps_{xy}/q-Qs_zs_{xy}/q^2\\ -Ps_{xy}/q-Qs_{xy}s_x/q^2 & -Ps_{xy}/q-Qs_{xy}s_y/q^2 & G_e-Qs_{xy}{}^2/q^2 \end{vmatrix} \times \begin{vmatrix}\Delta\varepsilon_x\\ \Delta\varepsilon_y\\ \Delta\gamma_{xy}\end{vmatrix} \quad (13)$$

where $s_x=\sigma_x-p$, $s_y=\sigma_y-p$, $s_z=\sigma_z-p$, $s_{xy}=\tau_{xy}$ are components of deviatoric stress tensor,

$$M_1=B_p+4G_e/3$$

$$M_2=B_p-2G_e/3$$

$$P=B_eG_e\gamma\ /\ (1+B_e\alpha+G_e\delta)$$

$$Q=G_e{}^2\delta\ /\ (1+B_e\alpha+G_e\delta)$$

$$B_p=\frac{B_e}{1+B_e\alpha}\left[1+\frac{B_eG_e\gamma^2}{1+B_e\alpha+G_e\delta}\right]$$

and

$$\alpha = (c_c - c_s)(M^2 - \eta^2) / (pM^2)$$

$$\beta = 4(c_c - c_s)\eta^2 / [pM^2(M^2 - \eta^2)] \qquad (14)$$

$$\gamma = 2(c_c - c_s)\eta / (pM^2)$$

$$\delta = \beta$$

In these equations p_r, M, c_c, c_s, and G_e are parameters. $B_e = p/c_s$ is elastic bulk modulus. p_r and M can be deduced from Mohr-Coulomb envelop, however, a straight line is use to yield the intersection $c*\cot\phi$ instead of the slight curved envelop for rock materials, thus

$$p_r = c*\cot\phi \qquad (15)$$

$$M_c = 6\sin\phi / (3 - \sin\phi) \qquad (16)$$

But Eq.16 is related to compression circle in π-plane, and a compromised circle

$$M = 2\sin\phi \qquad (16a)$$

is used in computation (Fig.4). J.M.Duncan et al.(1981) suggested to deduce c_c and c_s

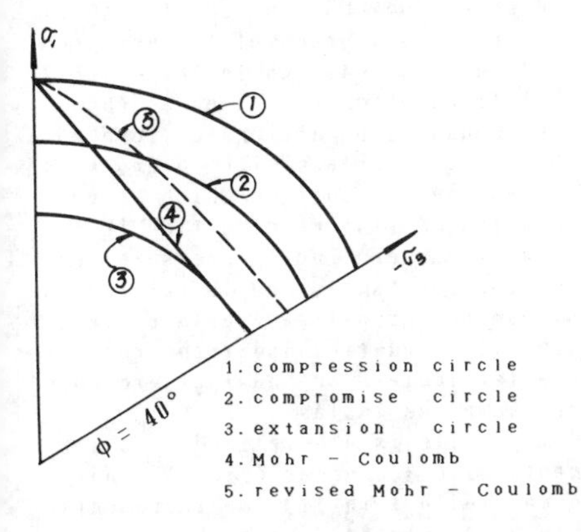

Fig.4 Yield locus in π-plane

also from data of triaxial test instead of isotropical compression test which is difficult to do for rock materials. However, because of nonlinearity of $\varepsilon_v \backsim \log p_0$ curve, a variable c_c is used as calculated by

$$c_c = c_{c1}(p/p_a)^{n_c} \qquad (17)$$

Finally, the elastic shear modulus G_e can be determined from elastic bulk modulus B_e as

$$G_e = 3(1-2\nu)p / [2(1+\nu)c_s] \qquad (18)$$

assuming Poison's ratio ν=const.

3.New elasto-plastic model

This model has the elasto-plastic matrix formally identical with the Cambridge model but its parameter determination is similar to that of Duncan's hyperbolic model. The model is based on double hardening concept and its two yield surfaces are recommended as follows (Fig.2b)

$$f_1 = p^2 + r^2 q^2$$

$$f_2 = q^s / p \qquad (19)$$

where r and s are two parameters. From multiple hardening theory the plastic strain increment can be calculated (Shen, 1984).

$$\Delta\varepsilon_{ij}{}^{p} = \sum_{n=1}^{l} A_n \partial f_n / \partial\sigma_{ij}$$

where l is total number of yield surfaces, f_n and A_n are the n-th yield surface and related plastic coefficient or its contribution to plastic deformation, and A_n=0 when f_n is inactive. Therefor, for double hardening model the increments of volumetric and deviator strain components can be written as

$$\Delta\varepsilon_v = \frac{\Delta p}{B_e} + A_1\frac{\partial f_1}{\partial q}\Delta p + A_2\frac{\partial f_2}{\partial p}\Delta p$$

$$\Delta\varepsilon_d = \frac{\Delta p}{G_e} + A_1\frac{\partial f_1}{\partial q}\Delta q + A_2\frac{\partial f_2}{\partial q}\Delta q \qquad (20)$$

where $\varepsilon_d = \sqrt{2}/3\,[(\varepsilon_1-\varepsilon_2)^2 + (\varepsilon_2-\varepsilon_3)^2 + (\varepsilon_3-\varepsilon_1)^2]^{1/2}$. Defining

$$E_t = \Delta\sigma_1 / \Delta\varepsilon_1, \quad \mu_t = \Delta\varepsilon_v / \Delta\varepsilon_1$$

and considering $\Delta p = \Delta\sigma_1/3$, $\Delta q = \Delta\sigma_1$, $\Delta\varepsilon_d = \Delta\varepsilon_1 - \Delta\varepsilon_v/3$ in triaxial state, the plastic coefficient A_1 and A_2 in Eq.20 can be expressed through E_t and μ_t as follows

$$A_1 = r^2[\eta(9/E_t - 3\mu_t/E_t - 3/G_e + s(3\mu_t/E_t - 1/B_e)] / [(1+\eta r^2)(s+\eta^2 r^2)] \qquad (21)$$

$$A_2 = [9/E_t - 3\mu_t/E_t - 3/G_e - r^2\eta(3\mu_t/E_t - 1/B_e)] / [(s-\eta)(s+\eta^2 r^2)] \qquad (22)$$

If the $(\sigma_1-\sigma_3)\backsim\varepsilon_1$ plot and the $\varepsilon_v \backsim \varepsilon_1$ plot are fitted with hyperbolic and parabolic function respectively then the expression of E_t is the same as Eq.3, and

$$\mu_t = 2c_d(\sigma_3/p_a)^{n_d} E_i S_1(1-R_d)\{1-S_1(1-R_d)/ / [(1-S_1)R_d]\} / [R_d(\sigma_1-\sigma_3)] \qquad (23)$$

where c_d, n_d, and R_d are 3 constants instead of G, F, D in the original Duncan-Chang's (1970) model.(Fig.3b)

If the Prandtl-Reuss flow law is used again, the same stress-strain relationship can be deduced as shown in Eq.13, but in this case in Eq.14

$$\begin{aligned} \alpha &= A_1 / r^2 + \eta^2 A_2 \\ \beta &= r^2 \eta^2 A_1 + A_2 \\ \gamma &= \eta (A_1 - A_2) \\ \delta &= \beta + B_e(\alpha \beta - \gamma^2) \end{aligned} \tag{24}$$

Because in hyperbolic model modulus E_t is calculated according to Mohr-Coulomb criterion, which may underestimate soil rigidity under plane strain state when the intermediate principal stress $\sigma_2 > \sigma_3$, an enhanced value of $\underline{\sigma}_3$ instead of σ_3 in Eq.5 and Eq.7 is used (Fig.4), i.e.

$$\underline{\sigma}_3 = \sigma_3 [\sigma_2 / \sigma_3]^{1/3} \tag{25}$$

The elastic bulk and shear modulus can be determined through E_{ur} (Eq.10) as follows

$$\begin{aligned} B_e &= E_{ur} / 3(1-2\nu) \\ G_e &= E_{ur} / 2(1+\nu) \end{aligned} \tag{26}$$

4. Compressibility and flow of fluid

In addition to stress-strain model of soil skeleton, the laws of compressibility and flow of pore fluid must be defined for the core zone. A compressible mixed fluid model is used instead of seperate consideration of water and gas flow. The coefficient of compressibility of this fluid is calculated by

$$c_p = n_p(1 - s_r + c_h s_r) / (p_w + p_a) \tag{27}$$

where p_w - pore pressure, s_r - degree of saturation, n_p - porosity and c_h - Henry's coefficient of dissolution. If s_{ro} is the degree of saturation at the compaction of soil, the variation of s_r with pore pressure can be estimated (Zaretski & Orhov, 1982) by

$$s_r = (p_w + p_a) / [\, p_a + (1 - c_h) s_{ro} p_w \,] \tag{28}$$

Besides, Darcy's law with constant coefficient of permeability is used in this computation.

METHOD OF ANALYSIS

If displacements u, v and potential of pore fluid h are accepted as the independant variables, finite element formulation of Biot's theory can be written as follows

$$\begin{aligned} &\sum_{j=1}^{n} (k_{ij}^{11}\Delta u_j + k_{ij}^{12}\Delta v_j + k_{ij}^{13}\Delta h_j) = \Delta F_i^1 \\ &\sum_{j=1}^{n} (k_{ij}^{21}\Delta u_j + k_{ij}^{22}\Delta v_j + k_{ij}^{23}\Delta h_j) = \Delta F_i^2 \\ &\sum_{j=1}^{n} k_{ij}^{31}\Delta u_j + k_{ij}^{32}\Delta v_j + \alpha \Delta t k_{ij}^{33}\Delta h_j) = \Delta F_i^3 \\ &\qquad - \Delta t \sum_{j=1}^{n} k_{ij}^{33}\Delta h_{j0} \\ &\qquad (i = 1, 2, 3, \cdots\cdots n) \end{aligned} \tag{29}$$

where Δu_j, Δv_j and Δh_j are increments of node variables in time interval Δt, ΔF_i^1, ΔF_i^2 and ΔF_i^3 are increments of horizontal and vertical loads and inflow in node i respectively, h_{j0} is existing pore fluid head in node j before the time increment, n is total number of nodes, α is time integration parameter and $\alpha = 2/3$ is used. Elements of matrix k_{ij}^{11}, k_{ij}^{12} ... k_{ij}^{33} can be deduced by using standard finite element technique.

Q4 element associated with substructure solution method is used in the computer program. The dam section is subdivided into 3 substructures, i.e. up- and down-stream shell and core zone. For the substructures of shells only two degrees of freedom are prescribed in each node, while there are 3 degrees of freedom for core zone as shown in Eq.29. The boundary condition for core zone are ; the contact surface with bedrock is unpermeable, the top surface which is variable in height during construction is assumed having zero pore pressure or having inflow constantly due to raining, and the upstream and downstream surface are equipotential boundaries under the respective water tables, but having zero pore pressure above the tables.

The dam profile is discretized into 302 Q4 elements with 348 nodes (Fig.5). Analyses are carried out in 17 time increments to simulate the planed 4 stage schedule of construction and reservoir operation.(Fig.6)

The parameters for 3 models used in computation are derived from data of triaxial test of $\phi\, 30^{cm}$ samples and are summurized in table 1. However, in the new model, parameters r=2 and s=2 for clay or s=3 for coarse materials are assumed empirically, computation is not sensitive to the variation of these two parameters.

RESULTS OF ANALYSIS

Fig.7 shows pore pressure distribution computed with the new model in clay core in different stage of construction and water level. Results with other two models are similar, but the maximum pore pressure at

Table.1 Soil parameters

	①	②	③	④	c_s	c_{c1}	n_c	M	c	ϕ	$\Delta\phi$	R_f	K	n	K_b	m	c_d	n_d	R_d
I	2.12	---	---	---	.0017	.007	1.33	1.50	---	54.0	10.	.75	900	0	200	-.24	.002	1.87	.63
II	1.91	---	---	---	.0025	.018	1.06	1.48	---	52.0	9.0	.64	600	.25	227	0.20	.004	1.21	.52
III	1.89	---	---	---	.0035	.013	0.62	1.44	---	50.4	7.3	.64	320	.43	210	0.38	.006	0.74	.50
IV	1.88	.51	.94	7.0	.0035	.021	0.76	1.04	41.	31.0	---	.64	250	.15	92	0.12	.018	0.47	.64

I - shell rock, II & III - stone & sand of transition zone , IV - core material
① - Density in t/m^3 , ② - porosity , ③ - Degree of saturation
④ - Permeability in 10^{-9} m/s , c - cohesion in kPa

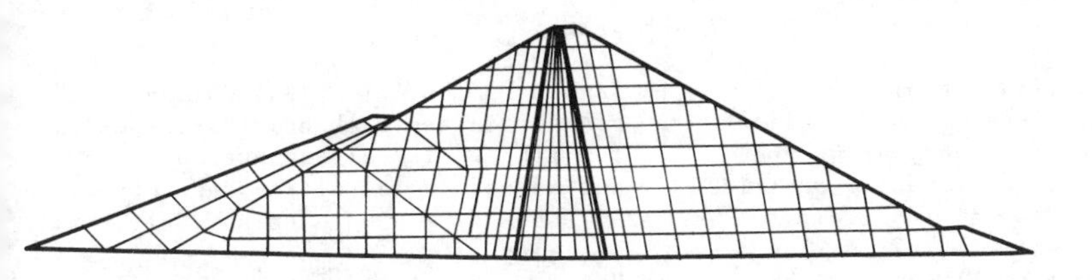

Fig.5 Finite element mesh

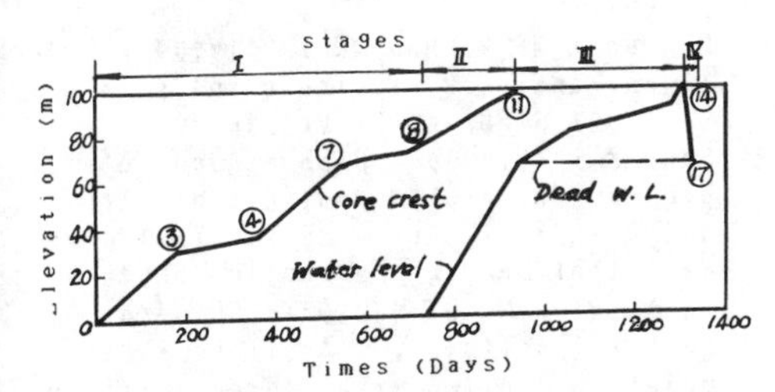

Fig.6 Schedule of construction

bottom at the end of first stage is highest (250 kPa) with Duncan's model, and lowest (170 kPa) with Cambridge model, and the new model gives a moderate value of 220 kPa.

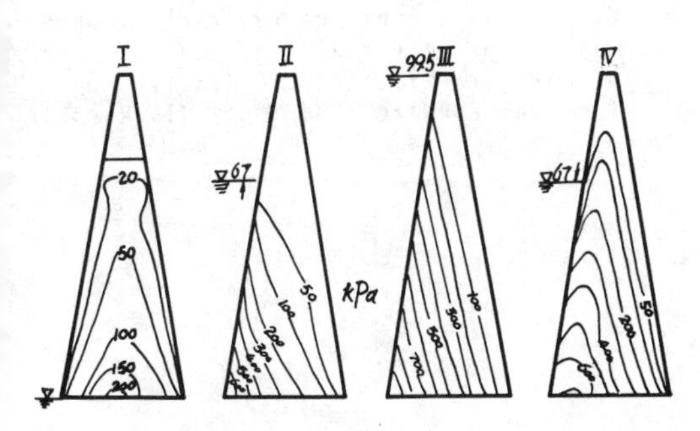

Fig.7 Pore pressure distribution in core

Because of quite high permeability of the core material, no excess pore pressure can be detected obviously after increase of water head in the front of core.

Fig.8a shows the settlement distribution along the center line of core at the completion of the dam construction. In addition to the results of finite element analyses with 3 models, also shown is a curve obtained by the conventional method used in settlement anaalysis of foundation. The later must overestimate the settelement because of assuming no load transfer from core to shell. The maximum value of settlement from these 4 estimations are : ① Cambridge model - 93^{cm}; ② new model - 100^{cm}; ③ Duncan's model - 149^{cm}; ④ conventional method - 171^{cm} .

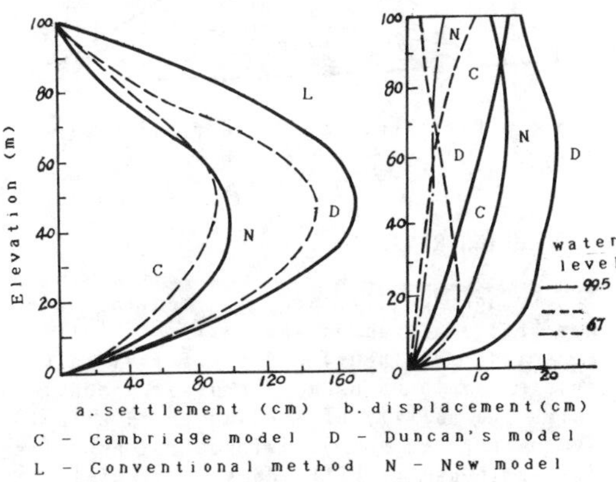

Fig.8 Movement of core center line

Fig.8b shows the horizontal movement of center line of core when the water level has rised from $\nabla 67^{m}$ to $\nabla 99.5^{m}$ and drawn down back to $\nabla 67^{m}$. The maximum displacements are : Cambridge model - 15.2^{cm} , new model - 15.2^{cm} , and Duncan's model - 22.3^{cm}, but according to later two models the maximum responses to the reservoir impounding are located at $^2/_3$ height of dam. After draw-down of water level, all 3 models give some amount of residual displacement.

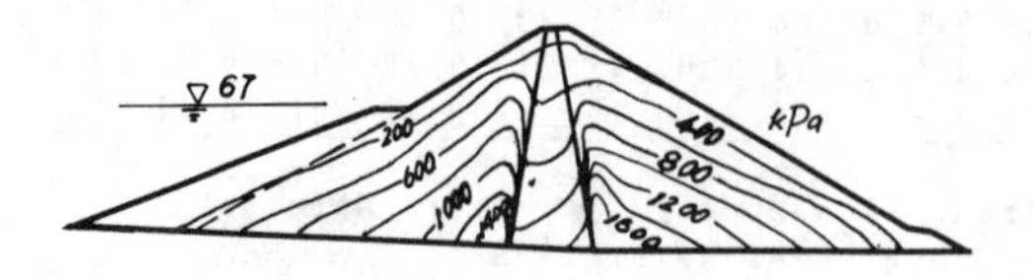

Fig.9 Distribution of major principle stress

Fig.9 shows the distribution of major principle stress at the end of construction. The load transfer effect is very pronounced and the stress distribution is not symmetric because of buoyancy effect of reservoir water which have been anticipated to rise to the dead water level of $\nabla 67$ ^{m} at this time. Fig.10 shows the distribution of minor principle stress at the high reservoir water level. Inspite of substantial drop of effective stress due to increase of water pressure, no tensile stress is obtained except for some zone near upstream slope, hence no danger of hydraulic fracture is predicted. The results obtained with the new elasto-plastic model, and with Cambridge's are similar, but Duncan's model yields much larger zone of tensile stress at the highest water level.

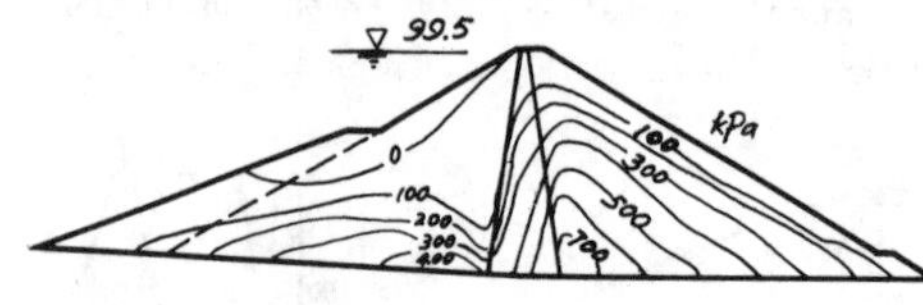

Fig.10 Distribution of minor principle stress

CONCLUDING REMAK

The Lubuge rockfill dam with central core has been analyzed on the basis of Biot's consolidation theory and 3 stress-strain models have been used. The paper demonstrates the ability of the suggested effective stress method of deformation analysis for earth and rockfill dams to simulate their response to the construction load and reservoir water pressure and give overall picture of movement and stress-strain and pore pressure distribution of dam body in the various stage of dam constuction and reservoir operation. The results of analysis with the new developed double hardening elasto-plastic model seem quite reasonable in comparison with the Cambridge and Duncan's model, though the final conclusion could be made only after getting the data of field measurement.

ACKNOWLEDGEMENT

This study is the part of the research program supported by the Science Foundation Council of Ministry of Water Conservancy and Electric Power.

REFERENCES

[1] Duncan J.M. et al.(1980) :Strength, stress-strain and bulk modulus parameters for finite element analyses of stress and movements in soil masses, Report No.UCB / GT / 80-01, Uni.of California, Berkeley

[2] Duncan J.M. et al.(1981) : CON2D - A finite element computer program for analysis of consolidation, Report No.UCB / GT / 81-01, Uni. of California, Berkeley

[3] Duncan J.M. & Chang C.Y. (1970) : Non-linear analysis of stress and strain in soils, J.SMFD, ASCE, Vol.96,SM5

[4] Shen Z.J. (1983) : BISS - A computer program for Biot's analysis of stress and strain of earth and rockfill dams, Report of Nanjing Hydraulic Research Institute, No.GT-8306.[In Chinese]

[5] Shen Z.J. (1984) : A stress - strain model for soils with three (triple) yield surfaces, Acta Machanica Solida Sinica, No.2 [In Chinese]

[6] Shen Z.J. (1987) : A new model for stress-strain analysis of soils, Paper presented to 5th Chinese Conf. SMFE [In Chinese]

[7] Zaretski U.K. & Orhov V.V. (1982) : Effect of schedule of reservoir impounding on the stress-deformation response of earth-rockfill dam, Hydraulic Engineering Construction, No.3 [In Russian]

Numerical Methods in Geomechanics (Innsbruck 1988), Swoboda (ed.)
© 1988 Balkema, Rotterdam. ISBN 90 6191 809 X

Effect of tectonoclastic zones in rock foundation on dam stress

J.C.Chen
Fourth Design and Survey Institute, Harbour Eng. of Ministry, Guangzhou, People's Republic of China

ABSTRACT: This paper deals with an engineering practice of an effect of tectonoclastic zones in rock foundation on dam stress analyzed by a finite element method. The dam is located at an upstream branch of the Yangtze River and is a 112m high concrete gravity dam with a crest spillway. Its foundation was composed of phyllite with two large dipping upstream tectonoclastic zones and an intrusive mass of igneous rock near the downstream of dam toe. Their elastic modulus varied in range of above ten times. In order to confirm the posibility of dam construction in the complex foundation the stresses of the foundation and dam were analyzed and the treatment measures for the tectonoclastic zones were studied.

1 GENERAL DESCRIPTION AND MAIN DATA

This paper is based on a report on dam stress study of a hydroelectric station in design stage. There is a typical geological section of a spillway dam section, the geometrical size of which was simplified and given in Fig. 1.

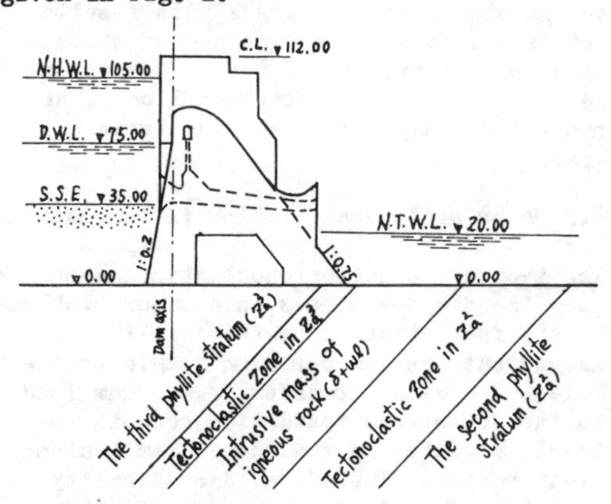

Fig. 1 A simplified geological section of spillway dam section

Water level(relative elevation):

Normal high water level at upstream	**105.00m**
Normal tailwater level	**20.00m**
Dead water level at upstream	**75.00m**
Silt sediment elevation at upstream	**35.00m**

The physical and mechanical properties of the dam concrete and rock in foundation are shown in Table 1.

Table 1. Adopted elastic modulus for the dam and rock in foundation

Positions	E(Mpa)	μ
Dam concrete	1.96×10^4	1/6
The third stratum of phyllite(Za^3)	10^4	0.25
Tectonoclastic zone in Za^3	3.92×10^3	0.27
Intrusive mass of igneous rock	1.96×10^4	0.2
The second stratum of phyllite(Za^2)	7.84×10^3	0.25
Tectonoclastic zone in Za^2	1.96×10^3	0.32

2 PURPOSE OF STUDY

The study is to prove an effect of the second and the third strata of phyllite(Za^2, Za^3) in foundation on the dam stress under combination condition of design load and check load, the safety of the dam with a tensile stress in its upstream heel and distribution of the foundation stresses as well as to take treatment measures for the tectonoclastic zones, e.g. observe the improvement of the dam stress after removing the tectonoclastic zone in Za^3 in the toe

downstream to the depth of 15m(equal to two times of width for the tectonoclastic zone in Za^3) and filling a concrete plug.

3 METHOD OF CALCULATION

The finite element method was used for the stress calculation, regarding it as a two-dimension problem of elasticity mechanics. The dam was calculated as a plane stress problem considering its variation in thickness, whereas the foundation was computed on basic of one dam section(19m) as a plane strain problem by using different moduli. Several alternatives of computation are shown in following Table 2.

Table 2. Several cases of stress computation under combination of loads for design and for check*

Cases	Condition of foundation
1	Homogeneous foundation(for dam and foundation E=1.96 x 10^4Mpa)
2	Complex foundation without tectonoclastic zone
3	Complex foundation with a tectonoclastic zone in Za^3
4	Complex foundation with tectonoclastic zones in Za^2 and Za^3
5	In case 4 tectonoclastic zone in Za^3 removed to depth of 15m with filling concrete plug
6	In case 4 fissured rock in the vicinity of the dam heel

* Combination of loads
For design: Water pressure +silt pressure +dead weight of dam
For check: Water pressure +silt pressure +dead weight of dam +water pressure due to earthquake +silt pressure due to earthquake +inertia force of earthquake

The triangular elements were used for element's division. Considering that element's nets were divided according to the dimensions and elastic modulus, the nets were suitably thickened at the places where stress was concentrated. It assured that boundary planes of the zero displacement were located at a distance of 1.5 times of the dam height respectively away from the upstream, downstream and the base of the dam. The element's division and axes were used as shown in Fig. 2.

The standardized program of computer was used for calculation.

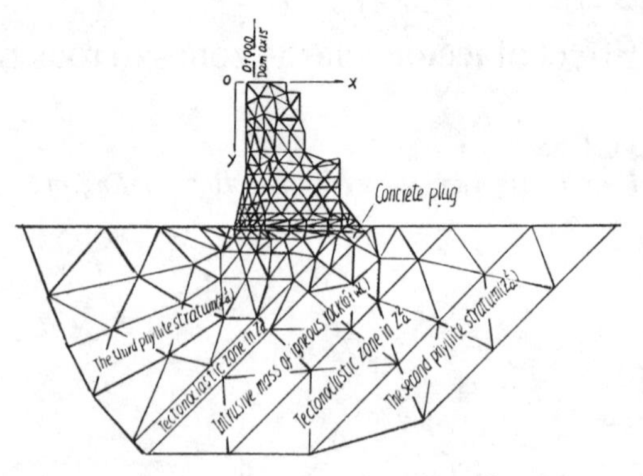

Fig. 2. Element's division for the dam and foundation

4 ANALYSIS FOR RESULTS OF CALCULATION

4.1 Effect of rock properties on dam stress

The results of computation indicate that the effect of physical and mechanical properties of the foundation rock and its geological structure on the stress of the concrete gravity dam is only limited on the area near the foundation, the height of which doesn't exceed 1/3 to 1/4 height of the dam(shown in Fig. 3). From stress values it is seen that a compression stress occurred in all positions except tensile stress in range of 2.5m height of the dam heel. Therefore it verified that the stress variation agrees with results obtained from a classic method of calculation for gravity dam. However, the stress in the foundation is more complicated and the failure may take place in its weak area.

4.2 Major geological problem in foundation

The two very wide tectonoclastic zones near the dam downstream is a major problem in the foundation. From the Fig. 1 it is shown that the dam rests on a thin triangular rock mass. Transfer of the dam load to the downstream foundation rock is seriously affected by the downstream tectonoclastic zones. Therefore, dam stability and stress conditions are very unfavorable.

Fig. 3 and 4 show the results of stress under design loads in the complicated geological condition. Results of calculation demonstrate that the priciple tensile stress near the dam heel increased due to existence of the tectonoclastic zone in the second stratum of phyllite(Za^2) (see Table 3). The tectonoclastic zone in the second stratum is wider and its elastic modulus is smaller, whereas the tectonoclastic zone in the third stratum is thinner and its elastic modulus is higher than that of second stratum by two times, so the effect of the tectonoclastic zone in the third stratum on the stress of dam heel is less (see Table 4). But the dam toe directly presses on the tectonoclastic zone in the third stratum so that the stress is concentrated.

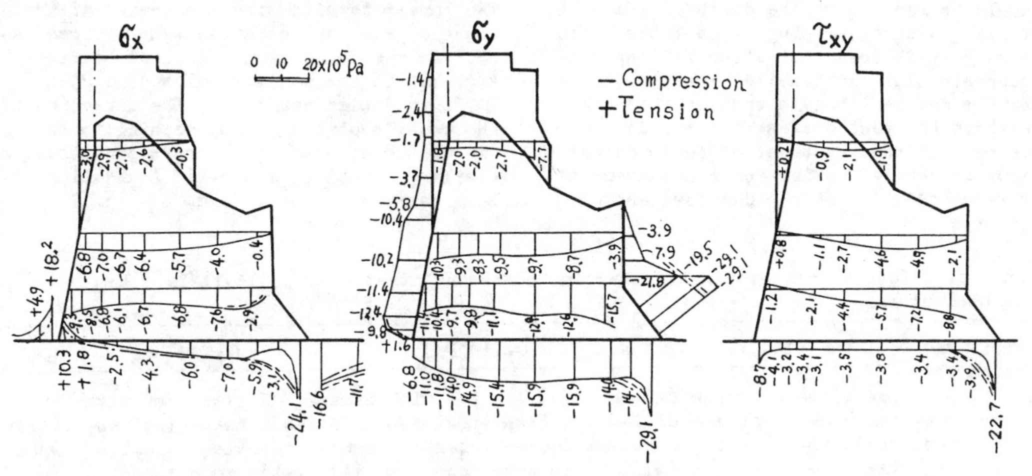

Fig. 3 Distribution of normal stresses(for design)

——— Za^2 and Za^3 with tectonoclastic zones

------ Za^3 with tectonoclastic zone

—.—. Without tectonoclastic zone

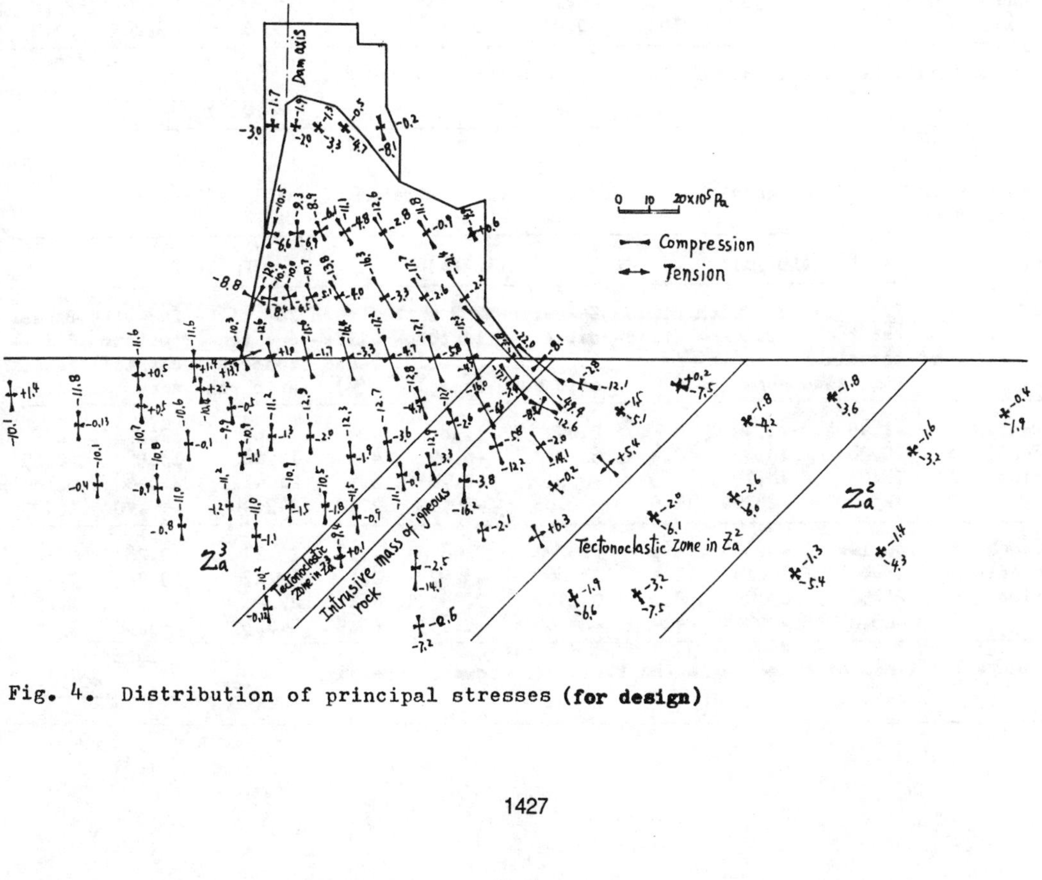

Fig. 4. Distribution of principal stresses **(for design)**

4.3 Improvement of stress in dam heel after treatment

In order to improve the stress condition at the upstream dam heel the tectonoclastic zone in the third stratum of phillite(Za^3) would be removed to the depth of 15m with filling a concrete plug. From Table 5 and Fig. 5 it is seen that after filling the concrete plug the tensile stress at the upstream dam heel became smaller than that without the tectonoclastic zone. It indicates that the treatment of tectonoclastic zone is very effective for improvement of stress condition at the dam heel and of great use for transfer of the compressive stress at the dam toe.

4.4 Tensile stress at lower edge of igneous rock mass

The larger tensile stress occurred at the lower edge of an intrusive mass of igneous rock in the foundation with weak bands(see Fig. 4). It has maximum value of 6.25 x 10^5Pa in design condition. The direction of the tensile stress is approaching to the dip of the stratum. Its stress condition is similar to that of a elastic foundation beam carrying a local load.

Table 3. Effect of tectonoclastic zone in the second stratum of phillite(Za^2) on dam heel stress

Combination of loads	Number of element	σ_x (10^5Pa)			σ_y (10^5Pa)			σ_1 (10^5Pa)		
		Za^2 without tectonoclastic zone	Za^2 with tectonoclastic zone	Stress difference (%)	Za^2 without tectonoclastic zone	Za^2 with tectonoclastic zone	Stress difference (%)	Za^2 without tectonoclastic zone	Za^2 with tectonoclastic zone	Stress difference (%)
Design condition	D_1	-2.99	-1.49	50	1.15	2.07	80	9.83	12.63	28.5
	D_2	6.70	10.28	53.4	-6.75	-6.42	5	10.39	14.08	35.5
	F	13.14	18.03	37.2	-3.63	-2.61	28	16.43	21.41	30.3
Check condition	D_1	-1.75	-0.14	92.0	6.50	7.62	17.2	16.15	19.21	19
	D_2	9.93	13.80	39.0	-4.27	-3.86	9.6	15.33	19.29	25.8
	F	17.31	22.56	30.3	-0.90	0.24	127	21.91	27.26	24.4

Remark Positions of elements in the table are shown in the Fig.

Table 4. Effect of tectonoclastic zone in the third stratum of phillite(Za^3) on dam stress

Combination of loads	Number of element	σ_x (10^5Pa)			σ_y (10^5Pa)			σ_1 (10^5Pa)		
		Za^3 without tectonoclastic zone	Za^3 with tectonoclastic zone	Stress difference (%)	Za^3 without tectonoclastic zone	Za^3 with tectonoclastic zone	Stress difference (%)	Za^3 without tectonoclastic zone	Za^3 with tectonoclastic zone	Stress difference (%)
Design condition	D_1	-1.49	-1.49	0	2.07	1.58	-24	12.63	12.27	-2.9
	D_2	10.28	10.32	0.4	6.42	-6.67	4	14.08	13.96	-0.85
	F	18.03	18.19	0.9	-2.61	-2.78	6.5	21.41	21.44	0.14
	D_3	-23.05	-24.08	4.5	-22.86	-29.09	27.2	-42.46	-49.40	16.34
Check condition	D_1	-0.14	-0.004	-97	7.62	7.48	-1.8	19.21	19.28	0.36
	D_2	13.8	14.13	2.4	-3.86	-3.97	2.8	19.27	19.54	1.4
	F	22.56	23.08	2.3	0.24	0.25	4.2	27.26	27.71	1.65
	D_3	-25.04	-26.14	4.4	-24.89	-31.33	25.9	-46.37	-53.58	15.5

Remark Positions of elements in the table are shown in the Fig.

Table 5. Effect of concrete plug on dam heel stress

Combination of loads	Number of element	σ_x (10^5Pa)			σ_y (10^5Pa)			σ_1 (10^5Pa)		
		Za^3 with tectonoclastic zone	Concrete plug	Stress difference (%)	Za^3 with tectonoclastic zone	Concrete plug	Stress difference (%)	Za^3 with tectonoclastic zone	Concrete plug	Stress difference (%)
Design condition	D_1	-1.49	-3.13	-110	1.58	-3.36	-330	12.27	6.27	-49
	D_2	10.32	6.76	-34	-6.67	-8.89	33	13.96	9.01	-35
	F	18.19	13.87	-24	-2.78	-5.39	94	21.44	16.04	-25
Check condition	D_1	-0.004	-1.81		7.48	1.89	-75	19.28	12.59	-35
	D_2	14.13	10.28	-27	-3.97	-6.36	60	19.54	14.08	-28
	F	23.08	18.41	-20	0.25	-2.57		27.71	21.78	-21
Remark	Positions of elements in the table are shown in the Fig. 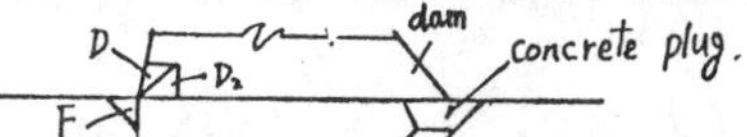									

4.5 Tensile stress in dam heel and safety of dam

The tensile stress occurred near the dam heel will change the stress condition of rock in this region. Therefore, considering that the redistribution of stress after rock cracking is significant to evaluate the safety of dam, two methods were used in computation. The first method is to reduce elastic modulus of rock elements in tension region to 1×10^9Pa. The results of computation(Fig. 5) indicate that the tensile stress at the dam heel is decreasing with reducing of the elastic modulus in the tension region. The second method is to assure E=o in the tension region. Then the tensile stress will turn into the compressive stress. It demonstrates that the rock cracking to a certain degree would be in a steady state and doesn't develop unlimitedly. So we can confirm that the tensile stress at the upstream dam heel is local and wouldn't affect the safety of the dam.

5 CONCLUSION

In the complicated rock foundation the tectonoclastic zones with low elastic moduli have an unfavorable effect on the dam and foundation stresses near the dam heel, but it is limited on the region of about 1/3 to 1/4 dam height near the foundation. After treatment of the tectonoclastic zone in the third stratum at the dam toe their stress condition would be improved. In further analysis and study of the tensile stress at the lower edge of the intrusive mass of igneous rock the nonlinear deformation of the tectonoclastic zones should be taken into consideration.

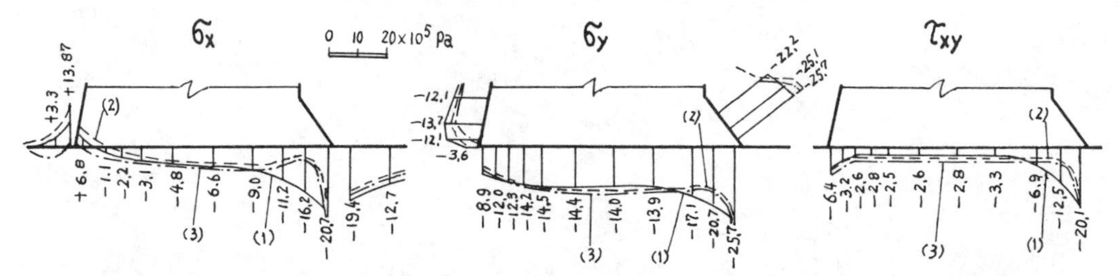

Fig. 5. Distribution of normal stresses under conditions of filling a concrete plug in Za^3 at dam toe and of rock cracking at dam heel
(1) Weak band in Za^3 to depth of 15m was removed and concrete plug was filled. (2) There are tectonoclastic zones respectively in Za^2 and Za^3. (3) There are tectonoclastic zones respectively in Za^2 and Za^3 and rock cracking in foundation at dam heel.

REFERENCES

1 Finite element method for elastic mechanics problems, written by Hydraulic Engineering College in Eastern China, Hydraulic-Electric Publication, 1978
2 Standardisation of programs for calculation of gravity dam with finite element method. Proceedings of conference on technical experiences exhange for usage of electronic computer in hydraulic and hydroelectric engineering in China 1974.

Numerical Methods in Geomechanics (Innsbruck 1988), Swoboda (ed.)
© 1988 Balkema, Rotterdam. ISBN 90 6191 809 X

Semi-analytical finite element analysis of laterally loaded piles

N.S.V.Kameswara Rao
Indian Institute of Technology, Kanpur
S.K.Prasad
M.S. Ramaiah Institute of Technology, Bangalore, India

ABSTRACT: The present study focusses on the behaviour of the Pile-Soil system, subjected to static lateral load within elastic limits using Semianalytical Finite Element technique. Straight, vertical piles in homogeneous soils, layered soils, soils with linearly varying modulus, as well as Battered and the Tapered piles subjected to lateral load have been studied. The results are presented in nondimensional form. The effect of provision of interface elements between the pile and soil media has been discussed. The results, from the above technique are compared with those of two and three dimensional Finite Element analyses and the relative computational efforts required for the above methods is presented.

1. INTRODUCTION

Modern structures (off-shore and onshore) exert complex loading - axial load, lateral load, biaxial moment, torsional moment on the foundation. Pile foundation becomes obvious choice in most cases. Piles transfer the forces to firm soil partly through end bearing and partly through side friction between soil and pile. The mechanism of transfer of these forces depends on the factors such as the magnitude and nature of force, depth of embedment, strength and stiffness of soil, interface characteristics and method of installation.

Piles derive their support against axial loading through base and/or shaft resistance. The lateral load is resisted by the normal earth pressure and the shear stresses acting on the shaft and base. Pile foundation may be socketed (in contact with rock, deriving resistance at base) or floating (deriving resistance only along shaft). Sometimes, the pile may be bent, battered or curved due to some reason or it may be tapered. Usually piles are used in groups.

Pile foundation, even a single pile is statically indeterminate to a very high degree. Hence, the chance of precise analysis of the pile is remote. Usually empirical knowledge and experimental results are being used in the design and analysis of Pile foundations.

In recent years many numerical techniques such as Finite Difference Method, Variational approach, Boundary Element technique, Finite Element method are being applied for the analysis of pile foundations. The present study is focussed on the behaviour of Pile-Soil system subjected to lateral load using special Finite Element technique called Semi-analytical Finite Element Method, as this method has established itself as a powerful numerical tool for solving problems involving complicated geometry, material behaviour and Boundary conditions.

Analysis of single pile subjected to static lateral load in elastic continuum has drawn the attention of many investigators over the years. This analysis can essentially be classified into the following four approaches.

1. the Subgrade Reaction or Winkler analysis in which the pile is idealised as an elastic transversely loaded beam supported by a series of unconnected linear springs representing the soil.

2. the Bounday Element or Integral Equation analysis in which the soil mass is modelled as semiinfinite elastic continuum. In this approach, Mindlin's equation for horizontal displacement due to horizontal load is applied to evaluate the displacements.

3. the Finite Element technique using either two or three dimensional soil pile idealisation, and

4. Semi analytical Finite Element technique in which, one dimension is taken care of by expressing the displacement field interms of Fourier series in the polar coordinate (θ) and the remaining dimensions by Finite Elements.

Simplest and the earliest idealisation for laterally loaded single pile behaviour is by using Winkler model for the foundation. Broms (1964) was among the first to use this model. Dispite the mathematical convenience and simplicity of this model, it does not account for the continuity of the soil mass and hence, the result cannot be directly related to any of the material properties of the soil.

More refined analysis using Boundary Element technique has been made by Poulos (1971), Banerjee and Davies (1978). In this model pile is idealised as a thin vertical strip and the effect of presence of pile has been neglected. The interface joint cannot be actually simulated and detailed three dimensional idealisation is difficult.

Using Finite Element model, Yegian and Wright (1973) have made Plane stress idealisation. Desai and Appel (1976) have used three dimensional idealisation for laterally loaded piles. However, a three dimensional treatment leads to computational complication and large storage space is required in a computer.

Kuhlemeyer (1979) and Randolph (1981) have attempted Semianalytical Finite Element technique for laterally loaded piles. The present study is based on the above approach. In this study, pile subjected to lateral load in homogeneous soil, layered soil and soil with varying modulus has been analysed. Battered pile and the tapered piles are also dealt with. The effect of use of interface elements at the joints between the pile and soil is studied. The results are compared with those of two and three dimensional higher order Finite Elements.

2. ANALYTICAL FORMULATION

A brief outline of Semianalytical Finite Element displacement formulation as applied to the problems in statics used in the present problem is given below. In case of laterally loaded piles, the geometry and the material property along one direction (θ direction in polar coordinate system) remain unchanged. However, load terms vary preventing the use of simplified assumption such as axisymmetric idealisation. Hence, Fourier series is applied to represent the solution in this direction, (Wilson (1965)). The soil-pile system in radial plane (r z-plane) is discretised using Finite Elements. Quadratic, eight noded, isoparametric elements (four corner nodes and four mid-side nodes) have been used for the analysis. Similar elements are used for two dimensional plane stress Finite Element formulation. Six noded interface elements are used to model the properties of the joint between the pile and soil medium. In three dimensional Finite Element idealisation soil pile medium is modelled by linear isoparametric elements each having eight corner nodes and quadratic isoparametric elements each having eight corner nodes and twelve midside nodes.

The shape function defining the variation of displacement 'u' in Semianalytical Finite Element formulation can be written as $u = N(r, z, \theta)\, a^e$

Here, a^e is nodal displacement & N is shape function.

$$u = \sum_{l=1}^{NH} [\bar{N}(r,z)\cos l\theta + \bar{\bar{N}}(r,z)\sin l\theta]\, a_l^e \quad -(1)$$

Here, l varies from 1 to NH where NH is the total number of harmonics.

The strain-displacement matrix [B] includes sine and cosine terms and is given by,

$$[B] = \begin{bmatrix} \frac{\partial N_i}{\partial r}\sin l\theta & 0 & 0 \\ 0 & \frac{\partial N_i}{\partial z}\sin l\theta & 0 \\ \frac{N_i}{r}\sin l\theta & 0 & \frac{lN_i}{r}\sin l\theta \\ \frac{\partial N_i}{\partial z}\sin l\theta & \sin l\theta & 0 \\ -\frac{lN_i}{r}\cos l\theta & 0 & \left(\frac{\partial N_i}{\partial r}-\frac{N_i}{r}\right)\cos l\theta \\ 0 & -\frac{lN_i}{r}\cos l\theta & \frac{\partial N_i}{\partial z}\cos l\theta \end{bmatrix} \quad (2)$$

The displacement components are described as

$$u^l = [N_1, N_2 \dots]\sin l\theta\, u^{le}$$
$$v^l = [N_1, N_2 \dots]\sin l\theta\, v^{le} \quad (3)$$
$$w^l = [N_1, N_2 \dots]\cos l\theta\, w^{le}$$

and the nodal loads are described as

$$R = \sum_1^{NH} \bar{R}^l \sin l\theta$$
$$Z = \sum_1^{NH} \bar{Z}^l \sin l\theta \quad (4)$$
$$T = \sum_1^{NH} \bar{T}^l \cos l\theta$$

where $\bar{R}$, $\bar{Z}$ and $\bar{T}$ are nodal forces per unit length of circumference in radial, axial and angular directions. The stress-strain matrix [D] is same as three dimensional case and two dimensional Jacobian matrix is used for shape function. In the present study, for the integration of the stiffness matrix [K], the numerical integration is carried out by Gauss-Legendre Quadrature method and the simultaneous equations are solved using Choleski Decomposition technique.

A two dimensional six noded interface element is used along with eight noded Pile soil system to represent the joint between Pile and soil.

3. RESULTS AND DISCUSSION

The above formulation has been adopted to analyse the behaviour of Pile soil system under lateral load and the results are compared with those of two dimensional plane stress idealisation and three dimensional idealisation using Finite Element technique.

The basic problem of a single laterally loaded pile is as shown in the Figure 1 wherein,

H = Applied horizontal force
P = Applied axial force
M = Applied axial moment
Θ = Applied torsional moment
L = Length of embedment of the pile
D = Diameter of the pile

E_p, ν_p = Young's modulus and Poisson's ratio of pile.

E_s, ν_s = Young's modulus and Poisson's ratio of soil.

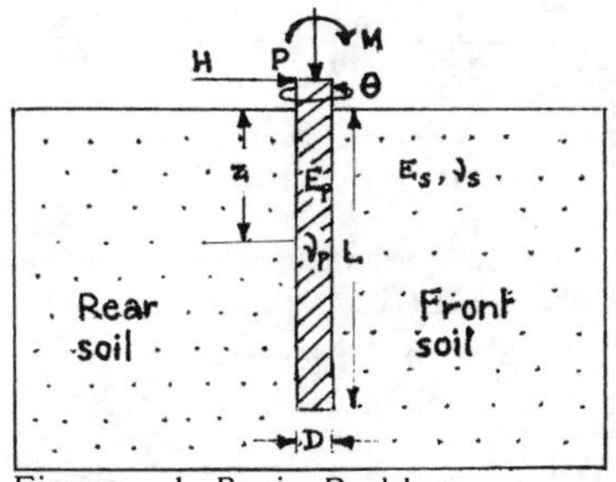

Figure 1. Basic Problem.

For the purpose of discussion, the portion of soil in the direction of horizontal force is called as Front soil and that on the other side is called Rear soil. In case of Battered piles, batter provided in the direction of the load is called Positive batter and that in the other direction is called negative batter. Similarly, in the Tapered piles the angle of inclination which the pile surface makes with the vertical axis is called Taper angle.

3.1 Comparison of results

The present results are comparable with the existing results. For illustration, the comparisons are presented in Figure 2 and figure 3. In Figure 2, lateral displacement of pile with L = 0.61m, D=0.013m and E_p = 2 X 10^{11} Pa in

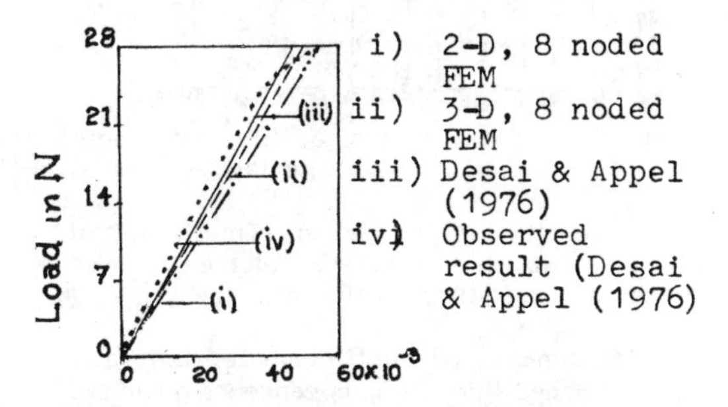

Figure 2. Comparison of lateral displacement results.

a homogeneous soil having E_s = 1.7 X 10^6 Pa at different loads is compared by different methods (Desai and Appel (1976)). In Figure 3, lateral displacement of pile with L = 25m,

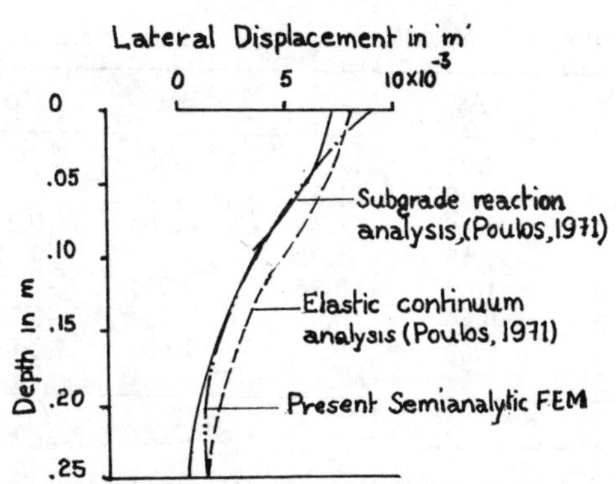

Figure 3. Comparison of lateral displacement results.

D=1m, E_pI_p = 3.9 X 10^5N-m^2 in homogeneous soil of E_s = 5 X 10^5 Pa at different depths, subjected to a horizontal load of 200N is compared by different methods (Poulos (1971)).

It has been observed that two dimensional Finite Element method generally gives higher displacements in comparison with those of three dimensional idealisation.

The displacements obtained by semi analytical analysis are slightly higher than those of three dimensional idealisation.

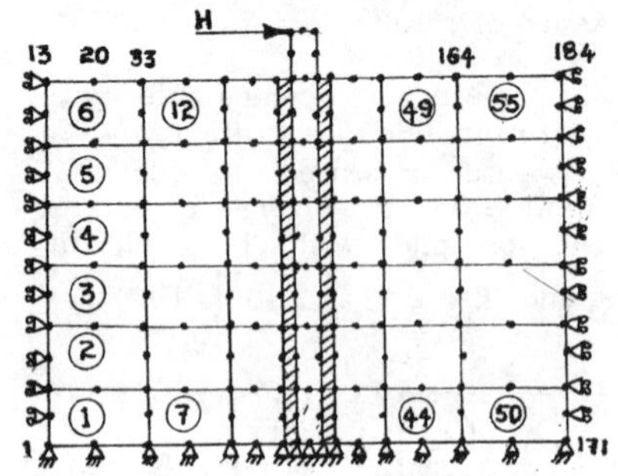

Figure 4. Typical Finite Element mesh with interface elements.

A typical Finite Element mesh detail along with the interface elements used in the above analysis is shown in the Figure 4.

3.2 Nondimensional coefficients for Laterally Loaded Piles in homogeneous soil mass.

Non-dimensional coefficients for lateral displacement, slope, moment and shear at different depths, in homogeneous soils are given in Table 1 in similar lines of Reese and Matlock (1956). The non-dimensional results allow for general and qualitative conclusions to be drawn.

Table 1. Non-Dimensional Co-efficients.

z	Ay	As	Am	Av
0.0	2.90	2.05	0.00	--
0.5	2.50	1.85	2.65	0.33
1.0	2.15	1.70	4.10	0.25
2.0	1.55	1.30	4.05	0.08
3.0	1.05	0.95	3.05	-0.12
5.0	0.40	0.45	1.30	-0.05
8.0	0.00	0.10	1.60	0.06

A characteristic length T_L is used instead of actual length of pile as lateral displacement will be negligible for most of the length. Non-dimensional parameters adopted in the present work are defined as follows:

Characteristic Length, $T_L = [E_pI_p/E_s]^{1/4}$

Depth coefficient, $Z = z/T_L$

Displacement coefficient, $Ay = (yE_pI_p)/(HT_L^3)$

Slope co-efficient, $As = (SE_pI_p)/(HT_L^2)$

Moment co-efficient, $Am = M_B/(HT_L)$ and

Shear co-efficient, $Av = V_s/H$ where

E_pI_p = Flexural stiffness of pile.

z = any depth varying from Zero to L.

E_s = Soil modulus, and at any depth

y = Lateral displacement of pile.

S = Slope of pile.

M_B = Bending moment of pile and

V_s = shear force in the pile.

3.3 Laterally loaded Piles in two layered soil system.

Following nondimensional parameters are used to study the behaviour of laterally loaded piles in two layered soil system and results are shown in Figures 5 and 6.

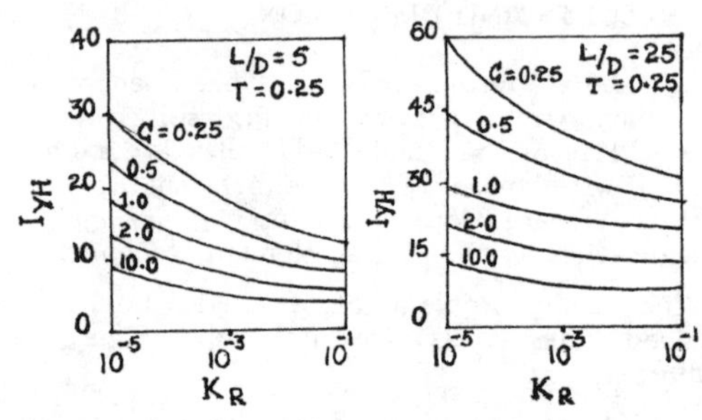

Figure 5. Variation of IyH with K_R and C.

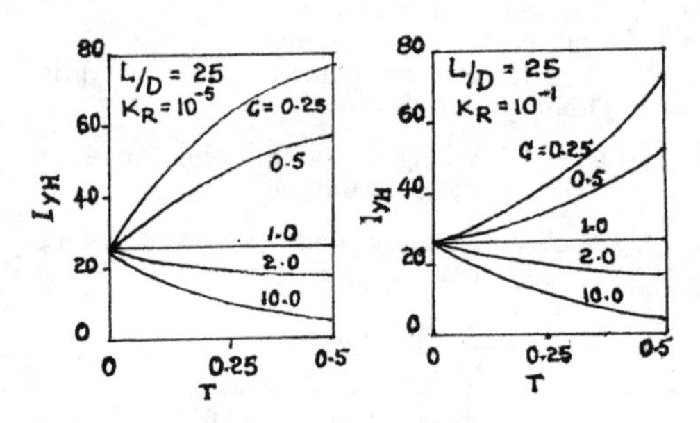

Figure 6. Variation of IyH with T and C.

Layer thickness ratio, T = Ls/L

Layer coefficient, C = Et/Eb

Pile flexibility factor, $K_R = E_pI_p/E_bL^4$

Lateral ground displacement of pile,

$Yg = I_{yH}\ (H/E_bL)$

Depth Factor, DF = z/L where

Ls = Depth of the surface (top) layer.

E_t, E_b = Soil modulus for top and bottom soil respectively.

IyH = Influence factor for lateral displacement,

In Figure 5, IyH is plotted against K_R for L/D = 5 and 25 at different C and T. It is seen that IyH at ground decreases as either K_R or C increases at constant L/D. At any K_R and C, IyH increases as L/D increases.

Figure 6 shows IyH plotted against T for typical L/D, K_R and C values. For C= 1, IyH remains horizontal. At $C > 1$, IyH will be below and at $C < 1$, IyH will be above the C = 1 line. Hence, with increase in C, IyH decreases considerably.

3.4 Laterally loaded Piles in soils with increasing modulus.

Idealisation of soils as modulus linearly increasing with depth gives a more realistic representation of cohesionless soils. In the present study, a socketed pile with $E_p = 2 \times 10^{11}$ Pa and L/D = 40 subjected to a lateral load of 100 N in soil having Es linearly varying from 0 to 4.2×10^6 Pa is compared with that in a homogeneous soil having Es = 2.1×10^6 Pa. At z/L = 0.15, the magnitude of laterial stress was 2450 Pa and that of shear stress was 700 Pa, less by about 60% and 100% respectively when compared with the stress in soil having constant modulus. However, lateral ground displacement was 0.69mm. In comparison it was 1.15mm for soil with constant modulus.

3.5 Battered Piles

Some times, the piles may have to be inclined to the vertical axis as in case of bearing piles below the dam. This inclination is called Batter angle. In the present study, piles with Batter angle of 15° and 30° (both positive and negative) have been analysed for lateral loads and the results are compared with those of vertical piles. Socketed piles with L/D = 20 and K_R = 5.2×10^{-3} in homogeneous soil medium are considered for the study. The stresses in the neighbouring soils are expressed in non-dimensional form as given below.

Lateral stress factor, $I_{\sigma r} = \sigma_r \times (LD/H)$
Shear Stress factor, $I_{\tau rz}$, $= \tau_{rz} \times (LD/H)$

It has been observed that the lateral deflection of the pile decreases with the increase in Batter angle, being maximum for vertical piles, remaining same for positive as well as negative Batter at a given batter angle. The magnitude of stresses on either side of battered piles will be non-uniform unlike those of straight piles. The typical results at z/L = 0.042 are illustrated in Table 2.

Table 2. Effect of Providing Batter in the pile.

	Batter Angle				
	-30°	-15°	±0°	+15°	+30°
IyH	5.03	5.63	5.73	5.63	5.03
$I_{\sigma r}$ (rear)	-0.56	-0.49	-0.40	-0.31	-0.21
$I_{\sigma r}$ (front)	0.21	0.31	0.40	0.49	0.56
$I_{\tau rz}$ (rear)	-0.25	0.19	-0.10	-0.04	-0.01
$I_{\tau rz}$ (front)	0.01	0.04	0.10	0.19	0.25

Sign Convention

3.6 Tapered Piles

Some times, tapered piles may be used having the cross sectional area decreasing gradually with depth, area at the base being minimum. In the present study, a tapered pile (taper angle = 5°) with L=0.25m and Ep = 1×10^{10} Pa is compared with straight pile having same volume as that of tapered pile in homogeneous soil mass having Es = 1×10^7 Pa. It is observed that the effect of tapering is to decrease the lateral displacement at ground level. At a lateral load of 100 N, the ground displacement for tapered pile was found to be 32×10^{-5}m as against 35×10^{-5}m. for straight pile.

3.7 Effect of Interface element

Often large relative movements occur between soil and pile, causing slip. It is necessary to allow for such movements and to transfer the shear stress across the interface. In the present study, normal stiffness Kn = 1×10^{11} N/m and shear stiffness, Ks = 3×10^6 N/m have been used. For interaction problems the value of normal stiffness is taken to be a very high value (Desai and Appel (1976)). A straight pile with L/D = 25 and K_R = 2.1×10^{-4} is analysed in homogeneous soil mass with Es = 1×10^7 Pa. It is observed that the provision of interface element results in larger lateral displacement at ground level. At a lateral load of 100N, the lateral displacement of pile at ground level was found to be 39×10^{-5}m with interface element, as against 35×10^{-5} for piles without interface elements.

3.8 Comparison of CPU time for the analysis by various methods.

The relative computational time required for different methods - two dimensional Finite Element method, three dimensional eight noded and twenty noded Finite element method and Semianalytical Finite Element method on DEC 1090 system of the Computer centre at I.I.T., Kanpur, India are shown in Table 3. Similar problems were solved by the above methods.

Table 3. Comparison of Computational times.

	2-D 8 noded FEM	3-D 8 noded FEM	3-D 20 noded FEM	Semi Analytical FEM
Elements	35	75	24	32
Nodes	130	144	193	121
CPU Time	7 Sec.	24 Sec.	49.5 Sec.	11.6 Sec.

4. CONCLUSION

The above investigation infers that the Finite Element technique can be conveniently used to analyse the problems of laterally loaded straight, battered, or tapered piles and the soil can be idealised to be homogeneous, layered or to possess linearly varying soil modulus.

A 2-D Finite Element Analysis is sufficient for the problems of laterally loaded piles. Semianalytical Finite Element Analysis may be adopted if 3-D idealisation is necessary. Complete 3-D analysis is required only for very accurate solution.

5. REFERENCES

Broms.B.B.,(1964), 'Lateral resistance of piles in cohesionless soils,' J. of S.M. and F.D.,A.S.C.E., V 90, SM3, pp 123-156.

Banerjee.P.K. and Davies.T.C.(1978), 'The behaviour of Axially and laterally loaded single pile embedded in homogeneous soil,' Geotechnique, V21, No.3, pp 309-326.

Desai.C.S., and Appel. G.C. (1976), 'Three dimensional analysis of laterally loaded structures,' Proc. II Int. Conf., Numerical methods in Geomechanics, Blackburg,Va.

Kuhlemeyer, R.L. (1979), 'Static and dynamic lateraly loaded floating piles,' J. Geotechl. Engg.Dvn, ASCE, V 105, No. GT2, pp 289-304.

Poulos.H.G (1971), 'Analysis of the displacement of laterally loaded piles, Part I Single Pile,' J. of S.M. and F.D., ASCE, V97, SM5, pp 711-731.

Prasad.S.K.(1985), 'Finite Element Analysis of laterally loaded piles,' M.Tech Thesis, I.I.T. Kanpur, India.

Randolph.M.F. (1981), 'The response of Flexible piles to lateral loading,' Geotechnique, V31, No.2, pp 247-259.

Reese.L.C. and Matlock.H.(1956), 'Non dimensional solution for laterally loaded piles with soil modulus assumed proportional to depth,' Proc. 8th Texas Conf. on SM and FE, Austin, Texas pp 71-74.

Wilson.E.I. (1965), 'Structural analysis of Axisymmetric solids,' J. of AIAA, V-3. pp 2269-2274.

Yegian.M. and Wright. S.G.(1973), 'Lateral soil resistance-displacement relationships for pile foundations in soft clays,' Proc. Offshore Technology Conf. Houston, Texas.

Numerical Methods in Geomechanics (Innsbruck 1988), Swoboda (ed.)
© 1988 Balkema, Rotterdam. ISBN 90 6191 809 X

Finite element analysis of certain aspects of deep anchors in cohesive soil medium

D.M.Dewaikar
Indian Institute of Technology, Bombay

ABSTRACT : The ultimate uplift resistance of a cohesive soil, in case of multiunderreamed deep anchor, is estimated for undrained loading conditions, by finite element method, on the basis of its tensile strength and anchor displacements. Multiunderreaming is effected by providing one similar additional anchor base in one case and two additional similar bases in another case. The computations show that multiunderreaming increases the ultimate uplift resistance as compared to the anchor with single base. Further, the computations show that multiunderreaming with increased number of bases may not be more advantageous.

1 INTRODUCTION

The distinctive feature in the analysis of anchor foundation is that the force acting on the soil is opposite to the force of gravity, and as a result, the material is subjected to tensile stresses.

For deep laid anchors, the ultimate uplift resistance in independent of embedment depth and a cavity is formed below the anchor base at failure.

Davie and Sutherland (1977), on the basis of model tests on deep laid anchors in cohesive soils, observed that, experimental uplift resistance values compared favourably with Vesic's (1971) deep anchor theory.

The author (1984), on the basis of model studies on deep anchors in cohesive soil under undrained loading conditions and simulated finite element analysis, has reported that the ultimate uplift resistance could also be estimated with a reasonable degree of accuracy, using finite element analysis, incorporating the influence of tensile strength of soil.

In the present investigations, the finite element analysis is extended to study the effect of multiunderreaming on the ultimate uplift resistance.

2 ANCHOR GEOMETRY

Model anchor with L/B (L = length of embedment, B = anchor base diameter) ratio of 4.17 was selected for analysis. For this anchor, ultimate uplift resistance was determined experimentally under undrained loading conditions and by simulated finite element analysis.

Multiunderreaming was effected in two ways. In the first case, another similar anchor base was provided above the first (lower) base, at a variable spacing as shown in Fig. 1(a).

In the second case, two similar additional anchor bases were provided above the first base, as shown in Fig. 1(b) and the spacing (S) was fixed at 0.67 B.

3 FINITE ELEMENT ANALYSIS

Keeping the other boundary conditions similar to that for anchor with single base (case DA-1), the anchor-soil interaction was

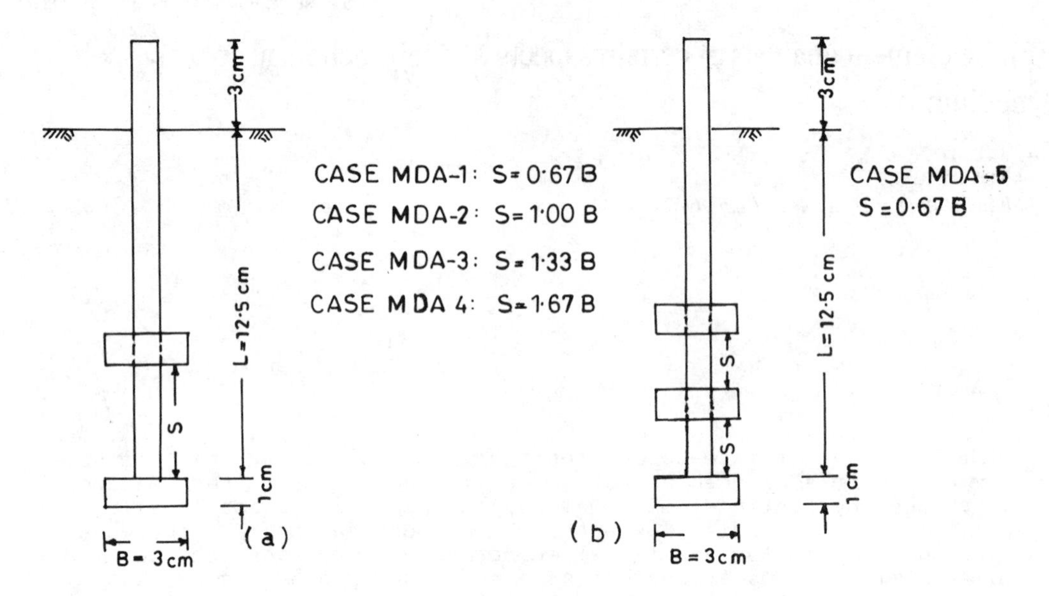

Fig. 1. Geometric details of multiunderreamed anchor

studied using axisymmetric elements of rectangular type. Interface elements were also used in the analysis.

3.1 Idealized material behaviour

The soil behaviour in compression for undrained loading conditions, was modelled to be elastic-perfectly plastic, following von Mises yield condition and associated flow rule.

For tensile stress condition in soil, it was specified that failure by cracking occurred when the principal tensile stresses reached the tensile strength of the soil.

Attempts were made to measure the tensile strength of soil by double punch test (Fang and Chen, 1975). However it was found that saturated soil samples could take very little amount of load. Therefore, consistent with the specified material behaviour, a very low value of tensile strength of 0.001 Kg/cm^2 was assigned to the soil.

The various parameters of anchor-soil system which were used in the analysis are reported in Table 1.

Table 1. Material and interface properties

Brass anchor model
modulus of elasticity - 0.97x10^6 Kg/cm^2
Poisson's ratio - 0.35
Cohesive soil
saturated density-0.0018 Kg/cm^3
cohesion - 0.16 Kg/cm^2
angle of shearing resistance-0.0
tensile strength - 0.001 Kg/cm^2
modulus of elasticity - 80.0 Kg/cm^2
Poisson's ratio - 0.48
Anchor-soil interface
normal stiffness - 10^7 Kg/cm^3
tangential stiffness - 0.0

3.2 Procedure of analysis

The analysis was carried out using initial stress method (Valliappan, 1968). After establishing initial stress conditions due to gravity, load was applied in increments, and for each increment, yielding in compression, failure in tension and tensile stress condition in interface elements were checked.

For yielding in compression, von Mises condition as expressed by the following equation was used.

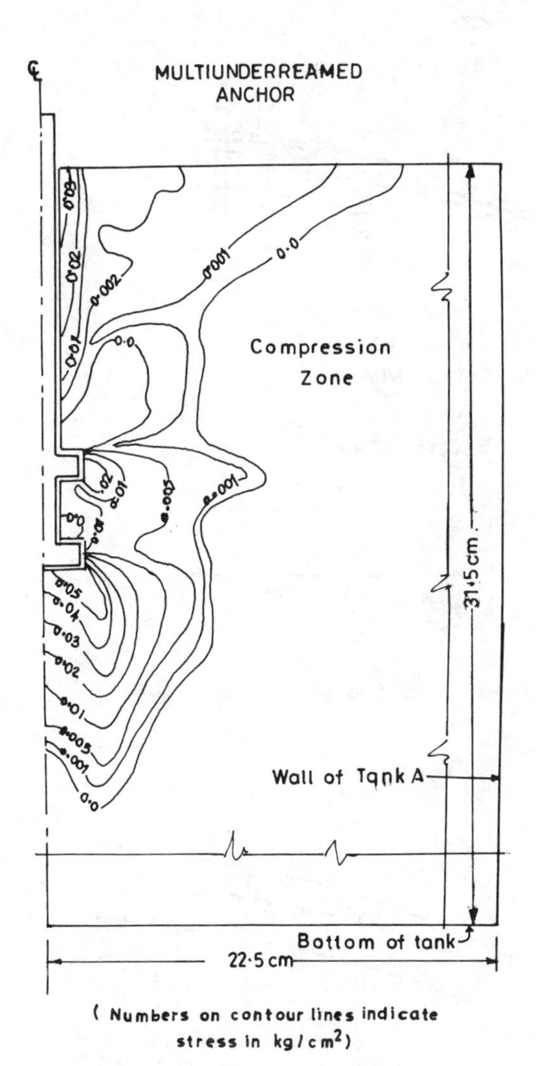

Fig. 2. Contours of maximum principal tensile stresses (case MDA-1)

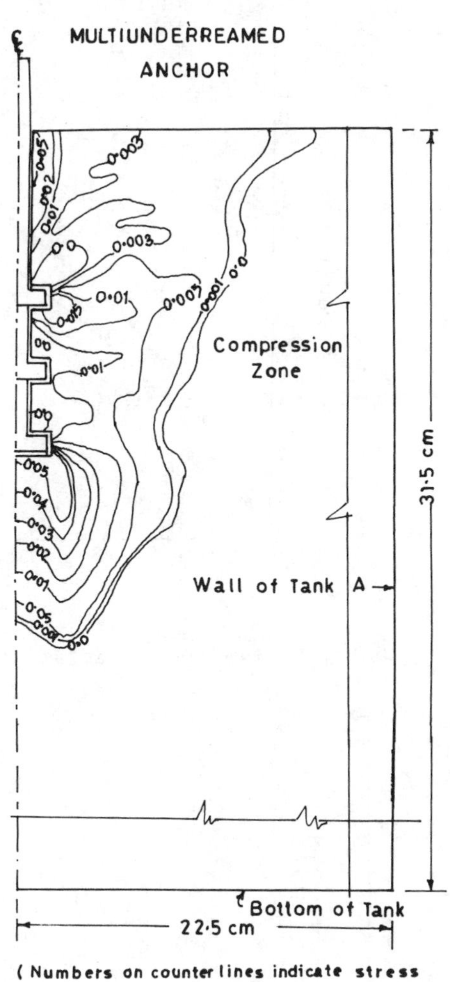

Fig. 3. Contours of maximum principal tensile stresses (case MDA-5)

$$\left\{\frac{1}{2}\left[(\sigma_r-\sigma_\theta)^2+(\sigma_\theta-\sigma_z)^2+(\sigma_z-\sigma_r)^2\right] + 3\tau_{rz}^2\right\}^{1/2} = \bar{\sigma} \quad (1)$$

In the above equation, σ_r, σ_θ, σ_z and τ_{rz}- coordinate stresses for axisymmetric stress condition and $\bar{\sigma}$ represents unconfined compression strength of the soil.

In case, there was an excess above the value of $\bar{\sigma}$, as calculated from the left hand side of Eq. 1, it indicated yielding of the element in compression and the excess was converted into nodal loads.

For elements wherein tension developed, the principle tensile stresses were compared with tensile strength of the soil. If these stresses exceeded the tensile strength, excess was converted into nodal loads and for subsequent iterations and load increments, the element was assigned zero stiffness.

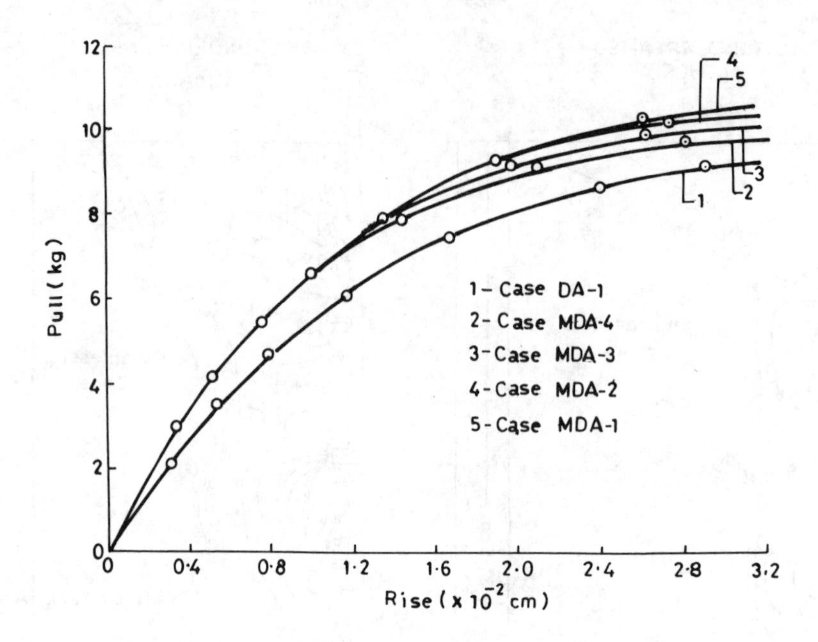

Fig. 4. Pull-rise characteristics (cases DA-1 and MDA-1 to MDA-4)

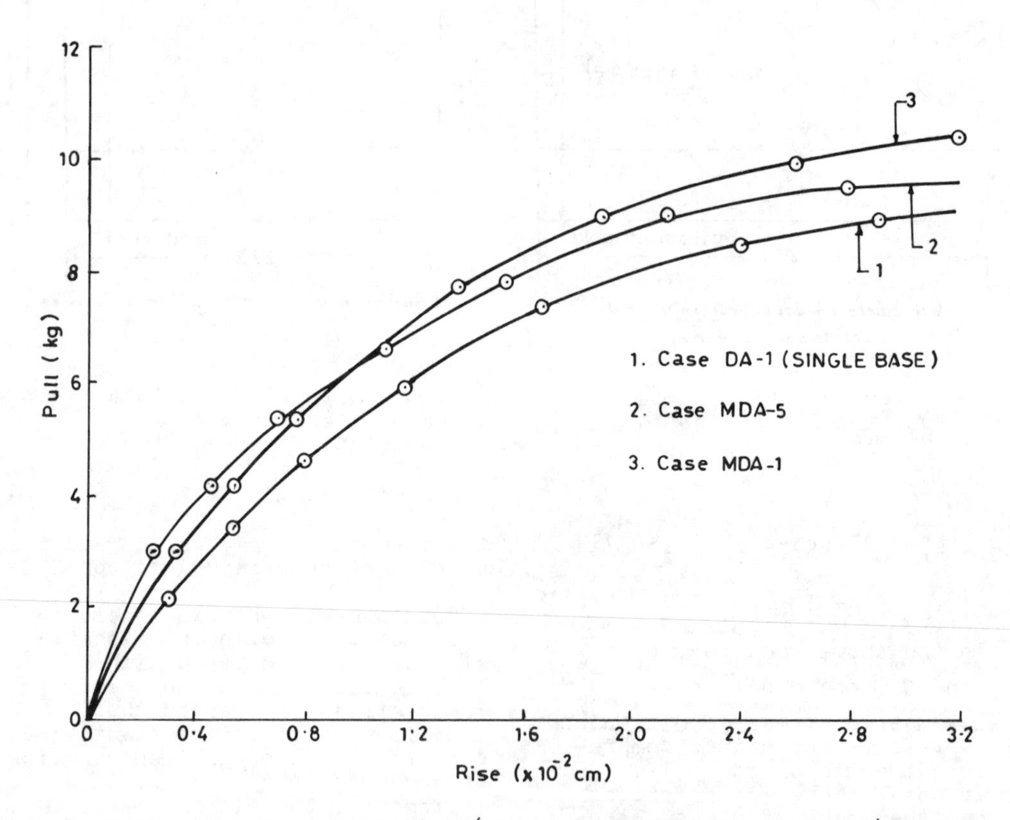

Fig. 5. Pull-rise characteristics (cases DA-1, MDA-1 and MDA-5)

The tensile stresses, if any, that developed in interface elements, were converted into nodal loads and for subsequent iterations and load increments, these elements were assigned zero stiffness.

The nodal loads due to above referred criteria were applied to the system and iterations were carried out till there was existence of negligible nodal loads as indicated by negligible incremental anchor displacements.

This procedure was repeated for other load increments. When the vertical anchor displacement continued to increase without any convergence under a particular load increment, it indicated a state of ultimate uplift resistance in the soil-anchor system.

4 DISCUSSION

For deep anchor with one additional similar base, the spacing S between the bases was varied at 0.67B, B, 1.33B and 1.67B (Fig. 1.a).

In Fig. 2, the contours of maximum principle tensile stresses are shown for the case MDA-1 (spacing S = 0.67B), at the state of ultimate uplift resistance. The computations did not indicate any yielding in compression and thus the failure in the soil-anchor system is seen to be due to development of tensile stresses. Similar trends were observed in respect of other spacings.

For deep anchor with two additional similar bases with a fixed spacing of 0.67B (Fig. 1.b, case MDA-5), the contours of maximum principle tensile stresses are shown in Fig. 3, for the state of ultimate uplift resistance. The failure pattern is seen to be similar to the one shown in Fig.2.

In Fig. 4, the pull-rise characteristics for deep anchor with single base (no multiunderreaming, case DA-1) and for multiunderreamed anchor (one additional similar base at variable spacing), are shown. It is seen that ultimate uplift resistance increases with multiunderreaming, with maximum increase occurring at closest spacing (S = 0.67B, case MDA-1) between the anchor bases.

In Fig. 5 are shown the pull-rise characteristics for the cases MDA-5, DA-1 and also for the case MDA-1. It is seen that provision of two additional similar bases (case MDA-5) also increases the ultimate uplift resistance. However, it is observed that multiunderreaming corresponding to case MDA-1, has proved to be more effective as compared to case MDA-5, thus indicating that, providing more number of anchor bases may not be advantageous.

5 CONCLUSIONS

The finite element analysis of deep laid anchor in cohesive soil medium, under undrained loading condition, incorporating the influence of soil tensile strength and anchor displacements shows that multiunderreaming increases the ultimate uplift resistance. The analysis further shows that providing more number of bases beyond a certain limit, may not be advantageous.

REFERENCES

Davie, J.E. and Sutherland, H.B. 1977. Uplift resistance of cohesive soils. J.S.M.F.E.Dn. A.S.C.E. GT-9. 103: 935-952.

Dewaikar, D.M. 1984. Study of deep anchors in cohesive soil medium. Indian Geotechnical Conference, Calcutta. 1: IV.1-IV.4.

Fang, F.Y. and Chen, W.F. 1975. Further studies of double punch test for tensile strength of soils. Proc. Third South East Asian Conference on Soil Engg. Hongkong. 1: 211-215.

Valliappan, S. 1968. Nonlinear finite element analysis of two dimensional problems with special reference to soil and rock mechanics. Ph.D. thesis, University of Swansea, U.K.

Vesic, A.S. 1971. Breakout resistance of objects embedded in ocean bottom. J.S.M.F.E. Dn. A.S.C.E. SM-9: 1183-1206.

Numerical Methods in Geomechanics (Innsbruck 1988), Swoboda (ed.)
© 1988 Balkema, Rotterdam. ISBN 90 6191 809 X

Analysis of unreinforced and reinforced embankments on soft clays by plasticity theory

G.T.Houlsby & R.A.Jewell
Department of Engineering Science, Oxford University, UK

ABSTRACT: The design of both unreinforced and reinforced embankments on soft clay is dominated by the clay strength, and can be treated as a bearing capacity problem. As well as exerting vertical stresses on the clay surface, unreinforced embankments exert outward shear stresses which result in a reduction of the bearing capacity factor. In reinforced embankments these shear stresses are carried by the reinforcement, and only vertical load is transferred to the clay below. For the important case of an embankment on a soil with an increasing strength with depth a futher benefit is gained due to inward shear stresses on the clay surface. Solutions from plasticity theory are presented which are of direct application to the design of embankments.

Keywords: analysis, clays, embankments, plasticity, reinforced soil

1 INTRODUCTION

The stability of an embankment on soft soil can be treated as a bearing capacity problem. The embankment fill applies vertical load to the foundation surface in combination with an outward shear stress caused by the horizontal stresses in the fill, Figure 1a. The influence of outward shear stress is to reduce the bearing capacity of the foundation soil, as illustrated in Figure 2 for a foundation with uniform strength (see Section 2 on Analysis).

Reinforcement at the base of an embankment can improve stability in two ways.

Firstly, the reinforcement can support the outward shear stress from the fill therby reducing the forces causing failure, Figure 1b. The foundation then only has to support the vertical stresses from the embankment.

Secondly, the reinforcement can provide inward shear stress on the surface of the foundation, Figure 1c. Inward shear stress increases the bearing capacity either when the foundation strength increases with depth, or when the foundation is on a clay of limited depth. Here the rein-forcement is increasing the forces resisting failure, (see Jewell (1987) for a full discussion).

Plasticity theory allows the calculation of the maximum vertical stresses that a foundation can support in combination with a surface shear stress. The existing solutions for a foundation with strength increasing with depth, relevant to deposits of normally or lightly over-consolidated clay, are shown in Figure 3, in terms of the overall bearing capacity factor N_c. The two solutions available are for zero surface shear stress and for full inward shear stress, which are the perfectly smooth and perfectly rough footing cases respectively (Davis and Booker, 1973).

The shaded regions in Figures 2 and 3 indicate the magnitude of the improvement in bearing capacity that could be expected from reinforcement acting in the two ways described above.

A complete set of results for a foundation with strength increasing uniformly with depth is now presented. The results include outward shear stress on the foundation surface, for the analysis of unreinforced embankments, and intermediate values of shear stress for the analysis of "partially" reinforced embankments. The results are presented in terms of the increase in the allowable vertical stress with distance from the edge of loading, which may be thought of as representing an ideal shape for embankment loading.

2 ANALYSIS

The problem of the calculation of the limiting pressure distribution on the surface of a clay foundation with strength increasing with depth can be solved using the lower bound theorem of plasticity theory. This approach is quite usual in the analysis for bearing capacity, and involves the calculation of a stress distribution which obeys the equilibrium equations and does not violate the yield condition. The relevant equations for a plane strain problem of bearing capacity on an undrained clay with strength increasing linearly with depth are as follows.

Equilibrium equations:

$$\frac{\partial \sigma_{xx}}{\partial x} + \frac{\partial \tau_{xz}}{\partial z} = 0 \qquad ...(1)$$

$$\frac{\partial \tau_{xz}}{\partial x} + \frac{\partial \sigma_{zz}}{\partial z} = 0 \qquad ...(2)$$

(Note that for problems with a horizontal ground surface it is not necessary to consider the self weight of the soil since this simply adds terms of the form γz to all stresses).

The yield condition, which is given by:

$$(\sigma_{xx} - \sigma_{zz})^2 + 4\tau_{xz}^2 = 4s_u^2 \qquad ...(3)$$

where s_u is the undrained strength which is a function of the depth z below ground level:

$$s_u = s_o + \rho z \qquad ...(4)$$

The above equations can be shown to form a hyperbolic set of equations in two variables which can be solved by the method of characteristics. Methods of solving such problems are discussed by Houlsby and Wroth (1982). Solutions to the bearing capacity problem on a soil with increasing strength with depth were published by Davis and Booker (1973) for the cases of a perfectly smooth footing and a perfectly rough footing, Figure 3, and some additional cases in axial symmetry were presented by Houlsby and Wroth (1983). In this paper solutions are presented to the full range of possible roughness values. Greater detail is provided to allow more precise calculation of pressure distributions, since these have been found extremely useful in the design of embankments on soft clays.

The results are shown in Figure 4 as the normalised vertical stress beneath the foundation σ/s_o against the normalised distance from the edge of the foundation $\rho x/s_o$.

Another way of presenting the data is as follows. The numerically calculated results may be compared with the approximate expression for the vertical stress distribution on the base of a foundation:

$$\sigma_{approx} = N_c s_o + \rho x \qquad ...(5)$$

where N_c is the bearing capacity factor for constant strength with depth:

$$N_c = 1 + \pi + (1 - \alpha^2)^{1/2} + \arcsin(\alpha) \qquad ...(6)$$

(see for example Houlsby and Wroth (1983)), where α is the roughness factor defined as:

$$\alpha = \tau/s_o \qquad ...(7)$$

and τ is the inward shear stress acting on the clay surface. The factor σ/σ_{approx} is exactly equal to unity for $\alpha = -1$ (full outward shear stress) and varies with $\rho x/s_o$ as shown in Figure 5 for other values of α. For very high and very small values of $\rho x/s_o$ the factor approaches unity, and for intermediate values it is always greater than unity, with a maximum value at $\rho x/s_o$ of approximately 8, i.e $\ln(\rho x/s_o)$ of approximately 2, but the exact position of the peak is dependent on the α value.

A typical network of characteristics is shown in Figure 6, which indicates that for any value of x the failure region extends to a maximum depth z_m. The magnitude of z_m for the solutions is shown in Figure 7. The maximum depth is relevant since it shows the range of applicability of the results. The soutions are valid as long as the strength variation is appropriate up to the depth z_m and that the strength does not decrease with depth below this level. If the foundation clay is underlain by a stronger layer, the depth z_m allows a check as to

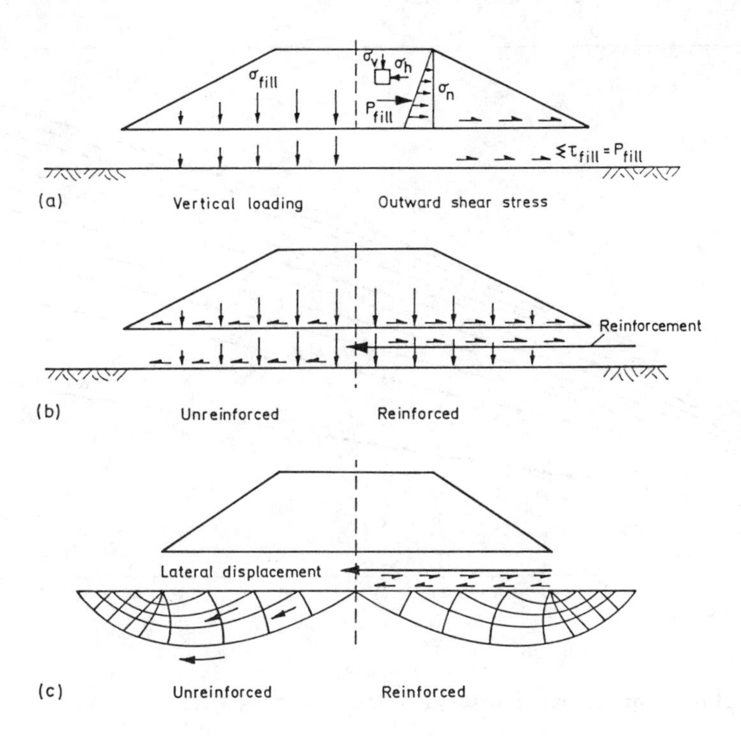

Fig. 1. (a) Loading from an embankment and the action of reinforcement (b) reducing the forces causing failure and (c) increasing the forces resisting failure

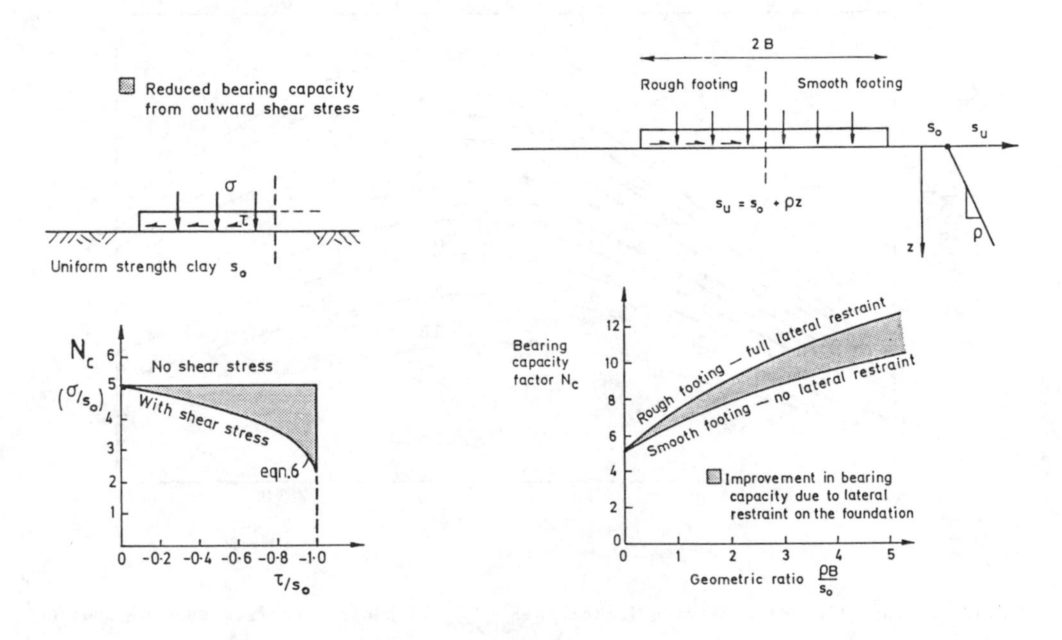

Fig. 2. Reduction of bearing capacity due to outward shear stress

Fig. 3. Improvement of bearing capacity due to inward shear stress (after Davis and Booker, 1973)

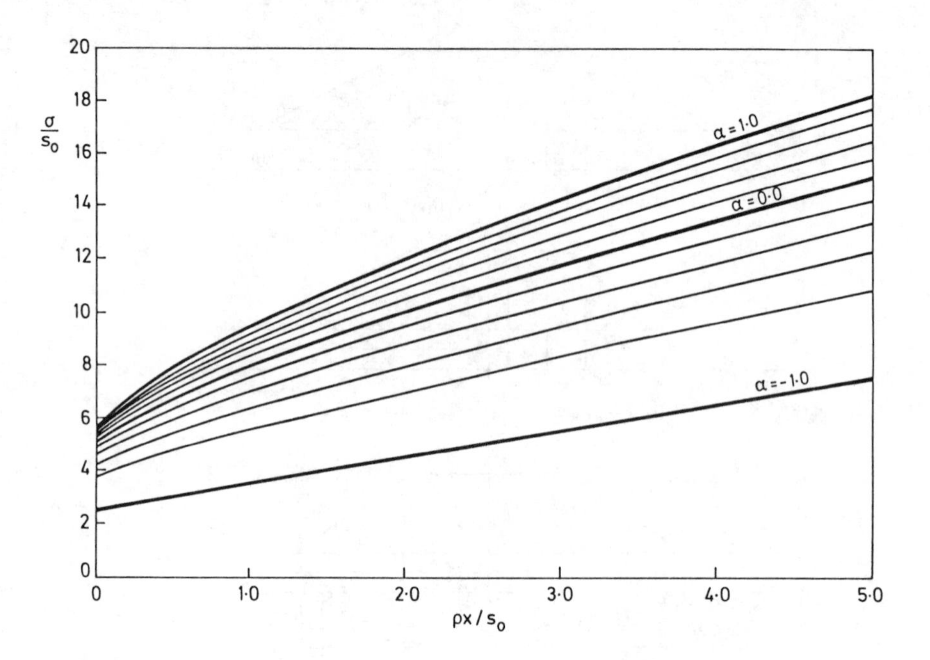

Fig. 4. Variation of normal stress with distance from edge of footing

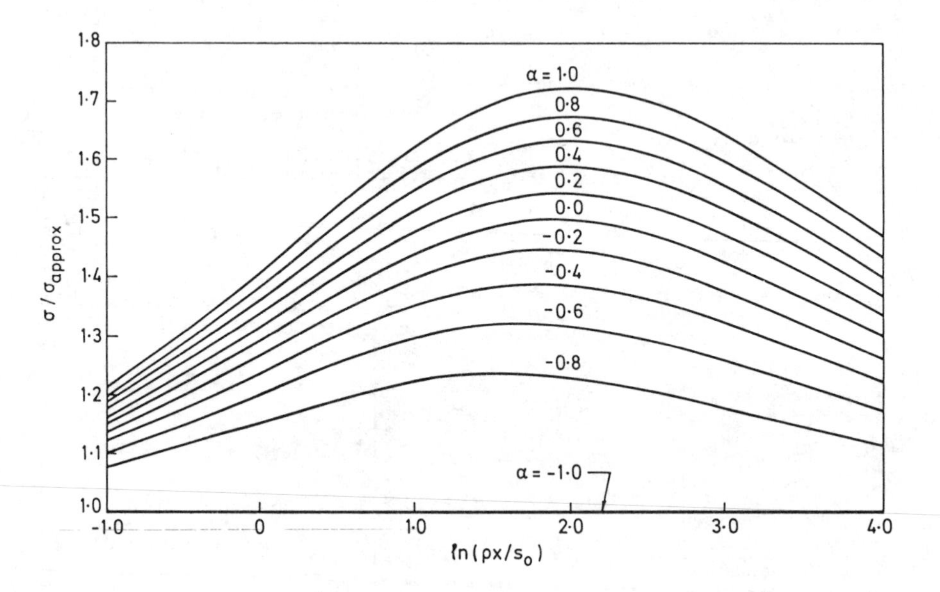

Fig. 5. Variation of modifying factor on σ_{approx} with distance from edge of footing

whether the stronger layer will provide additional benefit.

3 SLIP CIRCLE ANALYSIS

Slip circle analysis is a standard design tool for embankments. In this method allowance can simply be made for non-uniform variations of foundation shear strength. The slip circle analysis can be applied to reproduce the plasticity solutions shown in Figure 3. The procedure involves building up the vertical stress distribution incrementally from the edge of the loading. The result is a family of slip circles all with the same factor of safety, but with different positions for the circle centres. These circles each imply total moments about their centres from the applied loading, and can therefore be used to deduce a loading distribution. Figure 8 shows a comparison between the slip circle analysis and the plasticity solution for the fully reinforced case, Jewell (1987).

The finding is that the slip circle analysis provides results in good agreement with plasticity theory for the bearing capacity of a foundation with strength increasing with depth. Slip circle analysis may therefore be used with confidence for the analysis of real embankment profiles.

The agreement does not extend to the case of a foundation on a clay of limited depth when the slip circle analysis over-estimates the foundation bearing capacity, Jewell (1987). Also, for an unreinforced embankment, continuation of the most critical foundation circle through the embankment fill typically underestimates the disturbing lateral thrust from the fill, thereby overestimating the overall stability. To avoid this, the lateral thrust in the fill may be estimated either from a simple wedge analysis, using a log spiral through the fill, Leshchinsky (1987), or by use of appropriate earth pressure coefficients.

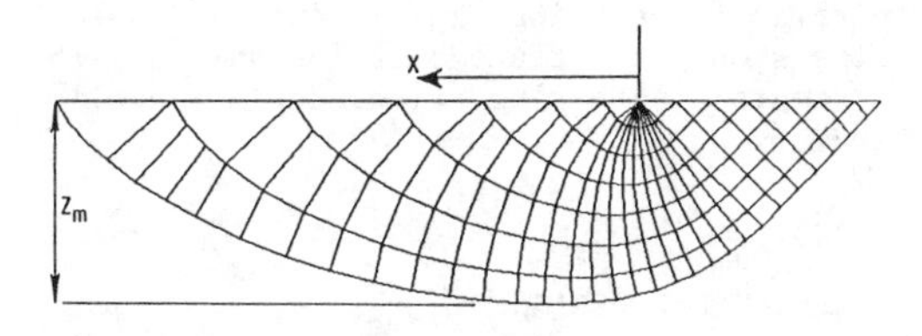

Fig. 6. Typical characteristic mesh for $\alpha = 0.5$

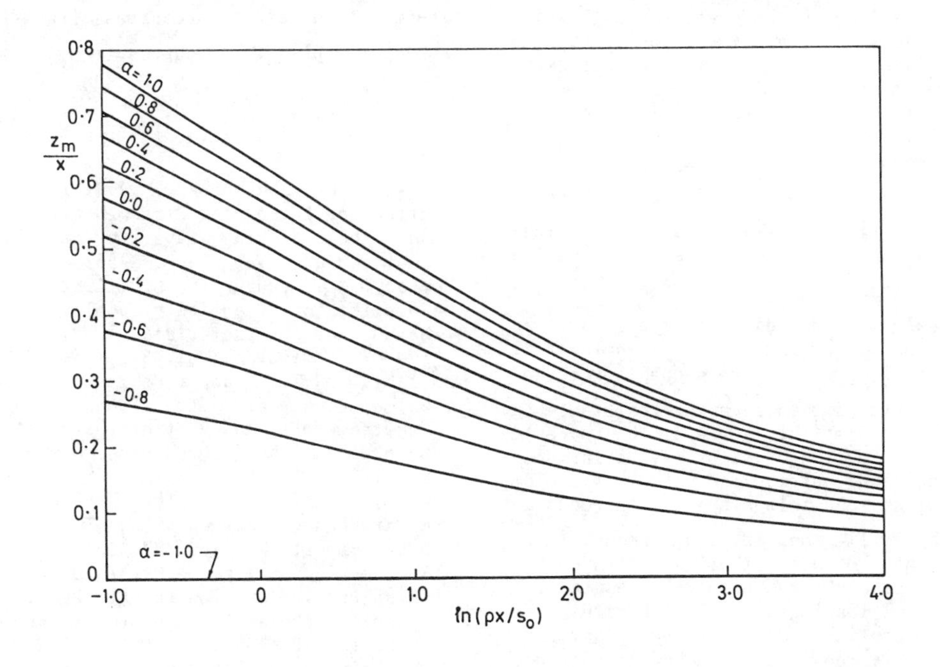

Fig. 7. Variation of z_m with distance from edge of footing

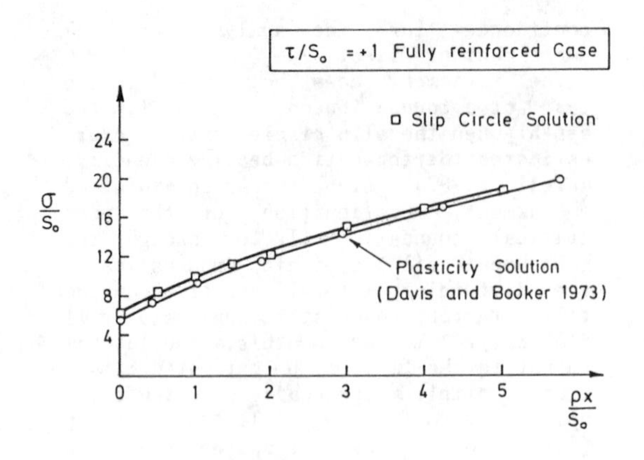

Fig. 8. Comparison of the slip circle and plasticity solutions for full inward shear stress on the surface of a foundation with strength increasing with depth (Jewell, 1987)

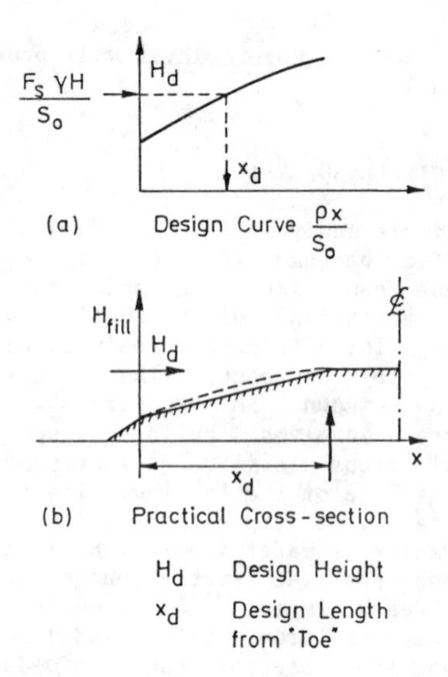

Fig. 9. Practical embankment shape to approximate the ideal vertical stress distribution

4 EMBANKMENT DESIGN

An important result is that the calculated foundation stability depends on the distribution of the applied surface stresses and not just on the highest value of the vertical stress. A satisfactory approximation to the exact loading, which will just bring the whole of the foundation to failure, for an embankment design is shown in Figure 9.

In the fully reinforced case, for example, the design curve used would be that with $\tau/s_o = 1.0$. Applying a factor of safety F_s to the foundation shear strength gives the desired vertical loading $\sigma/s_o = F_s\gamma H/s_o$ and the design length from the toe x_d, Figure 9a. The vertical load distribution may be approximated by a uniformly sloping fill over the length x_d. The slope is governed by the foundation shearing resistance (a topic outside the scope of this paper). A steeper section of slope may be included at the toe outside the uniform slope of length x_d, and this slope is limited only by the fill shearing resistance. It is important to note that a safe design cannot be achieved simply by maintaining the actual loading distribution within the ideal profile derived by plasticity theory. If too little loading is provided near the toe of the slope then the stability can be adversely affected.

For an unreinforced embankment, the outward shear stress τ/s_o depends on the fill height, shearing resistance and unit weight and the design length x_d. The determination of x_d requires iteration between the plasticity curves.

REFERENCES

Davis, E.H. and Booker, J.R. 1973. The Effect of Increasing Strength with Depth on the Bearing Capacity of Clays, Geotechnique, Vol. 23, 551-563

Houlsby, G.T. and Wroth, C.P. 1982. Direct Solution of Plasticity Problems by the Method of Characteristics. Proc. 4th ICONMIG, Edmonton, Vol. 3., 1059-1071

Houlsby, G.T. and Wroth, C.P. 1983. Calculation of Stresses on Shallow Penetrometers and Footings, Proc. IUTAM Symp. on Seabed Mechanics, Newcastle, 107-112

Jewell, R.A. 1987. The Mechanics of Reinforced Embankments on Soft Soils, OUEL Report No. 1694/87, Department of Engineering Science, Oxford University (Also Proc. Stanstead Abbotts Prediction Symosium, Thomas Telford Ltd (in press))

Leshchinsky D. 1987. Short-Term Stability of Reinforced Embankment over Clayey Foundation, Soils and Foundations, Vol. 27, No. 3, 43-57

Numerical Methods in Geomechanics (Innsbruck 1988), Swoboda (ed.)
© 1988 Balkema, Rotterdam. ISBN 90 6191 809 X

Analysis of polymer grid-reinforced soil retaining wall

Hidetoshi Ochiai & Shigenori Hayashi
Kyushu University, Fukuoka, Japan
Eiji Ogisako
Shimizu Construction Co. Ltd, Tokyo, Japan
Akira Sakai
Saga University, Japan

ABSTRACT: A method of analysis for polymer grid reinforced soil structures, which is capable of taking into account the displacement dependence property of the pull-out resistance of the polymer grid in soils, is proposed. The soil-polymer grid interaction is modelled by combining the joint element expressing the property of discontinuous plane with the truss element transmitting the axial force only. Application of the method to the analysis of retaining wall with polymer grid reinforced soil backfill is described.

1 INTRODUCTION

In the recent years interest has been shown in the use of polymer grids for reinforcement of soil structures. Much of the approaches for analysis of polymer grid reinforced soil have employed the limit equilibrium principle [1]. The limit equilibrium methods generally suffer from their inability to take account of the effect of finite displacement on soil-polymer grid interaction, although it is this effect which primarily controls the reinforcement mechanism [2].

In the paper, a distribution of the displacement of the polymer grids in soils and a process of mobilization of the pull-out resistance are presented from the labolatory pull-out tests on the polymer grids. Then, a method of analysis for polymer grid reinforced soil structures, which is able to consider the displacement dependence property of the pull-out resistance, is given. The method is applied to two types of retaining wall systems with polymer grid reinforced soil backfill, and the earth pressures and the wall deformations are discussed.

2 MECHANISM OF PULL-OUT RESISTANCE OF POLYMER GRIDS IN SOILS AND ITS ANALYTICAL PROCEDURE

When the polymer grid laid in soil is subjected to a pulling force, the polymer grid is pulled out of the soil with its deformation, and thus the pull-out resistance is mobilized on both grid junctions and ribs of the polymer grid as shown in Fig.1(a). In the case of the polymer grids through where the soil on either side of the reinforcing material is partially continuous, the resistance effect of the ribs at right angles with the direction of pulling is assumed to be transferred to the grid junctions in a

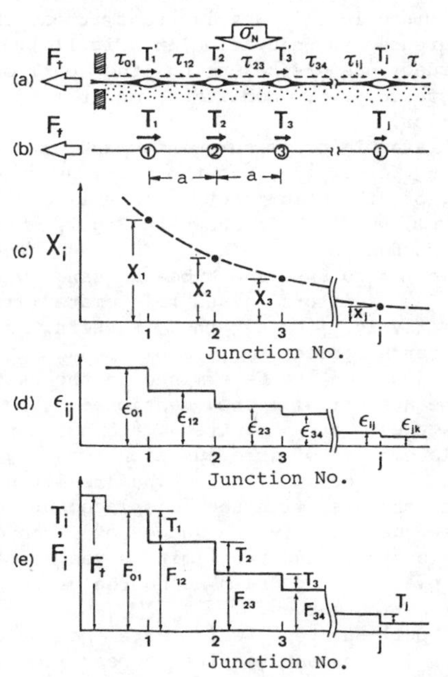

Fig.1 Analytical procedure of results obtained from pull-out test.

concentrated manners, thus the resistance mobilized on each grid junction playing a greater role than that mobilized on each rib. It may be therefore considered that the pull-out resistance will concentrate and act on each of the grid junctions as shown in Fig.1(b). The concept of the analytical procedure for determining the pull-out resistance mobilized on each grid junction in soil from the pull-out force, F_t, exerted on the front of the polymer grid and the displacement, X_i, of the grid junctions is as follows [3]: Fig.1(c) shows displacement, X_i, of each grid junction measured in a pull-out test under condition of constant vertical stress, from which the strain, ε_{ij}, of the grid between the junctions is calculated by the next equation;

$$\varepsilon_{ij} = (X_i - X_j)/ a \qquad (1)$$

in which a is length between each grid junction. As a result, Fig.1(d) is obtained. Then, the pulling force, F_{ij}, between each grid junction that corresponds to the strain, ε_{ij}, is determined by using a load-strain curve of the polymer grid. Although the load-strain curve is affected by the strain rate and temparature, the index curve [4], which was obtained by the tensile test under the conditions of strain rate of 2%/min and temparature of 20°C ± 2°C, is approximately used as a standard one. Fig.1(e) show how the pulling force as determined in this way is transmitted. The difference in elevation in Fig.1(e) is regarded as representing the pull-out resistance, T_i, mobilized on the grid junction.

An example of measurements made of the pulling force, F_t, and the displacements, X_i, of the grid junction in soil during the pull-out test is shown in Fig.2, which corresponds to Fig.1(c). In the test, polymer grid SR-1 and a beach sand were used as reinforcing and fill materials, respectively. Using the results, the resistance force, T_i, mobilized on the grid junction are determined on the basis of the analytical method mentioned above, and they are plotted against the displacements of grid junction at a given pulling force in Fig.3. The resistance force changes with the displacement of grid junction in a shape of convex distribution, and two limit curves, upper and lower, can be drawn along the group of distribution curves.

The pull-out tests which were performed under 4 constant vertical stress levels show that good linear relations between the pull-out resistance acting on the junctions and the vertical stress hold for any stages of the displacement on both upper and lower limit curves. This implies that for either limit curve there is a kind of friction law expressed by the following equation with the displacement of grid junction taken as a parameter;

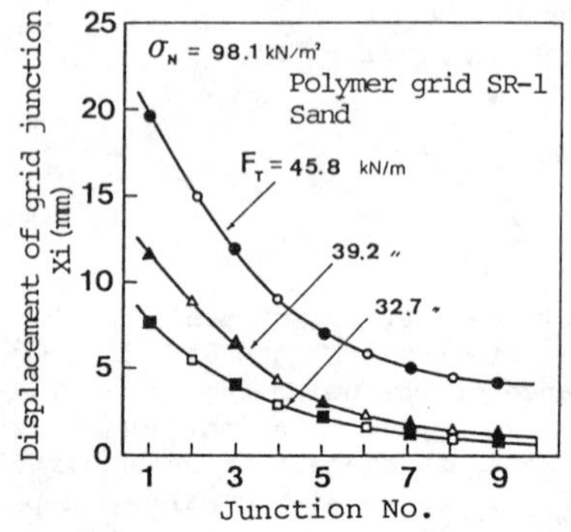

Fig.2 Measured displacement of polymer grid in soil during pull-out test.

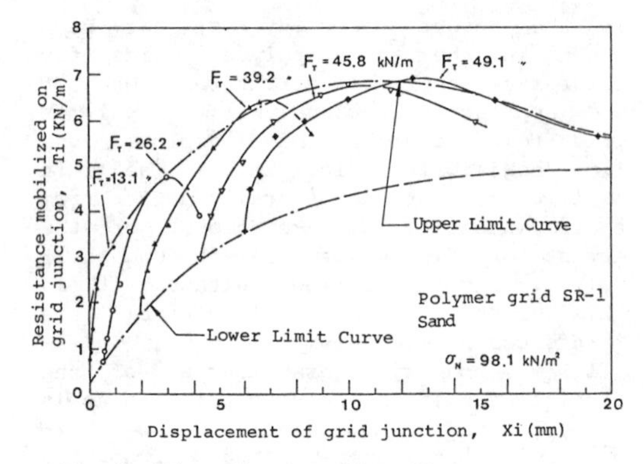

Fig.3 Distribution of pull-out resistance mobilized on grid junctions.

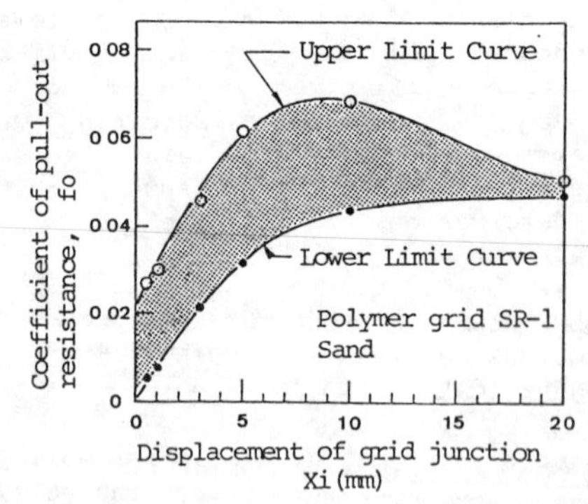

Fig.4 Process and range of mobilization of pull-out resistance.

$$PR = s_o + f_o N \quad (2)$$

where PR and N are pull-out resistance and vertical force per unit width, and s_o and f_o are coefficients corresponding to cohesion and friction components of the pull-out resistance, respectively.

The coefficient, f_o, for both upper and lower curves is plotted in Fig.4 with the displacement of grid junction as abscissa. The curves may be regarded as representing the process and range of development of the friction component of the pull-out resistance being mobilized on the polymer grid in soil as it is displaced. Similar result was also obtained for the coefficient, s_o.

3 ANALYTICAL METHOD FOR POLYMER GRID REINFORCED SOIL

3.1 Modeling of polymer grid reinforced soil

For the case of deformation analysis of a reinforced soil structure with a heterogeneous material such as the polymer grid, it is necessary to use an analytical method which is capable of expressing a behavior of discontinuous plane with a peculiar friction. Presented herein for a modeling of the polymer grid reinforced soil is a method of combining the joint element expressing the property of discontinuous plane with the truss element transmitting the axial force only. The polymer grid is modeled by the truss element whose ends are connected by the pin joint. Fig.5 shows the finite element model for polymer grid reinforced soil.

Using the joint element, the mechanism of pull-out resistance of the polymer grids in soils described in the previous section can be treated as a nonlinearity of the element. In particular, it is possible to express the change of coefficients of pull-out resistance with the shear displacement. Hence, the nonlinear behavior of the joint element may be evaluated by introducing the dependence of shear displacement into a shear stiffness [5].

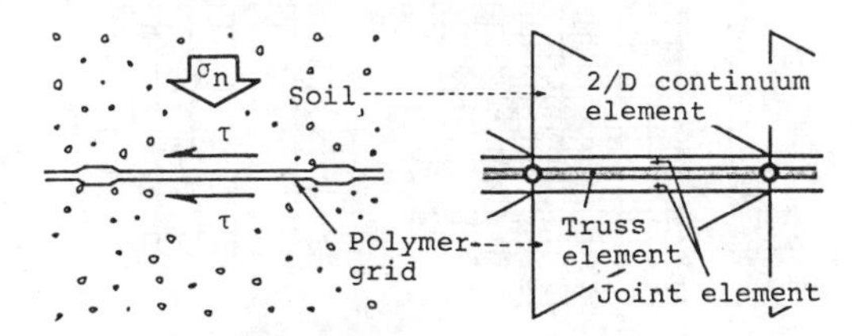

Fig.5 Finite element model for polymer grid reinforced soil.

The soil is assumed to be a nonlinear elastic medium (Duncan-Chang model) with the next tangent elasticity modulus.

$$E_t=\{1-\frac{R_f(1-\sin\phi)(\sigma_1-\sigma_3)}{2c\cos\phi+2\sigma_3\sin\phi}\}^2 K P_a(\sigma_3/P_a)^n \quad (3)$$

in which c and ϕ are cohesion and internal friction angle of soil, σ_1 and σ_3 are maximum and minimum principal stresses and K, n and R_f are experimental constants, respectively.

3.2 Nonlinearity of the joint element

The joint element has two unit stiffness, a normal stiffness, k_n, and a shear stiffness, k_s. The former expresses a transmission of compressive force only, and the latter a sliding against a shear displacement. In the pull-out tests on polymer grids in soils, the coefficients of pull-out resistance change with the displacement of grid junctions, as described above. Therefore, the nonlinear behavior of the polymer grid reinforced soil may be analized by using the shear stiffness which changes continuously with shear displacement, and not by dealing with the shear strength independently.

A determination of the shear stiffness, k_s, in the joint element is as follows: The mobilizing process and range of the cohesion and friction components of pull-out resistance are shown by broken lines in Fig.6, in which the coefficients, s and f, are values of s_o and f_o when Eq.(2) is represented in terms of stress

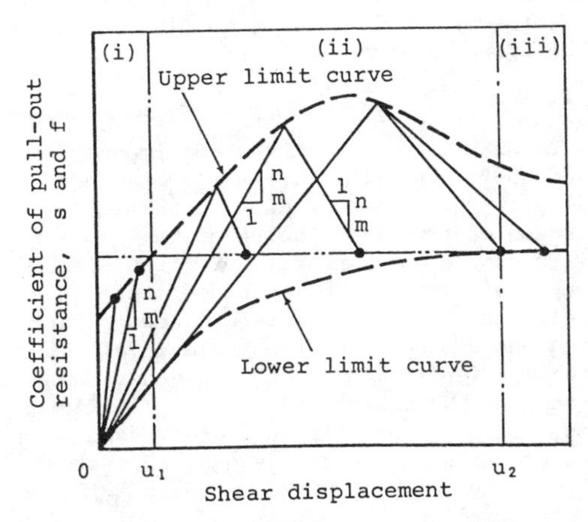

Fig.6 Determination of shear stiffness k_s in joint element.

unit. Assuming that the coefficients are linear function of the shear displacement u, s = nu and f =mu, shear stress τ may be represented by the next equation:

$$2\tau = s + f\sigma_n = (n + m\sigma_n)u \qquad (4)$$

Therefore, the shear stiffness k_s is given by

$$2k_s = n + m\sigma_n \qquad (5)$$

in which σ_n is the normal stress. A determination of values of n and m is classified into three regions by a maximum value, u_{max}, of displacements of the grid junctions at a given pulling force, as shown in Fig.6. The first classifing point is one on the upper limit curve according with the residual value of the lower limit curve, which corresponds to the shear displacement of u_1. The second is an intersection point of the residual value of the lower limit curve and the straight line which a line through origin being in contact with the lower limit curve is turned on the upper limit curve with the same absolute value of slope. The second point corresponds to the shear displacement of u_2.

(i) Region-1 ; $u_{max} \leq u_1$
Values of n and m are that of slope of a straight line which connects the origin to the point of u_{max} on the upper limit curve.

(ii) Region-2 ; $u_1 < u_{max} \leq u_2$
In the region, it is assumed that elements having arised u_{max}-value show the residual value of shear resistance. The coefficients, s and f, are given by two straight lines with the same absolute value of slope ($\pm m$, $\pm n$), where an intersection of them is a turning point on the upper limit curve.

(iii) Region-3 ; $u_{max} > u_2$
Values of s and f are given by two straight lines in the same manner as Region-2, but the line through origin is always the same one with $u_{max} = u_2$ independently of the value of u_{max}.

Thus, the nonlinearity of joint element may be expressed by means of the shear stiffness k_s which is determined in three regions depending on the maximum value of shear displacement. On the other hand, constant values of the normal stiffness in the joint element are used in compression and tension sides, respectively. The value of tension side, k_{n2}, is much smaller than that of compression side, k_{n1}, because of no-resistance to tensile force on discontinuous plane.

4 ANALYSIS OF POLYMER GRID REINFORCED SOIL RETAINING WALL

It is one of important programs to analize an earth pressure acting on and a deformation of a retaining wall with polymer grid reinforced soil backfill. The presented method, which is able to evaluate the mechanism of pull-out resistance of the polymer grids in soils, is applied to the finite element analysis of the polymer grid reinforced retaining wall.

4.1 Analytical model and procedure

Two types of the wall systems are analyzed in this study. One is a rigid wall which the lateral deformation is fixed perfectly, and it corresponds to the K_o-conditions. The other is a flexible wall which the lateral deformation is free except for the bottom end of the wall.

Fig.7 shows the cases to be studied, in which Model 1 is unreinforced one. The

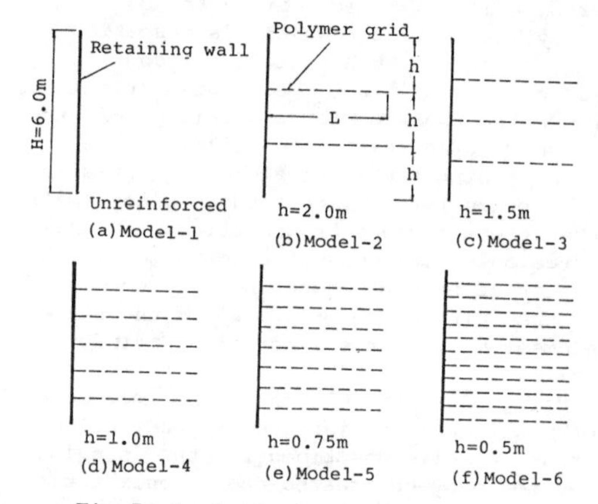

Fig.7 Analytical cases.
(H = 6.0m, L = 6.0, 4.5, 3.0m)

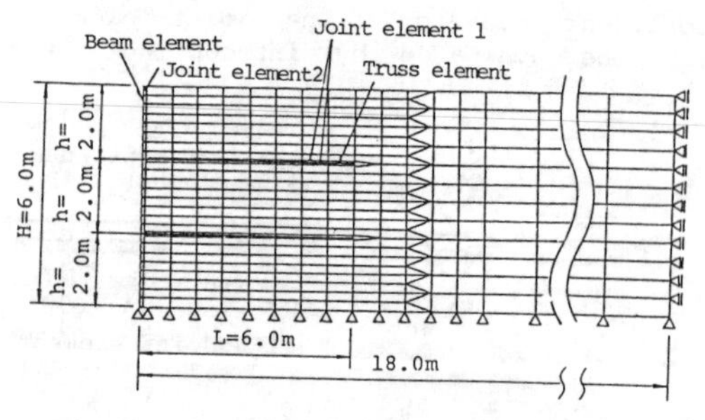

Fig.8 FEM mesh.

Table 1 Soil parameters

γ (KN/m³)	C (KN/m²)	φ (°)	K	n	R	ν
17.1	0	35	1000	1.0	0.9	0.333

Table 2 Material parameters

	E (KN/m²)	A (m²)	I (m⁴)
Polymer grid	1.65×10^{6}*	0.0012	——
Wall	2.45×10^{6}	0.18	4.86×10^{-4}

*initial value

height of retaining wall, H, is 6.0m. The spacing, h , of the polymer grid laid in the backfill are 2.0, 1.5, 1.0, 0.75 and 0.5m, and the length, L , of them are 6.0, 4.5 and 3.0m. The end of the polymer grids are anchored to the wall.

An example of analytical models for the case of h=2.0m, L=6.0m and flexible wall is shown in Fig.8. The flexible wall is modeled as beam element, and in the case of the rigid wall every node on the left side is fixed perfectly. Joint elements are used on the boundary of soil-wall as well as on that of soil-polymer grid, since the discontinuous plane is caused there.

The soil backfill is a beach sand. In the analysis, Duncan-Chang model is used as a soil model of which parameters are given in Table 1.

The polymer grid used is an uniaxially orientated grid. The properties of the flexible wall are determined on the basis of the assumption that the wall is constructed with concrete panels of about 1.0m × 1.0m × 0.18m connected by metal joint. The properties of the polymer grid and the wall used in the analysis are shown in Table 2. The values of normal stiffness, k_{n1} and k_{n2} , in the joint element between polymer grid and soil are 10^{6}KN/m³ and 10^{-2}KN/m³ respectively, and the shear stiffness in the joint element used on the soil-wall boundary is constant value of 10KN/m².

The procedure of analysis is to simulate the process of construction which the soil is backfilled in layers of 50cm step by step and after the completion of the backfill the uniform surcharge of 10KN/m² is applied. At each step of the backfilling stress-deformation analysis by self weight of soil is performed, and the calculation is iterated until the values of the shear stiffness k_s in the joint element used on the soil-polymer grid boundary and of the Young's modulus E in the truss element satisfy the convergency conditions.

4.2 Analytical results

Rigid wall

When any lateral deformation of the rigid wall is perfectly fixed, the earth pressure distributions acting on the walls after backfilling and surcharge are shown in Fig.9. In the case of unreinforced soil backfill,the earth pressure distribution caused by backfilling is almost trapezoidal one. Whereas for the reinforced soil one, when the spacing, h, of the reinforcement in the backfill is large, only the earth pressures in the positions of the reinforcement decrease. And, as the amount of the reinforcement increases and the spacing, h, becomes smaller, the earth pressures exerted on the whole height of the wall decrease.

Reductions of earth pressures resulted from the reinforcement are plotted against the spacing of the polymer grid in the backfill in Fig.10. Here, the reduction ratio of earth pressures, R_p , is expressed by the next equation;

$$R_p = 1 - (P_g / P_o) \quad (6)$$

in which P_g and P_o are total forces of

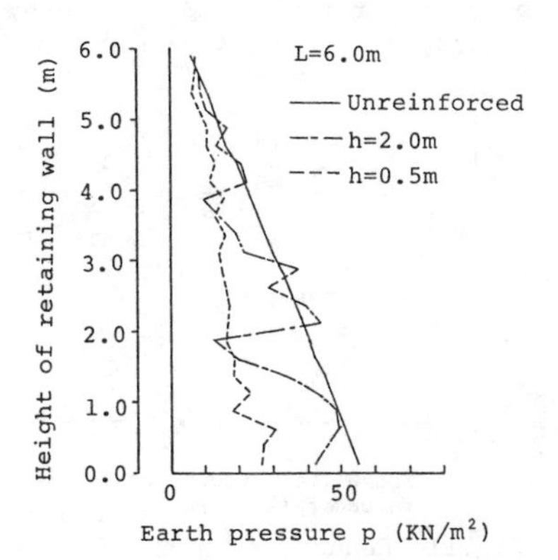

Fig.9 Distribution of earth pressure acting on rigid wall

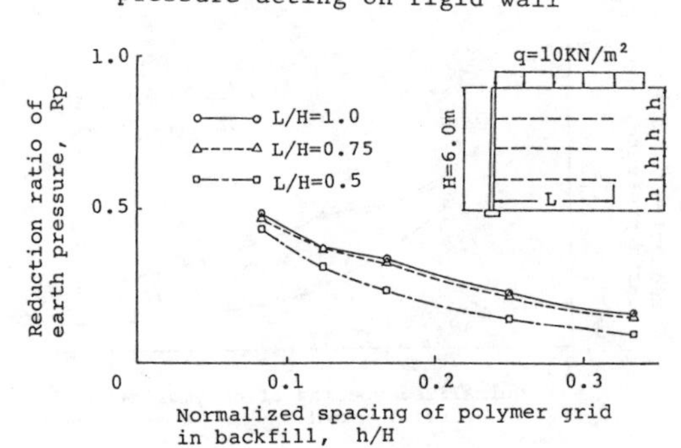

Fig.10 Reduction ratio of earth pressure for rigid wall.

earth pressures acting on the walls with and without reinforced soil backfills, respectively. And the spacing, h , is normalized by the wall height, H , and the ratio, h/H , is used as an abscissa in Fig.10. From this figure, it can be seen that the effects of the earth pressure reduction by the reinforcement become larger as the spacing of polymer grids becomes smaller and the length of them becomes larger. There is, however, less difference between the results of L/H=0.75 and L/H=1.0. It therefore seems that there exists a reasonable length of polymer grid to reduce earth pressures, and the value of L/H is about 0.75 in the case studied here.

Flexible wall

In the cace of the flexible wall, lateral deformations of the wall is caused by the backfilling and surcharge. It is considered that the reinforcement of the backfill may have a great effect on the deformation of the wall. Fig.11 shows that the relation between the reduction ratio of the lateral deformation of the wall, R_δ , and the normalized spacing of the polymer grid, h/H . Here, the reduction ratio, R_δ , is expressed by the following equation;

$$R_\delta = 1 - (\delta_g / \delta_o) \qquad (7)$$

where δ_g and δ_o are maximum values of the lateral deformations of the walls with and without reinforced soil backfill, respectively. From this figure, it is seen that small spacing and large length of the polymer grid back of the wall do reduce the lateral deformation of the flexible wall.

Reductions of earth pressures acting on the flexible wall by the reinforcement of backfill are shown in Fig.12. In the figure, the reduction ratio of earth pressure, R_p , is the same as Eq.(6), but P_g and P_o are maximum values of earth pressures exerted on the walls with and without reinforced soil backfill, respectively. The trends of the effects of the earth pressure reduction by the reinforcement are similar for both rigid and flexible walls, as shown in Fig.10 and Fig.12.

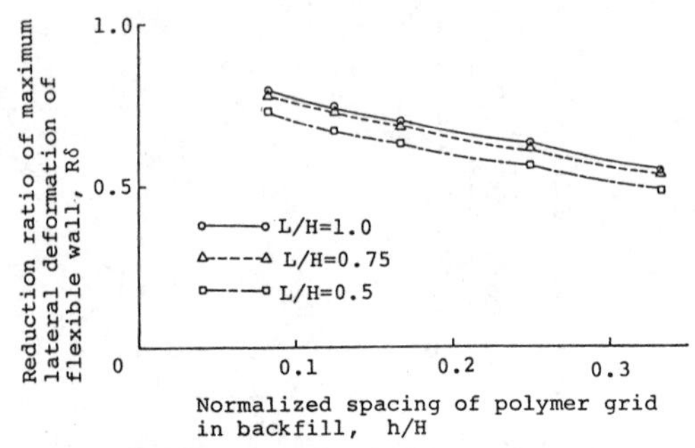

Fig.11 Reduction ratio of lateral deformation of flexible wall.

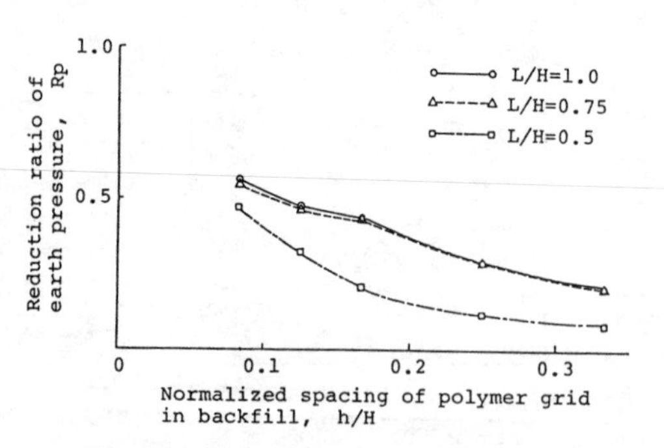

Fig.12 Reduction ratio of earth pressure for flexible wall

5 CONCLUSIONS

An analytical method for expressing the pull-out behavior of the polymer grids in soils is described and its application to the polymer grid reinforced soil retaining wall is discussed in this paper. It is shown from the analytical results that the polymer grid reinforcement reduces the earth pressures acting on the walls as well as the lateral deformation of the wall, and there exists a reasonable length of the polymer grids to reduce both the earth pressures and the wall deformation.

REFERENCES

[1] R.R.Berg et al: Design, construction and performance of two reinforced soil retaining walls, Proc. 3rd Int. Conf. on Geotextiles, 401-406, (1984).

[2] S.Frydman: Stability of slopes and embankments, Preprint of the 8th Asian Regional Conf. on SMFE, 197-225, (1987).

[3] S.Hayashi et al: Mechanism of pull-out resistance of polymer grids in soils, TSUCHI-TO-KISO, JSSMFE, vol.33, No.5, 21-26, (1985).

[4] Netlon Ltd: Test methods and physical properties of 'Tensor' geogrids, (1984).

[5] H.Ochiai and A.Sakai: Analytical method for geogrid-reinforced soil structures, Proc. 8th Asian Regional Conf. on SMFE, 483-486, (1987).

ANDERSONIAN LIBRARY
WITHDRAWN
FROM
LIBRARY
STOCK
UNIVERSITY OF STRATHCLYDE